Expotecnologia 2019 "Research, Innovation and Development in Engineering"

IOP Conference Series: Materials Science and Engineering
Volume 844

Cartagena, Colombia
30 October - 1 November 2019

ISBN: 978-1-7138-4728-1
ISSN: 1757-8981

Printed from e-media with permission by:

Curran Associates, Inc.
57 Morehouse Lane
Red Hook, NY 12571

Some format issues inherent in the e-media version may also appear in this print version.

Institute of Physics
Dirac House, Temple Back
Bristol BS1 6BE UK

Phone: 44 1 17 929 7481
Fax: 44 1 17 920 0979

techtracking@iop.org

Additional copies of this publication are available from:

Curran Associates, Inc.
57 Morehouse Lane
Red Hook, NY 12571 USA
Phone: 845-758-0400
Fax: 845-758-2633
Email: curran@proceedings.com
Web: www.proceedings.com

TABLE OF CONTENTS

IOP Conf. Series: Materials Science and Engineering **844** (2020) 011001 doi:10.1088/1757-899X/844/1/011001

Expotecnología 2019

Organized by

UNITECNAR
FUNDACIÓN UNIVERSITARIA ANTONIO DE AREVALO

FACULTAD DE CIENCIAS DE LA INGENIERÍA
UNITECNAR

Patrocinadores:

UNIVERSIDAD DEL SINÚ
Elías Bechara Zainúm
Seccional Cartagena

enel
codensa - emgesa

Asociacion Colombiana de Ingenieros
ACIEM Capitulo Bolivar

RENATA COLOMBIA
Red Nacional Académica de Tecnología Avanzada

SIEMENS
Ingenuity for life

CELSIA
La energía que quieres

IOP Conf. Series: Materials Science and Engineering **844** (2020) 011001 doi:10.1088/1757-899X/844/1/011001

Expotecnología 2019 "*Research, Innovation and Development in Engineering*", Expotecnología 2019, took place from October 30 to November 1, 2019, in the beautiful city of Cartagena de Indias (Colombia), with an agenda that joins to the academic dynamics through the promotion of activities that promote culture and research commitment, transferring knowledge and technological innovation, and the management year after year to create spaces updating the educational and professional community of all areas of engineering in Colombia.

Expotecnología is a scientific-academic event that comes from the Faculty of Design and Engineering of the University Foundation Antonio de Arévalo - UNITECNAR, in order to provide teachers, students and other professionals, a specialized meeting that seeks, through the creation of academic networking, create a new relationship between science, technology, and society.

The conferences took place in an environment of socialization of experiences, research results or discussions around several themes focused on 4 topics, which have been strategically selected to approach from the academic, industrial and social point of view, the present and the future of our profession in our country.

The general topics discussed were:
New Technologies: Information systems, Software Engineering, Information technology, Internet of Things and Mobile Applications.
Petrochemical and Process Control: Material science, Instrumentation and Control Systems, Robotics and Automation, Industrial maintenance.
Electricity, Electronics and Telecommunications: Renewable energy, Energy and Power Systems, Digital Signal Processing, Wireless networks, Mobile Communications, Bioengineering.
Construction Summit: Modern Construction Systems, Computational Modeling, Structural analysis.

For Expotecnología 2019, almost 200 participants attended, representing 21 national and international universities, including Italy, Cuba, Ecuador, Costa Rica, Mexico, Puerto Rico and the United States. 109 oral presentations took place on this occasion, of which 73 articles were selected for this volume of IOP Conference Series: Materials Science and Engineering and, and 34 articles for the book *Desafíos en Ingeniería: Investigación Aplicada* https://expotecnologia.unitecnar.edu.co/expo2019.html, after a peer review process. Three international speakers and three national speakers were invited.

Invited speakers
Dra. Mayra Rivera Anaya, Instituto Politécnico Nacional (UPN) (México).
Dr. Jorge Alejandro Miranda Ossandón, Universidad Católica de Temuco (UCT) (Chile).
Dra. Inés Analía Arriagada Puschel, Universidad Católica de Temuco (UCT) (Chile).
Dr. Osvaldo Novoa Arroyo, Head of Power Plant - ENEL – EMGESA (Colombia).
Dra. Mónica López Sánchez, Gerente Académica Renata (Colombia).
Dr. Ronald Gutiérrez Ríos, Universidad de la Sabana (Colombia).

Organizing Committee

President
Libis Valdez Cervantes, M.Sc, Fundación Universitaria Antonio de Arévalo

IOP Conf. Series: Materials Science and Engineering **844** (2020) 011001 doi:10.1088/1757-899X/844/1/011001

Vice-President
Mayerlis Vásquez Osorio, M.Sc, Fundación Universitaria Antonio de Arévalo

Coordinators
Luz Ahumada Torres, M.Sc, Fundación Universitaria Antonio de Arévalo
Yesid Tarriba Lezama, M.Sc, Fundación Universitaria Antonio de Arévalo
Alexander Catellon Arenas, M.Sc., Fundación Universitaria Antonio de Arévalo

Scientific Committee

President
John R Castro Suarez, Ph.D, Universidad Del Sinú, Seccional Cartagena.

Coordinator
Luis Rojas Martínez, M.Sc, Fundación Universitaria Antonio de Arévalo

Finally, we want to thank all the authors for submitting their research papers to Expotecnología 2019. We express our gratitude to all the peer reviewers for their time and dedication in reviewing the articles assigned to be published in IOP and Book chapters.

John R. Castro Suarez, Ph.D
Editor of the Proceedings of EXPOTECNOLOGIA 2019

IOP Conf. Series: Materials Science and Engineering **844** (2020) 011002 doi:10.1088/1757-899X/844/1/011002

Peer review statement

All papers published in this volume of *IOP Conference Series: Materials Science and Engineering* have been peer reviewed through processes administered by the proceedings Editors. Reviews were conducted by expert referees to the professional and scientific standards expected of a proceedings journal published by IOP Publishing.

Evaluating a new binder in the self-leveling mortar mixture

I Guzman [1], A F Porto [1 *], G Cervantes [2], and R I Porto[1]

[1] Industrial Engineering Department, Corporación Politécnico de la Costa Atlántica. Barranquilla, Colombia.
[2] Industrial Engineering Department, Instituto Tecnológico de Pachuca. Pachuca de Soto, México

*Email: aportos@pca.du.co

Abstract. The infrastructure of a country plays an important role in economic growth, contributing as a basis for other actors such as: production, transport and access to services that meet basic population needs. In this article we study the Sika-Colombia company dedicated to the construction material supplies, specifically self-leveling mortars, in relation to an improvement of physical properties (i.e. resistance and consistency) and production costs. The impact of new binder addition in a self-leveling mortar mixture Sika Grout 212 is analyzed experimentally. This approach is justified by the need to improve the granular skeleton of the study product, recover fluidity and maintain the resistance offered to customers. The research results conclude the physical and economic advantages of new binder adding to the mortar mixture as well as optimum composition.

1. Introduction

The infrastructure of a country plays an important role in economic growth, contributing as a basis for other actors such as: production, transport and access to services that meet basic population needs [1]. Some papers mention the relationship between infrastructure development with the economic growth of a country, therefore, a direct relationship is created between infrastructure quality and the country's gross domestic product, considering that they are reinforced. [1, 2].

In response to the relationship between infrastructure and economic development, has been to achieve a consensus among the major Latin American international economic organizations such as the Inter-American Development Bank (IDB), the Development Bank of Latin America (CAF) and the Economic Commission for Latin America and the Caribbean (ECLAC), to study and promote the improvement and infrastructure investment in Latin America and the Caribbean, through the provision and quality of official data about this topics. It is then that INFRALATAM was created, a platform in which data is shared for the general public, which allows "to orient the allocation of public resources, setting priorities, schedule projects, induce private participation and to allow progress monitoring of provision in infrastructure in Latin America and the Caribbean countries" [3].

The main data offered by INFRALATAM are on participation in investment (public and private) in the sectors like water, energy, telecommunications, transportation and other subsectors, for each one of 19 Latin American countries considered. In latest report (2008-2015) highlights that annual investment average in economic infrastructure as a percentage of GDP of the countries analyzed is 3.54% [3].

In this topic, Colombia does not stand out, in fact, your average investment is 3.96%, which, although it is above the Latin American average, is still distant from countries like Bolivia (5.79%), Panamá (5.32%), Honduras (4.9%), Paraguay (4.5%) and Costa Rica (4.14%). Infrastructure investment in Latin America is growing. In the last 10 years, it has been possible to demonstrate an investment around 6.2% of gross domestic product (GDP) per annum, however, to meet the needs of the sector, this investment should be more ambitious and receive a stronger support from private sector. As proof of above, its possible see reflected that by mid-2015 an investment of 1.5% of entries came from public sector, while the remaining 0.8% from private sector [1]. Figure 1 shows the behavior of GDP growth and construction of civil works.

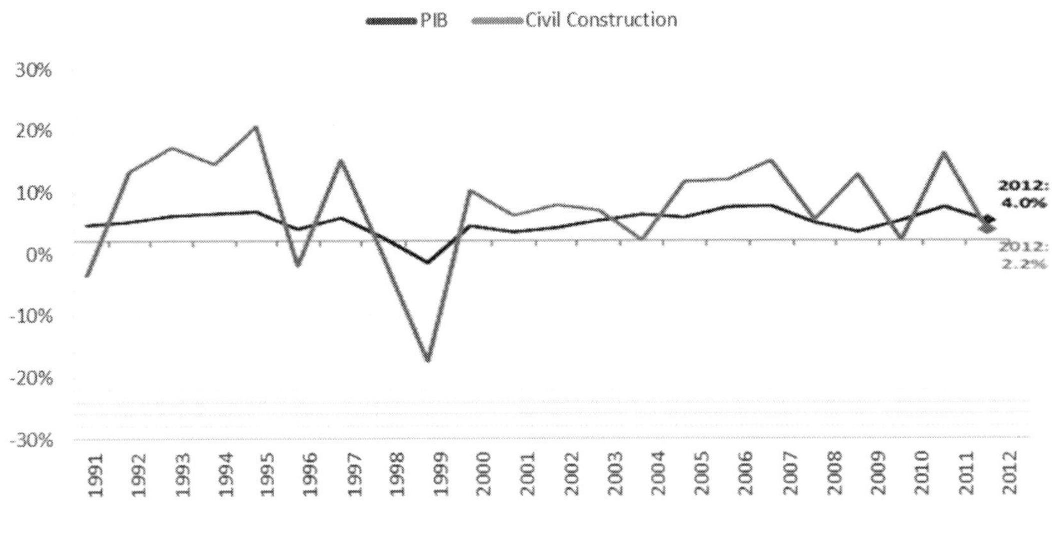

Figure 1. Economic indicator in Colombia of GDP and Civil construction. Serfínco DANE calculations (2013)

Figure 1 highlights the proportional trend relationship between annual growth rates in civil works construction respect to the GDP growth in Colombia.

When talking about infrastructure, reference is usually made to set of engineering structures and facilities, created from the techniques, methods, tools and materials that propitiate it [4]. Within this materials group, the most indispensable is Cement, which acts as a binder par excellence for the union bricks. According to the use of the structure and needs, cement is often combined with other elements such as water, sand, gravel and additives, to forming mixtures as grout, mortar and concrete [5].

Mortar is the main mixture with which bricks, stones or concrete adhere; and also for lifting and filling spaces of structures in civil works. The fluid mortar or "grout" is a material with properties, as the name implies, of fluidity, for self-leveling purposes. It is mainly used to fill bases, anchors and small spaces. The grout is able to perform a uniform and correct transmission of the efforts between supported structures and the foundation or anchors, making the structure more robust [6].

This paper presents an overview about importance and current state of infrastructure in Colombia, in addition to economic growth. Sika company case is presented, is a company dedicated to the manufacture and marketing of materials for civil construction, and the relationship with improvements in developments of mortars, aiming at better performance in characteristics of fluidity and resistance to compression. This approach is justified from the need to improve the mortar self-leveling Sika Grout 212 granular skeleton, regain fluency and maintain the resistance offered to customers stipulated in the data sheet, therefore, it is important to carry out studies and tests, with new materials with better granulometric characteristics, that make the mixtures more robust [6].

2. The problem

In 2014, Sika Colombia in its factory located in Barranquilla began manufacturing one of its most emblematic mortars in the construction sector, seeking to be the ideal support for engineering projects in all country. The following information is specified in the Sika Grout 212 mortar product data sheet:

> [...] It is a self-leveling ready for use in anchor fillings and leveling works. It has been specially designed to obtain the appropriate consistency and penetration for each type of application, it maintains its volumetric stability vertically, developing and reaching high initial and final resistances. (p. 1).

Around 2016, applicators on site or customers show that the Sika Grout 212 self-leveling had plastic consistencies, as it had to present fluid consistencies, taking into account the need to fill in confined areas. This problem has been evaluated by trained and trained R&D (Research and Development) personnel, concluding that this novelty is due to the granular composition of the raw materials or to the incompatibility of the cement with the other binders that are in the mixture. The main reason that promote the manufacturing of this mortar in Barranquilla is due to strategic geographical position and export possibilities in Central America, but this inconsistency in the product has led the company to lose the possibility of exporting this product to customers in other countries.

Currently, at a commercial level, there is a low demand for this product in the Caribbean region of Colombia. Among the 43 references of products manufactured by the Barranquilla factory, the participation of Sika Grout 212 was 15% in the periods 2014-2015 for months considered as good utility (e.g., October), while in 2017, in the same month, this mortar contributed only 0.45% of total manufactured units.

2.1 Sika grout 212 characteristics.

Sika Grout 212 self-leveling mortar is composed of Sands (coarse and fine), Cement and Pre-mix (i.e., Additive). Therefore, product improvements should be aimed at the reformulation of the mixture or the addition of a new compound that allows to reach the standard of properties and characteristics required by customers. In previous investigations within the company it was possible to verify that pre-mixing and cement were significant factors in the characteristics of consistency and resistance. Hence, the sands group are the main compounds to be reformulated (i.e., Change proportions) or partially replaced by a new compound.

Based on the unprofitable behavior that this product has had in the market due to low performance, it is vital that the company visualize strategies that regain the trust of customers, keep loyalty of the

Construction market and increase the competitiveness of the company's facing the foreign market. Therefore, the impact of new binder addition in the Sika Grout 212 self-leveling mortar mix is analyzed experimentally. Subsequently, the proportions of the compounds in the mortar mixture are determined. After all this, the economic benefit of using the new binder is estimated, showing the possible advantages of applicability and innovation of the binder.

3. Literature review

A detailed analysis was carried out, initially considering a background from different theoretical perspectives, and then recent mortars researches, which constitute one of the most frequent and important elements in the execution of masonry and construction works. For aforementioned, the materials used and the results that have been obtained from the different mixtures were taken into account for the analyzes.

Manuel Perez [7], Austral University of Chile, conducted a paper entitled "Characterization of mortars with addition of combinations of micro-silica and nano-silica". Although the goal of his work was to analyze the behavior of these materials, applied separately in the mortar together with concrete, they analyzed the results to mixed use. This research serves as the basis for analyzing the behavior of our self-leveling mortar Sika Grout 212 with the new binder.

Similarly, José Vásquez Romero [6] performed a job (applied Sika Madrid, Spain) that aimed to analyze the possibilities of optimizing particle size cementitious mortars by applying concepts of interference particles and the addition of filler in order to improve the physical and mechanical properties of these. It was focused on the analysis of commercial self-leveling mortars for floors, and was interesting for the development of this research because the changes in volume, density, fluidity and viscosity could be analyzed in detail due to addition of specific binder.

Tomás Escribano and Alba García [8] developed a working list showing the development of what has been the self-leveling mortar since its first appearance in the 70s, and how their leveling properties with little effort, using little staff and less time employment has led to self-leveling mortars to be based construction processes, positioning them as a technique that increases levels of productivity. The practical application consisted of making use of three types of additives: an aerating one, a setting retarder, and a cracking controller.

Petrement [9] has tested mortars replacing the sand by waste polyurethane foam. The physical and chemical characterization of mortars made shows that the mixtures obtained have properties that can make suitable for use in construction.

Janneth Torres, Ruby and Constanza Gutiérrez [10] presented results of performance evaluation of mortars added with a metakaolin (known by the acronym MK) of Colombian origin. The study showed that resistance of cement is higher in the presence of MK and that this effect is more noticeable by increasing the proportion of this substance in mortar. In another investigation by Bernal, S. Esguerra et al. [11] analyzed the mechanical properties of geopolymeric mortars based on Colombian steel slag (GBFS) and a high purity metakaolin, reinforced with carbon fibers (CF). Showing that the production of mortars with high mechanical resistance is possible

Puertas and Vázquez [12] in his work entitled "Behavior mortars alkaline cements reinforced with polypropylene fibers and acrylic" found that polypropylene fibers increases the impact resistance greater extent than acrylic fibers.

Furthermore, Izquierdo, Mejía de Gutierrez, and Torres Agredo [13] analyzed the deterioration of mortars containing catalyst residue catalytic cracking (FCC). Highlighting the increased chemical resistance of mortars with addition of this material.

3.1 Taxonomic summary

Table 1 shows a taxonomic classification of the previous works, which have incorporated and evaluated the granular composition of mortars indications for each of the following taxonomic characteristics are listed:

1. *Approach:* Characteristic in which the work is oriented to achieve the objectives. It is divided into Physical (F), which considers those physical and mechanical properties as resistance, fluidity, among others; Economic (E), which considers minimizing the cost of mortar production. Finally, it is also possible to consider both characteristics at the same time.
2. *Material type:* Rank the materials according to their properties and atomic structure. They are ceramic type (C) and polymers (PO).
3. *Addition:* The material added to the mortar mixtures is specified to achieve fluidity and mechanical strength. Here are considered: Silica (S), Limestone Filler (LF), Polyurethane (PLI), Aerating (A), Setting Retarder (SR), Cracking Controller (CF), Metakaolin (MK), Iron Slag (IS), Polypropylene (PR), Acrylic (AC) and Catalytic Cracking Catalyst (FCC)
4. *Consistency/Fluency:* Basic characteristic for study of mortars, related to the manageability of the mixture. It is presented in a Dry (D), Plastic (PL) and Fluid (FL) way.
5. *Resistance:* Ability to withstand and transmit the loads to which it is subjected mortar according technical standard conditions.
   5.1 *Pressure:* Maximum effort that a material can withstand under a crushing load. It is reported in megapascals (mpa).
   5.2 *Time:* Time period required to measure the resistance of material.
6. *Setting time:* Time period which a mortar has sufficient handleability to be used without further addition of water, in order to counteract hardening effects.
7. *Standard:* For measuring workability of the mortar flow test described in standard NTC 111 in Colombia used. The resistances are obtained from to the test results according to standard conditions UNE-EN 1015-11 [14]. Similarly, ICONTEC [14] suggest that the most widespread international standard is ASTM-270, where the setting times of a mortar are exhibited.

All of the above can be summarized as a qualitative explanation of taxonomy that will be presented in Table 1.

Table 1. Taxonomic table of literature review

Reference	Main goal	Focus	Material type	Material	Consistency/Fluency			Resistance		Setting Time	Standard
					D	PL	FL	Pressure	Time		
Puertas, F., Amat, T., & Vázquez, T. (2000)	Study the behavior of alkaline cement grout reinforced with fibers of different nature (acrylic and polypropylene).	F	PO	PR-AC	x			50 mpa	28 days	48 hours	UNE
Manuel Heraldo Pérez Bahamonde. Valdivia (2008)	Characterize the mechanical and physical properties of a mortar treated with additions of nanosilica, microsilica and combinations of both.	F-E	C	S	x	X	x	3.84 mpa	28 days	24 hours	ASTM -NTC-UNE
Torres Agredo, Janneth; Mejía de Gutiérrez, Ruby; Gutiérrez, Constanza (2008)	Present the results obtained from the performance evaluation in the presence of sodium sulfate from mortars added with MK produced from a kaolin of Colombian origin	F	C	MK		X		47 mpa	14 days	24 hours	ASTM
Bernal, S. Esguerra, J., Galindo, J., de Gutiérrez, R. M., Rodríguez, E., Gordillo, M., & Delvasto, S. (2009)	Present the results on the mechanical properties of geopolymeric mortars based on a Colombian steel slag and a high purity metakaolin, reinforced with carbon fibers	F	PO-C	MK-IS			x	40 mpa	28 days	24 hours	ASTM
José Vázquez Romero Madrid (2012)	Granulometric optimization of cement-based grouts through the application of concepts: particle interference, the addition of filler.	F	C	LF	x			46.9 mpa	28 days	24 hours	ASTM -NTC-UNE
Carlos Junco Petrement. Burgos (2012)	Obtain cement mortars with partial or total sand substitution by rigid polyurethane foam residues, in order to achieve viable products that improve the management of these wastes.	F-E	PO	PLI		x		23.6mpa	28 days	24 hours	ASTM -NTC-UNE
Escribano, y García (2014)	High conductivity grout	F	PO	A- SR-CF		x		5 mpa	8 days	24 hours	UNE
Izquierdo, S., Diaz, J., Mejía de Gutiérrez, R., & Torres Agredo, J. 2016	Analyze the displacement of grouts added with 10 and 20% of a catalytic cracking catalyst (FCC) residue	F	PO	FCC			x	56 mpa	60 days	24 hours	ASTM -NTC-UNE

From different materials used in mixtures analyzed, such as polymers and ceramics, it is possible to reach the normative values stipulated by NTC, ASTM and UNE as the case may be. The most frequent targets in the analyzed studies are aimed at physical focus, i.e., consider properties such as fluidity, resistance, among others.

4. Materials and methods

This section defines the stages of the methodology for the design of an experiment, which achieves the proposed physical and economic objectives.

4.1 Stage I: Selection and characterization of materials

This stage described, by a brief description, the materials used by Sika Colombia company for preparation of self-leveling mortar.

(a) *Fine sand:* Natural material mostly siliceous ratio, dry yellow, angular gradation with 85% between the meshes 50 and 70mm, classified by Sika Colombia specifications as sand 30/140, moisture absorption of 1.13% and wherein the mesh passes 200 mm does not exceed 5%.
(b) *Coarse sand:* Granular material, characteristic of large stones natural detachment in mines or by the result of a crushing process, 90% gradation between 16 and 30 mm meshes, 0.6% moisture absorption, rough texture, semi angular, resistant and where the mesh passes 200 mm does not exceed 1%.
(c) *Cement:* Binder resulting from the mixture of limestone and clay having a hardening property upon contact with water. This combination generates a chemical reaction with setting characteristics and high strength, good adhesion and cohesion. It is designed for use in mixtures, mortars and concrete.
(d) *New Compound:* Material with high calcium content. It has a purity of 90 to 94% CaCO3 clean, gradation 4 mm mesh pass 100%, textured, semi-angular with presence billed faces it helps good adhesion in mixtures.
(e) *Pre-mix:* Addition to mortars. Confidential use formula in Sika Colombia that improves the characteristics of adhesion manageability and resistance in mixtures.

4.2 Stage II: Selection and justification of experimental design

The experimental design selected is Simplex-lattice, in which k involved ingredients (factors) and each treatment is a particular combination including $(X_1, X_2, X_3, \dots, X_k)$. This design restricts that participation of ingredients (proportions) in a mixture must be greater than 0 and never greater than 1, at the same time, the sum participation of the ingredients must not exceed 1. In other words, it must be fulfilled:

$$0 <= X_i <= 1 \qquad (1)$$

$$X_1 + X_2 + X_3 + \cdots + X_k = 1 \qquad (2)$$

These restrictions cause that levels of ingredients X_i not to be independent of each other. Likewise, the increase in the participation of an ingredient in a mixture conditions or decreases the participation of other ingredients in it. For applications o more information about this type of design, it is recommended to review Gutiérrez & De la Vara [15].

4.2.1 Simplex lattice mixture design.

The restriction on experimental design selected mixtures is due to a tacit knowledge and previous trials by Sika company Laboratorians Colombia, regarding the involvement of the 5 compounds in the mixture of self-leveling mortar. Considerations about the design variables and factors are explained below:

(a) *Cement:* Fixed ratio 43% of the mixture by cost issues. Its cost is the highest among all compounds so it is not a design factor.
(b) *Coarse sand:* It has a granulometry that, in concatenation of others compounds and water, allows an increase in resistance. Its participation in the mix remains unchanged at 13%, value with which it is normally working in Sika Colombia.
(c) *Fine sand:* Participation in 41% of mixture. It presents a phenomenon called moisture absorption. ratios between 30% and 35% of the mixture were tested.
(d) *Pre-mix:* Previous experiments have yielded Sika Colombia that the optimal ratio of pre-mixture should be maintained to ensure a stable invariant resistance. To elaborate this conclusion, is possible review the paper by León [16].
(e) *New compound:* The new compound is selected by analysis and internal testing of the company Sika Colombia. Also, by financial issues and experimental intention, the proportion of new compound on total mixture is between 6% and 11%, as opposed to fine sand.

The input factors selected for the design of experiment are: fine sand and new compound, which, concatenated, represent 41% of the mixture. Table 2 shows a summary of treatments based on above restrictions.

Table 2. Percentage of self-leveling mortar compounds

% Self-leveling Mortar Compounds					
Treatment	*Cement*	*Coarse sand*	*Pre-mix*	*Fine sand (x_1)*	*New compound (x_2)*
1	43	13	3	30	11
2	43	13	3	32,5	8,5
3	43	13	3	35	6

Figure 2 shows the geometric representation of simplex lattice design restricted with two factors selected factors.

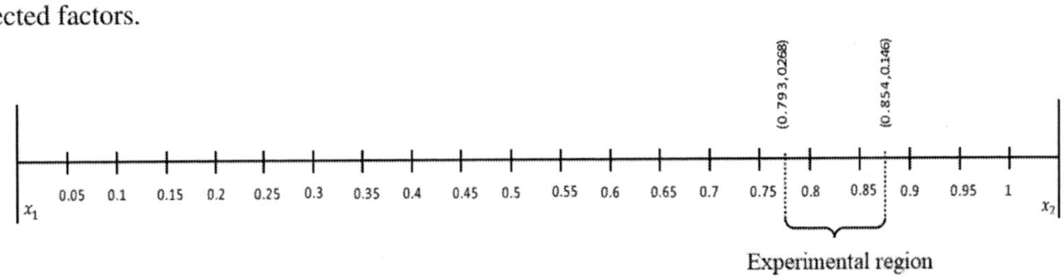

Figure 2. Geometric representation of mixtures experimental design with k = 2

4.2.2 Experimental runs.

Experimental runs based treatments described in Table 2 are performed using statistical software Statgraphics®. Three replications per treatment are considered, as some authors suggest [15]. The order of runs is completely randomized to minimize some kind of correlation between them. Table 3 shows experimental runs order.

Table 3. Experimental runs order

% Self-leveling Mortar Compounds						
Order	*Treatment*	*Cement*	*Coarse sand*	*Pre-mix*	*Fine sand (x_1)*	*New compound (x_2)*
1	2	43	13	3	32,5	8,5
2	1	43	13	3	30	11
3	3	43	13	3	35	6
4	1	43	13	3	30	11
5	3	43	13	3	35	6
6	3	43	13	3	35	6
7	2	43	13	3	32,2	8,5
8	2	43	13	3	32,5	8,5
9	1	43	13	3	30	11

4.3 Stage III: Evidence of physical behavior of mixtures at different ages of resistance failure

This subsection procedurally presents the steps taken to implement the experimental runs.

The mixtures were made under standardized procedures Sika Colombia, for preparation of product study. According to what indicated in technical specification, a sample of 3500gr, formed by the mixture proposal (i.e., fine sand, cement, coarse sand premix, new compound) it is taken. It is mixed in a water / product ratio at 0.17%. With the homogenized and fresh mixture state, 1000gr are taken, which a consistency analysis (i.e. fluidity) is performed, ensuring it complies with technical stipulations sheet (fluid consistency). With the remaining 2500gr, specimens are prepared in bronze cubes 50x50 mm in two layers with 54 strokes each, and finally each mold is obtained.

To measure resistance in the cured state at ages 24 hours, 14 days and 28 days, a universal device is used in compression (6 kg scale with accuracy of 0.01). Waiting for every age reached, the cubes are removed from the curing water. Subsequently, they are taken to test machine where it is sought to place the piston on the more uniform face of the hub. The press of test machine is activated and the device display shows data in Kg. / Cm2 at failure time.

5. Analysis of result

In this subsection, an analysis is performed in three stages of results. Initially in subsection 5.1. the results of experimental runs performed are presented, under the conditions described in subsection 4.2.2. Subsection 5.2. presents an analysis where the impact of the factors studied (in terms of consistency) is identified. In subsection 5.3. It is represented by a mathematical model the relationship between these factors, deriving an optimal latter operating condition. Finally, in subsection 5.4. shows a comparative analysis in terms of costs of the mixtures do not consider or the new compound.

5.1. Experimental results

The results presented in this subsection correspond to the values of consistency (Kg / cm2) of each of the treatments tested. Table 4 shows the results of the treatments at 24 hours, 14 days and 28 days.

Tabla 4. Results of experimental runs

Execution order	*% Fine sand (x_1)*	*% New compound (x_2)*	Consistency Kg /cm² (y)		
			24 hrs	**14 d**	**28 d**
1	32,5	8,5	183	511	541
2	30	11	141	426	477
3	35	6	175	430	465
4	30	11	177	439	478
5	35	6	176	433	488
6	35	6	198	432	470
7	32,2	8,5	169	523	551
8	32,5	8,5	208	513	546
9	30	11	181	429	465
		Mean	178,7	459,6	497,9
		Standard deviation	18,7	42,4	36,9

The average of data proves that the mean value of the fluid consistency (kg / cm2) for both ages 24 hours, 14 days and 28 days exceeded the lower limit of the sheet according to the resistance Sika compression Colombia (i.e., greater than 150, 340 and 430 respectively. See Annex 1). Only one experimental run at 24 hours of age yields a better value than allowed lower limit (141 kg / cm2 <150 kg / cm2). This usually happens by outside variables in the formulation and which has a certain permittivity because it failed the first age. The minimum values corresponding to ages of 14 and 28 days exceed the lower limit of compressive strength by a considerable margin.

Standard deviations of treatments 1 and 2 are similar. However, treatment 3 shows inconstant, especially 14 days. Taking this as a decisive factor, this treatment showed less variability to 28 days treatment was 2 standard deviations for treatment and age are presented in Table 5.

Table 5. Standard deviation by treatment and age of the self-leveling mortar

	Standard deviation		
Traetment	*24 hrs*	*14 d*	*28 d*
1	22,030	6,8068	7,2341
2	19,7565	6,42910	5
3	13	1,52752	12,0968

5.2 Analysis of Variance (ANOVA)

The tool par excellence for the analysis of experiments is the Analysis of Variance (ANOVA). The analysis stage is carried out almost entirely using the statistical software Statgraphics®. It should be mentioned that the data entered in Statgraphics® software proportions correspond to values between 0 and 1, not percentages.

A preliminary analysis shows the results of adjusting different data models resistance. The mean model is formed solely by constant. The linear model is first order terms for each of the components (i.e., fine sand for X_1 and X_2 New compound). The quadratic model adds cross between pairs of component products (i.e., their interrelationships). Each model is shown with a P-value, which tests whether that model is statistically significant when compared to the mean square of the term below. Typically, a model with a P value less than 0.05 is selected, assuming that one works at a confidence level of 95%. The standard error statistics of estimates and the R-square have been tabulated for each models. Some analysts prefer to select the model that maximizes R-Square adjusted.

(a) *ANOVA at 24 hours:* According to the P-value in Table 6, no model is suitable for data. Table 7 indicates the standard error statistics for the R-square fitting, the model that best describes the relationship between the factors studied is linear.

Table 6. Estimated effects of the full model for 24-hour resistance

Source	Sum of Squares	DF	Mean Square	F-Value	P-Value
Mean	287296,	1	287296,		
Linear	413,358	1	413,358	1,22	0,3067
Square	291,309	1	291,309	0,84	0,3957
Error	2089,33	6	348,222		
Total	290090,	9			

Table 7. Standard error of the 24-hour Full Model

Model	ES	R-Square	R-Square Ajd.
Linear	18,4416	14,79	2,62
Square	18,6607	25,22	0,29

(b) *ANOVA at 14 days:* According to the P-value in Table 8, the quadratic model is suitable for the data, as the result yielded Table 9.

Table 8. Estimated effects of the full model for 14-days Resistance

Source	Sum of Squares	DF	Mean Square	F-Value	P-Value
Mean	1.90E+11	1	1.90E+11		
Linear	0,0255036	1	0,0255036	0,00	0,9973
Square	14168,2	1	14168,2	472,27	0,0000
Error	180,0	6	30,0		
Total	1.92E+11	9			

Table 9. Results of the Complete Model at 14 days

Model	ES	R-Square	R-Square Ajd.
Linear	45,27	0,00	0,00
Square	5,477	98,75	98,33

(c) *ANOVA at 28 days:* The results in Table 10 and 11 allow further strengthen the conclusions generated in the analysis to 14 days. There is a significant relationship between the analyzed factors of quadratic type.

Table 10. Estimated effects of full model for 28-days Resistance

Source	Sum of Squares	DF	Mean Square	F-Value	P-Value
Mean	2.23E+11	1	2.23E+11		
Linear	0,544294	1	0,544294	0,00	0,9856
Square	10417,0	1	10417,0	139,72	0,0000
Error	447,333	6	745,556		
Total	2.24E+11	9			

Table 11. Results of Complete Model at 14 days

Model	ES	R-Square	R-Square Ajd.
Linear	393,961	0,01	0,00
Square	863,456	95,88	94,51

In general, the behavior to 24 hours was inconclusive and, somehow, is a result of observed variability between data in the descriptive analysis. However, the behavior of variability 14 and 28 days remained constant, as was seen descriptively also giving the conclusion fit a quadratic model.

5.3 Mathematical model to 28 days

Table 12. Results of Resistance Model Adjustment

Parameter	Estimate	Standard	T	P-Value
X_1: FINE SAND	474,3	4,985		
X_2: NEW COMPOUND	473,3	4,98		
X_1X_2	288,66	24,42	118,198	0,000

R-square = 95,88 %
R-Square (adjust g.l.) = 94,51 %
Standard Error = 8,63

Mean Absolute Error = 6,0
Statistical Durbin-Watson = 1,87 (P=0,4298)
Residual Autocorrelation Lag 1 = -0,11

The adjusted mathematical model that relates the new compound and fine sand:

$$Resistance = 474{,}3*X_1 + 473{,}3*X_2 + 288{,}66*X_1*X_2 \quad (3)$$

Table 13. Estimated results for Resistance

	Observed	Adjusted	Lower 95,0%	Upper 95,0%
File	Values (Y_i)	Values ($\hat{Y}_i$)	Mean	Mean
1	465,0	474,333	462,135	486,532
2	477,0	473,333	461,135	485,532
3	541,0	546,0	533,802	558,198
4	488,0	474,333	462,135	486,532
5	551,0	546,0	533,802	558,198
6	478,0	473,333	461,135	485,532
7	470,0	474,333	462,135	486,532
8	546,0	546,0	533,802	558,198
9	465,0	473,333	461,135	485,532

5.3.1 Testing statistical assumptions (Residual analysis)

Residual analysis of mathematical model may ensure suitability for estimation and representation of relationships between variables. Most cases can be validated analyzing residues ($e_i = Y_i - \hat{Y}_i$). These assumptions represent the distance of each observation to the fitted line model. To expand the issue of statistical validation of assumptions, it is recommended to review Montgomery [17]. Basically suitability assumptions can be validated in two ways: analytically and graphically.

i. *Normality:* Graphically it validates that the values follow a standard straight (i.e., From the probability distribution Normal) residual. This adjustment is measured in relation to the deviations of line, as evidenced in Figure 3, where the points are close to it. Analytically it could be validated by the Shapiro-Wilk test or the Kolmogorov-Smirnov test.

ii. *Homoscedasticity:* It assumes that, for each value of an independent variable, the variance of the residuals is constant. To validate this assumption, the residuals versus predicted values graph is used, no trend cloud of points should be presented, as can be seen in Figure 4. In an analytical way, Breusch-Pagan test could also be performed, if the data fit a normal distribution, or the Brown-Forsythe test, if they do not comply.

iii. *Independence:* This validates that there is no dependency between errors, that involve an error in taking measurements leading to a trend. Graphically, the residual dispersion, according to observation sequence is analyzed. In Figure 5 it can be seen that there is a marked tendency for the residual values for the adjusted model. Analytically it is possible to apply Durbin-Watson test.

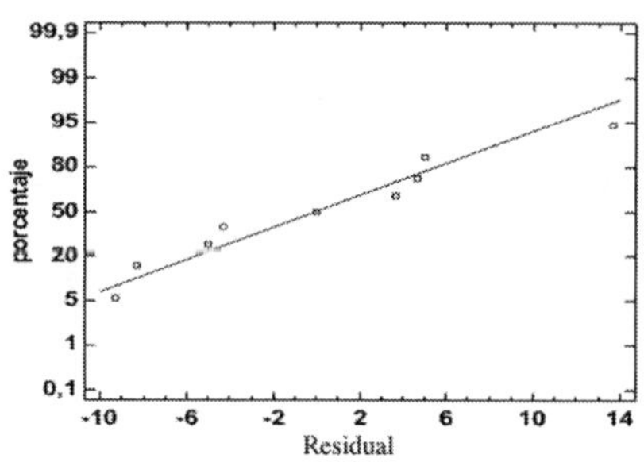

Figure 3. Probability graph normal-residual

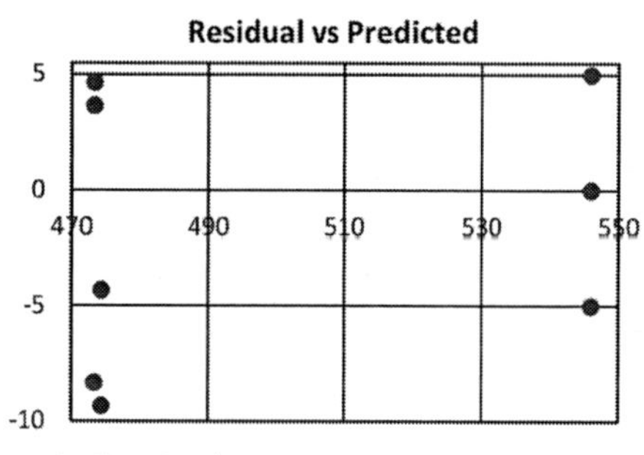

Figure 4. Graph of residuals vs. predicted values.

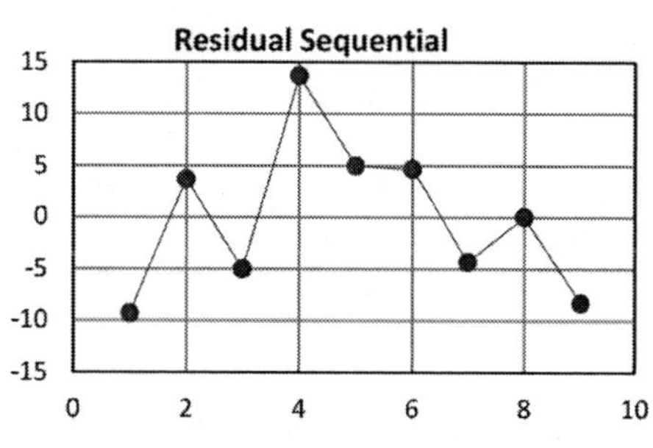

Figure 5. Graph of residual sequential

5.3.2 Optimum operating condition.

It is also important to know the kind of relationship between design factors, including conjecture that maximizes the response variable. Based on the proposed quadratic model (see equation 3) and using the statistical software Statgraphics®, it aims to maximize strength, resulting in an optimal value equal to 546.001 kg / cm2.

Table 14 shows the optimization resolved by Statgraphics® based on the proposed model. The optimum operating condition is .7932 .2067 fine sand and New compound. In terms of percentage it would be 32.5% and 8.5% respectively, which is equal to 2 experimental treatments.

Table 14. Optimum operating condition.

Component	*Low*	*High*	*Optimal*
Fine grit	0,7	0,8	0,793
New compound	0,147	0,267	0,206

5.4 Economic analysis of proposals

The proposed total of 12 samples prepared in the laboratory, either using the control specifications (QA / EPT 1221 edition 4 and AC / ME / Ed code 05.19.05), are compared with the results expressed in the data sheet version 13 / 02/2013 Sika Grout 212.

Table 15 breaks down the current formulation of Sika Colombia for manufacturing the self-leveling mortar, indicating the total cost per unit (Lump 30Kg). Furthermore, Table 16 shows the resistance in each of the different blocks of setting time

Table 15. Current formulation of mortar manufacturing.

Resourse	Consumption per Unit		Total cost per kilo	Total cost per Unit
Bag (Package)	1.00	Unit		833.00
Gray Cement	12.90	kg	402.94	5,197.93
Gross grit 8/30	3.90	kg	150.00	585.00
Fine grit	12.30	kg	75.00	922.50
PM	0.90	kg	2,700.00	2,430.00
Operator 1 - Product manufacturing	-	Hours		340.97
Operator 1 - Product packaging	-	Hours		340.97
Operator 2 - Product manufacturing	-	Hours		62.78
Operator 2 - Product packaging	-	Hours		62.78
Automatic machine operations (silos filling, dosing, mixing)	-	Hours		183.51
Laboratory (quality control, sample analysis)	-	Hours		19.76
Package 30kg Grout 212	**1.00**	**Bag**		**10,979.20**

Table 16. Compressive strength results of the original mortar formulation

Resistance 24 hrs (Kg\cm²)	Resistance 14 d (Kg\cm²)	Resistance 28 d (Kg\cm²)
194	547	571

The value of proportions of the compounds to the mortar mix and resistance in each time block for each setting proposal presented in Tables 17, 19 and 21. New compound is used at 2.55, 1.80 and 3.30 kg, these represent 8.5, 6 and 11% of total sample respectively. In Tables 18, 20 and 22 is added to the formulation New compound manufacturing self-leveling mortar, in order to meet the total unit cost.

(a) Proposal 1

Table 17. Proportions of mortart mix compounds according to proposal 1

Fine grit (%)	New Compound(%)	Resistance 24 hrs (Kg\cm²)	Resistance 14 d (Kg\cm²)	Resistance 28 d (Kg\cm²)
32.5	8.5	183	523	541
32.5	8.5	169	511	551
32.5	8.5	208	513	546

Table 18. Manufacturing including the new compound at 2.55 kg.

Resourse	Consumption per Unit		Total cost per kilo	Total cost per Unit
Bag (Package)	1.00	Unit		833.00
Gray Cement	12.90	kg	402.94	5,197.93
Gross grit 8/30	3.90	kg	150.00	585.00
Fine grit	9.75	kg	75.00	731.25
PM	0.90	kg	2700.00	2,430.00
New Compound	**2.55**	**kg**	**60.00**	**153**
Operator 1 - Product manufacturing	-	Hours		340.97
Operator 1 - Product packaging	-	Hours		340.97
Operator 2 - Product manufacturing	-	Hours		62.78
Operator 2 - Product packaging	-	Hours		62.78
Automatic machine operations (silos filling, dosing, mixing)	-	Hours		183.51
Laboratory (quality control, sample analysis)	-	Hours		19.76
Package 30kg Grout 212	**1.00**	**Bag**		**10,940.35**

(b) Proposal 2

Table 19. Proportions of mortar mix compounds according to proposal 2

Fine grit (%)	New Compound(%)	Resistance 24 hrs (Kg\cm²)	Resistance 14 d (Kg\cm²)	Resistance 28 d (Kg\cm²)
35	6	175	430	465
35	6	176	433	488
35	6	198	432	470

It is clear that proposal 3 does not meet the minimum specifications in one of its replicas. On the other hand, the proposal 1 present results that exceed the minimum values necessary for mortar studied (see Annex 1), variability between their results is a product unwanted feature. Finally, the proposed 2 (optimum condition found using Statgraphics®) present results with less variability among themselves and with values that meet product requirements.

In economic terms, the comparison of production costs shows that: a greater amount of new bond in the mix, the cost increases proportionally. Consistent with this proposal 3, with more of new compound, is the most economically attractive without considering the physical requirements necessary section in the product.

Table 20. Manufacturing including the new compound at 1.80 kg.

Resourse	Consumption per Unit		Total cost per kilo	Total cost per Unit
Bag (Package)	1.00	Unit		833.00
Gray Cement	12.90	kg	402.94	5,197.93
Gross grit 8/30	3.90	kg	150.00	585.00
Fine grit	10.50	kg	75.00	787.50
PM	0.90	kg	2700.00	2,430.00
New Compound	**1.80**	**kg**	**60.00**	**108.00**
Operator 1 - Product manufacturing	-	Hours		340.97
Operator 1 - Product packaging	-	Hours		340.97
Operator 2 - Product manufacturing	-	Hours		62.78
Operator 2 - Product packaging	-	Hours		62.78
Automatic machine operations (silos filling, dosing, mixing)	-	Hours		183.51
Laboratory (quality control, sample analysis)	-	Hours		19.76
Package 30kg Grout 212	**1.00**	**Bag**		**10,952.20**

(c) Proposal 3

Table 21. Proportions of mortar mix compounds according to proposal 3

Fine grit (%)	New Compound(%)	Resistance 24 hrs (Kg\cm²)	Resistance 14 d (Kg\cm²)	Resistance 28 d (Kg\cm²)
30	11	141	426	477
30	11	177	439	478
30	11	181	429	465

Table 22. Manufacturing including the new compound at 3.30 kg.

Resourse	Consumption per Unit		Total cost per kilo	Total cost per Unit
Bag (Package)	1.00	Unit		833.00
Gray Cement	12.90	kg	402.94	5,197.93
Gross grit 8/30	3.90	kg	150.00	585.00
Fine grit	9	kg	75.00	675.00
PM	0.90	kg	2700.00	2,430.00
New Compound	**3.30**	**kg**	**60.00**	**198.00**
Operator 1 - Product manufacturing	-	Hours		340.97
Operator 1 - Product packaging	-	Hours		340.97
Operator 2 - Product manufacturing	-	Hours		62.78
Operator 2 - Product packaging	-	Hours		62.78
Automatic machine operations (silos filling, dosing, mixing)	-	Hours		183.51
Laboratory (quality control, sample analysis)	-	Hours		19.76
Package 30kg Grout 212	**1.00**	**Bag**		**10,929.70**

6. Conclusions

It was assessed the impact that has the addition of a new binder maintaining the strength in self-leveling mortar Sika Grout 212. For this purpose, a restricted single lattice experimental design, low theoretical and tacit considerations of technical research and development is postulated company in which different ratios of fine sand (30% to 35%) and a new binder (6% to 11%) in the original mixture mortar tested.

The new compound maintains levels of consistency and fluidity required by self-leveling mortar Sika Grout 212. From the results obtained by the proposed experimental design, analysis of variance which shows significant effects at different ratios of fine sand is performed and New compound. The mathematical model that best fits the data (i.e., the relationship between the factors and resistance) was a square arrangement. This regression model met suitability tests on the residual analysis (i.e., Normality Homoscedasticity and Independence).

The optimum operating condition maximizing compressive strength corresponds to 32.5% of fine sand and 8.5% of New compound on the original formulation. As seen in estimating production costs, the formulation of proposal 2 (i.e., the statistically optimum) reduced by 0.24% manufacturing cost compared to the original formulation per package of 30kg. Mortar Sika Grout 212. The nominal economic margin of improvement is quite low, however, translated into an annual production (At least, 300 tons per year) it means a significant cost savings, helping to program development and innovation of company Sika Colombia.

Future research could be aimed at the implementation of experimental designs (Mixtures), based on inferential statistics to perform tests in the search for product improvements, and consider mixtures containing additions of Polymers and Ceramics at the same time. Likewise, future research could consider ecological material based on hemp, which some research assures that it improves physical properties of mortars (Collet et al., 2007; Hamzaoui et al. 2014).

7. References

[1] Sánchez, R., Lardé, J., Chauvet, P., & Jaimurzina, A. 2017 *Inversiones en infraestructura en América Latina: tendencias, brechas y oportunidades* p 187

[2] Sánchez, D. 2004 *Nuevas tendencias en la especificación y diseño de mezclas de mortero. Cartagena: Memoria técnica:Reunion del concreto p 33*

[3] Infralatam 2018 *Datos de inversión en infraestructura económica*

[4] Duarte, J., Gargiulo, C., Millán, G., Ruiz Pérez, L., Hinojosa, S. A., Moreno, M., & López, M. A. 2012 *Aprendizaje en las escuelas del siglo XXI: hacia la construcción de escuelas que promueven el aprendizaje, ofrecen seguridad y protegen el medio ambiente* p 40-42

[5] Riveral , G. 2013 *Manejabilidad del concreto* p 83-103

[6] Vásquez Romero, J. 2012 *Optimización granulométrica de morteros de base cemento* Vol 1 p 33-41

[7] Pérez, M. 2008 *Caracterización de morteros con adición de combinaciones de micro-silice y nano-silice. (*Valdivia: Universidad Austral de Chile) p 35

[8] Escribano T; García, A., 2014 *Morteros autonivelantes de elevada conductividad térmica.* (Tesis Doctoral) p 50

[9] Junco Pretement, C. 2012 *Morteros aligerados con residuos de espumas rígidas de poliuretano: obtención, caracterización y puesta en obra.* P 33

[10] Torres Agredo, J., Mejía de Gutiérrez, R., & Gutiérrez, C. 2008. *Desempeño de morteros adicionados con metacaolín frente a la acción de sulfatos Vol.* 28 p 117-122.

[11] Bernal, S., Esguerra, J., Galindo, J., de Gutiérrez, R. M., Rodríguez, E., Gordillo, M., & Delvasto, S. 2009 *Morteros geopolimericos reforzados con fibras de carbono basados en un sistema binario de un subproducto industrial* p 587-592.

[12] Puertas, F., Amat, T., & Vázquez, T. 2000 *Comportamiento de morteros de cementos*

alcalinos reforzados con fibras acrílicas y de polipropileno p 74-79
[13] Izquierdo, S., Diaz, J., Mejía de Gutiérrez, R., & Torres Agredo, J. 2016 *Durabilidad de morteros adicionados con FCC expuestos a sulfato de magnesio y sulfato de sodio* Vol.31 p 183-190
[14] ICONTEC. 1989 *Normas técnicas colombianas para el sector de la construcción*
[15] Gutiérrez, H., & De la Vara, R. 2008 *Análisis y diseño de experimentos p 42*
[16] León, G., & al, e. 2016 *Optimización de la pre-mezcla para la fabricación de un mortero auto-nivelante* p 1-8.
[17] Montgomery, D. C. 1991 *Diseño y análisis de experimentos p 18-22*

8. Annexes

In this subsection the annexes related to necessary information and cited in the document are presented.

Annex 1. Technical information of compressive strength of the self-leveling mortar without shrinkage for precision fillings and anchors

TECHNICAL INFORMATION

Compression Resistance	**Plastic consistency (kg/cm^2)**	**Semi-fluid consistency (kg/cm^2)**	**Fluid consistency (kg/cm^2)**
1 day	230	210	150
4 days	510	470	340
28 days	630	575	430

Setting time: 7 - 10 hours

Depending on the consistency of the grout and the temperature of the place this time may vary

Source: Product Data Sheet Sika Grout 212, Sika Colombia

IOP Conf. Series: Materials Science and Engineering **844** (2020) 012002 doi:10.1088/1757-899X/844/1/012002

Evaluation of triangle´s triple restriction on conventional construction and modular construction- Example: New building of the University of the Coast

S E de la Puente Molina[1], J L Espitia Ticora[1], O A Capera Rodríguez[1], Z A Palomeque Sánchez[1]

[1,] Universidad Católica de Colombia, Facultad de Ingeniería Civil

sedelapuente78@ucatolica.edu.co

Abstract. This document evaluates and compares the modular construction system against the conventional construction system, analyzing it from the triangle of the triple restriction of the PMBOK guide (5th edition), which corresponds to the analysis of the scope, time and cost that can be achieve in the execution, which is focused directly on a real building of higher education, as a case study was used the university corporation of the CUC coast located in the city of Barranquilla. The main purpose of the research is to demonstrate the strengths, advantages and disadvantages of each construction system and once the real results were obtained to be able to make the application in professional practice for the management of the works and the execution of higher education projects, in this way the new construction processes are potentialized as a contribution to the development of architecture and innovation in Colombia, which is a country that is seeking to enter new methodologies and construction systems which must be aligned with the management issues that are being applied in the construction guild.

1. Introduction

This document is structured on the basis of a research process based on the evolution of modular construction [1] and its relationship to traditional construction, this, in order to show the advantages and disadvantages that each one establishes, highlighting the physical, functional and economic characteristics, in order to promote a balance between both methods of construction.

Therefore the document is arranged based on the historical transcendence that has been transforming the way of seeing architecture and its construction systems, specifically in the evolution of industrialization processes that lead to the integration of a prefabricated modular construction [2] on a specific environment.

The objective of the research project is to know the limit of application of modular unconventional construction and how far it is possible to be implemented. In the other hand, it is necessary to know if it has the same resistance capacities as a conventional construction systems [3]; to know if it is possible with the modular construction, to reduce time and execution costs, besides verifying if the profitability margin is favorable. Finally, it will be determined if it is applicable in constructions for educational use with emphasis on higher education.

In the same way we want to show to what extent the unconventional modular construction can be counteracted or adapted in front of a conventional constructive system; in this way establish its constructive viability evaluated from the triangle of the triple restriction (scope, time and cost) [1] that is shown in the PMBOK Guide 5th Edition.

Part of the tangible scope of the document is that it has a significant contribution for project management [4] and encourages companies engaged in construction, to evaluate the possibility of applying the prefabricated modular construction as necessary and as much as possible as a tool for making decisions for the benefit of both business and those interested in this type of construction. The purpose of the intangible scope corresponds to giving value to modular unconventional construction and generating a social acceptance of the new trends in architecture so that, based on the comparative evaluation of the two types of construction, it could be possible to break the paradigm that the modular construction is inefficient.

2. Methodology

This research is a comparative and analytical type, where the advantages and disadvantages presented by conventional construction versus prefabricated modular construction were basically established, highlighting the potential that each offers for the management and application in construction projects of higher education institutions in Colombia. In addition, a qualitative and quantitative analysis of the two types of constructions is shown based on a case of study according to the triple restriction triangle. [5]

2.1. Project Development Phases

This research was carried out in four main phases, which are described below:

Phase 1: Overall evaluation based on the description and comparison of the two construction systems (Conventional and modular prefabricated).

Phase 2: Analysis of the parameters of scope, time and cost of each construction system, based on the triangle of the triple restriction of the PMBOK Guide 5th Edition in Chapter 1.

Phase 3: Determine the balance between the two constructive alternatives highlighting the one that represents greater efficiency and feasibility, establishing indicators that facilitate the management of educational projects within the companies of the infrastructure and building sector.

Phase 4: Emphasize on the construction system that represents the best processes within the stages for the Construction Management.

2.2. Tools to used

For data collection there were used several books, articles and monographs form which an analysis of information relevant to the research project was done. On the other hand, some regulations and normativity documents were applied during the research process, as well as websites associated with companies or corporations dedicated to these types of construction in Colombia.

Other important data was taken from architectural and structural 3D models that were develop for this investigation, in order to support the lists of activities, programming and costs of each type of construction, and, also to represent the different construction phases through more realistic images (renders).

3. Results and discussion

There were found that the modular system can be equated to a conventional constructive system, when evaluated from the triangle of the triple restriction (Scope, time and cost) of the PMBOK guide. Below it is possible to analyze the reasons that support the equity of both construction systems.

From the scope point of view, the modular construction system is equal to the conventional construction system, given that it has the same ability to meet technical, functional, structural and aesthetic needs, in addition to ensuring the spatial thermal comfort of the building, as shown by the project object of comparison Corporación Universidad de la Costa CUC [7], where it was proved that this modular construction process achieves the same scope of architectural compliance as the conventional process at the end of its execution as shown in figure 1.

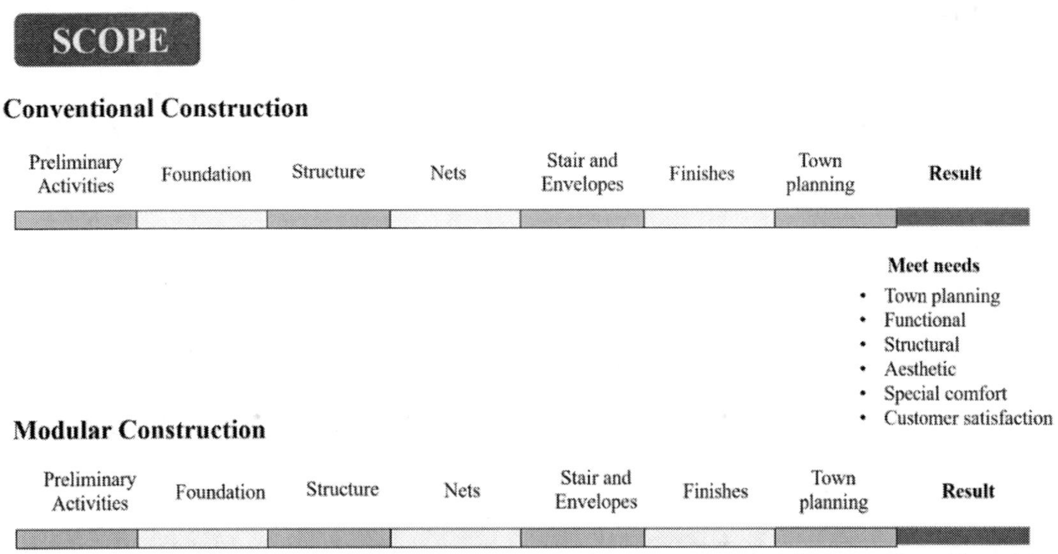

Figure 1. Comparative analysis of the scope management between conventional construction and modular construction. Own source

When analyzing the traceability of the project from the beginning to the end, it was observed that two large variables interfere with time and cost, since this is where the modular system highlights its own characteristics to show the benefits of optimization, performance and cost.

According to the results obtained, it can be guaranteed that when talking about time, the modular construction system equals and exceeds conventional construction, given that a project of this magnitude can be constructed by reducing 70% of the time compared to a conventional construction system executed under the same planned activities. As it is an industrialized and prefabricated system, it reduces on-site execution times and in turn grants certain advantages that make project planning more accurate, such as risk reduction, contingency 4reduction and resource optimization.

The building based on the modular construction, within its planning manages to execute the project under 91 days, while the conventional construction reaches 320 days, contemplating a significant difference of 229 days in favor of the modular construction system, executing the same programmed activities in a shorter time, as shown in figure 2.

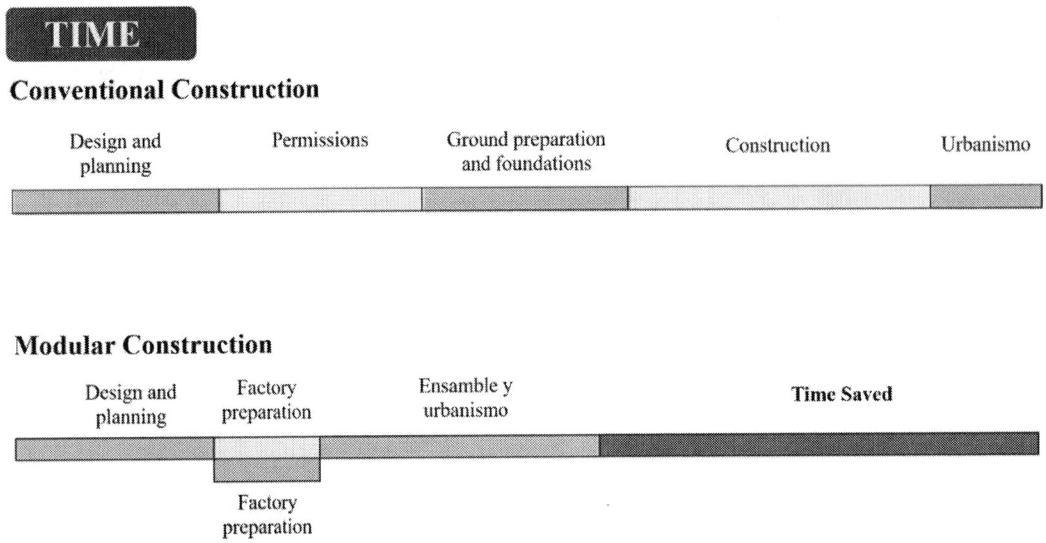

Figure 2. Comparative analysis of time management between conventional construction and modular construction. Own source

The latest comparative analysis of cost management, the modular construction system equals the conventional construction system, since the cost ratio between the same activities is clearly lower in the modular system simply because it requires fewer construction processes, less material handling and implementation and less labor recruitment compared to the number of activities and processes required by conventional construction.

A clear example of what was mentioned above, is the structural deliverable of the building which for the modular system has no greater involvement, since due to the standardization of this construction leads to minimal cost varies based on the daily change of the TRM, whereas in the conventional construction system the structural deliverable is one of the most relevant items because of the variation in costs it can have during the execution time. This can occur either by incidents of external agents such as location, delays in the work as a result of weather conditions or by delays in the work as a result of weather conditions or by delays characteristic of the materials because of their time of forging and stiffness.

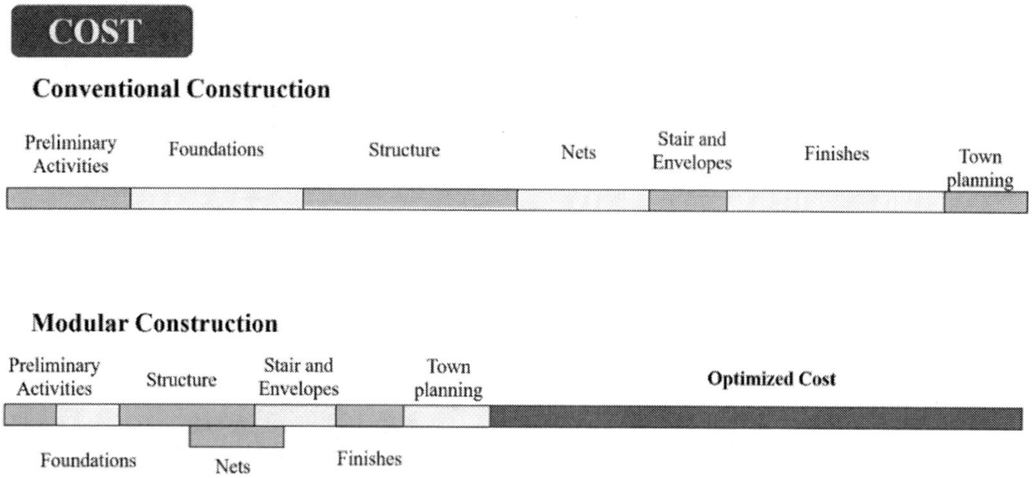

Figure 3. Comparative analysis of cost management between conventional construction and construction modular- Own source

The total value of the building based on modular construction is ($ 3,745,938,818) and the total value of the building under conventional construction is ($ 6,632,127,856), yielding a large difference corresponding to the total of ($ 2,886,189,038). Equivalent to a significant reduction of 43.51% in cost savings.

3.1. Impacts of the results

Based on the comparative analysis and the results obtained previously, it is considered that there are 3 primary impacts that directly affect the variation of time and costs, that in turn influence the total difference of each of the construction systems, showing the relevance that the modular construction can have on the conventional construction.

Based on the cost estimates and implementation schedules of the two educational buildings, three common activities of great relevance for the proper performance of the construction are identified below, considered as the differential impacts between the two systems.

3.1.1 Foundation

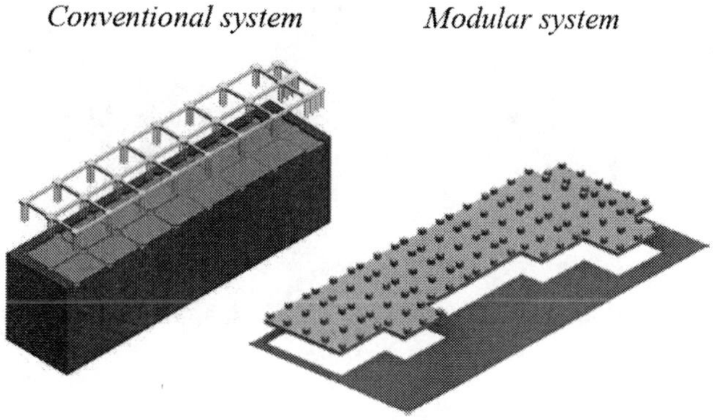

Figure 4. Conventional and modular foundation system. Own source.

Scope: Compliance	*Scope: Comply*
Time: 35 days	*Time: 23 days*
Cost: $811.904.020	*Cost: $71.925.029*

3.1.2. Structure

Conventional system *Modular system*

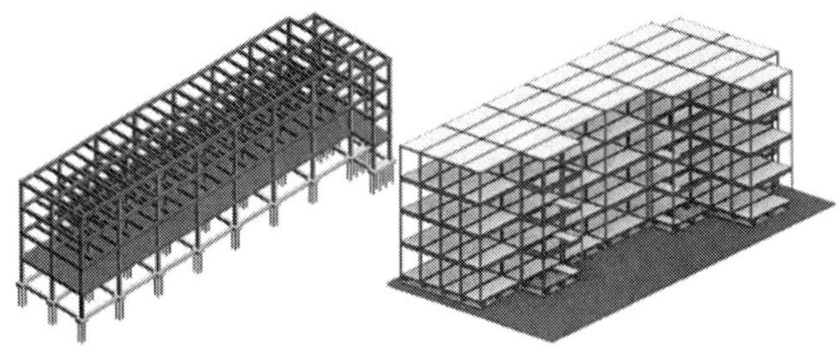

Figure 5. Conventional and modular system structure. Own source.

Scope: Compliance	*Scope: Comply*
Time: 148 days	*Time: 29 days*
Cost: $3.794.695.290	*Cost: $2.609.976.913*

3.1.3. Finishes

Conventional system *Modular system*

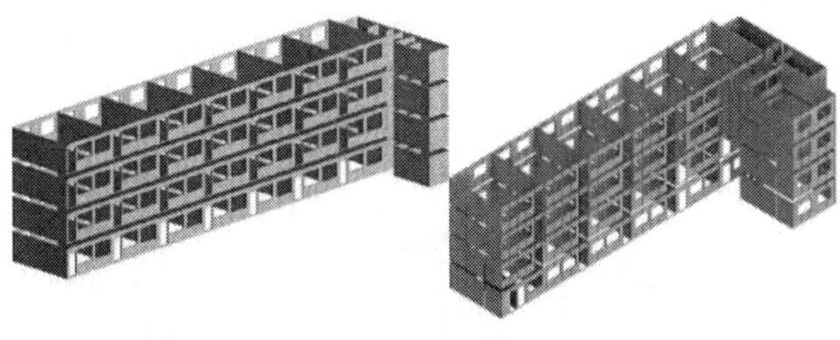

Figure 6. Conventional and modular system finishes. Own source.

Scope: Compliance	*Scope: Comply*
Time: 299 days	*Time: 30 days*
Cost: $1.338.631.000	*Cost: $114.307.614*

4. Conclusions

The conclusion of this research is the support tool for construction managers and their team, this reference will facilitate the running of construction projects specifically for higher education, since from the comparison of both methods in the same project, specific information can be obtained to measure the limitations and performance of execution of each system. In addition it is also possible to gather information about the financial traceability of the entire duration of the project that allows to estimate the cost of each of the systems, as well as determine which can become profitable and economically more attractive to the client.

In addition, with the incentive of promoting a starting point for pioneer construction companies, this document can encourage the implementation of new project construction alternatives, providing the same functional capacity, reducing execution and delivery time, obtaining more profitable and environmentally friendly projects, which for project management of any construction company is a plus of architectural and constructive innovation.

Given that Colombia is a developing country in all its aspects and more so when it comes to construction area, this is where the implementation of new construction systems can reduce traditional practices for the improvement and performance of the execution processes, as an added contribution from the construction management, the industrialization division can be boosted by the creation of companies dedicated to serial production and construction typing in order to enhance scope efficiency, time optimization and cost reduction of any type of construction project.

5. References

[1] Banco Interamericano de Desarrollo-Gestión de Proyectos- Guía de Aprendizaje Paraguay Bolivia y Salvados:2015
[2] Project Management Institute-GUÍA DE LOS FUNDAMENTOS Quinta ed Vols %1 de %2ISBN978-1-62825-009-1
[3] Universidad de la Costa CUC Universidad de la Costa avanza en Infraestructura 2015
[4] Ropero D-CONSTRUCCIÓN MODULAR DE VIVIENDAS Y ARQUITECTURA pp 77 2013
[5] Haas and Enshassi M and Walbridge S West JS -Integrated risk management framework for tolerance-based mitigation strategy decision support in modular construction projects Journal of Management in Engineering pp 35 2019
[6] López Del Puerto C and Shane-Keys to success in megaproject management in México and the united states: Case study -Claves del éxito en la gestión de Megaproyectos en México y los Estados Unidos- Un Estudio de Casos Journal of Construction Engineering and Management 2014
[7] Salas J and Oteiza -The basic construction materials industry and today's vast housing shortage-La industria de materiales básicos de construcción ante las ingentes necesidades actuales de edificación Materiales De Construccion, pp 58 2008
[8] Wong J M W and Chiang H and NgT-Construction and economic development- The case of hong kong,» Construction Management and Economics pp 813-824 2008

IOP Conf. Series: Materials Science and Engineering **844** (2020) 012003 doi:10.1088/1757-899X/844/1/012003

Evaluation of the coughing power of the *Opuntia ficus indica* for removal of turbidity in waters of the Guatapuri River (Colombian Caribbean)

C R Vidal-Tovar[1], E M Moscote-Vides[2], A J Vides-Redondo[2] and C A Severiche-Sierra[3]
[1]Grupo de Investigación Creando Ciencias CRECI, Universidad Popular del Cesar, Valledupar-Cesar, Colombia.
[2]Programa de Ingeniería Sanitaria y Ambiental, Universidad Popular del Cesar, Valledupar-Cesar, Colombia.
[3]Grupo de Investigación GITEC, Corporación Universitaria Rafael Nuñez, Cartagena de indias-Bolívar, Colombia.

carlosvidal@unicesar.edu.co

Abstract. The purpose of this article is to publicize the evaluation of the effectiveness of *Opuntia ficus indica* as a coagulant during the treatment of raw water from the middle basin of the Guatapuri River, in the Department of Cesar, Colombia. Tuna coagulant is purely organic; it has been shown by phytochemical studies performed on *Opuntia ficus indica* the presence of some minerals among some others; contains glycides or carbohydrates. The evaluation was carried out using the jug test equipment, which operated by varying the revolutions per minute (20, 30 and 40), keeping the mixture fast at 125 rpm and settling for 30 minutes. The samples were physically characterized (Turbidity, Alkalinity, pH and total suspended solids) taking into account the methods specified for parameter in the Standard Methods. With samples of 345.9 NTU and 2799.1 NTU at optimal doses of 658.75 mg/L and 1976.3 mg/L respectively, removals up to 97% and 99.5% were observed in the same order. It was shown that the variation of the slow mixture only influenced with the sample of 2799.1 NTU, being the most effective treatment at 30 RPM.

1. Introduction

Worldwide, the shortage of drinking water and consumption of water of poor quality is one of the main causes of diseases and death especially in the most vulnerable population [1], turbidity is one of the parameters widely applied as a criterion of water quality, both in the sources of supply and in the processes of potabilization and distribution systems [2,3], since it is a rapid and economic measurement. Currently, for the removal of this parameter have been using more coagulants of chemical origin as aluminum salts and iron [4,5], however, by using aluminum salts in the process may be low levels of aluminum in water, between 0.014 and 2.7 mg/l which is associated with diseases such as Alzheimer's disease [6]. On the contrary, the tuna coagulant is purely organic, where it has been demonstrated by phytochemical studies carried out on the *Opuntia ficus indica* the presence of some minerals among some others; also contains, carbohydrates or carbohydrates [7], is also a plant that has become raw material for the food and medicine industries, among the foods based on this plant are: pickles, juices, jams and food supplements rich in fiber or for the control of diabetes or obesity, among others [8].

On the other hand, one of the most important negative impacts found in the Guatapurí River basin is the loss of natural cover, which translates into erosion and sedimentation [9], the above generates high levels of turbidity in the receiving waters [10] especially in the winter season, thus making it difficult to treat the water, as water with low turbidity is easier to treat [11]. The objective of this research was to evaluate the effectiveness of an organic coagulant based on prickly pear for the removal of turbidity in raw water from the middle basin of the Guatapurí River.

2. Experimental

For the study, 2 simple samplings were made at the entrance of the water purification treatment plant located in Valledupar, Colombia, during the dry and rainy season, the samples were stored in high density polyethylene (HDPE) containers with a capacity of 20 liters, they were transferred to the fluid mechanics laboratory of the Universidad Popular del Cesar and refrigerated at 4 ° C for their preservation.

A completely randomized experimental design with three treatments was used, for this design preliminary tests were made with coagulant solution extracted from the *Opuntia ficus indica* using doses from 1 mL to 300 mL with slow mixing of 20 rpm in order to obtain an optimal range for each turbidity from which the 5 best doses were taken in the turbidity removal and control, this range was performed in triplicate in order to obtain average values to choose the dose of lowest concentration of coagulant that removed the highest value turbidity, obtained the optimal dose treatments were applied varying the revolutions per minute (20, 30 and 40), keeping the mixture fast at 125 rpm and sedimentation for 30 minutes. Each treatment was carried out in triplicate, taking as constant variables the dose of coagulant, temperature (25 °C ± 1 ° C), the volume of the sample, the time and speed of the rapid mixture and the time of sedimentation.

To obtain the coagulant, 795 g of *Opuntia ficus indica* buds were used, to which a series of processes ranging from peeled to macerated were applied, where 320 ml of prickly pear extract was obtained, which was used for the preparation of the 10% V / V coagulant solution.

The characterization of the raw water was determined by means of turbidity, total suspended solids, pH, temperature and alkalinity. For which the average values that corresponded to the initial physicochemical characteristics of the two samples studied were calculated. These parameters were measured before and after the coagulation treatment to evaluate the effectiveness of the coagulant. The parameters taken into account are:

Turbidity: measured by the nephelometer method, 25 mL of the sample was taken after agitation and taken to the Hatch 2100AN turbidimeter, selecting the 750 method, at a wavelength of 450 nm, calibrated with the target (distilled water), throwing the measure of turbidity.

pH. It was determined through the potentiometric method, using a pH meter, in which the electrode was introduced in the sample until obtaining the constant reading.

Total Suspended Solids (SST): The gravimetric method was used, using an analytical balance, filter paper with fiberglass and aluminum capsules. Suction filter in a vacuum pump 25 ml of sample through the pre-weighed filter paper which was the constant weight (P1), this was deposited in an aluminum capsule, then the capsules were introduced in an oven with a temperature range between 103 and 105°C for one hour. After this time, they were left to cool in a desiccator and weighed to obtain the weight (P2). The calculation of the Total Suspended Solids was made by means of the following formula:

$$SST\left(\frac{mg}{L}\right) = \frac{(P_2 - p_1) * 10^6}{VM} \quad (1)$$

Where:

P_1: Weight of the capsule, at room temperature, (g).

P_2: Weight of the residue + capsule, at 103-105°C, (g).

VM: Sample Volume (mL).

3. Results and discussion

3.1 Characterization of raw water samples

Table 1, presents the results obtained from the characterization of raw water from the middle basin of the Guatapurí River in dry and rainy season, according to the Regulation of the Drinking Water and Basic Sanitation Sector (RAS), the turbidity of both samples allows to categorize to the source as very deficient because it is higher than 150 NTU, and recommends a conventional treatment with coagulation, sedimentation, filtration and disinfection.

Table 1. Characterization of raw water samples.

Physical-chemical parameter	Standard Method	Units	Results Summer	Results Winter	Maximum acceptable values Resolution 2115 of 2007
Turbidity	Nephelometric	NTU	345,9	2799,1	2
Total suspended solids	Gravimetric	mg/L	400	1890	No register
pH	Potentiometer	pH	8,1	8,2	6,5-9,0
Total Alkalinity	Laboratory method	mg/L CaCO3	110	97,5	200
		°C	26	26	No register

3.2 Results of the turbidity parameter in treatments 20, 30 and 40 RPM for the turbidity of 345.9 NTU

In Figure 1, the results of each of the treatments are presented, which consisted of using a slow mixture of 20, 30 and 40 RPM, using the optimum dose for a turbidity of 345.9 NTU, which was 658.8 mg / l of tuna. According to Figure 1, with each of the treatments (20.30 and 40 rpm) the turbidity decreases significantly, obtaining removal percentages of 97.1, 97.4 and 97.4% respectively, indicating that the coagulant of tuna It is effective for the removal of turbidity, however the maximum acceptable value stipulated by resolution 2115 of 2007 is not reached, so a filtering process was carried out that managed to comply with the Colombian regulations.

On the other hand, the variation of the speed of agitation in the slow mixture during the coagulation process affected little the decrease of this parameter, where it is evident that the treatment that decreased more the turbidity was using the slow mixture at 40 RPM observing that the difference between this and the other treatments is minimal.

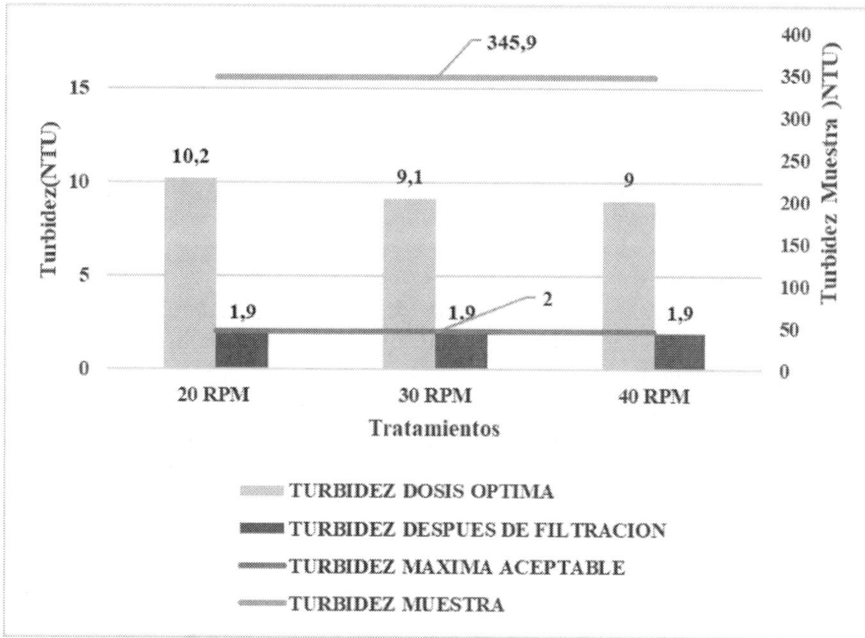

Figure 1. Turbidity results in each treatment.

Comparing these results with those obtained by Parra et al. (2011) [12] which worked with Opuntia goiana for the removal of turbidity in raw waters presenting removal percentages between 85.25 - 94.84% said percentage is high, however better results are presented with the Tuna Opuntia cochenillifera, on the contrary Lameda et al. (2014) [13] used the Tuna Opuntia ficus-indica for the removal of turbidity in synthetic waters obtained lower percentages between 17.22-82, 37%.

3.3 Results of the pH parameter in treatments 20, 30 and 40 RPM for the turbidity of 345.9 NTU
In Figure 2, the pH results of each of the treatments are presented.

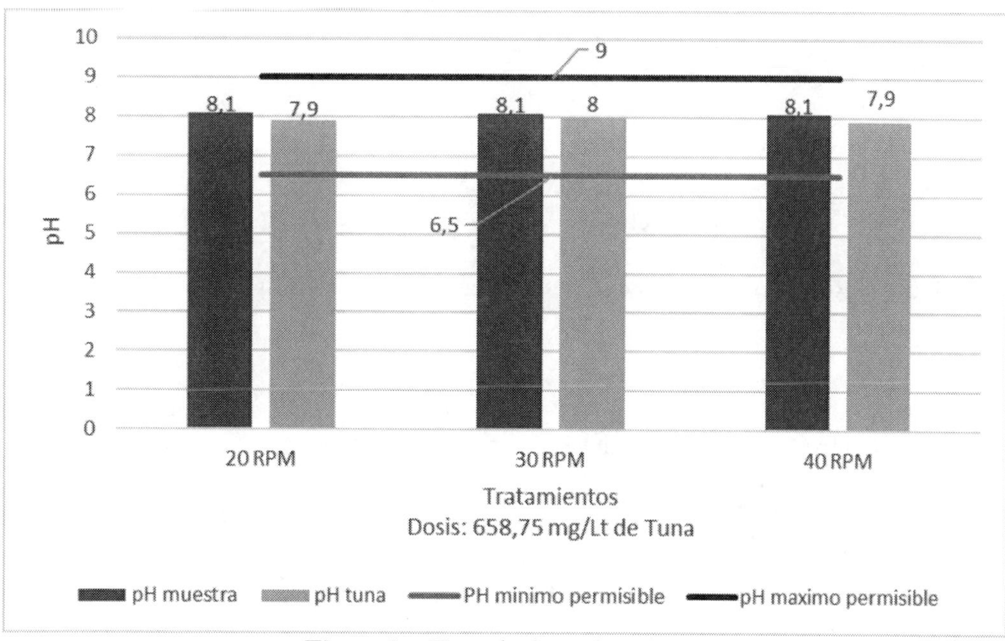

Figure 2. pH results in each treatment.

As it is observed, the pH does not present variations in each of the treatments, in addition it is evident that in none of the treatments the pH parameter exceeds the range established by Resolution 2115 of 2007. Comparing these results with those obtained by Villabona et al. (2013) [14] which used Opuntia ficus-indica in the removal of turbidity in raw waters, obtained a pH range between 7.5 - 7.9 which did not have a significant variation and also these values are within the limits established by the regulations Colombian, likewise Lameda et al (2014) [13] who worked with Opuntia ficus-indica in the removal of turbidity with synthetic water obtaining a range of 6.90 - 7.82 which presented greater variation however does not exceed the stipulated range by the norm mentioned above.

3.4. Results in the parameter of total suspended solids (TSS) in the treatments at 20, 30 and 40 RPM
In Figure 3, the results of the concentration of total suspended solids in the treatments at 20, 30 and 40 RPM are presented applying the optimal dose which was 658.75 mg / L of coagulant of prickly pear.

It can be evidenced that the treatment with the best behavior in the removal of the total suspended solids was the 20 RPM, with a remnant of 30 mg / L with a removal percentage of 92.5%, which is well below the maximum acceptable value by the Colombian technical standard 813, on the other hand the results of treatments at 30 and 40 RPM (35 and 40 mg / L respectively) also showed good results with percentages of removal greater than 90% and that are also found by below the maximum acceptable value stipulated for this parameter in said standard.

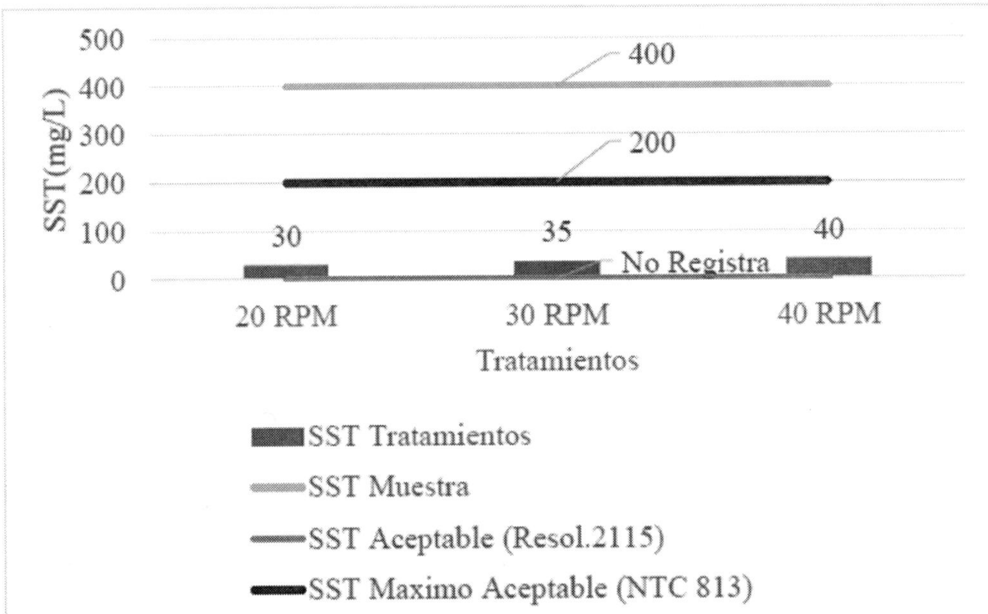

Figure 3. Results of the treatments in the parameter of Total Suspended Solids (SST).

3.5. Results in the turbidity parameter in treatments 20, 30 and 40 RPM for turbidity of 2799.1 NTU
In Figure 4, the results are presented with each of the treatments, with a coagulant dose of 1976.3 mg/ L of Tuna, which turned out to be the optimal dose for this turbidity.

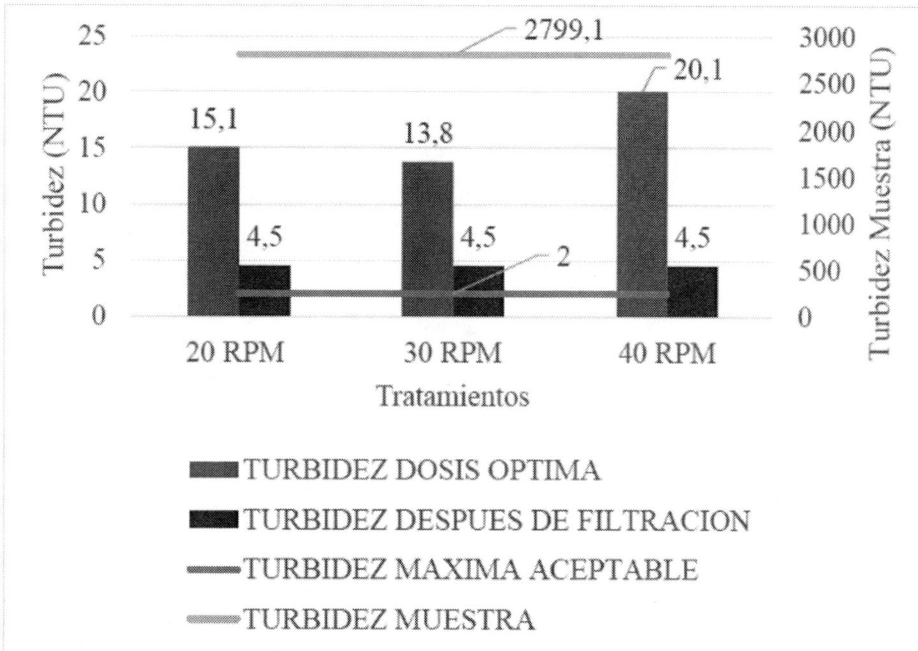

Figure 4. Turbidity results in each treatment.

As seen in Figure 4, with each of the treatments (20.30 and 40 RPM) the turbidity is significantly reduced, obtaining removal percentages of 99.5%, 99.5% and 99.3% respectively, indicating that the coagulant of Tuna is effective for the removal of waters with high turbidity, however the maximum acceptable value stipulated by Resolution 2115 of 2007 for this parameter is not reached, therefore a filtering process was carried out that did not comply with this norm, now it is necessary to take into account that the filtration carried out in the drinking water treatment plant that supplies the city of Valledupar carries out the process by means of mixed sand and anthracite filters, which suggests that with this Adequate filtration could reach the maximum acceptable value by Colombian regulations.

On the other hand, the stirring speed of the slow mixture influences the removal of turbidity, where it is evident that using the slow mix at 20 and 30 RPM, better removals were obtained, however the difference between these and the treatment at 40 RPM is minimal , the above can be explained that with high velocity gradients there is a greater probability of flocculation, while with very high gradients it generates the rupture of flocs already formed by the action of liquid cutting forces, that is to say, a shearing effect occurs. it produces the mechanical breakage of the floccules [15].

3.6. Results in the pH parameter in treatments 20, 30 and 40 RPM for turbidity of 2799.1 NTU

According to Figure 5, the pH results show that the coagulant of prickly pear does not alter this parameter in each of the treatments, also it is observed that none of the treatments exceeds the range established by Resolution 2115 of 2007. In addition, it can be affirmed that the natural coagulant based on Tuna has the property of not significantly altering the pH of the water when it is applied in high doses.

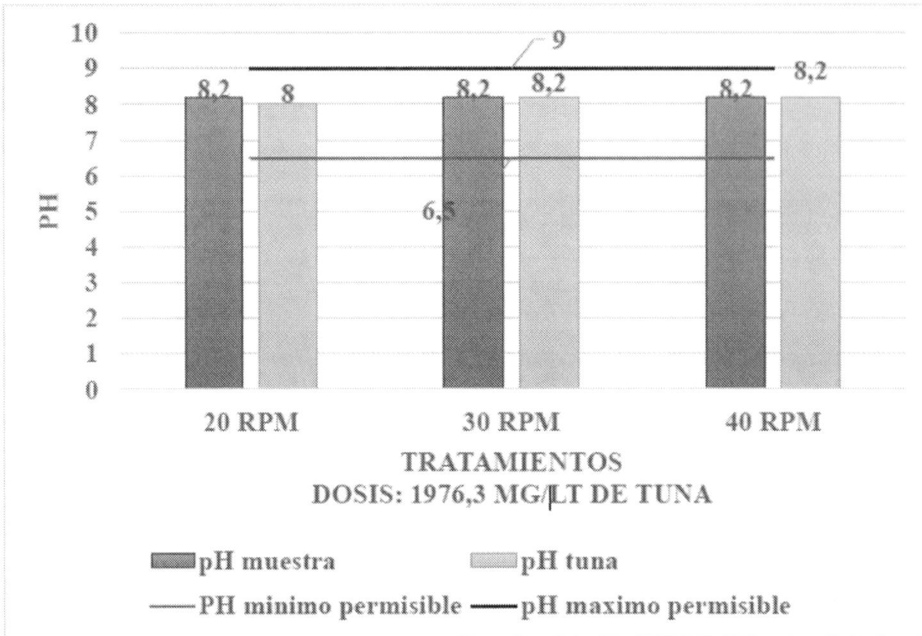

Figure 5. pH results in each treatment.

3.7. Results in the parameter of total suspended solids (SST) in the treatments at 20, 30 and 40 RPM
In Figure 6, we present the results of the concentration of total suspended solids in the treatments at 20, 30 and 40 RPM applying the optimum dose which was 1976.3 mg / L of coagulant of prickly pear.

As shown in Figure 6, there was a significant decrease in each of the treatments, decreasing from 1890 mg / l to 20 mg / l at 20 RPM with a removal percentage of 98.9%, while for treatments of 30 and 40 RPM showed a residual of 40 mg / l with a removal percentage of 97.9%, which indicates that the treatment at 20 RPM fuel is the most effective for the removal of SST, in addition with the three treatments a concentration is obtained well below the stipulated in the Colombian technical standard 813 that is 200 mg/ L, it should be noted that for resolution 2115 which indicates the characteristics for water for human consumption this parameter is not contemplated, so it is not possible to make a comparison with this parameter physical.

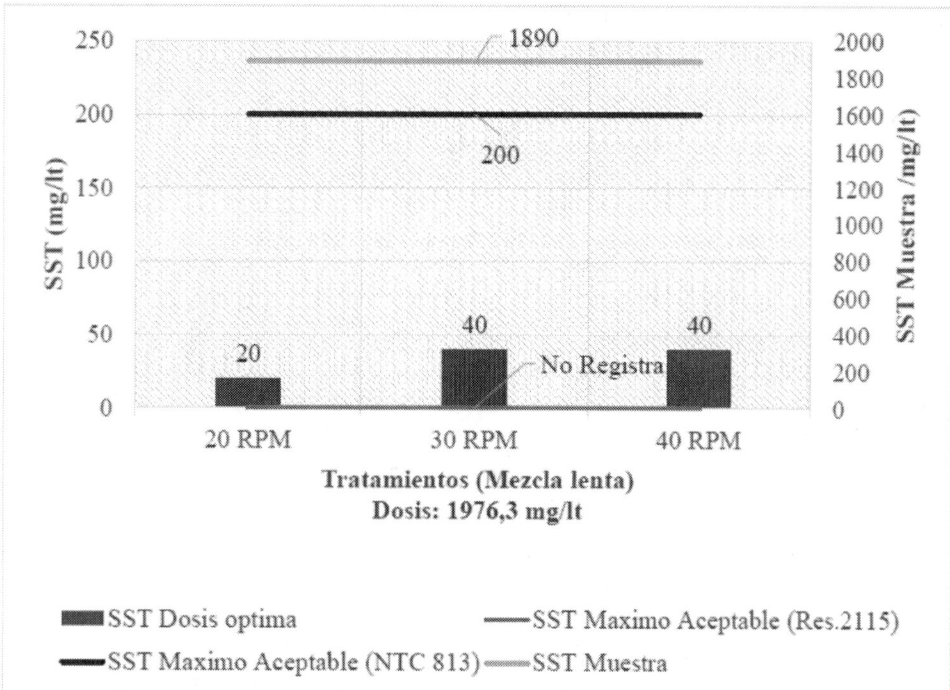

Figure 6. Results of the treatments in the parameter Total suspended solids (SST).

4. Conclusions

This study showed that the coagulant of Opuntia ficus indica is effective in the removal of turbidity, reaching percentages of removal between 97.1% to 99.5%, without affecting the pH greatly. The optimal doses obtained in the coagulation process using samples of raw water from the Guatapurí River with initial turbidities of 345.9 and 2799.1 NTU, under the conditions of rapid mixing with a speed of 125 rpm and a time of 1 minute, slow mixing with a speed of 20 rpm and time of 20 minutes and a sedimentation time of 30 minutes, were 658.75 and 1976.3 mg / liter respectively. Based on the above, the coagulant of tuna is considered as an alternative for the treatment of raw water of low and high turbidity, being more effective when it is high.

Conflict of Interest.

The authors report there are no conflicts of interest.

References

[1] Severiche C, Barreto A and Acevedo R 2013 Efecto de las Lluvias sobre la Calidad del Agua en La Ciénaga Grande de Santa Marta, Colombia. Avances: Investigación en Ingeniería 10 58- 67.

[2] Cho S, Nagendra K, Wu T, Raghunandan M and Ramanan R 2014 Utilization of plant-based natural coagulants as future alternatives towards sustainable water clarification. Journal of Environmental Sciences 26 2178-2189.

[3] Rodiño J, Feria J, Paternina R and Marrugo J 2015 Tratamiento del agua cruda del río Sinú por coagulantes naturales. Revista Facultad de Ingeniería Universidad de Antioquia 76 90-98.

[4] Bahman R 2014 Treatment of water turbidity and bacteria by using a coagulant extracted from Plantago ovata. Water Resources and Industry 6 36-50.

[5] Jimenez M, Mújica D and Welti J 2014 Caracterización fitoquímica de la tuna (Opuntia spp.) y de sus propiedades nutricionales y funcionales: una revisión. Current Nutrition & Food Science10 57-69.

[6] Arcila H and Peralta J 2016 Agentes naturales como alternativa para el tratamiento del agua. Revista Facultad de Ciencias Básicas 11 136-153.

[7] Olivero R, Florez A, Vega L and Villegas G 2017 Evaluación de una mezcla para coagulantes naturales, Opuntia ficus y Moringa oleíferaen clarificación de aguas. Revista Producción Mas Limpia 12 71-79.

[8] Asrafuzzaman M, Fakhruddin A and Hossain M 2011 Reduction of turbidity of water using locally available natural coagulants. ISRN Microbiology 632189

[9] Severiche C, Baldiris I, Acosta J, Bedoya E, Castro I and Pacheco H 2017 Multivariate Analysis of Water Quality in Rosario Islands National Park (Colombia). American Journal of Engineering Research 6 136-144.

[10]Maldonado W, Baldiris I and Díaz J 2011 Evaluación de la calidad del agua en la Ciénaga de la Virgen (Cartagena, Colombia) durante el período 2006-2010. Revista Científica Guillermo de Ockham 9 79-87.

[11]Oladoja N 2015 Avances en coagulantes poliméricos naturales en operaciones de tratamiento de agua y aguas residuales. Revista de ingeniería de procesos de agua 6 174-192.

[12]Lameda P, Negrilli C, Hidalgo C and Fuentes L 2014 Clarificación de aguas de baja turbidez empleandoel coagulante opuntia ficus-indica. REDIELUZ 4 44-48.

[13]Parra Y, Cedeño M, García M, Mendoza I, González Y and Fuentes L 2011 Clarificación de aguas de alta turbidez empleando el mucílago de Opuntia Wentiana (Britton & Rose) / (Cactaceae). REDIELUZ 1 27-33.

[14]Villabona A, Paz I and Martínez J 2013 Caracterización de la Opuntia ficus-indica para su uso como coagulante natural. Revista Colombiana de Biotecnología 15 137-144.

[15]Carrasquero S, Montiel S, Faría E, Parra P, Marín J and Díaz A 2017 effectivines of coagulants obtained from residues of potato (Sonalum tuberosum) and banana (Musa paradisiaca) in water purification. Rev. Fac. Cienc. Básicas 13 90-99.

Obtaining a sugar syrup from the use of the extract of the guácimo fruit (Guazuma ulmifolia lam)

C R Vidal-Tovar[1,2], C A Cadavid-Osorio[1], G E Angulo-Blanquicet[2], and C A Severiche-Sierra[3]
Grupo de Investigación Creando Ciencias CRECI, Universidad Popular del Cesar, Valledupar-Cesar, Colombia.
[2]Grupo de Investigación CIEMPIES, Universidad de Santander, Valledupar-Cesar, Colombia.
[3]Grupo de Investigación GITEC, Corporación Universitaria Rafael Nuñez, Cartagena de indias-Bolívar, Colombia.

carlosvidal@unicesar.edu.co

Abstract. The main objective of this work was to elaborate a sugary syrup from the use of Guácimo fruit extract (Guazuma Ulmifolia Lam) using the leaching method (solid - liquid), which can be used as a partial or total substitute for sugar. The experimental design that was applied for the extraction process was a factorial design 23; the data obtained were evaluated with the software Statgraphics. Once the extracts were obtained, physico-chemical tests were carried out and reductive sugars and dextrose equivalent were determined in each one. To obtain the syrup, the four extracts having the highest% ED were concentrated at constant temperature and time (50 ° C and 4 hours); The determination of the presence of reducing sugars was done using the techniques of colorimetric analysis (miller's method) and chromatography (HPLC). The products that were obtained during the process were extreme conversion syrups with a range of dextrose equivalents of 90.13% - 98.27%. The syrup that obtained the highest percentage of equivalent dextrose (J3) was subjected to a sensory hedonic test, which consisted of 3 samples of sweetened coffee at different concentrations of sweet sugar syrup and sugar (sucrose), where M1 contained 80% sugar syrup of guácimo and 20% of table sugar. M2 contained 50% guácimo sugar syrup and 50% table sugar. M3 contained 100% sugar guacimo syrup and 0% table sugar. The results showed that the mixture that obtained the highest organoleptic acceptance was M2, followed by M3, which states that the syrup can substitute partially the common sugar. The above data allow the guácimo fruit to be taken into account as an alternative raw material for the production of sugar syrups and other products aimed at human consumption.

1. Introduction

The use of sweeteners is important in products and markets around the world, because in most edible products and most of the pharmaceuticals are used sweeteners to increase the degree of acceptance of these by the consumer, this situation has become a necessity that has generated a great impact due to the chemical components that the sugars possess and that leads to the exploration of new substances or species [1].

In most industrial processes of foods, sweeteners of various types are used in a great proportion, such as sucrose, glucose, lactose, fructose, and others produced by the mixture of the aforementioned or another agent that produces a Pleasant palate sensation in ready-to-eat products. These sugars or sweeteners are used in order to improve the organoleptic characteristics of each product and also preserve their properties [2].

Nowadays, due to the high market cost of these sugars, it has been decided to investigate and focus their needs on the production, generation and development of substitutes for common sugars, which is why they have obtained positive results, among these are the natural sweeteners and artificial or synthetic. Within the natural sweeteners we find the honey of bee, the corn syrup rich in fructose; and the artificial ones are saccharin, Aspartame "Nutrasweet", among others [3].

In this way, the proposed study will provide a set of knowledge applied to obtaining sweeteners from a raw material that is cultivated naturally, artisanal and wild without any industrialization or technological application for its use. The guácimo, which predominates in the livestock areas of the department of Cesar, and is only used for feeding and shady livestock by means of live fences in the paddocks where they are present [4].

Likewise, the fruit of the guácimo is not taken into account in the region for the agro-industrial use or the production of food destined for human consumption, so it is intended with this work to provide a method, technique and procedure for its processing, in order that value can be added to the large number of fruits produced by guácimo trees in Valledupar since the extraction of a sweetening agent [5].

Being thus feasible to the markets and therefore would be benefiting directly with this product to people with metabolic diseases where sugar levels are very high, secondly, companies that produce low-calorie foods will benefit. As well as companies producing soft drinks, confectioneries, dessert creams, ice cream, etc [6].

In this sense, this research aims to encourage the use of guácimo as a key alternative for the production of natural sweeteners that can be used as a natural substitute for some sugars. In addition, motivate agro-industry researchers to use the guácimo fruit and its components as raw material for the production of products and by-products for direct human consumption, given its diverse physicochemical and nutritional composition [7].

Guácimo (guazuma ulmifolia Lam) is produced wild without any control, it has been used for animal consumption and traditional uses; similarly, no research is known in the region concerning or oriented towards the use of guácimo for the consumption of people. The investigations that have been carried out are aimed at the consumption of minor species [8]. In this regard it is convenient to say that, in the surrounding areas of Valledupar, the guácimo tree adapts easily thanks to the properties offered by the soils and the climatic conditions of the department.

At present the interest in sweeteners is gaining strength, it is opting for it as a substitute for common sugars, which have been causing damage to health. Sweeteners have been taken into account as a viable alternative in the food industry worldwide [9]. The existing interest for sweeteners opens space, and serves as a platform to share information on species never explored for these products, to take as alternatives for use and exploitation, as a form of sustainable development that can be consumed in Colombia and the entire world. This work is carried out in search of alternatives for the use of the guácimo fruit, for the obtaining of by-products, and the elaboration of food products aimed at human consumption. At the same time, this research is intended to encourage the use of guácimo as a key ingredient for the production of sugar syrups that can be used as a substitute for sugars and in turn give value to the crop.

2. Experimental

2.1 Experimental Design

To obtain guácimo extract, for the measurement of the causes and effects, and to be able to carry out the fulfillment of the objectives of this work, a factorial experimental design was carried out (23) because the factorial designs of two levels or 2k have the quality of proposing processes in all combinations of factor variables, which indicates that factor takes 2 different values, that is, when the number of factors is k, then the number of experiments will be 2k [10], where it was manipulated 3 independent variables such as: solvent ratio (water) - solid (guácimo), temperature and time; the

IOP Conf. Series: Materials Science and Engineering **844** (2020) 012004 doi:10.1088/1757-899X/844/1/012004

variations of the effects or consequences expressed as dependent variables were observed, which are: extract yield (%), concentration (Brix), and dextrose equivalent.

For the realization of this design, eight (8) runs were required each in triplicate of which the effects of each of them were measured at the end. In the process, time and temperature were strictly controlled, since the effects in each process could vary or change depending on the manipulation given to these variables. in Table 1, the experimental design model is shown.

Table 1. Factorial experimental design model 2^3.

Levels	Relationship Water: Guácimo	Temperature	Time (min)
G_1	[4:1]	Ambient	20
G_2	[5:1]	Ambient	20
G_3	[4:1]	37	20
G_4	[5:1]	37	20
G_5	[4:1]	Ambient	30
G_6	[5:1]	Ambient	30
G_7	[4:1]	37	30
G_8	[5:1]	37	30

The syrup was then obtained, taking into account the four best variables of the extract responses, and they were concentrated at a constant temperature of 50 ° C and a time of 4 hours, then the physicochemical characteristics of each of the syrups were determined. concentrated.

2.2. Sample / Population

Guacimo fruits were collected in the zones located between the road that goes from Valledupar to the dry river township, as well as between the output of Valledupar. These zones are similar in terms of climate and distribution of trees, so two lots were taken from each area (population); from each batch 8 trees were chosen and from each tree a kilogram of fruit, thus obtaining 8 kilograms per batch, obtaining a total of 32 kilograms of guacimo that were used deliberately in the process of obtaining the syrup.

2.3. Characterization of the fruit

Guácimo fruit harvest: Guacimo fruits were collected in places near the municipality of Valledupar, where two zones with similar characteristics regarding climate and distribution of trees were taken. Taking two lots from each zone, 8 trees were selected from each batch and 1 kg of fruit per tree, thus obtaining 8 kg per batch, for a total of 32 kg of guácimo.

Selection and classification of the raw material: For the selection and classification of the guácimo fruit, the objective was to discard the green and damaged fruits, in order to obtain a fruit in optimum conditions, which were used for the process. Once the fruits were selected, the classification was carried out, for this activity a small pulp was obtained, which is irregularly opened by 5 seed pores, which were removed as the heart of the fruit.

Storage of the fruit: already having the previously selected fruits, classified and cut, they were later stored to preserve the characteristics of the fruit and avoid any type of contamination.

2.4. Determination of the physicochemical characteristics of the fruit
Total soluble solids (Degrees Brix): the determination of Brix degrees of the guacimo fruit, because it is a dried fruit, which said composition is to have a low percentage of water. It was made by means of a material balance, for this, 20 grams of fruit were weighed and diluted in 200 ml of water, having a ratio of 1:10 and kept under constant agitation for 30 minutes, after this time it was filtered with the help of a vacuum pump thus obtaining a yield of 129 ml of extract (64.5%), then 250 ml of distilled water was added, the Brix degrees were measured in the aqueous solution with the help of a portable refractometer, mark ATAGO model PAL-1 with precision of 0.1 ° Brix, already known the values of SST, the balance of matter was made to obtain the Brix of the fruit.

Reducing sugars: Miller's method was used to determine reducing sugars. For which a glucose calibration curve (standard) was made, then the glucose standards.

For the determination of reducing sugars present in the fruit; the solution of 250 mL of extract mentioned in the previous point was started and dilutions were prepared in triplicate in a ratio of 1: 100 because the sample is quite dark, and it was subjected to react with the DNS at high temperatures to then measure absorbances at 540 nm, and thus be able to make the calculation and obtain the value of reducing sugars, in addition, a moisture test was carried out on the fruit to extract by difference the percentage of the dry substance, which serves to determine the dextrose equivalent.

Humidity and Dry Substance: To determine the humidity, different amounts of fruit were crushed, and dried in porcelain pots at 105 ° C. Once the percentage of humidity was obtained, it was possible to obtain, by difference, the% of dry substance (% SS = 100 -% Humidity).

2.5. Obtaining and characterization of the guácimo extract
Crushing the guácimo fruit: Initially, the fruit already selected was crushed, classified and stored under optimum conditions to reduce the particle size and facilitate the extraction process, taking into account the relationship or scale (water: guacimo).

Obtaining the extract: After the fruit was crushed, it was taken to a container that carried water (solvent) at a scale of 4: 1 and 5: 1 for each run respectively, to carry out the extraction method by solid-liquid leaching, each run was kept in constant agitation for different times of 20 and 30 minutes (according to the corresponding run), then it was filtered using lilycloth cloth, the extract was centrifuged for 20 minutes, to clarify and eliminate unwanted particles. Subsequently, samples were taken for the characterization of the extract. For which physicochemical tests were carried out (pH, Acidity, Brix), the percentage of reducing sugars and dextrose equivalent were also determined, these analyzes were carried out taking into account what is described in point 8.6.4, which deals with the methods of analysis physiochemical.

2.6. Characterization of the sugar syrup obtained
Obtaining the syrup: Obtained and analyzed the extracts, we proceeded to choose the ones with the best physicochemical conditions and the highest equivalent dextrose value, in this case were the extracts G1, G3, G4 and G7, then these were concentrated at a temperature of 45 ± 50 ° C and 4 hours' time, with constant agitation controlling the physicochemical characteristics every 2 hours; Once that time had passed, the syrup was centrifuged for 20 minutes to separate dissolved solids in the product, thus obtaining the product of interest, from which the samples were taken to perform the corresponding physicochemical tests (pH, Brix, acidity and organoleptic aspects) and the content of reducing sugars and dextrose equivalent was determined.

2.7. Methods of physicochemical analysis
Determination of pH: For the determination of pH, the requirements of the standards were taken into account: ISO 1842: 1991 (US ISO; 1991), NTC 4592 (ICONTEC; 1999) and AOAC method 981.12 (AOAC, 2005). pH meter OHAUS brand model STARTER 3100 with precision of 0.01. For the implementation of the pH-meter, a sample of 15 ml of each sample was taken in a Beaker, and the pH-meter was introduced into it, to finally read and record the pH value.

IOP Conf. Series: Materials Science and Engineering **844** (2020) 012004 doi:10.1088/1757-899X/844/1/012004

Determination of total soluble solids or Brix degrees: ISO Standard 2173: 2003. (US ISO; 2009), NTC 4624 (ICONTEC; 1999) and the AOAC method 932.14 (AOAC, 2005). For this test a portable ATAGO brand Refractometer model PAL-1 with precision of 0.1 ° Brix was used, where 5 drops of sample were placed in the refractometer lens to then star and start reading.

Determination of reducing sugars (RA), total (AT) and dextrose equivalent (ED): For the characterization of sugars, the method of MILLER (DNS), NTC 1779 (ICONTEC, 1997), Method of AOAC 923.09 (AOAC, 2005) was used. THERMO SPECTRONIC 4001/4 spectrophotometer at 540 nm using a calibration curve with a glucose standard. HPLC 1200 Agilent; Mobile phase [11].

Titratable acidity determination: The requirements of the Colombian standard NTC 4623 (ICONTEC, 2000) and NTC 4803 (ICONTEC, 2016) were implemented. One ml of extract was taken and 10 ml of water were added with distilled water, 3 drops of phenolphthalein were added and the mixture was stirred. Then a 0.1 N standardized NaOH base solution was added by means of a graduated burette, until the mixture turned to an opaque pink color. The titratable acidity was expressed as% citric acid.

Moisture Determination: The gravimetric method described by AOAC was used; 2005.

2.8. Application and measurement of organoleptic acceptability of sugar syrup as a partial or total substitute for sugar

Application of sugar syrup as a partial or total substitute for sugar: For the application the syrup produced in this project was used as a standard sample of sweetened coffee at different concentrations of syrup and table sugar (sucrose), as follows:

Sample No. 1 contained 80% sugar guacimo syrup and 20% table sugar.

Sample No. 2 contained 50% sugar guacimo syrup and 50% table sugar.

Sample No. 3 contained 100% guácimo sweetened syrup and 0% table sugar.

Obtaining thus, a partial and total substitution of sugar (sucrose) for sugary syrup obtained from the guácimo fruit; in this way to perform the calculations of the sweetener, they were made by the trial method, where guácimo syrup and sugar (sucrose) were weighed little by little, and then tested until an acceptable coffee sweetness was found; at the end a quantity of sugary syrup of 22.25 grams was weighed, equivalent to 79.6% and 5.7 grams of table sugar which is equivalent to 20.4% in the total sweetener mixture. Measurement of the organoleptic acceptability of sugar syrup as a partial or total substitute for sugar: To measure the degree of acceptance of sugar syrup, a sensory analysis was implemented, such as the hedonic, technical and analytical tasting. This test consisted of 3 samples of sweetened coffee to different concentrations of sugar syrup obtained from the guácimo fruit and sugar (sucrose), and they were given a survey format to each evaluator, so that this could give their opinion about the product, this person should observe, smell and test each of the samples, and then be able to evaluate the sensory attributes of the product (taste, smell, texture and acceptance) taking into account that the rating goes from 1 to 5, where "1 is equal to dislike me a lot" and "5 is equal to I like it a lot".

3. Results and discussion

3.1 Physicochemical characteristics of fruit

The fruits used in this work were collected on the road connecting Valledupar; for the characterization of the fruits, a selection and subsequent classification was carried out in order to proceed with the respective physicochemical tests (acidity, pH, Brix). In Table 2, the physicochemical characterization of the guacimo fruit is presented.

Table 2. Physicochemical characteristics of the fruit

Characteristics	Percentage (%)
Humidity	15,6
Brix	28,5
Reducing sugars	15,98
Equivalent dextrose	18,94
Dry material	84,4

It is evident that the guacimo fruit has a percentage of total soluble solids (Brix degrees) of 28.5%, and a percentage of reducing sugars of 15.98%; It should be clarified that in the bibliographic sources referring to the fruit of guácimo or other species belonging to the family of the malvaceaes, no information was found about the content of Brix degrees and the% of reducing sugars of interest for this research, however, there are jobs where they use fruits, tubers and / or other raw materials that can be very important to compare the data; Beltrán (2014) [12] reports a percentage of Brix degrees of 15%, whose value is lower than those shown in this research; and referring to reducing sugars, Gerena (2013) [13] reported higher values (16,876% - 19,277%) in different mixtures of orange and potato.

Regarding the MS% it can be said that those obtained in this research are very similar to those reported by Queiroz et al. (2019) [14], who obtained a dry matter% value of 85.17%, and in this work 84.4% ; which also states that the percentage of humidity are similar.

3.2. Characterization of the extract obtained from the use of the guácimo fruit

According to the applied methods, in Table 3, it is observed that a detailed characterization was made to the extract, where the variables are: yield, pH, total soluble solids, humidity%, reducing sugars., Thus obtaining the response variable of the extract (the Dextrose equivalent).

Table 3. Physicochemical and bromatological characteristics of the extract.

M	Solvent Relation ship: Guácimo	T °C	t(min)	% R	pH	Acidit y	Bri x	%H	%S S	%A R
G1	[4:1]	Ambient	20	61	4,55	0,098	9,16	94,37	5,63	4,62
G2	[5:1]	Ambient	20	66,8	4,58	0,084	7,34	94,37	5,63	2,72
G3	[4:1]	37	20	60	4,57	0,133	9,63	94,37	5,63	4,47
G4	[5:1]	37	20	67,5	4,59	0,098	8,27	94,37	5,63	4,85
G5	[4:1]	Ambient	30	60,1	4,5	0,091	9,65	94,37	5,63	3,82
G6	[5:1]	Ambient	30	64,7	4,49	0,067	7,75	94,37	5,63	2,51
G7	[4:1]	37	30	6,2	4,45	0,091	9,83	94,37	5,63	4,46
G8	[5:1]	37	30	65	4,57	0,094	8,39	94,37	5,63	3,16

Table 3 shows a higher% Reduction Sugars in samples G1, G3, G4 and G7 that range from 4 to 5, whose range is lower than those reported by Gerena (2013) [13]. It is intuited that the components of the guácimo fruit extract have the potential for the production of a sweetening agent.

Analysis of the dextrose equivalent obtained from the guácimo fruit extract: The percentage of dextrose equivalent (ED) of the guácimo fruit extract is shown in Table 4.

Table 4. Experimental matrix of the results obtained from dextrose equivalent.

Cumshots	Relationship Solvent: Guácimo	Temperature °C	Time (Minute)	Equivalent Dextrose (%)
1	4	25	20	92,04
2	5	25	20	49,05
3	4	37	20	91,75
4	5	37	20	76,93
5	4	25	30	74,32
6	5	25	30	50,5
7	4	37	30	85,94
8	5	37	30	48,47
9	4	25	20	92,33
10	5	25	20	49,05
11	4	37	20	87,1
12	5	37	20	94,94
13	4	25	30	70,54
14	5	25	30	54,86
15	4	37	30	89,42
16	5	37	30	47,89
17	4	25	20	92,33
18	5	25	20	51,08
19	4	37	20	86,81
20	5	37	20	81,58
21	4	25	30	72
22	5	25	30	54,57
23	4	37	30	87,39
24	5	37	30	51,37

For the dextrose equivalent, the statistical model obtained presented three significant terms: A, B and C with p values of 0.0000, 0.0191 and 0.0052 receptively, the correlation coefficient (R_2) can explain 80.6941% of variability in the dextrose equivalent in the response variable analyzed and a standard deviation of 9.86466. In addition, the adjusted R_2 was 70.3977%. This value is the most suitable to compare models with different numbers of variables.

The percentage of dextrose equivalent obtained in the different extraction conditions are influenced in the relationship of time and temperature, where it can be observed that thc bchavior of this is reflected in the combinations of levels that maximize the response variable as the water ratio: Guácimo 4: 1, time 20 minutes and temperature of 37 ° C to these conditions is where the optimum value of equivalent dextrose (95,7783%) is obtained, when applying a time of 30 minutes in the same temperature ranges it is obtained as result a lower% ED. And with regard to the 4: 1 and 5: 1 relationship it can be observed that at a lower (water: Guácimo) ratio, the extract will be more concentrated.

3.3. Physicochemical characterization of the sugar syrup

The syrup obtained in this work presented a brown color, flavor and sweet smell characteristic of guácimo, texture a little dense. The results obtained in the physicochemical characterization of the syrups with respect to yield, Brix degrees and reducing sugars are shown in Table 5; it is observed that the yield is between a range of 62% - 76.8%, also a total soluble solids concentration between 13.7 - 18.2 is evidenced, the previous Brix range is inferior to those reported by Kumar & Gurunani (2019) [15] who uses Stevia leaf, and concentrates at a temperature of 70 ° C; As for the yield of the syrup, this was superior to that of Stevia. Likewise, a production of reducing sugars (%) is observed from 6.89 to 7.52; The range discussed above regarding reducing sugars are lower than those reported by different authors Gerena (2013) [13] and Vidal (2011) [11] who use orange, potato and yam starch residues.

Table 5. Physicochemical characteristics of sugar syrup.

					Physicochemical and bromatological characteristics					
N°	Extraction	T°C	Time (horas)	Percent age (%)	Brix Ini	Brix 3 h	Brix 4 h	%H	%SS	% AR
J1	[4:1]	50	4	62,4	10,0	12,9	14,8	92,35	7,65	7,14
J3	[5:1]	50	4	62	10,3	15,3	18,2	92,35	7,65	7,52
J4	[4:1]	50	4	76,8	8,3	12,3	13,7	92,35	7,65	6,89
J7	[5:1]	50	4	68	10,6	15,4	17,3	92,35	7,65	7,14

Based on the above, the quantification of reducing sugars expressed as glucose (g/L) and fructose (g/ L) present in guácimo syrups, where it is evident that J3 was the one with the highest glucose concentration (5.99) % followed by J7 with 5.395%, J1 4.835% and J4 3.384%; and regarding the quantification of fructose, it is evident that J1 and J3 had values of 4.633% and 6.992% respectively, while J4 and J7 did not show fructose concentrations.

It can be seen that the syrups obtained have a% ED of 93.32 for J1, 98.27 for J3, J4 with 90.13 and 93.32 for J7; these values are higher than those reported by Gerena (2013) [13] in the hydrolysis of orange and potato residue (21.7 - 67.3% ED); the hydrolysis of yam starch that obtained a dextrose equivalent between 18.82% and 78.99% reported by Vidal (2011) [11]. But it is within the ranges of those reported by Salcedo et al. (2010) [1] who obtained 96.84 - 98.28%; Roncallo & Rodríguez (2013) [10], who reported 95.32 and 94.12 units of ED, before converting glucose to fructose.

From the above it can be inferred that the results obtained in the characterization indicate that guácimo contains a potential as a raw material for the production of sugar syrups or sweetening agents; it is evident that syrup J3 had the best values under all conditions and response variables (° Brix, reducing sugars, glucose, fructose and dextrose equivalent), showing difference compared to the other syrups. Thus, with respect to the dextrose equivalent it can be said that an extreme conversion syrup was obtained [2,3], and it can be used in industries for the production of beer, soft drinks, juices, jams, etc. [7].

3.4. Result of the hedonic test of the sugar syrup obtained

To determine the degree of acceptability of guácimo syrup, the attributes of the guacimo syrup were analyzed by means of a survey, which allows to have information of each one of the proposed attributes, as shown in Table 6.

Table 6. Organoleptic characteristics of sugar syrup

Attributes	Syrups Sample 1	Syrups Sample 2	Syrups Sample 3
Taste	$2{,}91^{b} \pm 1{,}08\ \sigma$	$3{,}91^{c} \pm 0{,}66\ \sigma$	$1{,}91^{a} \pm 0{,}79\ \sigma$
Odor	$4{,}0^{b} \pm 0{,}85\ \sigma$	$3{,}91^{b} \pm 0{,}66\ \sigma$	$2{,}75^{a} \pm 1{,}21\ \sigma$
Texture	$3{,}33^{ab} \pm 1{,}07\ \sigma$	$3{,}75^{b} \pm 0{,}62\ \sigma$	$2{,}83^{a} \pm 0{,}93\ \sigma$
General acceptance	$3{,}10^{b} \pm 0{,}83\ \sigma$	$3{,}91^{c} \pm 0{,}51\ \sigma$	$2{,}08^{a} \pm 0{,}79\ \sigma$

a-b-c Socks with the same letters between columns do not show significant differences. ($p \geq 0.05$).

The acceptance analysis for the sugar syrup in terms of taste showed statistically significant differences ($P < 0.05$), with a confidence level of 95.0%, in relation to syrups M1, M2 and M3, presenting M2 higher score 3.91c ± 0.66 σ, placing itself within the "like" scale, which reflects that the product has a suitable flavor for consumption.

For the smell attribute, when comparing the three samples, statistically significant differences were found ($p < 0.05$) with a confidence level of 95.0% for M3, in relation to samples M1 and M2, which did not present significant differences in between, the ratings were; 4.0b ± 0.85 σ and 3.91b ± 0.66 σ respectively, placing the samples within the "like" scale, which reflects that the product has a pleasant smell.

In the same sense, for the texture, the observation given by the tasters in this investigation shows that there is no statistically significant difference between the texture mean ($P \geq 0.05$) between one level of treatments and another, with a level of 95.0% confidence. Appreciating that the data obtained for this attribute, M1, is related to treatments, for samples M2 and M3 which are different, with M2 showing the highest grade 3.75b ± 0.62 σ.

For the general acceptance, lower grades were presented ($p < 0.05$) clarifying that there is a significant difference between the means, with a confidence level of 95.0% between the syrups M1, M2 and M3, see Figure 1, obtaining the syrup M2 the highest rating, these values are consistent with the previous characteristics where M3 presented the lowest rating with respect to the overall rating in the flavor, smell and texture attributes.

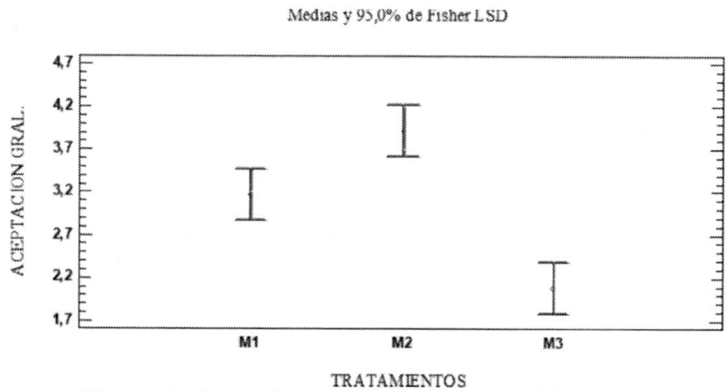

Figure 1. General acceptance diagram of the syrup.

For the syrup obtained from the guácimo fruit extract it is important to point out that of the points of the scale used for the evaluation, most of the qualifications that were obtained are between M1 and M2 higher points, which shows that in general These two types of syrups would have good acceptance by consumers.

4. Conclusions

Physicochemically characterizing the guácimo fruit (Guazuma ulmifolia lam), achieved a Brix degree value of 28.5%, 15.98% reducing sugars and 18.94% equivalent dextrose; which states that the fruit has basic components when producing sugar syrups or other products intended for human consumption.

The physicochemical analysis that was carried out on the different extracts allowed us to choose the four that have the highest percentage of dextrose equivalent, since they obtained values of 80.27% - 94.45%, which indicates that there is a product of high conversion and not it is necessary to concentrate to obtain a product with a high concentration of sugar, although its content of reducing sugars is between 4 and 5. However, it was concentrated to obtain higher dextrose equivalent values (90.13 ± 98.27).

The results obtained in the characterization indicate the potential of guácimo as a raw material for the production of sugar syrups; they reflect that the syrup J3 had the highest values in all the conditions and variables of responses (° Brix, reducing sugars, glucose, fructose and dextrose equivalent), showing difference before the other syrups. Thus, with respect to the dextrose equivalent (90.13 ± 98.27) it can be said that an extreme conversion syrup was obtained due to its high content of glucose and fructose, and it can also be used in the industries for the production of beer, soft drinks, juices, jams, etc.

The acceptance of the product was measured using a sensory hedonic scale, using a survey format, and the sugary syrup produced in a sample of sweetened coffee was applied at different concentrations of syrup and table sugar (sucrose), where it was determined that the mixture had greater organoleptic acceptability was M2 (50% syrup and 50% sugar), which indicates that the sugar syrup obtained from the guácimo fruit can be used partially in the diet of the person.

Conflict of Interest.
The authors report there are no conflicts of interest.

References
[1]Salcedo J, Montes E, Zapata J, Márquez D and Díaz M 2010 Obtención de jarabes de fructosa a partir de hidrolizados enzimáticos de almidón de ñame (Dioscorea Alata Y Dioscorea Rotundata). Vitae 17 243-251.
[2]Lafont J, Severiche C and Jaimes J 2018 Evaluation of the Starch Quantification Methods of Musa paradisiaca, Manihot esculenta, and Dioscorea trífida Using Factorial Experiments. International Journal of Food Science Article ID 5901930 7 pages.
[3]El Batal H, Hasib A, Ouatmane A, Dehbi F, Jaouad A and Boulli A 2016 Sugar composition and yield of syrup production from the pulp of Moroccan carob pods (Ceratonia siliqua L.). Arabian Journal of Chemistry 9 S955-S959
[4]Arrázola G, Alvis A and Herazo I 2015 Aprovechamiento tecnológico del almendro de india (Terminalia catappa L) para la obtención de productos alimenticios. ORINOQUIA 19 27-34.
[5]Quezada W and Gallardo I 2014 Obtención de extractos de plantas mucilaginosas para la clarificación de jugos de caña. Tecnología Química 34 114-123.
[6]Espitia J, Duran H, Fandiño J, Díaz F and Gómez H 2011 Química y biología del extracto etanólico del epicarpio de Crescentia cujete L. (totumo). Revista Cubana de Plantas Medicinales 16 337-346.
[7]Ramírez F, Campos R, Jiménez M and Brenes L 2016 Evaluación técnica, ambiental y económica de tres tipos de tratamiento para el cultivo de lechuga en huertas caseras de Guácimo, Limón, Costa Rica. Revista Tecnología en Marcha 29 14-24.
[8]Olivares J, Rojas S, Quiroz F, Camacho L, Cipriano M, Damián M, Ávila B and Villa A 2018 Diagnóstico de los usos, la distribución y características dasométricas del árbol cirián (Crescentia alata Kunth) en el municipio de Pungarabato, Guerrero, México. Polibotánica 45 191-204.
[9]Manríquez L, López S, Pérez P, Ortega E, López Z and Villarruel M 2011 Agronomic and forage characteristics of Guazuma ulmifolia Lam. Tropical and Subtropical Agroecosystems 14 453-463.
[10]Rodríguez G and Roncallo B 2013 Producción de forraje y respuesta de cabras en crecimiento en arreglos silvopastorales basados en Guazuma ulmifolia, Leucaena leucocephala y Crescentia cujete. Ciencia y Tecnología Agropecuaria 14 77-89.
[11]Vidal C 2011 Producción de Jarabes Edulcorantes por Hidrólisis Enzimática del Almidón de Ñame Variedad (Dioscórea rotundata). Publicaciones e Investigación 5 71-85.
[12]Beltrán J 2014 Elaboración de un edulcorante a base de cabuya y su efecto post-prandial en adultos diabéticos que asisten al centro cinco esquinas de la ciudad de quito durante el periodo diciembre 2013-abril. (Bachelor's thesis) Pontificia Universidad Católica del Ecuador. Quito, Ecuador.
[13]Gerena F 2013 Obtención de jarabes azucarados a partir de la hidrólisis química de residuos de cáscaras de naranja (citrus sinensis l var valencia) y papa (solanum tuberosum) variedad diacol capiro (r-12) para ser empleados como edulcorantes en la industria de alimentos. (Bachelor's thesis) Unibversidad Nacional Abierta y a Distancia UNAD. Duitama, Colombia.
[14]Queiroz R, Andrade K, Gomes L, Oliveira A, Ribeiro D, Ndiaye E and de Souza E 2019 Characterization of mutamba (Guazuma ulmifolia LAM.) fruit flour and development of bread. Biocatalysis and Agricultural Biotechnology 19 101120
[15]Kumar N and Gurunani S 2019 Guazuma ulmifolia LAM: A review for future View. Journal of Medicinal Plants Studies 7 205-210

IOP Conf. Series: Materials Science and Engineering **844** (2020) 012005 doi:10.1088/1757-899X/844/1/012005

Mobile application development to promote the sustainable transport of agricultural products

I A Vergara[1], M A Martínez[1] and L F Castañeda[1]
[1] Mechanical Engineering Department, GEMI Research Group, Universidad EAFIT, Medellín, Colombia

E-mail: inv_ivergara@eafit.edu.co

Abstract. Mobile applications today are present in most situations of daily life and in the productive sectors of the global economy. However, when talking about the rural areas of developing countries, the reality is different due to the difficulties for access, use and quality of the use of information and telecommunications technologies; the same problem of technological coverage occurs with the transport of agricultural products. This article presents a mobile application as a result of the research process, which makes the shared use of cargo vehicles easier and allows the integration of small and medium farmers with markets and transport companies. In this way, the costs of transport processes are reduced and agriculture becomes a more productive and competent activity with a view to fulfilling the sustainable development goals.

1. Introduction

The access and use of information and communications technologies (ICT) has consolidated as one of the trends of global growth in recent years, and Colombia is not unaware of this thanks to the growing expansion of cellular telephony and mobile internet. This has significantly favored the development of applications and digital media for different sectors such as urban transport, banking, food, commercial, among others. According to figures presented by the Ministry of Information and Communications Technologies (MinTIC), at the end of the third trimester of 2018, mobile telephony penetration was 127.3% with 63,427,667 subscribers, and mobile internet access (by subscription and demand) was 25,924,724 users [1].

In spite of the above, the field has been one of the productive sectors of the country in which there is a marked digital gap in terms of skills in the use of ICT, Internet connection and coverage of telecommunications and computing infrastructure [2]. The fact that small and medium farmers do not have tools that allow efficient and transparent access to markets, delays the growth of the rural economy and prevents their living conditions from improving. Likewise, the transportation of products is another factor that affects the development of the agricultural sector. According to the Food and Agriculture Organization of the United Nations (FAO), this process is distinguished by high complexity and heterogeneity, a fact that is aggravated by the poor condition of the roads and the conditions of the cargo vehicles. When agricultural products are grown far from the centers of consumption, export or industrialization, their transport becomes more important, which is reflected in the high costs that farmers must assume to the detriment of the earnings obtained from their work.

Undoubtedly, access to technology accelerates change and mobile devices can contribute to the development of rural areas by facilitating meetings between producers and traders [3]. ICT in this sector can be generalized as equipment, tools or applications that allow the rapid and accurate exchange of data and information through interaction and transmission processes [4], which generates a large number of possibilities to potentiate the field and make it more productive and efficient. Therefore, it is essential to increase the incorporation of ICT in each of the aspects related to the agricultural sector, from production and transport, to the commercialization of products; all in order to improve sustainability and competitiveness.

Thus, from the advantages offered by mobile applications it is possible to achieve the shared use of cargo vehicles, as well as the integration of small and medium farmers with transporters and marketers of agricultural products. In this way it is possible to reduce the chain of intermediaries, reduce the costs of transport processes, increase the profit of the producers and make the products competitive in the markets. In short, this contributes to the fulfillment of sustainable development goals (SDG), since it facilitates decent work and economic growth, reduces inequalities and leads to self-sustainable communities.

2. Methodology

The research process was developed taking into account the Design Science Research Methodology (DSRM), a trend to provide validity for research activity in engineering [5]. The application of this methodology was carried out following the process model suggested in [6], which consists of the five steps or stages shown in the figure 1.

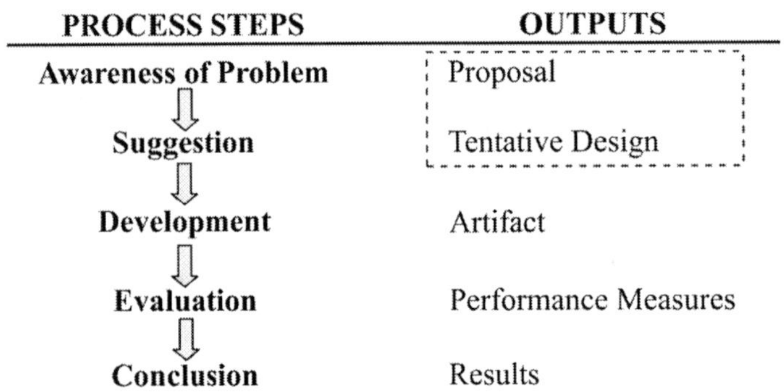

Figure 1. Design Science Research Process Model (DSRM) [6].

2.1. Awareness of Problem

According to the results of the last National Agricultural Census, conducted by the National Administrative Department of Statistics (DANE), 83.5% of the agricultural production units do not receive any type of technical assistance [7], besides this is the sector which presents the greatest delay in terms of incorporating ICT into productive, social and cultural activity [8]. The educational level and the lack of infrastructure such as networks, access points and equipment are the main constraints for the use of these technologies by farmers, which is demonstrated by the scarce presence of digital solutions in these areas. Moreover, the difficulty to quickly access the markets is another factor that impedes the development of rural areas. There are no tools that allow farmers to publicize their products or optimize the way they have been performing the transport processes in the field, a situation that increases production costs and decreases the income they receive as compensation for their work.

Taking into account the above, it is necessary to develop technological strategies that allow the effective integration of the actors involved in agricultural work, and that lead to the sector being more productive and profitable for all.

2.2. Suggestions

Considering the difficulties that small and medium farmers have to access the markets, as well as the high costs involved in transporting the products to the industrialization or consumption centers, the development of a multi-user mobile application, intuitive and easy to use, was proposed, capable of interconnecting producers, marketers and transport companies in a fast and direct way. Besides this, the application must allow the shared use of cargo vehicles as a strategy to reduce transport costs, since the value of the route is distributed among different users depending on the distance and the amount of product transported.

So, the application is constituted as an interaction space to know the available products and their location, generate and execute transport agreements, ease negotiations and bring historical relevant processes for each user.

2.3. Development

For the design and development process of the mobile application, the Software Life-Cycle model suggested in [9] was applied, which consists of the four phases shown in the figure 2.

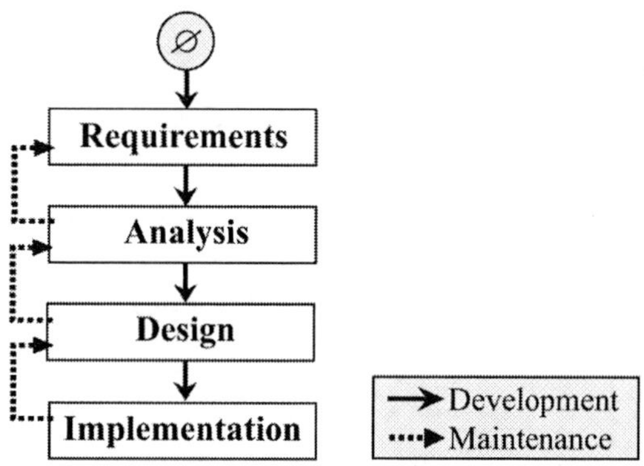

Figure 2. Phases of the Software Life-Cycle model [9].

2.3.1. Requirements. This phase structured in detail the functional and non-functional specifications required by the mobile application based on the awareness of problem and suggestions of the previous stages. These requirements are described below:

2.3.1.1. Functional Requirements. The functionality of the mobile application can be defined through the use cases, considering that these are a model of the interaction between actors and the software product itself [10]. The actor is a user who plays a specific role within the system and functional requirements are set by identifying the actors and actions [9].

As far as the actors are concerned, it is necessary to involve people in every marketing chain who produce, transport and buy the products. This is why producers, drivers and buyers were defined as actors of this mobile application; additionally, a coordinator was established for the management and supervision of the drivers of each transport company, as well as an administrator for the management of the software tool. The figure 3 represents the interaction of the actors with the system.

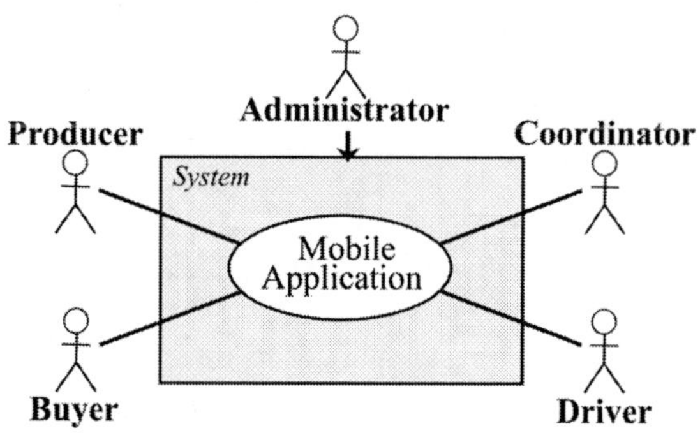

Figure 3. Actors established for the mobile application.

About the actions of the actors, it is established that the producer must make the publication of the products, the buyer must generate the purchase agreements and the transport requests, the coordinator must be in charge of the registration of drivers and the supervision of the agreements of transport, and the driver must execute the assigned route. Regarding the administrator, this one is responsible for the maintenance of the system and the linking of the transport companies. The use case diagram of the figure 4 shows in detail the interaction of the actors with the functional requirements established for this mobile application.

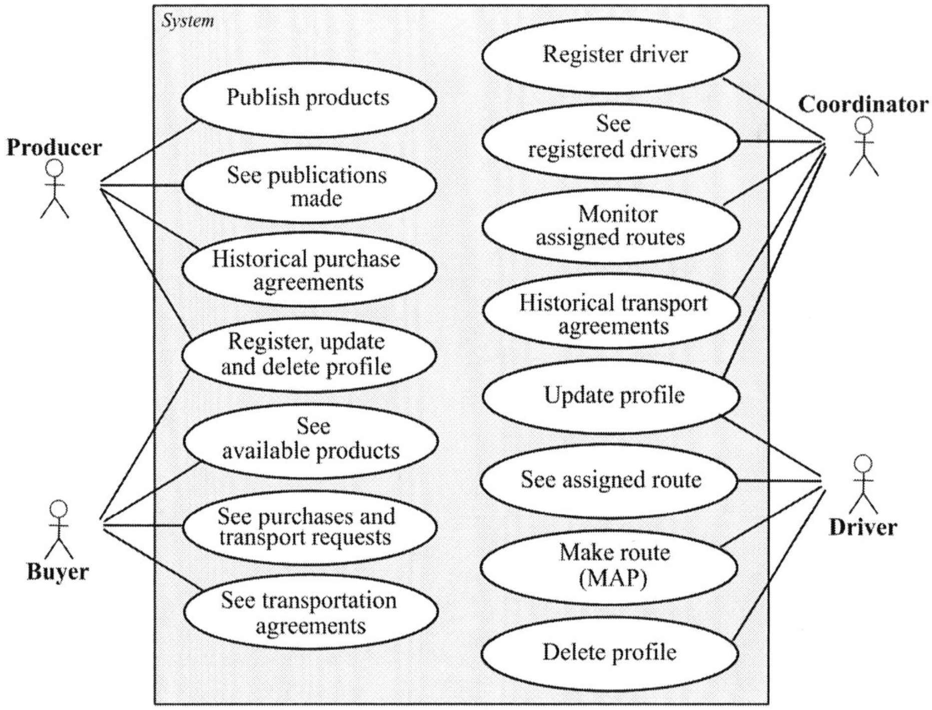

Figure 4. General diagram of cases of use of the mobile application.

2.3.1.2. Non-Functional Requirements. The non-functional requirements of a software tool describe the general aspects that do not include a direct relationship with the functional behavior of the system. These correspond to restrictions and demands for performance, availability, usability, among others [10]. As shown in the figure 5, the simple and intuitive design, the ease of use, the data validation, the quick response to the requests and the scalability of the number of users are examples of the non-functional requirements of this mobile application.

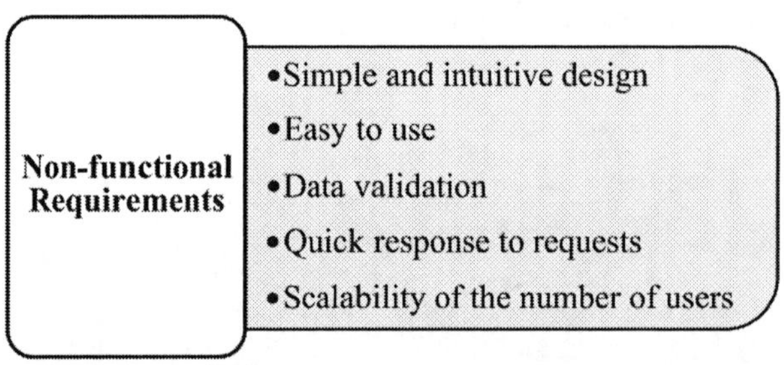

Figure 5. Non-functional requirements established for the App.

2.3.2. Analysis. In this phase the previously presented requirements were examined. Through this process it was determined that the functionalities established for the mobile application allow users to interact in an agile way and can share the cargo vehicles. Considering the above, these specifications were organized in the technical document that exposes the characteristics of the software tool.

2.3.3. Design. The established requirements were submitted to a process of architectural design and detailed design, to finally establish the architecture of the system.

During the architectural design, the mobile application was divided into the six components or modules of the figure 6, taking into account that these meet all their requirements.

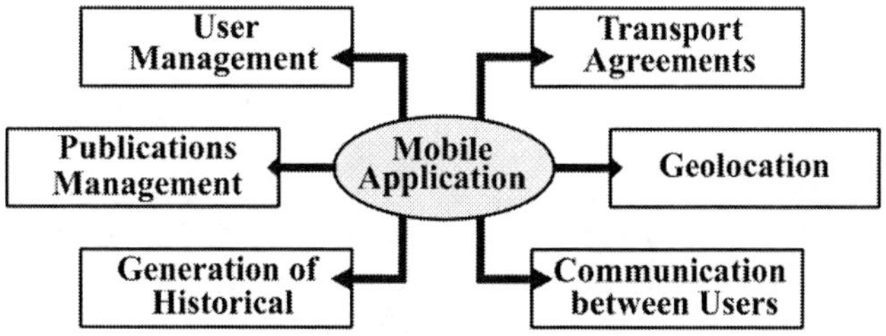

Figure 6. Modules of the architectural design of the App.

In terms of the detailed design, the characteristics of each particular module were defined and the programming algorithms were carried out. User management refers to the processes of registration, updating and elimination of profiles, as well as the beginnings and closings of sessions; the management of publications is associated with the basic functions in databases (CRUD - Create, Read, Update and Delete) to manage products, purchase agreements and transport requests; the generation of historical data is related to the consultations carried out to present the consolidated user on some aspect of interest; transportation agreements refer to the process of creation and allocation, after developing an analysis of the requests and the price calculations; geolocation involves obtaining the geographical coordinates of the driver's mobile device and its storage in order to show the location in real time to the other actors in the process; finally, communication between users is related to the making of phone calls as a means of consensus and negotiation.

On the other hand, the architecture of the system uses different Firebase products that are hosted in the cloud, among them, Authentication, Cloud firestore, Realtime database, Cloud storage and Hosting; in this way, scalability and performance problems can be mitigated in order to optimize the transfer of information and multimedia between the web server and the database [11]. The geolocation module and the memory of the mobile device are also used. The figure 7 represents the described architecture.

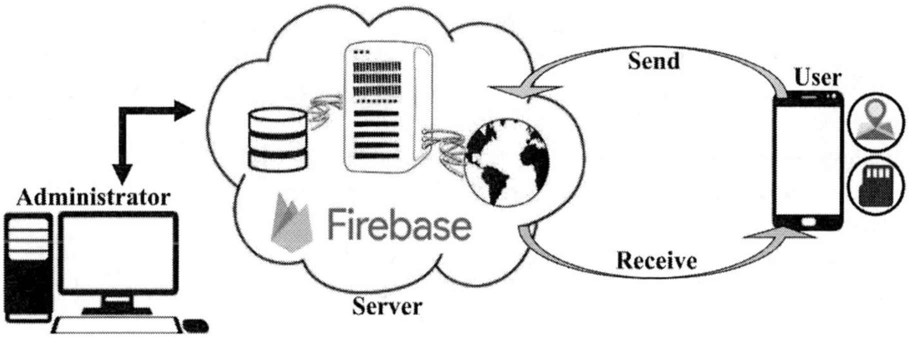

Figure 7. Architecture defined for the system. Based on genexus and firebase.google.

2.3.4. Implementation. The different modules of the previous phase were coded and tested. During coding, the algorithms designed in source code were converted, using for this Java, XML, HTML5, CSS5 and JavaScript as programming languages; the implementation was made in the integrated development environment of Android Studio and the Atom source code editor.

On the other hand, during the tests, the correct functioning of each module was checked separately, first making a revision of the source code in order to ensure the execution of the programmed methods, and then verifying that these generated the expected results; contrary situation the relevant modifications were made. The test cases were defined based on the functionalities specified in the use cases. At the end of the individual tests, the tests of all the combined modules of the application were made, that is, the integration of the software tool was carried out.

2.4. Evaluation

The evaluation of the developed mobile application was carried out through laboratory tests. In these, the registration of different users was made, who then made the publication and the purchase of distinct products. After analyzing the transport requests made by some of the buyers, and verifying their viability based on the quantity and places of origin and destination, transport agreements were generated that could be executed by the driver, and visualized in the options of each of the actors involved in the process. Based on these tests, the effectiveness of the software tool for the integration of users and the reduction of transport costs was evaluated.

2.5. Conclusion

In this last stage, the process of developing the mobile application was considered completed, considering that its operation is adequate and that the results of the evaluation are relevant to the proposal made. The information related to the software tool, its design, novelty and usefulness was communicated in the final report and in this paper.

3. Results

In the process to connect small and medium farmers to markets and allow the shared use of cargo vehicles, an intuitive and easy-to-use mobile application was developed that allows the integration of agricultural producers with buyers and transport companies, as well as, contributes to the reduction of costs due to the transport of agricultural products. Next, the data sheet of the application, its functionalities and a quality analysis are presented.

3.1. Data Sheet

In the data sheet of the table 1, the basic characteristics associated with the developed mobile application are presented.

3.2. Functions

From the preliminary tests carried out to the mobile application it was possible to establish that, within its functionalities, it allows:

- The management of the actors of the process, which includes the registration of new users, the validation and modification of data, the deletion of the profile and the resetting of the password.
- The logging in and out for each user, as shown in the figure 8.
- The publication of agricultural products and the updating of the data entered, as shown in the figure 9.
- The display of all published products available with its photo and features.
- The execution of the purchase agreement for the products of interest and the request for transportation if required, as shown in the figure 9.
- The analysis of transport requests and their grouping, in order to create and assign agreements through the execution of a backend hosted on the server in the Firebase cloud. This process includes the calculation of the price of transportation for each user that is part of the process.
- The visualization of transport agreements in the options of each of the actors involved.
- The implementation of the transport agreement assigned to the driver, as well as its follow-up by the coordinator of the transport company.

- The geolocation in real time, by the users grouped in the agreement, of the vehicle assigned to transport the products; as shown in the figure 9.
- The display of historical publications, purchase agreements and transport.
- Telephone communication between users without having to leave the application.

Table 1. Data sheet of the mobile application (TransportApp).

Feature	Description
Name of the application	TransportApp
Type of application	Mobile Application
Version	1.0
Description	TransportApp is a mobile application that facilitates the integration of small and medium producers with people interested in the purchase of their products and transport companies, which allows optimizing and reducing the costs of transport processes through the shared use of cargo vehicles.
Developer	Iván Andrés Vergara Morillo
Institution/Dependence	Universidad EAFIT Grupo de Estudios en Mantenimiento Industrial (GEMI)
Last update date	June 14 2019
Document creation date	June 15 2019
Integrated Development Environment (IDE)	Android Studio v3.2 and Atom
Programming languages	Java, XML, HTML5, CSS5 and JavaScript
Packaging format	.APK
Application size	16.2 MB
Database manager	Firebase (Cloud Firestore, Realtime Database and Cloud Storage)
Technical requirements for installation and use	- Android 4.4 (API 19) or higher - Built-in geolocation module - Minimum 512MB of RAM (2GB or more is recommended) - Network card (WiFi and Mobile networks)

Figure 8. Views of the App for the (a) login, (b) registration and (c) main interface.

Figure 9. Views of the App for the (a) publication and (b) display of products, (c) execution of the purchase agreements and (d) geolocation in real time.

3.3. Software quality

The quality of a software can be observed in its characteristics or attributes, that is, it refers to things that can be compared. In order to be understood in a better way, it must be expressed through terms that have more meaning for the user. Below are some elements to evaluate the quality of this mobile application.

3.3.1. Security and Data Integrity. Security and integrity are provided through Firebase Auth, which is integrated with the other products to control access to data and allow users to be managed by email and password stored in Firebase. To this end, security rules are established either that limit or admit operations on the database after the validation of a unique identifier (UID) during the authentication process.

3.3.2. Stability. The stability of the processes of the application is guaranteed since the functionalities offered have been verified during the evaluation stage.

3.3.3. Speed. The speed depends on the ability of the mobile device to execute the APK file, the internet connection and the speed of the server in which the websites are hosted for the terms and conditions, privacy policies and creation of transport agreements. On average, the login and publications have a response time of two seconds.

3.3.4. Usability. The mobile application has an intuitive interface in which most options are visible, the icons are simple and the labels are easy to understand.

3.3.5. Functionality. The mobile application performs a validation of the data entered before the execution of each process in order to avoid errors or inconsistencies. The result of the operations is also informed to the user, whether these are successful or unsuccessful.

4. Conclusions

The research process guided through the Design Science Research Methodology made it possible for the problem identification, design and development of the software tool to be a formal, controlled and rigorous activity in order to generate a technological solution for the agricultural sector.

A mobile application was designed and developed that eases the integration of farmers with buyers and transporters, while optimizing the transport process of agricultural products, that is, the tool allows the actors of the field to be connected in an agile way and reduce the costs due to transportation.

FAO, in its five principles for sustainable food and agriculture, exposes twenty action areas that describe approaches, practices, policies and tools that interrelate several SDGs and integrate its three dimensions, such as economic growth, social inclusion and environmental protection. The developed mobile application conforms to the second of these actions, connecting small farmers with markets, thus contributing to generating positive indicators in terms of compliance with these objectives and the incorporation of ICT in the agricultural sector.

Based on the actions generated by the mobile application in response to a specific event, preliminary tests performed on the tool indicate that its operation is adequate. That is why it is recommended to carry out its implementation in rural areas as a continuation of this investigation.

References

[1] Ministerio de Tecnologías de la Información y las Comunicaciones 2019 *Boletín Trimestral de las TIC Cifras Tercer Trimestre de 2018* (Bogotá: MinTIC) p 29 p 35
[2] Díaz J, Pérez A and Florido R 2011 Impacto de las tecnologías de la información y las comunicaciones (TIC) para disminuir la brecha digital en la sociedad actual *Cultivos Tropicales* **32** 83–5
[3] Organización de las Naciones Unidas para la Alimentación y la Agricultura 2018 *Transformar la Alimentación y la Agricultura para Alcanzar los ODS* (Roma: FAO) p 9 p 12 pp 18–9
[4] Flórez D and Uribe C 2018 *TIC para la Investigación, Desarrollo e Innovación del Sector Agropecuario* ed L Gaona (Mosquera: Agrosavia) pp 16–7
[5] Peffers K, Tuunanen T, Rothenberger M and Chatterjee S 2008 A design science research methodology for information systems research *J. of Management Information Systems* **24** 1–3 DOI: https://doi.org/10.2753/MIS0742-1222240302
[6] Vaishnavi V and Kuechler W 2015 *Design Science Research in Information Systems Int. Conf. on Design Science Research in Information Systems and Technology (Dublín)* pp 11–3
[7] Departamento Administrativo Nacional de Estadística 2016 *3er Censo Nacional Agropecuario* ed S M Escobar and E C Peñuela (Bogotá: DANE) p 109
[8] Nagel J 2012 *Principales Barreras para la Adopción de las TIC en la Agricultura y en las Áreas Rurales* (Santiago de Chile: CEPAL) pp 5–6 p 21
[9] Schach S 2011 *Object-Oriented and Classical Software Engineering* ed L Neyens (New York: McGraw Hill) p 41 p 577
[10] Forero H and Vega O 2015 *Sistema de Información Web de Apoyo al Concurso Docente de Planta en la Universidad de los Llanos* (Villavicencio: Universidad de los Llanos)
[11] Gómez U E, Pérez J P and Ramírez J L 2016 Sistema de información agrícola para la disminución de brechas entre oferta y demanda – AGROCRAFT *Información Tecnológica* **27** 217–9 DOI: http://dx.doi.org/10.4067/S0718-07642016000300020

Acknowledgments

The authors thank the Government of Cesar, COLCIENCIAS and the Universidad EAFIT for providing the necessary logistical and economic resources for the execution of the research work that allowed the development of the mobile application and this publication.

IOP Conf. Series: Materials Science and Engineering **844** (2020) 012006 doi:10.1088/1757-899X/844/1/012006

Learning styles for algebra concepts: The case of engineering students at the Universidad de Santander (Valledupar, Colombia)

C R Vidal-Tovar[1,2], H Valle-Fuentes[2], I C Rios-Dominguez[3], and A F Castro-Alfaro[4]

[1]Grupo de Investigación Creando Ciencias CRECI, Universidad Popular del Cesar, Valledupar-Cesar, Colombia.

[2]Grupo de Investigación CIEMPIES, Universidad de Santander, Valledupar-Cesar, Colombia.

[3]Maestria en Educación, Universidad Tecnologica de Bolivar, Cartagena de indias-Bolívar, Colombia.

[4]Grupo de Investigación CARTACIENCIA, Corporación Universitaria Rafael Nuñez, Cartagena de indias-Bolívar, Colombia.

carlosvidal@unicesar.edu.co

Abstract. The objective of this research was to evaluate the learning styles for algebra concepts of the students of the industrial engineering program of the University of Santander, Valledupar, Cesar, Colombia. A descriptive study was developed, consisting of 40 algebra students and 6 mathematics teachers, the questionnaire is self-administered elaborated with a dichotomous scale for the student and for interpretation of the means, a scale was designed based on a scale between zero and five that reveals the score obtained in said test, the styles studied were active, reflective, theoretical and pragmatic. Concluding that the predominant style is the pragmatic and theoretical, which allowed to select and apply strategies aimed at generating interactivity between students and teachers around the knowledge of algebra.

1. Introduction

Learning is achieved when there is a modification of the knowledge one has about something. Each being or individual has an exclusive way of learning and each one chooses his style to appropriate the new concepts [1,2,3]. In every teaching process in order to optimize it, the style of each participant in this process must be taken into account, an important input for the design of the teaching strategies to be undertaken in a learning session; because experience tells us that the student learns in a better way when instructed taking into account his predominant learning style [4,5].

But this activity is immersed in a complexity that encloses the human being as an individual that interacts with its environment. Learning Styles are defined as "cognitive, affective, physiological, preferences for the use of the senses, environment, culture, psychology, comfort, development and personality that serve as relatively stable indicators of how people perceive, interrelate and respond to their learning environments and their own methods or strategies in their way of learning "[6].

According to Gómez et al. 2017 [7], these features can be specified with instruments designed for this purpose, which guide how the student interacts with their environment. It is from there that he builds his own learning according to his style, which determines how he processes and perceives the information of his reality.

Each individual conceives his reality in a different way, this is how Safarin et al. (2013) [8], states that each student learns differently from others, uses different strategies, learns at different speeds, either more or less effectively and even if they have the same motivations, the same level of instruction, same age or are studying the same subject.

By analyzing learning styles over time and in the words of Waes et al. (2014) [9], the most outstanding and recognized studies by the academic community, are the models of Honey and Mumford in 1982 that establish four styles of learning: active, reflective, theoretical and pragmatic; and Kolb 1984 that classifies them into four stages: divergent, assimilative, convergent and accommodative.

Based on the above four types of learning styles are detailed according to how to organize and work with characteristics that clearly indicate the skill field of each of them as a result of a factorial analysis, which are shown in the Table 1:

Table 1. Types of learning styles.

Type of Style	Features
Active	These people get involved in the affairs of others and center around them all the activities they undertake with enthusiasm. They are open-minded, nothing skeptical. Your days are full of activity. They think that at least once you have to try everything. Spontaneous, creative innovators eager to learn and solve problems.
Reflective	They like to consider experiences and observe them from different perspectives. Collect data by analyzing them carefully before reaching any conclusion. They are prudent, observe well and consider all possible alternatives before making a move. They listen to others and do not act until they take ownership of the situation, they are pondered, patient, inquisitive, slow and detailed.
Theoretical	Adapt and integrate observations within logical and complex theories. They approach the problems vertically staggered, by logical stages. They tend to be perfectionists. They integrate the facts into coherent theories. They like to analyze and synthesize. They are profound in their system of thought, when establishing principles, theories and models. For them, if it is logical, it is good. They look for rationality and objectivity, fleeing from the subjective and the ambiguous.
Pragmatic	The practical application of ideas predominates in them. They discover the positive aspect of new ideas and take advantage of the first opportunity to experience them. They like to act quickly and safely with those ideas and projects that attract them. They tend to be impatient. They tread the earth when there is a decision to be made or a problem to be solved.

On the other hand, mathematics is present, to a greater or lesser extent, in each of the scientific advances and technological innovations of the contemporary world. There is a close correlation between the technological development in a society and the degree of insertion of Mathematics in its techniques [10]. In the training of engineers, the study of abstract sciences is preponderant, so much so that working with a contextualized mathematics is not an easy task neither for the teacher nor for the student since it is necessary to integrate mathematical knowledge with other areas of knowledge with the inherent implications of this process [11]. In the particular case of the concepts of algebra, when you want to use mathematics in the context of engineering, in its didactic phase, questions arise as to how to do it. This article seeks, without being the only way, to show how a teacher can work with Mathematics in the context of engineering to promote interdisciplinarity. The career of Industrial Engineer is taken as an example. With the above, it is expected to contribute to the evaluation of the learning styles for algebra concepts in engineering students of the Universidad de Santander, Valledupar headquarters in the Caribbean Colombiano.

IOP Conf. Series: Materials Science and Engineering **844** (2020) 012006 doi:10.1088/1757-899X/844/1/012006

2. Experimental

To obtain gu Population and Sample: For the present investigation, the population corresponds to the level of university undergraduate training in Industrial Engineering offered by the University of Santander, in the municipality of Valledupar, department of Cesar, Colombia. It is constitutedby 40 algebra students and 6 mathematics teachers.

Techniques and instruments for data collection: The data collection instrument will be the survey, the questionnaire is self-administered with closed type questions to apply to students and industrial engineering teachers of the Universidad de Santander. This was elaborated with a dichotomous scale for the student and for interpretation of the means, a scale was designed based on a scale between zero and five that reveals the score obtained in said test. The categories used in the scale for these scores are: total ignorance (DT), average ignorance (DM), neither know nor ignore (NC / ND), average knowledge (CM) and high knowledge (AC). To evaluate the learning styles of the students, it was designed with a Likert scale of 5-1, with the alternatives: Strongly Agree (TA), Medium Agree (MA), Neither Agree nor Disagree (NA / ND) , Medium Disagreement (MD) and Totally Disagree (TD).

Data analysis technique: The use of descriptive statistics was used with measures of central tendency, to carry out the treatment of the information obtained, in the processing of the results obtained through the application of the instruments. The average or arithmetic average was used, allowing the categorization of items, indicators, dimensions and study variables. Statistical packages such as Excel and SPSS were used for the statistical analysis of the information.

3. Results and discussion

The different learning styles of the engineering students of the University of Santander are presented next to the concepts of algebra.

Active learning style: This learning style could be evidenced in students of algebra because it involved being an animator, discoverer, risky and spontaneous. The tendency of the students regarding this style of learning was not to agree or disagree (NA/ND) with being animator, discoverer, risky and spontaneous, which can be verified in the scale with the scores respectively obtained in each indicator such as 3.7, 2.3, 2.6 and 3.4. In Table 2, the results obtained on the students' active learning style can be compared.

Table 2. Active learning style.

SUB DIMENSION	ítem	INDICATOR	TD	MD	D	MA	TA	CONCLUSION ACCORDING TO BAREMO
ACTIVE LEARNING STYLE	3	Animator	13%	0%	0%	31%	56%	The tendency is not to agree or disagree
	5		0%	13%	19%	25%	43%	
	6		18%	31%	13%	25%	13%	
		AVERAGE%	10%	15%	11%	27%	37%	
		SCORE	0,1	0,3	0,3	1,1	1,9	3,7
	4	Discoverer	6%	25%	38%	0%	31%	The tendency is not to agree or disagree
	8		75%	13%	6%	0%	6%	
	12		50%	18%	13%	13%	6%	
		AVERAGE%	44%	19%	19%	4%	14%	
		SCORE	0,4	0,4	0,6	0,2	0,7	2,3
	2	Risky	25%	19%	19%	25%	12%	The tendency is not to agree or disagree
	10		31%	13%	18%	25%	13%	
	11		31%	31%	13%	25%	0%	
		PROMEDIO %	29%	21%	17%	25%	8%	
		SCORE	0,3	0,4	0,5	1,0	0,4	2,6
	1	Spontaneous	13%	0%	6%	50%	31%	The tendency is not to agree or disagree
	7		13%	13%	6%	50%	18%	
	9		13%	25%	31%	25%	6%	
		AVERAGE%	13%	13%	14%	42%	18%	
		SCORE	0,1	0,3	0,4	1,7	0,9	3,4
HALF %			24%	17%	15%	25%	20%	The tendency is not to agree or disagree
AVERAGE SCORE			0,2	0,3	0,5	1,0	1,0	3,0

Finally, it can be seen that the average score of the active learning style is 3.0, which according to the scale gives the group a tendency to disagree or disagree (NA / ND) with this learning style.

Reflective learning style: This learning style is characterized in students of algebra who are weighted, conscientious, receptive and analytical. In Table 3, the results can be observed.

Table 3. Reflective learning style.

SUB DIMENSION	ítem	INDICATOR	TD	MD	D	MA	TA	CONCLUSION ACCORDING TO BAREMO
REFLECTIVE LEARNING STYLE	19	Weighted	38%	13%	18%	0%	31%	The tendency is not to agree or disagree
	23		0%	31%	19%	31%	19%	
	24		19%	13%	31%	31%	6%	
		AVERAGE%	19%	19%	23%	21%	19%	
		SCORE	0,2	0,4	0,7	0,8	0,9	3,0
	13	Conscientious	6%	25%	13%	25%	31%	The tendency is to be fairly in agreement
	14		0%	13%	6%	19%	62%	
	21		6%	6%	6%	63%	19%	
		AVERAGE%	4%	15%	8%	36%	37%	
		SCORE	0,0	0,3	0,3	1,4	1,9	3,9
	15	Receptive	6%	6%	26%	31%	31%	The tendency is not to agree or disagree
	17		6%	13%	25%	43%	13%	
	22		13%	37%	0%	0%	50%	
		AVERAGE%	8%	19%	17%	25%	31%	
		SCORE	0,1	0,4	0,5	1,0	1,6	3,5
	16	Analytical	0%	25%	25%	25%	25%	The tendency is to be fairly in agreement
	18		13%	13%	6%	31%	37%	
	20		0%	0%	19%	43%	38%	
		AVERAGE%	4%	13%	17%	33%	33%	
		SCORE	0,0	0,3	0,5	1,3	1,7	3,8
HALF			9%	16%	16%	29%	30%	The tendency is not to agree or disagree
AVERAGE SCORE			0,1	0,3	0,5	1,1	1,5	3,5

In the reflective learning style on algebra, students do not agree or disagree (NA / ND) with being weighted and receptive, which can be observed in the scale with the scores respectively obtained in each indicator such as 3.0 and 3.5. While they are fairly in agreement (MA) with being conscientious and analytical, which can be evidenced in the scale with the scores respectively obtained in these indicators such as 3.9 and 3.8. Finally, it can be seen that the average score of the reflective learning style is 3.5, which according to the scale gives the group a tendency to disagree or disagree (NA/ND) with this learning style.

Theoretical learning style: This learning style is observed in algebra students when it is demonstrated that they are methodical, logical, critical and structured. The students tendency towards the theoretical learning style about algebra, was to be in agreement (MA) with being logical and critical. And they do not agree or disagree (NA/ND) on being methodical and structured; which can be evidenced in the scale with the scores respectively obtained in each indicator such as 4.2, 4.1, 3.45 and 3.6. In Table 4, the results obtained with regard to this learning style are evident.

IOP Conf. Series: Materials Science and Engineering **844** (2020) 012006 doi:10.1088/1757-899X/844/1/012006

Table 4. Theoretical learning style.

SUB DIMENSION	ítem	INDICATOR	TD	MD	D	MA	TA	CONCLUSION ACCORDING TO BAREMO
ESTILO DE APRENDIZAJE TEÓRICO	31	Methodical	6%	38%	12%	6%	38%	The tendency is not to agree or disagree 3,45
	33		0%	13%	19%	25%	43%	
	36		19%	13%	19%	43%	6%	
		AVERAGE%	8%	21%	17%	25%	29%	
		SCORE	0,08	0,43	0,50	0,99	1,45	
	26	Logical	0%	0%	19%	38%	43%	The tendency is to be fairly in agreement 4,2
	27		0%	12%	19%	19%	50%	
	28		0%	6%	25%	13%	56%	
		AVERAGE%	0%	6%	21%	23%	50%	
		SCORE	0,0	0,1	0,6	0,9	2,5	
SUB DIMENSION	**ítem**	**INDICATOR**	**TD**	**MD**	**D**	**MA**	**TA**	**CONCLUSION ACCORDING TO BAREMO**
	25	Critical	0%	6%	13%	13%	68%	The tendency is to be fairly in agreement 4,1
	29		0%	0%	19%	31%	50%	
	30		6%	13%	19%	31%	31%	
		AVERAGE%	2%	6%	17%	25%	50%	
		SCORE	0,0	0,1	0,5	1,0	2,5	
	32	Structured	0%	0%	38%	50%	12%	The tendency is not to agree or disagree 3,6
	34		19%	6%	13%	37%	25%	
	35		0%	19%	31%	6%	44%	
		AVERAGE%	6%	8%	27%	31%	27%	
		SCORE	0,1	0,2	0,8	1,2	1,4	
HALF			4%	11%	21%	26%	39%	The tendency is to be fairly in agreement
AVERAGE SCORE			0,0	0,2	0,6	1,0	1,9	3,8

Finally, it can be shown that the average score of the theoretical learning style is 3.8, which, according to the scale, gives the group a tendency to agree moderately (MA) with this learning style.

Pragmatic learning style: This learning style is characterized by algebra students who are experienced, practical, effective and realistic. In Table 5, we can see the results in percentage and score that were obtained on the pragmatic learning style of the students.

Table 5. Pragmatic learning style.

SUB DIMENSION	ítem	INDICATOR	TD	MD	D	MA	TA	CONCLUSION ACCORDING TO BAREMO
PRAGMATIC LEARNING STYLE	38	Experimenter	0%	6%	13%	31%	50%	The tendency is not to agree or disagree
	41		13%	25%	25%	37%	0%	
	43		6%	25%	19%	13%	37%	
		AVERAGE%	6%	19%	19%	27%	29%	
		SCORE	0,1	0,4	0,6	1,1	1,5	3,5
	39	Practical	0%	0%	6%	44%	50%	The tendency is to be fairly in agreement
	44		0%	6%	19%	25%	50%	
	46		25%	25%	12%	19%	19%	
		AVERAGE%	8%	10%	12%	29%	40%	
		SCORE	0,1	0,2	0,4	1,2	2,0	3,8
	40	Effective	6%	0%	13%	38%	43%	The tendency is not to agree or disagree
	42		13%	6%	38%	43%	0%	
	47		0%	0%	38%	50%	12%	
		AVERAGE%	6%	2%	30%	44%	18%	
		SCORE	0,1	0,0	0,9	1,7	0,9	3,7
	37	Realistic	0%	13%	13%	43%	31%	The tendency is to be fairly in agreement
	45		13%	13%	19%	13%	42%	
	48		0%	0%	6%	25%	69%	
		AVERAGE%	4%	9%	13%	27%	47%	
		SCORE	0,0	0,2	0,4	1,1	2,4	4,0
HALF			6%	10%	18%	32%	34%	The tendency is to be fairly in agreement
AVERAGE SCORE			0,1	0,2	0,6	1,3	1,7	3,8

In the pragmatic learning style on algebra, students agree fairly (MA) with being practical and realistic, which can be observed in the scale with the respective scores of 3.8 and 4.0 obtained in these indicators. While in the group they do not agree or disagree (NA / ND) with being an experimenter and effective, which can be evidenced in the scale with the score for these indicators of 3.5 and 3.7. Finally, it can be observed that the average score of the pragmatic learning style is 3.8, which according to the scale gives the group a tendency to agree moderately (MA) with this learning style.

Students' predominant learning style: Based on the previous analyzes, it can be affirmed that the predominant learning styles in the students of algebra are the theoretical learning style and the pragmatic learning style, where the students obtained a score of 3.8 for each one of them, which according to the scale gives a tendency to the group to be fairly in agreement (MA) with these learning styles. Second, there is the tendency of students with the reflective learning style, where a score of 3.5 was obtained, which according to the scale gives a tendency to the group to disagree or disagree (NA / ND) with this style of learning. And finally the students identify with the active learning style where they obtained a rating of 3.0 which according to the scale gives a tendency to the group not to agree or disagree (NA / ND) with this style of learning.

4. Conclusions

The learning styles of the Engineering students of the University of Santander were identified, highlighting that it is very important to recognize which are the learning styles (active, reflexive, theoretical and pragmatic) that predominate most in the students, concluding that the predominant style is the pragmatic and theoretical, which allowed to select and apply strategies aimed at generating interactivity between students and teachers around the knowledge of algebra.

Conflict of Interest.

The authors report there are no conflicts of interest.

References

[1] Giuliano C, Moser L, Poremba V, Jones J, Martin E and Slaughter R 2014 Use of a unified learning style model in pharmacy curricula. Currents in Pharmacy Teaching and Learning 6 41-57.

[2] Romero L, Alarcón M and Gómez M 2014 Learning styles and web technology use in Business and Economics university students Carlota. Procedia-Social and Behavioral Sciences 141 1281-1290.

[3]Fong W, Severiche C, Jaimes J, Marrugo Y and Espinosa E 2017 Cognition and Its Relationship with Endogenous and Exogenous Factors in Engineering Students. International Journal of Applied Engineering Research 12 6929-6933.

[4] Shinnick M and Woo M 2015 Learning style impact on knowledge gains in human patient simulation. Nurse Education Today 35 63–67.

[5] Tocci A 2013 Estilos de aprendizaje de los alumnos de ingeniería según la programación neurolingüística. Revista Estilos de Aprendizaje 12 12-20.

[6] Fong W, Severiche C, Pitre R, Vargas L and Espinosa E 2017 Association Between Self-Regulation of Learning, Student Attitude, Provenance and Age in Engineering Students. Contemporary Engineering Sciences 10 665-672.

[7] Gómez E, Jaimes J and Severiche C 2017 Estilos de aprendizaje en universitarios, modalidad de educación a distancia. Revista Virtual Universidad Católica del Norte 50 383-393.

[8] Safarin M, Mohd N, Subari K and Zolkiflibin M 2013 Visualization Skills and Learning Style Patterns among Engineering Students at Universiti Teknologi Malaysia. Procedia - Social and Behavioral Sciences 93 1769-1775.

[9] Waes L, Weijen D and Leijen M 2014 Learning to write in an online writing center: The effect of learning styles on the writing process. Computers & Education 73 60-71.

[10] Abdul-Rahman S and Boulay B 2014 Learning programming vía worked-examples: Relation of learning styles to cognitive load. Computers in Human Behavior 30 286–298.

[11] Eishani K, Saa'd E and Nami Y 2014 The Relationship Between Learning Styles and Creativity. Procedia – Social and Behavioral Sciences 114 52-55.

IOP Conf. Series: Materials Science and Engineering **844** (2020) 012007 doi:10.1088/1757-899X/844/1/012007

Scientific reasoning skills and patterns of use of Information and Communication Technologies. The case of the students of the Universidad de Santander (Valledupar, Colombia)

G E Angulo Blanquicett[1], G E Angulo Blanquicett[2], C R Vidal-Tovar[3,4] and I C Rios-Dominguez[5]

[1]Dirección del Programa de Administración Financiera, Universidad de Santander, Valledupar-Cesar, Colombia
[2]Coordinación de Programas y Departamento, Universidad Jorge Tadeo Lozano Seccional del Caribe, Cartagena de indias-Bolívar, Colombia
[3]Grupo de Investigación Creando Ciencias CRECI, Universidad Popular del Cesar, Valledupar-Cesar, Colombia
[4]Grupo de Investigación CIEMPIES, Universidad de Santander, Valledupar-Cesar, Colombia
[5]Maestria en Educación, Universidad Tecnologìca de Bolivar, Cartagena de indias-Bolívar, Colombia

gangulo@valledupar.udes.edu.co

Abstract. This research establishes the relationship that has had the development of the skills of scientific reasoning and the patterns of use of tools and resources of information and communication technology (ICT). For that, it is necessary to use Lawson Test and to have into account ICT National Standards for students that USA International Society for Technology in Education (ISTE) suggest. Also, this paper is tackled from a non-experimental design featuring a correlational – transectional range. The population consists of 2035 students from a higher education institution of the Colombian Caribbean coast. Besides, stratified random sampling is used defining, defining strata and a representative sample for academic programs. Personal survey was used to collect information. The existence of relationship between scientific reasoning skills and patterns of use of information and communication technology was confirmed.

1. Introduction

When talking about scientific reasoning skills, reference is made to the existence of formal operations, which precisely, facilitate the elaboration of propositions in the individual, since to demonstrate scientific thought, abstraction, propositional logic, probabilistic notions are indispensable, among other factors, for its exercise. In this way the individual is able to use the eight formal operating schemes of combinatorial operations, proportions, combination of two reference systems, notion of mechanical equilibrium, notion of probability, notion of correlation, multiplicative compensations, forms of conservation, which are acquired in a solidary or homogeneous way from the domain of formal thought [1].

Lawson, in criticizing Piaget's postulates, argued that the development of scientific thought is associated with processes of instruction [2]. According to this author, the Piagetian approach to learning has several edges "[...] different levels of learning analysis, from behavior, information, representation and knowledge from the point of view of different psychological theories". Thus, Lawson places students in three categories of reasoning: concrete operations, transition operations and formal operations; evaluating six aspects of scientific reasoning: conservation of physical magnitudes, proportional thinking, identification and control of variables, probabilistic thinking, correlational thinking and combinatorial thinking [3].

On the other hand, the enormous accumulation of information, led to a series of practices that stopped being exclusive of science to become the whole of society. Thus, knowledge ceased to be something exclusive to a small group of people dedicated to science, and began to have an influence on each of the dimensions of human life: the everyday, the work, the family, etc. In this way, since the Century of Lights (eighteenth century) began to form what is known as the knowledge society. The notion of the Knowledge Society is characterized by the ability to make use of information through analysis or critical reflection to generate the knowledge necessary to promote individual and collective human development. In addition, this notion incorporates a network construction of society that promotes autonomy and encompasses the notions of plurality, integration, solidarity and participation [4].

In this sense, the development of digital technologies, added to the Knowledge Society, has made the relationship between man and his environment change according to the use given to information and communication technology. This relationship is deepened day by day according to technological advances, in such a way that the new generations are more integrated with these advances. They are known as digital natives. The digital natives are those people - mostly young people - who were born immersed in technology and live with it since they were very young, so they are used to it and are very skilled at using it [5].

For this, the Colombian state, occupies a position, not very privileged, in terms of indicators that measure, among others, the degree of preparation for the use of ICTs, with respect to other countries, which count and develop a clear policy against ICT. Those countries that have had policies that have explicitly pointed to the development of ICT and competitiveness, and that have been sustained over time, tend to see it reflected in these measurements. Chile, for example, is a country that has had a great commitment not only to economic policies aimed at increasing its competitiveness, but has also focused attention on the appropriation and use of ICT.

Likewise, Costa Rica has also had a very proactive approach in the use of these technologies. Although these indices constructed in this way have some deficiencies, as in many cases they are outdated, or based on perceptions and not on measurable variables, they still allow for comparisons between countries and draw attention to aspects that must be addressed. to get better.

Thus, the Ministry of Information Technologies and Communications (MINTIC) announced in December 2017, the first results of the Digital Economy Observatory: "The Observatory analyzed the adoption and management of mature digital technologies, such as broadband , mobile technologies and management platforms, and advanced digital technologies, such as cybersecurity, cloud computing, internet of things, robotics, big data, artificial intelligence, etc., taking into account the size of companies, the economic sector and the region [6].. This is how, in 2018, MINTIC is concerned about contributing with ICT to the transformation of education, since it considers as its objective 3 "Promote ICT as a platform for equity, education and competitiveness", trusting that what is established at the educational policy level for the use and access of ICT in society, contributes to improving the country's underprivileged position. In this sense, the objective of the work is to determine the relationship between the scientific reasoning skills and the patterns of use of information and communication technology of the students of a higher education institution of the Colombian Caribbean Coast.

2. Experimental

2.1. Method

The research was developed under the foundations of the quantitative, descriptive, field, transectional and non-experimental approach according to the proposal of Hernández and Others (2014) [7]. Some of its elements are described below.

2.2. Participants

The population is constituted by the students of a higher education institution of the Colombian Caribbean Coast, which amount to 2035 students. The students are distributed in 8 academic programs offered by the institution (7 professionals and 1 technology): Financial Administration, Physiotherapy, Industrial Engineering, Bacteriology and Clinical Laboratory, Psychology, Law, Veterinary Medicine and Zootechnics and Technology in Graphic Design and Advertising.

When using stratified random sampling, III Strata were defined (I: 1st to 3rd Semester, II: 4th to 6th Semester, III: 7th to 8th - 10th Semester.) Bearing in mind that the population is 2035 students and expecting a standard error of less than 0.04 and a confidence level of 0.95, the sample size is 464 students, distributed in the three strata, and Table 1 shows the distribution of the samples by cohort. At the beginning of the race, the distribution of the sample was 43%, in the middle of the race the distribution of the sample was 34% and at the end of the race the distribution of the sample was 23%.

Table 1. Distribution of the sample by cohort

	Frequency	Percentage
Early in the race	271	43%
In the middle of the race	209	34%
At the end of the race	146	23%
Total	626	100.0

2.3. Techniques

For the collection of information, the Survey was used, considered the most used quantitative technique for obtaining primary information; Among the different types of surveys, the personal survey was chosen for this research, where there is direct contact between the interviewer and the interviewee [7].

2.4. Instruments

Lawson questionnaire (standardized psychometric test). The scientific reasoning skills in the students were identified using a modified version of the Scientific Reasoning Test in the Lawson Classroom (1978). The validity of the original test was established in several studies (for example, Lawson, 1978, 1979, 1980, 1982, 1983, 1992, Lawson, Weser, 1990, Lawson, Baker, DiDonado, Verdi, Johnson, 1993) [9].

The modified test includes 24 multiple-choice questions that identify reasoning patterns related to: Conservation of physical magnitudes, proportionality thinking, identification and control of variables, probabilistic thinking, correlation and probabilistic thinking and combinatorial thinking. Similarly, the results of the Lawson Test allow students to be placed in three categories of reasoning: specific operations, transition operations and formal operations. This test was designed in such a way that the approaches formulated to the students do not require a complex understanding of contents, which reduces any type of interference and allows to improve the measurement of basic reasoning skills. The validity of the modified version was established by Lawson (1999) and Lawson, Clark, Cramer-Meldrum, Falconer, Seaquist, Kwon (2000) [9].

Questionnaire about the use of technology. To determine the patterns of use of technology by the students in the sample, a questionnaire of 24 multiple-choice questions was designed that gathers information related to: Type of resources and technological tools that it uses, hours and days of use of ICT to the week, places of access and use of technology, habits of use of ICT for learning, use of ICT in teaching-learning activities in the institution, assessment of learning and use of technology.

With this information will be reported the patterns of use of information technology and communication: Operations and concepts of computer science, knowledge of the Internet, use of ICT in the planning of learning, research and management of digital information for learning and

IOP Conf. Series: Materials Science and Engineering **844** (2020) 012007 doi:10.1088/1757-899X/844/1/012007

general ICT competence.

2.5. *Data analysis*

For the analysis of the following results we proceed first to make descriptive statistics such as frequency and percentages.

We used a Kolmogorov-Smirnov test, of goodness of fit, which serves to test the null hypothesis in which the distribution of a variable fits a certain theoretical distribution of probability. If the value of the criterion or level of significance is very small (less than 0.05), the hypothesis of normality is rejected and it is concluded that the scores of that variable do not adjust to a normal distribution. The results of the Kolmogorov-Smirnov test indicate that the hypothesis of normality is rejected with a critical level of $p < 0.005$, and we conclude that the scores of the variables do not conform to a normal distribution. That is, that nonparametric statistics must be used to analyze the data.

Then a Spearman correlation was made. The decimal number obtained by relating these variables indicates the strength of relationship and statistical significance of the same, thus from the numerical value of the correlation coefficient obtained, it is considered that values close to zero denote a weak relationship, while that approached + 1 or -1 indicate a stronger relationship. The correlation scores that showed a level of significance less than or equal to .05 were taken into consideration.

In order to check each of the hypotheses raised in this research, a Kruskal-Wallis H-test was used. It is an extension of the U Mann-Whitney test is the nonparametric analog of the one-way analysis of variance (ANOVA) and detects differences in the location of distributions. The statistical software SPSS was used to perform these analyzes.

3. Results and Discussion

Below is an analysis of the main results.

In determining the relationship between scientific reasoning skills and the patterns of use of information and communication technology of students who are beginning their careers at a higher education institution in the Colombian Caribbean Coast, it is observed that there is a significant relationship between the operations pattern and computer concepts and the proportionality thinking skills ($r = .183$, $p < 0.010$); correlational and probabilistic thinking ($r = .229$, $p < 0.010$); Combinatorial thought ($r = .127$, $p < 0.050$). In addition, there is a significant relationship between the internet knowledge pattern and the identification and control of variables ($r = .120$, $p < 0.050$); probabilistic thinking ($r = .143$, $p < 0.050$); Correlation and probabilistic thinking ($r = .190$, $p < 0.010$).

When determining the relationship between scientific reasoning skills and the patterns of use of information and communication technology of students who are in the middle of their higher education careers in the Colombian Caribbean Coast, it is observed that there is a significant relationship between the pattern of operations and computer concepts and the conservation abilities of physical magnitudes ($r = .191$, $p < 0.010$); Proportionality thinking ($r = .158$, $p < 0.050$); identification and control of variables ($r = .221$, $p < 0.010$); probabilistic thinking ($r = .274$, $p < 0.010$); correlational and probabilistic thinking ($r = .194$, $p < 0.010$); combinatory thinking ($r = .159$, $p < 0.050$). There is also a significant relationship between the Internet knowledge pattern and probabilistic thinking ability ($r = .247$, $p < 0.010$). Similarly, there is a significant relationship between the pattern of use of ICT in the planning of learning with the conservation skills of physical magnitudes ($r = .136$, $p < 0.050$); identification and control of variables ($r = .183$, $p < 0.010$); probabilistic thinking ($r = .184$, $p < 0.010$).

In determining the relationship between the scientific reasoning skills and the patterns of use of information and communication technology of students who are at the end of their careers at a higher education institution in the Colombian Caribbean Coast, it is observed that there is a significant relationship between the pattern of operations and computer concepts and the ability to conserve physical quantities ($r = .234$, $p < 0.010$). There is also a significant relationship between the knowledge pattern of the internet and the conservation skills of physical magnitudes ($r = .288$, $p < 0.010$); identification and control of variables ($r = .186$, $p < 0.050$). Similarly, there is a significant relationship between the pattern of research and management of digital information for learning and the ability to think correlational and probabilistic ($r = -. 164$, $p < 0.050$).

When determining the differences that exist between the scientific reasoning abilities among the students' scores in the different cohorts, there are significant differences in probabilistic thinking ($X2 = 6.039$, $gl = 2$, $p < 0.050$), because as students are advancing in their careers have a greater probabilistic thinking. That is to say that the hypothesis that there is a significant difference of the scientific reasoning ability "Probabilistic thinking" between the scores of the students in the different cohorts is approved.

When determining the differences that exist between the patterns of use of information and communication technology between the scores of the students in the different cohorts, there are significant differences in the use of ICT in the planning of learning ($X2 = 15.583$, $gl = 2$, $p < 0.050$), because as students progress in their careers they have a greater planning of their learning. That is, the hypothesis that there is a significant difference in the pattern of use of information and communication technology "Planning of learning" between the scores of students in the different cohorts is approved.

The results reveal that there is a significant relationship between some of the skills of scientific reasoning and some of the patterns of use of information and communication technology, coherently as it is contemplated by Prensky (2011) [8], the cognitive differences of the Digital Natives claim for new approaches in education, a better and more thoughtful "adjustment". This affirmation is made, since a relationship was found between the pattern of operations and computer concepts, and the thinking skills related to proportionality, correlational and probabilistic thinking and combinatorial thinking. Likewise, a relationship was found between the knowledge pattern of the internet and the skills of identification and control of variables, probabilistic thinking, correlational and probabilistic thinking.

In this way, when performing a correlation by cohort, according to the semester studied by the students in their training career, a significant relationship was found between some of the scientific reasoning skills and some of the patterns of use of information and communication technology. with students who are at the beginning, in the middle and at the end of their careers in a higher education institution of the Colombian Caribbean Coast. Similar result found by Benford, Russell; Lawson, Anton E. (2001) [9]. Relationships between Effective Inquiry Use and the Development of Scientific Reasoning Skills in College Biology Labs. National Science Foundation, Arlington, VA, in which the most important findings is that the ability to effectively use research in a university-level science lab It seems to be strongly associated with the formal and post-formal scientific reasoning skills of the instructor, and that the effective use of research promotes the acquisition of scientific reasoning skills in university-level students. To say, both the effective use of research and the use of ICT patterns facilitate the development of scientific reasoning skills in higher education students.

In that sense, it was found that the more students realize, whether early, mid or late in their careers, of the use of information and communication technologies, they will use more scientific reasoning skills. And in turn, while students use scientific reasoning skills, they will demonstrate patterns of use of information and communication technology. And precisely, Colombia, which in its history has not prevailed the provision of a clear policy for the use of information and communication technology [10], has led to the preparation of its inhabitants for the use of ICT is not the most appropriate and so has the use of information and communication technology, which does not facilitate the development of scientific reasoning skills, thus finding a significant relationship between scientific reasoning skills and the use that is made of it. makes of information technology and communication.

Likewise, it was found that, for the pattern of operations and computer concepts, there is a significant relationship for more scientific reasoning skills in students who are mid-career. Likewise, it was found that, for the pattern of Internet knowledge, there is a significant relationship for more scientific reasoning skills in students who are at the beginning of their careers. Likewise, it was found that, for the pattern of use of ICT in the planning of learning, there is a significant relationship for scientific reasoning skills, only in the mid-career students.

In turn, the Lawson Test allows students to be placed in three categories of reasoning: specific operations, transition operations and formal operations; In addition, it evaluates six aspects of scientific reasoning: a) conservation of physical magnitudes, b) proportional thinking, c) identification and control of variables, d) probabilistic thinking, e) correlational thinking and f) combinatory thinking [11].

In the same way, the categories of reasoning in which the students are located (specific operations, transition operations and formal operations), according to the results obtained with the Lawson Test, it was found that the predominant category is that of specific operations for all cohorts; However, the lowest value is presented to the students at the beginning of the career, followed by the value found for the students in the middle of the career and then the value found for the final-year students. It is observed with concern that the final year students do not show formal operations. It is highlighted that the mid-career students are those who show the most formal operations. Similar situation, found when applying Lawson's Test, to have a diagnosis of the degree of scientific reasoning of five groups of first-year students entering engineering degrees; and by applying a questionnaire on Force Concepts (Forcé Concept Inventory) to diagnose their level of conceptual knowledge in Mechanics, by Rodríguez, María D, Mena, Daniel A, Rubio, Carlos M. (2010) [11], in which the conclusion was obtained a significant percentage of students with a low level of scientific reasoning.

On the other hand, the students of the institution account for a medium or low level of scientific reasoning skills (medium level in conservation of physical magnitudes and low level in thought of proportionality, identification and control of variables, probabilistic thinking, correlational thinking and probabilistic and combinatory thinking). This is evidenced independently of the cohorts, that is, for the students at the beginning of their career, at the mid-career and at the end of their career, the medium or low level is presented for the same scientific reasoning skills.

Concerning the patterns of use of information and communication technology, it is concluded that the pattern of use of ICT in relation to learning planning [12] shows a high level; instead, both the Internet knowledge pattern and the overall ICT proficiency pattern have medium usage; On the other hand, the pattern of operations and computer concepts together with the pattern of research and management of digital information for learning present a low use. Ratifying as expressed by Garcia I, Gros B, Escofet A (2011) [13] in the XII International Congress of the Theory of Education, University of Barcelona: Profiles of use of technology and learning patterns among university students;

"Obviously, we can not assume that being a member of the" Internet generation "is synonymous with knowing how to use strategic technology-based tools to improve learning experiences in university environments."

With respect to the differences that exist between the different cohorts, it is concluded that the pattern of operations and computer science concepts gives an average level for the students of the beginning of the career, of a medium and low level (equally) in the students mid-career and a low level in the final-year students. It is evident that the level at which the patterns are presented: Internet knowledge, use of ICT in the planning of learning, research and management of digital information for learning and general tic competence; It is independent to the cohorts. This means that, for the students of career start, mid-career and final year, the high, medium or low level is presented for the same ICT use patterns. That is to say, it is observed that ICT usage patterns account for a better level achieved than scientific reasoning skills.

• Conclusion

No significant relationship was found between the pattern of research and management of digital information for learning and scientific reasoning skills. Likewise, it was found that, for probabilistic thinking ability, there is a significant relationship for more patterns of use of information and communication technology in students who are in the middle of their careers. Also, no significant relationship was found between the ability to conserve physical magnitudes and patterns of use of information and communication technology in students who are at the beginning of their careers. Likewise, no significant relationship was found between the abilities of: Proportionality thinking, probabilistic thinking and combinatory thinking; and patterns of use of information and communication technology in students who are at the end of their careers. In general terms, it is observed that the more students realize, at the beginning, in the middle or at the end of their careers, the use of information and communication technology, they will use more scientific reasoning skills. And in turn, while students use scientific reasoning skills, they will demonstrate patterns of use of information and communication technology.

Conflict of Interest

The authors report there are no conflicts of interest.

References

[1] Villanueva, G., & De la Luz Casas, M. (2010).e-competencias: nuevas habilidades del estudiante en la era de la educación, la globalidad y la generación del conocimiento. Signo y pensamiento, XXIX, (56), 124-138.
[2] Bedoya, E., Behaine, B., Severiche, C., Marrugo, Y. y Castro, A. (2018). Redes de Conocimiento: Academia, Empresa y Estado. Revista Espacios, 39(8), Pág. 16.
[3] Vargas L, Villalba V, Severiche C, Bedoya E, Castro A and Cohen H 2019 TICs y gestión de la innovación en MiPyMEs: Un análisis con experimentos factoriales para las utilidades. *Espacios* **40** 24
[4] Flores, Ó., & De Arco, I. (2012). La influencia de las TIC en la interacción docente y discente en los procesos formativos universitarios. Revista de universidad y sociedad del conocimiento, 9(2), 31-47.
[5] Sobrino-Morrás, Á. (2011). Proceso de enseñanza aprendizaje y web 2.0: valoración del conectivismo como teoría de aprendizaje post-constructivista. Estudios sobre educación, 20, 117-140.
[6] MinTIC (2017). MinTIC revela los primeros resultados del Observatorio de Economía Digital. Recuperado el 13 de noviembre de 2018, de https://www.mintic.gov.co/portal/604/w3-article-61929.html
[7] Hernández, R; Fernández, C; Baptista, P. (2014). Metodología de la Investigación. México: Mc Graw Hill.
[8] Prensky M (2011). Nativos e inmigrantes digitales. Madrid, España: SEK.
[9] Lawson A.E, Clark E, Cramer-Meldrum K.A, Falconer J.M, Sequist Y, Kwon (2000). "Development of scientific reasoning in college biology: Do two levels of general hypothesis-testing skills exist?". J. Res. Sci. Teach. 37, 81-101.
[10] MinTIC (2008). Plan Nacional de Tecnologías de la Información y las Comunicaciones. Recuperado el 21 de julio de 2016, de http://bit.ly/29Z8QNr
[11] Rodríguez M, Mena D, Rubio C (2010). Razonamiento científico y conocimientos conceptuales de mecánica: un diagnóstico de alumnos de primer ingreso a licenciaturas en ingeniería. Formación Universitaria, 3 (5), 37-46.
[12] Castro S, Guzmán B, Casado D. Las TIC en los procesos de enseñanza y aprendizaje
Laurus . 2007 13 (23): 213-234.
[13] Marulanda, Carlos E, Giraldo, Jaime, & López, Marcelo. (2014). Acceso y uso de las Tecnologías de la información y las Comunicaciones (TICs) en el aprendizaje: El Caso de los Jóvenes Preuniversitarios en Caldas, Colombia. Formación universitaria, 7(4), 47-56.

IOP Conf. Series: Materials Science and Engineering **844** (2020) 012008 doi:10.1088/1757-899X/844/1/012008

Traffic flow prediction model and probability of congestion

M Pilamunga Poveda[1*], J M Lavin[2], M Aldás Sánchez[3], D Ríos Insua[4]

[1] Design and Architecture School, Universidad Técnica de Ambato, Av. Los Chasquis and Rio Payamino, 180207, Ambato, Ecuador
[2] Cesine Centro Universitario, Calle Jose Simón Cabarga 6, 39005, Santander, España.
[3] Civil Engineering School, Universidad Técnica de Ambato, Av. Los Chasquis and Rio Payamino, 180207, Ambato, Ecuador
[4] CSIC, Serrano 117, 28040 Madrid, España

[*] Email: em.pilamunga@uta.edu.ec

Abstract. Investment in the management of integrated activities for better community development determines the progress of a city. Generate actions directed to change and evolution can promote greater satisfaction among its inhabitants. One of those activities is the one concerning transit, which is one of the most transcendental factors in the quality of life of a modern city. Thus, good road planning is of vital importance for their development. To do this, studies must be conducted that allow us to have a diagnosis of reality and thus make decisions would help to minimize or avoid problems that are occurring or that may appear in the future. However, to make a study of this type whenever it is required is expensive in terms of time and resources. This work proposes a model that allows predictions of traffic flows in different relevant sections of a city and, predict issues such as traffic saturation in them, thus determining possible solutions to road congestion, it can be presented at a given moment. These possible alternative solutions could consequently would reduce costs in time, fuel, pollution among others.

1. Introduction

Urban traffic is one of the problems that most influence the quality of life of residents in cities. This traffic has been accentuated in the last decades, due to the development of three simultaneous phenomena:

- The increase in the use of private cars to the detriment of public transport, due to the fact that it is insufficient to cover all sectors of the city;
- The generalization of the diffuse city model [1], which has reached its current development after a decentralization process; and,
- Economic growth.

Ambato is characterized for being a node of great commercial activity in the national context, existing an intense commerce. For this reason, in recent decades, the city has experienced a considerable increase in population and buildings, added to the geography of the city with many slopes and the growing presence of cars, the concentration of public and private entities in certain areas of the city, the good commercial activity of the marketplaces and industries, the great concentration around the educational centers, among others, complicate the traffic in several sectors of the city. For these reasons, it is very important to generate tools that help manage traffic and allow to determine a better road planning in a technical way, reducing costs depending on distances and waste of time, fuel expense and environmental degradation.

2. Related works

First, it is necessary to expose some definitions and basic principles of networks. A road network is a set of streets and avenues that intersect at intersections, which may occur at a level or slope. In general, the stretches of streets or avenues have a certain sense of circulation and, therefore, are designated as directed. Therefore, the basic model will use graphos. A grapho is a set, not empty, of objects called vertices (or nodes) and a selection of pairs of vertices, called edges, that can be oriented or no [2].
Mobility studies are based on the analysis and optimization of a traffic network through characterizing the infrastructure and its capacity include detailed field investigations necessary to know the origin / destination relationships and their main characteristics. This allows to adjust the functional characteristics of the different elements of the system, including a detailed analysis of the intersections and their different methods of regulation, Fernandez y Valenzuela [3].
Below are some existing vehicle traffic models that help to better understand the advantages of the proposed model. Each user defines how to reach a destination, choosing the one that considers the best route, considering factors such as cost, time, security and comfort. The estimated flows can serve both to describe traffic and to predict or recommend a pattern of vehicular flow in a network. For this, for example, can be used the models of Castillo et al. [4] or Tebaldi and West [5], which represent the macroscopic behavior of traffic in large urban areas or entire cities. Through these models it is possible to estimate the vehicular flows in a network where there is a certain demand for trips and the effects of congestion make the travel times in the arcs depend on the flows.
The urban and environmental impacts associated with traffic go, to a large extent, directly associated with the degree of saturation (x) of any road device (road, intersection, whereabouts, network). The degree of saturation is defined as the ratio between the flow used by the device (q) and its capacity (Q), that is, $x = q/Q$. Congestion is a consequence of the degree of saturation and usually manifests in:

- Additional delays to people beyond those required to make a trip at a safe and constant speed.
- Queues of vehicles or people, that are generated in sections of track, crossings or terminals of public or private transport like bus stops or parking lots.
- Involuntary stops of vehicles or users, which occur in the same places as mentioned before, due to the advance and stagnation of a queue (start-stop condition).

Different studies such as Akcelik [6], IHT [7], indicate that the manifestations of congestion begin to occur for values of (x) greater than 0,7 ó 0,8. During the last decades, traffic flow prediction models with different degrees of complexity have been proposed. The simplest are the combination of random paths, but only work well in specific situations [8].
A more complex approach [9], is based on time series models ARIMA, seasonal models ARIMA [8] and [10], Kalman filters [11], simulation [12] and [13], local regression models [14], combination of Kohonen maps and ARIMA models [15], and and Markov chains see Yu et al. [16]. Here dynamic linear models are used as a variation. Introductions to these models can be seen in West and Harrison [17] and Petris et al. [18], which include numerous applications in the economy and the business world. In comparison with the aforementioned proposal, the sample approaches that of Okutani and Stephanedes [11], but it has the advantage of allowing a priori information to be incorporated. Compared with the rest, it has the advantage of being a dynamic methodology, allowing predictions in real time.

3. Proposed work

To develop the study, it was necessary to analyze the information collected by the Municipality of Ambato, which contained the vehicular flows in different road intersections called gauging nodes, considered as nodes of high vehicular inflow.

IOP Conf. Series: Materials Science and Engineering **844** (2020) 012008 doi:10.1088/1757-899X/844/1/012008

To carry out an analysis in greater depth, we took as an example part of the Ambato road network, in the selected sector most of the traffic problems that exist in the rest of the city are presented, the gauging nodes are marked with orange color, and the red arrows represent the groups that took the data in the gauging node and the direction in which the vehicles were directed at the moment of counting, as can be seen in Figure 1.

Because, originally, the data was collected in the form of concentration of vehicles of each node, several considerations were taken into account for the allocation of data to each of the arcs of the network.

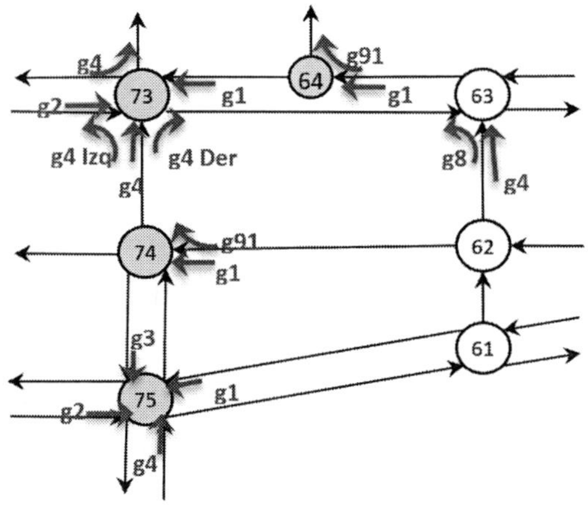

Figure 1. Scheme of the example network with the registered data groups

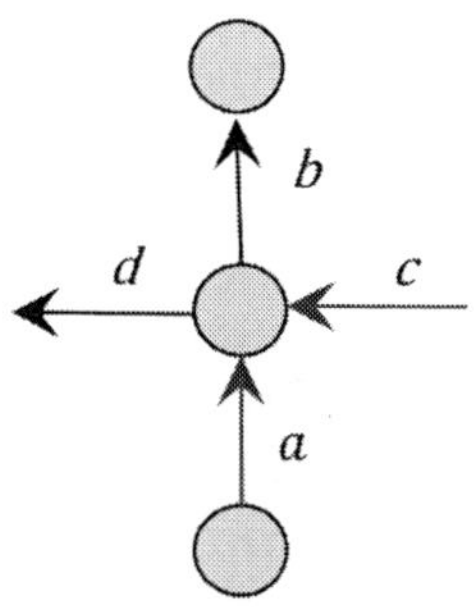

$$b = pa + qc \quad (1)$$

where p is the flow fraction a that goes in the direction of b; and q is the fraction of c that goes in the direction of b.

Figure 2. Scheme for the imputations formulation

The imputation of values for the arcs were made according to the format in Figure 2. For this reason, percentages have been assigned based on the knowledge of the sector and on the experience when traveling through such intersections, trying to bring them closer to the reality and keep relationship with the initial data. Thus, if the flows a and c are available and we want to predict the flow in b, we will use a prediction of the style $b = pa + qc$.

Basically, general rules of network flows are applied, such as the one that indicates that the sum of inflows to a node must be equal to the sum of outflows. Thus, according to the scheme of Figure 2., if information is available in the arcs a, c and d, it is easy to calculate the values for the remaining arc with the formula $b = a + c - d$.

In general, it has to estimate fractions p of traffic that adopt certain routes. For this, proceed as follows. We have data $((x_1, n_1), \ldots, (x_m, n_m))$ to the hours $h_1, \ldots, h_m$, where n_i is the total number of vehicles and x_i the number of vehicles by way of interest. A priori, it is believed that $p \sim \beta e(\alpha, \beta)$. It is easily verified that, under the model $x_i \sim Bin\,(n_i, p)$, it turns out that distribution posteriori is:

$$p|data \sim \beta e\left(\alpha + \sum x_i\,, \quad \beta + \sum (n_i - x_i)\right) \quad (2)$$

If we need to summarize the information, we can do it through

$$\hat{p}|data = \frac{\alpha + \sum x_i}{\alpha + \beta + \sum n_i}. \quad (3)$$

Under conditions of scarce information, we can do,

$$\hat{p}|data = \frac{1 + \sum x_i}{2 + \sum n_i}. \quad (4)$$

Next, it is shown in Figure 3, the arcs where it was possible to impute the data in the example network marked with red color, according to the format shown in Figure 2 and the exposed on the beta-binomial model, see French and Ríos Insua [19].

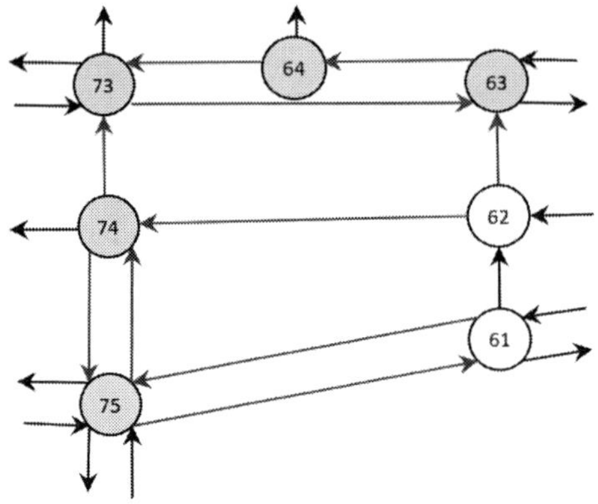

Figure 3. Scheme of the example network with imputed data

Once the data have been obtained in most of the arcs through the indicated method, the methods used in the problem are illustrated.

3.1. Principal Components Analysis in the Network Example

The Principal Components Analysis (PCA) [20] is a technique of exploratory data analysis, whose objective is the synthesis of information, or reduction of the number of variables. A main component is a linear combination of the original variables. In order to determine the nodes of greater influence within the example network, an PCA was made on the available data of the same. In this way, you can establish the most influential nodes in this network section and establish them as main nodes (components).

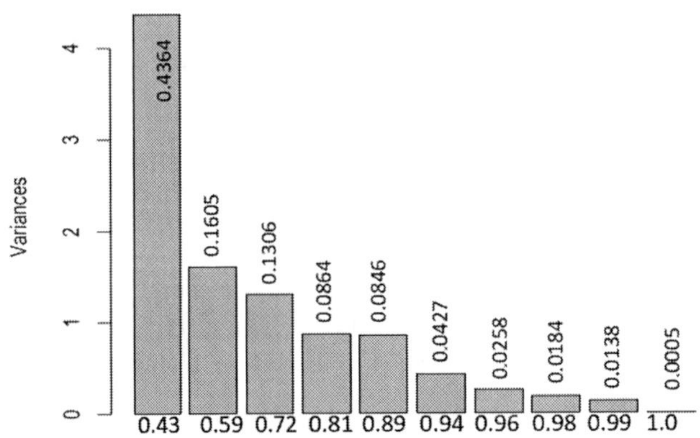

Figure 4. Bar chart of the variances explained by main components

Table 1. Matrix of results of the selected main components

Arco	PC1	PC2	PC3
X61.75	0,24	0,50	-0,11
X75.61	0,28	0,08	0,62
X75.74	0,39	0,18	0,31
X74.75	-0,32	0,24	-0,11
X62.74	-0,08	0,72	0,03
X74.73	-0,22	0,31	0,09
X73.63	-0,31	0,17	-0,33
X64.73	-0,40	0,01	0,44
X63.64	-0,41	-0,01	0,42
X62.63	-0,35	-0,10	0,12

Figure 4 contains the results of the PCA, which includes proportions of variance explained by the main components. In this case, it seems reasonable to keep the first three main components, since they explain 72.75% of the variance, the rest of the components contribute less than 10% each. The resulting charge matrix for the chosen components is shown in Table 1.
In order to be able to interpret them better, the following graph of the example network with blue arrows for negative charges, red for positive charges and those approximated to zero are not plotted, from the data matrix of **¡Error! No se encuentra el origen de la referencia.**.

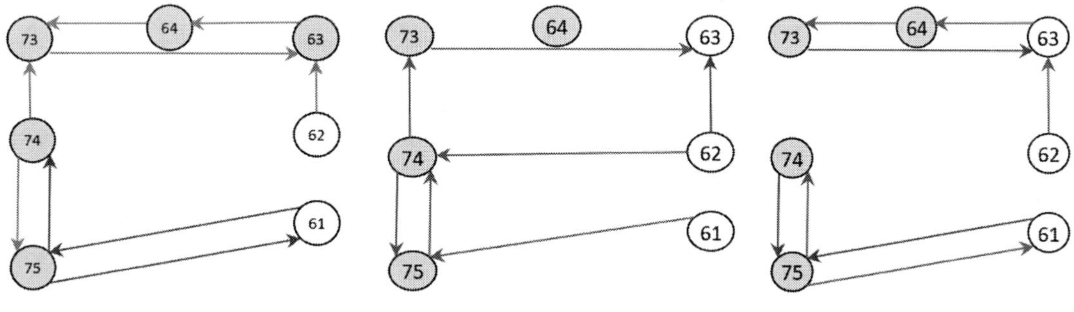

a) Component PC1 b) Component PC2 c) Component PC3

Figure 5. Graph of positive and negative charges of each component

As shown in Figure 5., the first component PC1 has a positive relationship with the arcs of node 75, which would mean that there is a high vehicular demand in that node, 3 of the 4 arcs have high vehicular circulation that influences the network and the remaining arc only acts as network output. In addition, it maintains a high negative relationship with the rest of the nodes, which would suggest that the vehicles that circulate through the rest of the nodes come from other inputs to the network and that node 75 acts as an exit node.
For its part, the second main component, according to the results matrix, has a high positive relationship with eight of the ten arcs. As far as negative correlations are concerned, they are values close to 0, as well as a pair of positive values, so they have not been taken into account when analysing the component. It could be said then that most of it has a positive relationship, especially with the arcs of entry and exit of nodes 74 and 75, indicating that there is a high traffic flow in these arcs, and potential trouble spots to consider. Finally, the third component has a high correlation with the data referring to movements towards node 73.
It could be concluded that the example network has been reduced to 3 variables that interpret what is happening within it, especially to and from nodes 75 and 73, which would be the most relevants from the point of view of traffic variability, suggesting that they are the areas that require more attention because they have greater influence within the network.

3.2. Prediction models

3.2.1. Time Series: The main problem in this article is the short-term forecast of traffic on the streets of Ambato. Figure 6 shows the trend and behavior lines of each of the arcs of the example network in the three-time bands, which has provided valuable information to help select the appropriate prediction models. For this reason, it has been considered convenient to apply dynamic linear models (DLM) to create a prediction model of vehicular flow in the arcs of the network under study. West and Harrison [17] and Petris et al. [18] provide introductions to the DLMs.

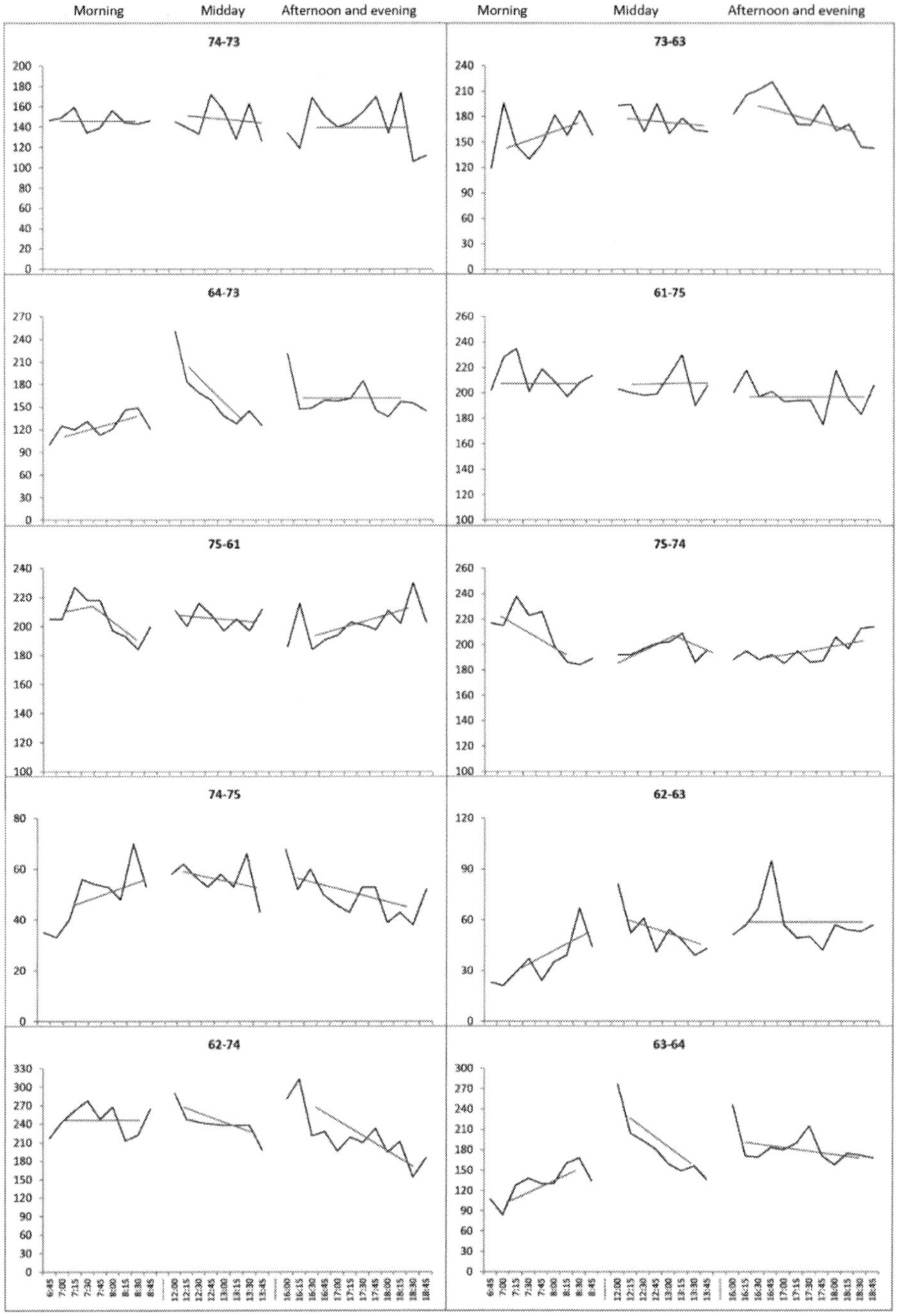

Figure 6. Time series graph of example network arcs

3.2.2. Dynamic Linear Models (DLM): The mathematical modeling of time series processes is based on dynamic models; the dynamic term refers to changes in those processes due to the passage of time as a fundamental driving force. Classic linear models are usually written as an equation:

$$Y = F'\theta + v \tag{5}$$

where Y is the response vector, F' this is the matrix of regressors or independent variables, θ this is a vector of unknown parameters, and v, it is the errors vector.

If you want to give a dynamic or evolutionary approach to the previous system, you must can be vary the elements of the model over time t and allow the set of parameters θ to evolve, in turn, with time.

IOP Conf. Series: Materials Science and Engineering **844** (2020) 012008 doi:10.1088/1757-899X/844/1/012008

Evolution takes the form of a Markov first-order process. In this way, a general normal dynamic linear model (DLM) could be written according to:

$$Y_t = F_t\theta_t + v_t, \qquad v_t \sim N_m(0, V_t) \tag{6}$$

$$\theta_t = G_t\theta_{t-1} + w_t, \qquad w_t \sim N_p(0, W_t) \tag{7}$$

$$\theta_0 \sim N_p(m_0, C_0) \tag{8}$$

which will be characterized by: $\{F_t, G_t, W_t, V_t\}$ (9)

where, for each instant of time t, F_t is a known vector of dimension n x 1, G_t is a known matrix $n\ x\ n$, V_t is a known variance and W_t a matrix of known variances $n\ x\ n$.

The specific model to use is a linear growth DLM, which is a state space model with:

$$Y_t = \mu_t + v_t, \qquad v_t \sim N(0, V), \tag{10}$$

$$\mu_t = \mu_{t-1} + \beta_{t-1} + w_{t,1}, \qquad w_{t,1} \sim N(0, \sigma_\mu^2) \tag{11}$$

$$\beta_t = \beta_{t-1} + w_{t,2}, \qquad w_{t,2} \sim N(0, \sigma_\beta^2). \tag{12}$$

When β is equal or very concentrated in 0, the linear polynomial model is of order 1 (as in the arc 74-73, in the morning). Otherwise, it is of order 2 (as in arc 75-61, afternoon schedule). In the DLMs they allow modeling typical features in temporary series as trends (polynomials), regression and seasonal effects, see Petris et al. [18].

Application of the DLM to the example network

In Figure 6 you can see the trend lines marked with blue, which suggest the level of the polynomial be first or second order. This helps to determine one of the main parameters to obtain values by applying the dynamic linear model (DLM) of prediction.

The data of the arch 75-61 represented in Figure 6 is taken as an example, since according to its trend lines in the morning schedule it is of second order, while in the hours of noon, and afternoon and night are of the first order. The values for W and V used in the DLM have been determined by maximum likelihood; the initial value of m_0 is equal to the mean of the data vector of the corresponding time band, and the C_0 initial, the variance thereof. The DLM mentioned has been applied in R, and the following models and graphs have been obtained:

Morning schedule (6:45 - 7:00 - 7:15 - 7:30 - 7:45 - 8:00 - 8:15 - 8:30 - 8:45)

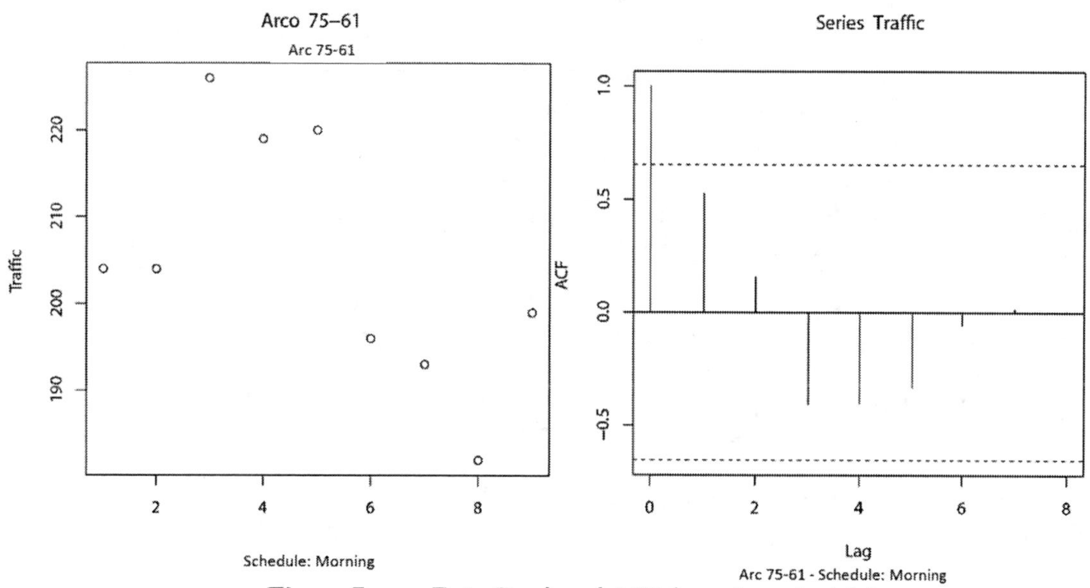

Figure 7. Data Graph and ACF (morning schedule)

In Figure 7 you can see, in the first table, the scatter plot of the morning data. The second table shows the autocorrelation function suggested by the trend line of order 2 as indicated in Figure 6 for the arc 75-61 in the morning schedule.

Figure 8 shows the predicted values marked in gray that follow a trend similar to the original values in black, with a good fit.

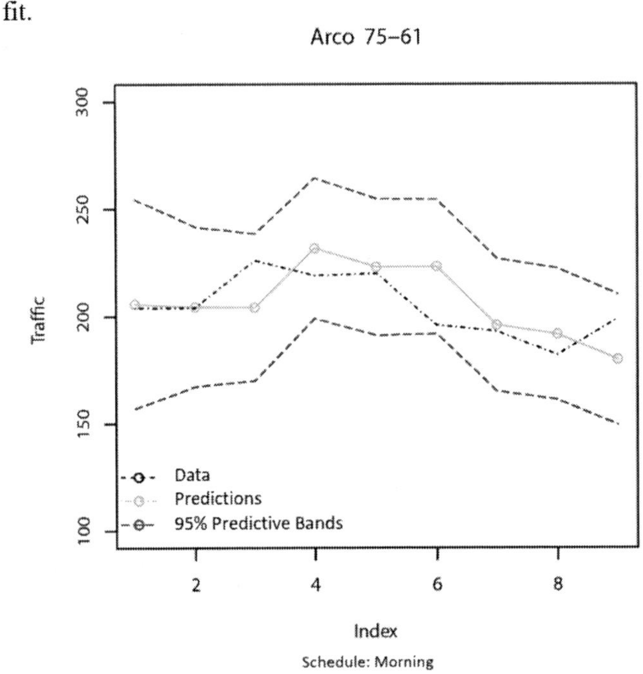

Figure 8. Graph of original and calculated data when applying the DLM (morning schedule)

In Figures 7 and 8, as can be seen, the generic DLM used is adjusted satisfactorily to predict the following quantities of vehicles that may circulate in the following periods of time. Similarly, under the same conditions, they apply to the rest of the arcs and thus predict 30, 45, 60, ..., n minutes.

3.2.3. Saturation Prediction: One of the most relevant uses of the previous DLM is a prediction of saturation in the corresponding network paths. It is defined as the capacity of a road at the maximum hourly intensity of people or vehicles when crossing a uniform stretch of a lane or roadway for a defined period of time under the prevailing road conditions, traffic and regulatory systems [21].

To make the calculation of maximum flow in each arc, considering that the capacity of a lane is the maximum number of vehicles that can pass through it, assuming a uniform speed.
Generally, it is expressed in vehicles per hour. Then the capacity in vehicles per hour would be:

$$C = \frac{v}{s}1000, \tag{13}$$

where C: maximum capacity of a lane, v: speed in km/h, s: minimum average separation in meters between the front or rear parts of two successive vehicles to determine the speed. The value of s is a function of the length of the vehicles, the reaction time of the drivers and the braking distance, can be expressed as a function of v using a formula of the type:

$$s = a + bv + cv^2 \tag{14}$$

The independent term (a) corresponds to the length of the vehicles, the term (bv) to conductors reaction time and (cv^2) to the braking distance. In a study linked in England by Smeed [22], he found the following formula for the most frequent separation based on speed:

$$s = 8 + 0.2v + 0.003v^2. \tag{15}$$

Replacing s in the formula of capacity C, we would have:

$$C = \frac{v}{8 + 0.2v + 0.003v^2} * 1000 \tag{16}$$

When the vehicle running speed is below 25km/h it becomes intolerable for the driver, so we have considered the use of the minimum tolerable to drive. Therefore, $v = 25Km/h$ and $C = 1681\ vehicles/hour$. Thus, we would obtain that for every 15 min for arcs with a lane, the maximum flow is 420, and 70% of its capacity would be 294, this information being necessary since when exceeding this percentage saturation problems begin to exist. The saturation probabilities are calculated by:

$$\Pr(Z > 0{,}7f), \tag{17}$$

where Z is the predictive distribution of the corresponding DLM.

After determining the probability values for each arc, it is observed in Figure 9, in the first table it indicates that in the arc 62-63 there is a very low probability of saturation, since the maximum value reached is 0.5 in the first two quarters of an hour in the morning and in the rest of the vehicle flow measurement periods, the probability is close to 0.

Regarding the arc 62-74, the opposite can be observed, in the first two periods of time the probability of saturation is higher than 50%, while in the rest of the periods it indicates that it is possibly always saturated, with a probability close to 1. This would suggest the need to interact in such traffic, through public service, limitations of traffic and other similar measures.

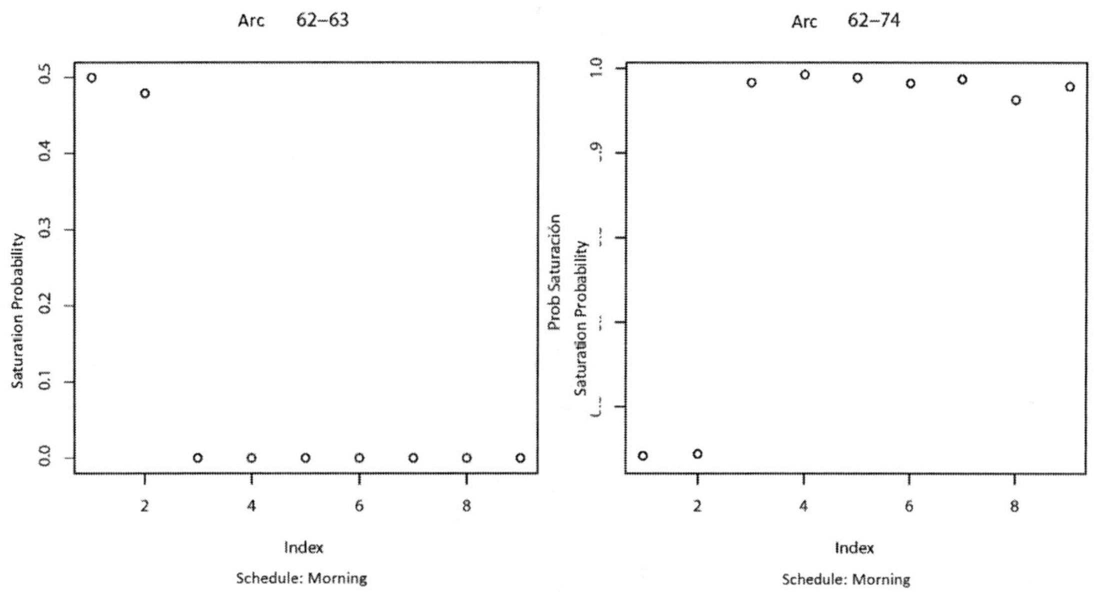

Figure 9. Saturation probability graph

4. Results

Finally, in Figure 10 you can see the results obtained after applying the methods described above in the example network. With the help of the PCA, the conflict nodes marked with yellow colour were determined, the same ones that require more attention within the network, therefore, improving the vehicular flow in these nodes would probably cause an improvement in the flow throughout the network. The arcs with greater probability of saturation determined with the help of the DLM, are marked with dashed lines in red, blue and green colors for each schedule respectively. This information can help to take the necessary corrective measures to avoid the probable problems of congestion.

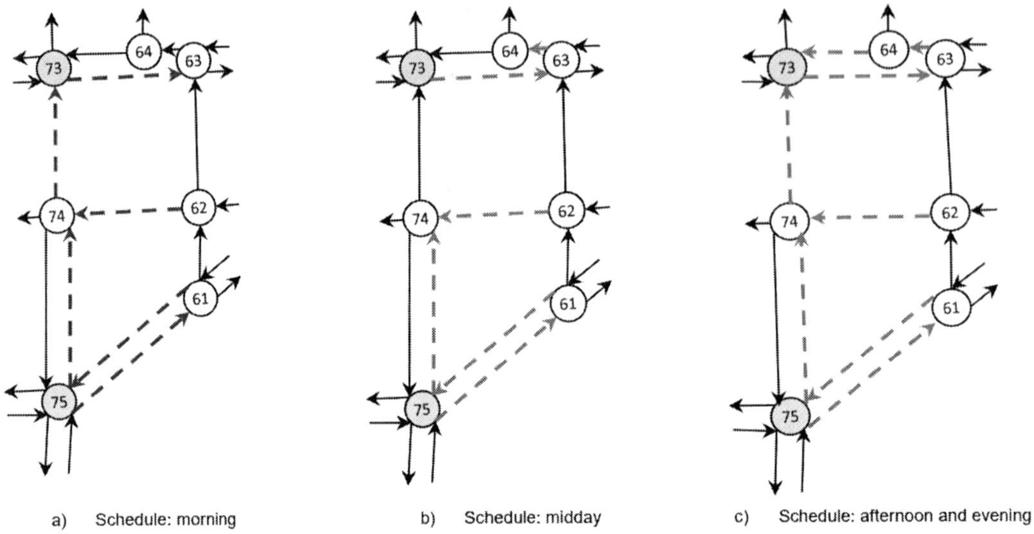

Figure 10. Saturation probability graph

Applying the indicated methodology to a larger road network in the city, the results are shown in Figure 11. were obtained, where the flows with the highest probability of vehicular congestion can be clearly seen, becoming priority places where measures must be applied to avoid saturation.

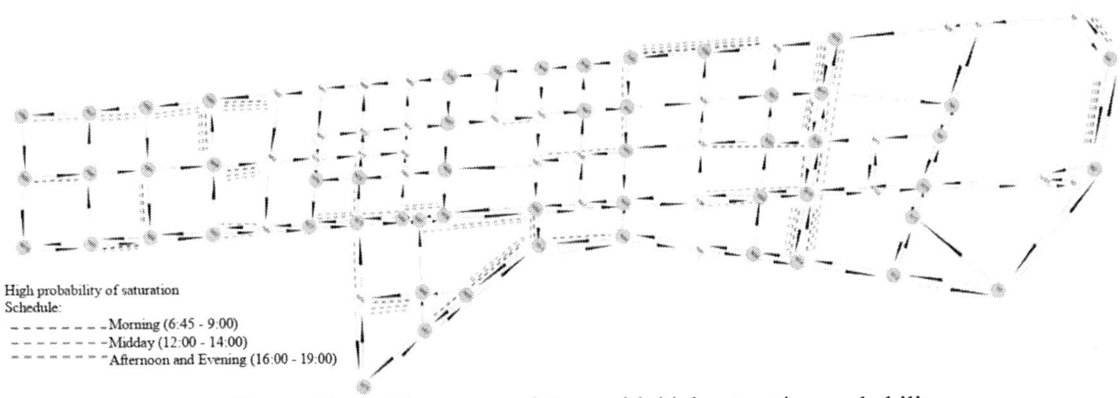

Figure 11. Hot spots and Arcs with high saturation probability

5. Conclusion

After applying the PCA to the study network, several of the so-called hot spots or with greater incidence in the vehicular flow are determined. Therefore, the nodes in which more care must be taken when making decisions regarding vehicle traffic planning. Another alternative offered by these points is determine the possible key locations where vehicle counting devices can be placed to obtain a traffic data history in the city and thus achieve traffic predictions with a lower margin of error by applying the DLMs.
The saturation probability values obtained helps to determine the possibility of congestion problems occurring in the different periods of time, with which the necessary corrective measures can be taken to avoid them. For example, to locate traffic lights, to eliminate parking lanes, to change the direction of tsome arcs in certain hours or permanently, etc., and thus allows the flow of vehicles is optimal and thus avoid congestion in strategic locations.

6. Future work

Based on the prediction model provided by the MDLs at the traffic measurement points, the concept of dynamic O-D matrices is intended to predict the main routes in the city. The concept of static O-D matrix can be seen, p. ex. in Vardi [23]. Having thus to fit to a model of simulation to experiment different measures to improve the vehicular traffic.

Acknowledgments
The work of Marcelo Pilamunga and Milton Aldás has been possible made by the project "1572-CU-P-2017 Models of management and optimization of traffic".

References

[1] Nel-lo O 1998 *Los confines de la ciudad sin confines. Estructura urbana y límites administrativos en la ciudad difusa* de *La ciudad dispersa. Suburbanización y nuevas periferias* ed Monclús J. (Barcelona: Centro de Cultura Contemporánea de Barcelona) p 35

[2] Taha H A 2004 *Investigación de Operaciones*, (México: Pearson Educación)

[3] Fernandez R and Valenzuela E 2004 Gestión Ambiental de Tránsito: Cómo la Ingeniería de Transporte *Eure* **29** 97

[4] Castillo E, Pedregal P, García R and Alguacil N 2002 *Formulación y Resolución de Modelos de Programación Matemática en Ingeniería y Ciencia* (México DF: Alfaomega Grupo Editor)

[5] Tebaldi C and West M 1998 *J American Stats Assoc* **93** 557

[6] Akcelik R 1981 *Analysis, Traffic Signals: Capacity and Timing* (Vermont: ARRB Group Limited)

[7] IHT 1997 *Transport in the Urban Environment* (London: The Institution of Highway and Transportation)

[8] Williams B M and Hoel L A 1999 *Modeling and Forecasting Vehicular Traffic Flow as a Seasonal ARIMA Process: Theoretical Basis and Empirical Results* Ph.D. Dissertation, Dept. Civil Eng., University of Virginia, Charlottesville, 1999.

[9] Head K L 1995 *Transp Research Rec* **1510** p 46

[10] Lee S and Fambro D B 1999 *Transp Research Rec* **1678** 179

[11] Okutani I and Stephanedes Y J, 1984 *Transp Research Part B: Method* **18** p 1.

[12] Banks J 1998 *Handbook of Simulation: Principles, Methodology, Advances, Applications, and Practice* (Atlanta, Georgia: EMP Books)

[13] Barceló J 2010 *Fundamentals of Traffic Simulation* (California: Springer)

[14] Willumsen L 1978 *Estimation of an O-D Matrix from Traffic Counts – A Review* (University of Leeds,Institute of Transport Studies)

[15] Van derVoort C, Dougherty M and Watson S 1996 *Transp Research Part C: Emerging Tech* **4** p 307

[16] Yu G QHu J M Zhang C S, Zhuang L K and Song J Y 2003 *Intelligent Vehicles Symposium, 2003. Proceedings IEEE*, p 208

[17] West M and Harrison J 2005 *Bayesian Forecasting and Dynamic Models* (New York: Springer)

[18] Petris G, Petrone S and P. Campagnoli P 2009 *Dynamic Linear Models with R* (New York: Springer)

[19] French S and Ríos Insua D 2000 *Statistical decision theory* (Oxford: Oxford University Press Inc)

[20] Hair J, Anderson R and Tatham R 1987 *Multivariate data analysis* (New York: Macmillan Pub. Co)

[21] National Academy of Sciencies 2000 *Highway Capacity Manual (HCM2000), Transp Research Board*

IOP Conf. Series: Materials Science and Engineering **844** (2020) 012008 doi:10.1088/1757-899X/844/1/012008

[22] Smeed R and Bennett G 1949 Research on Road Safety and Traffic Flow *Inst Civil Engineers - Road Paper* **29**

[23] Vardi Y 1996 *J American Stats Assoc* **91** p 365

IOP Conf. Series: Materials Science and Engineering **844** (2020) 012009 doi:10.1088/1757-899X/844/1/012009

Identification of critical variables in conventional transformers in distribution networks

I.A. Marriaga-Márquez[1a], K.Y. Gómez-Sandoval[1], J.W. Grimaldo-Guerrero[2], JR Nuñez-Álvarez[3]

[1] Electrical engineering student, Universidad de la Costa, Colombia.
[2] Department of Energy, Universidad de la Costa, Colombia, ORCID: 0000-0002-1632-5374.
[3] Department of Energy, Universidad de la Costa, Colombia, ORCID: 0000-0002-6607-7305.

[a] Corresponding author: imarriag2@cuc.edu.co

Abstract. Transformers are essential equipment to the operation of electrical power systems, a failure causes the lack of electricity supply to end-users, affecting the operating indicators of companies in the distribution sector. The investigation presents an identification of the faults in transformers through a fishbone diagram, an evaluation of the variables that cause the identified failure using the cross-impact matrix method and a proposal to improve the performance. The results will enable a plan to be developed for taking action with monitoring plans to avoid faults that could put the electrical asset at risk and achieve a better performance of the distribution network.

1. Introduction

The growth of cities requires the expansion of distribution systems to meet the growing demand for electricity [1]; one of the fundamental components are transformers, used to modify electrical voltage levels, reducing electrical losses and risks when it supplies electrical energy to the end-user [2].
The electricity service must be provided reliably to users, avoiding failures which may occur due to the malfunction of the electrical system or part of it [3]; for this purpose, standards NTC 2050 [4] and IEE C57. 12. 00-1987 [5] are established as requirements for the proper operation of distribution transformers in Colombia. Distribution systems have technical loss percentages of 15. 7% [6], due to oversized or overload of the circuit, resulting in an increase in the occurrence of faults and a decrease in the useful life of the equipment [7].
The costs associated with the maintenance and replacement of the transformer are high [8]; avoiding failures is a key factor for distribution companies, due to the impact on their service indicators. Knowing their causes allows companies to create preventive and corrective maintenance plans to extend the useful life of transformers. This study presents a characterization of failures presented in distribution transformers and the use of Cross Impacts Matrix-Multiplication Applied to a Classification (MICMAC) method, based on qualitative study between the variables identified in a fishbone diagram and their interactions, it allows us to identify key factors to the performance of the transformer, the results will allow to design monitoring plans to avoid faults that can put in risk the electrical asset and to achieve a better performance of the distribution network.

IOP Conf. Series: Materials Science and Engineering **844** (2020) 012009 doi:10.1088/1757-899X/844/1/012009

2. Methodology

The research begins with the parts of a transformer and the failures that they present during its operation; variables are identified which may be managed to the optimal performance of the transformer. The methodology is organized in three phases:

2.1. Identification phase

It involves the documentary review of failures in transformers and their components, the design of a fishbone diagram that relates the causes and effects, allowing the identification of variables that affect the performance of the electric asset.

2.2. Analytical phase

The identified variables and the cross-impact matrix method are used to identify the critical elements for the operation of the transformer; these variables are located in the conflict zone generated by the matrix.

2.3. Propositional phase

The variables located in the conflict zone of the structural analysis are identified, these variables are the strategic elements and their intervention will allow the adoption of strategies for the care of the transformers.

3. Parts of a distribution transformer

The transformer consists of two windings, one with high voltage and the other with low voltage, an iron core inside a tank full of dielectric oil; and bushings for high and low voltage, the latter allow the connection of the coil terminals to the medium and low voltage network [9]. Figure 1 shows a diagram of the distribution of the components:

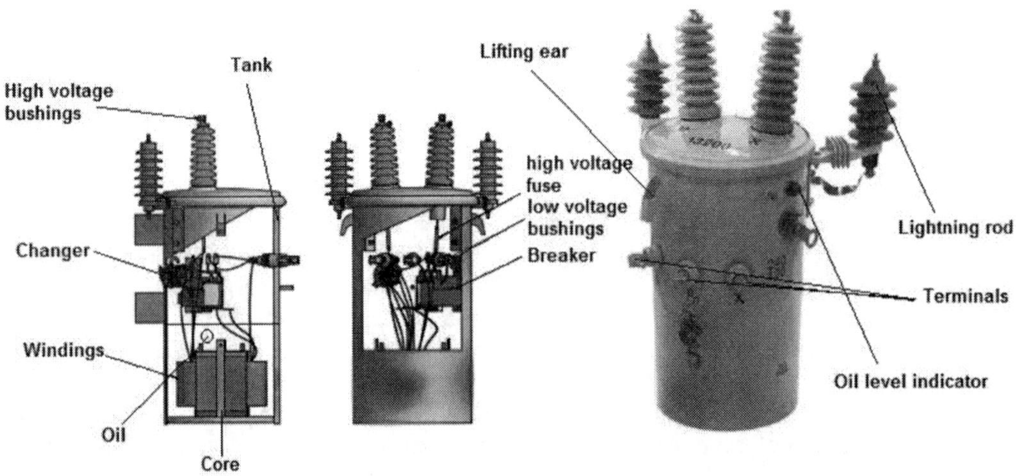

Figure 1. Parts of a conventional distribution transformer. **Source:** [10]

4. Distribution transformers faults

4.1. Overvoltage in distribution transformers

An overvoltage is caused by an increase in the grid voltage; it can be induced by an atmospheric electric shock [11] or caused by a decrease in the impedance into the transformers, preventing voltage regulation [12].

4.2. Ground fault
This fault is caused by the poor condition of the bushings or loss of dielectric rigidity in the transformer components, causing overloads and electric arcs to the grounded parts of the transformer. A bushing is damaged by the presence of hot spots due to poor maintenance of operators, reduced internal insulation, presence of plants or animals and deformation of the bushing [13].

4.3. Moisture in distribution transformers
Moisture is always present in a conventional distribution transformer in an acceptable quantity and is specified on the transformer data sheet. The initial moisture comes from manufacturing and increases due to atmospheric humidity or the ageing of the cellulose in the insulating paper [14].

4.4. Overload in distribution transformers
This fault occurs when the nominal transformer power value is exceeded, due to the connection of an additional load, a wrong dimensioning in the supply demand or a bad design of the operating temperature [15].

4.5. External short circuits
This fault is related with the environment where the transformer is located, affected by environmental factors such as the short circuit by contact with a tree, animal or object, vandalism, failures in the end-users circuits and wind gusts that cause contact between distribution lines [9].

4.6. *Internal problems*
These failures occur due to lack of tightness and moisture presence, false contacts in the switch, output terminals broken, low oil level, factory defects or damage due to improper in transport and assembly handling [16].

5. Results

5.1. Variables identification
According to the described faults a fishbone diagram is made to relate the faults and their causes [17], in figure 2 the diagram can be observed.

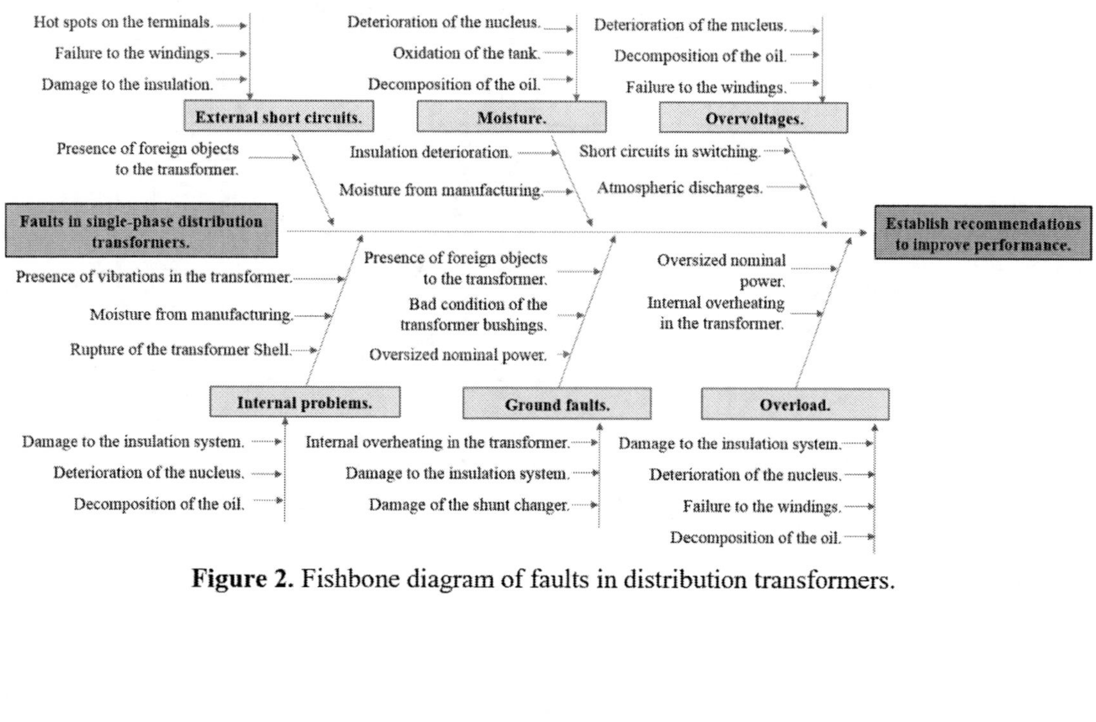

Figure 2. Fishbone diagram of faults in distribution transformers.

Figure 2 presents in a simplified way the relation of the causes and the effects due to the faults in the distribution transformers; table 1 lists ten (10) variables that affect the performance of these equipment, identified in Figure 2.

Table 1. Variables identified from the fishbone diagram.

Code	Causal
V01	Presence of foreign objects in the transformer
V02	Rupture of the transformer shell
V03	Internal overheating in the transformer
V04	Oversized nominal power
V05	Atmospheric electrical discharges
V06	Short circuits due to faults in the switching
V07	Moisture from manufacturing
V08	Presence of vibrations in the transformer
V09	Insulation deterioration.
V10	Bad condition of transformer bushing

5.2. Structural analysis: Identification of key variables

The method of structural analysis is composed of three steps, the first is the identification of a set of variables that characterize the system; Table 1 shows the variables identified; In the second step, the interaction matrix is made with the identified variables, a group of participants collaborated to evaluate the degree of influence between them, the variables identified were valued from 0 to 5 taking into account the following relationship, (0) null, (1) very low, (2) low, (3) medium, (4) high and (5) very high; the degree of dependence and driving power was calculated. Table 02 and Figure 03 presents the result of the matrix interactions and dependence vs driving power diagram. Finally, the identification phase of the key variables, the variables located in the conflict zone are selected due to their high dependence and high driving power; because they have a significant influence on the others.

Table 2. Cross-impact matrix with the variables in table 1

	V01	V02	V03	V04	V05	V06	V07	V08	V09	V10
V01	0	1	2	0	0	1	0	2	4	2
V02	0	0	4	0	0	1	0	0	2	1
V03	0	1	0	3	0	1	0	0	4	1
V04	0	2	5	0	0	2	0	1	2	2
V05	0	1	2	1	0	3	0	2	1	2
V06	0	1	1	1	0	0	0	1	2	2
V07	0	5	1	0	0	1	0	0	4	1
V08	0	1	1	0	0	2	0	0	1	1
V09	0	5	4	1	0	1	0	0	0	2
V10	0	0	1	0	0	3	0	0	2	0

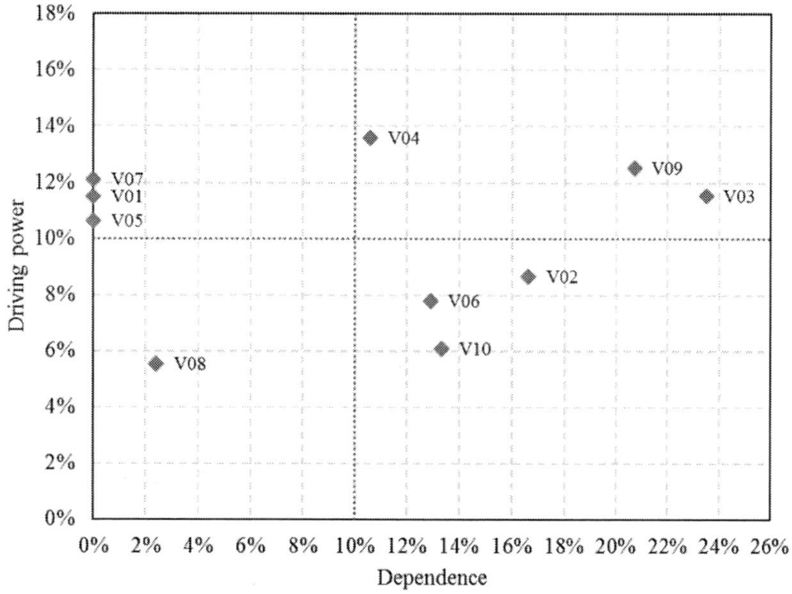

Figure 3. Driving power and dependence diagram.

Figure 03 shows variables V03 (Internal overheating in the transformer), V09 (Insulation deterioration) y V04 (Oversized nominal power) in the conflict zone; V03 and V09 are considered the critical variables to improve performance to decrease the failures occurrence due to its high driving power and dependence. The correct operation requires a proper use of transformers, do not oversize, check the internal overheating and the insulation; they can be affected by insulation failures caused by insulation gaskets damaged, moisture ingress conclude with the tank rupture due to oxidation.

5.3. Improve the operation of distribution transformers

The results presented the variables V03, V04 and V09 in the conflict zone, create plans for their care and prevention will help manage the care of the transformer affecting its useful life and optimal performance. The following are strategic to favor the control of these variables:

- ✓ Periodically check the installation area of the packages, keep the area free of sand or foreign particles from the outside.
- ✓ Verify the oil condition, by checking the color and condition or using the test dielectric breakdown voltage, in case of doubts; it is advisable to use accredited laboratories.
- ✓ Check the transformer paint and determine the degree of deterioration of the tank.
- ✓ Check the sealing of the transformer cover to prevent the entry of water.
- ✓ Evaluate the working conditions in front of the transformer data sheet, according to the connected load.

6. Conclusions

The work presented the components and faults of a transformer, a fishbone diagram was made in which the causes and effects of the failures were identified; allowing the identification and selection of ten variables that affect the operation of these equipment. The MICMAC method was used to identify critical variables, located in the conflict zone, and variables V03 (Internal overheating in the transformer) and V09 (Deterioration of the insulator) were the variables with the most influence.

The results suggest that preventive maintenances must be carried out, especially the checking of the external components of the transformer. It is suggested to check the system of isolation and verification of the oil condition, and look for hot spots; avoid connecting loads higher than the demand supported by the transformer.

The care of distribution transformers will allow to have a safe and reliable operation, improving operational indicators such as SAIDI and SAIFI, indicators evaluated by the Superintendence of Public Services of Colombia; if these indicators have a bad rating, they will represent financial penalties for network operators.

7. References

[1] Bracho CS 2005 Desarrollo de una metodología forense para el diagnóstico de fallas en transformadores monofásicos tipo poste en la red de distribución de 23.9 kV de ENELVEN *Universidad Rafael Urdaneta* Venezuela.
[2] Canasí Piñeiro M, Martínez Barbado JÁ 2012 Implementation of embedded system DIL/NETPG with Labview for test to transformers *Revista Científica de Ingeniería Energética* **1** pp 78-86.
[3] Campos FR. 1991, Análisis de las fallas en transformadores causadas por la operación del pararrayos ante sobretensiones externas *Ingeniería e investigación* **22** pp 34-46.
[4] NTC 2050. 1998, Instituto colombiano de normas técnicas y certificación. ICONTEC.
[5] C57.12.00-1987 IEEE Standard General Requirements for Liquid-Immersed Distribution, Power, and Regulating Transformers.
[6] UPME 2015 Unidad de planeación minero-energética. [Online] [Cited 2019 Feb. Available from: http://www1.upme.gov.co/Energia_electrica/Planes-expansion/Plan-Expansion-2015-2029/Plan_GT_2015-2029_VF_22-12-2015.pdf.
[7] Urrego, Acosta CA, Rendón RA, Isaza RH 2012, Reubicación óptima de transformadores de distribución para reducción de pérdidas técnicas *Scientia et Technica* **51** pp 28-34.
[8] Álvarez RE, del Pozo MD 2007 Mantenimiento de transformadores de potencia. Instituto de Investigaciones Tecnológicas para Redes y Equipos Eléctricos (IITREE) *XII Encuentro Regional Iberoamericano de CIGRE.*
[9] Montoya A, Murcia J 2014 Diagnóstico de fallas en transformadores de distribución monofásicos para la empresa Emcartago *Universidad tecnológica de Pereira* Colombia.
[10] RYMEL. Catálogo de transformadores. [Online]. [Cited 2019 Mar. Available from: http://www.rymel.com.co/CatalogoProductos.pdf.
[11] IEEE 2009 *IEEE Power & Energy Society* IEEE Std C62.22 pp 64.
[12] Griffel D, Leitloff V, Harmand Y, Bergeal J 1997 A new deal for safety and quality on MV networks *IEEE Transactions on Power Delivery* **4** pp 1428–1433.
[13] Jaimes JE, Ochoa JA, Camacho EA 2016 Análisis de criticidad y árboles de diagnóstico de fallas para transformadores de potencia *Revista Colombiana De Tecnologias De Avanzada (Rcta)* **27.**
[14] Araya G, Gómez GA 2016 Comportamiento de los aislamientos sólidos de transformadores de potencia en condiciones ambientales no controladas *Revista Tecnología En Marcha* **3** pp 99-116.
[15] Mago MG, Vallés L, Olaya JJ 2011 Análisis de fallas en transformadores de distribución utilizando ensayos no destructivos y pruebas de tensión mecánicas *Revista Ingeniería UC* **2** pp 15-26.
[16] Janer D, Areu O 2002 Un análisis sobre las fallas de transformadores de distribución en cuba *Revista Científica Ingeniería Energética* **2** pp 10-15.
[17] Zapata C, Villegas S, Arango F. 2006, Reglas de consistencia entre modelos de requisitos de UN-Metodo. Rev Univ Eafit *Revista Universidad EAFIT* **141** pp 40-59.

IOP Conf. Series: Materials Science and Engineering **844** (2020) 012010 doi:10.1088/1757-899X/844/1/012010

Drivers of Biomass Power Generation Technologies: Adoption in Colombia

W Reyes-Calle[1a], J.W. Grimaldo-Guerrero[2]

[1] Electrical engineer, Universidad de la Costa, Colombia.
[2] Department of Energy, Universidad de la Costa, Colombia, ORCID: 0000-0002-1632-5374.

[a] wreyes3@cuc.edu.co

Abstract. The share of non-conventional renewable energies in Colombia represents less than 4% of electricity generation; the government aims to increase the net installed electricity generation capacity to over 10% by 2028. A structural analysis was carried out for decision-making: international experiences was used to identify Social, Environmental, Technical and Economic indicators and with the MICMAC method, the key variables were identified to define improvement strategies. This research determined the driver elements that will allow Colombia to effectively include biomass as an electrical energy source.

1. Introduction

The essential need for energy supply has aroused global interest in having alternative energy sources [1, 2], mainly to replace fossil fuels due to the objective of reducing the emissions of greenhouse gases [3, 4] and the gradual depletion of oil reserves in the world [5]. Renewable energies are presented as an alternative for electrical and thermal energy sources [6]; to achieve these goals there needs to be an increase in projects that use these resources [7] and technological migration in the net installed electricity generation capacity (NIEGC) needs to be carried out [8, 9].

As a government plan, countries mainly encourage projects that use energy such as biomass, solar energy and wind energy [10, 11, 12]. Colombia aims to increase its share of Non-Conventional Renewable Energies (NCRE) in NIEGC to over 10% by 2028 [13]. Currently, this share represents less than 4% of electricity generation [14].

The government of Colombia has not developed projects with NCRE sources on a large scale [15, 16], its NIEGC is based on hydroelectric power (67.3%) and fossil-fuel power (27.1%) [17, 18]. Due to the climatic fluctuations suffered by the country, caused by the El Niño-Southern Oscillation (ENSO), the design of the NIEGC has led to energy crises, specifically in the years 1992 and 2015 [19, 20]; As a solution to the latest crisis, the law 1715 was proposed [21] as a fundamental pillar to create an incentive to renewable energy projects.

Biomass is used as an energy substitute for coal and gas in fossil-fuel power plants [22], in which technological adaptations have been made to use them for direct burning [23], lowering the acquisition costs of fossil fuels [24]. Among the international practices for the diversification and promotion of the use of biomass the following initiatives stand out: Japan proposed the concept of combustible garbage which achieved to promote the recycling and the use of all waste material [25]. The United States suffered an increase in the volume of produced garbage and in the consumption of electric energy due to the accelerated population and economic growth [26], as an alternative it was decided to generate

electricity and heat with urban biomass, which achieved the reduction of volume in landfills [27]. Sweden created laws to protect its forests and to use recycled materials to generate biodiesel; currently 20% of the fuel used for public transport is biodiesel [28].
Colombia should face changes for the inclusion of biomass energy due to its high potential [29]; Biomass power generation would be able to provide public electricity service to the populations of Non-Interconnected Zones (ZNI) [30, 31, 32] – the major part of these zones are located in remote areas. In spite of the efforts made by the Colombian government, no progress has been made and no large projects that influence the NIEGC have been launched, since the projects that were carried out were mostly projects for self-consumption and in most cases with capacities not exceeding 1MW [33]. In order to achieve this goal, it is necessary to carry out a restructuring that allows strengthening the system by means of the lessons learned from countries that have achieved technological migration.
This research presents a review of the strategies employed by some countries in America, Europe and Asia to produce electricity with the use of biomass in order to establish key factor using the MICMAC method. The results allow identifying current barriers and initiatives for the development of strategies to improve the incorporation and growth of technologies for the use of biomass in Colombia.

2. Methodology

The design of this research is oriented to analyse the characteristics applied for the development and use of biomass from a conceptual and methodological dimension, identifying best practices based on international experiences. The research is organized in the following three phases, namely:

2.1. Identification phase

A review of relevant sources, mainly documentation and regulations from certain governments of Europe, Asia and America, is used to promote the technological inclusion of biomass as an energy source for the production of electricity. The indicators and criteria that were identified for the incorporation of the power generation technology will be the basis for the formulation of key variables, which will make it possible to identify drivers.

2.2. Analytical phase

International experience shows the lack of initiatives and the inexperience about these processes in Colombia. The measures taken by other countries will help to create concepts and procedures for the development of a pertinent strategic vision to attain the objective that is pursued by the government. The Cross-Impact Matrix Multiplication Applied to Classification (MICMAC) method is used with the established variables; the results will present a conflict zone where the key elements for the promotion of technologies for the generation of electric power with biomass will be located.

2.3. Propositional phase

The analysis of the key elements will allow the design of medium – and long – term planning proposals in the electricity market. The results are a tool to evaluate the different factors according to their degree of relevance and to determine guidelines for the management and adoption of energy policies.

3. Results

3.1. Identification of factors for the implementation of biomass generation technologies

The European Union (EU) guidelines of 2017 were identified [34], these guidelines establish the goal of reducing the emission of greenhouse gases by 20%, increasing the use of renewable energies in the EU to 20% and improving energy efficiency by 20%; this commitment was set up due to the dependence on fossil fuels and the importing countries as well as to promote mitigation of related environmental problems [35]. Agents such as industries, governments and academia have developed alternatives to encourage the achievement of the goals; these alternatives are legal proposals to promote the use of biomass, the inclusion of generators in the grid, an increase of self-generation [36, 37] and an adaptation

IOP Conf. Series: Materials Science and Engineering **844** (2020) 012010 doi:10.1088/1757-899X/844/1/012010

of technologies in existing plants to use biomass [38]. A relationship between social and economic growth and the initiatives of renewable energy generation projects, especially with biomass has been identified in populations far away from the national electricity grid [39].
Figure 1 presents the identified elements from international experiences that were used to improve the implementation of technologies for the generation of energy with biomass; these elements are classified into four (4) groups according to their nature: Social, Environmental, Economic and Technical. Indicators belonging to the social group are identified as characteristics that benefit the individual and improve their quality of life; the environmental factors deal with the effects on the environment like the reduced pollution and improvements in air quality; the technical elements show the benefits in the NIEGC through technological advances; the economic ones establish movements of the market for the creation of new projects.

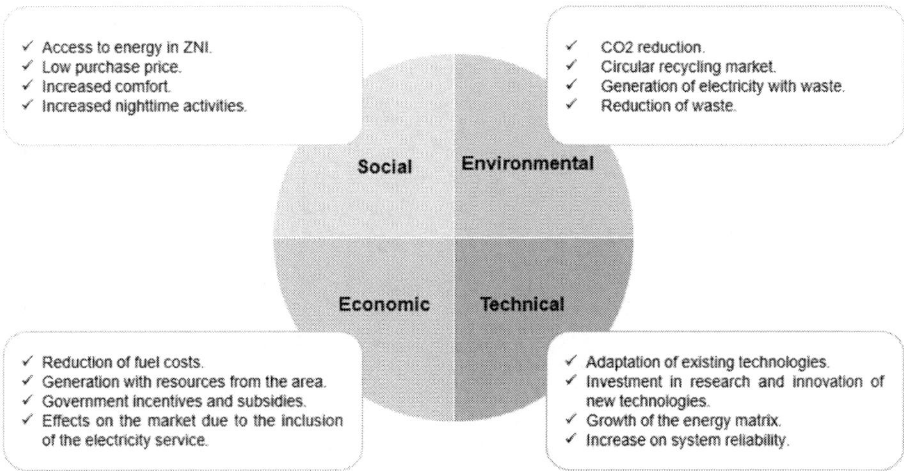

Figure 1. Indicators and criteria for the use of biomass. Sources: [40, 41, 42, 43, 44, 45, 46]

Figure 1 presents the groups of indicators and criteria that allowed obtaining an adequate management for the use of biomass and to benefit the different agents of the market, mainly the final users; these variables will promote the growth of biomass use in Colombia.

Table 1. Variables identified to improve the use of biomass in Colombia.

Code	Causal	Remark
V01	Energy use in urban, forest, agricultural and animal wastes	Garbage can be used as fuel in the generation of electric energy. [40, 44, 43]
V02	Lack of a garbage sorting system	Absence of a system structured by the government, in which it specifies how it should be recycled and in which it provides tools for doing so. [45]
V03	Energy model for the sale of energy and preferential tariffs according to the generation technology	It is seen as an incentive for private companies to generate energy with renewable sources and sell it to neighbouring communities. [28]
V04	Costs of acquiring technologies to generate energy with biomass	Currently, biomass energy is expensive because of the costs of technological investments. This is why many countries choose to adapt the existing technology.

V05	Lack of incentives for the implementation of new technologies	Colombia does not have incentives or initiatives from the government to acquire technologies or adapt existing thermal plants that do not work.
V06	Generation in non-interconnected zones	Biomass is a type of energy widely used for generation in non-interconnected areas, according to international experience. [42]
V07	Saturation of sanitary landfills	Saturation leads to the opening of a new sanitary landfills. Through the recycling and energetic use of the waste, this problem is reduced.
V08	Gas emissions from waste	The emission of gases into the atmosphere is reduced compared to fossil-fuel power plants. The environmental impact is lower. [26]
V09	Lack of awareness of recycling	Lack of training to classify and recycle the different type of garbage of the communities. [41]
V10	hydroelectric power and fossil-fuel power generation	Colombia depends mainly on these two energy sources. This has diminished the reliability in some moments of history. Biomass is an energy source that can be included in the NIEGC to provide versatility for the system. [18]
V11	Industrial sector initiatives	Agreements with industries concerning the generation and distribution of electric energy to neighbouring communities with their waste. [44]

3.2. *MICMAC Method for the identification of key variables*

Using the variables in table 1, the MICMAC method is applied to identify the interactions between them, and to determine the driving power and dependence of each of the variables. Table 2 presents the results of the cross-impact matrix.

Table 2. Cross-impact matrix with the variables in table 1

	V01	**V02**	**V03**	**V04**	**V05**	**V06**	**V07**	**V08**	**V09**	**V10**	**V11**
V01	0	0	5	0	0	5	4	3	0	2	4
V02	5	0	0	0	0	3	5	4	3	0	3
V03	5	0	0	0	3	5	0	3	0	4	5
V04	5	0	0	0	3	5	0	2	0	2	3
V05	5	0	2	5	0	5	0	3	0	2	3
V06	2	0	5	0	0	0	0	2	0	5	5
V07	4	2	0	0	0	0	0	4	5	0	2
V08	0	0	3	2	4	0	3	0	3	5	0
V09	5	4	0	0	0	4	5	4	0	0	0
V10	0	0	4	0	0	2	0	5	0	0	4
V11	5	0	5	3	5	5	0	4	4	3	0

Table 2 shows the driving power and the dependence of each variable according the cross-impact matrix, the conflict zone contains the variables with high driving power and high dependence. Due to their ability to influence the other variables, these variables are key elements for the performance of the object of study.

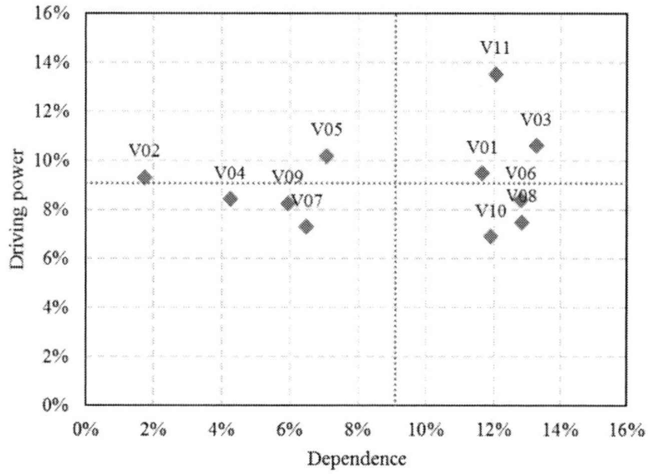

Figure 2. Results obtained from the cross-impact matrix.

The results in figure 2 focus on variables V01 (Energy use in urban, forest, agricultural and animal waste), V03 (Energy model for the sale of energy and preferential tariffs according to the generation technology) and V11 (Industrial sector initiatives) in the conflict zone.
V11 is the second variable with higher dependence and the highest driving power, it indicates the need to encourage the private sector to implement technologies for the generation of energy with biomass and for their integration into the electricity power system, and the possibility to sale the surplus. Variable V03 highlights that the government needs to launch initiatives in order to develop modifications to promote energy that is generated from biomass in the electricity market. V01 indicates that the country does not take advantage of the energy potential of its waste. The results present V02 as a variable with high driving power and low dependence which indicates that actions to promote waste classification are required.

3.3. Suggestions for the improvement of the identified key factors

The results show the lack of a system for the classification of waste, as it is for example employed in Japan [45]. A strict public policy which encourages recycling and stimulates the creation of new projects to generate energy is required to improve the use of waste and its energy potential; countries like Germany, Finland and Brazil have achieved great results by laws [28, 41, 44].
The government's commitment is required to establish duties and preferential tariff relating to the used technology and to provide clarity for the requirements for the sale of surplus electricity. These modifications in the Colombian energy model will encourage private companies that have the potential of generating energy with biomass to implement projects that are beneficial their financial accounts [10]. Colombia's mining and energy planning unit (UPME) estimated the amount of waste Colombia produced [47]. Furthermore, the UPME has a website that shows the technical, economic, financial and environmental viability of projects with biomass [48]. However, the energy department of Colombia should use the key factors (see table 1) in order to achieve the successful inclusion of biomass in the NIEGC of Colombia.
In every urban area, be it large or small, it is necessary to create waste reception centres in order to use the collected waste to power generation plants in the community. The installation of energy plants using biomass as fuel would generate energy and social development in accordance with the objectives set by

the Colombian government. This initiative includes the society and indirectly helps it to be aware of the environmental.

4. Conclusions

The article presented the identification of indicators and criteria in non-interconnected areas that are using biomass. Eleven (11) variables were to promote the growth in biomass use in Colombia were established, the MICMAC method was applied to identify strategic intervention elements and guidelines were provided that allow identifying the key elements identified.

Due to the flexibility of existing technologies and energy in the areas of influence, international experiences present biomass as a reliable alternative for the supply of electricity in areas that are isolated from the electricity grid. This highlights the need for tariff policies to promote the development of projects and to diversify the technologies already implemented.

A garbage classification system is required to optimize the use of waste and to reduce environmental impact. A strict compliance policy is recommended to achieve better results; countries such as Sweden and Japan impose fines to encourage compliance.

The electricity market should provide clarity in the guidelines concerning the sale of surplus electricity produced by companies that are not agents in the electricity sector. Currently the model allows for sale but the lack of knowledge and clarity discourage the participation of the private sector. Currently, Colombia has various means to determine the amount of biomass. However, the UPME should contemplate superior options in order to promote the use of biomass in Colombia, e.g. taxes, subsidies or incentives for the private sector.

For future work, it is recommended to study energy business models to encourage the implementation of projects and conduct a study on the key characteristics of energy models for the sale of energy with biomass.

References

[1] Gustavsson L, Börjesson P, Johansson B, Svenningsson P 1995 Reducing CO2 emissions by substituting biomass for fossil fuels. *Energy* **20**(11) pp 1097-113.

[2] Marques AC, Fuinhas JA, Pereira DA 2018 Have fossil fuels been substituted by renewables? An empirical assessment for 10 European countries *Energy policy* **116** pp 257-65.

[3] Panwar NL, Kaushik SC, Kothari S 2011 Role of renewable energy sources in environmental protection: A review *Renewable and Sustainable Energy Reviews* **15**(3) pp 1513-1524.

[4] Moriarty P, Honnery D 2016 Can renewable energy power the future? *Energy policy* **93** pp 3-7.

[5] Seifi S, Crowther D 2016 Managing with depleted resources *InCorporate Responsibility and Stakeholding* pp 67-86 Emerald Group Publishing Limited.

[6] Covert T, Greenstone M, Knittel CR 2016 Will we ever stop using fossil fuels? *Journal of Economic Perspectives* **30**(1) pp 117-138.

[7] Mathiesen BV, Lund H, Connolly D, Wenzel H, Østergaard PA, Möller B, Nielsen S, Ridjan I, Karnøe P, Sperling K, Hvelplund FK 2015 Smart Energy Systems for coherent 100% renewable energy and transport solutions *Applied Energy* **145** pp139-154.

[8] Owusu PA, Asumadu-Sarkodie S 2016 A review of renewable energy sources, sustainability issues and climate change mitigation *Cogent Engineering* **3**(1) 1167990.

[9] Sen S, Ganguly S 2017 Opportunities, barriers and issues with renewable energy development–A discussion *Renewable and Sustainable Energy Reviews* 2017 **69** pp 1170-1181.

[10] Abdmouleh Z, Alammari RA, Gastli A 2015 Review of policies encouraging renewable energy integration & best practices *Renewable and Sustainable Energy Reviews* **45** pp 249-262.

[11] Michelsen CC, Madlener R 2016 Switching from fossil fuel to renewables in residential heating systems: An empirical study of homeowners' decisions in Germany *Energy Policy* **89** pp 95-105.

[12] Urmee T, Md A 2016 Social, cultural and political dimensions of off-grid renewable energy programs in developing countries Renewable Energy **93** pp 159-67.

[13] UPME 2014 Plan de expansión de referencia generación–transmisión 2014-2028 *Unidad de Planeación Minero Energética (UPME)* Bogotá DC, Colombia.
[14] UPME 2015 Plan Energetico Nacional Colombia: ideario energético 2050 *Unidad de Planeación Minero Energética (UPME)* Bogotá DC, Colombia.
[15] Eras JJ, Morejón MB, Gutiérrez AS, García AP, Ulloa MC, Martínez FJ, Rueda-Bayona JG 2018 A look to the electricity generation from non-conventional renewable energy sources in Colombia *International Journal of Energy Economics and Policy* **9**(1) pp 15-25.
[16] Arias-Gaviria J, Carvajal-Quintero SX, Arango-Aramburo S 2019 Understanding dynamics and policy for renewable energy diffusion in Colombia. *Renewable energy* **139** pp 1111-1119.
[17] Castillo Y, Gutiérrez MC, Vanegas-Chamorro M, Valencia G, Villicaña E 2015 Rol de las Fuentes No Convencionales de Energía en el sector eléctrico colombiano *Prospectiva* **13**(1) pp 39-51.
[18] DNP 2017 Energy Supply Situation in Colombia *Departamento Nacional de Planeación (DNP)* Bogotá DC, Colombia.
[19] Valencia AC 2016 Crisis energética en Colombia *TIA Tecnología, investigación y academia* **4**(2) pp 74-82.
[20] Grimaldo J, Mendoza M, Reyes W 2017 Modelo para pronosticar la demanda de energía eléctrica utilizando los productos interno brutos sectoriales: caso Colombia *Espacios* **1**(2) pp 3.
[21] Colombia CD. Ley 1715. Por medio de la cual se regulala integracion de las energias renovables no convencionales al sistema electrico nacional. Colombia.
[22] Nogués FS 2010 Energía de la Biomasa (volumen I) *Universidad de Zaragoza*.
[23] Aslani A, Mazzuca-Sobczuk T, Eivazi S, Bekhrad K 2018 Analysis of bioenergy technologies development based on life cycle and adaptation trends *Renewable energy* **127** pp 1076-1086.
[24] Sanchez DL, Nelson JH, Johnston J, Mileva A, Kammen DM 2015 Biomass enables the transition to a carbon-negative power system across western North America *Nature Climate Change* **5**(3) pp 230.
[25] Greenpeace 2011 Nuevas tecnologias para el tratamiento de residuos urbanos *Greenpeace* Buenos Aires, Argentina.
[26] Weitz KA, Thorneloe SA, Nishtala SR, Yarkosky S, Zannes M 2002 The impact of municipal solid waste management on greenhouse gas emissions in the United States *Journal of the Air & Waste Management Association* 2002 **52**(9) pp 1000-1011.
[27] Ebers A, Malmsheimer RW, Volk TA, Newman DH 2016 Inventory and classification of United States federal and state forest biomass electricity and heat policies *Biomass and Bioenergy* **84** pp 67-75.
[28] Ericsson K, Huttunen S, Nilsson LJ, Svenningsson P 2004 Bioenergy policy and market development in Finland and Sweden *Energy policy* **32**(15) pp 1707-1721.
[29] Escalante H, Orduz JA, Zapata H, Cardona MC, Duarte M 2011 Atlas del potencial energético de la biomasa residual en Colombia *Unidad de Planeación Minero Energética (UPME)* Bogotá DC, Colombia.
[30] Instituto de Planificación y Promoción de Soluciones Energéticas para las Zonas No Interconectadas (IPSE) 2014 Soluciones energéticas para las zonas no interconectadas de Colombia. [Online]. Available from: https://www.minenergia.gov.co/documents/10180/742159/09C-SolucionesEnergeticasZNI-IPSE.pdf/2871b35d-eaf7-4787-b778-ee73b18dbc0e
[31] Bustos J, Sepulveda A, Triviño K 2014 Zonas no interconectadas Eléctricamente en Colombia.
[32] UPME 2016 Plan Indicativo de Expansion de Cobertura de Energía Eléctrica *Unidad de Planeación Minero Energética (UPME)* Bogotá DC, Colombia.
[33] Sistema de Información Eléctrico Colombiano (IPSE) 2019 Proyectos de Generación Inscritos. [Online]. Available from: http://www.siel.gov.co/Inicio/Generaci%C3%B3n/Inscripci%C3%B3ndeproyectosdeGeneraci%C3%B3n/tabid/113/Default.aspx

[34] European Commission Climate Action. [Online]. Available from: https://ec.europa.eu/clima/policies/strategies/2020_en

[35] Schallenberg J, Piernavieja G, Hernández C, Unamunzaga P, García R, Díaz M, Subiela V 2008 Energías renovables y eficiencia energética *Instituto Tecnológico de Canarias*.

[36] Von Roon S, Dossow P, Kern T, Hinterstocker M, Pellinger C 2019 Relevance and chances for industrial self-generation of electricity for high market shares of renewable energies *Proceedings of the 11th Internationale Energiewirtschaftstagung an der TU Wien* Vienna, Austria pp 13-15.

[37] Rodríguez-Monroy C, Mármol-Acitores G, Nilsson-Cifuentes G 2018 Electricity generation in Chile using non-conventional renewable energy sources–A focus on biomass *Renewable and Sustainable Energy Reviews* **81** pp 937-945.

[38] Ericsson K, Werner S 2016 The introduction and expansion of biomass use in Swedish district heating systems *Biomass and bioenergy*. 2016 **94** pp 57-65.

[39] González JM, Daza CA, Urueña CH 2008 Análisis del esquema de generación distribuida como una opción para el sistema eléctrico colombiano. *Revista Facultad de Ingeniería Universidad de Antioquia* (44) pp 97-110.

[40] Republic of the Philippines 2008 Promoting the development, Utilization and Commercialization of renewable energy resources and for other purposes. [Online]. Available from: https://www.doe.gov.ph/laws-and-issuances/republic-act-no-9513

[41] Allen & Overy LLP 2016 The German Renewable Energy Act 2017 – An overview for foreign investors/banks. [Online]. Available from: http://www.allenovery.com/SiteCollectionDocuments/The%20German%20Renewable%20Energy%20Act%202017%20%E2%80%93%20An%20overview%20for%20foreign%20investors%20banks.pdf

[42] Gobierno de El Salvador 2016 Plan maestro de desarrollo sostenible e inclusivo de la region oriental del el salvador 2015-2025 *Gobierno de El Salvador*, San Salvador.

[43] Energy Department 2009 Finland's national action plan for promoting energy from renewable sources pursuant to Directive *Filand government* Helsinki.

[44] IRENA 2015 Renewable Energy Policy Brief Brazil *International Renewable Energy Agency (IRENA)* Abu Dhabi.

[45] Japanese government 2001 Waste management and public cleansing law Tokio.

[46] Federal Ministry Republic of Austria 2018 Misson 2030 Austrian Climate and Energy Strategy Vienna.

[47] UPME 2011 Atlas del potencial energetico de la Biomasa Residual en Colombia *Unidad de Planeación Minero Energética (UPME)* Bogotá DC, Colombia.

[48] Unidad de Planeación Minero Energética (UPME) Modelo conceptual para el cálculo de la viabilidad de proyectos WTE. [Online]. Available from: http://www.upme.gov.co/ValoracionEnergeticaRSU/?fbclid=IwAR36POUW9GbvVCwaDb0VcSBRtjpMw2TFgVTUBZLuVbsrBql6e-Nx0DBYMrw

IOP Conf. Series: Materials Science and Engineering **844** (2020) 012011 doi:10.1088/1757-899X/844/1/012011

Kinetic Study of Cyanide Degradation from Gold mining wastewater Using Photocatalysis

R.R. Correa-Turizo[1], R.M. Mestre-Martinez[2], I Marrugo-Pautt[2], I Baldiris-Navarro[3], Carlos Severiche[4]

[1, 4]Faculty of Engineering, Comfenalco Process Research Center - CIPTEC, Fundacion Universitaria Tecnologico Comfenalco Cartagena. Cartagena de indias, Bolívar, Colombia.
[2]Deparment of quality control, Gestion Analitica SAS, Cartagena de indias-Bolívar, Colombia
[3]Environmental science program. SENA CINAFLUP Research Group.SENA, Cartagena de Indias, Bolivar, Colombia.

Email: ibaldirisn@sena.edu.co

Abstract— Highly toxic cyanides are among the most persistent chemical compounds in effluents, these chemicals are used in several industrial sectors and particularly in mining industry for the extraction of gold. Cyanides frequently appear at low levels in water, deteriorating its quality. In this paper a degradation of 55.76% and 62.99% was obtained using TiO_2 catalyst and TiO_2 doped with Cu heteropololybdate respectively in 360 min of reaction, applying photodegradation with a 125 watt mercury lamp source. Most of the trials showed that the degraded compound kinetics may be adjusted to a pseudo first order model.

1. Introduction

Industrial growth, agricultural exploitation and demographic expansion lead to the appearance of emissions and waste [1]. The presence of these wastes into the environment in quantities greater than their capacity to eliminate them, implies an imbalance in the biosystem. These changes may affect the quality and health of living beings that inhabit the environment and may even become impossible the life of existing species [2]. This situation demands an urgent study of the polluting forms in the environment that affect the habitat, and also the application of techniques and procedures for its monitoring, prevention and remediation [3].

The protection of is essential because it is vital for the development of life and for its scarcity [4]. Although about 70% of the Earth's surface is covered by water, only 2.5% of the planet's water is fresh, of which more than 99.5% is found as frozen water (polar ice caps, glaciers) or underground [5]. Due to the difficult access and the enormous technological and economic cost involved in the extraction of this fraction of water, man has historically resorted to the use and exploitation of rivers and surface lakes, which represent a low 0.01% of total water [6].

Cyanide is a chemical group made up of an atom of carbon bonded to a nitrogen atom by triple bonding [7].

Most cyanides, whether of natural or industrial origin, are powerful, fast-acting poisons [8]. The main cause of water contamination by cyanide is the industrial activity and includes activities such as: mining and chemical industry, wastewater treatment, incineration of municipal waste, use of pesticides, etc [9]. Water has the capacity to transport a large amount of substances either in dissolved form or as suspended material. Thanks to this characteristic, the contamination of a water source, even if it occurs in a specific location, has repercussions over long distances and therefore affects regions of great extension [10].

Exposure to cyanide for a short period may cause damage to the brain, heart, produce coma and finally death [11]. In aquatic ecosystems, the presence of cyanide concentrations above 5 and 7.2 μg/L reduce the swimming capacity and inhibit reproduction of many fish species [12].

Due to the high complexing capacity that cyanide ion possesses, it is able, for example, to form stable compounds with almost any heavy metal at low concentrations, while its compounds may fix chemical substances that are present in nature and materials produced by man [13]. In this way, there are more than 2,000 natural sources of cyanide, (different species of arthropods, insects, bacteria, algae, fungi and higher plants contain it), as well as numerous common organic products such as plastics, synthetic textiles, fertilizers, herbicides, dyes and pharmaceutical products [6, 10]. All this puts man in permanent contact with a large number of cyanide compounds, including food, that are removed from human body by the liver, not reaching in this way harmful concentrations to health [14].

Of all the processes to obtain and purify gold in operation, the cyanide leaching method developed at the end of the 19th century, is the most effective one [3, 9]. Although its application involves the use of cyanide compounds and the generation of by-products derived from said compounds that are dangerous for the environment and humans, the percentages of extraction of the precious metal are so high that they have converted it into the more used process today in different gold mining operations [12].

In recent years, there has been an increase in foreign investment in the mining sector in Colombia. The participation of this sector has increased its presence in the Gross Domestic Product and State policies are increasingly side by side with the exploitation of minerals through concession figures. According to the Ministry of Mines and Energy (MME), in 2014 the GDP of the mining sector without hydrocarbons represented 2.10% of the total GDP of the economy. Likewise, the sector received 43% of the direct foreign investment that arrived in the country [15]. Environmentally, the forms of free cyanide (HCN and CN^-) are not generally detectable in mining effluents. The concern for the toxicity of effluents lies in the complex forms of cyanide and the possibility of breaking down and generating free cyanide in the new system [16].

On the other hand, photo-catalysis is a useful process in the treatment and / or purification of water. Currently, this topic is of great interest to scientific community, as it has proven to be an adequate alternative treatment for the purification of waters with contaminant content of difficult removal like cyanide. The heterogeneous photo-catalysis, unlike most photochemical processes, is not selective and may be used to treat complex mixtures of pollutants [12]. Nowadays the methods used for dissolved cyanides removal are: physical, chemical, biological or mixed.

The photocatalytic degradation of cyanides is another interesting application since this technique does not produce sludge or highly toxic compounds, such as cyanogen chloride. The use of hazardous products such as chlorine is avoided and it is not necessary to store chemical reagents. Another additional advantage is the possible recovery of the metal normally complexed to cyanide [5]. Photocatalytic oxidation is capable of transforming CN- into products such as cyanate, OCN- (about 1000 times less toxic) with a careful choice of reaction conditions. Once this conversion is achieved, the OCN- is completely oxidized and the final products are mainly CO2 and NO3 [6, 12].

The physical methods are based on the ion exchange, the absorption on activated carbon and separation by flotation. Biological treatments (bio-remediation) are limited to low concentrations of cyanides. Chemical methods generally involve oxidation of cyanide with reagents such as SO2, air, chlorine, potassium permanganate, hydrogen peroxide and ozone, among others [13].

Some of the advantages of photo-catalysis are that the process may be carried out with solar energy, besides the organic pollutant is completely destroyed, solving the problem of subsequent disposal of industrial process waste. Additionally, it uses as basic input a relatively unexpensive and non-toxic substance that theoretically might be recovered in its entirety: titanium dioxide (TiO_2), which is used in crystalline form as a photocatalyst.[14].

An alternative to improve the quality of wastewater, is the heterogeneous photocatalysis mediated by different semiconductors such as TiO2, considering this an effective, simple, and economical technology in relation to bioremediation treatments [11]. The aim of this work was to evaluate the kinetic of the degradation of cyanide in wastewater of gold mining.

2. Materials And Methods

Kinetic studies of cyanide degradation in this system were performed at an initial contaminant concentration of 100 ppm, with pH and catalyst concentration obtained in previous experiments, during 50 min of reaction, taking 1 mL of sample at intervals of 10 min, under the following experimental conditions: Air, TiO_2, UV radiation, Air + TiO_2, UV Radiation + Air, UV radiation + TiO_2 and UV radiation + Air + TiO2.

To observe the effect of O_2, the intensity of radiation and the catalyst on the photocatalytic degradation, assays were made combining these parameters in order to make a more detailed analysis in the photocatalytic reaction. The conditions of pH and concentration of the catalyst were optimized. Table 1 shows the reaction targets. Tests with TiO_2 and TiO_2 doped with heteropololybdate were made under the same experimental conditions, taken as a model to compare the results. Figure 1 show the set up.

Table 1. Reaction targets for a concentration of CN- 100 ppm

Name	Air (O_2)	Light	pH	Catalyst concentration (TiO_2 g/L)	Catalyst concentration TiO2 with copper heteropololybdate
Aeration	X		X		
Balance of adsorption and desorption			X	X	X
Adsorption and aeration	X		X	X	X
Photolysis		X	X		
Photochemistry	X	X	X		
Photocatalytic without air		X	X	X	X
Photocatalytic	X	X	X	X	X

The degradation study of real samples was carried out in the reaction system taking into account the optimal concentration of catalyst (TiO_2, TiO_2 doped with copper heteropololybdate) and pH. The residual water from Gambote and the mining area of the South of Bolívar, had an initial concentration of 0.025 and 75.62 ppm of total cyanides respectively and a pH of 6.4 and 6.8 respectively, and they were conserved with NaOH, because the residual cyanide is in a very complex matrix, an analysis of the state of the samples was carried out, in order to eliminate interferences.

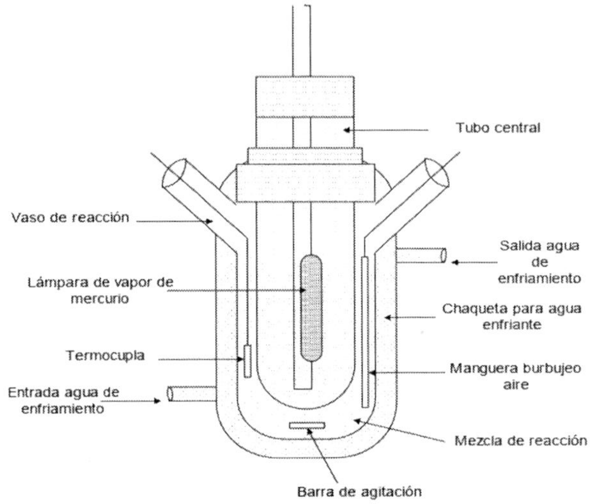

Figure 1. Photoreactor and its components

3. Results And Discussions

Kinetic study of cyanide degradation

The kinetic study of the degradation of cyanide under different experimental conditions is depicted in Figure 2, showing the contribution of the catalyst (TiO_2), air and type of radiation in the photocatalytic reaction.

In the CN^- + Cat curve, the equilibrium of adsorption - desorption of the cyanide on the catalyst is shown, observing a small adsorption of 0.33% on the catalyst since the concentration decreased by 4.11% in 50 min, due to the low adsorption of the cyanide ions on the surface of titanium dioxide. These data are in agreement with those reported in the literature by Chiang et al. 2002 [17] and Chiang et al. 2003 [18], reported an adsorption of 0.35% up to 3.71% with respect to the initial concentration of solutions. In the photolysis curve when only light is supplied, a concentration decrease was observed in 14.42% in 50 minutes, due to the lack of oxygen [19].

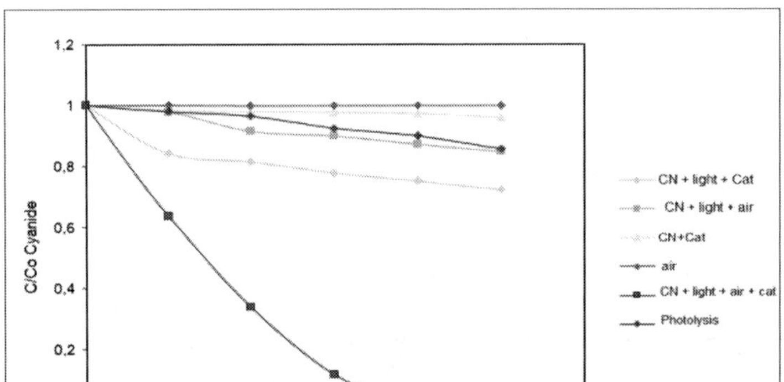

Figure 2. Kinetic study of the degradation of cyanide under different experimental conditions.
Co Cat. = 0.3 g/L TiO2 and Co cyanide = 100 ppm

When only air was supplied to the solution, it was observed that the concentration of cyanide decreased to 0.17% in 50 min, due to the lack of formation of C-radicals centered by absence of oxygen.

The curve CN + light + air, showed that the photolytic degradation of cyanide in the presence of air reaches 15.32% in 50 minutes of reaction, this degradation is contributed due to the presence of air in the system by the formation of species such as organic peroxides that contribute to the oxidation process, in the CN- + light + Cat curve, it was observed that photocatalytic degradation of airless cyanide is 27.78% in 50 min, although water dissolved oxygen was not enough due to the presence of the catalyst, that increases the degradation rate when irradiation provided to the system activated the catalyst generating electron-hole pairs.

The curve of CN- + Light + Cat + air, showed better results with a degradation of cyanide to 98.55% in 50 minutes, this evidenced that the presence of oxygen is essential for effective oxidation because the oxygen provided by the air acts as an acceptor of the electrons generated in the semiconductor when it absorbs the radiant energy, avoiding the recombination of the pair in the surface of the same, making the photocatalytic reaction more efficient [20].

The above is verified by the kinetic analysis shown in Figure 3, where the reaction constant is determined.

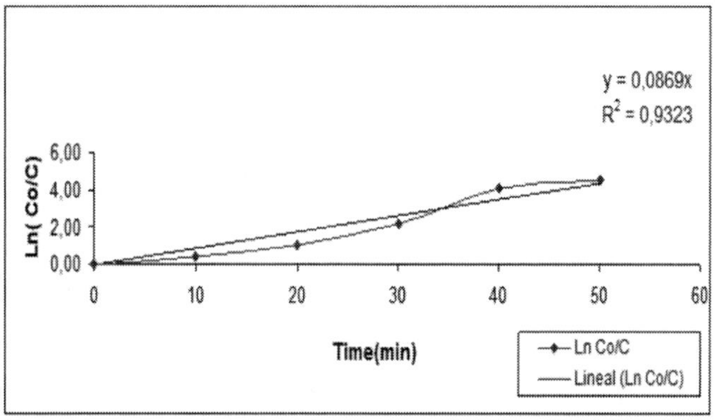

Figure 3. Kinetic adjustment of pseudo-first order for the photocatalytic degradation of cyanide

Kinetic study of the degradation of the real sample

When comparing the degradation of the stock solution with the synthetic one, it showed that there is an increase in time for cyanide degradation, which indicates that aggregated compounds produce interference.

Figures 4 and 5 show the variations in the concentration of total cyanides as a function of time for the real and synthetic samples, using as catalyst TiO_2 and TiO_2 doped with 1% copper heteropololybdate. Noting that the percentage of degradation is of 55.76% for the real sample and 98.08% for the synthetic with TiO_2 and 62.99% for the real one, 98.53% in the synthetic with TiO_2 doped with HPMoCu 1% in 390 minutes respectively, in which the real sample presents a slower kinetics of degradation, because not only are the interferences added to the synthetic sample, but a series of metal complexes with cyanide M (CN-) xy- that inhibit the photodegradation of cyanide in the presence of TiO2 by the competition between M (CN-) xy-complexes and free cyanides by the surface of the active sites of * OH-TiO2. These results are in accordance with those reported by Chiang et al. (2003) [18], and Carp et al. (2004) [21].

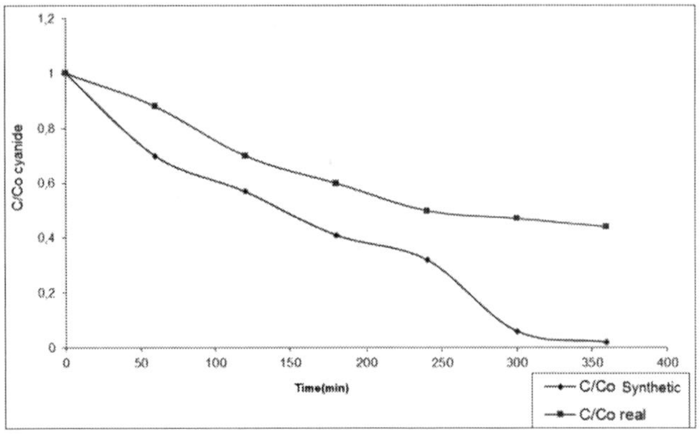

Figure 4. Photocatalytic degradation of cyanide in real and synthetic sample. C Cat: 0.3 g / L TiO2 and Co CN- total 75.62 ppm and 72.64 ppm of real and synthetic sample respectively

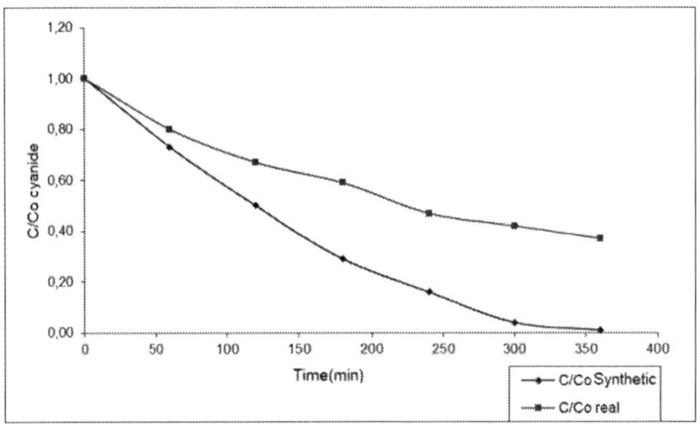

Figure 5. Photocatalytic degradation of cyanide in real and synthetic sample. Cat C: 0.3 g / L TiO_2 doped with HPMoCu 1% and Co CN- total 75.35 ppm and 72.64 ppm of real and synthetic sample respectively

In Figure 6, the influence of the type of catalyst on the degradation of cyanide in the real sample is shown, observing a small difference in the kinetics of degradation where the doped TiO_2 reaches 62.99%, while the TiO_2 55.76%. The degradation of cyanide reported, may be analyzed in terms of the langmuir-hinshelwood model:

$$r = -\, dc/dt = K_2\theta_{o2}\theta_{CN^-}$$

Where r is the reaction rate for cyanide oxidation in units mM / s and K2 (mM / s) is the kinetic constant of second order, the terms θo2 and θCN- refer to the surface fraction covered by oxygen and cyanate respectively.

The term θo2 can be assumed constant, considering the continuous bubbling of air to the reaction system, reducing the equation to:

$$r = -\, dc/dt = K_1\theta_{CN^-}$$

Where the kinetic constant of first order K1 is equal to K1 = K2θ02, the coated surface θCN- can be expressed as: θCN- = K * CCN- / 1 + K * CCN-, replacing in the equation we have that:

$$r = -dc/dt = K_1 K^* C_{CN^-} / 1 + K^* C_{CN^-}$$

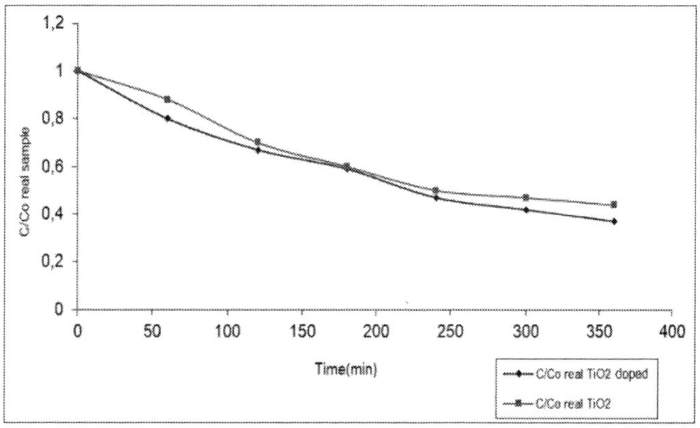

Figure 6. Kinetic photocatalytic degradation of cyanide present in real sample with respect to the influence of the catalyst 0.3g/L TiO_2 and 1g/L TiO_2 doped with HPMoCu 1%

A plot of the reciprocal of the initial cyanide photo-oxidation ratio r versus the reciprocal of the initial cyanide concentration shows a straight line as can be seen in Figure 7, with a value of r2 = 0.9607, which indicates a good approximation between the prediction of the equation and the experimental results.

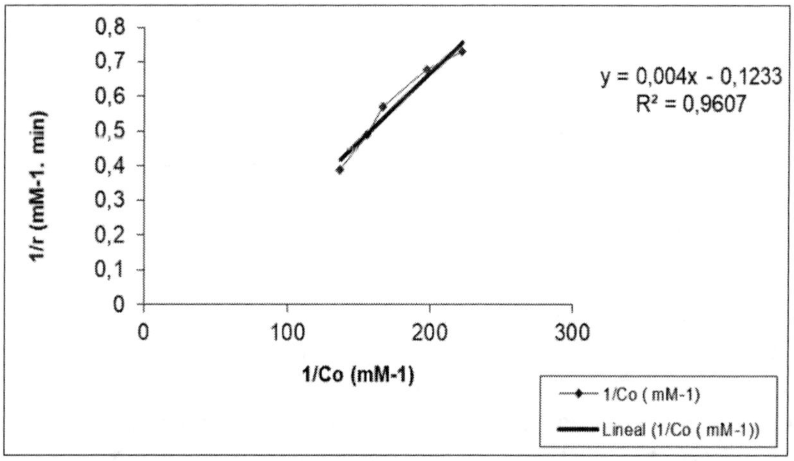

Figure 7. Reciprocal plot of the proportion of initial cyanide photo-oxidation (1/r) versus reciprocal of cyanide initial concentration (1/Ci)

In Figures 8 and 9, variations in the concentration of cyanide and cyanate during the photoreaction of cyanide degradation (75.72 mg / L) which is equivalent to 2.91 mM in real water, using TiO2 and TiO2 doped with HPMoCu 1% as catalyst are shown. In several photocatalytic oxidation studies of cyanide solutions, cyanate was the main oxidation product for both TiO2 and doped copper modified TiO2 [22, 23]. It may be observed that during the cyanide oxidation, the cyanate concentration increases in the same way as the cyanide concentration decreases. The non oxidation of cyanate was detected during the experimental assays, which is in agreement with the works of [22, 23]. Many reported results show that the percentage of CN- reaction with OH * are two orders of magnitude higher between the CON- and OH * reaction [24, 25, 26, 27].

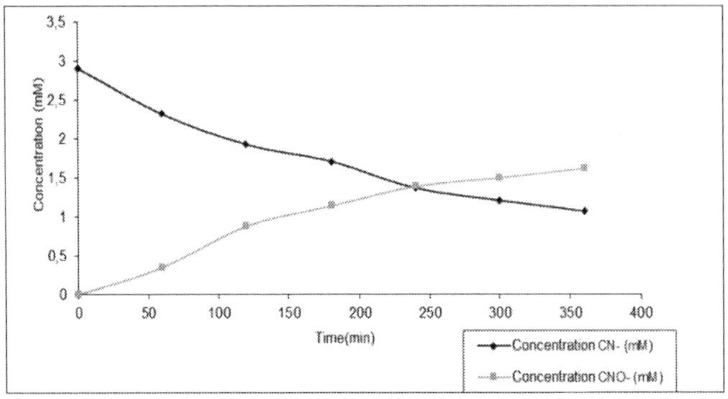

Figure 8. Photodegradation of cyanide (75.62 mg/L) to cyanate in real sample using TiO2 as catalyst

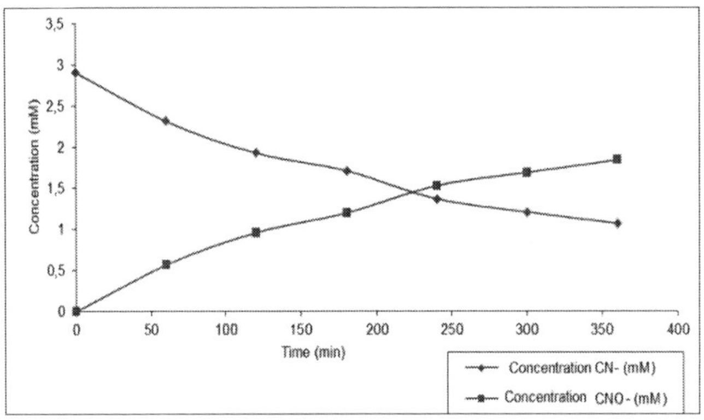

Figure 9. Photodegradation of cyanide (75.62 mg/L) to cyanate in real sample using TiO2 doped with HPMoCu 1% as catalyst

4. Conclusion

According to the review of the bibliography, the results and their discussion can be concluded that: The photocatalytic degradation when air was supplied was higher than that obtained when it was absent, therefore the oxygen supply is important because it increases the degradation rate of the substrate, since the lack of supply tends to deplete the dissolved oxygen in water.

Acknowledgment

The authors are grateful for the support provided by the Fundacion Universitaria Tecnologico Comfenalco Cartagena through the research vice-rectory for the funding of this work.

IOP Conf. Series: Materials Science and Engineering **844** (2020) 012011 doi:10.1088/1757-899X/844/1/012011

References

[1] N. Dwivedi, Ch. Balomajumder, P. Mondal, 2016 Water Resources And Industry,. 15, pp. 28,.
[2] L. Mekuto, S.K.O. Ntwampe, A. Akcil, 2016 Science of The Total Environment, 571 711
[3] T. Kim, T. Kim, A. Jo, S. Park, K. Choi, K. Zoh, 2018 Chemosphere, 208, pp. 441-449,.
[4] M.P. Guamán, D. Nieto, J. Of Water Process Eng, 2018 vol. 23, pp. 84-90,.
[5] A. Brüger, G. Fafilek, O. Restrepo, L. Rojas-Mendoza, 2018 Sc Of The Total Environment, vol. 627, pp. 1167-1173,.
[6] X. Yu, R. Xu, Ch. Wei, H. Wu, 2016 J. Of Hazardous Materials, Vol. 302, pp. 468-474,.
[7] M. Gebresemati, N. Gabbiye, O. Sahu, 2017 J. Of Appl Research And Tech, vol. 15, pp. 27-35,.
[8] A. Pala, R. Ruth-Politi, G. Kurşun, M. Erol, F. Bakal, G. Öner, and E. Çelik, 2015 Surface and Coatings Tech, vol. 271, pp. 207-216,.
[9] U. Ria-Irfan, , I. Nur, and M. Kasim, 2017 Int. J. on Adv Sc, Eng and Info Tech, vol. 7, pp. 2244-2250,.
[10] E.S. Baeissa, 2015 Chinese J. Of Catalysis, vol. 36, pp. 698-704,.
[11] G. Moussavi, M. Pourakbar, E. Aghayani, M. Mahdavianpour, 2018 Chem Eng J., vol. 350, pp. 673-680,.
[12] C. Feng, C. Aldrich, J.J. Eksteen, D.W.M. Arrigan, 2018 Minerals Eng, vol. 122, pp. 84-90,.
[13] D. Fortuna, A. Rahimsyah, and Y. Puspitasari, 2015 Int. J on Advanced Science, Eng and Info Tech, vol. 5, pp. 291-293,.
[14] L. Razanamahandry, H. Anderson, H. Karoui, K. Kouakou, H. Yacouba, 2016 Chemosphere, vol. 157, pp. 71-78,.
[15] J. Olivero-Verbel, K. Caballero-Gallardo, J. Negrete-Marrugo. 2011. Biol Trace Elem Res, vol. 144, pp. 118-132,.
[16] M. Pirmoradi, S. Hashemian, M. Shayesteh, 2017 Transactions Of Nonferrous Metals Society Of China, vol. 27, pp. 1394-1403,.
[17] K. Chiang, R. Amal, and T. Tran. 2002 Adv. In Environmental Res, vol. 6 , pp. 471-485,.
[18] K. Chiang, R. Amal, and T. Tran. 2003 Chemical, vol. 193,.
[19] L. Gómez, A. Urkiaga, M. Gutierrez, and L. De Las Fuentes, 2000 Revista De Ingeniería Química, pp. 211-216,.
[20] S. Parra, 2001 École Polytechnique Federale De Lausanne. Doctoral Thesis No. 2470.
[21] O. Carp, C.L. Huisman, and A. Reller, , 2004 Progress in Solid State Chemistry, . 32, . 33.
[22] V. Augugliaro, V. Loddo, G. Marcí , L. Palmisano, and M. J. López. 1997, J Of Catal,. 37,. 272.
[23] C.H Pollema, J.L Hendrix, E.B Milosaylievic, L. Solujic, and J.H. Nelson, 1992 J.Photochem. Photobiol.A: Chem, 66 235.
[24] D. Behar, 1974 J.Phys. Chem, 78,. 2660.
[25] H. Buchler, and R.E. Buhler, 1976 J.Phys.Chem,. 80, 1549.
[26] B.H.J. Bielski, and A.O. Allen, 1977 , J.Am. Chem.Soc, 99 5931.
[27] J.G. Leopold, and M. Faraggi. 1977 J. Phys. Chem. 81 803.

Degradation of Cyanide from Gold Mining Wastewater Using Photocatalysis

R M Mestre-Martinez[1], I Marrugo-Pautt[1], R R Correa-Turizo[2] C A Severiche Sierra[3]and I Baldiris-Navarro[4]

[1]Departamento de Control de Calidad, Gestión Analítica SAS, Cartagena de indias, Bolívar, Colombia
[2]Facultad de Ciencias Exactas y Naturales, Universidad de Cartagena, Cartagena de indias, Bolívar, Colombia
[3]Grupo de Investigación GITEC, Corporación Universitaria Rafael Nuñez, Cartagena de indias-Bolívar, Colombia

E-mail: rcuadrado019@gmail.com

Abstract: In recent years, water quality issues related with hazardous wastewater and toxic substances have attracted worldwide attention. Cyanide is a major toxin in wastewater resulting from a diversity of industries, including gold mining. Cyanide has adverse health effects on people as well as other living organisms. The toxic effects of cyanide are so important to cause nerve damage and thyroid glands malfunctioning. In this paper, the degradation of cyanide in waters resulting from gold mining activity was carried out in a batch system with two catalysts TiO_2 Degussa P-25 and TiO_2 doped with copper heteropolyimolybdate, that was synthesized in laboratory and characterized by FTIR and XRD. Assays showed a degradation of 98.55% with a concentration of 0.3 g/L of TiO_2 and 97.17% with TiO_2 doped with 1% of Cu heteropolyimolybdate (1 g/L) in 50 min of reaction. A real mining sample showed a cyanide degradation of 55.76% with TiO_2 in 240 min of reaction. All the assays were made applying photodegradation using a 125-Watt mercury lamp as source. These results are encouraging to use this technology for the removal of cyanide coming from highly contaminated aqueous effluents.

1. Introduction

The presence of dangerous chemical compounds in the environment is the result of anthropogenic activity. In the last years, environmental legislation has focused on demanding the protection of water bodies, due to the lack of it and the dramatic increase of diseases and dysfunctions of the nervous system in animals and humans [3]. Cyanide is one of the most prevalent toxic environmental pollutants, resulting from industrial processes such as coking plants, oil refineries, and several chemical industries processing resins, plastics, dyes, varnishes, pharmaceuticals and pesticides, etc. Cyanide is also present in chemicals that are used to reveal photographs. Cyanide salts are used for galvanization, metal cleaning and recovery of gold from mining activities [6,7]. Previous studies [11], showed the presence of cyanides at low concentrations in water that are of environmental and toxicological interest, because this may cause lethal effects on plants, animals and humans [12, 13]. In human health, cyanide avoids the body cells from receiving oxygen, causing tissue hypoxia and cyanosis. Cyanide enters the human body through inhalation, ingestion or absorption through the skin [14, 15]. Cyanide in acid water streams produces hydrocyanic acid (HCN), which is a highly toxic chemical; if the water pH is basic, it is maintained as cyanide. Due to this is necessary to reduce the cyanide levels in wastewater before discharging it into the water bodies, because of the social and public health problems that these involve [8]. In mines, gold containing waste is recovered using activated carbon. Once extracted from activated carbon, gold is concentrated by precipitation or galvanization [9]. Since gold is a noble metal, it is not soluble in water. To dissolve it, a substance like cyanide is needed, because it may form complexes and stabilize gold in solutions [10]. In Colombia, gold exploitation is done legally and illegally in rural conflict zones where authorities are not capable to protect the environment from these dangerous effluents. Most of the studies reviewed in the literature perform a quantification of cyanides present in water, others postulate absorbers that could recover or trap cyanide but would be transferring the problem of a pollutant in an aqueous effluent to a solid waste contaminated not giving a true solution [16]. So, the aim of this project is to remediate through Photocatalysis real mining water samples to provide a method to minimize cyanide effluents in conflict zones.

2. Experimental

2.1 *Preparation of Copper heteropolyimolybdate*

Copper heteropolyimolybdate synthesis was carried out by wet coprecipitation. The solutions of copper sulfate ($CuSO_4$) and ammonium heptamolybdate $(NH_4)_6Mo_7O_{24}.4H_2O$, were previously heated at 80 °C for 10 min., Mixed and then boiled in a Borosilicate glass reactor (at approximately 94 °C), forming a precipitate that after a certain time was filtered, washed and dried in an oven at 100 °C. The initial precipitate was a by-product of the synthesis where the copper heteropolyimolybdate was obtained. Copper heteropolyimolybdate ($(NH_4)_4$ $[CuH_6Mo_6O_{24}]$ $.5H_2O$,) was precipitated from the mother liquor which was obtained from the above filtration with an excess of acetonitrile.2.1.

2.2 *Collection and conservation of samples.*

The aim of the sampling procedure was to obtain representative samples of the parameters to be analysed, conserving the relative concentrations of all the components present in the original material. At each sampling site, enough water was taken at different times. For storage, a time of 8 hours was taken, due to the collection of the sample that took place on working days, from 8 a.m., because it is the time when the mining activity begins and ended at 4 p.m. approximately. Individual portions of the water under study were taken and mixed in a single bottle [17].

2.3 *Photodegradation experiments of cyanide.*

The cyanide photodegradation experiments were carried out in the reaction system at room temperature (30 ± 2 ° C) and at atmospheric pressure by bubbling air into the system, to avoid recombination of the electron-hole pairs (see figure 1). The temperature of the system was measured by a thermocouple and regulated by cooling water that removes the heat generated by the UV lamp. In order to optimize the photocatalytic degradation of cyanide, the following factors were taken into account, taking as a reference the work carried out by Sarla et al. (2004) [18]: Concentration of TiO_2 catalyst (0.2 - 3 g / L), Type of catalyst (TiO_2; TiO_2 doped with HPMoCu), Initial cyanide concentration (100 ppm), Reaction time (50 minutes) and pH (10.8 - 12),

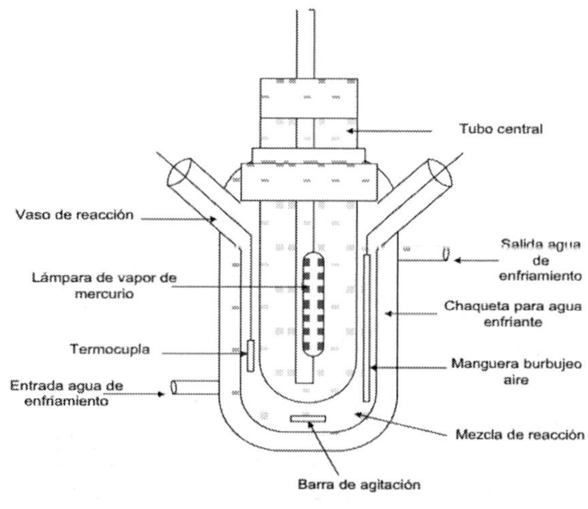

Figure 1. Photoreactor and its components

The above was done to obtain the optimum concentration of the catalyst. With this catalyst concentration, the initial pH was varied between 10.8 and 12, in order to obtain the value that maximized the percentage of degradation. The levels of catalyst concentration and pH were selected based on the literature review [19, 20]. The initial cyanide concentration was set taking into account the concentration of total cyanide present in the effluents of the mining area and the time of reaction in relation to a series of preliminary tests.

The degradation study of the real samples was carried out taking as reference a research done by Augugliaro et al. (1997) [21] in the reaction system taking into account the optimal fixed concentration of catalyst (TiO2 and TiO2 doped with copper heteropolyimolybdate) and pH. The residual water from the mining area of southern Bolívar, had an initial concentration of 75.62 ppm of total cyanides and a pH 6.8, and were preserved with NaOH, because the cyanide in Residual water is in a very complex matrix.

As the initial concentration of this effluent was 75.62 ppm than that fixed in the previous kinetic studies, a degradation test of a synthetic mixture with an initial concentration of 72.64 ppm of CN^- was previously carried out, adding the concentrations of the compounds found in the respective physical-chemical analysis of the water, very similar to the real sample, with the same concentration of catalyst and adjusting the pH of the initial solution to 10.8 adding NaOH, which allowed to estimate the reaction time necessary to degrade a solution with this initial concentration and have a reference study. For these tests, a time of 360 min was set and 4 mL of sample was taken at 1-hour intervals. The chemical oxygen demand (COD), total organic carbon (TOC) was analyzed for the corresponding samples from the mining area of southern Bolivar "Barranco de loba" at 60, 120, 180, 240, 300 and 360 minutes.

2.4 Determination of cyanide.
The samples taken from the reaction system were centrifuged for 7 minutes at 4200 rpm, then the potentiometric method was used to determine the final concentration in each of the samples, following the Standard Methods for the Examination of Water and Wastewater, 23rd Edition of the 2017 [17], using a cyanide selective ion electrode potentiometer, which allows us to measure the potential of the specific ion in the solution.

2.5 Physicochemical characterization of mining water.
Water samples from the mining area of Barranco de Loba, south of Bolívar (Colombian Caribbean), were subjected to physical-chemical tests for COD and TOC in accordance with the Standard Methods for the Examination of Water and Wastewater, 23rd Edition of 2017 [17].

3. Results and Discussion

3.1 Characterization of the catalyst
IR spectrum of TiO2 doped with $(NH_4)_4$ $[CuH_6Mo_6O_{24}].5H_2O$ is shown in figure 1, where a wide band between 3200-3400 cm-1 is observed which is attributed to the stretching vibrations of the hydroxyl groups from the hydration waters.

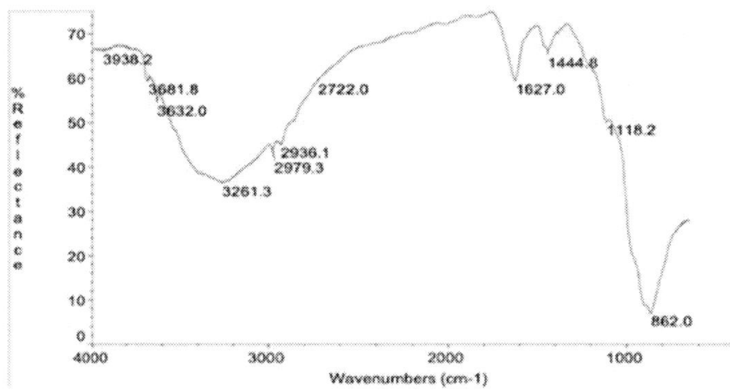

Figure 2. IR spectrum of TiO_2 doped with $(NH_4)_4$ $[CuH6Mo_6O_{24}]$ $.5H_2O$

The typical region of absorption of TiO_2 between 1200-1700 cm-1 is shown and they are physically assigned to water adsorbed on the surface, this behavior is similar to that reported by (Dvoranová D. et al., 2002). The signal at 862.0 cm-1 can be assigned to the Cu-O-H and MoO4 binding vibrations. The peak at 1118.2 cm-1 is due to asymmetric Ti-O stretches and octahedral Mo = O terminal oxygens.

The XRD pattern of titanium dioxide doped with 1% copper heteropolyimolybdate is depicted In Figure 3. The diffractographic analysis was performed based on the cards N ° 83-2243, 78-2485, and 22-0504 respectively, (XRD patterns).

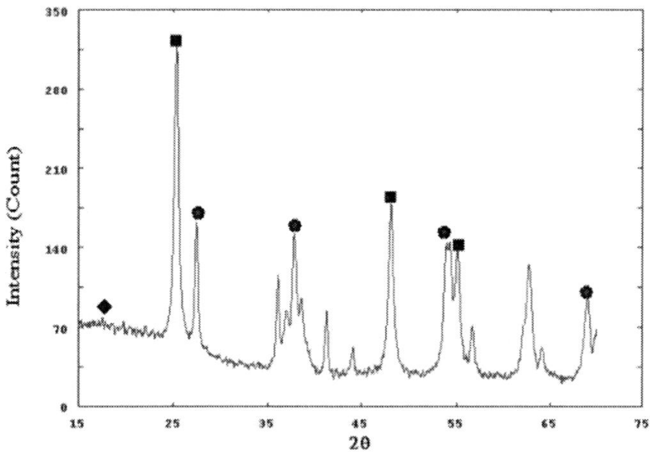

Figure 3. Diffractogram of TiO2 doped with (NH4) 4 [CuH6Mo6O24] .5H2O

The results showed a higher proportion in the phase of anatase-TiO2 (■) which showed a peak at 3.24688 Å, 1.89118 Å, 1.66578 Å which confirm the presence of TiO2 in all samples. Meanwhile the rutile phase-TiO2 (●) showed peaks at 3.24688 Å, 2.48552 Å, 1.69102 Å, 1.36051 Å, and the phase (NH4) 4 [CuH6Mo6O24] .5H2O (◆) showed a typical distance of 5.06805 Å.

3.2 Cyanide Photodegradation experiments.

The percentages of cyanide degradation obtained by varying the catalyst concentration (TiO_2 Degussa p-25) between 0.1 and 3 g / L are shown in Figure 1. The initial concentration of cyanide was 100 ppm, pH was 10.8 and reaction time was 50 min. Results showed an increase in degradation from 0.1 to 0.3 g / L, possibly due to the greater number of photons absorbed by the photocatalyst, which provides a high concentration of charge per volume unit. From 0.3 to 3.0 g / L there is a degradation reduction, since the penetration of the UV photons decreases, therefore, less TiO_2 particles may be activated. Therefore, the optimum catalyst point is 0.3 g / L with a degradation percentage of 98.55%. This was the selected concentration of TiO_2 catalyst used for the subsequent degradation tests.

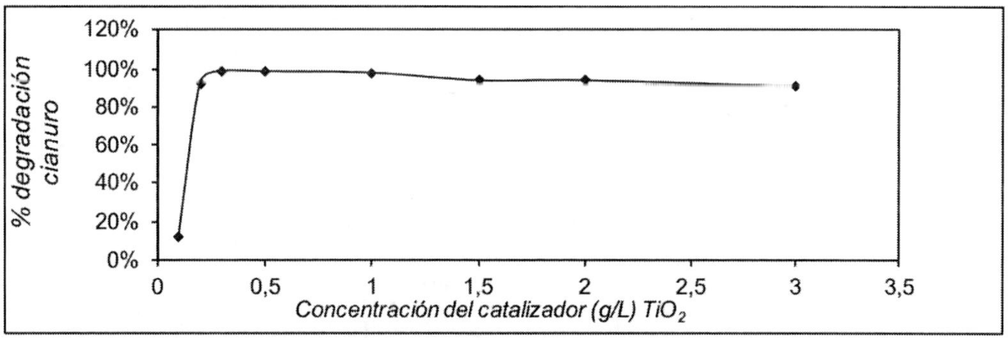

Figure 4. Influence of the concentration of catalyst TiO_2 degussa P-25 vs. percentage of cyanide degradation

Study of the influence of TiO_2 catalyst concentration doped with copper heteropolimolybdate Vs % Degradation of cyanide for the reaction system.

Percentages of degradation obtained from cyanide using doped catalyst are depicted in Figure 2. The concentration of the TiO_2 degussa P-25 catalyst were 0.3 and 1.0 g/L and Cu heteropolimolybdate concentration varied between 0.1% and 10% during 50 min of reaction. The best results were obtained with 1 g/L of TiO_2 and a doping of 1%, showing a degradation percentage of 97.17%. It may be seen that for the same catalyst concentration of 0.3 g/L of TiO_2 and doped TiO_2 there is a pronounced decrease in the degradation percentage for the case of modified TiO_2. The negative effect of the photooxidation of cyanide due to the presence of Cu (1) -CN- which are species that are formed in competition reactions with the hydroxyl radicals produced on the surface of TiO_2 under UV radiation.

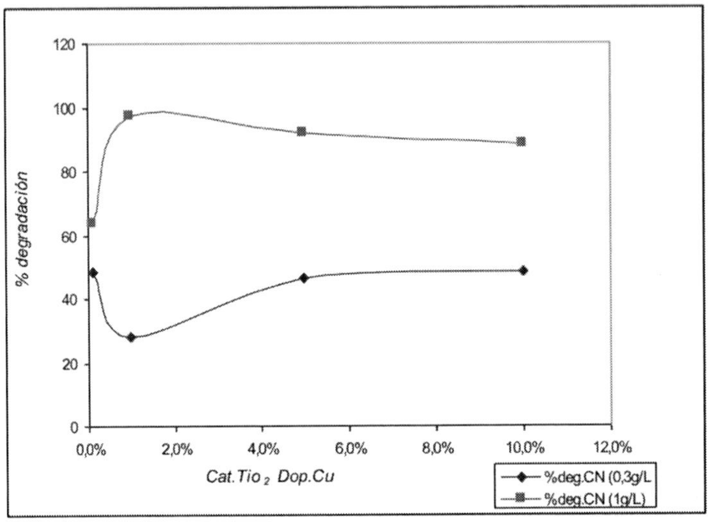

Figure 5. Influence of doped TiO_2 catalyst concentration on cyanide degradation

Influence of the catalyst TiO2 and TiO2 doped with heteropolimolybdate of Cu Vs% degradation of the cyanide for the reaction system.

In order to increase the uptake of the electrons and decrease the recombination of the electron - hollow pairs, strategies were followed in the modification of the structure and composition of the catalyst. These approaches include modifying the TiO_2 surface doped with metal ions and adding compounds that transfer charge to the semiconductor [22]. Table 1 shows the percentages of degradation obtained from cyanide of 98.55% and 97.16% in a time of 50 minutes, using TiO2 and TiO2 doped with copper heteropolimolybdate, respectively in which it was observed that the modified catalyst decreases the cyanide degradation in compared to the unmodified one, this can be attributed to the fact that during the preparation surface area has been lost.

Table 1. Influence of the catalyst on the degradation of cyanide

Time (min)	TiO_2 (%)	(TiO_2 – Cu) (%)	Without catalyst (%)
0	0	0	0.00
10	35,94	33,02	2.25
20	65,85	44,84	8.54
30	87,91	58,82	10.12
40	97,97	77,37	12.91
50	98,55	97,16	15.32

Study of the influence on the pH variation in the photocatalytic degradation of cyanide

Figure 3 shows the percentages of degradation of cyanide obtained by varying the pH between 10.8 and 12.0, starting from an initial concentration of cyanide 100 ppm and a concentration of catalyst 0.3 g / L, during 50 min of reaction. The results showed an increase in the degradation of cyanide by decreasing the pH, because at high pH values the adsorption of cyanide ions with negative charge by the hydroxide groups absorbed on the TiO2 surface is not favourable, allowing the permanence of free cyanide in solution. Therefore, the optimum point of pH is 10.8 with a degradation percentage of 98.55%, since the pH cannot be lowered below the pka of hydrogen cyanide which is equal to 9.21 at 25 ° C.

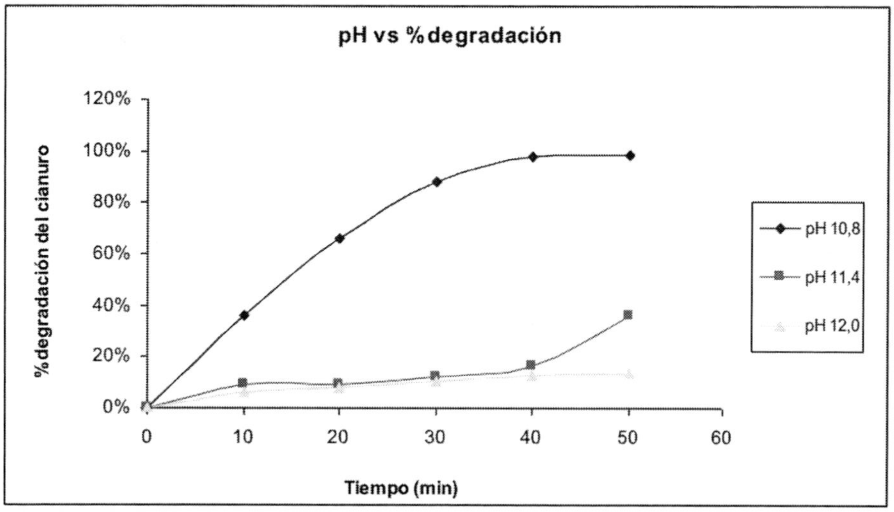

Figure 6. Influence of pH on the photocatalytic degradation of cyanide in the presence of 0.3 g / L TiO2, Co = 100 ppm

3.3 *Characterization of mining water.*
Generally, cyanides in industrial wastewater are found together with other contaminants that interfere with the photocatalytic degradation process. For this reason, the degradation of cyanide compounds presents in an industrial discharge of the mining area Barranco de Loba, Santa Cruz mine, Colombia was studied. DQO and TOC were applied to real samples and were analysed in laboratory according to the standard methods for water and wastewater analysis.

3.4 *Chemical oxygen demand (COD)*
The initial cyanide concentration of the sample was 75.62 mg/L and COD was 3911.6 mg/L, the behaviour of CDO and cyanide degradation process are shown in Figure 4. COD decreased 76.36%, while for the cyanide was 55.76% in 240 min of reaction, indicating the non-selectivity of the photocatalysis towards the destruction of other organic compounds present in the sample, which are also water pollutants.

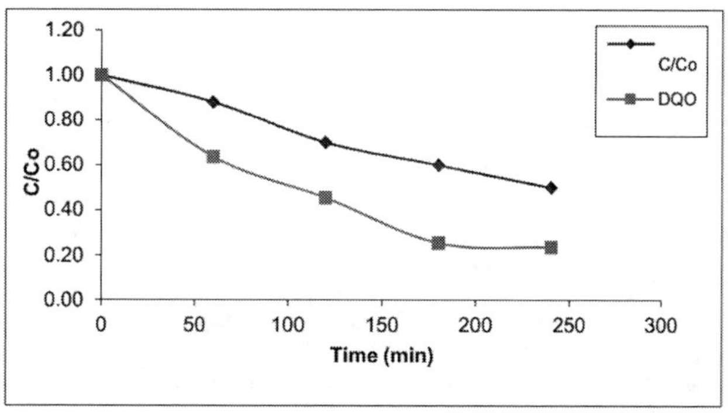

Figure 7. Curves of photocatalytic degradation of total cyanide and COD of mining water

3.5 *Total organic carbon (TOC)*
Curves of the photocatalytic degradation of the total cyanide and the total organic carbon of the real sample are depicted in figure 5. Initial concentration of TOC was 224 mg/L and experiments showed a decrease of the total organic carbon in 94.01 mg/L and a degradation of CN- in 55.76%, indicating that part the organic matter was also photodegraded confirming once again the non-selectivity of photocatalytic degradation.

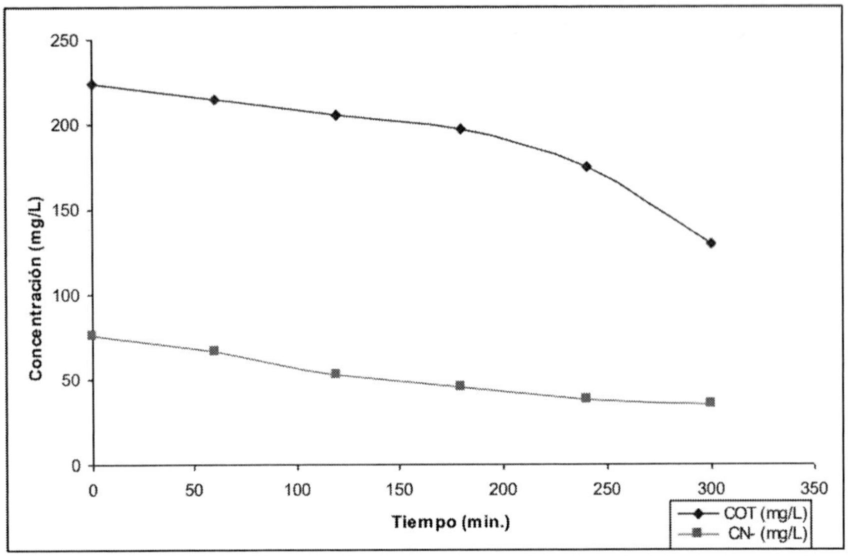

Figure 8. Curves of the photocatalytic degradation of total cyanide and the TOC of mining sample

4. Conclusions

According to the review of the bibliography, the results and their discussion can be concluded that: (i) The degradation system showed us that the total cyanides, present in the aqueous residual discharges, coming from the mining sector, may be degraded through heterogeneous photocatalysis, providing a viable alternative for the destruction of contaminants in mining effluents. (ii) In the photodegradation of the cyanide (75.62 ppm), there was a reduction of 76.36% of the COD, while for the cyanide it was 55.76% in 240 min. This means that cyanide was photodegraded and also another series of organic compounds that were present in the degradation, showing the non-selectivity of the photocatalytic process in the destruction of the organic contaminants. (iii) Cyanide compounds present in the real sample were degraded by heterogeneous photocatalysis in 55.76% with TiO_2, while the simulated synthetic sample reached 98.41% with TiO_2 %, which showed the effect of the interferences generated by the compounds present in the residual sample competing with the cyanides for the OH^- radicals produced by the excitation of the catalyst.

Conflicts of Interest

The authors declare that they have no conflicts of interest.

References

[1] Baldiris-Navarro I, Sanchez-Aponte J, Acosta-Jimenez J, Severiche-Sierra C, Correa-Turizo R 2017 Int J. Eng. and Technol (IJET) 9 4203
[2] Kuyucak N, Akcil A 2013 *Miner. Eng.* 50 13
[3] Factorovich M, Guz L, and Candal R 2011 *Adv. Phys. Chem.* 2011 8
[4] E.Y. Yazıcı E, Deveci H, Alp I 2011 *J. of Hazard. Mater.* 166 1362
[5] Yeddou A, Chergui S, Chergui A, Halet F, Hamza A, Nadjemi B, Ould-Dris A, Belkouch J, 2011 *Miner. Eng* 24 788
[6] Roshan Dash R, Balomajumder Ch, Kumar A 2009 *Chem. Eng. J.* 146 408
[7] Yu X, Xu R, Wei Ch, Wu H 2016 *J. Hazard. Mater.* 302 468
[8] Dwivedi N, Balomajumder Ch, Mondal P 2016 *Water Res. and Ind.* 15 28
[9] Gebresemati M, Gabbiye N, Sahu O 2017 *J. Appl. Res. and Technol.* 15 27
[10] Baeissa E.S., 2015 *Chin. J. Catal.* 36 698
[11] Olivero-Verbel J, Caballero-Gallardo K, Negrete-Marrugo J. 2011 *Biol. Trace Elem. Res.* 144 118
[12] Donato D, Nichols O, Possingham H, Moore M, Ricci P & Noller B. 2007 *Environ. Int.* 33 974
[13] Barakat M, Chen Y, & Huang C 2004 *Appl. Catal.* B. 53, 13–20.
[14] Parga J.R, Shukla S.S, Carrillo-Pedroza F.R 2003 *Waste Manag.* 23 183
[15] Acheampong M, Lens P, 2014 *Hydrometallurgy* 141 1
[16] Dash R, Gaur A, & Balomajumder C 2009 *J. Hazard. Mater.* 163 11.
[17] American Public Health Association (APHA), the American Water Works Association (AWWA) and the W E F (WEF). 1999 Standard Methods for the Examination of Water and Wastewater Part 1000 Standard Methods for the Examination of Water and Wastewater

[18] Sarla M, Pandit M, Tyagi D.K, Kapoor J.C. 2004 *J. Hazard Mater.* 116 49
[19] Gómez L, Urkiaga A, Gutierrez M and De las Fuentes L, 2000 *Rev. de Ing química* 211
[20] Doménech X, Jardim W.F y Litter M 1998 *CNEA*
[21] Auguglíaro V, Loddo V, Marcí G, Palmisano L, y López M. J. 1997 *J. catal.* 37 272
[22] Rasalingam S, Peng R, and Koodali R, 2014 *J. Nanomater* 2014, 42
[23] Ministry of agriculture 1984 Decreto 1594 vol 1984 (Bogota)

Risk assessment of PCP lift system using a fishbone diagram and MICMAC method

J. Osorio-Tovar[1a], J.W. Grimaldo-Guerrero[2]

[1]Petroleum engineer, Instituto Universitario Politécnico "Santiago Mariño", Maracaibo, Estado Zulia.
[2] Department of Energy, Universidad de la Costa, Colombia, ORCID: 0000-0002-1632-5374.

[a]joseosorio94@hotmail.com

Abstract. Hydrocarbons demand by industrialized and developed nations and the oil prices conditions last few years; many companies have turned again to fields that were not previously considered attractive to produce, to identify them as an alternative to meet that demand. This research shows the identification of a condition, which should be taken into account to produce a mature field by progressive cavity pumping (PCP) artificial lift system, an evaluation with the MICMAC method and the identification of the key variables to achieve the development in this practice. The results make possible to identify highest influence elements, which can guide intervention strategies and form the basis to formulate guidelines and policies for the PCP implementation. The results allow us to conclude that pressures and mechanical designs in field wells should be the guidelines for optimum production, the market rate improvement, and the reservoir productive life.

1. Introduction

The oil sector is a strategic asset due to its economic contributions [1]; for Colombia in 2018, this sector contributed 40% total exports [2]. There is a necessity to maintain an optimal and continuous production, to achieve a long productive life of the oil wells.

The main natural resources, which supply the growing energy demand, are oil and natural gas [3, 4]; the prices of these are volatile and they can generate problems the consumer countries economy, to avoid these volatilities the exploitation of these two natural resources is maintained [5]. To optimize extraction processes, improvements have been made in the Artificial lift systems (ALS) [6], but all of them are susceptible to fail.

The systems selection will depend on the conditions present in the reservoir, starting from natural production, then it uses an ALS and ending in secondary and tertiary recovery systems [7, 8]. During operation, the problems may occur in the electrical functioning and mechanical stresses in the production equipment [9], which may generate additional and indirect impacts on other components of the production system [28].

Failure management is effective when it begins prevention and ends with correction [10] these operations may involve high economic costs due to the equipment and maneuvers must be performed; identifying the occurrence frequency of failures allows the design of preventive maintenance plans which will have an impact on improving oil production and reducing incidents and accidents [12].

This research makes an evaluation of the failures which occur in a progressive cavity artificial lift system (PCP), to identify the key elements which can be used to propose preventive maintenance strategies and improve performance conditions.

2. Methodology

The methodology begins with the oil well production optimization scheme, it identifies failures in a well with PCP they are related through a fishbone diagram, to allow the variables selection which causes the failures. Using the Cross-Impact Matrix Multiplication Applied to Classification (MICMAC) method and the identified variables, the key elements for the prevention of system failures are identified. Procedural systematization is organized into three phases:

2.1. Identification phase

A documentary review was carried out in different information sources about the operating conditions of an oil well with a PCP, operation variables were identified and the effects they can generate in the good operation; a fishbone diagram was designed to relate and identify the key variables which can cause failures in the all mechanical system.

2.2. Analytical phase

The MICMAC method was used and the key variables selected to identify the riskiest variables in the production system. The results will be present these variables in the conflict zone, characterized by containing the variables with higher driving power and higher dependence; the identification of these variables will allow the creation of an operation strategic and maintenance plans.

2.3. Propositional phase

Strategies are proposed to manage and control oil wells with PCP, using the identified variables with the MICMAC method; these strategies will allow the design to follow-up plans to increase the reliability of the system and reducing the failure rate in the producing wells.

3. Production optimization in an oil field

Identify the characteristics in the good operation, it allows defining the production plan to optimize its useful life and quantity of oil produced. A production systems description using reservoir energy is presented below.

3.1. Natural flow

The first extraction stage uses the reservoir pressure; this method is known as Natural Flow. The energy of the fluids compressed and stored in the reservoir [13] are sufficient for the fluid to reach to the surface, due to the pressure differential between the reservoir and the surface production facilities; this method generates a reduction in the reservoir (Ps) and well (Pwf) pressure.

These pressure changes generate alterations in the Gas/Oil Ratio (GOR) and an increase in the production of waterreservoir [14]; when the energy of the fluids is not sufficient to produce a differential pressure to raise the fluids, equipment called the artificial lifting system [15], is implemented, which helps to release weight from the hydrostatic column.

3.2. Artificial lift systems

Downhole pumps are installed to suck and transport fluid to the surface, reservoir, rock and fluid conditions will define the best bottom hole assembly (BHA) design. The Gas Lift and Pneumatic Pumping are mechanisms that inject gas or water under pressure in the wells, to deliver energy and move the fluid to the surface, these systems are effective if the fluids do not have a high weight [16, 17]. Other mechanisms use a bottom pump, such as progressive cavity pumps (PCP), electric submersible pumps (ESP) and mechanical pumps (MP), driven by variable speed electric motors; these equipment allow better extraction management due to the ease changing operating speeds. Its greatest problems are in the designs and mechanical efforts generated by the movement [18]; during the operation of the pumping system, problems can arise related to the operation of the electrical equipment, designs of the production string, even in the mechanical efforts generated by the movement [17] this generates failures and operating outputs of the wells, negatively affecting the reliability of the system. this generates failures and operating outputs of the wells, negatively affecting the reliability of the system.

3.3. Progressive cavity pump (PCP)
The document presents a mechanical failure review in PCP systems, below is a diagram of the main components of this pumping system

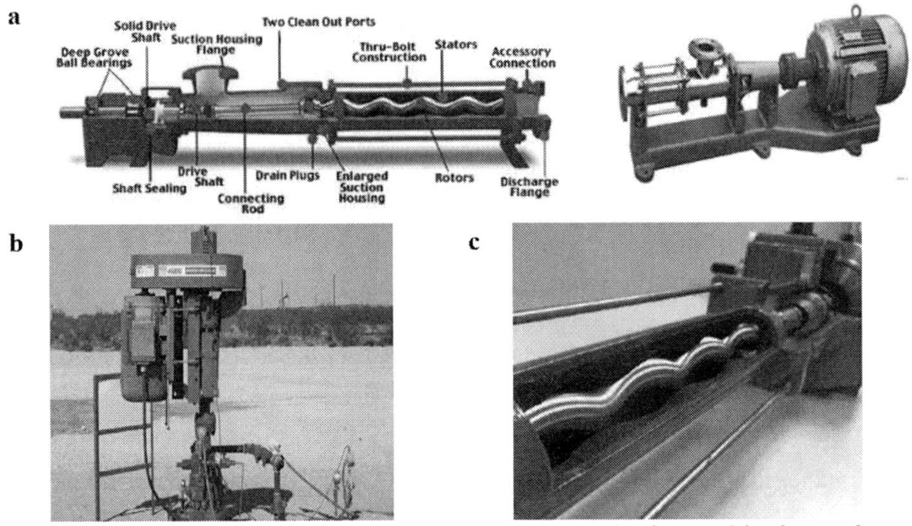

Figure 1.a) PCP system components b) PCP system head c) Progressive cavities internal structure

Figure 1 a, b and c show the design and components of a PCP system, the video in [18] shows the operation of the PCP system. Added to this are the main parts such as the surface equipment in image b and the fundamental bottom equipment in image c, which together constitute the elemental functioning of the progressive cavities

4. Failures in a progressive cavity production system

4.1. Design and Operation Failures
System size and configuration should consider depth, pipe size, fluid volume, and reservoir conditions to define variables such as geometry, stroke length, pump speed, and rod string. Poor design contributes to broke other components, such as tubing cut or rod failures as a result of compressive rod loads [19], presence of corrosive agents in fluids affecting [20] rods resistance producing premature failures [21] When one of the above conditions changes, the design of the artificial lifting system must be re-evaluated [15].

4.2. Fatigue Failure
These failures are of progressive type, identifiable by the fracture profile of multiple parallel marks; they do not present significant plastic deformation and it can start at any point where there is a concentration of stresses [22] The stress on the rod string should be distributed over the length to reduce workloads [23]. The fracture has ductile characteristics and may present edges at 45° [24].

4.3. Flex fatigue failures
Flex fatigue failures cause rods to bend in the failure point. New rods are manufactured with body straightness with no less than 1/16 inch along the overall length of the rod body; a greater degree of

flexion increases local stress. When the rod is bent and then tensioned straight during loading or axial stress, it will cause the material to break. The pumping cycle increased fluid weight and rod fatigue on the concave side of the flex are the causes of this failure type [23].

4.4. Connection Failures

These failures occur in connection rods with rotary load or also called API rods, due to loss of displacement may occur by improper lubrication, excess or lack of torque, wear by blow of pipes or some combination [23]; cause an adjustment loss and the total uncoupling of the connection [5]

5. Causes and effects related to the failures in a progressive cavity pumping system.

In accordance with the failures described, a fishbone diagram was drawn up relating their failures and their causes [25].

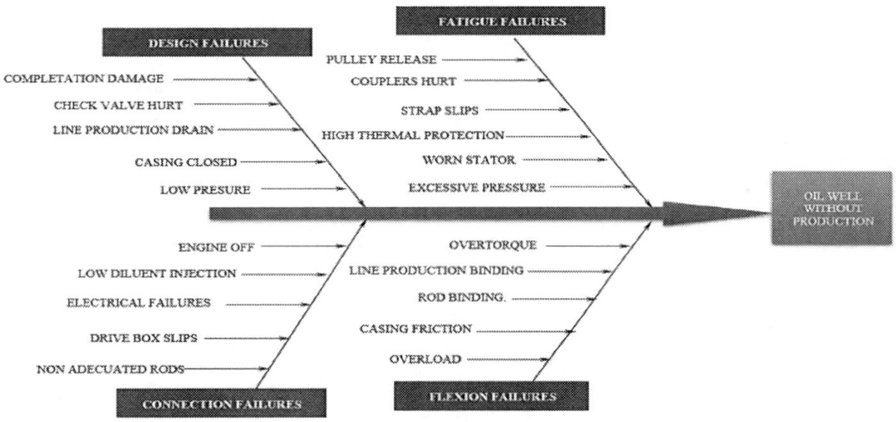

Figura 2.PCP failures in a Fishbone diagram

The figure 2 shows the causes and effects relationship due to failures in the PCP production system, causes allow recognizing the variables which affect the performance and trigger failure. The table selects eleven (11) variables that affect equipment performance.

Table 1. Variables identified from the fishbone diagram.

Code	Variable	Remark
V01	Acid presence	Production of acid compounds, such as H2S, cause changes in the physical properties of good components.
V02	Sand production	Production of reservoir sand, generated by unconsolidated sands or by operational actions; it agglomerate in different parts of the well
V03	Paraffins and asphalt presence	The presence of these compounds can generate scales reservoirs in different parts of the well.
V04	Mechanical design	Lack of consideration of good inclination, fluid petrochemistry, and reservoir petrophysics.
V05	% BSW	A high percentage of water and sediment production.
V06	Active and reactive electrical power	High consumption of electrical energy, generating a deficiency in its production process.
V07	Temperature rise	The temperature increase due to frictional energy losses in the pumping system.
V08	Torque	Increased torque to generate the rotary motion of the PCP system.
V09	Fluid level	Lack of consideration of the static and dynamic well levels.
V10	Gas presence	Increase of the Gas/Oil ratio in the fluid.
V11	PWh	Increased pressure at the wellhead.

IOP Conf. Series: Materials Science and Engineering **844** (2020) 012013 doi:10.1088/1757-899X/844/1/012013

6. Variables Identification by MICMAC

The structural analysis method offers the possibility to describe a system with the help of a matrix linking all its constituent elements [26]. Its objective is to show the main influential and dependent variables and therefore the essential variables to the system evolution.

This method, fundamentally is a tool to structure and organize the ideas, it is composed of three steps, the first is the identification of a variables set which characterize the system; in table 1 the identified variables are observed; in the second step, the matrix of interactions with the identified variables is carried out, a group of participants collaborated to evaluate the influence degree between them, the identified variables were valued from 0 to 5 taking into account the following relationship, (0) null, (1) very low, (2) low, (3) medium, (4) high and (5) very high; the degree of dependence and driving power were calculated. In table 02 and figure 03 you can see the result of the and the dependence vs Driving power diagram. Finally, the identification phase of the key variables, the variables located in the conflict zone are selected due to their high dependence and high driving power; because they have a significant influence on the rest.

Table 2. Cross-impact matrix with the variables in table 1

	V01	**V02**	**V03**	**V04**	**V05**	**V06**	**V07**	**V08**	**V09**	**V10**	**V11**
V01	0	0	5	4	0	0	0	0	3	5	5
V02	2	0	0	5	5	4	0	4	4	0	5
V03	3	4	0	1	0	4	3	4	0	4	4
V04	0	4	0	0	4	5	0	5	5	3	1
V05	0	5	0	4	0	1	4	4	4	0	5
V06	0	4	4	5	0	0	2	4	4	0	0
V07	4	0	5	1	2	4	0	3	2	5	5
V08	0	2	0	4	0	5	0	0	3	3	1
V09	0	3	0	5	5	3	4	0	0	0	5
V10	5	2	5	5	0	3	4	4	3	0	5
V11	5	5	4	5	4	4	1	5	5	5	0

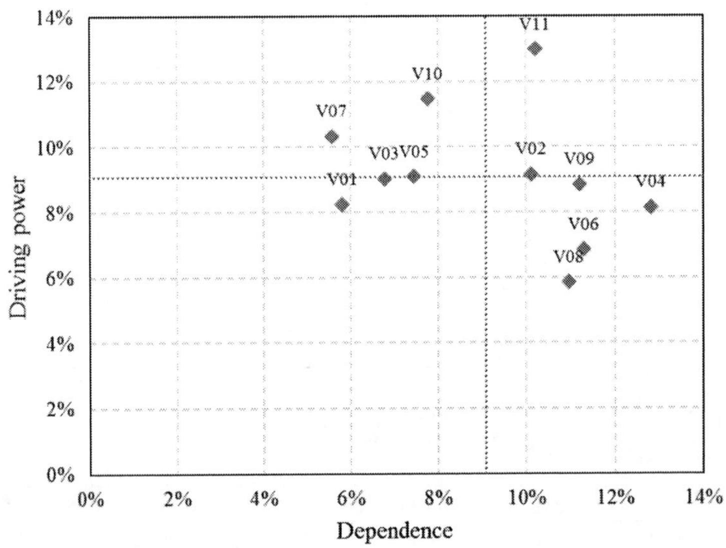

Figure 3. Results obtained from the cross-impact matrix.

Figure 03 shows the variables V11 (PWh Head pressure), V02 (sand production), located in the upper right zone, they are the key variables or challenge variables of the system, with high driving powers and dependence, which disturb the normal system operation. They are by nature unstable, where V02 is in the conflict zone; and is considered the critical variable to improve the performance and decreasing the occurrence of failures. Located at the top right of the driving power/dependency plane, they have a high level of driving power and dependency, which makes them extraordinary importance variables in the strategic axis. Correct operation requires wellhead pressure control, which is affected by failures caused by high sand production causing a chain of situations that affect the productive life of the well [27].

7. Strategies to prevent failures in progressive cavity systems

The results indicate the necessity to study on variables V02, V09, V11, well conditions, petrophysical variables, and reservoir fluid composition, to identify additional components to the design of the pumping system; a good diagnosis will predict operating conditions to minimize and control well variables, avoiding failures.

- ✓ Equipment implementation is necessary to make the proper system functioning. It can use a grating sand control, gravel pack, grooved Liner.
- ✓ Control the pumping speed, if an adequate speed is maintained, can extend the life of the equipment.
- ✓ Control pumping points when fluid levels are present, it is necessary by placing centering devices to reduce friction and knocking between the rod string and production tubing.
- ✓ Control the excessive presence of released gas.
- ✓ Keep the pump at an adequate level until you have better submergence.
- ✓ Install a gas separator under the pump.
- ✓ Make a correct analysis of fluid properties to optimize equipment subsurface parts selection and materials can prevent future corrosion problems.

8. Conclusions

The research showed the components and failures in the PCP system, one of the most used in shallow wells or those with low productivity. A fishbone diagram was made in which the faiñures causes and effects were identified; allowing the identification and selection of eleven variables which affect this equipment operation. The MICMAC method was used to identify critical variables located in the conflict zone, and the variable V11 (Pwh Head pressure) was the variable with the greatest influence.

The results suggest that analyses should be carried out in the control of production and operation variables which intervene in the monitoring and control process of the well-equipment system, allowing the optimization the production method behavior. Cause of it, the following topics are defined as variables related to the production of the well: production rate of fluids, fluid level, water, and sediments cut (%AyS), viscosity, depth, gas cut. which reflect the conditions that have of operation; with this, it is possible to detect problems in the well opportunely, and it allows to take corrective measures for more efficient operation.

The prevention of anomalies in the PCP system will allow safe and reliable operation, improving oil production and profitability indicators.

References

[1] López E, Montes E, Garavito-Acosta AL, Collazos-Gaitán MM 2013 *Capítulo 9. La economía petrolera en Colombia*. In Republica Bdl. Flujos de capitales, choques externos y respuestas de política en países emergentes pp 337-406.

[2] DANE 2019 Exports of Colombia/ coffee, coal, oil and its derivatives, ferronickel and non-traditional. [Online] Available from: https://www.dane.gov.co/index.php/estadisticas-por-tema/comercio-internacional/exportaciones.

[3] Grimaldo Guerrero W, Mendoza Becerra M, Reyes Calle WP 2017 Forecast Electricity Demand Model Using Predicted Values of Sectorial Gross Domestic Product: Case of Colombia. *Espacios* **22**.
[4] BP Statistical Review of World Energy. [Online] 2018. Available from: https://www.bp.com/en/global/corporate/energy-economics/statistical-review-of-world-energy.html.
[5] Robin B 2014 Pumping Oil: 155 Years of Artificial Lift. *Journal of Petroleum Technology*. October **10**.
[6] Flatern Rv 2015 Defining Artificial Lift. *Oilfield review*. Schlumberger..
[7] Carpenter C 2019 Best Practices for Waterflooding Optimization Improve Oil Recovery in Mature Fields. *Journal of Petroleum Technology* **01**.
[8] Alvarez JO 2018 Improving Oil Recovery in the Wolfcamp Reservoir by Soaking/Flowback Production Schedule With Surfactant Additives. *SPE Reservoir Evaluation & Engineering*. **04**.
[9] Gud B, Lyons WC, Ghalambor A 2007. Petroleum roduction Engineering a Computer-Assisted Approach. *Editorial Elsevier Science and Technology Books* .
[10] Dunn et al 1995 Progressing Cavity Pumping System Applications in Heavy Oil Production. *Society of Petroleum Engineers*.
[11] Noonan S. 2008 The Progressing Cavity Pump Operating Envelope. *Society of Petroleum Engineers* **01** pp 398-404.
[12] Gil E, Chamorro A 2009 Técnicas Recomendadas para el Aumento de la Producción en Campos Maduros[Online]Availablefrom:http://oilproduction.net/files/Aumento%20de%20produccion%20en%20campos%20maduros.pdf.
[13] Ezekwe N 2010 PETROLEUM RESERVOIR ENGINEERING PRACTICE *Massachusetts: Pearson Education Inc.*
[14] Nind TEW 1987 Fundamentos de Producción y Mantenimiento de Pozos Petroleros Mexico *Limusa*.
[15] Cam M 2015 Technology Focus: Heavy Oil. *Journal of Petroleum Technology* **03**.
[16] Babadagli T 2017 Technology Focus: Heavy Oil. *Journal of Petroleum Technology* **03.**
[17] Guo B, Liu X, Tan X 2017 Petroleum Production Engineering In Sucker Rod Pumping. (Second Edition) ed: *Gulf Professional Publishing* pp 515–548.
[18] National Oilwell Varco 2017 Youtube [Online] Available from: https://www.youtube.com/watch?v=UDr5_Bd4bnA.
[19] Chacín N 2008 Bombeo de Cavidad Progresiva Operaciones y Diagnóstico. *Manual para operadores Esp Oil*.
[20] Osorio Tovar, Grimaldo Guerrero W, Pacheco Torres J, Chaparro Badillo P 2018 Chemical Failure Analysis of Artificial Lift System in Petroleum Industry: A Review. *Journal of Engineering and Applied Sciences* **19** pp 8010-8015.
[21] Zou D 2008 Designe of PCP *SPE Journal*.
[22] D.L.Duan et al 2014 Failure mechanism of sucker rod coupling. *Engineering Failure Analysis* **36** pp 166-172.
[23] Norris 2008 Sucker Rod Failure Analysis.
[24] Castillo M 2001 Manual de equipos de superficie *PDVSA Exploración y Produccion*.
[25] Zapata J CM, Villegas S SM, Arango I F 2006 Reglas de consistencia entre modelos de requisitos de un método *Universidad EAFIT* **141** pp 40-59.
[26] Koontz H 2008 Administración: una perspectiva global y empresarial México, D.F *Mc Graw Hill Interamericana XXXVI* .
[27] Chamorro EG 2009 Técnicas Recomendadas para el Aumento de la Producción en Campos Maduros *oil production*.

IOP Conf. Series: Materials Science and Engineering **844** (2020) 012014 doi:10.1088/1757-899X/844/1/012014

Prospective towards implementation of electric vehicles in Colombia

E. Gómez-Mesino[1a], A.F. García-Quintero[1], J.W. Grimaldo-Guerrero[2],

[1] Electrical engineering student, Universidad de la Costa, Colombia.
[2] Department of Energy, Universidad de la Costa, Colombia, ORCID: 0000-0002-1632-5374.
[a] Corresponding author: egomez43@cuc.edu.co

Abstract. The effects of climate change have led to reduces fossil fuel in vehicle, developing new technologies such as the electric vehicles. The research presents an identification of barriers and inhibitors according to international experiences, an evaluation with the MICMAC method and the identification of key factors to achieve development of electric vehicles in Colombia. The results allow to identify the elements of high influence, which may guide intervention strategies and be the basis for the formulation of guidelines and policies. The results conclude that the business models for the purchase and sale of energy are the catalyst for the encouraging of the stakeholders, the improvement of the distribution network and the management of the electricity market.

1. Introduction

The increase of greenhouse gases and the volatility of the price of oil have allowed to develop alternatives for the reduction of fossil fuel consumption. Alternatives have been developed such as technological improvements in the injection system [1], efficiency in the combustion engine [2], use of biofuels [3] and migration to electric vehicles [4]. The latter achieves a reduction of CO2 produced by the combustion process and fuel demand for the urban transport sector, key factors for climate change and energy security.

The implementation of electric vehicle technology entails new challenges such as the management of existing infrastructure to allow changes and avoid bottlenecks; the study of the current state helps governments in strategic planning for future scenarios. With the lessons learned, it can identify characteristics that are necessary for the inclusion of new technologies; Colombia is a country that is doing its first investigations and pilot tests on this topic.

According to National Single Transit Registry (RUNT) [5], Colombia has a vehicle fleet of 14,671,694 vehicles, where 5,425 are electric vehicles; of which 872 are automobiles and 1,385 are vehicles such as vans and buses. The departments with the highest number of records are Antioquia (1,329), Bogotá D.C. (1,146), Cundinamarca (1,067), Valle del Cauca (743) and Bolívar (236) [5], These figures show the need and the existence of difficulties to encourage the integration of these technologies.

The research presents an identification of key elements that allow the development and integration of the technologies associated with electric vehicles in Colombia and the use of Cross Impacts Matrix-Multiplication Applied to a Classification (MICMAC) method, based on qualitative study between the variables and interactions identified between the Electric Vehicles and the Colombian Power System, these results will allow to be the base for the formulation of guidelines for the different agents of the market.

The work begins with section 1 where the motivation to carry out a qualitative study on the factors that affect the market of electric vehicles for their development in Colombia is presented, section 2 sets out the stages and the procedure of this analytical research. The legal framework for electric vehicles is presented in section 3. In section 4 the results of the methodological phases are presented, it begins with the identification of the variables and interactions identified between the Electric Vehicles and the Colombian Power System, then the MICMAC method is used to identify the key factors, and ends with the need to integrate users due the lack and ignorance in business models of electricity transactions. Section 5 denotes the limitations of the study and future research directions associated with the electric vehicle market.

2. Methodology

The research design involves the description of the Colombian legal framework and the identification of the interactions between the electric vehicle and the electricity market; through the lessons learned in different sources of information, it identified variables that allow defining strategic guidelines for the strengthening of the electric vehicle market. Procedural systematization is organized in three phases of research, namely:

Table 1. Systematization of the methodology.

Step	Remark	Phase
1	Review the regulatory framework.	*2.1. Identification phase* It involves the documentary review of lessons learned, which will allow identifying the factors that will affect the electric vehicle market. International experiences from one country to another show that ignorance of these processes can generate a chain of doubts or conceptual and procedural gaps regarding the strategic vision of greater relevance in correspondence with the intended objectives [6].
2	Identification of the interactions between the Electric Vehicles and the Colombian Power System.	
3	Identification of variables.	
4	Make the cross-impact matrix using the variables.	*2.2. Analytical phase* The identified factors and the Cross-Impact Matrix Multiplication Applied to Classification (MICMAC) method are used to identify the characteristics of the variables and define the strategic axes; these elements are catalysts for organizational development and allow the identification of facilitators, inhibitors and barriers. The analysis identifies strengths and weaknesses of the sector, regarding processes such as planning and policy formulation, which will be inputs for the definition of short, medium and long term plans.
5	Make the driving power and dependence diagram.	
6	Identification of key factors.	
7	Analyze the key factors.	*2.3. Propositional phase* The variables located in the conflict zone of the structural analysis are identified, these variables are the strategic elements and their intervention will allow the adoption of priority strategies for the development of the electric vehicle market in Colombia.
8	Identify and propose improvements.	

IOP Conf. Series: Materials Science and Engineering **844** (2020) 012014 doi:10.1088/1757-899X/844/1/012014

3. Colombian legal framework

The ministries of Mines and Energy [7] and Transport [8] oversee the electric vehicle market; the regulations have been made around the tariff measures for the importation of these technologies, in table 2 the regulations are presented.

Table 2. Regulatory framework for electric vehicles in Colombia.

Legal regulations	Remarks
Law 023 (2010)	Tariff reduction measures are established for the importation of transport systems that use electric power for their traction. The governors and mayors may contemplate the elimination of traffic restrictions.
Resolution 186 (2012)	Tax incentives are established for the technological reconversion of the vehicle fleet and the use of hybrid and electric transportation systems.
Decree 1116 (2017)	Electric vehicles will not pay duty and a sales tax of 5%. The hybrid vehicles will pay a tariff rate of 5% and a consumption tax of 8% or 16% in case of exceeding a Free On Board (FOB) price of US $ 30,000.

The measures adopted encourage the importation of electric vehicles, but the infrastructure, the electric and automotive market do not present the necessary conditions for the growth of this initiative. The prospective are to establish the exemption from vehicular taxes and traffic lights, subsidies for the purchase of electric vehicles and the replacement of conventional automobiles, and the installation of recharging systems in public and private places [9].

4. Results

4.1. Identification of factors

The understanding of the interactions between the object of study and the development environment allow identifying elements that can facilitate or inhibit their development in the market; Figure 1 shows the interactions that exist between an electric vehicle and the elements of the Colombian electric power system.

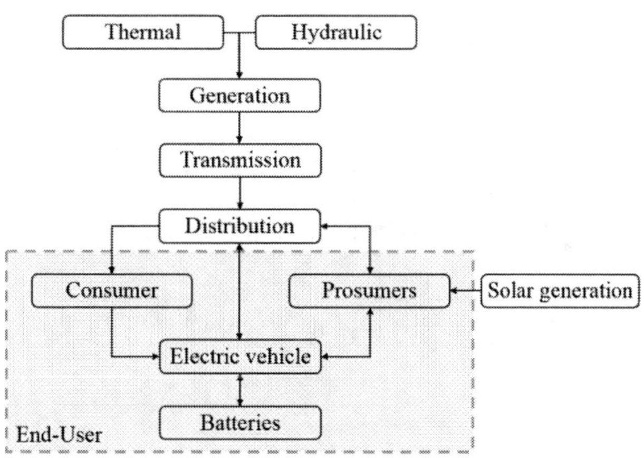

Figure 1. Interactions between the Electric Vehicle and the Colombian Power System.

The generation of electricity in Colombia comes mainly from hydroelectric power plants located in isolated and thermal power plants zones have a high cost due to the acquisition of fuels [10]; the energy is carried by the transmission and distribution system reaching the end-users. The inclusion of electric vehicles will lead to the participation of end-users, they may be consumers or prosumers. If the users own an electric vehicle, they will charge their batteries with energy from the grid or from their own production, mainly photovoltaic (PV) generation. The vehicle storage system is the key element for the interaction of users and the distribution system, allowing processes to purchase and sale of energy to the network. This indicates the need for Smart Grid, to give flexibility and to encourage users to a new energy storage business model. The variables presented in table 3 are proposed due the interactions identified.

Table 3. Variables identified for the implementation of electric vehicles in Colombia.

Code	Variables	Remark
V01	Displacement of electric power blocks.	Storage systems allow the displacement of generated energy to be used in times of high demand and low capacity generation [11, 12].
V02	Uncontrolled increase of the PV generation.	The increase in facilities and PV generation can generate an effect called "duck curve", which can affect electrical power systems [12, 13].
V03	Price of electric energy.	The model used to establish the price of electric power does not favor the reduction of prices, despite having a high contribution of hydraulic generation.
V04	Lack of infrastructure for intelligent measurement.	Currently there are no bidirectional measurement systems installed in most of the Colombian territory; there is the government's willingness to carry out the migration [12].
V05	Limited technological offer in the electric vehicle sector.	There is not a wide range of electric vehicles and components that incentivize users and generate free competition in the automotive market.
V06	Solar energy production by residential users.	There is not a high participation of residential users to install PV generation systems [12, 14].
V07	Lack of a system for the final disposal of electronic equipment.	Control for the final disposal of electronic equipment that has reached its useful life and reduce the environmental impact.
V08	Lack of charging stations.	The limited existence of charging stations for electric vehicles. 28 stations throughout Colombia and none in the Caribbean region [14, 15].
V09	Lack of energy business models.	Lack of knowledge and existence of business models for the purchase and sale of electric power for the end-user [16].
V10	Control in quality standards for electric power service.	The integration of PV generation technologies and storage systems can cause anomalies and failures in distribution systems [12, 13].
V11	Regulation for mobility and control of electric vehicles.	The existence of a special regulation can encourage use [14].
V12	Oil as competition	The price of oil generates uncertainty in the energy security of fuels, royalties and government plans.

Sources: [17, 18, 19, 20]

4.2. Structural analysis: identification of key issues

The method of structural analysis is composed of three steps, the first is the identification of a set of variables that characterize the system, in Table 3 the variables are observed; the second is a description of the relationship between the variables, and the interaction matrix is made with the identified variables. A group of participants collaborated to evaluate the degree of influence between them, 0 for a null influence up to 5 for a very high influence; the degree of dependence and driving power was calculated, in table 4 and figure 2 you can observe the result of the interaction matrix and the diagram between the dependence vs the driving power. Finally, the identification phase of the key variables, the variables located in the conflict zone are selected due to their high dependence and high driving power; because they have a significant influence on the others.

Table 4. Cross-impact matrix using the variables of Table 3

	V01	V02	V03	V04	V05	V06	V07	V08	V09	V10	V11	V12
V01	0	5	5	1	0	1	0	3	3	5	0	4
V02	4	0	5	3	1	2	4	1	4	5	0	0
V03	3	4	0	2	0	2	3	5	5	5	0	0
V04	5	5	3	0	2	4	2	5	5	5	0	0
V05	1	1	1	1	0	2	2	5	4	2	5	5
V06	4	4	4	2	2	0	4	0	5	5	0	3
V07	1	0	0	4	4	3	0	2	3	2	4	5
V08	5	1	5	3	5	3	1	0	4	4	5	4
V09	3	4	5	5	5	5	5	5	0	3	3	4
V10	3	4	5	5	3	3	0	3	3	0	4	0
V11	0	0	5	2	4	2	4	4	5	4	0	5
V12	2	4	5	2	3	3	0	5	2	0	3	0

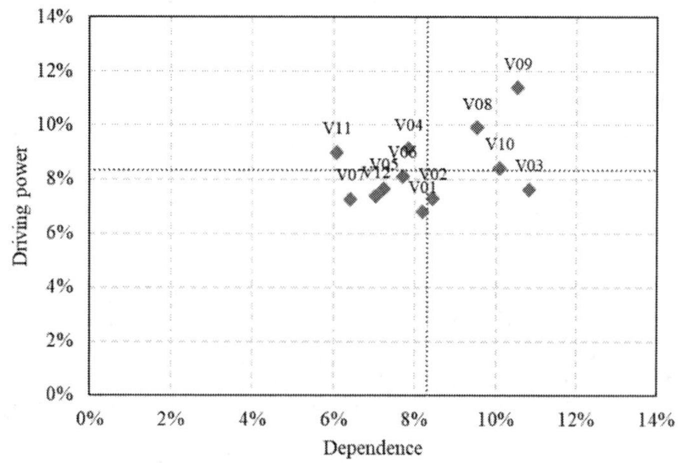

Figure 2. Driving power and dependence diagram.

Figure 2 shows the variables V08 (Lack of charging stations), V09 (Lack of energy business models) and V10 (Control in quality standards for electric power service) in the conflict zone; V09 is the furthest variable and refers to the mechanisms of participation in the electricity market, V08 and V10 are variables to make improvements in the infrastructure.

4.3. Energy models as a key factor for the implementation of electric vehicles

The lack and ignorance in business models of electricity transactions, presents the need to integrate users; this integration is weak due to the lack of knowledge and little incentive that exists now. Figure 1 shows end-users as a strategic element, their decision to own an electric vehicle can be incentivized to consider this purchase as an investment and not as an expense, you can have profits because energy transactions will favour the user in the maintenance and return of investment. By existing these models will increase the demand for electric vehicles and the growth of the necessary infrastructure; it will favour the installation of charging stations, the reliability of the electrical system and the technological update of the electric grids towards Smart grids.

Currently in Colombia, bidirectional meters and PV generation systems are installed, government entities have chosen net-metering as the mechanism for integrating prosumers; but the lack of incentives and clarity of the model has not allowed growth despite the high potential. There is a need to have electricity hourly rates, allowing the purchase of energy at times of low demand and high supply, and then be sold at times of high demand.

Energy models need policies to modify the regulations and methodologies used to define the price of electricity, define the requirements for the injection of energy into the grid, modify the current energy sales system to an intra-hour one and implement training to show the benefits of owning an electric vehicle

5. Conclusions

The work presented the identification of the interactions between the electrical system and electric vehicles, and the selection of twelve (12) variables; the MICMAC method was used to identify the variables located in the conflict zones and variable V09 was obtained as the most influential variable in the system.

The results suggest the restructuring of the electricity market to a model with hourly prices that allows to encourage the installation of generation and storage systems to move blocks of energy to areas of high demand and low supply. The acquisition of electric vehicles will be an attractive option to participate in the market; achieving growth in demand and supply. The end-user is an ally for the growth of the electrical infrastructure and in the support of generation and storage of electricity; the acquisition of electric vehicles will reduce the environmental impact due to CO2 emissions and reduce the demand for oil, strengthening energy security.

The best incentive for an end-user is their participation in the purchase and sale of electricity, motivating the purchase of equipment for generation and storage; the electric vehicle is an element that can improve the functioning of networks and ensure energy security due to fluctuations in the price of oil. For future research, technical studies and business models to improve the integration of electric vehicles and their integration into the network in Colombia.

6. References

[1] Arat HT, Baltacioğlu MK, Özcanli M, Aydın K 2015 Optimizing the Quantity of Diesel Fuel Injection by Using 25HHOCNG Gas Fuel Mixture *Advanced Engineering Forum* **14** pp 36-45.

[2] Sevik J, Pamminger M, Wallner T, Scarcelli R, Reese R, Iqbal A, Boyer B, Wooldridge S, Hall C, Miers S 2016 Performance, efficiency and emissions assessment of natural gas direct injection compared to gasoline and natural gas port-fuel injection in an automotive engine *SAE International Journal of Engines* **9**(2) pp 1130-1142.

[3] Velásquez-Piñas JA, Pacheco-Torres PJ, Calle OD, Mora-Higuera LM, Grimaldo-Guerrero JW, De-la-Ossa-Ruiz MP 2018 Production and Characterization of Biodiesel from Cotton Oil as an Alternative Energy in Substitution of Soybean Oil *Journal of Engineering Science and Technology Review* **11**(6) pp 182-186.

[4] Poullikkas A 2015 Sustainable options for electric vehicle technologies *Renewable and Sustainable Energy Reviews* **41** pp 1277-1287.

[5] RUNT Boletín de Prensa 003 de 2019 - Mercado de eléctricos en aumento [Online] Available from: https://www.runt.com.co/sites/default/files/Bolet%C3%ADn%20de%20Prensa%2003%20de%202019.pdf.

[6] Estrada Gasca CA 2013 Transición energética, energías renovables y energía solar de potencia *Revista Mexicana de Física* **59**(2) pp 75–84.

[7] MME Misión y Visión [Online] Available from: https://www.minminas.gov.co/mision-y-vision

[8] MT [Online] Available from: https://www.mintransporte.gov.co/publicaciones/33/quienes_somos/

[9] Alvarez CA, López LB 2017 Análisis normativo de vehículos eléctricos y sistemas de recarga para la adopción de soluciones de movilidad eléctrica en Colombia *Revista CIDET*

[10] XM Effective capacity by type of generation [Online] Available from: http://paratec.xm.com.co/paratec/SitePages/generacion.aspx?q=capacidad

[11] Agbonaye O, Vorushylo I, Keatley P, Huang Y, Bani Mustafa M 2019 Business models for the displacement of oil by heat pumps and energy storage in social housing *UKERC Annual Conference 2019: Local Energy Systems in National and International Contexts*

[12] Obi M, Bass R 2016 Trends and challenges of grid-connected photovoltaic systems–A review *Renewable and Sustainable Energy Reviews* **58** pp 1082–1094.
[13] Chandra Mouli GR, Bauer P, Zeman M 2016 System design for a solar powered electric vehicle charging station for workplaces *Applied Energy* **168** pp 434-443.
[14] Mersky AC, Sprei F, Samaras C, Qian Z 2016 Effectiveness of incentives on electric vehicle adoption in Norway *Transportation Research Part D: Transport and Environment* **46** pp 56–68.
[15] Electromaps [Online] Available from: https://www.electromaps.com/puntos-de-recarga/colombia
[16] Li Y, Zhan C, de Jong M, Lukszo Z 2016 Business innovation and government regulation for the promotion of electric vehicle use: lessons from Shenzhen, China *Journal of Cleaner Production* **134** pp 371–383.
[17] Brown MA 2001 Market failures and barriers as a basis for clean energy policies *Energy Policy* **29**(14) pp 1197-1207.
[18] Brown S, Pyke D, Steenhof P 2010 Electric vehicles: The role and importance of standards in an emerging market *Energy Policy* **38**(7) pp 3797-3806.
[19] Steinhilber S, Wells P, Thankappan S 2013 Socio-technical inertia: Understanding the barriers to electric vehicles *Energy Policy* **60** pp 531-539.
[20] Silva J, Ospino A, Balbis M 2015 Set of Elements, parameters and considerations to get the successful inclusion of the smart grids in colombian power systems

IOP Conf. Series: Materials Science and Engineering **844** (2020) 012015 doi:10.1088/1757-899X/844/1/012015

Offshore oil exploitation in the Caribbean Sea: Challenges for Colombia

J.W. Grimaldo-Guerrero [1a], Y.F. Contreras-Rueda[2],

[1] Department of Energy, Universidad de la Costa, Colombia, ORCID: 0000-0002-1632-5374.
[2] Universidad Nacional Sede Caribe, Colombia, ORCID: 0000-0003-0164-2032.
[a] Corresponding author: jgrimald1@cuc.edu.co

Abstract. Colombia began an expansion plan for oil exploration in the Caribbean Sea, this industry is recognized for the high impact it can generate, being necessary to identify mechanisms to improve management. This research presents a review of the experiences in the offshore oil industry, describes the case of oil exploration in the Arctic, a fragile ecosystem; methodologies and indicators used to strengthen the industry in order to achieve sustainability are identified. The results allow to conclude that it is necessary to strengthen the development plan from the institutional and political frameworks using a Strategic Environmental Assessment, also they indicate how this can improve the instruments and achieve improvement in the social, economic and environmental dimensions.

1. Introduction

The General Comptroller (2017) [1] presents a forecast of loss of oil self-sufficiency for the year 2021, to increase hydrocarbon reserves, the National Hydrocarbons Agency (ANH) has granted new exploration and exploitation blocks onshore and offshore [2, 3]. According to the Office of the General Comptroller (2017), it will not be enough to avoid the loss of national self-sufficiency if the prices of hydrocarbons are low in the market, or if policies are not formulated and adopted to encourage exploration and exploitation activities.

The drilling of an offshore well can cost on average $ 120MM, if it is a deposit in the Miocene and $ 230MM in the Jurassic one, while in an inner well it can range between $ 4.9 and 8.3MM, the production it is high and maintained for a period of 10 to 20 years [4, 5]. Offshore activities are highly sensitive and have a high impact on nature, due to the potential for spills and pollution that reaches a large scale, affecting marine ecosystems [6]; The regulations of the activities are much more rigorous and of permanent surveillance [7]. Oil activities on the high seas that are usually carried out are:

a. *Exploration* are carried out to locate areas with high potential, can use magnetic, gravimetric and seismic 2D and 3D, evaluates the seabed, winds and marine currents [8, 9].
b. *Drilling* involves the invasive activities to the maritime soil for the creation of a well that allows the extraction of the fluids from the deposit. For its proper operation, equipment is required that perforates the subsoil and controls the pressure of the fluids of the rock, fluids that are injected to lubricate the bit and remove the debris from the perforation [9, 10, 11].
c. *Production* is the extraction of the fluids contained in the previously drilled deposits, commonly the first years of production are oil or gas phase depending on the type of deposit, after a certain amount of time the production changes to the gas, oil and water phases; this last fluid will gradually increase its percentage of production, it will be necessary to install an

equipment for the treatment and its correct disposal [9, 12]. Other unwanted products that can be found within the production are the molecules of CO2 and H2S, corrosion generators in the equipment installed for the production management [9, 13, 14].

d. *Transport* is the activity where the fluids produced are transported to the coast, in order to give commercial value to the production. Large vessels called Tanker or by means of pipes are used depending on the distance to the coast and financial viability [9].
e. *Abandonment*, at the end of the productive life of the well or of some producing sands, the temporary or definitive closure is made; this operation installs expandable or inflatable seals (they are removable) or by means of pumping cement, plugs are created to seal the well and prevent the escape of reservoir fluids [15].

Offshore operations involve difficulties like land operations, added to others due to the wild environment and psychosocial risks, workers are isolated for long periods of time surrounded by water [16]. Among the risks of greatest occurrence are explosions, mechanical problems in the wells, equipment failures; which can have an impact on the ecosystem and implications for society and economy [9, 17].

To achieve viability and sustainability, Colombia must generate market conditions and implement policies that are attractive to foreign companies, giving greater confidence to investors due to the high investment values that are required, without neglecting the sustainability of the areas where develop the projects; so it is necessary to carefully evaluate the transformations that these projects will produce in the economy, society and the environment. This research presents a review of international experiences, methodologies and indicators used to carry out evaluations of oil activities, which can be the basis for formulating strategies, strengthening existing mechanisms and improving the growth of the offshore oil industry in the Caribbean Sea Colombian.

2. International experiences in the oil industry

The prevention of the different risks for the incursion in the projects is the motivation of the countries to adapt the policies and the guidelines of reference, to understand the actions carried out by other countries will allow to identify critical points for the later analysis and intervention. A review was made to identify aspects for the strategic planning in the oil industry.

Krupnick et al. (2014) [18] indicates that China's environmental standards are weak, ineffective or in non-existent cases, analysed the framework of the United States to develop shale gas exploitation projects to generate guidelines to strengthen China's framework. Aung (2017) [19] conducted a study in Myanmar; this is a region with petroleum resources, but backward on issues of environmental management; their policies and laws lack an administrative and legal structure. Its results propose the Environmental Impact Assessment (EIA) as an instrument to strengthen the prevention of risks and the preservation of the environment due to the activities of the oil sector. Shvarts et al. (2016) [20] proposes an instrument that can qualify as environmental responsibility of oil companies, with the purpose of generating motivation in Russia's oil and gas industry by disclosing information on its performance and economic and environmental impacts.

Ellis et al. (2017) [21] conducted a comparison of the instruments Environmental Risk Assessment (ERA), Environmental Impact Assessment (EIA) and Environmental Management Planning (EMP) in New Zealand in the offshore mining and oil industries, in the which results propose to carry out a Strategic Environment Assessment (SEA) to complement the instruments. Tait et al. (2014) [22] present the results obtained by the ExxonMobil company in the countries of Angola and Chad when using the Environmental, Socioeconomic and Health Impact Assessment (ESHIA), a variation of the EIA, its results exalt the success in identifying and assess the needs of the areas of influence.

The identification of risks and the integration of the social, economic and environmental dimensions are the basis of strategic planning, Rosenfeld et al. (1998) [23] conduct a study on the problems and challenges of an environmental and social nature, and what are the strategies to minimize the impacts that are generated in oil projects. Arora and Lodhia (2017) [24] use the case of the BP oil spill in the Gulf of Mexico and analyse how social and environmental relations can manage the reputation, their results identify five key elements: Financial Performance, Quality Management , the performance of social and environmental responsibility, the quality of the employee and the quality of the goods / services.

Raufflet et al. (2014) [25] identifies four different areas: ethics and governance, environmental management, relations with the community and social health and safety problems in Corporate Social Responsibility (CSR); the work of Hilson (2012) [26] compares CSR from different countries and identifies a great difference between environmental regulations, where developed countries have more comprehensive regulations. Linnen et al (2015) [27] presents the experience of Apache Corporation, an onshore oil company in Egypt, and the key social aspects treated with the Bedouin tribe on issues related to political risks, benefits in education, consultations and socializations.

The identification of environmental and social constraints are other lines of research, Touzel (2012) [28] describes China's plans for unconventional gas extraction, the prospect areas have environmental and social restrictions, such as the availability of water for the hydraulic fracturing and its future disposition, access routes to the zones and the relationship with the inhabitants, indicates that the social and environmental management will allow to achieve a success in the development thanks to the acceptance of the project. Sun and Wang (2015) [29] establishes a model for evaluating shale gas projects and their impact on the environment, comparing the impacts before and after the project taking as a variable the impact on surface water in the water underground, on land and in the air. Le (2018) [30] analysed the development of gas extraction activity in shale rocks, the environmental implications mainly the need for a large volume of fresh water (around 12 to 20 million litres of water per horizontal well). Hasheminasab et al (2018) [31] present a review of economic, environmental and social indicators to generate metrics and evaluate refinery projects in terms of sustainability.

The literature presents the SEA as an instrument that can improve the instruments and improve social, economic and environmental aspects; highlight the importance of CSR as a key instrument for companies to achieve communication channels with society and the need to identify indicators to assess the impacts generated in different dimensions by the activities carried out.

2.1. Oil exploitation in the Arctic

In the investigations carried out in [32, 33, 34, 35, 36] they present experiences in the Arctic area, an area belonging to eight (8) countries and considered a fragile ecosystem. Hasle et al (2009) presents a study on the management of environmental risk through the Matrix Risk Assessment (MRA) instrument for requests for block allocation and well drilling under the criteria of possible oil spills and contingency plans. Harsem et al (2011) reviews the capacities and activities carried out to improve oil policies and infrastructure in countries with arctic terrain and the challenges they have to face, such as market conditions, strong winters and little infrastructure The work of Noble et al (2013) presents the results of an SEA for the offshore hydrocarbon extraction activities in the Arctic in areas belonging to Canada, they conclude the process must be continuous and adaptive to changes in policy conditions, environmental, social or economic; whereas Dale (2016) shows the work of Norway for the promotion of the extractive activities of oil in the seas of Barent and Lofoten, historically dedicated to fishing, indicates that the implementation of instruments for the management of political, environmental and social networks allows for a change and approval in the change in the use of these areas, but these changes must be accompanied by techno-scientific knowledge to change political decisions.

Gulas et al (2017) presents the market conditions that have affected the industry such as the low prices of hydrocarbons and the political restrictions of the West; compares frameworks to prevent pollution in the Arctic of Canada, USA, Denmark, Norway and Russia, and only Norway uses the SEA as a risk management instrument, where there is a participation of the different stakeholders and an improvement of the response teams. From the USA highlights that the responsible for responding to a spill is the coast guard, in 2015 the regulation for drilling in Alaska was published, but the impacts and mitigation plans are not clear; Canada has a more complex system because it has national and state regulation, where specific EIA are requested for each project. Denmark is conducting a SEA by the government of Greenland and where EIA are additional requirements. Finally, it recommends that countries should be better prepared to control spills, improve well intervention and legal frameworks based on the SEA.

The investigations by Noble et al (2013) and Gulas et al (2017) highlight the SEA as an instrument that identifies the strong need of the oil sector to generate greater care for the environment and its effectiveness in establishing guidelines for the industry to generate greater security to the ecosystems where these activities are carried out.

3. Promoting sustainable development in the extractive industry

Talk about sustainable development in the extractive industry, such as the mining and oil sector, has a complex connotation to be defined due to the environmental and social impacts caused, due to changes in the traditional activities of society, the modification of the ecosystems and damages generated by process failures; seeking regulation and balance among the three dimensions proposed by sustainable development [37] creates a dilemma for the lack of compliance with the project. Wang et al (2014) [38] defines the concept of green image, where companies act in such a way that their corporate image delivers a concept of sustainable development through social and environmental responsibility. From a low index in its social and environmental dimensions, the companies carry out actions to increase it, among them there are actions such as the support of communities, improvement of access roads, creation of schools, among others [23, 24, 25, 26].

Table 1. Methodologies for the evaluation of environmental impacts.

Methodology	Remark
Strategic Environment Assessment (SEA)	It is the process by which environmental considerations are integrated completely into the preparation of plans and programs before their final adoption, its main objective is to provide a high level of environmental protection and promote sustainable development [39].
Environmental Impact Assessment (EIA)	It is a methodology for environmental management by forecasting the consequences that an activity can have on the environment, from the initial phase to the final phase. Its approach is practical and is guided in accordance with current legislation, allowing the adoption of preventive, corrective or compensatory measures [40].
Environmental Impact Statement (EIS)	It describes the effects of the proposed activities, considers and describes the positive and negative impacts, and the alternatives to mitigate the impacts. Laws and regulations require the federal government to evaluate the effects on the environment and consider alternative actions [41].
Risk Assessment	It used to characterize the nature and magnitude of health risks for humans and ecological risks of factors present in the environment; Data are collected to characterize the nature and extent of contamination, and thereby assess the frequency and magnitude of exposures that may occur as a result of contact. Its development is an iterative process [42].
Ecological Risk Assessment	It allows to evaluate the probability that the environment is affected by the agent exposure. Its evaluation starts from the collection of a characterization of the agent, then a determination of the effects and finally an estimation of the exposure risk, the results provide information to interpret the risks and identify a level of harmful effects [43].
Environmental Audit	It is a systematic, documented, periodic and objective process in the evaluation of the activities and services of an organization in relation to compliance with legal and internal requirements, promotion of good environmental management, maintaining credibility, sensitizing staff to compliance of environmental policy and create opportunities for improvement [44].
Regional Risk Screening	It generates standardized levels of detection for pollution levels. Detection levels are presented in tables with parameters and exposure conditions [45].

Identifying the mechanisms to guide the sustainable development of the projects is vital, to determine the baseline and evaluate the impact over time. The following subsections present methodologies used for the evaluation of impacts and a categorization of the indicators according to the dimension of sustainable development.

3.1. Methodologies for the evaluation of environmental impacts

Table 1 presents methodologies used for the evaluation of environmental impacts, ordered according to their management capacity.

These methodologies allow to evaluate and forecast the situation in the social, environmental and economic dimensions generated by a project, regardless of the business sector. The SEA was designed by Directive 2001/42 / EC of the European Parliament or SEA Directive as a legal evaluation procedure, to introduce the systematic evaluation on the environmental effects of the plans and programs related to the use of the territory. The SEA is not applied in all countries, instead the EIA applies whose purpose is totally different. The SEA must be carried out before carrying out a plan and will allow a global vision, while the EIA is focused on the prevention of impacts in the area of influence [46].

The SEA is used to estimate the medium and long-term consequences of the execution of investments of different types [36]; and it is internationally recognized for its potential to fill gaps by incorporating considerations of environment and sustainable development into strategic decisions, such as policies, plans and programs in different sectors such as mining, agroindustry, environmental health and fuels [47, 48].

3.2. Categories of indicators used in project management

Table 2 presents the category of indicators according to the dimensions of sustainable development.

Table 2. Categories of indicators according to dimension.

Environmental Dimension	Socio-environmental	Social Dimension	Socio-economic	Economic Dimension	Economic-environmental
Environmental compliance	Health and security	Impact to the communities	Employment and unemployment	Economic benefit	Resource efficiency
Biodiversity	Management of biodiversity	Equality in opportunities	Local economy/ entrepreneurship	Exploration and production	Energy efficiency
Strategies for the climate	Climate change	Human capital development	Development and training	Risk management	Demand for energy
State of the environment	Risk management	Retention of human talent	Sponsorship	Business governance	Sustainability
Emissions to the environment		Working practices		Attractive to the investor	

Source: [49, 50, 51]

The indicators allow the SEA to have a greater vision and provide better standards for the protection of the environment and promote sustainable development; these are based on some category of table 2. The integral selection of the indicators will allow a better follow-up of the projects and achieve greater control in the aspects to be improved.

4. Conclusions

A review of international experiences in the oil industry was presented for the evaluation of environmental impacts and the case of oil exploitation in the Arctic; the results indicate that comparisons must be made of the frameworks of other countries to strengthen management mechanisms, there must be collaborative work between the different state entities and the countries that make up the Caribbean Sea.

Performing an SEA is an excellent option to establish a route for sustainable development in offshore oil exploration and exploration activity in the Colombian Caribbean region, because it will allow institutions to be evaluated and processes to be improved, dialogue and participation structures to be established In the stakeholders for decision making, identify relevant policies and propose reforms or policies that contribute to decision making. The SEA will allow to have information that helps to make key decisions, and that the different actors can participate formally, opportunely and sufficiently for social, economic and environmental development in a sustainable manner.

This work is part of an ongoing doctoral research. The research development is expected to provide strategic energy planning and guidelines for the sustainable growth of the offshore hydrocarbons sector.

5. References

[1] Contraloría General Autosuficiencia petrolera en Colombia [Online] Available from: https://www.contraloria.gov.co/documents/463406/484739/Bolet%C3%ADn+Macrosectorial+No.+06+%28pdf%29/f01dfce0-493c-423a-9148-244fce46edc1?version=1.2.

[2] ANH Contrato E&P-TEA [Online] Available from: http://www.anh.gov.co/Asignacion-de-areas/Relacion-de-areas-asignadas/Paginas/Contrato-EandP-TEA.aspx.

[3] ANH Rendición de cuentas 2017 [Online] Available from: http://www.anh.gov.co/Atencion-al-ciudadano/Documents/Rendici%C3%B3n%20de%20Cuentas%202017%20-%20ANH.PDF.

[4] Oilscams The difference between Offshore and Onshore Oil Drilling [Online] Available from: http://www.oilscams.org/offshore-vs-onshore-oil-drilling.

[5] EIA. Trends in U.S. Oil and Natural Gas Upstream Costs [Online] Available from: https://www.eia.gov/analysis/studies/drilling/pdf/upstream.pdf.

[6] Bogotá Sanabria GH, Flórez García AC, Guzmán Manrique JA Sensibilidad ambiental ante un posible derrame offshore aplicando tecnologías geoespaciales, costa caribe colombiana *Revista AIDIS de Ingeniería y Ciencias Ambientales. Investigación, desarrollo y práctica* **11**(1) pp 95-109.

[7] Cusaría AA 2005 Petróleo, seguridad ambiental y exploración petrolera marina en Colombia *ICONOS. Revista de Ciencias Sociales* (21) pp 11-7.

[8] Alvarez C, Lupo C Evaluación de Localizaciones para la Perforación Costa Afuera. [Online] Available from: http://oilproduction.net/files/Evaluacion-de-Localizaciones-Offshore.pdf.

[9] Speight JG 2014 Handbook of offshore oil and gas operations *Elsevier*

[10] Martínez Rodriguez G 2005 Metodología para la evaluación de procesos de tratamiento y opciones de disposición de desechos de perforación generados costa afuera *Universidad Industrial del Santander*.

[11] Serrano MS, Munevar LD, Silva LPA, Palma JJÁ 2015 Modelo logístico para el apoyo a la perforación offshore como un aporte a la internacionalización del sector hidrocarburos de Colombia *Puente 9* **9**(2) pp 105-112.

[12] Fakhru'l-Razi A, Pendashteh A, Abdullah LC, Biak DRA, Madaeni SS, Abidin ZZ 2009 Review of technologies for oil and gas produced water treatment *Journal of hazardous materials* **170**(2-3) pp 530-551.

[13] Kermani MB, Harrop D 1996 The impact of corrosion on oil and gas industry *SPE Production & Facilities* **11**(03) pp 186-190.

[14] Osorio Tovar, Grimaldo Guerrero W, Pacheco Torres J, Chaparro Badillo P 2018 Chemical Failure Analysis of Artificial Lift System in Petroleum Industry: A Review *Journal of Engineering and Applied Sciences* **13**(19) pp 8010-8015.

[15] Guerrero Hernández JM Opciones de taponamiento para abandono de pozos petroleros 2014 *Universidad Nacional Autónoma de México*.

[16] Bergh LIVB 2017 Occupational health psychology and management: psychosocial risk management in the oil and gas industry *Doctoral dissertation, University of Nottingham*.

[17] Lindøe PH, Baram M, Renn O (Eds.) 2013 Risk governance of offshore oil and gas operations *Cambridge University Press.*
[18] Krupnick A, Wang Z, Wang Y 2014 Environmental risks of shale gas development in China *Energy Policy* **75** pp 117-125.
[19] Aung TS 2017 Evaluation of the environmental impact assessment system and implementation in Myanmar: Its significance in oil and gas industry *Environmental Impact Assessment Review* **66** pp 24-32.
[20] Shvarts EA, Pakhalov AM, Knizhnikov AY 2016 Assessment of environmental responsibility of oil and gas companies in Russia: the rating method *Journal of cleaner production* **127** pp 143-151.
[21] Ellis JI, Clark MR, Rouse HL, Lamarche G 2017 Environmental management frameworks for offshore mining: The New Zealand approach *Marine Policy* **84** pp 178-192.
[22] Tait RD, Moynihan KJ, Daetwyler NA, Young CJ, Scott-Wilson P 2014 ESHIA Hindcasts: Advancing Environmental, Social, and Health Impact Assessment and Risk Management *SPE International Conference on Health, Safety, and Environment.*
[23] Rosenfeld AB, Thomsen JB, Bowles IA 1998 Approaches to Minimizing the Environmental and Social Impacts of Oil Development in the Tropics S*PE International Conference on Health, Safety, and Environment in Oil and Gas Exploration and Production.*
[24] Arora MP, Lodhia S 2017 The BP Gulf of Mexico oil spill: Exploring the link between social and environmental disclosures and reputation risk management *Journal of Cleaner Production* **140** pp 1287-1297.
[25] Raufflet E, Cruz LB, Bres L 2014 An assessment of corporate social responsibility practices in the mining and oil and gas industries *Journal of Cleaner Production* **84** pp 256-270.
[26] Hilson G 2012 Corporate Social Responsibility in the extractive industries: Experiences from developing countries *Resources Policy* 2012 **37**(2) pp 131-137.
[27] Linnen LT, Noreldin AMA 2015 Social Sustainability & Stakeholder Engagement: An International Perspective *ASSE Professional Development Conference and Exposition.*
[28] Touzel P. 2012 Managing environmental and social risks in China *International Conference on Health, Safety and Environment in Oil and Gas Exploration and Production.*
[29] Sun R, Wang Z 2015 A comprehensive environmental impact assessment method for shale gas development *Natural gas industry B* **2**(2-3) pp 203-210.
[30] Le MT 2018 An assessment of the potential for the development of the shale gas industry in countries outside of North America *Heliyon* **4**(2) pp e00516.
[31] Hasheminasab H, Gholipour Y, Kharrazi M, Streimikiene D 2018 A novel Metric of Sustainability for petroleum refinery project *Journal of Cleaner Production* **171** pp 1215-1224.
[32] Hasle JR, Kjellén U, Haugerud O 2009 Decision on oil and gas exploration in an Arctic area: Case study from the Norwegian Barents Sea *Safety Science* **47**(6) pp 832–842.
[33] Harsem Ø, Eide A, Heen K 2011 Factors influencing future oil and gas prospects in the Arctic *Energy policy* **39**(12) pp 8037-8045.
[34] Noble B, Ketilson S, Aitken A, Poelzer G 2013 Strategic environmental assessment opportunities and risks for Arctic offshore energy planning and development *Marine Policy* **39** pp 296-302.
[35] Dale B 2011 Governing resources, governing mentalities. Petroleum and the Norwegian integrated ecosystem-based management plan for the Barents and Lofoten seas in 2011 *The Extractive Industries and Society* **3**(1) pp 9-16.
[36] Gulas S, Downton M, D'Souza K, Hayden K, Walker TR 2017 Declining Arctic Ocean oil and gas developments: Opportunities to improve governance and environmental pollution control *Marine Policy* **75** pp 53-61.
[37] UN Desarrollo sostenible [Online] Available from: http://www.un.org/es/ga/president/65/issues/sustdev.shtml.

[38] Wang D, Li S, Sueyoshi T 2014 DEA environmental assessment on U.S. Industrial sectors: Investment for improvement in operational and environmental performance to attain corporate sustainability *Energy Economics* **45** pp 254-267.
[39] EPA Strategic Environmental Assessment (SEA) [Online] Available from: http://www.epa.ie/monitoringassessment/assessment/sea/
[40] EPA Environmental Impact Assessment (EIA) [Online] Available from: http://www.epa.ie/monitoringassessment/assessment/eia/
[41] UF6 Environmental Impact Statement (EIS) [Online] Available from: http://web.ead.anl.gov/uranium/eis/whatiseis/index.cfm
[42] EPA Risk Assessment [Online] Available from: https://www.epa.gov/risk/about-risk-assessment#whatisrisk
[43] EPA Ecological Risk Assessment [Online] Available from: https://www.epa.gov/risk/ecological-risk-assessment
[44] EPD Environmental Audit [Online] Available from: https://www.epd.gov.hk/epd/english/how_help/tools_ea/audit_1.html
[45] EPA Regional Screening Levels [Online] Available from: https://www.epa.gov/risk/regional-screening-levels-rsls
[46] MARN Evaluación Ambiental Estratégica (EAE) [Online] Available from: http://www.marn.gob.sv/evaluacion-ambiental-estrategica-eae/
[47] MADS Evaluación Ambiental Estratégica. [Online] Available from: http://www.minambiente.gov.co/index.php/component/content/article?id=159:plantilla-asuntos-ambientales-y-sectorial-y-urbana-12#evaluaci%C3%B3n-ambiental-estrat%C3%A9gica-tabla-1
[48] Donaldson CV, Lichtenstein J 2013 Strategic Environmental and Social Assessment for REDD+: What they matter, and how to do them
[49] Scupholme PL, Bruney J, Armstrong K, Visser JP, Campbell JA 1998 Principles for Social and Environmental Assessment *SPE International Conference on Health, Safety, and Environment in Oil and Gas Exploration and Production.*
[50] Bell J. 2014 Interlinking Engineering and Social Performance into Sustainability Using the Triple Bottom Line Principal *Unconventional Resources Technology Conference* pp 2432-2440.
[51] Graveaud F, Verzat B, Jammes L 2014 Evaluating the Global Performance of Social Investment Programs: A Socioeconomic and Environmental Impact-Oriented Approach *SPE International Conference on Health, Safety, and Environment*

Communications in Flexible Supervisor for Laboratory Research in Renewable Energy

J. R. Nuñez-Alvarez[1a], I. F. Benítez-Pina[2], Y. Llosas-Albuerne[3]

[1] Department of Energy, Universidad de la Costa, Colombia, ORCID: 0000-0002-6607-7305.
[2] Department of Automatic Control, Universidad de Oriente, Cuba, ORCID: 0000-0003-2359-9768.
[2] Department of Electricity, Universidad Técnica de Manabí, Ecuador, ORCID: 0000-0002-5713-0565.

[a] jnunez22@cuc.edu.co

Abstract. In laboratories new systems are created where researchers hope to obtain improved functions at a reasonable cost. In them, specialists need efficient tools that offer better performance and flexible features. This paper designs and implements a SCADA system for GERA laboratory (Scientific Group for Applied Renewable Energy at the University of Oriente) adjusted to the inherent characteristics of flexibility and functionality required in research centers. All these features are supported in an efficient communication system with the implementation of various protocols and novel solutions for this type of systems for monitoring and control. It verifies and validates the system for formal methods.

1. Introductio

In today's highly competitive business environment, the industry is challenged by the demand for increased productivity, quality, safety and environmental protection, where the role of research centers in the development of new technologies is crucial in achieving these goals. The specialists who develop applications in these centers need efficient tools that offer better performance and flexible features [1]. On this issue is currently working on how to reuse existing software applications SCADA (Supervisory Control and Data Acquisition) [2, 3, 4]. This issue involves a major challenge for designers, since from the stages where software engineering is conceived it is necessary not only the use of diagrams and classical methodologies, but the use of formal mathematical tools is also imposed as Petri Nets (PN), which allow to model and simulate the behavior of systems prior to the relevant implementation [5, 6]. Although the use of formal methods a priori not guarantee a hundred percent correct result, you can however, greatly increase the understanding of the system revealing inconsistencies, ambiguities and inaccuracies that might otherwise go unnoticed [7, 8].

In the research [9], a test bench of a control and monitoring system for the discharge of drinking water in a well is designed and implemented. In the work, the graphical user interfaces (GUI) are designed to interact with the operator, as well as the human-machine interface (Human Machine Interface (HMI)) which was implemented in proprietary software and contemplates the rules for control and monitoring of system conditions. The HMI interface is connected to a programmable logic controller (PLC) where the process protection rules are implemented in ladder diagram (Ladder Diagram, LD).

The work of [10] develops a configurable SCADA system from a library of stationary, modular, parameterizable and reusable models. This tool is used in the energy diagnosis of sugar plants, taking process data and using a stationary process model, and using data reconciliation algorithms based on optimization techniques. It is also able to provide information about unmeasured variables, detect deviations from the measured variables and calculate indices related to energy efficiency. Since the application is going to be customized for different factories, tools have been developed that allow rapid reconfiguration and code reuse.

This paper designs and implements a SCADA system for GERA laboratory (Scientific Group for Applied Renewable Energy at the University of Oriente) adjusted to the inherent characteristics of flexibility and functionality required in research centers. All these features are supported in an efficient communication system with the implementation of various protocols and innovative solutions for such monitoring and control systems [11, 12].

In the next section II, the structure of SCADA is described, where the communication system (three fieldbuses) has the primary function. In III formal design details are explained in GHENeSys hierarchical extended PN [13] each of the three main software components. In IV implementation in Labwindows is explained [14] that allows its professional use in the GERA laboratory.

2. Methodology

2.1. General System Architecture

The GERA at University of Oriente has as main objective the development, study, evaluation and analysis of the different practical applications of renewable energy [15]. The beginning of the evaluating process of a system in the laboratory of the research center includes installation or assembly of technology in the polygon or center areas. The researcher together with center specialists in instrumentation and control will be responsible for installing sensors and actuators to be attached to experiment. Autonomously the server should detect the newly installed device and manager generate a software key where permissions and access to the physical I/O system is assigned; This will be delivered to the person responsible to develop the experiment. Finally, the researcher has a client software that will link with the server by a TCP-IP data network and have access to take measurements and/or implement control actions [16].

The flexible SCADA required differs from the standards that the concept of users is replaced by software keys that assign permissions to different copies of the client software running on the network. Permissions can be changed by the system administrator at runtime. Communication drivers allow link memory addresses located on remote equipment's with OPC servers that make up the application [17, 18]. The management of these are done automatically and completely transparent to users. The Synoptics of each experiment are created by different users at runtime, not by developers, where the desired graphical indicators are shown on the screen to analyze the information they really need in each experiment. By storing information associated with each process or experiment that is supervised or controlled in standardized relational databases facilitates systems of report generation to obtain the required information without investing great efforts in formatting and data preparation.

According with these requirements, the designed flexible SCADA system consists of distributed hardware support of three data buses and decentralized software support of three independent applications that interact through an Ethernet Data Network. The Figure 1 shows the three data buses of hardware block.

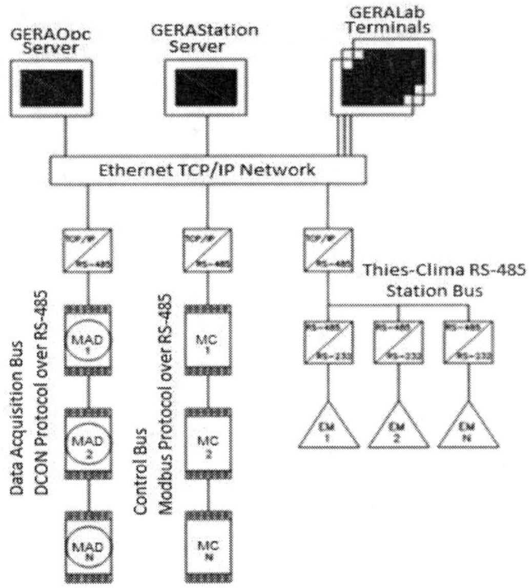

Figure 1. Schematic mono-linear SCADA Communications of GERA Laboratory.

2.2. System Components

The data acquisition Bus is based on the DCON protocol on the physical standard RS-485 [19]. Where multiple data acquisition boards that support this protocol are linked. The control Bus is supported on physical standard RS-485 where proprietary cards are installed with Modbus communication [20] and digital and/or analog outputs with standard current and voltage levels for controlling multiple actuators. The Bus of meteorological stations is based on the communication protocol of Thies-clima weather stations [19] on the physical standard RS-485. Through this bus can be linked to the system until ten of this type weathers stations.

On top of Figure 1 shows the three software applications that make up the system, two of which will ensure access to communication buses (GERAOpc and GERAStation) and the third application (GERALab) is the central SCADA software and communicates with the above two applications using the OPC standard on the TCP-IP protocol [2, 21]. The distribution of the communications system in three buses ensure efficient control of them from the first two applications and a safety information flow throughout the SCADA. The complexity of the interaction of all system components requires formal design method explained in the next section.

2.3. System Formalization using GHENeSys Hierarchical Extended Petri Nets

Following the design methodology on GHENeSys given by [13], the first thing to consider in first step is the study of process to control and its functional requirements, which corresponds to the informal specifications phase. Equally steps 2 to 5 of the methodology are closely related, and are associated with the phase of formalization of the applications, thereby ensuring obtaining the right model. Following these principles, each application that forms the system was decomposed into modular structures shown as the example in Figure 2, that permit a PN modeling. Figure 3 is the GHENeSys extended hierarchical Petri net model developed on Visual Object Net 2.7 [22, 23, 24] by means of Figure 2.

Attention to data acquisition and control bus (GERAOpc) are separated from the attention to station bus (GERAStation) because the sampling time and peculiarities of the data are different in each. Also, the functions of the data and control bus are different. Differentiated attention of the three buses allows the simplicity and efficiency of communication.

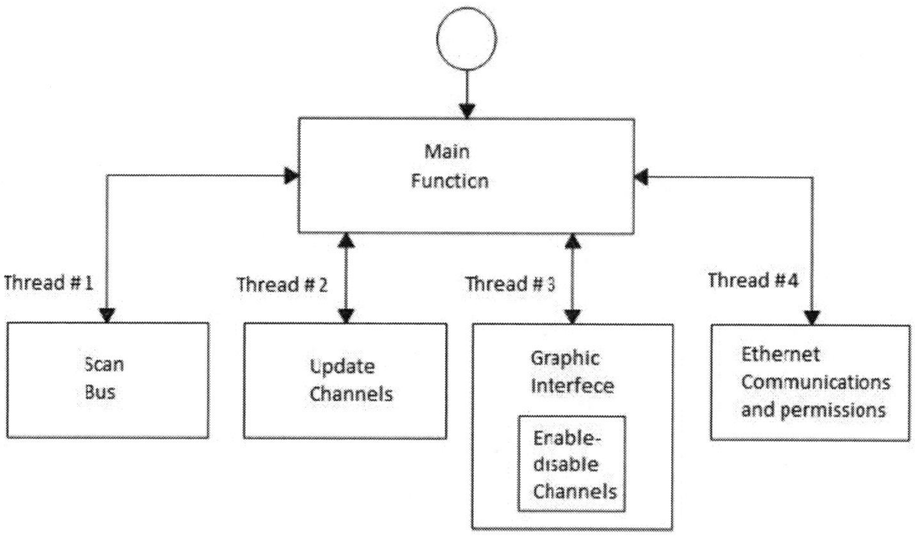

Figure 2. Hierarchical modular structure GERAOpc software.

The GERAOpc software (Figure 2) consists of two hierarchical levels and four concurrent threads to execute the program. At the top level is the functional unit, responsible for creating the four threads and manage access to global variables that will be used by the other functional blocks. At the second level the functions entrusted to scan and update channels in the Data Bus, to make communication in Ethernet network and to attend the graphical interface (GUI).

In the model on GHENeSys of this software (Figure 3) the P0 and P1 places correspond to the resource "Initial State" and "Main Function" respectively. By the T2 transition the provided four threads of concurrent execution are created. P2 and P6 places represent the "GUI" resource, P3 and P7 represent the "Communication Ethernet", P4 and P8 represent the "Scanning BUS" resource, P5 and P9 represent the "Update I/O" resource and P10 represents the "Final State" resource or close the application.

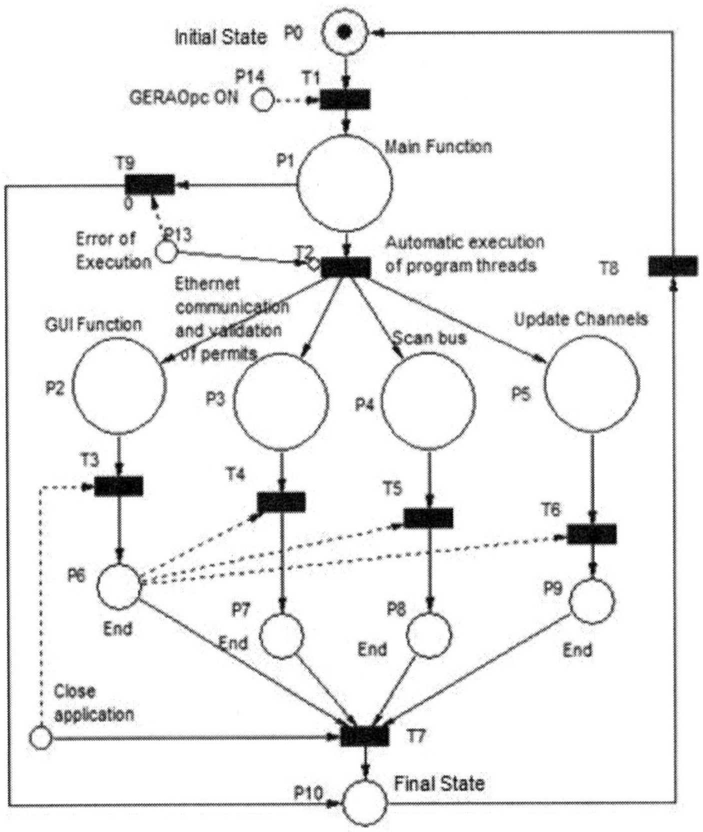

Figure 3. Formal Model in GHENeSys PN of GERAOpc software.

Using the same methodology, the models of other two applications and their subsystems were developed, can itemize their functions and interrelationships with other elements. Verification of the developed models were then performed using reduction methods and invariant [25]. Further validation of the models was performed by simulation to verify the functional behavior of each module in all its variations (normal operation, fail, etc.).

3. Results

3.1. Implementation in LabWindow/CVI

The implementation of the three software was developed under the LabWindows/CVI 10.0 environment. The GERAOpc (Figure 4) consists of three files, GERAOpc.uir where all information relating to the design of the graphical interface is stored and GERAOpc.hy GERAOpc.c files which were all functions are declared and implemented respectively.

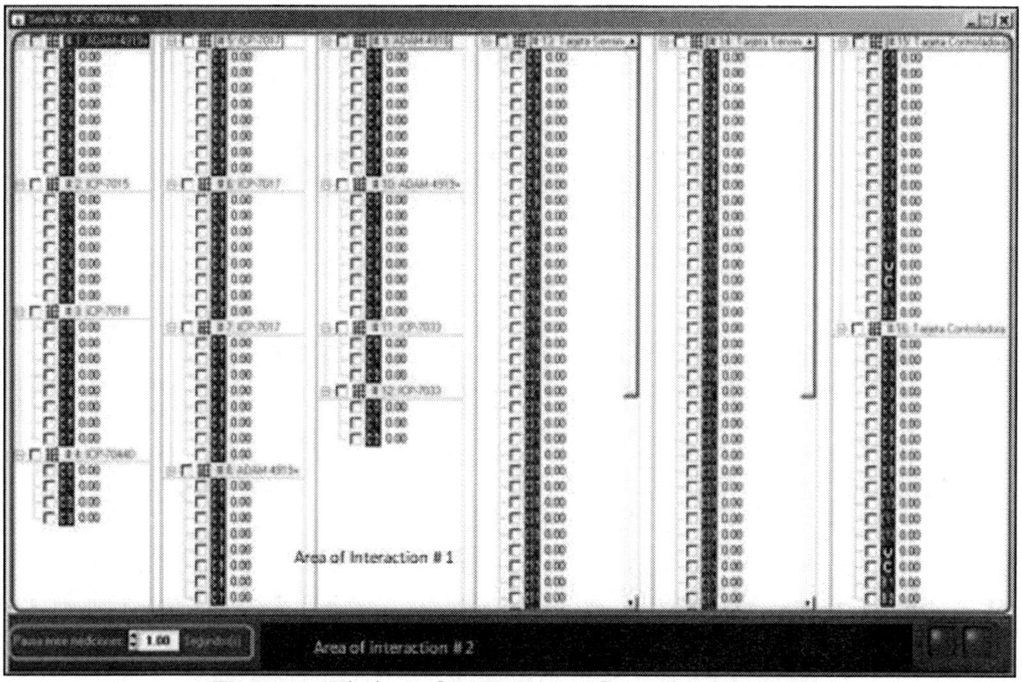

Figure 4. Window of GERAOpc software implementation.

Likewise, was created GERAStation (Figure 5) with files: Station.uir, Station.h and Station.c. The GERALab (Figure 6) includes GERALab.uir file where all the information relating to the design of the GUI is stored, asynctmr.h which is the library that includes the functions of asynchronous timers, inifile.h that enables automated reading of .ini file and movectrl.h that makes creating and editing of components at runtime.

The interface designed to GERAOpc includes a minimalist look with the aim of facilitating the characteristics of flexibility, but powerful auto-detection algorithms channels and visualization of data obtained using fieldbuses and control are included, providing the user only two areas of interaction and an information area (Figure 4). In the area of interaction # 1 (Figure 4, marked in red) all acquisition and control modules detected in the data buses are presented, and attached to each channel the user has a "Check Box" and a label displayed. By the first you can enable or disable the use of the channel in question and by the second can be displayed in real time the instantaneous values of each channel.

Through the interaction area # 2 (Figure 4, marked in green) can introduce a delay between each measurement cycle or channels scans, this in order not to overload excess resources available on the server that supports this software. And the display area (marked in blue) will be composed of two simple lighting indicators, one that flashes red to indicate when it is accessing data buses and other flash blue indicator to indicate when running storage routines in the database of real-time.

In the area of interaction (marked in red) GERAStation (Figure 5) all the weather stations linked to the network are displayed. By viewing area (marked in green) the program administrator can quickly check whether the values are logical.

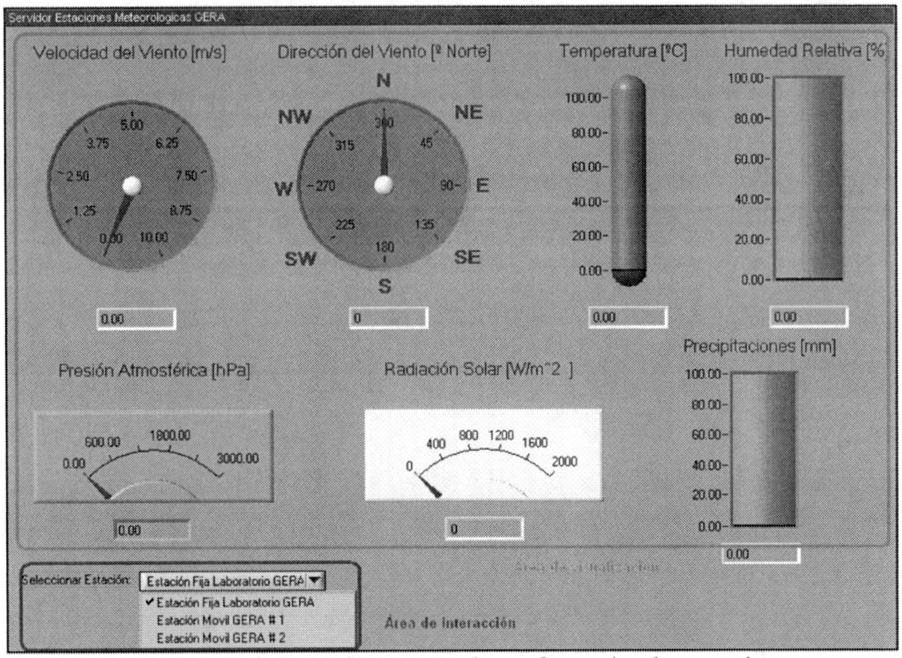

Figure 5. Window of GERAStation software implementation.

The designed interface to create synoptic (Figure 6) has a simple look with the objective of facilitating the characteristics of flexibility, but complex routines communication with OPC servers, data visualization, data storage and calculations of controllers are included, providing the user only initial interaction area at the top right of the application window (Figure 6). Through these options the user can start a wizard which will tell the system all the details of the experiment in question. Then the user can configure mimic the experiment, with the possibility of inserting five different types of components on-screen display. For each of the inserted components the user will edit options available to obtain a fast and safety interface for the required experiments (Figure 7).

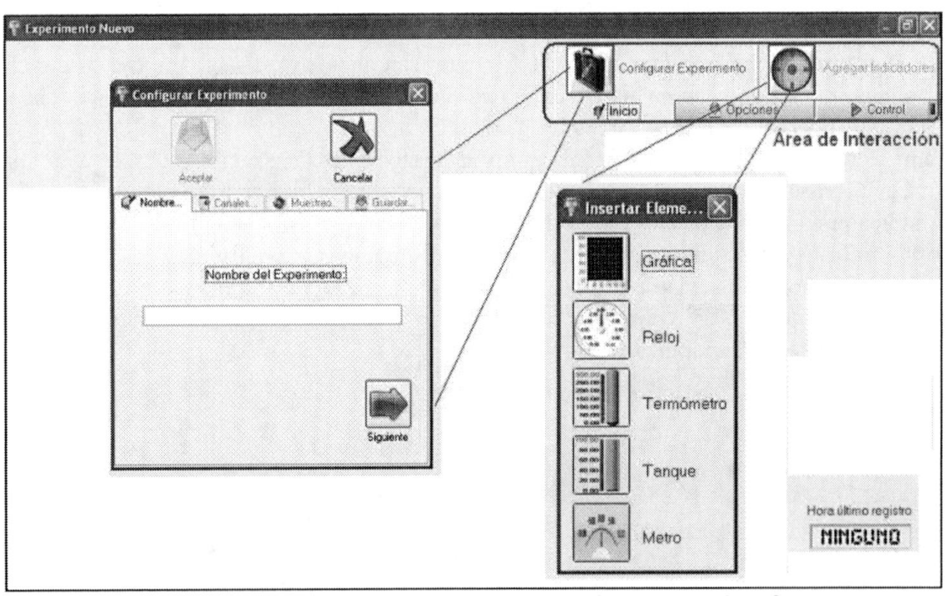

Figure 6. Configuration of a new experiment in GERALab software.

IOP Conf. Series: Materials Science and Engineering **844** (2020) 012016 doi:10.1088/1757-899X/844/1/012016

This Flexible SCADA system is being used in the GERA as the central tool for the development of experiments and data collection and national and international research projects, throwing good results.

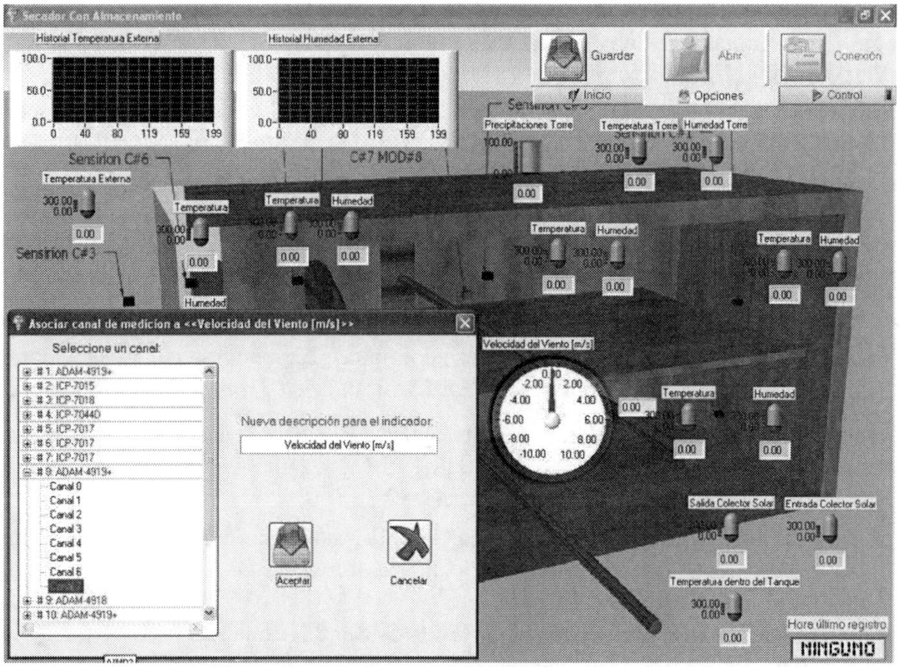

Figure 7. Example Window of GERALab software implementation.

Conclusions

The software implementation was developed under the LabWindows / CVI 10.0 environment. By means of the model in GHENeSys it was possible to evaluate the sequence of operation and the conditionals that generated changes of state.

The designed interface includes a simple aspect in order to facilitate the flexibility features, but it includes complex communication routines with the OPC servers, visualization of the data obtained, data storage and controller calculations, facilitating the user an area of interaction. The functional behavior of each implemented module was proven by demonstrating compliance with the requirements demanded by its physical interpretation.

The main contribution of the GERALab software is that it executes in a synchronized way the control and data collection functions with which a database is obtained that allows researchers to analyze the behavior of the studied object based on the same unit of time to represent the measured parameters and the control actions executed.

This Flexible SCADA system is being used in the GERA as the central tool for the development of experiments and data collection and national and international research projects, throwing good results.

References

[1] Valencia G, Nuñez J and Duarte J 2019. Multiobjective Optimization of a Plate Heat Exchanger in a Waste Heat Recovery Organic Rankine Cycle System for Natural Gas Engines. *Entropy*. 21(7), 655; https://doi.org/10.3390/e21070655

[2] Pérez López E 2015. Los sistemas SCADA en la automatización industrial. *Tecnología en Marcha*. vol. 28, no. 4, pp. 3–14

[3] Romero A, Marín A and Jiménez J. A 2014. SCADA System for Detection of Explosive Atmospheres in Underground Coal Mines Through Wireless Sensor Network. *IEEE Latin America Transactions*. vol. 12, no. 8, pp. 1398-1403

[4] Minchala L. I, Sánchez C, Yungaicela N, Mora A and Mata J. A 2016. Comparative Study Of Black – Box Models For Cement Fineness Prediction Using SCADA Measurement of a Closed Circuit Grinding. *IEEE Latin America Transactions*. vol. 14, no. 2, pp. 673-680

[5] Vega-de la Cruz L, Lao León Y, and Pérez Pravia, M 2016. Redes de Petri en la Determinación de Puntos Críticos para el Control Interno. *Revista Universidad y Sociedad*. 8(4), 219-226

[6] Castellanos J. S and Solaque L. E 2010. Modeling with Petri Nets and Implementation with Grafcet of a Flexible Manufacturing System with Concurrent Processes and Shared Resource. *Ciencia e Ingeniería Neogranadina*, Vol. 20-1, pp. 61-75

[7] David R and Alla H 2010. Discrete, Continuous and Hybrid Petri Nets. Editorial Berlin Springer. Book. 337 pages. ISBN 978-3-642-10669-9

[8] Silva M and Recalde L 2007. Redes de petri continuas: Expresividad, análisis y control de una clase de sistemas lineales conmutados. *IEEE Latin America Transactions* vol 4, no. 3, pp. 5-33

[9] Quesada J. Carlos, Flores E, Bautista J and Quezada V 2014. Diseño e implementación de un sistema de control y monitoreo basado en HMI-PLC para un pozo de agua potable. *Ingeniería, Investigación y Tecnología*. vol.15 no.1, pp.41-50

[10] Acebes L. F, Merino A, Alves R and Prada C 2009. Análisis en Línea del Estado Energético de Plantas Azucareras. *Revista Iberoamericana de Automática e Informática Industrial*, [S.l.], v. 6, n. 3, p. 68-75. ISSN 1697-7920

[11] Rendón A, Fuerte C. R and Calderón J. G 2015. State Estimation of Electrical Power Grids Incorporating SCADA and PMU Measurements. *IEEE Latin America Transactions*. vol. 13, no. 7, pp. 2245-225

[12] González I, Calderón A. J, Figueredo J and Sousa J. M 2019. A Literature Survey on Open Platform Communications (OPC) Applied to Advanced Industrial Environments. *Electronics*. 8, 510; doi:10.3390/electronics8050510

[13] Sasián F, Theron R and Gachet D 2016. Protocolo para comunicación inalámbrica en instalaciones de energías renovables. *Revista Iberoamericana de Automática e Informática Industrial RIAI*. vol 13. Issue 3. pp 310-321

[14] Silva J. R, Benítez I, Villafruela L, Gomis O and Sudriá A 2008. Modeling extended Petri nets compatible with GHENeSys IEC61131 for industrial automation. *The International Journal of Advanced Manufacturing Technology*. vol 36, Isuee 11-12, pp. 1180-1190

[15] Valencia G, Nuñez, J., Acevedo, C. (2019). Research Evolution on Renewable Energies Resources from 2007 to 2017: A Comparative Study on Solar, Geothermal, Wind and Biomass Energy. International Journal of Energy Economics and Policy. 9(6), 242-253. DOI: https://doi.org/10.32479/ijeep.8051

[16] Salazar C. A and Correa L. C 2011. Buses de campo y protocolos en redes industriales. *Ventana Informática*. 25. 83-109

[17] Pedraza L, Hernández-Suárez C. A, Alfonso D and López-Sarmiento 2013. Sistema de comunicación TCP/IP para el control de una intersección de tráfico vehicular. *Ingeniería, Investigación y Tecnología*. 2. 10.1016/S1405-7743(13)72268-6

[18] Caicedo-Erasoa J. C, Varón-Sernab D. R and Díaz Arango F. O 2012. Redes Industriales. *Vector*. 7, 12-17

[19] Custodio A, Urdaneta E, Barreto A and Oviedo A 2014. Sistema Electrónico de Adquisición y Entrega de Datos para Laboratorios de Enseñanza que puedan incorporarse a Entornos Virtuales de Aprendizaje. *Congreso Iberoamericano de Ciencia, Tecnología, Innovación y Educación*. ISBN: 978-84-7666-210-6 – Artículo 371

[20] Quiles F, Ortiz M. A, Moreno D and Brox M 2009. ADQPCI: Placa de Adquisición de Datos con Fines Docentes. *Formación universitaria*. 2(3), 25-30. https://dx.doi.org/10.4067/S0718-50062009000300005

[21] Núñez Alvarez, José Ricardo et al. Metodología de diagnóstico de fallos para sistemas fotovoltaicos de conexión a red. Revista Iberoamericana de Automática e Informática industrial, [S.l.], sep. 2019. ISSN 1697-7920. Disponible en: <https://polipapers.upv.es/index.php/RIAI/article/view/11449>

[22] Leal D. L, Celi M. R and Alvarez E 2016. Sistema integral para el diseño e implementación de control asistido por computadora. *Revista Iberoamericana de Automática e Informática Industrial RIAI*. 2016. 13. 228–237

[23] Minchala I, Ochoa S, Velecela E, Astudillo D and González J 2016. An open source SCADA system to implement advanced computer integrated manufacturing. *IEEE Latin America Transactions*. 14. 4657-4662. 10.1109/TLA.2016.7816994

[24] Benítez I. F, Lamar S, da Silva M, Miyagi E, Silva J. R 2017. Design of automatic control system based on unified timed hybrid Petri net. *DYNA*. Volumen 84, Número 200, p. 80-89. ISSN electrónico 2346-2183. ISSN impreso 0012-7353

[25] Munir A, Gordon-Ross A and Ranka S 201. Modeling and optimization of parallel and distributed embedded systems. Wiley-IEEE Press. 400 pages. ISBN: 978-1-119-08639-0

IOP Conf. Series: Materials Science and Engineering **844** (2020) 012017 doi:10.1088/1757-899X/844/1/012017

Design of a Fuzzy Controller for a Hybrid Generation System

J. R. Nuñez[1*], J. Mestre[2], J. J. Cabello[3], H. Dominguez[4], J. Fong[5], L. Peña[6], I. Benítez[7], D. De Oliveira[8]

[1,3]Departamento de Energía, Facultad de Ingeniería, Universidad de la Costa, Calle 58 No 55-66, 080002, Barranquilla, Colombia.
[2,5,7]Departamento de Automática, Facultad Ingeniería Eléctrica, Universidad de Oriente, Ave Las Américas, s/n, 90300, Santiago de Cuba, Cuba.
[4]Departamento de Informática, Facultad Ingeniería Informática, Telecomunicaciones y Biomédica, Universidad de Oriente, Ave Las Américas, s/n, 90300, Santiago de Cuba, Cuba.
[6]Departamento de Desarrollo, UEB Santiago de Cuba, Empresa de Hidroenergía, Ave Las Américas, s/n, 90300, Santiago de Cuba, Cuba.
[8]Departamento de Electricidad, Campus Reitor Joao David Ferrer Lima s/n. 88040-970. Universidad Federal de Santa Catarina, Florianápolis. Brasil.

* jnunez22@cuc.edu.co

Abstract. This paper presents the design of a control system for the automatic connection/disconnection and distribution of load, between an asynchronous alternator and a generator, in a hydroelectric central that works in isolated mode. The design of a control algorithm based on fuzzy logic is exposed, as this is a flexible method to be used in different installations with a variety of technology. The controller is supported on the Arduino Mega 2560 platform, in order to develop a low-cost system with its own technology, it is tested by computer simulation using the professional software Proteus v7.7, which guarantees that once validated the correct operation of the controller can be migrated to another system, say for example a PLC. The results obtained are shown and the simulations performed to the different blocks of the system are explained.

1. Introduction

Factors such as climate change and the increase in energy demand make a change in global energy policy increasingly urgent [1, 2]. The exploitation of Renewable Sources of Energy (RSE) is the main alternative for this change, since they are clean energies by not emitting CO_2 and can be applied both in networks interconnected to electrical systems and in isolated areas [3, 4, 5, 6].

The application of RSE in isolated areas has as main drawback its instability, which makes it necessary to use hybrid systems that guarantee permanent power supply [4, 6]. The operation of hybrid power generation systems in isolated areas of the National Electric System (NES) has as one of its main challenges to ensure the parameters of energy quality and permanent supply regardless of the climatic conditions and the RSE that can replace [7, 8].

In this work the design of an automatic control system for the hybrid generation system integration is made, to be used in an isolated area of the NES, using a hydraulic installation and a Diesel Generator Set (DGS), supported on the platform Arduino Mega 2560. Using artificial intelligence techniques, the operation of the system is simulated to guarantee the electrical supply in an isolated community, verifying the correct functioning of the algorithm developed based on fuzzy logic for the connection and disconnection of the generating sources, as well as the correct synchronization between these sources [9, 10]. Developing this work, the design and validation of a low-cost controller that is part of an electrical generation system using hydraulic energy in mountainous areas of the eastern region of Cuba, which are totally isolated from the National Electro Energetic System (SEN), is obtained.

2. Methodology

The implementation of the controller is developed through the XP software development methodology adapted, which consists of the following phases: Planning, Design, Coding and Testing.

2.1. System proposal

During the work development, the following parameters were considered: frequency, power and the phase shift between the two generating sources. Both have voltage control and the micro hydroelectric plant also has a frequency controller [11, 12]. The generation system that is proposed to automate is composed of:

- A Small Hydroelectric Plant (SHP) with a synchronous generator, and an asynchronous motor which is used as a generator. For the asynchronous motor, a speed control is required, which consists in varying the opening of the water inlet to the turbine that puts it in motion, in order to bring the motor to synchronism with the synchronous generator and then generate up to 30% of the nominal power of the synchronous generator to help generate power when necessary.
- A DGS with frequency automatically regulated to the frequency of the PCH.

For the measurement of frequency and electrical power, standardized output transducers from 0 to 5 V and 0 to 10 V are used, respectively, which allow the controller to make the measurements and make decisions about the generating sources (DGS, SHP or both) to satisfy the energy demand, as well as that of establishing an order of priority for the connection and disconnection of these sources in case the resource of all is needed. Other parameters controlled are the phase shift, the water inlet valve to the turbine of the PCH and the frequency of the GE. The controller used is the Arduino Mega 2560 because it is economical and has the required capabilities. At this stage of the work, and due to the complexity involved in the design of the controller, the study focuses only on the control signals that the Arduino receives and sends to carry out the whole process of connecting and disconnecting the generating sources, by what the electrical machines involved in the process are analyzed as a black box. It is expected that in future research work the dynamic behavior of each of them will be studied. Figure 1 shows a block diagram of the different parts of the system to be automated.

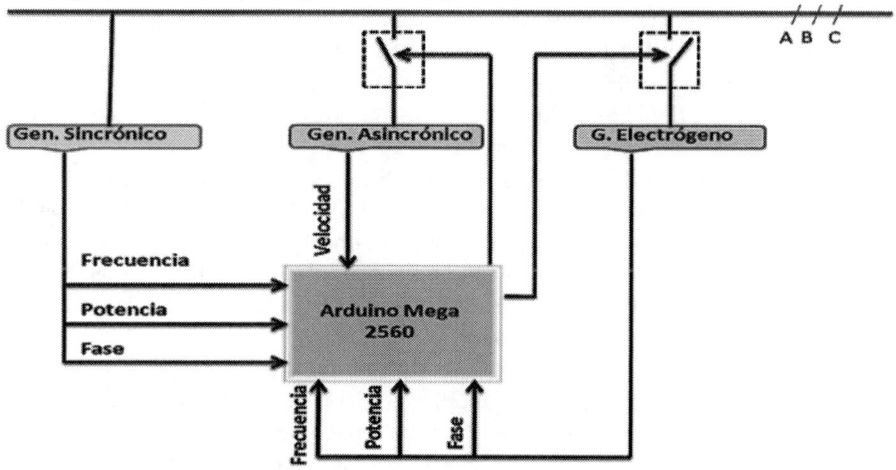

Figure 1. System Blocks Diagram. Source: The authors

2.2. Functional requirements

For the design of the controller, two modes of operation, manual and automatic, are defined, with the following requirements:

- Manual Mode:
  - RF1- Show the variables measured and controlled (Power and frequency of the SHP and DGS, asynchronous generator speed of the SHP and phase angle);
  - RF2- Control asynchronous generator;
    - RF2.1- Power on / off asynchronous generator;
    - RF2.2- Connect asynchronous generator to the network.
  - RF3- Control the power of the SHP through the water inlet valve;
    - RF3.1- Turn in positive direction the three-phase reversible motor for opening the valve;
    - RF3.2- Turn in a negative direction the three-phase reversible motor for closing the valve.
  - RF4- Control the DGS;
    - RF4.1- Power on / off the DGS;
    - RF4.2- Connect DGS to the network;
    - RF4.3- Sort increase / decrease the speed of the DGS generator.
  - RF5- Calculate phase angle.
- Automatic Mode:
  - RF6- Select mode according to the availability of sources and demand;
    - RF6.1- Operate only with the SHP and the asynchronous generator;
    - RF6.2- Operate only with SHP and the DGS;
    - RF6.3- Operate with the SHP, the asynchronous generator and the DGS.
  - RF7- Indicate flow level;
  - RF8- Activate alarm produced by low flow;
    - RF8.1- Eliminate alarm caused by low flow.
  - RF9- Provide some degree of intelligence to the controller, so that he can make correct decisions about auxiliary sources when operating in automatic mode.

2.3. Fuzzy Logic

Fuzzy control consists of establishing a series of rules to control actions that allow combining input variables, and by means of a relation of rules they result in one or more output values [13]. To satisfy the RF9, the designed controller is equipped with a certain degree of intelligence through fuzzy logic and a set of fuzzy rules so that it can make decisions when working in automatic mode and execute them with its switching system. For this, the fuzzy logic library (eFLL) for Arduino was used [14]. This library makes use of the Maximum and Minimum algorithms and Mamdani for inference and composition as well as the Area Center for defuzzification in a continuous universe [3].

Making use of this library defines the input set of the electric power values with the linguistic terms (low, medium, high), which are created using the FuzzySet function (w, x, y, z). The system is constantly measuring the power of the network, which is blurred through the function fuzzify(), and, according to the degree of belonging that has the value read power, to one of the defined linguistic terms, it is taken the decision to connect or disconnect the asynchronous generator and / or the DGS. For example: if the measured power has a degree of belonging greater than the term "high" than the term "average", then, according to the availability of the asynchronous generator and the DGS, and according to the water level, the decision is made to connect each one or both to help in the generation. Therefore, when the degree of belonging is greater than the term "average" or "low" that at the end "high", then the decision is taken to disconnect one or both auxiliary sources that are connected [15, 16].

The fuzzy set of power input was created through an object of the FuzzyInput class called power, this object contains three objects of the FuzzySet class called low1, medium1 and high1, leaving the set defined as shown in figure 2.

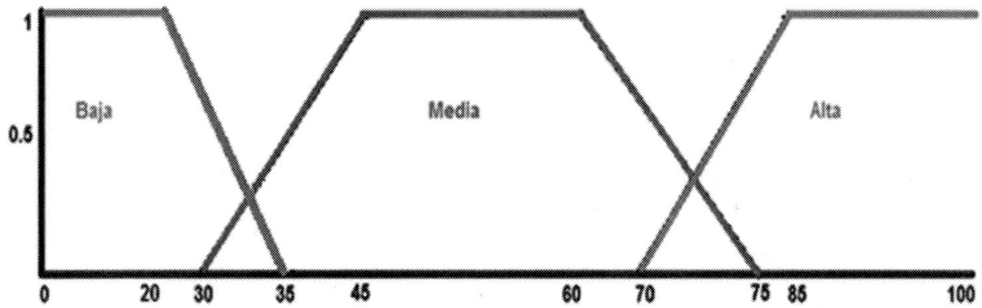

Figure 2. Input membership function (Power). Source: The authors.

To obtain the behavior of the different variables, a trapezoidal function model is established that represents each set of system variables denoted by ranges within the universe of discourse.

$$\mu_{trapezoidal}(x, a, b, c, d) = \begin{cases} 0; & x < a \\ \frac{x-a}{b-a}; & a \leq x \leq b \\ 1; & b \leq x \leq c \\ \frac{d-x}{d-c}; & c \leq x \leq d \\ 0; & d \leq x \end{cases} \quad (1)$$

Equation for the low diffuse set:

$$\mu_{low}(x,0, 0, 20, 35) = \begin{cases} 1; & 0 \leq x \leq 20 \\ \frac{1}{25}(-x + 35); & 20 < x < 35 \\ 0; & 35 \leq x \end{cases} \quad (2)$$

Equation for the average diffuse set:

$$\mu_{average}(x, 30, 45, 60, 75) = \begin{cases} 0; & x < 30 \\ \frac{1}{15}(x - 30); & 30 \leq x < 45 \\ 1; & 45 \leq x \leq 60 \\ \frac{1}{15}(-x + 75); & 60 < x \leq 75 \\ 0; & 75 < x \end{cases} \quad (3)$$

Equation for the high diffuse set:

$$\mu_{high}(x, 70, 85, 100, 100) = \begin{cases} 0; & x < 70 \\ \frac{1}{15}(x - 70); & 70 \leq x < 85 \\ 1; & x \leq 85 \end{cases} \quad (4)$$

From the above equations it is obtained that, e.g., for a power value (read from the power transducer) equivalent to 73% after performing the blurring of this value through the functions fuzzy→ setInput (1, pot) and fuzzy → fuzzify(), belonging to the eFLL library. It reaches a degree of belonging to the diffuse set average = 0.13 and a degree of belonging to the diffuse set high = 0.2. Therefore, as the power begins to be "high" because the degree of belonging to the "average" group is less than the degree of belonging to the "high" group, then the asynchronous generator or the DGS is put into operation depending on the availability and water level, and subsequently connects the generator to the grid to help with cogeneration.

This behavior is obtained from the set of fuzzy rules defined for the controller:

- If power is high, flow is enough, asynchronous motor is available and asynchronous motor is off, then ignite asynchronous motor;
- If power is high, DGS is available and off, then turn on generator set;
- If power is high, asynchronous motor is available and off, has enough flow, generator set is available and generator set is off, then turn on asynchronous motor [17];
- If power is high, asynchronous motor is available, motor is off, flow is insufficient and DGS is available and off, then turn on DGS;
- If power is high, asynchronous motor is available, has enough flow rate, asynchronous motor is on and DGS is available and off, then turn on DGS;
- If power is medium or low and asynchronous motor is on, then shut off asynchronous motor;
- If power is medium or low and DGS is on, then turn off the DGS;
- If power is medium or low, asynchronous motor is on and DGS is off, then shut off asynchronous motor;
- If power is medium or low, asynchronous motor is on and DGS is on, then turn off DGS.

The use of this controller makes the installation of the PCH system more flexible with different hydraulic and technological characteristics because it is based on fuzzy logic for decision making, which makes it possible to simplify the system since it is not necessary to know the mathematical model of the plant, unlike, for example, PID controllers. In addition, the use of the Arduino as a controller is an economically feasible option to acquire, develop and implement.

2.4. Controller's work sequence

In the controller programming and considering that always it should be the SHP who defines the voltage and frequency parameters of the network, two modes of work are required, the manual and the automatic. In addition, the controller follows a strategy of cyclical planning in which it supervises that all the tasks that are executed within the main cycle comply with the times defined for each one. At the same time, for the work in automatic mode, the diffuse rules described above are defined, which allow a normal operation regime, and which depends on the following conditions:

2.4.1. Condition 1: Abundant flow and only the asynchronous generator available as auxiliary source

Connection process: The controller is measures and always blurs the power of the SHP. When this reaches a value belonging to the fuzzy set "high", the system sends a signal to start the asynchronous generator which starts to open the water inlet valve until the synchronous speed is reached. Then it connects with the SHP, it waits for a time of approximately 2 seconds and it takes it up to 30% of the generation power of the SHP, thus achieving cogeneration. Figure 3 shows the behavior that the controller follows in this condition.

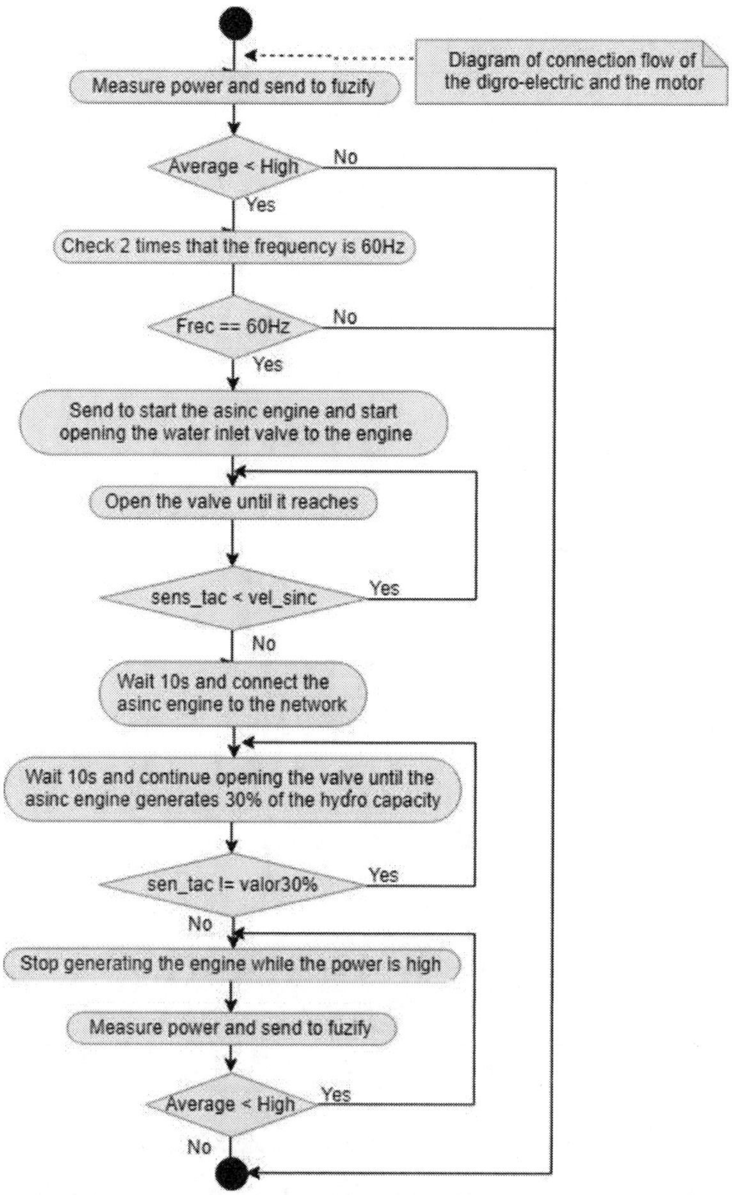

Figure 3. Flow diagram of the connection sub-process of the micro hydroelectric plant and the asynchronous generator. Source: The authors.

Disconnection process: The controller permanently measures and blots the power of the SHP. When it reaches a value of belonging to the fuzzy set "average" greater than a value of belonging to the fuzzy set "high", or a value greater than zero of the set fuzzy "low", begins to close the water inlet valve to the turbine of the asynchronous generator and waits for a certain time, until it reaches the synchronous speed. Then, disconnects it from the SHP and continues closing the valve until the speed of the asynchronous generator is less than 5% of the maximum value. Figure 4 shows this behavior.

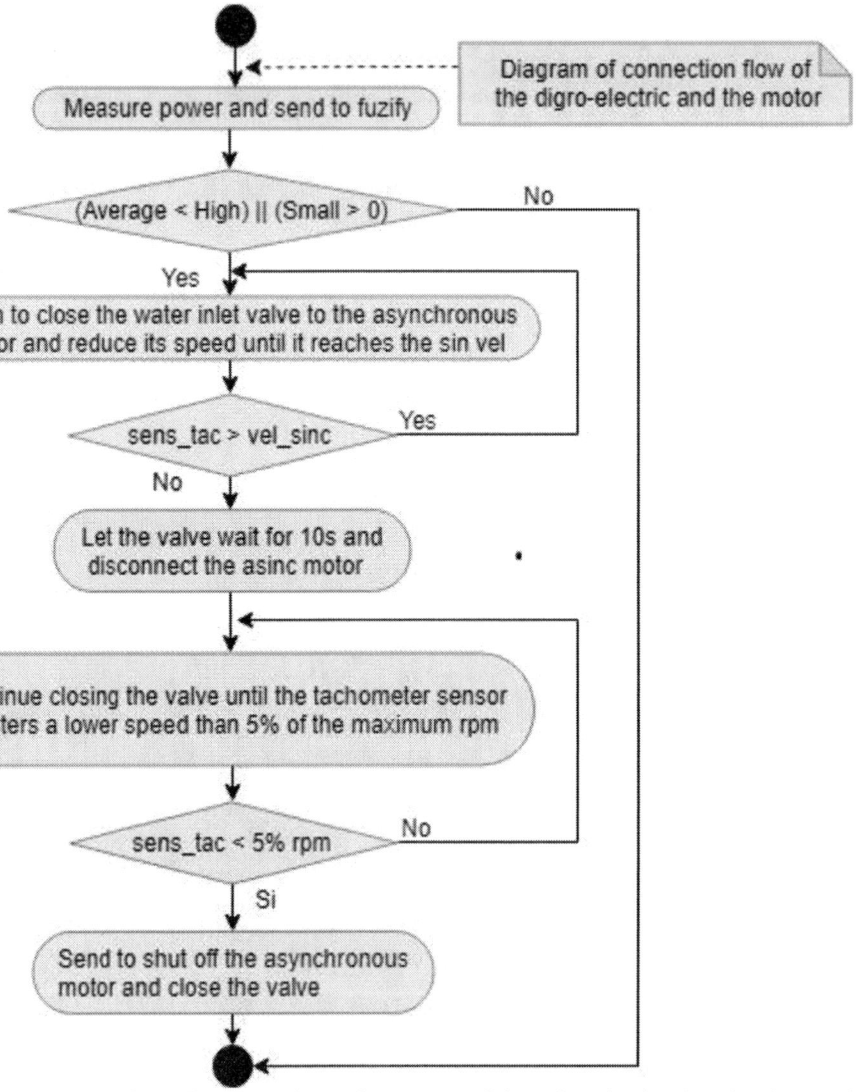

Figure 4. Flow diagram of the disconnection sub-process of the micro hydroelectric plant and the asynchronous generator. Source: The authors.

2.4.2. Condition 2: Only the generator set (DGS) available as an auxiliary source

Connection process: the controller permanently measures and blots the power of the SHP (PotCH). When it reaches a value of belonging to the fuzzy set "high", greater than a value of belonging to the set fuzzy "average", then the system turns on the DGS, verifies that its frequency is equal to that of the SHP in a range of ± 0.2 Hz according to [18]. If they are not equal, the system increases or decreases the speed of the DGS to adjust its frequency. If they are equal, it is verified that the phase shift is less than 15°, and if it is fulfilled, the DGS is connected in vacuum with the SHP, 10 seconds are expected and the generation power of the DGS is increased up to 100% of its capacity. The flow chart of Figure 5 shows the behavior of the controller in this case.

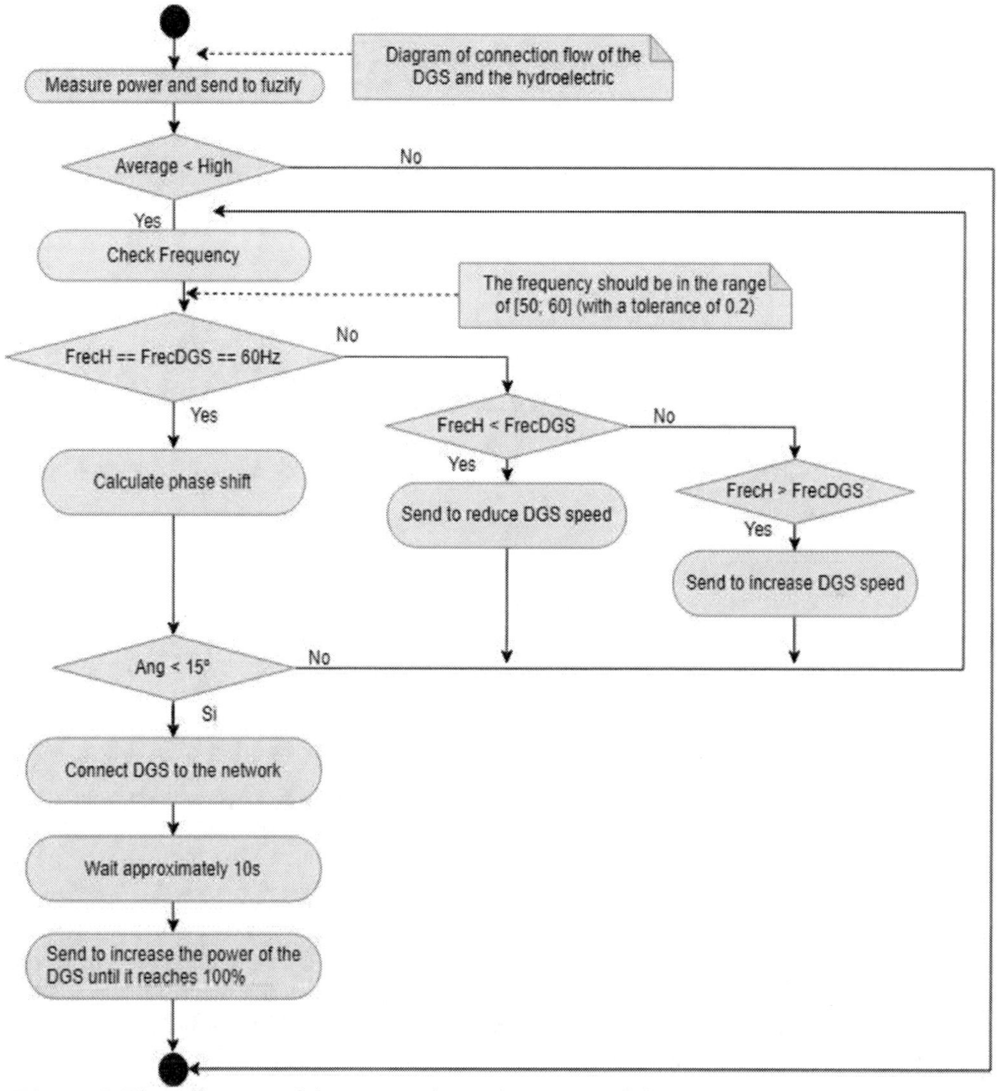

Figure 5. Flow diagram of the connection sub-process of the micro hydroelectric plant and the DGS. Source: The authors.

Disconnection process: the controller permanently measures and smudges the power of the SHP (PotCH). When it reaches a value belonging to the fuzzy set "average" greater than a value of belonging to the fuzzy set "high" or a value greater than zero of the fuzzy set "low", the system decreases the power of the DGS to 0%, waits for 10 seconds disconnects it from the SHP and turns off the DGS (See figure 6).

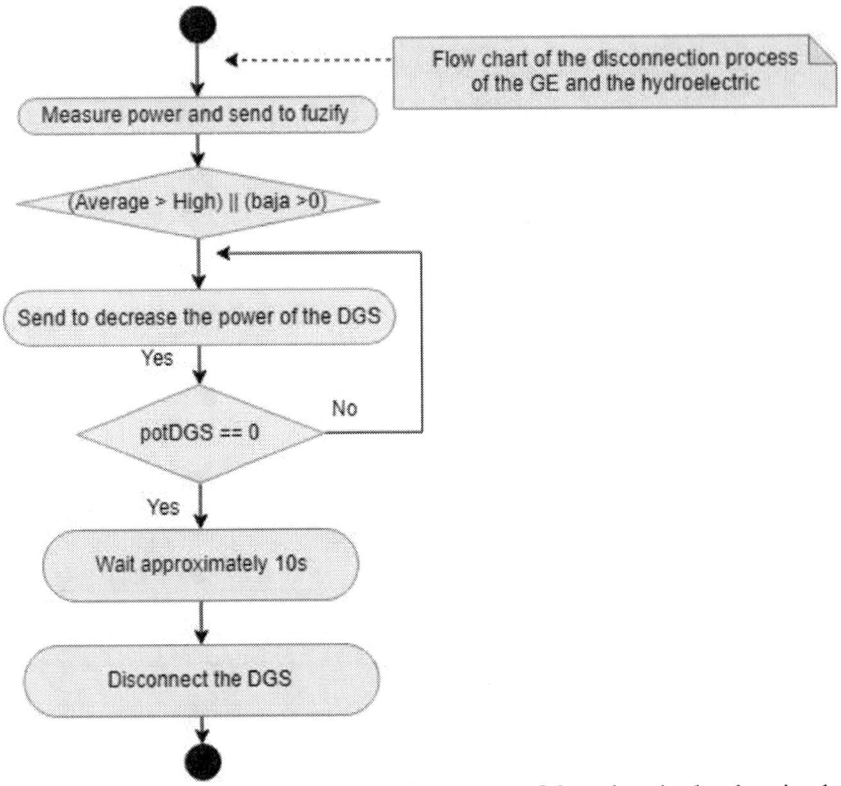

Figure 6. Flow diagram of the disconnection sub-process of the micro hydroelectric plant and the DGS. Source: The authors.

2.4.3. Condition 3: Abundant flow and asynchronous generator and DGS available as auxiliary sources
Connection process: the controller measures and permanently blots the power of the SHP (PotCH). When it reaches a value belonging to the fuzzy set "high" greater than a value of belonging to the fuzzy set "average". The system starts the asynchronous generator and begins to open the water inlet valve to the turbine of the asynchronous generator until it reaches the synchronous speed and then connects it with the SHP, waits 2 seconds and takes it up to 30% of the SHP generation power, remaining these cogenerating. Wait 10 seconds. If the power continues to be high, the DGS is connected and it is verified that the DGS frequency is equal to that of the SHP in ± 0.2 Hz according to [18]. If they are not equal, the system commands to increase or decrease the speed of the DGS to adjust this frequency. If they are equal, it is verified that the phase shift is less than 15° and the DGS is switched on with the SHP, 10 seconds are expected and the power of the DGS is increased to its nominal capacity, thus leaving the asynchronous generator and the DGS cogenerating with the SHP.

Disconnection process: the controller measures and blots always the power of the SHP (PotCH). When it reaches a value belonging to the fuzzy set "average" greater than that of the fuzzy set "high" or a value greater than zero of the set fuzzy "low", the system decreases the power of the DGS to 0%, waits 10 seconds, disconnects it from the SHP and turns off the DGS. Wait 10 seconds. If the power continues to be low, then the system reduces the opening of the water inlet valve to the turbine of the asynchronous generator until it reaches the synchronous speed and disconnects it from the micro SHP and continues closing the valve until the speed of the asynchronous generator is less than 5% of the maximum, leaving only the SHP working.

In manual mode, the operator has a panel through which he can make decisions about the two auxiliary sources.

If it operates on the asynchronous generator through several actuators, it can:

- Turn on / off the asynchronous generator.
- Control the water inlet valve to the turbine, to vary the speed of the asynchronous generator and bring it to synchronous speed and 30% of the generation power of the SHP.
- Connect / disconnect the asynchronous generator in parallel with the hydroelectric power plant.

In the case of the DGS:

- Turn on / off the DGS.
- Measure the phase shift between the HPH and the DGS.
- Increase / decrease the power of the DGS.
- Connect / disconnect in parallel the DGS with the SHP.

3. Results and Discussion

In the coding stage several programmed functions were defined in C language that satisfy all the established functional requirements. Among the main functions can be mentioned:

- void Fuzificar_Potencia: This function receives as a parameter a value of type double, which is equivalent to the power value read from the power transducer and performs the blurring of this value to start making decisions from it.
- void connect_Sinc_Asin: This function receives as parameter an integer value that reflects the state (synchronized / non-synchronized) of the asynchronous generator. (See figure 3).
- void disconnect_Sinc_Asinc: This function receives as parameters two variables of type bool that allow knowing if the asynchronous generator is on and connected to the network, or if it is only on, as well as the third parameter an integer value is passed, which allows making decisions jointly with the function connect_Sinc_Asinc, this variable allows to divert the flow of the program from one point to another. (See figure 4).
- void conecta_Hidro: is responsible for connecting the synchronous generator and the DGS. (See figure 5).
- void disconnect_Hidro_GE (): it oversees disconnecting the synchronous generator and the DGS. (see figure 6).

To carry out the simulation tests of all the functions performed by the controller, the PROTEUS 7.7 professional program is used, which provides the possibility of simulating them in real time and, in addition, programming the Arduino Mega 2560 and representing the tasks it performs.

The system (see figure 7) is composed of the Arduino Mega 2560, an alphanumeric LCD LM044L of 20x4, potentiometers to simulate the output signal of the transducers and the tachometer, 8 LEDs to indicate the status of the relays and the status indicators both asynchronous generator and the DGS. It also has 4 logicprobe to indicate the status of the control relays for closing / closing the water inlet valve to the turbine of the asynchronous generator and to indicate the increase or decrease in the speed of the DGS generator. It also has a switch to clear the alarm if it occurs and 14 logicstate to manually control the different parameters, both the asynchronous generator and the DGS, and to select manual / automatic mode, simulate the flow level and the availability of different sources.

IOP Conf. Series: Materials Science and Engineering **844** (2020) 012017 doi:10.1088/1757-899X/844/1/012017

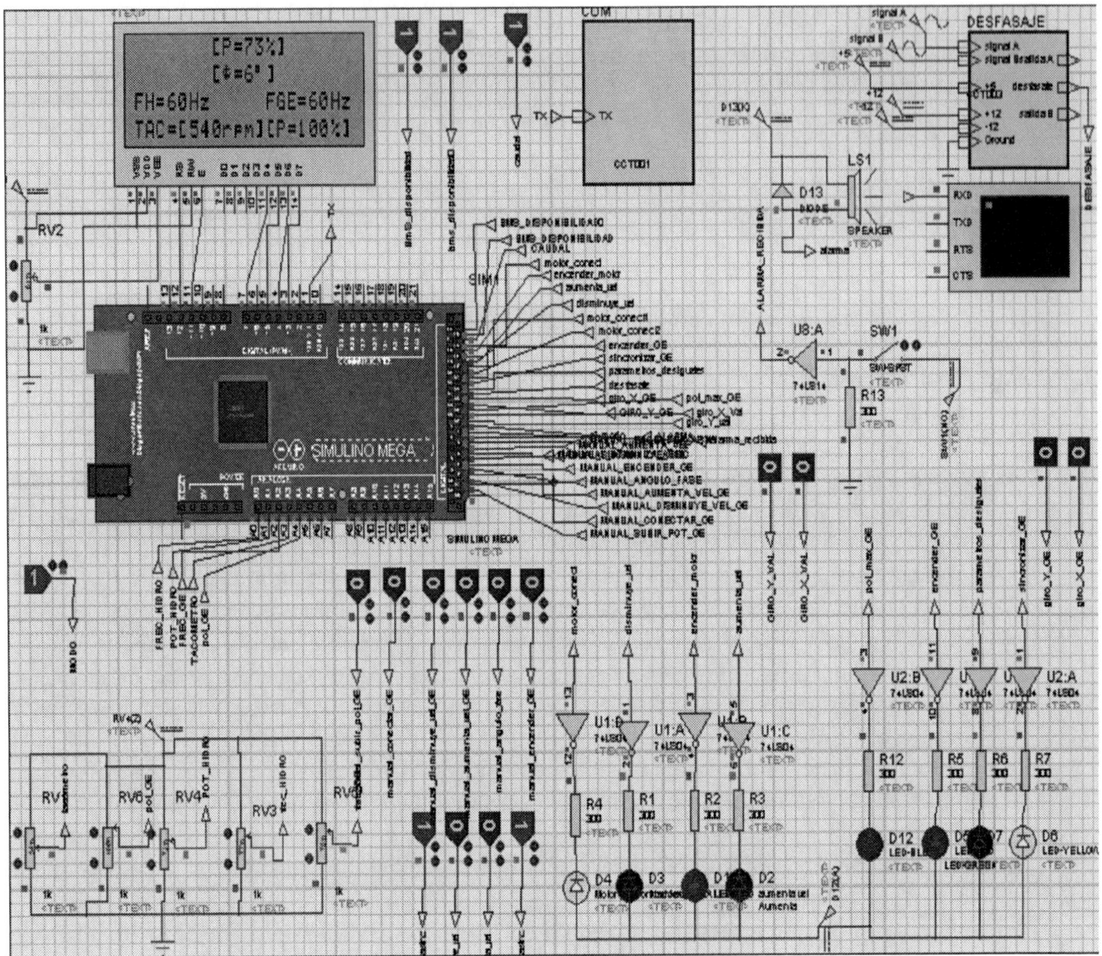

Figure 7. Simulation of the system in Proteus. Source: The authors.

3.1. Simulation of the transducers and the tachometric sensor and visualization of the controlled variables

The simulation of the frequency and power measurements of the DGS and SHP transducers are made with potentiometers, as is the simulation of the tachometric sensor measurement of the asynchronous generator speed. The voltage outputs of these are connected to 5 analogue inputs of the Arduino Mega 2560 (A0- frequency of the SHP, A1- power of the SHP, A2- frequency of the DGS, A3- tachometric sensor and A4- power of the GE).

Measurements of the variables: SHP frequency (FH), DGS frequency (FGE), power (P) of the SHP and DGS power, phase shift (φ) and speed (TAC) are displayed on a 20x4 LM044L LCD.

3.2. Work simulation with the asynchronous generator

For the work with the asynchronous generator, the on/off simulation is simulated, using a red LED connected to pin 26 of the Arduino. To simulate the connection in parallel with the SHP, a yellow LED connected to pin 25 is used. To indicate the increase/decrease of the speed of the motor, two LEDs are used, one green connected to the pin 27 which indicates if it is necessary to increase the speed and a blue LED connected on the pin 28 which indicates that it is necessary to decrease the speed of the generator, to bring the asynchronous generator to the synchronous speed and to increase or decrease the power generated by it. These two indicators work in conjunction with 2 logicsprobe connected to pins

38 and 39, these logicsprobe are used to simulate the relays that control the opening and closing of the water inlet valve to the turbine of the asynchronous generator.

Figure 8 shows the simulation test of the work with the asynchronous generator. The red LED indicates that the asynchronous generator is on, the yellow LED indicates that the generator is connected to the network. The green LED indicates that the speed of the asynchronous generator must be increased to take it to 30% of the generation power of the micro hydroelectric plant and the logicsprobe with the combination -01- indicate the activation of the relay, which allows the opening of the valve of water input to the turbine to increase the power of the asynchronous generator.

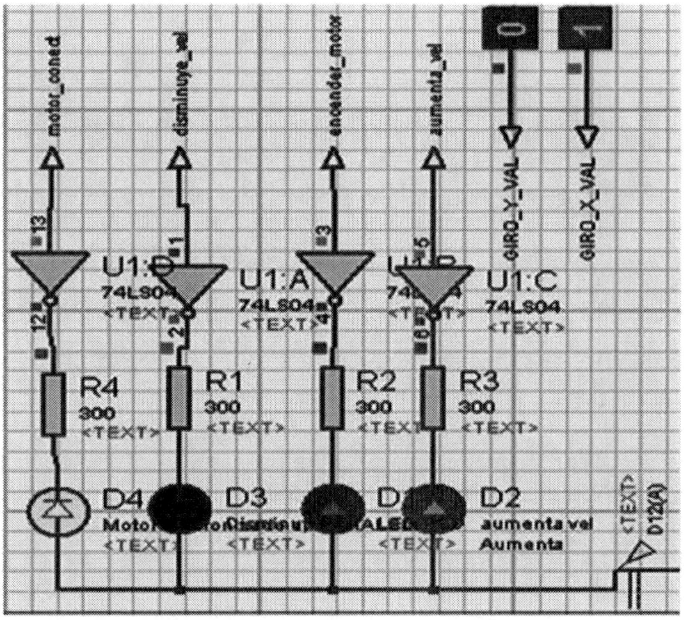

Figure 8. State of the indicators and relays that control the asynchronous motor. Source: The authors.

3.3. Connection of the auxiliary relays with the arduino for the parallel connection of the asynchronous generator with the synchronous generator

For the connection of the asynchronous generator with the synchronous generator, the use of power relays is necessary. The modules admit high intensity peaks, and consequently, have enough reserves to activate the connected fuses. They also resist capacitive and inductive interferences without problems.

3.4. Connection of relays for three-phase reversible motor drive

Figure 9 shows the connection diagram used in the drive control of the three-phase reversible motor that regulates the valve that controls the flow of water entering the turbine that moves the asynchronous generator.

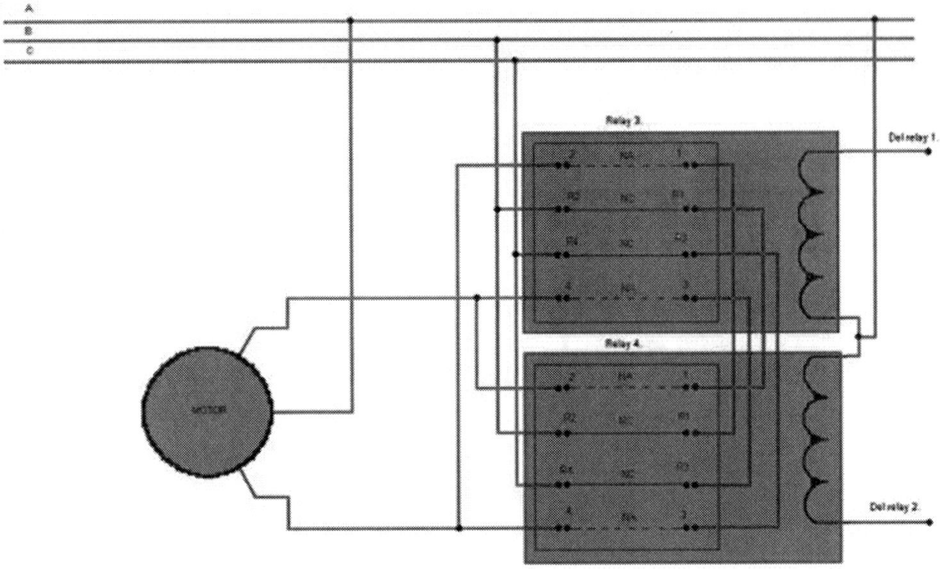

Figure 9. Connection diagram of the relays for control of rotation of the three-phase reversible motor. Source: The authors.

The operating principle of the motor drive is as follows:

The Arduino has two digital outputs (pins 38 and 39) responsible for connecting the relay system to control the rotation of the three-phase reversible motor.

To achieve a correct performance of the motor operation, two relays, model MULTI9, were selected as final stage, which have four contacts, two normally open and two normally closed. Its connection is shown in figure 9, where it is observed that the motor drive is only carried out if one of the relays is activated and the other is not, that is, if one of the coils is energized and the other is not. In case the coils of relays three and four, respectively, are energized or de-energized, the motor remains at rest. For the motor to rotate one way or the other, only one of the coils, relay coil three or relay coil four must be energized, so that the normally open contacts close, and the normally closed contacts open, allowing the start and start-up of the engine.

To activate relays three and four (see figure 9), the drive coils are connected to lines B and C of the network, and through two micro relays (see figure 10), relays three and four are activated. These micro relays are activated by the control signal coming from the Arduino (pins 38 and 39), according to the state (code) programmed according to the following rule:

- Code 00: Micro relays one and two are deactivated, therefore relays three and four are in normal state. Three-phase motor at rest.
- Code 11: Activated micro relays one and two, therefore in activated state relays three and four. In this state the normally open contacts close, and the normally closed contacts open. Three-phase motor at rest.
- Code 01: Micro relay one is activated, therefore the coil of relay three is energized and relay four is kept deactivated, therefore the normally closed contacts of relay three are opened and the normally open contacts are closed. In this condition the three-phase motor is energized and rotates in one direction.
- Code 10: The micro relay two is activated, therefore the coil of relay four is energized and relay three is kept deactivated, therefore the normally closed contacts of relay four are opened and the normally open contacts are closed. In this condition, the three-phase motor is energized and rotated in the opposite direction.

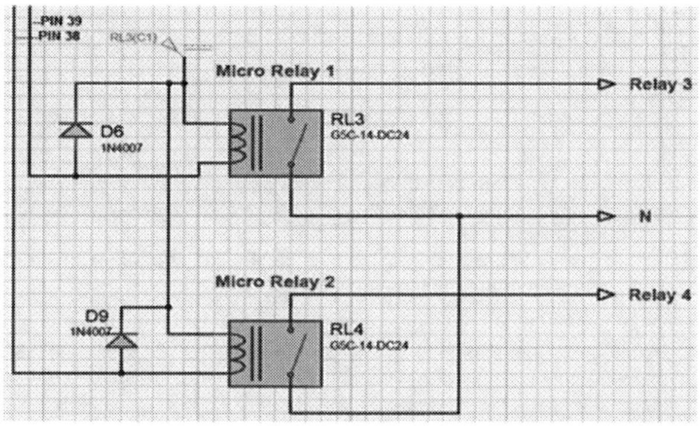

Figure 10. Connection diagram of micro relays 1 and 2. Source: The authors.

3.5. *Simulation of work with the DGS*

For the work with the DGS, the on / off simulation is simulated using a red LED connected to pin 31 of the Arduino; To simulate the connection in parallel with the SHP, a yellow LED connected to pin 32 is used. To indicate the inequality between the parameters of frequency and phase angle measured from the SHP and DGS, a green LED connected to pin 33 is used, when being on reflects the inequality between these parameters. Furthermore, when the blue led connected to pin 37 is on it indicates that the DGS is generating at its maximum power, and the 2 logicsprobe indicate the increase or decrease of the DGS speed to increase its power generation.

Figure 11 shows the simulation performed of the DGS operation. The red led indicates that the DGS is on, the green LED off indicates that the frequencies measured in the central micro and the DGS coincide and that these are between 60 ± 0.2 Hz and that in addition the measured phase shift between the DGS and the micro central does not exceed 15° as recommended [18]. The yellow LED indicates that the DGS is connected to the network. The logicsprobe with the combination -01- indicate the increase of the speed to bring the DGS to its maximum power generation and as this is not yet at the maximum of its power, then the blue LED remains off, because this it only turns on when the DGS is generating at maximum power.

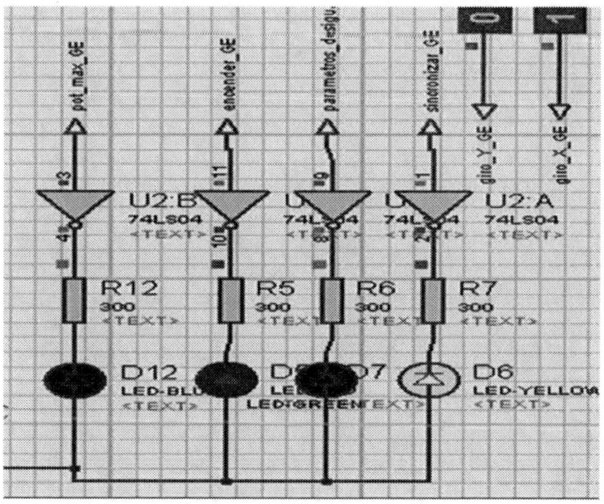

Figure 11. State of the indicators and relays that control the generator set. Source: The authors.

3.6. *Calculation of the phase angle*

For the phase measurement, two sinusoidal signals, whose phase shift between them is to be measured, are converted into rectangular pulses. The separation between the pulses of the signals is equal to the phase shift between them. In figure 12 two sinusoidal waves are shown offset by an angle Δt.

IOP Conf. Series: Materials Science and Engineering **844** (2020) 012017 doi:10.1088/1757-899X/844/1/012017

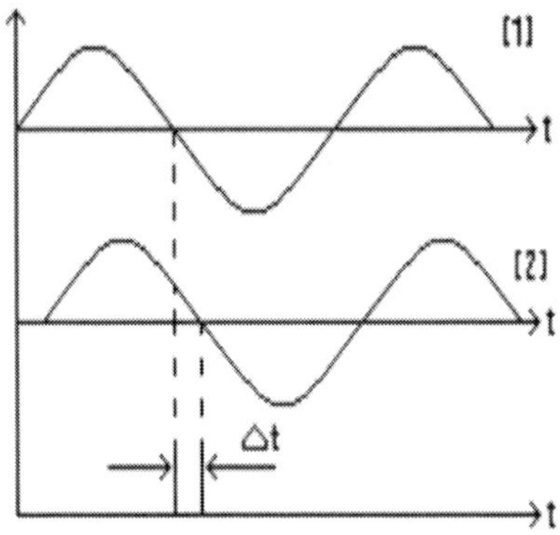

Figure 12. Graph of two signals offset in time. Source: [19].

The phase can be expressed as:

$$\varphi = w\Delta t \qquad (5)$$

Where w is the angular velocity, in rad-sec, and Δt is the time, in seconds, between the zero crossing of signals 1 and 2.

$$\Delta t = \frac{\varphi}{w} \qquad (6)$$

The time difference Δt is directly proportional to the phase shift φ and inversely proportional to the angular velocity w, so that an electric circuit is formed that generates a rectangular pulse of duration Δt [20, 21]. The circuit shown in Figure 13 is used to convert sine waves into rectangular waves.

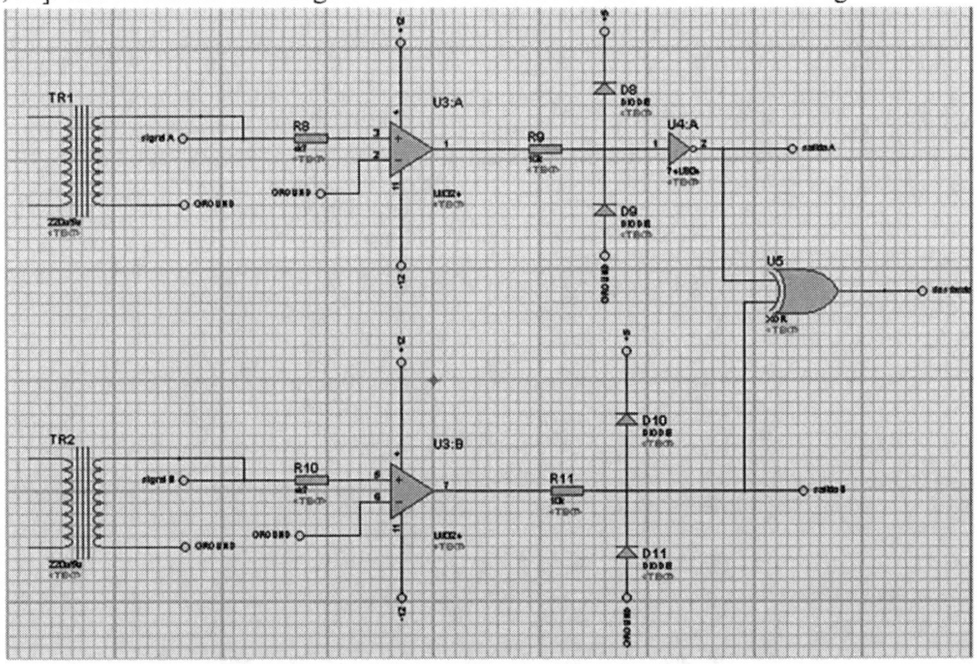

Figure 13. Circuit for phase measurement. Source: The authors.

Firstly, the voltage value of a phase of the micro hydroelectric plant and one phase of the DGS are transformed to voltage values of 5V by the transformers TR1 and TR2. Then, the signal is sent to the comparators that are formed by the operational amplifiers LM324, and in the comparators the sinusoidal waves are converted into rectangular waves. These square waves maintain the same phase shift as the sinusoidal waves.

IOP Conf. Series: Materials Science and Engineering **844** (2020) 012017 doi:10.1088/1757-899X/844/1/012017

The signal offset at a given angle is taken to a NOT gate, integrated circuit 7404, where it is inverted and then an exclusive sum is made by an XOR gate with the reference signal, at the gate exit a proportional rectangular pulse is obtained at the phase shift, see figure 14. This pulse is taken to the Arduino through pin 34 and is captured by the PulseIn function where the time of the impulse generated by the circuit is obtained and it is taken to the value of the phase shift.

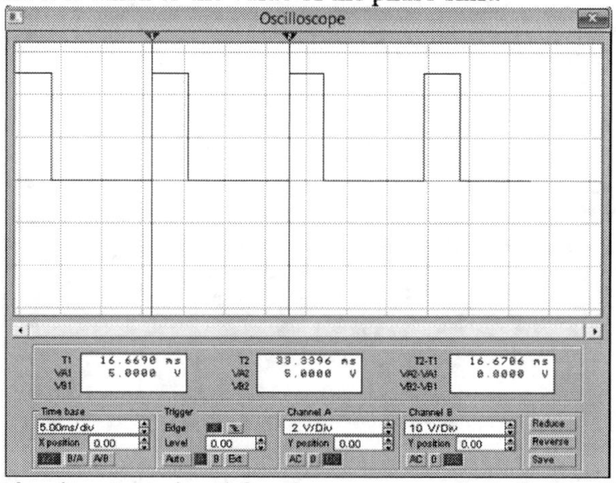

Figure 14. Circuit output signal for phase measurement. Source: The authors.

3.7. *Work mode selection*

The designed controller allows two modes of operation: manual and automatic. These modes can be selected through a connected logicstate on pin 42, a value of "0" written on this pin selects manual mode and a value of "1" written selects automatic mode.

For the operation in manual mode, there is a set of logicstate that allows the control of both the asynchronous generator and the DGS. For control of the asynchronous generator, there are four logicstates that allow the following operations:

- Turn the generator on/off (pin 43).
- Increase the generation power of the asynchronous generator and therefore the opening of the water inlet valve to the turbine (pin 44).
- Decrease the generating power of the asynchronous generator and therefore close the water inlet valve to the turbine (pin 45).
- A combination on the pins 44 and 45 of -01- indicates increasing the opening of the valve and a combination of -10- indicates decreasing the opening of the valve. A combination of -00- or -11- on these pins would leave the valve at rest, and a generator connected to pin 46 is used to connect the generator to the grid.

For control of the DGS there are five logicstates that allow the on / off (pin 47), calculate the phase shift (pin 48), increase / decrease the speed of the DGS to control its frequency or take it to the maximum / minimum of its power, pin generation 49 and 50, a combination of -01- on these pins indicates the increase in the generator speed of the DGS and a combination of -10- indicates the decrease in speed, any other combination would have no effect. To connect the DGS to the network, the logicstate connected to pin 51 is used.

Figure 15 shows the simulation of the operation in manual mode of the asynchronous generator and the DGS, in it the state of the indicators described above for the control of each of the auxiliary sources of power generation is reflected.

IOP Conf. Series: Materials Science and Engineering **844** (2020) 012017 doi:10.1088/1757-899X/844/1/012017

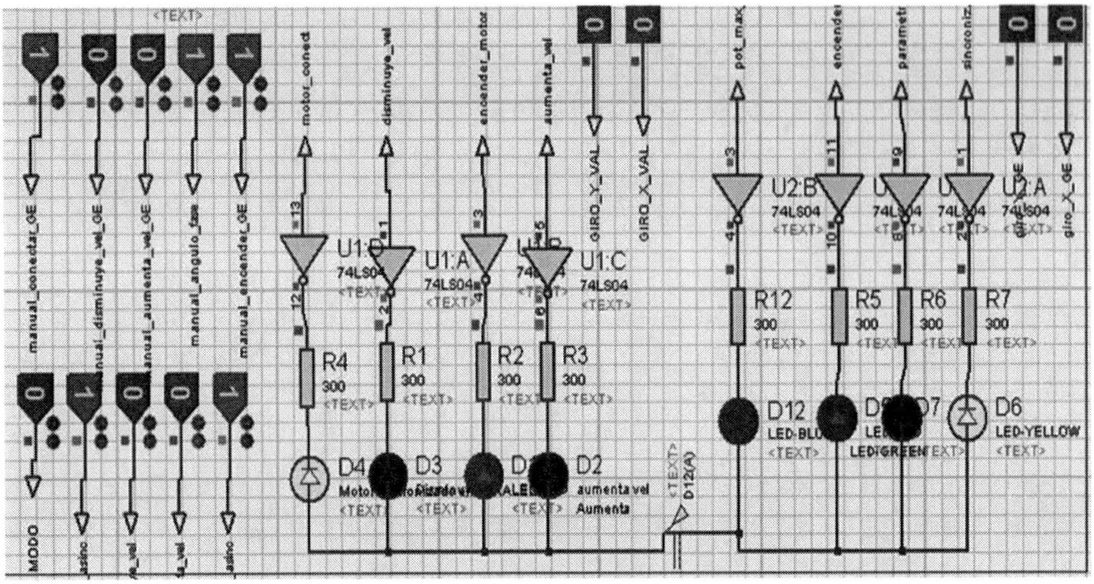

Figure 15. Operation in manual mode of the asynchronous generator and generator set. Asynchronous generator on (red led on the left) and connected to the grid (yellow led on the left) and DGS on (red led on the right), connected to the grid (yellow led on the right) and generating at maximum power (blue led). Source: The authors.

For operation in automatic mode, there are three logicstates that are used to indicate the flow level (pin 24). A value of "0" written on this pin indicates insufficient flow, which makes the use of the asynchronous generator as an auxiliary source impossible of generation and a value of "1" written on this pin indicates that the flow is enough which allows the asynchronous generator to be used as an auxiliary generation source.

The logicstate connected to pins 22 and 23 are used to select the three working modes that the controller allows in automatic mode. These modes are derived from the auxiliary sources available to assist in the generation, so a combination on these -00- pins indicates that neither the asynchronous generator nor the DGS are available to cogenerate with the SHP. A combination of -01- indicates that it can only cogenerate with the SHP the asynchronous generator; a combination of -10- indicates that only the DGS can cogenerate with the SHP, and a combination of -11- indicates that both the asynchronous generator and the DGS can be used as auxiliary sources for generation. Working in automatic mode, the controller makes use of all the rules defined using fuzzy logic for the decision making of the sources that are needed in the generation.

In the event that the operator selects the automatic mode and indicates that the flow is insufficient and selects a combination of -01- (on pins 22 and 23), it would select the asynchronous generator as an auxiliary source and because the reservoir flow rate it is not abundant, then an alarm is activated (pin 40) indicating to the operator that the asynchronous generator cannot be used because the flow is low.

4. Conclusion

In this work, the simulation of the operation of an electrical generation system is obtained using the hydraulic energy that guarantees the supply of energy to an isolated community of the National Electrical System, verifying the proper functioning of the algorithm developed based on the fuzzy logic for the connection and disconnection of the generating sources, as well as the correct synchronization between these sources.

An automatic control system is designed for a hybrid electric generation system that operates in an isolated operation regime, consisting of a hydroelectric central (SHP) with synchronous and asynchronous alternators and a generator, taking advantage of the potential offered by the Arduino Mega platform 2560, which simplifies the hardware.

A control algorithm is selected that makes system installation flexible to SHP with different hydraulic and technological characteristics.

The tests are performed by simulation using Proteus 7.7, which shows the correct functioning of the design.

The design made allows the use of other controllers such as PLCs or PIC controllers.

References

[1] Noa A and Montero R. 2017. Comportamiento operacional de grupos electrógenos: particularidades del índice de consumo específico de combustible. *Ingeniería Mecánica*. vol. 21. No. 1. pp. 19-27

[2] Núñez Alvarez, José Ricardo et al. 2019. Metodología de diagnóstico de fallos para sistemas fotovoltaicos de conexión a red. *Revista Iberoamericana de Automática e Informática industrial*, [S.l.], sep. 2019. ISSN 1697-7920. Disponible en: <https://polipapers.upv.es/index.php/RIAI/article/view/11449>

[3] Andújar J. M and Barragán A. J. 2014. Hibridación de sistemas borrosos para el modelado y control. *RIAI-Revista Iberoamericana de Automática e Informática Industrial*. vol. 11, núm. 2, pág. 127-141

[4] Beleño K. J and Berrio J. S. 2013. Diseño de una smart grid para un sistema híbrido de energía. *Prospect*. vol. 11, No. 2, págs. 94-101

[5] Guerrero J. W, Toscano A, Pacheco L. V, Tovar J. O. 2018. Analysis of the Energetic and Productive Effects Derived by the Installation of a Conveyor Belt in the Metal-mechanic Industry. *International Journal of Energy Economics and Policy*. 8(6), 196-201. DOI: 10.32479/ijeep.7066

[6] Valencia G, Nuñez, J and Acevedo C. 2019. Research Evolution on Renewable Energies Resources from 2007 to 2017: A Comparative Study on Solar, Geothermal, Wind and Biomass Energy. *International Journal of Energy Economics and Policy*. 9(6), 242-253. DOI: https://doi.org/10.32479/ijeep.8051

[7] Colectivo de autores. 2017. Cogeneración de energía, eléctrica y térmica, mediante un sistema hibrido biomasa-solar para explotaciones agropecuarias en la isla de Cuba. *Editorial CIEMAT*, Madrid, España

[8] Fong J, Domínguez H, Benítez I and Nuñez J. R. 2018. Smart Grid proposal in communities of Guama Municipality of Santiago de Cuba Province. *Journal of Engineering and Technology for Industrial Applications*. 2018. Vol. 04. N° 13, pp 61-69

[9] Real Calvo, Rafael et al. 2017. Sistema Electrónico Inteligente para el Control de la Interconexión entre Equipamiento de Generación Distribuida y la Red Eléctrica. *Revista Iberoamericana de Automática e Informática industrial*. v 14, n. 1, p. 56-69. ISSN 1697-7920

[10] Marín J. D. 2103. Estudio del Control de una Pequeña Central Hidroeléctrica para la Operación por Micro Redes en el Sistema de Distribución Local. *Facultad de Ingeniería y Arquitectura*. Universidad Nacional de Colombia. Manizales. Colombia

[11] Barrero F, Milanés M. I, González E, Roncero C and González P. 2015. El Control de Frecuencia y Potencia en los Sistemas Eléctricos Multitarea. Revisión y Nuevos Retos. *Revista Iberoamericana de Automática e Informática Industrial*. vol. 12, núm. 4, pág. 355-500. DOI: 10.1016/j.riai.2015.07.001

[12] Salazar A and Fong J 2004. Mediciones Eléctricas. Empresa Editorial Poligráfica Félix Varela. Edición 2004-06-06. ISBN 978-959-258-758-8

[13] Asprilla D. B. 2016. Estudio de Sistemas Híbridos de Energía Renovable (solar – gasificación de biomasa) como alternativa para satisfacer necesidades energéticas en Zonas no Interconectadas del Departamento del Chocó. Tesis de maestría para optar al grado de Máster en Ingeniería Mecánica. Facultad de Minas. Universidad Nacional de Colombia. Medellín, Colombia

[14] eFLL - A Fuzzy Library for Arduino and Embeded Systems. Disponible en https://blog.zerokol.com/2012/09/arduinofuzzy-fuzzy-library-for-arduino.html

[15] Espino A. 2017. Control de Temperatura con lógica difusa para un sistema de espectroscopia láser. Trabajo de diploma en opción a título de Ingeniero Electrónico. Facultad de Ingeniería. Universidad Nacional Autónoma de México
[16] Barragán A. J, Al-Hadithi B. M, Andújar J. M and Jiménez A. 2015. Metodología formal de análisis del comportamiento dinámico de sistemas no lineales mediante lógica borrosa. *Revista Iberoamericana de Automática e Informática Industrial RIAI*. 12 (4), 434–445
[17] Valencia G, Nuñez J and Duarte J. 2019. Multiobjective Optimization of a Plate Heat Exchanger in a Waste Heat Recovery Organic Rankine Cycle System for Natural Gas Engines. *Entropy*. 21(7), 655
[18] Rodríguez S and Sarmiento A. 2015. Competitividad de los sistemas híbridos eólicos-fotovoltaicos para la electrificación rural. *Ingeniería Mecánica*. vol.18 no.1. ISSN on-line 1815-5944
[19] Mikati M, Santos M and Armenta C 2012. Modelado y Simulación de un Sistema Conjunto de Energía Solar y Eólica para Analizar su Dependencia de la Red Eléctrica. *Revista Iberoamericana de Automática e Informática Industrial-RIAI*. vol. 9, núm. 3. pp. 267-281
[20] Bordons C, García F and Valverde L. 2015. Gestión Óptima de la Energía en Microrredes con Generación Renovable. *Revista Iberoamericana de Automática e Informática industrial*, [S.l.] vol. 12, n. 2, p. 117-132
[21] Guerrero J. W, Toscano A, Pacheco L. V and Tovar J. 2018. Analysis of the Energetic and Productive Effects Derived by the Installation of a Conveyor Belt in the Metal-mechanic Industry. *International Journal of Energy Economics and Policy*. 8(6), 196-201

IOP Conf. Series: Materials Science and Engineering **844** (2020) 012018 doi:10.1088/1757-899X/844/1/012018

A proposal for updating the Response Spectrum of San Jose de Cucuta, Colombia

C. H. Florez[1] J F Marquez[2] C F Lozano[3]

[1]Universidad Francisco de Paula Santander. Grupo de Investigación en Geotecnia Ambiental, GIGA. carloshumbertofg@ufps.edu.co
[2]Universidad Francisco de Paula Santander. Grupo de Investigación en Estructuras, POLIMATAS. jorgefernandomp@ufps.edu.co
[3]Universidad de Santander. Grupo Ambiental de Investigación Aplicada, GAIA. ca.lozano@mail.udes.edu.co

Abstract. An analysis of seismic signals captured in three stations of the network of accelerographs of the Colombian Geological Service (SGC, by its acronym in Spanish) was developed for the city of San Jose de Cucuta, Colombia. Earthquake Equivalent, Response Analysis (EERA) and DEGTRA4 software were used for analysis, processing and calculation of signal response spectra during the characterization and propagation modelling of the signal by profiles. For the accelerograms family, a normalization factor computed with respect to the Peak Ground Acceleration (PGA) was obtained by superposition against the typical design spectrum of the Colombian earthquake resistant code (NSR10). It was found that there is a band of frequencies above the maximum acceleration zone of the design spectrum of the current standard. This suggests a revision of the seismic threat of the city of Cucuta and the urgent need to develop the seismic microzoning, which would allow a substantial reduction of the seismic risk.

1. Introduction

In 1875 San Jose de Cucuta (in this work, Cucuta city) was destroyed by a strong earthquake and since then has endured successive periods of seismic crisis. This event killed almost 400 people and is considered as one of the most powerful in Colombia's seismic history. It is of paramount importance to study the seismic threat of Cucuta city because of its history and the high probability that a new crisis can come from the activity of the nearby field sources fault lines that surround the city. Cucuta city and its Metropolitan Area have a population around one million inhabitants and is the main border crossing between Colombia and Venezuela [1]. High hazard and high exposure are naturally combined with the high vulnerability of buildings to produce a high seismic risk. Then, to improve knowledge about seismic risk is necessary to study seismic hazard. To obtain information of the local seismic hazard the response spectra are very useful. They can be used by the engineers to determine the frequencies range that could demand a structural system during the occurrence of a random or deterministic load. The displacement, velocity or acceleration values of an oscillatory system subjected to the action of an external load are analytically assessable if the function that represents the load is a mathematical function [2] for any possible system of a degree of freedom [3]. If the external load is random (such as an earthquake) the displacements must be computed using numerical techniques. Once response spectra are obtained, digitized signals consisting of a pair of values (time, acceleration) are available. The solution to the problem of dynamic equilibrium of an oscillator is given in terms of a homogeneous solution. The ground movement imposes a displacement to the oscillator mass and an inertial force is generated because of the system stiffness. Such force is continuously modified due to the system damping. Response spectra obtained from real earthquakes are highly irregular and their shapes reflect details of the specific frequency contents and phasing [4]. The concept of response spectrum developed by Beniof (1934), Biot (1941) and Housner (1952) lead to a better evaluation of the (random) seismic response on the simple oscillator. Combination of multiple response spectra lead to the construction of the so-called "elastic spectra" which are commonly used in the Resilient Earthquake Construction Standards and are valid for a given structural damping value (critical fraction) and a certain soil profile. In a typical elastic spectrum, the acceleration level for short periods is determined by the seismic zone while for long periods it is determined by the soil profile. Its construction is usually derived from the effective amplitudes of ground movement and pseudo values [5] such as: a) maximum acceleration, velocity and terrain displacement; b) period of the structural system; c) maximum displacement of the mass relative to the base; d) maximum pseudo velocity and maximum pseudo acceleration. In this paper, analysis and conclusions related to response spectra obtained for three specific sites of Cucuta city are presented. Their dynamic characteristics have been determined from widely accepted empirical correlations in modern engineering.

2. Materials and methods.

2.1. Input information for the model

The information of seismic signals was captured using three accelerograph stations located in three different points of Cucuta city: Profile 1 (CUCC1), profile 2 (CUCC2) and profile 3 (CUCC3). Profiles 1 and 3 correspond to a deposit of rock (weathered arcillolite) of Oligocene tertiary age with a power or depth of 30 meters. Profile 2 corresponds to a very dense soil or soft rock with an approximate depth of 180 meters located on the limnological line of the Pamplonita River, old quaternary. Shear wave velocity varies from 360 m/s to 760 m/s for Profile 2 and from 180 m/s to 360 m/s for Profiles 1 and 3. EERA program was used to determine the dynamic parameters of each profile. DEGTRA4 program was used to determine the response spectra for each profile using the signal recorded in the respective accelerographs.

2.2. Determination of the response spectrum

The general solution of the mathematical system of a simple oscillator in an elastic state can be expressed as two parts: that associated to a homogeneous response, $u(t)_h$ plus a complementary part, $u(t)_c$. The external load is considered as an effective inertial load, so the solution corresponds to that of a homogeneous equation of the form established in equation [2]

$$u(t)' = e^{-\beta t'} \int p(\tau) \sin \omega_d t' d\tau \qquad (1)$$

Where $p(\tau)d\tau$ represents an impulse force that acquires the mass of the system for a succession of impulses. $\omega_d = \omega(1-\xi^2)^{1/2}$ is the damped frequency that will depend on the damping with respect to the critical, ξ; $\beta = \xi\omega$; Equation (1) can be written as:

$$u(t') = -\left(\frac{1}{\omega_d}\right)\int e^{-\xi\omega(t-\tau)}\, \ddot{u}_g(\tau)\left[\sin\omega_d t \cos\omega_d \tau - \cos\omega_d t \sin\omega_d \tau\right] d\tau \qquad (2)$$

Once the seismic signal is digitized, pairs of values $\left[\ddot{u}_{gi}, \tau_i\right]$ are obteined. Suchs values will conform the basis of the numerical integration of the equation, that is, the acceleration that is a random value and the time that is a deterministic value. Let $t' = t - \tau$ a convolution condition that allows to express the equation (2) as equation (3):

$$S_d = \left| \frac{1}{\omega_d} \int_0^t e^{-\xi\omega(t-\tau)} \ddot{u}_g(\tau) \sin \omega_d (t-\tau) d\tau \right|_{\max} \quad (3)$$

Where S_d is the spectral displacement. By deriving the equation (3) the pseudo-spectral velocity, S_v is obtained:

$$S_v = \left| \int_0^t e^{-\xi\omega(t-\tau)} \ddot{u}_g(\tau) \cos \omega_d (t-\tau) d\tau + \frac{\xi}{(1-\xi^2)^{1/2}} \int_0^t e^{-\xi\omega(t-\tau)} \ddot{u}_g(\tau) \sin \omega_d (t-\tau) d\tau \right|_{\max} \quad (4)$$

By deriving equation (4) the spectral acceleration, S_a is obtained:

$$S_a = \left| \omega(1-2\xi^2 / (1-\xi^2)^{1/2} \int_0^t e^{-\xi\omega(t-\tau)} \ddot{u}_g(\tau) \sin \omega_d (t-\tau) d\tau + 2\xi\omega_d \int_0^t e^{-\xi\omega(t-\tau)} \ddot{u}_g(\tau) \cos \omega_d (t-\tau) d\tau \right|_{\max} \quad (5)$$

The above results can be summarized as follows, equation (6):

$$S_d = \left| \frac{1}{\omega_d} \int_0^t e^{-\xi\omega(t-\tau)} \ddot{u}_g(\tau) \sin \omega_d (t-\tau) d\tau \right|_{\max}$$

$$S_v \approx \left| \int_0^t e^{-\xi\omega(t-\tau)} \ddot{u}_g(\tau) \cos \omega_d (t-\tau) d\tau \right|_{\max} \quad S_a = \left| \omega_d \int_0^t e^{-\xi\omega(t-\tau)} \ddot{u}_g(\tau) \sin \omega_d (t-\tau) d\tau \right|_{\max} \quad (6)$$

Summarizing:

$$S_v \approx \omega S_d$$
$$S_a = \omega^2 S_d = \omega S_v$$

The record of earthquakes constitutes, unequivocally, the most important tool in seismic engineering to study the energy released from a seismogenic source and its effects on civil structures [6, 7, 8]. Equations (6) allow to relate the earthquakes records with the structural features and with the soil deposits properties. Such equations are useful to compute the maximum response values in the oscillators (structures) produced by the land movements which are in turn caused by the propagation of seismic waves in the soil [9,10].

3. Results and Discussion

According to the Colombian earthquake resistant code (NSR10) the local effects of seismic response must be evaluated based on the soil profiles. To do so, NSR10 states some procedures and parameters to use. In this work, a comparison between maximum values obtained from NSR10 and those resultant from the application of the procedure described in the precedent section has been made. Figures 1, 2 and 3 show the result obtained for the normalization factor relative to Peak Ground Acceleration (PGA) for stations CCUC1, CCUC2 and CCUC3 normalized response spectra. The black band in the figures corresponds to the design spectrum of the NSR10.

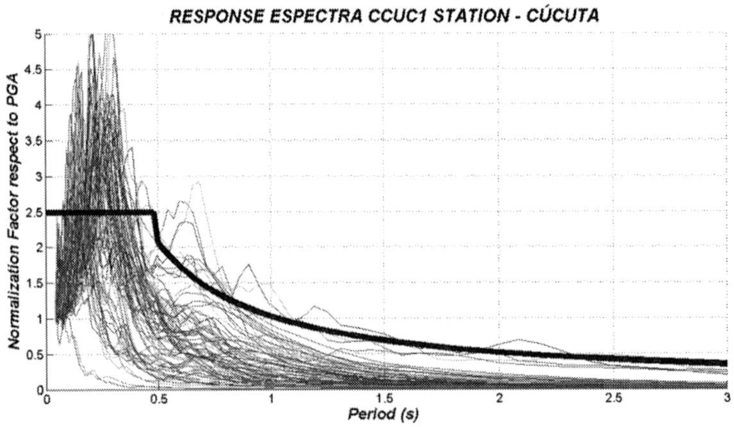

Figure 1. Normalized response spectra from CCUC1 data.

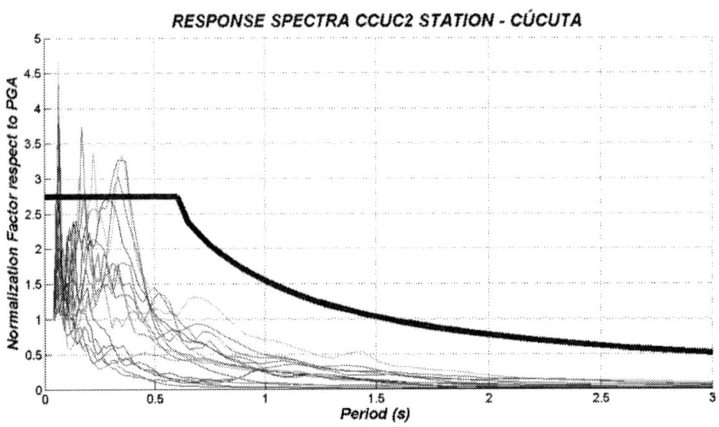

Figure 2. Normalized response spectra from CCUC2 data.

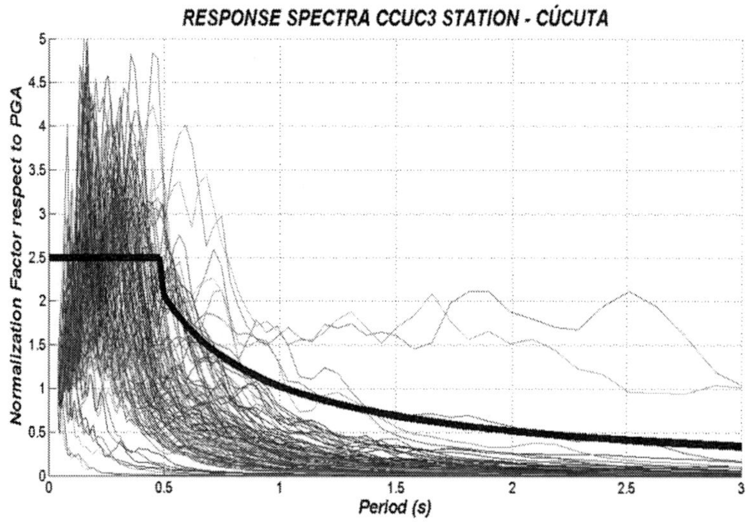

Figure 3. Normalized response spectra from CCUC3 data.

4. Conclusions

In the three studied sites, a high concentration of frequencies within a range from 5 to 6 Hz (periods from 0.17 to 0.20 s) is observed in figures 1, 2 and 3. Indeed, this range delimits the frequency characteristics of each earthquake family. The dominant frequency content in the signal can be linked to a frequencies band delimited between 6 and 7 Hz. This range is probably highly influenced by the soil profile characteristics, i.e. is mainly defined by the geomechanical characteristics and geometry of rigid soil deposits. The response spectra show an important amplification when period is low ($0.1 < T < 1.0$ s). This suggests that for buildings with less than five floors could be in high risk if an earthquake with magnitudes between 6.0 and 7.0 occurs. When comparing the family of response spectra for the three soil deposits against the profile type of the elastic design spectrum of the NSR10 a band of spectral accelerations is clearly shown that are above the range of short periods of the spectrum. This suggests that a rigorous review should be made of the site effects that would be occurring in the city's soil profiles. Hence the imperative need to conduct a seismic microzoning study.

References

[1] La Opinión 2018 En: *https://www.laopinion.com.co/economia/703-mil-personas-hay-en-cucuta-dane-165484#OP*
[2] Sarria, A. 1995 *Ingeniería Sísmica.* (Segunda Edición, Ediciones Uniandes) p 569
[3] Paz M and Leigh W 2004 *Structural Dynamics Theory and Computation* (Kluwer Academic Publishers) p 812
[4] Kramer S L 1996 *Geotechnical Earthquake Engineering Prentice-Hall international series in civil Engineering and Engineering mechanics* (Upper Saddle River, NJ: Prentice-Hall) 653 p.
[5] Hernández Julio, J 2010 *Desempeño estructural y Espectros de Respuesta*
[6] C. Ruiz Esparza M. G. y Aguirre G. J. 2002 *Obtención de espectros de respuesta considerando efectos de sitio obtenidos de registros por medio de la utilización de algoritmos genéticos. Instituto de Ingeniería* (UNAM. México, DF. C.P. 04510)

[7] Florez G, C. y Lozano L C 2007 *Análisis y tratamiento de señales de fuentes sismogénicas de campo cercano a San José de Cúcuta, Colombia* (Revista Respuesta. Año 12 No 2 ISSN 0122820X)
[8] García R L E 1998 *Dinámica estructural aplicada al diseño sísmico* (Bogotá: Universidad de los Andes) p 574
[9] Das Braja M 1993 *Principles of soil dynamics* (Brooks/Cole) p 570
[10] Chopra A K 2014 *Dinámica de estructuras* (Pearson Educación de México) p 944

IOP Conf. Series: Materials Science and Engineering **844** (2020) 012019 doi:10.1088/1757-899X/844/1/012019

Potential Energy savings in compressed air systems in industrialized cities. A case study in Barranquilla and Cartagena

Hernan Hernandez Herrera[1], Jorge Iván Silva-Ortega[2], Vicente Leonel Martínez Diaz[1], Aliana Gomez Pons[3], Zaid García Sanchez[4].

[1]Universidad Simón Bolívar, Facultad de Ingenierías, Barranquilla, Colombia.
[2]Research Group in Energy Optimization GIOPEN - Universidad de la Costa, Barranquilla Colombia.
[3] Accounts Department, Universidad de Cienfuegos, Cuba- e-mail:
[4]Center of Energy and Environmental Studies Department, Universidad de Cienfuegos, Cuba.

hernan.hernandez@unisimonbolivar.edu.co

Abstract. The increase of energy consumption, global warming, resource depletion and the rise of policies focused on climate change and greenhouse gas emissions reductions, have promoted to countries and industries the implementation of strategies focused on increase energy efficiency and reduce GHG emissions. Compressed Air Systems (CASs) are one of the most widespread systems used in industry. In countries such as China, USA, Australia, France and Italy, CASs accounts around 10% of the overall electricity costs. In Colombia the energy used in the industrial sector, rise the 33 % of the total energy consumption, equal to 481.429 TJ/year; the electricity consumption is a 13 % of this value, equivalent to 13,3 TWh/year. This paper determine the potentials energy saving of CASs for two industrial cities of the Colombian Caribbean Coast region, showing that there is a high energy saving potential, around 50 GWh/year and a reduction of CO_2 emissions of 10,702 tons of CO_2/year, which can be taken in consideration by the government and organizations to develop projects focused on reduce energy consumption and mitigate CO_2 emissions.

1. Introduction

Worldwide energy consumption will increase 1.2% per year to 2040, being the industrial sector the biggest consumer, with about 54% of the total energy consumption [1, 2]. This growth is led by gas consumption and electricity uses as it is shown in Figure 1.

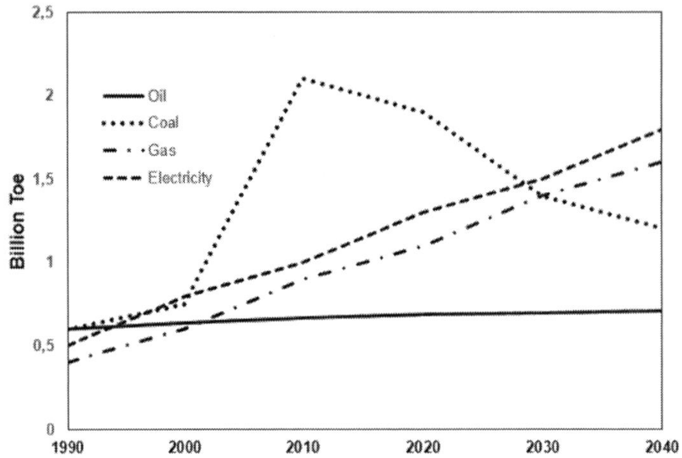

Figure 1. Worldwide energy consumption in industry by fuel

In Colombia the energy used in the industrial sector, rise the 33 % of the total energy consumption, equal to 481.429 TJ/year; the electricity consumption is a 13 % of this value, equivalent to 13,3 TWh/year [3]. In Colombia is estimated that the electricity uses will have a growth rate of 52% between 2016 and 2030 [4] highlighting the Cost-Caribbean region as the fastest growing. Figure 2 shows this trend by region.

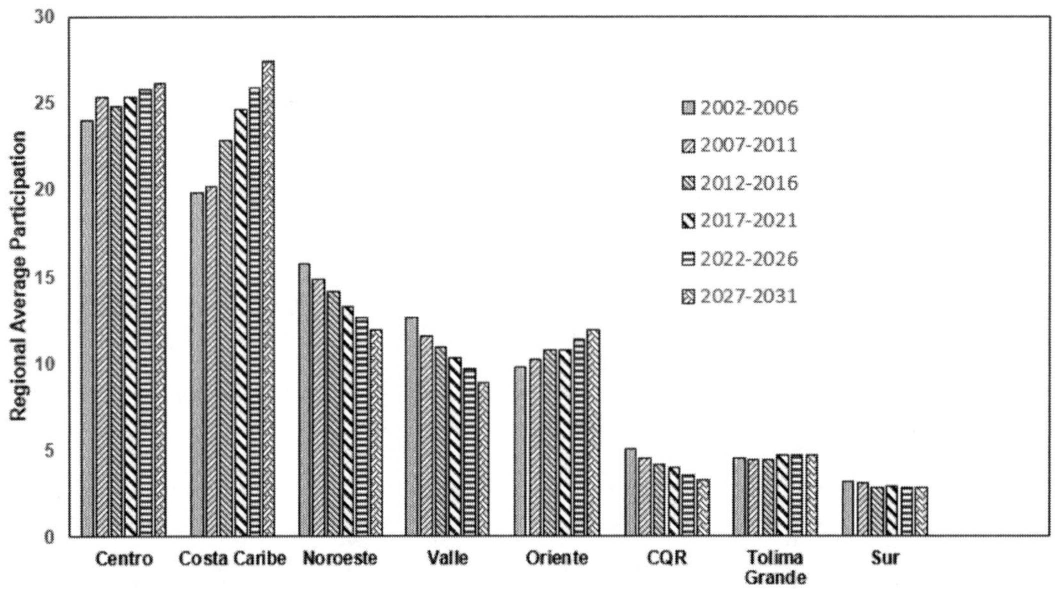

Figure 2. Electricity consumption in Colombia by region from 2002 to 2031.

This increase in energy consumption, the global warming, the resource depletion and the rise of policies focused on climate change and greenhouse gas emissions reductions, have promoted to countries and industries the implementation of strategies focused on increase energy efficiency and reduce GHG emissions [5-7]. Colombia is committed to this, and the government has proposed a group of policies which includes: The Indicative Action Plan (PROURE), Law 1715, National Development Plan (PND) and the Colombian Low Carbon Development Strategy (ECDBC) [8]. These policies focus on reducing 20% of GHG by 2030 considering an inertial scenario, a commitment proposed by Colombia at COP 21 [9]. The implementations of good practices in compressed air systems are outstanding strategies to develop the indicative action plan for energy efficiency 2017-2022 [4].

Compressed Air (CA) is one of the most widespread systems used in pneumatic systems to operate valves and cylinders, transporting products, cleaning, among others uses, due to factors such as good transportability, speed, cleanness, and riskless [10-16]. In many countries as China, USA, Australia, France and Italy Compressed Air Systems (CASs) accounts around the 10% in the overall electricity costs for industries [17, 18]. In Colombia this value reaches the 8 %, equivalent to 1.39 TWh/year [19-21]. However, CASs has been recognized as one of the most expensive forms of energy, only about the 19 % of the consumed energy reaches the end use point [22, 23]. Energy losses can be caused by heat, inefficient uses, pressure drops, inadequate storage, leaks, among others [24, 25]. According with [26] the energy cost required to operate a CASs can be up 75 % of total life cycle cost, higher than investment and maintenance which represents the 13 % and 12 % respectively [27, 28].

The energy efficiency measures applied in CASs such as: reducing the compressors pressure, matching the supply and demand, adequate storage capacity, reducing the air intake temperature, optimizing the air compressor location and to reduce leaks, may give energy savings between 20-60 % with a usual payback lower than two years [29-31]. Furthermore, energy savings, increase energy efficiency in CASs, may ensure other Non-Energy Benefits (NEBs) such as: maintenance cost reduction, capital avoidance, productivity improvement [32, 33]. These are reasons why CASs can be considered as one of the main targets in the implementation of energy efficiency actions at country, region and industry level [11].

In some industries, staff are quick to identify energy losses from pump systems, hot surfaces, but do not pay same attention to the losses involved in the generation of compressed air, because they do not produce dirt, residues or accidents; also, the wrong approach considering compressed air as cheap, or even it is considered a free source of energy [34-36]. Another aspect is that CASs is often considered as a common utility and in most cases, they are not assigned to a specific area and therefore they become difficult to manage [37]. However, the only moment where CASs gets all the attention is when the air losses and pressure drops interfere with the normal operation of the plant [26]. This paper analyzes the potential savings in CASs for two important cities in the Caribbean Coast Region of Colombia if the different energy efficiency measures are taken there is a potential energy savings around 50GWh/year and reduction the CO2 emission in 10 702 tonnes of CO2/year.

2. Potential energy savings in Compressed Air Systems.

CASs are composed by two fundamental areas: Supply and Demand. In the Supply side the inlet air is converted in compressed air, this side include compressors, controls, air dryers, air receivers, filters and drains; in the demand side the air is carry out to the end use equipment and it is comprises of distribution lines, filters, pressure and flow controls and end-use consumers as is shown in Figure 3.

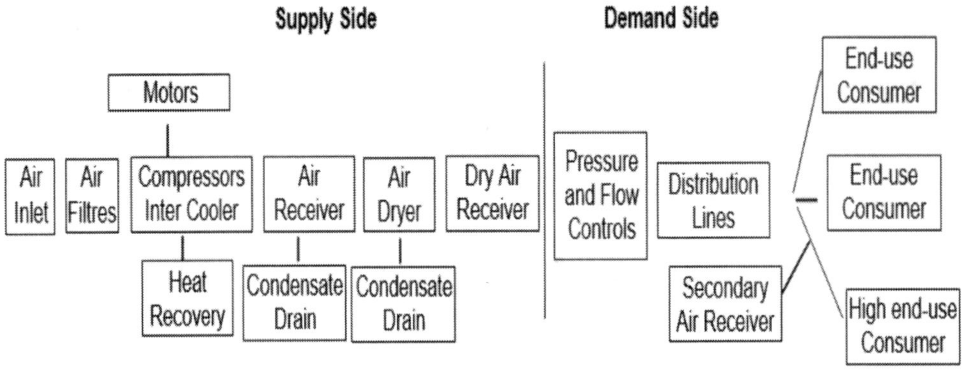

Figure 3. CASs structure and elements divided according to supply and demand sides. Source of data: Prepared by the authors based on data from: [16, 32].

In [6, 26, 37] have found that the 50-70% of the overall energy efficiency measures are related to demand side, the main measures are: air leaks reduction, inappropriate uses, reduction artificial demand, recover waste heat, inefficiencies in air system distribution and an inadequate maintenance. On the supply side the main measures are: to improve compressor operation and control, to reduce intake air temperature and to install new technologies based in efficiency concept. Table 1 shows these strategies, energy savings, applicability, payback period and costs.

Table 1. Strategies for saving Electricity in CASs. Source of data: Prepared by the authors based on [6, 26, 37].

Energy savings strategy	Annual Energy savings %	Applicability %	Potential Contribution %	Payback Period (Months)	Annual Cost ($$)
Using cooler intake air	10	40	4	5	1400
Compressors Controls	10	30	3	8	7900
Compressors pressure reduction	15	50	7,5	4	2800
Reducing air leaks	30	53	15,9	3	3900
Eliminating or reducing compressors air used	25	10	2,5	6	7300
Air compressor Waste Heat Recovery	20	20	4	10	2700
Improving filter maintenance	2	40	0,8	-	-
Use variable speed drives	15	25	3,8	-	-
Total			41,5		

3. Energy savings in Barranquilla and Cartagena Cities.

3.1 Structure of manufacturing sector.

In Colombia, the manufacturing sector is formed by 23 divisions, as product of the revision 3.1 A.C updated by the Statistics Administrative Department (DANE), adopting the revision 4 A.C which corresponds with the fourth revision of the International Standard Classification of All Economic Activities [38, 39]. This sector is made up by more than 8000 companies. Tables 2 shows a detailed

IOP Conf. Series: Materials Science and Engineering **844** (2020) 012019 doi:10.1088/1757-899X/844/1/012019

structure of the manufacturing sector, the number of companies by division and the electricity consumption in Barranquilla and Cartagena cities respectively.

3.2 Energy savings by divisions.

A study developed by [19, 20, 21] in more than five hundred manufacturing companies in the country established the percent of electricity consumption in CASs by divisions of total electricity consumption. The CASs electricity share costs in all divisions are around 8%, highlighting divisions such as textile manufacturers, where it was reached a 24%. Table 3 shows these values.

Figure 4 shows the energy savings potential in the CASs in Barranquilla and Cartagena for all divisions. In most of the divisions the saving potentials in the city of Barranquilla are greater than Cartagena, only in the sectors of Manufacture of chemicals and chemical products, Manufacture of rubber and plastics products, Manufacture of other non-metallic mineral products and Manufacture of coke and refined petroleum products, the saving potential is greater in Cartagena.

Table 2. A Detailed structure of manufacturing section in Barranquilla and Cartagena: Data source: Prepared by the authors based on [38, 39].

Section	Division	Description	Number of companies in Barranquilla	Electricity consumption (Kwh/year)	Number of companies in Cartagena	Electricity consumption (Kwh/year)
C. Manufacturing	10	Manufacture of food products	70	164.761.051	16	34.030.948
	11	Manufacture of beverages.	9	90.604.089	4	33.773.345
	12	Manufacture of tobacco products	-	-	-	-
	13	Manufacture of textiles	10	37.070.428	-	-
	14	Manufacture of wearing apparel	22	2.505.818	-	-
	15	Manufacture of leather and related products	6	9.760.246	-	-
	16	Manufacture of wood and of products of wood and cork, except furniture; manufacture of articles of straw and plaiting materials	3	154.376	4	189.657
	17	Manufacture of paper and paper products	5	48.111.729	-	-
	18	Printing and reproduction of recorded media	21	6.167.848	-	-
	19	Manufacture of coke and refined petroleum products	7	886.154	9	124.953.234
	20	Manufacture of chemicals and chemical products	27	85.756.298	21	606.271.720
	21	Manufacture of pharmaceuticals, medicinal chemical and botanical products	7	30.993.542	-	-
	22	Manufacture of rubber and plastics products	22	47.623.089	13	161.469.858
	23	Manufacture of other non-metallic mineral products	23	98.529.920	8	162.296.034
	24	Manufacture of basic metals	10	75.179.118	4	1.158.365
	25	Manufacture of fabricated metal products, except machinery and equipment	23	10.631.964	10	2.191.413
	26	Manufacture of computer,	-	-	-	-

Section	Division	Description	Number of companies in Barranquilla	Electricity consumption (Kwh/year)	Number of companies in Cartagena	Electricity consumption (Kwh/year)
		electronic and optical products				
	27	Manufacture of electrical equipment	-	-	-	-
	28	Manufacture of machinery and equipment N.E.C.	14	4.861.090	3	1.490.521
	29	Manufacture of motor vehicles, trailers and semi-trailers	3	138.526	-	-
	30	Manufacture of other transport equipment	-	-	-	-
	31	Manufacture of furniture	14	8.318.401	-	-
		Total	296	722.053.687	92	1.127.825.095

Table 3. CASs electricity share costs of total electricity consumption in all divisions.

Divisions	% of electricity Consumption	**Divisions**	% of electricity Consumption
Manufacture of food products	10	Manufacture of pharmaceuticals, medicinal chemical and botanical products	3.2
Manufacture of beverages.	11	Manufacture of rubber and plastics products	12.8
Manufacture of tobacco products	2	Manufacture of other non-metallic mineral products	4.2
Manufacture of textiles	24	Manufacture of basic metals	2
Manufacture of wearing apparel	12	Manufacture of fabricated metal products, except machinery and equipment	4.3
Manufacture of leather and related products	15	Manufacture of computer, electronic and optical products	6.8
Manufacture of wood and of products of wood and cork, except furniture; manufacture of articles of straw and plaiting materials	17	Manufacture of electrical equipment	2.7
Manufacture of paper and paper products	3	Manufacture of machinery and equipment.	2.8
Printing and reproduction of recorded media	9	Manufacture of motor vehicles, trailers and semi-trailers	4.1
Manufacture of coke and refined petroleum products	5.1	Manufacture of other transport equipment	7.7
Manufacture of chemicals and chemical products	4.1	Manufacture of furniture	6.8

IOP Conf. Series: Materials Science and Engineering **844** (2020) 012019 doi:10.1088/1757-899X/844/1/012019

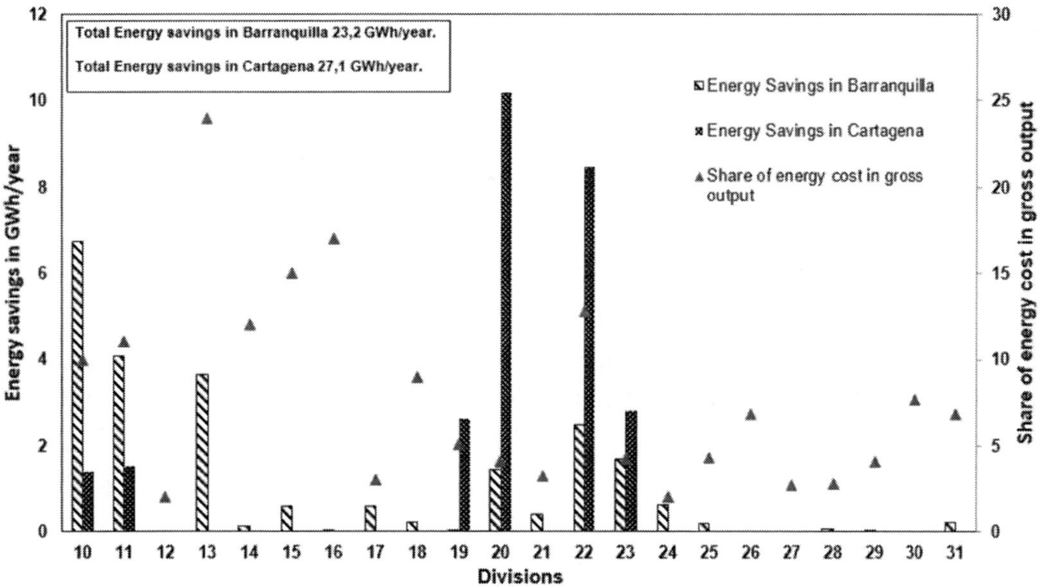

Figure 4. Savings potential in the compressed air system in the city of Barranquilla and Cartagena for different divisions.

Although in Cartagena there are three times fewer industries than in Barranquilla, the energy savings potential are greater due to the fact that the existing industries are large consumers of electricity.

3.3 Potential of CO_2 reduction.

Figure 5 shows the impact in the reduction of CO_2 emissions, considering the specific emission factor for power generation in Colombia of 0.2129 kgC0$_2$/kWh (Roa and Castellanos, 2018). The total reduction is equivalent to 10.702 tonnes of CO_2/year, 4.939.169 in Barranquilla and 5.763.157 in Cartagena.

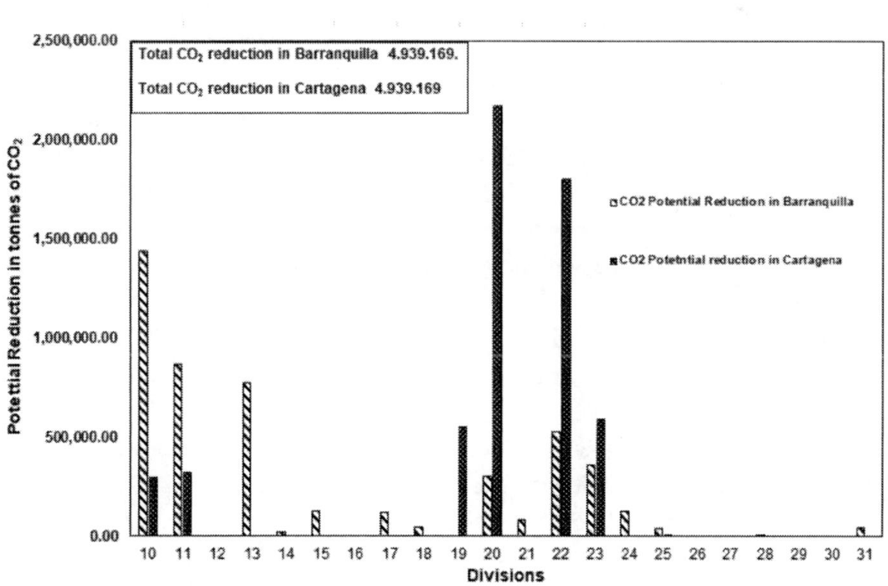

Figure 5. Potential of CO_2 reduction in energy efficiency measures are applied.

4. Conclusions.

Compressed Air Systems (CASs) are one of the most widespread systems used in industry. In countries such as China, USA, Australia, France and Italy CASs accounts around 10% of the overall electricity costs. However, CASs has been recognized as one of the most expensive form of energy, applied energy efficiency measures may give between 20-60 % of energy savings with a usual payback lower than two years. These are reasons why CASs can be considered as one of the main targets systems for the implementation of energy efficiency actions in many country, regions and industry levels.

The implementation of good practices in CASs is one of the strategies to be developed by governments focused on reduce energy consumption. This paper shows the potential energy savings and CO_2 reduction in CASs for two cities of the Colombian Caribbean Coast region, which are around 50GWh/year and 10 702 tonnes of CO_2/year, these results can be used by the government and organizations to implement energy efficiency projects in the identified energy savings potentials, which is related with the Colombia commitment focused on reduce a 20% of GHG by 2030, considering an inertial scenario, as it was proposed during COP 21.

References.

[1] Outlook, B. E. 2019 edition. London, United Kingdom2019. Available From: https://www.bp.com/content/dam/bp/business-sites/en/global/corporate/pdfs/energy-economics/energy-outlook/bp-energy-outlook-2019.pdf. [Last accessed on 2019 Jun 28].

[2] Chikunov S O, Gutsunuk O N, Ivleva M I, Elyakova I D, Nikolaeva I V and Maramygin M S, 2018, Improving the economic performance of Russia's energy system based on the development of alternative energy sources. *IJEEP*, 8(6), 382-391.

[3] UPME (2018), Balance Energético Colombiano. BECO. Available from: http://www1. upme.gov.co/ InformacionCifras/Paginas/BalanceEnergetico.aspx.[Last accessed on 2019 Jun 24].

[4] UPME (2016), Plan de acción indicativo de eficiencia energética 2017-2022, una realidad y oportunidad para Colombia (PAI Proure 2017-2022), *Available from:*http://www1. upme.gov.co*/DemandaEnergetica/MarcoNormatividad/PAI_PROURE_2017-2022.pdf.*[Last accessed on 2019 Jun 12].

[5] Faizah S I and Husaeni U A, 2018, Development of Consumption and Supplying Energy in Indonesia's Economy. *IJEEP*, 8(6), 313-321.

[6] Benedetti M, Bertini I, Introna V, and Ubertini S, 2018, Explorative study on Compressed Air Systems' energy efficiency in production and use: First steps towards the creation of a benchmarking system for large and energy-intensive industrial firms. *Applied Energy*, 227, 436-448.

[7] B Gopalakrishnan, K Ramamoorthy, E Crowe, S Chaudhari and H. Latif, 2014, A structured approach for facilitating the implementation of ISO 50001 standard in the manufacturing sector, *Sustainable Energy Technologies and Assessments*, 7, 154-165.

[8] Ocampo N, Garcia J, Ghazoul J and Etter A, 2018, Quantifying impacts of oil palm expansion on Colombia's threatened biodiversity. *Biological Conservation,* 224, 117-121.

[9] Berruezo J A and Jiménez J D, 2017. Situación del Convenio Marco de Naciones Unidas sobre el Cambio Climático. Resumen de las Cumbres de París, COP21 y de Marrakech, COP22. *Revista de Salud Ambiental,* 17(1), 34-39.

[10] Taheri K, and Gadow R, 2017, Industrial compressed air system analysis: Exergy and thermoeconomic analysis. *CIRP-JMST*, 18, 10-17.

[11] Bonfà F, Salvatori S, Benedetti M, Introna V and Ubertini S, 2017, Monitoring compressed air systems energy performance in industrial production: lesson learned from an explorative study in large and energy-intensive industrial firms. *Energy Procedia*, 143, 396-403.

[12] Annegret C and Radgen P, 2003, Efficient compressed air a successful campaign for energy efficient compressed air systems in Germany, *ECEEE 2003 summer study proceedings; 2 - 7 June 2003*, Saint-Raphaël, France.

[13] Dos Santos Mascarenhas J, Chowdhury H, Thirugnanasambandam M, Chowdhury T and Saidur, R, 2019, Energy, exergy, sustainability, and emission analysis of industrial air compressors. *Journal of Cleaner Production*, 231, 183-195.

[14] Yin Y, Zheng B, Yang C, and Zhang X, 2015. A proposed compressed air drying method using pressurized liquid desiccant and experimental verification. *Applied Energy*, 141, 80-89.

[15] Zahlan J, Asfour S, 2015, A multi-objective approach for determining optimal air compressor location in a manufacturing facility, *Journal of Manufacturing Systems*, 35, 176–190.

[16] DoE US, 1998, Improving Compressed Air System Performance, a Sourcebook for Industry. Prepared for the US Department of Energy, Motor Challenge Program by Lawrence Berkeley National Laboratory (LBNL) and Resource Dynamics Corporation (RDC).

[17] Šešlija D, Ignjatović I, Dudić S and Lagod B, 2011, Potential energy savings in compressed air systems in Serbia. *African Journal of Business Management*, 5(14), 5637-5645.

[18] Viholainen J, Grönman K, Jaatinen-Värri A, Grönman A, Ukkonen P, and Luoranen M, 2015, Centrifugal compressor efficiency improvement and its environmental impact in waste water treatment. *Energy conversion and Management*, 101, 336-342.

[19] UPME INCOMBUSTION, 2013, Determinación del potencial de reducción del consumo energético en los subsectores manufactureros códigos CIIU 10 a 18 en Colombia. Available from: http://www1.upme.gov.co/DemandaEnergetica/INFORME_III_Caracterizacion_energetica_Ver Pub.pdf. [Last accessed on 2019 Mar 6]

[20] UPME CORPOEMA, 2014 a, Determinación y priorización de alternativas de Eficiencia Energética para los subsectores manufactureros informe final códigos CIIU 19 a 31. Vol. I. Colombia. Available from: http://www.upme. gov.co/Estudios/2014/Informe_Final_Volumen_1 .pdf. [Last accessed on 2019 Mar 8].

[21] UPME CORPOEMA, 2014 b, Determinación y priorización de alternativas de Eficiencia Energética para los subsectores manufactureros informe final códigos CIIU 19 a 31. Vol. II. Colombia. Available from: http://www1.upme.gov.co/DemandaEnergetica/ DeterminacionEficiencia/Informe_Final_Volumen_2.pdf. [Last accessed on 2019 Mar 8].

[22] Kriel C J, Marais J H and Kleingeld M. 2014, Modernising underground compressed air DSM projects to reduce operating costs. *In 2014 International Conference on the Eleventh industrial and Commercial Use of Energy*, 1-6.

[23] Shaw W, Mathews M and Marais J, 2019, Using specific energy as a metric to characterise compressor system performance. *Sustainable Energy Technologies and Assessments*, 31, 329-338.

[24] Corsini A, De Propris L, Feudo S and Stefanato M, 2015. Assessment of a diagnostic procedure for the monitoring and control of industrial processes. *Energy Procedia*, 75, 1772-1778.

[25] Mousavi S, Kara S and Kornfeld B. 2014, Energy Efficiency of Compressed Air Systems, *21st CIRP Conference on Life Cycle Engineering, Procedia CIRP,* 15, 313-318.

[26] Saidur R, Rahim N A and Hasanuzzaman, M, 2010, A review on compressed-air energy use and energy saving, *Renewable and Sustainable Energy Reviews,* 14, 1135-1153.

[27] European Commission, 2009, Reference Document on Best Available Techniques for Energy Efficiency. Available from: http://eippcb.jrc.ec.europa.eu/. [Last accessed on 2019 Feb 22].

[28] Yang M, 2009, Air compressor efficiency in a Vietnamese enterprise. *Energy Policy*, 37(6), 2327-2337.

[29] Abdelaziz E A, Saidur R and Mekhilef S, 2011, A review on energy saving strategies in industrial sector, *Renewable and Sustainable Energy Reviews*, 15, 150-168.

[30] Vittorini D and Cipollone R, 2016, Energy saving potential in existing industrial compressors. *Energy,* 102, 502-515.

[31] Cloete S, le Roux D and Buhrmann T, 2013, Reducing Compressed Air Wastage By Installing New Technology in Underground Mines. *In: 2013 Proceedings of the Conference on the Industrial and Commercial Use of Energy, ICUE,* 1-6.

[32] Nehler T, 2018, Linking energy efficiency measures in industrial compressed air systems with non-energy benefits–A review. *Renewable and Sustainable Energy Reviews*, 89, 72-87.

[33] Fleiter T, Hirzel S and Worrell E, 2012, The characteristics of energy-efficiency measures a neglected dimension. *Energy Policy*, 51, 502–513.

[34] Broniszewski M, and Werle S, 2018, The study on the heat recovery from air compressors. In *E3S Web of Conferences,* 70, p 03001, EDP Sciences.

[35] Neale J R, and Kamp P J, 2009, Compressed air system best practice programmes: what needs to change to secure long term energy savings for New Zealand, *Energy Policy*, 37,3400–3408.

[36] Kaya D, Phelan P, Chau D and Ibrahim H, 2002, Energy conservation in compressed-air systems. *International Journal of Energy Research*, 26(9), 837-849.

[37] Radgen P and Blaustein E, 2001, Compressed air systems in the European Union: energy, emissions, savings potential and policy actions. *LOG_X Verlag GmbH*, Stuttgart, Germany.

[38] United Nations, 2018, International Standard Industrial Classification off all Economics Activities (ISIC) Revision 4, ISBN: 978-92-1-161518-0, United Nations, New York.

[39] DANE, 2018, Clasificación industrial Internacional uniforme de todas las actividades Económicas, Revisión 4 adaptada para Colombia CIIU Rev. 4 A.C.

IOP Conf. Series: Materials Science and Engineering **844** (2020) 012020 doi:10.1088/1757-899X/844/1/012020

Evaluation of trace-metal concentrations in water consumption of streams near Guajira (Colombia)

N. Garzon[1], E Espinosa[2], L. A. Sarmiento-Rubiano[3], R. Manzoli R[4], R. Cantero[2]and J. R Castro-Suarez[5]

[1] Grupo de Investigación en Calidad de Aguas, Modelación Hídrica y Ambiental (CAMHA), Schools of Engineering and Architecture. Faculty of Sanitary and Environmental Engineering, Universidad Pontificia Bolivariana (UPB), Montería, Colombia

[2] Department of Engineering, Universidad Libre-Barranquilla, Carrera 46 No. 48-170, 080002, Barranquilla, Atlántico, Colombia.

Barranquilla, Atlántico, Colombia, Calle 58 # 55 - 66, Barranquilla, Colombia

[3] Grupo de Investigación en Alimentación y Comportamiento Humano. Universidad Metropolinana. Programa Nutrición y Dietética. Cl. 76 #42-78, Barranquilla, Atlántico.

[4] Grupo de Investigación en Gestión y Sostenibilidad Ambiental (GESSA) , Department of Civil and Environmental. Universidad De La Costa (CUC),

[5] Ciencias basicas exactas, Universidad del Sinu, Cartagena, Colombia.

eduardoespinosafuentes@gmail.com

Abstract. The spatial distribution of metals in the water for consumption in the different supply sources, natural sources, family storage tanks and deep wells were studied. 16 samples were made for different municipalities belonging to the department of La Guajira upper part. The water samples were analyzed in 2016 and analyzed physically and chemically and microbiologically. The results were compared with national standards and those established by the WHO. The results found show some traces of metals in some of the sampling sites. The concentrations of Calcium, Magnesium exceeded the concentrations established by the standards. The results reinforce the need for better control of the physicochemical and microbiological parameters of the water consumed by the population.

1. Introduction

The analysis of the water resource is of great importance for human health [1,2] as well as concentrations of metals in surface and groundwater has been of great research interest [3-5]. Various studies have shown that metals can reach the bottom of water bodies causing alterations in ecosystems and communities. Therefore, the metals in the basins will contaminate the drinking water system, can cause impacts on human health, on the soils and on the aquatic biota [5]. The effects of metals in drinking water have been reported by different researchers [6-11]. According to the WHO, most consumers do not have the means to judge for themselves the safety of the water they consume, but their attitude towards drinking water and their water providers will be greatly affected by aspects of water quality that you are able to perceive with your own senses. Additionally, different studies have been carried out in order to evaluate the quality of groundwater, especially from the United States, through the evaluation of water quality, which is designed to evaluate water quality conditions, determine spatial trends and temporal, and to identify the Physical, chemical and biological factors that affect the underground resource (Gilliom., et al 1995). In some cases, the spatial and temporal focus of metals is analyzed, including an integrated water evaluation that includes a combination of physicochemical, biological and microbiological analyzes with chemical aspects [3]. Similarly, Abraham J et al. [5] analyzes some of the main factors that affect the metal cycle and the dynamics in hydrographic basins such as forest fires, soil and water pH, hydrology, biological processes, events of precipitation, organic matter in the soil and water. Finally, another important aspect in the analysis of drinking water is hardness. According to Dahl, C et al. [12], drinking water provides small doses of calcium and other essential minerals for bones throughout life. When a water is referred to as "hard" water it simply means that it contains more minerals than normal water. There are especially calcium and magnesium minerals. The presence of calcium and magnesium in very high proportions in the water, generates what is known as hard water [1]. For these reasons, it has been considered of great importance to carry out a study of the concentration of metals, chemical parameters that determine hardness, and microbiological parameters in order to determine if the concentrations may represent a risk for daily use and consumption in the area. study.

2. Materials and methods

2.1. Study area

The Guajira peninsula, which corresponds to the northernmost part of the country, made up of a series of low hills of marine origin, extensive plains and, in the extreme north, a more prominent relief with heights of up to 864 meters above sea level (masl) in the Macuira mountain range. In a department where the least rainfall occurs in relation to the other departments of the country with annual totals of 500 mm or less. The mangrove precipitation ranges from 150 mm per year in La Guajira. The highest values of the average air temperature are between 28 °C and 32 °C, which occur in the high and medium Guajira. In relation to runoff, the plain of La Guajira in contrast presents values ranging from 25 mm (in the middle and high Guajira) to 200 mm in the foothills of the Sierra Nevada. Finally, it has one of the largest populations of indigenous communities [13]. It is the department in which it has the largest number of ecosystems from perpetual snow to the sea, and in Manaure the largest exploitation of sea salt in the country is carried out.

La Guajira is home to the Cerrejon mines that make up the largest open pit coal mine in the world, likewise it owns 36% of the Sierra Nevada de Santa Marta territory where the highest elevation in the country is located with 5775 masl, Depto del Magdalena) considered the highest coastal mountain in the world [14]. The department of La Guajira has 15 municipalities [14]. Municipalities such as San Juan del Cesar, Albania, Barrancas and Fonseca are considered with ecoepidemiological conditions [15].

The present study was designed to evaluate the temporary concentration of the metals Aluminum, Arsenic, Boron, Barium, Beryllium, Calcium, Cadmium, Cobalt, Chromium, Copper, Iron, Lithium, Manganese, Molybdenum, Nickel, Phosphorus, Lead, Antimony, Selenium , Tin, Titanium, Thallium, Vanadium and Zinc, as well as microbiological analyzes at 16 sampling points in the waters that are used for consumption by the inhabitants of the communities located in the municipalities of Papayal, Chancleta, Patilla, Hato Nuevo, Guairmarito , Albania, Cuestecias and Matitas (Figure 1).

The study area constituted Media Guajira which is uniform, straight and regularized, with alternation of sectors of erosion that cut the central plain in low cliffs, and sectors of coastal accumulation that isolate salty or brackish lagoons, in addition to the Ranchería river that builds a delta cut by coastal drift and lower Guajira made up of gallery forests, pastures and crops, main urban centers and coal mining [14].

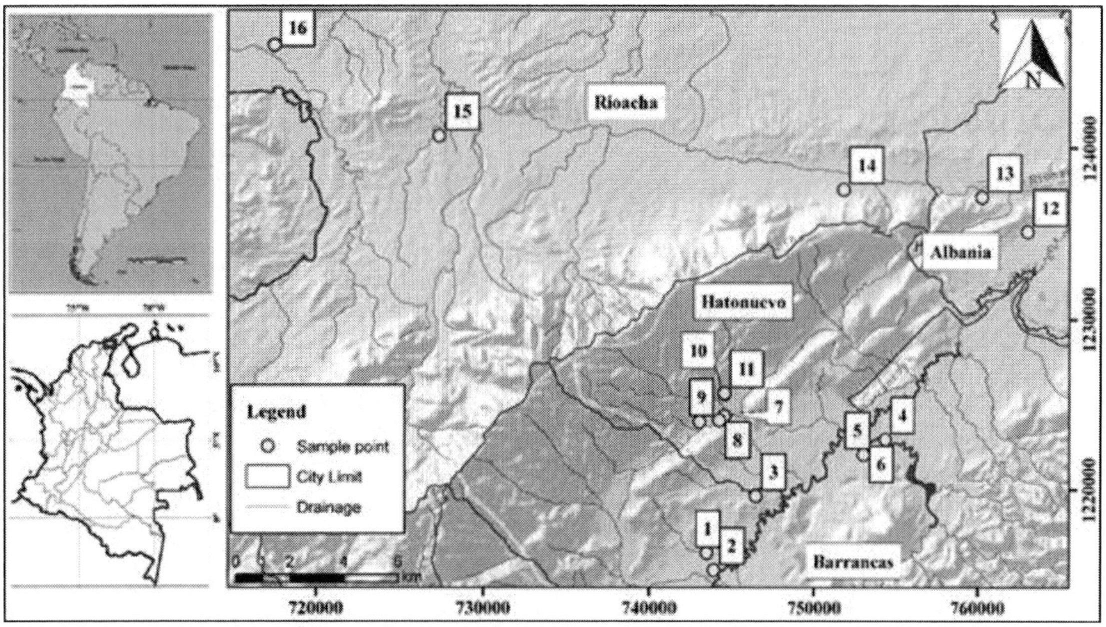

Figure 1. Location of the study area and sampling sites in the Department of La Guajira.

2.2. Sampling and analysis method

2.2.1. Sampling. A total of 16 water samples collected for chemical and microbiological analysis in June 2016 from supply sources, see Table 1.

Table 1. Location of study sites near Guajira State.

Site	Municipality	Sample point	Longitude	Latitude
1	Papayal	Underground storage tank	743471,08	1216302,16
2	Papayal	Rancheria River	743912,55	1215301,79
3	Papayal- Serrito 2 Sector	Deep well	746426,38	1219630,64
4	Chancleta	Cerrejoncito River	754284,89	1222953,11
5	Patilla	Storage tank	752943,92	1222066,24
6	Patilla	car tank	752943,92	1222066,24
7	Hatonuevo	kitchen faucet restaurant	744540,37	1224334,85
8	Hatonuevo wellspring " El Pozo"	kitchen faucet family home	744247,16	1224081,47
9	Hatonuevo wellspring " El Pozo"	Hatonuevo wellspring " El Pozo"	743027,19	1223999,70
10	Guaimarito	Storage tank open	744551,52	1225616,33
11	Guaimarito	Storage tank open	744551,19	1225660,59
12	Albania	kitchen faucet family home	763012,79	1235099,67
13	Cuestecitas	Hosepipe	760242,41	1237151,15
14	Homestead "La Isla"	Deep well	751753,22	1237589,48
15	Homestead "Villa Martin"	Deep well	727368,32	1240724,30
16	Matitas	Deep well	717546,04	1246050,00

The sampling period corresponded to June where the sea breeze and the northeast trade winds act during most of the year and influence the climate regime in this area of the country, with rainy periods when their intensity decreases (June-November and periods dry when it increases (December-May). The climate in the department of La Guajira is warm and dry at sea level, where the annual average temperature is 29 °C, with maximums of 39 °C in the desert area from Uribia [14]. All the material, including the sampling bottles, used for the determination of heavy metals, were subjected to washing with neutral soap and then immersed in HNO_3 diluted to 10%. In the field, the pH parameters were determined and temperature for each of the samples.

2.2.2. Chemical analysis. The analyzes of the reported metals were carried out on an optical emission spectrometer by Inductively Coupled Plasma (ICP-OES), brand VARIAN VISTA-MPX, with a water recirculation system, argon gas cylinder, extractor for vapors and fumes generated by the plasma, located approximately 30 cm above the plasma generating chamber, Schott Ceran heating plate, Thermolyne Cimarec 2 or equivalent.

The elements analyzed were Silver, Aluminum, Arsenic, Barium, Beryllium, Calcium, Cadmium, Cobalt, Copper, Chrome, Iron, Potassium, Lithium, Magnesium, Manganese, Molybdenum, Sodium, Nickel, Phosphorus, Lead, Antimony, Selenium, Silicon, Tin, Strontium, Titanium, Thallium, Vanadium, Zinc and Sulfur. Argon gas AP grade 4.8. Ultrapure water. Nitric acid, HNO_3, 65% ultrapure. Acidulated water: Dissolve 2.5 mL of ultrapure 65% nitric acid in 2 L of ultrapure water. Hydrogen peroxide, H_2O_2, 30%. Standard solutions traceable to SRM of 1000 mg / L of the analyzed metals.

2.2.3. Microbiologic analysis. Aliquots of water were collected in 15 of the 16 sampling points in 250 ml capacity amber plastic bottles contained with a volume of water of approximately 200 ml with an air chamber corresponding to a volume of 50 ml. The samples were kept in refrigeration from their origin and until the moment of analysis. The microbiological parameters evaluated were total Coliform count (CT) and Fecal or thermotolerant Coliforms (CTT) by two methods; (I) Most Probable Number method (NMP per 100 / ml of water) according to the protocols of the (American Public Health Association) APHA29 and in accordance with Decree 1594 of 1984 of the Colombian Ministry of Health. (II) membrane filtration with a 0.45 micron pore and subsequent seeding in Cromocult Agar according to the protocols established by IDEAM. In both cases the strain used for quality control of the protocols was E. coli ATTC 25922.

The reference values According to decree 1594/1984, the microbiological quality of the water intended for primary contact activities, such as swimming and diving, allows a maximum count of 200 NMP / 100 mL of thermotolerant Coliforms and 1000 NMP / 100 mL Total coliforms; for secondary contact activities such as fishing, the total Coliform count should not exceed 5000 NMP / 100 mL. And According to Decree number 1575 of 2007 "By which technical standards of drinking water quality are issued." ARTICLE 25. The water for human consumption must comply with the following values admissible from the microbiological point of view by the MEMBRANE FILTRATION method: Total coliforms 0 CFU / 100mL and E. coli 0 CFU / 100mL [16].

3. Results and Discussion

3.1. Chemical parameters

Figure 2 presents the concentrations obtained in the laboratory for different metals analyzed and according to this, the most impacted points. Most of the monitored points presented higher values than those stipulated by the drinking water standard for Calcium (60 ppm) in points 2, 3, 10, 13 and 14 that correspond to the townships of Papayal, Guaimarito, Cuestecitas and the sidewalk of "the island" with an average value of (52.60 ppm) in 16 samples an interval of (4.1 to 105 ppm), see Figure 2a and Magnesium (36 ppm) were found an average of 13.8 and an interval in the 16 samples (1.8 to 56.6 ppm), see Figure 2b. In the study area (5) points exceed the norm for Calcium and 1 point the norm for Magnesium. Kippler M et al. [17] evaluated and found mean calcium concentrations of 44 ± 24 in 40 samples an interval of (12 - 86 ppm) and magnesium an average of 24 ± 4 in 22 samples (96.1 and 46 ppm), in drinking water in Bangladesh. This indicates that the water has characteristics of hardness, derived from the presence of calcium and magnesium, which is generally evidenced by the precipitation of soap residue and the need to use more soap to achieve the desired cleaning (WHO 2006) [16]. Studies have shown the relationship between calcium and magnesium on mortality from stroke and ischemic diseases. This relationship is stronger in cerebrovascular disease than in ischemic heart disease, is more pronounced for women than for men, and is more evident with magnesium than with calcium concentration levels [18]. Other studies indicate a relationship of the thyroid size by waters with high calcium content [19]. Chandra et al. [19] conclude in their study that chronic exposure to high calcium content progressively modify the activities of the thyroid hormone by synthesizing enzymes, possibly acting on the cellular and molecular levels that are reflected in the altered profile of the thyroid hormone.

This situation is also conditioned by factors of acceptability of drinking water for consumers, which is subjective and can be affected by various components, that is, the concentration of these components that is unpleasant for consumers is variable, and depends on individual factors. and local, such as the quality of water to which the community is accustomed and various social, environmental and cultural considerations (WHO 2006) [16]. In the case of the study area and in the same way, tourism, trade and exploitation of mining-energy resources have also been considered some of the main drivers of population concentration in the coastal area of La Guajira (Corpoguajira and Invemar 2012, [14]) as factors in the increase of the population and its location in areas that have little guarantee in the access of drinking water.

For the studied regions, high concentrations were found with respect to the levels accepted by Resolution 2215/07 of the following metals: calcium (Ca), strontium (Sr), silicon (Si), Sodium (Na) and potassium (K) . In particular cases, concentrations over the norm value of the following metals were found: Barium (Ba) Figure 2c, Lithium (Li). Figure 1e shows the concentration values per sampling point with respect to the accepted average value for Silicon are the corregimientos denoted by the numbers 2, 3, 4, 6, 14, 15 and 16.

Metals also alter the health of the environment, in fact, the increase in heavy metals due to industrial activity has seriously unbalanced and polluted many natural ecosystems. Once these heavy metals are released by human action, they can remain in the environment for hundreds of years [1]. Likewise and according to Corpoguajira and Invemar [14] during 2009, the four main rivers of La Guajira (Cañas, Jerez, Palomino and Ranchería) released nearly 60 m^3 / s of water loaded with various pollutants, which allowed calculating how About 26,215 t / a of solids, 212 t / a of inorganic nitrogen, 67 t / a of phosphates and about 918 MPN of microorganisms with faecal contamination. Where the main polluting sources are Mining, agriculture, livestock, industries, port activity and transportation of fuels and domestic origin (wastewater and solid waste).

According to Abraham J et al. [5], the water resources and watersheds contaminated with metals have the potential to represent a significant risk for downstream communities and ecosystems due to the effects of these metals on the health of ecosystems and human beings. The use of deep wells has been reported as a protection factor against some diseases including diarrhea (Wistons et al 2013), however this situation shows the high concentrations of metallic elements such as Manganese. The townships of Papayal, Chancleta, Patilla, Guaimarito and Cuestecitas mostly impacted by high levels of metals are those denoted by numbers 2, 3, 4, 6, 10, 14 and 15, which is a worrying finding for the health of the communities that are supplied from the water resource. In the study area, the highest concentrations of Potassium were found at the sampling points: 2, 3, 4, 6, 14 and 15. Strontium (Sr): 2, 3, 4, 6, 10, 11, 12, 13 , 14 and 15. Sodium (Na): 2, 3, 4, 6, 10, 11, 12, 13, 14 and 15. Barium (Ba): 2, 3, 4, 6, 10, 11, 12 and 13.

The highest concentrations of lithium were found at points: 3, 7, 12, 13 and 14. Harari F et al. [10] conducted a study of exposure to lithium through drinking water during pregnancy, the authors suggest that exposure A high lithium through drinking water during pregnancy can affect calcium homeostasis, particularly vitamin D. The danger of metals is that, once they enter the body, they accumulate and are not easily self-eliminated. These compounds are very dangerous for human health and negatively affect vital organs. We are continually exposed to metals, whether through food, water, or the air we breathe.

IOP Conf. Series: Materials Science and Engineering **844** (2020) 012020 doi:10.1088/1757-899X/844/1/012020

The main organs highly impacted by the accumulation of heavy metals in the body are the kidneys, liver, lungs and also the nervous system (central and peripheral). According to Kippler M et al. [17]iron-rich water can cause an unpleasant taste or odor, as well as brown precipitation in stored water before use. An advantage with such precipitation is that arsenic can be co-precipitated with it, resulting in less than estimated exposure from the concentration in the sampled water. When relating the sampling points with the water supply sources to the population of the study area, it was observed that the points that presented the values above the norm correspond to underground wells, domestic storage tanks that do not have protection (without cover) and surface sources such as the River of Rancheria.

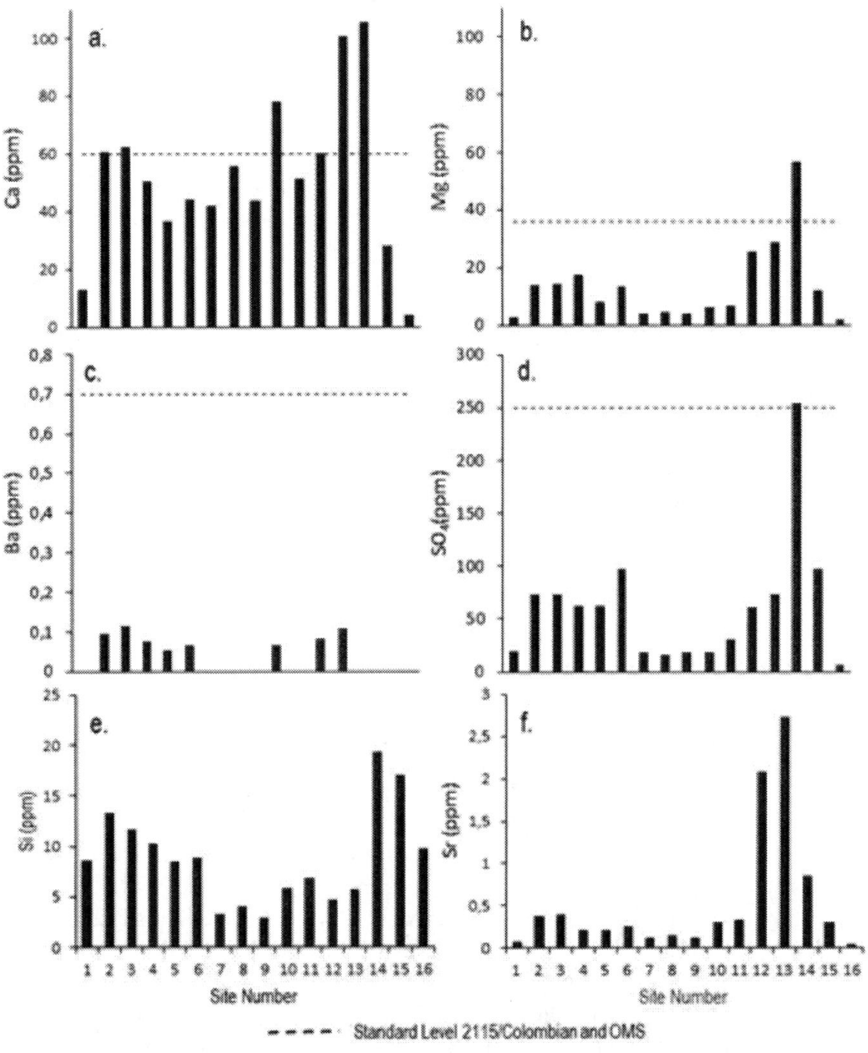

Figure 2. Concentration of metals: Ca, Mg, Ba, SO4, Si and Sr in parts per million per sampling point

The high reactivity and therefore retention of metals in any biological matrix, is due to the favourable acid-base properties of the metallic elements, which allows it to interact strongly with any organic and inorganic electro-donor or electro-acceptor species. This property causes the metallic elements to be almost irreversibly retained in complex molecular structures, such as proteins and nucleic acids. In most cases, when a metal enters the internal structure of proteins, these structures lose or change their

IOP Conf. Series: Materials Science and Engineering **844** (2020) 012020 doi:10.1088/1757-899X/844/1/012020

physiological functionality, altering the vital processes for which they were designed. Therefore, metals have a highly toxic effect on vital systems and the concentrations allowed in water for human consumption allowed by current legislation should be very low, for the preservation of public health.

The most common sources of aluminum in drinking water are naturally occurring aluminum and aluminum salts used as coagulants in water treatment. Sampling point 6 is the only point that presents higher values of aluminum 0.532 ppm, this sampling point corresponded to the sample taken in a tank car which is supplied from the Rancheria River. This can be attributed to the fact that (WHO) the final water can be contaminated with chemicals used in water treatment and substances from materials in contact with water. This indicates the need for mechanisms to deliver water to the population so that water access guarantees its quality.

3.2. Microbiologic analysis

The determinations of Total Coliforms (CTT) and Thermotolerant Coliforms (CT) evaluated by the NPM / 100 mL method, were below the maximum limit allowed for primary contact (200 NMP / 100 mL for CTT and 1000 NMP / 100 mL for CT) , but not in all cases the potability conditions established in Decree number 1575 of 2007 (0 UFC / 100mL for CT and CTT) are met, see figure 3.

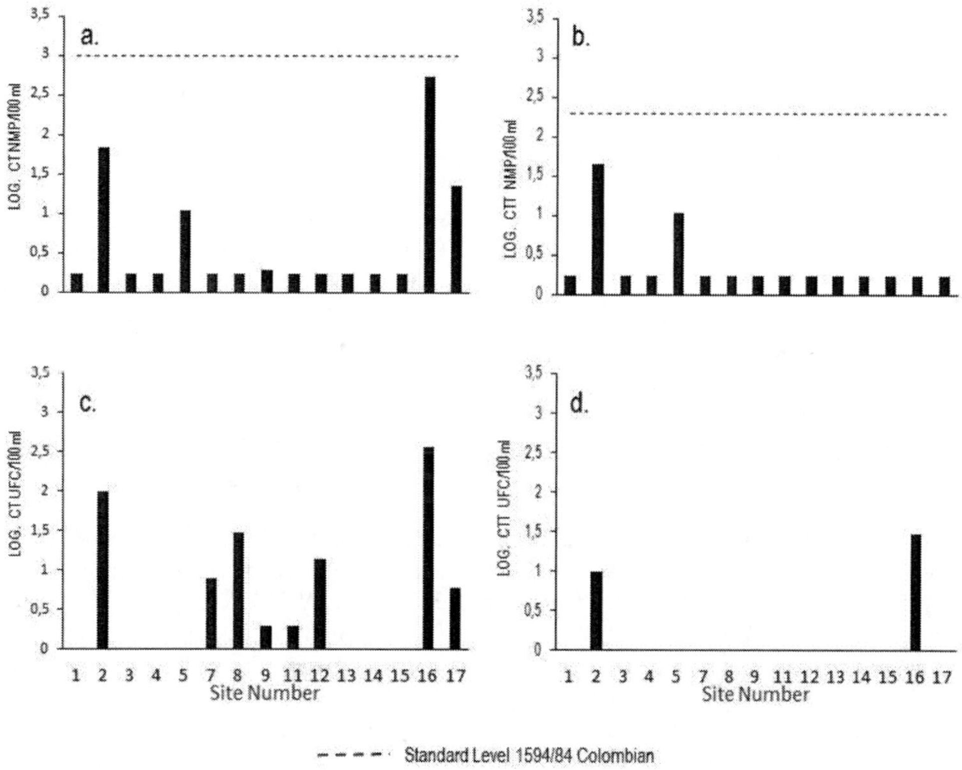

Figure 3. Determinations of Total Coliforms (CTT) and Thermotolerant Coliforms (CT) in the sites tested.

4. Conclusions

Assessment and monitoring of metals in local drinking water could provide effective regional prevention programs including mitigation strategies and prevention of contamination of the water resource. It is essential that additional mitigation strategies are implemented and effectively evaluated in the long term for water monitoring throughout the Guajira region. Additionally, the need to assess the exposure of the population by biomarkers is established, reflecting the exposure of all sources, not just drinking water.

References

[1] Zamora, J.R., 2009. Parámetros fisicoquímicos de dureza total en calcio y magnesio, pH, conductividad y temperatura del agua potable analizados en conjunto con las Asociaciones Administradoras del Acueducto,(ASADAS), de cada distrito de Grecia, cantón de Alajuela, noviembre. *Pensamiento Actual*, *9*(12), pp.125-134.

[2] Jobbágy, V., Altzitzoglou, T., Malo, P., Tanner, V. and Hult, M., 2017. A brief overview on radon measurements in drinking water. *Journal of environmental radioactivity*, *173*, pp.18-24.

[3] Giddings, E. M., Hornberger, M. I., Hadley, H. K. 2001. Trace-metal concentrations in sediment and water and health of aquatic macroinvertebratecommunities of streams near park city, summit county, Utah U.S. *geological survey water-resources investigations report* 01-4213 national water-quality assessment program salt lake city, utah.

[4] Bhuyan, S. Md., Bakar, A. M., Akhtara, A., Hossain, M. B., Alid, Mir M., Md., Islam, S., 2017. Heavy metal contamination in surface water and sediment of the Meghna River, Bangladesh. *Environmental Nanotechnology, Monitoring & Management*. 8, 273–279.

[5] Abraham, J., Dowling. K., Singarayer, F. 2017. Risk of post-fire metal mobilization into surface water resources: A review. *Science of the Total Environment*. 1740–1755.

[6] Ebrahimian, F., Asl Zahedi, S., 2005. The effect of high calcium on activity of thyroid gland in rat. *Endocr. Abstr.* 89. 11 P775

[7] Chandra, A.K., Tripathy, S., Debnath, A., Ghosh, D., 2007 . Bioavailability of iodine and hardness (magnesium and calcium salt) in drinking water in the etiology of endemic goiter in Sundarban delta of West Bengal (India). *J. Environ. Sci. Eng.* 49, 139–142.

[8] Broberg, K., Concha, G., Engstrom, K., Lindvall, M., Grander, M., Vahter, M. 2011. Lithium in drinking water and thyroid function. *Environ. Health Perspect.* 119 (6), 827–830.

[9] Vita, A., De Peri, L., Sacchetti, E. 2015. Lithium in drinking water and suicide prevention: a review of the evidence. *Int. Clin. Psychopharmacol.* 30 (1), 1–5

[10] Harari, F., Bottai, M., Casimiro, E., Palm, B., Vahter, M., 2015. Exposure to lithium and cesium through drinking water and thyroid function during pregnancy: a prospective cohort study. *Thyroid* 25 (11), 1199–1208.

[11] Harari. F., Åkesson. A., Casimiro, E., Lua, Y., Vaht, M. 2016. Exposure to lithium through drinking water and calcium homeostasis during pregnancy: A longitudinal study. *Environmental Research* 147. 1-7.

[12] Dahl, C., Søgaard, J. A., Tell S.G., Forsén, L., Peder, F. T., Hongve, D., Kristin, O. T., Holvik. K., Meyer, H. E., Aamodt, G. 2015. Population data on calcium in drinking water and hip fracture: An association may depend on other minerals in water. *A NOREPOS1 study*. (81), 292-299.

[13] Cabrera, L. M., Ortega, D. M., Quiroga, L.P. S., Lozano, P. R. J., Orenzon G. 1999. *República de Colombia segunda comunicación nacional ante la convención marco de las naciones unidas sobre cambio climático*. Capitulo 1. p 94.

[14] Corpoguajira e Invemar. 2012. *Atlas marino costero de La Guajira*. Serie de Publicaciones Especiales de Invemar No. 27. Santa Marta, Colombia. 188p. ISBN: 978-958-8448-45-9

[15] Rodriguez, M. A. J., García, L. C. J., Galindo, M. M. L., Sabogal, R. J. A., Marin, L. S., Lozada, R. C. O.; Diaz, Q. F. A. 2016. Zika infection GIS-based mapping suggest high transmission activity in the border area of La Guajira, Colombia, a northeastern coast Caribbean department, 2015 - 2016: Implications for public health, migration and travel. *Travel Medicine and Infectious Disease* .14, 286 - 288.

[16] Organización Mundial de la Salud. 2006. *Guías para la calidad del agua potable* [recurso electrónico]: incluye el primer apéndice. Vol. 1: Recomendaciones. Tercera edición. Versión electrónica para la Web. 1. Normas sobre el agua potable. 2. Normas sobre el agua. 3. Normas

sobre la calidad del agua. 4. Directrices. I. Título. ISBN 92 4 154696 4 (Clasificación de la NLM: WA 675).

[17] Kippler. M., Skrödera, H., Moshfiqur. S., Fahmida, R., Vahter. M.T. 2016. Elevated childhood exposure to arsenic despite reduced drinking water concentrations — A longitudinal cohort study in rural Bangladesh. *Environment International.* (86) 119-125.

[18] Ferrándiz. J., Abellán, J. J., Gómez, R. V., López, Q. A., Sanmartín. Pilar., Abellán, C., Martínez. B. M. A., Melchor, I., Vanaclocha, H., Zurriaga, Ó., Ballester. F., Gil, J. M. Pérez, H. S., Ocaña, R. 2004. Spatial Analysis of the Relationship between Mortality from Cardiovascular and Cerebrovascular Disease and Drinking Water Hardness. *Environmental Health Perspectives.* 112 (9) 1037-1043.

[19] Chandra Amar K., Haimanti Goswami, Pallav Sengupta. 2012 Dietary calcium induced cytological and biochemical changes in thyroid. *Environmental toxicology and pharmacology.* (3 4) 454–465.

IOP Conf. Series: Materials Science and Engineering **844** (2020) 012021 doi:10.1088/1757-899X/844/1/012021

Use of technological tool for the consolidation of mathematical pre-knowledge in higher education

D. Diaz[1] and K. Rueda [2]
1. Unidades Tecnológicas de Santander, Bucaramanga, Colombia
2. (GICBA) Research group of basic and applied sciences, Unidades Tecnológicas de Santander, Bucaramanga, Colombia

marceladiaz312@hotmail.com, krueda@correo.uts.edu.co

Abstract: Currently, information and communication Technologies (ICT) have permeated the different educational, political, economic and sociocultural contexts, contributing to the inclusion and development of society by optimizing time, efforts and resources. On the other hand, (ICT) are considered powerful facilitators of the teaching-learning process at all training levels because they visibly improve the way of acquiring knowledge, becoming essential tools for educational transformation and innovation. Therefore, the purpose of this present study is to analyze the results of the first implementation of the PREIN - UTS virtual course developed in the web Khan- Academy tool. The sample was formed by new students who entered the first semester of the technological level in the academic period 2018-II of a higher education institution located in Bucaramanga Santander (N=801). The methodological approach was quantitative of exploratory type, with the use of descriptive statistics and SPSS software for data analysis. Among the research results it was evidenced that the PREIN course is an innovative and easily accessible strategy, that contributes to the assurance of mathematical skills of the upcoming students that will enter proximally.

1. Introduction

Information and communication technologies (ICT) are important within the public policies of each country because they promote the accurate development of society through the improvement of the quality of life and reduction of large inequality gaps [1]. Precisely, in the educational field, (ICT) have become an essential facilitators tools and methods that boost knowledge since they allow the proper decentralization of knowledge and the creation of teaching methodologies in innovative learning spaces [2-3]. Nonetheless, the ICT incorporation within the formative process is limited due to the lack of training by teachers and it's also presented in several cases because of the high costs they generate [4].

Now, it can be appreciated and consider that despite the different alternatives offered by (ICT) to improve learning, the most clear results of the Report of the International Program for Student Evaluation or PISA Report reveal the assumption that the percentage of Colombian students who achieve high performance in the area of Mathematics is minimal with 1.8% [5]. Similarly, the State tests Saber 11, implemented to measure the educational quality of Colombian people, show that students reach higher education with low levels of performance in the area of mathematics, implying high rates of desertion and repetition in the first academic semesters [6]. The above is one of the problems that Higher Education Institutions must address in order to train mathematically competent professionals, so that a possible way of tackling the problem is through the (ICT) incorporation in training processes, mainly in the area of mathematics, since they generate different ways of learning and several of them are freely accessible. For example, digitals tools such as: Geogebra, Moodle, Mathway, Khan- Academy and others similar tools in between are able to benefit the learning of math's because they facilitate the making of didactics activities where the mathematical concepts are presented in a virtual and interactively way, allowing to relate the mathematics with real current living situations in accessible and attractive learning spaces that are used as support material to the developement to the development of each student self learning.

Indeed, Khan-Academy is an innovative technological tool where different areas of knowledge can be learned online: mathematics, biology, chemistry, physics, computers, humanities, economics, finance and history. Khan-Academy courses are aimed at all educational levels and are offered free of charge [8-9]. The formative process at Khan-Academy is based on in an artificial learnign system that generates one unique training path for each student, depending from the reasults obtained int the diagnostic test, the system determinates the weaknesses, the strengths and skills in which the students are getting prepared to the learning processWhile the students Works on the platform, Khan- Academy periodically evaluates them through "Domain Challenges " to ensure that the themes learned will be remainded and retained over the time is passing through. Khan-Academy courses include several study material like videos, exercises, clues and instant feedback, which makes the different skills in the student to be formed and strengthened. Likewise, Khan-academy allows the student to go at their own pace and self-regulate their own learning resulting in people with extensive knowledge and excellent academic performance [10-11]. In Chile [12], the Khan- Academy massive incorporation results in the math classrooms highlight that the use of Khan-Academy improves the academic performance, it encourages the students participation, increases the collaborative learning and develops the self-learning, besides outstanding references has been said about what it is highly relevant to the Khan-academy and that is a universally adaptable tool for different types of teachers, students and countries.

Finally, assuming the primordial case in which the Technological Units of Santander, a Colombian Higher Education Institution, is involved; the previous problem is enlarged by the inclusionary component, where the Institution allows to receive all the students who are bachelor and who have presented the Saber 11 test without having Consider a minimum score to get admitted in the institution. Therefore, in the second semester of 2018, the PREIN-UTS course was implemented as a solution alternative in order to improve academic levels and favor student retention in the subjects of the area of mathematics. The PREIN-UTS course is an online training, of free use, implemented in the Khan-academy technology platform. The course addresses the thematic axes of: fundamentals of algebra, algebraic expressions, quadratics and polynomials, linear equations and inequalities, geometry and equations, line and slope graphs, systems of equations and expressions with exponents.

2. Methodology

The methodological approach was quantitative of exploratory type. For the analysis of data, descriptive statisticians were used in the SPSS software. The random sample was made up of students from all the academic programs, previously inscribed in the actual modality, admitted or enrolled in the Technological Units of Santander during the second academic period of 2018. The research was divided into the following phases:

2.1 First Phase.

Implementation of the PREIN-UTS course. Via e-mail and through the institutional Contact Center, the invitation to the first semester students was extended to carry out the free virtual course PREIN-UTS, in order to level the mathematical pre-knowledge. The activities to be executed on the platform were the thematic axes of "Fundamentals of Algebra", which collect on a large scale the competences of the educational levels of basic and secondary education according to the curricular guidelines of the Ministry of National Education MEN.

2.2 **Second Phase.**
Collection of performance reports in the PREIN course generated by the Khan-Academy platform. After the first implementation of the PREIN course, continuously in this stage it gets proceeded where reports were generated from the Khan-Academy platform regarding the thematic axes of the "Fundamentals of Algebra" course, composed of 104 skills.

2.3 Third Phase.
Data analysis. The results obtained were analyzed considering the following study variables:

- Descriptive variables: gender, headquarters, academic program, faculty and group.
- Khan-Academy Variables: skills, activities and gamification.

The classification of the Skills variable was calculated using the qualitative scale: with difficulty, it needs practice, practiced, level 1, level 2 and dominated. While the variable Activity was assessed taking into perspective the total minutes, minutes of videos and skills minutes. Finally, as the evidence for the Gamification variable, the energy points and the won medals have been taken into account.

3. Results
In the second academic semester of 2018, the first implementation of the PREIN course was taking form properly alluding with the considerable participation of 801 students distributed as follows: 5.9% corresponding to students inscribed or admitted, but in fact not enrolled in the Institution, which are considered within the extension factor that the Institution contributes to society. 94.1% corresponding to students enrolled in the Institution and who attended the first academic semester in 2018-II, (see Figure 1).

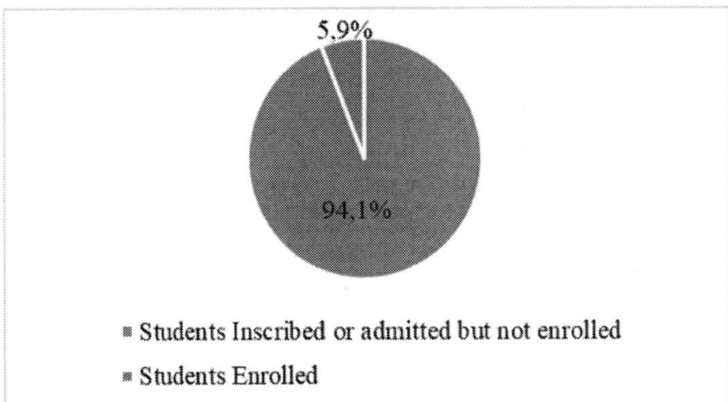

Figure 1. Sector graph of percentage distribution of students PREIN-2018-II according to their state

On the other side, 94.1% of the students enrolled in the PREIN2018-II course, equivalent to 754 students, obtained the following results: of the 1786 new students who entered the Institution to attend the first academic semester in 2018 -II, 42% completed the PREIN course. On the other hand, Figure 2 shows the percentage of students per headquarters, standing out a greater participation of students from the main campus. However, in spite of evidencing a minimum percentage of participation of the students of the regional headquarters, the accessibility of the course is highlighted since it made possible the decentralization of the institutional academic processes.

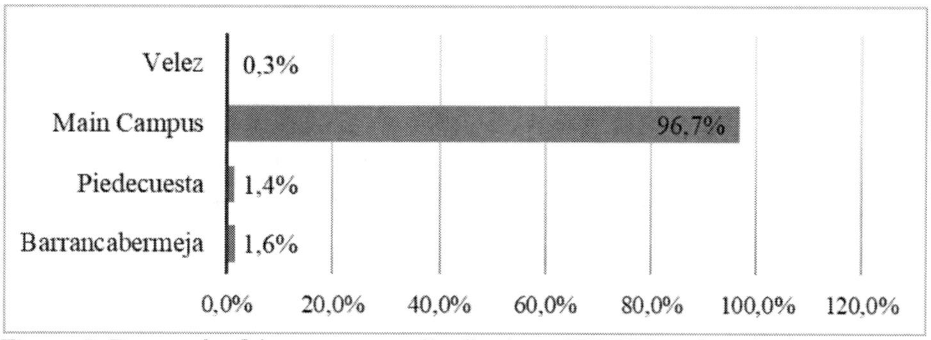

Figure 2. Bar graph of the percentage distribution of PREIN students by headquarter

Likewise, Figure 3 shows the percentage of participation of PREIN students by academic faculties, evidencing greater participation by students of the Faculty of Socioeconomic and Business Sciences with 61%.

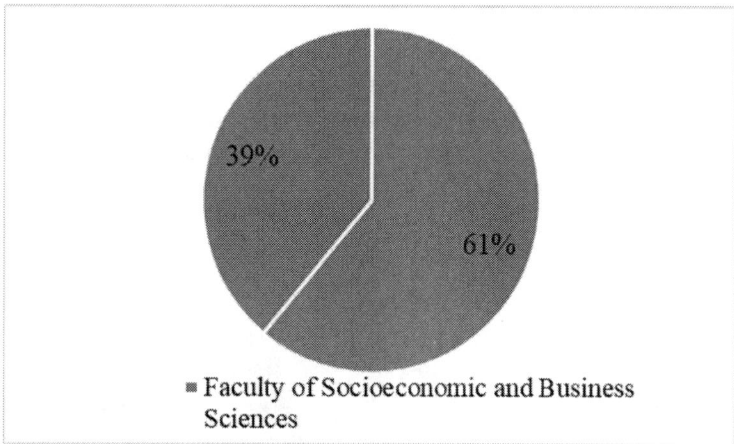

Figure 3. Sector graph of the percentage distribution of PREIN students by Faculties

Now looking at the participation of students by academic programs, Figure 4 shows that Technological program in Financial Accounting (N = 137), Technological program in Business Management (N = 104) and Technological program in Sports (N = 93), programs belonging to the Faculty of Socioeconomic and Business Sciences, they obtained a greater number of participants in the course.

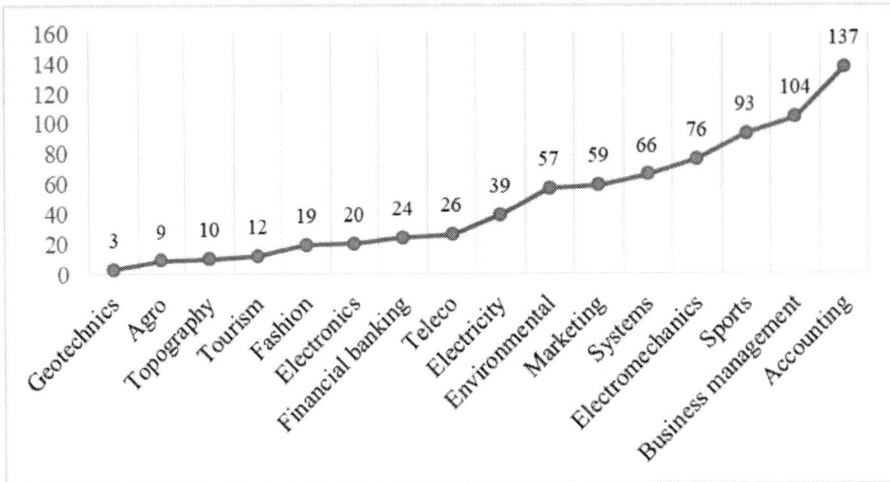

Figure 4. Line graph of the distribution of the number of PREIN students by academic programs

Regarding the time invested in the development of the PREIN course, it is evident that the students spent an average of 468.53 minutes on the platform, distributed as follows: 51.24 minutes watching video tutorials and 417.29 minutes to practice missions (see Figure 5). From the above, it is concluded that students prefer to spend time solving exercises than observing videos.

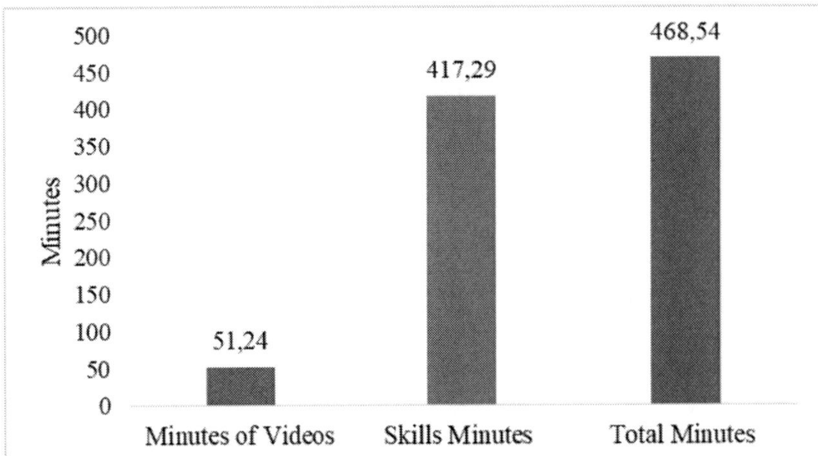

Figure 5. Histogram of the average time spent by students according to PREIN activities.

From referential point in the PREIN course of 104 skills measured in the qualitative scale: With difficulty, needs practice, practiced, level 1, level 2 and dominated, it is evident that 66.46% of students managed to position in the levels: practiced, level 1, level 2 and mastered more than 40 skills (see Figure 6).

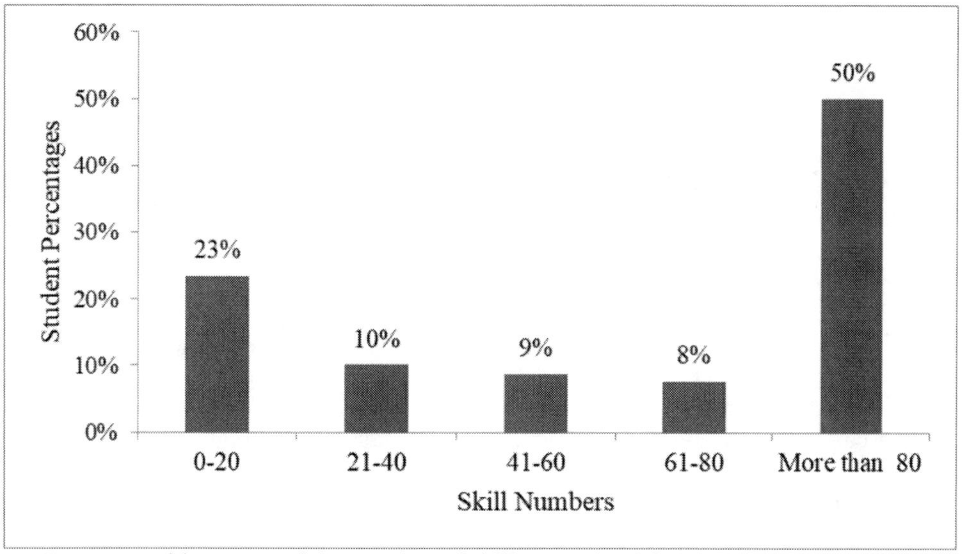

Figure 6. Percentage histogram of the number of skills: practiced, level 1, level 2 and mastered in the PREIN course.

Regarding the academic performance of the first cut in the subjects of the area of mathematics, the results show that the students who made the PREIN course obtained a higher percentage of approval of the subjects with 60%, compared to the students who did not Take the course, which obtained an approval percentage of 42%. In the same way, the average of the grades was higher for the students who took the PREIN course 3.26, unlike those who did not complete the course, which was 2.64. Finally, we will see that during the first academic cut none of the students who made the PREIN course defected from the subjects of the area of mathematics.

The technological tools may affect the math's learning, however, Khan- Academy has a ultimate adding value in referent to the others technological tools due by the fact that maintains the usage of an artificial learning system, the high quality training process that conform the different courses and the easily adaptability in different teachers methodology type, students variable and countries.

4. Conclusions

The primary fact that Khan-Academy is a free software makes the implementation of the PREIN-UTS course generate minimal costs to the institution contributing to the use of (ICT) in improving student learning in the area of mathematics and implementation of didactic and innovative strategies easily accessible in higher education. The PREIN-UTS course contributes to the promotion of educational quality by strengthening and ensuring the mathematical pre-knowledge of new students, thus helping to mitigate the decline and repetition in the first academic levels with respect to the subjects of the area of mathematics.

Accordingly to several studies made in Chile [12] we positively can conclude that the Khan-Academy Platform is an easy implementation system and can be really usefull to start the most profounds pedagogical procces in the students mathematical competences contribuying to mitigate the academic challenges that The Colombian HEIs are facing.

References

[1] Martínez P, Pérez J and Martínez M 2016 Las TICS y el entorno virtual para la tutoría universitaria *Educación XXI: Revista de la Facultad de Educación*, **19**(1) 287-310
[2] Pabón J 2014 Las TICs y la lúdica como herramientas facilitadoras en el aprendizaje de la

matemática *Eco matemático*, *5*(1) 37-48
[3] Utomo H, Bon A and Hendayun M 2018 The integrated academic information system support for education 3.0 in higher education institution: Lecturer perspective *Journal of Physics: Conference Series* **1049** 012102
[4] Valdés A, Angulo J, Urías M, García R and Mortis S 2011 Necesidades de capacitación de docentes de educación básica en el uso de las TIC. *Pixel-Bit Revista de Medios y Educación* ***39*** *211-223*
[5] Murcia M and Henao J 2015 Educación matemática en Colombia, una perspectiva evolucionaria *Entre Ciencia e Ingeniería*, **9**(18) 23-30
[6] Rodríguez G, Gómez V and Ariza M 2014 Calidad de la educación superior a distancia y virtual: un análisis de desempeño académico en Colombia *Investigación & Desarrollo* ***22***(1) 79-120
[7] Casales R, Rojas J and Paulí G 2008 Algunas experiencias didácticas en el entorno de la plataforma Moodle *Revista de informática educativa y medios audiovisuales* *5*(19) 1-10.
[8] Thompson C 2011 Cómo Khan Academy está cambiando las reglas de la educación. *Revista alámbrica* ***126*** 1-5
[9] Rueda K and Serrano O 2019 Online training for the strengthening of mathematical pre-knowledge mediated by Khan-Academy platform *J. Phys.: Conf. Ser.* **1161** 012019
[10] Cardeño J, Muñoz L, Ortiz H and Alzate N 2017 La incidencia de los objetos de aprendizaje interactivos en el aprendizaje de las matemáticas básicas, en Colombia *Revista Trilogía* ***9***(16) 63-84
[11] Rueda K and Rey M 2018 The learning of differential calculus mediated by the platform Khan-academy *J. Phys.: Conf. Ser.* **1126** 012047
[12] Light D and Pierson E 2014 Increasing student engagement in math: The use of Khan Academy in Chilean classrooms International Journal of Education and Development using ICT **10**(2) 103-119

Methodology to define an integration process between frameworks SCRUM, Django REST framework y Vue.js, implemented for software development, from quality management approach and agility

J G Arévalo[1, 2], L Viecco[1] and L Arévalo[1]
[1]University of La Guajira, Riohacha, La Guajira – Colombia.
[2]ENAPROT, Investigation Group of the University of La Guajira.

E-mail: jarevalo@uniguajira.edu.co

Abstract. At present, the industry has great challenges for the construction and maintenance of software, in terms of quality, costs and time, for this it is significant to have a methodological framework, which allows the possibility of having effective and practical methods, applying programming techniques from an agile approach, which guarantees the construction of quality code from the beginning of software development. This study analyzes the integration between frameworks SCRUM [1], Django REST framework [2] and Vue.js [3], which are commonly implemented in isolation for the execution of software development projects, and components are also incorporated of the IEEE 830 standard defined for the specification of software requirements [4], and of the ISO 25000 standard for software quality assessment [5]. Likewise, the characteristics of each study framework within the software architecture are analyzed, establishing of traceability between them and obtaining as a result a methodology for the definition of a process map; where each framework is integrated and it is described the procedure for structuring the planning, development and delivery phases. Thus providing a guide for the application of programming techniques from the concept of agility and quality management, from the process of integration of the frameworks analyzed in this study.

1. Introduction

There is a tendency for the use of agile methodologies for software development projects; because they are included in frameworks where different layers of the architecture of a project are structured, planned and controlled significantly reducing the execution time of the same. Currently, a large number of technologies and frameworks are available, however, in a relative way they are directed to medium-scale implementations down. In this study, the analysis of various frameworks is presented, to define the architecture of an integrated process, having a practical guide for the development of scalable software projects, based on their use.

In the software architecture the different levels for the development of a system are grouped, from the design, data storage, views for the validations and data processing, security aspects, quality, among others. Beyond the programming and definition of data schemas, a guide or methodology is required to keep track of the activities that are carried out throughout the architectural process. Several studies have presented cases focused on the integration of work frameworks, however on the one hand they focus on the permanent search to ensure software quality [6] [7] [8]; and on the other hand they are grouped into having agile tools [9] [10]. The referenced research's guides this study to obtain the methodology to define the process of integrating SCRUM frameworks, Django REST framework and Vue.js for software development from the agility and quality management approach.

2. Theoretical bases

The following sections present related concepts, used as theoretical bases in this study, including: Software architecture components (section 2.1), Development languages and tools (section 2.2) and frameworks (section 2.3).

2.1. Software architecture components

2.1.1. Software architecture. It is part of the fundamentals of software engineering and is also a broad term, so it is necessary to specify its definition for this particular study, this definition refers to the stages of software development throughout the execution of a project, where each main component is included, its behavior within the system, the iterations between the components and the coordination in search of reaching each objective of the system. It is important to clearly have every stage of the development of the architecture, because in the search to guarantee the quality of the product, in some stages the standards recognized by the Institute of Electrical and Electronic Engineers are implemented and developed - IEEE, world association of engineers dedicated to Standardization in technical areas; and by the International Organization for Standardization (ISO), dedicated to the creation of international standards; In addition to establishing the traceability of each frame for each architectural stage, in the application of the agile manifesto.

The Standard ISO/IEC/IEEE 42010: 2011 [11], addresses the creation, analysis and sustainment of architectures of systems through the use of architecture descriptions. A conceptual model of architecture description is established. The required contents of an architecture description are specified. Architecture viewpoints, architecture frameworks and architecture description languages are introduced for codifying conventions and common practices of architecture description.

2.1.2. Integrated Development Environment (IDE). Corresponds to the logical components or applications, which, depending on their use, allow, on the one hand, the definition of the database schemas and, on the other, the software programming. The IDE integrates tools, libraries or components that allow to speed up the development process. An alternative not to use the IDE and that is used with great frequency are the text editors, which, like integrated environments, allow the creation of digital files in the formats of the different programming languages. They are used for the development of server-side code and for user-side interfaces.

2.1.3. Hardware. Corresponds to computer equipment or input or output devices, used to perform activities related to data processing and software development.

2.1.4. Database. It is the set of structures or schemes, defined logically for data storage. The definition of these schemes, for the most part, must respond to the data generated from the definition of the functional requirements. These schemes must also allow the consultation of data for proper processing and have the required reports. There are several managers that allow you to manage the content of the databases and are defined as database management systems DBMS. On the other hand, the database engines are responsible for providing the protection of the schemes, as well as allowing the storage and processing of the data.

Finally, it is important to indicate that the schemas can be defined in different structures, the most currently implemented are the Relational Database - BDR, BDOR Relational Object Database and Non-Relational Database - BD NoSQL.

2.1.5. Software Requirements Specification. The description of the requirements, is made up of a list of functionalities that the software to develop must comply with, also includes the use cases of the iterations that the user would have in the software. The IEEE 830 standard [4], has a structure for the specification of software requirement, where it contains the requirements at a level of detail sufficient to allow designers to design a system that meets the needs of the client and allow the team plan and perform tests that meet or not the established requirements.

2.1.6. Life cycle of a project. It is significant to have clarity about the life cycle for the execution of the projects, depending on the type of project they can have several stages, however, in the vast majority for software management projects, the stages are constituted as the specific guide of the fundamentals for project management [12]: start, planning, execution, monitoring-control and closure; Thus, in this way they have been established in various methodology for software development, however, difficulties have arisen in relation to the current challenges of the industry for the construction and maintenance of software, in terms of quality, costs and time.

However, for its part SCRUM encompasses the life cycle stages of a software development project in: Product Backlog, Sprint Planning, Sprint Backlog, Sprint, Sprint Burndown, Release or Product Burndown, Impediments Backlog, Product and Retrospective Increase, throughout these stages the requirements are specified, each work day is established, the list of tasks is made, each task is executed with a time limit, the progress is monitored, the rhythm of delivery of functionalities tested, deliveries are made to the final customer and the process is retrospectively and how the team works, including techniques, practices and development tools.

2.2. Development languages and tools

2.2.1. Structured Query Languag – SQL. It is a language designed for the administration of transactional database management systems. Based on the use of SQL statements, statures or schemas are defined and grouped into data definition language (DDL); records management in relation to the basic actions to create, read, update and delete records known by the acronym CRUD, are grouped into data manipulation language (DML); to manage and control access to data, the data control language (DCL) statements are used; and to control the processing of transactions in the database the transaction control sentences (TCL) are used.

2.2.2. Python. It is an interpreted programming language, with an intuitive syntax that favors readable code. It is a multi-paradigm programming language, since it supports object orientation, console programming, web programming, and mobile programming and commonly used for the analysis of large volumes of data and artificial intelligence.

2.2.3. JavaScript. Like Python, it is an interpreted programming language, dialect of the ECMAScript standard. It is defined as object oriented, prototype based, imperative, weakly covered and dynamic.

2.3. Frameworks

2.3.1. Methodology of development software. It is a framework used to structure, plan and control the development process in information systems.

2.3.2. Scrum. It is a very complete agile framework for project development. Scrum is an adaptable, iterative, fast, flexible and efficient framework that is designed to deliver value to the client throughout the project development. The primary objective is to meet the needs of the client through an environment of transparency in communication, collective responsibility and continuous progress [1].

2.3.3. Django rest framework. It is a powerful and flexible toolkit for creating web API [2]. Some of the reasons that define in its documentation to use this REST framework are: i) The web browsing API is a great usability gain for developers, ii) Authentication policies that include packages for OAuth1a and OAuth2, iii) Serialization compatible with ORM and non-ORM data sources, iv) Customizable to the end: only use regular function-based views if you don't need the most powerful functions, v) Used and trusted by internationally recognized companies, such as Mozilla, Red Hat, Heroku and Eventbrite.

2.3.4. Vue.js. It is defined as the progressive framework for JavaScript [3]. Integrates HTML, CSS and JavaScript, reference languages used in the development of client-side views. Similarly, an ecosystem that can be adopted incrementally and that can be scaled between a library and a full-featured framework is defined as versatile.

3. Methodology for defining the integrated process map

3.1. Framework features

From the analysis of the characteristics of each study framework, they can be classified within the architecture of a software development project. The SCRUM structure is selected for the project execution process. The processes implemented in SCRUM are structured in time boxes as one of the principles of the framework, and that allows to effectively manage the planning and execution of a project, where you can have several iterations throughout in order to increase the product in the deliveries to the end user, these iterations are divided into the planning, development and delivery phases.

Django REST framework, is made up of a powerful and flexible toolkit to create web APIs, therefore it is used for the development of server-side software or Backend, which is the part that connects to the database and the server. For its part, Vue.js is an open source JavaScript framework for creating single-page user interfaces and applications (SPA), its implementation will be on the client side or Frontend, which is the part of a website that interacts with the users. Table 1 shows the characteristics of each framework.

Table 1. Framework features

Items	Framework	Phase Scrum	Features
1	Django REST Framework	Development	• Powerful and flexible toolkit for building Web APIs. • Serialization that supports both ORM and non-ORM data sources. • Customizable all the way down - just use regular function-based views if you don't need the more powerful features. • JSON Web Token Authentication, Hawk HTTP Authentication and HTTP Signature Authentication. • Djoser and Django-rest-knox.
2	Vue.js	Development	• Build APIs for it frontend • Is incrementally adoptable, scalable between a library and a full-featured framework. • Integrates HTML, CSS and JavaScript. • Minimal Optimization Efforts.

3.2. Selection of standards

As described above, is selected the structure SCRUM for the planning and execution of a project is maintained, in relation to this, then in Table 2, the criteria for the incorporation of standards in the process of integration of the frameworks are presented. The selection of standards and integrate them into the process, is done in search of providing a guide for the definition of requirements and the evaluation of the software product, however, requirements can be defined with another method that allows to standardize the process, but it is recommended the continuation of the guide. The selected standards apply in different phases of software development.

On the one hand the selected criteria of the IEEE 830 allow to clearly specify the software requirements to have an overview of the product perspective, to define the restrictions that are imposed on the development, the assumptions and dependence that represent the factors that, if change, can affect the requirements specify, finally the functional and external requirements can be defined as design and performance, it is significant to highlight the standard is integrated in the planning phase. On the other hand, the selection of the ISO 25000 standard, allows to establish the quality requirements and the evaluation guide for independent developers, acquirers and evaluators, the standard is integrated into the delivery phase of the software product.

Table 2. Criteria for selection of standards

Items	Standard	Phase Scrum	Selection Criteria
1	IEEE 830	Plannig	• Overview of the software product perspective. • Restrictions that are imposed on development. • Assumptions and dependence. • Definition of specific requirements at the level of functionalities. • Definition of external requirements at the level of design and performance.
2	ISO 25000	Delivery	• Quality requirements. • Evaluation guide for developers, acquirers and independent evaluators.

3.3. Methodological integration

At this point, once the characteristics of the frameworks and the criteria for selecting the standards to be incorporated have been described, the process of their integration can be graphically represented, where the traceability of the frameworks is observed in the stages of Planning, development and delivery. In Figure 1, the defined process map can be observed, it should be noted that the structure of SCRUM is maintained, and the other study frameworks and / or standards are integrated as complementary mechanisms for the development of the execution phases of the project. During the execution of a project, several iterations can be presented, in order to increase the product for the end user.

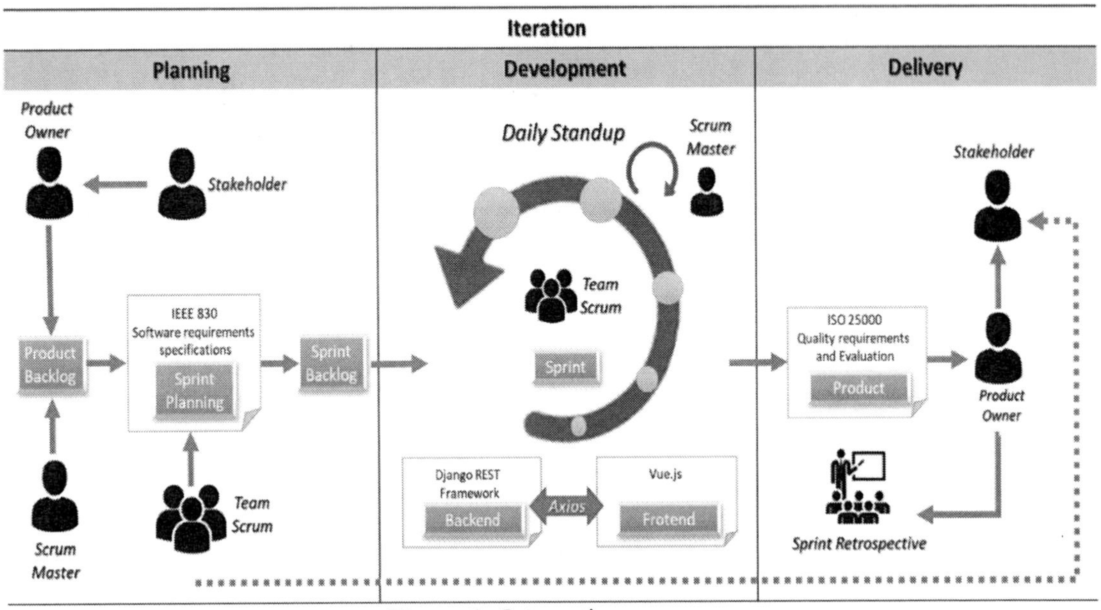

Figure 1. Integration process

The integrated model focuses on the unification of the planning, development and delivery phases, where the methodological components of SCRUM, Django REST framework and Vue.js are identified, as well as the complement of the IEE 830 and ISO 25000 standards. A brief description of the stages of the representative model of the integration process follows.

3.3.1. Plannig. In this phase the iteration of a project begins, from the proportion of requirements of the Stakeholder to the Product Owner, this turn, in the Product Backlog in establishes and communicates to the Scrum Master the prioritized business requirements. For their part, the Team Scrum and the Scrum Master, are responsible for the planning of each sprint, supported by the specifications of the IEEE 830 standard, obtaining as a result the Sprint Backlog, where the general vision of the product perspective is defined, the restrictions that are imposed on the development, the specific requirements at the level of functionalities, design and performance, and also the criteria of acceptance of the product. For the use of the IEEE 830 standard, the following sections are established.

3.3.1.1. Model overview

- Functionalities: This subsection will show a summary, broadly speaking, of the functions of the future system.
- User characteristics: This subsection describes the general characteristics of the users of the product, including educational level, experience and technical experience.

3.3.1.2. Requirements

- External interface requirements. Requirements that affect the user interface, interface with other systems (hardware and software) and communications interfaces will be described.
- Functional requirements. This subsection must specify all those actions (functions) that the software must carry out. In table 3, the form to specify the functional requirement is structured.

Table 3. Functional requirements

Identifier		:	FN-0X
Description		:	
Dates	Entry	:	
	Process	:	
	Exit	:	
Conditions	Initials	:	
	Finals	:	

- Design requirements. The design requirements are listed, in table 4 a simple structure is available for this.

Table 4. Design requirements

Requisite	Description
DI-01	short description
DI-02	short description
DI-0N	short description

3.3.2. Development. In this phase, each of the activities that were planned for each Sprint is developed, direct responsibility of the Team Scrum, where the deliverables are developed incrementally in each Sprint. For which it is necessary that on the backend side implement Django rest framework, and its three fundamental components corresponding to routers, views and sterilizers; i) the routers are the tools that allow to define the URLs of the API, arranging in an orderly manner a set of URLS for the methods based on HTTP such as POST, GET, PUT, among others; ii) the views are extensions of the Django class-views, which allow you to easily render an html in response, in a Json, XML or other data structure that interests us to return our API; and iii) the serializers allow to define in detail how the responses that our API will return will be and how we will process the content of the requests that come to us.

To install Django Rest framework, you need to install Python, from the command console you enter *"virtualenv -p python env"*, and then pip your dependency handler, induced in console *"pip install"*. The project is created with the command *"django-admin startproject"* + *"project name"*. Finally create an application (module) with the command *"python manage.py startapp"* + *"app name"*.

In each application (in terms of modularity) created in Django, the data schemas that are subsequently migrated to a database engine are established. SQL allows data management in several engines such as the database Oracle, PostgreSQL, MySQL, MariDB, among others. For a practical case, it is recommended to use one of the most common engines today and, which is compatible with Oracle, we refer to MySQL, for which it is necessary to implement the connector whit the command *"pip install mysql-connector-python"*; and general settings such as database name such as user, password, engine and auto-commit. Once the schemes are defined, migrations to MySQL can be performed. The files views, serializer and router are configured and in this way, the definition of the migrated database schemas in the respective engine, and the workable Api Web, would be used to consolidate the data in Json or other formats allowed by Django REST framework.

For the development of the frontend, the use of vue.js is required, implementing the *npm* your dependency manager, in console enter the command *"npm install"*, equally it is implemented the Vue CLI tool that allows the creation of user interfaces is installed with command *"npm install -g @vue/cli"*, from components where HTML, CSS and JavaScript are integrated, in addition to building and sharing reusable solutions for needs common. These interfaces will consume the processed data of the API built with Django rest framework, for which it is necessary to use Axios which is an HTTP client based on promises it is installed with the console command *"npm install axios"*. Beyond consuming and displaying API data, it also communicates with the functions of the server to execute the main actions of CRUD, after configuring the access for it.

In this same phase of development, in order to monitor the progress of each sprint, daily meetings of 15 minutes of obligatory nature are held to discuss the progress and inconveniences that arise, these meetings correspond to the Daily Standup, where the Team Scrum and the Scrum Master. From now on, several Vue components can be implemented to improve the design of the interfaces, such as bootstrap-Vue, which allows the implementation of templates developed with bootstrap for the vue.js components, another recommended library is sweetalert for the visualization of the notifications or alerts on the interfaces, among other libraries that will allow a better experience in terms of use by the end user.

3.3.3. Delivery. At the end of each Sprint, the application of product quality tests begins, defined in the ISO 25000 standard, where quality requirements are evaluated, for verification and incremental delivery of the product to the Product Owner and in cases where relevant Stakeholders are required. Throughout this phase, at the end of each Sprint, the functional requirements, design, performance, and the acceptance criteria of the product are also verified, in addition the progress and the rhythm of delivery the functionalities tested are monitored and the retrospective of the process and the way in which the team works, including techniques, practices and development tools.

The Quality Measurement Division corresponds to ISO/IEC 2502n, these standards include a reference model of product quality measurement, definitions of quality measures (internal, external and in use) and practical guidelines for their application. Table 5 shows the divisions of the selected standards for the methodology.

Table 5. Divisions of the standards ISO 25000

Items	Division ISO / IEC	Description
1	25022	Measurement of quality in use: specifically defines the metrics to measure the quality in use of the product.
2	25023	Measurement of system and software product quality: specifically defines the metrics for measuring the quality of products and software systems.
3	25024	Measurement of data quality: specifically defines the metrics for measuring data quality.

4. Evaluation test and resulted.

The methodological guidelines presented solve a problem of integration of frameworks that act in different layers of the software development architecture. The description of each integrated component automates the process of selecting and analyzing the characteristics of the frameworks, as well as their integration, and establishes the procedure to structure, plan and control the software development process. To evaluate the suitability of the proposed methodology to integrate the frameworks, a panel of experts in the area of engineering, architecture and software development was convened, consisting of teachers assigned to the systems engineering undergraduate of the University of La Guajira, academic program It has a high quality accreditation certificate from the Ministry of National Education, a component that provides reliability in the evaluation by the experts convened, and is also a national reference program.

To the panel convened, they were presented with the methodological proposal and also the evaluation criteria, defined in Table 6. Evaluation criteria of the methodological guideline. Subsequently, in Table 7. Results obtained in the evaluation, the consolidation of the evaluation carried out by each invited expert can be observed, in relation to each evaluation criterion.

Table 6. Evaluation criteria of the methodological guideline

Items	Evaluation Criteria (EC)
1	The descriptive level of the components of the methodological guidelines is adequate.
2	It considers the use of methodological guidelines for integration with the frameworks of study work viable.
3	An adequate solution is represented, for the integration of frameworks for backend and frontend development.
4	The technologies implemented represent modern or innovative technological tools.
5	The basic foundations presented are in accordance with the technologies implemented.

Table 7. Results obtained in the evaluation

Items	Results				
Developer Experts (DE)	*Score by evaluation criteria (EC)*				
	EC # 1	*EC # 2*	*EC # 3*	*EC # 4*	*EC # 5*
DE # 1	*4.5*	*4.6*	*5.0*	*4.2*	*4.8*
DE # 2	*4.7*	*4.4*	*4.6*	*4.9*	*4.8*
DE # 3	*4.5*	*4.3*	*4.5*	*4.4*	*4.9*
DE # 4	*4.6*	*4.4*	*4.8*	*4.4*	*4.5*
DE # 5	*4.8*	*5.0*	*4.4*	*4.6*	*5.0*
AVERAGE	***4,6***	***4,5***	***4,7***	***4,5***	***4,8***

The results of the evaluation reflect that, the descriptive level of the components that integrate the proposed methodological guidelines is adequate, in addition the use of the defined process with the frameworks of study is considered viable, also, and it allows to show that the process represents a solution convenient, for the integration of innovative technologies implemented in software development. For the rest, in this study the analysis of the characteristics of the theoretical foundations, which allowed to determine the process specifications of each study framework, within the software development architecture is left in this study.

5. Conclusion

In this study, developing effectively methodology to define an integration process between the frameworks implemented for software development, from the agility and quality management approach. In the process, the criteria for selecting the implemented standards and the characteristics or components of each framework for its integration were defined for each stage of the software architecture. Finally, a process map is presented that leads to the result of the integration of the frameworks SCRUM, Django REST framework and Vue.js.

The integration of SCRUM, Django REST framework and Vue.js, has a guide to positively impact the planning, development and delivery of software, providing for each iteration incrementally a product of greater value for the client. The above from the use of the efficient tool that supports the process with greater precision because it allows through the definition of three phases to manage and control each activity of the project; where IEEE standard are integrated to define specifications that are required, the use of framework for development significantly reducing the times of the same and applying methodologies for the agility of the project, but still does not guarantee that the product is of quality; for which they are complemented with ISO standards, which allows to assess compliance with the parameters that are stable to measure the quality of the software product.

References

[1] Platzi. 2017 , Qué es SCRUM y los roles en SCRUM. [En línea]. Available: https://platzi.com/blog/que-es-scrum-y-los-roles-en--scrum/?utm_source=google&utm_medium=paid&utm_campaign=intention&utm_content=&adgroup=71299386940&gclid=EAIaIQobChMI8IGTt_OP4wIVx4SzCh0H3giVEAAYAiAAEgIZefD_BwE.

[2] Django REST framweork 2011. [En línea]. Available: https://www.django-rest-framework.org/.

[3] E. You, 2014 VUE.JS The Progressive JavaScript Framework, [En línea]. Available: https://vuejs.org/.

[4] Mendez G 2008 Especificacion de Requisitos segun el estandar IEEE 830. [En línea]. Available: https://www.fdi.ucm.es/profesor/gmendez/docs/is0809/ieee830.pdf

[5] ISO - Organización Internacional de Normalización, Portal ISO/IEC 25000, 2019. [En línea]. Available: https://iso25000.com/index.php/normas-iso-25000.

[6] Holguien Barrera J A 2015 Integración de marcos de trabajo para desarrollo de software: Scrum, PSP e ISO 25000, *Ventana Informatica,* nº 32, pp. 151-164.

[7] Herrera Uribe E and Valencia Ayala L E 2007 Del manifiesto ágil sus valores y principios, *Scientia et Technica,* vol. XIII, nº 34, pp. 381-385.

[8] Mercado Ramos V H, Zapata J and Fernado Ceballos Y 2015, Tools and good practices for software quality insurance under skilful methodologies, *Revista de invetigacion desarrollo e innovacion: RIDI,* vol. 6, nº 1, pp. 73-83.

[9] Sihuay M, Dávila A and Pessoa M, Factores en la Adopción de Métodos Ágiles en el Proceso de Desarrollo de Software: Revisión Sistemática de la Literatura.

[10] Vivas L, Cambarieri M, Garcia Martinez N, Petroff M and Muñoz Abbate H 2014 Un Marco de Trabajo para la Integración de Arquitecturas de Software con Metodologías Ágiles de Desarrollo, *ResearchGate,* pp. 1288-1297.

[11] International Organization for Standardization 2019, "IEEE 42010-2011 - ISO/IEC/IEEE Systems and software engineering -- Architecture description," 31 10 2011. [Online]. Available: https://www.iso.org/standard/50508.html

[12] PMBOK® 2017 Guide, Guide to the Project Management Body of Knowledge, Pennsylvania: Project Management Institute.

Acknowledgments
The authors thank the help of colleagues from the research group ENAPROT, engineering faculty and to the research Center of the University of La Guajira, and the organizing members of EXPOTECNOLOGÍA 2019 of the Antonio de Arévalo Technological Foundation.

IOP Conf. Series: Materials Science and Engineering **844** (2020) 012023 doi:10.1088/1757-899X/844/1/012023

Design of a library using Matlab software for modeling and simulation of parabolic cylindrical solar collectors (CCP)

S A Sanchez [1], M A Quiroz[1], LR Sierra [1], A D Morales [1], J M Palencia[1]

[1]Corporación Universitaria Antonio José de Sucre, Sincelejo, Colombia, Faculty of Engineering Science, GINTEING Research Group

sergio_sanchez@corposucre.edu.co

Abstract. Much of the energy available to the earth comes from the sun, it is the largest source of radiation, emitting approximately 1353 W/m2 of energy into the Earth's atmosphere, in the form of electromagnetic waves, with wavelengths (λ) between 0,3 and 3 μm. Within this spectrum we find the following radiations: Gamma (1%), Ultra violet (4%), Luminous (49%) and Infrared (46%). Only 51% of the solar radiation that enters the atmosphere reaches the earth's surface. This radiation represents a main source for renewable energy, presenting great applicability in solar thermal energy and specifically in the use of parabolic cylindrical solar collectors; that take advantage of this radiation to raise the temperature of a fluid, located in a receiving tube. The great problem of these systems is to generate a good design, which allows to reach high temperatures and good yields, therefore, many investigations have generated multiple design software, which mitigates the margin of error, but generally this software are limited; With licenses, they are complex and do not specify all the features for processing. The main objective of this research is an open source library that allows obtaining the highest efficiency of parabolic cylindrical collectors, based on the theoretical references Duffie, Beckman and Kalogirou. For this, the Matlab software and the Soltrace tool were taken into account, in order to validate the sizing and optical, thermal, geometric and global performance. Within the results, the maximum concentration ratio of the parabola is checked, which is obtained when the edge angle is 90 °. A functional library can be generated, where multiple measurements and shapes of the parabola were simulated, obtaining the highest overall performance, when the focal length is equal to the maximum height of the parabola, obtaining a performance of 62.64%.

1. Introduction

The sun is a sphere of gas from which a large amount of light and heat emanates. Life on Earth depends closely on the amount of solar radiation, is the largest star in the solar system, emits a plurality of electromagnetic waves, considering that energy as solar radiation, varies between ultraviolet rays far to gamma rays.

The sun is a black body whose surface temperature is 5,500 ° K, emits radiation continuously throughout the space, radiating approximately $4x10^{26}$ W. Due to this, the radiated solar energy is distributed as follows; Gamma (1%), Ultra violet (4%), Luminous (49%) and Infrared (46%), Gamma and ultraviolet radiation have a higher energy level, due to their low wavelength, and represents 5% of total solar energy.

The energy of the visible and infrared spectrum represents 95% of the total, and is the one that must be taken into account for its use. Of this large portion of energy that enters the atmosphere, only 51% reaches the surface of the earth, because 4% is reflected on the surface, 20% is reflected in the clouds, 6% is reflected in the atmosphere, and 19% is absorbed by the atmosphere and clouds. The radiations that reach the earth can be due to the albedo effect, direct or diffuse.

There are different types of energy sources; non-renewable and renewable; the non-renewable sources that are found in nature in a limited quantity and once consumed in their entirety, cannot be substituted, since there is no production or extraction system, for example; fossil fuels (coal, oil, natural gas) and nuclear fuels (uranium, plutonium, among others). On the other hand, there are renewable sources, which are those that are obtained from virtually inexhaustible natural sources, either because of the immense amount of energy they contain, or because they are capable of regenerating by natural means, for example; solar, wind, geothermal, hydroelectric, tidal, wave, biomass, among others.

The sun is one of the virtually inexhaustible renewable source, so it has multiple uses and applications, within the collection of solar energy can be divided into passive thermal collectors, active thermal collectors and photovoltaic collectors. For the particular case of this investigation, thermal energy will be studied in active thermal collectors, specifically in parabolic cylindrical solar collectors.

For a long time, solar collectors have been objects of study within the use of solar thermal energy, in the 19th century the first investigations were initiated to apply and develop technologies that would allow this energy to be used as a caloric and electrical source on a commercial scale and massive, [1] qualifying it as a new alternative of renewable energy, solar thermal energy defines the release of heat in a system, this heat is transferred from a hotter body to another that has less heat, so that, while more heat be introduced, the higher the temperature until a phase modification is reached, this is widely used to change from liquid to gaseous state and for any state of matter [2].

Such investigations caused the invention of various systems for the use of solar energy from the sun, obtained by means of solar panels and mirrors, so that the mirrors, concentrate the light in a receiver that reaches temperatures of 1,000 ° C, heat produced is used to heat a fluid that generates steam, important for multiple uses and derivative systems [3], however, for the collection and concentration, not only enough with common mirrors, so the design of these solar collectors has changed, highlighting the most used today; cylindrical parabolic collectors, central tower collectors, stirling, composite, semi-parabolic collectors, among others.

The parabolic cylindrical collectors are linear-focus solar collectors, which transform direct solar radiation into thermal energy, thanks to the heating of a working fluid that can reach 400 ° C in very favorable cases, within their applications they are found, drinkable water, disinfection, evaporation drying and distillation, among others.

Taking into account that, in 1866, Auguste Mouchout used parabolic trough collectors (CCP) to move a steam engine, that same year the first patent of a solar collector was granted and in subsequent years John Ericsson or Frank Shuman designed systems and plants complete for irrigation, refrigeration, and locomotion. Between 1984 and 1991, in the Mojave Desert (California), 9 plants known as SEGS (Solar Energy Generating Systems) with an installed capacity of 354 MW were built and connected, with these plants being the only commercial centers built until 2007 [4] in Spain, in 2008 the construction of Andasol 1 began, one of three 50MW power plants very similar to each other, being considered as the true beginning of the rebirth of the generation of solar thermal energy worldwide, since 2013 the SENER trough is the collector that is most frequently built today; However, for the modeling of these solar collection systems it includes the study of different designs, the geometric, optical and thermal parameters being relevant, which have a detailed and specific study related to theoretical-analytical thermodynamics, optical principles and geometric analysis [5].

Currently, there are different software that facilitate the design of libraries for solar collectors, including SPSS, Staca, SAS, R, model, soltrace among others. The relevance of this research is to achieve an open source library, scalable, functional, versatile, simple to use and mainly to obtain efficient prototypes.

2. Fundamentals

Solar collectors are elements that capture solar radiation and convert it into thermal energy, that is, they generate heat [6], they are composed of a flat or curved plate, of reflective material that direct and concentrate the light beams, in addition, of a receiver that receives concentrated energy by adsorbing the greatest amount of energy to convert it into heat, generating high temperatures [7].

To reach high temperatures in the collectors, it is important to modify the collection system, so that the concentration is increased, which leads to different types of collectors for each application, in the following section, we will observe the types of solar collectors in function of temperature ranges and solar concentration ratio.

			Name	Sketch	Concentration reason "C"	Temperature range reached [C°]
Stationary		Flat adsorber	Flat water collector	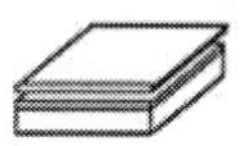	$C \leq 1$	$30<T<250$
Stationary		Flat adsorber	Empty tubes		$C \leq 1$	$50<T<200$
Followers	Track an axis	Tubular adsorbents	Compound parabolic collector CPC	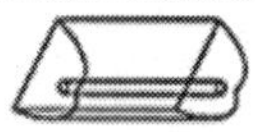	$1<C<15$	$70<T<300$
Followers	Track an axis	Tubular adsorbents	Parabolic cylindrical collector PCC	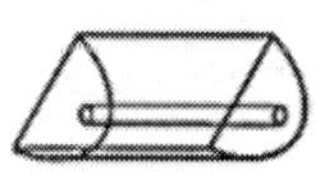	$15<C<40$	$70<T<350$
Followers	Two axis tracking	Point adsorbents	Parabolic disc		$100<C<1000$	$70<T<1500$
Followers	Two axis tracking	Point adsorbents	Central tower		$100<C<1500$	$150<T<1500$

Figure 1. Classification of collectors according to the concentration factor.

As can be seen in the image, the classification of solar collectors, where static collectors and followers, their respective subclassifications and the concentration ratio are more clearly denoted. On the other hand, the figure defines the attainable temperature range, where it is evident that the central tower collectors are the ones that can achieve higher temperatures, but their installation and sizing are more expensive compared to the others.

The performance of the systems is given by the typology of the concentrator, but also by the fluid used. However, it can be determined that the yield increases with concentration [8].

IOP Conf. Series: Materials Science and Engineering **844** (2020) 012023 doi:10.1088/1757-899X/844/1/012023

2.1. Parameters of cylindrical parabolic collectors or CCP.

Parabolic cylindrical collectors have shown in real systems efficiencies of between 10 and 15%, reaching maximum yields of up to 21.5% in any of the SEGS plants, currently the fluids used in the receiving tube are synthetic oils that can exceed 400 ° C. The CCP are composed of different parts, one of them is the geometric shape of a parabola, so that all the rays affect the focus of it, taking advantage of a much larger area where the sun's rays would be reflected, with the capacity of absorbing heat, where thermal energy is obtained that will be transformed later.

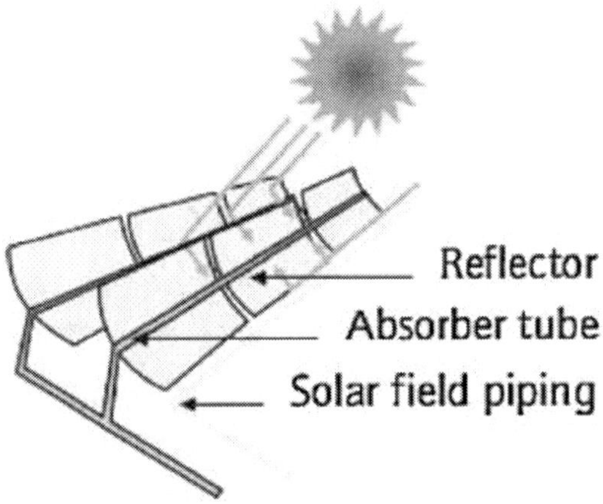

Figure 2. Parts of the parabolic cylindrical parabolic.

In Figure 3, the different parts of a parabolic cylindrical solar collector are appreciated; the parabolic reflector, the receiver tube and the pipes; All these elements play an important role in increasing thermal energy. When observing the collector transversely, we observe a parabola that is characterized by certain parameters, defined below.

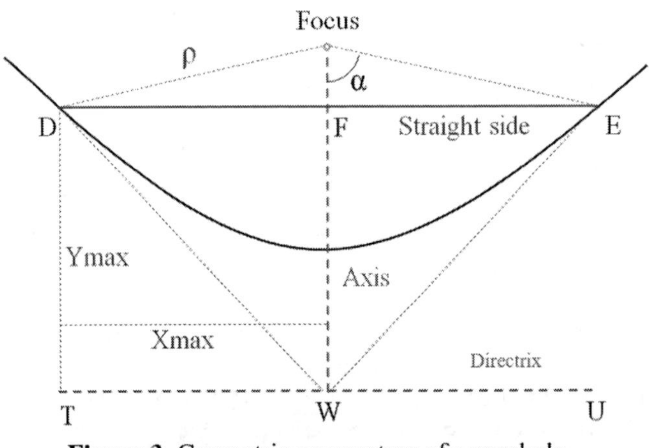

Figure 3. Geometric parameters of a parabola.

In Figure 3, the main parameters of a solar collector are specified, which defines ρ as the distance from the focus to the edge of the collector, α as the angle of incidence, the straight side that corresponds to the width of the collector, Ymax to the maximum height of the collector and Xmax the maximum width of half a parabola.

The conditions of the CCP models implemented for the design of the proposed library favor collectors with an incidence angle of 90 ° (α), to reach the highest solar concentration, according to the theoretical reference Kalogirou, the geometry of the CCP it is calculated based on the focal length, so that the straight side (Fig. 5) is 4 times the focal length.

$$Lrecto = 4 * Dfoco \qquad (1)$$

The parabola defines an Xmax, described as the maximum distance of half a parabola, therefore.

$$xmax = \frac{L\,recto}{2} \qquad (2)$$

To corroborate the parameters of the parabola, the manual geometric method was analyzed, by method of guidelines and the crossing point of perpendicular points on the directrix, through the focal length and any point of a directrix [9].

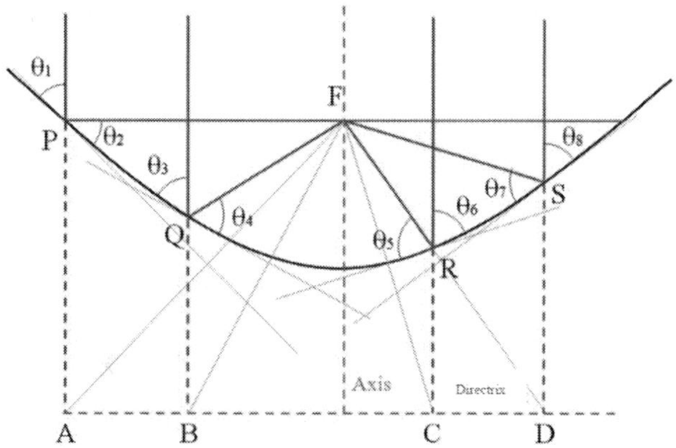

Figure 4. Principle of bisector and reflection.

In the red lines A, B, C and D, the principle of reflection on a flat surface is observed, where, the same angle of incidence θ1 is equal to the angle of reflection θ2, so the reflections are directed to the focus, at any point of the curved line. Likewise, with the bisectors (blue lines) are the points P, Q, R, S, which draw the curved line, which as it is constructed, the parameters of Figure 4 are observed.

For the construction of the parabola by means of the library, equation 3 was used, taking samples from –Xmax, to Xmax, finding Y in the Cartesian plane, such samples end until reaching Ymax in this case.

$$Y = \frac{x^2}{4*Dfoco} \qquad (3)$$

It is also necessary to find the distance of the curved side of the parabola from its vertex to its Xmax, so that:

$$S = \int_0^{X_{max}} \sqrt{1 + \frac{{X_{max}}^2}{4*Dfoco}} \qquad (4)$$

The incident angle is not always 90 °, nor is the distance ρ equal to Xmax, these parameters are defined in equation 5 and 6.

$$\alpha = \tan^{-1}\left(\frac{Xmax}{Dfoco-Ymax}\right) \qquad (5)$$

$$p = \sqrt{X_{max} + (Dfoco - Y_{max})^2} \qquad (6)$$

With the defined equations of the reflective area, we will design the receiver tube.

2.2. Receiver tube

For the design of the receiver or adsorbent tube it is necessary to take into account that the sun does not emit its energy in a timely manner, observed from the earth's surface, it is known that the solar disk corresponds to the projected area for an opening angle of 32 ° and does not emit its rays in parallel, however, they are distributed over a cone of solid angle directions of 16 °, which means that the rays do not reach the diameter of the larger receiver. [10], this is known as the cosine effect, as evidenced in Figure 5.

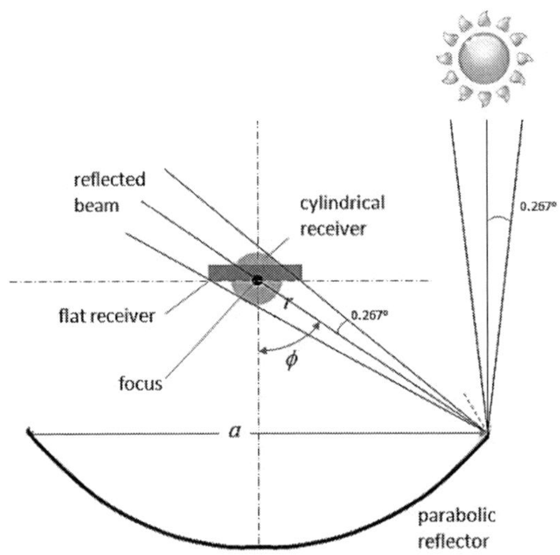

Figure 5. Solar projection.

The linear CCP receiver is the element that converts concentrated solar radiation into thermal energy, through a fluid. This is located on the focal length of the parabolic trough concentrator, being decisive for the overall performance of the collector [11]. Within the receiving tube, the working fluid is important, depending on the temperature that is intended to be reached and can be reached, as indicated in Table 1 below.

Table 1. Fluid temperatures in the manifold.

Fluid	Temperatures
Demineralized water	T < 200 °C
Synthetic oils	200 °C < T < 450 °C
Santotherm 55	T. Máx a 300 °C
Monsanto VP-1	T. Máxa 400 °C
Syltherm 800	T. Máxa 425 °C

There are different types of receiver tube design, the most efficient are those that work under the greenhouse effect principle, because of the temperature they can reach. This research focuses on this type of tubes.

IOP Conf. Series: Materials Science and Engineering **844** (2020) 012023 doi:10.1088/1757-899X/844/1/012023

This type of receiver tube consists of two tubes: one metallic interior (through which the fluid that heats up circulates) and another glass exterior (Figure. 6).

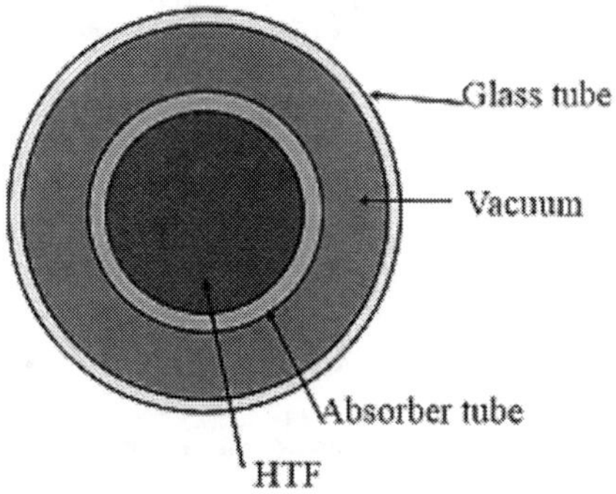

Figure 6. Parameters of a receiver tube.

To calculate both diameters it is necessary to apply the following equations with respect to Figure 6.

$$D_{min} = X_{max} * \frac{\sin(0.267)}{\sin\left(\frac{\alpha}{2}\right)} \tag{7}$$

$$D_{max} = X_{max} * \frac{\sin\left(0.267+\frac{1.15}{2}\right)}{\sin\left(\frac{\alpha}{2}\right)} \tag{8}$$

To calculate the thermal concentration factor for the adsorbent tube it is necessary to know the areas of both cylinders, like this:

$$FC = \frac{A\ capt}{A\ total} \tag{9}$$

Being, (Acapt) the catchment area and (Atotal) the area of the receiving tube.

In addition, it is also necessary to calculate the amount of accumulated soles (Cmax) and the volume (Vol) of the receiver area, defined as follows:

$$C_{max} = X_{max} * \frac{2*L}{Dtubo^{*}*L} \tag{10}$$

$$Vol = 1000 * \pi * \left(\frac{D_{max}}{2}\right)^2 * L \tag{11}$$

Being (L) the length of the collecting tube and (Dtubo), the diameter of the receiving tube of the market.

On the other hand, it is important to take into account in the receiver tube the transfer of heat by radiation, where some surface properties such as; emissivity, absorptivity and reflectivity. There is another property, transmissivity, which is usually of little importance. This can be corroborated in Figure 7.

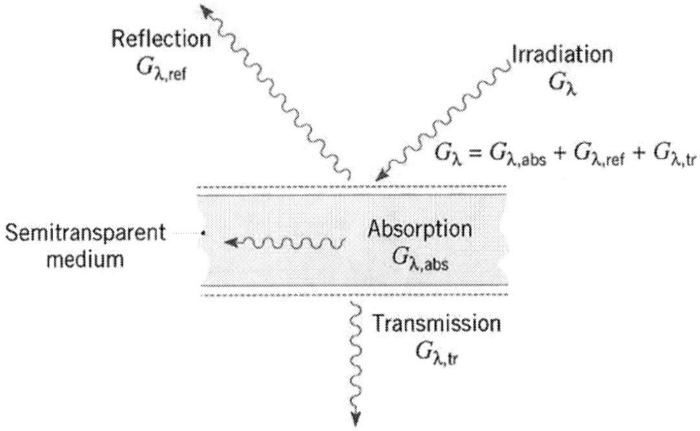

Figure 7. Radiation heat transfer.

The metal inner tube has a selective coating that has a high absorptivity (~ 94%) and a low emissivity in the infrared spectrum (~ 15%). Most of the coatings degrade in contact with the air, when they are hot, so it is required that there is a high vacuum in the chamber that is between the inner metal tube and the glass cover, in addition it usually takes an anti-reflective treatment on both sides, to increase its transmissivity, consequently, the optical performance of the collector. In Figure 8, the parts of a receiver tube are observed.

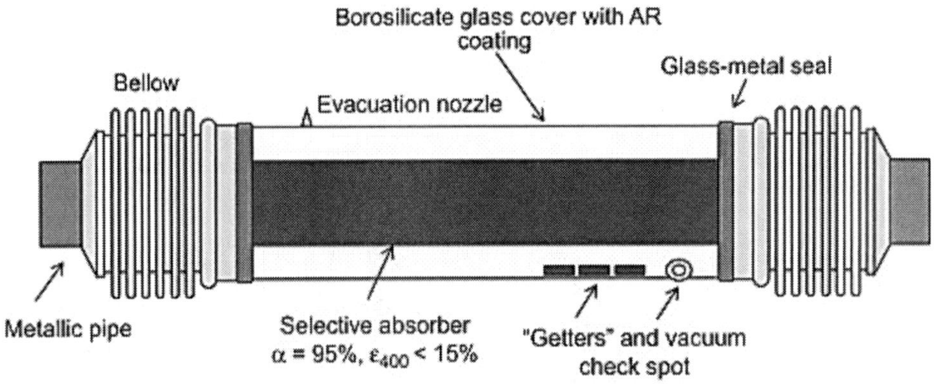

Figure 8. Parts of a receiver tube.

2.3. CCP requirement.

Normally, CCPs are installed in such a way that their axis of rotation is oriented in the East-West or North-South direction, although intermediate orientations could also be used, and thus obtain the greatest possible solar utilization and increase their overall performance [12] [13]. As seen in figure 9.

IOP Conf. Series: Materials Science and Engineering **844** (2020) 012023 doi:10.1088/1757-899X/844/1/012023

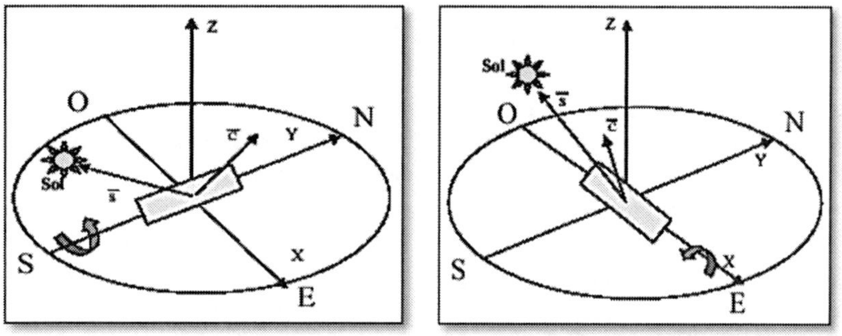

Figure 9. North-South and East-West Orientation.

In the process of thermal use of direct solar radiation by the CCP, a series of losses appear. These losses can be divided into three groups: optical losses, thermal losses and geometric losses. On the other hand, with losses, yields can be determined.

2.4. Losses of solar collectors

When solar radiation reaches the earth's surface, a large amount of this energy is lost due to different aspects, these losses can be divided into 3 groups; optical, geometric and thermal losses.

Optical losses occur when not all the energy that reaches the area of the reflector (parabola) reaches the fluid that circulates inside the receiver tube, these losses originate for different reasons; initially because the area of the reflector is not a perfect mirror, producing losses due to reflectivity (ρ), transmissivity (τ), absorptivity (α) and losses that occur due to intrinsic imperfections of the material, called the interception factor (γ).

The geometric losses occur due to the angle of solar incidence on the collector, within these losses is the cosine effect, lost by the end of the collector, lost by shadows and lost by the purely optical properties of the collector.

Thermal losses are caused by the difference in ambient temperatures and the average temperature of the fluid that travels through the receiver tube, within these losses is found by radiation, by convection and conduction of the materials that make up the collector.

2.5. Performance of solar collectors

Not all incident solar power will be transformed into useful thermal power in the fluid that travels through the receiving tube. Due to the different losses experienced by solar collectors. There are 4 types of returns; optical (ηópt), geometric (ηgeom), thermal (ηtérm) and global (ηglobal). In the following equations the different yields are described.

$$\eta_{opt} = \rho * \gamma * \tau * \alpha * F_{limp} \quad (12)$$

In equation 12 the optical performance (η_{opt}), is described as the product of the reflectivity (ρ) of the reflector area, transmissivity (τ) of the glass cover, the absorptivity (α) of the receiving tube and fouling factor (F_{limp}) of the installation.

$$\eta_{geo} = K(\theta) * F_{sombras} \quad (13)$$

In equation 13, the geometric performance (η_{geo}), is described as the product of the angle of incidence modifier (K (θ)) and the shadow factor ($F_{sombras}$).

$$\eta_{term} = \frac{Q_{util}}{DNI * A_{aper} * \eta_{opt} * \eta_{geo}} \quad (14)$$

In equation 14, the thermal efficiency (η_{term}) is described as the ratio between useful heat (Q_{util}) received by the fluid that travels through the receiving tube and the product of the solar irradiation of the place where the collector is located (DNI), the reflector aperture area (A_{aper}), optical performance (η_{opt}) and geometric performance (η_{geo}).

$$\eta_{global} = \frac{Q_{util}}{DNI * A_{aper}} \quad (15)$$

3. Methodology

The proposed library for the design of parabolic cylindrical solar collectors, is characterized according to the factors that affect the sizing, conditions and limitations of these, based on theoretical references that propose designs of solar collectors, in addition to the mathematical relationships that affect in the calculation of its performance, applying the optical, geometric and thermal principles that interact, which make up the overall efficiency of the collector based on the estimated solar radiation during the day.

The Matlab Software version 8.5 R2015b with professional license and as a complementary tool Soltrace was used for the design of the library, which is a software developed at the National Renewable Energy Laboratory (NREL) to model solar concentration systems (CSP) and analyze its optical performance. It is ideal for solar applications, the code can also be used to model and characterize many general optical systems.

Table 2. Shows a comparative chart of different software that is used for the design of parabolic cylindrical solar collector libraries.

Software	Learning curve	Data management	Gráficos
SPSS	Gradual	Moderate	Good
Stafa	Moderate	powerful	Good
SAS	Steep	Very powerful	Very good
Modelica	Steep	Very powerful	Very good
R	Steep	Very powerful	Excellent
Matlab	Steep	Very powerful	Excellent

Different software was validated, according to its versatility and functionality, so Matlab was chosen due to its scientific approach, the use of matrix expressions and vector arrangements, the different tools (toolbox) that facilitate complex analyzes, its ease to integrate flows of works with other hardware, with other environments, its analysis in the cloud and finally, for being reliable and fast.

The library has all the equations listed in the previous sections, it should be noted that the measures are handled in the International System of Units.

When executing the library, the focal length, the length of the reflective area and the adsorbent tube must be specified with their respective diameter based on the technical characteristics. Then, there is the straight side and therefore, Xmax, Ymax and the curved side, based on equations 1-4.

Subsequently, there is the area of the reflecting surface and the collection area, the opening angle (α) and distance p (Figure 4) of the CCP, corresponding to equations 5 and 6.

The minimum and actual diameter of the absorbent tube is calculated, based on equations 4 and 5, of Figure 6.

It is very important to know other parameters, such as the thermal concentration factor and the amount of accumulated soles, defined by equations 7, 8 and 9, based on figures 1, 5, 6, and 7. On the other hand, it is necessary validate the solar irradiation in the place where the solar collector is going to be located, and check the solar behavior during the 12 months of the year. In order to identify which are the most favorable and least favorable months.

The design of the collector was conditioned to the dimensions of the standard vacuum receiver tube, which has a structural specification of 1.8 m long and a diameter of 5.6 cm.

For the validation of the library, different tests were performed using the initial parameters described in Table 3, with the purpose of reviewing the mathematical equations mentioned above, based on the theoretical references Duffie, Beckman and Kalogirou.

Table 3. Parameters obtained by the library in Matlab.

Variable	Valor	Unidad	Descripción
Dfoco	0.5	m	Focal distance
Largo	1.8	m	Receiver tube length
Dtubo	0.056	m	Tube diameter of the market.
Largo2	2	m	Collector Length
Lrecto	2	m	Distances straight side of the parabola
Xmax	1	m	Distance from the vertex to the side of the collector
Ymax	0.5	m	Manifold height
Lcurvo	2.1560	m	Curved side distance of the parabola
Asurf	4.3112	m	Reflective surface area
Acapt	4	m	Collection area or collector opening.
P	1	m	Distance from the edge of the parabola to the focus.
A	90	Grados	opening angle
Dprop	0.0066	m	Minimum tube diameter
Dprop2	0.0207	m	Real tube diameter
Dpropcm	0.65	cm	Minimum tube diameter
Dprop2cm	2.07	cm	Real tube diameter
Atotal	0.1182	m^2	Actual collector tube area
Fm	0.5104	m	Average distance between surface and receiver
Io	-1.0184	m	Lost distance in the absorber tube
Ae	-2.0368	m^2	Lost area absorbing tube
FC	33.84	Adi.	Concentration factor
Cmax	35.7143	Adi	Amount of accumulated soles
Vol	0.6106	litros	Tube and liquid volume
Emisv	0.09	decimal	Emissivity of the reflective surface
Absrt	0.96	decimal	Absorptivity
Rflec	0.935	decimal	Reflectivity of the reflective surface
Transm	0.96	decimal	Transmissivity of the absolve tube

IOP Conf. Series: Materials Science and Engineering **844** (2020) 012023 doi:10.1088/1757-899X/844/1/012023

Based on the result obtained by table 1, after varying the focal length between (0.4 - 0.6 m), maintaining the initial parameters, the author Kalogirou's theory is fulfilled; when the focal length is equal to the height of the collector, a higher overall efficiency is obtained. As shown in Figure 18. Regarding geometric, thermal and optical losses, it was taken into account that the receiver tube had a smaller distance based on the length of the collector, to avoid losses due to the end of the collector, in addition to the characteristics of Emissivity, absorptivity and reflectivity of the reflective area material are theoretically good. The following figure shows the behavior of the light beams reflected in the area of the reflector, over the focal length.

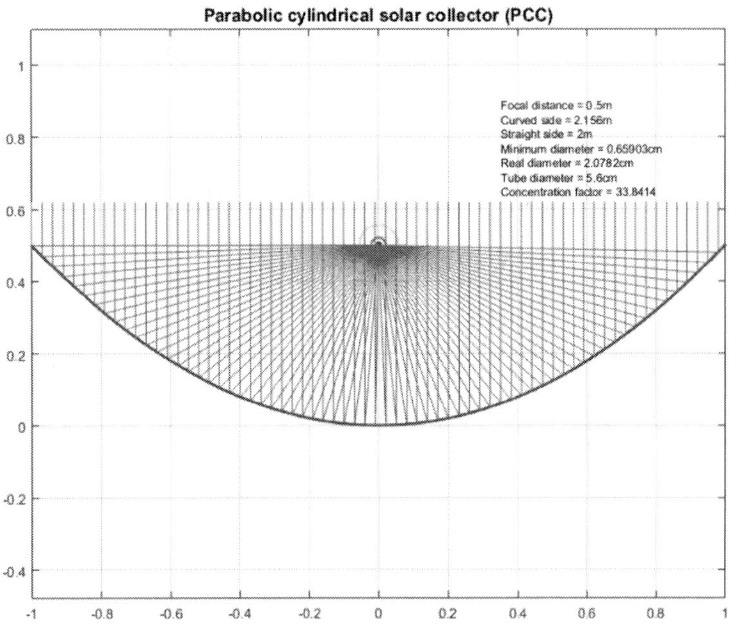

Figure 10. 2D image solar collector with 0.5m focus.

Based on the results obtained in Figure 10, it is observed that all the light beams that are reflected in the area of the reflector, all are redirected to the focus of the parabola, this principle is fulfilled because the focal length e is equal at the height of the collector, that is, the angle of solar incidence is 90 °.

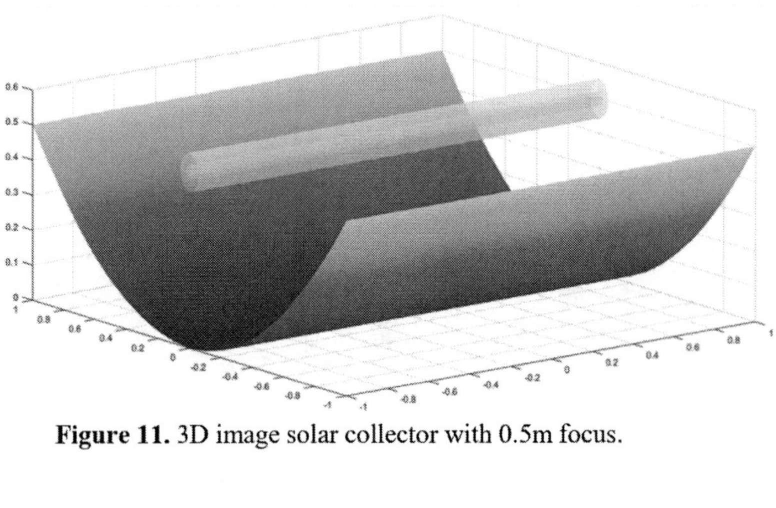

Figure 11. 3D image solar collector with 0.5m focus.

IOP Conf. Series: Materials Science and Engineering **844** (2020) 012023 doi:10.1088/1757-899X/844/1/012023

In figure 11, it is observed that theoretically all the light beams are concentrated in the receiving tube, because it is located exactly in the focus of the parabola, where it is evidenced that there is a higher concentration of heat, observing a yellow color, and in the parts where the blue color is observed, it is due to shadows caused by the receiving tube.

In Figure 12, we observe a focal length of 0.6 m and a maximum height of 0.5 m.

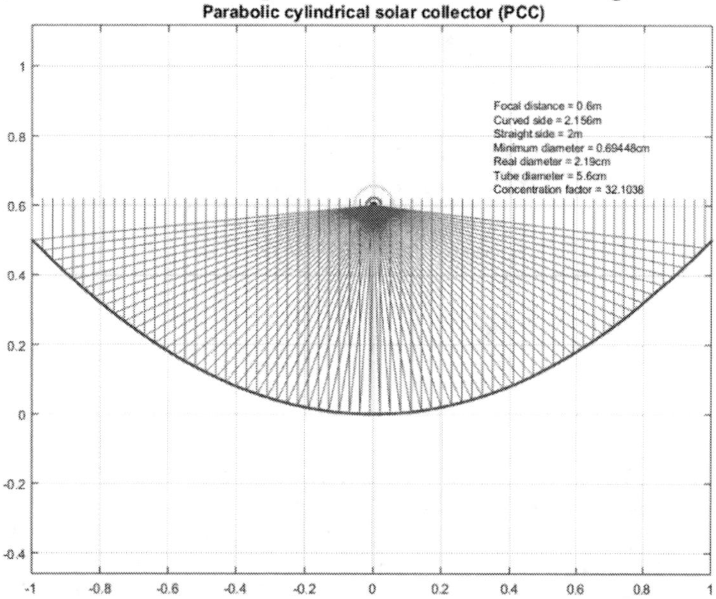

Figure 12. 2D image solar collector with focus higher than agreed.

With respect to Figure 12, it can be seen that most of the light beams are not satisfactorily reflected towards the focus, which affects the concentration factor. In Figure 13, the focal length is 0.4 m compared to the maximum collector height that is 0.5m.

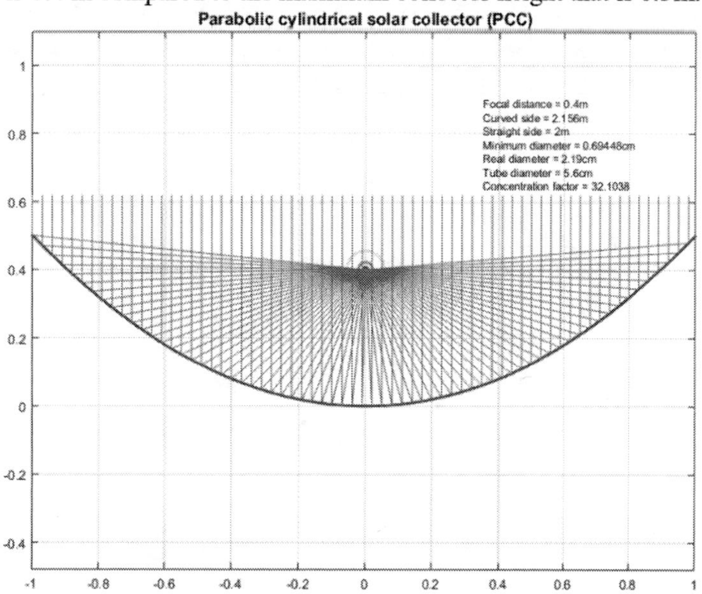

Figure 13. 2D image solar collector with lower focus than agreed.

IOP Conf. Series: Materials Science and Engineering **844** (2020) 012023 doi:10.1088/1757-899X/844/1/012023

Similarly, in Figure 13, the same phenomenon in Figure 12 occurs, all the light beams do not reach the focus, that is to say the receiving tube, which directly affects the concentration factor.

With the SolTrace software, the solar behavior of the incident beams and the flow map is illustrated in more detail, where it is observed that, with a solar acceptance angle of 90 degrees, it is theoretically guaranteed that all the light beams that project in focus. As illustrated in Figure 14 and 15.

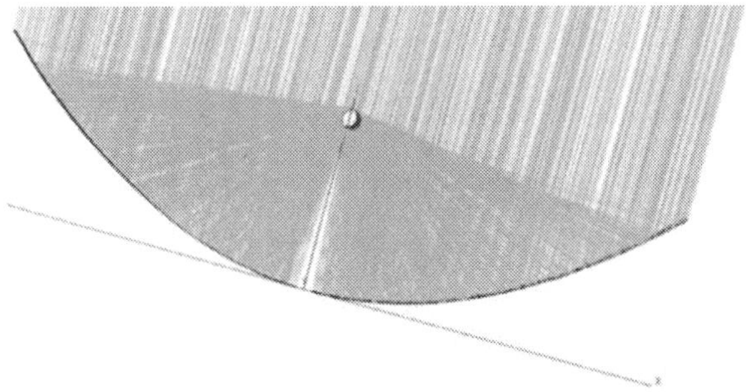

Figure 14. Solar Analysis in SolTrace software with a focal length of 0.5 m.

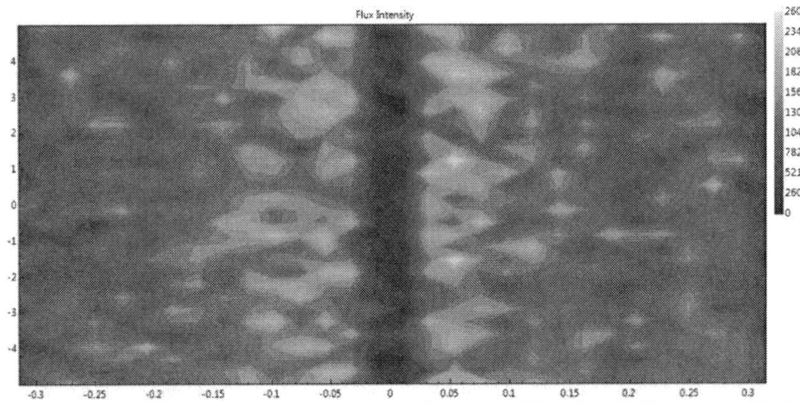

Figure 15. Flow map in the SolTrace software with a foval distance of 0.5m.

All the incident light beams are satisfactorily reflected, obtaining the highest solar collection by the receiving tube, therefore, the maximum performance. In the figure of the flow map, it is evident that the points with blue color are caused by shadows and in the more yellow ones, where solar radiation is more reflected.

In Figure 16, a focal length is shown, higher compared to the height of the collector.

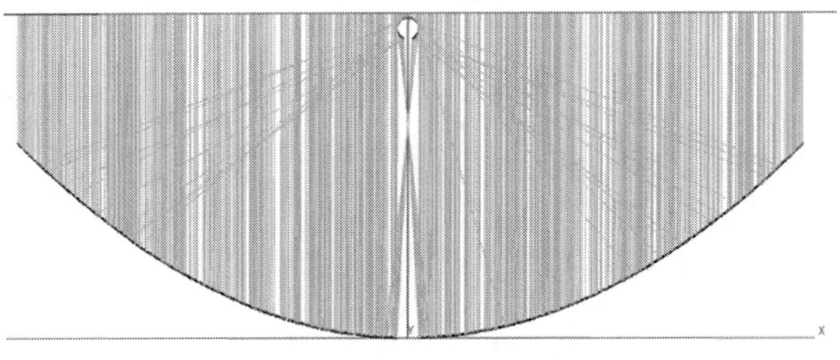

Figure 16. Solar Analysis in SolTrace with focus greater than defined.

IOP Conf. Series: Materials Science and Engineering **844** (2020) 012023 doi:10.1088/1757-899X/844/1/012023

It is observed in Figure 16, that many of the solar rays that affect the area of the reflector do not reach the focus, due to the cosine effect and the location of the focus. Which allows the overall collector performance to decrease. Figure 18 shows the flow map of Figure 17.

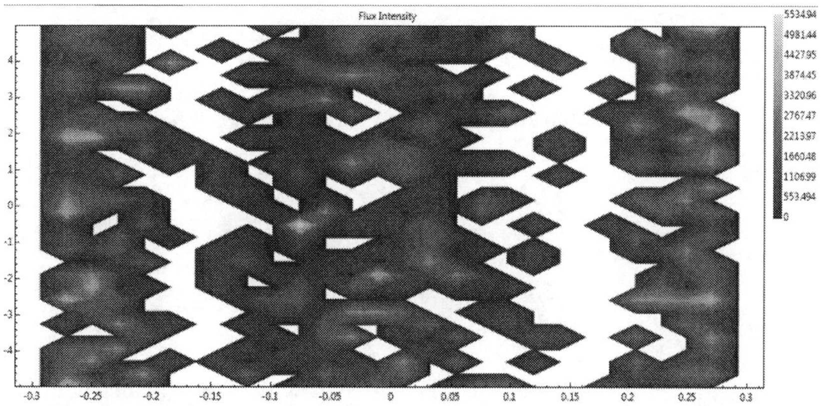

Figure 17. Flow map in SolTrace with focus greater than defined.

When observing the previous image, it shows the lack of solar concentration in the receiving tube, in addition there is little solar concentration.

In figure 18 and 19 a focal distance is observed that is located inside the solar collector, presenting a phenomenon similar to figures 15 and 16, where the rays reflected in the area of the reflector, do not reach to focus on the focus.

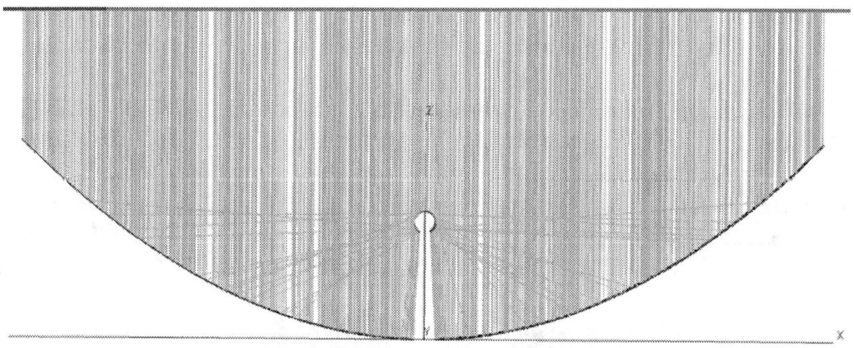

Figure 18. Solar Analysis in SolTrace with focus less than defined.

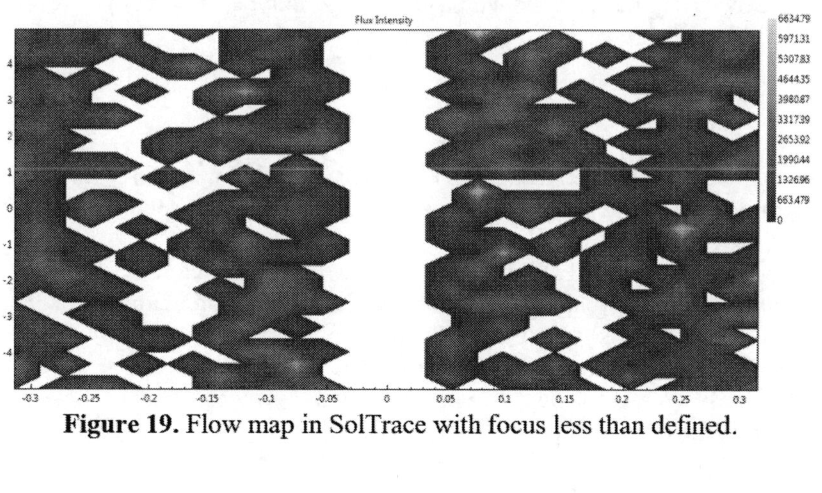

Figure 19. Flow map in SolTrace with focus less than defined.

It shows more spaces with shadows and low heat concentration, which causes the overall efficiency of the collector to decrease considerably.

Table 4 shows the optical, thermal, geometric and global yields of a solar collector test with the library, varying the focal length.

Table 4. Performance of the solar test collector, with different focal distances.

	0.4 m focal length	**0.5 m focal length**	**0.6 m focal length**
Variable	value	value	value
Nopt	0.8617	0.8617	0.8617
Ngeo	0.5823	0.7270	0.6354
Nterm	0.9556	1	0.9732
Nglobal	**0.5099**	**0.6264**	**0.5564**

In table 4, it can be corroborated that a solar collector with an acceptance angle of 90 degrees, allows to increase the overall efficiency of the collector, it is important to take into account that these systems are exposed to different losses, so it is essential to define a Good structural design, relevant materials and good software that facilitates calculations and reduces physical tests that increase costs.

4. Conclusions

The library developed in Matlab software and verified with the complementary tool Soltrace, allowed to simplify the process of design of parabolic trough collectors, creating a functional model that allows simulating the relevant variables to find the performance based on solar radiation.

The development of this methodology allowed to analyze in detail the variability of some factors, allowing to select the appropriate components and obtain yields adjustable to the local environmental conditions of the implementation. In addition, the expected behaviors for each of the variables were obtained, taking into account the theoretical references, where it was validated that the focal length should have the same measure as the maximum height of the collector, so that the angle of solar incidence is 90 degrees. This allows the overall collector performance to be the most efficient.

The dimensions must be adjusted to the measurements of the receiver tube and the most optimal focal length, to achieve maximum solar concentration in the absorber tube fluid. As future work, this library is expected to be validated with a practical prototype, in order to better optimize losses and yields in different contexts.

[References

[1] Ideam.gov.co. (2019). RADIACIÓN SOLAR - IDEAM. [online] Available at: http://www.ideam.gov.co/web/tiempo-y-clima/radiacion-solar [Accessed 14 Jul. 2019].

[2] Como-funciona.co. (2019). ENERGIA TÉRMICA | Características, ejemplos y funcionamiento. [online] Available at: https://como-funciona.co/la-energia-termica/ [Accessed 14 Jul. 2019].

[3] Acciona.com. (2018). ¿Qué beneficios tiene la energía solar? | ACCIONA. [online] Available at: https://www.acciona.com/es/energias-renovables/energia-solar/ [Accessed 10 Jul. 2019].

[4] Villasante, C. (2010). [online] Sc.ehu.es. Available at: http://www.sc.ehu.es/sbweb/energias-renovables/temas/termoelectrica/revision/Articulo_Revision_Tecnologias_Solares_Termoelectricas_.pdf [Accessed 8 Jul. 2019].

[5] Taylor & Francis. (2007). Object oriented modelling and simulation of parabolic trough collectors with modelica. [online] Available at: https://www.tandfonline.com/doi/abs/10.1080/13873950701847199 [Accessed 8 Jul. 2019].

[6] Solar-energia.net. (2019). Colector solar. [online] Available at: https://solar-energia.net/definiciones/colector-solar.html [Accessed 9 Jul. 2019].

[7] VELASCO, C. (2012). DISEÑO DE CAPTADOR SOLAR CILÍNDRICO PARABÓLICO. [online] Available at: http://oa.upm.es/14011/1/PFC_CARMEN_PAREDES_VELASCO.pdf [Accessed 6 Jul. 2019].

[8] Laenergiasolar.org. (2018). Colector Cilindro Parabólico: ¿qué es y cómo funciona? - La Energia Solar. [online] Available at: https://www.laenergiasolar.org/energia-termica-solar/colector-cilindro-parabolico/ [Accessed 15 Jul. 2019].

[9] Sastre Alfaro, A. (2016). ANÁLISIS DE COLECTORES SOLARES DE MEDIA TEMPERATURA. Leganés. [Accessed 15 Jul. 2019].

[10] Cherian, R. (2016). Performance Enhancement of Solar Water Heaterusing Compound Parabolic Reflector and Numerical Simulation of Thermal Losses. [Accessed 7 Jul. 2019].

[11] Busso, MSc. Arturo (2019). Estudio y caracterizacion de un colector plano. [online] Available http://exa.unne.edu.ar/investigacion/energia_solar/PUBLICACIONES/informe_PI455.pdf

[12] J. Roviras, V. Sarrablo, M. M. Casanovas (2016). Integración arquitectónica de colectores solares térmicos cerámicos para clima mediterráneo.

[13] Rodriguez, Juan. (2015). Energía solar, Medio Ambiente y desarrollo. [online] Available at: http://www.guzlop-editoras.com/web_des/ener01/enersolar/pld0524.pdf. [Accessed 15 Jul. 2019].

IOP Conf. Series: Materials Science and Engineering **844** (2020) 012024 doi:10.1088/1757-899X/844/1/012024

A review: Comparison of performance metrics of pretrained models for object detection using the TensorFlow framework

S A Sanchez [1], H J Romero [1] y A D Morales [1]

[1]Corporación Universitaria Antonio José de Sucre, Sincelejo, Colombia,
Faculty of Engineering Science, GINTEING Research Group

sergio_sanchez@corposucre.edu.co

Abstract. Advances in parallel computing, GPU technology and deep learning facilitate the tools for processing complex images. The purpose of this research was focused on a review of the state of the art, related to the performance of pre-trained models for the detection of objects in order to make a comparison of these algorithms in terms of reliability, ac- curacy, time processed and Problems detected The consulted models are based on the Python programming language, the use of libraries based on TensorFlow, OpenCv and free image databases (Microsoft COCO and PAS-CAL VOC 2007/2012). These systems are not only focused on the recognition and classification of the objects in the images, but also on the location of the objects within it, drawing a bounding box around the appropriate way. For this research, different pre-trained models were re- viewed for the detection of objects such as R-CNN, R-FCN, SSD (single- shot multibox) and YOLO (You Only Look Once), with different extractors of characteristics such as VGG16, ResNet, Inception, MobileNet. As a result, it is not prudent to make direct and parallel analyzes between the different architecture and models, because each case has a particular solution for each problem, the purpose of this research is to generate an approximate notion of the experiments that have been carried out and conceive a starting point in the use that they are intended to give.

1. Introduction

Increased Industry 4.0 or better known as the fourth industrial revolution promises great advances and technological challenges, the concept of artificial intelligence is the central protagonist of this transformation, related to the analysis of large volumes of data (Big Data) and use algorithms for processing. Generating great expectations due to the impact that this technology can conceive that promises to change the paradigm of machines. Today the words artificial intelligence, machine learning and learning Deep are widely used, but there is no clear difference between each, to the point of believing that is the same concept.

Artificial intelligence is simulating machine behavior and human reasoning for this, use different techniques including machine learning, for example; Siri voice assistants and Alex, on the other hand, Machine Learning is the ability of computers to learn by itself from data and experiences or subsets of techniques using statistical methods to enable machines to learn by itself. In its most complex use uses Deep learning, for example, predictive analytics in autonomous vehicles, face detection of persons with judicial positions, among others.

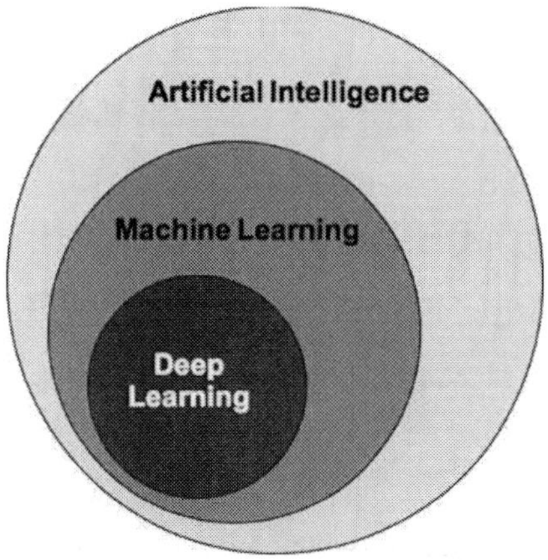

Figure 1. Relationship of the concepts of artificial intelligence, learning and deep learning machine.

With the emergence of the cognitive computing era, many researchers seek to incorporate human processes such as learning and communication with the purpose of reproducing them in machines. For many years, this technology has been surpassing different challenges, achieving rapid computation times (real time) and data processing as a human. Today we are talking about a new technology that is coming into great boom, which is Deep Learning, based on a series of neural networks that resemble the human brain. It consists of a set of unsupervised algorithms that form layers of artificial neurons to determine hidden features in a data set. In Figure 2, we observed some artificial intelligence techniques for the analysis of large volumes of data, from the most traditional to the most complex used today [2].

1.1. Using neural networks in Deep Learning

Neural networks have been crucial to teach computers to think and understand the world the way humans do, retaining innate advantages such as speed, ac- curacy and lack of bias. Artificial neural networks are widely used for their properties of nonlinear character, adaptability, generalization, fault tolerance, and scalability task decomposition. But today have disadvantages such as complexity in the design of architecture, lots of parameters to adjust difficulty and train networks.

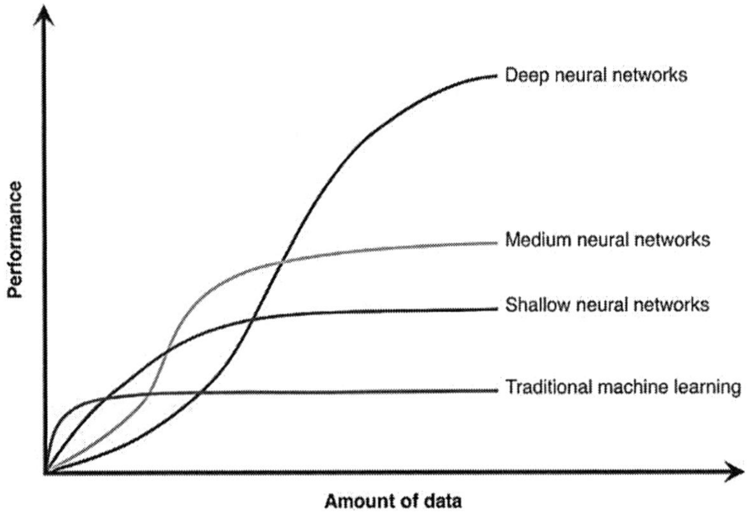

Figure 2. Artificial intelligence techniques vs. data volume.

Artificial neural networks can be classified according to their network topology and by learning algorithm. Depending on the topology can be the type of the (hidden or visible) layers, input or output and the directionality of the connections of neurons. According to the algorithm can be monitored, unsupervised, competitive or reinforcement. Furthermore, neural networks can be classified according to its architecture, pre-fed Neural Networks, convolutional and recurrent [3] networks.

The pre-fed neural networks were the first to be deployed, for its simple model. In these networks, the information is shifted in one direction, in this architecture include pioneers were perceptron multilayer perceptron and [4].

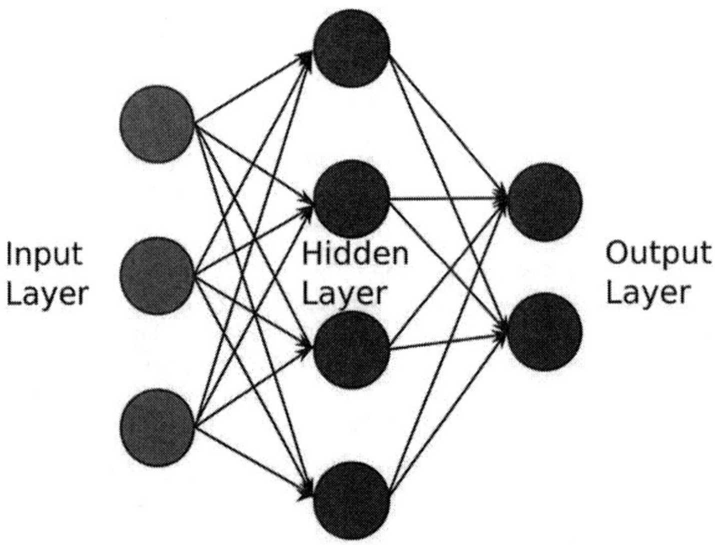

Figure 3. Topology of an artificial neural network.

The convolutional neural network (CNN) are very similar to the multilayer perceptron, are made up of neurons having weights and biases, which can be punished and learn from the entries. All this supported in a loss function or cost over the last layer. Some specific implementations of such networks are: ResNet, VGG16, Xception, Inception V3, among others [5].

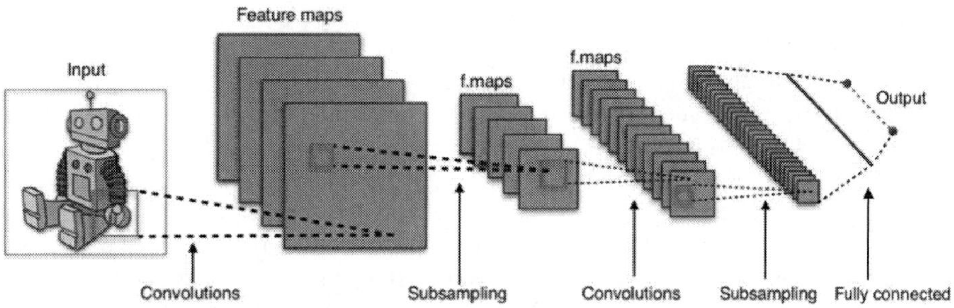

Figure 4. convolutional neural networks.

Recurrent Neural Network (RNN) using feedback loops, which allow information to persist. These apply the same operation to each element of a sequence of input data (text, voice, video, etc.), whose output depends on the input data as past information, characterized by memory capacity. The RNN conventional networks have problems in their training because backward gradients tend to grow or fade over time, because the gradient of the error depends not only present but also past mistakes. This causes the difficulty of memorizing dependencies for a long time. That is why architectures and learning methods have been developed that avoid these problems such as LSTM, neurological Turing machines and memory networks [6].

There are several types of training neural networks, such as manual tuning, learning algorithms and genetic algorithms. The most common problems during training are underfitting, overfitting and local minimum.

2. Methodology

In recent years, the architecture of the deep networks has been a significant progress for the moment Keras and TensorFlow dominates with different-pre- trained models already included in libraries, among these include VGG16, VGG1, ResNet50, Inception V3, Xception, MobileNet. The VGG and AlexNet 2012 networks follow a typical pattern of classical convolutional networks. MobileNet is a simplified architecture Xception architecture, optimized for mobile applications. The following architectures; ResNet, Inception and Xception have become a reference point for subsequent studies of artificial vision and learning for its versatility Deep [7].

There are many factors that explain the revolution of deep learning, among these factors is highlighted; availability of sets of huge data and quality, parallel computing (GPU) features efficient activation for backpropagation, new architectures, new regularization techniques that allow train more extensive networks with less danger of overshooting, robust optimizers and software platforms with large communities like TensorFlow, Theano, Keras, CNTK, PyTorch, Chainer and Mxnet. All this has allowed solving problems easier. Today the Python programming language has great importance in Machine Learning compared to other languages because of its support for Deep learning framework.

Within this framework include TensorFlow which is a library of open source software for machine learning that allows you to deploy computing in CPU or GPU, developed by Google, using graphs flow data, PyTorch uses Python language and has the support of Facebook, Theano is a Python library that supports mathematical expressions involving tensioners, CNTK are a set of tools developed by Microsoft, open for Deep learning code, Keras is a library of neural networks

high level created by Francis Chollet, member of Brain google equipment that lets you choose whether the models that are built will be executed in Theano, TensorFlow or CNTK. Keras and TensorFlow can construct models of three different ways; using a sequential model, a functional API and pre-trained models.

Earlier we talked about the different architectures (MobileNet, Inception, ResNet, among others), now we discuss models for object recognition and Keras TensorFlow; Faster-CCN R, R-FCN, SSD and YOLO. These models are classified based detectors in the region (Faster R-CNN, R-FCN, FPN) and single shot detectors (SSD and YOLO), start from different paths, but they look very similar now fighting for title faster and more accurate detector.

There are different metrics that can improve object detection algorithms based on more accurate positioning, faster speed and more accurate classification; metrics that stand out are: Intersection over Union (IoU), mean average precision (MAP) and rendered frames per second (FPS). Intersection over Union (IoU)It is an indicator that determines how close the predicted picture of the real picture [9]. The average metric average accuracy (MAP) is the product accuracy and recovery detection bounding boxes. It's a good combined measure of how sensitive the network to objects of interest and how well it avoids false alarms. The higher the score the map, the more precise the network, but this has a cost of execution speed [10]. Processed frames per second (FPS) is used to judge how fast is the system [11].

a. Datasets

Architectures and above models need data lately have focused on free data sets posted on the web, like Microsoft COCO (common objects in context) and PAS- CAL Visual Object Classes (VOC). Microsoft COCO is a dataset of 300,000 images with common objects 90 supported on an API that provides different models of object detection, which compensates for the speed and accuracy based on bounding boxes suitable objects [12]. PASCAL Visual Object Classes (VOC) is a reference point in the visual recognition of object categories and detection. It consists of a set of standard image data, annotations, and evaluation procedures [13]. Organized since 2005 to the present.

b. Comparison Between Deep Learning Algorithms for Object Detection

It is difficult to define a fair feature of different object detectors, each case of real life can have different solutions to reach a decision concerning the accuracy and speed, it is necessary to know other factors that affect performance; the type of feature extractor, steps out of the extractor, income resolutions images, strategy coincidence and threshold (as predictions are excluded when calculating the loss), Threshold IOU no maximum suppression ratio of positive anchor and negative, number of proposals or predictions of frame coding limit, increased data set of training data, using multi-scale images training or testing (with clipping), map layer features for object detection, It is important to note that technology is constantly evolving, any comparison can become obsolete quickly.

A comprehensive review of scientific papers and academics on the performance of different models object detection with TensorFlow framework where investigations found the following was performed; The authors Shaoqing Ren Kaiming He, Ross Girshick, and Jian Sun in his research entitled "Faster R-

CNN: Towards Real-Time Object Detection with Region Proposal Networks" analyzed several test set where different results were found based on the as medium accuracy (mAP) and the number of predictions using the R-CNN Faster method evidenced in table 1 [15].

Table 1. Performance Model Faster R-CNN based metrics speed, accuracy and num- bers predictions using database COCO MS [16].

Method	# proposals	data	mAP (%)
SS	2000	12	65.7
SS	2000	07++12	68.4
RPN+VGG, shared	300	12	67.0
RPN+VGG, shared	300	07++12	70.4
RPN+VGG, shared	300	COCO+07++12	75.9

Better average precision measurement (MAP) of the model is evidence Faster R-CNN based on RPN 300 + VGG methods predictions using data bases MS COCO + PAS-CAL VOC 2007+ PAS-CAL VOC 2012. In the following table, focused tests were performed on a K40 GPU with the test set PASCAL VOC 2007.

Table 2. Performance Model Faster R-CNN function metrics speed, accuracy and numbers predictions database PASCAL VOC 2007 [17].

model	system	conv	proposal	region-wise	total	rate
VGG	SS + Fast R-CNN	146	1510	174	180	0.5 fps
VGG	RPN+ Fast R-CNN	141	10	47	198	5 fps
ZF	RPN+ Fast R-CNN	31	3	25	59	17 fps

Can be determined in Table 2, the models made by RPN + Fast R-CNN and RPN + Fast R-CNN systems were the most efficient, with a rate of rendered frames per second of 5 and 17 fsp and a rate prediction 10 and 3 respectively. On the other hand, the authors Jifeng Dai, Yi Li, Qinghua He, Jian Sun in his re- search entitled "R-FCN: Object Detection via Region-based Fully Convolutional Networks" compared with detectors based on faster R-CNN, consumption com- puter is expensive, in contrast to NGF-R technique is based on convolutional calculations are shared across the image [18].

In Table 3, the results obtained according to the medium accuracy mea- surement (MAP) and the response time, which worked with the database MS COCO test and PAS-CAL VOC 2007, 2012. The code is evidenced available at the following link: https://github.com/daijifeng001/r-fcn.

Table 3. Result sobtained from the R-R-CNN and FCN models based metrics speed and accuracy.

	training data	mAP (%)	test time (sec/img)
Faster R-CNN	07++12	73.8	0.42
Faster R-CNN	07++12+COCO	83.8	3.36
R-FCN multi-sc train	07++12	77.6	0.17
R-FCN multi-sc train	07++12+COCO	82.0	0.17

better average precision measurement (MAP) is evidenced by the R-CNN method using the data bases MS COCO + PAS-CAL VOC 2007+ PAS-CAL VOC 2012, with 83.8 mAP and a testing time of 3.36 sec/img.

Another method was analyzed SSD, where the authors Wei Liu, Dragomir Angelov, Dumitru Erhan, Christian Szegedy, Scott Reed, Yang Cheng-Fu, Alexan- der C. Berg conducted a study entitled "SSD: Single Shot MultiBox"; presenting a method with a deep neural network, where the output space defined by a set of predetermined tables on different aspect ratios and scale is discretized.

The network generates scores for each category of object in each frame and produces default settings for that box delimiter closely matches the shape of the object. In addition, the network combines multiple predictions feature maps with different resolutions to naturally handle objects of various sizes. This makes it simple to train SSD with different databases [20].

In Table 4, the results based on the measurement of average precision (MAP) and the response time in databases PASCAL VOC 2007, 2012 and MS COCO shown using input images with a resolution of 300 300 and 512 512 [21].

Table 4. Performance SSD method using databases PASCAL VOC 2007, 2012 and MS COCO.

Method	VOC2007 test 07+12 07+12+COCO 0.5 0.5		VOC2012 test 07++12 07++12+COCO 0.5 0.5		COCO test-dev2015 trainval35k 0.95 0.5 0.75		
SSD300	74.3	79.6	72.4	77.5	23.2	41.2	23.4
SSD512	76.8	81.6	74.9	80.0	26.8	46.5	27.8
SSD300*	77.2	81.2	75.8	79.3	25.1	43.1	25.8
SSD512*	79.8	83.2	78.5	82.2	28.8	48.5	30.3

In the above table it shows that the SSD method has better performance using input images with resolution of 512 x 51 databases PASCAL VOC 2007, 2012 + MS COCO.

In Table 5, we find the results based on the measurement of average precision (MAP) and FPS with databases PASCAL VOC 2007, 2012 and MS COCO using input images with a resolution of 300 300 and 512 512 and Faster methods R- CNN, YOLO and SSD [22].

Table 5. Results in the test set PASCAL VOC 2007, 2012 and MS COCO using input images with different resolutions, implementing the YOLO, Faster R-CNN and SSD [23] methods.

Method	mAP	FPS	batch size	# Boxes	Input resolution
Faster R-CNN (VGG16)	73.2	7	1	6000	1000X600
Fast YOLO	52.7	155	1	98	448X448
YOLO (VGG16)	66.4	21	1	98	448X448
SSD300	74.3	46	1	8732	300X300
SSD512	76.8	19	1	24564	512X512
SSD300	74.3	59	8	8732	300X300
SSD512	**76.8**	**22**	**8**	**24564**	**512X512**

In the above table the yield of YOLO, Faster R-CNN and SSD methods where it validates that each method has its advantages and disadvantages, concerning MAPs, FPS parameters, the number of predictions and image resolution.

Finally, the authors Joseph Redmon and Ali Farhadi University of Washing- ton and Allen Institute for AI, conducted a study entitled "YOLO9000: Better, Faster, Stronger" YOLO is a method of detecting objects that uses the latest technology in time real, which can detect more than 200 classes and 9,000 different categories of objects. This novel method has various improvements, as is YOLOv2 and YOLOv3, using a training method multiscale, is a relatively new and very efficient technology standard detection tasks databases and MS COCO PASCAL VOC. Exceeding methods as faster and SSD RCNN with ResNet [23].

In Table 6, the results evidence function as medium accuracy (mAP), FPS and databases PASCAL VOC 2007, 2012 and MS COCO using input images with different resolutions, implementing methods Faster R-CNN, Yolo and SSD Table 6. Performance of YOLO, Faster R-CNN and SSD methods works metrics speed and accuracy using the test set PASCAL VOC 2007 [24].

In Table 7, the results evidenced in function of the measured average precision (MAP) and databases PASCAL VOC 2007 and 2012, using the methods Faster R-CNN, YOLO and SSD.

It is not wise to make a parallel analysis of the above article, these experiments are performed in different environments. But the purpose of this article is to have a general notion about these methods.

In Figure 5, the results of input images is shown with dimensions of 300 x 300 and 512 x 512 using free databases IMAGEnet, PASCAL VOC 2007, 2012 and MS COCO for different methods making comparisons based on the metric precision. The YOLO method has different results for input images of 288 x 288, 416 x 461 and 544 x 544. The higher resolution images for the same model have better Map but are slower to process [26].

Table 6. Performance of YOLO, Faster R-CNN and SSD methods works metrics speed and accuracy using the test set PASCAL VOC 2007 [24].

Detection Frameworks	Train	mAP	FPS
Fast R-CNN	2007+2012	70.0	0.5
Faster R-CNN VGG-16	2007+2012	73.2	7
Faster R-CNN ResNet	2007+2012	76.4	5
YOLO	2007+2012	63.4	45
SSD300	2007+2012	74.3	46
SSD500	2007+2012	76.8	19
YOLO v2 288 x 288	2007+2012	69.0	91
YOLO v2 352 x 352	2007+2012	73.7	81
YOLO v2 416 x 416	2007+2012	76.8	67
YOLO v2 480 x 480	2007+2012	77.8	59
YOLO v2 544 x 544	2007+2012	78.6	40

Table 7. Results in test set PASCAL VOC 2007, 2012 using the R-CNN, and SSD YOLO [25] method.

Method	data	mAP	aero	bike	...	bird	boat
Fast R-CNN	07++12	68.4	82.3	78.4	...	70.8	**52.3**
Faster R-CNN	07++12	70.4	84.9	79.8	...	74.3	**53.9**
YOLO	07++12	57.9	77.0	67.2	...	57.7	**38.3**
SSD300	07++12	72.4	85.6	80.1	...	70.5	**57.6**
SSD512	07++12	74.9	87.4	82.3		75.8	**59.0**
ResNet	**07++12**	**73.8**	**86.5**	**81.6**		**77.2**	**58.0**

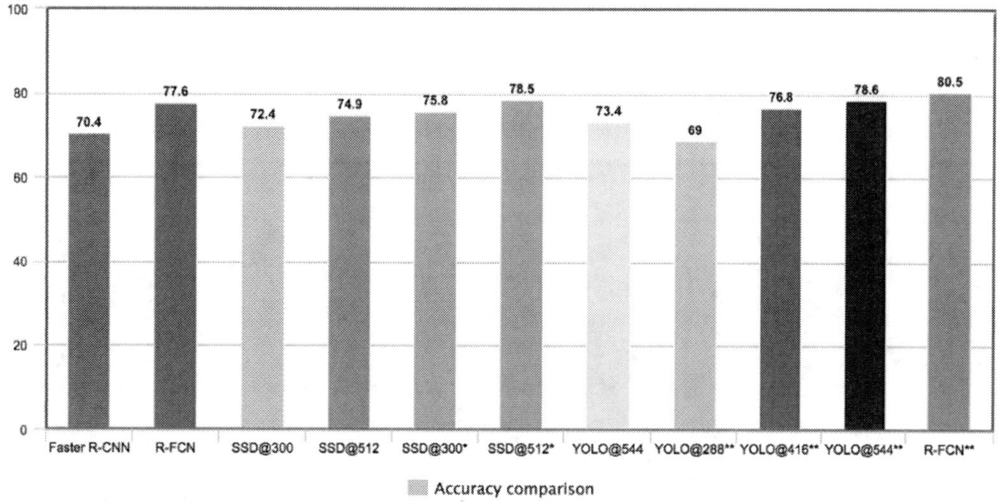

Figure 5. Comparison accuracy Faster R-CNN, R-FCN, SSD and YOLO models using input images with different resolutions.

In Figure 6, one can observe the resolutions of the input images and the feature extractors vary the speed of the detectors. Then the metric FPS highest and lowest reported by the above information is displayed, however, the result can then be very biased, in particular when measured at different map [27].

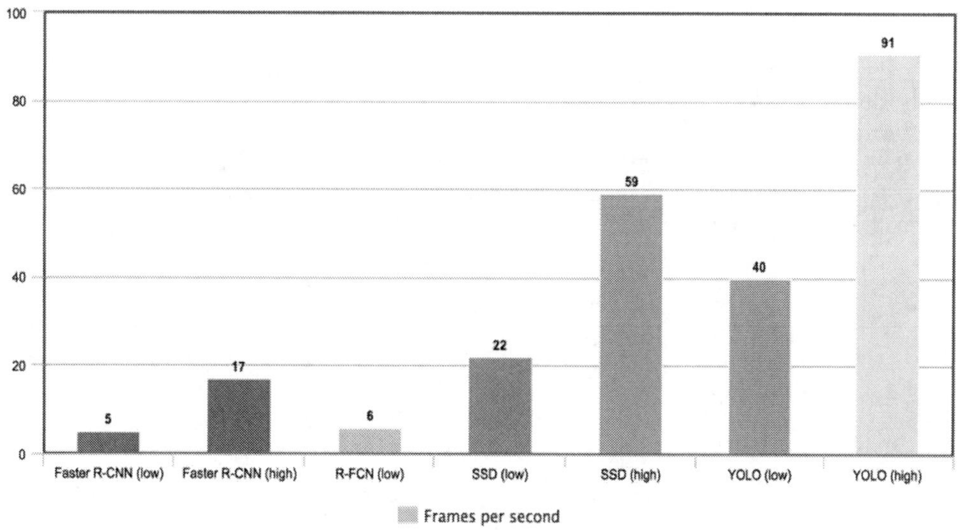

Figure 6. Comparison of frames processed per second (FPS) implementing the Faster R-CNN, R-FCN, SSD and YOLO models using input images with different resolutions.

In recent years, many results are measured exclusively with the data set MS detection COCO objects. This data set is more difficult to detect objects and, generally, detectors reach a much lower mAP. To measure the accuracy of these models has been used AP, where 0.5 denoting a fair and detection 0.95 indicating a very accurate detection. Here are some comparisons of key detectors. This can be seen in Figure 7 [28]. Table 11.

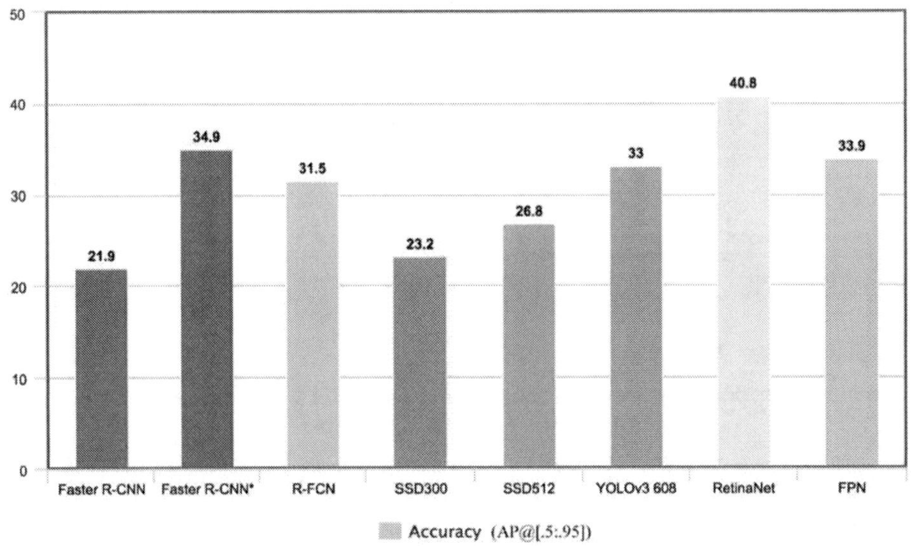

Figure 7. Comparison of accuracy Faster R-CNN, R-FCN, SSD and YOLO models using the database MS COCO.

The above items are studying how the resolution of the input images and feature extractors affects accuracy and speed detector. Overall, Faster R-R-FCN CNN and compared with SSD and YOLO models are slightly slower but more accurate. YOLO SSDs and methods have difficulty detecting small objects, but are faster.

In Table 9 the advantages and disadvantages of methods Fast R-CNN, Faster R-CNN, R-FCN, SSD and YOLO for detecting objects in images will be high- lighted by various experiments conducted by a variety of authors investigated in Deep Learning area.

IOP Conf. Series: Materials Science and Engineering **844** (2020) 012024 doi:10.1088/1757-899X/844/1/012024

Table 8. Advantages and disadvantages of some methods for detecting objects in images.

Method	Authors	Advantage	Disadvantages
Fast R-CNN	(Girshick, 2015)	The calculation of the characteristics of CNN is done in a single iteration, achieving that the detection of objects is 25 times faster than the RCNN method (it requires 20 seconds on average to analyze an image).	The use of an external candidate region generator creates a bottleneck in the detection process.
Faster R-CNN	(Renet al., 2015)	The RPN method allows object detection to be almost real-time, approximately 0.12 seconds per image.	Despite the efficiency of the algorithm, it is not fast enough to be used in applications that require real time, as would be the autonomous vehicles.
R-FCN	(Dai et al., 2016)	The test time of R-FCN is much faster than that of R-CNN	R-FCN has a competitive mAP but lower than that of Faster R-CNN.
Mask R-CNN	(He et al., 2017)	The location of the objects is more precise, when making a segmentation of the objects in the images.	Its execution time is greater than that used by the Faster-RCNN method, therefore, it can not be be implemented in applications that require real time.
YOLO	(Redmon et al., 2015)	The location of objects is very efficient, allowing its use in real-time applications.	The method has difficulties to correctly detect small objects.
SSD	(Liu et al., 2016)	The use of a single network, makes the location of the objects faster than the Fast-RCNN and Faster-RCNN methods.	The detection accuracy of the objects is lower compared to the Fast-RCNN and Faster-RCNN methods.

3. Conclusion

This research has been conducted a systematic review other than 50 scientific papers, where the recent progress of deep learning networks object detection is evident, deep models have significantly improved performance, but there are still many challenges and challenges. Parallel computers and GPU have greatly reduced training time of artificial neural networks and pre-trained models they have succeeded in reaching loaded into the devices reduced computation times, without having to use a GPU. Allowing to have a starting point for researchers seeking to venture into this line and not have to build and train an object detector from scratch, which would require a long time.

The difference between the detectors is shrinking. Single shot detectors use more complex designs in order to be more precise and regions based detectors accelerate the operations to be faster. The detector Yolo single shot, for example; associated features of other detectors, the specific difference may not be the basic concept but the details of implementation.

Detectors based on regions such as CNN and Faster R-R-FCN show a small advantage if speed precision is needed in real-time, single-shot detectors are here for real-time processing. But applications must verify if it meets your requirements accurately. The difference YOLO and SSD model is simply that the model does not include YOLO feature maps of varying size that makes SSD, and also uses its own custom base architecture. The YOLO and SSD methods have difficulty detecting small objects. In the case of these methods, the accuracy of detection of objects is smaller as compared to Faster-RCNN and R-FCN methods.

The R-FCN, YOLO and SSD models are faster on average, but cannot beat the R-CNN faster in accurately if speed is not a concern. feature extractor and resolution of the input image significantly affect the accuracy of the models. For large, SSD and YOLO objects can beat CNN and Faster R-R- FCN in precision with lighter and faster extractors. Although many researchers have made great strides in detection methods of objects, there are still many challenges that must be overcome. Deep learning will have a prospective future in a wide range of applications.

References

[1] Wang, L., & Sng, D. (2015). Deep Learning Algorithms with Applications to Video Analytics for A Smart City: A Survey 1-8. Retrieved from. http://arxiv.org/abs/1512.03131

[2] Orera Floria, JM. (2015). Development of a system for detecting people in indoor environments using fish-eye camera in overhead and algorithms based on Deep Learning plane. pp 5.

[3] Chaves, D., Saikia, S., & Fern, L. (2018). A Review of Methods Sistem tica to Locate Objects in Images Automatically, 15.

[4] Auer, P., Burgsteiner, H., & Maass, W. (2008). A learning very simple universal rule for approximators Consisting of a single layer of perceptrons Chair for Information Technology. Theoretical Computer Science, 015 879 1-29.

[5] Zhao, B., Li, X., Lu, X., and Wang, Z. RNN-CNN Architecture for Multi-Label Weather Recognition, pp-5. (2019).

[6] Manuel, U., Alcocer, R., Tello-loyal, E., Bertha, A., & Alvarado, R. (2018). NEURAL NETWORK BASED MODEL NEURAL NETWORK BASED rcurrent LSTM- MODEL FOR summary, 40 (130) 962-974.

[7] Sundar, KVS (2018). Evaluating Training Time of Inception-v3 and Resnet-50, 101 Models using CPU and GPU across TensorFlow. Second International Confer- ence on Electronics, Aerospace and Communication Technology (ICECA), (Iceca).

[8] Sun, C., & Murphy, K. (2017). Speed / accuracy trade-offs for modern convolutional object detectors, pp-11.

[9] Tsoi, N., Gwak, J., Reid, I., & States, U. (2019). Generalized Intersection over Union: A Metric and A Loss for Bounding Box Regression.
[10] Li, Y., Huang, C., Ding, L., Li, Z., Pan, Y., & Gao, X. (2019). Deep learning in bioinformatics: introduction, application, and perspective in big data was. White-filgueira, B., Garc, D., Fern, M., Brea, M., & Paula, L. Deep Learning- Based Visual Tracking on Multiple Object Embedded System for Mobile Edge IoT and Computing Applications 1-8.
[11] T.-Y. Lin, M. Maire, Belongie S. J. Hays, P. Perona, D. Ra- mannan Dolla'r P., and C. Lawrence Zitnick. (2014). Microsoft COCO: Common objects in context. In ECCV, 1 May.
[12] M. Everingham, L. Van Gool, CK Williams, J. Winn, and A. Zisserman. (2010). The visual object classes pascal (voc) challenge. International Journal of Computer Vision, 88 (2): 303-338,
[13] Wang, L., & Sng, D. (2015). Deep Learning Algorithms with Applications to Video Analytics for A Smart City: A Survey 1-8. Retrieved from http://arxiv.org/abs/1512.03131.
[14] Ren, S., He, K., Girshick, R., & Sun, J. (2017). Faster R-CNN: Towards Real- Time Object Detection with Region Proposal Networks. IEEE Transac- tions on Pattern Analysis and Machine Intelligence, 39 (6), pp 1. https://doi.org/10.1109/TPAMI.2016.2577031
[15] Ren, S., He, K., Girshick, R., & Sun, J. (2017). Faster R-CNN: Towards Real- Time Object Detection with Region Proposal Networks. IEEE Transactions on Pattern Analysis and Machine Intelligence, 39 (6), 1137-1149. p.8 https://doi.org/10.1109/TPAMI.2016.257703.
[16] Ren, S., He, K., Girshick, R., & Sun, J. (2017). Faster R-CNN: Towards Real- Time Object Detection with Region Proposal Networks. IEEE Transactions on Pattern Analysis and Machine Intelligence, 39 (6), 1137-1149. p 9. https://doi.org/10.1109/TPAMI.2016.2577031.
[17] Dai, J., Li, Y., He, K., & Sun, J. (2016). R-FCN: Object Detection Region- based via Convolutional Networks Fully. Retrieved from. P1. http://arxiv.org/abs/1605.06409.
[18] Dai, J., Li, Y., He, K., & Sun, J. (2016). R-FCN: Object Detection Region- based via Convolutional Networks Fully. Retrieved from. p7. http://arxiv.org/abs/1605.06409.
[19] Liu, W., Angelov D., Erhan, D., Szegedy, C. Reed, S., Fu, C., & Berg (2016). AC SSD: Single Shot MultiBox Detector, pp-1
[20] Liu, W., Angelov D., Erhan, D., Szegedy, C. Reed, S., Fu, C., & Berg (2016). AC SSD: Single Shot MultiBox Detector, pp-7.
[21] Liu, W., Angelov D., Erhan, D., Szegedy, C. Reed, S., Fu, C., & Berg (2016). AC SSD: Single Shot MultiBox Detector, pp-11.
[22] Liu, W., Angelov D., Erhan, D., Szegedy, C. Reed, S., Fu, C., & Berg (2016). AC SSD: Single Shot MultiBox Detector, pp-15.
[23] Redmon, J., & Farhadi, A. (2016). YOLO9000: Better, Faster, Stronger. Pp-1.
[24] Redmon, J., & Farhadi, A. (2016). YOLO9000: Better, Faster, Stronger. Pp-4.
[25] Redmon, J., & Farhadi, A. (2016). YOLO9000: Better, Faster, Stronger. Pp-5.
[26] Sun, C., & Murphy, K. (2017). Speed/accuracy trade-offs for modern convolutional object detectors .
[27] K. He, X. Zhang, S. Ren, and J. Sun. Deep residual learn- ing for image recognition. arXiv preprint arXiv: 1512.03385, 2015.two, 4,5.
[28] Sundar, (2018). KVS. Evaluating Training Time of Inception-v3 and Resnet-50, 101

Models using CPU and GPU across TensorFlow. 2018 Second International Conference on Electronics, Aerospace and Communication Technology (ICECA), (Iceca), 1964-1968.
[29] Sun, C., & Murphy, K. (2017). Speed / accuracy trade-offs for modern convolutional object detectors, pp-11.
[30] Chaves, D., Saikia, S., & Fern, L. (2018). A Review Sistem atic M ethods for Automatically Locate Objects in Images, p15.
[31] Tsoi, N., Gwak, J., Reid, I., & States, U. (2019). Generalized Intersection over Union: A Metric and A Loss for Bounding Box Regression.
[32] Li, Y., Huang, C., Ding, L., Li, Z., Pan, Y., & Gao, X. Deep learning in bioinformatics: introduction, application, and perspective in big data.
[33] White-filgueira, B., Garcia, D., Fern, M., Brea, M., & Paula, L. Deep Learning-
[34] Based Visual Tracking on Multiple Object Embedded System for Mobile Edge IoT and Computing Applications 1-8 (2019)
[35] T.-Y.Lin, M. Maire, Belongie S. J. Hays, P. Perona, D. Ra- mannan Dolla'r P., and C. Lawrence Zitnick. Microsoft COCO: Common objects in context. In ECCV, 1 May (2014)
[36] TYLin and P. Dollar. Ms coco api.https: // github. com / pdollar / coco, 2016.5

IOP Conf. Series: Materials Science and Engineering **844** (2020) 012025 doi:10.1088/1757-899X/844/1/012025

Growth Optimization of *Chlorella Vulgaris* in Mixotrophic Culture Enriched with Nutrients using Experimental Design

Ildefonso Baldiris-Navarro[1], Jorge Sanchez Aponte[1]

Environmental Science Programs, Sena-Cinaflup, Mamonal km4 Cartagena, Colombia.

E-mail: ibaldirisn@sena.edu.co

Abstract: In this research, statistical optimization designs were applied to evaluate the growth of the *Chlorella vulgaris* in mixotrophic cultures enriched with nitrogen and phosphorus. The influence of each factor was evaluated at three levels, so a multifactorial experimental design was created 32. The results obtained showed that microalgae growth improves 32% by adding 50 mg of nitrogen and 25 mg of Nitrogen to the culture. This project demonstrates the relevance of the design of experiments for the optimization of algal biomass production in order to obtain bioproducts.

1. Introduction

The increase in water pollution worldwide due to the discharge of domestic and industrial wastewater, have caused an impact on the quality of water bodies, proven by the increase in the concentrations of nitrogen in the form of ammonium, nitrites, nitrates, and phosphorus in the form of phosphates. The eutrophication of aquatic systems is caused by an excessive concentration of nutrients which alter the characteristics of the food chain and increase entropy in the ecosystem as a consequence of reduced biodiversity, with opportunistic species occupying places previously occupied by other species. Over time, the eutrophication processes might turn a swamp or lake into the mainland, because with the increase in nutrients the amount of plants and animals that live on the surface also increases, which when they die accumulate in the bottom and they are not completely decomposed [1].

In recent years, sewage treatment with microalgae has gained great interest due to its impressive properties. Microalgae are of great ecological importance due to their biodiversity, high growth rates and adaptability to different environmental conditions. They are present in any aquatic environment where that contains a source of carbon, nutrients, sufficient light and appropriate temperatures according to the species. They provide the biosphere with a considerable amount of oxygen for fixing more than 40% of the earth's coal, they are very attractive for producing many products of commercial interest since they can use any residual effluents as a source of nutrients [2]. The use of microalgae, in the treatment of wastewater, has been the subject of numerous investigations due to its ability to remove significant amounts of Nitrogen (N) and phosphorus (P), heavy metals, fats during its growth and at the same time may accelerate the inactivation of pathogenic bacteria [3]. The wastewater bioremediation capacity of microalgae has already been demonstrated, relevant aspects such as the conversion of nutrients present in algal biomass are described in several researches [4] [5].

The microalgal biomass generated may be used in the production of bioenergetic alternatives such as biodiesel, biohydrogen or methane, in addition animal food, beauty products and pharmaceuticals may also be obtained from it [6] [7] For this reason the production of microalgae using wastewater as cultivation media, may be an alternative for an environmental problem, such as the deterioration of water bodies and also may be a source of different byproducts of great commercial interest in agriculture and industry.

In microalgae cultures, nitrogen is an essential component of many key biomolecules, such as amino acids and nucleotides. Protein synthesis usually depends on an adequate supply of nitrogen to the culture. On the other hand, phosphorus plays an important role in most cellular processes, especially those that are involved in the generation and transformation of metabolic energy, so it is essential for the growth and reproduction of microalgae [8].

Design of experiments (DOE) is a technique that consists of carrying out a series of experiments in which deliberate changes in the variables of a process are induced, in order to observe and identify the causes of the changes in the output response [1]. With this technique may be achieved, for example, to improve the performance of a process and reduce its variability or production costs. Its application in industry includes fields such as chemistry, industrial processes, quality control, etc [9] [10] [11]. The statistical technique of experiment design has been used by several authors in different fields of microalgae use. Skorupskaite et al (2014) used experiment design to optimize a mixotrophic culture of *chlorella sp.* for the production of biofuel [12]. Priyadharshini et al (2016) used a Plackett-Burman design for the treatment of phenol with *Chlorella pyrenoidosa* microalgae [13]. Jaafari et al studied the effect of the interaction between the initial concentration of metals, the reaction time and the microalgae dose for the chromium, cadmium and cobalt biosorption by *colonial chlorella* microalgae [14]. Ido et al applied experimental designs to optimize lipid production of *oblique Scenedesmus* microalgae assisted with solvent extraction and ultrasound [15].

The aim of this research was to evaluate the growth of *Chlorella vulgaris* in a medium similar to wastewater enriched with nutrients and determine the concentration of nitrogen and phosphorus for optimal cell growth.

2. Methodology

2.1. *Organism and cultivation*

Freshwater microalgae *chlorella vulgaris* was obtained from the biotechnology laboratory of SENA CINAFLUP. *Chlorella* cells were maintained in modified Conway medium, which contained $FeCl_3 \cdot 6H_2O$ (26 g), $MnCl_2 \cdot 4H_2O$ (0.72 g), H_3BO_3 (67.2 g), EDTA (90 g), $Na_2HPO_4 \cdot 12H_2O$ (40 g), $NaNO_3$ (200 g), Na_2SiO_3 (40 g), H_2O (2 L), traces of metals (2 mL) and solution of vitamins (100 mL). The solution with metal traces contained $ZnCl_2$ (2.1 g), $CoCl_2 \cdot 6H_2O$ (2 g), $(NH_4)_6Mo_7O_{24} \cdot 4H_2O$ (0.9 g), $CuSO_4 \cdot 5H_2O$ (2 g) and distilled water (100 mL), and the vitamin solution was composed of decamyl (210 mg) and distilled water (100 mL). The sources of nitrogen and phosphorus for the bioassays were potassium diacid phosphate (KH_2PO_4) and sodium nitrate ($NaNO_3$).

2.2 *Experimental design*

To evaluate the microalgae growth cultured in excessive nitrogen and phosphorus, a factorial design 3^2 was implemented. This factorial design consisted of two factors (added nitrogen and phosphorus) and 3 levels for each factor (0 - 25 - 50 mg, respectively). Table 1 shows the factors and their levels in the experimental design. The design was used to determine the joint effects of several factors on a response variable. It was also used to determine the individual and cumulative effects of these variables and the mutual interactions between them. The complete design consisted of 9 runs, which were performed in duplicate to optimize the levels of selected variables. Data processing and calculations were carried out using a statistical software, Minitab, to estimate the coefficients of the regression equation. The equations were validated by analysis of variance (ANOVA) to determine the significance of each term in the adjusted equations and to estimate the adjustment in each case.

Table 1. Experimental design Factors and levels

Factors	levels
Nitrogen	0 - 25 - 50 mg
Phosphorus	0 - 25 - 50 mg

The statistical model for the design may be written considering the individual effect of each factor and the interaction between them, and is as follows:

$$Y_{ijk} = \mu + \gamma_i + \delta_j + (\gamma\delta)_{ij} + \varepsilon_{ijk}$$

With $i= 1,2,3$ $j=1,2,3;$ $k=1,..,n$

Where γ_i is the effect of the concentration of Nitrogen at its level i, δ_j represents the effect of the phosphorus concentration factor at its level j, $(\gamma\delta)_{ij}$ is the interaction effect of both at levels ij and n is the number of repetitions of each treatment. Consequently, the hypotheses to be tested are: H_0: $\gamma_i= 0$ (there is no significant effect of the Nitrogen factor over Growth), H_0: $\delta_j = 0$ (there is no effect of the Phosphorus factor over Growth) and H_0: $(\gamma\delta)_{ij} = 0$ (there is no interaction effect of factors A and B on the response variable). These hypotheses will be tested with the ANOVA; for this, the sums of squares for the three effects included in the previous equation are given by:

$$SC_A = \sum_{i=1}^{3} \frac{Y_{i**}^2}{3n} - \frac{Y_{***}^2}{n3^2}$$

$$SC_B = \sum_{j=1}^{3} \frac{Y_{*j*}^2}{3n} - \frac{Y_{***}^2}{n3^2}$$

$$SC_{AB} = \sum_{i=1}^{3}\sum_{j=1}^{3}\frac{Y_{ij*}^2}{n} - \frac{Y_{***}^2}{n3^2} - SC_A - SC_B$$

Graphically, the design of experiments for the optimization of microalgae growth may be seen in figure 1.

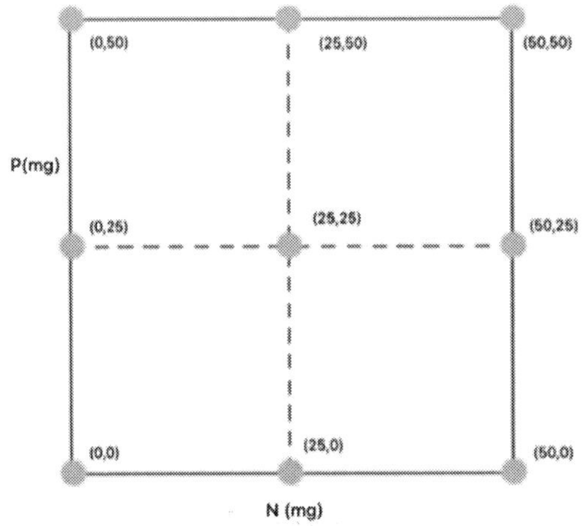

Figure 1. Experimental design for microalgae growth

For this model the total sum of squares is calculated with:

$$SC_T = \sum_{i=1}^{3}\sum_{j=1}^{3}\sum_{k=1}^{n} Y_{ijk}^2 - \frac{Y_{***}^2}{n3^2}$$

and the random error is calculated with the difference:

$$SC_E = SC_T - SC_{AB} - SC_A - SC_B$$

2.3 Determination of the microalgae biomass and growth rate

Biomass concentration was determined by cell count using neubauer chamber and measuring the optical density at 680 nm. In addition, the dry weight per liter of microalgae was measured day by day, using the total dissolved solids method. A linear relationship was found between dry weight and absorbance at 680 nm.

The specific growth rate was calculated as follows:

$$\mu(d^{-1}) = \frac{(lnX_2 - lnX_1)}{(t_2 - t_1)}$$

Where X_2 y X_1 are the biomass concentration in (gL^{-1}) in t_2 y t_1(days), respectively. Duplication time of the cell concentration was calculated using:

$$t_d = \frac{ln2}{\mu}$$

3. Results

The biomass concentration (dry weight per liter) of the microalgae cultures was calculated by measuring the solids and the optical density daily. The cell count showed that it was closely related to the optical density at 680 nm and a relationship was found between the two variables according to the following formulation:

$$Concentración\ de\ Biomasa(gL^{-1}) = 0.42Abs_{680nm} - 0.22,\ R^2 = 0.974$$

This relationship is depicted in figure 2

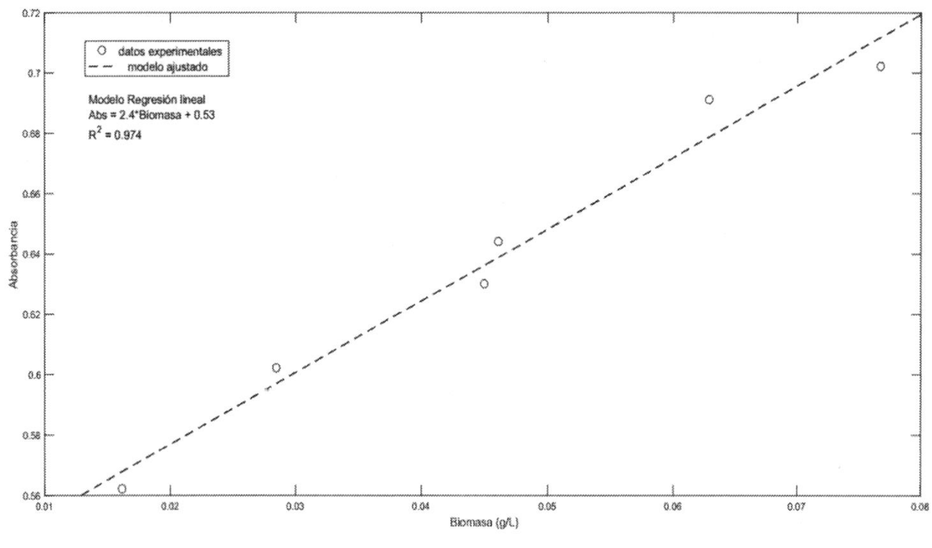

Figure 2. dry weight vs. absorbance for *chlorella vulgaris*

The growth of the microalgae with nutrients was made for a period of 8 days, during this time the optical density of the microalgae at 680 nm was measured as well as the dry weight every 24 hours. The variation in cell concentration is shown in Figure 2.

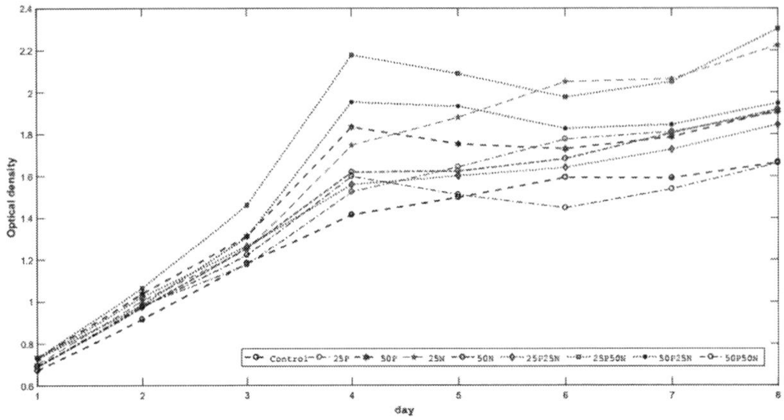

Figure 3. Microalgae growth profiles at different concentrations of N and P.

Population growth of the *Chlorella vulgaris* showed differences in the bio assays carried out with the nutrients. While *Chlorella vulgaris* with 50N-25P concentration reached a population close to 11'800,000 cells / mL, with 25N-50P concentration slightly exceeded 10'000,000 cells / mL (Figure 1) and in the control bioassay did not exceed 8'600,000 cells / mL, which implies a 37.2% improvement in the growth process. The effect of these concentrations may be seen reflected in the growth rates and doubling times (see figure 4). This shows that the best performer was obtained with the mixture of 25 mg of phosphorus with 50 mg of nitrogen.

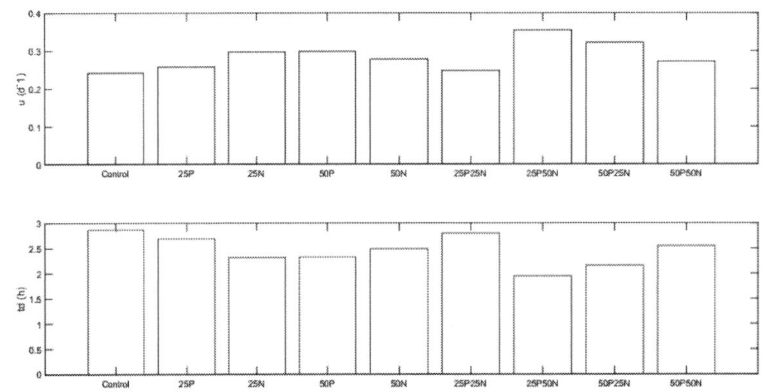

Figure 4. Growth kinetics for *Chlorella vulgaris*

3.1 Experimental Design

Pareto chart shows the main effects that are those passing the dotted line, in this case are the quantity of nitrogen and phosphorus-nitrogen interaction. These factors affect the growth of the microalgae at the experimental level (see figure 5).

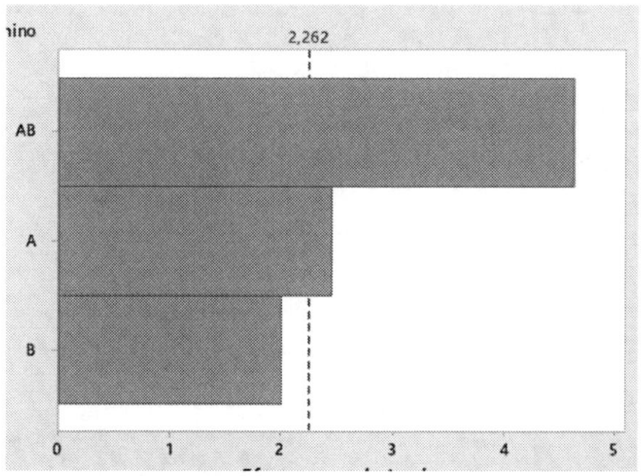

Figure 5. Pareto chart for *Chlorella vulgaris* growth. Factor A= N, Factor B= P

Table 2 shows the coefficients of the effects, as well as the analysis of variance. The main effects are those lower than the 0.05 significance level, p-Values (column P), these are shaded in the table.

Table 2. Variance analysis of factors and interaction.

Source	GL	SC Ajust.	MC Ajust.	F-value	p-value
Model	8	0,62724	0,078406	8,04	0,003
Lineal	4	0,16397	0,040992	4,20	0,034
Nitrogen	2	0,09569	0,047845	4,91	**0,036**
Phosphorus	2	0,06828	0,034139	3,50	0,075
interaction of 2 terms	4	0,46328	0,115819	11,88	0,001
NxP	4	0,46328	0,115819	11,88	**0,001**
Error	9	0,08774	0,009749		
Total	17	0,71498			

According to these values, the Nitrogen-Phosphorus interaction has a greater significance in the cultivation of chlorella. The mathematical model obtained has an R-quad of 87.73%, which fits well with the experimental data obtained.

a. Influence of individual factors on cell growth

As may be seen in Figure 6, the nitrogen factor creates an optimal effect at the second level (25 mg) and decays by adding 50 mg. Phosphorus has a similar effect when 25 mg of this is added, but as the dose rises, the growth becomes null and when the is not added, microalgae also has a moderate growth. This indicates that the effect of the added phosphorus is not significant in cell growth as shown by the pareto chart.

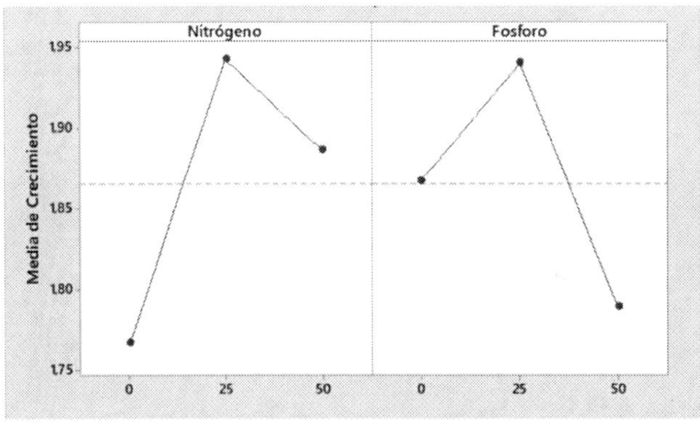

Figure 6. Chart of main effects for *Chlorella vulgaris* Growth

The effect of the interactions of the factors on cell growth may be seen in Figure 7. Here it shows that growth is optimal when 50 mg of nitrogen and 25 mg of Nitrogen are added. It may also be noted that as the added phosphorus concentration increases, cell density decreases, which shows that the phosphorus factor has a limit in its concentration, since otherwise the number of microalgae cells decreases.

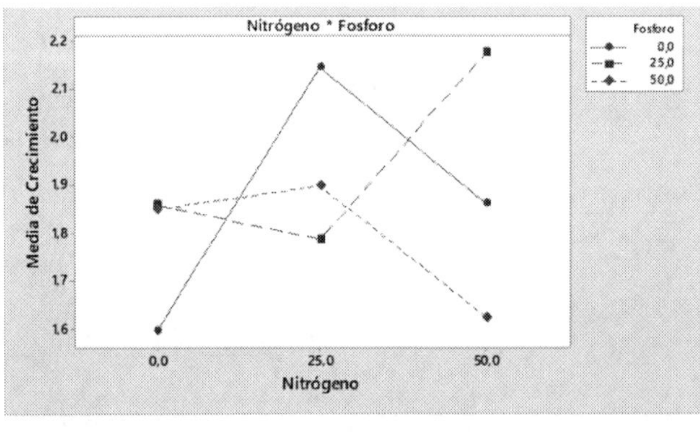

Figure 7. Interaction plot for *Chlorella vulgaris* growth

The responses of the population of *Chlorella vulgaris* in the different treatments with nutrients of the experimental design reveal that this species is particularly susceptible to the concentration of nitrogen compared to the concentration of phosphorus, since increasing the concentration of the former significantly increases the growth rate, while, although there is also a statistically significant decrease in the values of these variables due to phosphorus deficiency compared to the control treatment, the results are not as marked and have a limit value of nutrient acceptance, which generates a decrease in the population of the microalgae when it is reached.

Nitrogen deficiency affects growth as it limits the formation of aminoacids that are the building blocks of proteins, which are involved in many aspects of growth, including structural and metabolic processes. If the nitrogen supply is not enough, not enough proteins may be formed to maintain optimal growth levels.

4. Conclusions

The growth of *Chlorella vulgaris* was studied based on the addition of phosphorus and nitrogen using multiple factorial design. It could be observed that the growth is optimal when 50 mg of nitrogen and 25 mg of Nitrogen are added, with this treatment an improvement of 37.2% in the process was obtained. The factor with the greatest influence on the growth variable was the addition of nitrogen and the interaction Nitrogen-Phosphorus.

Acknowledgments

Authors want to thank SENA CINAFLUP y SENNOVA for financing this research.

References

[1] Baldiris-navarro I and Sanchez-aponte J 2018 *Cont. Eng. Sci* 11 1339
[2] Samorì G, Samorì C, Guerrini F and Pistocchi R. 2013 *Water Res 47* 791
[3] Lananan F, Abdul Hamid S. H, Din W, Ali N, Khatoon H, Jusoh A and Endut A 2014 *Int. Biodeterioration and Biodegradation 95* 127
[4] Roa-Parra A and Cañizares-Villanueva, R. 2012 *Bistua 10* 71
[5] Wang X, Wu Y, Zhang T, Xu X, Dao G and Hu H 2016 *Water Res. 94* 215
[6] Yesica I, Luis A. Ignacio A, Martha P, Paula N, Elvira L, and Hector M 2015 *African J. of Biotech. 14* 1710.
[7] Baldiris-navarro I, Sanchez J and Torres-Virviescas M 2017 *Int. J. of Chemtech* 10 411.
[8] Beuckels A, Smolders E and Muylaert K 2015 *Water Res, 77*, 98.
[9] Miller J N and Miller J C 2010 *Statistics and Chemometrics for Analytical Chemistry*
[10] Montgomery D C, Runger G C and Hubele N F 2009 *Engineering statistics.*
[11] Gutierrez H and De la Vara R 2008 *Análisis y diseño de experimentos.*
[12] Skorupskaite V Makareviciene V and Levisauskas D 2015 *Algal Research, 7*, 45.
[13] Dayana Priyadharshini S and Bakthavatsalam A K 2016 *Bioresource Tech 207* 150
[14] Jaafari J and Yaghmaeian K 2019 *Chemosphere 217* 447
[15] Ido A L, de Luna M D G, Capareda S C, Maglinao A L and Nam H 2018 *Energy 157* 949

IOP Conf. Series: Materials Science and Engineering **844** (2020) 012026 doi:10.1088/1757-899X/844/1/012026

Benefits of the instrumentation, monitoring and control systems (IMC) of bridges in Colombia, from an economic point of view through a case study

J D Castro[1], L Suarez[1] and I C Cerón[1]

[1] Catholic University of Colombia, Faculty of Engineer. Diagonal 46 A # 15B -10 Bogotá, Colombia

E-mail: jdcastro53@ucatolica.edu.co

Abstract. An instrumentation, monitoring and control (IMC) system is the result of the application of different branches of engineering such electrical, systems and telecommunications, applied to solve a problem in civil engineering, and achieving a radical innovation in the construction and the maintenance of bridges. These different branches come together to carry out activities such as the development of measurement instruments, applications for the storage and analysis of data, and the transmission thereof. This article presents the state of art about road infrastructure controls in Colombia, particularly on bridges. It summarizes some important cases of collapse with their main causes, establishing as an effective solution IMC of the bridges throughout the structures' life cycle, showing the benefits associated with the initiation of the system. In addition, an economic analysis of the installation of an IMC system is presented through a case study of one bridge located in Colombia, which has enabled us to identify and find relevant information related to those responsible for the project, the required instrumentation and the investment percentages regarding this particular type of project. The text consists of seven sections, which address the problems, instrumentation, software and personnel requirements for this type of project in Colombia, and finally to examine the associated costs.

1. Introduction

In the last few years, there has been a regular increase in the amount of problems related to the stability of the bridge structures in different countries [1]. Regardless of whether the structure is under construction or is in operation, the need to monitor structures throughout their life cycle has been demonstrated [2]. In a large number of cases, the stability of the structure is so affected that its collapse is inevitable. The following table shows several cases of bridge collapse (See Table 1).

It is necessary to mention that the causes of collapse of bridges vary, depending on whether the structure is in use or not. However, it is common to mention at the beginning of the investigations the following factors for bridges that are under construction: problems in design, foundations, material quality and geology [3]. In the case of bridges in operation: aging, corroded concrete, and lack of investment in maintenance [4] [5] [6]. Other cases may be due to the particular use of the structure [7]. Generally, stability problems generate various cost overruns in projects that can be avoided, within the different repercussions we have: questions in engineering, professionals [8], companies [3], public entities [9] and in general to the state of intervention within the project. On the other hand, economic costs are generated for projects [10], insurers and reinsurers [11] [12] [13], mobility inconveniences [14] [15], environmental impact [16] and people injured or deceased [1] [17] [18].

Table 1. Cases of bridge collapse.

Bridge	Date	Country
Morandi	08 - 14 - 2018	Italy
Cancura	06 - 23 - 2018	Chile
Chirajara	01 - 15 - 2018	Colombia
Charte	08 - 22 - 2016	Colombia
Calcutta	03 - 31 - 2016	India
Chalma	06 - 25 - 2014	Mexico

The alternative to reduce and mitigate the problems associated with stability of bridge structures is a bridge instrumentation, monitoring and control system, consisting of an instrumentation system, an information analysis and management system, and an alarm system. The IMC system is made up of computers that allow data acquisition, which will later be analyzed and stored, establishing alerts that generate alarms about the behavior of the structure, all in order to avoid its collapse and, therefore, avoid or mitigate the aforementioned repercussions [19]. Some researchers have been developing associated technology that has been collected in the literature [20].

After the collapse of the Chirajara bridge on January 15, 2018, the controls on structures under construction were strengthened, especially in this type of megaproject. This generated in the initial contract of the case study an addition for the instrumentation, monitoring and control of the bridge, and in turn, the personnel for the execution of the different activities related to the IMC system.

It is important to indicate that in Colombia there is no regulation, state directive, or technical requirement that requires systematic inspection and control of a structure through IMC systems during the life cycle of the bridge.

2. Who exercises control in Colombia

This section is divided into two parts: the first section presents the controls exercised by the main parties involved in the supervision of bridges during their life cycle.

2.1. Those who exercise control

In this order of ideas, two possible cases are described, classified depending on whether the road over which the bridge is located is under the charge of INVIAS, or it is part of a road concession. The following table identifies those involved in the first case.

2.1.1. Road under the charge of INVIAS. INVIAS, as a state authority, formulates and performs the selection process for the building of the structure of a bridge and then opens a selection process designating a contractor and auditor for the execution of the contract. Once the construction is finished, they deliver the work to INVIAS to carry out its administration and maintenance.

2.1.2. Concessioned road. On concessioned roads, depending on the terms agreed to in the contract, after delivery to INVIAS (1.1.1 Road under the charge of INVIAS), the road or bridge is delivered to ANI, which in turn delivers it to the Road Concession in charge of administration and maintenance until its final delivery to INVIAS after the end of the concession contract.

Table 2. Those involved in supervision in Colombia.

Involved	Function according to the stage
National Institute of roads - INVIAS	Contracting authority of the construction and auditing contractor, and state supervisor of the contract
Building Contractor	Constructor and responsible during the validity of policies of compliance, quality, extracontractual civil liability and stability of work.
Auditing	Supervisor and responsible by default.
National Agency of Infrastructure- ANI	State authority that receives the work from INVIAS and delivers them road concessions.
Road Concession	Private company or semi-public association, which performs the maintenance and operation of a road during certain times.

2.2. Controls performed
Depending on the stage of the life cycle, bridges in Colombia are under the supervision of different stakeholders. In the next paragraphs, we will describe the following stages:

2.2.1. Building contractor. The building contractor performs a basic instrumentation during the construction of the bridge, in order to verify its behavior.

2.2.2. INVIAS. The Bridge Administration System of Colombia (SIPUCOL), assigned to INVIAS, is in charge of the bridges department of the national road network. Currently, it monitors active bridges through visual inspections of the structural elements. This is done based on the professional criterion linked to INVIAS which assigns a quantitative and qualitative grade depending on the current state of the bridge structure. A report is then delivered to contract the necessary activities.

2.2.3. Road concession. Depending on the terms agreed to in the road concession contract, they can be charged with inspections, maintenance and / or adaptation of the infrastructure in their respective road corridors. It is worth mentioning that some road concessions already have structures in place with IMC systems.

3. Overview about controls in the region
In other countries, such as Mexico, the IMC is in charge of the monitoring center of bridges and intelligent structures (CEMPEI), which is responsible for monitoring the infrastructure of the main road corridors in the country. One of the most prominent cases is monitoring and load testing of El Carrizo bridge (a project monitored by CEMPEI), in which a road accident occurred. Thanks to the IMC system, it was possible to perform a quick evaluation of the structure in order to put it into service with the necessary repairs, thus demonstrating the benefit of the system[1].
In other countries, such as Chile, after the last collapse events (Table 1), bridge revision and monitoring projects were initiated, highlighting that there are already monitored bridges there[2]. It is necessary to expand IMC programs in order to increase the infrastructure covered and avoid fatal incidents.

[1] Some conclusions are shown in https://www.gob.mx/imt/es/videos/monitoreo-y-pruebas-de-carga-en-el-puente-el-carrizo. Mexican Institute of Transport (IMT) leads other examples about their practices. https://www.gob.mx/imt.

[2] An important local-newscasts published this information: https://www.t13.cl/noticia/politica/video-pinera-destituye-al-director-obras-del-mop-caida-puente-osorno

In the USA, an article determined that they have a worn infrastructure, with lack of maintenance, and with a completed time of useful life [21]. This emphasizes the importance of an IMC, which allows establishment of the necessary maintenance for bridges.

In conclusion, it is necessary that in Colombia the bridge administration system (SIPUCOL) starts with the implementation of an IMC system for the bridge infrastructure connected to INVIAS, according to the importance and relevance of the structures along with their respective road corridors. Other factors to take into consideration are vehicular traffic, old age, lighting, budget for instrumentation, level of project information, location and typology.

4. Instrumentation, software and staff for an IMC System

4.1. Instrumentation

This section presents the status of instrumentation equipment, software and staff for bridges' IMC systems. Table 3 describes instrumentation equipment together with measurement data.

Table 3. Equipment.

Equipment	Measurement data
-Anemometers	Wind speed and direction
-Deformeters	Deformation
-Load cells	Tension
-Scouring sensor	Scouring
-Settlement sensor	Settlement
-Temperature multipoint FBG sensor	Deformation and temperature
-Tiltmeter	Tilt
-Triaxial accelerometers sensor	Acceleration
-Type K thermocouple	Temperature
-Uniaxial accelerometers sensor	Acceleration

4.2. Software

The necessary software must analyze the data and compare it with the ranges established by professionals, thus issuing an alarm when the bridge does not behave optimally. Alonso et al [22] recently presented a review of internet-based technologies of things (IoT) that are currently used. In summary, they can originate in three ways: through the purchase or rental of software developed by companies in the sector; through own developments of data analysis programs and / or the use of free license software oriented to data analysis.

4.3. Minimum professional staff

The personnel require strong qualifications for the project. The main staff is described as follows:

Table 4. Staff.

Staff	Experience	Specialist
Civil engineer	10 years	Structures and bridge design
Telecommunications or electrical engineer	5 years	N/A
Systems engineer	10 years	Programming or software development
Installation technicians	2 years	N/A

5. Equipment, typology and benefits of case study

5.1. Equipment

The following table describes the equipment of instrumentation used in the case study, the measurement data and its location in the structure. Later in the chapter, the typology of the case study is addressed, before comparing the costs related to the instrumentation and the final cost analysis.

Table 5. Equipment.

Equipment	Measurement data	Location
-Load cells	Tension	Cable
-Temperature multipoint FBG sensor	Deformation and temperature	Deck
-Type K thermocouple	Temperature	Deck and cable
-Deformeters	Deformation	Deck, towers and anchorage block
-Uniaxial accelerometers sensor	Acceleration	Cable
-Triaxial accelerometers sensor	Acceleration	Anchorage block and point out of the bridge
-Anemometers	Wind speed and direction	Towers and deck
-Tiltmeter	Tilt	Top of the tower

It should be noted that no equipment was used for scour control, however in [23] relevant information for the selection of the scour sensors was obtained. Additionally, and in accordance with the equipment installed on site, the author concludes: "*...Both the shallow foundation experiment and the deep foundation experiment give promising results that the accelerometer can be used to predict bridge failure as well as tilt-meter and sonar sensor.*" ... "*approach is proved to be effective to analyze the accelerometer data as they showed significant change when the scour depth reached the bottom of the column, and the column started to settle and rock*" [23, p. 87] ... "*the scour monitoring system will result in a program that is successful to ensure the safety of the bridge and of the traveling public*"... [23, p. 271].

5.2. Typology

The approximate typology of the case study corresponds to a cable-stayed bridge 400 meters in length, with a main span of 250 meters and entrance and exit spans of 80 meters. The deck has a useful width of 15 meters and a height of 1.40 meters. The bridge is a harp and fan type. The separation of the points of support of the deck is 10 meters. Two beams every 4.50 meters support the cross section of the deck. The deck is supported by two towers and at its ends by two counterweights, to which some braces are anchored. The bridge has an orthotropic deck formed by two lightened main longitudinal ribs to which the braces are anchored, joined by transverse beams every 4.50 meters and an upper slab as a board. The batteries have a type H geometry; each column is responsible for a cable plane. The stack has two transverse braces 1.50 meters high and 1.30 meters wide, with concentric post-tensioning. The lower brace supports the deck with two insulators. The bridge is not compensated in its construction, so it has two counterweights at its ends that anchor six braces.

5.3. Benefits

The main benefit of an instrumentation, monitoring and bridges control system lies in having the data on the behavior of the structure in real time, analyzing and storing it, so it can be known if the structure is behaving properly and in the right ranges, and when it stops behaving in that way. It is advisable to begin the IMC from the construction of the structure, in order to acquire the data of its behavior, and perform the installation of embedded equipment in the concrete. Even during its construction, stage it is possible establish the ranges of the structure's behavior, determining the possibility of an imminent collapse, and solving the problems described in the introduction. In that order of ideas, with the data of the structure it is also possible to verify the state and behavior of isolated elements, which may be susceptible to maintenance or reinforcement, even warning of the malfunction of those elements with the goal of making preventive corrections, avoiding invasive maintenance and additional expenses that could close road corridors. The instrumentation also generates early warnings of scour, as mentioned in previous chapters, which allows us to know a bridge's state after earthquakes or road accidents and geological movements. The IMC of bridges takes on special relevance considering that each year the country's infrastructure sees greater aging, increasing the need to perform maintenance where the structure requires. Additionally, climate change increases the scour factors on the structure. Along those lines, it is important to highlight the following Table 6, which shows 63 real cases of collapse registered in Colombia since 1986.

Table 6. Causes of collapse of bridges in Colombia [24].

Main Cause	Number of bridges	Percentage
Structural and design deficiency	6	10%
Scouring	15	24%
Overload and impact	3	5%
Terrorist attacks	20	32%
Landslides, growth, etc.	15	24%
Lack of maintenance	1	2%
Deficiencies in construction or auditing	3	5%
Total	63	

In this way, the need for an instrumentation, monitoring and control system of bridges is clear, particularly in considering the investment in the infrastructure of the country, in that it would allow us to know, anticipate, and facilitate the maintenance of the structures and therefore mitigate or reduce the collapse events that may happen in the future. [25].

The IMC system has another benefit associated with the clarification of investigations and responsibilities in cases where the structure was not built with the established materials, due to the ease of the analysis of the structure. It is also possible to share the data analysis with other entities like national risk and disaster management unit UGRD.

Below, a cost and risk graph shows the IMC level and the level of IMC efficiency. Note the point "IMC adequate," which represents the appropriate point between the level of efficiency and level of the IMC (see Figure 1).

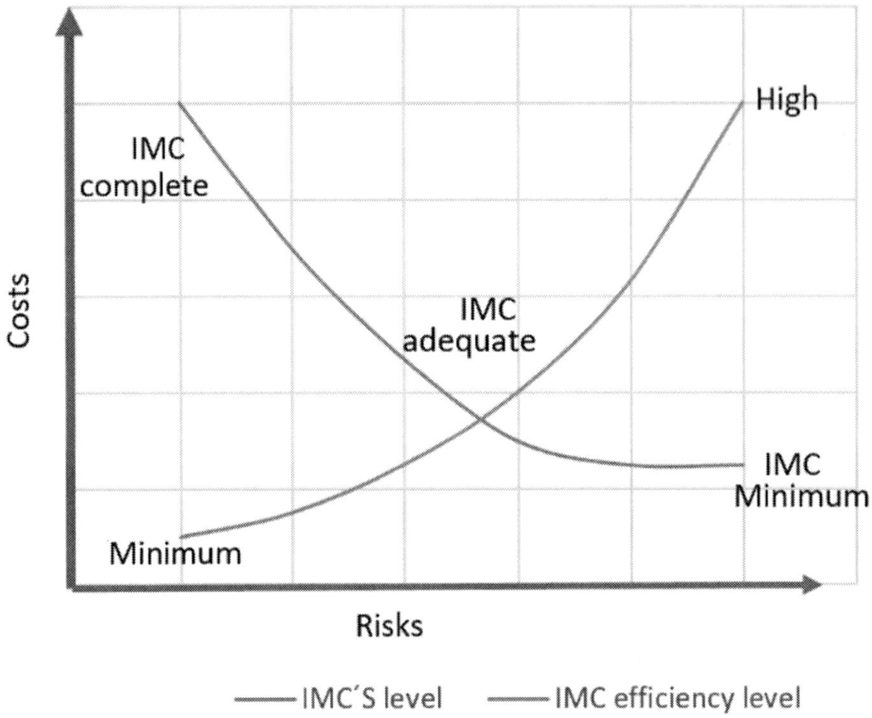

Figure 1. Costs vs Risks

6. Interest groups Colombia

Interest groups in Colombia are divided in two: state entities and private organizations, which are described in the table below.

7. Costs

Through a financial exercise on the real costs of a bridge in recent construction in Colombia, the following values for the typology described above were found. The cost of a bridge IMC system for this typology corresponds to a percentage around 6.8% to 7.1% of the total value of the bridge's construction, with a high level of instrumentation. However, if we take into account the total value of the contract, which includes social, property and environmental management, the percentage of the IMC system is around 4.4% to 4.6%. The percentage that corresponds to the instrumentation of the bridge is around 58.9% to 60.9% of the total value of the IMC contract and the remaining percentage of 39.1% to 41.1% corresponds to staff.

Table 7. Interest groups in Colombia.

State entities	Because	Due to
INVIAS	Formulation and estimation of costs of IMC systems of bridges that make or will be part of the network in charge of the entity. In the programming of the maintenance, monitoring and checking of the state of the infrastructure of bridges that are attached to the entity.	In compliance with its functions according to Decree 2056 and 2067 of July 24, 2003.
National Infrastructure Agency, ANI	Formulation, estimation of costs of projects that make or will be part of road concessions.	In compliance with its functions according to Decree 4165 of November 3, 2011 In case of any irregularity or investigation of the project.
Control entities, Prosecutor's Office, Procuracy Office and Comptroller's Office	Clarification of investigations and controversies in case of collapse of a structure that uses the IMC system.	In case of any irregularity or investigation of the project.
UGRD National risk and disaster management unit	System of early warnings and verification of the state of the structure after natural disasters or road accidents that involve structures with IMC.	Coordination of activities in the event of a disaster.
Private Organizations	**Because**	**Due to**
Construction contractors	Start-up of the IMC system of bridges during the construction stage, with early warnings in case of the structure's collapse.	Due to the responsibilities acquired in the contract.
Road Concessions	Start-up of the IMC system of bridges during the construction stage, with early warnings in case of the structure's collapse.	Due to the responsibilities acquired in the contract.
Insurers	Start-up of the IMC system of bridges during the construction stage, with early warnings in case of the structure's collapse. Offering the opportunity to reinforce the structure, or safeguard the lives of workers or third parties who may be involved, avoiding the payment of policies Provision of data due to the action of repetition after the payment of policies, in order to avoid costly and / or delayed investigations to establish responsibilities.	Payment of indemnities associated with policies of responsibility, quality, stability, being those affected due to the payment of claims. Legal defense of the insurer.

8. Conclusions

The instrumentation, monitoring and control of bridges is necessary, in order to avoid and mitigate the problems related to the collapse of the structures, since identifying the deformations that are generated in early phases allows the generation of alerts to the community in general, fundamental information for the authorities in charge i.e. INVIAS and ANI, control authorities, and insurers in the case of Colombia, thereby allowing the necessary repetition actions due to action or omission.

Additionally, this article teaches the benefits related to the maintenance and control phase of structures due to the specific technical information obtained on damages caused by earthquakes, road accidents, overloads, scour or geological events. Further, in an eventual investigation, information from the IMC allows the easy solution of contractual problems related to the technical behavior of the infrastructure, in such a way that the responsibility of anyone involved can be identified.

The IMC is not restricted only to the behavior of the substructure and superstructure, but can also implement slopes, retaining walls and other elements of civil works. However, a particular pre-feasibility study is required for each project due to the factors specific to each type of project. It is also advisable to plan the IMC from the formulation of the building project, since it is advisable to generate a budget and schedule of technical and administrative tasks that fit the needs of the project. Therefore, it is necessary to have the collaboration of the professional structural designer of the bridge.

The main factor that determines the direct cost of the IMC systems is the type of bridge; additionally, the analysis and maintenance of the resulting data throughout the life cycle of the bridge must be considered as an operational cost of the system. Currently these costs are not considered in the budget, and it is recommended that they be done through independent contracts. For the typology described it conclude that the investment cost of IMC system with a high level of instrumentation is close to 7% of the construction value of the bridge.

Finally, the cost vs. risk is outlined where, at a higher investment cost in the IMC system, the risk in the structure is lower. In the same way, the efficiency level function indicates that an excessive level of instrumentation generates financial inefficiency in the investment. The graph shows a point called ADEQUATE IMC, the point where the two functions intersect and is the level of instrumentation suitable for the bridge. It must be said that this chart applies to any bridge that requires an IMC system.

References

[1] El Nuevo Diario 2018 Los derrumbes de puentes más mortíferos de los últimos 20 años Available: https://www.elnuevodiario.com.ni/internacionales/472161-ultimos-derrumbes-mortiferos-puentes/

[2] Farrar C and Worden K An introduction to structural health monitoring 2006 *Philosophical Transactions of the Royal Society A: Mathematical, Physical and Engineering Sciences* vol 365 no 1851 pp 303-315

[3] Hernández J M 2018 Las empresas que responderían por el colapso del puente de Chirajara *El Espectador* Available: https://www.elespectador.com/economia/las-empresas-que-responderian-por-el-colapso-del-puente-de-chirajara-articulo-733921

[4] Gómez Fuentes Á 2018 Diez puentes se han desplomado en Italia en los últimos cinco años *Diario ABC* Available: https://www.abc.es/internacional/abci-diez-puentes-desplomado-italia-ultimos-cinco-anos-201808142109_noticia.html

[5] El Nuevo Herald 2018 Diseño y mantenimiento del puente de Génova, Italia, están bajo sospecha Available: https://www.elnuevoherald.com/opinion-es/trasfondo/article217070475.html

[6] García J M, Ospina J and Graciano E A 2014 Evaluación técnica de los puentes en la infraestructura vial del departamento de Antioquia *Ingeniería Solidaria* vol 10 no 17 pp 49-54

[7] Parkasiewicz B, Kadela M, Bętkowski P, Sieńko R and Bednarski Ł 2007 Application of structure monitoring systems to the assessment of the behaviour of bridges in mining areas

IOP Conference Series: Materials Science and Engineering vol. 245 no 3 p 032018
[8] Palomino C 2018 Ética y corrupción de la ingeniería estructural Asociación Colombiana de Ingeniería Estructural Available: www.aciescolombia.org
[9] El Espectador 2018 Fiscalía abre investigación por desplome del puente Chirajara, en vía al Llano Available: https://www.elespectador.com/noticias/judicial/fiscalia-abre-investigacion-por-desplome-del-puente-chirajara-en-al-llano-articulo-733619
[10] Caracol Radio 2018 Lo que quedo del puente Chirajara será dinamitado, según Coviandes *6 AM Hoy por Hoy* Available: http://caracol.com.co/radio/2018/02/23/nacional/1519383146_948387.html
[11] Benítez K 2018 Gisaico contrató con QBE la póliza "Todo Riesgo Construcción" *La Republica*. Available: https://www.larepublica.co/economia/gisaico-contrato-con-qbe-la-poliza-todo-riesgo-construccion-2589525
[12] Caracol Radio 2018 Coviandes tenía todos los seguros exigidos para obras como el puente de Chirajara Available: http://caracol.com.co/radio/2018/01/17/nacional/1516188907_972678.html
[13] Caracol Radio 2018 Pólizas son suficientes para cubrir el siniestro del puente de Chirajara Available: http://caracol.com.co/radio/2018/11/07/nacional/1541553265_786059.html
[14] El Telégrafo 2017 Nuevo puente une a más 12 comunidades de Napo Available: https://www.eltelegrafo.com.ec/noticias/regional/1/nuevo-puente-une-a-mas-12-comunidades-de-napo
[15] Caracol Radio 2017 En riesgo de colapso, se encuentra viaducto entre Boyacá y Casanare *Caracol Radio* Available: http://caracol.com.co/emisora/2017/06/14/tunja/1497439053_572341.html
[16] García R 2018 Megaproyectos y desastres ambientales: ¿qué está pasando? *UN Periódico Digital* Available: http://unperiodico.unal.edu.co/pages/detail/megaproyectos-y-desastres-ambientales-que-esta-pasando/
[17] Diario ABC 2018 Ya son 43 muertos en el puente Morandi en Génova Available: https://www.abc.es/internacional/abci-43-muertos-puente-morandi-genova-201808191815_video.html
[18] El Universo 2014 Puente colapsa en Napo; doce personas resultaron heridas Available: https://www.eluniverso.com/noticias/2014/04/04/nota/2596646/puente-colapsa-napo-12-personas-resultaron-heridas
[19] Instituto Mexicano de Transporte 2017 Monitorean Puentes y Estructuras de México Available: https://www.gob.mx/imt/articulos/monitorean-puentes-y-estructuras-de-mexico?idiom=es
[20] Klikowicz P, Salamak M and Poprawa G 2016 Structural health monitoring of urban structures *Procedia engineering* vol 161 pp 958-962
[21] Infrastructure report card 2017 Infrastructure report card Available: https://www.infrastructurereportcard.org/wp-content/uploads/2017/01/Bridges-Final.pdf
[22] Alonso L, Barbarán J, Chen J, Díaz M, Llopis L and Rubio B 2018 Middleware and communication technologies for structural health monitoring of critical infrastructures: A survey *Computer Standards & Interfaces* vol 56 pp. 83-100
[23] Briaud J L, Hurlebaus S, Chang K A, Yao C, Sharma H, Yu O Y, Darby C, Hunt B E and Price G R 2019 Realtime monitoring of bridge scour using remote monitoring technology Available: https://static.tti.tamu.edu/tti.tamu.edu/documents/0-6060-1.pdf
[24] Muñoz Díaz E E 2002 Estudio de las causas de colapso de algunos puentes en Colombia *Ingeniería y Universidad* vol 6 no 1 pp 33-48
[25] Correa M V 2017 Así avanza la construcción de cinco grandes puentes *El Colombiano* Available: http://www.elcolombiano.com/colombia/obras-de-cinco-grandes-puentes-en-colombia-LI6695278

IOP Conf. Series: Materials Science and Engineering **844** (2020) 012027 doi:10.1088/1757-899X/844/1/012027

Causal Analysis of the Pinion Teeth Failure in a Limestone Reclaimer

L M Castellanos[1*], J J Cabello[2], A Sagastume[2], V L Martínez[1] and A E Ortiz [3]

[1*] Facultad de Ingeniería, Universidad Simón Bolívar, Carrera 59 No. 59-65, Barranquilla ,Colombia.
[2] Departamento de Energía, Universidad de la Costa, Calle 58 # 55 – 66, Barranquilla, Colombia.
[3] Universidad Tecnológica de Bolívar, Parque Industrial y Tecnológico Carlos Vélez Pombo, Km 1 Vía Turbaco, Cartagena, Colombia.

E-mail: * luis.castellanos@unisimonbolivar.edu.co

Abstract. This paper presents a causal analysis of an unexpected failure of the pinion teeth of a large reducer installed in a limestone reclaimer which is part of a raw material transportation system of a cement factory. The Montano methodology is applied. It includes a collection of documentation, evidences, evaluation of failure mechanism by visual inspection and some hypothesis and their validation based on the analysis of the material resistance. It is demonstrated that although the teeth of the pinion were made of a suitable material and cemented correctly, they worked overloaded due to the presence of large rocks in the raw material. A resulting increase of contact tensions in the teeth over the allowed limits led to a fatigue failure process which is described in this work. The results of the analysis can be of great interest in order to develop preventive measures and avoid future undesired plant stoppages and economic losses.

1. Introduction

Mechanical elements are designed and manufactured for a specific purpose taking into account the loads, mode of loads application, planned operational conditions and environmental factors in order to guarantee a long useful life. However, mechanical failures during service are relatively frequent. When a failure occurs, a thorough analysis is necessary to establish its causes and then avoid its repetition [1].

Gears are the most widely used mechanical transmissions in the industry and to the study of their failures have been dedicated numerous research which had classified and characterized the failures according to its characteristics and causes in order to facilitate their study [2][3], also a methodology to gear failures study have been stablished [4] although each gear failure case have own particularities and need specifically analysis [5].

In the cement industry, gearboxes drives are massively used to drive most material handling and transportation equipment, working in hard conditions usually 24 hours by 7 days, inside a high powder content environment and usually with significant dynamics loads [6].

The significant economic losses as a result of mechanical failures in the cement industry are highlighted by Mayer [7], also Yanusa et al. [8] point the importance of analyzing each one in order to solve it correctly avoiding its repetition.

This paper is focused on the analysis of a gears critical failure in the gearbox drive of limestone rocks reclaimer in a Colombian cement factory, identifying the kind of failure, its causes and solutions to avoid repetition.

2. Limestone reclaimer

The limestone reclaimer has two main functions: drag the raw material from the pile where it is stored to the push system by using the rake arms showed in figure 1. The second function is to push continuously the material to a conveyor by means of an endless rolling mechanism with steel tablets as seen in figure. 2. Finally, the limestone is transported by a conveyor to the crude mill.

Figure 1. Drag system in the limestone reclaimer

Figure 2 shows the steel tablets that push the limestone through a bottom path of the pusher. These tablets return from the upper part doing a cycling process. A chain-driven mechanism connected with a gear reducer is used for the mechanical power transmission. This work is motivated by an unexpected failure of the pinion teeth of this reducer.

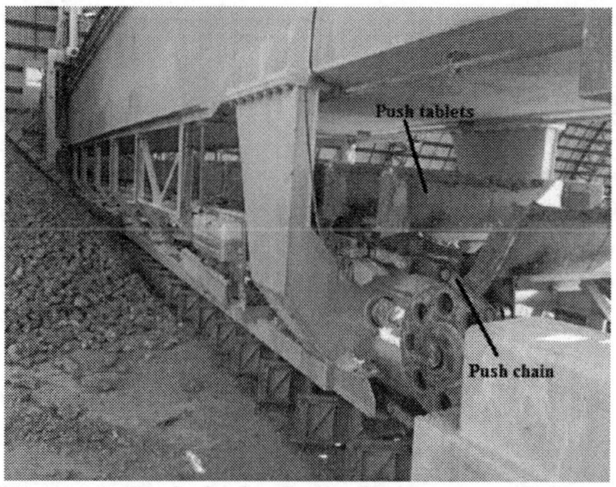

Figure 2. Push device in the limestone reclaimer.

3. Failure Analysis

The methodology used in this work is based on the applied by Montano [9] and proposed by Errichelo [4]. The steps are shown next

1. Collection of documentary evidence.
2. Analysis of service conditions.
3. Visual inspection and evaluation of the failure mechanism
4. Preliminary hypothesis.
5. Material evaluation
6. Mechanical resistance assessment.

3.1 Collection of documentary evidence.

At this stage, data about gearbox X4FH210/T model was collected. Although in the cement factory there are other seven similar reducers operating different equipment, there are no records about failures or incidents in their use. The gearbox has two steps, the failure occurred in the output gear. The nominals data of gearbox are shown in table 1.

Table 1. Drag arm´s gearbox operational data

Nominal motor power	93 kW
Motor speed	1750 rpm
Output speed of gearbox	15,53 rpm
Input speed to the output gear pair	372,7 rpm
Total transmission ratio	112,6
Nominal gearbox power	80 kW

Since there is no available data about the dimensions of the gears, a methodology to geometric decryption of damaged cylindrical gears [10] in order to get the geometrical parameters of the faulted gear pair was used. The results are shown in Table 2.

Table 2. Geometric parameters of damaged gear pair.

Number of teeth of the pinion	19
Number of teeth of the crown	63
Transmission ratio	3,31
Module	9,5 mm
Distance between centres	400 mm
Helix angle	13,15 o

3.2 Analysis of service conditions.

The service regime of the push mechanism is very hard because of frequently big rocks in the limestone as can be seen in figure 3 (a) provoking dynamics loads which deform tablets and structure of the rake arm as are shown in figure 3 (b) and (c). Also the chain wear increases the dynamic loads.

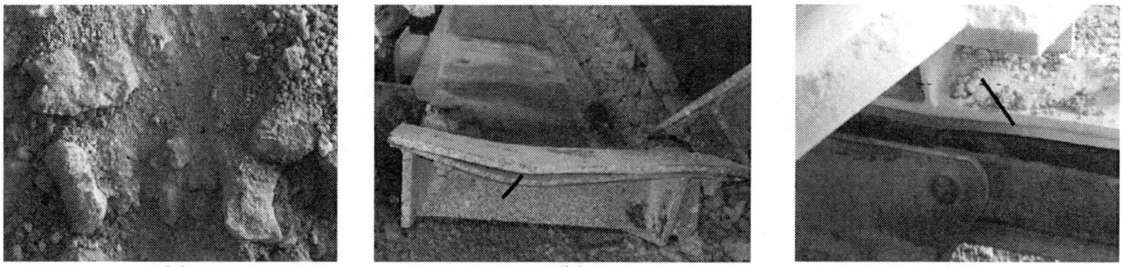

(a) (b) (c)

Figure 3. Operational conditions of push device

3.3 Visual inspection and evaluation of the failure mechanism

In visual inspection, the damaged gears were carefully examined in order to identify the failure mechanism. Large stones are observed in the raw material and plastic deformation in the steel structure and tablets

In figure 4 some pictures of the failure in the pinion are shown, the failure of the gears pair is mainly in the pinion. In figure 4 (a) the circle marked teeth surfaces show pitting and micro-pitting failure mode with scales removed, on teeth surfaces mark by box, which are on the teeth back, some fragile fractures are observed through the surface. In figure 4 (b) the highly destructive character of the failure is shown and two teeth fully removed from the pinion are shown, evidenced the effects of the high overload level produced by the progressive spoilage the teeth profile and also by the external dynamic load action. The tooth before the completely removed teeth has a large significant loss of material. Figure 4 (c) show a best picture of the mass loss in the work tooth flanks are shown. No significant evidence of wear or plastic deformation was found on any affected teeth.

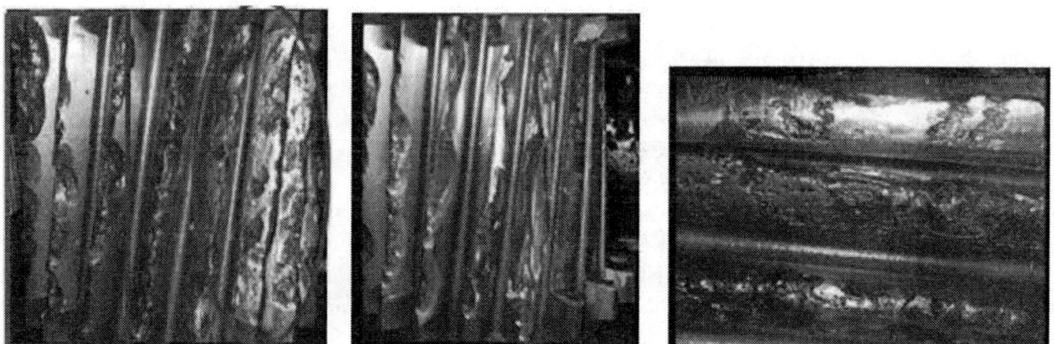

(a) (b) (c)

Figure 4. Visual inspection of the pinion teeth.

The fractographic study was carried out on both sides of the tooth as are shown in figure 5 (a). The behaviour is typical of this kind of failure with micropitting at the beginning and progressive pitting and lost large part of the surface with the consequent increase of surfacing stress and removing scales from the surface until its destruction. In figure 5 (b) a fractographic study of the tooth back surface is shown, the high level of destruction of work surface and the dynamics loads cause the extension of fragile fracture to this surface destructing.

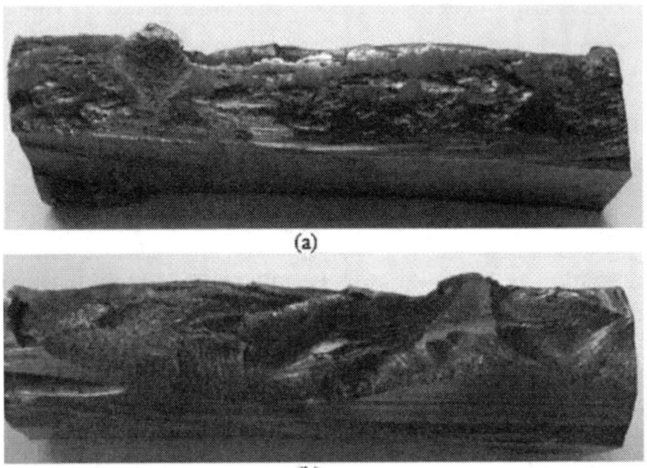

(a)

(b)

Figure 5. Macrographic study (25 X) on the tooth surface, (a) work surface y (b) back surface

Surface fatigue in the closed gears is not easily detected and have a trend to worse because of the growth of contact stresses and the presence of metal´s fragments detached from teeth surfaced which introduce additional loads which contribute to fracture by the base of other teeth. When a tooth is broken by a sudden shock or an overload, the fracture usually has a fibrous appearance, even when the tooth is fully hardened, the fracture will look like the fibers of a plastic material that has been twisted [4]. When consecutive teeth are broken, it usually happens that one or two are broken by fatigue; As the gear continues to rotate under torque, the hit of the wheel that meshes when jumping into the gap left by the fatigued tooth will break other additional teeth. Looking at several fractured teeth, one can define which one failed due to fatigue (it will present a smooth appearance) and which failed later due to overload (fibrous appearance). This phenomenon is manifested in figure 6 where a detail of removed teeth are shown, in which the tooth-marked No 1 was fractured by fatigue and those on the right of this failed due to overload, in the absence of tooth No 1.

Figure 6. Removed teeth

3.4 Preliminary hypothesis.

There are six generic causes for gear failure: wrong design, inadequate material selection, faulty thermal treatment, defective manufacture, faulty assembly, operation and poor maintenance [9]. Since the gearbox is a successful commercial product and there are in the factory several of the same kind without technical problems the hypothesis associated with design, manufacturing or materials are refuse. The preliminary hypothesis is that the dynamics load and overloads provoked by operational conditions with the usual presence of oversized stones y in limestone pile and the push chain wear are the main cause of the gear failure and if these conditions are corrected the gearbox could work reliability.

3.5 Material evaluation

Since the material and thermal treatment of the gears were not known several tests were carried out in order to identify the material and its characteristics. The results of the chemical analysis indicate that the steel is slightly alloyed to the chromium - nickel, with very low percentage of carbon and without the presence of grain refiners. Steel corresponds to an 18NiCrMo 7 steel, by German and European standard. In table 3 the results of chemical analysis are shown.

Table 3. Chemical analysis

Element	C	Si	Mn	P	S	Cr	Mo
%p/p	0.159	0.313	0.609	0.023	0.025	1.560	0.264
Element	Ni	Cu	Al	As	B	Co	Nb
%p/p	1.513	0.211	0.019	0.013	0.00055	0.017	0.0051
Element	Pb	Sb	Sn	Ta	Ti	V	W
%p/p	0.0002	0.0033	0.017	0.005	0.0025	0.0071	0.001
Element	Zr	Ca	-	-	-	-	-
%p/p	0.0005	0.003	-	-	-	-	-

In figure 7 (a), a specimen from a pinion tooth taken from lateral side which not contact con crown tooth is shown, used for macrographic analysis, hardness measurement and micrographic analysis. The macrographic analysis of polished area evidenced two zones, one peripheral, around 1,3 to 1,6 mm, where the surface is darker as a result of the chemical attack with Nital at 3 % and other inside the tooth clearest. The structure is typical of cemented steel which is corresponding with hardness measures the 60 HRC in surface and 45 HRC below the cemented layer. Additionally, the micrograph of the surface layer in figure 7 (b) shows martensite with low tempering, typical of the cemented and tempered. It should be noted that the steel is of very good quality due to the degree of cleaning, there is a low level of non-metallic inclusions, in the form of oxides.

(a) (b)

Figure. 7. Tooth section (a) macro structural analysis and (b) microstructure showing martensite

The results let concluded that that the Pinion was cemented, with posterior tempering, with a core (inside the tooth), for its chromium and nickel levels the inside of the tooth hardens enough (45 HRC), limited by the Low carbon inside (0.16 % C), giving rise to a tempered martensitic of medium hardness. In contrast, the surface has high hardness (60 HRC) due to the martensitic rich in carbon and chromium. In addition, small grain size is evident, which is desirable.

3.6 Mechanical resistance assessment

In order to verify the results of the visual inspections, which evidenced the occurrence of destructive pitting destructing the surface and the profile of the tooth increasing the dynamic loads until the full fracture of the teeth in its base is also appreciated; a verification of the surface resistance and breakage by bending of the teeth for different levels of dynamic loading is carried out using the software KISSOFT (http://www.kisssoft.ch/castellano/home/index.php) applying the calculation methodology established by ISO 6336: 2006 [11]

The input data for the calculation are taken from tables 1 and 2 and material analysis. In order to take into account dynamics loads, the recommended values for the load application coefficient of ISO 6336:2006 [11] shown in Table 4 are used.

Table 4. Regimen of work machine driven

	Steady	Light shock	Medium shock	Heavy shock	Very heavy shock
Steady	1	1.25	1.5	1.5	2
Light shock	1.1	1.35	1.6	2.5	3.5
Medium shock	1.25	1.5	1.8	2.8	4
Heavy shock	1.5	1.75	2.5	3.5	4.5
Very heavy shock	2	2.5	3.7	4	5

As the operating regime of the chain and the push plates driven by the reducer is heavy collisions, the verification calculation consisted in calculating the contact and bending stresses at the foot of the tooth and the admissible tensions for values of the Load application coefficient between 1.5 and 5.5. The results are shown in figure 8.

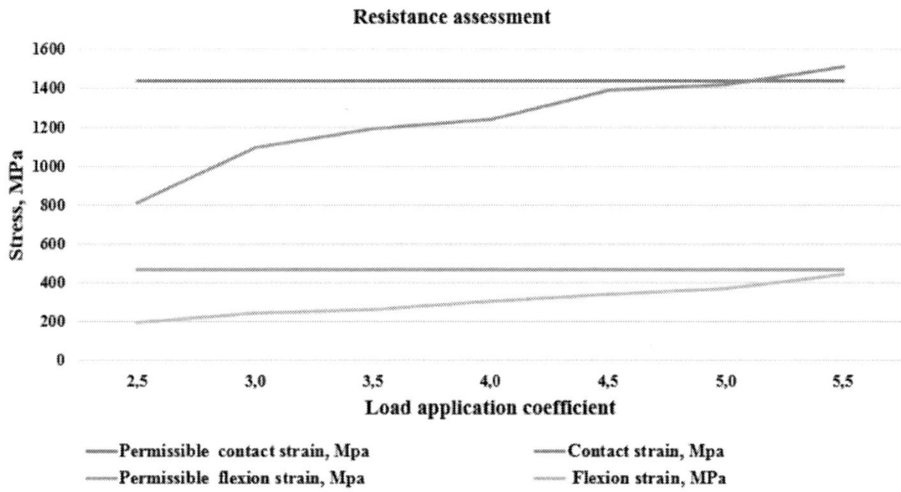

Figure 8. Mechanical resistance assessment results

In figure 8 can be seen that values of the load application coefficient below 3, the safety coefficient to the pitting is greater than 1.5; nevertheless, with values close to 5 the superficial stress exceeds the admissible stress of the material and the useful life of the gears must be shortened significantly. The bending stresses at these coefficient levels are remarkably close to the allowable stresses even without taking into account the effects of the destructive pitting on the profile of the teeth.

From the analysis, is possible concluded that the pair of gears is well designed and can have a useful life of the order of 20 000 hours considered in the verification calculations, the cause of the failure is the conditions of severe dynamic loads to which is subjected to the operating conditions of the mechanism that triggers.

4. Conclusions

The pinion steel of the gears is well selected and corresponds to the group of the best steels for cementing, slightly alloyed chromium, nickel and molybdenum, type 17CrNiMo-7 by European standard. The mechanical properties and the microstructures indicate that it was correctly treated with cementation, more tempering and tempering, presenting very few impurities in the form of oxides.

The gearbox is well designed for normal working conditions according to the resistance calculations. The verification of the contact and bending stresses in the tooth and the admissible stresses for values of the Coefficient of Application of the Load of between 1.5 and 5.5 show that up to values of the Coefficient of Application of the Load of 3, the safety coefficient at the pit is greater than 1.5; nevertheless, already with values close to 5 the acting superficial stress exceeds the admissible stress of the material and the useful life of the gears must be shortened significantly.

The more probable cause of failure in the pinion is the overloads and high contact stresses, which increase significantly with the degree of deterioration. The failure begins with superficial fatigue, progressive pitting and scaled; then the volumetric fatigue fracture is caused in the base of one or two teeth and the lack of these, the inertia and the torque of the motor cause the fracture by the base of several contiguous by overload and impact.

It is concluded that the working conditions of the reducer are extremely severe, with shocks and strong dynamic loads resulting from the oversized stones in the limestone pile and push chain wear.

References

[1] Asi O. 2006. Fatigue failure of a helical gear in a gear box. *Engineering Failure Analysis*. **13** 1116-1125.

[2] Yadav A. 2012. Different types Failure in gears-A Review. *International Journal of Science, Engineering and Technology Research*. **5** 82-92.

[3] Jotram Patel J, Sahu G, Kumar P. 2015. A study on common failure of gears. *International Journal Of Innovative Research In Technology*. **6** 301-307

[4] Errichello R. 2002. How to analyze gear failures. Journal of Failure Analysis and Prevention, **2, 6** 8-16.

[5] Peng C. 2011. Failure analysis of a gear tooth fracture of a rolling mill decelerator. *Engineering Failure Analysis*. **18,1** 25-35

[6] Goytisolo R, Noa J, Hernández H, Cabello M, Alemán G, Consuegra M, Alfonso R, Sagastume P and Morffi A. 2007. Análisis y soluciones de la avería de la criba de la planta de trituración y secado de la empresa "Cementos Cienfuegos SA. Proceeding of 8º Congreso Iberoamericano De Ingeniería Mecánica. Cusco, Perú. Octubre 2007. Available from: http://congreso.pucp.edu.pe/cibim8/pdf/32/32-07.pdf

[7] Mayer C. 1981.Torsional Vibration Problems and Analyses of Cement Industry Drives. *IEEE Transactions on Industry Applications*. **17,1** 81-83

[8] Yunusa A, Kermani M and Labib A. 2017. Investigation of critical failures using root cause analysis methods: Case study of ASH cement PLC. *Engineering Failure Analysis*, 73, 25-45.

[9] Montano E. 2014. Sistema experto prototipo para análisis de falla en engranes. Universidad Nacional Autónoma De México. Master Thesis. Available on: http://www.ptolomeo.unam.mx:8080/xmlui/bitstream/handle/132.248.52.100/4221/Tesis.pdf?sequence=3

[10] González G. 1997. Descifrado Geométrico y Cálculo de la Capacidad de Carga de Engranajes Cilíndricos. Proceeding of 9no Congreso Iberoamericano de Mantenimiento. Lima, Perú. 245-250

[11] International Standard (ISO) 6336. 2006. Calculation of load capacity of spur and helical gears

IOP Conf. Series: Materials Science and Engineering **844** (2020) 012028 doi:10.1088/1757-899X/844/1/012028

Agriculture 4.0 the advance in models and new knowledge to improve production.

F González-Salazar
Electronic Engineering. Seedling leader in Agriculture 4.0 – GiosTIC Research Group. Services Technician at Netux Tecnología. University Institution of Envigado

Email: ferdinan7@gmail.com

Abstract. This article presents Research on the design and application of a model that allows controlling agroecological variables and conditions based on inductive-deductive methods supported by qualitative and quantitative tools such as metadata, trend tools and comparison graphs. This Model represents the knowledge of the farmer in a different kind of applications that allows the continued progress of the agriculture. The model has been tested with some agricultural, plant and animal species in controlled in real environments, where the quality of the cultivated species can be measured to determine improvements in nutritional quality, this measure is made by sensors of temperature, pH, lumens and humidity then we have a mathematical model that allows the simulation of the variables that we want to control that includes monitoring and control times. From here, concepts of agriculture 4.0 are associated, which consists in mitigating the risk in different in the environments. Currently, the integration of concepts and technologies associated with the fourth industrial revolution is evaluated through the simulation of variables and comparison between ideal and real models.

1. Introduction

To contextualize the space of the research carried out, some aspects of the agricultural plans that Colombia has and the elements of interest for the agriculture 4.0 project, its development and the possible elements of application are described.

"In Colombia, the National Development Plan proposes as a transversal strategy the transformation of the field and green growth to strengthen the competitiveness of the agricultural sector, seeking to modernize processes and projects" [1]. Colombia, like most Latin American countries, is a country of regions, so its diversity of climates benefits it for the growth of various plant and animal species. However, climate changes has caused problems of drought for long periods and instability in climatic conditions, in addition, the erosion caused by deforestation and the little technification implemented in agricultural practices, complicate the greater expansion of this type of crops Precision agriculture (PA) discourses this problem, "the term occurs in the early 1980s, with the use of sensors and positioning systems, the use of data from all types of sensors, both in-situ and remote, together with positioning via GPS / GNSS offers the possibility of generating useful recommendations for farmers to make informed decisions about what actions to carry out in their plots "[2], since then developments have been generated where agriculture concepts are applied 4.0, which has managed to introduce not only electronics, but the convergence of different knowledge as if stemas, mathematical modeling and agrobiotechnology, seeking a better production process through the joint innovation of technological and agronomic sciences.

"All the techniques are already being introduced in our territory as evidenced by the fact of the number of companies that offer different products and services of precision agriculture"[2], the implementation of technologies to improve agricultural production and efficiency processes, together with the diversification of Colombian and foreign species, establish the initial conditions that allow developments in agriculture 4.0.

"Now, with information and communication technologies (ICT), unprecedented ways have emerged to generate, store, transmit and distribute information, causing important changes not only in formal and non-formal education, but also in social relationships, work, economy, politics, culture and daily life" [3], the electronic signal conditioning models establish ideal growth conditions; Agriculture 4.0 manages to concatenate the information that nature gives us and converts it into data that allows its management, in the same way it includes metadata analysis that determines trend lines and behavioral graphs that make constant comparisons with the actual measurement for maintain ideal conditions as much as possible.

"The application of optimal quantities at the right time and in the right place allows a reduction in time, costs and impact on the environment, as well as an increase in yields obtained in crops" [2]. Although Colombia It is a country with the optimal agro-ecological conditions for cultivation, it is located below the world average of agricultural production, due to the insufficiency of land for agriculture. Agriculture systems 4.0, generate optimal conditions regardless of the availability of the land or external environmental conditions, the only thing required for its implementation is a space dedicated to cultivation.

Information and communications technologies (ICT) have begun to be used and implemented throughout the crop's life cycle" [1]. "It is a strategic initiative for the industrial sector, triggered by the development of communication and information technologies (ICT), which results in a transformation of traditional production systems" [4]. Food is defined as one of its strategic focuses. It is from here that the objective of this research project is inspired, which seeks to generate sustainable production models for the cultivation of agricultural species, innovating in its growth process and final result for its consumption and/or transformation, improving the productivity of Colombian agriculture.

2. Methodology

Agricultural production processes require models and processes that that allow their reproduction, regardless of changing agroecological conditions and the different environments associated with them, for this the model of agricultural production applied to the technification of their processes, starts in agriculture models precision (AP), which have the ability to increase the productive efficiency of a crop in turn managing in a more efficient way the resources required for it. In the agriculture of precision, different technologies are observed such as satellite geo-positioning systems, both physical and digital geographical processes of agronomic conditions, performance monitors, sampling and fertilization management, automatic nutrient management, agroecological conditions sensors such as temperature, humidity , lumens and pH, the convergence between these technologies establishes a more precise production model but lacking human conditions that allow decisions to vary according to the terrain, and types of planting.

"In agriculture, water quality is of great importance because it is a factor to be considered for obtaining high agricultural yields" [5], precision agriculture models are today a combination of technologies applied in order to increase, preserve and evolve agricultural management techniques, it is there that the convergence of current schemes nested in a single production model based on concepts of agriculture 4.0, manage to develop disruptive models and schemes of intakes of Critical decisions for agricultural production, through manual and semi-automatic processes, human expertise is included according to the ideal production conditions.

"Due to the impacts of climate change (increase in temperature, increase in CO_2 and variation in the acidity of the soil) agriculture has been affected during the cultivation process" [6] current models establish technological models while Agriculture 4.0 establishes conditions of convergence that can be scaled over time. These allow a number of applicable solutions, which in the Colombian planting process generate conditions of evolution towards new exploitation processes and, above all, the conservation of our environment.

3. Developing

"A third of the global population derives its livelihoods from agriculture, and in emerging economies it can represent up to 30% of GDP" [1]. Conventional exploitation processes in agriculture make it difficult for them to be replicable and used in time and space independent of the agro-ecological conditions to which it is involved, that is where agriculture 4.0, with each of the technological systems at the service of the controlled exploitation of agriculture, allow "ideal" and "variable" conditions of production, always within the framework of conservation and quality of the final product.

"Precision Agriculture means the differentiated application of inputs, taking into account the characteristics and needs of each specific area of the field or land, through the use of geo-computer techniques and other technologies" [7]. Agriculture 4.0, in all its definition applied to our developments, manages to concatenate technology at the service of nature, where it gives us valuable information that, through mathematical models and microelectronic control and automation schemes, generate a formula for the acquisition of information that will become metadatas, which will be adjusted in time by the agronomist of the present and future, a person with a high knowledge of the environment and agro-industrial with a great sense and competence of textual information analysis, which will allow guiding their knowledge towards controlled equipment, establishing more demanding and efficient agricultural administration models according to environmental and market needs.

Apply mathematical models and prospective techniques that control the conditions of the environment and its variables such as humidity, light, temperature, pH among others; It achieves environments adjusted to agricultural needs.

The incorporation of vibro sensor systems and microelectronic devices are a fundamental part so that the conditions of the plant are optimal and the climate is not a determining factor for its growth obtaining efficient and effective results in agriculture 4.0.

"We want to describe the possibilities of applying these tools to decision making in agriculture" [8], it is ideal to test management tools under the "Block Chain" scheme that allow the implementation of models using tools of risk management and determination of critical points in the process.

These tools are based on a production model, the systems are reflected in the low productivity and quality in most crops, "perhaps due to the lack of a system of instruments that allow the farmer to know when and how to produce according to the climate" [6], a method applied to agriculture 4.0 was generated, which has been consolidating throughout the world since 2016 a few years after the 4.0 industries boom exploded, of which it is part . For the integration of these two branches such as systems and electronics to control the variables of each system, it is necessary to be clear about which are the variables to be controlled, that is, which of the thousands of variables will be the selected ones, In this case, the project team considered that the lumens, relative humidity, pH and temperature are variables specific to each system and that in some variation of it can affect the entire system and cause imbalances with dire consequences and potential loss of this one.

Measuring points are specified, which indicate that parts of the process are considered more vulnerable, after having clarity on this, a mathematical model based on statistical concepts such as trend lines, graphs of behavior of agroecological variables that are fed by sensors was generated which in turn feed the metadata. With this, a 4.0 integration model is generated based on a "PHVA" (plan to verify), and the recurring need for models that allow greater efficiency in the management and quality of agricultural products. This means that in Colombia and in much of Latin America, it is not easy to find models that generate greater efficiencies in the final product and that also offer a higher quality in agricultural production. This creates the need to generate their own production model for agriculture 4.0.

IOP Conf. Series: Materials Science and Engineering **844** (2020) 012028 doi:10.1088/1757-899X/844/1/012028

The design plan begins, based on knowing in detail the agricultural process that is to be intervened, which generates an impact mitigation plan, in addition to a change management system to ensure that it is effective. At this stage of the process, the sensors that will be used for the data collection of the different variables are specified, the way in which the data is quantified and its processing, based on metadata and data science, in order to submit this information under a continuous improvement system, which indicates that the system can always be improved in each of its versions, until it is considered to comply with the necessary standards of ISO 9001/2015.

When this level was reached, they proceeded to implement, which resulted in a demo version of a functional and viable system and by repeating the process several times the productive version was reached.

Research is complemented with simulation of automatic control for irrigation systems. Methodologically, the Penman-Monteith and water balance models were used to analyze the water requirement of the crop" [9].

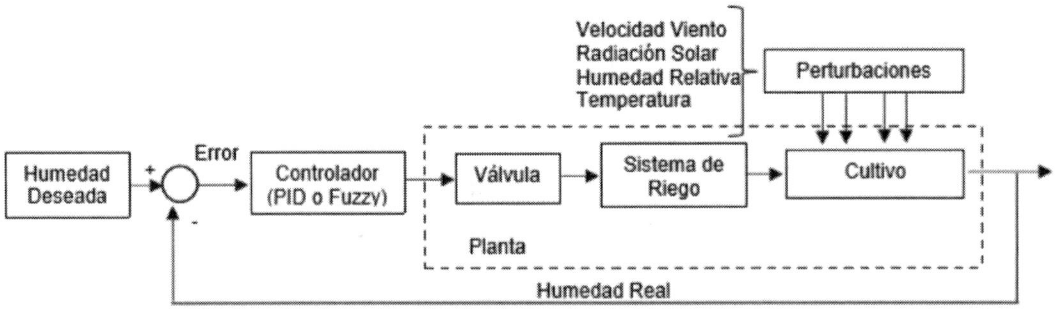

Figure 1. Block diagram of the controller. [9]

As a control strategy, the water systems could be associated according to the type of controller according to the system designed proportional, integral and derivative (PID) and diffuse type controllers were used to compare comportments and establish the best option in the face of crop water resource difficulties under climatic conditions. See Figure 1.

"Adopting the concept of 4.0 is the opportunity to make the country's production more competitive, mainly in exportable goods" [4]. For the generation of this model, the inductive-deductive method was applied, which allows transcending from a general idea to a final product, applying quantitative tools that facilitated decision making and data mining, and the use of qualitative tools that allowed the differentiation of each of the systems and how optimal was their intervention in the processes. With this, improvements were achieved in growing times in both crops and animal species and in quality standards to make it faster in growth, without losing nutritional quality.

The uses of micro-electronic systems on planting allow cultivation in unproductive or very small spaces, where it would normally not be possible to cultivate.

Finally, the last component of the model, the administration and management of the integrality of production or what was called traceability according to ISO 28001. It is a strategic initiative for the industrial sector, triggered by the development of communication technologies and information (ICT), which results in a transformation of traditional production systems" [4]. "An agricultural company is made up of the natural, human and capital resources it has and the external and internal factors that condition production; where management plays a very important role in the efficient control of internal factors" [10]. Inspired by these concepts, the administrative component for the management of all the systems implemented and the decision making of the projects is carried out, it was defined at what moment the possible vulnerabilities, risks and the identification of the critical points can be presented, control points at each step of the system. Eight fundamental schemes of an organization were evaluated:

Human, Economic and Financial Resources, Information, Corporate Image, Environmental, Operation, Customers and Suppliers.

MATRIZ "JOIBEM" VALORACIÓN DEL RIESGO						VERSION: 01
SEVERIDAD O CONSECUENCIA		PROBABILIDAD O POSIBILIDAD				
NIVEL		1	2	3	4	5
	FACTOR	RARO	IMPROBABLE	POSIBLE	PROBABLE	CASI CIERTO
5	CATASTRÓFICO	5	10	15	20	25
4	MAYOR	4	8	12	16	20
3	MODERADO	3	6	9	12	15
2	MENOR	2	4	6	8	10
1	INSIGNIFICANTE	1	2	3	4	5
CONVENCIONES						
MUY BAJO	BAJO	MODERADO	ALTO	EXTREMO		

Figure 2. 5x5 matrix called (Joiben matrix)

The Joiben matrix is shown where the risk is analyzed with respect to its severity and its probability of occurrence and how each of these factors can be managed or mitigated its impact so that the severity is lower or minimum after its management. This is achieved by means of a matrix which is responsible for making these qualitative variables take a quantitative value with respect to the severity and the probability of occurrence after each of these risks are outlined in some technical sheets which will mitigate this risk from foresight, prevention, protection, attention, control, transfer, recovery; this to attack the risk before it occurs and implement. A contingency plan for this so that its probability of occurrence is low.

PLAN DE RIESGOS EN SEGUIRDAD PARA:								
RECURSOS IMPACTADOS POR LOS RIESGOS								
R. HUMANO	ECONOMICO Y FINANCIERO	INFORMACION	IMAGEN CORPORATIVA	AMBIENTAL	OPERACIÓN	CLIENTES	PROVEEDORE S	VULNERABILIDAD DE LA ORGANIZACIÓN

S	P	V	S	P	V	S	P	V	S	P	V	S	P	V	S	P	V	S	P	V	S	P	V	S	P	V

Figure 3. Guidance table for the creation of the technical sheets under the seven parameters considered.

To replicate the model it is necessary to make a study of the natural phenomena of the space that is desired to intervene since all the areas do not have the same conditions, therefore for each intervention their own indicators or natural values of each variable are generated; This implies that this condition must be added to the model every time one wishes to implement.

"While additive manufacturing presents greater precision in the terms that define it, this is not the case with Industry 4.0, as it does not have a standard terminology as such, rather it refers to a wide spectrum of technologies, processes and applications in the framework of digitalization, autonomy and networking" [11]. The strategies allow us to have a large number of information available to the industry because of the influence of the 4.0 industries and in turn the data mining and the possible applications of these tools.

4. Results

Once the production model has been established, the work team applied the control of agro ecological variables to potato crops and the clean production of pigs whose processes and results generated two patents: (Ref. File No. NC2016 / 0005923) entitled "irrigation system by aeroponics for controlled production in potato variables" which in its productive model manages to increase starch production per unit of potatoes by 35%, decreasing its final production from 4 months to 3 months, and the second patent (Ref. File No. NC2017 / 0004296) entitled "automation of the livestock production process" which mitigates the proliferation of epidemic swine diarrhea in each stage of semovient growth, thus generating optimal traceability and prolificacy conditions for its production, and commercial development

The patents were evaluated by the Superintendence of Industry and Commerce of the Republic of Colombia under resolution No. 37606 and No. 37358.

These patents attest that the model is functional and meets the academic and productive objectives set for the development of this. Currently working on new projects such as the incorporation of new species and aquaponics systems among others in the field of agriculture, focused on systems that optimize growth times in both animal species and plant species under the generation of microclimates and Microelectronics tools that act on agricultural species.

5. Discussion of the results

"The benefits of agricultural production for society and its interaction with the environment; they are subjective and inaccurate, so it is suitable for the application of mathematical models of fuzzy or fuzzy logic "[8], this as a cause has a social component under agribusiness projects and the export of these raw materials to other countries since around 12 million people in Colombia have a livelihood through the countryside. This model of Agriculture 4.0 has a significant impact as it improves the efficiency and productivity of these areas, which will help economically the people who join these processes.

"The technification of crops, although it improved productivity against crop pests and diseases, deteriorated the environment, increased deforestation, decreased wildlife and damaged the soil" [12]. The author states that the techniques they are currently using do not favor the environment, which is why it is important to make planting systems in favor of the environment with the use of clean energy and good water management Residual crops, so a solution for deforestation is the conditioning of agricultural spaces and tend to be more productive, thus minimizing the interaction with wildlife is where the microelectronics application model proposed when controlling Irrigation systems reduce water consumption and prevent damage to the soil. In this way, agriculture 4.0 generates a solution to the problem posed by Osorio & Alberto by implementing the technification of agriculture to reduce space, harvest time and water consumption and not to deforest areas with traditional crops and make a revolution technological technology that can meet the demand for food without affecting it irreparably.

"Some research has included different aspects of inter-organizational relationships in their models or empirical work, including areas of organizational behavior"[13]; This depends on the level of expertise of him who is in the execution or usually group different types of models in order to set the parameters that best fit and thus generate their own models. The innovative thing about this implementation is that it generated its own model that has been tested in several species and is currently advancing the planting of a non-native species such as the blue tomato of European origin and whose habitat was replaced by a space which does not It is suitable for agricultural production.

Currently, the new crops are tested in real environments in the La Heliodora Linear Nature Reserve in the Municipality of Envigado, where it is expected to observe the effectiveness of the model by growing the blue tomato in a controlled space.

Figure 4. Germination of the Blue Tomato in the nature reserve Parque Lineal La Heliodora. Photo taken by the author.

6. Conclusions

The processes of Agriculture 4.0 in Colombia and Latin America are at a fundamental moment so that the acquisition of 100% organic products is accessible to all socio-economic strata. Innovation in production, where the amount of land is not a requirement for the development and production of quality agricultural products, will allow countries with "ideal" planting conditions, position their processes and planting models of generations in enduring models in time thanks to the technological convergence between agriculture and technical sciences.

Multidisciplinary work generates more robust models; this model is the result of combining microelectronics, mathematics, the new concepts of the fourth revolution and the standards that allow continuous improvement.

Thanks to the application of mathematics, it was possible to strengthen the statistical techniques and the analysis of agroecological variables that determine the trend lines and expose behavior graphs of these variables, allowing more precise and reliable adjustments in the application of the Agriculture 4.0 model.

The administrative management model is a complement that will strengthen the industry and the adverse factors that are related to the production, times and quality of the agricultural industry.

It is possible to grow crops in unfit or unproductive spaces through the generation of controlled conditions so that the species to be cultivated has optimal growth, in addition to reducing production times and achieving the introduction of non-native species that diversify food for consumption, better quality of life and export opportunities.

7. References

[1] Orozco Ó and Llano G 2015 Sistemas de información enfocados en tecnologías de agricultura de precisión y aplicables a la caña de azúcar, una revisión. Medellín: Revista Ingenierías Universidad de Medellín.

[2] Garitano X 2018 Servicios de posicionamiento GNSS y usos de la teledetección en la agricultura de precisión. 27th ed. (p. 2,3,5). Mapping.

[3] Gaviño G, González C and Sánchez Á 2019 Sistema para la enseñanza de control y monitoreo de variables a través de entornos virtuales de enseñanza aprendizaje (EVEA) En Cu Uaem Valle de México. (40th ed., p. 4). Valle de México: Revista Investigación Operacional.

[4] Alejandro E and Henderson S 2019 Industria 4.0 ¿qué se debe conocer del concepto para subsistir en el mercado? (pp. 2,3). La Alimentación Latinoamericana.

[5] Mola B, Bonet C, Rodríguez D, Guerrero P, Avilés G and Noa Y 2019 Calidad del agua en la producción de organopónicos (9th ed., pp. 3,4). Cuba: Revista Ingeniería Agrícola.

[6] Ossa S 2017 Monitoreo y control de variables ambientales mediante una red inalámbrica para agricultura de precisión en invernaderos (1st ed.). Vector 12.
[7] Hernández P 2006 Obtención del mapa de rendimiento georreferenciado del cultivo de la papa mediante el empleo de técnicas de agricultura de precisión (17th ed.). Revista Ciencias Técnicas Agropecuarias.
[8] Martínez Y and R E 2019 La toma de decisiones en la agricultura con empleo de modelos matemáticos difusos (28th ed.). Cuba: Revista Ciencias Técnicas Agropecuarias.
[9] Cohen C, Rodríguez J and Salgado R 2018 Modelado del Microclima de un Cultivo de Sandía (Citrulluslanatus) en la Sub-región Sabana del Departamento de Sucre, Colombia (p. 2). Información Tecnológica.
[10] Cabrera D, Sánchez M and Ramos R 2012 Factibilidad económica del empleo de las herramientas de agricultura de precisión en la Empresa Pecuaria "Niña Bonita" (21st ed., pp. 2,3,7). Revista Ciencias Técnicas Agropecuarias.
[11] Rodríguez M and Mancilla J 2017 Presencia de la Industria 4.0 en la fabricación aditiva: análisis de tendencias tecnológicas (14th ed., pp. 3,4). Monterrey: colaboración.
[12] Osorio N and Alberto O 2019 La evolución tecnológica y la percepción de la calidad ambiental de los caficultores de El Águila, Valle del Cauca, Colombia (31st ed., pp. 1.3,4). Valle del Cauca, Colombia: revista internacional de fronteras, territorio y regiones.
[13] Sosa J, Svensson G and Mysen T 2011 La calidad de la relación en la cadena de suministros (2nd ed., pp. 2,5). Gerencia.

Acknowledgments
To the University Institution of Envigado, its teachers and especially Julián Andrés Yepes for his support that was fundamental for the creation of the research hotbed and his accompaniment allowed this innovation to be reached. To the Environmental Linear Park the Heliodora de Envigado for facilitating the land, infrastructure and technical support of its staff. To the work team of this project and for their perseverance and commitment.

IOP Conf. Series: Materials Science and Engineering **844** (2020) 012029 doi:10.1088/1757-899X/844/1/012029

Influence of thermal-varying material properties on stress calculation for creep life assessment of seamless medium-carbon steel boiler tubes

L M Castellanos-González [1], H Hernández Herrera [1], R Goytisolo Espinosa[2†]
L M Castellanos Molina [3], E E Vergara Verbel [4] and J A Alcalá Ballestero [4]

[1] Facultad de Ingeniería, Universidad Simón Bolívar, Carrera 59 No. 59-65, Barranquilla ,Colombia
[2] Facultad de Ingeniería, Universidad de Cienfuegos, Cienfuegos, Cuba
[3] Department of Mechanical and Aerospace Engineering, Politecnico di Torino, Turin, Italy
[4] Escuela Naval de Cadetes Almirante Padillas, Cartagena, Colombia.

E-mail: luis.castellanos@unisimonbolivar.edu.co

Abstract. Seamless medium-carbon steel boiler tubes from a thermoelectric plant are studied to predict their remanent life. An alternative for stresses calculation in creep life-assessment method is presented. The proposed stress calculation incorporates not only material dimensions and working pressure but also thermal-varying properties of the material and the radial temperature gradient in the tube wall. The methodology for creep life prediction established by ISO / TR 7468-1981 is used. It is compared with the API std 530 to show that the predicted creep rupture times can result longer when thermal-varying properties and the radial temperature gradient in the tube wall are considered. This enhanced creep failure prediction method may favour a cost reduction avoiding premature maintenance and unnecessary plant shutdowns.

1. Introduction

In the Power-Generation industry where components failure can cause injury, societal hardships, and economic losses, life-assessment technology has been a major focus of research. The creep failures of the tubes of steam generators are very common in almost all power plants, therefore, much research is conducted on the improvement of creep life prediction methods in order to minimize damage and reduce expenses.

At high temperatures, metal components deform slowly and continuously under load below the yield stress. This time dependent deformation of stressed components is known as creep. Creep occurs due to atomic movement of the crystal structure under high temperature and stress, and it results in the creep strain. The creep damage has definite relation with stress and temperature, and rate of creep damage can be calculated based on empirical formulas [1] and microstructure examination [2], [3].

One of these empirical approaches to calculate remaining life against creep is through Larson and Miller Parameter (LMP) which correlates tensile stress to operating temperature and time to exhaust remaining life. The LMP curves data is generally produced by the alloy manufacturers and taken into account at the design stage. However, these curves are equally important for in-service life estimation based on records of operating parameters and accelerated stress rupture test. Figure 1 shows a typical creep failure in a boiler tube.

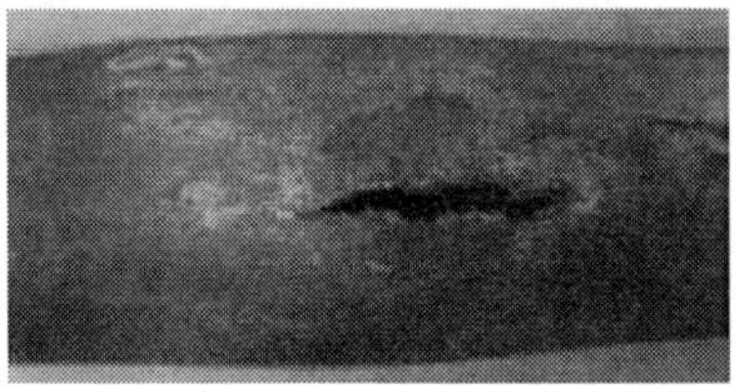

Figure 1. Creep failure in a boiler tube.

Finite element (FE) modeling is increasingly used as an integral part of creep analysis for the integrity assessment of high temperature structures. An important consideration in such Finite element simulations is the constitutive model used to represent the creep strain response of the component material as a function of temperature, stress and time [4]–[6].

This work can be outlined as follows. Firstly, dimensional and chemical characteristics of the case study tube are presented. In Section 3, the procedure for calculating the creep time to rupture is detailed based on two different norms: ISO/TR 7468 and API std 530. A formulation for stress calculation that incorporates thermal-varying material properties and the difference of temperature across the wall of tubes is also presented. In Section 4, the creep life prediction is analyzed based on a comparison of the norms mentioned before and the last Section concludes the work.

2. Case study material specification

The case study material in this research are seamless medium-carbon steel boiler tubes corresponding with SA-210 A1. It chemical composition based on ASTM standard is detailed in Table 1.

Table 1. Chemical composition of SA-210 A1 (%)

C	Mn	P	S	Si
0.27, max	0.93, max	0.035, max	0.035, max	0.10, min

The physical dimensions for the case study are presented in Table 2.

Table 2. Material specification, dimension and operating condition.

Material specification/grade of steel (ASTM)	A-210 A1
Design water pressure [MPa]	14.3
Design inner surface temperature [°C]	480
Operating inner surface temperature [°C]	440
Operating outer surface temperature [°C]	448
Tube thickness [mm]	4.9
Inner diameter [mm]	53.7
Service exposed [years]	6

3. Creep life prediction based on Larson Miller Parameter

The Larson Miller Parameter is a model based on Arrhenius rate equation used for prediction of creep rates or creep rupture times to longer times that have been measured. It also enables rating comparison to be made for different materials. It is empirically obtained based on creep testing procedures. The results for different steels are stated in API std 530. LMP correlates tensile stress to operating temperature and time to exhaust remaining life. The general model is presented typically as follows:

$$P = T(C + \log t_r) = f(\sigma) \tag{1}$$

where P is the Larson Miller Parameter, T is the work temperature (°K), t_r is the time before rupture ($hours$), C is a material specific constant and $f(\sigma)$ is a logarithmic polynomial depending on the equivalent stress (σ). From reference [7], equation (2) represents the Larson Miller Parameter for SA-210 A1:

$$T(20 + \log t_r)10^{-3} = f(\sigma) \tag{2}$$

After calculating the stress, the time to rupture can be obtained as follows:

$$t_r = 10^{\left[\frac{10^3 f(\sigma)}{T} - 20\right]} \tag{3}$$

3.1 Creep life prediction based on Norm ISO/TR 7468-1981
According to ISO/TR 7468-1981, the master curve ($P(\sigma)$) for SA-210 A1 is represented by equation (4) as follows:

$$P(\sigma) = \frac{\log t_r - 10.656877}{T-500} = a + b(\log\sigma) + c(\log\sigma)^2 + d(\log\sigma)^3 + ec(\log\sigma)^4 \tag{4}$$

where a, b, c, d, e are given constants and their values are presented in Table 3.

Table 3. Constants values of the correlation equation.

Constants SA-210 A1	SA-210 A1
a	-0.66639
b	1.418113
c	1.152629
d	0.413428
e	-0.05584

After knowing the equivalent stress, and working on equation (4), the time to rupture can be found using equation (5).

$$t_r = 10^{[P(\sigma)(T-500)+10.656877)]} \tag{5}$$

3.2 Calculation of tensile stress for creep life prediction
It is well known that stress calculation has an important influence in creep life predictions. There is an extensive flexibility for stress calculation being the mean-diameter equation, one of the most used. The mean-diameter equation for stress is as follows [7]:

$$\sigma = \frac{p}{2}\left(\frac{D_0}{\delta} - 1\right) = \frac{p}{2}\left(\frac{D_i}{\delta} + 1\right) \tag{6}$$

where p[MPa] is the operating point gauge pressure, D_0[mm] and D_i[mm] are outside and inside diameters, respectively, and δ[mm] is the thickness. Equation (6) provides good correlation between the creep rupture of a pressurized tube and a uniaxial test specimen.

On the other hand, the Larson Miller approach could be responsible for non-conservative predictions because it does not account for strain enhanced thermal ageing affects which could influence creep deformation and rupture mechanisms at long times. Hence, in order to attain a more accurate prediction, a new formulation for stress calculation is proposed [8]:

$$\sigma = \frac{(\epsilon+1)^2}{4\epsilon}\sqrt{3p^2 + 3pm_1\Delta t + (m_1\Delta t)^2} \quad (7)$$

in which $m_1 = \frac{E(T)\cdot\alpha(T)a_1}{1-\mu(T)}$, $a_1 = \frac{2\epsilon}{(\epsilon+1)^2\ln(\frac{\epsilon+1}{\epsilon-1})} - 1$ and $\epsilon = D/\delta$,

where E[MPa]: modulus of elasticity of pipe material, α [°K^{-1}]: coefficient of thermal expansion of pipe material, μ [−]: Poisson's ratio of pipe material. The functions $E(T)$, $\alpha(T)$ and $\mu(T)$ depend on the working temperature (T) and type of material. D[mm] is the mean diameter and ΔT[°C] is the difference between the inner and outer surface temperature (according to [8], $\Delta T < 0$ for external heating). The proposed stress calculation in equation (7) is an extension of Lamé Problem equations for triaxial stress calculation. It incorporates not only material dimensions but also includes thermal-varying material properties and the radial temperature gradient in tube wall. By inspecting equation (7), can be deduced that the analysis of the effect of thermal-varying material properties does not make sense when it is assumed no difference between inner and outer surface temperature ($\Delta T = 0$).

From reference [9] and using least-squares method, linear polynomial models for the materials properties (i.e., $E(T)$, $\alpha(T)$ and $\mu(T)$) were obtained for the case study. Accordingly, the equivalent stress for different ΔT values and working temperatures is shown in Figure. 2.

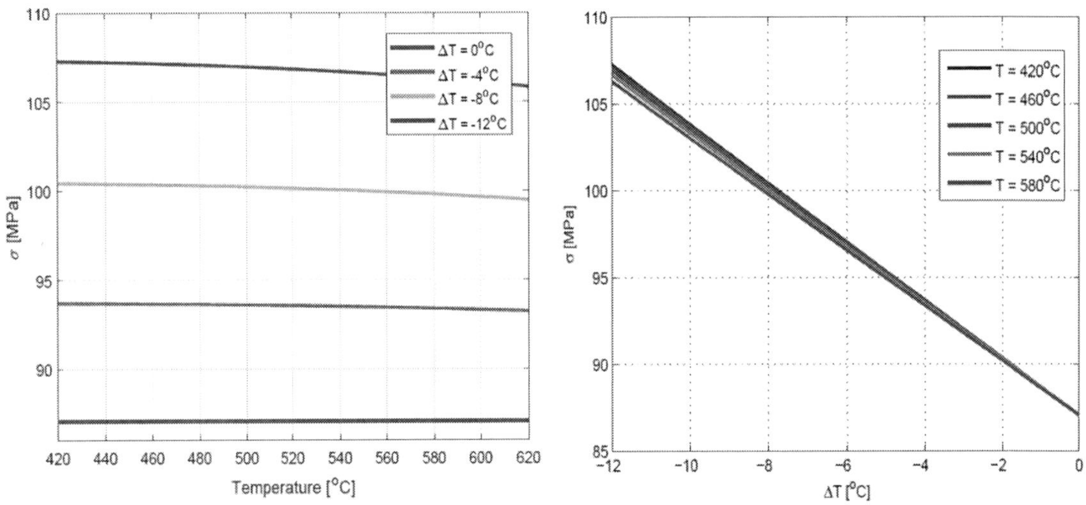

(a) (b)

Figure 2. Case study equivalent stress a) vs working temperature, b) vs ΔT.

It is appreciated a slight decrease of stress in Figure. 2.a when working temperature increases but a considerable increase for large values of ΔT (Figure. 2.a). When $\Delta T = 0$, the resulting stress value is equal to the triaxial stress calculation from Lamé equations and slightly bigger than the ones obtained from equation (6).

4. Results and discussion

A comparison of the ISO/TR 7468-1981 and API std 530 is carried out by considering thermal-varying material properties on creep time to rupture for the case study. The tensile stress is calculated in two different ways: (i) using equation (7) to apply the ISO/TR 7468-1981, and (ii) based on the API std 530 by using equation (6) as recommended in the API. An overview of the time to rupture at working temperature, is given in Table 4. The t_r values from API std 530 were obtained using minimum rupture strength curve which is the lower 95% confidence limit, (i.e., 95% of all samples, should have rupture strength greater than this value). It means that creep time to rupture values obtained from API std 530 and using minimum rupture strength curve guaranties conservative predictions. On the other hand, using ISO/TR 7468 based on the stress calculation method and considering the thermal-varying material properties, the time to rupture is less conservative.

Table 4. Time to rupture comparison for SA-210 A1.

Working temperature [°C]	ISO/TR 7468 t_r[years] /ΔT[°C]	API std 530 t_r [years]
420	59.66/ 0	39.67
	46.84/-4	
	36.93/ -8	
	29.19/-12	
440	18.36/ 0	7.62
	14.07/ -4	
	10.83/ -8	
	8.36/ -12	
460	5.65/0 1.6	1.6
	4.23/-4	
	3.18/-8	
	2.40/-12	

The times to rupture obtained from both norms are presented in Figure. 3 for a large range of working temperature. It is appreciated that at any working temperature, for smaller ΔT values, the time to rupture increases above the resultant t_r values from API std 530. Therefore, when the effect of thermal-varying material properties and working ΔT estimation are considered, the resulting exploitation time results longer. It is also observed that, for larger gradient of temperature in the tube wall, the time to rupture gets closer to the results from API std 530 using minimum rupture strength curves.

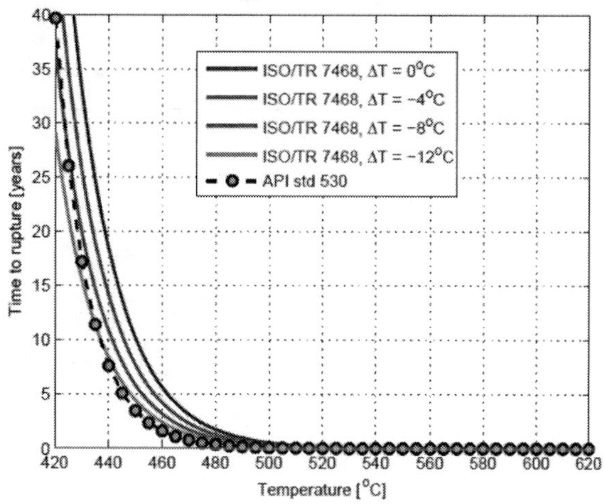

Figure 3. Time to rupture, SA-210 A1.

5. Conclusions and future work

Seamless medium-carbon steel boiler tubes from a thermoelectric plant are studied to predict their creep time to rupture. The proposed stress calculation incorporates not only material dimensions and working pressure but also includes thermal-varying material properties and the gradient of temperature across the tubes. From comparison of ISO/TR 7468 and API std 530, all results indicate that taking into account the effect of thermal-varying material properties and working ΔT estimation, less conservative prediction times are obtained. Hence, when using the enhanced prediction method, an extra amount of exploitation time is justified before plant maintenance or parts replacement need to be carried out. The proposed methodology can be also extended to any other seamless medium-carbon steel tubes of the power plant that work under creep failure risk. Nevertheless, it is worth mentioning that having measurements of actual ΔT values is a hard task, almost impossible indeed, but an estimation can be done in future works based on thermal and mass balances and indirect measurements. The implication of wall-tubes slimming due to corrosion can also be considered for future works.

References

[1] Kimura K. , H. Kushima, and K. Sawada, "Long-term creep deformation property of modified 9Cr-1Mo steel," *Mater. Sci. Eng. A*, vol. 510–511, no. 0, pp. 58–63, 2009.

[2] Mazarheri M. , F. Djavanroodi, and K. M. Nikbin, "Creep life assessment of an overheated 9Cr-1Mo steel tube," *Int. J. Press. Vessel. Pip.*, vol. 87, no. 12, pp. 746–752, 2010.

[3] Sugiura R. *et al.*, "Effects of Alloying Additions and Material Microstructure on the Accuracy of the Predictive Law of Creep Crack Growth for W-Strengthened 9–12%Cr Ferritic Heat-Resistant Steel," *Mater. Trans.*, vol. 48, no. 11, pp. 2928–2936, 2007.

[4] Hosseini E. , S. R. Holdsworth, and E. Mazza, "Experience with using the {LICON} methodology for predicting long term creep behaviour in materials," *Int. J. Press. Vessel. Pip.*, vol. 92, no. 0, pp. 70–76, 2012.

[5] Hosseini E. , S. R. Holdsworth, and E. Mazza, "Creep constitutive model considerations for high-temperature finite element numerical simulations," *J. Strain Anal. Eng. Des.*, vol. 47, no. 6, pp. 341–349, 2012.

[6] Hosseini E. , S. R. Holdsworth, and E. Mazza, "Stress regime-dependent creep constitutive model considerations in finite element continuum damage mechanics," *Int. J. Damage Mech.*, 2013.

[7] "API Standard 530, ISO 13:704:2001 (E), Petroleum and natural gas industries - Calculation of heater tube thickness in petroleum refineries," no. 530. Jan-2003.

[8] Jusmatulin E. R., *Handbook of vessels and pipings at high-pressure*. Mashinostroenie, 1990.

[9] Maslienkov and Maslienkova, *Steels and alloys for high temperatures*. Metalurgiya, Moscow, 1991.

IOP Conf. Series: Materials Science and Engineering **844** (2020) 012030 doi:10.1088/1757-899X/844/1/012030

Thermal comfort evaluation in an educational building with air conditioning located in the warm tropical climate of Colombia.

M Balbis-Morejon[1], A Noya-Sambrano[1]

[1] Department of Energy, Universidad de la Costa, Calle 58 No. 55-66. Barranquilla, Colombia.

E-mail: mbalbis1@cuc.edu.co

Abstract. The objective of this work is to evaluate the perception of thermal comfort that students have in building 10 of the Universidad de la Costa located in the city of Barranquilla. For the investigation the methods for the thermal comfort evaluation are analyzed, selecting the Steady-State methodology of Fanger according to the ASHRAE 55 Standard. With the application of the surveys, the thermal sensation that the students had in the building was evaluated, which was preferably fresh and thermal acceptability where 87% responded satisfactorily to indoor climate conditions in classrooms. It was also obtained that the perception of inner thermal comfort is between 21.1 ° C and 23 ° C of temperature. With the results obtained, recommendations are issued to ensure that students feel thermal comfort and the proper regulation of air conditioning systems.

1. Introduction

Thermal comfort is constituted as the person's mental perception of the thermal environment [1]. Therefore, when evaluating the perception of the occupants, the thermal comfort state of a place can be determined. It is also defined as a neutral sensation of the person with respect to a certain thermal environment without the need to sweat or develop their activities without disturbance [2–4]. The expectations of thermal comfort depend on the place where the person is and the climatic conditions inside and outside the enclosure [5–8]. The psychophysiological variability of the human being makes it impossible to get everyone to feel comfortable with the thermal environment, whatever the reference conditions. According [9], the discomfort related to the sensation of cold or heat has a negative effect on the well-being of the participants however, the welfare decreases to a greater extent when the environments are colder.

Due to the size and dynamic environment of buildings, reducing energy while providing the desired comfort for users is a challenge. The cost of energy is based on the non-linear cost relation with the energy consumed, which is modeled consecutively in terms of the specific heat load of the building [10]. A study carried out in Chile showed that the state of a building depends to a large extent on the functionality of the air conditioning systems; therefore, insufficient thermal comfort and high energy consumption are the result of inefficient construction. Correcting these deficiencies, means great benefits, it is possible to raise the standard, increase the thermal comfort inside and lower the expenses in air conditioning [11–14]. In a study conducted at the University of Bordeaux in France, it was determined that among the factors that influence the variability of energy consumption, are the climatic factors and the thermal comfort sensation of the users [15], [16]. As well, [17] evaluated the thermal comfort and its behavior from the energy point of view in a teaching educational center in Valencia - Spain.

The studies listed above indicate thermal discomfort may be due to misuse, design errors and inadequate operation of air conditioning systems, hence the importance of studying the thermal comfort state of a building and its causes. The greater the amplitude and the greater adaptation to the outside temperature of the comfort zone, the less active air conditioning systems must be activated and, therefore, there will be less energy consumption [18].

The study of thermal comfort has been applied in several universities and schools with the purpose of achieving a balance between the energy efficiency of the building and the thermal comfort of the students in the classrooms. In Taiwan, an experiment was carried out to measure thermal comfort in the classrooms of seven Universities and the result was to acquire the ideal range of thermal acceptability and climate adaptation for students [19]. An evaluation on thermal comfort was also carried out in a building of the National University of San Juan in Argentina; that allowed identifying opportunities for improvement in the building [20]. Another study was conducted in Universities of Zulia in Venezuela [21]. Research on the importance of thermal comfort in the performance of learning have also been carried out in schools and universities by different authors [9], [22–24].

Considering that the behavior of the conditions that are generated in buildings to ensure thermal comfort, may be significantly affected by parameters associated with climate and energy technologies that are used, this work aims to assess the perception of thermal comfort they have students in block 10 of the Universidad de la Costa. With the results obtained, recommendations are issued to ensure that students feel thermal comfort and the proper regulation of air conditioning systems.

2. Methodology

2.1. Selection of the thermal comfort measurement model.

For the development of the project, a search was made of the thermal comfort measurement methodologies and their characteristics, in addition to the measurement variables used for a climate-controlled thermal environment that allowed knowing the degree of influence of comfort on the welfare of building occupants. The most used methodologies were compared and the standard model Steady-State of Fanger was chosen for its practical value, based on the ASHRAE 55: 2013 standard. This methodology is ideal for measuring thermal comfort in closed environments, where people are exposed to long working or academic hours, allowing knowledge of the thermal state of the place to propose alternatives for improvement [25], [26].

2.2. Selection of the reference building for the study of perception of thermal comfort and calculation of the sample of students for the application of the survey.

It is selected for the study and application of the survey, building 10 of the Universidad de la Costa, of figure 1, located in the city of Barranquilla, which has an area of 480 m², 26 classrooms that measure on average 43.16 m² and maximum capacity of 40 students. And a laboratory with an area of 41.99 m² and capacity for 15 students

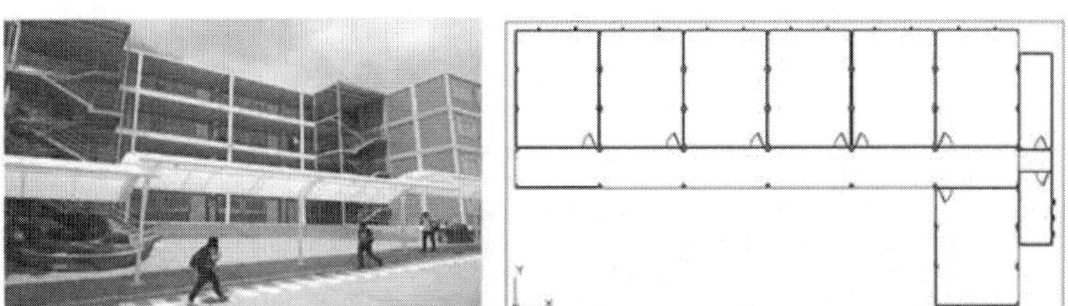

Fig 1. Building 10 of the Universidad de la Costa.

To determine the number of students to be surveyed, the sample size was determined with a confidence level of 95%, applying equation (1):

$$n = \frac{Z\alpha^2 \cdot N \cdot (p \cdot q)}{e^2 \cdot (N-1) + Z\alpha^2 \cdot (p \cdot q)} \quad (1)$$

Where, n-Population size of 1080 corresponding to the maximum capacity allowed by the building,$Z\alpha^2$- Designated value of 1.96, α-Confidence interval of 0.95, e-Acceptable error limit of 0.05, p.q-Standard deviation of 0.5

The sample size calculated was 283 students, and 584 students were surveyed. The number of students surveyed in each floor was 133 on the first floor, 125 on the second floor, 181 on the third floor and 140 on the fourth floor. On average, 22 students were surveyed per classroom. And of the total number of respondents, 303 were male and 281 were female.

2.3. Development of the survey for the measurement of the thermal perception of the occupants of building 10, of the University of the Coast.

The survey was designed to evaluate psychological, physiological and physical aspects of the user regarding the environment in which the users were and the measure corresponded to a scale of value suggested in the standard. The questions asked mainly considered the location of the occupant in the classroom, the level of insulation of the clothes, level of activity in the last instance of sitting time and the criterion of sensation and thermal acceptability. The process of evaluation of surveys was taken into account the following conditions: The surveys were applied in all classrooms of building 10 of the Universidad de la Costa, only students with a minimum time of 15 minutes in the classroom as indicated by the ASHRAE 55: 2013 Standard.

Studies carried out by different researchers have shown that one of the main factors influencing thermal comfort are the environmental conditions present in the city where the building is located his is due to the direct proportionality that exists between the external climate and the interior climate of the building, for the case of Barranquilla, which is characterized by having a hot-humid climate, especially during the summer, and average temperatures that range from 28°C to 33°C, with a higher incidence at midday. Therefore, measurements of temperature and relative humidity were made in each classroom at the time of the surveys.

3. Results

The main factors analyzed in the project to determine the perception of thermal comfort by students in building 10, are: the external climate, the indoor climate (air conditioning system), the amount and type of clothes, the location of the person and activity level of the students within the classroom.

3.1. Analysis of the general characteristics of the students and their location in the classrooms.

The survey was conducted in the months of July to September 2018, from 10:30 am to 12:30 p.m. The distribution of students in the classrooms was identified considering the location of the air conditioning, windows and doors of each room, where it was identified that criteria of sensation and thermal acceptability were different depending on the location. The average height of the students surveyed was 1.70 meters. The clothes most used by the respondents, was light of characteristic use for the type of climate, with a low Clo value.

The metabolic rate of the occupants is estimated based on the data provided by the ASHRAE 55 standard, for a variety of common activities, such as sitting, standing, walking, office activities and leisure activities, among others. The standard allows averaging the metabolic rate of the occupants, when the level of activity developed is common. For the study, the metabolic rate of the students was estimated for the activity of sitting and writing, as shown in table 1.

Table 1. Metabolic table according to activity carried out

Activity	Metabolic rate W/m²
Sitting (office, home, school, laboratory)	60
To write	60

3.2. Analysis of the thermal sensation identified in the students in building 10.
Figure 2 shows the results obtained when the analysis was performed to assess the thermal sensation that the students had in building 10, where it was possible to identify that 34% felt the environment cool, although there was a minimum percentage of 12% that had a feeling of heat inside the classroom. In general, a cool thermal sensation predominates in the building.

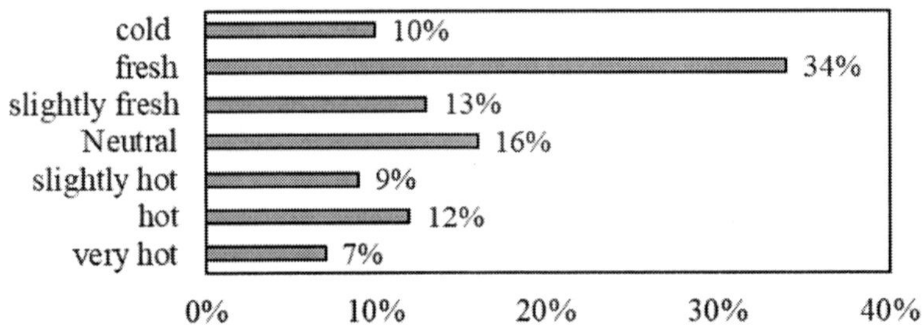

Fig 2. Thermal sensation of the students in the building.

When each floor is analyzed, represented in Figure 3, it is observed that the highest percentage of students agree that their thermal sensation is cool in all the floors, being the floor 2 identified as the hottest and the 1st floor the coldest for the most critical conditions of thermal sensation evaluated.

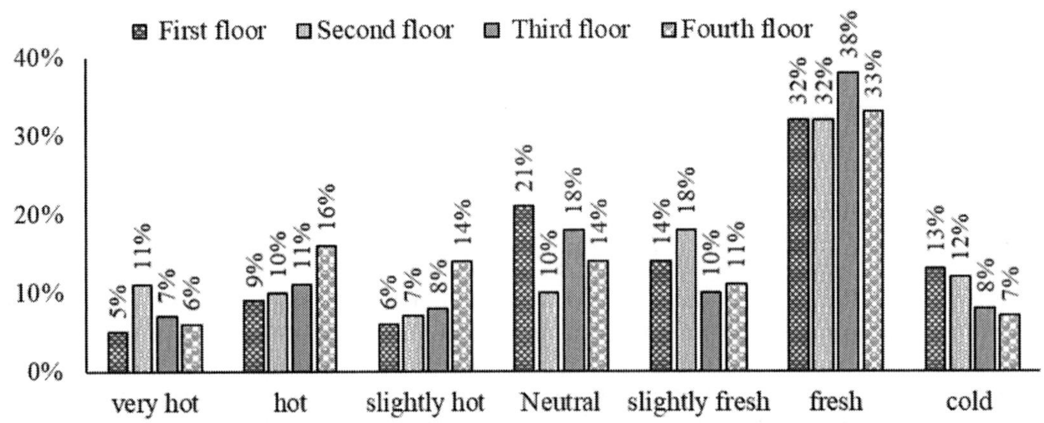

Fig 3. Thermal sensation of the students on each floor of the building

The results for each classroom were found that the two hottest rooms in the building were on the second floor and the coldest room was located on the first floor. This, product of the technical condition and temperature selection in the air conditioners.

3.3. Analysis of thermal acceptance identified in students in building 10
The thermal acceptability was also analyzed, as shown in Figure 4, obtaining as a result that the students in the highest percentage found in the range of acceptable thermal comfort in building 10. The unacceptability had a very low percentage of only 14%.

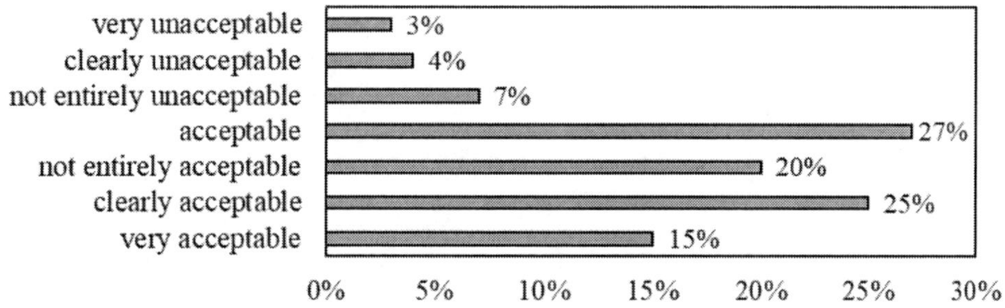

Fig 4. Thermal acceptability of the students in the building.

Figure 5 shows the results of thermal acceptability on each floor of building 10. It can be seen that the second floor obtained the best percentages in acceptability and the fourth floor was the one with the least acceptability on the part of the students. Regarding the classrooms analyzed, the opinion of clearly acceptable and acceptable predominates in most of the classrooms surveyed in all the floors.

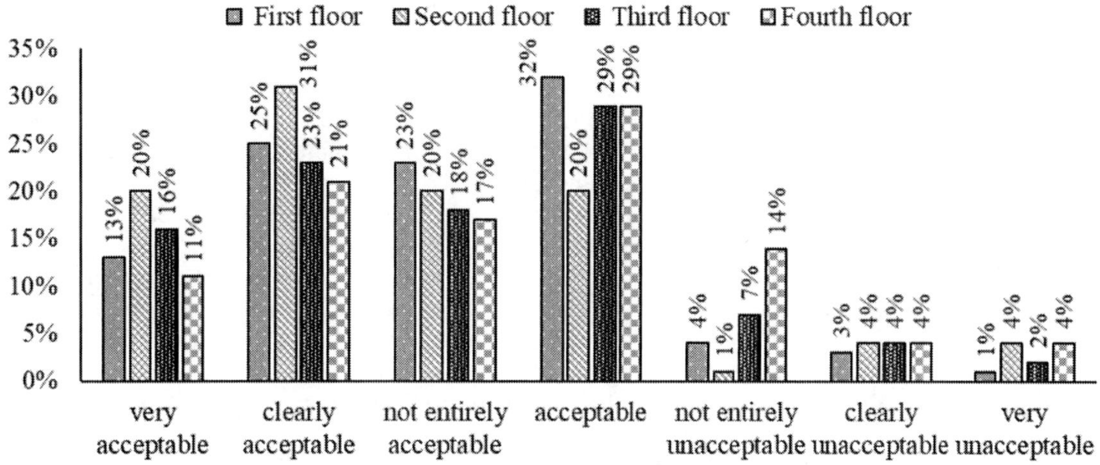

Fig 5. Thermal acceptability of students on each floor of the building.

3.4. Analysis of the thermal sensitivity according to the temperature inside the classroom.
At the time of the survey, the temperature and relative humidity were measured outside and inside the classrooms, observing, in Figure 6, that there is a thermal sensation of coolness that predominates.

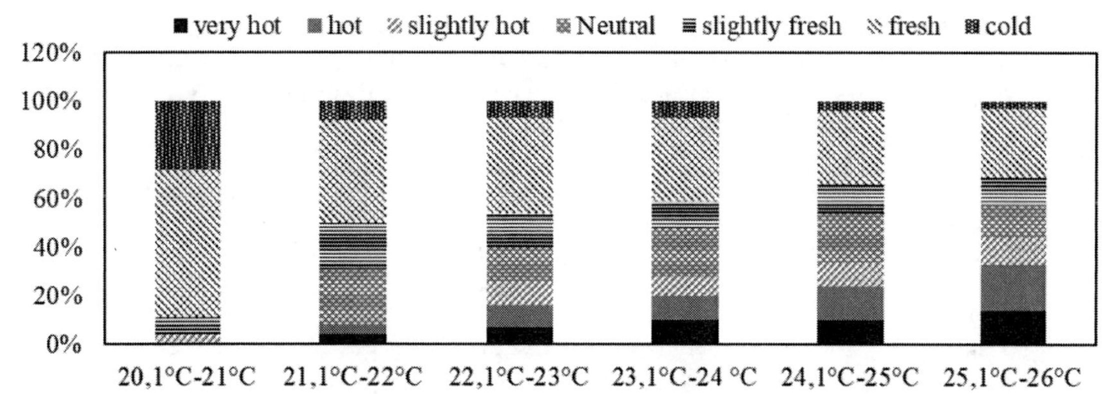

Fig 6. Thermal sensation of the students according to the interior temperature of the classroom.

From Figure 6, it is concluded that, although in the range of 20.1 to 21 ° C the sensation of cold increases and from 25.1 to 26 ° C that of heat, the temperature range of 21.1 to 23 ° is preferred by students. C. Therefore, it can be considered to establish set point temperatures of 22 ° C in the air conditioners of each room of the building.

Studies have shown that neutral thermal sensations are not always the ideal or preferred thermal state for people, demonstrating adaptive behavior in air-conditioned classrooms [27]. This is reflected in the students' preference for low temperatures, where they express cool to cold thermal sensation, but they show acceptability to the indoor climate in classrooms and, therefore, perception of comfort in the building.

4. Conclusions

The most well-known methodologies to evaluate the thermal comfort perception are: the standard model, the adaptive model and the adaptive variable model. These models seek to assess thermal comfort inside buildings in order to establish well-being for their occupants. But it also allows to establish adequate parameters for air conditioning systems, mainly when the Fanger model adopted by the ASHRAE 55 standard is applied, because it has been shown to be ideal for evaluating closed environments when occupants are exposed to long hours within the building.

For the study, Block 10 of the Universidad de la Costa was selected, for having its classrooms with standard measurements in the four floors that make up the building. The results showed that the thermal sensation for block 10 was fresh for 72% of the students surveyed, not reaching 80% as established by the ASHRAE 55 standard. The floor with the best results in terms of thermal sensation of comfort was the third floor, and in those that showed greater discomfort were the first and second floor with 67% and 33% respectively. Regarding thermal acceptability, 87% responded satisfactorily to indoor climate conditions in the classrooms. The highest thermal acceptability was on the first floor with 92%, followed by the second floor with 91%.

The measurement of the interior temperature was key in the analysis of perception of thermal comfort that students had in the building. 84% of the students found the greatest thermal sensation of comfort in the range of 21.1 to 23 ° C. Which indicates that the Universidad de la Costa can establish as a measure that the temperature of the set point is set in the air conditioners of the halls of building 10 between 21 and 23 ° C depending on the class schedule and the time of year.

References

[1] Schoen LJ, Alspach PF, Arens EA, Aynsley RM, Bean R 2013 *ASHRAE/IES standard 55-2013. Thermal Environmental Conditions for Human Occupancy* vol. 2013 Atlanta

[2] Höppe P 2002 Different aspects of assessing indoor and outdoor thermal comfort, *Energy and Buildings* vol. 34 no. 6 pp 661–665

[3] Mondelo P, Gregori E, Comas S, Castejón, E 2001 ERGONOMIA 2 - Confort y estrés térmico, 3rd ed. Alfaomega

[4] Jara P 2015 Thermal comfort and its importance for the architectural design and environmental quality of indoors space *Arquitectura y cultura* vol. 7 pp 106–121

[5] Vera M, Ordenes S 2011 Thermal and Energy Performance Evaluation of a Social Housing in Chile, Using a Building Energy Simulation Software *Revista Ingeniería de Construcción* vol. 17 no. 3 pp. 133–142

[6] Djamila H 2017 Indoor thermal comfort predictions : Selected issues and trends *Renewable and Sustainable Energy Reviews* vol. 74 pp. 569–580

[7] Djamila H, Chu C M, Kumaresan S 2013 Field study of thermal comfort in residential buildings in the equatorial hot-humid climate of Malaysia *Building and Environment* vol. 62 pp. 133–142

[8] Castilla M M, Álvarez J D, Berenguel M, Pérez M, Rodríguez F, Guzmán J L 2010 Técnicas de Control del Confort en Edificios *Revista iberoamericana de automática e informática industrial* vol. 7 no. 3 pp. 5–24

[9] Jiang J, Wang D, Liu Y, Xu Y, Liu J 2018 A study on pupils' learning performance and thermal comfort of primary schools in China *Building and Environment* vol. 134 pp. 102–113

[10] Gaonkar P, Bapat J, Das D 2018 Location-aware multi-objective optimization for energy cost management in semi-public buildings using thermal discomfort information *Sustainable Cities and Society* vol. 40 pp. 174–181
[11] Hatt T, Saelzer G, Hempel R, Gerber A 2012 High indoor comfort and very low energy consumption through the implementation of the passive house standard in Chile Revista de la Construccion vol. 11 no. 2 pp. 123–134
[12] Barros-Alvarez M, Balbis-Morejon M, Tovar-Ospino I, Castro-Peña J, de Leon-Siado L, Silva-Ortega J I, Rosales D 2017 Energy consumption comparison between air conditioning system Mini-Split and Variable Refrigerant Flow in an educational building *Espacios* vol. 38 no. 43 pp 19
[13] Fernández F and Martilli A 2012 El clima urbano: aspectos generales y su aplicación en el área de Madrid *Índice* pp. 21–24
[14] Ré I, Blasco M 2014 A hygrothermal and energy evaluation of a school building belonging to the '700 schools' national program, in the metropolitan area of San Juan, Argentina *Hábitat Sustentable* vol. 6 no. 2 pp. 40–51
[15] Valderrama C, Cohen A, Lagiere P, Puiggali J R 2011 Análisis del comportamiento energético en un conjunto de edifi cios multifuncionales. Caso de estudio Campus Universitario *Revista de la Construccion* vol. 10 no. 2 pp. 26–39
[16] Cabello J J, Sousa V, Sagastume A, Vandecasteele C 2019 Data supporting the improvement of forecasting and control of electricity consumption in hotels *Data in Brief* vol. 25 pp. 104-147
[17] Campos A 2017 Confort Térmico Y Eficiencia Energética En Un Centro Docente *Universidad Politécnica De Cataluña* pp. 188
[18] Madrigal J A, Cabello J J, Sagastume A, Balbis M. 2018 Evaluación de la Climatización en Locales Comerciales , Integrando Técnicas de Termografía , Simulación y Modelado por Elementos Finitos Evaluation of Air Conditioning in Commercial Buildings , Integrating Thermography Techniques , Simulation and Modeling *Información Tecnológica* vol. 29 no. 4 pp. 179–188
[19] Ruey-Lung H, Tzu-Ping L, Nai-Jung K 2006 Field experiments on thermal comfort in campus classrooms in Taiwan *Energy and Buildings* vol. 38 no. 1 pp. 53–62
[20] Kuchen E, Fisch M N, Gonzalo G E, Nozica G N 2009 Predicción del indice de disconformidad térmica en espacios de oficina considerando el diagnóstico de usuarios *Avances en Energías Renovables y Medio Ambiente* vol. 13 pp. 15–22
[21] González E M, Claret G Morales B 2009 About thermal comfort: neutral temperatures in the humid tropic *Revista de Investigación Científica en Arquitectura Journal of Scientific Research* pp. 33–38
[22] Mishra A K, Derks M T, Kooi L, Loomans M G, Kort H S 2017 Analysing thermal comfort perception of students through the class hour, during heating season, in a university classroom *Building and Environment* vol. 125 pp. 464–474
[23] Puteh M, Ibrahim M H, Adnan M, Che'Ahmad C N, Noh N M 2012 Thermal Comfort in Classroom: Constraints and Issues *Procedia - Social and Behavioral Sciences* vol. 46 pp. 1834–1838
[24] Corgnati S P, Filippi M, Viazzo S 2007 Perception of the thermal environment in high school and university classrooms: Subjective preferences and thermal comfort *Building and Environment* vol. 42 no. 2 pp. 951–959
[25] Pacheco J C, García R, García C A 2016 Methodological guide for determining the effect of thermal discomfort on standardized industrial operations *Tecnura* vol. 20 pp. 122–131
[26] Ciuha U, Tobita K, Mcdonnell A C, Igor B 2019 The effect of thermal transience on the perception of thermal comfort *Physiology & Behavior*
[27] Kwok S S and Lee E W 2011 A study of the importance of occupancy to building cooling load in prediction by intelligent approach *Energy Conversion and Management* vol. 52 no. 7 pp. 2555–2564

IOP Conf. Series: Materials Science and Engineering **844** (2020) 012031 doi:10.1088/1757-899X/844/1/012031

Energy performance analysis between two air conditioning systems used in an educational Building in warm-climate

M Balbis-Morejón[1], J I Silva-Ortega[1], J Castro-Peña[1]

[1] Department of Energy, Universidad de la Costa, Calle 58 No. 55-66. Barranquilla, Colombia.

Email: mbalbis1@cuc.edu.co

Abstract. Energy saving measures, in the design air conditioning systems, are crucial in the development of energy schemes with rational energy consumption. Traditionally, integrated buildings systems have been assessed individually to optimize the energy performance, however they have different parameters that affect energy performance that demands the use of detailed analysis using dynamic simulation. This paper is focused on compare an air conditioning system to be implemented in educational buildings in warm-climate, considering energy schemes provide for a constant air volume (CAV) flow system with a water chiller, while the other integrates a variable refrigerant flow (VRF) system. Adding in each case dedicated outdoor air System (DOAS) units. Energy consumption achieved by each AC system is obtained considering the configuration achieving energy savings of 40% of the annual electricity demand for cooling. Finally, the use of DOAS represents an increase of 20% of total electricity consumption.
Keywords: constant air volume; variable refrigerant flow; outdoor air System; energy efficiency.

1. Introduction

According to the Energy Information Administration (EIA), estimations, energy consumption will increase approximately in 56% in all regions during the following three decades, however this percentage is led by residential and commercial sector with 51 %, industry with 47%, and transportation with a 2 % [1]. It is expected that in dynamic regions energy consumption will falls in relation to final energy consumption related with emerging economies, due to population growth, lives standard improvement such as countries located in South-East Asia and Latin America region. Otherwise, Colombia posts an annual and sustained growth of 3.7% of electric consumption, being the commercial and public services, the fastest-growing sector (5.6%) [2], which is characterized for the high energy consumption of heating, ventilation and air conditioning (HVAC) systems, which is approximately between 40% to 60% of the total electricity consumption of buildings [3,4].

The HVAC systems have been identified as one of the largest energy consumers in developed countries, it is necessary to know the energy efficiency requirements of these equipment, which can be classified into: minimum efficiency of equipment, fluid distribution systems, HVAC control, ventilation, heat recovery and free cooling [5]. As a result of this behavior, mature economies have developed programs that have allowed the establishment of good practices in the construction sector, without compromising the comfort conditions. These good practices require verification through a detailed analysis of factors that affect the final consumption of energy in a building [6–9].

The method used for this analysis corresponds to the application of the heat transfer function, developed by the American Society of Heating, Refrigerating and Air-Conditioning Engineers (ASHRAE), that allows to calculate with precision the profiles of energy consumption of all the components which integrate the building such as: climate, location, loads, occupation profiles, construction characteristics, end-use technologies and variations in the configurations of the integrated systems [10,11].

These models require the use of computational tools due to their level of complexity since it considers the interaction of all the components of the building with the dynamic flows derived from the climate and levels of occupation. Therefore, this type of analysis contemplates a holistic approach based on dynamic simulation, which is precise and efficient because it allows to quantify the impact of important factors in the energy consumption that can have the components of the buildings [12]. For an accurate simulation, four factors that play a vital role in the energy consumption of a building are taken into account: (a) physical properties such as location, orientation and materials, (b) equipment installed to search internal comfort conditions, C) external environmental conditions, such as temperature, humidity, solar radiation, among others, and (d) the behavior of its occupants such as their schedules, activities and ways of using energy [13].

The literature reports the evaluation of savings measures focussed on improve the quality of the enclosure by limiting the thermal transmittance of the envelope in a dynamic regime [14]. Results showed that the surround with superior quality can reduce the energy consumption by approximately 35% [15]. Likewise, the incidence of ten constructive parameters in final energy consumption in commercial buildings in fourteen cities in Brazil was evaluated in [16]. In 2009 a comparative analysis was developed between variable air volume (VAV) and variable refrigerant volume (VRF) technologies in an existing office building [17]. These highlights the energy advantages that the VRF system can achieve, compared to the VAV system, which variates from 32.3% to 37% for the primary system and from 81.4% to 83.4% for the secondary system. In 2010, various energy saving measures combining construction alternatives, internal loads, and low-and high-efficiency air conditioning systems applied to a medium-sized non-residential building were studied in [7].

In 2013, it was characterized the energy performance of different systems and configurations of air conditioning systems in an existing educational building through using computational tools [18]. The results show a reduction in the energy consumption of auxiliary equipment from 15 to 35% and from 5 to 15% of the total energy consumption of the building when a VAV system is used versus a VRF. Likewise, it was demonstrated that the use of heat to recover and economize in the proposed air conditioning schemes, reducing the final energy consumption based on the refrigeration concept between 33 and 65% respectively.

Also, in 2017 an energy consumption comparison between air conditioning system Mini-Split and VRF in an educational building was realized in Barranquilla city (Colombia), obtaining a 30% of energy savings after the replacement of the air conditioning system technology. A 26% of savings were obtained as energy consume reduction [19].

Likewise, the comparative advantages of the three typical configurations for water chillers are described in [20]. Similarly, the strategies in the configuration in the design of chillers have demonstrated potential savings of approximately 10% in subtropical climates [21,22]. Where, the cooling load profiles of the buildings have been simulated using detailed analysis tools, allowing in this way to establish strategies in the design of these technologies. Also, the use of DOAS mixed with cooling has been studied its energy behavior for different climatic conditions. The results showed increases in energy efficiency between 10 and 40% depending on the climate [23-26].

2. Materials and methods.

This study is focused on the analysis of variable refrigerant volume (VRF) and chiller technologies, in the configurations of each of the systems contemplating the energy efficiency. The methodology used makes it possible to verify in detail the influence of the installation of a dedicated aerial unit (DOAS), construction parameters, internal loads, among others, in the energetic demand that reaches by the air conditioning system for a typical year of operation. The computational tool for the detailed analysis considered dynamic simulation using Energy Plus V.8.4 (E +), which provides excellent reference points in the modelling of energy demand in buildings and mechanical equipment that admit the detailed configuration of the system of generation, transportation and final distribution; The modelling of the educational building of the University was carried out to simulate the energy performance of the technological variants as viable alternatives to cover the energy needs that the building can demand to achieve thermal comfort.

2.1 Building Characteristics

The simulated building has a total area of 2.695 m^2 spread over eight floors. Between the second and the seventh floor the building's characteristics do not change, a zone multiplier factor was used, so that in the graphic representation of the building only 4 of the 8 floors are shown. The height of the ground floor is 5 m, plus a plenum of 1.2 m in height. The long wide ratio is approximately 1.4 m, with glazing proportions regarding the envelope of 6.7%, the building is intended for classrooms, offices and two auditoriums as is shown in **¡Error! No se encuentra el origen de la referencia..**

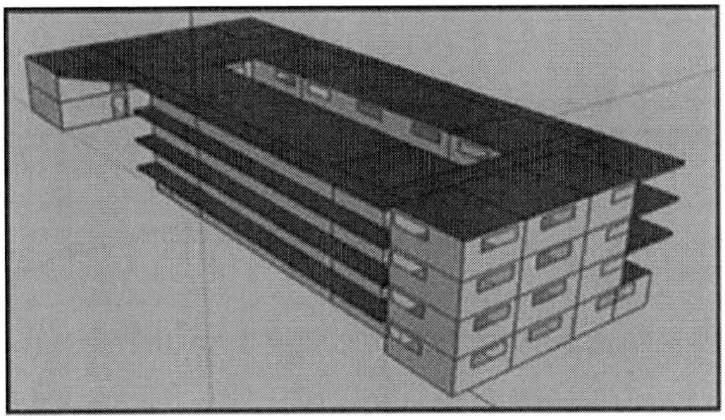

Figure 1. Educational building 3D view.

During the different simulations, the specific characteristics of the building envelope and of internal loads were kept constant. In Table 1, the characteristics of the the building are listed.

Table 1 Characteristics of base building envelope model.

Item	Material	Value	Unit	Property
Walls	Brick	150	mm	Case 1
Glazing	Standard clear glass	8	mm	Case 2
Heat Transfer	---	3.56	W/m^2K	Case 3

The typical work schedules of the building were considered, usually in the full operation from 7:00 am to 10:00 pm with slight variations between 12:30 pm to 2:00 pm. for a load level of 80% [5]. The operating levels for lighting and electrical equipment correspond to 100% during the working day, and for the rest of the time, it remains with a load of 20%. The internal charge levels for electrical equipment and illumination correspond to 10 and 12 W/m^2 respectively.

2.2 Air conditioning system specifications

The installed air conditioning system is a Chiller with a Screw compressor (COP 3) condensed by water. The configuration is based on VRF. Figure 2 describes HVAC operation scheme. Also, it can be observed a pump control flow and a bypass valve [19].

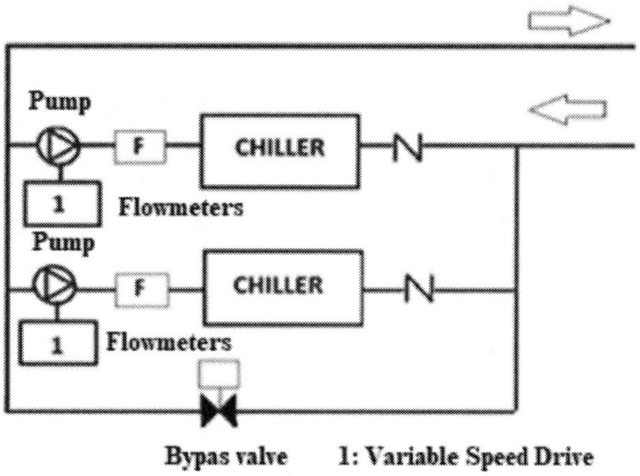

Figure 2. Configuration of the air conditioning system of the educational building.

2.3 Simulation model

The first simulated air-conditioning scheme corresponds to a constant air system (CAV), considering in some cases air units (DOAS) and forty-one Fan Coil type terminal units for the same number of air-conditioned zones. The cold production system corresponds to a water condensing chiller and an average performance coefficient (COP) of 3.0 and a screw type compressor. The temperature of the water was set at 7 ° C and return 12 ° C. The thermostat control is independent in each zone at 25 °C.

Also, the second technological variant analyzed for the building under study corresponds to a variable refrigerant volume system (VRF), capable of providing air to multiple zones. This was configured to operate at a drive temperature of approximately 14 ° C, equally, this type of system was simulated considering the use of dedicated air units (DOAS) in many scenarios. The average yield coefficient (COP) is 3.3. The system has a master control for the 41 climate zones. The technological characteristics of simulated air conditioning systems are shown in table 2.

Table 2. Air conditioning system specifications and configurations.

Features/Components	Description
Air Conditioning System	Constant air volume (CAV), with and without dedicated air units (DOAS) with 41 Fan Coil for an equal number of air-conditioned zones. Chiller with an average performance coefficient (COP) of 3.0 and compressor type Screw.
	Variable coolant volume (VRF), with and without dedicated air units (DOAS) with 41 direct expansion terminal units and 5 air condensing units, with an average efficiency coefficient (COP) of 3.3.
Thermostat	25°C
Ventilation Airflow	10 / person

3. Results.

Figure 3 shows the percentage distribution of the electrical demands for end-use equipment of the electric energy integrated into the educational building. This shows the significant impact of air conditioning systems on final energy consumption with a 49% for VRF and 64% for CAV, with an increase in energy consumption with the addition of DOAS of 5% and 7% respectively. The impact of these systems is not only due to the climatic severity but also as a product of the great internal load generated in the interior.

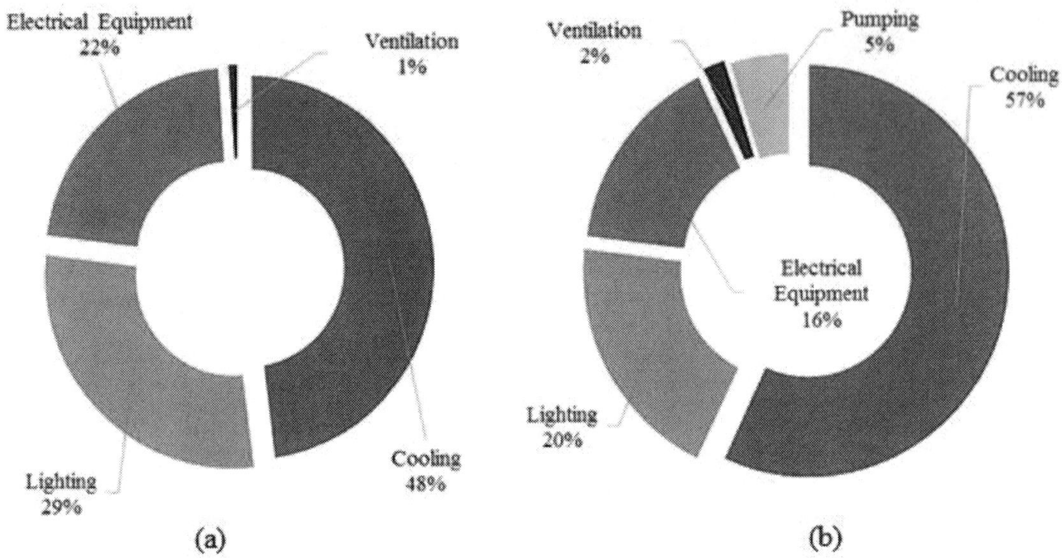

Figure 3. Distribution of energy final consumption in the education building. a) VRF b) CAV

For its part, Table 3 shows the energy impact derived from the scenarios of the four technological variants of the air-conditioning system. It stands out for its high performance the behaviour of the air conditioning system configured with the VRF technology, it shows a reduction of 40% in the electric consumption by refrigeration.

Table 3. Consumption of electrical energy per system or subsystem in the educational building.

System and Subsystem	VRF	VRF-DOAS	CAV	CAV-DOAS
Refrigeration	335016.88	422066.39	560990.25	675212.74
Lighting	197762.74	197762.74	197762.74	197762.74
Electrical Equipment	152557.57	152557.57	152557.57	152557.57
ventilation	9682.46	41454.58	16905.75	37261.57
Pumping	0	0	48758.3	68090.15
Total	695019.65	813841.28	976974.61	1130884.77

Likewise, the total electric energy impact of the VRF system compared to the system configured in VAC mode is 30%. On the other hand, the employment impact of dedicated air units (DOAS) represents an approximate increase of 20% for the two technologies analysed. The advantage of the VRF system lies in the possibility that present of this technology, in the adjustment of the system load to the instantaneous thermal demand.

4. Conclusion

The results showed the potential reduction of 30% in total electricity consumption when the air conditioning system is configured with VRF technology. This is due to the master control system present in this technology used, thus achieving the adjustment of the load of the system of generation of cold to the thermal demands of the zones for a greater diversity. It should be noted that although the total electric energy consumption is increased by approximately 20% when the dedicated air units

(DOAS) are added for both technologies, these are guarantees for compliance with minimum air requirements per standard, and humidity control in each room.

Acknowledgments
The authors acknowledge the support provided by COLCIENCIAS and Universidad de la Costa whose supports the project thought the open call No. 543 approving the project titled: knowledge network Consolidation on energy efficiency and its impact on the productive sector under international specifications. The authors acknowledge the support to the Colombian Energy Efficiency Network RECIEE.

Reference

[1] EIA 2018 Monthly Energy Review U.S. Energy Whashington, DC
[2] Universidad Nacional, Fundación Bariloche Política Energética 2010 Análisis y Revisión de los Objetivos de Política Energética Colombiana de Largo Plazo y Actualización de sus Estrategias de Desarrollo Bogotá
[3] Grupo de Demanda Energética 2006 Caracterización Energética de los Sectores Residencial, Comercial y Terciario Bogotá D C
[4] UPME 2015 Plan Energetico Nacional Colombia: Ideario Energético 2050. Unidad Planeación Min Energética, Repub Colomb 2015:184.
[5] Pérez-lombard L, Ortiz J, Coronel JF, Maestre IR 2011 A review of HVAC systems requirements in building energy regulations. *Energy Build* 2011 vol. 43 pp. 255–68. doi:10.1016/j.enbuild.2010.10.025.
[6] Ministerio de Vivienda 2006 Código Técnico de la Edificación Partes I y II Madrid España
[7] Barros-Alvarez M, Balbis-Morejon M, Tovar-Ospino I, Castro-Peña J, de Leon-Siado L, Silva-Ortega J I, Rosales D 2017 Energy consumption comparison between air conditioning system Mini-Split and Variable Refrigerant Flow in an educational building *Espacios* vol. 38 no. 43 pp 19
[8] ASHRAE Sandards Committee 2006-2007 Energy Standard for Buildings Except Low-Rise Residential Buildings Atlanta ASHRAE
[9] Iwaro J, Mwasha A 2010 A review of building energy regulation and policy for energy conservation in developing countries. *Energy Policy* vol. 38 pp. 7744–55. doi:10.1016/j.enpol.2010.08.027.
[10] Luo C, Moghtaderi B, Page A 2010 Modelling of wall heat transfer using modified conduction transfer function , finite volume and complex Fourier analysis methods. *Energy Build* vol 42 pp. 605–17. doi:10.1016/j.enbuild.2009.10.031.
[11] Li XQ, Chen Y, Spitler JD, Fisher D 2009 Applicability of calculation methods for conduction transfer function of building constructions. *Int J Therm Sci* vol 48 pp. 1441–51. doi:10.1016/j.ijthermalsci.2008.11.006.
[12] Kim H, Stumpf A, Kim W 2011 Automation in Construction Analysis of an Energy Efficient Building Design Through Data Mining Approach. *Autom Constr* vol. 20 pp. 37–43. doi:10.1016/j.autcon.2010.07.006.
[13] Madrigal J A, Cabello J J, Sagastume A, Balbis M. 2018 Evaluación de la Climatización en Locales Comerciales , Integrando Técnicas de Termografía , Simulación y Modelado por Elementos Finitos Evaluation of Air Conditioning in Commercial Buildings , Integrating Thermography Techniques , Simulation and Modeling *Información Tecnológica* vol. 29 no. 4 pp. 179–188. doi: 10.4067/s0718-07642018000400179
[14] International Organization for Standardization ISO. ISO 6946:2007. Building components and building elements — Thermal resistance and thermal transmittance — Calculation method. Switzerland: 2007. doi:10.1109/IEEESTD.2010.5733835.
[15] Chan KT, Chow WK 1998 Energy impact of commercial-building envelopes in the sub-tropical climate. *Appl Energy* vol 60 pp. 21–39. doi:10.1016/S0306-2619(98)00021-X.

[16] Signor R 1999 Análize de Regressao do Consumo de Energía Eléctrica Frente a Variáveis Arquitetonicas para Edifícios Comerciais Climatizados em 14 Capitais Brasileiras. Universidade Federal de Santa Catarina
[17] Aynur TN, Hwang Y, Radermacher R 2009 Simulation comparison of VAV and VRF air conditioning systems in an existing building for the cooling season. *Energy Build* vol 41:1143–50. doi:10.1016/j.enbuild.2009.05.011.
[18] Alvarez O, Sanjuan M, Bula A, Amaya F 2017 VAV System Operating in an Educational Building Under Tropical Conditions: Energy Analysis. ASME 2013. 7th Int. Conf. Energy Sustain. ES, Minneapolis, USA: ASME; 2017 pp. 1–8.
[19] Balbis-Morejon M, Tovar-Ospino I, Castro-Peña JJ, Cárdenas-Escorcia Y del C 2017 Energy assessment of the system pumping a climate control scheme with water coolers for an educational building using dynamic simulation. *Espacios* vol 38 pp19–32
[20] Hubbard RS 2011 Energy Impacts of Chilled-Water-Piping Configuration. *HPAC Eng* pp. 20–6
[21] Wang F, Lin H, Tu W, Wang Y, Huang Y 2015 Energy Modeling and Chillers Sizing of HVAC System for a Hotel Building *Procedia Eng* vol 121 pp. 1812–1818. doi:10.1016/j.proeng.2015.09.161.
[22] Yu FW, Chan KT 2007 Part load performance of air-cooled centrifugal chillers with variable speed condenser fan control. *Build Environ* vol 42 pp. 3816–3829. doi:10.1016/j.buildenv.2006.11.029.
[23] Mumma S 2001 Ceiling Panel Cooling Systems. ASHRAE vol. 2 pp. 28–32.
[24] Tian Z, Love JA 2009 Application of Radiant Cooling in Differente Climates: Assessment of Office Buildings Through Simulation. Elev. Int. IBPSA Conf., Glasgow, Scotland: Building Simulation pp. 2220–2227.
[25] Kwok SSK, Lee EWM 2011 A study of the importance of occupancy to building cooling load in prediction by intelligent approach. *Energy Convers Manag* vol 52 pp. 2555–2564. doi:10.1016/j.enconman.2011.02.002.
[26] Rahman MM, Rasul MG, Khan MMK 2010 Energy Conservation Measures in an Institutional Building in Sub-Tropical Climate in Australia. *Appl Energy* vol. 87 pp. 2994–3004. doi:10.1016/j.apenergy.2010.04.005.

IOP Conf. Series: Materials Science and Engineering **844** (2020) 012032 doi:10.1088/1757-899X/844/1/012032

Study of linear strain gauges at micron scale for the reduction of external factors

C T Villamizar[1], W Y Arevalo [1] and C C Castiblanco [1]

[1] RFMEMS seedbed, Catholic University of Colombia, Sede Claustro, Bogota Colombia

Abstract. In the present article, the design and simulation of a strain gauge is shown, which serves to measure deformation or pressure in different applications of the industry. In this way, the respective simulation was carried out in the COMSOL Multiphysics 5.2 software, where the 3 linear meander gauge was subjected to external factors such as temperature and pressure, with this, the different results obtained were shown the mechanical behavior and of Graphene and Silver grid materials and substrate Paper. On the other hand, in front of the design of the three-way linear strain gauge, M.E.M.S (Microelectromechanical Systems) parameters technology was used to reduce its dimensions in micrometers to reduce the effects of the respective external factors where these produce errors in the measurements.

Keywords: M.E.M.S (Microelectromechanical systems), GF (Gauge Factor), Linear Strain Gauges, simulation, temperature, Strength.

1. Introduction

In the mechanical industry (e.g. Vehicular, aeronautical) a measurement system is required to maintains continuous monitoring of the various mechanical parts. Due to this, the strain gauges appear as a transducer device, which can obtain precise measurements of the variations of the parameters (as is the case of the deformations that the mechanical components) and in turn detect possible ruptures in such as components.

The strain gauges are defined as a transducer based on the extensometric principle, which when are used to some mechanical effect (e.g. tension, pressure, etc.) vary their resistance, being able to convert small deformations of a material in quantifiable measures. On the other hand, the manufacture and miniaturization of strain gauges have shown a visible breakthrough; since the impact of this type of transducers have had on the different fields, (e.g. mechanics) contributing to an improvement in deformation control. Furthermore, the diversity of the parameters within which the strain gauges can work, as well as the structure of these gauges and what are the characteristics that should be taken into account when are used to the design and simulation of strain gauges.

In the Figure 1, the results obtained from the search in some databases such as Science Direct ® and ProQuest ®, as well as processed with a Bibliometric Tool such as Vantage Point ® and VosViewer ®. Then as result can be found keywords used as the main input of this research, which indicates that the work done is located within the global trends. Similarly, it can be seen that in Latin America it is not an area with a high level of research.

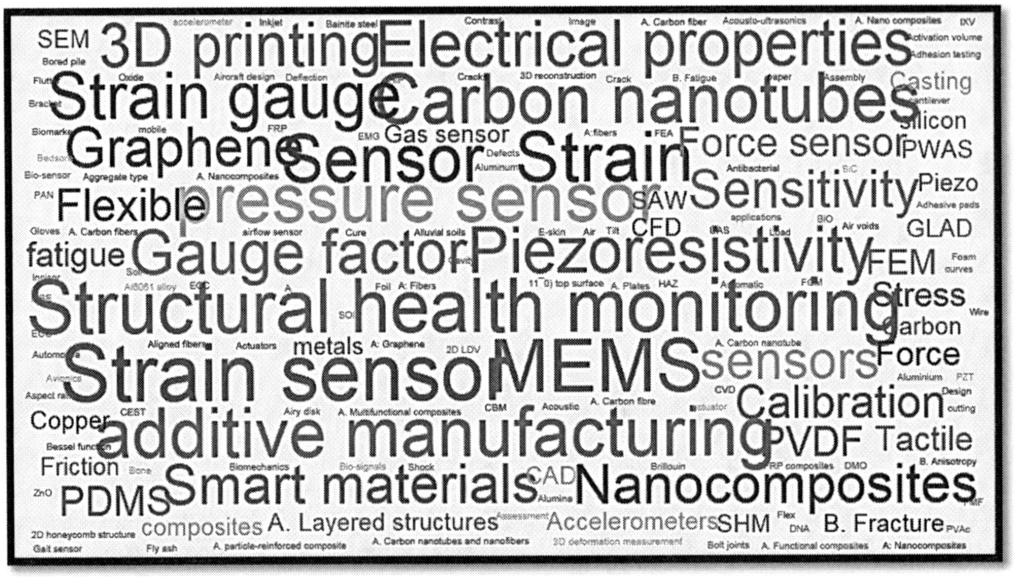

Figure 1. Vantage Point word map.

Additionally, in this article we shown the process of design and simulation of linear strain gauges, starting with the basic concepts of this type of transducer, such as the materials used and the parameters to be taken into account, also the equations and the results obtained when implementing these equations used to find the values of the electrical and mechanical parameters of these.

2. Design of strain gauges

2.1 History of strain gauges

In 1856 Lord Kelvin in his research, could applied a force on a conductive wire or a semiconductor, which generated a variation in its electrical resistance. This principle allows measurements of forces of very small magnitude to be taken, which are generated by slight deformations in the conductor material. These precepts were essential for working with transducer devices, among which are strain gauges. Meanwhile, strain gauges were invented in 1938 by Edward E. Simmons, who defined them as passive transducers, which, when adhered to a surface, help to measure the deformation that this suffers when a force is applied on this measure is given in ranges of variation of the strength of the strain gauge. At the same time, the simplest type of gauge is made of a flexible insulating line that is supported by a flat metallic pattern. Therefore, such as gauges are attached to the surfaces by using a special adhesive. In addition, the resistance variations are of very small values, so it is necessary to use a Wheatstone bridge, which results in a measurement of an electrical signal [1].

However, within the main design and simulation conditions of an ex-tensile gauge, there are the materials used, which must have piezo resistive properties, which refer to the ability to transform any deformation into a variation of its resistance and the type of substrate, which is the base on which the sensors are manufactured. Likewise, in a subsequent process such as the manufacture of some tensometric gauges, factors such as sensitivity or gauge factor must be taken into account at the time of being implemented, without neglecting, that have a short time of useful life..

IOP Conf. Series: Materials Science and Engineering **844** (2020) 012032 doi:10.1088/1757-899X/844/1/012032

2.2 *Materials*

Among the materials used in the manufacture of strain gauges are the following:

- *Polymers.* In nature there are huge molecules called macromolecules. Hundreds of thousands of atoms form these molecules, so their molecular weights are very high. Polymers are a particular type of macromolecule, which is characterized by having a unit that repeats throughout the molecule [2].
- *Fiber.* It is a crystalline structure that have their chains ordered and are composed of natural and synthetic, organic and inorganic materials, such as acetate and cellulose [3].
- *Graphene.* It is defined as a thin flat sheet of carbon atoms with sp2 hybridization in two dimensions (2D), forming a structure similar to a bee panel [5].
- *Silver. It* is a lustrous whitish precious metal that is found throughout the world, but most of it is extracted from mines in Mexico, the western United States, Bolivia, Peru, Canada and Australia [6].

2.3 *Parts of a strain gauge.*

The materials (shown above) are necessary for the Strain Gauge construction, the union of its three main parts, such as [4]:

- *Grid.* It is the metallic part, which varies its resistance as a compression, or tension is applied.
- *Base.* It is the carrier support of the grid and is manufactured in different insulating materials.
- *Tags.* They are used to connect the signal strain.

The material is located between the surface of the object of study and the grid. This provides firmness and support to the grid for the welding of cables to the labels, also provides a manageable surface for gluing the gauge on the object of study and finally providing electrical insulation between the grid of the gauge and the object of study.

It is important to mention that the strain gauge has a relationship between tension that is applied to the object of study and resistance variations, which is called tension sensitivity or gauge factor; which is defined as the change ratio of the electrical resistance of the conductor by varying the relative length. This parameter is dimensionless as mentioned G, Ruiz "... *The gauge factor is a function of the tensile sensitivity of the alloy of the metals which the gauge grid was manufactured and other factors such as: Temperature, size, grid configuration and how to measure it*" [4].

At the edges of the grid (final curves), a transverse sensitivity is presented. This sensitivity is observed when excessive tension is applied to the gauge, which generates the in-swelling (compression) or stretching (tension) of the gauge, causing the grid edges to vary in length and with this, the gauge has a change in its resistance. In the figure. 2 the parameters of the linear strain gauge are observed.

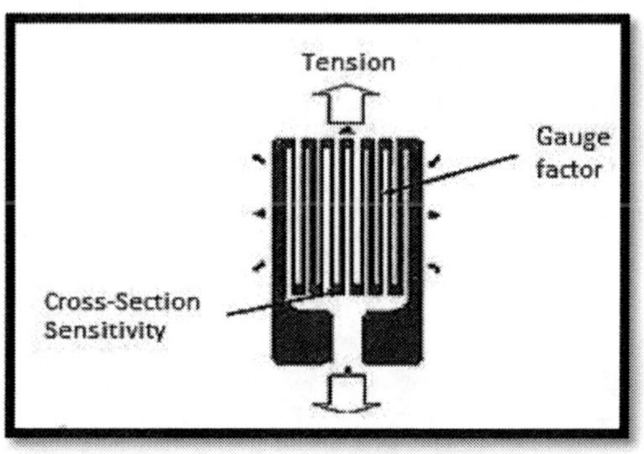

Figure 2. Parameters of the linear strain gauge [4].

To know what the practical values for the design of the strain gauge are; we proceed to formulate equations 1,2,3,4, for its development within the proposed work. Equation 1 shows the variation of the resistance based on the multiplication of the resistivity of the material and the division of the changes in the length of the gauge, the resistance and the area of the conductor.

Eq 1. Change of electrical resistance

$$\Delta R = \rho \frac{\Delta L}{A} \quad (1)$$

Where:

- ΔR = Changes in electrical resistance (Ω).
- ρ = Resistivity of the material (Ωm).
- ΔL = Change in effective length of the conductive material (m).
- A= Conductive material area (m).

Equation 2 shows the gauge factor which is obtained by dividing the variation of the electrical resistance, the resistance of the gauge and the unit deformation.

Eq 2. Gauge factor equation

$$GF = k = \frac{\frac{\Delta R}{Rg}}{\frac{\Delta L}{L}} = \frac{\frac{\Delta R}{R}}{\varepsilon} \quad (2)$$

Where:

- GF=K= Gauge factor [dimensionless].
- Rg= Initial gauge resistance (Ω).
- L= Gauge Length (m).
- ε = Unitary Deformation [dimensionless].

Equation 3 shows the value of the stress applied to the object, which is obtained by dividing the tension, force, weight, load between the cross-sectional areas of the structure.
Eq 3. Equation of strain applied to an object [4]

$$\sigma = \frac{P}{A} \quad (3)$$

Where:

- σ = Strain applied to the object [N/m^2].
- P= Tension, force, weight, load [N].
- A= Cross- Sectional structure area [m^2].

In Equation 4, the value of the deformation is obtained by dividing the change in the length of the object by the initial length of the object.

Eq 4. Unitary Deformation Equation

IOP Conf. Series: Materials Science and Engineering **844** (2020) 012032 doi:10.1088/1757-899X/844/1/012032

$$\varepsilon = \frac{\Delta L}{L0} \tag{4}$$

Where:

- L_0= Initial object length [m].

3. Calculation of mechanical and electrical parameters of strain gauges

The equations presented previously, are proposed for the design, taking into account the proposed geometries characteristics, which can be evidenced in Table 1.

Table 1. Measurement of the designs of linear strain gauges.

Name	Number	Description
h_sust	1 [cm]	Substrate height
a_sust	0.5 [cm]	Substrate length
p_subst	50 [μm]	Depth of Substrate
h_grid	0.65 [cm]	Grid height
a_grid	197.36 [μm]	Grid width
p_grid	50 [μm]	Grid depth
posh_grid	2.4 [mm]	Grid position
d_grid	625 [μm]	Grid distance
a_electrode	300 [μm]	Electrode width

Related to table 1, different variations were made, among which are the number of meanders of 1, 2,3 and 5 with round, square and 5 meander edges with triangular edges. It is also important to mention the use of substrate in paper fiber materials and grating material of graphene and silver type material.

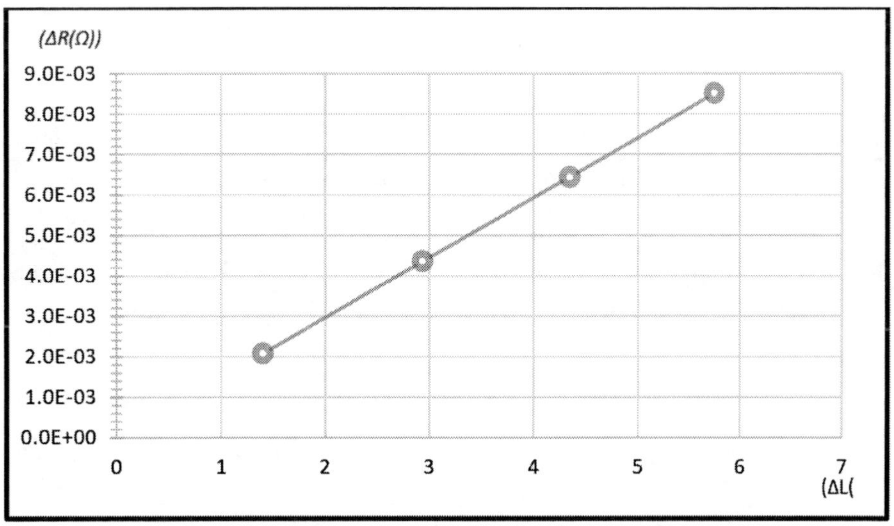

Figure 3. Variation of the electrical resistance of graphene

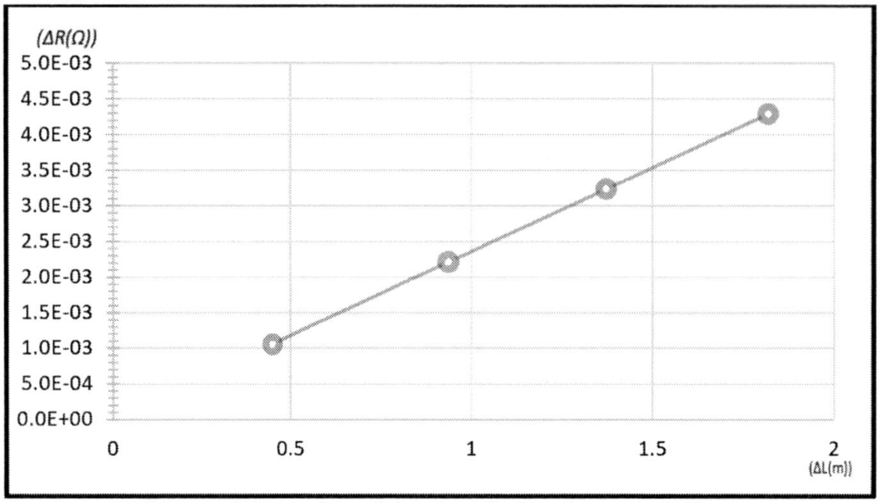

Figure 4. Variation of the electrical resistance of silver

In the figure. 3. Its shown a variation of the electrical resistance of graphene. On the other hand, the figure. 4, shown the variation of the electrical resistance of the silver, which it is dependent on the value of the variation of the length of the material.

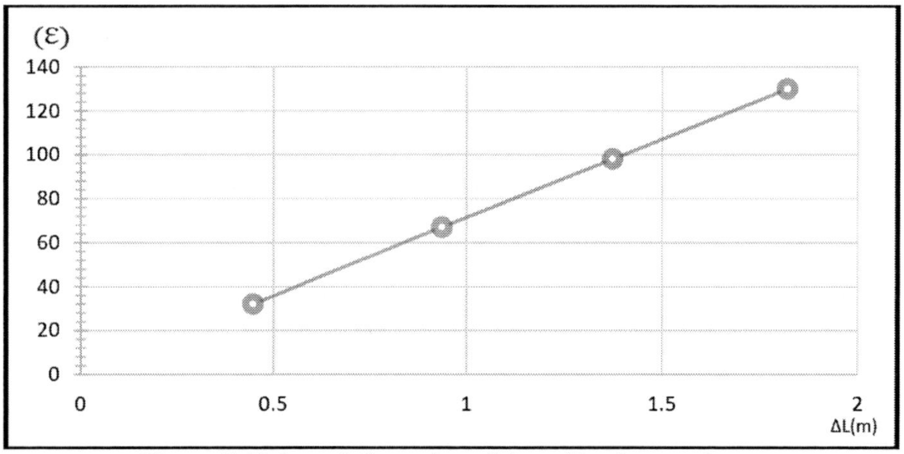

Figure 5. Unitary deformation of the grid using graphene.

In the figure 5 shows the slope of the unitary deformation, which indicates the value of the deformation suffered by the material (graphene) as a function of the variation in length.

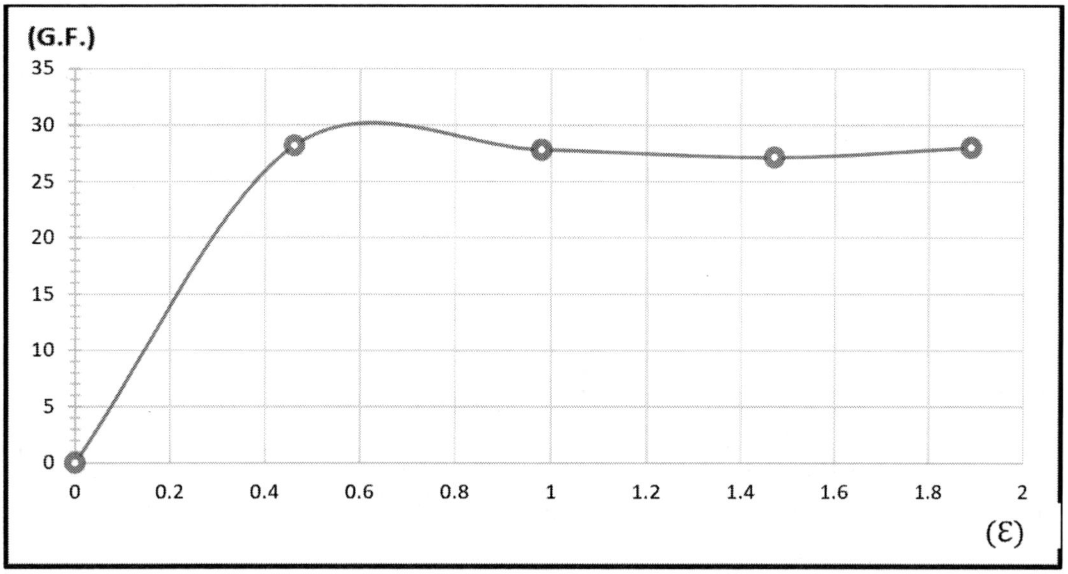

Figure 6. Silver Gauge Factor.

In Figures 6 and 7, the behavior of the Gauge Factor can be observed depending on the variation in the length of graphene and the theoretically obtained silver. The theoretically obtained values of the Gauge Factor are observed for the silver and graphene materials used in the designs, these materials behaved in very different ways, with graphene having higher values compared to silver.

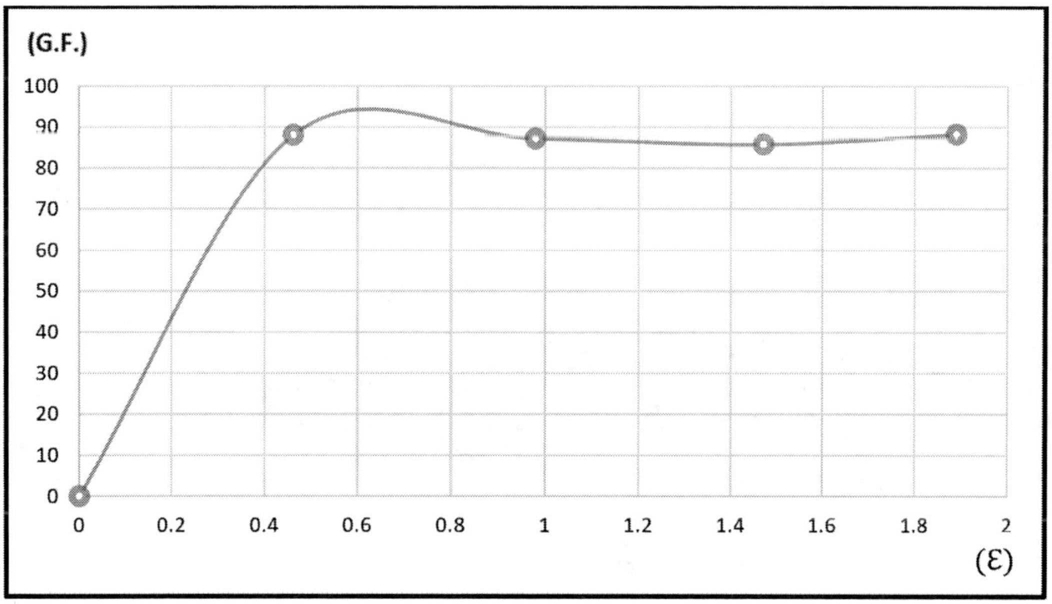

Figure 7. Graphene Gauge Factor.

IOP Conf. Series: Materials Science and Engineering **844** (2020) 012032 doi:10.1088/1757-899X/844/1/012032

4. Analysis of the results in the simulations performed

The simulations were worked on in the COMSOL MULTIPHYSICS 5.2 ® program where various geometries (mentioned above) were generated, which underwent changes in temperature and in the force applied to the linear strain gauge. In Figures 8 and 9, the 3-gauge linear strain gauge with square edges, paper substrate and graphene grid material is observed, which was subjected to various forces of 2N and 8N.

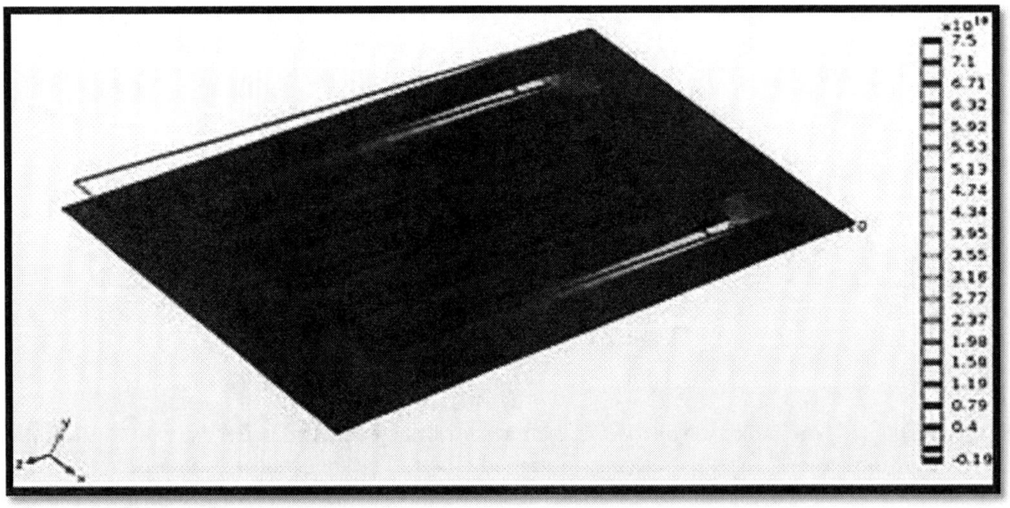

Figure. 8. Linear strain gauge of 3 meanders with square edges, paper substrate and graphene grating material subjected to 2N..

The results of the deformations obtained, respectively, were the graphene, which showed the best performance compared to those of silver grid material. Similarly, variations in color that indicate the concentration of stresses in the material are observed

Figure 9. Linear strain gauge of 3 meanders with square edges, paper substrate and graphene grating material subjected to 8N.

The design of the linear strain gauge also underwent temperature variations that started at 26.85°C up to a maximum temperature of 226.89°C. The results are illustrated in the following figures 10 and 11, the linear strain gauge is observed which It was subjected to a temperature of 26.85 ° C and 226.89 ° C and it is evidenced that the material is not affected in large quantities, which indicates that it will not suffer large variations in the measurements as its temperature is increased in the measurement environment.

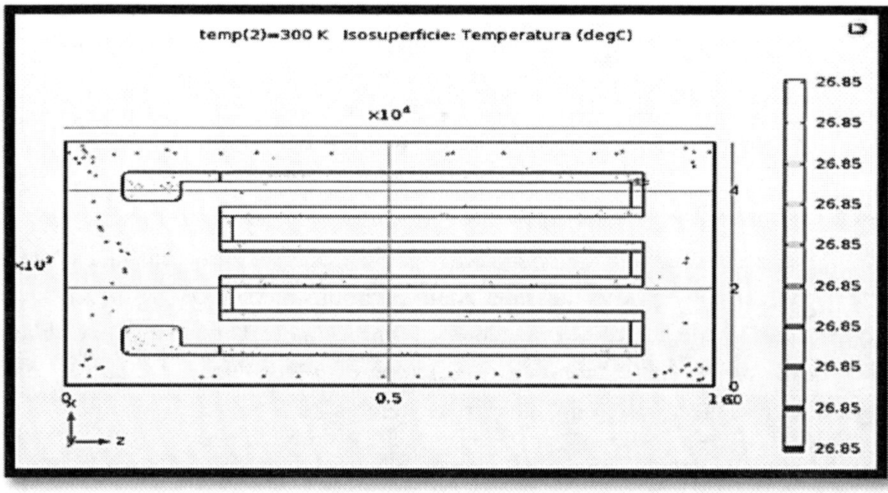

Figure 10. Linear strain gauge 3 meanders with square edges (temperature range 26.85°C)

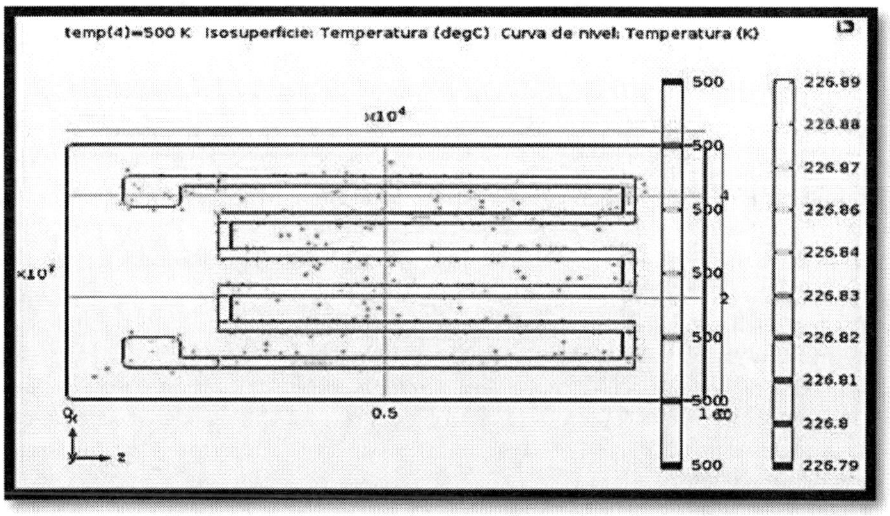

Figure 11. Linear strain gauge 3 meanders with square edges temperature range 226.89°C

The temperature variations in the designed linear strain gauges, present a behavior that was observed in the temperature of 26.85°C or 300°K. They had no major effects on their materials on both the paper substrate and in the silver or Graphene meanders, when is increasing this temperature flow to 226.85°C or 500°K temperature (which is maximum point near supported by the paper before its combustion of 233°K).

5. Conclusions

This article showed that simulations were performed in which variations were generated in the materials used in the designs of linear strain gauges, paper in the substrate, silver and graphene in the meanders, in addition to the different geometries evaluated and the variables to which these designs were subjected to temperature and force.

In the tensile gauge of 5 meanders with square edges it has a value of 6.0444x1017 unlike the gauge of 3 meanders that is of a tension value of 7.90812 x1010, this maximum tension decreases exponentially implying that the greater number of meanders, the strain gauge is more rigid. In the same way, it can be seen that the greater the force to which the strain gauge is subjected, there is a greater tension, and from another point of view, it is evident that at a lower number of meanders there is a decrease in the maximum tension.

It was found to vary in different ways the affectation generated and this depending on the type of geometry and materials used in the design, with the geometry of 3 meanders with square edges being the one that suffered the least damage as the temperature increased.

Acknowledgment

We appreciate the collaboration of Dr. German Álvarez Botero, who, during his time at the University, supported the technical processes in the simulation environment of COMSOL®. We also want to express our gratitude to the RFMEMS Seedbed of the Catholic University of Colombia for the opportunity of development and exploration in this type of topics, which are susceptible for future research projects such as the one presented in Expotecnologia 2019.

References

[1] V. Ciscar. *Galgas Extensométricas*. In: Info PLC. Net- Automatización Industrial, Robótica e industria 4.0, s, f. [en línea]. Available in: http://www.infoplc.net/files/documentacion/instrumentacion_deteccion/infoPL C_net_Galgas_extensiometricas.pdf

[2] F, López. *Fundamentos de los polímeros*. Trabajo de grado. Mérida: Universidad de los Andes Venezuela. Facultad de Ciencias. Departamento de Química, 2004. 1-64, p. Available in: https://www.researchgate.net/publication/262639386_FUNDAMENTOS_DE_ POLIMEROS

[3] J, González. *Efecto del Contenido de Fibra de Vidrio sobre las Propiedades Mecánicas en Probe tas de Polibutilentereftalato (PBT) Reforzado*. Trabajo de grado. Ingeniero Industrial. Cartagena: Universidad Politécnica de Cartagena. Facultad de Ingeniería. Programa de Ingeniería Industrial, 2013.1-120, p. Available in http://repositorio.upct.es/bitstream/handle/10317/4002/pfc5725.pdf;jsessnid=AE29C449F559A91944F1EEA0D1DFE832?sequence=1

[4] G, Ruíz. *Diseño, construcción y caracterización de un sensor para medir cargas a tensión por extensométrica*. Tesis. Ingeniero en Electrónica. Oaxaca: Universidad Tecnológica de la Mixteca. Facultad de Ingeniería, 2005. 1-108 p. Available in: http://jupiter.utm.mx/~tesis_dig/9676.pdf

[5] A, Rodríguez. *Grafeno: síntesis, propiedades y aplicaciones biomédicas*. Trabajo de fin de grado. Farmacéutico. Madrid: Universidad Complutense. Facultad de Farmacia, 2016. 1-18 p. Available in: http://147.96.70.122/Web/TFG/TFG/Memoria/ALBA%20RODRIGUEZ%20VILLALON.pdf

[6] E, Coutiño & A, Pérez. *Los compuestos de la plata y la salud*. En: Revista Altepepaktli. 2007, Vol. 3, no. 5. Available in: http://132.248.9.34/hevila/Altepepaktli/2007/vol3/no5/5.pdf

IOP Conf. Series: Materials Science and Engineering **844** (2020) 012033 doi:10.1088/1757-899X/844/1/012033

Application of Business Intelligence in studies management of Hazard, Vulnerability and Risk in Cuba

C Milanés-Batista[1a], H Tamayo-Yero[2], D de Oliveira[3], J R Nuñez-Alvarez[4a]

[1]Civil and Environmental Department. Universidad de la Costa, Colombia, ORCID: 0000-0003-2560-8859.
[2]Computer Engineering Department, Universidad de Oriente, Cuba.
[3]Departamento de Electricidad, Universidad Federal de Santa Catarina, Brasil.
[4]Department of Energy, Universidad de la Costa, Colombia, ORCID: 0000-0002-6607-7305.

[a]Corresponding author: jnunez22@cuc.edu.co; cmilanes1@cuc.edu.co

Abstract. In Cuba, the state invests considerable resources in the establishment of preparation plans to mitigate and to minimize the negative impacts of natural threats. As a sample of it, since the year 2010, the country carried out the studies of hazard, vulnerability, and risk (HVR), however, the form in that results of these studies are analyzed present serious limitations for the excessive quantity of data that are dispersed and not very understandable for users belonging to the Centers of Risks' Management make decisions in an agile way. The present work, exposes the management's pattern of the studies of Hazard, Vulnerability and Risk in Cuba alongside the occurrence of hydrometeorological extreme events in vulnerable territories; and the structuring process of the computer proposal that modifies and computerizes the current analysis procedure of these studies in the country. As contributions of the research and by means of the use of techniques and tools of Business Intelligence, is designed and implemented, for the case of Santiago de Cuba province, the data warehouse that centralizes the results of the studies for hydrometeorological extreme events in 2011 and 2016. A decision support system (DSS-HVR) is built, and integrated at the developed data warehouse in this work, it allows the analysis of the studies in a holistic manner and from several perspectives, to obtain outstanding and better-represented information through interactive and dynamic reports, both tabulate and graph representation, all as a support for a fast and effective decision making.
Keywords. Business Intelligence, decision support system, data warehouse, hazard, vulnerability and risk.

1. Introduction

The current increase in the frequency and the destructive force of extreme hydrometeorological events in Cuba, and its main manifestations, such as: coastal flooding due to heavy rains, sea penetrations, and the effects of high winds [1] have contributed to the fact that in Cuba perfects the political, economic, social, scientific-technological and environmental approach to risk management through the implementation of a set of strategies, which are the result of multidisciplinary projects and research, aimed at preventing, reducing and controlling risk factors of disasters, and thus reduce vulnerabilities to the negative impact of these hazards [2].

The Santiago de Cuba province is located south of the Island of Cuba, in direct contact with the Caribbean Sea [3], a position that increases its fragility in the face of extreme hydrometeorological events, which already occur with a high frequency and greater destructive power [4].

Coupled with this situation, the progressive economic development and population growth that exists in towns near the Santiago coast stand out, which is multiplying the risk of affecting economic and social infrastructure and the loss of the most precious of society, "life of man", so it is necessary to establish more efficient preparedness and risk management plans, in order to mitigate and minimize the conditions of vulnerability caused by the main damages.

In 2005, the first Directive No. 1 of the Vice President of the National Defense Council for the organization, planning and preparation of the country in the event of disasters [5] emerges, which is perfected in 2010 [6]. Since then, the Ministry of Science, Technology and Environment (CITMA) was responsible for conducting studies of hazard, vulnerability and risk (HVR) in different types of situations. At the same time, CITMA delegated to the Environmental Agency (AMA) the responsibility for compliance with this directive, with the help of the participation of the country's scientific potential, [1, 7]. These studies allow to identify, measure, quantify, analyze and understand the risks associated with extreme natural phenomena, establish an action model that places great emphasis on preventive aspects, since it is not only about responding to the phenomena, but about anticipating by identifying hazards and their risks, how to manage them, transform them and modify them with the objective of reducing susceptibilities [8]. The studies also offer an overview of the coastal and marine environment and the socio-economic conditions to act proactively in the face of natural disaster risk [1] to, from their results, contribute to the development of disaster reduction plans, management territorial and the design of sustainable economic and social development policies. However, despite the advantages obtained in the implementation of the HVR studies for the province of Santiago de Cuba, the following difficulties can be seen as fundamental:

I. Significant amount of data that is scattered or not accessible. Not all decision makers can access the information resulting from HVR studies in the province. The results were stored in abundant maps, spreadsheets and hard-to-understand text documents for quick and timely decision making;
II. For different actors to access the required information, it must be duplicated, which does not guarantee the reliability and integrity of the data [9]. This may result in documents with conflicting information and does not ensure access by specific levels of users;
III. There is no homogenous format that allows presenting the data properly, nor the ability to analyze them according to the different perspectives and types of danger. Neither does the way in which the information is presented today allows us to analyze the effects that the interactions between different frontal systems would have;
IV. There is no analysis or reporting system that generates useful information for planning, neither does it consider the new information needs of risk managers and decision makers;
V. The information requires systematic updating (through censuses, inventories, frontal systems that have occurred over time, etc.), however, there are currently no information systems in the country that are considered true monitoring systems for the management of HVR studies. The follow-up processes that are carried out are often manual.

The information above justifies the following question: How to contribute to improve the decision-making process in reducing risk to hydrometeorological events from the analysis of the current results of the Danger, Vulnerability and Risk studies?

The current research describes the design and implementation of a decision support system (DS) with support in a data warehouse, which optimizes the processing of the results of HVR studies in the province of Santiago de Cuba before extreme hydro meteorological events, to be integrated for analysis and obtain relevant information with the desired level of detail for a quick and effective decision making in disaster risk reduction.

2. Methodology

Through the review of the methodologies that support the studies of HVR in the face of the exposed hazards and the results collected from the spreadsheets, it was possible to identify which parameters were taken into account to calculate the vulnerability, to the different categories of Hurricane (I, III, V), in each Popular Council of the nine municipalities that make up the province. For high winds and heavy rains, all municipalities were analyzed. In the case of the danger of the sea penetration, only the two coastal municipalities that the study region has are valued, which correspond to the municipalities of Guamá and Santiago de Cuba [3].

Vulnerability is classified according to its type as: structural, non-structural, functional, economic, social and ecological, according to different criteria and expressed from the mathematical point of view as a value limited between zero and one. It takes value $\{1\}$, when the damages are total and, zero $\{0\}$ when these are null [1] Table 1 show intervals for assessment different level of susceptibility used in this research. Table 2 reflects the parameters that define the calculation of vulnerabilities to the hazards of coastal flooding due to sea penetration (1), heavy rains (2) and effects due to high winds (3) respectively.

In the case of coastal flooding due to heavy rains, physical vulnerability is also calculated as the sum of structural, non-structural and functional vulnerabilities [1]. From the results, the total vulnerability (V) is obtained as a sum of the previous ones and the estimated level of susceptibility for each popular council is evaluated by intervals as shown in Table 1.

Considering the value of the danger of a potentially damaging event with a certain intensity, which is estimated through topographic, geomorphological, permeability criteria, among others, the total risk (R) for each hydrometeorological hazard was calculated as:

$$R = P \times \sum V \tag{1}$$

And is valued as shown in Table 3.

To estimate the risk range of the coastal flood hazard due to sea penetration, a comparison of this value is established between the popular councils studied in each municipality or province, the maximum risk value obtained within them is considered for the same intensity of danger and the highest obtained is divided by three to establish the range and breakdown at high, medium and low risk [1]. This procedure is performed for each hazard intensity analyzed.

2.1. Business Intelligence in the management of HVR studies in Cuba.

In the investigation the appropriate techniques and tools proposed by the Business Intelligence (BI) process were used to exploit and analyze structured information on a specific area [10], regularly contained in data stores and which have as Aim to develop perspectives that lead to better and better decision making [11]. Among the features provided by these components are:

I. Accessibility to information: Data is the main source of this concept. It is a priority to guarantee users' access to them regardless of their origin [12];
II. Increase efficiency: With BI all the information can be centralized and visualized on the same platform and converted into useful and organized information, saving time and making decision-making more efficient;
III. Support in decision-making: It seeks to go further in the presentation of information, so that users have access to analysis tools that allow them to select and manipulate only those data that interest them [13].

Table 1. Intervals for the assessment of the level of susceptibility. Source: Modified from AMA 2014

Hazard \ Vulnerability Level	Low	Average	High
Coastal flooding due to heavy rains	0-0.33	0.34-0.66	0.67-1
Coastal flooding by sea penetration	0-0.33	0.34-0.66	0.67-1
Affectations due to high winds	0-0.3	0.4-0.6	0.7-1

Table 2. Definition of the parameters for the calculation of Vulnerability to the different hazards. Source: Modified from AMA 2014.

(a)

Vulnerability Type	Parameters (1)	Parameters (2)	Parameters (3)
Structural	Permeability	Permeability	Construction damage
	Pending	Pending	Housing affectation index
	Ground level	Ground level	Housing quality index
	Location	Location	Construction height index
	-	-	Wooded index and affected buildings
Non- structural	Affectation of transport routes	Affected roads	Access roads blocked
	Sewage system	Drainage network	Aerial power networks and high voltage towers affected
	Aqueduct system	Damaged electrical networks	-
	Other vital lines	Gas lines affected	-
	-	Communication lines affected	-
Functional	Availability of generator sets	Availability of generator sets	Availability of generator sets
	Health system preparation	Health system preparation	Health system preparation
	Accommodation capacity	Accommodation capacity	Accommodation capacity
	Access to isolated areas	Access to isolated areas	Reserve supplies
	Reserve supplies	Reserve supplies	-
Social	Population Affectation	Population Affectation	Population Affectation
	Risk perception	Risk perception	Risk perception

(b)	Population Preparation	Population Preparation	Population Preparation
	Unhealthy Neighborhoods Solid waste	Unhealthy neighborhoods Solid waste	Unhealthy Neighborhoods -
Economic	Budget Execution Level	Budget Execution Level	Budget Execution Level
	Industrial zones	Industrial zones	Industrial zones
	Accounting cost response.	Accounting cost response	Accounting cost response
	Cultivated areas	Cane cultivated areas	Cane cultivated areas
	Animals risk	Cultivated tobacco areas	Cultivated tobacco areas
	-	Coffee cultivated areas	Coffee cultivated areas
	-	Cultivated areas of other crops	Cultivated areas of other crops
	-	Animals at risk	Animals at risk
Ecologic	Ecological sensitive areas	Ecological sensitive areas	Ecological sensitive areas
	Protected Areas Affected	Protected Areas Affected	Protected Areas Affected

Table 3. Intervals for risk level assessment. Source: Modified from AMA 2014

Risk Level / Hazard	Low	Average	High
Coastal flooding due to heavy rains	0-0.33	0.34-0.66	0.67-1
Affectations due to high winds	0-0.11	0.12-0.43	0.44-1

2.2. Use of multidimensional modeling technique for data storage.

As part of the design of the computer proposal, the multidimensional model was used to store the data obtained from the HVR studies carried out. This model is a technique of designing database structures that takes into consideration essential elements of the standard conceptual model "entity-relationship", and which as advantages it offers, simplicity of navigation, survey and allows the rapid analysis of information based on of several indicators to support decision making [14]. It consists of two types of tables: dimension tables, each formed by tuples of attributes of the dimension and a fact table, composed of tuples, one for each recorded fact. This fact contains some variable or variables measured or observed [15]. The model proposes three multidimensional schemes, and because of this research the two employees are exposed: 1) The star scheme, which consists of a central fact table and several dimension tables related to it through their respective keys. This model must be totally denormalized, that is, it cannot be presented in the third normal form (3rd FN) [13] and the 2) constellation scheme: a model composed of a series of star schemes, formed by a main fact table and by one or more tables of auxiliary facts, which may be summaries of the principal. These tables lie in the center of the model and are related to their respective dimension tables [16].

2.3. Data Warehouse usage to support the DS-HVR.

The variety of formats, types of storage and the permanent difficulty in presenting reports derived from the analysis of HVR studies in the province turns necessary the structural transformation that modifies the current process that manages the results. As an option to change, analytical tools and techniques provided by business intelligence were selected, which among its solutions is evidenced the creation of Data Warehouses: data repositories related to the activities of a company and recorded in a database designed specifically to obtain strategic information and support the decision-making process [17]. These repositories also enable the extraction of data from operational systems and external sources, to then be integrated, homogenized and subsequently analyzed by the indicated users, allowing timely, reliable and integrated information [16]. The proposal for the implementation of a data warehouse, made it possible to develop a design focused on responding efficiently to what are the elements, taken from all external sources, which most affect the levels of susceptibility and the likelihood of being affected by exposure to a certain hazard and how to be analyzed from different approaches and in the period determined. The above responds to the fundamental characteristics expressed by [18] about data stores: theme-oriented, integrated, time-variable and non-volatile.

2.4. Data warehousing architecture

In the proposal for the selection of a data warehouse as a solution to the problems raised, the need to build repositories that individually house data related to the parameters and indicators that define the state and behavior of the province in the face of confrontation with each of the mentioned hazards, to later compose the general warehouse on which its subsequent analysis can be carried out in an integrated way. To reach it the Data Mart resource is used, which is no more than the implementation of a Data Warehouse with scope restricted to a functional area, problem, issue or group of needs. According to the operations that are desired or required to be developed, the data mart is expressed through the following architectures: Top-Down: first the warehouse is defined and then the data mart is developed, constructed and loaded from it and Bottom - Up: where the warehouse is built centrally and incrementally over time from the data mart independently developed [19, 20, 21].

The construction of the data warehouse was conducted by a process called Data Warehousing: responsible for extracting, transforming, consolidating, integrating and centralizing the information obtained from operational sources and defining the set of tools necessary to consult, analyze and present the information [16] so that the managing staff does not require advanced computer skills for access.

The stages that are carried out during data warehousing, adapted to the Bottom-Up architecture defined above, are used in the present investigation and consisted primarily of examining and selecting the external sources from which the data resulting from the studies will be obtained. In this case, spreadsheets referring to the two stages 2011 and 2016, to then be extracted, transformed to solve problems of inconsistency between them and guarantee their optimum quality, and once purified they are loaded to the warehouse.

The data warehouse is managed in terms of fact tables: precisely containing fact, numerical values or result of mathematical expressions of high utility. In this context are those referring to the parameters and indicators defined for each hazard; and the tables of dimensions: perspectives for estimating the facts, as they are, by popular councils, municipality, among others; and they are organized around a multidimensional database or cube. It is important to note that these procedures were performed for each hazard and that once completed they are visualized and explored in an integrated manner. Finally, through the Online Analytical Processing (OLAP) operation, the DS-HVR is formed, which allows, through an intuitive graphical interface, that decision makers can generate multidimensional queries to the data and obtain desired reports in real time and in a familiar format that support the decision-making process to minimize the negative impact of these events.

2.5. Methodology and tools used during data warehousing.

The Pentaho open source suite is defined as a BI platform and provides the set of utilities for the construction of data warehouses. Pentaho Data Integration (KETTLE) V5.3 was used, which includes a set of tools, which, in a graphic way, speeds up the process of extracting, transforming and loading (ETL) the data contained in the spreadsheets towards the multidimensional database. Data sources were accessed through visual components and prepared for analysis, relevant transformations were defined by interconnecting blocks and with the adjustment of various functions they were programmed for execution.

Still, Mondrian Schema Workbench V3.8, is the mechanism that from a visual interface provided connection to the defined schemas, corresponding to the behavior of the data for each hazard and that make up the warehouse, to make the multidimensional scheme or cube in terms of dimensions, hierarchies and facts, and subsequently perform the OLAP operation and finally through the BI-server V6.0. The cubes are published, accessible through the browser, and with the use of various add-ons such as Saiku Analytics and Pentaho Audit, among others, the processes of exploration, analysis and information security are improved and pertinent reports are created dynamically in support of the role of risk managers in the event of the different hazards.

The methodology used for the construction of the Data Warehouse was the HEFESTO methodology. It was selected for its many benefits which correspond to the following characteristics: a) The objectives and expected results in each phase are easily distinguished and are simple to understand; b) is based on the user's requirements, so its structure is able to adapt easily and quickly to changes in business and; c) punctually specify the steps to follow in each phase, unlike other methodologies that mention the processes but do not explain how to explain them [16]. Figure 1 shows the steps that guide this methodology and were applied to develop the results that this work presents.

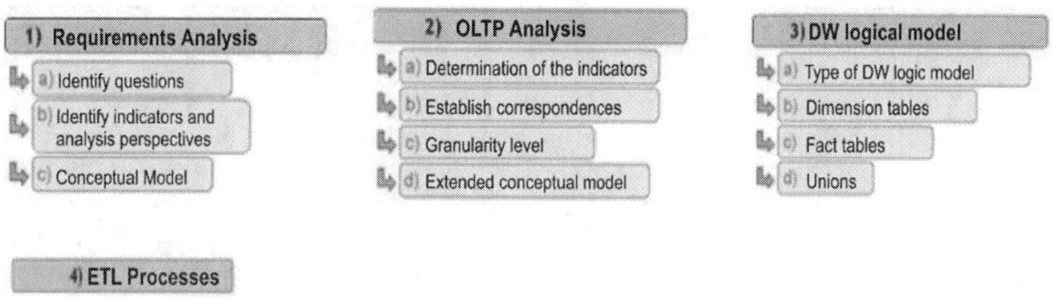

Figure 1. Steps oriented in the Hephaestus methodology. Source [17].

3. Discussion and Results

The spreadsheets from which the results of the HVR studies are obtained have a design that can hardly be interpreted by the exposed analysis tools, so it became necessary to create intermediate sheets in the document itself that respond to the format requirements, making links to the data without duplication.

According to the provisions of the Hephaestus methodology, these documents were examined and the correspondence with the indicators and analysis perspectives that define the scope of the Data Mart that make up the Data Warehouse was determined (Figure 2).

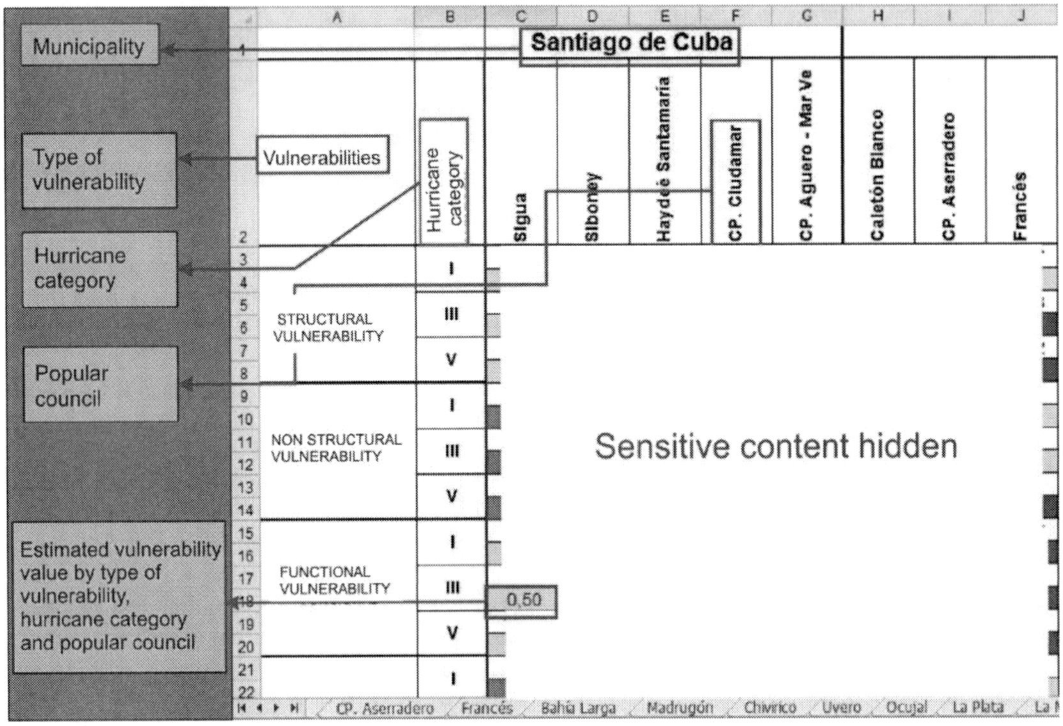

Figure 2. Establishment of correspondences for the control of values by vulnerability, for each type of hazard.

For the correspondence between the data sources and the determination of the indicators and analysis perspectives, the need to form a new secondary fact table for the estimation of the risk associated with the corresponding hazard is established, since it is independent of the type of vulnerability, and therefore the Constellation scheme, described in (2.1), is incorporated into the design of the multidimensional database [22, 23].

To design the data mart that make up the data warehouse, the Star and Constellation schemes were used. The database design is shaped from the logical models defined in Figures 3 and 4.

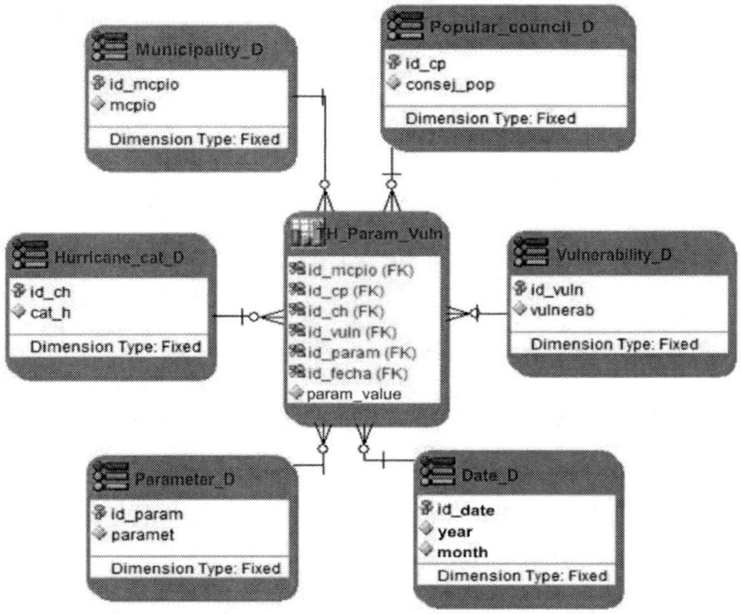

Figure 3. Logical model of the Data Mart for each type of danger.

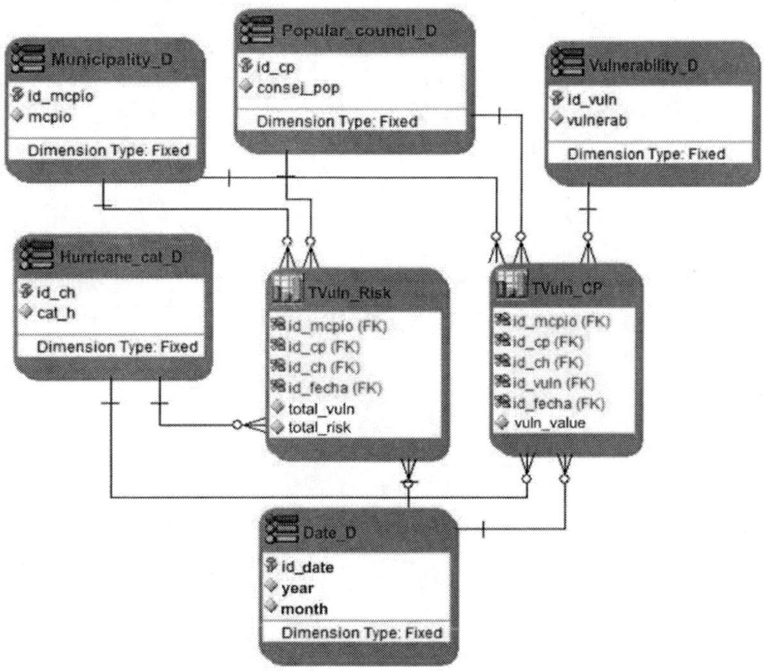

Figure 4. Logical model of the Data Mart for each type of hazard.

Once the multidimensional database was created, the dimension tables and the fact table of the data mart defined for each hazard were loaded through the ETL process (Extraction, Transformation and Load) as part of the premises that BI incorporates to eliminate inconsistency and heterogeneity in the data contained in the operational sources.

Once the database is loaded, the multidimensional cubes are built. The structure of the cubes is carried out in correspondence with the data mart defined, with the aim of representing the flat data found in the rows and columns of the database in a matrix of N dimensions [17], in which for each data mart, the values of the parameters and indicators that act as facts and that are determined from predefined or customized aggregation functions are included for subsequent evaluation criteria, and the analysis perspectives represented by the dimension tables.

Finally, the DS-HVR is formed, accessible through the browser, and in which the created schemes are selected, corresponding to the distribution of the developed data warehouse, and visualized the dimensions, measures or facts that compose them and with only to carry out drag and drop operations of these components within visual containers, the query or report is constructed in a tabular or graphic format dynamically, with improved representation and export options to different formats such as PDF, Excel and PNG.

Within the system, different lists and option buttons are displayed that provide navigability to the interactive and dynamic analysis of the data contained in the warehouse.

Figure 5 shows the multidimensional consultation carried out by the user in tabular format, corresponding to the result of the multi-vulnerability and multi-risk analysis of the popular councils belonging to the coastal municipalities Santiago de Cuba and Guamá, before the hydrometeorological hazards, simultaneously. The query is constructed by dragging and dropping operations, giving the manager the possibility to modify and display their results dynamically as it is structured. Figure 6 shows the behavior in vulnerabilities of the popular councils of the coastal municipalities of the province for the year 2016, specifying the differences before the manifestation of the three dangers in unison. In this way, the changes occurred in the territories before exposure to events and alert about the level of attention they should receive to reduce fragility conditions are verified.

Hazard	High Winds				Heavy Rains				Sea Penetration			
Category	III				III				III			
Year	2011		2016		2011		2016		2011		2016	
Popular council	Vulnerability	Risk	Vulnerability	Risk	Vulnerability	Risk	Vulnerability	Risk	Vulnerability	Risk	Vulnerability	Risk
Aserradero												
Bahía Larga												
Caletón Blanco												
Chivirico												
El Francés												
La Magdalena												
La Plata												
Madrugón												
Ocujal												
Uvero												
Aguero - Mar Verde												
Ciudamar												
Haidee Santamaría												
José Martí Norte												
Siboney												
Sigua												

Sensitive content hidden

Figure 5. DS-HVR. Tabular Report obtained from multi-vulnerability and multi-risk analysis in coastal municipalities of the Santiago de Cuba province (2011 and 2016).

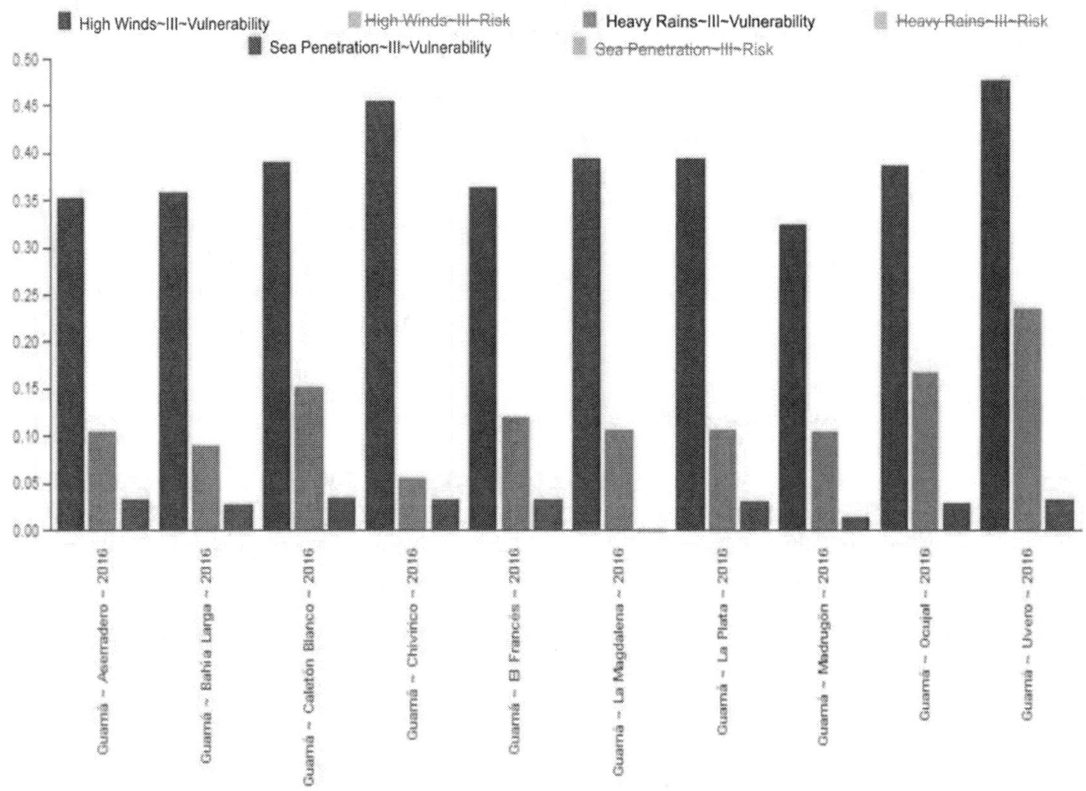

Figure 6. DS-HVR. Graphic Report on Bars of multi-vulnerability analysis in coastal municipalities of the province (2016).

3.1 Information security management from DS-HVR.

Access to the information contained in multidimensional cubes can be restricted based on the different permissions granted to each role defined in the system. So, users belonging to a specific role will have specific access to content defined only for their level.

From DS-HVR it is possible to configure this functionality at the same time as it is enhanced with the inclusion of the Audit V2.0 add-on, to manage the entire record of activities occurred within the platform. It means that the user in charge of managing the contents, may Verify all the historical actions carried out by all the users that have accessed the system, guaranteeing the correct operation and security in the management of the stored information, given its vital importance. The present research takes into account the analysis of threat, vulnerabilities and risks and data bases to consider in the study area [3,24,25].

4. Conclusions

The Hazard, Vulnerability and Risk (HVR) studies before hydrometeorological events, are among the first of their kind elaborated and implemented in Cuba. Because of their importance, they constitute instruments for disaster risk management and timely decision making for governments at their different levels of action.

In this article we present a large-scale model that optimizes and computerizes the management of HVR studies. With the result achieved, all the hydrometeorological hazards information is centralized in a single database, which was designed in a multidimensional way when evaluating historical trends in a comparative way. The decision support system (DS-HVR) is formed, which, integrated into this database, allows studies to be analyzed in an integrated way and from various perspectives, to obtain relevant and better represented information through dynamic reports, which guarantees fast and effective decision making.

Our work also includes the following contributions:

- It allows obtaining information through interactive consultations with the desired level of detail, performing simple operations without this implying to have advanced programming or database knowledge;
- It provides data representation in 11 types of graphics that provide dynamism, flexibility and high power of intuition, with style options and real-time filtering, adapting easily to the information needs of decision makers;
- It is immediately accessible through the network, which allows the information to be available to several users at the same time;
- It guarantees information security, through access restricted by specific roles where only the actors with defined permissions can access the information.

In future work we intend to:

- Extend the data set integrating other hazards of natural origin such as seismic threats, severe drought, forest fires, technological threats, among others;
- Develop appropriate probabilistic and statistical models, through the use of specialized computer tools such as Weka, CAPRA and CRISIS, which allow estimating, quantifying or predicting disaster risk in objective terms, given the relevance of these aspects as necessary components to be incorporated into territorial planning processes to face the impact of these hazards;
- Build a geospatial module integrated to the platform that allows representing the indicators highlighted by type of threat, spatially distributed through different shades of color, where the intensity gradation expresses the intervals of the indicator in territorial units.

Acknowledgments

The authors like to thank the Master's program in Integrated Coastal Zone Management coordinated by CEMZOC at the Universidad de Oriente, Cuba, as well as the project Project Index No. INV.1106-01-007-12 named "Resilient cities: minimization of vulnerabilities to phenomena extreme weather and climatic changes in coastal communities".

References

[1] Serrano J, Pedroso I and Pérez O. 2014. Metodologías para la determinación de riesgos de desastres a nivel territorial. Parte 1. Grupo de Evaluación de Riesgo de la Agencia de Medio Ambiente del Ministerio de Ciencia, Tecnología y Medio Ambiente (CITMA). Cuba. pp- 1-67

[2] Botero, C and Milanés C. 2015. *Aportes para la gobernanza marino-costera. Gestión del riesgo, gobernabilidad y distritos costeros*. Libro de investigación. Fondo de publicaciones de la Universidad Sergio Arboleda, Bogotá, Colombia. ISBN: 978-958-8866-67-3. ISBN: 978-958-8866-68-0. 554 p

[3] Milanés C, Suárez A and Botero C. 2017. Novel method to delimitate and demarcate coastal zone boundaries. *Journal Ocean and Coastal Management*. 144 (2017) 105-119p.http://dx.doi.org/10.1016/j.ocecoaman.2017.04.021
[4] Núñez Alvarez, José Ricardo et al. 2019. Metodología de diagnóstico de fallos para sistemas fotovoltaicos de conexión a red. *Revista Iberoamericana de Automática e Informática industrial*, [S.l.]. ISSN 1697-7920. Disponible en: <https://polipapers.upv.es/index.php/RIAI/article/view/11449>
[5] Pérez Montero y Milanés B Celene. Social perception of coastal risk in the face of hurricanes in the southeastern region of Cuba. Ocean & Coastal Management. Available online 10 October 2019, 105010. https://doi.org/10.1016/j.ocecoaman.2019.105010
[6] Planas, F.J.A., Milanés, B.C., Fanning, L.M. and Botero, C.M. (2016) Validating Governance Performance Indicators for Integrated Coastal and Ocean Management in the Southeast Region of Cuba. *Open Journal of Marine Science*, 6, 49-65. doi: DOI: 10.4236/ojms.2016.61006
[7] Milanés Batista, Celene; Galbán Rodríguez, Liber y Olaya Coronado,Nadia J. 2017. *Amenazas, riesgos y desastres: Visión teórico-metodológico y experiencias reales*. Libro de investigación. 306 p. ISBN: 987-958-8921-44-0 (Digital). Disponible en http://repositorio.cuc.edu.co/xmlui/handle/11323/927
[8] Milanés C, Pereira C and Botero C. 2019. Improving a decree law about coastal zone management in a small island developing state: The case of Cuba. *Marine Policy*. doi.org/10.1016/j.marpol.2018.12.030
[9] Devece C. A, Lapiedra R and Guiral H 2011. Introducción a la gestión de sistemas de información en la empresa. Universitat Jaume I. pp 7-10
[10] Lluís Cano J. 2008. Business Inteligence: competir con información, Esade. Depósito Legal: M-41185-2007. pp. 21-37
[11] Kopáčková H and Škrobáčková M. 2006. Decision support systems or business intelligence: what can help in decision making. Scientific Papers of the University of Pardubice. Series D, Faculty of Economics and Administration, 10
[12] Huamantumba R. 2007. Manual para diseño y desarrollo de Datamart
[13] Calzada L and Abreu J. L. 2009. El impacto de las herramientas de inteligencia de negocios en la toma de decisiones de los ejecutivos. *International Journal of Good Conscience*. 4(2). pp. 17-52
[14] López N. M, Vela J. P and Mondejar J. C. 2010. Diseño y explotación de almacenes de datos: Conceptos básicos de modelado multidimensional. Editorial Club Universitario, pp 28-29
[15] Kimball R. 1997. A dimensional modeling manifestó. *Dbms*. 10(9), 58-70
[16] Dario B. R. 2009. Data Warehousing: Investigación y Sistematización de Conceptos–HEFESTO: Metodología propia para la Construcción de un Data Warehouse. Licencia de Documentación Libre de GNU, Versión 1
[17] López C. P 2007. Minería de datos: técnicas y herramientas. Editorial Paraninfo. pp. 74-77
[18] Inmon W. H. 2005. Building the data warehouse. John wiley & sons. pp. 31-32
[19] Oketunji T and Omodara O. 2011. Design of Data Warehouse and Business Intelligence System: A case study of Retail Industry. pp. 12-13, 21-22
[20] Barrio J, Abreu H, Pina I and Alvarez J. 2018. SmartGrid proposal in communities of Guamá Municipality of Santiago de Cuba Province. Journal of Engineering and Technology for Industry Applications, 4(14), 66-74. https://doi.org/10.5935/2447-0228.201831
[21] Valencia G, Nuñez J and Acevedo C. 2019. Research Evolution on Renewable Energies Resources from 2007 to 2017: A Comparative Study on Solar, Geothermal, Wind and Biomass Energy. International Journal of Energy Economics and Policy. 9(6), 242-253. DOI: https://doi.org/10.32479/ijeep.8051

IOP Conf. Series: Materials Science and Engineering **844** (2020) 012033 doi:10.1088/1757-899X/844/1/012033

[22] Botero C. M, Arrizabalaga M, Milanés C and Vivas O. 2017. Indicadores de gobernabilidad para la gestión del riesgo costero en Colombia. *Revista Luna Azul*, 45, 227-251. DOI: 10.17151/luaz.2017.45.12 Disponible en: http://200.21.104.25/lunazul/downloads/Lunazul45_12.pdf

[23] Pereira C. I, Carvajal A. F, Milanés Batista C and Botero C. M. 2019. Regulating human interventions in Colombian coastal areas: Implications for the environmental licensing procedure in middle-income countries. Environmental Impact Assessment Review 79,106284. https://doi.org/10.1016/j.eiar.2019.106284

[24] Milanés B, Celene., 2012 Unidades costeras ambientales para el manejo en Santiago de Cuba: delimitación y prioridades de actuación. Revista Arquitectura y Urbanismo. Volumen XXXIII No. 3. p. 83-97. Consultar en http://revistascientificas.cujae.edu.cu/Revistas/Arquitectura/Vol-XXXIII/3-2012/08_au_03_2012.pdf

[25] Milanés B, Celene y Pacheco M. Alicia., 2011 Asentamientos costeros en la bahía de Santiago de Cuba: estudio de su vulnerabilidad urbana. Revista Arquitectura y Urbanismo. Vol. XXXII, No 3. 18-26 pp. Disponible en http://es.scribd.com/doc/86054921/Revista-Arquitectura-y-Urbanismo-3-2011

Characterization of the current electric energy needs in subsidized residential users

S F Umaña-Ibáñez[1], J I Silva-Ortega[2], Y Gomez-Charris[1], K M Berdugo-Sarmiento[2] and S J Carbonell-Navarro[2]

[1] Universidad de la Costa CUC, Facultad de Ingeniería, Departamento de Gestión Industrial, Agroindustrial y Operaciones. Grupo de Investigación PRODUCOM. Calle 58 N° 55-66, Colombia.

[2] Universidad de la Costa CUC, Facultad de Ingeniería, Departamento de Energía. Grupo de Investigación GIOPEN. Calle 58 N° 55-66, Colombia.

Sumana1@cuc.edu.co

Abstract. In this paper it is considered the characterization and behaviour of the residential energy demand of users located in socioeconomic sectors of the Department of Atlántico, also, the participation of communities to declare the typical energy consumption at the residential level. The results are based on direct measurement sources which allows the development of energy efficiency projects, renewable energies that respond to the needs of the communities and seek to encourage the proliferation of renewable energy projects by adequately sizing the installed capacities of the resources, with the possibility of establishing grid connection as well as systems for energy management in residential users.

1. Introduction

In the different world debates on sustainable development and environmental care, influential aspects are highlighted such as the generation and consumption of energy in countries due to the accessibility, quality, technological level and state of the infrastructure in a nation with regard to power, has direct effects on the quality of the inhabitants and the productive processes of the same [1].

As a result, different countries have studied the trend and relationship of variables such as the development of a nation and the generation and distribution of energy, an example of which is Turkey, where studies show that an effect of the use of renewable energy helps to supply the demand for electricity, which has increased annually by 6% and 8% in recent years and also includes the need for a sustainable energy policy to meet the demand and use of Renewable Sources (RS) [2]. Another country that targets the use of RS is the Republic of Azerbaijan, a nation whose economy is significantly based on oil production, demonstrating that the transition to renewable energy and sustainable energy development diversifies its economy through a green economy by addressing the environmental and social concerns of that country.

Vidadili, Suleymanov, Bulut, y Mahmudlu [3], Japan, for its part, is studying the inclusion of RF in the generation of energy through the cultural development of a country. Shortall y Kharrazi (2017), some African countries, promotes the financing of renewable energies considering that it is a crucial challenge of the sustainable development objectives of this continent, Nigeria, analyzes the barriers to sustainable energy development, where it is highlighted in costs and prices, legal and regulatory and market performance, where it was identified that government policies have an impact on overcoming these barriers [4]. In general, the development of studies for the inclusion of RF in a country's power generation has been of global relevance [6].

On the other hand, the trends in economic models that examine the behavior of energy demand are structured in two large blocks, the demand management model or "demand-driven," which emphasizes the concepts of supply and demand. The operating principle of this model applied to the electricity sector encourages demand to schedule its consumption during the day to reduce its energy costs. On the other hand, the supply-driven model is based on the control of generation prices and the search for alternatives to cut them, through the participation of renewable and non-conventional resources.

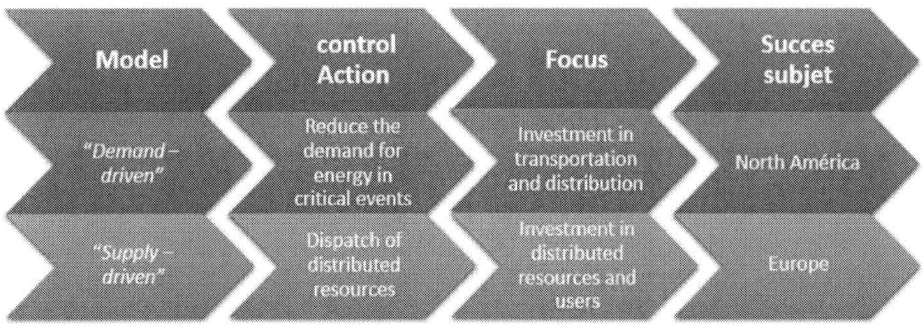

Figure 1. Models of operation and control of an electrical network in general.

The demand management model is based on controlling demand through the application of efficient energy management techniques to obtain appropriate and rational use of available resources, which has led to the inclusion of intelligent devices in the electricity network. As an initial measure in the 1970s, the "virtual utilities" proposed as computing and consumption management techniques are shown, which are capable of generating control overloads and focus on reducing maximum levels during critical hours. To solve the problems of high consumption that began with the proliferation of electronic technology, communications, the presence of household appliances and equipment in residential and administrative buildings (FEDIT & Electrónica, Tecnologías de la información y Telecomunicaciones, 2011).

The successful inclusion of the model is attributed by many researchers to the maturity of the capacity to respond to demand behavior [8]. At the same time, the inclusion of technologies based on original concepts and the Internet of Things (IOT) has made it possible to guarantee the control of consumption by end users.

According to Giordano and others [9], the energy policies of the United States of America encourage investments that improve the grid operation scheme to a value of 476 billion US dollars before 2030, where 24% will be directed to energy transmission, and substations, 71% to distribution and the rest of the funds will be allocated to users. The North American electricity system is focused on supply reliability, so more than eight million smart meters will be installed by network operators in the United States (Giordano et al., 2011; Physical Measurement Laboratory, Office of the National Coordinator for Smart Grid Interoperability, & Information Technology Laboratory, 2012). Among the projects aimed at demand management are the objectives of the EPRI project (Electrical Energy Research Institute), which considers the inclusion of the concept of a virtual plant for remote user dispatches on the North American electricity grid [11].

On the other hand, supply-driven models consider the use of renewable energies as one of the fields of action [9], [12]. This model encourages the development of research into non-conventional sources, with emphasis on research into the control of microgrids, distributed energy resources, evaluation of energy potentials and high penetration participation from renewable energies. The monitoring of energy storage technologies is contemplated with the objective of reducing the intermittency of non-conventional sources to guarantee reliability, new technologies are considered on an individual basis, and intelligent devices are contemplated to allow connection to the network of users [9], [13].

However, they do not focus on solutions that strengthen the power grid, and it is possible that the high penetration of new technologies in automation will generate a new approach as mentioned in [14].

Additionally, the scheme considers projects in the areas of smart metering, grid automation in energy transport, integrated systems, application to end users and development of energy storage technologies.

Some researchers believe that within ten years, the penetration of new technology will be underway due to the increase in current automation trends at the residential and industrial levels, which will allow the rapid deployment of technology related to the management and uses of electricity.

It is essential that structured incentives be put in place to replace current business opposition and to boost user programmers. These technologies will create the potential to radically change the electrical system in a supply-centered order to a highly reliable "perfect" demand-driven scheme (p. 160). As presented above, there are different trends and models for operating an electricity system. However, they agree that the inclusion of new intelligent tools and technologies will favor the integration of small-scale generation and user management.

This is evidenced by the outstanding work and developments in the areas of market models [15], [16], cogeneration technologies [17], renewable energies [18], smart metering [19], energy storage [20] and a set of topics related to power quality [21].

Advances in China, North Korea, India, and Japan are considering schemes focused on tracking demand growth. This model requires substantial investments in the grid and control to transfer high power capacities through a robust interconnected system, which is integrated with the energy policies established to complement the implementation of the model:

- China: China's electricity system is less developed compared to European and North American operations. However, China's economic growth has led it to integrate different infrastructures in intelligent networks [22]. So China has focused on including 360 million smart meters in its transportation and distribution systems by the end of the year. 2030 [10].
- South Korea: Model has additional commitment to mitigate greenhouse gas emissions [10].
- Japan: Establishing new technologies due to its crisis with the Fukushima nuclear power plant [10].
- India: India's electricity system scheme focuses on mitigating losses and controlling energy theft [10], [23].
- South American countries: Given the emerging economy in South America whose case tends to be similar to that of India, the models have been based on reducing energy theft. Brazil is the country with the most significant development in this area, and its focus is directed towards the replacement of conventional meters by intelligent systems. [24].

At present, the energy demand on the Atlántico coast between regulated and unregulated market is between 107.47 kWh and 205.81 kWh in 2017 with an unmet need of 20.13 GWh, which implies the use of non-conventional energy sources [25]–[27], in this way, Colombia's energy policy is directed towards satisfying the needs of economic agents and the population, making use of available resources with criteria of economic, social and environmental sustainability [28], [29]. For this reason, the National Energy Plan (PEN) considers the ecological component as a strategy that cuts across all its objectives: the conservation and improvement of environmental quality at all decision-making levels, production processes and future investments in the energy sector are primary objectives of the national energy strategy.

2. Materials and Methods

The aim of this study is to evaluate the behavior of the residential consumption of socioeconomic level 1, 2 and 3 in the different municipalities of the Department of Atlántico to determine the current energy needs of the communities and to propose the formulation of alternative energy and energy efficiency projects. In this way serving research as a reference to replicate similar studies in other regions looking to consolidate a national map showing consumption habits of its inhabitants.

IOP Conf. Series: Materials Science and Engineering **844** (2020) 012034 doi:10.1088/1757-899X/844/1/012034

2.1. Estimated consumption
Consumption estimates are made through surveys that precisely determine the daily consumption of the population per dwelling. Therefore, a community was chosen 442,341 customers, taking into account a 5% margin of error and a 95% confidence level. The survey is carried out with the support of the Association of Municipalities of the Department of Atlántico ASOATLANTICO and the community action boards 50% of the population.

It was carried out in person (50%) and by telephone (50%) during September 2018 for standardized users with individual measurement of ELECTRICARIBE S.A. E.S.P. and ENERGÍA SOCIAL DE LA COSTA. The survey asked about the inventory of electrical equipment and household appliances and their hourly use and asked about the consumption of the last three months recorded in the invoices of energy. It was considered different socioeconomic levels: level 1 (48,657), level 2 (203,476), level 3 (115,008), level 4 (61927), level 5 (4423), level 6 (8850).

2.2. Characterization of energy consumption in the municipalities of the Department of Atlántico
In the second measure, the average residential energy consumption of socioeconomic level 1, 2, and three were identified, where the consumption habits and trends by area and socioeconomic level were determined. For this purpose, the data of the Superintendence of Public Domiciliary Services is taken into account, which indicates that in Colombia the number of residential users corresponds to 89%, who consume 11% of the country's total electrical energy, with the Department of Atlántico representing 5% of residential and non-residential users [30].

Table 1. Study characteristics in the Department of Atlántico

Description	Studied item
Public Service to which you belong	Energy
Region covered	Interconnected Zone
Time range	2010 - 2018
Entities providing information	Utility companies with marketing activity registered with the Single Register of RUPS Service Providers.

3. Results
Initially, differences of up to 30% were found between the departmental average of consumption for residential level 1, 2 and three published by SSPD [30]. This is because the survey does not include elements causing technical or non-technical losses of the low-voltage network, nor does it include small commercial premises associated with the residential sector or public lighting that may be registered by these same community systems. [32]. The values corresponding to the consumption of the departmental average (kWh month) for the volumetric energy according to the survey is 193.22, for the reported power it is 248.90 according to the SSPD Superintendency.

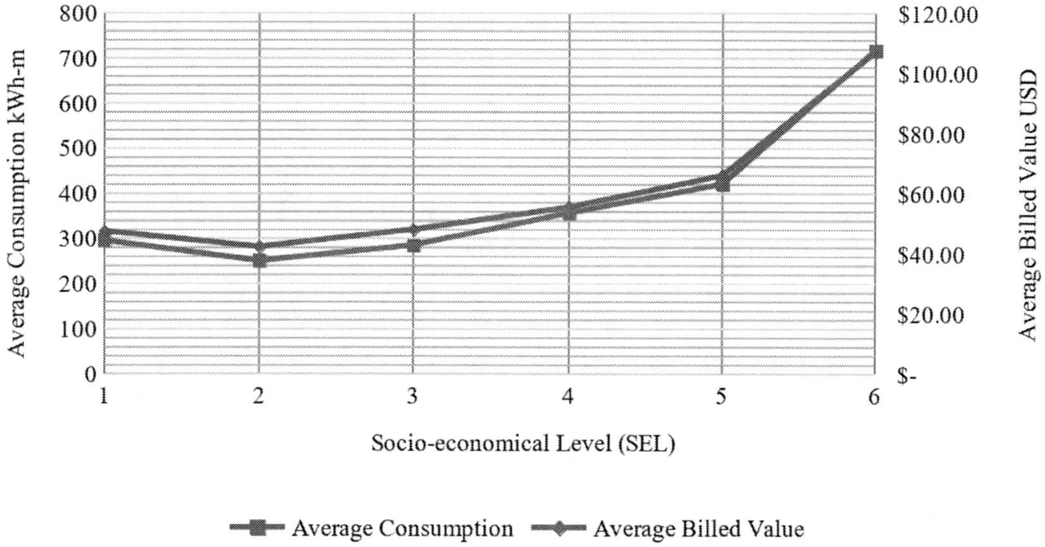

Figure 2. Average Consumption Vs Average Billed in the Department of Atlántico

On the other hand, energy consumption is directly related to the socioeconomic situation of the inhabitants as shown in Figure 2. In this case, the higher consumers were identified in the urban district that make up the metropolitan area.

The other municipalities have lower consumption rates and have an economy with an agricultural or livestock tradition. Another important indicator is the average per capita consumption. In the case of the department, according to the survey is located at 42.64 kWh month, while in the studies of energy consumption published by the Ministry of Mines and Energy which characterizes energy consumption in the residential, commercial and tertiary sectors [33] it is established a range of consumption per capita of 40 kWhm for stratum 1, 38 kWhm for stratum 2 and 62 kWhm for stratum 3. The average consumption for the Barranquilla district is 43.74 kWhm.

Figure 3 describes the Growth in users and energy consumption in the Department of Atlántico 2010-2018. The average energy consumption of homes in the socioeconomic department considering the average typical use among the three socioeconomic level. You identify yourself The levels of energy consumption in the Department of Atlántico are directly related to the socioeconomic level of the inhabitants and the proximity they have to the capital of the department. The city of Barranquilla and its surrounding municipalities such as Soledad, Puerto Colombia, Galapa, and Malambo, which make up the metropolitan area of the department, represent a higher energy consumption due to their population and economic activities developed in them; the municipalities bordering free trade and industrial zones represent an average energy consumption. The cities with less waste are far from the industrial zones and have a more significant population and connected users; these have an agricultural, fishing and livestock economy.

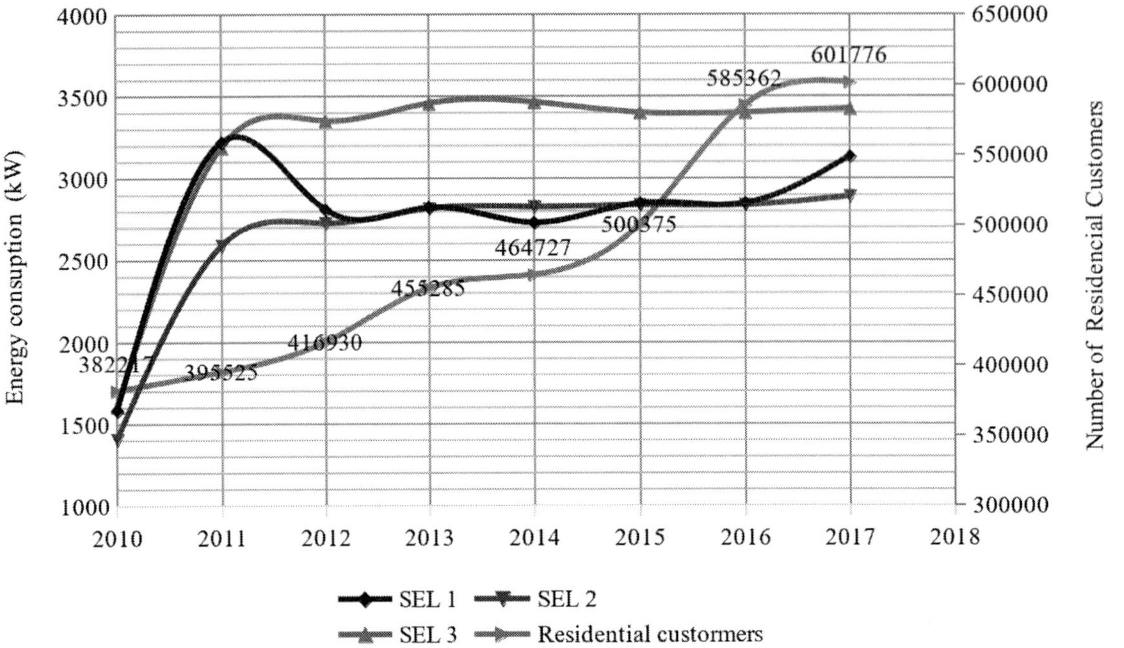

Figure 3. Average annual consumption in the Department of Atlántico by socio-economic level (SEL) 2010-2018

In Figure 4, it can be seen that the highest consumption is from 5 pm to 10 pm, while the lowest consumption is from 12 m to 10 am, so from 10 am to 5 pm there is an average consumption. On Saturdays, Sundays and public holidays there may be a 20% increase. The consumption habits surveyed show that the lowest demand is between 4 am and 10 am, while in weekend, the energy consumption can increase by 20%.

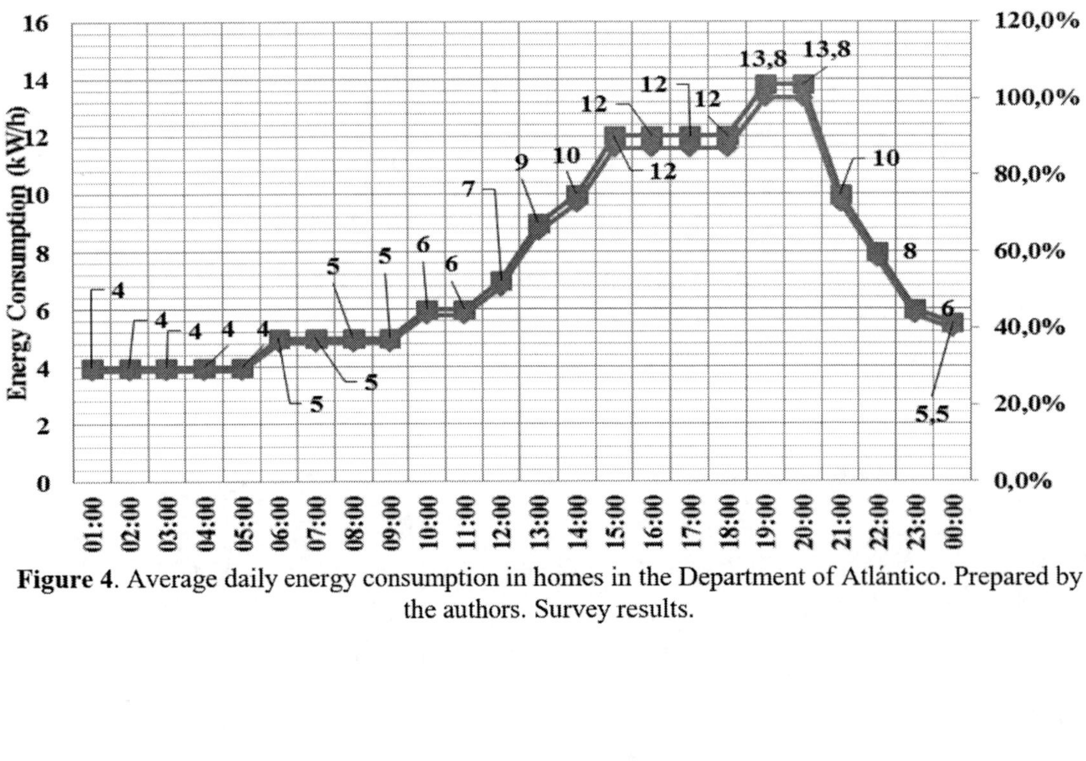

Figure 4. Average daily energy consumption in homes in the Department of Atlántico. Prepared by the authors. Survey results.

4. Discussion

The use of electricity for residential uses in rural and urban areas increases related to technological development and modern needs. According to the Colombian statistics established in by UPME, residential energy demand in Colombia is typified at 44%, characterizing the country with a residential consumption activity. However, utilities in Colombia, as mixed-economy (private-public) entities within the energy distribution activities, use to focus their strategies for customer service such as unregulated users, those users with an installed capacity greater than 0.1 MW, being then, the highest economic incomes recorded in this type of user. For other hand residential users are often regulated, who do not exceed this installed capacity and whose individual consumption does not represent a strong income for utilities.

Also, in rural areas, energy needs and consumption are lower, and these types of users are less attractive to utilities and energy retailers. However, according to Law 142 of 1992, public services must be guaranteed by the state and their supply quality by utilities is a responsibility to accomplish by rural users. However, rural projects electrification often considers sub-normal, or non-standard users which difficult electricity connection, which difficult to connect them to the available electricity networks. For another part, the nation usually supports these projects through standardization electricity network projects and funds for the integration of non-interconnected areas with the purpose to guaranteeing connection to final users and guaranteeing access to the networks as is established in current regulations.

On the other hand, subnormal and non-standard users do not have payment culture and often have negative experiences with quality of electricity supply in the department, making difficult the utilities activities such as energy reads, repair electrical grid problems and changes of technologies. At the same time, in many cases they do not have the internal electrical installations able to connect to the electrical power system, causing electrical failures inside homes, which compromises the security of the electrical circuit supply to which they belong and affecting other users.

The above suggests that electricity connection projects for rural and urban users in socioeconomic strata 1, 2 and 3 should have a management and direction plan based on ensuring proper electrical connections and access to the electricity networks in a regulated and organized manner. The development of rural communities and the overcoming of conditions of extreme poverty begin with actions that promote the proper management of public services to which communities have access as the fundamental axis of urban and rural development.

With the opening of Law 1715 of 2014 and the new models of commercialization for renewable energies, there are opportunities to guarantee that users can access renewable energy projects as an alternative for self-sustainability and consumption management. However, these projects must consider the joint participation of the state, utilities and sector-based committees to provide an integrated service that satisfies the needs of the communities, which is why this work is a reference point to identify the growth and monitoring of consumption of rural and subnormal communities in the department of Atlántico

5. Conclusion

The economic models established to achieve sustainable energy development that has been validated and are currently has been implementing can have been classified into two categories, an energy efficiency-based model, demand management models, and one based on the use of renewable energy, supply-driven model. Demand management proposes a robust model capable of supplying and meeting all the needs of its users with high efficiency and the reduction of demand in peak hours; Similarly, the supply drive model proposes a decentralized system where users who are capable of generating energy to meet their needs with renewable sources, however, although it contemplates an interconnection system, its primary objective is not to reduce losses or guarantee operating conditions that optimize the system; these models focus on reliability in the same way that they stimulate the development of new technologies and methodologies.

Colombia is developing a renewable energy inclusion model over a period of no more than 25 years, evaluating the probable fluctuations of a generation with renewable energy sources, with the assumption that the country's Caribbean coast can produce 21GW, which would supply twice the country's current demand. In the case study, the conditions for solar and wind farms in the department of Atlántico are evaluated, while establishing the consumption patterns of the population, with the result that stratum 1 in most cases is the cause of the highest consumption that occurs from 5 pm to 10 pm, while the lowest consumption occurs from 12 m to 10 am, so that from 10 am to 5 pm an average consumption occurs.

On Saturdays, Sundays and public holidays there can be an increase of 20% in the municipalities, of which the district and its metropolitan area represent the most considerable consumption, as evidenced by the results of the study, transversally the potential of the department for wind and solar exploitation is good guilty.

References

[1] Livas-García A 2015 Análisis de insumo-producto de energía y observaciones sobre el desarrollo sustentable, caso mexicano 1970-2010 *Ing. Investig. y Tecnol.* Vol **16** no 2, pp. 239–251
[2] Kok B and Benli B 2017 Energy diversity and nuclear energy for sustainable development in Turkey *Renew. Energy* vol **111** pp 870–877
[3] Vidadili N Suleymanov E Bulut C and Mahmudlu C 2017 Transition to renewable energy and sustainable energy development in Azerbaijan *Renewable and Sustainable Energy Reviews* vol **80** pp 1153–1161
[4] Shortall R and Kharrazi A 2017 Cultural factors of sustainable energy development: A case study of geothermal energy in Iceland and Japan *Renewable and Sustainable Energy Reviews* vol **79** pp 101–109
[5] Edomah N 2016 On the path to sustainability: Key issues on Nigeria's sustainable energy development *Energy Reports* vol **2** pp 28–34
[6] Caspary G 2009 Gauging the future competitiveness of renewable energy in Colombia *Energy Econ.* vol 31 no **3** pp 443–449
[7] FEDIT and Electrónica Tecnologías de la información y Telecomunicaciones 2011 Smart Grids y la Evolución de la Red Eléctrica FEDIT p 82
[8] Kranz E Broadband Networks 2013 Smart Grids and Climate Change. New York: *Springer*
[9] Giordano V et al 2011 Smart Grid projects in Europe : lessons learned and current developments European U.
[10] Physical Measurement Laboratory Office of the National Coordinator for Smart Grid Interoperability and Information Tecnology Laboratory 2012 NIST Framework and Roadmap for Smart Grid Interoperability Standards
[11] EPRI Smart grid demostration initiative final update 2014.
[12] Kieny C Berseneff B, Hadjsaid N Besanger Y and Maire J 2009 On the concept and the interest of Virtual Power plant : some results from the European project pp 1–6
[13] Awerbuch S and Preston A 1997 The Virtual Utility: Accounting, Technology & Competitive Aspects of the Emerging Industry 1st ed New York: *Springer US*
[14] Gellings C 2014 The Smart Grid: Enabling Energy Efficiency and Demand Response
[15] Hatziargyriou N 2014 Microgrids: Architectures and Control
[16] Chakrabortty A and Ilić M D 2012 Control and Optimization Methods for Electric Smart Grids, 1st ed New York: *Springer-Verlag*
[17] Kaestle G 2006 Virtual Power Plants as Real CHP-Clusters: A New Approach To Coordinate in *Energy 2nd International Conference on Integration of Renewable and Distributed Resources* 2006.
[18] Zare M and Niknam T 2013 A new multi-objective for environmental and economic management of Volt / Var Control considering renewable energy resources *Energy* vol **55** pp 236–252
[19] Aghaei J and Alizadeh M I 2013 Demand response in smart electricity grids equipped with renewable energy sources: A review *Renew. Sustain. Energy Rev* vol **18** pp 64–72
[20] Ausubel L M and Cramton PN 2010Virtual power plant auctions vol **18**

[21] Carmeli M S Dezza F C Faranda R Marchegiani G and Mauri M 2006 Universal Digital Controller for Power Quality and Distributed Generation Systems pp 17–22
[22] International Energy Agency 2011 Technology Roadmap Smart Grids Paris
[23] MIT 20101 The Future of the Electric Grid: An Interdisciplinary MIT Study
[24] Guasch Murillo D 2003 Modelado y análisis de sistemas fotovoltaicos p 223
[25] UPME 2017 Mercado regulado y no regulado
[26] Castro-Ospino A 2010 Análisis del potencial energético solar en la Región Caribe para el diseño de un sistema fotovoltaico *Inge Cuc*, vol **6** no 6 pp. 0–8
[27] Silva-Ortega J I Castro-Ospino A and Balbis-Morejon M 2013 Set of Elements, Parameters and considerations to get the successful inclusion of the Smart Grids in Colombian power systems in IMETI 2013 - *6th International Multi-Conference on Engineering and Technological Innovation* pp 166–169
[28] Ojeda Camargo R Candelo J E and J. Silva-Ortega J I 2017 Perspectives of Native Community in La Guajira Facing Sustainable Development and Energy Supply *Rev Espac* vol **38** no 11 pp 25–32
[29] Ojeda Camargo E, Candelo Becerra J E, and Silva-Ortega J I 2017 Solar and wind energy potential characterization to integrate sustainable projects in native communities in La Guajira Colombia *Espacios* vol **38** no 37 pp. 11–27
[30] SUI 2018 Consolidado de energía por empresa y departamento
[31] González J M Daza C A D and Ureña C H G 2008 Análisis del esquema de generación distribuida como una opción para el sistema elécctrico colombiano *Rev. Fac. Ing.* no. 44 pp 97–110
[32] M de M y Energía CREG 1994 (Comisión Reguladora de Energía y Gas Resolución 108 de la Ley 142 de 1994) pp 1–36
[33] Ministerio de Minas y Energia 2006 Caracterización energética sectores residencial, comercial y terciario Colombia p 135

IOP Conf. Series: Materials Science and Engineering **844** (2020) 012035 doi:10.1088/1757-899X/844/1/012035

Automatic detection of number-plate and traffic infractions of motorcyclists by Intelligent Transportation Systems

J F Valencia[1], T Ramírez-Guerrero[1], L F Castañeda[1] and M Toro[2]

[1] Mechanical Engineering Department, GEMI Research Group, Universidad EAFIT, Medellín, Colombia

[2] Informatics and Systems Department, GEMI Research Group, Universidad EAFIT, Medellín, Colombia

E-mail: inv_jvalencia@eafit.edu.co

Abstract. Public Transportation Strategic Systems are a digital transformation for the urban transportation service in the Colombian medium-size cities, from 200,000 to 600,000 inhabitants. The implementation of the urban public transportation services has not started in many cities; as a result, this cities in Colombia have an increase of illegal motorcycle taxis, which cause a high number of road accidents. The main method for controlling illegal motorcycle taxis is by giving the drivers transit fines. In order to give such transit fines, transit police have several control points in medium-size cities. In this article, we developed a mobile application to detect three types of transit infractions usually committed by motorcycle taxi drivers. The purpose of this application is to be a tool for the transit police agents to disincentivize the use of motorcycle taxis. The mobile application detects a possible transit infraction, saves the photo, motorcycle plate number, date, and time. Afterwards, the application generates a support for the transit infractions that will be issued by the transit police officer. We performed an experiment using this mobile application in the city of Valledupar, Colombia. The mobile application captured 40 possible transit infractions.

1. Introduction

The lack of a public transport system that meets quality requirements –such as, efficiency, coverage, accessibility, and sustainability– is one of the main reasons for the rise of illegal transport providers in medium-size cities in Colombia [1]. Gwilliam defines illegal transportation (also known as paratransit) as a transportation service for passengers that is not part of the public transportation [2]. The most common types of paratransit are: buses, collective taxis, particular cars and motorcycle taxis. Motorcycle has become the principal transport medium in Thailand, India, and Colombia, where most of the population use this medium for daily travels. In India, 11.7 million motorcycles were sold in 2016 [3]. Motorcycle Accident in Depth Study (MAIDS), made by France, Italy, Netherlands, Spain, and Germany, mentions that in 68.7% of the motorcycle reported accidents, the helmet could have prevented or reduced the amount of accidents [4]. For this reason, in many countries the use of helmet is mandatory by law.

In Colombia, paratransit has caused the elimination of bus routes in medium-size cities. This has caused a decrease of coverage of the bus routes, and paratransit providers took this opportunity to provide substitute alternatives for public transportation [5]. The use of paratransit alternatives, motorcycle taxis in particular, has caused a reduction of road traffic safety. In 2018, more than 3,000 people died in traffic accidents in Colombia, according to the *Colombian National Road Safety Agency* (ANSV, in Spanish). The Colombian Ministry of Transportation confirmed that there is a correlation between paratransit and deadly road accidents [6]. The main control that transit officers do to reduce paratransit is giving transit fines to motorcycle drivers that are not wearing helmets, driving in prohibited places in the city (for instance, downtown), or are driving with a passenger when it is prohibited (depending on the city, this could be fixed days or permanently). This kind of motorcycle monitoring is inefficient because transit police officers are limited, and human senses are limited to tracking every motorcycle driving in the area along their number-plates.

The control measurements performed by transit police officers are previously established by the local authorities previously. Valledupar city, in Colombia, is an example of these measurements. The city mayor created the following policies by law. (1) Do not allow the driving of motorcycles, during weekdays, with passengers, inside the city center area. (2) Do not allow the driving of motorcycles on Wednesdays from 8:00am to 6:00pm. (3) Do not allow to transport passengers in a motorcycle during Saturdays. (4) Both passenger and motorcycle driver must wear a helmet. The infringement of these policies can be punished with a transit ticket by the transit police officers [7].

The hypothesis of this article is: using a mobile phone application, is it possible to detect plate number and transit infractions caused by motorcycle drivers? We developed a mobile application that could serve as a tool for transit police officers to automatically detect, using their cellphones, the motorcycles driving in a certain area and the transit infractions that their drivers perform. The mobile application uses machine learning and artificial intelligence to recognize 3 types of infractions: not wearing a helmet, driving in prohibited places, and transporting a passenger.

In what follows, we present Intelligent Transportation Systems in Section 2. After, in Section 3, we describe automatic number-plate recognition. In Section 4, we explain object detection using computer vision. In Section 5, we explain materials and methods. In Section 6, we show the obtained results. Finally, in Section 7 we give some conclusions and future work.

2. Intelligent Transportation Systems

An *Intelligent Transportation Systems* (ITS) refers to the use of *Information Communication Technologies* (ITCs) to solve problems in conventional public transportation systems using technology to increase safety, effectiveness, efficiency, accessibility, and sustainability of the transportation network without increasing the capacity of the network [8]. The Colombian Ministry of Transportation defined an intelligent transportation system as the combination of several engineering areas (transportation, telecommunications, environmental, financial, among others) to increase the efficiency of public transportation, reduce the environmental impact, and safe human lives (road safety) [9].

The incorporation of edge technology in transportation systems is one of the strategies used in the main cities of the world to improve mobility and road safety. Singapore, London, New York, Barcelona, among others have implemented this kind of systems, which are physically and operationally integrated, using traffic management centers. In doing so, the have paid less attention to big road construction projects proposed in the second half of the twentieth century. The benefits of the operation of an ITS are traffic reduction, travel time, cost, emissions, noise, and accidents [10].

2.1. Technologies in Intelligent Transportation Systems

ITS include a wide range of functions. These functions range from a simple mobile phone alert to sophisticated traffic control systems. It uses several technologies available such as those shown in Table 1 [11].

IOP Conf. Series: Materials Science and Engineering **844** (2020) 012035 doi:10.1088/1757-899X/844/1/012035

Table 1. Technologies, implementation and examples of ITS [11,12].

Technology	Implementation and Examples
Data processing and management	Archived Data Management Systems (ADMSs)
Detection technologies	Traffic detector, automated vehicle detection, speed detection, environmental sensors
Communication technologies	Communication architecture
Information transmission technologies	Dynamic Message Sign (DMS)
Positioning and location technologies	Global Navigation Satellite System (GNSS)
Vehicle and Transit Control Technologies	Motorway Control Center
Electronic Payment Technologies	Front-end, Back-end

To control and manage incidents, one solution is to use a television closed circuit. These systems are composed of digital cameras. Control and management incident systems are composed by intelligent digital cameras placed in strategic places to monitor traffic, speed, and other characteristics of traffic. As an example, autoscope cameras and cetrac system [13].

3. Automatic number-plate recognition

In the first years of the twenty-first century, there has been an increased interest to develop safety mechanisms. Technologies such as biometric identification (for instance, voice, face, and signature) appear as new options replacing fingerprints. *Automatic Number-Plate Recognition* (ANPR) has been studied since 1976 by the UK police [14]. In all the cities of the world, there has been vehicle traffic problems, accidents, and drivers that break the law. This kind of problems led to the creation of identification and following of vehicles and has been extended to the identification of stolen vehicles [15].

The vehicle control systems using number-plate recognition is among the new technologies that stands out. Nonetheless, its efficiency was not good at the beginning, and only recently, it has improved with the use of infrared systems and refined algorithms. In London, since 2003, there is a governmental program to reduce traffic during peak hours in downtown. To achieve this, cameras were installed in downtown, and using an ANPR software, it was possible to know which vehicle entered downtown, compared the number against a databased of prepaid vehicles, and in case that the vehicle was not registered in the database, issued a payment form. Although this system was originally planned to be a temporal measurement, the British authorities claimed to have made it permanent and extent it through out all the United Kingdom because they say that it helps to reduce crime [16]. As another example, in Italy, the company Advanced Technologies implemented an ANPR system to control access to restricted places in the city [17].

4. Person and object detection using artificial vision

Automatic object detection starts with the acquisition of an image in a three-dimensional scene. Afterwards, the image is processed to improve quality and remove imperfections. The next step consists to separate the object from the background and obtain characteristics from the object such as color, texture and geometry. Finally, the characteristics of the object are compared with other objects in a knowledge database to determine the kind of object [18].

The characteristics used in artificial vision are *Hough transform, local binary patterns* (LBP), *histogram of oriented gradients* (HOG) and *Scale Invariant Form Transformation* (SIFT) [19].

Chiverton developed a system to detect helmet on motorcycle drivers [20]. Chiverton's system used a *Support Vector Machine* (SVM) model trained with characteristics obtained from data of the images in the region where the motorcycle drivers appear.

The selected characteristics capture the form and reflecting properties of the helmets, as the first half of the helmet is brighter than the second half. This system also considers the circular shape of the helmet. The system uses a circular shape detection using Hough transformation [21]. The main inconvenient with Chiverton's approach was the high classification error because other objects in the scene are classified as helmets and some helmets are not detected. Another limitation is that the system does not detect first the motorcycle and its driver before detecting the helmet; therefore, the system may detect other helmets in the scene that are not related to motorcycle drivers.

To improve the classification problems of Chiverton's system, Silva [22] developed a system that first identify the motorcycle divers in the scene using an SVM classifier that uses characteristics extracted from the LBP descriptor. After that, the helmet classification is made using another SVM model on the characteristics obtained from combining three descriptors: Circular Hough, HOG, and LBP.

5. Methods and Materials

In this article we use an incremental version of the design science research methodology [23]. The methodology has 5 steps: awareness of problem, suggestion, development, evaluation, conclusion.

Awareness of the problem: We made a literature review to understand the problem that motorcycle taxis are causing in Colombia. We also found out that the medium-size cities have not had enough budget to implement strategic transport systems, even though a law requiring this was passed 10 years ago.

Suggestions: We studied three designed alternatives to solve the problem of automatic plate number and transit infraction detection by considering the constraint that the solution needs to be low cost and respect the Colombian law. The first idea was to use a mobile application in a moving vehicle to detect the plate numbers and detect the transit infractions. We discarded this solution because the Colombian law does not allow automatic detection systems in moving vehicles [24]. The next solution was a fixed-place automatic detection system. This was discarded because the Colombian law requires that such a device comply with the following criteria: high accident rate, history of transit infractions, calibration mechanisms, and announce the detection system 500 meters before its location. These requirements will increase significantly the cost of the solution, Table 2 shows compliance with the requirements by type of tool. Finally, we designed a system that uses a mobile phone over a camera tripod and a mobile application as a tool for transit police officers to monitor motorcycle infractions. According to the Colombian law, transit police officers can use this kind of technological tools to detect transit infractions, in the same way they can use a speed detector or alcohol meter.

Table 2. Compliance of the requirements of the designed solutions according the Colombian law for automatic detection tools.

Type of tool	Solution requirements				
	Technical criteria	**Calibration mechanisms**	**Software tests**	**Authorization from the Colombian Ministry of Transportation**	**Signalization 500 meters before**
Fixed-place device	✗	✗	✗	✗	✗
Mobile device	✓	✓	✓	✓	✓

IOP Conf. Series: Materials Science and Engineering **844** (2020) 012035 doi:10.1088/1757-899X/844/1/012035

Development: The mobile application was developed using open source tools, in Figure 1 shows the design of our system. We used OpenCV, a computer-vision library, to process the images and locate the plate number area. We also used *Tesseract*, an *Optical Character Recognition* (OCR) library to recognize the characters of plate numbers.

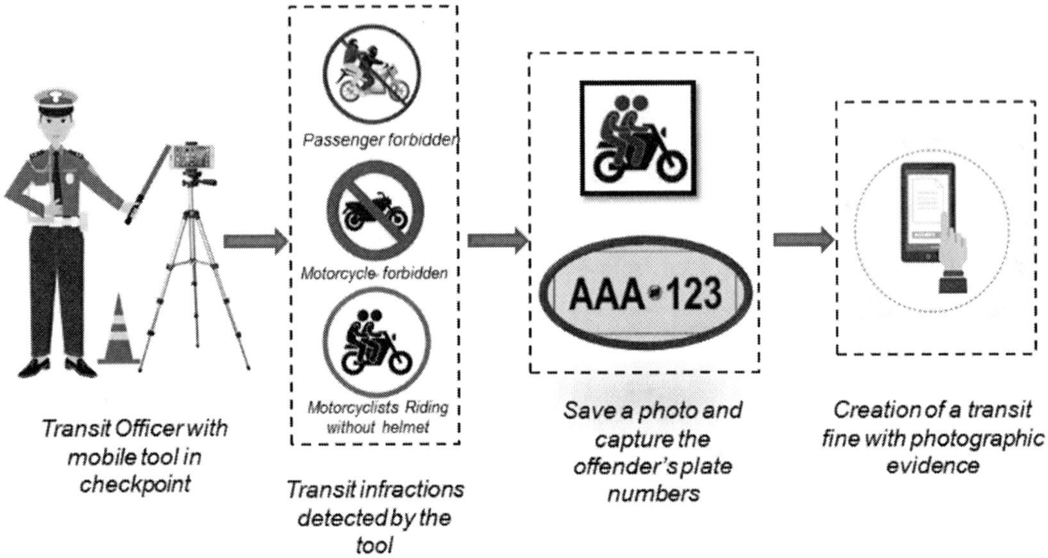

Figure 1. Design of the automatic detection system composed of a mobile phone, mobile application, portable cell phone power bank and camera tripod.

Figure 2. System used for data collection in Valledupar, Colombia.

For helmet detection, we used MATLAB 2018 function *imfindcircles* that uses Hough circular transformation to detect circles in the analyzed images. We calibrated the parameters for sensibility and luminosity to detect the motorcycle helmets.

The experiment was performed for an hour, on a Sunday, in Valledupar City, placing the designed system at a traffic light with high vehicular traffic. For the tests, we used a Motorola Moto G4 mobile device that has a 13Mp camera. The mobile phone was installed on a tripod that reaches 70 cm in

IOP Conf. Series: Materials Science and Engineering **844** (2020) 012035 doi:10.1088/1757-899X/844/1/012035

height as shown in Figure 2. The system allows to have a good visibility of the number-plates and people on motorcycles using the camera. Figure 3 shows the detection and recognition of a number-plate by the mobile application. In this case, the characters of the number-plate are correctly recognized.

Figure 3. License Plate Recognition by the App.

Evaluation: The metric to assess this experiment were the accuracy and precision [25]. The metrics were calculated using the Equation (1) and (2), respectively

$$Accuracy = \frac{p}{N} \quad (1)$$

$$Precision = \frac{p}{N+q} \quad (2)$$

where p is the *number of samples correctly classified*, q is the *number of samples incorrectly classified*, and N is the *total number of samples*.

Conclusions: To conclude this research we considered the problem in the context of Colombian sociocultural aspects, Colombian law, and the hypothesis of the research.

6. Results

During the experiment, 40 number-plates were detected. The accuracy was 85.5% and the precision was 77.7%. Figure 4 shows a case in which the number-plates could not be detected, analyzing the image shows that the angle of number-plate is not ideal for detection.

The system could not recognize 5 of the 40 number-plates captured by the camera during the experiment. This error was caused by environmental factors such as overexposure of the license plate or the angle of the camera with respect to the motorcycle. Figure 4 and Figure 5 shows an example of this problem.

The tests done in MATLAB for the detection of helmet and passenger showed us a precision of 53% and an accuracy of 67%, respectively. An example of the tests can be seen in Figure 6 and Figure 7.

Figure 4. Number-plate with an angle that does not allow recognition.

Figure 5. Detection of number-plate and helmet with overexposure.

Figure 6. Helmets detection using MATLAB. Left side shows a case in which detection was not correct. Right side shows a case in which the detection was correct.

Figure 7. Passengers detection using MATLAB. Left side shows a case in which detection was not correct. Right side shows a case in which the detection was correct.

In what follows, in Table 3 and Table 4, we show the confusion matrices of the helmet detection problem and the passenger detection problem, respectively. In the case of the helmet detection, the main error arises when detecting people without a helmet as if they had a helmet and in the case of passenger detection it happens that in many cases it does not detect the passenger.

Table 3. Confusion matrix of the helmet detection.

		Predicted value	
		With helmet	**Without helmet**
Real value	**With helmet**	**32**	**7**
	Without helmet	**15**	**10**

Table 4. Confusion matrix of the passenger detection.

		Predicted value	
		With Passenger	**Without passenger**
Real value	**With passenger**	**9**	**11**
	Without passenger	**1**	**19**

7. Conclusions

The local authorities of medium-size cities in Colombia have a huge interest in increasing road security. In particular, they want to reduce traffic accidents caused by motorcycles because it has been proven that they play a big role in this problem. In this article, we propose a new system to monitor and control motorcycle illegal transportation which is highly correlated with road accidents. Through our experiment, we confirmed the initial hypothesis: Using a mobile phone application, is it possible to detect number-plates and traffic infractions caused by motorcycle drivers? Using our mobile application, we were able to recognize the number-plates of the motorcycles and using MATLAB we were able to detect the motorcyclists who were not wearing a helmet. The precision was of 77.7% and the accuracy was of 85.5% for the case of number-plate recognition. The limitations of our system as in the case of excessive sunlight, in which the number-plates are overexposed. Another limitation is that the system is not designed to work 24 hours, 7 days a week. In order to do it, a different hardware is needed. The accuracy for helmet and passenger detection were 53% and 67%, respectively. These results can be improved because other circles were detected in the images that do not correspond to a helmet and or passenger. A solution to these problems in the detection of the helmet, is to first detect the motorcycle, and only after analyzing the area of interest of the image to detect the circles. For future work, we recommend improving the training of the algorithms for *Automatic Number-Plate Recognition* to increase its accuracy. We also recommend implementing algorithms in the mobile application for object detection in order to detect driver or passengers not wearing a helmet and when the motorcycle has a passenger in addition to the driver. In this way, MATLAB will not be required and the mobile application will work stand alone.

Acknowledgments
This study was funded by Departamento Administrativo de Ciencia, Tecnología e Innovación-COLCIENCIAS, Colombia, under the "Jóvenes Investigadores e Innovadores" programme [Contract No. FP44842-301-2018]. Authors extend their thanks to Vice-rectory of Discovery and Creation of Universidad EAFIT, and to Cesar Governorate.

References

[1] Ministerio de Transporte 2018 *Unidad de movilidad urbana sostenible: Informe Ejecutivo 2010-2018*
[2] Olarte T, Carvajal A M, Villamizar O and Castex C 2015 Estudio del comportamiento de los usuarios de medios de transporte informales en la ciudad de Bucaramanga *Ensayos: Revista de Estudiantes de Administración de Empresas* **8** 86–100
[3] Doval P 2017 India is now the world's biggest two-wheeler market *Times of India*
[4] Association des Constructeurs Européens de Motocycles 2009 MAIDS: Motorcycle Accident In Depth Study *In-depth Investigations of Motorcycle Accidents*
[5] Tobón A and Galvis D 2009 Análisis Sobre La Evolución Reciente Del Sector De Transporte En Colombia *Perfil de Coyuntura Económica* **13** 147–163
[6] Leal K 2019 Autoridades le ponen el ojo al "mototaxismo" en Colombia *RCN Radio*
[7] Alcaldía de Valledupar 2018 *Por medio del cual se adopta medidas tendientes a reglamentar el transporte en motocicletas, se promueve la seguridad vial y se dictan otras disposiciones, Decreto 00558* (Colombia)
[8] Mfenjou M L, Abba Ari A A, Abdou W, Spies F y Kolyang 2018 Methodology and trends for an intelligent transport system in developing countries *Sustainable Computing: Informatics and Systems* **19** 96-111
[9] Ministerio de Transporte 2015 *Decreto 2060 de 2015* (Colombia)
[10] Peláez Valencia A M 2016 *Modelo de cuantificación económica para implementación de Sistemas Inteligentes de Transporte (ITS) en Medellín* (Universidad Nacional de Colombia)
[11] Sadek A 2016 Tecnologías ITS *Manual Explotación de la Red Vial & Sistemas Inteligentes de Transporte*
[12] PIARC 2016 *Manual Explotación de la Red Vial & Sistemas Inteligentes de Transporte*
[13] Sayeg P y Charles P 2006 *Módulo SUTP 4e - Sistemas de Transporte Inteligente* (Eschborn, Alemania: Deutsche Gesellschaft für Technische Zusammenarbeit (GTZ))
[14] Nguyen T, Nguyen D and Nguyen P 2014 UIT-ANPR: Toward an open framework for automatic number plate recognition on smartphones *Proceedings of the 8th International Conference on Ubiquitous Information Management and Communication, ICUIMC'14* (USA: ACM) Article 113
[15] Espinoza D and Salinas C 2015 *Desarrollo de un Sistema de Reconocimiento de Placas Vehiculares* (Universidad del Azuay)
[16] Toledo L E 2005 *Reconocimiento automático de matrículas de automóvil* (Centro de investigaciones en Óptica) p 168
[17] BBC Inside Out 2014 Car cloning *BBC News*
[18] Avil C 2008 *Detección y clasificación de objetos dentro de un salón de clases empleando técnicas de procesamiento digital de imágenes* (Universidad Autónoma Metropolitana, Honduras) p 158
[19] Dalal N and Triggs B 2005 Histograms of oriented gradients for human detection *IEEE Computer Society Conference on Computer Vision and Pattern Recognition, CVPR'05* (USA: Centre pour la Communication Scientifique Directe) **1** 886-893
[20] Chiverton J 2012 Helmet presence classification with motorcycle detection and tracking *IET Intelligent Transport Systems* **6**(3) 259-269
[21] Pei S C and Horng J H 1995 Circular arc detection based on Hough transform *Pattern recognition letters* **16**(6) 615-625

[22] Silva R, Aires K, Santos T, Abdala K, Veras R and Soares A 2013 Automatic detection of motorcyclists without helmets *XXXIX Latin American Computing Conference (CLEI)* (Venezuela: IEEE) 1-7

[23] Vaishnavi V y Kuechler B 2004 Design Science Research in Information Systems (Last Updated: November 15, 2015)

[24] Ministerio de Transporte and Agencia Nacional de Seguridad Vial 2018 *Por la cual se reglamentan los criterios técnicos para la instalación y operación de medios técnicos o tecnológicos para la detección de presuntas infracciones al tránsito y se dictan otras disposiciones, Resolución 0000718* (Colombia)

[25] Kulkarni Y, Kamthe A, Bodkhe S and Patil A 2018 Automatic Number Plate Recognition for Motorcyclists Riding Without Helmet *International Conference on Current Trends towards Converging Technologies, ICCTCT* (India: IEEE) 1-6

IOP Conf. Series: Materials Science and Engineering **844** (2020) 012036 doi:10.1088/1757-899X/844/1/012036

Approach to professional practice from simulators and virtual laboratories

C J Peña , L Vargas and J C Murcia

Corporación Universitaria Minuto de Dios, Vicerrectoría Regional Bogotá Sur, Departamento de Investigaciones, Unidad de Ingenierías y Unidad de Ciencias empresariales, Calle 1 # 9-50, Bogotá – Colombia.

cjpena@uniminuto.edu

Abstract. The professional practice and student experimentation have undergone changes in the mode of the instruction processes, which has had an impact not only on the adaptation of business processes but also in the reorientation of academic dynamics around the use of simulators, due to the development of new learning technologies and their high influence in educational environments. Therefore, higher education institutions are responsible for designing complementary learning situations to the future professionals' training of the disciplinary area through qualified practices. That is why, through this research and the use of mixed methods, the perception of students and teachers of the support that new technologies have been providing just as a teaching resource for the transfer of knowledge, boosting the experimental classroom and its implementation in the professional practice; Simulators and virtual reality have allowed the improvement of science as an active part of academic processes and business development. The results highlight the importance of continuing to work on the implementation and familiarization with these tools.

1. Introduction

New information and communication technologies (ICT) have allowed higher education institutions to implement differentiating technological tools in their teaching-learning environments. [1] Virtual laboratories represent a creative, modern and cheap option for universities. Not only non-face, but also face-to-face educations require these spaces within their training processes. Consequently, classes will accomplish two didactic objectives simultaneously: a) carry out practices related to the subject by expanding the availability of laboratories and b) training students in the use of ICT.

Engineering education, is one of the fields of greatest theoretical-practical development, seeks that engineers can gain the necessary skills by applying their knowledge in practical experiences, specifically in those situations that, due to their high costs, it is not possible to implement through a traditional laboratory, which generates a window for the implementation of virtual laboratories.

The huge transformations that we experience today in highly changing professional contexts are typical of globalization. The prevailing paradigms on professional learning in universities indicate that in today's world the real question must be whether students receive at universities is according to the needs of the environment. Therefore, students can solve or tackle real problems in the future. Then is almost an obligation for future professionals and its preparation in this field not only a personal commitment on the part of the student but also that the educational infrastructure should gradually foster its development.

Implementation of these elements as a preface to the exercise and professional practice is essential for students. This incursion develop skills at the technological level in the various fields of application according to the basic training, which will allow not only a change in the teaching thought and pedagogical instruction, but will also provide different tools in students during the process of their professional practice. Those previous abilities associated with the globalization components they face during their first approach to the work life that will set an important guideline for their professional development.

Many academics considered that conventional or traditional classrooms, that means face-to-face education have been the only spaces where students can develop practices and do direct experimentation with reality, it is a space where their interactivity allows the student to observe what happens in their experiments, which help with the development of their cognitive abilities [2]. However, it must be consider that face-to-face method entails high costs of infrastructure investment, adaptation, support and that in some cases they carry great risks of not being able to achieve completely the learning objectives because it is not possible to do a certain amount of experimentation either. Most of the time it is due to the impossibility of accessing materials such as those that are considered dangerous, or due to the limited use times of these. However, the link between private companies and study institutes have managed in many of these cases to be able to solve, through sponsorship, the demand for their use and benefit different zones that expect results and advances in fields such as research.

Figure 1. Neos VR, Virtual Lab Software used by students in some areas

Figure 2. Use of Oculus VR virtual reality technology

A virtual laboratory or simulator is computer software that allows experimentation from a computer of different situations in almost any field of engineering, whereby manipulating different environments, variables, parameters, and input constants, As seen in figure 1 and figure 2 as an example of experimentation with these, both in software and hardware. It is allowed to run a model in which is usually based on mathematical algorithms to obtain output data. One of the advantages of this type of program is that it allows almost immediately obtaining mathematical, logical, and statistical results and data, which in face-to-face method would take a little longer to obtain. Other advantages are that a greater number of students can have simultaneous access to the experimental processes while improving the self-learning processes, as well as allowing the repetition of the experimental situations in a very short time and with fewer potential risks that could have reality; however, this practice does not replace real experimentation in any case.

2. Related Works (Literature)

For engineering education, throughout history, it has been evolving in the didactics used for the teaching-learning process, taking into account the dynamics of the markets and the inclusion of ICT in the various fields of engineering action. Likewise, the learning methods developed by new generations, which have grown with a greater link to technology and self-learning processes, require that educational institutions provide them with diverse and innovative tools for these processes.

As a consequence, the inclusion of tools such as LV and LR in the teaching of some of the topics of the subjects related to engineering programs have been increasingly nowadays, offering students different alternatives for facing problems on subjects as mathematics, physics, chemistry, robotics, electronics, programming, soil studies, among many other engineering applications. This innovation allows that the theoretical concepts acquired be used in such a way that future engineers not only keep the memorial exercise of theories but also can apply them through the development of laboratories.

For example in The University of Alicante (Spain), particularly for the teaching of robotics in which a virtual laboratory was designed, allows simulating the handling of a robot arm, and also the teleoperation of the equivalent real arm by several students through the Internet [3]. This practice generated as a result greater conservation and useful life of the arm by not executing erroneous commands that could damage it and also saving on Purchase of materials, time spent for practice and possibility of repetition of functions for clarity of the concepts used by students.

A similar application made at the Universidad Europa in Madrid, in which was used a remote laboratory based on internet for teaching control engineering: LABNET, assured the acceptance of the system by students. It dues to the experience acquired demonstrates that a remote experimentation environment is feasible and students adapt quickly to an educational setting that involves the use of remote laboratories [4].

On the other hand, a study with students of the IV semester of Electronic Engineering of the National University of Rosario (Argentina) in 2011 in relation to the use of the Remote Laboratory of Electronic Physics, evidenced the appropriation of this tool by the students. In addition to this, transcending the objectives of the study; meanwhile, the reports showed progress in the presentation of other tests that the students saw possible, resulting in a complementary use of the remote and traditional laboratories [5].

Talking about non-face-to-face engineering teaching, virtual and remote laboratories represented a significant role in theoretical-practical processes. For instance, in the study done at the University of Extremadura (Spain), whereby a simulator was developed for the study of hydraulic power networks, the students' found out that the user interface was easy to understand by most of students. Also, they considered better understanding the operation of hydraulic networks considering that many of them are not possible to develop in a traditional laboratory environment [6] (p. 66).

The available information on the internet related to applied cases of the use of virtual laboratories is quite robust. It allows us to infer that the application of these in engineering classrooms has generated a long history throughout history. Moreover they have become a tool of greater use in higher education institutions, which also leads to the presentation of various studies related to strategies for the development of virtual laboratories adapted to the needs of the courses, as well as others focused on the implementation of virtual laboratories of free access or licensed that acquire these institutions.

Another research done [7] at the University of the Coast (Colombia), the design of a model of a remote laboratory applied to the creation of microprocessor systems, in which fundamental stages such as the identification of requirements are established. The methodology for the project of the laboratory and its implementation, while [8] exposes in its research a compilation of outstanding resources on the web related to the area of Systems and Automation Engineering, categorizing them in automation and industrial instrumentation, architecture and computer networks, robotics, simulation and dynamics of systems, plants and software and other resources.

As mentioned [4], several projects have been deployed worldwide, some of them financed by the European Union, as CYBERLAB has culminated in the creation of a service provider for the integration of network laboratories. PEARL is a project focused on infrastructure development that allows remote access to laboratory equipment and investigates its pedagogical impact to validate its development. MARVEL developed a mixed environment that combines remote and virtual laboratories on the website of the Spanish Committee Automatics. It presents a compilation of several virtual and / or remote laboratories related to the area of Systems Engineering such as robotics, computer vision, automation or automatic control (p. 19), which allows viewing a wide range of options for the implementation of this type of tools in engineering education.

The RV teacher [9] belonging to the National University of Colombia, in an interview as part of this research, argued that today many teachers are interested in knowing other teaching methods. He exposes some researches and among them the National Institute of Educational Technologies and Teacher Training (INTEF) of Spain, estimates that every year about 10,000 teachers seek training opportunities to acquire digital skills to be able to transfer these experiences to the students. There are currently technological advances by companies such as Hewlett-Packard-HP for Spanish classrooms such as some computers and monitors specially designed to convey this type of experience in educational environments. Another contribution is the improvement in the understanding of the curricular contents, remembering the knowledge in the short and long term and makes the students more attentive and cooperative, achieving a better personalization of the class.

The fact that direct work in a traditional laboratory is still the optimal method to train to learn in students should not be ruled out, noting that such work, complemented with practice in virtual laboratories, has a greater impact on the teaching-learning process. This is reflected in the study developed by [10], which concludes that the use of the simulator as a complement to the dictated topics helps the student to pose challenges and goals in the development of the practices; since the simulator there is always in capacity to show the final result and verification of the proposed workshop, while in practical work, the detection of the fault and the verification is not so immediate and effective. Likewise, it was shown that by working in a combined way the simulator and physical assemblies, the level of drop out of the subject by the students was reduced to 10%, associated with teamwork, due to the distribution of functions and activities among the participating students.

3. Theoretical framework

The simulation of processes has evolved in parallel with computers and electronic devices. In the past with the entrance of the first machines there was a computer program for each process; however, as these were improving and other elements such as graphic interfaces, audio, multimedia, the development of more complex algorithms, object-oriented programming and complementary programming languages such as Java, Dreamweaver, Flash, among others, it was then possible to create general purpose tools with which different processes of reality can be simulated.

Simulation is one of the most powerful tools available to decision makers, design and operation of a system regardless of its complexity. Nowadays engineers, designers, analysts, administrators, among other professions make use of this resource because it allows the study, analysis and evaluation of situations and phenomena that would not otherwise be possible to analyse in a short time [11]. In several of these cases, the use of pre-decision simulation tends to be a factor to be taken into account as a requirement because the risk of making a bad decision decreases.

Since a simulator and a virtual laboratory are based on mathematical models that run on computers, their configuration and commissioning are much simpler than the configuration and commissioning of real laboratories. Also, they have a much higher degree of robustness and security since there are no real devices that cannot cause problems in the environment [12], considering that the majority of these experimental cases arise in ideal scenarios free of affectations and external factors outside the experimentation presented in the real world.

In relation to this [13], they express the advantages and disadvantages that arise with the implementation of virtual laboratories. Emphasizing positive aspects such as a greater number of students experiencing asynchronously, with flexible schedules for their practice, and the possibility of experiencing freely until understanding the concepts in a clearer way, promoting self-learning environments, as well as breaking the space-time barriers in the use of traditional laboratories.

IOP Conf. Series: Materials Science and Engineering **844** (2020) 012036 doi:10.1088/1757-899X/844/1/012036

Despite all the advantages offered by new technologies, simulators and virtual laboratories, these can also have their disadvantages, as is the fact that to be successful the experimentation processes, prior training of both the teacher is required as of the student that allows the development of skills in their management. Another aspect is that the simulation needs many highly reliable input data, so the acquisition of this data can be very expensive in time and money and if you do not have control over it. The simulation can never compensate for the poor quality of the data input and its results will then be very different from what is necessary. Since the input data is wrong then the output data will also be wrong, or if the proposed model is not well described, the output will not be adjusted to reality, so that a bad model can lead to wrong decisions in reality [11].

In the LV there is also the risk that the student will be totally abstracted from reality and tend to behave like a simple spectator, so that the instructional design of the educational experiences, must contemplate that the activities in the LV are accompanied by guides instructions that allow the student to be an active part of the experimental process. They must contain clear learning goals that allow one or another way to assess the competence in the subject. Not everything should be simulable, some experiences must be done in the real world so as not to lose the complete vision of reality, which implies that it is the teacher who must determine and analyse the relevance of the issues that should be simulated and which not [two]. In addition to all previous information, we must add the fact that we must tend to break the resistance that may exist among teachers regarding the inclusion of ICTs in their classes. For that reason, it is necessary to change the thinking in teaching methodologies by the teacher, since Students leave the classroom and be supported by autonomous student practices as the main input to independent work [14-15], as well as the need for support staff for the resolution of aspects related to technical support, software development and installation.

Another aspect that has been opening up spaces in the field of experimentation is the simulation based on the WWW or online simulation, which are considered as an effective means for the distribution and unification of criteria regarding the simulation process. In which the simulation process progresses continuously and dynamically, obtaining the results of the simulated time in each period of sampling in the form of a continuous flow of numerical values or graphs evolving in a sustained way; however these types of simulators depend exclusively of being connected to the network all the time [16]. Therefore, you can limit its effectiveness and use to the presence of the connection, however one of its advantages is that you will not have to have powerful computer equipment or servers to its operation, which implies a considerable reduction in operating costs as well as allowing access to your information from any Place of the world.

Virtual tools can be found on the internet and can be classified in other denominations. As proposed [17] in: blogging, chat tools, charts & graphs, file sharing and mashups, microblogging, notebooks and notation tools, online office suite, Online presentation , personal web portals, photo editing / sharing, search engines, social bookmarking, social network, survey and polling tools, task and group management, video chat, video editing / sharing, website creation, wikis and finally group template page. Some of them are available for free to the teacher.

4. Approach and Methodological Design

This research is exploratory and a mixed methodology used, between interpretative and praxeological. From the interpretative point of view the hermeneutic that the different social actors make of their "reality", this means studying it from people and emphasizing the process of understanding on the part of researchers to try to discover the meaning of their Actions. From the view of phenomenology that allows us to distinguish the form of things from how people think it is a reality [18].

4.1. Population and Sample

4.1.1. Target population of the study. Students of the Engineering Faculty from Minuto de Dios University Corporation of South branch of Bogotá, enrolled in the first semester of 2018 (380 students, which corresponds to 100% of the potential students), belonging to the Informatics Technology programs , Electronics Technology, SR Management Technology, Logistics Technology, Network Technology and IT security. Like 35 teachers belonging to the different programs mentioned above

4.1.2. Sample design. The sample design is probabilistic, in two stages, stratified in the first stage, the selection method is Simple Random Sampling (MAS) in the two stages. First sampling stage selection. They are selected by the program according to the different semesters obtained from each program. The stratification criteria considered is semester, for which six semesters and number of students registered for each semester were formed. In this first stage, 6 programs distributed in 20 semesters were selected. In the second stage of sampling, students over 18 were selected with a minimum of 20% of students per program.

4.1.3 Sample size. Total population 380 students, Total students surveyed, qualified (sample): 191
Total respondents who expressed and signed the consent to use data: 191.
For a total population of 380 and a margin of error of 5%, the sample should be 191 people.

4.1.4. Sample distribution. Regional South branch of Bogotá, "San Camilo headquarters", disaggregation into six technology programs: IT, Electronics, Market Management, Network Security, Business Logistics.

4.2 Methodological Design
4.2.1 Sample frames. Registration of enrolled students given by the Regional Information System Coordination of Regional South branch of Bogotá.
Collection date: from May 28 to June 5, 2018.
The number of pollsters: 4 pollsters, 1 field supervisor.
The margin of error: the error is 5%, with a confidence level of 95%, and heterogeneity of 50% is assumed.
4.2.2 Collection technique. A virtual survey format, applied face-to-face in classrooms assistants.
4.2.3 Instruments for collecting information. The instruments in research in the semi-structured interview and surveys, to make a diagnosis of what is known by learning with augmented reality in both students and teachers. And what is marketed in the market on resources or virtual devices for learning, for this purpose, questionnaires are used, according to [19] "The questionnaire is a set of questions designed to generate the necessary data, in order to achieve the objectives of the research project", This is why this instrument is chosen, to reach one of the objectives of the research, these instruments are chosen, to reach one of the objectives of the research.

5. Results

According to the information collected in the research through interviews with teachers and surveys of students from different academic engineering programs, it is evident that using virtual laboratories, simulators and virtual reality in the classroom, allow traveling and experimenting without leaving class. Students can meet, visit and study without leaving the place, enriching their learning in a fun and practical way, while removing economic and geographical barriers; some institutions of higher education, due to lack of spaces to rehearse or build, activities or workshops cannot be easily developed, so these pedagogical resources become an alternative to direct the classes and facilitate teaching. The usage of this technology allows they can implement for example laboratories, as mentioned throughout this article, recreating scenarios for the student to explore and interact with knowledge.

From the teaching perspective, it is expressed that these resources become an appropriate tool to create new educational environments, diversifying the options and allowing the teacher to meet the needs of each student without neglecting the group dynamics. Neuroeducation experts argue that this technology, due to its ability to create scenarios within the classroom, allows for more vivid and emotional experiences. That is why accessing digital resources can benefit the educational experience for both students and educators, although it is not an essential requirement to be a teacher at present, including the transformed learning in higher education institutions, which seek the preparation of its students within a digital environment to develop their disciplinary skills.

For the interviews, analysis categories were taken into account, such as: knowledge and interaction with simulation technology, virtual laboratories, and virtual reality, their use in education and devices such as VR, for the respective analysis.

On the other hand, intervention was made with the students of first semesters through surveys, whose categories to analyse were: emerging competencies, virtual laboratories, simulators and virtual reality in education, virtual spaces, in order to analyse the perception regarding, the use and relevance of these in class and the skills they have with this technology, so they were asked the following questions:

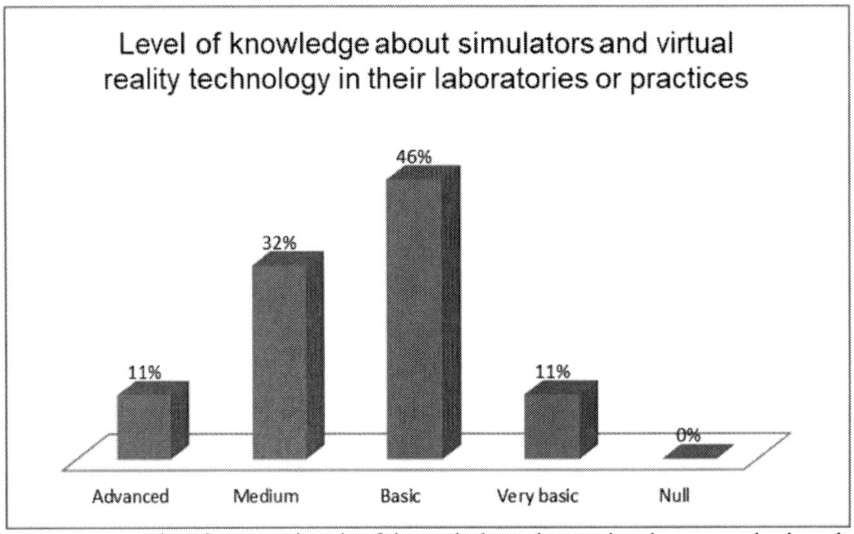

Figure 3. What is your level of knowledge about simulators and virtual reality technology in your laboratories or practices.

The answers to this question were: 42.9% basic, that means, that commercially recognize the devices and through internet consultations; 31.4% average, at some point they have identified programs or devices but without thoroughly exploring their operation; 11.4% very basic, only the specificity of what is worked on in some classes and nothing; more and 8.6% advanced, some students who have been interested in investigating this technology remain a very small population that knows it and 5.7% null (blue segment), do not recognize these neither experience nor by literary review, which represents little interest in this area. In the face of being a generation that has grown hand in hand with technology, they have little management and knowledge of these tools, they recognize it through online consultations and videos, they do not know that it can be used for their learning; they only identify it for use of training.

Therefore, it is important to relate the question about frequency in the use of virtual reality simulators and devices, with the answers on the level of knowledge to contrast them and verify if they have been curious to perform exploration with them beyond the technical review and documentary film.

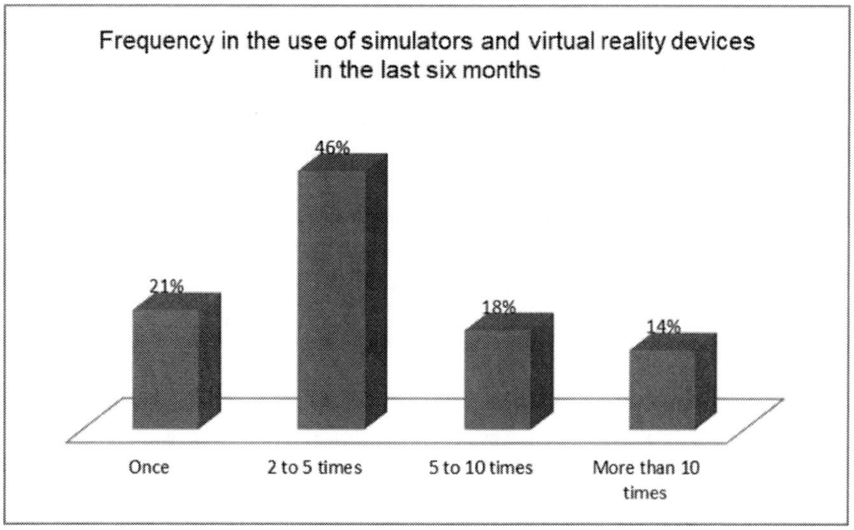

Figure 4. At what frequency have you made use of simulators and virtual reality devices in the last 6 months

According to the answers, it was found that 21.4% at least once they have used these resources, which may represent that they were once interested in knowing, but without generating greater impact that induces them to deepen this technology; 46.4% two to five times, which means that they are students who have used them occasionally for academic or entertainment use; 17.9% have used it 6 to 10 times, indicating that they are students who are interested in technology and finally 14.3% have used it more than 10 times become fans of technology and discovering new applications, making them research on new trends in versions of devices and software.

Finally, within the questions asked, the inquiry about hardware devices that students have frequently used and how they have been implemented in their classroom and laboratory processes is highlighted.

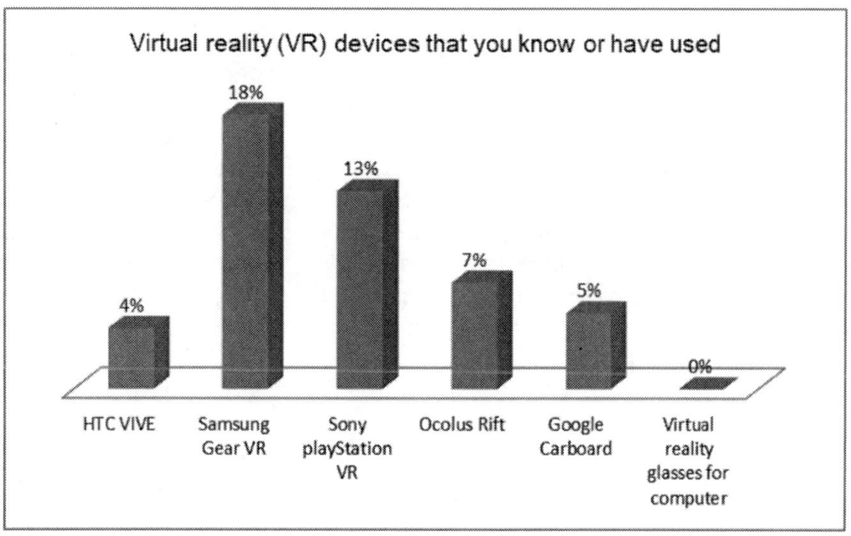

Figure 5. Virtual reality (VR) devices that you know or have used

It is identified that in 18% Samsung Gea VR is the most used for virtual glasses; 13% use Sony Play Station; 6% Oculus Rift; 5% Google Cardboard and 3% HTC Vive. The factors that are into account when selecting a device are costs and technical specifications of the same; for example: the peripherals Samsung Gea VR facilitate immersion with cell phones; Sony Playstation VR is for exclusive use on PS4 console, which makes it a game element and allows a high immersive experience for video game fans.

As evidenced by the VR regardless of the devices and software used, they are used for entertainment. According to survey results, students are unaware of the application of this technology in the academic subject; but it is important as a teacher to know what they handle. Students have digital skills, in this way the teacher characterizes the study groups, selects the virtual investment devices and software, as well as the preparation of the curricular contents for the implementation of laboratories in classes.

When consulting teachers about the use of resources in the classroom as simulators or virtual laboratories in cases such as laboratory practices in electronics, computer science, logistics processes among others, there is a clear trend towards the frequent use of these resources

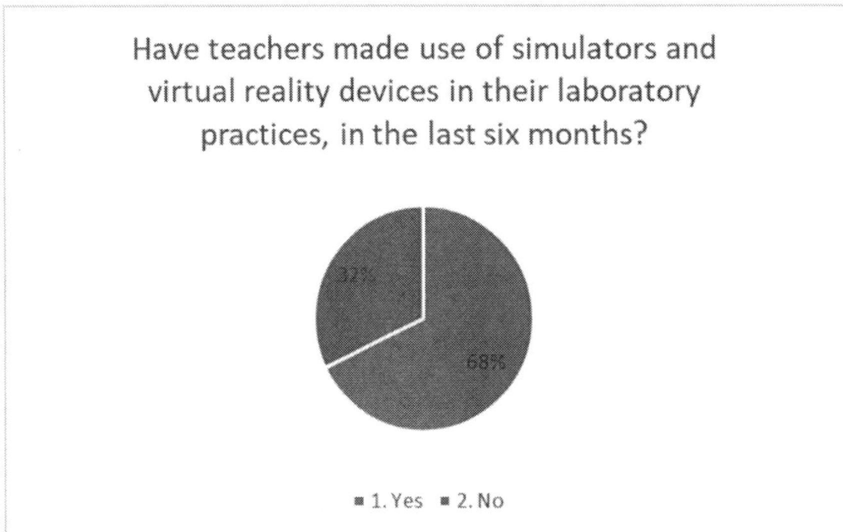

Figure 6. As a teacher he has made use of simulators and virtual reality devices in his laboratory practices in the last 6 months

The great majority affirm that they are very practical to be able to advance in the achievement of learning objectives or goals; most recurrent themes are for the development of Computational Algorithms, circuit simulation, computer network simulation, computer security, logistic models, study of physical and chemical phenomena, as well as study of the operation of systems and machines.

Both teachers and students were asked to evaluate their experience with the use of simulators, virtual reality elements and the implementation of virtual laboratories, and the vast majority as seen in Figure 7 expresses that their experience it was very satisfying

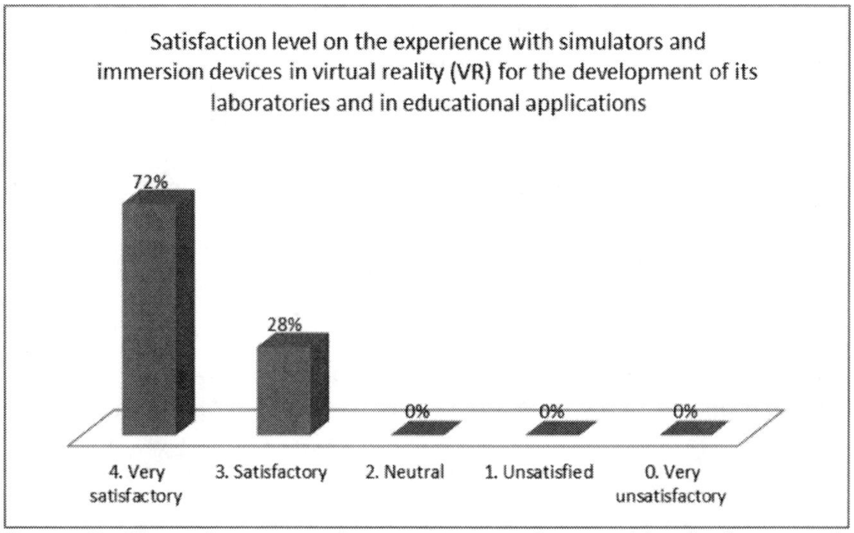

Figure 7. How do you evaluate your experience with simulators and immersion devices in virtual reality (VR) for the development of your laboratories and in educational applications on the scale of 0 to 4, where 0 is very unsatisfactory and 4 is very satisfactory.

6. Conclusions and future work

There are some elements to be considered in the implementation of virtual laboratories in the teaching-learning process of Engineering: for choosing or development should be based on aspects associated with resources, the installed capacity, the intentionality of the laboratory, the skills to be developed, the branch of knowledge, language, among many others that can generate both positive and negative results in the use of these tools. Therefore, the importance of constantly monitoring the impact that generates in students the implementation of virtual laboratories in the Engineering chairs.

It is important to highlight that Virtual Reality is in full evolution in the world [20] uses the immersion technique, where the user immerses himself through a viewfinder or other device to another reality or different space from the one he is in, being in a mode of "Teleportation", to interact with virtual elements in environments that can be laboratories, museums, universe among others; Virtual reality is a form of learning in which interacting with a computer, a virtual space is created where tasks are developed, test theories or live simulated experiences. This technology has two challenges to get into the academic and work environment, for example, the skills gap represents difficulties within education organizations and institutions to have enough technological experts to manage the digital transformation and another challenge is ensure that the institutional changes highlight the importance of integrating new technologies into their assets.

The proposal to implement virtual laboratories in the classroom becomes a useful tool for education because it improves attention in students, due to it is an experience in which they get in touch with the senses [21], allowing the assimilation of the concepts efficiently and recreating a real context reducing learning times, since "living" the situation directly, understanding what happens and solving the problems; It is a more dynamic process. Remember that the main instrument in the case of virtual reality glasses is the image, this is the format to which digital natives and young students react better and use more in their study and leisure.

By using laboratories, students can experiment, rehearse, make mistakes and correct, without being exposed to risks or without exposing real resources to economic or physical losses. In this way, it is possible to contribute as a teacher to a development of life facing new changes to do less traumatic and worked in advance in virtual environments. The RV helps the personalization of learning, because this technological application expands the resources to be used in classes, so that the preparation of future professionals is according to the real life where they are going to work. Like the case of architecture students or engineering when they can see their plans come true; the study of the land and the necessary rectifications, they visualize it prior to the execution and materialization, because a 3D recreation of the infrastructures is made and additionally they run them as if they were there, improving the effectiveness and quality of their works.

In the same way, in the case of artistic subjects and cultural management, virtual reality offers animated versions of museums and galleries in which you can tour and study an environment or a work of art and its context. Finally, as a demonstration from the need for virtual laboratories, in medical subjects, in the exploration with simulations students learn about the prevention of occupational hazards, dissections, identification of traumas or diseases, by virtual inspections. The theory to virtual scenarios enriches knowledge, more participation and teamwork is generated and what is learned in class is preserved in the short term, the teacher's mission is very important at the time of implementing it, because this way Students either take advantage of the spaces or lose teaching time, without achieving the objectives of the subject.

It is proposed as part of future work to deepen the issue of applications of hardware devices such as HDMI for laboratory practice processes, as well as deepening the different programs that are currently free of charge to improve processes of simulation in subjects of the area of physics, mathematics, general theory of systems as well as in the administration part of computer systems. While it is true that not everyone is competent implementing virtual laboratories, their study could lead to the development of other information technologies in support of curriculum development within the different academic programs.

References

[1] Infante C 2014 Propuesta pedagógica para el uso de laboratorios virtuales como actividad complementaria en las asignaturas teórico-prácticas *Revista mexicana de investigación educativa* **19** (62) pp 917-937.

[2] Medina A, Saba G, Silva H., and de Guevara E 2011 Los laboratorios virtuales y laboratorios remotos en la enseñanza de la ingeniería *Revista internacional de educación en ingeniería* **4**(1) pp 24-30.

[3] Torres F et al. 2002 Laboratorio virtual remoto para la enseñanza de robótica (Alicante III Jornadas de trabajo: EIWISA'02: enseñanza vía Internet-Web de la ingeniería de sistemas y automática: Alicante, 18-19 de abril de 2002. X Editorial Club Universitario ISBN 84-8454-159-2) pp 65-69

[4] Aliane N, Fernández J, Martínez A, and Ortiz J 2007 Un Laboratorio de Ingeniería de Control Basado en Internet *Información tecnológica* **18**(6) pp 19-26

[5] Marchisio S, Lerro F and Von O 2011 Empleo de un Laboratorio Remoto para Promover Aprendizajes Significativos en la Enseñanza de los Dispositivos Electrónicos Pixel-Bit *Revista de Medios y Educación Universidad de Sevilla España* núm 38 pp 129-139.

[6] Barrio R, Parrondo J, Blanco E and Fernández J 2011 Introducción de laboratorios virtuales en la enseñanza no presencial mediante entornos de trabajo propios *Revista de Formación e Innovación Educativa Universitaria* **4** pp 55-67.

[7] Zamora R 2010 Laboratorios remotos: Análisis; características y su desarrollo como alternativa a la práctica en la Facultad de Ingeniería *INGE CUC* **6** pp 281-290

[8] Candelas F and Sánchez J 2005 Recursos didácticos basados en Internet para el apoyo a la enseñanza de materias del área de ingeniería de sistemas y automática *RIAI: Revista Iberoamericana de Automática e Informática Industrial* **2** pp 93-101.

[9] Solórzano L 2018. Implmentación de Realidad Virtual en espacios académicos. (C. Murcia, Entrevistador)

[10 Vargas-Puentes L and Gonzalez S 2016 Análisis de la percepción de un grupo de estudiantes de tecnología en informática e ingeniería de sistemas de la Corporación Universitaria Minuto de Dios sobre el uso de un simulador en la materia de electrónica digital *Virtu@lmente* **3** pp 102-123

[11] Cataldi Z, Lage F and Dominighini C 2013 Fundamentos para el uso de simulaciones en la enseñanza *Revista de informática educativa y medios audiovisuales* **10** pp 8-16

[12] Calvo I, Zulueta E, Gangoiti U, López J, Cartwright H and Valentine K 2009 Laboratorios remotos y virtuales en enseñanzas técnicas y científicas *Ikastorratza* **3** pp. 1-21.

[13] Lorandi A, Hermida G, Hernández J and Ladrón de Guevara E 2011 Los Laboratorios Virtuales y Laboratorios Remotos en la Enseñanza de la Ingeniería *Revista Internacional de Educación en Ingeniería* 4 pp 24-30

[14] Rojas W and Fagua A 2013 Laboratorios remotos y virtuales: una herramienta para el desarrollo de prácticas en ingeniería *Ciencia, Innovación Y Tecnología* **1** pp 71-80

[15] Lapeira C 2015 Las nuevas tecnologías de aprendizaje y su participación en la exclusión social *Cooperativismo & Desarrollo* pp 107.

[16] Dormido S, Sánchez J and Morilla F 2000 Laboratorios virtuales remotos para la práctica a distancia de la automática *Sevilla: XXI Jornadas de Automática, Conferencia plenaria*

[17] Gross M 2010 165 Herramientas Web 2.0 distribuidas en 20 categorías

[18] Husserl E and Boehm, R 1996 Erste Philosophie (1923/24): Theorie der Phanomenologischen Reduktion (xliii, 592 S.) Martinus Nijhoff.

[19] Bernal C 2010 Metodología de la investigación, administración, economía, humanidades y ciencias sociales (Bogotá: Pearson)

[20] Cabero J and Fernández B 2018 Emerging digital technologies come into the University: AR and VR. RIED *Revista Iberoamericana de Educación a Distancia* pp 119-138

[21] García C, Valdovinos R, Salgado M. and Muñoz R 2014 Realidad virtual y entornos virtuales como apoyo al acercamiento universidad-comunidad: el caso de la Facultad de Ingeniería de la UAEMex| *Apertura* pp 76-85.

IOP Conf. Series: Materials Science and Engineering **844** (2020) 012037 doi:10.1088/1757-899X/844/1/012037

Automatic Classification System by Color of Plastic Bottle Caps (SISAC)

H. A. Rodriguez-Arias[1], A. Mendoza-González[2]

[1] Research Group in Robotics, Automation, Telecommunications, Electricity, Electronics and Computation. (GIRATEC), Faculty of Engineering Sciences, Antonio de Arevalo University Foundation, Cartagena, Colombia, e-mail:
[2] Research Group in Architecture Design and Engineering (GIADI), Faculty of Engineering and Architecture, Colegio Mayor de Bolivar Technological Institution, Cartagena, Colombia.

harold.rodriguez@unitecnar.edu.co

Abstract. The following project presents an autonomous system of classification of objects by colors which it is applied to the caps of plastic bottles, emulating an automated work cell, which serves to students to know the operation of industrial processes, from construction, programming and even maintenance point of view. The entire project consists of the design of a conveyor system which has a sensor to identify colors, and actuators which allow the caps are correctly classified in the containers according to its color. The work cell was built in a modular way in such a way that in the future it can be adapted to other work cells and simulate flexible manufacturing cells, since it can be reprogrammed not only to identify by colors but also by sizes. Besides, the pieces were built using the 3D printer and the laser cutter in order to use the tools of the laboratories of our institution in the construction of elements for future laboratory equipment which will strengthen the electromechanical program.

1. Introduction

The next work shows the development of a system to select objects by color, using free software and hardware, reducing construction costs.

The system comprises three main parts, the distributor of covers, the conveyor belt and the selector, in the case of the distributor it is a mechanism that ensures that one cover passes at the same time, the conveyor belt carries the covers to the selector, and in that sensing process is carried out; Then, it reaches the selector that is responsible for locating the lid in the corresponding place according to the color.

There are multiple similar works, with some differences compared to ours, such as Padilla Magaña J, [9] which builds a system for selecting lemons by color for the University of La Salle in Bolivia, here uses a CMyCam5 PixyCam vision sensor, which with the design that identifies, allows the identification of three colors, for the case lemon green, yellow and light green, leaving many restrictions for new jobs on the same platform, another work is that of Arenas Campoverde Alfredo, [10] which also used vision classifies soda caps as well as this work, but it only does it for 4 colors and the system is closed to the possibility of new colors, since they deliver a module of use of hardware and software closed, in the work of Zhaokun Wang, Binbin Peng, Yanjun Huang and Guanqun Sun [11] of the School of Mechanical Engineering of Nanjing University of Science and Technology of China and the Department of Mechanical and Mechatronics Engineering, University of Waterloo of Canada, in December 2018 they present a classification of 7 colors for the recycling of plastic bottles, also using artificial vision, which increases the costs with the use of the cameras, and also leaving a closed job since it is for the solution of a particular problem, other works that use the TCS3200 as a sensor [2] such as that of Efraim Gregorio and Benny Achmad of the Faculty of Information and Information Technology of the Gunadarma University of Indonecia [12] that uses this sensor to compare colors in a quality process of products, it is very good although you have to feed the system initially and it will always compare color to color, it means you don't select the objects, but compares them with a single predetermined color; and thus a sweep continues and multiple works aimed at solving specific problems will be found, the work of Amitesha Sachdeva, Mahesh Gupta, Manish Pandey and Prabham Khandelwal [13] of Department of Mechanical Engineering SRM University, Modinagar Ghaziabad, can also be observed India, who built a robot arm that selects objects of various colors, however make it clear that the system works for the colors red, green and blue, leaving the system limited.

The objective of this work is to leave a platform on which the student can interact with it, that can program it and detect as many colors as possible, based on groups of 6, since the conveyor belt is designed to separate 6 colors to the At the same time, the work shows how the TCS3200 sensor signal is received and there is an example of programming for some colors, giving the necessary tools for the student based on the graph and the example algorithm, to develop new algorithms that carry it to the detection of new colors.

Finally it is worth saying that professional stations [3], cost up to 4000 US dollars, while this platform does not exceed 500 US dollars.

2. Materials y methods

2.1. Problem statement

Without a doubt, as the industry grows, it requires new technologies, automatic processes and technological development. For this reason, it is necessary to carry out investigations and constructions with technology that allow speeding up and reducing costs in industrial processes. This project aims to achieve the aforementioned through automation, in a process that is usually done manually in small industries, the classification of objects by colors.

For the Technological Institution Colegio Mayor de Bolívar, this project becomes a work in accordance with the policies of Colciencias that seeks to change the priorities of the traditional policy in which the emphasis is on the generation of knowledge rather than on its transfer and application [5], therefore, it is a project which does not generate new knowledge, but it applies that existing knowledge with autonomy and own criterion.

On the other hand, this project allows and promotes the generation of own technology in the institution which is translated into a low-cost endowment for the electromechanical laboratory. Likewise, it can positively impact the development of the city and the region by training students with the example of making their own tools tailored to the needs. Hence, the question arises: Can industrial solutions be given to the problem of selecting materials by colors with the tools that the institution has, such as Solid Works, 3D printer and laser cutter,?

This is a technological development project; therefore it is an applied research, where acquired knowledge and existing theories will be put into practice.

To build an automatic system for classifying objects by colors at low cost as an initiative to equip the ITCMB's Electromechanics laboratory with useful tools for teaching-learning and generating own technology, the following methodology is followed (see figure 1), in item 3 each of the steps of the methodology is described.

IOP Conf. Series: Materials Science and Engineering **844** (2020) 012037 doi:10.1088/1757-899X/844/1/012037

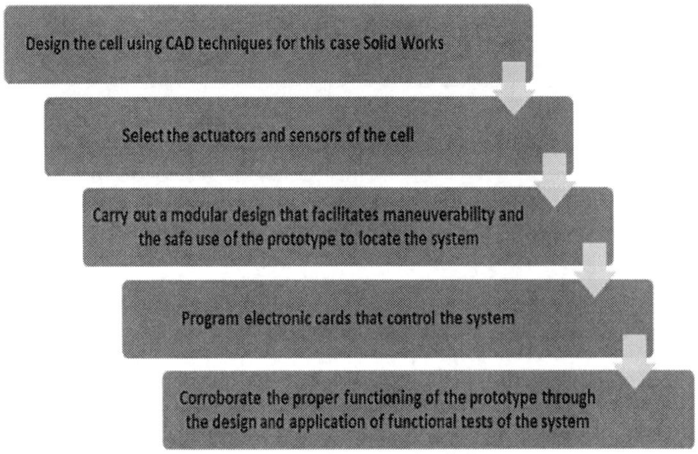

Figure 1. Project methodology

3. Phases applied to the construction the prototype

3.1. Designing the cell using CAD techniques, for this case Solid Works.

At this stage, the detail of the morphology of the cell was done. The computer-aided design program Solid Works [1] was used for this work, software with which the institution has an academic license, and hence the construction plans resulted.

Basically, the system is composed of three fundamental parts, which are: the cap distributor, the conveyor belt where the color sensor is located, and the selector [3].

The distributor (see figure 2a) is responsible for providing the conveyor belt covers ensuring they are separated and lying down. Also, it controls the time between cap and cap and it has a rotating system that ensures that one cap passes at a time, and an internal mechanism that prevents the caps from jamming when the storage part is full. Until now, it works very well, however, a new design is being worked on, which facilitates the feeding of the distributor, since at this time, the prototype is done manually.

Figure 2. Distributor designed in Solid Works (a.), Real Distributor (b).

Each piece was machined in a laser cutter and it was thought that some materials were transparent for academic reasons, since in this way we could visualize the mechanisms and their operation. The distributor works with truncated standard servomotor, so that it has a 360 degree turn, here it is not of interest to control angle, but speed.

The next fundamental part is the conveyor belt (Figure 3). In this case it was not manufactured, although this is possible and as a result, costs are lowered. This project used a band which we had, for this reason there is no emphasis on the construction of it.

However, it is highlighted that the system where the sensor is located that serves for the identification of colors is on the band, also it is a presence sensor, which helps control the entire system. The band is controlled with a simple PWM, and in this way it avoids jams when sensing the color of the cap.

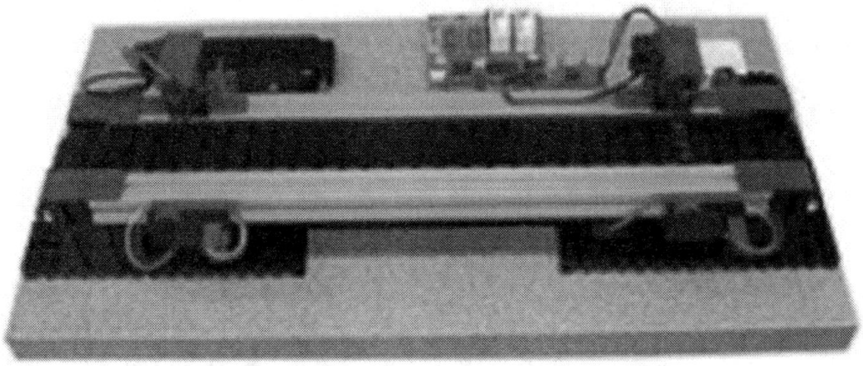

Figure 3. Conveyor belt

Last but not least, it is the selector, whose job is to get the cap of the identified color to the corresponding container (Figure 4), and in this way the process is finished.

Figure 4. Rotary selector.

3.2. Sensors and actuators

In this stage the actuators, such as motors, servomotors, and other elements necessary for the operation of the cell were determined. They will not be detailed, since it is not the focus of this article. However, there is a lot of literature on these devices. Basically, the system consists of a motor reducer, which is responsible for moving the conveyor belt; a standard micro servo which is responsible for stopping the cap on the conveyor belt at the time of sensing; a standard servomotor that gives mobility to the distributor and a servomotor of 9Kg * cm that is in charge to give movement to the selector. In this case the position control is necessary to be able to locate each cap of each color in the allocated compartment.

Also, the most suitable sensors for the task were selected, since the system is autonomous, the basis of its operation are the sensors, and moreover the associated electronics was added, in which Arduino technology was used for the specific case.

After several tests, it was decided to identify the color of each cap by using the RGB color sensor module TCS3200 [2] (Figure 5), compatible with Arduino technology. It is an economic sensor, easy to use and above all that, it worked well for the assigned task.

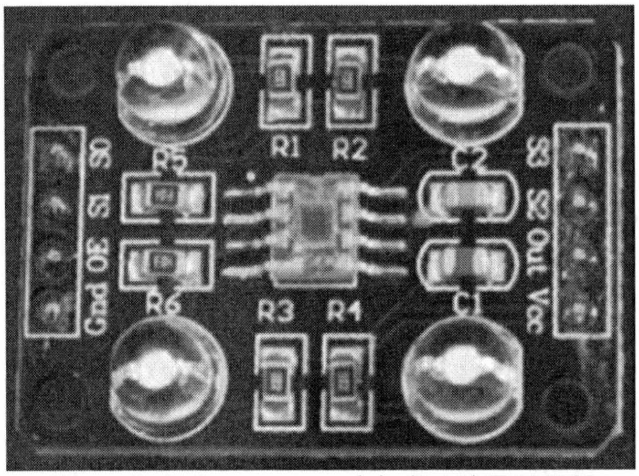

Figure 5. Color sensor module (RGB) TCS3200

The other sensor used is an infrared presence sensor reference E18-D80NK [6] (Figure 6), and this is used when the cap starts to enter the sensing area. The order can be given to the micro servo that stops the cap for 300 milliseconds so that the TCS3200 can read and perform the processing on the Arduino.

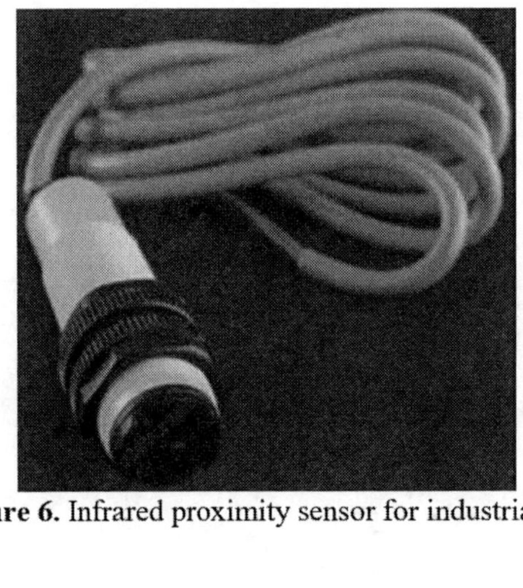

Figure 6. Infrared proximity sensor for industrial use

3.3. Developing a modular design that facilitates the maneuverability and safe use of the prototype to locate the system

Having already the design and the planes of the pieces together with the location of the actuators and sensors, as well as the associated electronics, we proceed to the construction of each of the elements that make up the cell. It is intended that the vast majority of elements can manufacture using the 3D printer. However, the drawings are not shown in this document because the number of pages is limited but if any design is required, the author can be contacted by email. Consequently, only figure 7 is shown as evidence.

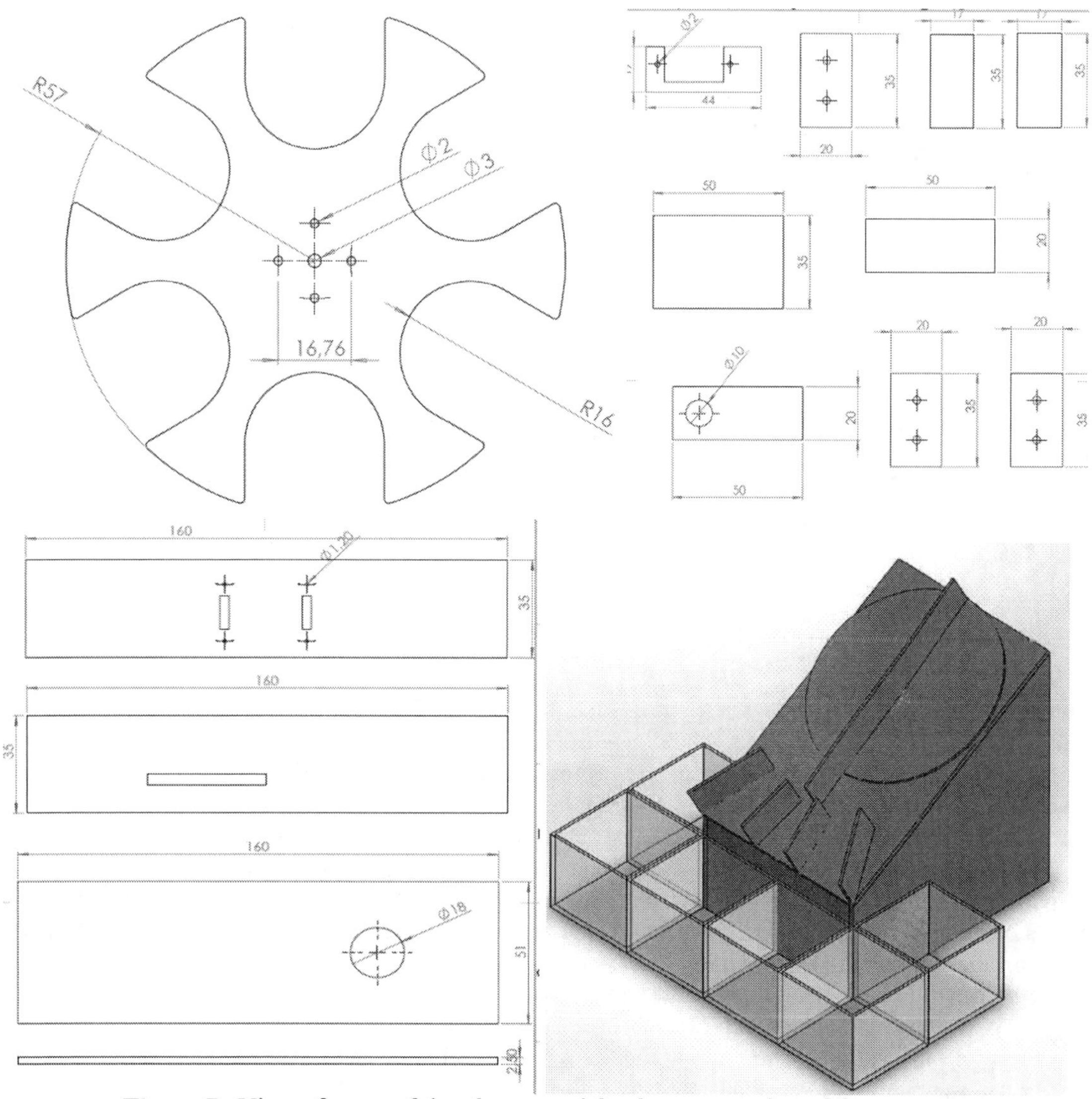

Figure 7. View of some of the planes used for the construction of the prototype.

3.4. Programming the electronic cards that control the system

Having already assembled all the elements of the cell, and each sensor and each actuator being in place, we proceed to the programming of the system, which when using the Arduino control cards as a base, free software will be used for its programming. The result is already the cell working as a system for selecting objects by colors.

Basically, the system starts to work and when the cap is detected by the infrared sensor, it is the signal to activate the micro servo that is located to stop the cap for a period of time and the color sensor can take the information, which will be processed on the Arduino card, and since the algorithm has determined what color the cap is, the order is given to the servomotor that is in the selector to guide the cap to the assigned tank. Currently, there are no security systems as interpreting that there are no more caps, or that there are bottlenecks in the band. However, this work is underway for the next version.

3.5. Corroborating the proper functioning of the prototype through the design and application of functional tests of the system

The functional tests prepared to check the operation of the built system are below.

The results of the application of the tests are presented in the results section.

Table 1. Operating tests of the cell

Test design
Collecting data with TCS3200 sensor exposed to ambient light and without any objects present in the band
Gathering data with TCS3200 sensor exposed to ambient light and objects present in the band
Collecting data with sensor TCS3200 located on a tunnel that hides the caps from the ambient light when they are sensed and any object presents in the band
Gathering data with sensor TCS3200 located on a tunnel that hides the cap of the ambient light when they are sensed and objects present in the band

4. Results

In this section it will be shown in more detail how the TCS3200 sensor was able to take the information more effectively, thus achieving close to 100% efficiency at the time of selecting the caps.

The main result is the operation of the prototype, making the selection of each of the caps and placing them in their assigned deposit according to the color: red, blue, green, yellow, white or black. These colors were selected for this specific case, showing the great variety of colors it can face.

This prototype, unlike the commercial ones at the educational level, gives the possibility of simultaneously selecting up to 6 colors, but leaving open the possibility of including new selection of colors by adjusting the algorithm. In this way, the possibility for the student to interact with the system is open, since it can be reprogrammed.

Then, it will be shown in more detail how we worked with the TCS3200 sensor, the data that was taken and how the input data was improved.

It was selected as a sensitive element the light, in the visible range (400 nm - 700 nm) the TCS3200 that transforms the perceived light intensity into frequency. This element can be configured to work with red, green, blue or any filters [7]. The microcontroller, in this case an Arduino UNO card, configures the sensor and takes 100 measurements in approximately 1.2 seconds in order to make a statistic because it was found that with a single reading there were errors of up to 10%, since with the taking of the 100 measurements it was reduced to 0.

The first tests were done exposing the sensor, being exposed to ambient light, and the results showed a great variety of interference. The sensor takes RGB readings; this means the range of red, green and blue light of the element that will measure, or the light that is affecting the sensor. For the case shown in Figure 8, the sensor is without the presence of an object. The result would be that there is no object, but it can be seen how the signals are intertwined. However, the tendency is that the strongest signal is green, followed by red and finally blue.

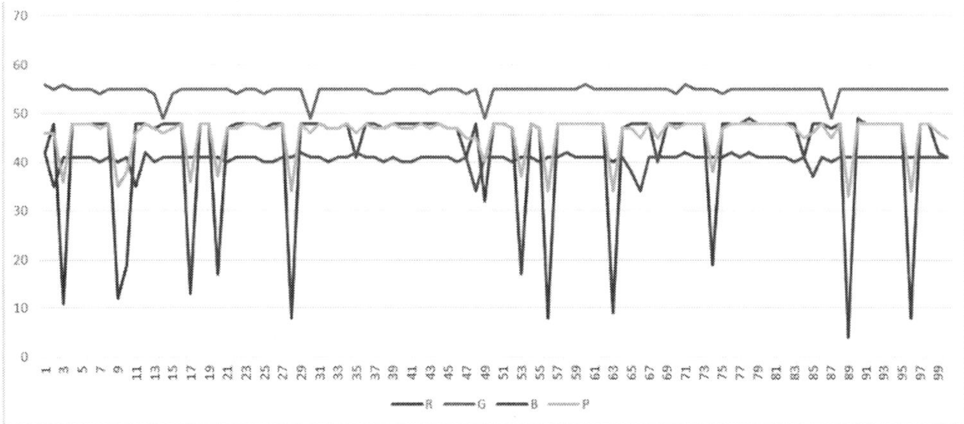

Figure 8. Data collected with the TCS3200 sensor in a natural environment.

On the other hand, it is not fulfilled throughout the course of the 100 samples; in fact it occurs in 79% of cases. These variations mean that if only one reading is taken, there is a 21% chance of error. In order to improve this situation, it was decided to control the environment, which is nothing more than tunnelling on the conveyor belt (see figure 9) to avoid external light as much as possible, and thus improve the sensor data collection.

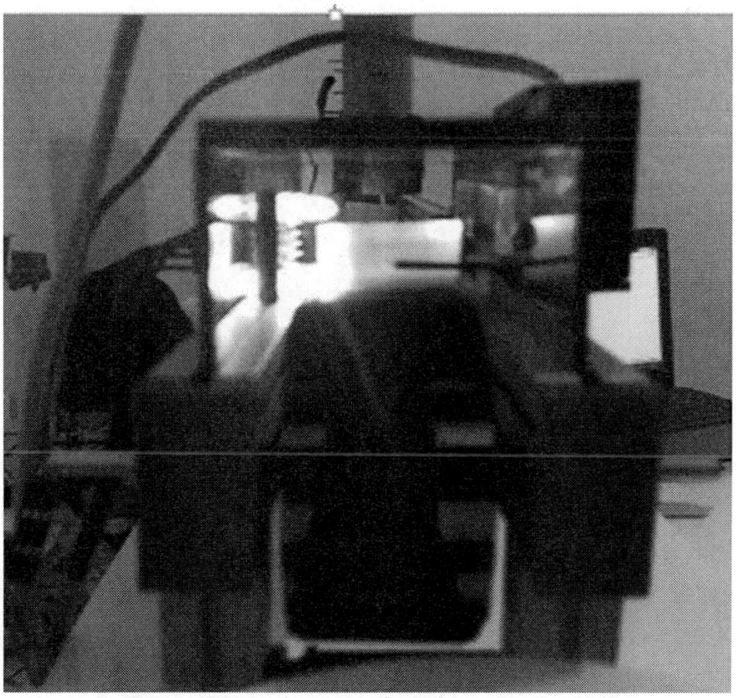

Figure 9. Sensing tunnel on the conveyor belt

After several tests, the improvement in the signal for the 100 samples was noted, going to 96% of hits (see figure 10), both without the presence of objects, and with the presence of the caps.

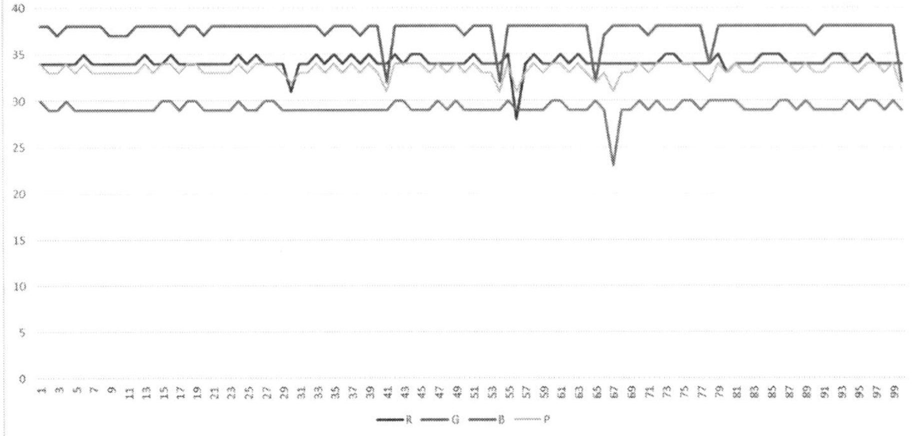

Figure 10. Data collected with the TCS3200 sensor in a controlled environment

Having already eliminated most of the noise, we proceeded to do the same for the identification of each of the colors, since with the improved signal, the selection algorithm could be programmed, passing for the case of blue 85% effectiveness to 100%, as can be seen in the comparison of the graphs in figure 11. The upper graph is in natural environment and the lower graph in controlled environment, even for some colors the signals are much more constant and stable.

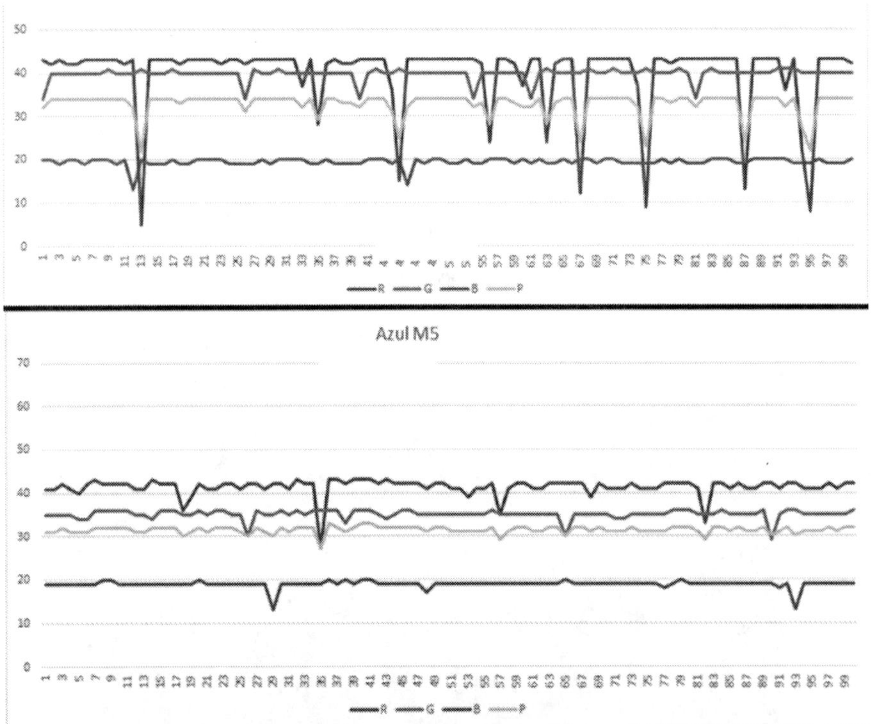

Figure 11. Data comparison for blue color, superior in natural environment, lower controlled environment

It is valid to add that the program was made in a modular way, or with subroutines, which makes the main program simply have the correct sequence of the call of these subroutines, as can be seen below.

```
//*******************main program*******************
void loop(){
presencia (); //the presencia subroutine is called that identifies if there is an object. (IR sensor)
 if (pre==1){ //If pre equals 1, it is because there is an object. and continues to identify the color.
   p=60;// The mini servo is sent at 60 degrees for the color sensor data capture.
   servomini();//With p = 60, take the mini servo to this position.
  delay(1200);// data collection time, and is the maximum time required for the selector to go to its destination.
   calculo (); // Color selection algorithm
   p=20;// The mini servo is sent at 20 degrees to let the object pass.
  servomini();//With p = 20, take the mini servo to this position.
  servo();//The selector servo motor goes to the position that the calculo subroutine sends, according to the identified color.
 }
//*********************************************************
```

With this data collected by the TCS3200 sensor, the algorithm for the selection of colors was carried out, for example, observing the behaviour of figure 11 at the bottom where the tendency is clear that red is greater than green and green greater than blue, it can be determined for this case that the object is blue, in the same way the analysis is done for each of the covers and the result is shown below, it should be noted that for reasons of space only will show the part of the algorithm that was used to identify the color, since turning on the conveyor belt, adjusting the PWM, giving the order to the servo motor of the positioning angle for the final classification of the covers, is a simpler job that even arrives to be routine for those who work with microprocessors.

```
//*******************calculation*******************
void calculo(){
r=0;// variables started at 0. Have already been declared in the setup.
g=0;
b=0;
h=0;
n=0;
w=0;
s=0;
may=0;
for (c=1; c<=100; c++){ // 100 data taken by the TCS3200 sensor
//************TCS3200 sensor data socket*******************
//****This is the routine to capture the red, green and blue data*****
//*****on the TCS3200 sensor according to the filters applied*******
//*******************************************************
digitalWrite(s2, LOW);
digitalWrite(s3, LOW);
rojo = pulseIn(out, digitalRead(out) == HIGH ? LOW : HIGH);
digitalWrite(s3, HIGH);
azul = pulseIn(out, digitalRead(out) == HIGH ? LOW : HIGH);
digitalWrite(s2, HIGH);
verde = pulseIn(out, digitalRead(out) == HIGH ? LOW : HIGH);
//*******************************************************
```

```
//********************Color selection***********************
//*********red*********
 if (rojo < azul && verde > azul && rojo < 25 && verde>20) //conditions for red
{
 r=r+1;
 }
 //*********blue*********
 else if (rojo>verde && Verde>azul && azul<25) //conditions for blue
 {
  b=b+1;
 }
 //*********green*********//
 else if (rojo > verde && abs(azul-verde)<10 && rojo>20) //conditions for green
  {
  g=g+1;
 }
 //*********white*********//
 else if (azul < 15 && rojo < 15 && verde < 15 ) //conditions for white
  {
  w=w+1;
 }
 //*********black*********//
 else if (azul<rojo && rojo<verde && verde>40 ) //conditions for black
  {
  n=n+1;
 }
 //*********yellow*********//
 else if (rojo<verde && verde<azul && rojo< 18) //conditions for yellow
  {
  h=h+1;
 }
 else //otherwise it is not one of the colors to select
{
  s=s+1;
 }
 //delay(10);
//******************************************
//***Here it is defined which one obtained greater****
//***successes, that determines the detected color****
//******************************************
if (may<r){
 may=r;
}
if (may<g){
 may=g;
}
if (may<b){
 may=b;
}
if (may<h){
 may=h;
```

IOP Conf. Series: Materials Science and Engineering **844** (2020) 012037 doi:10.1088/1757-899X/844/1/012037

```
}
if (may<n){
  may=n;
}
if (may<w){
  may=w;
}
if (may<s){
  may=s;
}
//*****************************************************************
//****Here the times of successes are compared with the correct data of each******
//****color and the detected color is printed as well as the value is given to********
//**given to the motorcycle servo to place the lid on the corresponding container***
//*****************************************************************
}
if (may>10){
if (may==r){
  Serial.print(may);
  Serial.println("rojo");
  p2=50;//servo position for red
  servo ();//Call the servo subroutine to place the selector in the corresponding container.
  }
if (may==g){
  Serial.print(may);
  Serial.println("verde");
  p2=65; //servo position for green
  servo ();//Call the servo subroutine to place the selector in the corresponding container.
}
if (may==b){
  Serial.print(may);
  Serial.println("azul");
  p2=80; //servo position for blue
  servo ();//Call the servo subroutine to place the selector in the corresponding container.
}
if (may==h){
  Serial.print(may);
  Serial.println("amarillo");
  p2=115; //servo position for yellow
  servo ();//Call the servo subroutine to place the selector in the corresponding container.
}
if (may==n){
  Serial.print(may);
  Serial.println("negro");
  p2=105; //servo position for black
  servo ();//Call the servo subroutine to place the selector in the corresponding container.
}
if (may==w){
  Serial.print(may);
 Serial.println("blanco");
 p2=130; //servo position for white
```

IOP Conf. Series: Materials Science and Engineering **844** (2020) 012037 doi:10.1088/1757-899X/844/1/012037

```
    servo ();//Call the servo subroutine to place the selector in the corresponding container.
  }
  }
  }
//***************************END***********************************
```

It is in the hands of the students other applications, such as selecting sizes or other objects, since the industry has experienced a quantitative growth in terms of brands and products in a considerable way [8], and in this way there is a prototype that approaches the concept Flexible manufacturing.

5. Conclusions

The main conclusion is the achievement of the proposed objective, since a solution was achieved to the question posed. Moreover, it is clear that from the technological programs can be given solutions to real problems of the industry.

As a laboratory team, students were brought into direct contact with the equipments, not only for maintenance work, but also for programming and design.

A relatively inexpensive solution was achieved, since the most economical equipment in the market, for the selection of colors costs for about US $ 4600 and only selects two colors (Recreational Didactic Technology EIRL, 2013) which brings its own card already programmed, while this prototype did not reach USD 1500 (see figure 12).

Regarding the TCS3200 sensor, its efficiency was evidenced for a project like this one. It is very fast since 100 samples are taken in less than 1.2 seconds, which allow to make a statistic to ensure the color selection with an effectiveness of 100% as long as it is in optimal working conditions, for example, that there are no external lights among others.

The fact that the Technological Institution Colegio Mayor de Bolívar has equipment such as a 3D printer, laser cutter, among others, facilitates the learning and development of technology-based equipment. These products are typified in COLCIENCIAS, similar to the generation of new knowledge; making groups of research are strengthened. Moreover, they are generating and appropriating technologies that are normally brought from abroad.

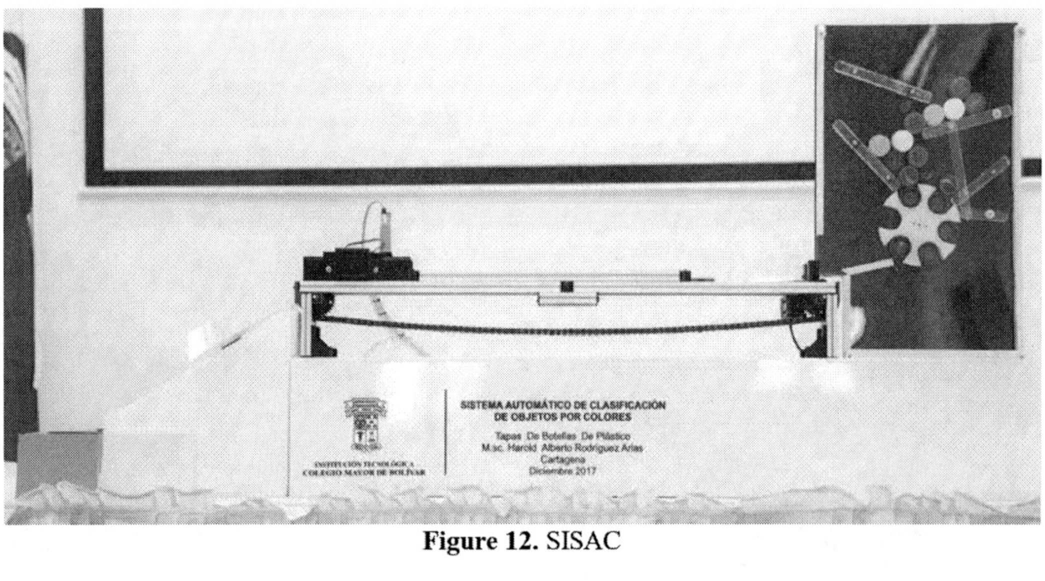

Figure 12. SISAC

Acknowledgments
Our gratitude to the Technological Institution "Colegio Mayor de Bolívar" because it supported and financed the project. Moreover, it is the facilitator of the institution's technological development. Likewise, thanks are expressed to the students who take part in the research seedbed of the electromechanical program (A&Sel) because of their dedication and performance to carry out this project.

References

[1] Dassault Systèmes SolidWorks Corporation, «Solid Works Education: Learn. Create. Succeed,» 2010. [En línea]. Available: https://www.solidworks.com/sw/docs/Student_WB_2011_ESP.pdf. [Último acceso: 2018].
[2] DF Robot, «DF Robot: Drive the Future,» 2017. [En línea]. Available: https://media.digikey.com/pdf/Data%20Sheets/DFRobot%20PDFs/SEN0101_Web.pdf. [Último acceso: 2018].
[3] FESTO AG & CO. KG, «Estación de Clasificación: Final,» 2018. [En línea]. Available: https://www.festo-didactic.com/es-es/productos/mps-sistema-de-produccion-modular/estaciones/estacion-de-clasificacion-final.htm?fbid=ZXMuZXMuNTQ3LjE0LjE4LjYwNi4zOTQ4&basket=add&vid=5918.
[4] Tecnología Didáctica Recreacional EIRL, «TECNODIDÁCTICA. Educación STEM,» 2013. [En línea]. Available: http://www.tecnodidactica.pe/index.php/productos/entrenamiento-industrial/linea-de-transporte-y-clasificacion-st00220024.html.
[5] COLCIENCIAS, «COLCIENCIAS: Ciencia, Tecnología e Innovación,» 2016. [En línea]. Available: http://www.colciencias.gov.co/sites/default/files/ckeditor_files/politiciadeactores-snctei.pdf. [Último acceso: 2018].
[6] 61mcu, «Adjustable Infrared Sensor Switch Manual,» 2016. [En línea]. Available: http://dl.btc.pl/kamami_wa/e18-d80nk-ds.pdf.
[7] H. A. Rodriguez-Arias, D. Ruiz y J. R. Castro-Suarez, «Design And Implementation Of A Photocolorimeter Using Microcontrollers To Determine Analyzes In Aqueous Solutions,» 16 th LACCEI International Multi-Conference for Engineering, Education, and Technology: "Innovation in Education and Inclusion", 2018.
[8] L. Solís and M. Carpena, «Prototipo de un selector automático de materiales y colores,» Escuela Profesional de Ingeniería Industria, Universidad Ricardo Palma, Lima, 2010.
[9] J. Padilla-Magaña, I. Sánchez-Suárez and P. Oseguera-Espinoza «Control Untomático de una máquina seleccionadora por color mediante la PixyCam CMUCAM5 para aseguramiento de la calidad,» Revista de Tecnología e Innovación, Vol 3, No. 8 Pag 35-44, Septiembre 2016, Bolivia.
[10] A. Arenas, «Diseño e implementación de prototipo a escala de máquina clasificadora de color por visión,» Universidad Católica de Santiago de Guayaquil, Septiembre 2018, Ecuador
[11] W. Zhaokun, P. Binbin, Y. Huang and G. Sun, «Classification for plastic bottles recycling based on image recognition,» Waste Management, Vol 88, Pag, 170-181, Abril 2019, Elsevier Ltd.
[12] G. Efraim and A. Benny, «Development of Robotic Arm for Color Based Goods Sorter in Factory Using TCS3200 Sensor with a Web-Based Monitoring System,» 2018 Third International Conference on Informatics and Computing (ICIC), Agosto 2018, DOI 10.1109/IAC.2018.8780461, IEEE, Indonesia.
[13] A. Sachdeva, M. Gupta, M. Pandey and P. Khandelwal, «Development Of Industrial Automatic Multi Colour Sorting and Counting Machine Using Arduino Nano Microcontroller and TCS3200 Colour Sensor,» The International Journal of Engineering and Science (IJES), Vol. 6, Pag. 56-59, 2017. India

IOP Conf. Series: Materials Science and Engineering **844** (2020) 012038 doi:10.1088/1757-899X/844/1/012038

Parameter estimation of a second order system via nonlinear identification algorithm

Patricio Ordaz*,1, Liliam Rodríguez-Guerrero[1], Omar Santos[1], Carlos Cuvas[1], Hugo Romero[1], Mario Ordaz-Oliver[2], Pablo López-Pérez[3]

[1]Electronic and Control Academic Group, Research Center on Technology of Information and Systems, Universidad Autónoma del Estado de Hidalgo, Pachuca de Soto, Hidalgo, México.
[2] Universidad La Salle-Pachuca, engineering School, Hidalgo, Pachuca de Soto, México
[1] Escuela Superior de Apan, Universidad Autónoma del Estado de Hidalgo. Apan, Hidalgo, México
* Corresponding author.

E-mail: jesus_ordaz@uaeh.edu.mx

Abstract. In this paper, a parameter estimation for a class of second order nonlinear system is presented. The considered system, can be represented in such way that it is linear respect its parameters. Since the system output is given, only, for the second integrator a state estimation of unmeasured state variables is reconstructed via nonlinear observer based on the terminal sliding mode observer. Therefore, the main contribution of this paper deals with a full order nonlinear observer algorithm design to enhance parameter identification of a nonlinear second order system. Finally, to illustrate the theoretical performance of the proposed identification algorithm, an experimental result of a mechanical system is presented.

1. Introduction

In system dynamics theory, study of trajectories performance is essential to job to analyze its performance. However, in deterministic systems, not all information is available to be measured or it is an estimate value of a real one. This is the case of mechanical, electric and electronic devices [1, 2, 3]. For this reason, successful parameters identification algorithms and/or trajectory estimation approaches of unmeasured state space variables, are very common problems in data processing and control theory [4, 5, 6, 7]. Today, several parameter identification algorithms were designed where the considered system fulfills the linearity respect its parameters [4, 5, 7, 8]. Actually, the Least-Square Method (**LSM**), attracts a lot attention of engineers by the easiness of implementation of this algorithm in real time processes and its applications on adaptive-control and neural-network systems [4, 5, 7, 9, 8, 10, 11]. In fact, parameter identification approaches use the regressor-vector scheme into a recursive algorithm which gives a numerical solution of a differential equation to update the current parameter estimation [9, 12, 13, 14, 15]. However, almost all algorithms to identify system parameters, algorithms like instrumental variables (**IV**), **LSM**, autoregressive moving average model with exogenous input model, **IV** with forgetting factor, among others need all variable states [9, 10, 11, 16].

In this work, it is assumed that only one of the state variables is available to be measured. Since, the vast majority of algorithms to identify parameter require an accurately measured state variable. For this reason, the reconstruction of high quality state signals is key piece before to identify the system parameters. In this context, the state estimation may be given by different procedures as; signal processing with the use of advanced filters and state estimation via full order or reduced type observers [17, 18, 19, 12]. Due that, the **LSM** needs accurately measured state variables to work efficiently, it is necessary to have a workable state estimation. This problem can be attached by using Sliding Mode Observer (**SMO**), this one uses the concept of finite time convergence to guarantee correct state estimation [20, 21]. Actually, the concept of terminal sliding mode control guarantees the trajectory convergence of second order systems [22, 23]. Thus, if the finite time convergence of the estimated states to real one is guaranteed, then the parameter identification via **LSM** could be suitable. In this way, a state estimation based on terminal sliding mode observer (**TSMO**) was successfully implemented [24]. In fact, the designed observer function supplies of relevant information to the **LSM** for the parameter identification.

The outline of this paper is as follows. The problem statement, system description and preliminaries are presented in section 2. Next, Section 3 provides the main paper contribution, parameter identification where the no available state variable is obtained via **TSMO**. Subsequently, in Section 4, the identification-state estimation algorithm is tested in numerical example. Finally, concluding remarks are displayed in Section 5.

2. Problem statement and preliminaries

Consider the system

$$\begin{aligned} \dot{x}_1 &= x_2, & x(0) &= x_0, \\ \dot{x}_2 &= f(x,t) + g(x_1)u, & y &= x_1, \quad \text{a. e. } t \in \mathbb{R}^+, \end{aligned} \tag{1}$$

where $x \in \mathbb{R}^2$ is the state space, $f : \mathbb{R}^2 \to \mathbb{R}$ is a nonlinear vector function which fulfills

$$\|f(x,t)\| \leq \delta_0 < \infty, \quad 0 < \|g(x_1)\| \leq \delta_1 < \infty \text{ and } \|x_2\| \leq \delta_2 < \infty. \tag{2}$$

The control input is defined as $u \in \mathbb{R}$, $g \in \mathbb{R}$ can be a linear or nonlinear function. Since, x_2 is not available to be measured and system dynamics contain unknown parameters. The problem here, design a full order observer in order to obtain the estimate $\hat{x}$, which in finite time converges to the real one. Furthermore, in same time that the state estimation algorithm works, identify the system parameters. In this way, the state observer is proposed in next format

$$\begin{aligned} \dot{\hat{x}}_1 &= \hat{x}_2, \\ \dot{\hat{x}}_2 &= \varphi(\sigma), \ \hat{x}(0) = \hat{x}_0, \end{aligned} \tag{3}$$

where the estimation error is defined as $e = x - \hat{x}$, $\varphi \in \mathbb{R}$ is the observer function, $\sigma \in \mathbb{R}$ is the sliding manifold, defined below. Moreover, nonlinear functions $\hat{f}$ and $\hat{g}$ has known dynamics with unknown but estimate parameters. Thus, the dynamic error function follows as

$$\begin{aligned} \dot{e}_1 &= x_2 - \hat{x}_2, & e_1(0) &= e_1^0, \\ \dot{e}_2 &= f(x,t) + g(x_1)u - \varphi(\sigma), & e_2(0) &= e_2^0, \end{aligned} \tag{4}$$

Notice that, systems (1)-(3) can be written as

$$\dot{x}_1 = x_2, \quad \dot{x}_2 = \mathbf{\Theta}\psi(x,u), \tag{5}$$

where $\mathbf{\Theta} \in \mathbb{R}^{1\times p}$ is a matrix which contains "p" parameters of system in linear form, $\psi(x,u) \in \mathbb{R}^p$ is the system regressor vector and it contains all state functions.

Summarizing, the problem covered by this work, is design a full order nonlinear observer such that the state estimate error function converges to zero solution in finite time; at the same time joins the designed observer function in such way that it be used in the **LSM** to identify the system parameters. According to the above, the following definition is introduced.

Definition 1 (Least-Square estimation) *For a given joint uncertainty* $\Delta(\hat{x}, u)$, *the function*

$$\hat{\Theta} = \arg\min_{\hat{\Theta}} \int_{t_0}^{t} \|\Delta(\hat{x}, u)\|^2 d\tau, \tag{6}$$

denotes the Least-Square estimate $\hat{\Theta}$, *obtained from the measurement of variables* $(\hat{x}, u)$.

3. State estimation and parameter identification algorithms

Here, the estimate dynamic (3), is based on Sliding Mode Observer function $\varphi(\sigma)$. The sliding manifold is stablished, in such way, that the estimation error e_1 converges to zero in finite time. To this end, consider next quadratic storage function

$$V(\sigma) = \frac{1}{2}\sigma^2, \quad \sigma = e_2 + Re_1|e_1|^{-\frac{1}{3}}, \; 0 < R \in \mathbb{R},$$

where $\dot{e}_1 = e_2$, and $\sigma \in \mathbb{R}$ is the sliding manifold. Notice that, its time derivative along the trajectories (4) fulfills next relation

$$\frac{d}{dt}V(\sigma)\bigg|_{(4)} = \sigma\left\{\dot{e}_2 + R\dot{e}_1|e_1|^{-\frac{1}{3}} + Re_1\left(-\frac{1}{3}|e_1|^{-\frac{4}{3}}\frac{d}{dt}|e_1|\right)\right\}. \tag{7}$$

Further, time derivative of $|e_1|$ yields to $\frac{d}{dt}|e_1| = \text{Sign}(e_1)\dot{e}_1$. Thus, (7) yields to

$$\tfrac{d}{dt}V(\sigma)\big|_{(1)} \quad \sigma\left\{f(x,t) + g(x_1)u - \varphi(\sigma) + \tfrac{2}{3}R\dot{e}_1|e_1|^{-\frac{1}{3}}\right\}.$$

By introducing the nonlinear observer function $\varphi(\sigma) = \frac{2}{3}R\dot{e}_1|e_1|^{-\frac{1}{3}} + \phi(\sigma)$, previous equation yields to

$$\tfrac{d}{dt}V(\sigma)\big|_{(1)} = \sigma\left\{f(x,t) + g(x_1)u - \phi(\sigma)\right\},$$

without loos of generality, $\dot{V}(\sigma) \leq |\sigma||f(x,t)| + |\sigma||g(x_1)||u| - \sigma\phi(\sigma)$ holds. Using assumption (2), and form previous inequality, $\dot{V}(\sigma) \leq \delta_0|\sigma| + \delta_1|\sigma||u| - \sigma\phi(\sigma)$ is satisfied. In this way, if the function $\phi(\sigma)$ is given by

$$\phi(\sigma) = (\rho_1 + \rho_2|u|)\,\text{Sign}(\sigma), \tag{8}$$

the storage function time derivative yields to $\dot{V}(\sigma) \leq -\alpha_1|\sigma| - \alpha_2|u||\sigma| \leq -\alpha_1|\sigma|$, then $\dot{V}_1(\sigma) \leq -\alpha_1|\sigma| = -\alpha_1\sqrt{2}\,V_1^{\frac{1}{2}}(\sigma)$. Previous differential inequality holds next solution

$$V_1^{\frac{1}{2}}(\sigma(t)) \leq V_1^{\frac{1}{2}}(\sigma(t_0)) - \frac{\alpha_1}{\sqrt{2}}(t - t_0),$$

which means that the sliding variable converges to the origin in finite time

$$t_r = \frac{\sqrt{2}}{\alpha_2}V_1^{\frac{1}{2}}(\sigma(t_0)) + t_0, \tag{9}$$

where $\alpha_2 = \rho_2 - \delta_1 > 0$. After this time, the sliding motion $\sigma = 0$ for all $t \geq t_r$. On the other hand, under assumption that $\dot{e}_1 = e_2$, so where $\sigma = 0$ is fulfilled, then

$$\dot{e}_1 = -Rx_1|e_1|^{-\frac{1}{3}}, \; e_1(0) = e_1^0,$$

solving, for all $\tau \in [t_r, t)$, next relation is obtained

$$|e_1(t)|^{\frac{1}{3}} = |e_1(t_r)|^{\frac{1}{3}} - \frac{R}{3}(t - t_r).$$

Thus, trajectory $e_1(t)$ converges to zero in finite time $T = \frac{3}{R}|x_1(0)|^{\frac{1}{3}} + t_r$. Once fixed the sliding manifold, it is evident that, after time T, the estimate state $\hat{x}_1 = x_1$, so $\hat{x}_2 = x_2$ and $\psi(x,u) = \psi(\hat{x},u)$. In this sense

$$0 = \mathbf{\Theta}\psi(\hat{x},u) - \varphi(\sigma) \;\; \forall\, t \geq T,$$

However, matrix $\mathbf{\Theta}$ is unknown, and the aim of this paper, is to identify the parameters of matrix $\hat{\mathbf{\Theta}}$. Additionally, $\tilde{\mathbf{\Theta}} = \mathbf{\Theta} - \hat{\mathbf{\Theta}} \in \mathbb{R}^{1\times p}$, defines the estimate error matrix, and it can be estimated form previous statement. By adding and subtracting $\hat{\mathbf{\Theta}}$, $\tilde{\mathbf{\Theta}}\psi(\hat{x},u) = \hat{\mathbf{\Theta}}\psi(\hat{x},u) - \varphi(\sigma)$ is fulfilled.Thereby, it can be defined the adjoint uncertainty $\Delta(\hat{x},u)$ as

$$\Delta(\hat{x},u) = \tilde{\mathbf{\Theta}}\psi(\hat{x},u) = \left(\hat{\mathbf{\Theta}}\psi(\hat{x},u) - \dot{\hat{x}}_2\right). \tag{10}$$

From this statement, the joint uncertainty allows to apply the identification of considered parameters $\hat{\mathbf{\Theta}} \in \mathbb{R}^{1\times p}$. Using Definition 1, and selecting the following performance index

$$\begin{aligned}\mathcal{J}(\mathbf{\Theta}) &= \textstyle\int_{t_0}^{t} \|\Delta(\dot{\hat{x}},u)\|^2 d\tau = \int_{t_0}^{t} \left\|\left(\hat{\mathbf{\Theta}}\psi(\hat{x},u) - \dot{\hat{x}}\right)\right\|^2 d\tau \\ &= \mathrm{tr}\left\{\hat{\mathbf{\Theta}}\left(\textstyle\int_{t_0}^{t} \psi(\hat{x},u)\psi^{\mathsf{T}}(\hat{x},u)d\tau\right)\hat{\mathbf{\Theta}}^{\mathsf{T}}\right\} + \textstyle\int_{t_0}^{t} \dot{\hat{x}}_2^2 d\tau - 2\mathrm{tr}\left\{\left(\textstyle\int_{t_0}^{t} \dot{\hat{x}}_2\psi^{\mathsf{T}}(\hat{x},u)d\tau\right)\hat{\mathbf{\Theta}}^{\mathsf{T}}\right\}.\end{aligned} \tag{11}$$

The derivative respect to the parameters $\hat{\mathbf{\Theta}}$ yields to

$$\frac{\partial}{\partial\hat{\mathbf{\Theta}}}\mathcal{J}(\mathbf{\Theta}) = -2\int_{t_0}^{t} \dot{\hat{x}}_2\psi^{\mathsf{T}}(\hat{x},u)d\tau + 2\hat{\mathbf{\Theta}}\int_{t_0}^{t} \psi^{\mathsf{T}}(\hat{x},u)\psi(\hat{x},u)d\tau.$$

The minimization of the estimate $\hat{\mathbf{\Theta}}$ can be achieved as $\frac{\partial}{\partial\hat{\mathbf{\Theta}}}\mathcal{J}(\mathbf{\Theta}) = \mathbf{0}$, then

$$\hat{\mathbf{\Theta}} = \mathbf{Z}(t)\bar{\mathbf{\Gamma}},\;\; \mathbf{Z}(t) = \textstyle\int_{t_0}^{t} \dot{\hat{x}}_2\psi^{\mathsf{T}}(\hat{x},u)d\tau, \;\;\; \bar{\mathbf{\Gamma}} = \left[\textstyle\int_{t_0}^{t} \psi(\hat{x},u)\psi^{\mathsf{T}}(\hat{x},u)d\tau\right]^{-1}. \tag{12}$$

Notice that, for $t \geq T$, the solution $\hat{x} = x$ is fulfilled, consequently $\dot{\hat{x}} = \dot{x} = (\rho_1 + \rho_2|u|)\,\mathrm{Sign}(\sigma)$. Introducing a small enough $0 < \varkappa \in \mathbb{R}$, such that

$$\mathbf{\Gamma}^{-1}(t) := \int_{t_0}^{t} \psi(\hat{x},u)\psi^{\mathsf{T}}(\hat{x},u)d\tau + \varkappa\mathbf{I}_n,$$

the matrix $\mathbf{\Gamma}^{-1}$ is a non singular matrix. Hence, it is evident that $\frac{d}{dt}\left(\mathbf{\Gamma}^{-1}(t)\right) = \psi(\hat{x},u)\psi^{\mathsf{T}}(\hat{x},u)$. Accordingly with $\mathbf{\Gamma}^{-1}\mathbf{\Gamma} = \mathbf{I}_n$, the time derivative of this expression yields to $\dot{\mathbf{\Gamma}}(t) = -\mathbf{\Gamma}(t)\psi(\hat{x},u)\psi^{\mathsf{T}}(\hat{x},u)\mathbf{\Gamma}(t)$, thus, (12) is rewritten as

$$\tfrac{d}{dt}\hat{\mathbf{\Theta}}(t) = -\varphi(\sigma(t))\psi^{\mathsf{T}}(\hat{x},u)\mathbf{\Gamma}(t),$$

which defines the identification algorithm based on estate estimation (3)-(8).

4. Illustrative example

The mathematical model of the pendulum is given by the next ordinary differential equation

$$\ddot{q} = -\frac{g}{l}\sin(q) - \frac{b}{Ml}\dot{q} + \frac{1}{Ml}\tau, \tag{13}$$

Description	Notation	value	Units
Mass 1	M	0.37	kg
Pole length	l	0.25	m
Gravity constant	g	9.81	$\frac{m}{s^2}$
Damping coefficient	b	unknown	$N\frac{s}{m}$

Table 1. Parameters of the pendulum system.

where the measured output is the angular position as $y = q$. Introducing the change of variable $x_1 = q$, $x_2 = \dot{q}$ and $u = \tau$, next nonlinear system control afine, in state space representation, is reached as

$$\dot{x} = f(x) + g(x)u,\ y = Cx,\ C = [1,\ 0]$$
$$f(x) = \begin{bmatrix} x_2 \\ -\frac{g}{l}\sin(x_1) - \frac{b}{Ml}x_2 \end{bmatrix}, g(x) = \begin{bmatrix} 0 \\ \frac{1}{Ml} \end{bmatrix},$$

Previous system has the format as (1). System parameters are given in Table 1. The observer algorithm is implemented in Matlab-Simulink R2016a. The numerical solver is the fixed step 4th-order Runge Kutta with step size $h = 0.001$. The experimental results were obtained on the PendCon system, see Fig. 1. A National Instruments Multifunction DAQ NI USB-6343 is used to acquire signals and to transmit the control input to a 24 volt DC motor with 3600 Cnt/Rev optical encoder from Maxon Inc. Appropriately, a 24VDC, 2.1AMP power supply is used.

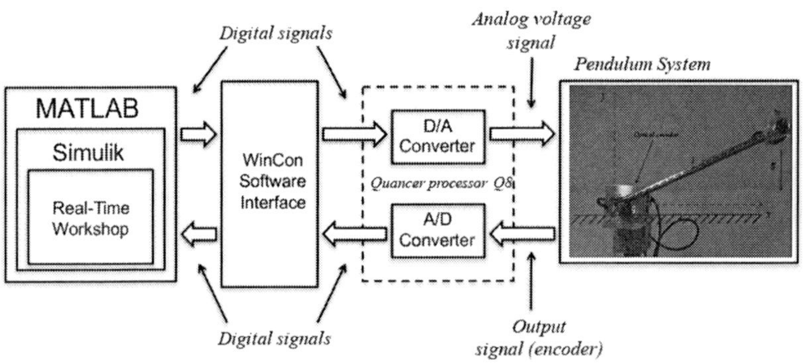

Figure 1. Realtime pendulum system.

The initial conditions of the real and estimate system are fixed as $x_0^{\mathsf{T}} = [1.3,\ 0]$ and $\hat{x}_0^{\mathsf{T}} = [-0.5,\ 0.5]$ respectively. In a sake to clarify the results of this algorithm, the estimation of system (13) is addressed for dynamics free of control, $u = 0$. Initial time is selected as $t_0 = 0$, final simulation time is $t_f = 20$ seconds, and parameters of observer are

$$R = 17,\ \delta_1 = 20,\ \delta_2 = 10.5,\ \rho_1 = 31,\ \rho_2 = 83,\ \alpha_2 = 72.5.$$

Thus, sliding manifold convergence time is $t_r = 1.1166$ and error convergence time $T = 6.6198$ seconds. For identification estimation algorithm, its parameters are selected as

$$\hat{\mathbf{\Theta}}_0 = [-0.1,\ 0.34,\ 0.25],\ \varkappa = 0.0001,\ L = 5.$$

Real parameters of considered system are $\mathbf{\Theta} = [39.2400,\ 0.1330,\ 10.8108]$. The experimental results are depicted on figures 2, 3, and 4.

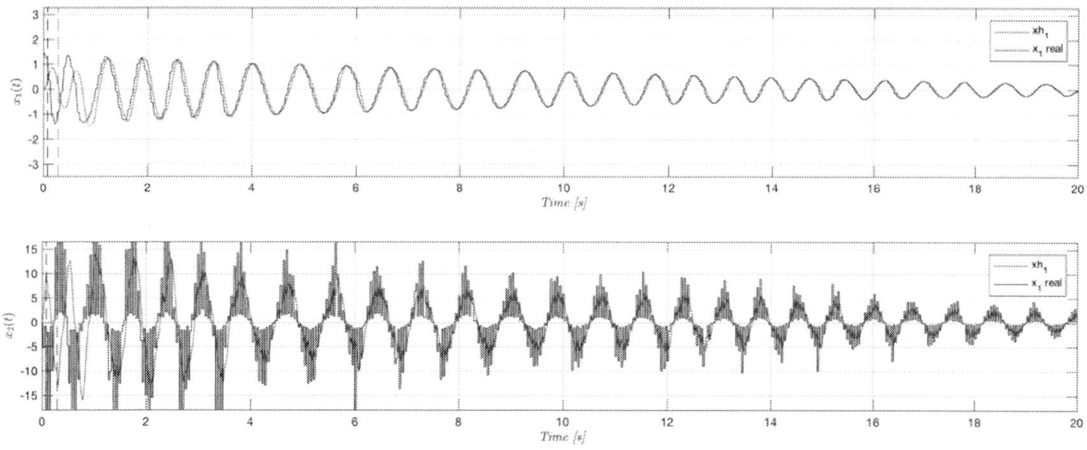

Figure 2. Numerical result for state estimation and real state, where pendulum velocity is not available.

Figure 2 depicts system estimate and real position and velocity. Figure 3 shows the observer function φ and the sliding manifold σ. Finally, Figure 4 presents the estimation of system parameters. In this figure, solid line refers to the estimated parameter and dashed one deals with the real parameter.

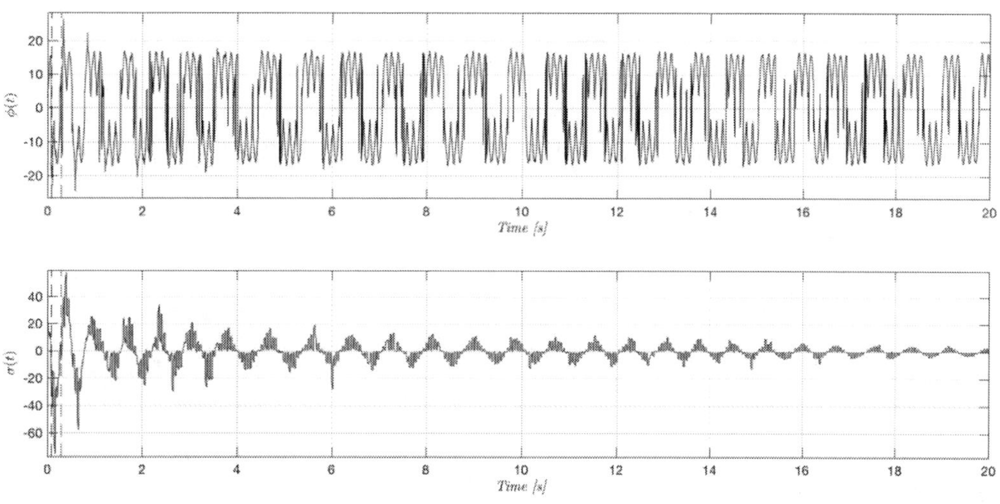

Figure 3. Experimental result for state estimation and real state, where pendulum velocity is not available.

5. Concluding remarks

In this paper, a full order nonlinear observer algorithm is combined with Least square method to parameter identification of a nonlinear second order system. Further, by using terminal sliding mode observer concept, the full order observer guarantees the finite time convergence of the estimated state of real one. Finally, to validate the effectiveness of this method, the proposed algorithm is tested in a real-time process.

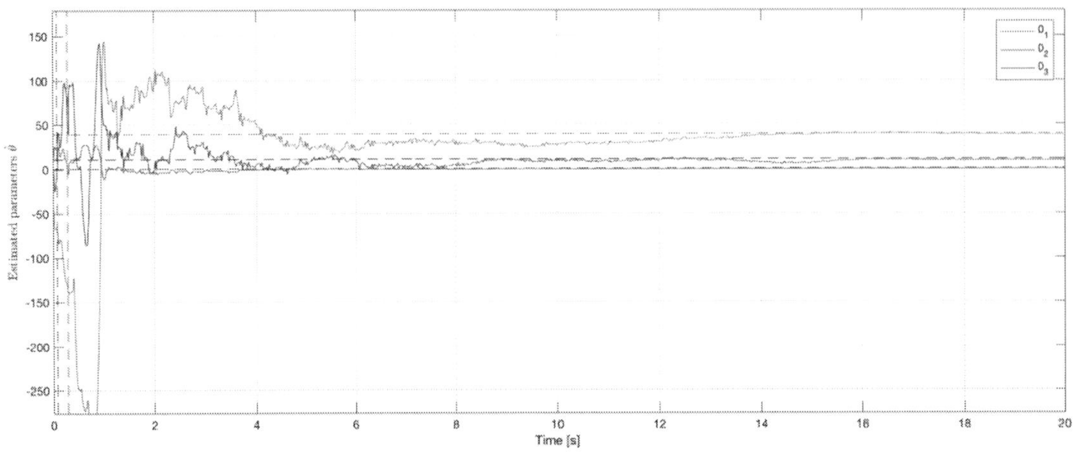

Figure 4. System parameter estimate and real one.

References

[1] Utkin, V., Guldner, J., and Shi, J., *Sliding mode control in electro-mechanical systems,* CRC press, 2017.
[2] Spong, M. W. and Vidyasagar, M., *Robot dynamics and control,* John Wiley & Sons, 2008.
[3] Jamshidifar, H., Askari, H., and Fidan, B., Asian Journal of Control (2017).
[4] Ljung, L., System identification, in *Signal analysis and prediction,* pages 163–173, Springer, 1998.
[5] Poznyak, A. S., Advanced mathematical tools for automatic control engineers, vol. 2: Stochastic technique, 2009.
[6] Bitmead, R., IEEE Transactions on Information Theory **30** (1984) 183.
[7] Chen, B., Zhu, Y., Hu, J., and Principe, J. C., *System parameter identification: information criteria and algorithms,* Newnes, 2013.
[8] Poznyak, A., International Journal Of Systems Science **11** (1980) 577.
[9] Janot, A., Vandanjon, P. O., and Gautier, M., Control Engineering Practice **25** (2014) 85.
[10] Astrom, K. J. and Wittenmark, B., *Adaptive control,* Addison-Wesley, NY., 1989.
[11] Narendra, K. S. and Annaswamy, A. M., *Stable adaptive systems,* Dover Publications Inc., 2005.
[12] Wu, J., Wang, J., and You, Z., Robotics and computer-integrated manufacturing **26** (2010) 414.
[13] Jin, J. and Gans, N., Robotics and Computer-Integrated Manufacturing **31** (2015) 21.
[14] Gao, G., Sun, G., Na, J., Guo, Y., and Wu, X., Mechanical Systems and Signal Processing **113** (2018) 145.
[15] Davila, J., Fridman, L., and Poznyak, A., International Journal of Control **79** (2006) 1251.
[16] Ioannou, P. and Fidan, B., *Adaptive control tutorial,* SIAM, 2006.
[17] De Silva, C. W., *Sensors and actuators: Engineering system instrumentation,* CRC Press, 2015.
[18] Liu, G., On velocity estimation using position measurements, in *American Control Conference, 2002. Proceedings of the 2002,* volume 2, pages 1115–1120, IEEE, 2002.
[19] Merry, R., Van de Molengraft, M., and Steinbuch, M., Mechatronics **20** (2010) 20.
[20] Edwards, C., Fossas Colet, E., and Fridman, L., *Advances in variable structure and sliding mode control,* Springer Verlag, 2006.
[21] Shtessel, Y., Edwards, C., Fridman, L., and Levant, A., *Sliding mode control and observation,* volume 10, Springer, 2014.
[22] Zhihong, M., Paplinski, A. P., and Wu, H. R., IEEE transactions on automatic control **39** (1994) 2464.
[23] Tang, Y., Automatica **34** (1998) 51.
[24] Tan, C.-P., Xinghuo, Y., and Zhihong, M., Automatica **46** (2010) 1041.

IOP Conf. Series: Materials Science and Engineering **844** (2020) 012039 doi:10.1088/1757-899X/844/1/012039

Evaluation methodology for Speech To Text Services similarity and speed characteristics focused on small size computers

J.E. Aguilar-Chacon[1] and D.A. Segura-Torres,[1]

[1]Facultad de Ingeniería Electrónica, División de Ingenierías, Universidad Santo Tomás, Carrera 9 No. 51-11, Bogotá, Colombia

jorgeaguilar@usantotomas.edu.co, dariosegura@usantotomas.edu.co

Abstract. The currently small size computers use and the high tendency to use human – voice interfaces to interact with any systems lead to merge these technologies. But implements strong Speech To Text algorithms on a small micro controlled system will reduce its computational capacity getting lower the system performance to execute other tasks. This work presents some selection criteria and a methodology to evaluate the Speech To Text services, using similarity and speed as metrics and the results obtained from an experiment deployed on a Raspberry Pi 3 using the proposed methodology with Spanish speakers. The application of the proposed methodology helped to detect the Google's STT system as a best option over IBM and Microsoft STT systems because shows a faster conversion and better similarity measures.

1. Introduction

This work was developed to find the better option to implement a Speech To Text (STT) system for Spanish speakers into a small size computer like Raspberry Pi. Before to explore the wide world of the STT systems the possibilities were limited by the following criteria. First, to reduce the development time was decided to find a STT API. Second, there are two types of STT systems according to the internet connection needs approach, some systems need an internet connection, called online, and other didn't, called offline. The offline STT systems process the audio information into the micro-controlled system, these has a program to capture, analyze and process audio signals. These systems manage all information locally, therefore, the system should be trained before their use, and a dictionary must be fed with new words to improve the STT system performance. The online STT systems usually invokes a web service, the audio is locally captured, but the audio is processed on remote server, the system on the server is trained by every audio processed, and many collaborators around the world feeds its dictionary. In this way, the online STT systems are better than the offline because the online systems are more trained and have a bigger dictionary, both properties give to the online STT systems a wide recognition range, can recognize more speakers and return a more precisely text. Then to get better results was selected an online STT System. Also, an online STT system reduce the computational load to the small size computer [1], because the process is made on a remote machine. Finally, the options were reduced by select the three most STT systems commercially used, these are: Google Web Speech API, Watson, and Azure Speech Service.

¿Which of the three is the best Speech To Text System to be implemented on a small computer like Raspberry Pi with Spanish speakers? to answer this question a methodology was proposed, an experiment has done using the proposed methodology, the experiment was developed over a Raspberry Pi and Python as programing language. Two metrics was selected to measure and compare the STT systems.

The following, it showed the proposed methodology, the work development and the obtained results by the experiment.

2. About commercial Speech To Text Services

In the field of technology there is a wide variety of voice-to-text services, the most commercially used are developed by recognized software companies. These services are called "Google Web Speech API" created by Google, "Watson" created by IBM and "Azure Speech Services" created by Microsoft. Python has a library called "SpeechRecognition", this library supports different APIs for speech recognition and audio to text conversion, including Google, IBM and Microsoft Services. These APIs provide real-time speech to text conversion, allowing systems to implement features such as voice commands, conversation transcripts and call centers. In addition, the voice-to-text services mentioned above have the ability to adapt to a different voice accent or voice pattern without changing the text resulting from a voice conversion.

"Google Web Speech API" and "Google Cloud Speech" are APIs implemented by some web pages, home automation systems and other devices that use voice recognition. The accuracy of the APIs is increasing due to the large neural networks that support them, the improvements in each version and the updates for each language and region.

"Google Web Speech API" is not recommended for an arduous production environment, its use is limited to fifty requests per day and it is not possible to increase this number of events, but "Google Cloud Speech" allows you to create a trial version that gives sixty minutes of audio conversion, or creates a paid account. When purchasing any of the packages offered by the company, the package will have features such as automatic speech recognition, more than one hundred and twenty languages adapted to different regions, support for pre-recorded files or live stream, filters for different noise conditions or inappropriate vocabulary, and selection of deployment models.

"Azure Speech Services" is the Microsoft API cloud service that allows the conversion from voice to text. In contrast with "Google Web Speech API", "Azure Speech Services" has two integration options, a device can use the Software Development Kit (SDK) or the REST-API. The SDK allows an easy way to integrate "Azure Speech Services" into a device, using traditional programming languages such as C# or JAVA. "Azure Speech Services" offer the protection of captured data and integration with other language services like cognitive services provided by Microsoft. This API has a limitation when using the account, the free version lasts sixty days, after this time, you must make a paid subscription.

The IBM Speech to Text service, better known as "Watson Speech to Text" has conversion time limitation like Google's, with this service it is possible convert one hundred minutes audio, to increase this time it is necessary to realize a payment subscription.

3. Proposed methodology

The selected metrics for the methodology, the task that conform it, and the algorithm used to implement the methodology are described in this section.

3.1. Metric used to evaluate speed and similarity

The currently experiment aims to know the faster and more reliable Speech To Text commercial service for male and female speakers, this way, the experiment will measures the conversion time for each of three systems to convert an audio and the percentage similarity comparing the original text and the output text of the STT system.

The conversion time its measure with a difference between two timestamps, one taken just before calling a service and the second when the service return the text.

The similarity between two text strings is determined with Levenshtein's algorithm, this is a dynamic programming algorithm devised by Vladimir Levenshtein in 1965. Levenshtein's distance is a measure of similarity between two text sentences, this measure is the number of elementary operations (insertion, substitution and deletion of a character) performed to the original string to obtain the target string. The number of operations performed indicates the similarity between the two words, so if Levenshtein's distance is a very large, it means that the two texts are very different [2]. Then the similarity precision percentage is the ratio between the Levenshtein's distance and the length of the longest of the two analyzed text strings.

3.2. Proposed methodology
The usual flow for a Speech To Text system has the following steeps [3-5], see ***Figure 1***, first the speaker talk or read in front of a microphone, this audio is taken by a digital system using an analog to digital converter, the digital system process the audio using an algorithm, and the algorithm return a text. This flow represents some problems to compare two STT systems.

The first problem is the speaker training, that means the speaker will improve his reading skills while the experiment is repeated, the second reading of a text made by someone will be better than the first read, therefore repeat an experiment many times with a speaker will falsely improve the result on a STT system. The second problem is the experiment repeatability, it is about to repeat a STT experiment under same conditions with a speaker will be difficult, because in the second iteration the speaker may be will be tired, the environment noise will change, the speaker cannot be available for the experiment all the time, and other considerations.

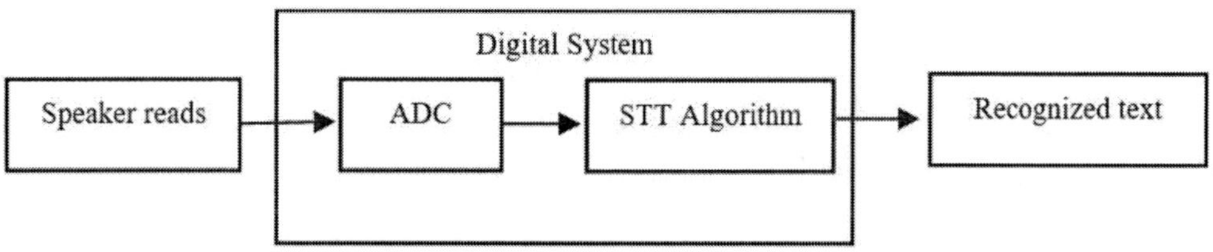

Figure 1. Flow for a Speech To Text system.

Record the speaker voice when reading once is the solution for these problems, the speaker read or speak only one time, this reading is recorded and the audio can be processed may times, giving to the digital system always the same information.

Then, the experiment methodology selects a text, record some female and male voices reading the text, process each voice file many times, store the measured results for conversion time and percentage similarity, then these results are analyzed by hand.

The selected text was taken from the fairy tale "Pinocho" because it is well known for the speakers and do not have difficult words, the text has 39 word divided in two sentences. The experiment uses Spanish speakers, 4 female and 6 male different voices were recorded into a file using a standard cellphone. The files were processed by an algorithm into a micro processed system, this algorithm execute STT algorithm, calculate the metrics and stores the results. Finally, the information was analyzed using a commercial spreadsheet.

3.3. Algorithm for the methodology
As mentioned before, the methodology has steps made by hand, these are capture voice and information analysis. Then, the automatized steps are the reading voice file, the calling STT service, getting the identified text from the STT, calculate the time taken by the service and the percentage, and finally store these results.

The developed algorithm allows automatically process many files into selected STT service or SST system. The STT algorithm flowchart used is showed in the Figure 2. The files to be processed should be copied into specified folder into the system. A list with the files names is preloaded by code, the system loads the first file in the list. Just after loading the audio an initial time mark is stored and the audio file is sent to the selected STT service. This STT system will begin to perform the conversion of the audio and at the end of its process return the identified text, at this point a final time mark is added, the value of the difference between the final mark and the initial mark will indicate the total time taken by a system to perform the speech to text conversion, the next step is obtain the text returned from the system, this text is compared with the original text using the Levenshtein's distance and calculating the similarity precision percentage.

The time and precision data obtained are stored in a CSV file with additional information like date and time experiment, file name, and STT service name. The system reads the next file on the list and repeat the process until file names on the list was ended. In order to perform the cycle again with a different system, the experiment is reloaded but calling a different STT system and storing these results.

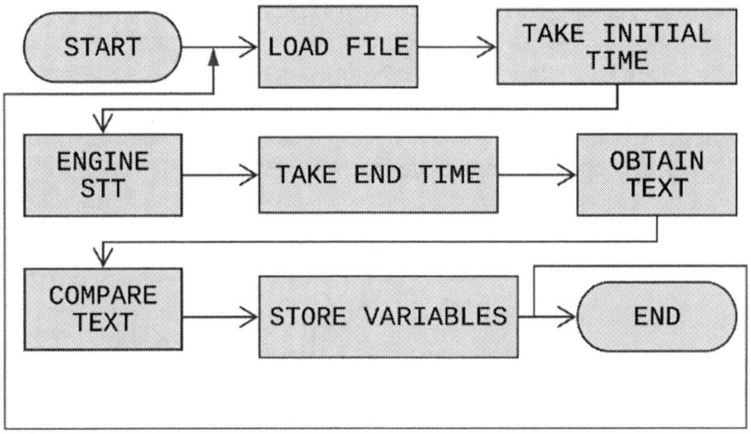

Figure 2. STT Algorithm Flowchart.

3.4. Comparison with previous work

Prior to experimental development, a bibliographic consultation on similar experiments was carried out. The most important difference is the language used in the tests, for this experiment the language is Spanish, in other experiments the language used is English.

The methodology used for the comparison of different recognition systems in the article " Comparing Speech Recognition Systems (Microsoft API, Google API And CMU Sphinx)" [6] does not use the time variable as a metric, and they have another algorithm different from "Levensthein's distance" for the comparison of the received audios with respect to the original audios, another difference is the implementation of different sentences for the tests, we only use the same text but recorded by different people. In another article " A Comparison of Online Automatic Speech Recognition Systems and the Nonverbal Responses to Unintelligible Speech" [7], a difference was found with respect to the methodology used for the experimental phase, the methodology used for the insertion of the audio to the different conversion services is manual, this methodology can be convenient if the experiment is carried out only once, but if in the experiment several tests are needed, it is better to automate the process and create an algorithm.

IOP Conf. Series: Materials Science and Engineering **844** (2020) 012039 doi:10.1088/1757-899X/844/1/012039

4. Experimental tests for the proposed methodology

4.1. Environment and system description

Small computers have become a useful ally for those who in recent years have decided to be immersed in the world of development, learning and use of new technologies for different functionalities. These devices have many specifications and services which can be used to implement an application according to the system requirements. This experiment uses Raspberry Pi 3 board, the most successful minicomputer in the market, because its powerful processor, different ports and connections, and a good purchase price in the market. Some specifications for the used board are listed in Table 1.

Table 1. Raspberry Pi 3 Specifications.

Specifications	Raspberry Pi 3	Specifications	Raspberry Pi 3
Price	$35 USD	Connections	HDMI
System on Chip	BCM2837		4xUSB
Manufacturer	Broadcom		10/100 Ethernet
CPU	ARM		40 GPIO pins
Instructions	ARMv8 64bits		MIPI Camera Connector
Cores	Quad-Core		MIPI Display DSI
Speed	1200MHz		Composite video (PAL and NTSC) via 3.5mm TRRS jack shared with stereo audio
RAM	1GB (1024MB)		Wi-Fi 2.4GHz
Storage	MicroSD Slot		Bluetooth 4.1
GPU	400MHz Broadcom VideoCore IV	Size	Length: 8.6 cm Width: 5.7cm
Power supply	5V, 3A microUSB		Weight:45g
OS	Raspbian V4.19		

The algorithm was implemented in using Python scripts. For the experiment, it is necessary to install certain libraries, libraries that allow updating the system to its latest version and the inclusion of libraries for microphone uses and speech recognition. In addition, an audio coding command was installed and all the requirements for the proper functioning of the STT systems. Finally, the "Speech Recognition" library was installed, this library implements the voice recognition interfaces to invoke some STT services or STT systems.

Commands used in the Raspberry Pi console application to get the system ready are:

- *sudo apt-get update*
- *sudo apt-get install python*
- *sudo apt-get install python-pyaudio python3-pyaudio*
- *sudo apt-get install portaudio19-dev python-all-dev python3-all-dev && sudo pip install pyaudio*
- *sudo apt-get install flac*
- *pip install SpeechRecognition*

The "SpeechRecognition" library is installed by default with "Google Web Speech API", this API is ready for testing. The installation process was done following the instructions of a guide found on the book called Building a virtual assistant for Raspberry Pi: The practical guide for constructing a voice-controlled virtual assistant, where the process is explained step by step. [8-9]

5. Results

After runs the experiment using the 10 recorded files, using three systems, and repeated it several times, the average conversion times are showed in the Table 2. The faster system was Google's, and IBM's was the slowest. The difference between Microsoft's and Google's takes around 33.7% less time than Microsoft's STT system, and around 48% less time than IBM's, its means significative differences between IBM's system and Google's.

Table 2. General average of time, with respect to STT systems.

STT systems	Average conversion time (mS)
Google	6101,579
Microsoft	9205,645
IBM	11744,616
Grand Total	9017,279822

The results on the **¡Error! No se encuentra el origen de la referencia.**, shows the average time by STT system and gender, all the engienes perform the convertions in similar time for both genders, but curiosly for all STT systems the female voice convertions takes less time than male voice, but its require other experiment or diferents studies beyond this work.

Table 3. Average conversion time respected to genders and STT systems.

Average conversion time	Gender		
STT systems	Male Voice (mS)	Female Voice (mS)	Grand Total (mS)
Google	6592,209	5365,630	6101,579
Microsoft	9308,767	9050,961	9205,645
IBM	12141,361	11149,497	11744,616
Grand Total	9347,446	8522,031	9017,280

According to the Table 4, the best similarity percentage was achived by Google's service, but Microsoft's and IBM's services are bery similar. Nevertheless, all services are over the 90% similarity percentage.

Table 4. General average of precision in percentage, with respect to STT systems.

STT Systems	Similarity percentage Average
Google	97,116%
Microsoft	90,825%
IBM	90,337%
Grand Total	92,759%

Also for the percentage similarity, the STT systems are gender independent, accordin to the results on the Table 5.

Table 5. Average similarity, with respect to gender and STT systems.

Similarity percentage Average	Gender		
STT Systems	Male Voice	Female Voice	Grand Total
Google	96,566%	97,940%	97,116%
Microsoft	90,893%	90,722%	90,825%
IBM	91,056%	89,258%	90,337%
Grand Total	92,838%	92,640%	92,759%

6. Conclusions

The conclusions obtained after applying the proposed methodology are:

- The proposed methodology allows the experiment repeatability with just one reading of speaker.
- With the proposed methodology the experiment can be replicated several times in a short time.
- According to the selected metrics, Google's STT system is the faster and presents the higher percentage similarity compared with Microsoft's and IBM's systems.
- According to the obtained results, the STT services are gender voice independent for the percentage of similarity.
- The used STT services perform faster conversion with female voices than male voices but is necessary a deeper test to confirm or deny this conclusion.

References

[1] Rosales J, Castillo J, Mansilla F and Boza G 2014 *A Prototype of a Semantic Platform with a Speech Recognition System for Visual Impaired People* (Journal of Intelligent Learning Systems and Applications) pp 87-92.

[2] Ene A and Andrei E 2017 *An application of Levenshtein algorithm in vocabulary learning* (2017 9th International Conference on Electronics, Computers and Artificial Intelligence (ECAI)) pp 1-4.

[3] Kishore, Kumar M, J Jayalakshmi and S Karthik 2018 *A Python base Virtual Assistant using Raspberry Pi for Home Automation* (SSRG International Journal Of Electronic And Communication Engineering) pp 22-27

[4] Al-Amin Md and Syeda Zinath Aman 2016 *Design of an Intelligent Home Assistant* (2016 7th International Conference on Intelligent Systems, Modelling and Simulation (ISMS)) pp 69-71.

[5] Surinder, Kaur, Sanchit S, Utkarsh J and Arpit R 2016 *Voice Command System Using Raspberry Pi* (Advanced Computational Intelligence: An International Journal (ACII)) pp 43-49.

[6] Kepuska, Veton and Gamal Bohouta 2017 *Comparing Speech Recognition Systems (Microsoft API, Google API And CMU Sphinx)* (Journal of Engineering Research and Application) pp 20-24.

[7] Kim Y, Joshua and others 2019 *A Comparison of Online Automatic Speech Recognition Systems and the Nonverbal Responses to Unintelligible Speech* (AiXiv) pp 1-13.

[8] Pant T 2016 *Building a virtual assistant for Raspberry Pi: The practical guide for constructing a voice-controlled virtual assistant* (Berkeley: Apress) volumen I

[9] Pedraza J, Ramos J, Vargas J, Aceves M and Martinez L 2018 *Prototipo de Asistente Virtual para su Aplicación en Edificios Inteligentes* (Robotica y Mecatronica) chapter 4 pp 37-46

IOP Conf. Series: Materials Science and Engineering **844** (2020) 012040 doi:10.1088/1757-899X/844/1/012040

Thermal energy capture by pyroelectric effect

F Cárdenas[1] and J Rodríguez[1]

[1]Facultad de Ingeniería Electrónica, División de Ingenierías, Universidad Santo Tomás, Ak 9 N 51-11, Bogotá, Colombia

E-mail: brayan.cardenas@usantotomas.edu.co

Abstract. Taking advantage of the physical nature of the pyroelectric materials, in this investigation a characteristic of these materials known as the pyroelectric effect was studied and corroborated, which allows this type of materials are sensitive to infrared electromagnetic radiation, and whose response can be seen in the form of electric potential difference. In this way, in this document the theoretical basis of operation of this effect is presented, and in turn, the results of experiments carried out in laboratories of the University of Santo Tomás Bogotá D.C. The experiments were performed using the "PIR325" sensor, which is composed of two thin films of pyroelectric material. In the first instance, the response of the sensor to thermal electromagnetic radiation produced by the human body is obtained, in the second instance, the same experiment is repeated, However, on this occasion, a Fresnel lens is used to ensure the highest infrared electromagnetic energy collection. Finally, the capture of the signals obtained in each experiment is performed and the results are contrasted, where the theoretical basis of the pyroelectric effect is demonstrated, and in turn, the importance of using a Fresnel lens in applications to carry out this type of sensors.

1. Introduction

Pyroelectricity, is a quality that some materials called pyroelectric have, it has been used by engineering to partly forge the current technological world. Pyroelectric materials have important physical characteristics for the industry, for example, they are used in the manufacture of electronic devices, sensor manufacturing, radiation measurement, infrared ray analyzers, automatic domotic systems, etc. The importance of these materials in the world is such that it has been decided to study their theoretical and experimental behavior, in order to characterize its principle of operation and physical properties of operation.

2. Pyroelectric effect

In terms of closed circuit with two electrodes, the current generated by a material pyroelectric is defined by equation (1), which expresses the relationship between the intensity of electric current I, and the variation of electric charge Q with respect to time, In addition, this current represents the amount of electrical charge that crosses the section of the pyroelectric material on a time interval.

$$I = \frac{dQ}{dt} \tag{1}$$

IOP Conf. Series: Materials Science and Engineering **844** (2020) 012040 doi:10.1088/1757-899X/844/1/012040

Taking into account the electric dipole moment, for pyroelectric materials, the equation (2) defines the relationship between the temperature change ΔT as a function of time, the charge generated Q, the pyroelectric current I, the surface area of material A and the pyroelectric coefficient of the material p under conditions of short circuit with electrodes that are oriented normal to the direction polar[1].

$$\frac{dQ}{dt} = pA\frac{dT}{dt} \tag{2}$$

Taking into account that the pyroelectric coefficient and the surface area of the pyroelectric material are constant, then, integrating on both sides of the equation (2) is obtained:

$$\Delta Q = pA\Delta T \tag{3}$$

Equation (3) expresses the directly proportional relationship between the electric charge and the temperature variation in a pyroelectric material, given that behaves like a dielectric material, the capacitance of the pyroelectric element is defined according to the equation(4), the above, taking into account a thin film of material as shown in figure (1), which behaves as a parallel plate capacitor, where the absolute electrical permittivity is expressed by equation (5), where E_0 is the electric permittivity under vacuum and E_r the relative permittivity of the material. [1] [4] [5]

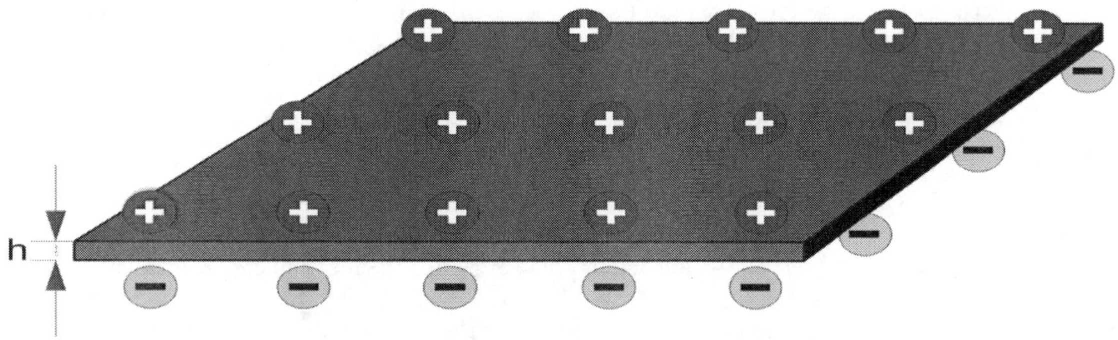

Figure 1. Thin film form of pyroelectric material.

$$C = A\frac{E}{h} \tag{4}$$

$$E = E_r E_0 \tag{5}$$

In open circuit conditions, the pyroelectric charge generates a potential difference between the faces of the material, the above based on the fact that the surface charge is proportional to the potential difference, therefore, the equation (6) expresses the relationship existing between capacitance C, the potential difference V and the electric charge Q. Accordingly, equation (7) defines the potential difference V in a pyroelectric material under open circuit conditions. [1]

$$Q = CV \tag{6}$$

$$V = \frac{\rho}{E}h\Delta T \tag{7}$$

3. Pyroelectric Infrared Sensor

Radiation is a component of the electromagnetic spectrum, but since the length of this radiation is greater than that of the visible region, it is not possible to observe it with the human eye, but it is feasible to detect it. According to black body radiation a hot object emits electromagnetic waves from the infrared region, but in addition to hot objects animals and the human body also emit them, these have the strongest radiation emission present in a wavelength of 9.4 u[m]. [2]

A pyroelectric sensor is a device sensitive to temperature changes, these based their operation in thin films of pyroelectric materials, which in accordance with the relationship expressed in equation (7), generates a potential difference directly proportional to the temperature change, therefore, this type of devices behave as transducers, transforming a response thermal energy into electrical energy as can be seen in figure (2).

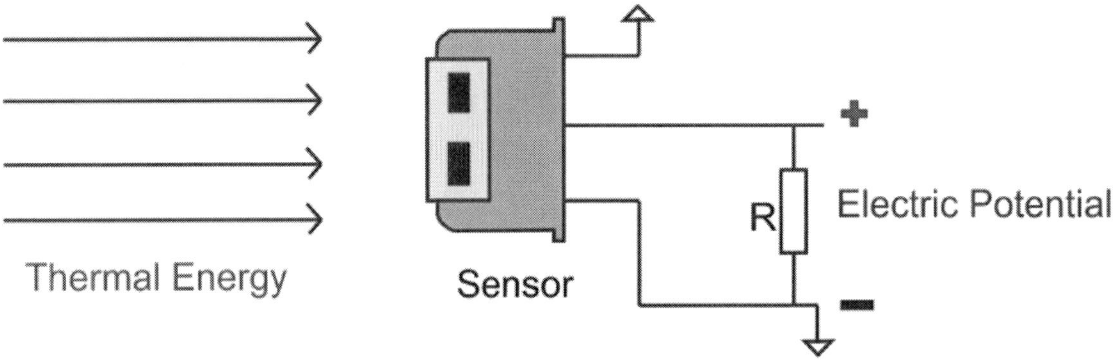

Figure 2. The infrared light with which the sample is irradiated, changes the temperature of this, resulting in a variation in the electric field proportional to the electrical potential.

4. Experiment of capture of thermal energy

In order to obtain the response of a pyroelectric sensor to temperature variation, we used the pyroelectric sensor PIR325 shown in Figure (3a), which is made of a crystalline material that generates an electrical charge on the surface when exposed to heat in the form of infrared radiation. When the amount of radiation inciding the crystal changes, the amount of charge also changes and can be measured with an integrated FET device in the sensor. The sensor elements are sensitive to radiation over a wide range, so a filtered window is added to the TO5 package to limit incoming radiation in a range of 8μ[m] to 14μ[m], being the wavelength range most sensitive to radiation of human body. [2]

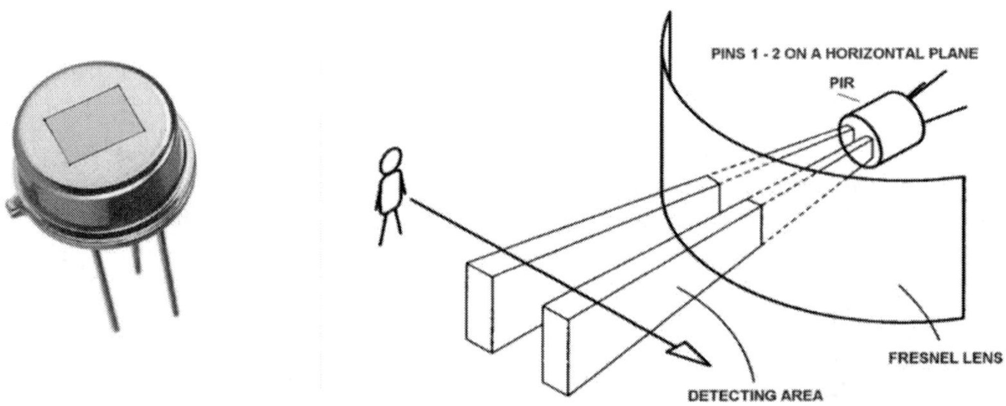

Figure 3. (a) Pyroelectric Sensor PIR 325, (b) PIR 325 pyroelectric sensor operation.[2]

The PIR sensor325 has two pyroelectric elements connected in a setting of voltage variation. This arrangement cancels the signals caused by vibrations, temperature changes and sunlight. A body that passes in front of the sensor will first activate one and then the other element as shown in figure (3b), while other sources will affect both elements simultaneously and be canceled. [2].

To obtain the sensor response signal, the circuit shown in figure (4) was implemented, where the signal coming from the sensor is passed through a stage of amplification and filtering, this way, the circuit is composed of a low pass filter active of the first order, this was designed to have an input-output gain of 10 [V/V], in turn, a cutoff frequency of 10 [Hz], taking into account that these types of sensors present their response in the frequency spectrum not exceeding 10[Hz].

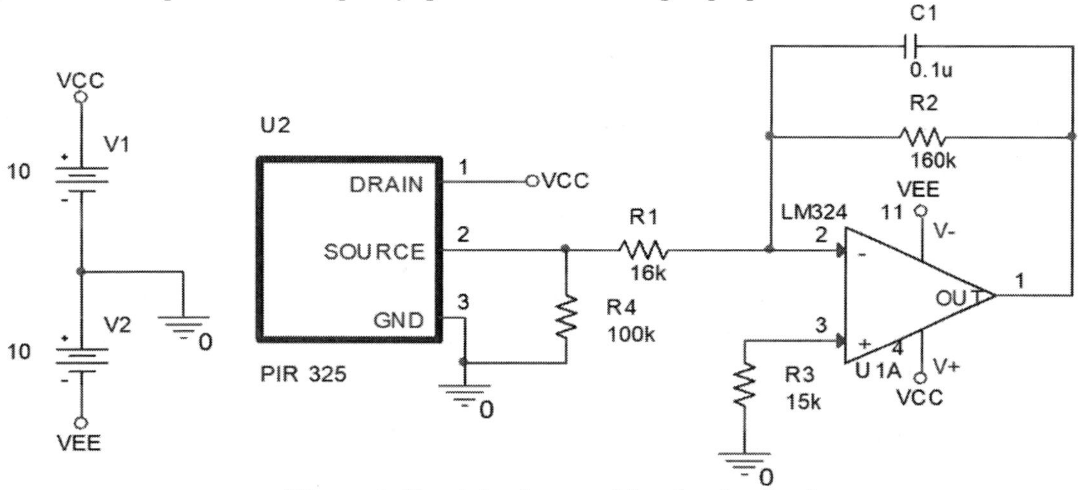

Figure 4. Circuit implemented for signal collection

In order to obtain the response signal from the sensor to temperature variation, the sensor was exposed to the infrared electromagnetic radiation produced by the human body according to the operation shown in figure (3b), exciting the two elements of pyroelectric material present in the sensor, from the left and right positions respectively, the above to demonstrate the excitation of the two elements of pyroelectric material, thus achieving by passing the hand to excite first one pyroelectric element and then the other respectively. Therefore, the signals are obtained as shown in figures (5) and (6), in which the response signals of the sensor can be seen from the positions described above.

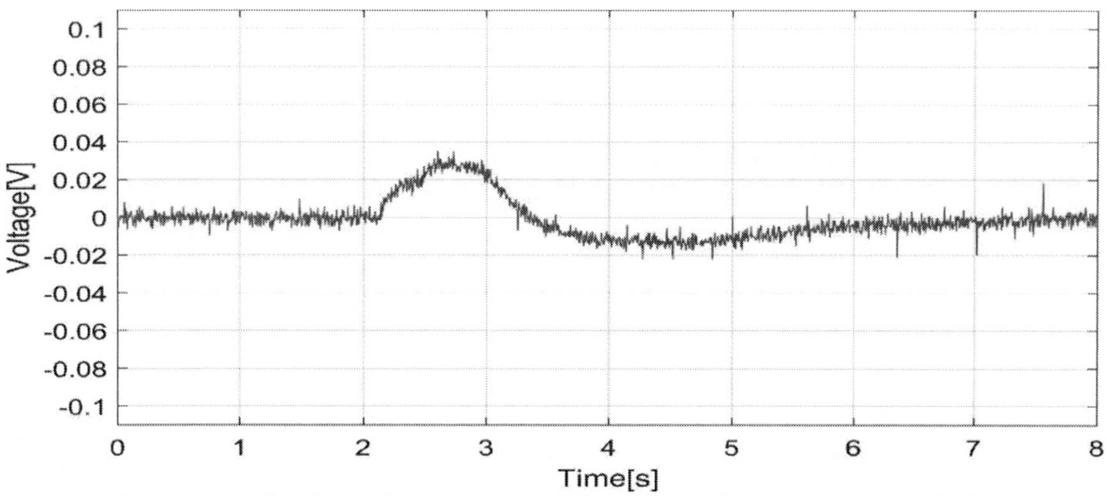

Figure 5. Response obtained from the sensor to temperature variation by the human body from the Left position.

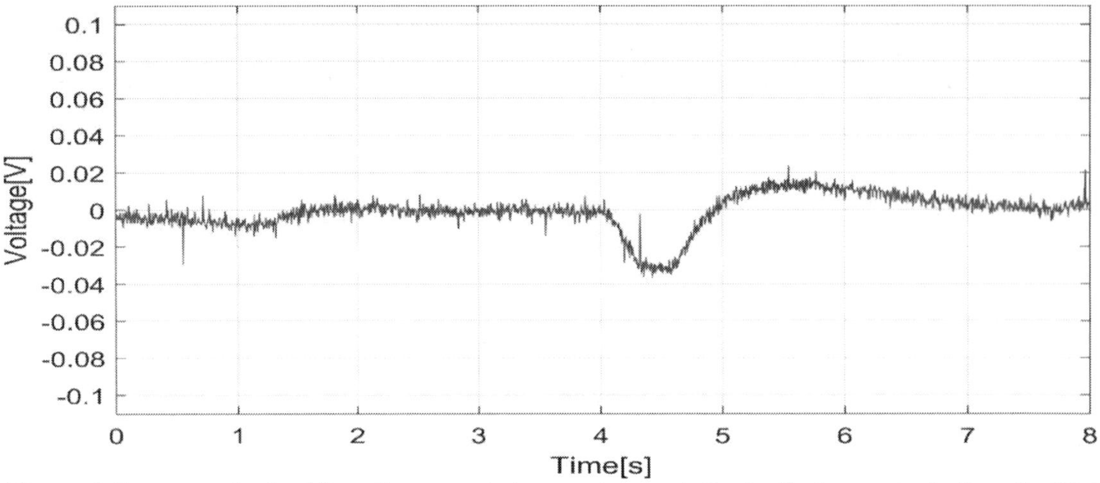

Figure 6. Response obtained from the sensor to temperature variation by the human body from the Right position.

The signals shown in figures (5) and (6), show the electric potential is generated according to the relationship of equation (7) in response to the change in temperature caused by the radiation of the human body, in turn, it is possible to observe how the shape of the signal obtained changes when passing the hand from the left and right positions, remembering that the pyroelectric elements present in the sensor are in voltage variation configuration, the response signal depends on the pyroelectric material that is excited in the first instance, thus, when passing the hand from left to right, the response of the pyroelectric element located on the left side of the sensor is evidenced first and then that of the element located on the right side of the sensor, it works the same way if it is passed from the right position, activating first the right pyroelectric element and then the left. This can be observed in the signals obtained.

5. Thermal energy capture using Fresnel lens

In applications of pyroelectric effect, an optical system used to ensure the greatest amount of infrared electromagnetic energy collection is composed of a Fresnel lens, whose function is to focus the radiation on the area of pyroelectric material. Fresnel lenses have behaviors similar to convex lenses, however, the cost of manufacturing a Fresnel lens is much lower, which makes them effective in some of the low-cost applications, its operation is shown in figure (7b), in which it can be seen as an electromagnetic wave that hits the surface of the lens, diffracts and converges to the central point P_o and at a focal length d, what causes that based on the principle of conservation of the energy, all the irradiated energy is concentrated in the angle of vision of the lens in the central point P_o.

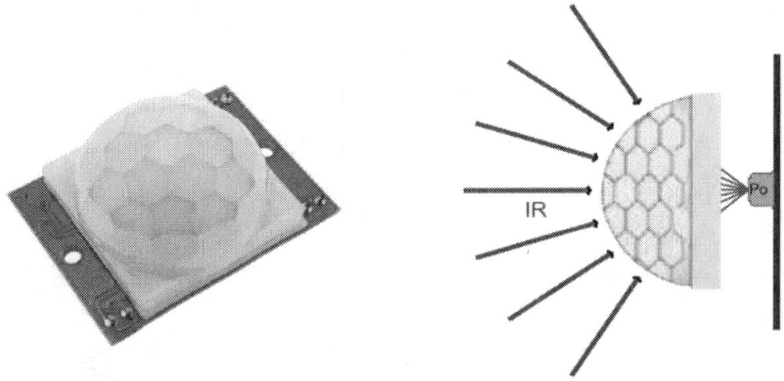

Figure 7. (a). Fresnel lens used (b). Fresnel curve lens operation.

The lens shown in figure (7a) was used, which, is made of plastic and has a low cost in the market. In addition, it is implemented on other various sensors passive infrared motion detectors, therefore, the experiment was performed from the left and right positions by adding the Fresnel lens, being so, the figures (8) and (9) show the signals captured, corresponding to the two positions raised previously.

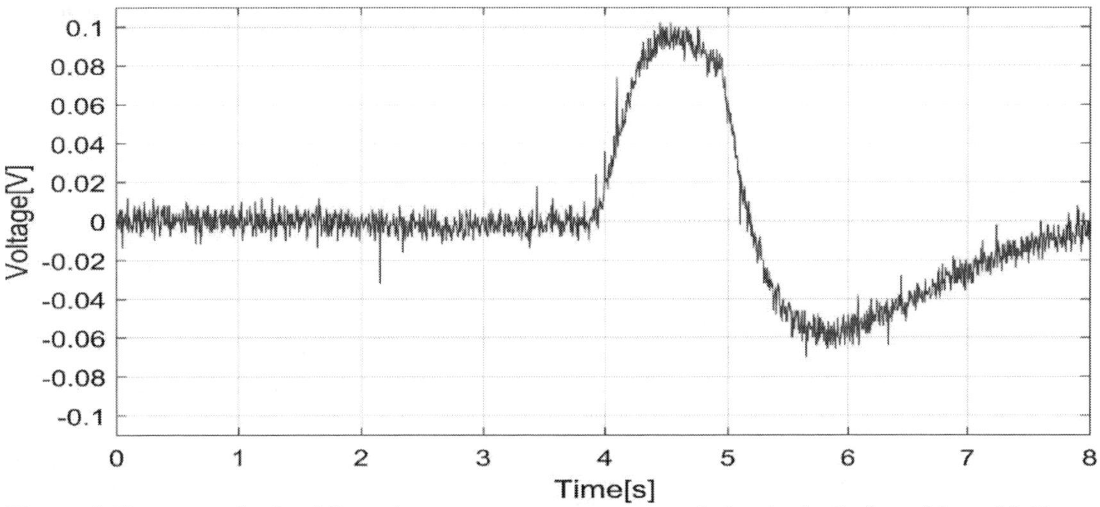

Figure 8. Response obtained from the sensor to temperature variation in the Left position with Fresnel lens.

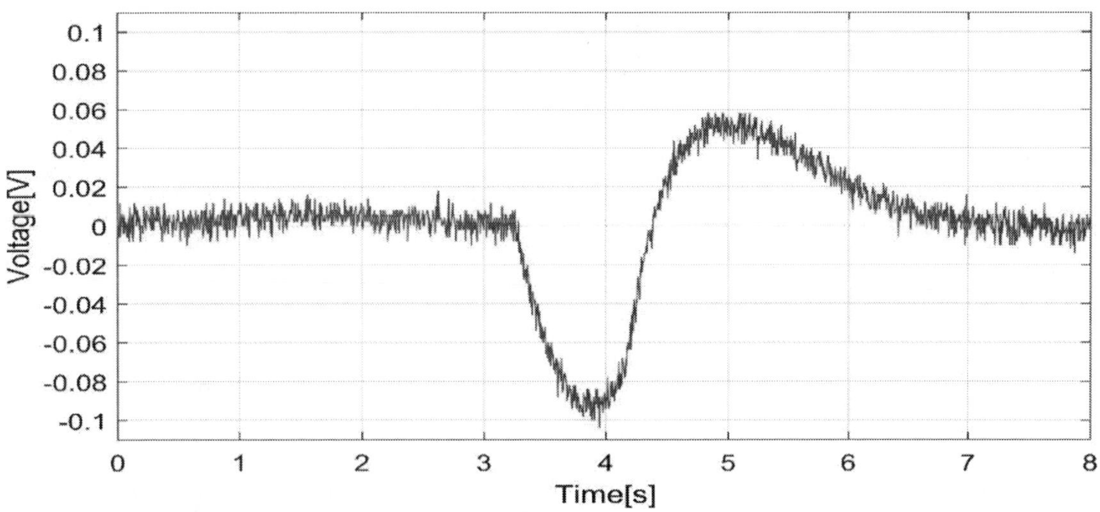

Figure 9. Response obtained from the sensor to temperature variation in Right position with Fresnel lens.

Figures (8) and (9) show a considerable increase in electrical potential when the Fresnel lens is added, this is because all the energy irradiated on the lens converges to the point where the elements of pyro electrical material are located, in such a way, the importance of this type of lens can be seen experimentally when wanting to capture as much infrared radiation as possible, in turn, in the same way as in figures (5) and (6) the excitation of the two pyroelectric elements located on the left and right side of the sensor, respectively, can be evidenced.

6. Conclusion

Performing the experiments presented above, the operation of the pyroelectric effect could be corroborated, thus managing to capture the electrical potential generated on the surface of each of the pyroelectric material elements present in the sensor, the above, when the infrared electro- magnetic radiation coming from the human body affects the surface of these elements, leaving in evidence that the infrared light with which this type of materials is irradiated, produces a variation in its temperature, which, translates into a variation of electric field proportional to the electrical potential on the surface of the material.

Finally, when supporting the experiment using the Fresnel lens, it is found that the amount of electrical potential that is generated by irradiating the elements of pyroelectric material is considerably greater, mainly because the field of vision is increased to 120°, and all the energy radiated on the Fresnel lens converges to the location point of the pyroelectric material elements, this demonstrates the importance of using this type of optical systems in infrared thermal energy capture.

References

[1] Xie M, Dunnb S, Le Boulbar E, and Bowen C 2017 Pyroelectric energy harvesting for water splitting International Journal of Hydrogen Energy **42** 23437-23445
[2] Corporation G 2003 Infrarred Manual Parts (New York: Glolab Corporation)
[3] Zhang X, Huang R, Senthil A and Liu K 2018 High-efficiency swinging-rotating diamond shaping of Fresnel lenses on roller molds CIRP Annals **67** 121-124
[4] Aline A and Bendawih M Impact of crystallization on ferro-, piezo- and pyro-electric characteristics in thin film P(VDF–TrFE) Organic Electronics **25** 92-98
[5] Chin H, Mao S, Huang C, Ohemeng K, Wagner S, Purohit P and McAlpine M 2015 Pyro-paraelectricity Extreme Mechanics Letters **2** 20-27
[6] Bhowmick S et al 2017 Investigation of pyroelectric fields generated by lithium niobate crystals through integrated microheaters Sensors and Actuators A: Physical **261** 140-150

IOP Conf. Series: Materials Science and Engineering **844** (2020) 012041 doi:10.1088/1757-899X/844/1/012041

NDT Non-Destructive Test for Quality Evaluation of Concrete specimens by Ultrasonic Pulse Velocity measurement

L Pedreros [1], **F Cárdenas** [2], **N Ramírez** [1] **and E Forero** [2]

[1] Facultad de Ingeniería Civil, División de Ingenierías, Universidad Santo Tomás, Ak 9 N 51-11, Bogotá D.C, Colombia

[2] Facultad de Ingeniería Electrónica, División de Ingenierías, Universidad Santo Tomás, Ak 9 N 51-11, Bogotá D.C, Colombia

laura.pedreros@usantotomas.edu.co

Abstract. Knowing the quality of concrete is very important, because this type of material is part of the design and implementation of most structures such as bridges, buildings, tunnels, among others, in this investigation non-destructive tests were carried out on test specimens of using ultrasound, which were designed and developed in the concrete laboratory at Santo Tomas University in Bogotá D.C. taking into account for its preparation a dosage of the estimated concrete mixture in order to obtain specimens with a compressive strength of 3500psi, 2500psi and 1500psi, in such a way, the specimens were allowed to cure for a period of 28 days, ensuring that they achieved their maximum compression strength.

As mentioned above, non-destructive ultrasound tests were applied in order to determine the quality of the manufactured concrete specimens, thus, an ultrasound system is proposed that makes it possible to pass an acoustic wave through a concrete specimen, this was proposed with the purpose of measuring the velocity with which the wave passes in the volume of this material, known as the ultrasonic pulse velocity. Therefore, the respective velocities for each of the concrete specimens were measured, thus finding a relationship between the ultrasonic pulse velocity and the compressive strength, which finally makes possible the evaluation of the manufactured concrete specimens.

1. Introduction

In the construction area, concrete acts as a main material when developing any type of structure, this is because it provides security, stability and is comfortable to work being adapted to any means, taking advantage of its malleability characteristic. For the realization of the mixture that composes the concrete materials that are easy to obtain are used because they are extracted from nature, among these materials is sand, gravel and water. The dosage of them has an important role, such that the amount of each component used in the mixture directly affects the final strength of the concrete.

To evaluate the quality of the concrete, the compressive strength is usually measured in a machine with hydraulic press, so that the press compresses the concrete to its breaking point, thus determining the strength of the concrete according to the applied force, however, at present there are other non-invasive methods, such as non-destructive tests by means of ultrasound, which allow to estimate the concrete strength without having to alter its condition or the properties it possesses, this is achieved by measuring the ultrasonic pulse velocity with which an acoustic wave travels through the concrete specimen.

IOP Conf. Series: Materials Science and Engineering **844** (2020) 012041 doi:10.1088/1757-899X/844/1/012041

Therefore, it is decided to manufacture concrete specimens with specific characteristics and therefore measure the ultrasonic pulse velocity with the help of ultrasound equipment available in the laboratories of the Santo Tomás University. In this way, the methodology shown in figure (1) is proposed and implemented, which is composed of three stages, in the first instance, the design of the concrete specimens, in the second instance the preparation of the same, taking into account as a design criterion, three different strengths, 3,500 *psi*, 2,500 *psi* and 1,500 *psi*, lastly, the measurement of the ultrasonic pulse velocity at each of the concrete specimens.

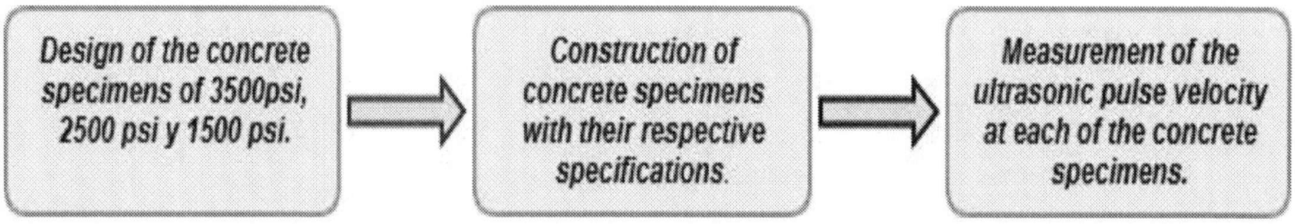

Figure 1. Methodology used.

2. Concrete

Concrete is the most useful material in structures such as retaining walls, bridges, foundations, columns, beams, among others, since it is economical, easy to melt, durable and resistant to the moment of setting, that is, when the material passes from a plastic state to a solid state.

Concrete is made of cement, a fine aggregate such as sand, a coarse aggregate such as gravel, water and, if necessary, additives are added. In figure (2), the proportion of the components within the concrete mixture [1], which form a manipulable paste, uniform due to the chemical reaction of the water with the cement, to form part of the solid body can be seen of concrete [2] and with excellent compressive strength when the concrete is hardened, that is, the ability of the material to withstand a flattening load. It is also important to highlight that the concrete has some weaknesses such as low tensile or tensile strength, this property refers to the ability of the material to oppose being elongated by forces in different directions.

Hydraulic cement is a finely ground powder, which is produced by burning limestone together with clay in an oven, at a temperature of 1,450 C [3], the union between these two elements is known as a clinker, which It is mainly composed of calcium silicates and, to a lesser extent, calcium aluminates, which, mixed with water, combines, sets and hardens at room temperature, both in air and underwater [4]. In turn, the cement is a conglomerate, that is, a material capable of joining elements and that when hydrated with good quality water it is transformed into a paste or dough [4], so that it obtains stiffness and strength over time, which is mostly acquired by the cement used in the concrete, which is why the greater the amount of cement content in the concrete, the greater the strength, however, in order to achieve an adequate and effective concrete, it is necessary to use of a single type of cement, since if several types are used in a single mixture, the components and characteristics of each one can alter the results that are desired [5].

To reduce costs and contribute to the concrete strength, a large percentage of aggregates that are clean of waste are used, which contribute significantly to the volume in the concrete mixture. This material is originated by igneous, sedimentary and metamorphic rocks, either by crushing or disintegrating them [6], which are classified into two groups such as the thin ones, where the sand and the coarse ones such as gravel, these are named as such because of the difference in size they have.

Due to the texture of the particles, they make up the rigid concrete body when it is in the fresh state, which when mixed adhere to the other elements being completely covered by the cement paste with water, this satisfies the results of hardened concrete and also providing smooth surfaces, suitable for developing different tests [7]. Within the characteristics that the aggregates possess, granulometry is important since it determines the sieve, that is, the size of the particles in which the sand must be found, is classified as having a minimum particle size of 0.075 mm and a maximum of 4.76 mm, in turn, the gravel is classified into a minimum particle size of 4.76 mm and maximum of 75 mm.

Finally, in the large environment of the constriction there is a wide variety of additives either in liquid or powder form, which are incorporated in small parts either before or while preparing the mixture [8], that is, in the fresh state in order to improve the original condition of the concrete, this occurs due to the various characteristics and reactions that the use of each additive possesses, taking into account the function that is required such as accelerating or retarding the setting, reducing water, accelerate strength, among others. The above is used to improve strength, avoid failures and improve conditions to work better concrete.

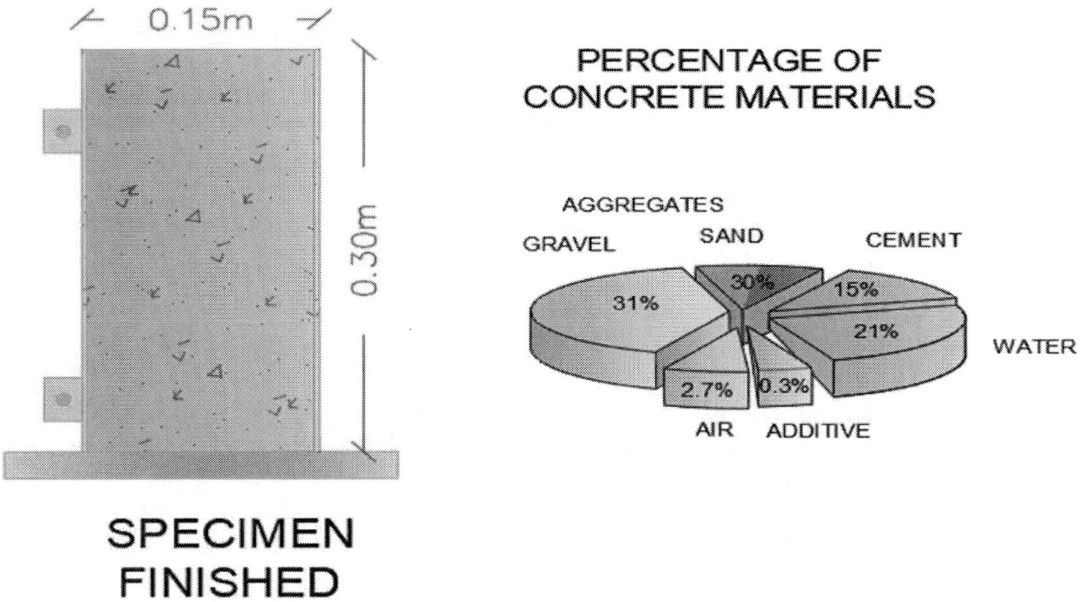

Figure 2. Percentage of the volume occupied by each material in a concrete mixture.

3. Measurement of the ultrasonic pulse velocity in concrete specimens.

Among the specific techniques to perform non-destructive tests on concrete, those that use ultrasound are included, these are used to determine the quality of concrete strength, without altering its physical properties, therefore, the use of these techniques helps to determine the state of the structures made with said material.

The measurement of the ultrasonic pulse velocity in concrete structures is based on the measurement of the return time it takes for an ultrasound wave to travel through a concrete volume. For this measurement one of the most used methods is the direct method, where the transducers are located on the opposite surfaces of a concrete specimen, in this case cylindrical as shown in figure (3). For the correct implementation of the method, sufficient acoustic coupler must be used on the joint surfaces between the transducers and the concrete, the above, in order that the acoustic impedance present due to the change of medium decreases and allows the acoustic wave to travel through concrete without reflecting.

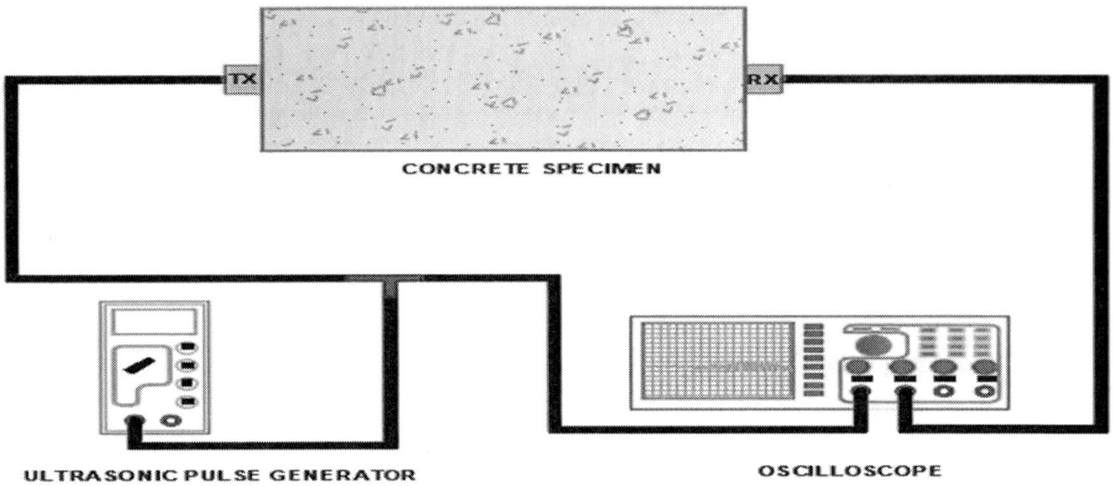

Figure 3. Ultrasound system posed by measuring the ultrasonic pulse velocity, using direct method.

In order to generate an acoustic wave and measure your flight time in passing through the test specimen of concrete, it makes use of an ultrasound system like the one that can see in figure (3), it is composed of a Ultrasonic pulse generator, which is responsible for exciting an ultrasonic transmitter transducer (Tx), which generates pressure waves that travel through the concrete, this generated signal can be seen through an oscilloscope. The transducer receiver (Rx) picks up the acoustic wave when it reaches the end of the specimen, and fulfils the function of generating an electrical signal corresponding to the received ultrasonic wave, this electrical signal is shown in an oscilloscope. Therefore, by synchronizing the oscilloscope channels, the time between the pulse and the received signal can be evidenced, this time corresponding to the flight time (t) of the acoustic wave through the distance (d) of the concrete specimen, in such a way that given the length of the specimen and the flight time, the ultrasonic pulse velocity in the concrete is determined, as shown by the expression in equation (1).

$$V = \frac{d}{t} \quad (1)$$

4. Design and development of concrete specimens.

As discussed above, are designed and produced concrete test specimens of three different strength, which are 3500*psi*, 2500*psi* and 1500*psi* as can be evidenced in figure (4), in order to verify and have a greater accuracy of the ultrasonic pulse velocity, two specimens are made for each strength since it is established that it is the minimum number of specimens to be tested.

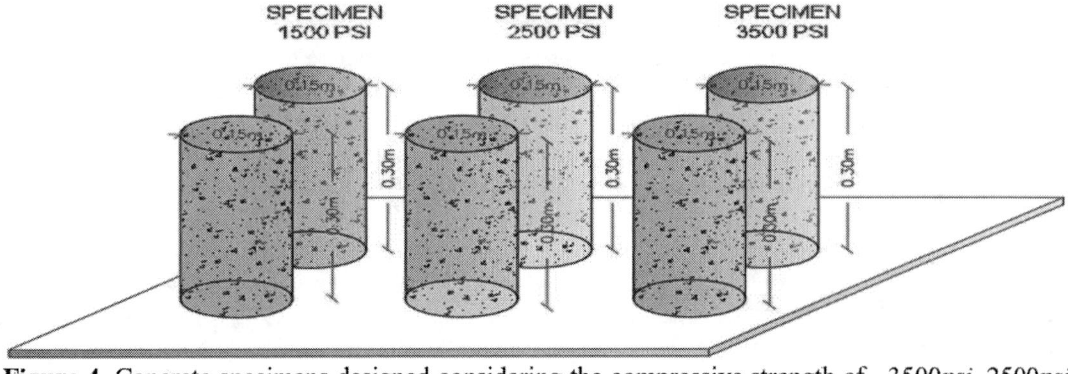

Figure 4. Concrete specimens designed considering the compressive strength of 3500*psi*, 2500*psi* and1500*psi*.

For the composition of the concrete, the elements used were the following: First, cement as the main element to acquire strength; which for the present case was used the gray cement type one of Argos, this is the one of general or normal use [8] and is used in any structure or civil works.

On the other hand, in the mixture, arid elements such as sand and gravel were used; where the gravel has a density of 1500 kg/m^3 and a particle size of 3/4", This material, to ensure excellent quality, it must be clean of any fine like clay, well graded and with a rough texture, because if the material is round, elongated and smooth [7], it affects the strength that want to acquire. Likewise, the sand used in the mixture is that obtained from the river, because it is of high quality and is often used in the concrete mixture, especially because of its high quartz content [6]. This type of sand is characterized by a density of 1450kg/m^3, and in order to be used in construction it must meet certain conditions as well as gravel, since both must have high durability and be clean, without having a coating of other materials, which interfere with the adequate hydration of the particles.

Finally, to the elements mentioned above, the necessary amount of water must be added, in order to form the concrete mixture; because water is the essential component that is responsible for the cementation process, on the one hand, it hydrates the cement molecules, transforms the physical and mechanical properties of this material and also providing malleability and the fluidity necessary for the development of the concrete that you want to perform. It is important to keep in mind that the water that must be used to obtain a quality concrete, must be that which is also suitable for human consumption, that is, it must not have other substances such as: oils, acids, garbage, matter organic or other fluids that alter its properties or molecular structure [7].

However, in order to acquire the desired strength, an excellent dosage is needed, that is, the necessary quantity of each previously named material, which must contain the concrete mixture, this quantity must be appropriate to provide an easy working mixture and economical. From the water content, the water / cement ratio and volume of the aggregates [11], the dosage for a concrete specimen can be obtained, however, table (1) presents an accepted dosage example for a volume of $1m3$.

Table 1. Dosage materials for a concrete mixture with a volume of 1 m^3

Resistance (psi)	cement (Kg)	Sand (m^3)	Gravel (m^3)	Water (l)
3500	420	0,67	0,67	220
2500	300	0,48	0,96	170
1500	210	0,5	1,0	160

According to the specifications of the materials, and according to the quantity required by each of these, the value of the dosage is calculated on the basis of the data found in the table (1) for the volume of the concrete specimen (Vp) of 0.0053 $m3$, as shown in the table (2); this value corresponds to the dimensions of the specimen pieces of 0.15m diameter (D) by 0.30m height (h). Considering the above situation, for to obtain the volume was determined using the expression in the equation [2].

$$Vp = \frac{\pi}{4} \, x \, D^2 \, x \, h \tag{2}$$

Table 2. Dosage of concrete for one volume 0,0053m^3

Resistance (psi)	cement (Kg)	Sand (m^3)	Gravel (m^3)	Water (l)
3500	2,226	0,003551	0,003551	1,166
2500	1,59	0,002544	0,002544	0,901
1500	1,113	0,00265	0,00265	0,848

When determining the correct volume that is required for a concrete specimen, proceeds to convert the amount of sand and gravel into kilograms, this is done in order to facilitate the work of measuring each of the materials in the mixture, for this, using the expression in equation (3), the data of the mass variable (m) is obtained, because the known data are the densities (ρ) of each aggregate and the volume (V) of them found in table (2). After these calculations, the necessary dosage of the concrete mixture for the six specimens is found, as shown in the table (3); in the same way, it should be considered that each quantity is doubled, because in the design the realization of two specimens was proposed for each strength.

$$\rho = \frac{m}{V} \tag{3}$$

Table 3. Concrete dosage for two specimens with dimensions 0,15 x 0,30 *m*

Resistance (psi)	Cement (kg)	Sand (kg)	Gravel (kg)	Water (l)
3500	4,4532	10,656	10,301	2,3326
2500	3,1809	7,6341	14,759	1,8025
1500	2,2266	7,9522	15,374	1,6965

Once the quantity of material required is established, the mixture is developed for the three concrete strengths as explained below:

For the construction of concrete specimens, suitable molds must be used to shape and resist the cylinders; the molds used are made of solid iron, with dimensions of 15*cm* in diameter and 30*cm* in length, it should be noted that the height of the specimen must correspond twice the diameter. Before melting the concrete, it must be verified that the cylindrical specimens, are in good condition and properly covered, by a thin layer of mineral oil in its interior, this component prevents the concrete from sticking to the instrument.

The molds to be used must be covered in three layers in a progressive manner by the concrete mixture, in each phase, these layers must be applied proportionally to the size of the specimen, that is, occupying the same volume as shown in the figure (5). To perform this procedure, it is necessary to use a smooth rod 5/8 "in diameter, which will be embedded in each layer, emitting 25 strokes evenly; likewise, the use of a rubber hammer is required, to that in each of the layers of the procedure, between 10 and 15 additional hits are generated around the molds, in order to suitably compact the material.

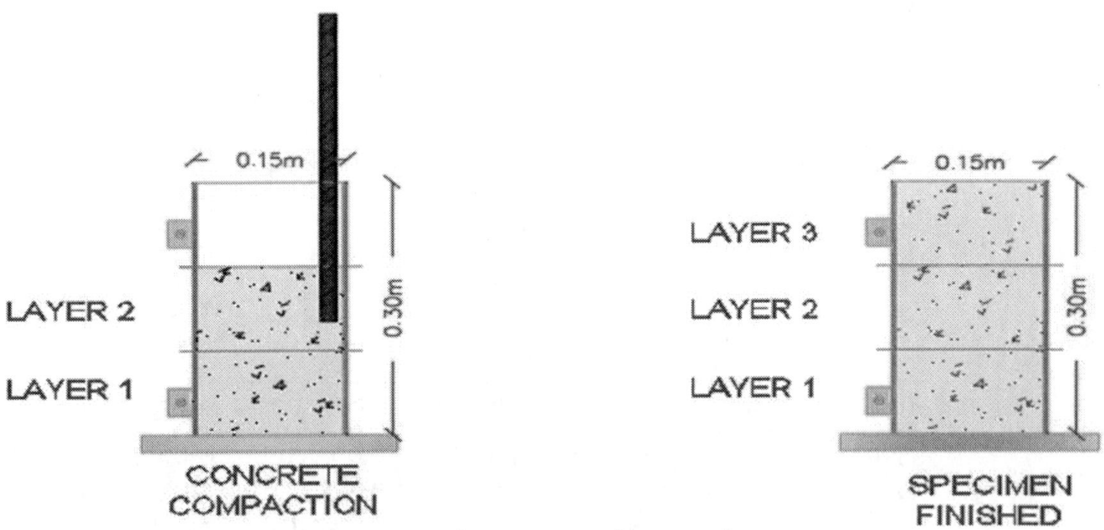

Figure 5. Specimen construction procedure.

Once the concrete specimens have been made, they must be unmolding over a period of 18 to 24 hours and transferred to a humid area, thus initiating the curing process, this is a process by which chemical hydration reactions occur of the cement components, mainly silicates and aluminates. Finally, in the final phase of the process; the specimens must be constantly hydrated with water for a period of 28 days, so that the concrete can reach the maximum strength with which it was designed. Figure (6) shows the specimens made.

Figure 6. Concrete specimen.

5. Measurement of the ultrasonic pulse velocity for each one the concrete specimens.

After 28 days of curing time of the concrete specimen, the ultrasonic pulse velocity is measured for each, so the experiment shown in figure (3) was performed, using the ultrasound equipment with reference to HUMBOLDT HC-6390, the generates an ultrasonic pulse signal as shown in figure (7), allowing to vary the maximum amplitudes of the signal, between 250[V], 500[V], 750[v] and 1000 V. In addition, 55 k[Hz] resonance frequency transducers were used. In this case the amplitude of the chosen ultrasonic pulse was 500[V], in turn, white vaseline was used acting as an acoustic coupling between the transducers and the surface of the concrete.

The respective test is performed for each specimen, and the signal generated by the receiving transducer is captured, Figure (8) shows a captured signal for a 3500psi specimen, in this one can observe the approximate flight time of the ultrasonic wave, taking into account that the equipment produces a small exciting on the receptor transducer that reference the initial time, in turn, it can be seen how some time later voltage variations begin to be noticed corresponding to the sound wave received by the transducer, thus the time between the reception of the signal and the initial time, corresponds to the flight time of the ultrasonic wave. For figure (8), the approximate flight time is 73u[s].

Finally, the table (4) shows the corresponding measured flight time values, and the ultrasonic pulse velocity calculated for each of the concrete specimens.

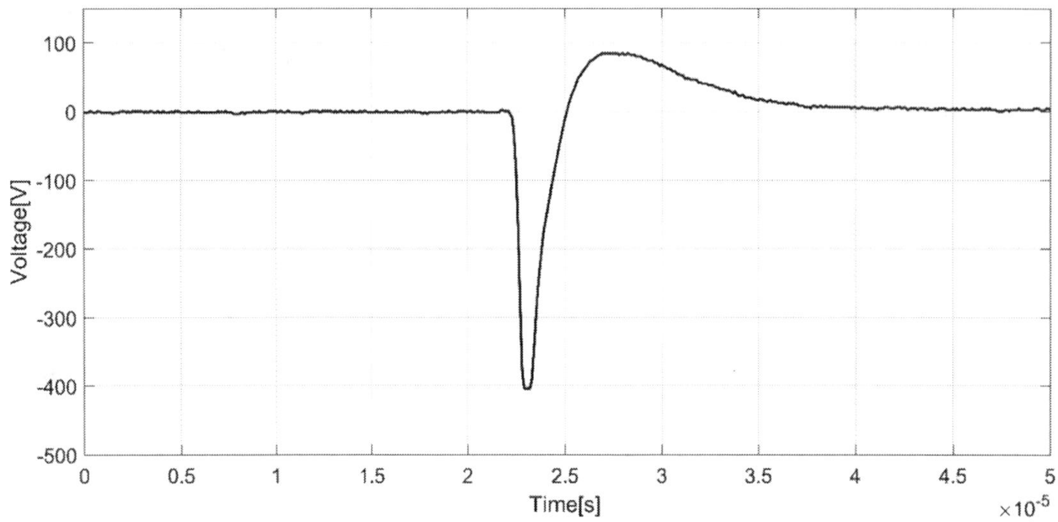

Figure 7. Ultrasonic pulse signal generated by the HUMBOLDT HC-6390 with a maximum amplitude of 500 [V].

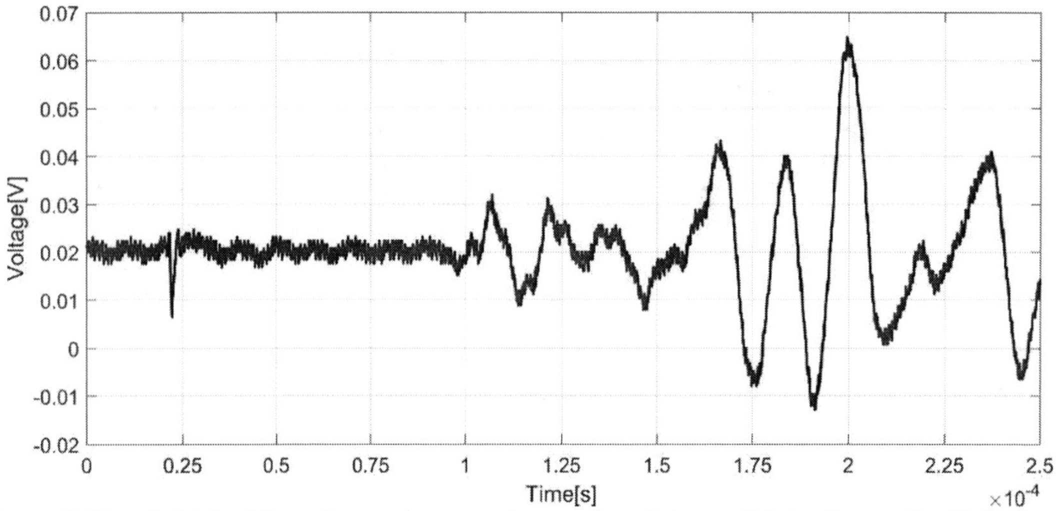

Figure 8. Signal obtained from the receiver transducer, where it is possible to observe the flight time of the ultrasonic wave, through one of the 3500psi specimens.

Table 4. Flight times measured and ultrasonic pulse velocity calculated for each of the specimens

Concrete Specimen	Flight time u[s]	Ultrasonic pulse velocity [m/s]
P1 3500psi	73,53	4080,0
P2 3500psi	72,76	423,11
P1 2500psi	81,30	3690,0
P2 2500psi	80,31	3735,
P1 1500psi	82,43	3639,5
P2 1500psi	82,72	3626,7

6. Conclusion

Taking into account the results obtained from table (4), it can be seen that the measured ultra- sonic pulse velocity is related to the compressive strength of concrete, because higher ultrasonic pulse velocity were obtained in the 3500psi specimens, lower in the 1500psi specimen, and intermediate in the 2500psi specimen, thus demonstrating that if the compressive strength of the concrete specimen increases, therefore the measured ultrasonic pulse velocity will also do this, this implies that the ultrasonic pulse velocity measurement technique implemented gives a notion about the quality of the concrete tested.

With the obtained results it is possible to classify the concrete specimen according to the ultrasonic pulse velocity, so based on [10] and [11], if a velocity greater than 4575[m/s] the concrete is considered excellent, if the velocity is between 3660 [m/s] and 4575 [m/s] the concrete is considered good, if it is between 3050 [m/s] and 3660[m/s] the concrete is considered to be questionable, if it is between 2125 [m/s] and 3050 [m/s] considered to be poor concrete, and if the velocity is less than 2125 [m/s] it is considered very poor quality concrete. Thus, according to the Classification mentioned Above, it is found that the specimen made of 3500 psi and 2500 psi respectively are in the quality range of good concrete, However, the Ultrasonic pulse velocity Measured on the 1500 psi specimen, suggests that their quality is questionable, which is reasonable due to their low compression strength.

References

[1] Culma A andRojas F 2018 Caracterización mineralógica y física de los agregados de la cantera Rodeb y Acopios, aplicada a concretos y filtros (Bogotá D.C.: Universidad Catolica de Colombia)
[2] Terreros L and Carvajal I 2017 Análisis de las propiedades mecanicas de un concreto convencional adicionando fibra de cañamo" (Bogotá D.C.: Universidad Católica de Colombia)
[3] Imbabi M, Carrigan C and McKenna S 2013 Trends and developments in green cement and concrete technology **1** 194-216
[4] Sanjúan M and Chinchón S 2018 Introdución a la fabricación y normalización del cemento portland (España: Universidad Alicante)
[5] Saov 2018 Norma de construcción concretos (Medellín: Epm)
[6] Gutiérrez L 2003 El concreto y otros materiales para la construcción (Manizales: Universidad Nacional de Colombia)
[7] Gerardo A and Rivera L Concreto simple (Cauca: Universidad del Cauca)
[8] Babilonia I and Urango S 2015 El uso de aditivos de origen natural integral a masas de concreto para la protección contra la corrosión del acero estructural embebid (Cartagena: Universidad de Cartagena)
[9] Culman A, Herrera P, Rocha D and rojas F 2017 Cartilla dosificación de mezclas de concreto: Agregado delpunto acopio rodeb (Bogotá D.C.: Universidad Santo Tomás)
[10] Quintana E, Quintana 2016 ULTRASONIDO:Aplicación para la determinación del modulo de elasticidad dinamico **61** 39-48
[11] Feldman R 1977 Non-Destructive Testing of Concrete (Canadá:NRC)

IOP Conf. Series: Materials Science and Engineering **844** (2020) 012042 doi:10.1088/1757-899X/844/1/012042

Automation of the Conductive Ink Synthesis from Recycled Parts from Metallurgical and Chemical Residues Joined to Polymer Resin and Solvent

Jhon Lara[1], Sergio Cantillo[1], Karolina Castillo[1], Jorge Montalvo[2] and Karen Lara[3]

[1]Ingeniería Mecatrónica, Universidad Autónoma de Occidente, Cali, Colombia.

[2]Departamento de Automática y Electrónica, Universidad Autónoma de Occidente, Cali, Colombia.

[3]Departamento de las TIC's, Universidad ICESI, Cali, Colombia.

E-mail: jhon.lara@uao.edu.co

Abstract. Conductive ink is a modern method for making printed circuit boards (PCB's), with new and somewhat different properties such as electrical conductivity, adhesion on non-conventional surfaces (glass, wood and paper) and the printability it possesses. However, such technology is cost and limited access given the machinery needed for the synthesis of ink and the tax on our currency. This situation leads to an increase in the technological gap between developed and developing countries. Therefore, the objective and motivation of this work is to present InkCo, a machine that automates the process of synthesis of conductive ink from metallurgical and chemical waste of the region. Understanding that through automation, virtual CAD modeling and materials science it is possible to develop the objectives of sustainable development (ODS) 4, 9 and 12, which seek innovation in industry and responsible consumption. Based on these ideas, InkCo is a mechatronic technology product that seeks to reduce this technological gap, through the implementation of rapid prototyping technologies (3D printing, laser cutting and prototyping of controlled circuits), for its structural construction, and thus generate an input (conductive ink) that can be used for the creation of more unconventional technology (3D printing of amorphous circuits on dielectric surfaces).

1. Introduction

In the context of the construction of electronic and some electrical devices, different types of physical arrangements have been developed for their connection and subsequent fastening. The product of this development has been called PCB (Printed Circuit Board). The development process of these boards is very complex. This complexity lies in the need for very specific knowledge, for example, mechanics, in cases where the process is by removal of chips, or chemistry, if lithographic methods are used for burning and developing circuits. On the other hand, the inputs required to produce a PCB are geometrically formal elements of a material similar to copper sheets on a normally thermostable polymeric layer such as Bakelite. Therefore, the new ways of creating electronic or electrical devices are of little variety in their geometric arrangement due to the options that exist. Currently, in order to design and produce an innovative electronic device, we are looking for elements that help to achieve physical arrangements for much more flexible connections and in shorter times. In addition, bearing in mind that in order to make each of these so-called PCB's, resources are needed, such as copper as a conductive element, polymers, or other inputs according to the manufacturing method, such as, for example, machinery or chemicals in the case of the developing method. It is imperative to seek new ways to manufacture PCB's with a low cost in raw materials and meeting what the current environment demands.

However, there are alternative processes that allow the production of PCB's with better characteristics than traditional processes, one of them is conductive ink. This process allows the realization of rapid prototypes [1] on any type of surface, as long as it is possible to permeate with this ink, giving solution to the needs of flexibility and time. The knowledge required for the use of this good is little, and for this reason it is a good ally. On the other hand, this ink opens a door to innovation in the environmental field, given that the elements for the creation of this ink are a waste of the metallurgical industry and these can be taken from different sources such as steelworks or metal-mechanic workshops [2, 3]

Bearing this in mind, it is proposed to create a machine designed to obtain a product that emulates the conductive environment of electrical signals, which is no more than the emulation of the copper layer in conventional PCB's. For the emulation of the polymer we can use any surface that allows it to be permeated by the ink and create a bond as does the Bakelite (thermostable) with the copper layer.

Therefore, we will present the parameters under which InkCo must work in order to obtain a conductive ink synthesis, the general requirements of InkCo for its operation where its materials and work inputs will be explained, the design that was made in computational design [4] according to the defined parameters required for operation will be presented, and finally the results obtained from InkCo with its conductive ink synthesis process.

2. Parameters for InkCo's Work

InkCo is a machine that automates the process of conductive ink Synthesis from recycled parts from metallurgical and chemical waste, together with a polymeric resin and a solvent. Therefore, it is necessary to keep a check on the parameters, depending on these needs, the variables that affect said synthesis are presented, taking into account that later these variables will become parameters that can be varied from the software (app development) that has InkCo. Therefore, the variables that affect the automated process will be defined by, Amount of material [gr], mixing time [s], hierarchy in the mixture [list], Temperature [C°] and relative humidity [%]. Taking these variables into account, the reasons for the importance of each one will be explicitly explained, taking into account that some items have already been dealt with in other InkCo publications.

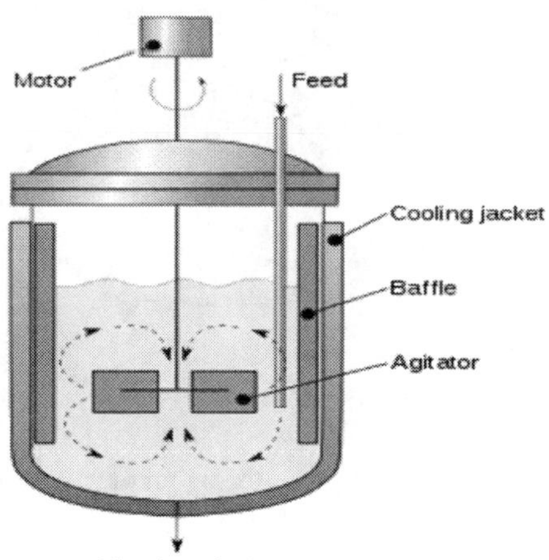

Figure 1. Mixer Tank.

(i) Quantities of material [5], this variable allows us to govern the final characteristics of the conductive ink, such as the max conductivity, adhesion, and viscosity of the conductive ink, these characteristics can be obtained by mixing the four elements that are part of the conductive ink, since each one can be graded according to the desired quantities [gr] for a defined composition or to be experienced.

(ii) Mixing time, this variable allows us to carry out the homogeneous mixing of the conductive ink since this variable governs the InkCo subsystems that are between (dosing, mixing, grinding and dosing). which are the automated processes in InkCo, this automation makes use of the variable time for mixing.

(iii) Hierarchy in the mixture [6], this variable is of type list since the order at the time of the mixture if it affects the properties that can be obtained in 30%, therefore, it is advised that the order is governed from liquids to the ceramics is (solvent, base, metallic and ceramic powder). However, the user can make an experimental change to obtain new or better mechanical properties [7] that he considers appropriate.

(iv) Temperature and relative humidity, are variables to be taken into account given by the inputs contained in InkCo that are classified from solvents to ceramics, that is, we need a controlled environment to obtain a conductive ink with a high repeatability every time you want to make said mixture. Therefore, sensors are located around the machine that monitors the environment and control it by means of actuators and keeps it within a working range suitable for inputs and for InkCo.

2.1. *General Requirements*

Within this section, we will analyse the InkCo systems, understanding what are the needs for automation in terms of actuators and sensors, as well as the requirements for controls that allow you to comply with the InkCo operating parameters, so we will start with the subsystems of InkCo, these subsystems are divided in feeding, mixer, grinding and dosing, which has the function of containing and carrying out the process of conductive ink synthesis through actuators, event-controlled mechanisms and signals sent from a micro-controller. These subsystems have the following functions.

(i) Feeding: contain the necessary supplies to carry out the conductive ink in four containers designed for this function.

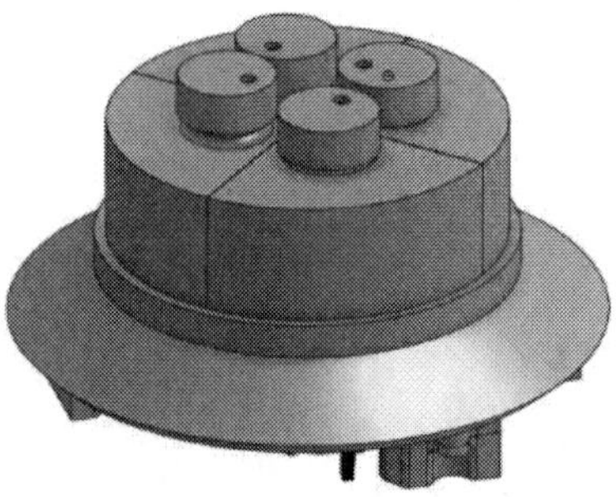

Figure 2. Feeding System.

(ii) Mixer [8]: make the first mixture of the inputs to join the different elements in an aqueous state that contains the four elements.

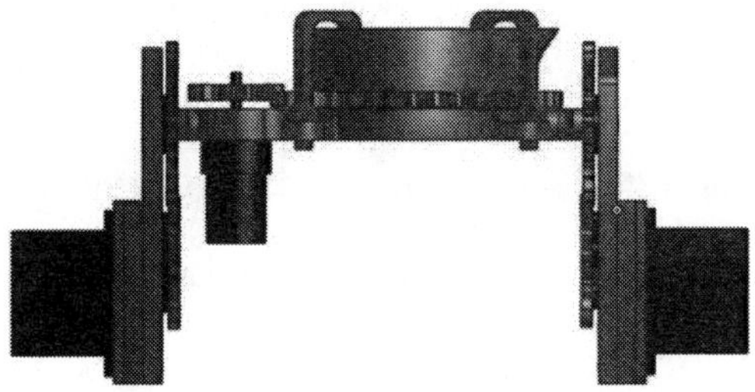

Figure 3. Mixing System.

(iii) Grinding [9, 10]: it has the function of homogenizing the mixture through the crushing and compaction of the four elements thus obtaining a homogeneous conductive ink.

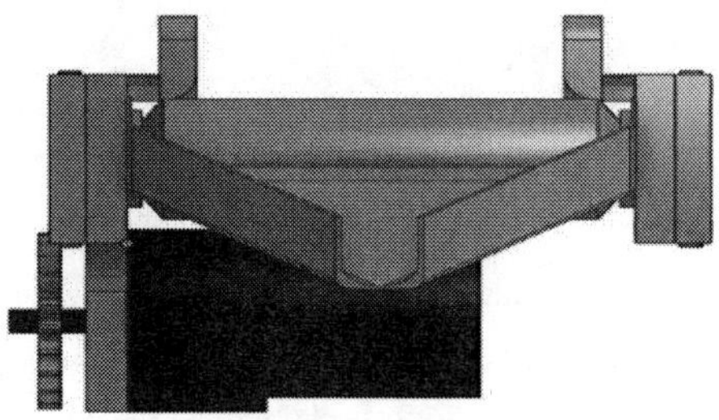

Figure 4. Grinding System.

(iv) Dosing: this subsystem helps us to continue mixing the homogeneous mixture preventing it from drying out and at the same time delivering an amount in [ml] desired by the user.

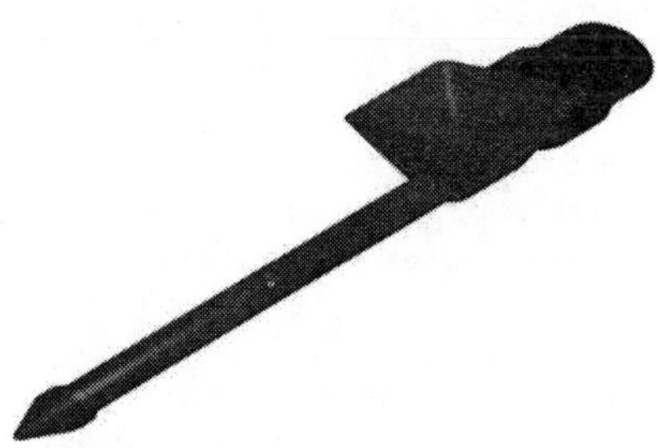

Figure 5. Dosing System.

On the other hand, the conductive ink mixture has infinite form of realization, however InkCo is based on the concept of common ink that is made with a base (resin) and a pigment (colour) that are mixed until obtaining a homogeneous mixture with the desired viscosity. Therefore, it is clear that the actuators and mechanisms are fundamental for this function of conductive ink synthesis, mainly the actuators used are DC motor reducers [11], which are controlled by a relative encoder to control the rotational speed and angular position of the same. Finally, the sensors used are constituted between end of stroke, temperature sensors and finally a time sensor.

IOP Conf. Series: Materials Science and Engineering **844** (2020) 012042 doi:10.1088/1757-899X/844/1/012042

2.1.1. Materials This section covers the materials necessary for the construction of InkCo and conductive ink with emphasis on InkCo since there is documentation on the conductive ink of InkCo [12]. InkCo is built mainly on the basis of rapid prototyping technologies such as 3D printing and laser cutting [13]. Therefore, most InkCo parts are made of PETG (a material that possesses the properties of PET and ABS that are considered as engineering polymers), which has mechanical properties suitable for the realization of mechanisms that transmit mechanical energy or that contain liquids such as the nearest similar to water bottles made of PET.

With respect to this the pieces of InkCo are made in 70% of PETG, and in 20% in laser cut with Inputs from MDF to aluminum 6061 sheets that has a hardening given that there are pieces that require a rigid structural element for grinding[14], and the remaining 10% in steel elements such as bearings, screws and motion transfer coupling joints[15].

Figure 6. Interaction of the different materials in InkCo.

In the case of InkCo, a mixture of four elements defined as follows was used:

(i) *Base:* It is the medium that allows the combination of the elements that make up the ink, as well as the subsequent homogenization of the mixture; for this element was used an input that has the properties of a resin, regardless of whether it is natural or synthetic[13]; for this case, a fast-drying nail varnish was used, since it is a highly commercial and affordable synthetic resin.

(ii) *Powder* [16]: Metallic input that is in a pulverized state, responsible for providing the electrical properties to the mixture, is usually obtained in highly expensive processes such as sintering, however there are more economical and sustainable sources of this input as the waste generated by processes such as the volumetric conformation of iron foundries made in steel mills, waste generated by processes of removal of chips (chips or metallic dust) to this same type of foundries, and the waste generated in the polishing process in metallic materials, the latter 2 sources usually seen in workshops of metal-mechanics, the latter 2 sources usually seen in workshops of metal-mechanics, however there are more economical and sustainable as the waste generated by processes such as the volumetric conformation of iron foundries made in steel, waste generated by processes of removal of chips or metal dust to this same type of foundries, and the waste generated in the polishing process in metallic materials, the latter 2 sources usually seen in workshops of metal-mechanics.

(iii) *Mineral:* Sedimentary rock, preferably ceramic, which provides hydroplasticity to the mixture, as well as the viscosity of an ink itself, given that its mechanical properties such as density and adhesion correspond; In the same way this mineral establishes other properties (mechanical, thermal, electrical and physical) inherent to its constitution, for this reason, the ink has 2 carbon inputs thanks to its good properties for the purpose of the ink, the first obtained from the transformation of CO2 pollution from gasoline vehicles into carbon (KAALINK)[12] and the second obtained from the nucleus of conventional alkaline batteries.

(iii) *Solvent:* It is the substance that allows the union between the base, the powder and the mineral that allows to obtain a "paste" that will be later ground by InkCo's grinding rollers.

3. InkCo Automation

Automation of InkCo is the most important point of development and methodology, given that this point depends on the whole process and therefore developed a scheme that allows us to observe the actions performed by InkCo for the synthesis of conductive ink. In function of this scheme we developed, all the dynamic models and of movement necessary for the temporal control and by variables previously raised for the correct synthesis of the ink, in function of this we will announce a few points.

In this subsystem we have elements that require dosing the four materials therefore, we designed massive division systems such as the ratchet and a bite valve for liquids, so that the volumetric control of the liquid is for the opening time of the valve through a servo motor, and the volume control of the ceramics is through the angular position of the ratchet as shown.

On the other hand, the subsystem of mixture is the main one in receiving the inputs in first instances therefore this system must realize angular displacements to fulfill this task, in function of this idea it is necessary to understand that it is realized by means of a train of gears controlled by the angular positioning of a servo motor, which has very clear events that are governed by final sensors of race. With the objective of indicating which is the rotation limit given that he must execute two movements at the same time one rotating before Z and the other rotating before Y.

Figure 7. Mixing Subsystem.

3.1. Systems Dynamic Control

Therefore, the design of the state machine that performs the InkCo automation is governed by binary configurations, which usually ask if a process has finished to continue with the other, although there are also tasks in parallel to optimize the processing time.

The automation of InkCo is the most important point of development and methodology, since this point depends on the whole process therefore developed a scheme that allows us to observe the actions of InkCo for the synthesis of conductive ink.

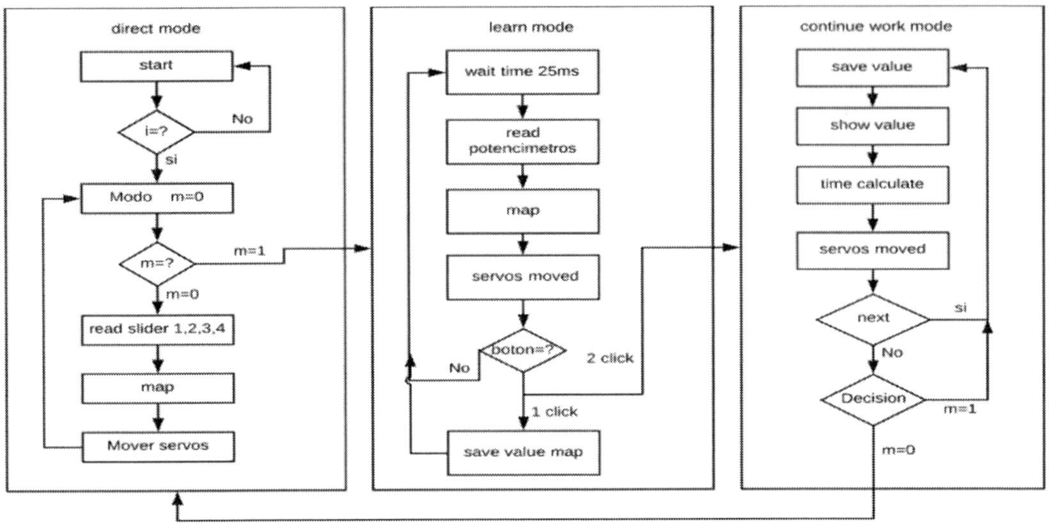

Figure 8. System's state machine.

In function of this scheme we developed, all the dynamic models and of movement necessary for the temporal control and by variables previously raised for the correct synthesis of the ink, in function of this we will announce a few points.

In this subsystem we have elements that require dosing the four materials therefore, we designed massive division systems such as the ratchet and a bite valve for liquids, thus remaining that the volumetric control of the liquid is for the opening time of the valve through a servo motor, and the volume control of the ceramics is through the angular position of the ratchet as shown.

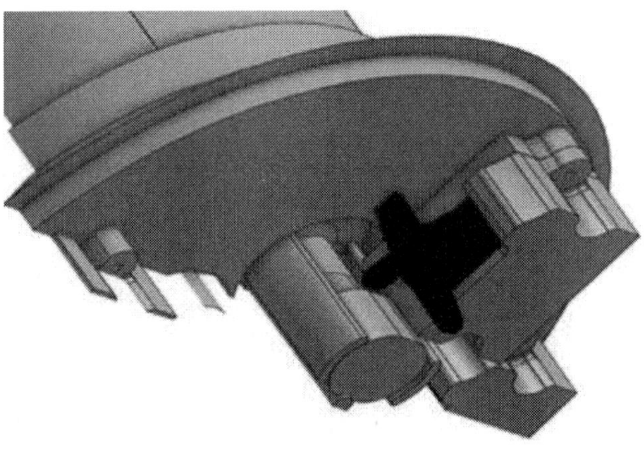

Figure 9. Material's Receptacle.

On the other hand, the subsystem of zirconium [9] mixture is the main one in receiving the inputs in the first instance therefore this system must perform angular displacement to fulfill this task, depending on this idea is necessary to understand that it is done by a train of gears controlled by the angular positioning of a servo motor, which has very clear events that are governed by final sensors of race, in order to indicate which is the limit of rotation given that the must execute two movements at once one rotating before Z and the other rotating before Y.

4. Results and Discussion

The conductive ink obtained from InkCo, is an ink that meets the basic requirements that are encountered in the market, for example that for 30cm does not have more or less 2Ω, such data and measures are exposed with greater clarity in the documentation related to InkCo ink [6, 7] [11, 12].

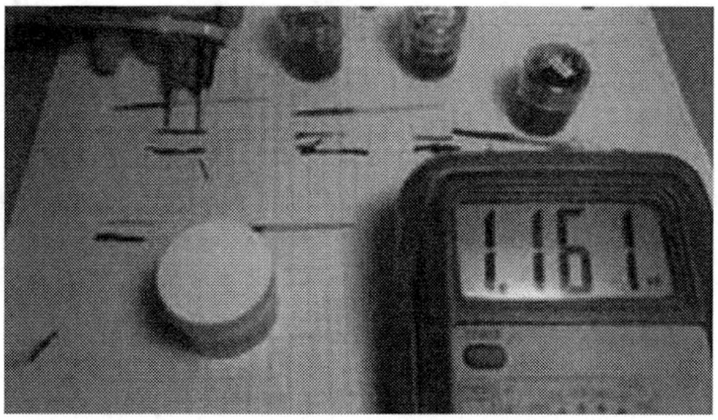

Figure 10. Test of resistance measurement in the application of conductive ink InkCo.

5. Conclusion

The development of Colombian technology that consumes inputs from our region, improve the innovation process that our country must face, through the use of rapid prototyping that helps us convert computer models into physical products that solve needs and in this case, give solution to the objectives of sustainable development since the innovation does not lie in the process but in how we get the same product making an analysis of the materials held in the region, in addition InkCo can help the equal knowledge among equals is a reality given that the commercial value of InkCo is zero because it is an open source development.

References

[1] Council N R 2005 *Linkages: Manufacturing Trends in Electronic Interconnection Technology* (Washington, DC: The National Academies Press) ISBN 978-0-309-10034-2

[2] Gulbinska M K 2014 *Lithium-ion battery materials and engineering: current topics and problems from the manufacturing perspective* (Springer)

[3] Li J, Zhou S and Han Y 2016 *Advances in battery manufacturing, service, and management systems* (John Wiley & Sons)

[4] Mills C B 2011 *Designing with models: a studio guide to architectural process models* (John Wiley & Sons)

[5] Sopicka-Lizer M 2010 *High-energy ball milling: mechanochemical processing of nanopowders* (Elsevier)

[6] Wion C Capture pollution and turn into ink URL https://www.youtube.com/watch?v=cZtF5TPQ0Gs

[7] LaSexta Como se hacen las pinturas URL https://www.youtube.com/watch?v=LYWJhRwI-ws

[8] Ronetika Tinta conductiva URL https://www.youtube.com/watch?v=DFd-X3u3s2U

[9] Mott M and Evans J 1999 *Materials Science and Engineering: A* **271** 344–352

[10] Li J, Monaghan T, Masurtschak S, Bournias-Varotsis A, Friel R J and Harris R A 2015 *Materials Science and Engineering: A* **639** 474–481

[11] Zozaya J Build hardware faster voltera v-one. prototyping tool URL https://www.voltera.io/docs/curing-conductive-ink/

[12] Sabán A Kaalink, un dispositivo que convierte contaminación en tinta URL

https://hipertextual.com/2017/02/este-dispositivo-convierte-la-contaminacion-en-tinta
[13] Air-ink: The world's first ink made out of air pollution URL https://www.kickstarter.com/projects/1295587226/air-ink-the-worlds-first-ink-made-out-of-air-pollu
[14] Aydelotte B and Thadhani N 2013 *Materials Science and Engineering: A* **570** 164–171
[15] Abolhasani D, Seyedkashi S M H, Hwang T W and Moon Y H 2019 *Materials Science and Engineering: A* **763** 138–161
[16] Totenart Como se fabrica la tinta URL https://www.youtube.com/watch?v=Y-YOPRdFp68

IOP Conf. Series: Materials Science and Engineering **844** (2020) 012043 doi:10.1088/1757-899X/844/1/012043

Smart Institution Model

M Pilamunga Poveda[1], J L Santamaría Aguirre[1] and E Viera Alulema[1]

[1] Faculty of Design and Architecture, Technical University of Ambato, Av. Los Chasquis and Rio Payamino, 180207, Ambato, Ecuador

Email: em.pilamunga@uta.edu.ec

Abstract. An institutional management model focused on the optimization of processes and services that emulates a Smart City on a smaller scale is the proposal that the Technical University of Ambato must direct its development towards a vision of efficient connectivity, reduction of resources and the approach to a Green workflow. The current problem of bureaucracy existing in public institutions in developing countries is a factor that leads to unsuccessful expenditure of time, money and personnel that must generate several steps within the value chain and due process to achieve a procedure that results in an action or service.. The model proposal focuses on the connectivity between teachers, students and the institution for the automation of activities that were performed manually, generating a virtual model that reduces time and processes; the use of security devices for the identification of people within the university environment and the communication carried out mainly through wearables with support of touch screens, mobile devices and apps; these tools will serve as a support for fast and fluid results under a solid data platform, that allows the collection of relevant information of diverse nature and its application to solve specific needs of the institutional and academic environment.

1. Introduction

Smart Cities is the new concept that appear as a result of the use of Information and Communication Technologies (ICT), integrating in an intelligent way the economy, population, mobility, the environment and administration, with the objective of guaranteeing an increase in the quality of life of the people and increasing the efficiency and effectiveness of the processes, services, infrastructure and resources of a city. For this reason, a greater capacity for innovation and better technologies are the key to demonstrate the viability of Smart Cities through the active participation of its residents [1]. Under this same concept, a development approach in the activities that take place within an educational institution are based on two visions: a). Emulate the environment of the university campus with a city giving way to the Smart City approach, where activities and interactions are developed that promote good living, learning and the sustainability of their processes. b) Also, we can look for a complementary perspective that approaches the Smart Industry, where the strategic operations and the business model are routed and optimized through technology [2]. Both environments present in the contemporary discussion of technological progress are based on connectivity as an indispensable requirement for their development and productivity, giving way to a Smart Campus.

2. Related works

Currently there are multiple examples of Smart Cities, orient at solving real problems of cities and automate them in a technical way, through the deployment of communication infrastructures. Proper information management is the key to the development of future services offered by the model Smart

City at competitive costs [3], [4]. Smart cities are considered if they have the following elements: Smart economy, Smart mobility; Smart environment; Smart people; smart living; Smart governance [5]. These elements are based on the theory of competitiveness of cities and regions, social capital, governability and new public management, inscribing the use of advanced technologies among the different elements [6].

Smart environments can derive other activities complementary to the connectivity of all their processes, Service Design as a strategy for improving and creating new services adapted to the interactive environment has great importance, its application is increasingly accepted within business environments that seek to generate positive experiences in users, furthermore of a positioning of the institutional image [7], [8]. Other directly related topics are the user experience (UX) and user interface (UI) that must be present and optimized in digital environments related to the provision of services and useful information for the user [9], [10].

Classroom use extends to student connectivity through mobile devices that, linked to the educational platform of the university, the teachers maintain contact with the student and upload to the cloud all the necessary material for the activity in the classroom and learning inside and outside of this. Another aspect that should be highlighted is the possibility of flexible workstations that can be adapted to different activities and teaching methods in the classroom. Each student can interconnect and develop content through portable devices with charging stations via the web or via USB, besides that the mobile device can be the online connection center to download information or activities on the network through technologies such as the rapid communication system and low-cost Li Fi (light fidelity), practical in small spaces of interaction without interference for transmitting data at high speed. In this context, users would have the possibility to move large files that are especially developed in class workshops and when you have image and video editing software, animation and 3D. This interaction is based on continuous and constant improvement that creative, professional and productivity apps have had in recent years [4].

A productive and for knowledge exchange ecosystem is the basis for development of quality education and application of innovative strategies in the classroom, Freedom is a key factor for dynamic processes and creative methodologies, in addition to the information of instantaneous access, it gives priority to the development of knowledge and gives fluidity to the learning process through unfettered flows in knowledge [11] .

3. Proposed work

The process is based on an input represented by the needs and users with their queries and institutional procedures as main actors; whereas, the output focuses on user satisfaction and the generation of documents and responses that meet those needs [12], [13]. Activities such as the registration of a course, consultations of schedules or tasks, the request for legal documents or certificates are covered by the platform that centralizes the user's activity through an assigned code, which in turn receives a response from the process through a document with electronic signature that generates validity to the process. **¡Error! No se encuentra el origen de la referencia.** shows a general scheme of the workflow of the platform that automates the processes.

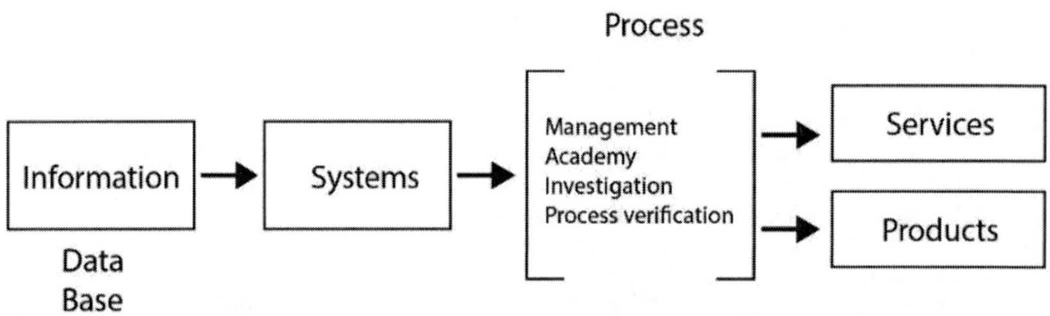

Figure 1. General scheme of the workflow of the platform

The profile of the people behind the digital platform should be suitable for process control, information management, user service and resolution of technical problems, so that it would become an environment of human talent with knowledge to give user support; and on the other hand, the personnel specialized in legal activities, secretarial or other administrative activities would be free to carry out their own activities and not lose time in simple procedures and transfer of documentation from one place to another.

An ecosystem that is closed and controlled by the institution also allows classroom activities to focus on achieve their objectives, that the connection not to external links that do not contribute to class activity is necessary to avoid interruptions and delays. Learning workshops with electronic equipment would be connected through the network and students would manage the equipment and develop prototypes and projects with the ease of aerial interconnection and a control system adapted for the classroom [14].

In order to implement this automated set of services, it is necessary to take into account several factors that will help in the interconnection, execution of processes and generation of products or services. In **¡Error! No se encuentra el origen de la referencia.**, it plans the process flow of the platform taking into account the software and hardware necessary for its operation.

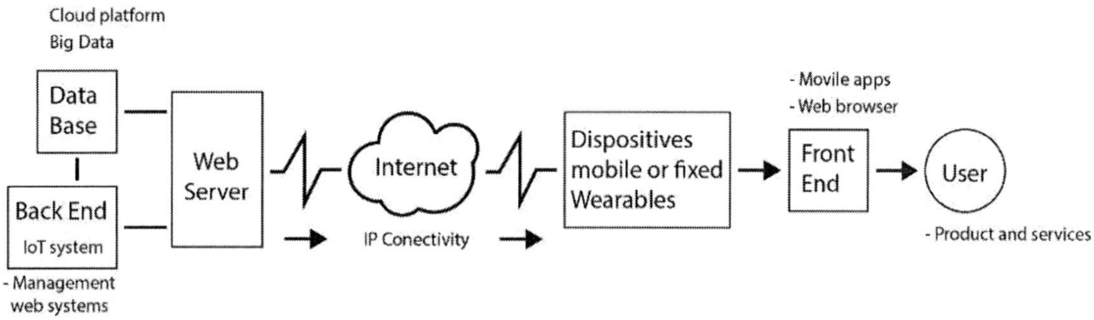

Figure 2. Process flow of the platform integrating software and hardware

- **Database:** The database service is essential for this type of systems, and depending on the volume and type of information with which it is going to work, you can use traditional databases on a server or CLOUD platform service. In the same way you can use different methods of database administration depending on the volume and type of information handled, which could be through traditional SQL managers, or in turn, Big Data techniques [14].
- **Back End:** The set of applications, web systems of information management or IoT systems that manage information through a set of algorithms, which help to provide a service or product to the end user (students, teachers, administrators, etc.). **¡Error! No se encuentra el origen de la referencia.** shows the process schema for the development and implementation of applications and systems [14].

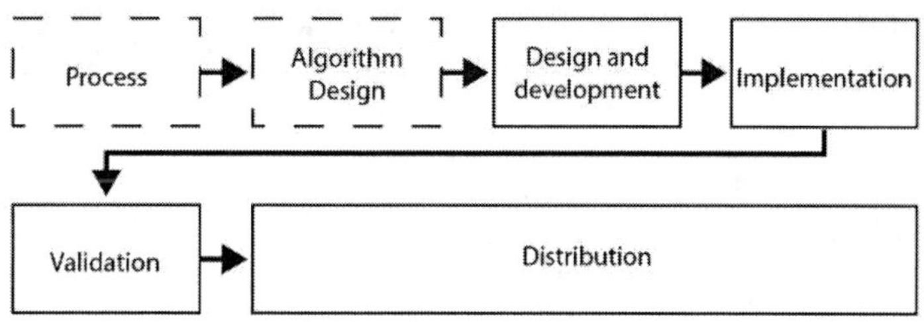

Figure 3. Process diagram for the development and implementation of systems and applications

- **Web Server:** Similarly, it is necessary to use web servers in order to effectively manage the information stored and directed to the end users [11].

- **IP Connectivity (Internet Protocol):** The internet is a network whose base is the IP network protocol. The same, that allows to interact with the connected objects (mobile or fixed devices). The interconnection can be of two types: connection through structured wiring and wireless connection [15].
- **Wearables Devices:** refers to the set of devices and electronic devices that are incorporated in some part of our body, interacting continuously with the user and other devices, in order to perform some specific function, for example, smartwatches, shoes with GPS, etc. These devices can be acquired or manufactured within the campus [2]. **¡Error! No se encuentra el origen de la referencia.** shows the process schema for the design and production of a wearable device.

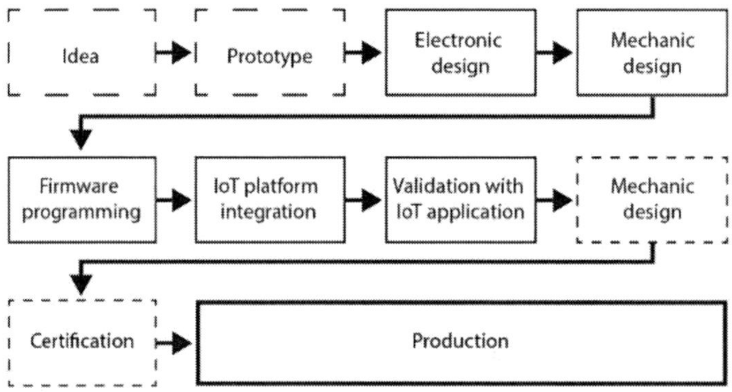

Figure 4. Process diagram for the design and production of a wearable device

- **Front End:** refers to the set of applications that the user has access to request a product or service to the institution. Generally, access is made through the web browser of the devices that are connected to the network, however, you can also develop mobile apps that can be installed different mobile devices such as smart phones, tablets, wearable devices, etc. [16], these applications offer simple interfaces and integration standards with new applications on them.

4. Conclusion

The automation of processes is very important in the development of any field, having as a consequence the minimization of time, resources and costs in the administrative and economic part of the institution, optimizing human and economic resources. In addition, the policy of "zero paper" can be adopted by supporting the environmental aspect. From the academic point of view, it helps to improve the quality of education through tools to support learning, such as the generation of virtual classrooms, which combine different multimedia resources such as images, audio and video to generate greater understanding of the students.

Initially the cost to implement this type of platforms is high due to the technological resources that are needed, however, practically all institutions of higher education have the software and hardware resources, as well as the human resources necessary for this task, therefore , the costs would be substantially reduced.

5. Future work

This project reveals other important aspects that must be analyzed in detail and developed, for example, the design and development of information management systems for Big Data or traditional databases; web systems; management of virtual classrooms and mobile apps to request services and products focused on users; and the design, development and production of wearable devices according to your needs.

6. Acknowledgments
The work has been made possible thanks to funding from the Directorate of Research and Development (DIDE) of the Technical University of Ambato.

References
[1] Ernst E, Ferrovial Y and Madrid Network 2012 *Libro Blanco, Smart Cities* (Madrid: Imprintia)
[2] FME-CWM, Chamber of Commerce, Ministry of Economic Affairs, TNO and VNO-NCW 2017 *Smart Industry, Dutch industry fit for the future* (Netherlands)
[3] Cruz V M, Oliete V P, Morales R C, González L C, Cendón M B and Hernández S A 2015 *Las tecnologías IOT dentro de la industria conectada 4.0* (Madrid: EOI Esc. Organiz. Industrial)
[4] Schaffers H, Komninos N, Pallot M T B, Nilsson M and Oliveira A 2011 Smart cities and the future internet: Towards cooperation frameworks for open innovation *The future internet assembly* pp 431-46
[5] Sikora-Fernández D 2017 Factores de desarrollo de las ciudades inteligentes *Revista Universitaria de Geografía* **26** 1 pp 135-52
[6] Anttiroiko A V, Valkama P and Bailey S J 2014 Smart cities in the new service economy: building platforms for smart services *(AI & Society)* **29** 3 pp 323–34
[7] Sleeswijk-Visser F 2017 Service design by industrial designers (Delft: TU Delft)
[8] Hassan QF, Khan A R and Madani S A 2017 Internet of Things: Challenges, Advances, and Applications (CRC Press)
[9] Kuniavsky M 2010 Smart things: ubiquitous computing user experience design (Elsevier)
[10] Hassenzahl M 2011 User Experience and Experience Design *The Encyclopedia of Human-Computer Interaction, 2nd Ed.* (The Interaction Design Foundation)
[11] Buyya R and Dastjerdi A V 2016 Internet of Things: Principles and paradigms (Elsevier)
[12] Mercado G, Álvarez L, Bocaccini L, Ledda M, Membrives J, Muros M, López L, Juárez D, Guillén L, Gómez F, Favaro G, Ciperiani G, Dumé S and Berra G 2017 Diagnóstico y Metodología para la Implementación de" Internet of Things" en el Planeamiento y Desarrollo de Ciudades Inteligentes *XIX Workshop de Investigadores en Ciencias de la Computación (WICC 2017)* pp 171-75
[13] Kamal R 2017 Internet of things: architecture and design principles (McGraw-Hill Education)
[14] Ontiveros E, Vizcaíno D and Sabater V L 2016 *Las ciudades del futuro: inteligentes, digitales y sostenibles* (Madrid: Ariel)
[15] Dordoigne J 2015 *Redes informáticas-Nociones fundamentales* (5ª edición):(Protocolos, Arquitecturas, Redes inalámbricas, Virtualización, Seguridad, IP v6...), (Barcelona: Ediciones ENI)
[16] Garrett J J 2010 *Elements of user experience, the user-centered design for the web and beyond* (Pearson Education)

IOP Conf. Series: Materials Science and Engineering **844** (2020) 012044 doi:10.1088/1757-899X/844/1/012044

Design of an Information System for optimizing the Programming of Nursing Work Shifts

F Osores[1], G Cabrera[2], R Linfati[3], S Umaña-Ibañez, J Coronado-Henández[4] and G Gatica[1]

[1] Universidad Andres Bello. Engineering Faculty, Chile.
[2] Pontificia Universidad Católica de Valparaíso. Engineering Faculty, Chile
[3] Universidad del Bío- Bío. Engineering Faculty, Chile
[4] Universidad de la Costa. Departamento de Gestión Industrial, Agroindustrial y Operaciones. Colombia.

jcoronad18@cuc.edu.co

Abstract. Health institutions operate twenty-four hours a day, seven days a week. They face a demand that fluctuates daily. Unlike jobs with fixed hours and obligatory days off, in health, operational continuity is required. The allocation for nursing shifts generates a rotation of people for health services according to legal and casuistic guidelines. Assigning and planning shifts results in a workload that takes an average of five to six extra hours. Existing applications offer a partial solution because they do not consider the news and contingencies of a health service. A web application is presented that, given a list of nurses, historical shifts and restrictions, a work shift planning is generated. This application comes to support the current shift allocation method based on electronic spreadsheets. The development consists of two modules. The first module has a shift allocation algorithm developed in C ++ and the second module has a graphical interface. As a case study, a set of health services from Chile and Colombia was used. The services have a defined number of nurses, who work different shifts according to the role and need of the institution. The results obtained are similar to a historical one. The proposed system takes less time and delivers various files and parameters that can be useful for nurses, the service and the health institution.

1. Introduction

Health institutions are facing various challenges, in which they increase economic pressure and demand for care challenges. One of the outstanding problems in health care operations is the problem of the allocation of work shifts [1].

The Nurse Rostering Problem (NRP) is a problem of the NP-Hard class [2].The problem involves making the allocation of work shifts, days off and vacations for nurses [3] [4]. The NRP faces a variety of constraints that hinder the goal of optimizing shift allocation [5]. Among them are number of personnel versus required personnel; maximum monthly work hours; maximum number of workable hours and days off [6].

There are several restrictions that vary according to the country and corresponding labour legislation [7]. Therefore, there are various solutions to the NRP, each according to the needs of each institution and country [8] [9].

NRP has been the focus of study since the 1970s when the first studies on the problem emerged [10]. In [10] after studying a hospital unit, they propose a solution focused on minimizing an objective function that balances compensation of personnel coverage and personal time preferences. That same year, [11] proposes a solution to the problem, whose objective function to quantify the preferences of the staff versus the time of work estimation, with rotation patterns and requests for free days, takes as restrictions the number of personnel and the change of rotation [12].

For NRP, new algorithmic approaches and developments of computer systems that facilitate the task of assigning shifts to arise. In spite of the above, the personnel in charge of assigning shifts prefer to perform this task manually [13]. The task of assigning shifts takes five hours on average, this scenario is affected when a nurse must be absent [14] [15]. Thus, an immediate reprogramming is required without overloading the remaining staff [16]. Based on the foregoing, it can be assumed that a quick way to deal with this situation is to offer extra shifts to the rest of the staff, but this sometimes exceeds the legal limit and may even generate additional costs for the institution [2] [17].

The research focus of this study is focused on contributing to improving the efficiency of the work shift programming and scheduling tasks through the development and implementation of a web application that automates the process and significantly reduces the time spent on this task.
This document presents in its structure a case study used for the analysis, the solution technique used, then the graphical interface and the technical architecture used are presented, finally, the results and conclusions are presented.

2. Case Study

The base scenario for the development of the web application, establishes a nursing service with five nurses, each with their own preferences for the request of free days and vacations [18] [19] [20] [21] [22][23][24]. It must be considered that any nurse at any time may present a medical license of an indeterminate length. External constraints define labor laws and the number of nurses available versus the minimum number of nurses for the service to function normally [6] [25] [26]. These restrictions will serve as input to develop an algorithm that performs a shift assignment taking into account the available constraints [27]. Figure 1. shows an example of how the work shift programming was carried out without any computational help. This manual assignment included Day Shifts (D), Night Shifts (N) and days off (-).

TURNOS DE ENFEREMERÍA 2018
ENERO

ENERO	1	2	3	4	5	6	7	8	9	10	11	12	13	14	15	16	17	18	19	20	21	22	23	24	25	26	27	28	29	30	31
	L	M	M	J	V	S	D	L	M	M	J	V	S	D	L	M	M	J	V	S	D	L	M	M	J	V	S	D	L	M	M
Francisca Torres	D	N			D	N			D	N			D	N			D	N			D	N			D	N			D	N	
Africa Fuentes	N			D	N			D	N			D	N			D	N			D	N			D	N			D	N		
Inés Taplas		D	N			D	N			D	N			D	N			D	N			D	N			D	N			D	N
Mariela Capurro			D	N			D	N			D	N			D	N			D	N			D	N			D	N			D
Francisca Torres	D	N	-	-	D	N	-	-	D	N	-	-	D	N	-	-	D	N	-	-	D	N	-	-	D	N	-	N	-	-	-
Africa Fuentes	N	-		D	N	-	-	D	N	-	-	D	N	-	-	D	N	-	-	D	N	-		D	N	-	-	D	N	N	-
Inés Tapias	-	-	N	N	-	D	N	-		D	N	-	-	D	N	-	-	D	N	-	-	D	N	-	-	D	N	-	-	D	N
Mariela Capurro		D	D	-	-	-	D	N	-	-	D	-	-	-	D	N	-	-	D	N	-	-	D	N	-	-	D	-	D	N	D
Margarita Vásquez																															
Vanessa Quispe												N																			

Figure 1. Example of a manual work shift programming.

3. Methodology

3.1 Solution Architecture

A three-layer architecture was defined as shown in Figure 2, the presentation layer corresponds to the user interface, displayed in the web application format. The second layer corresponds to the logical layer, where the allocation algorithm generates the monthly and weekly planning for the nurses of a service. The first two layers communicate with a third layer, where the data of an Access control list is stored according to the case study, this ensures that users cannot see information and plans from other users of other health institutions.

We chose to use this type of architecture because it offers extensible benefits to the needs of health services. The three-layer architecture allows developers to modularize and expand the application, in addition, the costs of implementing new functionalities are reduced, this is because the implementation of three independent layers between them allows making changes on one of them without unnecessarily affecting the others [28].

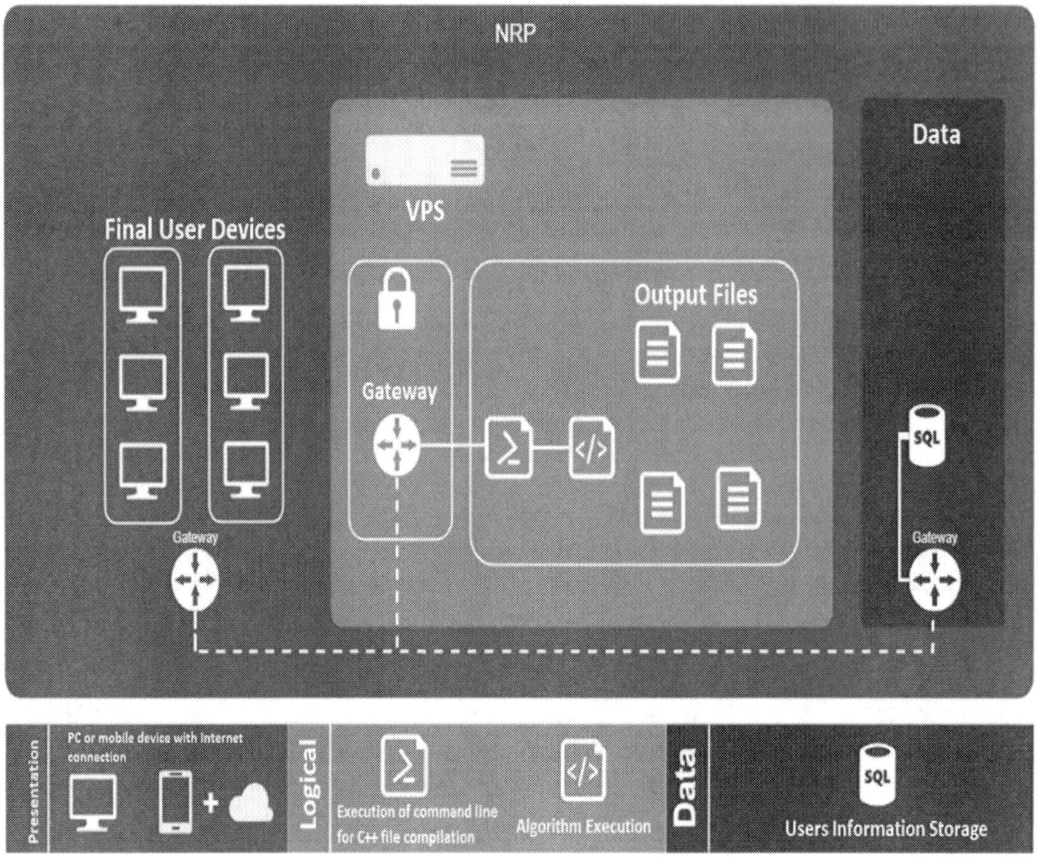

Figure 2. Technological architecture.

3.2 Solution Algorithm – Logical Layer
The proposed solution consists of an algorithm developed in C ++, whose pseudocode is shown in Figure 3, which receives 2 files as input, one of them corresponds to the nurses list with their respective preferences and the second, a configuration file, the which defines the internal variables of each service as shown respectively in Table 1.

3.3. User interface – Presentation Layer
A web module was developed (figures 4 and 5) to facilitate the use of the system by final users.
The web application is completely in Spanish. The interfaces were designed for data acquisition the content and functionality, complying with the quality standard established by ISO 25000 [29].

As shown in Figure 3, the option is shown when selecting a file to upload, there is the option to modify it, in a more intuitive way than to modify the input file directly, prioritizing the usability of the system, in addition to this way the nurse in charge of the assignment of shifts does not make mistakes when entering the desired information.

Attributes of functionality, maintainability, efficiency, usability, and portability stand out as attributes of quality. This in relation to the non-functional requirements developed.

IOP Conf. Series: Materials Science and Engineering **844** (2020) 012044 doi:10.1088/1757-899X/844/1/012044

```
1  Inputs: personal.csv, config.
2  Output: planification.csv
3  Start
4  While Last day of the month condition is not met Do
5        While Unassigned nurses in a daily work shift > 0 Do
6        If the nurse is available for assignment Then
7           If the nurse has no assignments Then
8              Assign a work shift
9              Go to Line 5
10          End If
11          If the nurse has assignments Then
12          Analyze if assignment is possible according hours and consecutive work shifts.
13          Assign work Shift
14          Go to Line 5
15          End If
16       End If
17       If Number of nurse and the work shift have no changes Then
18       Assign an extra work shift to the nurse with less hours and less extra works shifts.
19       End If
20       If No extra shifts was assigned Then
21       Update the list of required float nurses
22       End If
23       If there is remaining nurses with no assignments Then
24       Schedule the nurse as Day Off.
25       End If
26       End While
27       Go to the Next day of the month
28 End While
29 End
```

Figure 3. Proposed Solution Algorithm

Table 1. Available configurations

Type of configuration	Data
Maximum number of hours per day	Integer number between 0 and 24
Maximum number of hours per week	Integer number between 0 and 168
Types of shifts	List of elements separated by comma, in the following order: ShiftID (integer), the name of shift (String), Duration in hours (integer)
Shifts that cannot be preceded by other shifts (combinations not valid)	List of elements separated by comma, in the following order: ShiftID, ShiftID which cannot precede.
Combinations of valid turns	List of elements separated by comma, in the following order: new ShiftID, ShiftID 1, ShiftID 2, ..., ShiftID n
Requirements of shifts per day	List of elements separated by comma, in the following order: ShiftID, integer number indicating the number of nurses for shift on Monday, integer number indicating the number of nurses for shift on Tuesday, integer number indicating the number of nurses for shift on Wednesday, integer number indicating the number of nurses for shift on on Thursday, integer number indicating the number of nurses for shift on Friday, integer number indicating the number of nurses for shift on Saturday, integer number indicating the number of nurses for shift on Sunday

IOP Conf. Series: Materials Science and Engineering **844** (2020) 012044 doi:10.1088/1757-899X/844/1/012044

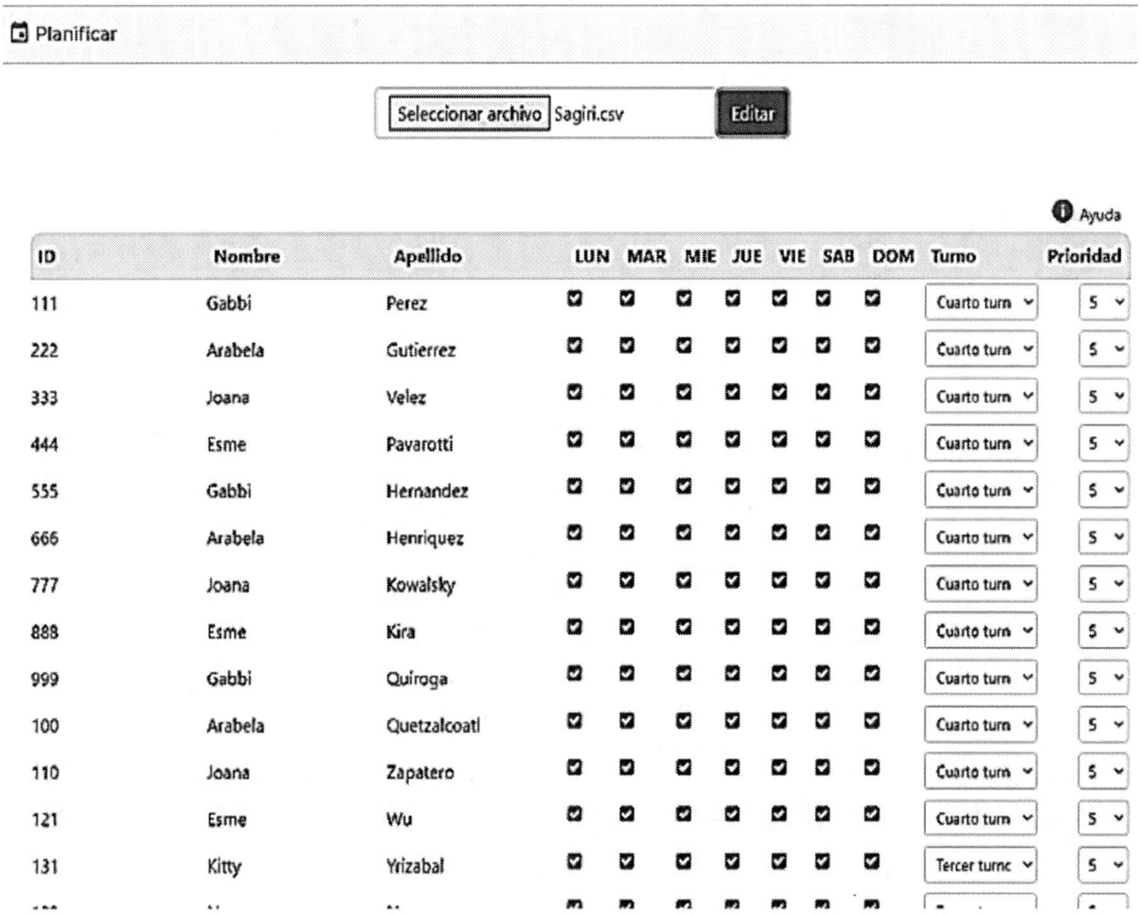

Figure 4. Input file example.

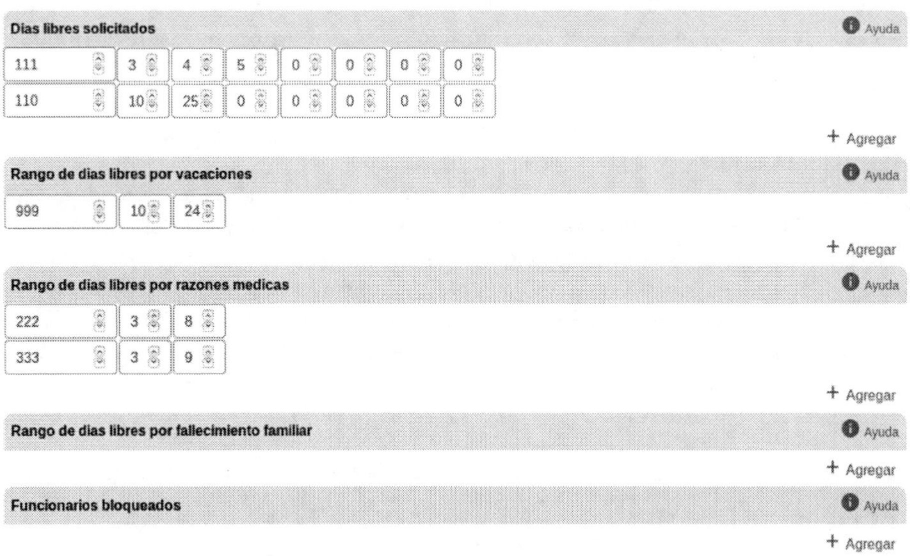

Figure 5. User interface example.

4. Results

To carry out a validation of the system, a real nursing service was used, the objective was to obtain results similar to a manual assignment.

Figure 1 shows hand-made planning, the top part shows what was planned at the beginning of the month and in the part below it shows the result at the end of the month.

An input file was defined with the list of nurses and some constraints like the one shown in figures 2 and 3 to obtain a monthly shift plan.

ID	Nombre	VIE	SAB	DOM	LUN	MAR	MIE	JUE	VIE	SAB	DOM	LUN	MAR	MIE	JUE	VIE
2		NOCHE	LIBRE	LIBRE	LARGO	NOCHE	LIBRE	LIBRE	LARGO	NOCHE	LIBRE	LIBRE	LARGO	NOCHE	LIBRE	LIBRE
3		LIBRE	LARGO	NOCHE	LIBRE	LIBRE	LARGO	NOCHE	LIBRE	LIBRE	LARGO	NOCHE	LIBRE	LIBRE	LARGO	NOCHE
4		LARGO	NOCHE	LIBRE	LIBRE	LARGO	NOCHE	LIBRE	LIBRE	LARGO	NOCHE	LIBRE	LIBRE	LARGO	NOCHE	LIBRE
5		LIBRE	LIBRE	LARGO	NOCHE	LIBRE	LIBRE	LARGO	NOCHE	LIBRE	LIBRE	LARGO	NOCHE	LIBRE	LIBRE	LARGO
6		DIA	LIBRE	LIBRE	DIA	DIA	DIA	DIA	DIA	LIBRE	LIBRE	DIA	DIA	DIA	DIA	DIA
7		LIBRE	LIBRE	SOLICITUD	SOLICITUD	LIBRE	LIBRE	SOLICITUD	SOLICITUD	LIBRE	LIBRE	SOLICITUD	SOLICITUD	LIBRE	LIBRE	SOLICITUD
8		LIBRE	LIBRE	LIBRE	LIBRE	LIBRE	LIBRE	LIBRE	LIBRE	LIBRE	LIBRE	LIBRE	LIBRE	LIBRE	LIBRE	LIBRE

Figure 6. Output file (1/2).

16	17	18	19	20	21	22	23	24	25	26	27	28	29	30
SAB	DOM	LUN	MAR	MIE	JUE	VIE	SAB	DOM	LUN	MAR	MIE	JUE	VIE	SAB
LARGO	NOCHE	LIBRE	LIBRE	LARGO	NOCHE	LIBRE	LIBRE	LARGO	NOCHE	LIBRE	LIBRE	LARGO	NOCHE	LIBRE
LIBRE	LIBRE	LARGO	NOCHE	LIBRE	LIBRE	LARGO	NOCHE	LIBRE	LIBRE	LARGO	NOCHE	LIBRE	LIBRE	LARGO
LIBRE	LARGO	NOCHE	LIBRE	LIBRE	LARGO	NOCHE	LIBRE	LIBRE	LARGO	NOCHE	LIBRE	LIBRE	LARGO	NOCHE
NOCHE	LIBRE	LIBRE	INCAPACIDA	INCAPACIDA	INCAPACIDA	INCAPACIDA	INCAPACIDA	INCAPACIDA	INCAPACIDA	LIBRE	LARGO	NOCHE	LIBRE	LIBRE
LIBRE	LIBRE	DIA	DIA	DIA	DIA	DIA	LIBRE	LIBRE	DIA	DIA	DIA	DIA	DIA	LIBRE
LIBRE	LIBRE	LIBRE	LARGO	NOCHE	LIBRE	LIBRE	LARGO	NOCHE	LIBRE	LIBRE	LIBRE	LIBRE	LIBRE	LIBRE
LIBRE	LIBRE	LIBRE	LIBRE	LIBRE	LIBRE	LIBRE	LIBRE	LIBRE	LIBRE	LIBRE	LIBRE	LIBRE	LIBRE	LIBRE

Figure 7. Output file (2/2).

When uploading the generated file to the system, resulting in a plan as shown in illustrations 6 and 7, which, when comparing both schedules, the system delivered similar results to the planning done in Figure 1.

When performing a more detailed analysis, it can be seen in figure 1, that the assignment for the first nurse of the list is that corresponding to the nurse with ID 2 of figures 4 and 5, it is exactly the same, with the difference that Figure 1 shows an anomaly in the structure of shifts between days 28 and 31, where the nurse performs a night shift when it does not correspond and is assigned an extra day off. Another case to analyze is that of the nurse with ID 5 of figures 6 and 7, which corresponds to the fourth nurse of the list in figure 1, by way of proof this nurse was assigned a medical license between days 19 and 25 (figure 6) this medical license was specified by the system as shown in figure 5 , the objective of this was to observe the behavior of the allocation algorithm to adapt to the minimum personnel requirements required daily, in this case it was not necessary allocation of extra work shifts to other nurses.

5. Conclusions and Future Work

The software architecture is portable, manageable, with reusable components, and allows modifications and extensibility, as well as being scalable. This is due to a set of layers interacting with each other, which have interfaces to communicate with other services.

The developed product is extensible and can be implemented in similar areas. As future work, it is planned to include new restrictions such as the time an nurse takes from his home to his place of work, work with historical records to avoid repeating the same shifts from one year to the next in an nurse and implement new functionalities to the Web Application.

References

[1] Fügener A, Pahr A and Brunner J O 2018 Mid-term nurse rostering considering cross-training effects *Int. J. Prod. Econ.* vol **196** pp 176–187

[2] Azaiez M N and Al Sharif S S 2005 A 0-1 goal programming model for nurse scheduling, *Comput. Oper. Res.* vol **32** no 3 pp 491–507

[3] Burke E K, Curtois T, Post G, Qu R, and Veltman B 2008 A hybrid heuristic ordering and variable neighbourhood search for the nurse rostering problem *Eur. J. Oper. Res.* vol **188** no 2, pp 330–341

[4] Burke E K, Curtois T, Qu R, and Vanden Berghe G 2010 A scatter search methodology for the nurse rostering problem *J. Oper. Res. Soc.* vol **61** no 11 pp 1667–1679

[5] De Causmaecker P and Vanden Berghe B 2011 A categorisation of nurse rostering problems, J. Sched vol **14** no 1 pp 3–16

[6] Dowsland K A 1998 Nurse scheduling with tabu sGGGarch and strategic oscillation *Eur. J. Oper. Res.*vol **106** no 2–3 pp 393–407

[7] Valouxis C and Housos E 2000 Hybrid optimization techniques for the workshift and rest assignment of nursing personnel *Artif. Intell. Med.* vol **20** no 2 pp 155–175

[8] Vanhoucke V and Maenhout B 2009 On the characterization and generation of nurse scheduling problem instances *Eur. J. Oper. Res.* vol **196** no 2 pp 457–467

[9] Komarudin, M A, Guerry T and Vanden Berghe G 2013 The roster quality staffing problem – A methodology for improving the roster quality by modifying the personnel structure *Eur. J. Oper. Res.* vol **230** no 3 pp 551–562

[10] Miller H E, Pierskalla W P and Rath G J 1976 Nurse Scheduling Using Mathematical Programming *Oper. Res.* vol **24** no 5 pp 857–870

[11] Warner D M 1976 Scheduling Nursing Personnel According to Nursing Preference: A Mathematical Programming Approach *Oper. Res.* vol. **24** no 5 pp 842–856

[12] Berrada I, Ferland J A and Michelon P 1996 A multi-objective approach to nurse scheduling with both hard and soft constraints *Socioecon. Plann. Sci.* vol **30** no 3 pp 183–193

[13] Clark A, Moule P Topping A and Serpell M 2015 Rescheduling nursing shifts: scoping the challenge and examining the potential of mathematical model based tools *J. Nurs. Manag.* vol **23** no 4 pp 411–420

[14] MJC2 2018 Workforce Planning & Scheduling

[15] Bester M J, Nieuwoudt I and Van Vuuren J H 2007 Finding good nurse duty schedules: a case study *J. Sched.* vol **10** no 6 pp 387–405

[16] Cheang B, Li H, Lim A, and Rodrigues B 2003 Nurse rostering problems—a bibliographic survey *Eur. J. Oper. Res.*, vol **151** no 3 pp 447–460

[17] Erhard M, Schoenfelder J, Fügener A and Brunner J O 2018 State of the art in physician scheduling *Eur. J. Oper. Res.* vol **265** no 1 pp 1–18

[18] Jaumard B, Semet F and Vovor T 1998 A generalized linear programming model for nurse scheduling *Eur. J. Oper. Res.* vol **107** no 1 pp 1–18

[19] Burke E K, De Causmaecker P, Vanden Berghe G and Van Landeghem H 2004 The State of the Art of Nurse Rostering *J. Sched.* vol **7** no 6 pp 441–499

[20] Bard J F and Purnomo H W 2005 Preference scheduling for nurses using column generation *Eur. J. Oper. Res.* vol **164** no 2 pp 510–534

[21] Constantino A A, Landa-Silva D, de Melo E L, de Mendonça C F X, Rizzato D B and Romão W 2013 A heuristic algorithm based on multi-assignment procedures for nurse scheduling *Ann. Oper. Res.*

[22] Lin C C, Kang J R, Chiang D J, and Chen C L 2015 Nurse Scheduling with Joint Normalized Shift and Day-Off Preference Satisfaction Using a Genetic Algorithm with Immigrant Scheme *Int. J. Distrib. Sens. Networks*, vol **11** no 7 p. 595419

[23] Romero Conrado A, Coronado-Hernandez J, Rius-Sorolla G, and García-Sabater J 2019A Tabu List-Based Algorithm for Capacitated Multilevel Lot-Sizing with Alternate Bills of Materials and Co-Production Environments *Appl. Sci.* vol **9** no 7 p 1464

[24] Vélez Vargas V, Henao Baena V and Calvo Salcedo 2018 A Design and implementation of a wired intercommunication prototype for hospital care *INGE CUC* vol **14** no 1 pp 101–112
[25] Ernst A, Jiang H, Krishnamoorthy M and SierD 2004 Staff scheduling and rostering: A review of applications, methods and models *Eur. J. Oper. Res.*vol **153** no 1 pp 3–27
[26] Lin C C, Kang J R, Liu W Y and Deng D J 2014 Modelling a Nurse Shift Schedule with Multiple Preference Ranks for Shifts and Days-Off *Math. Probl. Eng.* vol **2014** pp 1–10
[27] Wu T H,Yeh J Yand Lee Y M 2015 A particle swarm optimization approach with refinement procedure for nurse rostering problem *Comput. Oper. Res.* vol 54 pp 52–63
[28] Wijegunaratne I and Fernandez G 1998 The Three-Tier Application Architecture pp. 41–78.
[29] International Organization for Standardization 2005 *ISO/IEC 25000: Systems and software engineering - Systems and software Quality Requirements and Evaluation (SQuaRE).*

IOP Conf. Series: Materials Science and Engineering **844** (2020) 012045 doi:10.1088/1757-899X/844/1/012045

Information Technology Governance Model, Based on Risk Management and Information Security for Colombian Public Universities: Case on Study University of La Guajira

L R Viecco[1] **and J G Arevalo**[1, 2]
[1]University De La Guajira, Riohacha, La Guajira – Colombia.
[2]ENAPROT, Investigation Group of the Universidad De La Guajira.

E-mail: lviecco@uniguajira.edu.co

Abstract. Nowadays Universities require the implementation of practices in (IT)information technology governance issues where it involves structures, processes, mechanisms and measurement indicators for the good performance of Information Technology. In order to make a strategic alignment that defines the generation of value, and that is why this research shows the design of an IT governance. Based on model risk management and information security for Colombian public universities, for which its strategic alignment, its strategy is analysed business. Its institutional educational project in order to consolidate a model adjusted to the particular needs of each academic faculty.

1. Introduction

Technologies are now a reality in all processes of organizations, allow to store, manipulate, transport and manage information in universities and represents a valuable asset. It is increasingly common for companies to manage your information through electronic means, relying on various information technologies to comply with the processing objectives of information demanded by business processes today [1].

Undoubtedly, the support of Information Technology (IT) in the academic sector represents many benefits in the short and long term. Many of the largest universities in the world make extensive use of information technology to achieve their institutional objectives. IT support has allowed organizations around the world to be more competitive, but it has also created new scenarios of threats and risks, many of which are inherent to electronic or digital systems. Since the emergence of the first digital and electronic computing systems, there have been risks that have threatened three main attributes of high-quality information management: availability, integrity and confidentiality. Right now, information security has become crucial for any University and represents the protection of one of their most important assets: information. This important asset is stored and managed through the institutions hardware and software systems, which represent their base platform for information management. Unfortunately, there are serious risks that can threaten the continuity of their operations, and due to the fact that these risks can be characterized with having a high probability of occurrence and a potential for generating catastrophic damage; it necessitates that the organization adequately mitigate risks through a defined plan.

The threats and computer risks are of various kinds and day out they increase in number and probability of occurrence, such as the loss of data intentionally or not, information theft, loss of information caused by computer viruses, loss of information due to natural disasters such as electrical storms, earthquakes, fires etc. Due to the great variety of threats and risks to the security of information, the management of information security is a task that demands the commitment of the top management of the organization and for each of the management levels. The panorama of risks and computer threats information technologies have brought, has been the subject of studies by various agencies and companies of the largest in the world. Such Organizations as ISACA of the United

States, which is mostly integrated by computer auditors, the government of England, companies such as IBM, General Electric, American Telephone and Telegraph AT&T, and many others, have carried out research for decades. With the aim of collecting recommendations and best practices regarding the use and IT management, and its value for the achievement of business objectives, generating as results of their research, documents that formalize the methodologies of governance and management of information technology as best practices in organizations. [2].

The concept of IT governance was a subject of debate since the 90´s, and many definitions have been introduced by academics and professionals in the field. It has been noted that the various definitions differ in some aspects; however, most all of them share a commonality relating to the integration between companies and IT, usually known as strategic alignment and is considered one of the main elements of IT governance. The definitions also explicitly state that IT governance is an integral part of the company's operations and incorporates processes and structures for decision making and best use of IT performance. It has also been observed that there is a clear difference between IT governance and IT management. IT governance is much broader and focuses on present and future demands for IT and the business while the latter maintains a management approach providing services, IT products, and operations. [3]

IT governance is a framework for an organization's leadership, organizational structures, business processes, standards, and compliance with these standards, which ensure that the IT of the organization supports and allows for the achievement of their strategies and objectives. In the future, IT governance will be even more important than corporate governance today. Information and IT are critical to business survival, and organizations that do not" direct and control" their IT with a competitive advantage should expect to be left as roadkill on the information superhighway [4].

As previously discussed, the main purpose of IT governance is to align the strategic IT objectives with the purpose of the organization [5]; It may seem that this is only a matter of strategic planning and that the responsibility of this falls only the institutional managers, but in reality, this is not the case. Generally, in all organizations the IT areas are subject to different impositions, which can develop into impediments that can threaten the proper functioning of the organizational processes. Problems can also arise from succumbing to regulatory pressures and focusing too heavily on quick responses and not sustainable solutions, which can easily lead to losing alignment with the organization and leading them to dedicate themselves to solving lots of small, specific problems instead of focusing on the bigger picture [6]. According to the author, "it is intended to put the government and risk management of information technology in context, as an integral part of government and information technology management." Based on this concept, organizations will not only contemplate management, control, operation, and administration systems, but also involve within their systems the risks that may exist in an organizational and technological structure from the dynamism that yields under the regulatory criteria [7]. The success stories and recommendations collected over the research on information technology management have concluded that companies must elevate IT management to top upper management applying the top-down methodology. This implies that IT business decisions must depart of filter down from the organization's board of directors, thus elevating IT professionals to corporate governance and in this way, it is therefore, making it common to hear today the term, government of information technologies, applied to within the business world.

The governance of IT, ultimately involves the alignment of IT with business strategy, involving all processes of the organization, to promote better use of technology and organizational structures in order to achieve the organization's objectives. The Colombian national government, aware of the need to continue advancing the construction of an inclusive and equitable that demonstrates adherence to the global standards of quality of life, has been promoting, through the ministry of IT. Important initiatives related to the government of the same and its management, including the launch of its IT4+ model demonstrating a great contribution to the advancement of a sector that is so vital to the further development of the nation. In short, it is crucial that a company use IT to streamline their processes, achieve a competitive advantage, and manage their most important asset: information.

However, it will be necessary to have IT governance in order to align the objectives of information technology with the business strategy.

2. Literature Review

Nowadays, higher education institutions, both public and private, have been stimulating a technological transformation with the incorporation of information technologies as a key element for generating value and seeking competitive advantages through focusing on achieving strategical alignment in order to determine which technological models should be implemented. Once this is determined, this is where IT governance comes into play and as a practice, has been strengthening itself through existing reference frameworks elevating technologies from a tactical level to a strategic level. Due to its constant innovations and the complexity of its planning process and execution, and producing positive reactions in top management a result of achieving favourable outcomes. When talking about governance, we are focusing on its primary objective, the creation of value, which allows us to enhance the resources and minimize the risks of IT, therefore obtaining benefits through the development of optimal processes.

When you reach this point in the process, you can address a governance model based on good practices to improve the efficiency and effectiveness of the fundamental resource of any organization: information and these in turn day by day are constantly updated, becoming an asset of great importance in them. As is well known, information and communications technology (ICT) used in the university are usually conceptually different, depending on whether they are focused on teaching, research, or community service. There are often important differences in ICT trainings among students as well as some manifestations of heterogeneity among the student's training experiences depending on the area of the university that is taken into consideration [8].

In many universities, the relationship between IT and the business is totally disconnected; it is even more frequent that the IT infrastructure does not support the concept of an agile organization. The proliferation of information technologies, the increase in complexity, and the increase in costs are common features of many IT infrastructures. It is necessary to know the framework of concepts on which the research is based, so the following sections include a discussion of relevant findings from the literature review:

2.1 Strategy

They are the means by which long term objectives are achieved. Business strategies include geographic expansion, diversification, acquisition, product development, market penetration, cost reduction, disposal, liquidation and joint ventures. Strategies are potential actions that require decisions on the part of the management and resources of the company. In addition, the strategies affect the long-term finances of a company, at least for five years, thus guiding the future. The strategies produce effects on the functions and divisions of the company, and require that both external factors and internal factors facing the company be taken into account [9].

2.2 Strategic Management

Strategic management is defined as the art and science of formulating, implementing and evaluating decisions through functions that allow a company to achieve its objectives. According to this definition, the strategic direction focuses on the integration of management, marketing, finance, accounting, production, operations, research and development, and computer information systems to achieve the success of the company. The term strategic direction is used in this text as a synonym for the term strategic planning. The latter term is used more often in the business world, while the former is used in the academic environment.

2.3 Corporate governance
It is the set of mechanisms used to manage the relationship between stakeholders and determine and control strategic direction and performance at its core, corporate governance is concerned with identifying ways to ensure that strategic decisions are taken effectively. As a means of Corporations use it to establish order between the owner parties and their high-level managers whose interests may conflict. In modern corporations, especially the United States and the United Kingdom, a primary objective of companies is to ensure that the interests of senior managers are aligned with the interests of shareholders. Corporate governance implies supervision in areas where owners, managers and members of the board of directors may have conflicts of interest [10].

2.4 Business Governance
Corporate governance is the set of responsibilities and practices exercised by the Board and executive management, with the aim of providing strategic direction, ensuring that plans and objectives are achieved, assessing that risks are managed proactively and ensuring that Company resources are used responsibly. Corporate governance deals with the separation of ownership and control of an organization, while corporate governance focuses on the direction and control of the company, and IT governance focuses on the direction and control of IT [11].

2.5 University Government
It is a set of responsibilities and practices exercised by the higher university council and executive management in order to provide strategic direction, ensure compliance with objectives, establish an adequate management of risks, verify the responsible and efficient use of resources University. Compliance is a pattern of systematic behaviour of stakeholders, administrators and staff of the university, who are focused on achieving sustainable social and financial results. This behaviour must be directed towards the achievement of the four main assets of the organization: Infrastructure, customers and interested third parties, internal personnel, processes and the creation of value [12].

2.6 IT Government
Governance formalizes and clarifies oversight, accountability and decision rights for a wide range of IT strategy, resources and control activities. It is a collection of policies, practices and processes of management, planning and performance review; controls and metrics of performance on investments, plans, budgets, commitments, services, important changes, security, privacy, business continuity and compliance with laws and organizational policies [13]. An IT Government provides structures, processes, resources, and information to align with the company's strategic objectives. As well as integrating and institutionalizing internationally recognized good practices in Organizational Planning, acquiring and implementing, delivering service and support, monitor the performance of IT [14]. There is no doubt that more and more organizations depend on Information Technology as the engine to generate value in the business. The implementation and administration of an IT government according to its line of business would generate a competitive advantage in the market. IT plays a role not only tactical or operational but also strategic, enabling new models, products and business services that ensure the leadership, growth and sustainability of the organization in the long term, transforming even its industries.

In the process of implementing GTI in any type of organization, it is necessary to take into account five fundamental areas as the main dimensions of the business, which in turn frame its development and its generation of value for its shareholders [15]. These Areas are listed below: <u>Strategic Alignment</u>: Focuses on securing the link of business and IT plans; in defining, maintaining and validating the IT value proposition and in aligning IT operations with the operations of the company. <u>Delivery of Value</u>: It refers to executing the value proposition throughout the delivery cycle, ensuring that IT delivers the agreed benefits aligned with the strategy, concentrating on cost optimization, and demonstrating the intrinsic value of IT. <u>Resource Management</u>: It refers to the optimal investment and the proper management of critical IT resources such as applications, information, infrastructure and data. Performance Measurement: Follows and supervises the implementation strategy, the completion of projects, the performance of processes and the delivery of service. If there is no way to measure and evaluate IT activities, it is not possible to govern them or ensure alignment, delivery of value, risk management and effective use of resources.

3. IT Governance Framework

There is a growing concern in the top management of the entities about the activities of the IT function. The role that IT plays in organizations is vital, not only to maintain competitiveness, but to guarantee daily operations. This situation has led to the emergence of standards, models, methodologies and practices such as ISO / IEC 38500 own IT Government, COBIT 5, model GTI4U among others; aimed at ensuring better governance or a more optimal performance of IT in organizations. Although frameworks and guides such as COBIT and ITIL have been widely adopted, there is no absolute standard for Government IT [16]. Following is a description of each of the reference frames:

3.1 Joint Information Systems Committee (JISC)

The first initiative to design an IT governance model that serves as a reference for an entire university system was the Joint Information Systems Committee (JISC) for the universities of the United Kingdom. The JISC designed a reference model, and a self-assessment toolkit, which have become a starting point that helps universities in the process of identifying and defining the role of IT within the planning and governance of your organization. This framework was designed to be very flexible and can be used by different types of universities: large or small, old or modern and to take into account the different cultures that prevail in the institutional governance of universities. The reference model for IT governance of the JISC is based on five perspectives: government, administration, resources, organization and services.

3.2. ISO 38500:2015

The ISO / IEC 38500 standard, standardized and published by the ISO (International Organization for Standardization) and IEC (International Electrotechnical Commission), is the first internationally recognized standard in IT Governance. It was published in June 2008 based on the Australian standard AS8015: 2005, provides a set of definitions related to the corporate governance of IT. Principles on which a task model is based and a task model focused on informing and guiding to all personnel involved in the design and implementation of the management policy system, processes and structures that support IT governance [17]. It is an international standard for corporate governance of information and communication technology. It is a "principled, high-level consultative standard", provides "broad guidance on the role of the governing body and encourages organizations to use appropriate standards to support their governance of IT." ISO / IEC 38500 does not replace, in others words, those norms and frameworks (such as COBIT ™, ITIL, ISO 27001, etc.) that an organization has already implemented for the better governance of its IT, what it does do is provide a coherent framework to ensure that the council appropriately participates in the effective governance of IT[18].

The objective of this standard is to provide principles, definitions, and a model for governing bodies when using, evaluating, directing and monitoring the use of information technology (IT) in their organizations. This International Standard provides principles, definitions and a model for good governance of IT, to help those at the highest level of organizations to understand and comply with their legal, regulatory and ethical obligations related to the use of IT. Your organizations [19].

ISO 38500 establishes 6 principles for Corporate Governance of IT which are effective, efficient and acceptable, ensuring that organizations that follow these principles will help to balance risks and foster opportunities for the use of IT. They also express the desired behaviour that guides the decision making in the organization. According to the ISO / IEC (2015) these principles are: Responsibility, Strategy, Acquisitions, Performance, Conformity and Human behaviour. ISO 38500 divides IT governance activities into three main tasks: Evaluate the current and future use of IT, Direct the

Preparation and execution of plans and policies to ensure that the use of IT satisfies the objectives of the organization and Monitor the fulfilment of the policies and the performance in relation to the planned thing.

3.3 COBIT 5.0

It is the system that helps organizations create value to optimize risk levels within the organization, allowing information technologies to be managed holistically at all levels of the organization [20]. Provides of a comprehensive framework that helps companies achieve their goals for the governance and management of corporate IT. COBIT 5 enables IT to be governed and managed in a holistic manner for the entire company, encompassing the entire business from start to finish and the functional areas of IT responsibility.

COBIT 5 describes a set of good practices for Senior Management and administration. This set establishes a set of controls on information technologies and organizes them around a logical framework of the processes related to IT. COBIT is not an isolated framework, but is aligned with other existing standards, such as the older COBIT versions, some standards such as: ISO 38500, essentially complementing the principles of (Evaluate, Direct and Monitor) (Isaca, 2015); ISO 15504 aligning with it in the concept to process capacity [21].

The process reference model of COBIT 5 divides the governance and management processes of business IT into two main process domains:

Government: Contains five government processes; within each process, evaluation, guidance and supervision (EDM) practices are defined.

Management: It contains five processes of government; within each process, evaluation, guidance and supervision (EDM) practices are defined.

The names of these domains have been chosen according to these designations of main areas, but contain more verbs to describe them: Align, Plan and Organize (Align, Plan and Organize, APO), Build, Acquire and Implement (Build, Acquire and Implement, BAI), Deliver, give Service and Support (Deliver, Service and Support, DSS) and Supervise, Evaluate and Evaluate (Monitor, Evaluate and Assess, MEA).

3.4 Model GTI4U

The GTI4U Model has been proposed by a group of Spanish researchers and developed by Fernadez (2009) and modified in the year (2011). This model is based on the ISO 38500 standard and has been developed to be implemented in the university environment and is also used to evaluate the overall level of maturity of the Spanish university system (SUE) [22]. The GTI4U Model is composed of three levels, which are mentioned below:

The first level: Includes all the elements of the ISO 38500 standard: IT governance model, principles, good practices and dictionary of terms and this in turn establishes that IT must be governed through three proposed actions such as assess, direct and monitor. Likewise adopts its principles of responsibility, strategy, acquisition, performance, compliance, and human factor. The principles express what are the behaviours that must be adopted when making decisions, and each of them establishes what should happen, but does not indicate how, where or who should implement these principles. These aspects will depend on the nature of the organization.

The second Level: It is composed of a Maturity Model (MM) using benchmarking tools for each principle, which will be used to establish at what level of IT governance maturity is the organization, MM usually establish several levels or states which each it has its level of demand, from a level of non-existent (0) to an optimized level (5), which helps the organization to self-evaluate and one level can't surpass the other but has complied with the requirements in its entirety.

The Third Level: It includes the indicators that will serve to measure the extent to which the criteria presented in the standard are met and if good practices of the IT Government are being satisfactorily carried out. At this level, a broad catalogue of government indicators has been designed, divided into three groups: Maturity Model (CMM), Government Evidence Indicators (IEG) and Government Quantitative Indicators (ICG).

4. Results

The Colombian public universities in some particular cases, their technologies have been used for the support of their estates, which avoids doing the activities manually and that they are efficient in each established process. Resulting in an important organizational development supported by the exponential change of the technological development that has been experienced over time and the latter has increased the need for IT governance. Such university faculties obtaining as results two important variables to be addressed, such as the generation of value according to organizational objectives and the mitigation of the risks associated with IT services, allowing to achieve a correct strategic alignment. There are models on which they have been based to implement the IT Government in Universities such as: COBIT, JISC (Committee of Joint Information Systems) in the United Kingdom, MGTIU (Model of Government IT for Universities) in Spain, and the ISO standard / IEC 38500: 2015 (International Organization for Standardization), the latter as it focuses for public entities, it is necessary to know more details and integrate its model that proposes to govern IT through three fundamental actions such as Evaluate, Direct and Monitor.

The Responsibilities in the development of an IT Government planning in the universities should fall on the highest directives of the universities such as the Superior Council, Rectors, Vice-Rectors, Deans. For that reason, it is important to convince them technically of the convenience of adopting a Good governance of IT that focuses on establishing strategies and that these are aligned with those of the Universities. Determine the responsibilities of strategic IT planning where the current level is identified and clearly show where we want to go, proposing prioritized projects and these in turn must be monitored by an evaluation and monitoring of the performance of the processes through appropriate indicators.

According to the above, an IT Governance model is presented below, which is adapted to Colombian public universities where it is intended to provide guidance for managers, area managers and institutional support personnel on the acquisition and use of technology information. In this regard, the following model of IT governance is proposed, which brings together the guidelines for implementation in higher education institutions, which highlights the main processes for institutional improvements such as information security and risk management, which are identified in figure 1.

5. Macroprocesses of the Model and Government of IT

The following defined processes are presented below:

5.1 IT Strategic Planning

It focuses on determining an IT governance framework that ensures the strategic alignment of IT with institutional objectives by integrating its policies, processes and projects. The following are the Processes and Define Activities:

5.1.1 IT Strategic Management: Defined by the following activities:

PET1.1: Establish the current situation of the organization; identify its strategic direction to establish itself as an input for the construction of a good IT governance scheme.

PET1.2: Define the IT strategic plan where the desired level of the organization is defined and the definition of projects aimed at achieving the stated objectives, and define, in cooperation with relevant stakeholders, how IT will contribute to the strategic objectives of the company (goals) as well as the related costs and risks.

PET1.3: Analyse and know the initial situation of the organization in relation to IT governance, through the use of maturity models.

IOP Conf. Series: Materials Science and Engineering **844** (2020) 012045 doi:10.1088/1757-899X/844/1/012045

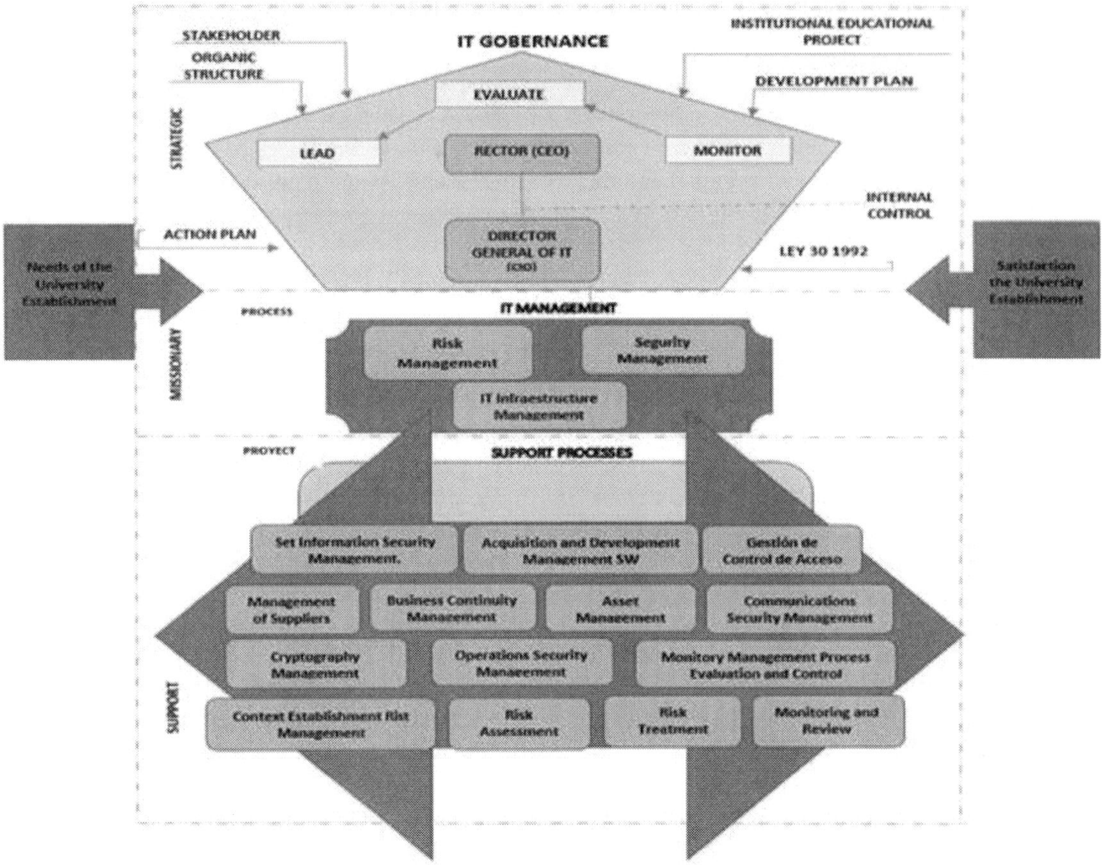

Figure 1. IT Government Model Colombian Public Universities

5.1.2 Organizational Structure of IT: Defined by the following activities:

PET2.1: Understand the business culture of decision making and determine an optimal model in decision making for IT.

PET2.2: Define the policies, guidelines and guidelines that are part of the IT Governance strategy, in accordance with institutional policies.

PET2.3: Conformation of an IT committee where functions are attributed with respect to aspects related to IT that have participation in all ICT decision activities and make sure to provide to the senior management performance reports on the performance of IT plans, policies and activities.

PET2.4: Define the roles and responsibilities in the IT structure, which have controls in IT decision making.

5.1.3 Evaluate and Supervise the IT Governance Model: Defined by the following activities:

PET3.1: Determine the relevance of IT and its role with respect to the business.

PET3.2: Periodically assess whether the mechanisms for IT governance agreed upon (structures, principles, processes, etc.) are established and operating effectively.

PET3.3: Supervise routine and regular mechanisms to ensure that the use of IT complies with relevant obligations (regulatory, legislation, common laws, contractual), standards and guidelines.

PET3.4: Evaluate the effectiveness of government design and identify actions to rectify any deviations.

PET3.5: Evaluate the effectiveness of the integration and alignment of IT strategies with institutional objectives to ensure that this contributes value.

5.2 Information Security Management
The objective of this process is to try to determine good practices through an ISMS in order to protect information assets in any of its states in the face of a series of risks or gaps that violate its fundamental principles of confidentiality, integrity and availability of information. through the implementation of processes and projects that allow managing and reducing the risks to which it is exposed and maximizing the return on investments in business opportunities. The following are the Processes and Define Activities:

5.2.1. Establish the SGSI: Defined by the following activities:

GSI1.1: Define the organizational understanding and its context, analysing the current situation of the entity in relation to the management of information security.

GSI1.2: Determine the scope of the SGSI.

GSI1.3: Define a diagnosis of the level of compliance of the entity in relation to the control objectives and controls established in Annex A of the ISO 27001 standard, and the action plans aimed at closing the gaps found.

GSI1.4: Determine the Maturity Level in which the entity is located for its information security model.

GSI1.5: Determine the needs and requirements of the stakeholders of the entity in relation to the Information Security Management System.

GSI1.6: Define and document information security policy that covers a defined scope and that is in the public domain for all officials of the institution.

GSI1.7: Establish the organizational structure, roles and responsibilities regarding Information Security.

GSI1.8: Define the Methodology of Analysis, Evaluation and treatment of Risks.

GSI1.9: Obtain authorization and support from senior management in the implementation of the SGSI.

5.2.2 Asset Management: Defined by the following activities:

GSI2.1: Assets associated with information and information processing facilities should be identified and an inventory of these assets should be developed and maintained.

GSI2.2: Each information asset must have its owner, who is responsible for the use throughout the life cycle.

GSI2.3: Rules for the acceptable use of information and assets associated with information and information processing facilities should be identified, documented and implemented.

GSI2.4: The assets of the organization that are under their charge must be returned, once the employment contract is finished, the delivery must be formalized.

GSI2.5: The information must be classified according to the legal requirements, value, criticality and susceptibility to disclosure or authorized modification.

GSI2.6: A set of procedures should be developed and implemented for the labeling of the information according to the information classification scheme adopted by the organization.

GSI2.7: Procedures for asset management should be developed and implemented according to the information classification scheme adopted by the organization.

IOP Conf. Series: Materials Science and Engineering **844** (2020) 012045 doi:10.1088/1757-899X/844/1/012045

GSI2.8: Procedures for the management of removable media should be implemented as these could store confidential information.

GSI2.9: Media containing information must be protected against unauthorized access, misuse or corruption during transport.

5.2.3 Communications Security Management: Defined by the following activities:

GSI3.1: Networks must be managed and controlled to protect information in systems and applications.

GSI3.2: The security mechanisms, service levels and management requirements of all network services must be identified.

GSI3.3: Groups of information services, users and information systems must be separated from the networks.

GSI3.4: Policies, procedures and formal transfer controls should be in place to protect the transfer of information through the use of all types of communication facilities.

GSI3.5: The information included in electronic messaging must be adequately protected.

GSI3.6: The requirements for confidentiality or non-disclosure agreements that reflect the needs of the organization to protect information must be identified, regularly reviewed and documented.

5.2.4 Access Control Management: Defined by the following activities:

GSI4.1: An access control policy must be established, documented and reviewed based on business requirements and information security.

GSI4.2: Users should only be allowed access to the network and to the network services for which they have been specifically authorized.

GSI4.3: A formal process of registration and cancellation of user registration must be implemented to enable the allocation of access rights.

GSI4.4: A formal user access provision process must be implemented to assign or revoke access rights for all types of users for all systems and services.

GSI4.5: The allocation and use of privileged access rights must be restricted and controlled.

GSI4.6: The allocation of secret authentication information must be controlled through a formal management process.

GSI4.7: The access rights of all employees and users external to the information and information processing facilities must be withdrawn upon termination of employment, contract or agreement, or adjusted when they are made changes.

GSI4.8: Users must be required to comply with the organization's practices for the use of secret authentication information.

GSI4.9: Access to information and system functions of applications must be restricted in accordance with the access control policy.

GSI4.10: The use of utility programs that may have the ability to override the system and controls of applications should be strictly restricted and controlled.

GSI4.11: Access to program source codes must be restricted.

GSI4.12: A policy on the use of cryptographic controls for the protection of information should be developed and implemented.

GSI4.13: A policy on the use, protection and time of life of the cryptographic keys must be developed and implemented throughout its life cycle.

5.2.5 Supplier Management: Defined by the following activities.

GSI5.1: Information security requirements should be agreed upon and documented to mitigate the risks associated with provider access to the assets of the organization.

GSI5.2: All relevant information security requirements must be established and agreed upon with each provider that can access, process, store, communicate or provide IT infrastructure components for the organization's information.

GSI5.3: Agreements with suppliers should include requirements to address the information security risks associated with the supply chain of information and communication technology products and services.

GSI5.4: Organizations must regularly monitor, review and audit the provision of services by providers.

GSI5.5: Changes in the supply of services by suppliers must be managed, including the maintenance and improvement of existing information security policies, procedures and controls, taking into account the criticality of information, systems and business processes. involved and the re-evaluation of risks.

5.2.6 Continuity Management: Defined by the following activities.

GSI6.1: The organization must determine its requirements for information security and the continuity of information security management in adverse situations, for example, during a crisis or disaster.

GSI6.2: The organization shall establish, document, implement and maintain processes, procedures and controls to ensure the level of continuity required for the security of information during an adverse situation.

GSI6.3: The organization must verify at regular intervals the information security continuity controls established and implemented, in order to ensure that they are valid and effective during adverse situations.

5.2.7 Management of Acquisition, Development and Maintenance of Systems: Defined by the following activities.

GSI7.1: Requirements related to information security should be included in the requirements for new information systems or for improvements to existing information systems.

GSI7.2: The information involved in the services of the applications that pass over public networks must be protected from fraudulent activities, contractual disputes and unauthorized disclosure and modification.

GSI7.3: The rules for the development of software and systems should be established and applied to the developments within the organization.

GSI7.4: System changes within the development life cycle must be controlled through the use of formal change control procedures.

GSI7.5: Modifications to software packages should be discouraged, which should be limited to the necessary changes, and all changes should be strictly controlled.

GSI7.6: Principles for the construction of secure systems should be established, documented and maintained, and applied to any information systems implementation activity.

GSI7.7: Organizations must establish and adequately protect secure development environments for system development and integration activities that comprise the entire systems development life cycle.

GSI7.8: The organization should supervise and monitor the development activity of externally contracted systems.

GSI7.9: During the development, tests of security functionality must be carried out.

GSI7.10: Security tests must be performed based on the security requirements of the organization.

5.2.8 Cryptography Management: Defined by the following activitie.

GSI8.1: A policy on the use of cryptographic controls for the protection of information should be developed and implemented.

GSI8.2: A policy on the use, protection and time of life of the cryptographic keys must be developed and implemented throughout its life cycle.

GSI8.3: Key management must be used to support the use of cryptography techniques in the organization.

5.2.9 Process Monitoring, Evaluation and Control Management: Defined by the following activities.

GSI9.1: Internal audits must be implemented at defined times.

GSI9.2: Periodic reviews of the ISMS should be carried out by Management to ensure that the scope is still adequate and that improvements have been identified in the ISMS process.

GSI9.3: The internal control office must ensure strict compliance with the improvement plans of the processes.

GSI9.4: Establish the methods to track, measure, analyze and evaluate the security processes and controls of the ISMS.

GSI9.5: Define an internal audit plan to measure the state of information security based on the ISO 27001: 2013 standard.

GSI9.6: Determine a continuous monitoring plan for the processes that lead the technology areas based on information security to identify compliance or otherwise implement corrective actions necessary for its good performance.

GSI9.7: Determine and document the causes of nonconformities with the ISMS and implement corrective actions identifying the vulnerability.[22]

5.3 Risk Management of IT

Its function is to determine the risk management applied to Colombian public universities and four processes are proposed such as an establishment of the context of the risks, then a risk assessment, then a treatment of risks, then a stage of monitoring, which will try to mitigate the risk of a possible materialization. The following are the Processes and Define Activities:

5.3.1 Establishment of Context Defined by the following activities:

GRTI1.1: Understand the organization and its context. The organization must articulate its objectives and define the internal and external criteria and the evaluation factors to be considered in IT risk management.

GRTI1.2: Define the scope and appropriate depth of risk analysis activities, taking into account all the risk factors and the business criticality of the assets.

GRTI1.3: Establish and maintain policies for classification and analysis related to IT risks, with capacity for multiple types of events, multiple categories of IT risk and multiple risk factors.

GRTI1.4: Define and communicate the roles and responsibilities of IT risk management.

GRTI1.5: Define resources for risk management.

GRTI1.6: Establish mechanisms for internal communication and reporting.

5.3.2 Risk Assessment: Defined by the following activities.

GRTI2.1: The organization must identify the sources of risks, the areas of impacts, the events and the causes and potential consequences. The identification should include risks regardless of whether or not their origin is under the control of the organization, even though the origin of the risk or its cause may not be evident.

GRTI2.2: The organization must apply tools and techniques to identify risk, which are appropriate to its objectives and capabilities and the risks faced.

GRTI2.3: Factors that affect consequences and probability must be identified because the risk analysis involves the consideration of the causes and sources of the risk, its positive and negative consequences and the probability of that such consequences may occur.

5.3.3 Risk Treatment: Defined by the following activities:

GRTI3.1: Design and implementation plan that includes priorities in the treatment of risks.

GRTI3.2: Establish priorities in the implementation of risk treatment measures.

GRTI3.3: Apply security controls of Annex A of ISO 27001: 2013 to reduce the risk.

GRTI3.4: Allocate the necessary resources to carry out the implementation plan.

GRTI3.5: Give applicability to the most appropriate response plan in the event of occurrence of risk incidents and achieve the minimization of the impact that may be generated.

5.3.4 Monitoring and Revision: Defined by the following activities:

GRTI4.1: Measure the performance of risk management in relation to performance indicators, which are periodically reviewed for their suitability.

GRTI4.2: Periodically review if the risk management framework, the policy and verify whether the plan is still adequate, taking into account the internal and external context of the organization.

GRTI4.3: Evaluate the effectiveness of the risk management framework.

6. Roles and Responsibilities of the G&G Model for Colombian Public Universities

The following structure is proposed that defines roles and responsibilities according to the Macroprocesses and defined processes which have specific functions and we relate them below:

6.1 CEO- Rector

The highest executive (Chief Executive Officer - CEO), is responsible for making IT governance work in the organization, this in turn must have the organization committed to the vision of the business in terms of information. IT ensuring that executives and business users understand through a realistic internal communication strategy that is what is expected of IT, its contribution to the business in the future and that is expected from IT users. Finally, it has the following functions:

- Establish the current situation of the organization, identify its strategic direction that allow establishing itself as an input for the construction of a good governance scheme of you.
- Define the IT strategic plan where the desired level of the organization is defined and the

definition of projects aimed at achieving the stated objectives. Define in cooperation with relevant stakeholders, how IT will contribute to the strategic objectives of the company (goals) as well as the related costs and risks.

- Define the policies, guidelines and guidelines that are part of the IT Governance strategy, in accordance with institutional policies.
- Determine the relevance of IT and its role with respect to the business.
- Supervise routine and regular mechanisms to ensure that the use of IT complies with relevant obligations (regulatory, legislation, common laws, contractual), standards and guidelines.
- Periodically assess whether the mechanisms for IT governance agreed upon (structures, principles, processes, etc.) are established and operating effectively.

6.2 CIO- IT General Manager

The role of the general director of IT (Chief Information Officer - CIO) is strengthened, becoming strategic and fundamental support of the business strategy as responsible for the management of information in the company, among the functions in charge are established the following:

- Analyze and know the initial situation of the organization in relation to IT governance, through the use of maturity models.
- Define the roles and responsibilities in the IT structure, which have responsibilities in IT decision making.
- Evaluate periodically if the mechanisms for IT governance agreed upon (structures, principles, processes, etc.) are established and operating effectively.
- Evaluate the effectiveness of government design and identify actions to rectify any deviations.
- Evaluate the effectiveness of integrating and aligning IT strategies with institutional objectives to ensure that this adds value.

6.3 Information Security Leader

Its objective is to lead implementation projects in the information security management system in the entity and its defined functions are as follows:

- Define the organizational understanding and its context, analyzing the current situation of the entity in relation to information security management.
- Determine the Maturity Level in which the entity is located for its information security model.
- Determine the needs and requirements of the interested parties of the entity in relation to the Information Security Management System.
- Define and document information security policy that covers a defined scope and that is in the public domain for all officials of the institution.
- Define an internal audit plan to measure the state of information security based on the ISO 27001: 2013 standard.
- Determine and document the causes of nonconformities with the ISMS and implement corrective actions identifying the vulnerability.
- Design a system that allows continuous improvement of the ISMS through a systematic process.
- A formal process of registration and cancellation of user registration must be implemented to enable the allocation of access rights.
- All relevant information security requirements should be established and agreed upon with each provider that can access, process, store, communicate or provide IT infrastructure components for the organization's information.
- All relevant information security requirements should be established and agreed upon with each provider that can access, process, store, communicate or provide IT infrastructure components for the organization's information.

6.4 Risk Management Leader

It is responsible for establishing all the risk management margins and is in charge of the following functions:

- Define the scope and appropriate depth of the risk analysis activities, taking into account all the risk factors and the criticality of the business assets.
- Establish and maintain policies for classification and analysis related to IT risks, with capacity for multiple types of events, multiple categories of IT risk and multiple risk factors.
- Establish mechanisms for internal communication and reporting.
- Design and implementation plan that includes priorities in the treatment of risks.
- Establish priorities in the implementation of risk treatment measures.
- Give applicability to the most appropriate response plan in the event of the occurrence of risk incidents and achieve the minimization of the impact that may be generated.
- Measure the performance of risk management in relation to performance indicators, which are reviewed periodically in terms of their suitability.
- Evaluate the effectiveness of the risk management framework.

6.5 IT Infrastructure Leader

It is in charge of proposing and managing the IT infrastructure and the operations that are given in it and is in charge of the following functions:

- The rules for the development of software and systems must be established and applied, to the developments within the organization.
- Establish and adequately protect secure development environments for system development and integration activities that encompass the entire systems development life cycle.
- Control changes in the organization, business processes, facilities and information processing systems that affect information security.
- Plan and obtain the homogeneity of equipment and base software in order to maximize the availability of information services.

6.6 Internal Control Leader

Its functions are the following:

- Define the policies, guidelines and guidelines that are part of the IT Governance strategy, in accordance with institutional policies.
- Evaluate periodically if the mechanisms for IT governance agreed upon (structures, principles, processes, etc.) are established and operating effectively.
- Verify that the planning, management, operation and support of the technological infrastructure, guarantee the availability, quality and operation of the electronic computer services, ready to fulfill the mission of the entity

7. RACI Matrix

Taking into account the roles defined in the government and management model for public universities, the matrix of responsibilities of each of the roles in the different macroprocesses is defined with their processes. Using the responsibility allocation matrix (RACI), they are named by the four letters with which the type of relationship with a process that each agent has is encoded:

- R: Responsible / Responsible. He is the one in charge of doing the task or activity.
- A: Accountable / Person in charge. It is the person who is responsible for the task being done. It is not the same as the R, since it does not have to be the one who performs the task,

you can delegate it to others. However, if he is the one who must ensure that the task is done, and done well.

- C: Consulted / Consult. The resources with this role are the people with whom one must consult data or decisions regarding the activity or process that is defined.
- I: Informed / Inform. These people are informed of the decisions that are made, results that occur, states of service, degrees of execution.

Table 1. RACI Matrix

MACROPROCESS	PRCESS ID	PROCESS	CEO - RECTOR UNIVERSITY	CIO - DIRECTOR GENERAL OF IT	RISK MANAGEMENT READER	LEADER INFORMATION SECURITY	IT INFRAESTRUCTURE LEADER	INTERNAL CONTROL LEADER
IT STRATEGIC PLANNING	PET1	IT STRATEGIC ADDRESS	A-I	R-I	C	C	C	C
	PET2	ORGANIZATIONAL STRUCTURE OF IT GOVERNANCE	A-I	R-I	C	C	C	C
	PET3	EVALUATE AND SUPERVISE THE IT GOVERNANCE MODEL	A-I	R-I	C	C	C	C
INFORMATION SECURITY MANAGEMENT	GSI1	ESTABLISH THE INFORMATION SECURITY MANAGEMENT SISTEM	I	A-I	R-I	R	C	C
	GSI2	ASSENT MANAGEMENT	I	A-I	C	R	C	C
	GSI3	COMMUNICATIONS SECURITY MANAGEMENT	I	A-I	C	R	C	C
	GSI4	ACCESS CONTROL MANAGEMENT	I	A-I	C	R	C	C
	GSI5	SUPPLIER MANAGEMENT	I	A-I	C	R	R	C
	GSI6	CONTINUITY MANAGEMENT	I	A-I	R-I	R	C	C
	GSI7	MANAGEMENT OF ACQUISITION, DEVELOPMENT AND MAINTENANCE OF SISTEMS	I	A-I	R-I	R	C	C
	GSI8	OPERATIONS SECURITY MANAGEMENT	I	A-I	C	R	C	C
	GSI9	CRYPTOGRAPHY MANAGEMENT	I	A-I	C	R	R	C
	GSI10	FOLLOW-UP MANAGEMENT, EVALUATION AND PROCESS CONTROL	I	A-I	R-I	R	C	C-I
IT RISK MANAGEMENT	GRTI1	ESTABLISHMENT OF THE CONTEXT	I	A-I	R	C	C	C
	GRTI2	RISK ASSESSMENT	I	A-I	R	C	C	C
	GRTI3	RISK TREATMENT	I	A-I	R	C	C	C
	GRTI4	MONITORING AND REVIEW	I	A-I	R	C	C	C

8. Conclusions

IT governance implemented with a top down management system allows upper management to have direct control of IT investments, so that they are able to contribute directly to the fulfilment of institutional goals. This method is achieved through the distribution of roles and responsibilities that support decision making, as well as the implementation of rules and procedures that allow for strategic monitoring of IT decisions, ensuring the strategic alignment of IT with the institutional plans.

As it has been demonstrated evidenced throughout the article, it is clear and undeniable that information is the most important asset of an organization, that is why the importance of beginning starting from with this premise is vital when one is applying the framework to for public universities. The power to have Having an IT governance Model of Government and IT Management model of IT where, in which an Information Security Management System (ISMS) is established and a Risk Management strategy is established focusing on the needs of the business and based on standards and good practices. Such as ISO/IEC 27001:2013 and ISO/IEC 31000, allows enables the directing, aligning, and monitoring of policies, processes and structures that support the corporate governance of information technologies in alignment with the institutional government within the framework of strategic, mission, and support processes.

The proposed model has characteristics of Colombian Public Universities, however to the extent that private universities want to adopt them, they will have to align it with their specific institutional needs and characteristics. At the time of design, senior management is taken into account who is the visible head of the organization and a general manager of IT who ensures that IT objectives are aligned with the strategy of the particular university and a good agreement of its resources. Manage security information, risk management and in addition to the performance of the aforementioned to generate added value to the company.

It is important to encourage public universities in the need to implement the IT Governance and Management model that in order to generates maturity in IT processes, where there are including its strategic, tactical and operational components. It is clear to determine that today's organizations are aware of the need to govern IT, taking into account especially considering that there are critical mission processes and operations which would indicate that the model would achieve the goal of minimizing risks and strengthening the security of the organization. knowledge in this particular case. It is for this reason that it is important to implement the model it because it clearly establishes strategic guidelines to successfully carry out the processes involved for continuous improvement, which would mean that Therefore, in order to achieve success in public universities. It is necessary to have knowledge is needed of standards and norms with clear methodologies and good IT governance that leads, organizes, and defines the guidelines to follow, and maintains with a view a vision for to sustaining their processes under an organizational culture.

Finally, from the results of the article, some recommendations have been determined on aspects that one should to take into account when implementing these IT governance model in public universities:

- It is important that the board of directors and senior management are involved in matters of governance related to the Government in order to ensure the commitment and alignment of between the IT area strategies with and the strategies defined by the University. This should be done in such a way that it will be able to generate obtains the maximum value through from the investments made.
- Those implementing the model should carry out trainings and maintain awareness of the importance of implementing good quality IT governance in Universities. Where highly relevant standards are present involved, such as information security to mitigate risks of attacks of all kinds.
- The figure of the CIO should be created in order to have a direct communication with the Superior Council and the upper management. Develop responsible decision making regarding the acquisition of the technology and understand that this is of great help in working towards achievement of the decided objectives.
- The Information Security Management System and risk management must be implemented to protect the most valuable asset of the organizations, the information, and for this, it is necessary to have professionals specialized in the aforementioned areas.

References

[1] Berciano J 2016 La importancia y la necesidad de proteger la información sensible. Obtenido de: RedSeguridad.com:http://www.redseguridad.com/especialidades-tic/proteccion-de-datos/la-importancia- y-la-necesidad-de-proteger-la-información-sensible.

[2] Isaca 2016 Incident Management and Response. Obtenido de http://www.isaca.org/Knowledge-

[3] Janahi L, Griffiths M, Al-Ammal H 2015, A conceptual Model for IT Governance in Public

Sectors.

[4] Calder A 2008, *ISO/IEC 38500 The IT governance standard.*

[5] Webb P and Pollard C 2006, Attempting to Define IT Governance: Wisdom or Folly. Proceedings of the 39th Hawaii Ridley G, International Conference on System Science, IEEE Computer Society.

[6] Weill P and Ross J 2004, *IT Governance: How Top Performers Manage IT Decision Rights for Superior Results*. Harvard Business School Press.

[7] Butler B 2016, The cloud computing subnet. Obtained from the cloud computing industry.: https://www.networkworld.com/article/3016926/cloud-computing/ibm-will-manage-atandt-s-hosted-and-cloud-services-as-part-of-partnership.html

[8] Fernádez V E 2008, UNITIL: Gobierno y Gestión de TIC basado en ITIL.

[9] Fred D R 2003, *Conceptos de Administración Estratégica.*

[10] Hitt M A, Ireland D and Hoskisson R 2007, Strategic Management: Competitiveness and Globalization (Concepts and Cases) Seventh Edition

[11] ISACA, IT Governance Developing a Successful Governance Strategy, 2005.

[12] Gómez V H, Tejada R A and Giraldo L M 2011, Un modelo preliminar de gobierno de tecnologías de información para universidades colombianas

[13] Selig G J PMP COP 2008, Implementing IT Governance A Practical Guide to Global Best Practices in IT Management.

[14] Isaca 2015, COBIT 5: A Business Framework for the Governance and Management of Enterprise IT. Rolling Meadows, IL: ISACA

[15] Muñoz I and Ulloa G 2011, Gobierno de TI. Revista S&T, **9**(17), 23-53. Cali: Universidad Icesi.

[16] Bayona S, Ayala M 2017, IT governance: Progress and challenges on public administration.

[17] De la Cámara Delgado Ms, Sáenz F J, Calvo J & Fernández V 2002, Project Management and IT Governance. Integrating PRINCE2 and ISO 38500.

[18] Calder A 2008, ISO/IEC 38500 The IT governance standard.

[19] International Organization for Standardization (ISO) / International Electro technical Commission (IEC), 2015.

[20] ISACA. COBIT 5 for Information Security. Illinois, Michigan, Estados Unidos de América: ISACA, 2012. p. 220.

[21] De Haes, S, Van Grembergen, W and Debreceny, R S 2013, COBIT 5 and Enterprise Governance of Information Technology: Building Blocks and Research Opportunities. Journal of Information Systems., **27**(1), p307-324.

[22] Norma Iso 27000, Sistema de Gestion de Seguridad de la Informacion 2013.

IOP Conf. Series: Materials Science and Engineering **844** (2020) 012046 doi:10.1088/1757-899X/844/1/012046

Technological surveillance as element for the positioning of public universities

F O Moya[1], S L Villero[1] and F P Perez[1]
[1]Universidad de La Guajira, Riohacha, Colombia.

Email: fmoya@uniguajira.edu.co

Abstract. This research work is primarily aimed at Technological Surveillance as an Element for the Positioning of Public Universities. This research is characterized by being descriptive, non-experimental design of transectional-descriptive type. The population was made up of the National Open and Distance University and the University of La Guajira, the units of informants made up of the managers responsible for the technologies in the mentioned universities. For the data collection, a simple questionnaire composed of 57 items and constituted by the response alternatives was always, almost always, sometimes, never and almost never. In turn, for the validation of the data collection instrument, it was necessary to submit it to 5 an expert and to apply a reliability test through the Alpha Cronbach statistics formula, giving a result of 0.83 for the Technological Surveillance variable. An analysis of the results obtained in the research was carried out, the presence in public universities is reiterated, however, there is evidence of weakness in the activities for the positioning of the Technology Watch.

1. Introduction

Technological advances establish the momentum of organizations due to the operational advantages that are possible in the immediacy that is achieved in the processes, in addition to the ability to minimize response times and quality of activities that are managed in this scenario. Therefore, generating opportunities to sustain this alternative determines a scenario of high chances of success for the objectives set.The Colombian Higher Education system is organized by Law 30 of 1992 issued by the Presidency of the Republic, according to this regulation, the system is made up of three types of institutions: professional techniques; Universities or technology schools. In addition, the system is constituted by the set of educational programs, which seek to achieve the aims of higher education. Its direction and control corresponds to the national government, which delegates to the Colombian Institute for the Promotion of Higher Education (ICFES) the action of promoting and coordinating the development of the sector. For the theoretical bases, contributions from authors such as [1][2][3] others were used

On the other hand, a quick analysis of the scenario allows us to gather that although our country has had a slight rebound in the appropriation of technologies there is still a marked underutilization of them, Regarding the figures, it can be said that in Colombia the technological gap has narrowed over time, an example of this is that the number of Colombians who have access to a computer has passed 48% in the year 2005 to 54% in 2006. Another important fact, which ratifies this reduction in the technology gap, is that more and more Colombians use computers in their homes, it indicates that the number of computers in homes has increased, and the most interesting thing is that 84% use it to connect to the Internet. According to the seventh report of the Economist Intelligence Unit (EIU), a unit of the company that publishes The Economist magazine, Colombia has evolved very slowly compared to the progress made worldwide.

Taking into account the above, the importance generated by this research work is aimed at Technological Surveillance As an Element for the Positioning of Public Universities, presenting the optimization of activities where the experience of university institutions will be determined, the academic offer they generate will be valid for the group to the extent that these are viable to their interest and professional alternatives. Undoubtedly, this is also stipulated with the plans and programs that the countries lead according to the current technological development plan or program.

2. Technological surveillance and public universities in Colombia in the department of La Guajira.

At present, organizations are influenced by the great competition that exists in their various regional, national and international markets, in which not only low prices or quality are decisive to survive on stage, but the effort they make to materialize a position in these that can be durable and scalable [1], says that technological surveillance is an innovative tool available to any organization which allows it to capture information from outside, analyze it and convert it into knowledge, so that it can be oriented to make decisions with lower risk and be able to anticipate changes This is a process that must be carried out systematically (capture, analysis, dissemination and exploitation of information) to alert the organization about innovations that are likely to create opportunities or threats and thus increase its competitiveness.

In this order of ideas, the technology surveillance resource will allow organizations to make a diagnosis in the scenario in which they are located, to know how their competitors have developed a sustainable work scheme based on details and definitions which can be determinants of its success, which is why this data is relevant for the performer in the definition of key decisions in the business. When considering the previous theorization, it is assumed that the positioning will allow organizations to trigger an internal and external analysis process to achieve the ideal image in the consumer's mind against the competition. That is why, this dynamic of work the reality of organizations determines a high possibility of transit in this space in which technological surveillance and the positioning of universities in Colombia are linked.

For the development of the research, the person in charge has managed to hold conversations with the technology managers at the National Open and Distance Universities and the University of La Guajira, so that I get to know some elements that characterize the situation of these educational institutions, which It can mean a possibility of problems, given the sequence of some symptoms. In the same way, in the absence of the manager or director of the institution as a promoter of the dissemination and management of technological surveillance, staff assume that attitude, therefore considering that their responsibilities do not integrate it into a methodology that may qualify them as Suitable for high technology specialists. Such facts raised, achieve a greater gap between university staff and the work of technological surveillance when no training and education plans have been conducted in this regard, causing a functional illiteracy towards this key work component within the institutions which disadvantages intellectual capital in its organizational and personal value.

This research raises the need to generate a study, which allows to know in detail the reality of this problem, thus obtaining relevant and objective information to drive decision-making in the construction of a viable alternative, to achieve monitoring or technology surveillance in the sector of these public universities in Colombia in the department of La Guajira.

3. Technological surveillance dimensions

3.1. Requirements of the technology surveillance system

The Technology Surveillance and Competitive Intelligence units are responsible for the implementation of a technological surveillance system, in which, through a set of information methods and resources, it is systematically captured, analyzed and disseminated with intelligence the managers and staff of the organization who can act from it. Technology surveillance units can basically provide two types of services, passive surveillance or active surveillance. As for the products, some of a regular nature and others of a specialized nature can be offered, the former are characterized by their greater frequency of elaboration (alert messages, technical bulletins, generic newsletters and profiles of customers and competitors), the latter are distinguished due to the depth of the contents to be treated (in-depth evaluations, strategic forecast, situation analysis). In order to comply with each of these products, it is necessary to comply with a series of requirements to achieve the effectiveness of the Technology Surveillance system which appear to be established by standardized guidance standards or those guidelines established by organizations.

3.2. Management responsibilities in technological surveillance

At this point, emphasizes that internal controls have limitations, and managers should know that even the best performance will also provide deficiencies and errors in the transfer of information. Although the level of protection can be determined by the priority of the supervision carried out, when management takes the time to find out if the commitments and revisions are being regulated as planned, there must be a continuous mechanism that reiterates a message to all analysts that internal controls are important. On the contrary, when the administration does not pay attention to the responsibilities that they think are in place, it is likely that these controls are not working effectively; it is then a clear responsibility of the management to assume the controls that in the matter of technology surveillance have to be arranged for the organization. That for this reason, in addition to performing some type of continuous monitoring, the administration must communicate its expectations for internal controls to all employees, as well as establish a form of dissemination that transmits knowledge tacit from the base of the organization to the top and vice versa.

3.3. Resource management in technological surveillance

Says that the design of resource supply must be linked to the development of exploitation maneuvers, which anticipate future needs, and focus on strategic objectives. With the support of a central computer platform to support the mission of the organization, effective management planning immediately becomes a vital axis, and therefore, guaranteeing the viability, relevance and value of the infrastructure of said platform also requires adequate financial approach. On the other hand, considers that the benefits proactively against the capital requirements can help to minimize the risks given by the probable leverage and activities outside the operating budgets. In this way, the long-term vision of financing programs, when an organization extends to the acquisition and support costs, aligns the synchronization of the real profits of the Technological Surveillance searches. Subsequently, when observing the procedures from a perspective of the modeling analysis of the technological life cycle based on patents, a serious congruence can be appreciated to ensure that the resources are adequately focused on the transverse axes. In this way, through an exhaustive technical evaluation taking the opportunity cost as a financial control, an organization can offer realistic life cycle scenarios that mark the user's expectations and allow the generalities of the human components and infrastructure to be managed as a service.

3.4. Performance of the Technology Watch (standard UNE 166006:2011)

- Have the most current scientific and technical information and facilitate its dissemination in the organization.

- Identify and track the main trends for one or several technologies in order to anticipate changes.
- Have relevant information for use in the decision-making process of R & D & I: i) Define an R&D strategy ii) Analysis and selection of R&D ideas and iii) Planning of the portfolio of technological projects.

3.5. Mechanisms for monitoring technological surveillance

According to Standard UNE 166006:2011, the organization must plan, program and implement the processes of monitoring, measurement, analysis and improvement of the process of Technological Surveillance and Competitive Intelligence. The organization must track the perception of interested parties regarding the satisfaction of their needs and expectations. This section of the standard includes the following requirements: i) Follow-up of satisfaction and fulfillment of expectations of interested parties, ii) Internal audits, iii) Follow-up and measurement of the R & D & I process. iv) Monitoring and measurement of the results of the R + D + i process, that is, of the R + D + i products, v) Control of deviations, vi) Data analysis, vii) Corrective actions and viii) Preventive actions.

3.6. Positioning of public university institutions

The positioning of a product is the way in which consumers define the products based on their important attributes; the place that the product occupies in the minds of consumers, in relation to competing products define positioning as a term that refers to the place that a product or brand occupies in the market and even In the consumer's mind, this order of preference is established with respect to the other competitors. Similarly, the Standard UNE 166006:2011, this tool deals with the strategic positioning and the evaluation model of the positioning actions of the company / organization itself against its rivals taking into account financial power, competitive advantages, strengths of the industry sector and turbulence in the external business environment.

4. Performing Technological Surveillance methodology

4.1. Population object of study.

The spatial delimitation points to the operational universe, otherwise the process would become very complex. The population is, according to [5], the set of all the data that match a series of specifications, the relevant population depends on the objectives of the research, that is, it is the processes, things or people, which are intended to generalize the results. The population for the case study was selected taking into account those responsible for technology of public universities, which are detailed below (see table 1). In view of the fact that the population is made up of a finite group of subjects that correspond to those responsible for technology at public universities, it is not necessary to go to sampling criteria to define the group of informants and in view of them, the term of census type population.

Table 1. – Population distribution

Public universities	Responsible for technology in public universities
Open and distance national university	3
University of la guajira	5
Total	**8**

4.2. Data collection technique and instrument

According to Bavaresco (2006), data collection techniques lead to the verification of the silver problem. Each type of investigation determines what techniques should be used and at the same time

IOP Conf. Series: Materials Science and Engineering **844** (2020) 012046 doi:10.1088/1757-899X/844/1/012046

establishes the tools, instruments or means that should be used. According to Blanco (2002), observation by means of the survey consists in obtaining information from a representative sample of a population, using an interview or questionnaire.

The simple questionnaire will be available, which constitutes a technical form of observation, will be prepared in accordance with the objectives, variable, dimensions and indicators proposed. For the purposes of this study, it was made up of a total of 57 items and it was arranged as stated [6], a Likert type response scale consisting of five alternatives; where the highest scores are: Always (5) and almost always (4); referring to the positive responses, followed sometimes (3), indicative of a neutral posture of the respondent, culminating with almost never (2) and (1) never, related to the negative trend responses. (See table 2).

Table 2. - Abbreviation Scale Score

Escala	*Abreviación*	*Puntuación*
always	(S)	5
Usually	(CS)	4
Sometimes	(AV)	3
Rarely	(CN)	2
Never	(N)	1

Elaborated by: Hernández, Fernández y Baptista (2006)

5. Results and discussion

The data collected through the instrument applied in this research are presented, indicated by dimension to interpret the research objectives. The information obtained was arranged to analyze the variables through absolute, percentage, standard and average deviation frequencies. These results were analyzed descriptively through the design of tables detailing the input (frequency and percentage), of which the study was presented, as well as the interpretation of the theory confrontation with the reality assessed by the researcher.

5.1. Dimension: technological surveillance system requirements.

Table 3. - Technological surveillance system requirements.

INDICATOR	(S)		(CS)		(AV)		(CN)		(N)		Totals		MEDIUM INDICATOR	STANDARD DEVIATION INDICATOR
Frequencies	**Fa**	**Fr (%)**	**Fa**	**Fr (%)**	**Fa**	**Fr (%)**	**Fa**	**Fr (%)**	**Fa**	**Fr (%)**	**Fa**	**Fr (%)**	**X**	**X**
Requirements Generals	0	0,0	3	37.5	5	62.5	0	0	0	0	8	100	4	1.0
Requirements of the documentation	0	0,0	0	0,0	3	37.5	3	37.5	2	25	8	100	2,6	0,47
Confidentiality requirements legality and ethical aspects	0	0.0	1	12,5	4	50	2	25	1	12.5	8	100	2	1,22
STANDARD DEVIATION DIMENSION	**1,0**													

Regarding the particular indicator, it is possible to show that 62.5% of respondents say that sometimes in the Institution there are general requirements corresponding to the technological surveillance system, however, 37.5% of respondents say that almost always in the institution there are general requirements corresponding to the Technology Watch system, while the other 0% of those consulted always refer, almost never and never respectively; In addition, the standard deviation of the indicator is 1, which indicates a medium level of dispersion and therefore a high reliability index. Consequently, the UNE 166006: 2011 standard establishes that the areas of performance of the Technology Surveillance system must be identified. Be clear in what areas, activities, structures of the department will operate the Technology Watch system. If the Technology Surveillance system has a performance in all departments of the analyzed area / unit ("Full performance"), in the majority ("Enough performance"), in a minority ("poor performance"), in any department ("Null performance").

The previous indicator is consistent with what was stated by the authors [3]), where it raises the aspects that must be treated with a high sense of responsibility, ethicism in addition to the respect of each organization, the forecasts are not a coincidence of The International Association of Professionals of this field, Society for Competitive Intelligence Professionals, SCIP, with more than 5,000 members has developed a deontological code based on professional ethics and strict respect for legality, one of its foundational pillars.

5.2. Dimension: Management responsibilities in technological surveillance

Table 4. - Management responsibilities in technological surveillance

INDICATOR	(S)		(CS)		(AV)		(CN)		(N)		Totals		MEDIUM INDICATOR	STANDARD DEVIATION INDICATOR
Frequencies	Fa	Fr (%)	Fa	Fr (%)	Fa	Fr (%)	Fa	Fr (%)	Fa	Fr (%)	Fa	Fr (%)	X	X
Commitment to management	0	0,0	4	50	1	12.5	3	37,5	0	0,0	8	100	2.66	1.24
Technology Watch Policy	0	0,0	0	0.0	2	25	5	62.5	1	12,5	8	100	2,66	1.69
Responsibility authority and communication	0	0,00	1	12.5	1	12.5	4	50	2	25	8	100	2,0	1.22
STANDARD DEVIATION DIMENSION			0,38											

Regarding the particular indicator, it is possible to show that 50% of those consulted said that there is almost always a commitment to management in the Institution, corresponding to the Technology Watch system and almost never 35.7%, in turn 12.5% of the consulted refers that sometimes; while 0% of those consulted said that there is always and never respectively in the institution commitment to the direction corresponding to the Technology Watch system, in addition the standard deviation of the indicator is 1.24; which indicates a low dispersion level and therefore a high reliability index. In relation to the aforementioned, the results are in accordance with the provisions of Standard UNE 166006: 2011, the standard sets out the guidelines for the organization in which the importance of the Technology Surveillance system can be visualized. Likewise, meetings must be held, at least quarterly, between the person in charge of the project and the workers to be aware of the development and implementation of the Technology Watch service and ensure that they establish the objectives of the system.

5.3. Dimension: Resource management in technological surveillance

Table 5. - Resource management in technological surveillance

INDICATOR	(S)		(CS)		(AV)		(CN)		(N)		Totals		MEDIUM INDICATOR	STANDARD DEVIATION INDICATOR
Frequencies	Fa	Fr (%)	Fa	Fr (%)	Fa	Fr (%)	Fa	Fr (%)	Fa	Fr (%)	Fa	Fr (%)	X	X
Human Resources	0	0,0	0	0,0	2	25,0	3	37,5	3	37,5	8	100	2.33	0,47
Means materials and infrastructure	0	0,0	0	0,0	2	25	5	62.5	1	12.5	8	100	2.66	1.69
STANDARD DEVIATION DIMENSION	0,23													

First, the human resources indicator exposes an average 2.33; tending to the category of negative response, according to the scale developed. In addition, the standard deviation of the dimension was 0.23; what allows to interpret a high reliability of the answers and very low dispersion on the particular indicator, allows to show that 37.5% of the consulted ones say that almost never and never exists in the Institution the human resources corresponding to the Technological Surveillance system. Therefore, 25.0% of those consulted said that sometimes there are human resources in the institution corresponding to the Technology Watch system, while the other 0% of those consulted always and almost always refer respectively; In addition, the standard deviation of the indicator is 0.47; which indicates a low dispersion level and therefore a high reliability index.

The particular indicator is consistent with what is established in Standard UNE 166006: 2011 in relation to the number of training courses related totally or partially to specialized databases; The staff must have a minimum of 3 courses related to the previous theme, each of them must have a minimum duration of 4 hours. Management and exploitation of specialized databases. Prepare a report on "tools and information resource available on the internet" on the area that corresponds to the entity, in which the main tools for information search are found. It will be necessary to keep it updated (minimum update every 3 months) Tools and resources to search for information available on the internet.

5.4. Dimension: Performance of technological surveillance.

Table 6. - Performance of technological surveillance.

INDICATOR	(S)		(CS)		(AV)		(CN)		(N)		Totals		MEDIUM INDICATOR	STANDARD DEVIATION INDICATOR
Frequencies	Fa	Fr (%)	Fa	Fr (%)	Fa	Fr (%)	Fa	Fr (%)	Fa	Fr (%)	Fa	Fr (%)	X	X
Process of identification of needs, sources and means of access to Information	0	0,0	0	0.0	1	12.5	5	62.5	2	25	8	100	2.66	1.69
Technology Watch Products	0	0,0	1	12.5	2	25	3	37.5	1	12.5	8	100	1.75	0,82
Results of the Technology Watch	0	0,0	0	0.0	2	25	1	12,5	5	62.5	8	100	2	0,816
STANDARD DEVIATION DIMENSION	0,47													

Regarding the particular indicator, it is possible to show that 62.5%, 25% and 12.5, of those consulted, state that almost never, never and sometimes there is a process of identification of needs, sources and means of access to Information in the Institution , corresponding to the Technology Surveillance system, however, 0% of those consulted said that there is always and almost always the process of identifying needs, sources and means of access to Information; In addition, the standard deviation of the indicator is 1.69; which indicates a high level of dispersion and therefore a high reliability index.

According to the previous indicator, the results agree with the UNE 166006:2011 Standard, which states that a procedure must be established by the organization for the identification of information needs. A preliminary document must be prepared that includes a minimum of information (specified above). Identification of information needs to perform searches establishing previous search strategies and actions in the selected sources (descriptors, terminology, keywords, operators used, geographical and temporal segmentation). In turn, the results are in accordance with the statement by the author Winter, (2010). In this sense, the author states that one can aspire to full satisfaction; in the previous fulfillment of the R + D + i profile in order to improve the competitiveness of the organization and act as a promoter for scientific-technological monitoring. Therefore, it is a commitment to be aware of the innovations and technological changes required in the market, as well as the demand of the clients, strengthening the instruction and motivation of the staff of the Technology Watch Committee.

5.5. Dimension: Mechanisms for monitoring technological surveillance.

Table 7. - Mechanisms for monitoring technological surveillance.

INDICATOR	**(S)**		**(CS)**		**(AV)**		**(CN)**		**(N)**		**Totals**		MEDIUM INDICATOR	STANDARD DEVIATION INDICATOR
Frequencies	**Fa**	**Fr (%)**	**Fa**	**Fr (%)**	**Fa**	**Fr (%)**	**Fa**	**Fr (%)**	**Fa**	**Fr (%)**	**Fa**	**Fr (%)**	**X**	**X**
Internal audits	0	0,0	1	12.5	4	50	2	25	1	12.5	8	100	2,0	1.22
Monitoring and measurement of the Technological Surveillance process	0	0,0	0	0,0	2	25,0	4	50,0	2	25,0	8	100	2.66	0,94
Monitoring and measurement of the results of the Technology Watch process	0	0,0	0	0,0	2	25	2	25	4	50	8	100	2.66	0.94
Improvement	1	4,17	4	16,6	4	16,6	8	33,3	7	29,1	8	100	2,33	1,00
STANDARD DEVIATION DIMENSION	**0,31**													

Regarding the particular indicator, it is possible to show that 50% and 25% of respondents state that, sometimes, almost never, there are internal audits corresponding to the technological surveillance system, however, 12.5% of respondents say that almost always and never; while 0% of those consulted said that there are always internal audits corresponding to the technological surveillance system, in addition the standard deviation of the indicator is 1.22; which indicates a low dispersion level and therefore a high reliability index. From the particular indicator the results are in accordance with the approach of Standard UNE 166006: 2011. According to indicator, Monitoring and measurement of the technological surveillance process, it exposes an average of 2.66; This indicator allows us to show that 50.0% of respondents say that almost never and 25.0% say that sometimes and never, while 0% of respondents always and almost always refer, there is monitoring and measurement of Technological surveillance process, in addition the standard deviation of the indicator is 0.94; which indicates a moderate level of dispersion and therefore a high reliability index.

These results coincide with the provisions of Standard UNE 166006: 201 Corrective action to eliminate "non-conformities", in the VT / IC system in order to prevent them from happening again should be established. The area coordinator will be in charge of performing these tasks. A report will be produced that will collect the requirements. The elaboration of the same in its employees, as long as it continues supervising it. The Preventive Action in the same way specifies Standard UNE 166006: 2011, which to eliminate the causes of "nonconformities", among other potentials in the VT / IC system in order to prevent them from happening again, a procedure must be established.

5.6. Dimension: Positioning of university institutions.

Table 8. - Positioning of university institutions.

INDICATOR	(S)		(CS)		(AV)		(CN)		(N)		Totals		MEDIUM INDICATOR	STANDARD DEVIATION INDICATOR
Frequencies	**Fa**	**Fr (%)**	**Fa**	**Fr (%)**	**Fa**	**Fr (%)**	**Fa**	**Fr (%)**	**Fa**	**Fr (%)**	**Fa**	**Fr (%)**	**X**	**X**
Positioning according to technological capacity	0	0,0	1	12.5	1	12.5	5	62.5	1	12.5	8	100	2	1.73
Durability of the competitive advantage	1	12,5	0	0,0	5	62,5	1	12.5	1	12.5	8	100	2	1.73
Return	0	0,0	1	12.5	3	37,5	2	25	2	25	8	100	2	0.70
Uncertainty and risk	1	12.5	2	25,0	3	37.5	2	25,0	0	0.0	8	100	3,08	0,70
STANDARD DEVIATION DIMENSION	**0,54**													

First, the indicator Positioning according to technological capacity, exposes an average 2; tending to the low response category, according to the scale developed, in addition the standard deviation of the dimension was 0.54; what allows to interpret a high reliability of the answers and very low dispersion. Regarding the particular indicator, it shows that 62.5% of respondents almost never refer; and 12.5% of respondents said that almost always, sometimes and never. However, 0% of respondents say that there is always positioning according to technological capacity university institutions corresponding to the technological surveillance system. In addition, the standard deviation of the indicator is 1.73; This indicates a high level of dispersion and a high reliability index. The results are in accordance with the statement by the author [6] where he states that technologies in the face of their interest in technological innovation processes can be classified with generic criteria and that they have an endogenous root, typical of technology, is say, they value their difficulty, their grade.

These results coincide with the author [6], where he postulates the analysis of the risk that has been mentioned before, provides, at least, two other indicators of attractiveness. Project One is the probability of technical success, in the sense that the project meet the intended objectives and other, the probabilities of commercial success, that is, the probability that, in addition, the result of the project will be translated into new products, processes and services or improvements for those currently offered by the company that are valued by customers and consequently, bought by them. In many cases a single risk factor calculated from the product of the previous two is used

IOP Conf. Series: Materials Science and Engineering **844** (2020) 012046 doi:10.1088/1757-899X/844/1/012046

6. Conclusions

After carrying out the analysis and discussion of the results, taking into account the objectives of the investigation, the following conclusions were reached: In relation to the first specific objective, which refers to identifying the requirements of the technological surveillance system as an element for the positioning of public universities. It was possible to establish that in the components of the technological surveillance the identification, the object of study or critical factors prevail, because the processes of the technological surveillance include all kinds of documentation that can be used for the analysis and reflection on business management strategies. In this sense, the organization must focus its efforts on mission objectives.

Regarding the second specific objective, which refers to describing the responsibilities of management in technological surveillance as an element for the positioning of public universities, it was established that the Center for Technological Development or Systems Department plays a predominant role, as opposed to To the character of the external development of the Information and Communication Technologies of an aggressive nature, the most obstacle-free alternative in the short term has to do with the capacity of these centers, by quickly identifying technological advances, making a strategic reading of they, in terms of identifying the opportunities and / or threats that are generated on a day-to-day basis. The institution must creatively take advantage of these opportunities; To this end, technological capacities must be developed to identify and evaluate the competitive impacts of their environment.

Regarding the third specific objective, to characterize the management of resources in technological surveillance as an element for the positioning of public universities, it was established that the absorption of the technological changes identified requires, in some cases, a deep technological and productive reconversion. Which requires senior management some capital investment commitments and training of technology area personnel, in the opinion of the experts consulted they must be an essential condition in the solution of the characterization in the management of technological surveillance in university positioning public.

In relation to the fourth specific objective, which refers to describing the performance of technological surveillance as an element for the positioning of Public Universities. It could be established that it is a systematic and organized process of searching, capturing and filtering information, preferably automated, of which a number of documents are generated, which must be organized and systematized to serve as an aid to decision making anticipating to external threats and opportunities that affect business strategy and research in science and technology. Regarding the fifth specific objective, it refers to establishing the monitoring mechanisms in technological surveillance as an element for the positioning of public universities. It could be established that everything.

The process requires feedback, which helps to enrich the exercise itself, in this sense of the technological surveillance projects, they are not alien to the audit processes and monitoring the information processed by the technological surveillance systems, in the era of knowledge management, where information abounds, is relevant and of extreme importance the processes of analysis of the results for its success. Regarding the sixth specific objective, the elements that characterize the positioning of university institutions were analyzed based on the efforts of technological surveillance. It was established that positioning is one of the most managed concepts in the world of current administration, and yet it is one of the worst understood when defining and putting it into practice.

In this sense, technological surveillance will support the management of corporate identity, as well as communications, which represents what public universities intend to convey to the target audience, in this same order of ideas is the perceived image, which is how the environment looks at the academic community. In accordance, with what is indicated in the first conclusion, it is the stamp or imprint that public universities want; To be recognized and to alternate what the competition is doing, in the continuous or systematic exercise of the technological surveillance process, it is key to enduring in time and in the opinion of the experts consulted it must be an essential condition within the

IOP Conf. Series: Materials Science and Engineering **844** (2020) 012046 doi:10.1088/1757-899X/844/1/012046

institutions. The solution of the characterization in the management of technological surveillance for the positioning of public universities.

References

[1] P. Escorsa 2007, INTEC: La inteligencia competitiva, factor clave para la toma de decisiones en las organizaciones. Madrid: Comunidad de Madrid. Fundación Madrid para el Conocimiento.

[2] F. PALOP y J. VICENTE 1999, Vigilancia tecnológica e inteligencia competitiva: su potencial para la empresa española. Madrid: Fundación Cotec para la Innovación Tecnológica, p. 116.

[3] F. SANCHEZ SELLERO y M. CRUZ GONZALEZ 2012, Desarrollo de Sistemas de Vigilancia Tecnológica en la Acuicultura Española. Journal of Technology Management & Innovation, *SCIELO,* vol. 7, nº 3, pp. 214-226.

[4] M. Fernández Fernández 2014, El control, fundamento de la gestión por procesos,.

[5] R. Hernandes , C. Fernandez y P. Bastidad 2010, Metodología de la Investigación. México: Ed. McGraw Hill.

[6] F. Casini, La Compatibilidad Estratégica del Proyecto de I+D.

Acknowledgments

The authors thank the help of the organizing members of EXPOTECNOLOGÍA 2019 of the Antonio de Arévalo Technological Foundation.

Analysis of the water quality risk index for human consumption in urban areas of the department of Bolivar

Emilia Polo-Carrillo[1], Geraldine Morales-Simancas[2], Yudis Cabarcas-Ariza[2] and Jorge Valle Rodríguez [3]

[1]Gitac: Grupo de investigación de turismo, administración y comercio. Colegio Mayor de Bolívar, Cartagena, Colombia.
[2]Facultad de Ciencias Económicas, Fundación Universitaria Antonio de Arévalo Bolívar, Cartagena, Colombia.
[3]Gobernación del Departamento de Bolívar, Cartagena, Colombia

c.admon@colmayorbolivar.edu.co

Abstract: Objective: To analyze the results of the Water Quality Risk Index for Human Consumption in Urban Areas of the Department of Bolívar. Materials and methods: A descriptive investigation was developed, with the objective of evaluating the compliance of the physical, chemical and microbiological characteristics of water for human consumption. The direct observation technique was used whose results are based on the application of the standardized measuring instrument authorized by the Ministry of Social Protection in Colombia - IRCA for estimating compliance with acceptable parameters established in current regulations. Results: It was found that of the 46 municipalities in the department of Bolívar, study focus 18 presents water that is not suitable for human consumption that requires special surveillance and 3 municipalities have sanitary waters that require maximum, special and detailed surveillance. Conclusions: The need to establish alternatives by the competent entities that improve the quality of water for consumption in some municipalities of Bolívar and implement strategies for the prevention of waterborne diseases was evidenced.

1. Introduction

United Nations Human Rights [1] Affirms that water is the essence of life. Drinking water and sanitation are essential for life and health, and fundamental for the dignity of every person [2] Expresses that the right of all persons was mentioned worldwide for the first time to have access to drinking water in sufficient quantity, and the evaluation and use of water resources in the world for the assurance of the well-being of the people was discussed, as the states were encouraged to carry out internal reviews on water matters to establish plans and policies focused on meeting the needs of drinking water and sanitation. The United Nations General Assembly explicitly recognized the human right to water and sanitation, reaffirming that drinking water and sanitation are essential for the realization of all human rights. [3]. Health and water quality are essential for the development and well-being of humans. Providing access to safe water is one of the most effective instruments to promote health and reduce poverty. (WHO, 2017).

In the same way according to statements of [4], Water is essential for life. No living being on Earth can survive without water. Water is essential for human health and well-being as well as for the preservation of the environment. Despite this, four out of ten people in the world lack access to a simple latrine and almost two out of ten do not have access to a safe source of drinking water. Every year, millions of people, mostly children, die from diseases related to inadequate water supply, sanitation and hygiene.

In accordance with the above [3] it indicates that water is at the epicenter of sustainable development and is essential for socio-economic development, energy and food production, healthy ecosystems and for the very survival of human beings. Similarly [5] expresses that the supply of drinking water constitutes a public service that states must provide in an obligatory manner to individuals, since access to the resource is a collective, basic or fundamental need, said supply has not could be entirely satisfied by many governments in various countries, especially in those underdevelopment conditions prevail [5].

According to figures from the World Health Organization (WHO), it is estimated that 10% of the world population consumes irrigated food with untreated sewage, and that 32% of the world population does not have access to adequate basic sanitation services, generating 280,000 deaths associated with water diseases. It is estimated that 4% of total deaths in the world are related to water quality, hygiene and sanitation. In Latin America and the Caribbean, acute diarrheal diseases (ADD) are one of the ten leading causes of deaths per year (5), due to problems in water quality, mainly due to improper wastewater management [6].

The Political Constitution of Colombia establishes as one of the main purposes of the State's activity, the solution of unsatisfied basic needs, among which is access to the drinking water service, which is fundamental for human life. The adequate supply of quality water for human consumption is necessary to avoid cases of morbidity due to diseases such as cholera and diarrhea [7]. Referring to the importance of water quality in health [5] Most intestinal parasites are transmitted by environmental pollution and in this aspect, water and food play an important role. If the feces are not disposed of properly, the cysts, oocysts and eggs of the intestinal parasites can remain in the home environment or contaminate water sources or crops irrigated with sewage. Therefore, it is estimated that many of the deaths in the world are due to problems related to water, drainage and hygiene.

The framework for which a water sample can be considered adequate or "safe" is a drinking water quality standard. An appropriate standard is, then, the reference that will ensure that water is not harmful to human health. 9 Most of the countries in Latin America, North America, the European Union and other regions and communities have adopted in their quality standards of drinking water, WHO guide values as specific indicators for its national standards [8].

According to the latest data published for the department of Bolívar on waterborne diseases (EVA) and water quality risk index (IRCA) in Colombia 2015. [9] the incidence rate of diseases directly related to the water in Colombia shows that for that last report year 2015, the mortality rate due to acute diarrheal disease (ADD) in children under 5 years and shows a general incidence rate of ADD of 49.5 with very low data in the rate of incidence of hepatitis A "of 0.6 paratyphoid fever incidence rate of 0.5 and incidence rate of leptospirosis of 1.7 showing an average risk IRCA with a 22.13.

This investigation takes into account the provisions of Decree 1575 by [10]. By which the system for the protection and control of water quality for human consumption is established, in which the IRCA is defined as the degree of risk of occurrence of diseases related to non-compliance with the physical, chemical and microbiological characteristics of water for human consumption. In this sense, the present study aims to analyze the results of the Risk Index of Water Quality for Human Consumption in Urban Areas of the Department of Bolívar in order to establish strategies for the prevention of waterborne diseases.

2. Materials and Methods

A descriptive investigation was developed, with the objective of evaluating the compliance of the physical, chemical and microbiological characteristics of water for human consumption according to Resolution 2115 of 2007.

It is very important to express that the data bank used for this investigation on water quality was obtained from the Information System of the Monitoring of the Quality of Water for Human Consumption (SIVICAP), reported to the Government of Bolívar, the data recorded in This base is the result of the report on the monitoring of water quality for human consumption carried out by health authorities in the different territories of Colombia.

This surveillance circumscribes the collection of samples, the analysis and interpretation of results, the provision of information and the use of this information for public health. For the collection of samples, the water quality monitoring plan established in Resolution 2115 of 2007 is developed, and in which the frequency and number of samples are established according to the characteristics and the population supplied. In the analysis the characteristics of color, pH, turbidity, free residual chlorine, total coliforms and Escherichia coli were considered, as they are the most representative indicators of the microbiological and physicochemical quality of the water and mandatory requirement for all 46 municipalities of the country taken for the study. Likewise, the water quality risk index was considered, an instrument established by the regulations to evaluate the quality of drinking water in the country.

It is important to state that in this study the information of the water quality monitoring system for human consumption was taken into account with the Interlaboratory Program for Quality Control of Potable Water (PICCAP), whose objective is to guarantee the reliability of the results of the analysis of water quality governed by the parameters established in Colombian technical standards: NTC-ISO / IEC 17025: 2005 NTC-ISO / IEC 17043: 2010 This program continuously evaluates the analytical quality of the different laboratories in charge of physical analysis , chemical and microbiological water for human consumption.

2.1. Collection of Information

For data collection, the direct observation technique was used, the results of which are based on the application of the standardized measuring instrument authorized by the Ministry of Social Protection in Colombia - IRCA in which the risk score is assigned to each physical, chemical and microbiological characteristic, for estimating compliance with acceptable parameters established in current regulations. For the interpretation of the results of the inspection of the Water Quality Surveillance for Human Consumption in the Urban Areas of the Department of Bolívar, work was carried out in accordance with the guidelines of Resolution 2115 of June 2007 in which the He developed the application "Information System for the Monitoring of Water Quality for Human Consumption-SIVICAP", which indicates the characteristics, basic instruments and frequencies of the control and surveillance system for the quality of water for human consumption that labels ranges of classification according to the IRCA to assess the level of water risks, therefore the rating ranges are established as follows (see table 1).

The measurement ranges detailed above were applied in the samples taken by municipalities to generate surveillance, inspect and analyze the risk levels of water for human consumption in our Department of Bolívar; Forty-six (46) municipalities taken as a study sample, only 6 of them reached a range of (0 -5) that is equivalent to NO RISKS, which shows that these waters are suitable for human consumption as the evidenced (table 2), but monitoring should continue. However, it is worth clarifying that this number of municipalities that approved the Water Quality Monitoring Process for Human Consumption is equivalent to 13% of the analyzed sample, the Number of samples chosen, the municipality and the level of risks.

IOP Conf. Series: Materials Science and Engineering **844** (2020) 012047 doi:10.1088/1757-899X/844/1/012047

Table 1. Classification of the level of health risk according to the IRCA.

CLASSIFICATION IRCA (%)	RISK LEVEL	CONSIDERATIONS
70.1 – 100	INVITABLE SANITARIATELY	Water not suitable for human consumption and requires maximum, special and detailed surveillance
35.1 – 70	HIGH	Water not suitable for human consumption and requires special monitoring
14.1 – 14	MEDIUM	Water not suitable for human consumption, direct management of the provider
5.1 – 14	LOW	Water not suitable for human consumption, susceptible to improvement.
0 - 5	WITHOUT RISKS	Water suitable for human consumption. Continue monitoring

Table 2. Municipalities without Risk Level in the process of Surveillance of Water Quality of the Department of Bolívar.

MUNICIPALITIES	N°.	% IRCA	RISK LEVEL	% CLASSIFICATION	WATER SUITABLE FOR HUMAN CONSUMPTION
El Carmen De Bolívar	4	0%	Without Risks	0 – 5	X
Montecristo	8	0.90%	Without Risks	0 – 5	X
San Pablo	8	0%	Without Risks	0 – 5	X
Santa Rosa Del Sur	12	0%	Without Risks	0 – 5	X
Simiti	12	3.12%	Without Risks	0 – 5	X
Turbana	3	0%	Without Risks	0 – 5	X

3. Results and discussions:

The study results were the result of the analysis of 393 water samples reported in 2017, derived from 46 municipalities of the Department of Bolívar.

The characterization was carried out by means of descriptive statistical techniques, considering the dispersion and the normal distribution of the data, and calculating the range between quartiles, taking as reference in this study the compliance with the norm regarding the potability values by means of the calculation of the percentage of samples that presented results among the maximum values allowed [10] And subsequently, the results of the physicochemical characteristics were analyzed using descriptive statistics techniques, calculating the measurements of position (maximum value, minimum value, median, first quartile, third quartile and average) and dispersion (standard deviation). The above is detailed in Table 3.

Table 3. Water potability characteristics and values by calculating the percentage of samples

PARAMETERS	MAXIMUM VALUES ALLOWED
Colour	15 units of cobalt platinum
Turbidity	5 nephelometric turbidity units (UNT)
Ph	6.5-9.0; turbidity
Liquid Residual Chlorine	0,3 a 2,0 180 Guzmán BL 35(Supl.2):177-90 mg/L
Total Coliforms	0 UFC/100 cm3
Escherichia Coli	0 UFC/100 cm3
Colour	15 units of cobalt platinum

For the analysis of the water quality risk index, the average for the total samples per year and by municipality was calculated, the values were descriptively analyzed using historical series, and the municipal geographic distribution and the collection area were considered (rural or urban). The interpretation of the results of the inspection of Water Quality Surveillance for Human Consumption in the urban areas of the Department of Bolívar, was worked according to the guidelines of the regulations in force in Colombia that contains the development of the application "System of Information on the Monitoring of Water Quality for Human Consumption-SIVICAP ", which indicates the characteristics, basic instruments and frequencies of the control and surveillance system for the quality of water for human consumption, having this classification ranges according to the IRCA for evaluate the level of water risks, therefore the rating ranges are established , see table 4.

The measurement ranges detailed above were applied in the samples taken by municipalities to generate surveillance, inspect and analyze the risk levels of water for human consumption of the Department of Bolívar; where of the 46 municipalities taken as a study sample, only 6 of them reached a range of (0 -5) that is equivalent to NO RISKS, which indicates that in the samples taken excellent results were achieved in the characteristics of color, pH, turbidity, free residual chlorine, total coliforms and Escherichia coli, therefore these waters are suitable for human consumption but monitoring should continue. However, it is clear that this number of municipalities that approved the Water Quality Monitoring process for Human Consumption is equivalent to 13% of the analyzed sample, of which the Number of selected samples, the municipality and the level will be detailed below of risks.

IOP Conf. Series: Materials Science and Engineering **844** (2020) 012047 doi:10.1088/1757-899X/844/1/012047

Table 4. Classification of the level of health risk according to the IRCA by sample and actions to be taken.

CLASSIFICATION IRCA (%)	RISK LEVEL	CONSIDERATIONS
70.1 - 100	INVITABLE SANITARIATELY	Water not suitable for human consumption and requires maximum, special and detailed surveillance.
35.1 - 70	TALL	Water not suitable for human consumption and requires special monitoring.
14.1 - 14	MEDIUM	Water not suitable for human consumption, direct management of the provider.
5.1 – 14	LOW	Water not suitable for human consumption, susceptible to improvement.
0 - 5	WITHOUT RISKS	Water suitable for human consumption. Continue monitoring.

Table 5. Municipalities without risk in the process of Water Quality Surveillance of the Department of Bolivar, 2017.

MUNICIPALITIES	N°. SAMPLE	IRCA %	RISK LEVEL	% CLASSIFICATION	WATER SUITABLE FOR
El Carmen De Bolívar	4	0%	Without Risks	0 - 5	X
Montecristo	8	0.90%	Without Risks	0 - 5	X
San Pablo	8	0%	Without Risks	0 - 5	X
Santa Rosa Del Sur	12	0%	Without Risks	0 - 5	X
Simiti	12	3.12%	Without Risks	0 - 5	X
Turbana	3	0%	Without Risks	0 - 5	X

The results of the remaining 40 municipalities of the Department of Bolívar, taken as objects of study where the water samples taken for human consumption by SIVICAP, obtained different levels of risks, resulting in of water not suitable for consumption, are analyzed below not suitable for human consumption Human and requires special vigilance. Results were confirmed with a LOW risk level for 6 municipalities with an IRCA between 5.1 - 14, where its description is Water not suitable for human consumption, susceptible to improvement; results with a MEDIUM risk level for 10 municipalities with an IRCA between 14.1 - 14, with a description Water not suitable for human consumption, direct management of the provider; results with a HIGH level of risk for 18 municipalities with an IRCA between 35.1 - 70, with a description Water not suitable for human consumption and requires special monitoring; and it should be noted that 3 municipalities obtained results INVITABLE SANITARIATELY, which means Water not suitable for human consumption and requires a maximum, special and detailed surveillance, where the IRCA obtained is between 70.1 – 100, see table 5.

Table 6. Municipalities with Risk in the process of Water Quality Surveillance of the Department of Bolivar, 2017.

Municipalities	N° SAMPLE	IRCA %	Risk Level	% CLASSIFICATION	WATER NOT SUITABLE FOR HUMAN CONSUMPTION
Achí	12	45.52%	High	35.1 - 80	X
Altos Del Rosario	12	10.15%	Low	5.1 - 14	X
Arenal	7	45.51%	High	35.1 - 80	X
Arjona	8	24.33%	Mean	14.1 - 35	X
Arroyohondo	0	No report	-	-	-
Barranco De Loba	12	43.17%	High	35.1 - 80	X
Calamar	12	49.70%	High	35.1 - 80	X
Cantagallo	12	34.97%	Mean	14.1 - 35	X
Cicuco	12	38.54%	High	35.1 - 80	X
Clemencia	9	76.55%	High	35.1 - 80	X
Córdoba	7	76.78%	High	35.1 - 80	X
El Guamo	8	28.12%	Mean	14.1 - 35	X
El Peñon	8	25.23%	Mean	14.1 - 35	X
Hatillo De Loba	12	86.63%	I. Sanitary	80.1 - 100	X
Magangué	12	16.66%	Mean	14.1 - 35	X
Mahates	10	5.11%	Low	5.1 - 14	X
Margarita	11	68.51%	High	35.1 - 80	X
Marialabaja	0	No report	-	-	-
Mompox	9	29.56%	Mean	14.1 - 35	X
Morales	12	50%	High	35.1 - 80	X
Norosí	12	22%	Mean	14.1 - 35	X
Pinillos	12	50.12%	High	35.1 - 80	X
Regidor	8	51.35%	High	35.1 - 80	X
Rio Viejo	8	59.62%	High	35.1 - 80	X
San Cristobal	12	47.04%	High	35.1 - 80	X
San Estanislao	4	9.40%	Low	5.1 - 14	X
San Fernando	7	44.64%	High	35.1 - 80	X
San Jacinto	0	No report	-	-	
San Jacinto Del Cauca	12	92.44%	Not Viable Sanitary	80.1 - 100	X
San Juan Nepomuceno	4	85.27%	Not Viable Sanitary	80.1 - 100	X
San Martin De Loba	8	22.78%	Mean	14.1 - 35	X
Santa Catalina	10	71.45%	High	35.1 - 80	X
Santa Rosa	10	51.97%	Alto	35.1 - 80	X
Soplaviento	8	18.75%	Mean	14.1 - 35	X
Talaigua Nuevo	12	9.37%	Low	5.1 - 14	X
Tiquisio	12	75%	High	35.1 - 80	X
Turbaco	8	5.13%	Low	5.1 - 14	X
Villanueva	6	36.62%	High	35.1 - 80	X
Zambrano	8	9.37%	Low	5.1 - 14	X

Therefore, it is consistent to establish that the parameters shown in this study relate the importance of verification by government entities to accurately identify the quality of water for human consumption, based on the basis that it must have ranges of values that are within the parameters established by the WHO and by the applicable legislation in Colombia regarding the water quality standards established by the Ministry of Social Protection and Environment and Housing of Colombia. Regarding the conduct of frequent tests to determine if water contains indicator organisms, it is still the most sensitive and specific way of estimating water quality from the point of view of hygiene. In countries where there is an integrated surveillance system, most outbreaks of water origin have been linked to treatment failures, inclusion of contaminated water in the network or to problems of bacterial regrowth in the distribution system [11].

The most recent data from the [12] Confirms that the quality of the air we breathe and the water we drink are the main environmental risk factors that affect the health of the Colombians.

However, [9] Waterborne Diseases (VAS) were classified according to their direct or indirect association with water and their priority for health. Among the directly related and high-priority VAS, there are acute diarrheal diseases (ADD), hepatitis A, cholera, typhoid / paratyphoid fever, Leptospirosis, cryptosporidiosis, giardiasis.

This study presents evidence that shows that the water quality in the municipalities of the Department of Bolivar from the analyzed samples. Finding that there are alarm data such as 18 municipalities in the department of Bolivar with water not suitable for consumption that require space surveillance and even more alarming is to find that 3 municipalities of the department have sanitary non-viable waters that require maximum, special and detailed surveillance see table 6 Among several studies is the one carried out by the water treatment plant in the municipality of El Rosal - Cundinamarca Colombia, showed that 100% of the untreated water and 90% of the treated wastewater were contaminated with microorganisms considering intervention necessary urgent of regulators and manifesting these waters as a potential risk to public health [13]. Other studies such as that of [14] where an analysis was conducted based on the IRCA of drinking water, the result showed that the provider is not complying with the residual Chlorine parameter, when calculating for the risk index From the quality of water for human consumption, the household tap water samples in both samples showed a medium level of risk, which classifies them as unfit for human consumption; therefore it is essential that the provider carries out corrective actions, since, the breach of a single parameter can lead to the rejection of water intended for human consumption. Another study, such as the one carried out in the urban and rural area of the municipality of Guatavita, Cundinamarca, Colombia, focused like other studies in the country in Resolution 2115 of 2007, analyzed the water for human consumption in the urban area of the municipality of Guatavita. of this the inhabitants of the paths consume the water not suitable for human consumption, since although it is true that in the Bocatomas there is a basic pretreatment that consists of a sand trap and filter, it does not fulfill the appropriate functions, since it allows the water resource reaches homes in non-potable conditions.

This study confirms what was expressed in the research carried out by [15] on water quality indices in surface sources used in the production of water for human consumption; the greatest impact on public health occurs through water supply systems; The alteration of the organoleptic, physical, chemical and microbiological characteristics of the supply source directly affects the level of sanitary risk present in the water. They define as the risk of transporting pollutants that can cause diseases of water origin to man and animals or alter the normal performance of work within the home or industry. [16] They argue that safety is synonymous with impeccability and this is necessary in terms of nutritional quality, since the only guarantee to prevent diseases is the protection, conservation, cleaning, disinfection and other operations committed to the preparation process, it is done respecting the parameters quality.

Being consistent with the above, the results of this study allow to express that for the low, medium and high levels it is essential that direct intervention by the state through the authorized entities in Colombia and in turn in each Department to control and ensure the water quality for human consumption, as well as the level that shows unfeasible sanitary, the concentration of the state and the community is important for the prevention of waterborne diseases as a public health problem which requires special, maximum and detailed attention, water quality being even a determinant of poverty in a community. Being the optimum water quality critical aspect of development and social welfare and is the starting point to educate poverty. WHO asserts that WHO estimates that 94% of diarrhea cases could be avoided through environmental modifications, for example, through interventions that increase the availability of clean water and improve sanitation and hygiene.

This study considers that if the prevention of waterborne diseases is taken as a strategy and sensitize the inhabitants of the municipalities, they will act with a preventive culture, even if the correct procedure is that there is drinking water from the state entities Suitable for human consumption with the physical, chemical, microbiological and organoleptic characteristics of a quality water accompanied by an optimal service that comes directly from the taps of homes with the required temperature and pressure and does not need to be stored as currently occurs in some places. Conduct disease promotion and prevention campaigns where they can guarantee the population the acquisition of a storage tank that meets the conditions to guarantee the non-contamination of the liquid.

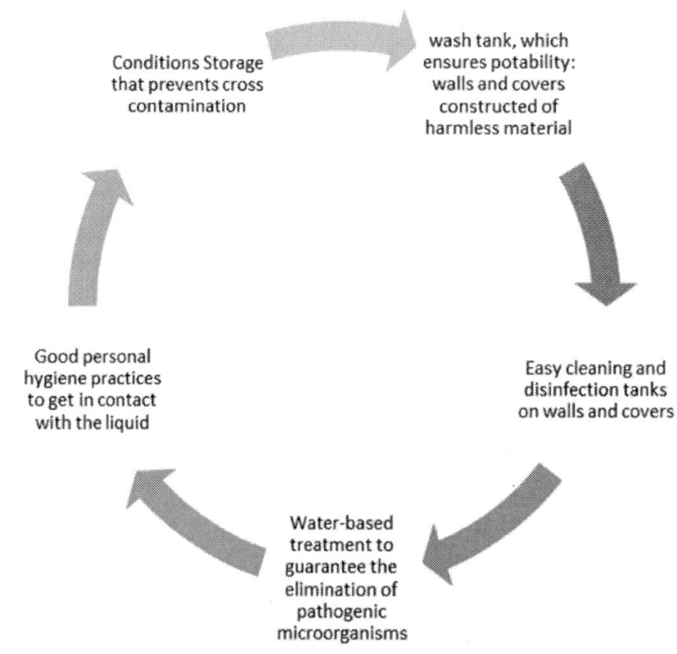

Figure 1. Own elaboration.

The present study has limitations. The most important is due to the characterization and costs of water-borne diseases, as well as the occurrences of water-associated diseases in the municipalities where the samples that present high-level risks in sanitary areas are not taken. However, it was possible to establish the Risk Index of Water Quality for Human Consumption, so that it serves as the basis of other studies that involve the quality of water for human consumption in Urban Areas of the Department of Bolívar.

4. Conclusions

Finally, after analyzing the results on the Risk Index of Water Quality for Human Consumption in Urban Areas of the Department of Bolívar, it must be concluded that even when there is no evidence of contamination or physical, chemical and microbiological alterations in all of the 46 Municipalities studied that greatly alter the quality and logically the potability of water, it is alarming that only 6 of them produce results "without risks", taking for granted that their characteristics for human consumption are suitable; However, the most worrying alarm is to find a large number of municipalities with water not suitable for human consumption and much more impressive still, is that there are some, although a minority, that are not sanitary.

As a consequence of the above, In view of the evidence collected and despite the worldwide concern by the WHO, and all the nations of the world for having quality water, both in developed and developing countries, the importance is stressed that the healthiness of water has for the health of populations and how this is the fundamental pillar for the proper functioning of a nation, as an important part of the axes that determine government development plans in which water is always as a point heading for the basic sanitation of the municipalities and cities. The importance of having water for society is such that every day the sick population grows due to not having this liquid, with the established quality indices based on IRCA. This study leads us to see clearly the reality of water quality in the municipalities of the department of Bolívar in Colombia, concluding this opens a great space and opportunity for improvement and that the results of this study can be the input and the tool for the adequate decision-making related to the design of robust public policies aimed at the prevention of diseases caused by water, which is ultimately the great problem when there are populations in which water does not meet the criteria and whose results will always go target a significant number of women, men and children who are constantly getting sick from the use and consumption of water under unsanitary conditions.

Reference

[1] Naciones Unidas Derechos Humanos. (2011). Derecho del agua Vol 35. Recuperado 16 de junio de 2019, de https://www.ohchr.org/Documents/Publications/FactSheet35sp.pdf

[2] Valdés De Hoyos, E. I. P., & Uribe Arzate, E. (2016). EL DERECHO HUMANO AL AGUA. UNA CUESTIÓN DE INTERPRETACIÓN O DE RECONOCIMIENTO. *Cuestiones Constitucionales*, *34*, 3-25. https://doi.org/10.1016/j.rmdc.2016.07.001

[3] ONU. (2010). Decenio Internacional para la Acción «El agua, fuente de vida» 2005-2015. Áreas temáticas: Derecho humano al agua y al saneamiento. Recuperado 16 de junio de 2019, de https://www.un.org/spanish/waterforlifedecade/human_right_to_water.shtml

[4] González, G., & Isabel, M. (2013). Un futuro a favor de la protección del agua. *Revista Cubana de Higiene y Epidemiología*, *51*(2), 126-128.

[5] Moreno, L. F. T. (2012). *El acceso al agua potable, ¿un derecho humano?* 23.

[6] Rodríguez Miranda, J. P., García-Ubaque, C. A., & García-Ubaque, J. C. (2016). Enfermedades transmitidas por el agua y saneamiento básico en Colombia. *Revista de Pública*, *18*, 738-745. https://doi.org/10.15446/rsap.v18n5.54869

[7] UNICEF – Colombia, & Procuraduría General de la Nación. (2016). El agua potable y el saneamiento básico en los planes de desarrollo. Recuperado 4 de julio de 2019, https://www.unicef.org/colombia/pdf/Agua3.pdf

[8] Robert Pullés, M. (2014). Microorganismos indicadores de la calidad del agua potable en cuba. *Revista CENIC. Ciencias Biológicas*, *45*(1). Recuperado http://www.redalyc.org/resumen.oa?id=181230079005

[9] Muñoz, N. J. M., & Gómez, F. R. (2016). *ENFERMEDADES VEHICULIZADAS POR AGUA (EVA) E ÍNDICE DE RIESGO DE LA CALIDAD AGUA (IRCA) EN COLOMBIA*. 101.

[10] Ministerio de la Protección Social. (2007a). Decreto-1575 pdf. Recuperado 14 de julio de 2019, de http://www.minambiente.gov.co/images/GestionIntegraldelRecursoHidrico/pdf/Disponibilidad-del-recurso-hidrico/Decreto-1575-de-2007.pdf

[11] Fernanda;, C., María Alejandra; Del Coco, Valeria. (2010). Agua y Salud Humana. Recuperado 21 de julio de 2019, de https://www.redalyc.org/pdf/863/86315692002.pdf

[12] Instituto Nacional de Salud INSDirección de Vigilancia y Análisis de Riesgo en Salud Pública. (2019). Informe Carga de Enfermedad Ambiental en Colombia. Recuperado 14 julio de 2019, de https://www.ins.gov.co/Noticias/Paginas/Informe-Carga-de-Enfermedad- Ambiental-en-Colombia.aspx

[13] Ortiz, C., López, M. C., & Rivas, F. A. (2012). Prevalencia de helmintos en la planta de aguas residuales del municipio El Rosal, Cundinamarca. *Revista de Salud Pública*, *14*, 296-304. https://doi.org/10.1590/S0124-00642012000200010

[14] Torres, S. M. E., & Navia, S. L. A. de. (2010). Calidad físico-química y microbiológica del agua del municipio de Bojacá, Cundinamarca. *NOVA*, *8*(14).

[15] https://doi.org/10.22490/24629448.451

[16] Lozada, P. T., Vélez, C. H. C., & Patiño, P. (2009). Índices de calidad de agua en fuentes superficiales utilizadas en la producción de agua para consumo humano. Una revisión crítica. *Revista de Ingenierías: Universidad de Medellín*, *8*(Extra 15), 3.

IOP Conf. Series: Materials Science and Engineering **844** (2020) 012048 doi:10.1088/1757-899X/844/1/012048

Algorithm for the dynamic compensation of Chromatic Dispersion in XGS-PON network architecture

A Trujillo[1], G Gomez[1] and A Zemanate[1].

[1] GNTT – (*Grupo de Nuevas Tecnologías en Telecomunicaciones*) Departament of Telecommunications, Faculty of Electronic Engineering and Telecommunications, University of Cauca, Colombia.

angievti@unicauca.edu.co

Abstract. Nowadays, communication system by 10 gigabit capable symmetric passive optical network (XGS-PON), provide the highest bandwidth and speed of all modern communication networks, however, these networks present difficulties on responding to degradations in transmission in optical fiber communication systems. Effects such as dispersion, attenuation and nonlinear effects lead to the degradation and/or lossing of information in the link. In the case of optical fiber, chromatic dispersion (CD) is the degradation that most affects communication systems, this effect happens because of the pulses emitted by the source are made up of different wavelengths, which in turn travel at different speeds, which generates different arrival times to the receiver, causing a temporary widening of the pulse and giving rise to the emergence of intersymbolic interference [1]. The CD impact is highly significant in the quality of the transmitted optical signal, since it restricts the transmission speed and link length. Therefore, it is necessary to implement compensation techniques that contribute to reduce the degradations caused by chromatic dispersion This article presents the elaboration of a chromatic dispersion compensating module implemented and co-simulated in Matlab of MathWorks and OptSim of Rsoft, in order to mitigate the degrading effects that the CD brings, and in this way to find solutions in the field of digital processing of signals that support the demands of modern architectures.

1. Introduction
The demand by more users and better services, promotes the study of methods that allow a more efficient use of the resources present in the communication systems. In this context, fiber optic is proposed as an alternative in the design of wired networks due to the handling of high data transmission rates, low attenuation and implementation in long distance links [1]. XGS-PON, provides solutions based on the handling of high access capacities, long transmission distances and high bandwidth. However, when an optical signal is transmitted on the fiber, it suffers the effects of linear and non-linear degradations that may significantly affect the performance of the system. The CD is the linear phenomenon that most affects the quality of the signal, which makes required the implementation of advanced compensation techniques in order to counteract this effect.

2. XGS-PON (ITU-T G.9807.2)
The International Telecommunications Union (ITU), defines the G.9807.2 standard, that specifies an XGS-PON type network, where a symmetric speed of 10Gbps is reached, obtaining higher bandwidths. The figure

1, represents an XGS-PON technology where a simultaneity of various types of channels on an only network is admitted, allowing to extend the usability of the network, in terms of speed, capacity and bandwidth, improving the spectral efficiency without needing to intervene in the physical network infrastructure. This implies that more information can be transported on the same bandwidth, obtaining that four optical channels of 10Gbps occupy the same bandwidth that four optical channels of 40Gbps [2].

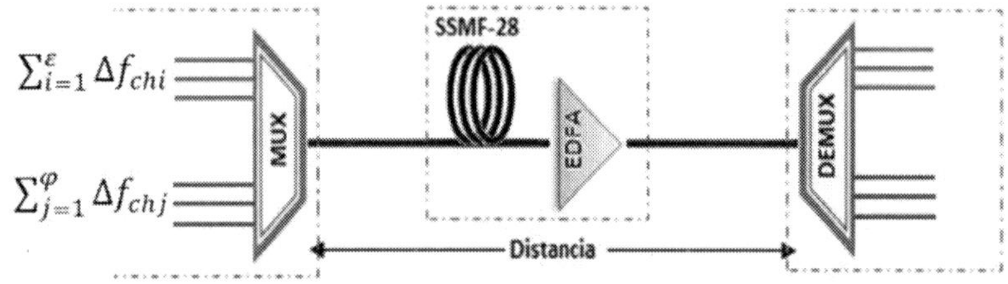

Figure 1. Architecture XGS-PON [2].

Despite the efficiency of this type of networks in the transmission of information, they must face to physical characteristics of the fiber and the quality of the optical components that could eventually be limiting in their performance and implementation costs can also be increased [3].

3. Dispersion

The dispersion defines the effect in which pulses of the signal transmitted through optical fiber travel with different speeds, therefore, the arrival at the receiver is at different times. Aspects as diameter core, refractive index, wavelength, influence in the behavior and state of the transmitted signal, causing the pulse to widen. The effect of the dispersion affects the performance of the optical link implemented, causing the pulses to overlap, thus generating intersymbolic interference (ISI) [4].

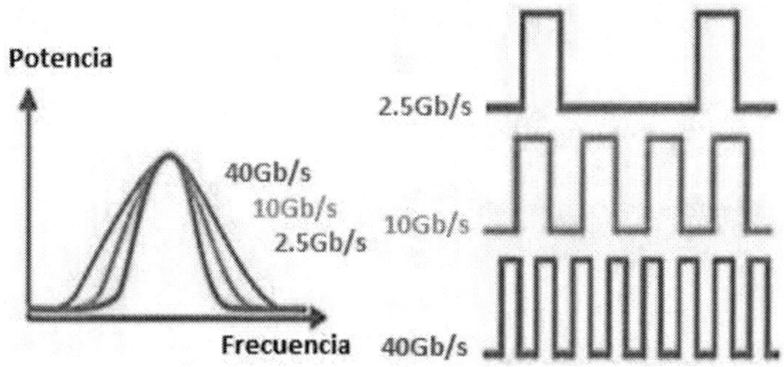

Figure 2. Dispersion at different speeds [5].

The figure 2 shows that the signal is more vulnerable to dispersion if it is transmitted at high speeds, therefore the dispersion is proportional to the length of the link and transmission speed, thus, these aspects are limiting in an optical communication link.

3.1. Chromatic Dispersion.

The CD is caused, mainly, because different spectral components of the transmitted signal are propagated at different speeds over the fiber length, as a result, these components have different times of arrival at the receiver. In figure 3 it can be seen how the pulse width increases and the space between bits is reduced. The result is that in reception it isn't possible to correctly recognize if the data received at a given moment corresponds to a logical value of one or zero [1].

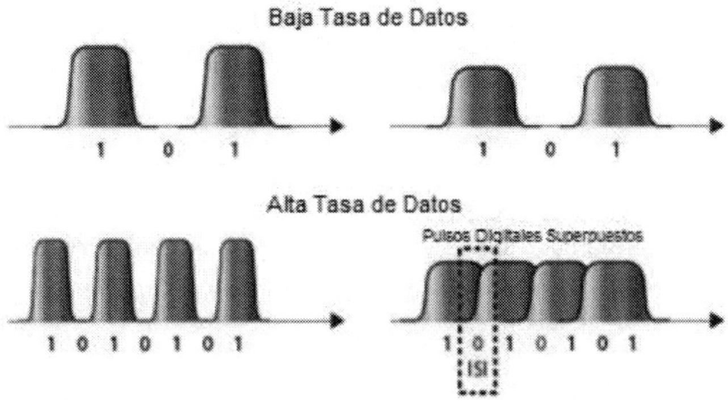

Figure 3. Pulse dispersion due to wavelengths [5].

As consequence of this effect, the pulse in the receiver arrives distorted, causing intersymbolic interference. CD is a cumulative effect, where the longer link length, the greater the total accumulation will be, therefore, the distance that a digital signal can travel before needing regeneration or compensation is limited [6]. Chromatic dispersion occurs due to two components: material dispersion and waveguide dispersion.

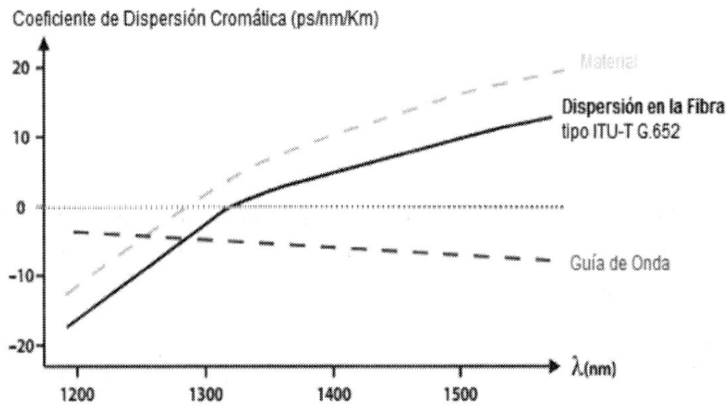

Figure 4. Total CD. Material Dispersion + Waveguide Dispersion [1].

The material dispersion involves the refractive index, which consists in the ratio between the speed of light in the material and the speed of light in vacuum. This type of dispersion is caused by the variation in the refractive index of the material used to make the fibers with respect to the wavelength [4]. Furthermore, the waveguide dispersion depends on several parameters [1] such as: i) the difference of the refractive index between the core and the coating (Δn), ii) reduction in the diameter of the core and, iii) the manufacture of the fiber. As a result, the material dispersion and waveguide dispersion, produce the same effect on the signal transmitted by the fiber, therefore, the concept chromatic dispersion is used in a general way. Figure 4 illustrate how the total CD in a fiber results from the sum of material dispersion and waveguide dispersion.

IOP Conf. Series: Materials Science and Engineering **844** (2020) 012048 doi:10.1088/1757-899X/844/1/012048

3.2. Chromatic Dispersion Compensation.

The impacts of the effects of the CD are highly significant in the quality of the transmitted optical signal, since they restrict the transmission speed and link length. Therefore, in the fiber optic communication links it should be considered the implementation of a compensating module that allows a proper operation of the optical link [7] [8].

3.2.1. DCF: Dispersion Compensating Fiber.

These are fibers that have a very high negative CD value and are capable of compensating the total dispersion in an optical link. The compensation is done by inserting sections of standard transmission fiber with DCF compensating fibers. The length of these sections depends on the total length of the standard fiber link, thus, the effects caused by the cumulative total dispersion are compensated compensated and minimized. The insertion of the DCF fiber in the links can be done in different ways: at the beginning of the link (Pre-compensation), at the end of the link (Post-compensation), or inside of the link (Symmetric compensation) [9] [10].

However, despite the great utility of this type of optical compensator, it has drawbacks [11] since it adds high attenuation (0,6 dB / km of DCF), requiring more complex optical amplifiers, where these can cause additional dispersion by polarization mode.

3.2.2. Electronic compensation.

Electronic Dispersion Compensation (EDC), is composed of techniques that make use of Digital Signal Processing (DSP) and electronic equalization [12], which makes it possible to eliminate the total accumulated CD at the end of the link and allows to reduce installation costs compared to DCFs. The EDC is described in recommendation ITU-T G.959.1 [13], where is contained information about the long distance networks with standard G.652 fibers [14].

4. Characterization of the simulation environment.

The tools used for signal processing and the integration into modules are Matlab [15] and OptSim® [16], respectively. These tools are characterized by their usefulness in integrated development, simulation and modeling environments, providing precision in the obtained results. One of the functionalities that OptSim presents is co-simulation with the Matlab tool. This interface allows OptSim users to run Matlab simulations that interact completely with the OptSim data structure and leads to a joint simulation. The step-by-step tutorial of its interconnection is reported on the article "Methodology of co-simulation environments for Matlab/OptSim tools" [17], where environments that allow the interaction of modules and processes based on algorithms are generated, to increase the collective and individual capacities of the mentioned tools.

For the simulation environment, XGS-PON technology is defined at a distance of 100Km according to the definition of the standard G.691 [3] [18], the optical fiber used along the link is single-mode Corning SMF-28e+ standard, the dispersion coefficient is obtained from the datasheet provided by the manufacturer; being in this case around 17 ps/nm*Km and the attenuation value around 1550 nm corresponds to 0.275 db/Km. The general characteristics of the model are defined in table 1 [14] [19].

Table 1. Characteristics of the simulation scenario.

SECTIONS	DEVICES
TX	Mach-Zehnder Optical modulator Bessel Filter of order 6 Laser CW, 2,5MHz of FWHM, 0dBm 100Ghz channel spacing C Band, arround of 1530nm 2.5Ghz Transmission speed
CHANNEL	ITU-T G.652 standard SSMF-28 fiber EDFA, 15dBm EDFA BOOSTER, 20dBm DCF fiber
RX	-28 dBm sensitivity Bessel electric filter of order 6 Cosine optical filter

5. Algorithm design.

The elaboration of the compensating code starts from the theoretical formulas that generate the dispersion [12], which include waveguide dispersion (D_G) and material dispersion (D_M) defined in equation (1):

$$CD_T = D_G + D_M \tag{1}$$

Initially, the size of the vector coming out of the modulator should be considered, because this will be the signal from OptSim to be treated in Matlab, for this procedure, the Matlab command window is used, since it allows to see values, size and orientation of the vector, in this case 1 x 55296 values corresponding to the size of the vector. The minimum and maximum wavelength value (lambda) expressed in nanometers is taken according to the single-mode fiber working windows recommended by ITU-T G.652 [14]. The values present in figure 5 characterize the optical fiber mathematically, these are necessary in the development of this algorithm.

```
lambda = [1000:0.0081136067708499999999:1997]*1e-9
G=0.6961663; %Coeficientes de Sellmeier para Germanio - Para determinar dispersión de luz en el medio
lambda1=68.4043*1e-9; %Longitudes de Onda para Germanio dopado de Silicio
c = 299792458;
pi = 3.1415926;
a = 4.1e-6; %Radio del núcleo
delta = 0.003; %Diferencia de índice entre el núcleo y el revestimiento.
n1=1.3024 %indice refractivo en el núcleo
```

Figure 5. Parameters that characterize the fiber [12].

5.1 Material dispersion calculation.

The calculation concerning to the material dispersion; equation (2), is determined in terms of the wavelength vector (λ) of size 1x55296 and the vacuum speed of light (c), which for conversions between frequency and wavelength must be equal to 2,99792458*10^8 m/s, according to ITU G.694.1 [12]:

$$D_M = -\frac{\lambda}{c} * \frac{d^2n}{d\lambda^2} \tag{2}$$

The material dispersion occurs due to the variation in the refractive index, depending on the wavelength [12]; this is equation (3):

$$\frac{d^2n}{d\lambda^2} = \frac{d^2\sqrt{1+\left(\frac{G*\lambda^2}{\lambda^2-\lambda_1{}^2}\right)}}{d\lambda^2} \tag{3}$$

Figure 7 contains the total material dispersion obtained, getting the expected behavior according to [1].

```
Matdisp = -(lambda./c).*p1
%Conversión a ps/nm.Km
Matdisp = Matdisp*1e-3
B=Matdisp.';
figure(1)
clf
hold
xlabel('nm')
ylabel('ps/nm.km')
title('Dispersión de Material')
plot(lambda, Matdisp, '.-')
grid on
```

Figure 6. Total Material Dispersión Calculation.

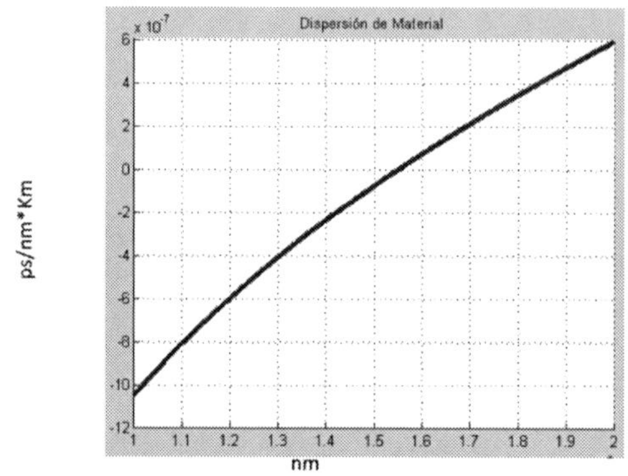

Figure 7. Material Dispersion.

5.2. Waveguide dispersion calculation.

According to **[12]**, considering that for Corning smf-28e + single mode optical fiber, for the respective simulations, it is determined how many propagation modes the fiber supports, it must be less than 2.5, to ensure that the fiber propagates a single mode, however, the line of code necessary for multimode fibers has been commented, for general interest. In order to facilitate and guarantee a correct obtaining of calculations in the Matlab tool, it is decided to split the total formula in two [12]; remaining as evidenced in equations (4) to (6) and in figure 8.

$$D_{\lambda 1} = -\frac{\sqrt{1+\left(\frac{G*\lambda^2}{\lambda^2-\lambda_1{}^2}\right)}*\Delta}{c*\lambda} \tag{4}$$

$$D_{\lambda 2} = 0.80 + 0.549 * (2.83 - V)\hat{}2 \tag{5}$$

Now, the calculation of the total waveguide dispersion is performed, resulting

$$D_G = D_{\lambda 1} * D_{\lambda 2} \tag{6}$$

```
    % % ------------------------------------
    % % Cálculo de la Dispersión de Guía de Onda
%V = (2 * pi * a * n1 * sqrt(2 * delta)) % Frecuencia Normalizada.Determina cuantos modos soporta
Dlambda1 = -(noflambda .* delta)./(c * lambda);
Dlambda2 = 0.080 + 0.549 * (2.834 - 2.3).^2;
Dlambda = Dlambda1 .* Dlambda2;
%Conversión a ps/nm.Km
Dlambda = Dlambda*1e-3;
C=Dlambda.';
figure(2)
plot(lambda,Dlambda, '-')
title('Dispersión de Guía de Onda')
```

Figure 8. Total waveguide dispersion calculation.

Figure 9 shows the result of this dispersion.

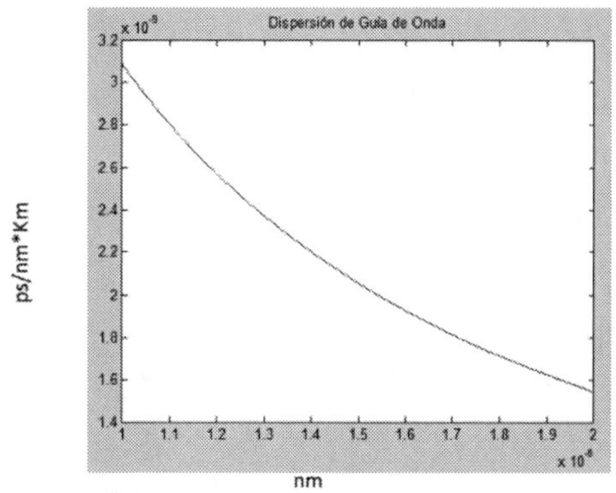

Figure 9. Waveguide dispersion.

5.3. Total Dispersion Calculation (Chromatic Dispersion)

Having the two types of dispersions that compose the chromatic dispersion, the total dispersion value is obtained. It can be seen how the slopes of the mentioned dispersions complement each other, adding point to point and thus generating; the total dispersion, as shown in figure 10.

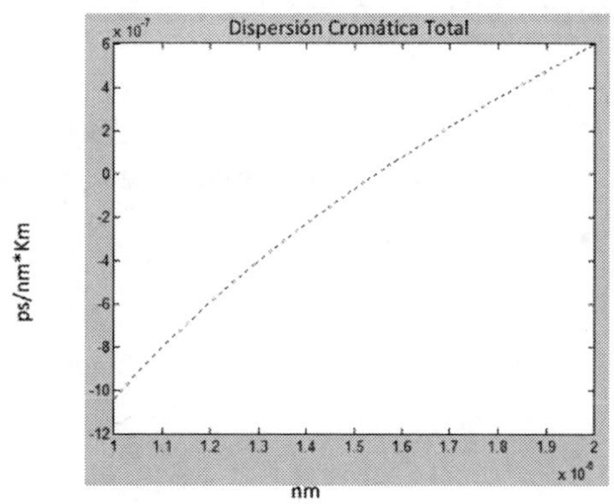

Figure 10. Total Dispersion. (CD).

An important and fundamental part for the development of the work, and that allows to reach the objective of reducing the CD immersed in the link, is presented in figure 11, in the framed code line. When changing the slope of the total dispersion obtained, dispersion is subtracted from the signal from the OptSim® modulator, that is that a dispersion is being generated that is counteracts point to point to the accumulated in the simulation link. Figure 12 shows the result of this change of slope.

```
    % % ------------------------------
    % % Cálculo de la Dispersión Total
TotDisp = B+C;
TotDisp=TotDisp*-1;
figure (3)
plot(lambda, TotDisp3, ':')
```

Figure 11. Change of slope of total dispersion obtained.

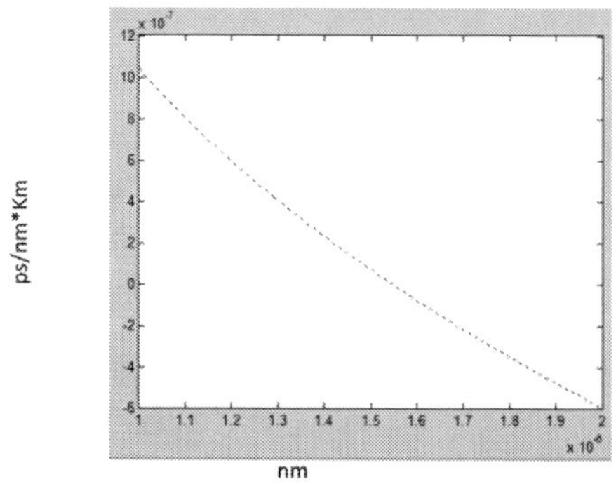

Figure 12. Total dispersion with negative slope.

Through the development and implementation of the chromatic dispersion compensating algorithm, it was found that it is possible to modify some configuration variables, which leads to the generation of different degrees of liberty, focused on the control of the core radius, the refractive index and the working wavelength, thus allowing to change the dispersion slope to achieve different compensation percentages in the digital treatment of the signal, without any need to make physical changes in the communication link.

6. Implementation and analysis

The compensations are usually made in the form of: zero, normal, anomalous, but due to the characteristics and objectives of the present work, it is carried out in a normal and anomalous way, with percentages of 25, 50, 75 and 100% to perform the respective analysis, and thus generate the relevant case studies.

As presented in Figure 13, and as mentioned in the development of the algorithm, the processing of the signal that is processed in the designed compensator module is carried out in the electrical domain, and for this reason requires prior information of the accumulation of the dispersion and the effects that it produces in the output signal generated by the transmitter, allowing the implementation only of the post compensated mode, due to the high information processing limitations of the electro-optical conversion modules of the OptSim tool, generated by the high data transmission speeds, however, it is possible to implement higher processing capacity equipment in the future, which will mitigate these limitations. A back to back link is generated, in order to obtain ideal values of optical monitoring parameters due to the absence of elements that interfere in the course of the signal. The above is evidenced in table 2.

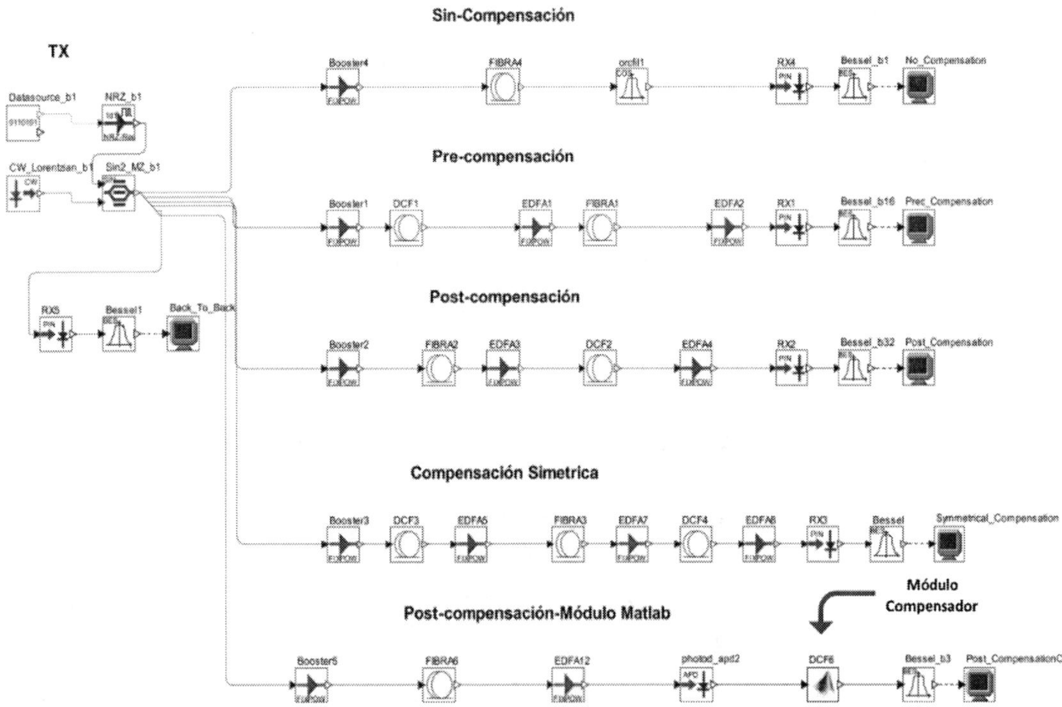

Figure 13. Simulation scenario.

Table 2. Optical monitoring parameters dependent on system degradations. Back to back link.

Distance [Km]	BER	Q Factor [dB]	Jitter[ns]
100	1e-40	40	0.0170

In the same way, a link without compensation is implemented, referring to the case in which the worst OPM values would be obtained, given that there are not devices that recover the signal that arrives at reception. The above is corroborated in table 3.

Table 3. Optical monitoring parameters dependent on system degradations. Link without compensation.

Distance [Km]	BER	Q Factor [dB]	Jitter[ns]
100	4.35 e-6	13.091	0.0282

From tables 2 and 3, the significant difference in OPM values can be seen, noting how the network needs regeneration of the transmitted signal. However, in terms of compensation percentages, the method used by the module consists in the variation of a multiplier factor (Matlab®) that interacts with the data vector that represents the input signal in terms of the delay and widening of the pulse introduced by the medium due to the CD accumulated in the system (OptSim®). This factor is related to the possibility of delaying the signal (jitter) allowing to generate a decision interval, for a minimum delay compensating in a 100% the dispersion of the input signal; multiplier factor of 1; and a null compensation for a multiplier factor of 35.

IOP Conf. Series: Materials Science and Engineering **844** (2020) 012048 doi:10.1088/1757-899X/844/1/012048

```
ele_out=ele_in +26.25*(TotDisp);
```

Figure 14. Configuration factor for 25%.

```
ele_out=ele_in +17.5*(TotDisp);
```

Figure 15. Configuration factor for 50%.

```
ele_out=ele_in +8.75*(TotDisp);
```

Figure 16. Configuration factor for 75%.

```
ele_out=ele_in +(TotDisp);
```

Figure 17. Configuration factor for 100%.

When implementing the simulation scenario with each configuration factor, it was analyzed that only for a 100% compensation percentage, the OPM values were within the ranges established by the ITU; Due to this, the results presented below correspond to the data thrown by the communication link, compensated to 100%.

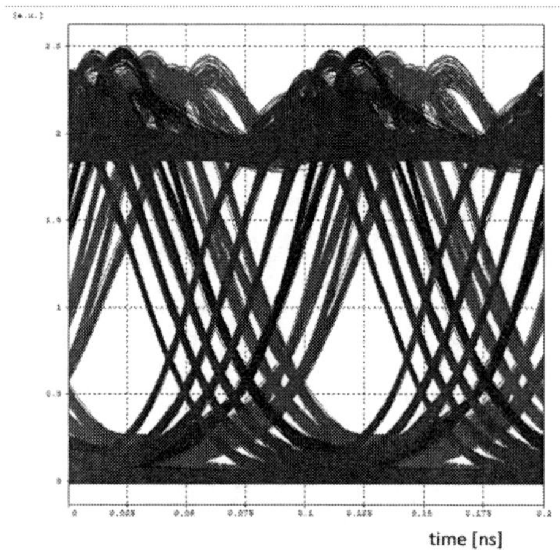

Figure 18. Pulse shift given increased CD compensated system Matlab module.

In order to analyze the behavior of the CD and the vulnerability of the module against it, it is decided to see the shift of the pulses in reception. It is necessary to see the eye diagrams that are generated once the CD increases, therefore, the CD is used as a variable to modify, this results are shown in the figure 18.

17 different CD values have been generated, starting at 100 ps/nm*Km and ending at 1700 ps/nm*Km, with jumps of 100 ps/nm*Km. This is because the maximum accumulated CD value in a 100Km link with SSMF-28 CD fiber = 17 ps/nm*Km is 1700 ps/nm*Km. Figure 18 illustrates the pulse shift once the CD increases, indicating that the expected jitter value in reception will increase. This is done in order to show how the CD increases directly proportional to the increase in link length. To avoid redundancy, the 17 cases are not presented, but only those that generate significant analyzes. These are presented below:

- 100 ps/nm*Km

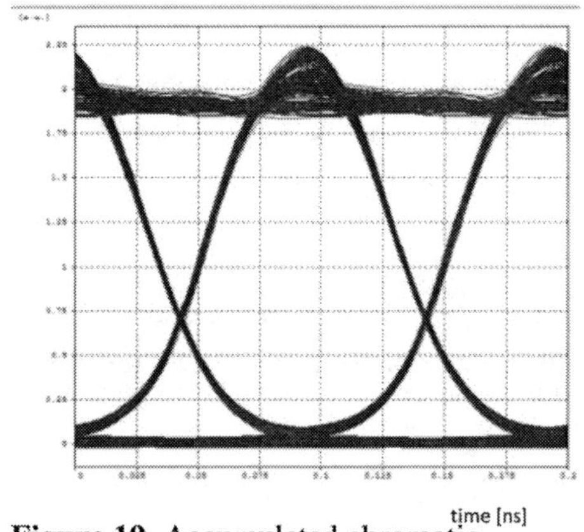

Figure 19. Accumulated chromatic dispersion of 100 ps/nm*Km.

Since it is known that the total accumulated CD is 1700 ps / nm * Km for a total length of 100Km, by equivalence it has to be 100 ps / nm * Km accumulated, they correspond to a length of 5.88Km of fiber. The parameters obtained from this configuration are presented below:

Table 4. Optical monitoring parameters dependent on system degradations. Accumulated link of 100 ps/nm*Km.

Distance [Km]	BER	Q Factor [dB]	Jitter[ns]
5.88	1e-40	27.81	0.0167

As evidenced in Figure 19 and Table 4, the accumulated CD does not affect the performance of the link, therefore, the CD value is increased, in order to determine the limiting CD for the generated module.

- 1200 ps/nm*Km

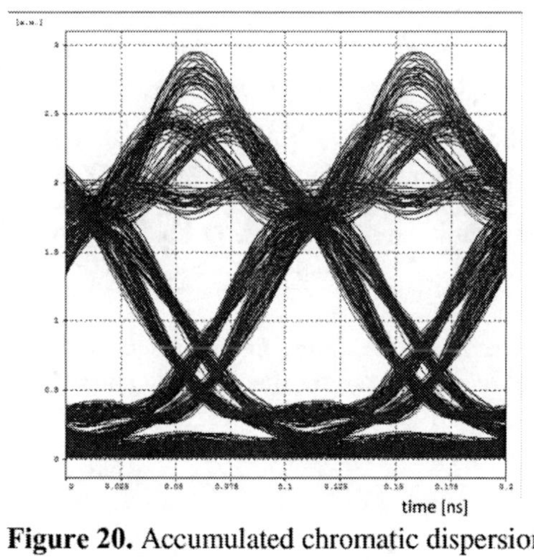

Figure 20. Accumulated chromatic dispersion of 1200 ps/nm*Km.

IOP Conf. Series: Materials Science and Engineering **844** (2020) 012048 doi:10.1088/1757-899X/844/1/012048

In the same way, by equivalence, for 1200 ps/nm*Km, 70.58 Km are necessary. The parameters obtained from this configuration are presented below:

Table 5. Optical monitoring parameters dependent on system degradations. Accumulated link of 1200 ps/nm*Km.

Distance [Km]	BER	Q Factor [dB]	Jitter[ns]
70.58	6.37e-18	18.74	0.0275

The change in the parameters of optical monitoring, indicates how to increasing the length of the link, causes more CD to be generated, causing the quality and performance of the link to decrease considerably. Since the BER still complies with the standards established by the ITU, the accumulated CD is increased again, in order to find the limitation.

- 1300 ps/nm*Km

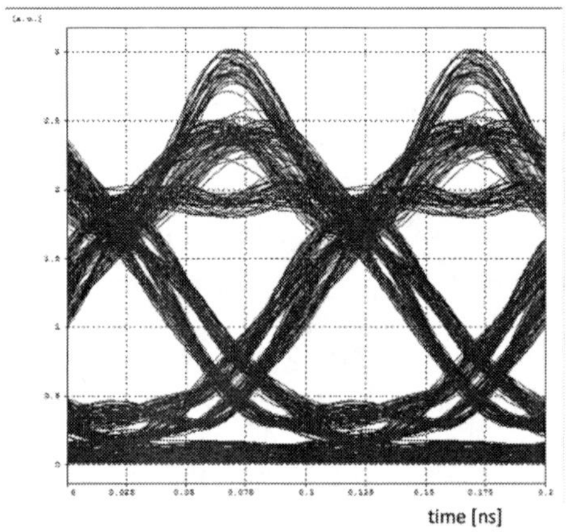

Figure 21. Accumulated chromatic dispersion of 1300 ps/nm*Km.

Therefore, 1300 ps / nm * Km accumulated, correspond to a length of 76.47Km of fiber. The parameters obtained from this configuration are presented below:

Table 6. Optical monitoring parameters dependent on system degradations. Accumulated link of 1300 ps/nm*Km.

Distance [Km]	BER	Q Factor [dB]	Jitter[ns]
76.47	4.85e-12	16.53	0.0279

Finally, it has been possible to obtain the minimum value accepted by the ITU in terms of BER, that is, until this link length is capable to compensate for the module generated in the so-simulation environment, complying with international standards. By increasing the distance, the jitter value is increased, however, lower values are achieved to the link without compensation, thus confirming that the compensator is doing its job.

- 1700 ps/nm*Km

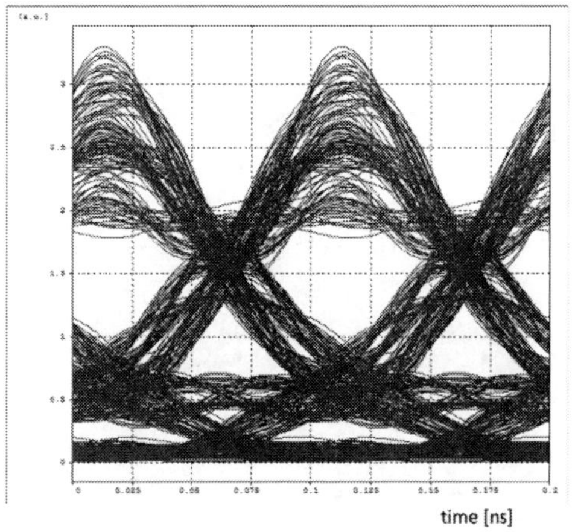

Figure 22. Accumulated chromatic dispersion of 1700 ps/nm*Km.

Table 7. Optical monitoring parameters dependent on system degradations. Accumulated link of 1700 ps/nm*Km.

Distance [Km]	BER	Q Factor [dB]	Jitter[ns]
100	¡Error!	11.49	0.0285

As shown in Table 7, this configuration is not acceptable, because the quality of the signal that reaches the receiver is poor. Due to this analysis, and in order to check the maximum length of the link obtained, it is decided to implement the link with 76Km of fiber, where it was obtained (see figure 23).

Given the above, it is concluded that the module designed in the present work, at a speed of 10Gbps, works in systems in which its maximum link length is 76Km; being this the limitation of this compensating module; otherwise, the system would not comply with international standards. Below are the values that were generated from the 17 CD variations, which were mentioned above. Table 8 contains this information.

Case studies were generated around the behavior of the DCF immersed in the communication link, this in order to obtain a base behavior of the compensated system and thus be able to generate a comparative analysis of the behavior of the compensated link by means of DCF and designed module. The information obtained from the above is shown in table 9.

Table 8. Results Matlab module configured to 100%.

CD [ps/nm*Km]	Jitter [ns]	BER	Q Factor [dB]
100	0.0167	1e-40	27.81
200	0.0183	1e-40	24.44
300	0.0169	4.10e-39	22.51
400	0.0143	9.50e-23	19.92
500	0.0152	2.27e-18	18.75
600	0.0156	2.31e-19	19.17
700	0.0159	5.44e-22	19.88
800	0.0171	1.29e-32	21.46
900	0.0210	1.49e-35	22.04
1000	0.0244	2.58e-29	21.10
1100	0.0271	2.19e-25	20.49
1200	0.0275	6.37e-18	18.74
1300	0.0279	4.85e-12	16.53
1400	0.0280	1.21e-7	14.58
1500	0.0281	2.57e-6	13.17
1600	0.0281	2.89e-5	12.10
1700	0.0285	¡Error!	11.49

Table 9. Abstract table.

<table>
<tr><th colspan="3">DCF - COMPENSATED 100%</th></tr>
<tr><th>POST</th><th>PRE</th><th>SYMMETRICAL</th></tr>
<tr><td>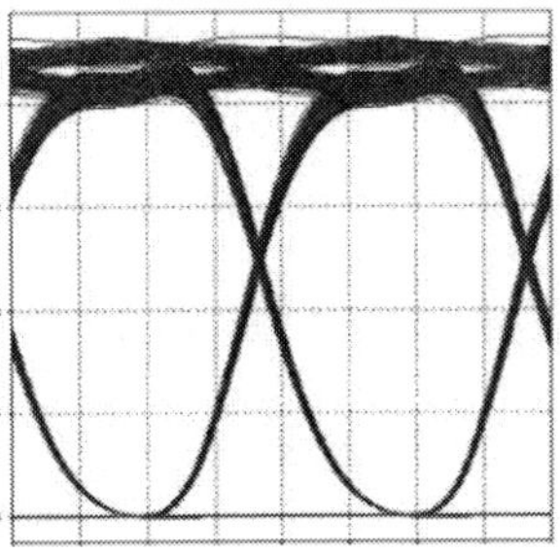</td><td>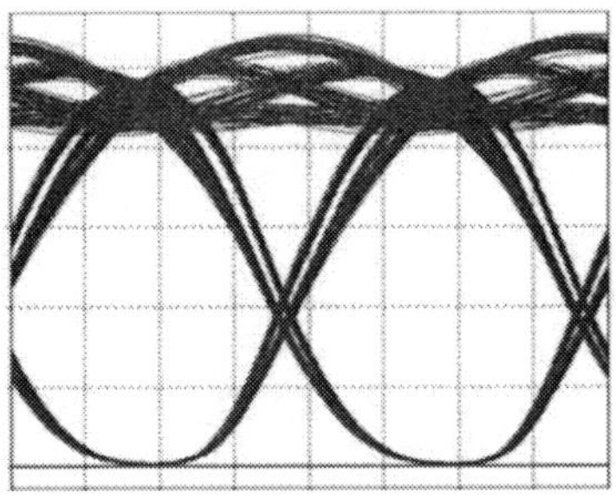</td><td>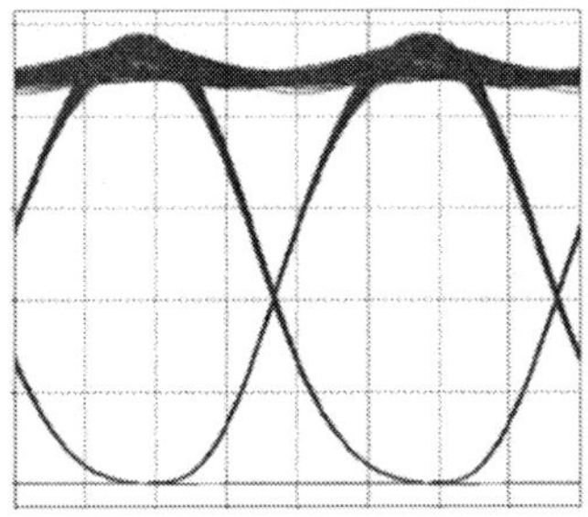</td></tr>
<tr>
<td><table><tr><th>BER</th><th>Factor Q [dB]</th><th>Jitter [ns]</th></tr><tr><td>1e-40</td><td>28.34</td><td>0.0180</td></tr></table></td>
<td><table><tr><th>BER</th><th>Factor Q [dB]</th><th>Jitter [ns]</th></tr><tr><td>1e-40</td><td>27.89</td><td>0.0257</td></tr></table></td>
<td><table><tr><th>BER</th><th>Factor Q [dB]</th><th>Jitter [ns]</th></tr><tr><td>1e-40</td><td>32.55</td><td>0.0093</td></tr></table></td>
</tr>
<tr><th colspan="3">MATLAB MODULE – COMPENSATED 100%</th></tr>
<tr><th>100ps/nm*Km</th><th>1300ps/nm*Km</th><th>1700ps/nm*Km</th></tr>
<tr><td>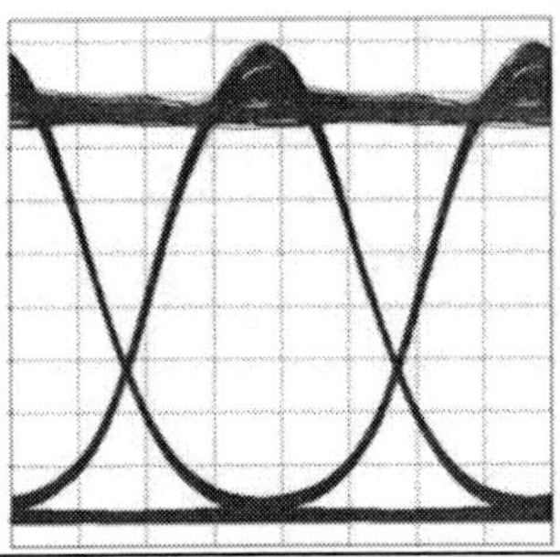</td><td>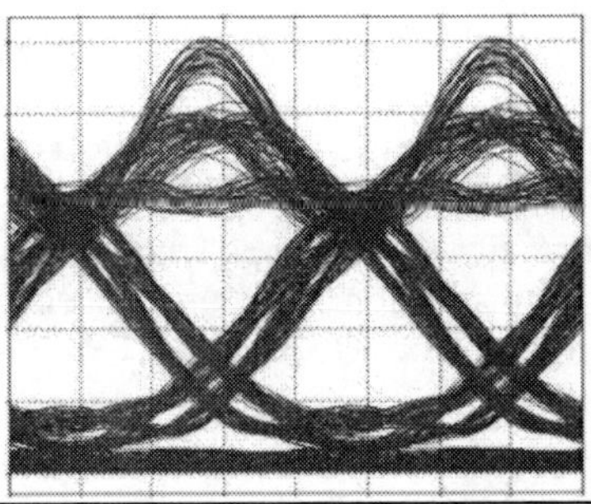</td><td>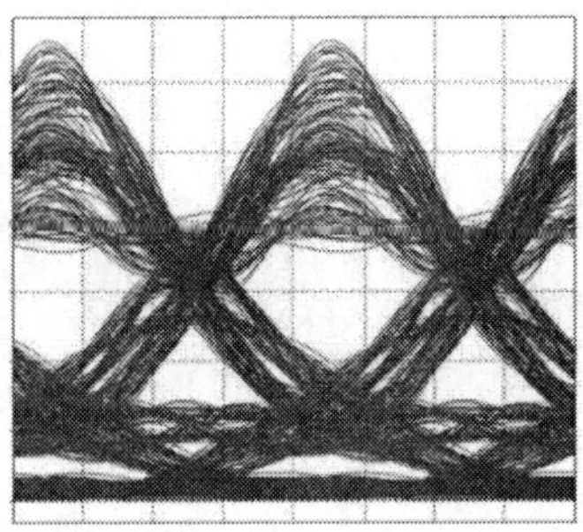</td></tr>
<tr>
<td><table><tr><th>Distancia [Km]</th><th>BER</th><th>Factor Q [dB]</th><th>Jitter [ns]</th></tr><tr><td>5.88</td><td>1e-40</td><td>27.81</td><td>0.0167</td></tr></table></td>
<td><table><tr><th>Distancia [Km]</th><th>BER</th><th>Factor Q [dB]</th><th>Jitter [ns]</th></tr><tr><td>76.47</td><td>4.85e-12</td><td>16.53</td><td>0.0279</td></tr></table></td>
<td><table><tr><th>Distancia [Km]</th><th>BER</th><th>Factor Q [dB]</th><th>Jitter [ns]</th></tr><tr><td>100</td><td>✖</td><td>11.49</td><td>0.0285</td></tr></table></td>
</tr>
</table>

7. Conclusions

Through the realization of this article, it was determined the incidence that the linear and non-linear effects generate in relation to the increase of the speed of data transmission was determined, observing the maximum operating limitation for the designed algorithm, for a compensation of 100% of the CD and lower propagation distances. In the same way, by simulating the different scenarios, it was identified that additional amplification stages are required so that the designed modules can work within the target parameters. A dynamic compensation of the Chromatic Dispersion was achieved in an XGS-PON network architecture.

References

[1] Collings B, Heismann F and Lietaert G 2010, Reference Guide to Fiber Optic Testing, *JDS Uniphase Corporation* **2**.
[2] Gómez G 2019, Eficiencia espectral de formatos de modulación avanzados en redes XGS-PON, *Visión Electrónica, ISSN: 1909-9746.*
[3] Herrera J and Toledo J 2016, Análisis comparativo del desempeño de los formatos de modulación RZ-DQPSK y RZ-PDPSK frente a técnicas de compensación ópticas de la dispersión cromática en redes WDM a 10Gbps, *Universidad del Cauca*, Colombia.
[4] Pawan D and Vibha S 2014, Dispersion in Optical Fiber Communication, *International Journal of Science and Research* (*IJSR*), ISSN (Online): 2319-7064 **3** Issue 10.
[5] Stepanek L 2012, Chromatic dispersion in optical communications, *Journal of Electrical and Electronic Engineering* **7**, No. 2, 142–151.
[6] Shevgaonkar R 2005, Fiber Optics, *Dept. of Electrical Engineering, Indian Institute of Technology*, Bombay.
[7] Ramaswami R, Sivarajan K and Sasaki G 2009, Optical networks: a practical perspective.
[8] Sethi R and Goel A 2015, Dispersion compensation in optical communication system by employing 16-QAM modulation using OFDM, *IMPACT: International Journal of Research in Engineering & Technology (IMPACT: IJRET)* ISSN (E): 2321-8843; ISSN (P): 2347-4599 *3, Issue 2.*
[9] Florez J 2016, Compensación de los Efectos de Propagación en Enlaces Ópticos, *Universidad Pontificia Bolivariana*, Medellín.
[10] Agrawal P 2012, Fiber-optic communication systems, *John Wiley & Sons* **222**.
[11] Kahlon N and Kaur G 2014, Various dispersion compensation techniques for optical system, open journal of communications and software **1**, Number 1.
[12] Binh L 2009, Optical Fiber Communication Systems, *CRC* **Press 1**.
[13] ITU-T Rec G.959.1 2012, Optical transport network physical layer interfaces, Telecommunication Standardization Sector of ITU.
[14] ITU-T Rec G.652 2016, Characteristics of a single-mode optical fiber and cable, Telecommunication Standardization Sector of ITU.
[15] Matlab 2014, User Guide, *Mathworks.*
[16] RSoft, Optsim user guide physical layer division 400 executive boulevard, *Suite 100 Ossining,* NY 10562.
[17] Trujillo A, Zemanate A, Gómez G and Toledo A 2018, Metodología de entornos de co-simulación para las herramientas Matlab/OptSim, *Universidad del Cauca.*
[18] Kartalopoulos S 2000, Introduction to DWDM technology: data in a rainbow, *SPIE Optical Engineering Press.*

[19] UIT-T Rec. G.691 2006, Interfaces ópticas para los sistemas monocanal STM-64 y otros sistemas de la jerarquía digital síncrona con amplificadores ópticos, *Telecommunication Standardization Sector of ITU*.

IOP Conf. Series: Materials Science and Engineering **844** (2020) 012049 doi:10.1088/1757-899X/844/1/012049

Business model for the optimization of energy consumption by VAC equipment in service sector buildings

Y Gómez-Charris[1], S Umaña-Ibáñez[1,], J Silva-Ortega[1,2], J Caicedo-Ortiz[1,3] and G Camargo-Acuña[1,2]

[1] Industrial, Agro-Industrial and Operations Management Department. Universidad de la costa CUC. Calle 58 N°55-66, Colombia
[2] Energy Department
[3] Computer Science and Electronics Department
[4] Economic Sciences Department

Abstract. Electricity consumption has been growing in recent decades, and it is estimated that air conditioning accounts for 30% of a building's total electricity consumption. However, in hot climates this percentage may be even higher, which has led to the development of energy management techniques aimed at reducing the high electric bills paid by end users. This article proposes a business model for the commercialization of intelligent solutions for the optimization of electricity consumption by air conditioning equipment (VAC) in service sector buildings. The study used the Canvas methodology to validate the proposed model, in combination with other techniques such as SCAMPER, interviews, and discussion groups, among others, to validate the proposed business model aimed at the introduction of new techniques for the commercialization of air conditioning. Service-sector buildings exposed to the climate conditions of the Colombian Caribbean region were used as case study. The results summarize the characteristics of the business model, including the respective validations of the nine "Canvas" modules, which enabled reviewing two possible scenarios offering different alternatives for implementation of this initiative.

1. Introduction

Research has found that energy consumption is primarily concentrated in three sectors: industry, transportation and others (agriculture, services and residential). Buildings account for between 20 and 40% of total final energy consumption [1,2]. Authors such as Lemmet [3], Lin y Liu (2015) believe this percentage may be higher, at between 40 and 50% of global energy consumption in buildings due to heating, ventilation and air conditioning (HVAC) systems. Additionally, the electricity consumed by buildings for the operation of ventilation and air conditioning system is close to 60% [4].

The energy policies of several countries take into consideration the development of energy efficiency and savings strategies, due to the high levels of energy consumption and of CO^2 emissions [2]. The impact is two-fold: on the economy and on the environment. Financially, it involves high costs of facilities, of equipment maintenance and high energy consumption, which translates into higher costs. Environmentally, CO^2 emissions have been found to produce the greenhouse effect and the consequent global warming [5]. In Spain, according to the Ministry of Industry, energy consumption by air conditioning systems accounts for approximately 2.5% of the country's energy consumption. Efforts to reduce this consumption level have focused on three techniques: the reduction of demand for heat by buildings, energy recovery, and more efficient use of air conditioning systems [6].

In the United States, the three main energy consumption factors in the retail industry are heating, lighting and air conditioning of stores, which account for close of half of energy consumption at the buildings [7]; in the residential sector, heating and air conditioning account for 54% of energy consumption by households [8].

In Colombia, the residential sector accounts for approximately 20% of the country's final energy consumption. Some of the major sources of energy inefficiency in this sector have been found to be: the high use of electricity for air conditioning, due to the use of energy-intensive equipment due to factors such as technology (low energy efficiency), insufficient maintenance and age [9]. According to a study carried out by UPME in 2013, the business, public and services sectors account for close to 6% of total energy demand in the country [9].

As part of global initiatives aimed at addressing the above, the Kyoto Protocol includes agreed commitments by participating nations to reduce greenhouse effect gases [10]. However, such commitments are often merely rhetorical, and in practice few actions have been implemented in this regard due to the non-binding nature of the signed agreements, which reduces it to a compendium of good intentions. For this reason it is important to view new technologies not only positive for the planet, but also beneficial for the finances of the business sector, by generating new business ideas and strategies to commercialize new products focusing on the reduction of energy consumptions, without loss of efficiency or quality of living [11].

From an energy perspective, the main option for addressing these challenges and the alternative for reducing the propagation of CO^2 emissions are energy products and services that consume less energy. In the climatization and air conditioning industry, in response to these energetic and environmental issues, in recent years VRF (Variable Refrigerant Flow) equipment has been installed in different types of building, including residential and business units, because of their high energy saving potential [12]. This system offers significant advantages, such as: low-noise operation, use of integrated controls, design and installation of compact units, flexibility, lower life cycle and maintenance costs [13].

Additionally, it has been found that an excellent complement for cooling systems are non-invasive monitoring systems incorporated into HVAC (Heating, Ventilation and Air Conditioning) equipment, which produce consumption statistics in real time, thereby providing data for energy savings decision-making. A recent trend in HVAC systems is the development of models that predict energy consumption in different types of buildings [3], [14–16]. A noteworthy example is the work performed by the firm *Johnson Controls* at the El Dorado International Airport, which was contracted to provide HVAC equipment through the implementation of a technology contracting model that reduces high front-end costs, eliminates the duplication of systems and infrastructure, assures quality and manages innovation, while at the same time maintaining and preserving integration between the various subsystems and technologies used at the airport [17]. **¡Error! No se encuentra el origen de la referencia.** displays business models that take into consideration the inclusion of various technologies to reduce energy consumption in buildings, incorporating monitoring, energy management and energy rationalization systems that affect cooling and heating consumption.

Table 1. Business Strategies for Reducing Energy Consumption.

Country	Reference	Business strategy	Contribution
United States	[18]	*Power Purchase Agreements* (PPAs).	PPAs are performance-based contracts whose purpose is to achieve fair payment for the purchase and sale of energy between the public utilities company (the buyer) and a generator (the seller). Use of this strategy has been increasing worldwide and is commonly used in Europe, United States and Latin America.
Italy	[19]	*Vertical Greening System* (VGS).	VGS produces a favorable impact in terms of thermal performance. Studies have demonstrated potential energy savings on air conditioning of up to 40-60% in the Mediterranean region.
Iran	[20]	*Green Roof*	Green Roof's main advantage is that it reduces a building's energy consumption costs on cooling and heating by increasing the thickness of the roof's insulation, providing natural shade against direct sunlight. Green Roofs are most suitable for the European and North American regions.
China	[21]	*Smart Home System*	The intelligent home system enables more systematic management of household appliances and facilitates energy management at the home. It provides information in real time of the home's energy consumption.
United States	[22]	*Demand Response* (DR)	Clients hire them to reduce demand during emergency periods. The final use of energy in the residential and business sectors is allocated to cooling, heating and ventilation, among others.

In this study, based on the objective of designing, developing and validating a Business Model for a smart solution for the optimization of energy consumption by VAC (Ventilation and Air Conditioning) in service-sector buildings, the Canvas method was used to develop a business model. The selected object of study was service-sector buildings in the city of Barranquilla; additionally, interviews were also carried out in the cities of Bucaramanga and Cartagena.

2. Description of the canvas model.

Market Segment: In this module we define "our customers", i.e., the various groups of people or companies to which the product or service is to be sold. To this end, [23] consider that market segmentation is appropriate because each customer group has differences in terms of:

- Their specific needs;
- The distribution channel required to reach them;
- Profitability index;
- Willingness to pay for different aspects of the offering.

The market segment was defined as service-sector buildings that use VAC equipment. In order to understand the potential market, a search was performed on the sectors that account for the largest shares of energy consumption. The standard categories for energy consumption are three sectors: industry, transportation and others (agriculture, services and residential). Buildings account for between 20 and 40% of total final energy consumption, and its main uses are: lighting, use of electrical appliances, water heating, building heating, air conditioning, etc. [24]

Value proposition: The initial hypothesis was pivoted through interviews and the development of empathy maps between the customer segments and the value proposition, in order to identify the customers' real and potential needs, i.e. the real "pain", which is associated primarily with "the cost of energy consumption associated with the use of air conditioners."

In order to take advantage of the opportunities identified in the market niche, a value proposition was formulated [23], adjusted to the needs and obstacles they currently face, by proposing modern technologies and innovative solutions that produce benefits and gains for potential customers:

"Intelligent solution to achieve savings in electricity consumption by air conditioning equipment".

Channels: In developing a business model, definition of the distribution channel is particularly important after having defined the market segment and the value proposition, because of the benefits provided to customers in terms of time savings for not having to travel long distances to find the product or service they require; and for setting prices for the products or services based on the cost of distributing products in the market [25].

The shorter it is, the more it can be controlled, even though coverage will be limited, and the price will be higher; instead, a longer channel offers less control because there is greater product coverage, which in turn has lower costs [26].

Customer relations: In the business world, customer relations are of utmost importance and are often the company's most valuable asset; the business does not depend on the company, but quite the contrary, it depends on the customers, because they are the ones who produce revenues for the organization. But it is essential to identify customers as humans, with both reason and feelings, who understand, have emotions, preferences and tastes, among others [27].

Due to the above, understanding the customer enables focusing the company's actions on satisfying the customers' needs through strategies that help build sustainable relations over time and increase their loyalty in favor of buying the company's products and/or services. However, achieving this is not an easy task, given the ever-changing characteristics of markets, which make customers increasingly demanding; new technologies, innovation and competitiveness lead customers to search for products with added value.

Revenue sources: In this module, the cash flow generated by the company is analyzed for each segment. The company needs to ask itself: how much is each market segment willing to pay? It is important to understand the various price-setting mechanisms [23]. According to Osterwalder and Pigneur [23], most business models have two different types of revenues: Revenues from individual transaction payments and recurring revenues of periodic payments made in exchange for the supply of a value proposition or customer care post-sales services. In this study, validation was performed based on forecast demand for the services in the market, interviews and trend analysis of VAC equipment sales and their energy consumption.

Key results: All business models require key resources that enable companies to create and offer a value proposition, reach markets, establish relations with market segments and earn revenues. Each business model requires different key resources. A microchip manufacturer requires capital-intensive production facilities, whereas a microchip designer depends more on human resources [24], [28-30].

These resources are directly associated with the key activities, because their transformation into activities is what enables the delivery of the company's value propositions. In the Osterwalder business model, the key resources are classified as physical, intellectual, human and financial. Additionally, the company may either own them, lease them or obtain them from its key partners [23], [31].

Key activities: In a business model, an activity can be defined as a commitment of human, physical and capital resources to be used to fulfill a given purpose that leads to the achievement of an overarching goal [32], through the company's core activities, i.e. all the things a business does to achieve customer satisfaction. A business model must clearly define this module, because all the efforts carried out are aimed at achieving the company's corporate purpose. Often, when an inventory is made of all the potential activities that a business model can have, it becomes conceptually challenging, because the number of activities is often very large, and their technical or strategic identification is complex [33].

The activities associated with the business model developed in this project involve resolving issues and the platform/network. This is because the nature of the value proposition defined as "Intelligent solution to achieve savings in electricity consumption by air conditioning equipment" focuses on solving a problem for potential customers, which requires having a service platform available that offers an attractive value proposition based on the use of a device that monitors different variables involved in electricity consumptions by VAC systems in order to achieve costs savings in electricity consumption by air conditioners. This device operates through an Arduin platform, which is considered a disadvantage for the model, given the scalability of its main functionality.

Key partnerships: This module presents a company's relationships in terms of strategic suppliers and partners that have direct influence on the operation of our business model. It is important to understand the partnerships to be worked with; an entrepreneur must be aware that in the business context he/she must work with others in order to publicize the business. Companies partner up for a variety of reasons; a widely used technique is networking. "It is difficult to find a general definition for networking; however, this technique, strategy or discipline points to the same objective: increasing the network of professional contacts" [34]. Effective networking requires working on establishing and maintaining relationships. The key partnership considered in a business model is that with the suppliers of payment and technology [35].

Cost structure: This module describes all the costs incurred in working with a given business model, both for creation and delivery of value, once a clear picture is available of the key resources, partnerships and activities [23]. Two scenarios were considered in this study: The creation of a company that provides energy consumption monitoring and control services, or the development and sale of a device that enables monitoring a building's energy consumption. In other words, either the total or partial transfer of the rights to use the system's intellectual property. In this module, the value are in Colombian pesos.

3. Methods

In the development of this business model and the validation of its relevance as an intelligent solution to optimize energy consumption in VAC equipment, this study was structured in three stages: the first was a diagnosis of the target audience, during which characterization and segmentation of the market was performed in terms of existing needs for the use and management of technological climatization alternatives. The second stage involved designing the operating structure of the model to be used, product commercialization and generation of the value chain. Lastly, the model was validated in order to assess its relevance.

3.1. Market characterization and segmentation

The proposed business model used as case study companies of the service sector located in the city of Barranquilla, to which end a specific query was made on the companies registered under the "Uniform International Industrial Classification of all economic activities adapted to Colombia" (CIIU, by its acronym in Spanish). Based on this query, 4 companies were selected from Barranquilla and the department of Atlántico, where there are 49,696 registered companies, approximately 40% of which engage in activities closely related to the tertiary (service) sector. However, for the effects of this project, only companies grouped under the sections displayed in table 2 were selected as potential clients, because of their close relationship to the model's business objective.

3.2. Model Definition

The conceptualization of the business model took into consideration the development and validation of each of its components, using the Canvas methodology as guiding strategy to develop and assemble the model. Table 3 summarizes the conceptualization of the initial business model, which was the starting point for its development.

Table 2. Companies registered in the city associated with the service sector. Source: [36]

COMPANY TYPE	NO. OF COMPANIES
Food and lodging services	3074
Administrative and support services	2713
Other service activities	2943
Human healthcare and social assistance activities	1490
Information and telecommunications	1306
Financial and insurance activities	941
Education	546
Public administration and defense	144
TOTAL	**13157**

Table 3. Model definition

No.	Segmentation based on CANVAS	Description of activities to be considered
1	Market segments	Identification of potential customers
2	Value proposition	Development of the value proposition for the customer, i.e. the product or service
3	Channels	Identification of the means for distribution to the customer
4	Customer relations	Development of strategies to establish good customer relations
5	Revenue sources	Explore and propose revenue generation alternatives for the proposed initiative
6	Key resources	Identify the resources required to carry out the initiative
7	Key activities	Identify the business model activities that contribute to growth of the value proposition.
8	Key partnerships	Identify strategies to achieve successful relations with partners, suppliers and others.
9	Cost structure	Based on the project's development, identify the cost structure to support subsequent implementation of the initiative.

3.3 *Model validation*

During the process of validating the customer segmentation and value proposition modules, group sessions were scheduled in three hot-climate cities, distributed as follows: sixteen (16) hotels, ten (10) service sector companies, six (6) universities and two (2) hospitals. Additionally, interviews were held with two (2) subject-matter experts on energy efficiency, in order to validate the model's effectiveness.

The model validation process with the companies took into consideration the following components, included in the assessment form: presentation of the model's structure and segmentation, description of activities, willingness to pay, characterization of energy consumption, dimensioning of the technologies to be used.

4. Results

It was found that according to the National Statistics Administration (DANE, by its acronym in Spanish, 2016), 66% of the companies registered in the city are in the tertiary (services) sector, and the following services: Food and lodging, Administrative and support services; financial and insurance companies, education, public administration and defense account for 12% of the records provided by the entity.

The above information indicates that a business opportunity exists in this sector, since there is a potential target audience with existing needs relating to the use and management of climatization technological alternatives, taking into consideration market situations at the global, national and regional level.

In order to validate this possibility, the methodology proposed by Osterwalder and Pigneur [23] was used, which is based on an initial working hypothesis as starting point, a need of a customer segment associated to the costs of maintaining, repairing, acquiring and installing air conditioning equipment for service sector buildings in Table 4.

Table 4. Initial business model (Hypothesis)

CANVAS SEGMENT	HYPOTHESIS OF THE STUDY
	H_1
VALUE PROPOSITION	Customer's pain: Associated with the maintenance, repair, acquisition and installation costs of HVAC equipment rather than consumption
	H_2
CUSTOMER SEGMENT	Customers exist for the proposed business idea
	H_3
SOURCES OF REVENUE	Charging for cool air consumption (Ton) / % over savings
	H_4
COST STRUCTURE	The company acquires, installs and maintains the equipment.

We thereby obtained a business model that enabled the interaction of specific segments to generate an attractive value proposition for companies in this sector.

Figure 1 shows the final Canvas obtained as a result of the validation process of each of the components of the proposed business model. The validation exercises enabled making several adjustments to the initial business model, by identifying some specific customer needs that could be turned into functionalities to provide additional value to the final solution. In order to validate the business idea, a service was designed using the minimum amount of resources achieved from the initial to the final iteration.

IOP Conf. Series: Materials Science and Engineering **844** (2020) 012049 doi:10.1088/1757-899X/844/1/012049

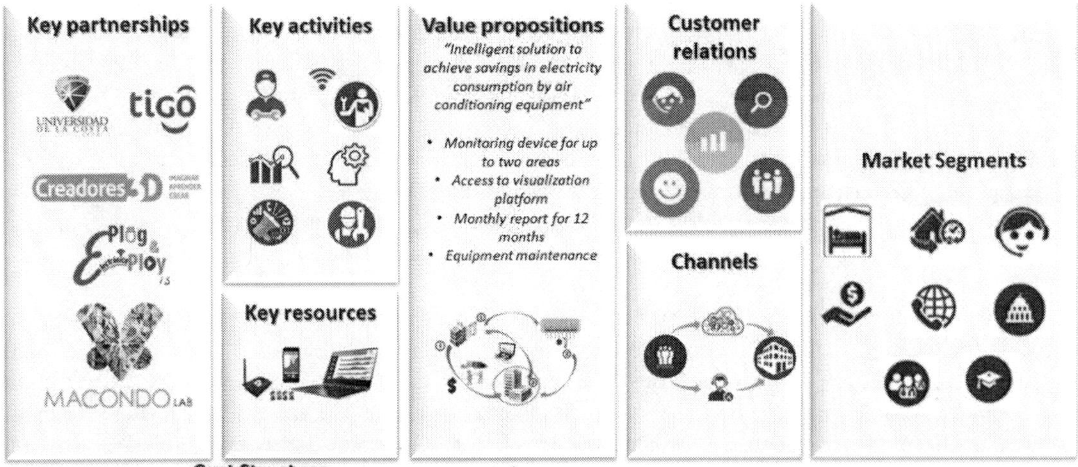

Cost Structure

ITEM	AMOUNT	%
RAW MATERIALS AND SUPPLIES	$119,351,733.33	31.50%
MACHINERY AND EQUIPMENT	$52,969,225.00	13.98%
INVESTMENT IN INTANGIBLES	$8,050,000.00	2.12%
PERSONNEL	$177,909,840.00	46.96%
SALES EXPENSES	$20,600,000.00	5.44%
TOTAL	$378,880,798.33	100.00%

Revenue Sources

	Year 1	Year 2	Year 3	Year 4	Year 5
Customers	66	72	80	68	96
Devices installed	197	217	239	263	289
Revenues from devices	$394,710,000.00	$455,890,050.00	$526,553,007.75	$608,168,723.95	$702,434,876.16
Service revenues (1 per year)	$0	$39,471,000.00	$45,589,005.00	$52,655,300.78	$60,816,872.40
Total annual revenues	$394,710,000.000	$495,361,050.000	$572,142,012.75	$660,824,024.73	$763,2551,748.56

Figure 1 Final Business Model
Note: The cost structure and revenue sources have its value in Colombian pesos.

Based on the nine modules of the Canvas model, the main finding was that the target market segment are buildings in the services sector (retailing, transportation and communications, business services and social services). The value proposition of the monitoring and control service on energy consumption by air conditioning equipment displays a clear differentiation from the model currently used for energy consumption management.

It was also found that the distribution channel for the service should be direct marketing (Company to Customer). Similarly, the marketing strategy is to manage the Green Consumption brand for the energy sector.

Based on the defined market niche, sustainable relationships with customers are to be based on dedicated personal relations, and self-service: the customer serves itself and effective communications are achieved through ICT. All the key resources required to develop the service were defined, the most important of which are human talent and technological resources (monitoring devices, website, among others).

Delivery of the value proposition implies carrying out the following key activities: technical and/or sales visits, installation of transmission, monitoring and information analysis equipment, maintenance of equipment and platforms, marketing and customer service. Such activities are actively involved in reinventing and updating the energy consumption monitoring and control service for air conditioning equipment, because adequate operation of the business model involves the constant search for individual solutions to each customer's problems.

Regarding key partnerships, there are two strategic alliances for adequate development of the proposed business model: firstly, with suppliers of equipment and supplies to develop an energy consumption monitoring device, and secondly, with universities and technology-based business accelerators to reduce risks and acquire key resources.

5. Conclusions

This study presents a business model based on the Canvas methodology proposed by Osterwalder y Pigneur. The model was designed and validated with the purpose of finding an intelligent solution for the optimization of energy consumption by VAC equipment in service-sector buildings. The proposal calls for using VRV technology, which is considered profitable based on a feasible economic investment for a model of this type, taking into considerations that service-sector buildings need to assure suitable infrastructure for application of this technology.

Development of the model enabled the incorporation of existing buildings with a common need, which is to save energy; the first step was to review energy consumption in buildings and existing energy business models worldwide.

This study shows that VRF technology is not considered the most profitable, taking into consideration its high cost and the delicate handling required for installation. Validation of the model was important in order to understand customer needs and their most important considerations that would lead them to buy into the model. For this it is necessary to create an attractive value proposition that would provide greater benefits to both parties. Initially customer surveys were made to get to know their "pain" and "wishes", which was valuable information for understanding how the value proposition should work and the characteristics it should have in order for it to be attractive for end users, finding as a result the need to achieve considerable energy savings at their companies.

Two alternative scenarios were created for model validation: 1. The creation of a company responsible for monitoring and control, and 2. The development and sale of an energy consumption monitoring device for VAC in service-sector buildings. The main challenge for future work of this study will be to implement this business model in different market segments that seek energy efficiency not only for air conditioning equipment, but also in other types of equipment connected to the electricity grid.

References

[1] Lin B and Liu H, 2015 China's building energy efficiency and urbanization, *Energy Build.*, vol. 86, pp. 356–365

[2] Pérez-Lombard L, Ortiz J, and Pout C 2008 A review on buildings energy consumption information, *Energy Build.*, vol. 40, no. 3, pp. 394–398

[3] Lemmet S 2009Buildings and Climate Change. Summary for Decision-Makers, *UNEP SBCI*

[4] ASHRAE 2009 Chapter 18, 'Nonresidential cooling and heating load calculations, in *ASHRAE Handbook*

[5] Coloma J.F and García M, CO2 Emissions Savings Produced by the Construction of an Upgraded Freight Rail Corridor. Application to Extremadura, *Transp. Res. Procedia*, vol. 18, no. June, pp. 156–163, 2016.

[6] Renedo C, Ortiz A, Pérez S, Delgado F, Fernández I, and Carcedo J, Improving the efficiency of an air conditioning system using a fire water tank as thermal accumulator, *Build. Serv. Eng. Res. Technol.*, vol. 36, no. 3, pp. 386–405, 2015.

[7] U.S. Department of Energy, Buildings Energy Data Book, *Chapter 3: Commercial Sector*, 2012. [Online]. Available: http://buildingsdatabook.eren.doe.gov/ChapterIntro3.aspx. [Accessed: 27-Jul-2016].

[8] U.S. Department of Energy, Buildings Energy Data Book, *Chapter 2: Residential Sector*, 2012. [Online]. Available: http://buildingsdatabook.eren.doe.gov/ChapterIntro2.aspx. [Accessed: 27-Jul-2016].

[9] UPME, Plan Energetico Nacional Colombia: Ideario Energético 2050, *Unidad Planeación Min. Energética, Repub. Colomb.*, p. 184, 2015.

[10] Ballester F, Díaz J, and Manuel Moreno J, Cambio climático y salud pública: escenarios después de la entrada en vigor del Protocolo de Kioto, *Gac. Sanit.*, vol. 20, no. Supl 1, pp. 160–174, 2006.

[11] Pinnola C. F, Vargas J. V. C, Buiar C. L, and Ordonez J. C, Energy consumption reduction in

existing HVAC-R systems via a power law controlling kit, *Appl. Therm. Eng.*, vol. 82, pp. 341–350, 2015.

[12] Aynur T. N, Hwang Y, and Radermacher R, Simulation comparison of VAV and VRF air conditioning systems in an existing building for the cooling season, *Energy Build.*, vol. 41, no. 11, pp. 1143–1150, 2009.

[13] Kwon L, Lee H, Hwang Y, Radermacher R, and Kim B, Experimental investigation of multifunctional VRF system in heating and shoulder seasons, *Appl. Therm. Eng.*, vol. 66, no. 1–2, pp. 355–364, 2014.

[14] Fumo N, A review on the basics of building energy estimation, *Renew. Sustain. Energy Rev.*, vol. 31, pp. 53–60, 2014.

[15] Suganthi L and Samuel A. A, Energy models for demand forecasting - A review, *Renew. Sustain. Energy Rev.*, vol. 16, no. 2, pp. 1223–1240, 2012.

[16] Zhao J and Ma T, Optimizing layouts of initial AFV refueling stations targeting different drivers, and experiments with agent-based simulations, *Eur. J. Oper. Res.*, vol. 249, no. 2, pp. 706–716, 2015.

[17] Johnson Controls, Controlling Building Systems, 2017. [Online]. Available: http://www.johnsoncontrols.com/es_latinamerica. [Accessed: 02-Jun-2017].

[18] Bruck M, Sandborn P, and Goudarzi N, A Levelized Cost of Energy (LCOE) model for wind farms that include Power Purchase Agreements (PPAs), *Renew. Energy*, vol. 122, pp. 131–139, 2018.

[19] Perini K, Bazzocchi F, Croci L, Magliocco A, and Cattaneo E, The use of vertical greening systems to reduce the energy demand for air conditioning. Field monitoring in Mediterranean climate, *Energy Build.*, vol. 143, pp. 35–42, 2017.

[20] Refahi A. H and Talkhabi H, Investigating the effective factors on the reduction of energy consumption in residential buildings with green roofs, *Renew. Energy*, vol. 80, pp. 595–603, 2015.

[21] Fan X, Qiu B, Liu Y, Zhu H, and Han B, Energy Visualization for Smart Home, *Energy Procedia*, vol. 105, pp. 2545–2548, 2017.

[22] Stoll B, Buechler E, and Hale E, The value of demand response in Florida, *Electr. J.*, vol. 30, no. 9, pp. 57–64, 2017.

[23] Osterwalder A and Pigneur Y, *Generación de Modelos de Negocio*, 13th ed. Centro Libros PAPSF, S.L.U., 2015.

[24] Ovallos-Gazabón D, Pacheco-Torres P, and Gómez-Charris Y, 5° Congreso internacional de gestión tecnológica y de la innovación COGESTEC., in *Hacia la construcción de un modelo de negocios innovador para el suministro, comercialización y venta de frío en edificaciones en el sector terciario mediante el uso de tecnologías limpias y eficientes.*, 2016.

[25] Vásquez G, Los canales de distribución y el valor para el consumidor, *Temas Manag. - Cent. Investig. en Manag. Entrep. e Inversión (CIMEeI)2*, vol. 2, pp. 10–15, 2009.

[26] Velázquez Velásquez E, *Canales de distribución y logística*, Primera. Mexico D.F.: Red Tercer Milenio, 2012.

[27] Ortiz Ospino L. E and Miranda Peralta P. E, Análisis de los Factores que Influyen en la Toma de Decisiones de los Habitantes de la Ciudad de Barranquilla para Realizar Compras en Tiendas O Supermercados, *Sotavento*, vol. 24, pp. 72–87, 2014.

[28] Ricart J. E, Modelo de Negocio: El eslabón perdido en la dirección estratégica, *Universia Bus. Rev.*, vol. 3, no. 23, 2009.

[29] Zott C and Amit R, Innovación del modelo de negocio: creación de valor en tiempos de cambio, *Universia Bus. Rev.*, vol. 3, no. 33, 2009.

[30] Mejia-Trejo J and Sánchez-Gutiérrez J, Factores Determinantes De La Innovación Del Modelo De Negocios En La Creación De Ventaja Competitiva (Business Model Innovation Determinants to Get Competitive Advantage), *Rev. del Cent. Investig.*, vol. 11, no. 42, pp. 105–128, 2014.

[31] Osterwalder A, Pigneur Y, and Tucci C. L, Clarifying business models: Origins, present, and future of the concept, *Commun. Assoc. Inf. Syst.*, vol. 16, no. 1, p. 1, 2005.

[32] Zott C and Amit R, Business Model Design : An Activity System Business Model Design : An Activity System Perspective, *Long Range Plann.*, p. 11, 2009.
[33] Porter M. E, *Competitive Advantage: Creating and Sustaining Superior Performance.* New York: The Free Press, 1985.
[34] Cortés M, ¿Qué es Networking? Todo lo que un Emprendedor Debe Saber, *SerEmprendedor*, 2016. [Online]. Available: https://www.seremprendedor.com.mx/que-es-networking/. [Accessed: 24-Aug-2017].
[35] Zolnowski A, Weiß C, and Böhmann T, 47th Hawaii International Conference on Systema Science, in *Representing Service Business Models with the Service Business Model Canvas - The Case of a Mobile Payment Service in the Retail Industry*, 2014, pp. 718–727.
[36] DANE, Geoportal, 2016. .

Impact through digital marketing tools in social networks and web analytics in higher education institutions

C J Gómez Sierra

Fundación universitaria Antonio de Arévalo, UNITECNAR, Cartagena-Colombia

cesar.gomez@unitecnar.edu.co

Abstract. Any commercial activity that is implemented today must have the innovation and growth component. Under this premise, the education business cannot be left behind and that is why it is necessary to implement strategies and models to match and overcome the closest competition. This is where digital marketing appears as a profit strategy for the customer, displacing traditional marketing now almost disappeared.

There are different ways to develop good campaigns and digital marketing plans to obtain competitive advantages, among the most important we have: content marketing, inbound marketing, relational marketing, conversational marketing and permission marketing. In an academic business unit, such as a Higher Education Institution, it is essential to conduct a study of the behavior of social networks, know in which aspects it is very weak and be able to aggressively attack the target market. For this reason, the behavioral investigation of Facebook, Twitter, Instagram of the Antonio de Arévalo Unitecnar University Foundation, will be carried out with the following tools: Facebook statistics, Twitter statistics and Instagram statistics. The methodology implemented will be quantitative research, since all variables will be measured and analyzed to draw conclusions.

1. Introduction

Today, with the rise and consolidation of information and communication technologies, strategies to do a good job or good digital marketing campaigns are very important and companies are very concerned and invest time and money for this purpose. The arrival of the Internet caused a change in the way in which human beings communicate and relate. The development of Web 2.0 and social networks brought with it a communication dynamic where each user ceased to be a receiver of information and became a co-creator. Personalization and immediacy of content became the new dynamic in the use of the Internet [1].

It is important for any company to know how to communicate with their customers and, above all, to maintain them. Currently, higher education institutions are affected by the low income index of students and through good digital marketing campaigns on social networks, you can get a perfect complement to increase the number of people interested in entering the higher education.

Social networks are on everyone's lips. Facebook, Twitter, Instagram, LinkedIn, YouTube, Google+ and many other platforms do not take away sleep only to the people who use them: companies strive to develop effective and innovative strategies that allow them to increase their visibility, improve their public image and Optimize your relationship with consumers. Social media marketing poses new challenges for brands. We refer only to technical issues, knowing how to use web platforms is only one of the points necessary to win in the competitive world of online marketing. But this is not the only thing that is required, and not even the most important. In a dynamic market, where change is something natural, instead of knowing the tools, the essential thing is to understand the logic of the new interaction processes [2].

Digital marketing campaigns in social networks have reached a very high point of interaction and impact with end users, so it is very important that this research is known and in what state of acceptance are the social networks of Antonio University Foundation from Arévalo Unitecnar. The objective of this article is to present an analysis of the areas involved and the impact generated in the community.

We understand the pressure and desire to quantify the performance generated by investing in social networks, but we believe that most digital marketing specialists are addressing the problem incorrectly. The effective measurement of social networks should begin by turning the traditional approach of return on investment. That is, instead of emphasizing their own investments in marketing and calculating returns in terms of customer response, managers should start considering the motivations of consumers to use social networks and then measure the investments in social networks they make customers when they relate to the vendor's brands [3].

2. Methodology implemented

The methodology implemented in this article is quantitative research, with which we can measure related variables, analyze their results and express conclusions. Information technologies have experienced spectacular growth since the 1950s, at a rate at which the power of information technology grows exponentially every year. This natural growth of information technology has been accompanied by information, whose volumes make it indecipherable by itself. This has forced specialists in this branch to resort to analysis systems to obtain the greatest value [4].

3. Web Analytics

For-profit companies are accessing social networks as a rich source of information and a commercial execution platform for product design and innovation, customer and stakeholder relationship management and marketing. For them, social networks are an essential component of the next-generation business intelligence platform [5].

Web analytics is the process of collecting, measuring, evaluating and analyzing all the data found on the Internet in different channels, such as social networks and websites. With this technique, the visibility and impact of any company in this digital age is optimized. This research was carried out on the social networks of the Antonio de Arévalo University Foundation in the period from July 03 to July 31, 2019. The strategies used for web analytics were:

3.1. Facebook Statistics

It is a powerful tool that allows us to make an analysis and know the impact of a Facebook account on the users who interact (number of people who have seen publications, users who interact, engagement, etc.). It is important that the community manager knows how to write the news correctly, as this will help increase the visits of the publications and gain more followers. The official website of Unitecnar - @ Unitecnar.edu.co has been used. Figure 1.

The advantages offered by social networks for the construction of value and communication of corporate values, have allowed new marketing strategies to occupy more space in the working day of the communication departments, which have allowed to air with a social breath of fresh air traditional sales channels, without losing their importance. The integration of the Social Media Plan into the marketing strategy and the incorporation into the workforce of specific professionals constitute clear signs of investment and trust in Social Media [6].

Figure 1. Official Unitecnar Facebook Page

3.2. Twitter Statistics

It is also a very robust tool to analyze the behavior of our accounts. Some of the most important aspects of writing a good tweet are: The use of hashtags to reach more people, use a maximum of 100 characters to make room for hashtags and comments, use shortened URLs to save more characters and get more clicks, always mention the source of information, do not make mistakes and use all spelling signs.

Retweeting is the key mechanism for the dissemination of information on Twitter. It emerged as a simple but powerful way to disseminate information on the social network of Twitter. Although much information is shared on Twitter, little is known about how and why certain information is disseminated more than others [7].

3.3. Instagram Statistics

It is important to know about the analysis carried out in our account, if the plan we have designed is achieving the objectives and if it is reaching the people we need to reach through our actions in this social network.

Many companies are now commonly using social media to promote their brands and to maintain a friendly relationship with their customers (Saravanakumar & Lakshmi, 2012). Aside from having an account in Facebook and Twitter, new social media platform is coming in to provide another platform for many marketers. Lately, there are many businesses that starting to integrate their business with the Instagram experience (Tekulve & Kelly, 2013). According to an article published on BBC News (2012), study has said that a large amount of the top brands around the world are using Instagram as one of their marketing strategy. Instagram can help a company to promote their products or services (Bevins, 2014). The point that makes Instagram different from other social media is that Instagram is applying a visual based strategy (Hird, 2013). Everything about Instagram is about photograph (Linaschke, 2011). Instagram marketing is an effective way to advertise a product as it is said that a picture speaks a thousand words (Silva & de Melo & Almeida & Salles & 4 Loureiro, 2013). Additionally, Instagram can help one in saving cost for brand designing. Every image taken for a product can be edited and filtered using the Instagram functions (Herman, 2014) [8].

4. Results

Facebook:

In Figure 2, we can see that there is a total of 329 likes on the page during the month of July (measurement period), which have been decreasing by 1%.

The Reach obtained by publication is 530,021 people, which have been increasing by 28%. It is necessary to clarify that this data refers to the people who saw the publications of the page on its walls. This does not specifically mean that the person liked it, that they agree with what is being published, that they have read the article or that they have taken any action.

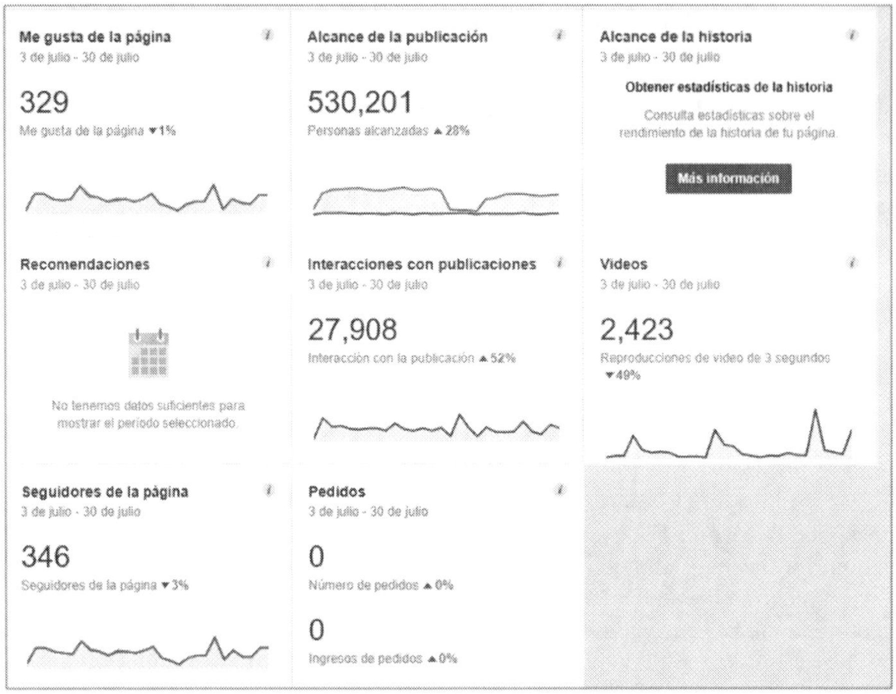

Figure 2. Facebook Unitecnar statistics

We can also observe that 27,908 people have interacted with the publications of the page during the month of July 2019. Here it is necessary to highlight that the interactions correspond to the fact that people have clicked on the publications or liked, that is, they had participation (Engagement) and a significant increase of 52%.

The followers who had the page was 346 during the period analyzed, with a decrease of 3% and the number of video views during the first 3 seconds was 2423, with a decrease of 49%. The total number of followers as of July 31, 2019 was 24,204 as shown in Figure 3.

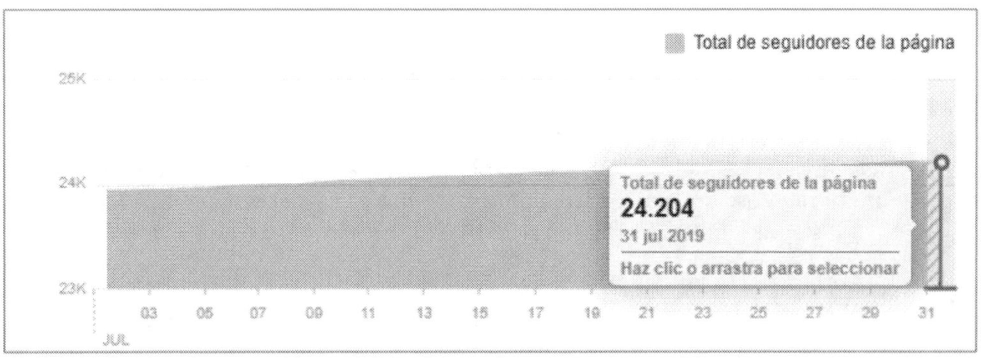

Figure 3. Total followers of the page

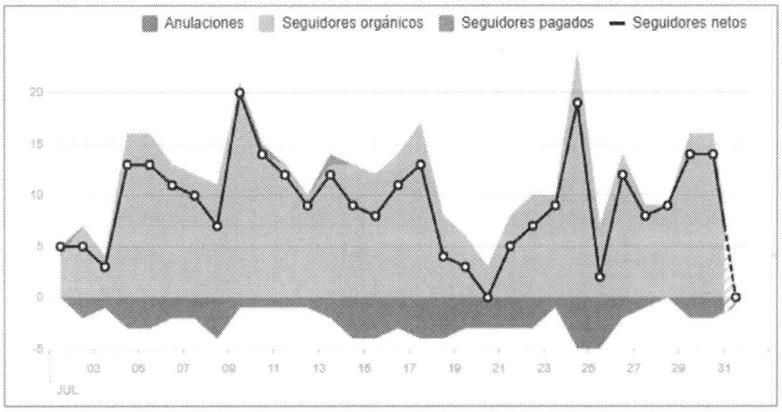

Figure 4. Net followers

The graph of figure 4, shows that the net follow-ups show us the number of new followers minus the number of people who stopped following the page, and for this particular case there are some peaks on July 9 and 25, 2019 with 20 registered followers.

Twitter:

Figure 5. Official Twitter Account Unitecnar

During the analysis made to the official Twitter account, it was observed that it serves to know the frequency of Tweets that are being achieved and if by increasing the frequency you can earn more impressions or get more visits.

In an annual study conducted by Stelzner (2010, 2011), Facebook and Twitter were the top two social media tools used by marketers for the past 2 years. In the hotel industry, about 75% of hotels have used social media for marketing purposes (Hotelmarketing.com, 2011). Facebook and Twitter were also found to be the two most useful social media channels for marketing purposes in the hospitality industry (Friebe & Campbell, 2010; Hotelmarketing. com, 2011) [9].

IOP Conf. Series: Materials Science and Engineering **844** (2020) 012050 doi:10.1088/1757-899X/844/1/012050

Figure 6. Twitter Unitecnar statistics

During the investigation of the account in the period of July 2019, it is observed that 17 was the total Tweets, the impressions reached were 6846, the visits to the profile reached 297, there were a total of 26 mentions and 4 new users followers, Figure 6.

Figure 7. Follower and main multimedia content

The Tweet with main multimedia content was sent with the Hashtag #Gastronomy and had 375 impressions, in addition the main follower was Emprendecultura @emprendecultura, as can be seen in figure 7.

Instagram:

Figure 8. Official Unitecnar Instagram Account

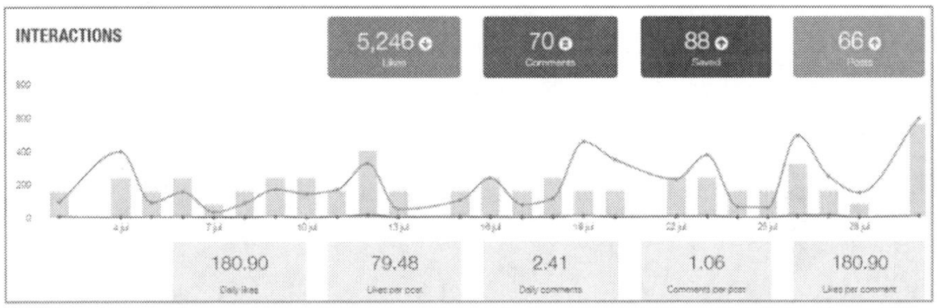

Figure 9. Instagram Unitecnar interactions

The interactions or activities of this account are all the times that have been touched on the profile, visits to the profile and website. The number of likes is 5246 during the period analyzed, with a daily average of 180.90. Here what is counted is the number of times it has been seen, even if it is the same person. It can be seen in figure 9.

Regarding the scope, we can also observe that 3400 unique people have seen the post of the account, the visits to the profile amount to 3774, a total to the website of 89, as can be seen in Figure 10.

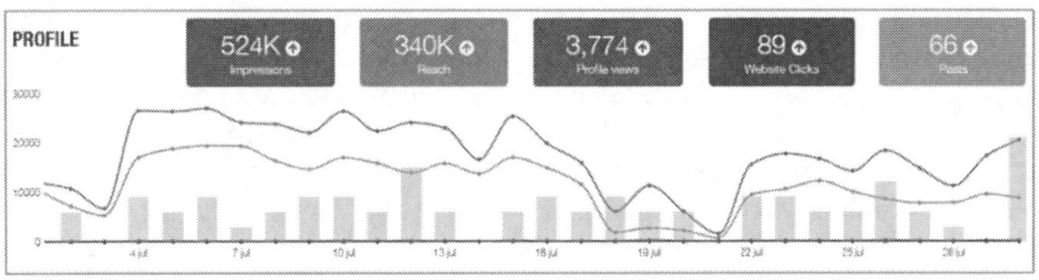

Figure 10. Instagram Profile Unitecnar

Speaking of segmentation, we analyzed a very valuable information on how are the followers, in which you can see the number of male and female population to structure future posts. In the case of the period under investigation, it is observed that the largest population is female and their ages range between 25 and 34 years of age. As well as the followers by city, where we observe that the highest percentage is in the city of Cartagena de Indias with 51.67%, followed by the city of Barranquilla with 15.90%. All this can be seen in figures 11 and 12 respectively.

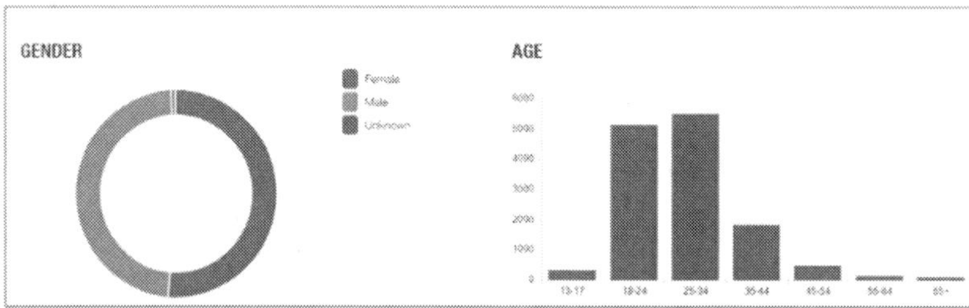

Figure 11. Instagram Unitecnar Gender Statistics

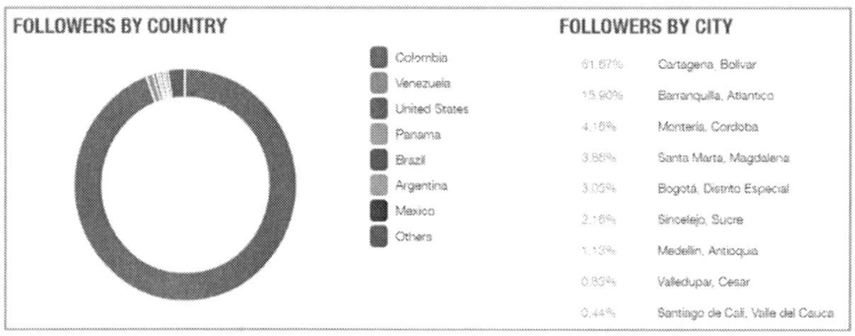

Figure 12. Followers by city Instagram Unitecnar

5. Conclusions

Once the statistics of the Unitecnar social networks have been analyzed and to gain many more followers on the Facebook page they share, comment, interact, the implementation of the 4 "C" Strategy is suggested. The first C has to do with Content, where the success of the marketing campaign depends on how much content is uploaded, how much it solves a problem and how much it becomes interesting for followers. The second C corresponds to the Constancy, where even uploading valuable content you have to be constant and upload the contents based on a schedule (create an interesting content plan) previously stipulated and not with disorder in time. The third C corresponds to Calm, where you have to keep in mind that to grow the page is not achieved in a short time and you have to do a hard job, where growth is done slowly and patiently. The fourth C corresponds to the Community, where it should be taken into account that the page is not created thinking of a business of its own but for the followers, it is something for the community, which is what really suggests the issues that should be tartar

As for the Twitter tool, it is necessary to have better Tweets quality, where the main Tweet is shown as a small guide to know what followers like most and is a tool to improve the percentage of interactions. It is also necessary to make more Tweets of multimedia content, which serve to improve the percentage of interactions and with a good main follower it is important to generate many enriching retweets.

In the case of the Instagram account, it is suggested to develop more multimedia content (videos) and the most important thing is that they are of maximum 1 minute, but taking into account that the attention should be achieved during the first 5 seconds, in addition not to use only images for the contents. It is also noted that the publications do not have as many Hashtags or some are without them, it is recommended to use many and related to important dates. At least 20 Hashtags are recommended in the comments and not in the description of the photos. Finally, it is recommended to publish very often, that is 2 photos or images and about 10 stories per day, because the Instagram algorithm will qualify better and thus more followers will be gained.

References

[1] Maqueira, J. M., & Bruque, S. (2009). Marketing 2.0: el nuevo marketing en la web de las redes sociales (pp. 106-110). Ra-Ma.

[2] Moschini, S. (2012). Claves del marketing digital. La nueva comunicación empresarial en el mundo, 3.

[3] Hoffman, D. L., & Fodor, M. (2010). Can you measure the ROI of your social media marketing?. MIT Sloan management review, 52(1), 41.

[4] Sarduy Domínguez, Y. (2007). El análisis de información y las investigaciones cuantitativa y cualitativa. Revista cubana de salud pública, 33.

[5] Zeng, D., Chen, H., Lusch, R., & Li, S. H. (2010). Social media analytics and intelligence. IEEE Intelligent Systems, 25(6), 13-16.

[6] Clavijo, I. G. (2016). Facebook para empresas y emprendedores. IC Editorial.

[7] Suh, B., Hong, L., Pirolli, P., & Chi, E. H. (2010, August). Want to be retweeted? large scale analytics on factors impacting retweet in twitter network. In 2010 IEEE Second International Conference on Social Computing (pp. 177-184). IEEE.

[8] Huey, L. S., & Yazdanifard, R. (2014). How Instagram can be used as a tool in social network marketing (Doctoral dissertation, Master's thesis. Retrieved from http://www. academia. edu/8365558/How_Instagram_can_be_used_as_a_ tool_in_social_network_marketing).

[9] Leung, X. Y., Bai, B., & Stahura, K. A. (2015). The marketing effectiveness of social media in the hotel industry: A comparison of Facebook and Twitter. Journal of Hospitality & Tourism Research, 39(2), 147-169.

IOP Conf. Series: Materials Science and Engineering **844** (2020) 012051 doi:10.1088/1757-899X/844/1/012051

Design and implementation of an intelligent thermal chamber to produce plantain hills

J López Hurtado[1] and H A Gallego Becerra [2]

[1]Master in Physical Instrumentation, DICOPED research group, Faculty of Basic Sciences, Technological University of Pereira, Colombia.
[2]Master in Physics, Headline Professor, Director of the DICOPED research group, Senior Researcher, Faculty of Basic Sciences, Technological University of Pereira, Colombia.

jholopez@utp.edu.co

Abstract. The project is aimed on the design and construction of a prototype device which allows the measurement, visualization and control of the temperature and relative humidity, and soil moisture inside a thermal chamber. This prototype consist mainly of two sensors, one of them to measure the temperature and humidity in the air, and the other one to measure the humidity of the soil; In addition, an Arduino development card, which be used to process the data and control the different measured Physical variables. A liquid crystal display (LCD) to display the measured data and the process it takes; In addition to other electronic devices that complement the processing and control in real time, the backup or monitoring of the thermal chamber through the days, the automatic turning on and off of a misting irrigation system and the automatic opening or closing of two windows used to let circulate oxygen inside the place.

1. Introduction

Although it is not a secret to say that any country in the world moves economically for what it produces, in Colombia one of the most important economic sectors is the agricultural sector, where different types of vegetables, fruits, plants, among others are produced. It is perhaps, the countryside one of the most vulnerable areas of Colombia and, without technological and industrial investment to help technify the agricultural processes that moves this nation.

It is worth bringing to the discussion that raised by the former president "Juan Manuel Santos", within the framework of the Fourth International Cocoa Seminar, based on the agreements made in the peace process with the guerrillas of the "FARC-EP", where he mentioned "The central axis of this agreement is in the field and it is the one that will benefit most from peace" [1]. Where he explained that this is stipulated, which will mean more investment, more public goods, more productive projects and more financing. Then, he said "This is what we are going to take to the countryside, especially those regions affected by war, which are almost all" [1].

According to President Santos, almost all of Colombia rural areas has been affected by the armed conflict presented for many years in this country, which has left thousands of people dead, entire families displaced, families that have remained in these areas that live with very little sources and, above all, that do not have the possibility of generating a greater income.

The implementation of this intelligent thermal chamber to produce plantain hills seeks to generate a positive social impact on the population of the "Villanueva Village Center", in El Águila - Valle del Cauca, Colombia; since this project encourages a new form of employment in this place, including and using electronic equipment as a work tool. It is also intended to improve the coexistence of the community, because it is expected that their economic income will improve by having greater quantity and quality in the production of plantain hills and the fruit of the plantain plant; and thus, having a better quality of life.

The project is implemented from the adaptation of a land and construction of a greenhouse with metal structure attached to the ground by concrete, where it is sought to control the temperature and humidity of this place, for accelerating the production of plantain hills, and with this achieve more uniform plants and fruits, and with a minimum probability of infection by fungi and pests. This is achieved by setting the temperature and humidity in a reference range within the greenhouse.

Also, it seeks to train a group of students of the Educational Institution Justiniano Echavarría, in different academic subjects concerning Mathematics, Physics, Electronics and Programming, in order to bring them closer to the project and make them understand the operation of the intelligent stage of the thermal chamber.

2. Methodology

For the execution of this project, different phases or stages were established to carry an orderly sequence and arrive at an accurate result. Initially, several activities were carried out in the first phase.

2.1. *first phase*

A bibliographic review was carried out in specialized journals of national and international order. On the other hand, we searched regional publications on the web and meetings with agronomists, in order to determine the progress of instrumentation in this area and the use of technology as a tool to technify this type of processes or other similar, in the Colombian rural area [2] [3] [4] [5].
It was also reviewed all the documentation requested from the Mayoralty Office of El Águila town and on web pages of the state, everything concerning the demographic situation of the aimed Municipality; all this with the purpose of knowing more details and establishing a geographical framework with the main territorial, population, economic and social difficulties of the habitants where this project is developed [6] [7].

In this same stage, the current government regulations related to the creation and permanence of nurseries in Colombia were reviewed. This documentation was collected from the Colombian Agricultural Institute (ICA) [8] [9].

At the same time, it began with the training of a group of students of higher degrees, of the Educational Institution Justiniano Echavarría, in consolidating concepts related to the environment on physical variables which are very important in the germination of crops; variables such as: temperature and relative humidity and soil moisture. In addition, a course of Basic Electronics, Algorithmia and Programming in C language begins with them, all this with the aim of teaching them how to program on development cards such as Arduino and use different peripherals and sensors from the world of electronics. It should be noted that this stage of the project is executed throughout all its implementation.

2.2. *second phase*

Through the use of rectangular metal tubes, different work tools in the field and work techniques developed by the same people in the rural area of Colombia, the land was adapted and vertical supports fixed to the ground with cement, and other in horizontal way which were fixed to the vertical supports with welding, to give shape to the space of thermal chamber. This work was carried out in collaboration with students who were part of the work group and some habitants of the Villanueva district. Figure 1 shows the beginning and ending of the architectural construction work of the thermal chamber space.

Figure 1. a) Beginning of construction of the thermal chamber, b) Ending of construction of the thermal chamber.

2.3. third phase

Adaptation and implementation of the irrigation system. Figure 2 shows the pressure pipe system and installed nebulizers. In addition, a series of stopcocks and an electrovalve were installed to control the flow of water.

At this point it is important to emphasize that the irrigation system by fogging becomes important, since the plants need a minimum part of water in the form of mist for a correct growth [3].

Figure 2. Misting irrigation system.

2.4. fourth phase

Different ways of implementing with high-performance electronic techniques, an intelligent system that was minimally operated by the human being, all this for making more technic the production process of plantain hills in a thermal chamber and thus be able of designing a team which will execute this task with the least possible error.

Already having cleared the electronic techniques necessary for the design of the prototype, software was created in which the studied techniques could be implemented. After having created the necessary software, the appropriate hardware was established for implementing the realized program. With the hardware modules developed, the software operation of the entire set of devices was interconnected and simulated. Figure 3a and 3b shows the designed equipment, in which it was also necessary to power it by solar energy, because in the Villanueva area there are constant interruptions of the electricity supply.

IOP Conf. Series: Materials Science and Engineering **844** (2020) 012051 doi:10.1088/1757-899X/844/1/012051

Figure 3. a) Electronic system for data processing and control of an intelligent thermal chamber, b) Electric power supply by solar energy.

2.5. fifth phase
Having already designed and implemented this hardware, it began with the functional tests of the equipment in the field to verify its proper functioning and make adjustments. Then we adapted a box or cabinet which was fitted outside the thermal chamber to keep the prototype.

With the verification of the proper functioning of the prototype of the designed equipment, the sowing and accelerated reproduction of the plantain hills proceeded. Figure 4 shows the process of sowing the mother seed of plantain inside the thermal chamber.

Figure 4. Planting of plantain mother seed inside the thermal chamber.

3. Results and discussion
Practical comparative analyzes in the field showed a delay of 4 months in the outbreak of plantain hills in plants exposed to the environment (figure 5) and 2 more months, to be extracted and transplanted to another place.

Figure 5. Plantain hills after 4 months in plants free exposure in the environment.

Otherwise, it occurred inside the thermal chamber, as shoots of baby seedlings of the mother seedlings were observed, at approximately 18 days and 10 more days, they took time to be extracted from the thermal chamber (figure 6a) and transplanted into bags outside this place. Then, for 1 to 2 more months, plantain hills were ready to transplant to the field (figure 6b).

Figure 6. a) Plantain hills of 18 day inside the thermal chamber, b) Plantain hills of 45 day outside the thermal chamber.

This is how it is possible to obtain plantain hills ready to transplant to the field with a reduction in time greater than 40%, through the use of the Technique of Accelerated Reproduction of Seed or Planting Material (TRAS) [2], than with the Traditional method to free exposure.

The best seed reproduction results within the thermal chamber were obtained at temperatures between 30 and 60 ° C and relative humidity levels between 30 and 100%; which was possible to control by means of the misting irrigation system and the automatic opening and closing of 2 windows (figure 7a) and, using the digital temperature and humidity sensor SHT10 “SEN0148” shown in figure 7b. In addition, to maintain a soil moisture level (rice husk) between 30 and 80% (figure 7c). Emphasizing that all site control is performed with the "Arduino Uno" Microcontroller (figure 7d).

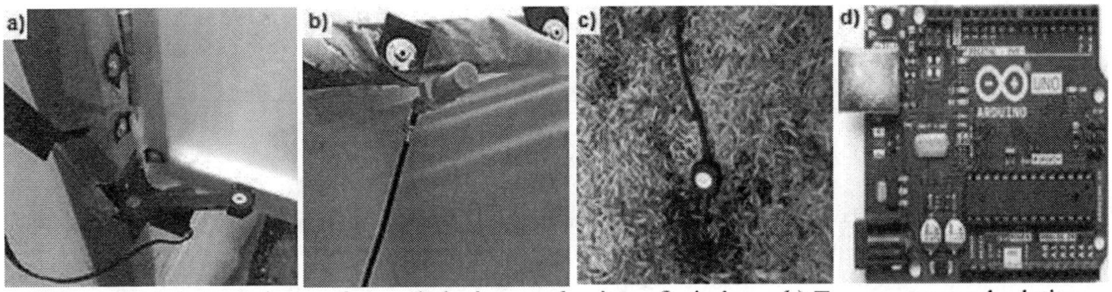

Figure 7. a) Automatic opening and closing mechanism of windows, b) Temperature and relative humidity sensor, c) soil moisture sensor, d) "Arduino Uno" microcontroller.

Figure 8 shows, after 5 months, plantain plants transplanted to the field, extracted from the thermal chamber; Here you can see healthy plants with an average height of 3.5m. Evidence that this method of seed reproduction is ideal for the production of plantain hills in a shorter time and of good quality.

Figure 8. Plantain plants extracted from thermal chamber and planted 5 months ago.

4. Conclusions

With the implementation of this project and especially using a device built to control various physical variables, it has been obtained a product of very good quality in less time than traditional techniques. This is important for generating a changes in the traditional way of producing plantain hills in the countryside and also motivating both the national government and farmers to promote and invest in technification for agricultural processes in rural areas of Colombia. In addition, it can be impacted a population whose has suffered with the armed conflict in Colombia and with limited economic sources, since new form of work was generated for improving their budget income.

References

[1] Ministerio de Agricultura y desarrollo Rural de Colombia 2016 El eje central del acuerdo de paz está en el campo [online].
Available in: https://www.minagricultura.gov.co/noticias/Paginas/El-eje-central-del-acuerdo-de-paz-esta-en-el-campo.aspx

[2] Aguilar M, Reyes G and Acuña M 2004 Guía técnica No 1 Métodos alternativos de propagación de semilla agámica de plátano (Musa spp.) *Universidad Nacional Agraria* (Managua, Nicaragua) p 18.

[3] Stanley M 2014 Los sistemas de nebulización se utilizan cada vez más en cualquier clima – Control de la temperatura y la humedad con un sistema de nebulización *Freshplaza* [online].

Available in: http://www.freshplaza.es/article/85147/Control-de-la-temperatura-y-la-humedad-con-un-sistema-de-nebulizaci%C3%B3n

[4] Álvarez E, Ceballos G, Gañán L, Rodríguez D, González S and Pantoja P 2013 Producción de material de siembra limpio en el manejo de las enfermedades limitantes del plátano *Centro Internacional de Agricultura Tropical* (Cali, Colombia).

[5] Ardiles Hurtado J, Caparachin Paredes E T and Clavijo Turpo J C 2013 Folletos de cámaras térmicas para plantaciones de plátano *Ministerio de Educación* (Oxapampa, Perú) [online]. Available in: https://issuu.com/prea-pasco/docs/folleto_platanos_finalizado

[6] Cadavid Palacio L, Cadavid Palacio M, Aguado L E and Mejía F J 2000 Esquema De Ordenamiento Territorial Municipio de El Águila Valle del Cauca *Alcaldia Municipal de El Águila Valle del Cauca* p 416.

[7] Departamento Nacional de Planeación Gobierno de Colombia 2018 El Águila Valle del Cauca *TerriData* **Ficha 76243** p 23 [online].
Available in: https://terridata.dnp.gov.co/index-app.html#/perfiles/76243

[8] Instituto Colombiano Agropecuario (ICA) 2009 Resolución 003180 de 26 de Agosto de 2009 Requisitos y procedimientos para la producción y distribución de material de propagación de frutales en el territorio nacional y se dictan otras disposiciones p 11.

[9] Instituto Colombiano Agropecuario (ICA) Manual técnico para viveristas en musáceas (plátanos y bananos) [online].
Available in: https://sioc.minagricultura.gov.co/Platano/Documentos/005%20-%20Documentos%20T%C3%A9cnicos/D.T.%202009%20Agosto%20-%20Manual%20Tecnico%20para%20Viveristas.pdf

Acknowledgement
Special thanks for Technological University of Pereira from Colombia, for sponsoring the research project through the Vice Rectory of Research, Innovation and Extension project number E3-19-1. Also, to the Educational Institution Justiniano Echavarría of El Águila - Colombia, for facilities provided and their great collaboration in the execution of this project.

Diagnostic tool for radical improvement in business processes

A Sanchez-Comas[1], L E Vasquez Osorio[1], M Pérez-Vargas[1], M Caicedo-García[1], D Neira-Rodado[1] and A Troncoso-Palacio[1]

[1]Department of Industrial Management, Agroindustry and Operations, Universidad de la Costa CUC, Barranquilla, Colombia.

E-mail: asanchez@cuc.edu.co

Abstract. The present work presents a diagnostic tool applicable in any business sector; it works as a support to managers for decision-making towards the need for a radical improvement of processes. A literary review was conducted to collect case studies in order to identify the problems it addresses and the impacts generated by the process of reengineering application in organizations. Through conceptual modeling and synthesization, some diagnostic questions were formulated related to the problems that reengineering has addressed in the literature, and an evaluation model was built that allows organizations to identify what type of problem the organization has, its processes or departments, which should be approached from a radical improvement approach. Finally a diagnostic tool with an evaluation method was formulated, to recommend initiate process improvement either organization, departments o process, formulated fully integrable with PDCA cycle and evaluation or monitoring phases for systems like business activity monitoring based on IT systems, relevant in today's contexts when most of the nowadays organization are technology-based, and progressive improvements do not become a pragmatic and agile alternative in the business context.

1. Introduction

The organizations should design differentiation policies, be cost leaders in order to obtain competitive margins. In order to achieve these objectives, it is essential that all the personnel of the organization get involved, manage and execute the processes to contribute in this way to the success of the institution. It is widely known that a large number of organizations have ceased to be competitive due to the low productivity that is generally linked to inefficient management of their processes, failures in the traceability of information, processes that hinder the functioning of the organization. This scenario has created the need for organizations to achieve a global change in their processes, by directing them based on market demands through the application of radical improvement methodologies [1], where Process Reengineering takes place seeking to restructure organizations for more effective results. Process reengineering follows the philosophy of taking a process and completely restructuring it in a way that generates a totally new and improved process through actions like decrease in process times, greater knowledge and control of processes, and a better flow of information within the organization [2]. Reengineering seeks results of great impact unlike other approaches that characterized by seeking incremental and continuous results. However, a study conducted found that reengineering is one of the three options to achieve Continuous Process Improvement (CPI) [3]. According to the research of [4], reengineering improves activities and seeks to make the work faster and easier by radical and drastic changes on the existing processes. For a reengineering of processes to be effective, it should been based on the reexamination of the current process and its objectives, with the aim of achieving spectacular improvements in its realization.

In response to the previous approach, this research takes Process Reengineering as a base methodology for the creation of a diagnostic tool, which will allow organizations to decide whether to the redesign radically their processes, improving the results associated with the performance of the processes.

2. Methodology

Based on an analytical approach from the global to the particular, a three-phase methodology was determined as shown in Figure 1, as followed by [5]. The first phase of the project focused on consolidating a compendium of case studies through a Review of Scientific Literature, which allowed us to see the practical approach to the implementation of Process Re-engineering, the problems it covers, and the impacts it obtain. In the second phase, the studies were modeled taking into account information of interest: problems, activities, tools, and results. In the third phase, the modeled information of the problems covered was simplified and normalized, and the impacts generated by the process reengineering to turn it into evaluation inputs for the diagnostic tool.

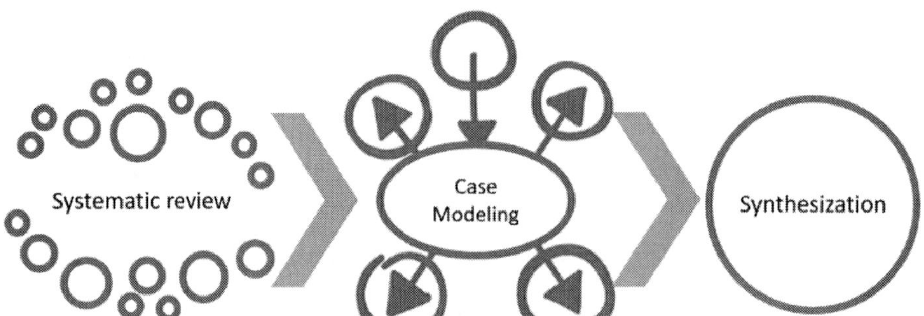

Figure 1. Applied methodology.

The Systematic Review phase (Figure 1) was conductd according [6], the questioning was geared towards identifying "what cases of Process Reengineering application had been documented in the scientific literature and if they had been applied to quality management". The review conducted in both Spanish and English, with the focus of information being on case studies documented in scientific articles or proceedings, so that information could be extracted from business experiences that enriched information support. The databases used were IEEExplorer, Proquest, Google Scholar, Ebsco and Science Direct. A methodology very similar to that applied by [7], which is evidenced in Figure 2.

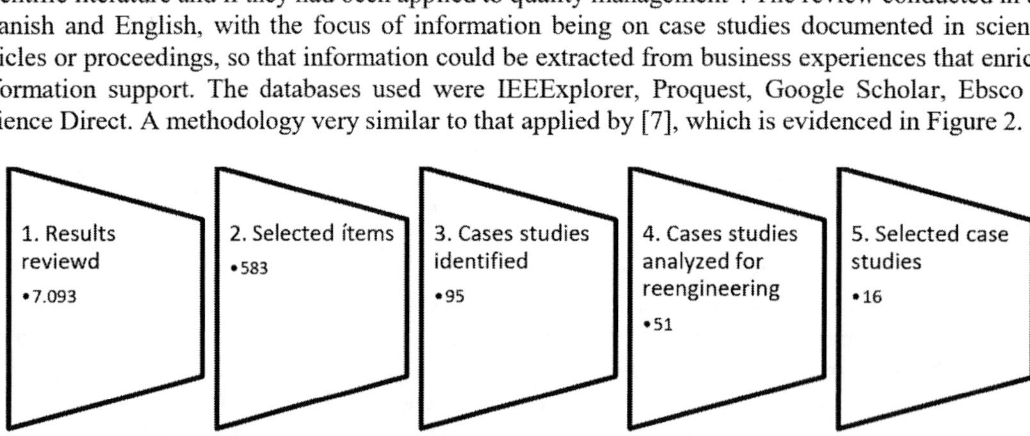

Figure 2. Flow of information selection in the Systematic Review of the sources.

The exclusion of articles done throughout the review process to arrive at the selected case studies following the methodology implemented by [6], as evidenced in the previous figure in five stages. The first stage proceeded from the accounting of the results of the 24 search strings inserted in each of the six databases, seeking to identify which of them were effective to review. The second stage consisted of reading the title, and keywords seeking to identify wether or not the result was relevant to the study in a preliminary way. The third stage consisted of identifying, selecting and downloading case study documents. In the fourth stage, the case studies were analyzed in depth, in order to identify that they were effectively related to process improvement through the concept of reengineering. The fifth stage consisted of a deeper review of the case studies, selecting for the present work those that described the information of interest: the application process, tools, problems and results/impacts obtained. This debugging method, together with reading and navigation methods, helped to reduce the information considerably. The amount of results classified in Figure 2 it established, allowing us to find 21 documented case studies of the Reengineering implementation with the study variables needed for the construction of the tool.

In the Case Modeling phase (Figure 1) in order to abstract and relate the information of interest from the case studies (problems, activities, tools and results), Cmaptools was used as a conceptual modeling tool that would allow us to have a global a visual and centralized overview of all the information to generate the variables of interest. An example of the modeling of a case study can be seen below in Figure 3.

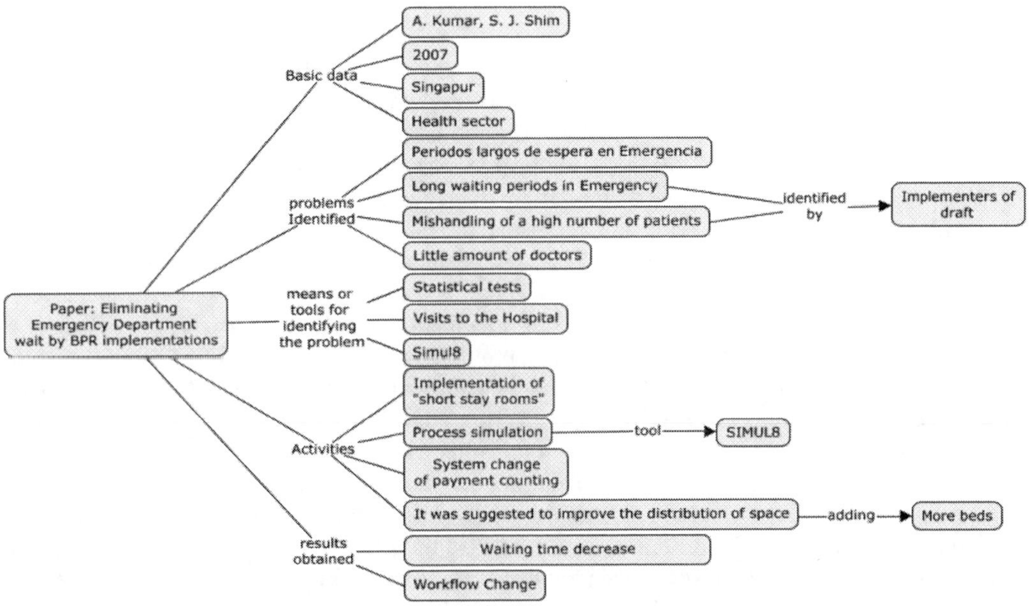

Figure 3. Conceptual case model of a study and information of interest: problems, activities, tools, and results.

In the Synthesization Phase (Figure 1), each case study modeled on the tool, initially simplified into a model by sector (Figure 4), the simplification consisted of methodically grouping the information of interest (problems, activities, tools, and results). Afterwards, the terms or concepts commonly found between the results of the different cases standardized so that a general and standardized overview of the types of problem, tools, activities, and results obtained could obtained. Once the information of interest standardized by sectors. They were synthesized into a single model in order to proceeding to identify a deeper level of classification and detail of the problems and impacts of the Process Reengineering application, so that the resulting info allows the transformation of the information of interest extracted from the cases of studies analyzed, into diagnostic variables for the construction of the diagnostic tool.

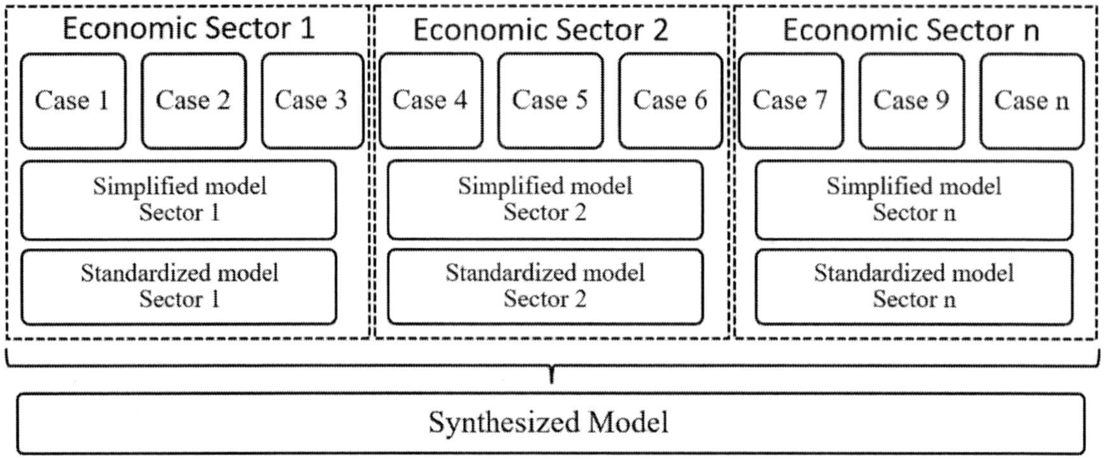

Figure 4. Simplification and standardization process in the characterization of the case studies.

3. Cases of re-engineering process studies

The literature review showed that Process Reengineering is widely focused on business restructuring in many sectors, due to the impact in productivity as well as serves as a starting point for looking excellence in business process [8, 9]. In the industry sector, [10] published an article in which he implemented a Reengineering process in the Rank Xerox (in the printing sector), in response to customer complaints regarding its products at a high price and its inefficient management. [11] Built collaborative teams, one of it, specialist in Process Reengineering and another in the management of the company. They carried out the modeling of the client management process, a subsequent analysis for improvements application, and then a completely new and improved model design; it applied as a pilot test and adjusted according to the feedback received. In India [12], decided to document the case study of Piped Natural Gas, a company within the oil sector that presented critical problems such as the slow billing to its customers, which led to the possession of a very inefficient small customer payment system. In order to address the problem, it decided to implement Process Reengineering and in this way, interviews first carried out with the company members and clients of the firm to gather as much information as it could contribute to the project objective. After that, the organization's payment system was completely changed, an executive committee in charge created and a device called Handled Billing Device implemented, with all this leading to reduction in the time it took to complete the process from 35 days to approximately 13 days, the billing was almost instantaneous and the error was minimized during the process. The application of reengineering in a construction company in China was documented [13], they addressed problems with payment in general, many repetitive jobs and problems with prevalent measures. As part of the Reengineering implementation process, the cause and effect figure carried out in previous steps, was analyzed and the process matrix was reorganized based on a lot of theory and the GB50500 standard of 2003, with this the processes were redesigned and unnecessary work was eliminated. The investigation of [14], a cereal company where the greatest inconveniences arose in customer dissatisfaction (low credibility of the company), insufficient marketing, poor strategic processes, and late billing. In the [15], documented the process of implementing Reengineering in an automotive company in Ecuador with lack of a system of information to control the inventory and to monitor the processes, in addition to the absence of indicators in the operation with many bottlenecks. To tackle these problems, the organization entered a stage of complete redemption of the activities, an indicator called OITFWQ, implemented. That allowed not only control the processes, but also to improve their quality, complementing with the application of an information system that allowed the restructuring of the company's information but in particular that of the suppliers, resulting in the elimination of bottlenecks and workload growth by 39%.

In the health sector, the investigation of [16] obtained a 44% improvement in the delivery of results by making changes in the assignment of shifts to the workers of a pharmaceutical laboratory. In the work of [17], the organization and its processes completely restructured, bringing the company to a more customer focused approach and implementing information technologies that would make it easier to manage the amount of data that should entered every day, in addition to improved internal communication in the organization. Even has been applied in public's hospitals [18].

In the financial sector, an insurance company [19], implemented reengineering in internal process; problems related to inefficiency of the development of information technologies; during the implementation, an information strategy planning project was developed to revolutionize all the processes that were up to that moment, this was carried out with excellent results due to the commitment of management during the process. Also [20], implemented reengineering within a process in an insurance company to be able to change internal problems such as the inefficiency of the development of the technologies of information; During the implementation, an information strategy planning project was developed to revolutionize all the processes that were up to that moment; This was carried out with excellent results due to the commitment of management during the process. In a trading company in Singapore, process reengineering is applied in response to the need to increase competitiveness and the elimination of such complex processes.

In the education sector, [21] reports the application of Process Reengineering in a private school due to the need for restructuring by departments and lack of documentation of the processes. Prior to the start-up of the Process Reengineering, it was ensured that all departments were committed to the project since they considered it of the utmost importance, after this, the organization was restructured in an integral way, and through the feedback, the policies were reviewed from the institution to not only eliminate unnecessary expenditure of resources but to improve the processes of each of the departments, reduce operating costs and make processes simpler.

Within the telecommunications sector, several articles on the implementation of Process Reengineering were published in a variety of organizations. [22] published a case study of a Korean company facing an economy changing at a dizzying pace at that time, and the costs of maintaining were increasingly high. In addition, the competition lurked and the staff was not committed to management decisions; during the implementation period of Reengineering, the organizational culture was redesigned, and efforts were focused on employee welfare, schedules were changed, employees were given training and family time was encouraged. [23] Documented a reengineering project in an Irish company that was a victim of quality policy failures and was facing a new type of market in which its competition was much stronger. The way in which the application of the Reengineering was approached was through the document drafting of the company's processes. The company was reorganized and the ISO 3001 and BS5750 certification was achieved. In addition, a quality council was created to guide the processes of company improvement. Many processes were automated and a management layer was eliminated, a release scheme was created for employees to leave the company voluntarily, roles were eliminated and a team was created to attack organizational problems in the future. With all this, this Irish company achieved a masterful adaptation to the market. In the

Logistics Sector in Aeronautics, [24], published an article describing the implementation of Process Reengineering in response to the need to improve the process of loading and unloading packages since it was a process too late. As part of the implementation, the limits of the process were defined due to the nature of the business; the process was restructured through data collection, observation, information analysis, and a subsequent improvement design. And as an added bonus a customer service thread was created generating considerable simplification in the company's activities. Continuing with the same economic sector [25], documented the application of Process Reengineering in a Spanish company that faced customer service problems and poor management of claims. To address the situation, the process to work was broken down and diagrammed separately, then a brainstorm was carried out with the members of the organization to materialize the improvement. After that, were documented 577 ideas for the process improvement. Then, a technique called "Knowledge transfer" was applied as a method to communicate the results to the entire organization. An overview of the case studies of Process Reengineering application reviewed can be seen in Figure 5 below.

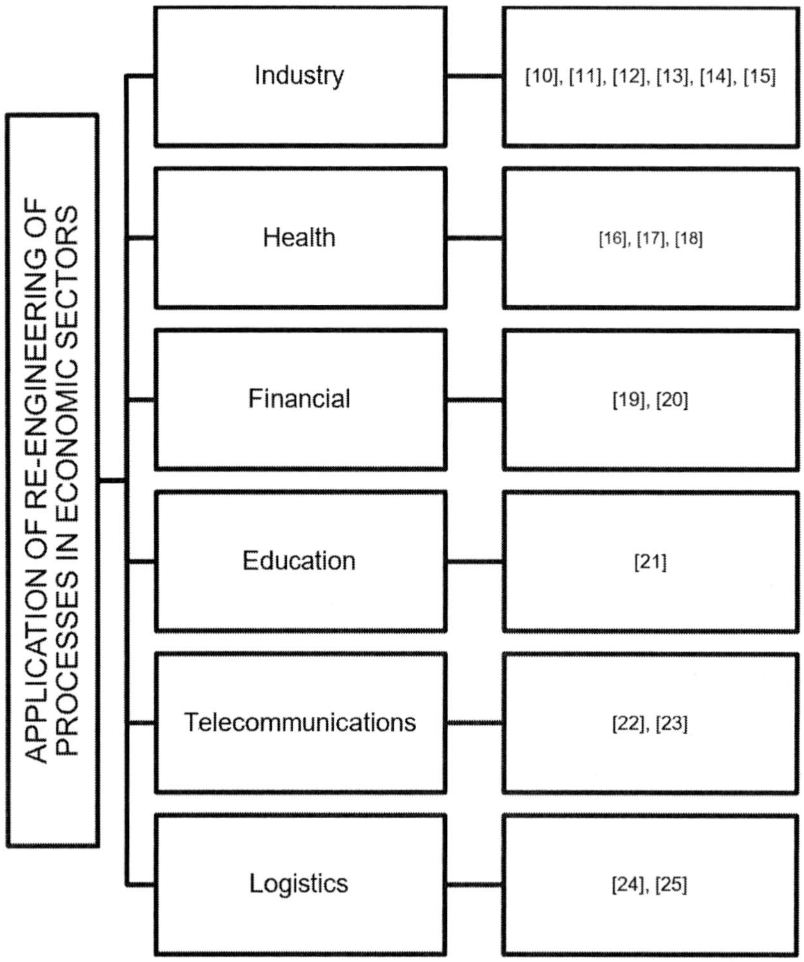

Figure 5. Economic sectors in which Process Reengineering has been applied, and its documented case studies.

4. Results and Discussion

The process of simplification and synthesization of the modeled cases classified by sectors, allowed to identify the problems that through the Process Reengineering were addressed in the case studies (Table 1), likewise, the overall positive impacts that the Reengineering implementation had overall in the organizations were identified (Table 2).

Table 1. Problems addressed in each of the sectors.

PROBLEMS ADDRESSED BY THE RE-ENGINEERING PROCESSES IN DIFFERENT ECONOMIC SECTORS	Health	Financial	Telecommunications	Industry	Logistics	Education
Customer service / equipment warranty			X			
Lack of quality policies			X			X
High operating costs	X	X	X	X	X	
Staff commitment			X			
Lack of organizational culture			X			
Information systems			X			X
Inefficient processes				X		
Inefficient resources	X	X		X		
Obsolete resources	X	X	X	X		
Complex processes	X	X		X	X	
Reprocesses	X	X		X	X	
Excess staff	X	X	X	X		
Lack of staff	X	X		X		
Indicator failure	X			X	X	
Little follow up processes	X	X				
Poor documentation	X	X		X		
Poor communication with the client	X	X			X	
Delayed processes		X		X		
Uncoordinated processes	X	X			X	
Little organizational culture		X				
Lack of quality policies / manuals		X		X		X
Breach of customer requirements	X	X			X	
Changes in the market		X				
Weak information system / Failures in physical infrastructure		X				
Supplier administration				X		
Market transformation			X			
Strengthening competition			X			
Changing economy			X			

IOP Conf. Series: Materials Science and Engineering **844** (2020) 012052 doi:10.1088/1757-899X/844/1/012052

Table 2. Positive impacts reported in the case studies by economic sector.

POSITIVE IMPACTS ACHIEVED BY THE REENGINEERING OF PROCESSES IN DIFFERENT ECONOMIC SECTORS	Health	Financial	Telecommunications	Industry	Logistics	Education
Improvements in service quality	X	X	X	X		
Customer increase	X	X	X	X		
Generation of customer benefits		X	X	X		
Staff reduction	X	X	X	X		
Less work each	X	X	X	X		
Efficient staff	X	X	X	X		
Increased motivation and commitment of staff	X	X	X	X		
Shorter response times	X					
Improve operating times		X	X	X	X	X
Process automation	X	X	X	X	X	
Effective and quality processes	X	X	X	X	X	X
Less complex processes	X	X	X	X	X	X
Shorter waiting times for the customer	X					
Reduction of operational costs	X	X				X
Better information and documentation management	X	X				
Occupational risk reduction	X					
Information traceability	X		X	X		X
Improvement in the indicators	X		X	X		X
Competitiveness and productivity improvements			X			

The problems were synthesized in eight groups: problems of processes, quality, costs, documentation, competitiveness, organizational culture, and resources, as can be seen in Table 3, and the documented causes of each in the case studies, which laid the foundations for the construction of this diagnostic tool.

Table 3. Classification of the problems addressed.

Type of problem addressed by process reengineering	Cause
Processes	• Delayed • Expensive • Inefficient • Uncoordinated • Complexes • Reprocesses • With little follow-up • Problems with the indicators
Quality of service Customer satisfaction	• Poor communication • Breach of requirements
High costs	• Organizational • Operation
Documentation	• Lack of information systems • Function's manual • Processes / Requirement • Quality Policies
Competitiveness	• Changes in the market
Lack of Organizational Culture	• Lack of staff commitment
Resources	• Obsolete • Inefficient • Lack or excess of staff

Likewise, the positive impacts that in general the reengineering implementation achieved in organizations are identified fields like: service, organizational structure, processes, and management. Table 4 belowallows seeing a panorama of results that any organization might be in need, and use it's as guidelines for the decision-making of process improvement projects within the organization, for purposes of continuous improvement.

Table 4. Classification of results, positive impacts obtained.

Impacted area	Type of impact achieved by process reengineering
Service improvements	• Customer Increase • Generation of customer benefits • Increase in the perception of quality • Decreases in customer waiting times
Organizational structure	• Staff reduction • Decrease in workload • Efficient staff • Increased motivation and commitment of staff
Process improvement	• Decrease in response times • Improvement in operational flow • Effective and less complex processes • Reduction of operational costs • Improvement in information and documentation management
Management improvement	• Occupational Risk Reduction • Information traceability • Improvement in the flow of information • Indicator improvements

Based on the problems addressed and their causes (Table 3), diagnostic questions formulated based on the opinion of the members of an organization about the processes, or departments of a company. In the same way, said diagnostic questions can be applied to an organization. During the correlation, we sought to synthesize the relationships in order to reduce redundancy and facilitate the analysis of the respondent, finally was generated three categories of diagnostic questions: Management and Operation, Resources, and Response Capacity. This resulted in a total of 15 diagnostic questions, which can be seen in Table 5 below. Each of these questions is correlated with the identified problems that an organization may present, and that were documented throughout the scientific literature.

Table 5. Diagnostic questions for the implementation of process reengineering, and its relation to problems.

	Diagnostic question	Associated problem
MANAGEMENT AND OPERATION	1. In the processes, there are moments in which results are not delivered in the required time?	Delayed processes
	2. Do you consider that the processes have problems with the planning of the activities?	Uncoordinated processes
	3. Do you consider that studies are carried out in the processes to see the effectiveness of the results according to the initial planning?	Processes with little follow-up
	4 Do you consider that activities are repeated, or that many of them do not add value to the processes?	Reprocesses in the organization
	5. Do you consider that the processes ask many requirements to carry out your activities?	Complex processes
RESOURCES	6. According to what is observed in the market, do you consider that the processes have a high expenditure of resources such as: money, time, personnel or machinery for example?	High operating costs
	7. Does management commit sufficient resources to the development of processes?	Insufficient resources in the processes
	8. Does the company have the resources required for the development of all its activities?	Insufficient resources in the processes
	9. Do you consider that the documentation of quality policies, function manuals, procedures and instructions for the development of daily activities in the organization, is deficient?	Bad documentation in the processes
	10. Do you think you have obsolete and/or inefficient resources?	Obsolete resources
	11. Do you think there is a lack of organizational culture and staff commitment?	Staff Commitment
ANSWER'S CAPACITY	12. How often are complaints or claims, made in the organization's processes?	Breach of requirements
	13. How often do you fail to comply with the requirements requested in the processes?	Breach of requirements
	14. Do you think there are failures in the internal communication of the processes?	Poor communication
	15. Do you consider that complaints and claims in the processes generally managed?	Low quality of service

Diagnostic questions can handle multiple answers such as Always, Sometimes, Never, Yes, No and N/A, which should be applied to all concerned according to the domain of the analysis: The organization, or processes or departments. When evaluating the entire organization globally, given that each question diagnoses a type of problem present in the organization that can be addressed from a reengineering approach, for each answer is gived a weight. As can be seen in Figure 6, for "always" is given a weight of 1, for "sometimes" is given 0.75, for "never" is given 0, and for N/A is given 0,5. This weighting of each percentage of response is totaled and compared with a reference point that will indicate whether or not to do reengineering, as is ilustraded in Figure 6.

Table 6. Illustrated case of diagnostic evaluation for an organization.

	Results Distribution								Weights				Reference point: 50%	
Diagnostic Question	Always		Sometimes		Never		N/A		1	0,75	0	0,5	Total	Recommendation
1. In the processes, there are moments in which results are not delivered in the required time?	5	25%	5	25%	6	30%	4	20%	25%	19%	0%	10%	54%	APPLY RADICAL IMPROVEMENTS

	Results Distribution								Weights				Reference point: 50%	
Diagnostic Question	Always		Sometimes		Never		N/A		0	0,75	1	0,5	Total	Recommendation
3. Do you consider that studies are carried out in the processes to see the effectiveness of the results according to the initial planning?	7	35%	3	15%	4	20%	6	30%	0%	11%	20%	15%	46%	ANY

Responsible for the analysis, should parameterize as they consider the weights for each answers, the importance of deciding the percentages of each question is because, although the types of answers may be common, these do not apply in the same way for each diagnostic question. As evidenced in the illustrated case in Figure 6, the weights of "never" turn to 1 due to is negative this answer considering the question 3, and the weigh of "always" turns to 0 due the positive consequence that always evaluates the process checking effectiveness in the process. The Reference Point evaluates if it is recommended to apply or not Radical Improvements, serving as a measuring gate of the level of a certain problem in each diagnostic question. Varying reference point, is due to a continuous evaluation strategy over time into the organization, where the diagnostic tool is applied periodically, improvements are achieved, so the problems should be less and less through time, making every time harder to identify some problem. So the requirement for initiate reengineering should decrease, increasing the sensibility to explore the need of process improvement (Figure 7). In this projected scenario, just start a deep study of the problem identified should be considered to activate a reengineering improvement process.

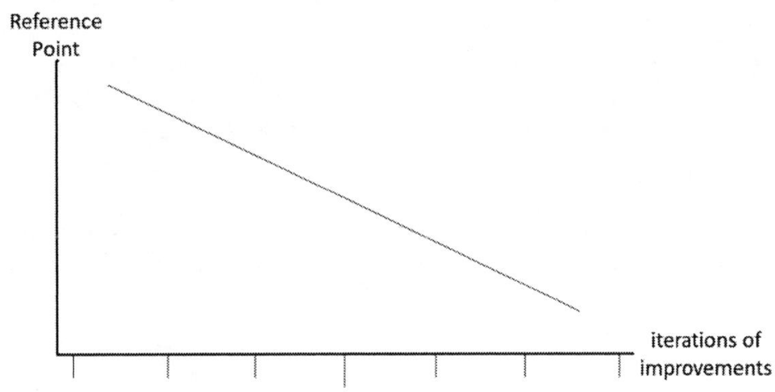

Figure 6. Behavior of the reference point over time for evaluating reengineering implementation.

IOP Conf. Series: Materials Science and Engineering **844** (2020) 012052 doi:10.1088/1757-899X/844/1/012052

If necessary, that the diagnostic tool be applied to processes or departments, executing the evaluation model explained above respondents must associate the responses to one or more processes or departments as they are being evaluated, in order that the application recommendation of reengineering be associated process through a Pareto Analysis as illustrated in Figure 7 below, where purchases and storage are the candidate departments to start a reengineering processes, seeking to address a possible problem of uncoordinated processes according to the relationship of the diagnosed question and the associated problem (Table 5). In this way, through this tool product of the described research, it can be determined if the processes or departments of an organization need to be intervened or at least analyzed under the optics of process improvement according to need.

2. Do you consider that the processes have problems with the planning of the activities?

PROCESSES/DEPT.	FREC-Q2	% Accumulated		80-20
PURCHASES	8	38%	8	80%
MAINTENANCE	6	67%	14	80%
WAREHOUSE	5	90%	19	80%
MANAGEMENT	2	100%	21	80%
ACCOUNTING	0	100%	21	80%
HUMAN RESOURCES	0	100%	21	80%

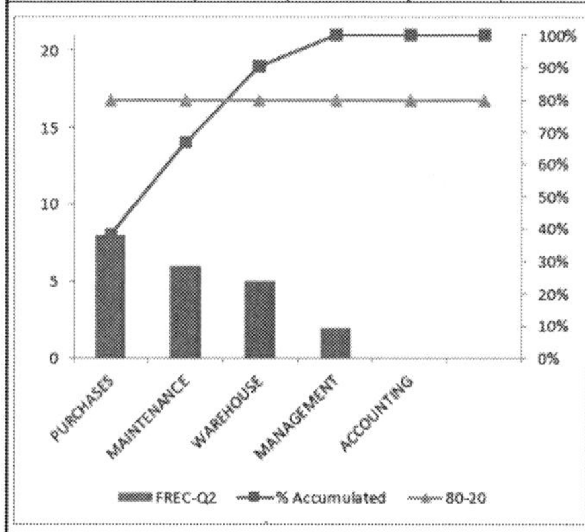

Figure 7. Illustrated case of diagnostic evaluation for an organization.

Tools such these represent a contribution to the business process management since new technologies and information systems have facilitated the capture and evaluation of information for making effective decisions. The authors of this paper propose process diagnosis tools use in improvement strategies which perceive organizations as a chain of events organized in process, whose performances are causes of bad or good results, determinants to reach business goals [26]. Quality Management Systems (QMS) or disciplines as Busines Process Management (BPM) are fully integrable with tools like these. This diagnostic tool can also assist QMS based on a PHVA cycle, with a point of view external assessment in order to identify aspects to make an improvement beyond just the process assurance. BPM looks at the entire organization as a process network that must be managed to impact the organization aims; because of the magnitude, complexity, and quantity of tasks that it must handle, BPM must rely on IT technology resources that support an organization, or software platforms such as Business Activity Monitoring, to identify risks and improve opportunities in which diagnostics tools play an important role (Figure 8).

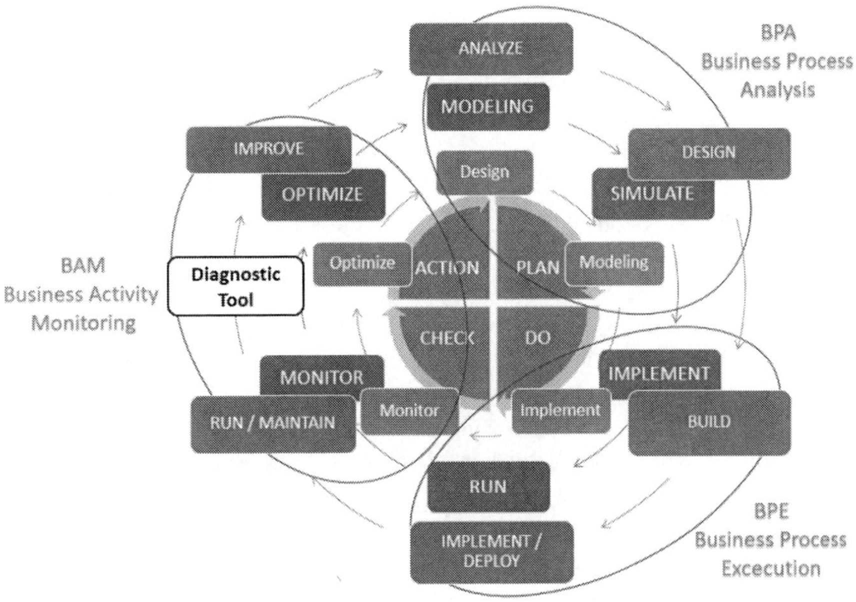

Figure 8. Diagnostic tools

5. Conclusions.

This research generated a diagnostic tool that will allow organizations in decision-making for the implementation of strategies for process improvement. The tool seeks to contribute to align processes performance with organizational goals and customer satisfaction, both internal and external, especially in those cases where incremental improvements are not enough and a radical change in the processes of organizations is necessary. The tool is based on 15 diagnostic questions that evaluate according to the level of the opinion of the members of an organization if it or its processes or its departments present some kind of problem, which has been documentarily addressed by a radical improvement of processes.

The problems on which the diagnostic tool is based were identified from a review of the literature that sought to collect case studies of the application of the methodology. Through the review, it was also possible to identify that reengineering has been applied in the financial, health, telecommunications, industrial, logistics, and education sectors and that many of the problems were common in all sectors. The review also allowed the identification of a consolidated benefit that was commonly achieved in many case studies in different economic sectors.

Reengineering is based on documenting as little as possible and focusing its objectives on redesigning one or more processes by completely recreating them, in order to achieve processes that (product of the redesign), truly tune the organization to its growth. Based on this, the proposed diagnostic tool indicates the need to improve processes from a radical approach either throughout the organization, departments or processes. Also is fully integrable with PDCA cycles, and disciplines like BPM which has IT-based systems like Business Activity Monitoring. This is relevant in today's contexts when progressive improvements do not become a pragmatic and agile alternative in the business context, and most of nowadays organization are technology-based.

Acknowledgement

This research was executed within the Process Quality of Process Research Program, as part of the —QoProcess Framework technology package, from Universidad de la Costa CUC.

References

[1] Alvarado F 2018 Interviewee, *BPM: metodologías para lograr un cambio radical en los procesos.* [Entrevista]. 04 09 2018.

[2] Johnson M E, Cramer and Cross Aimin R L Sources of Fidelity in Purposive Organizational Change: Lessons from a Re-engineering Case (*Journal of Management Studies,* vol. 40, nº 7) pp. 1837-1870, 2003.

[3] Suárez-Barraza M and Ramis-Pujol J 2008 Aplicación y Evolución de la Mejora Continua de Procesos en la Administración Pública (*Journal of Globalization, Competitiveness & Governability,* vol. 2, nº 1) pp. 74-86, 2008.

[4] Villasmil Molero M and Crissien Borrero T 2015 Cambio de paradigma en la gestión universitaria basado en la teoría y praxis de la reingeniería, (*Journal Económicas CUC,* vol. 36, nº 1) pp. 245-274, 2015.

[5] Sánchez-Comas A 2018 *QoProcess: framework para el mejoramiento de procesos durante la implementación y certificación de sistemas de gestión de calidad,* Barranquilla: Universidad de la Costa

[6] Sanchez-Comas A, Neira-Rodado D and Cabello-Eras J J 2016 Marcos aplicados a la gestión de calidad - Una revisión sistemática de la literatura *(Journal Espacios vol 37)* pp. 17-34.

[7] Caicedo Garcia M and Pérez Vargas M 2018 *Marco para la aplicación de mejoras radicales en los procesos durante la implementación y certificación de sistemas de gestión de la calidad, (*Barranquilla: Universidad de la Costa).

[8] Bancoldex, 2016 10 tips para mejorar la productividad de su empresa (*Revista Dinero)*

[9] ISOTools 2017 www.isotools.com.co. [En línea]. Available: https://www.isotools.com.co/3-estrategias-para-lograr-la-excelencia-de-los-procesos-empresariales-de-hse. [Último acceso: 20 07 2019].

[10] Bounds G and Hewitt F 1995 Xerox: Envisioning a corporate transformation *(Journal Review of Financial economics,* vol. 4, nº 1) pp. 3-17

[11] Moreno-Garcia R R and Parra-Bofill S 2017 Methodology for the reengineering of processes. Validation in the company Cereals Santiago (*Ingeniería Industrial,* vol. 38, nº 2) pp. 130-142

[12] Paliwal S, Seth D, Yadav D, Yadav R and Paliwal S 2011 Development of a robust QSAR model to predict the affinity of pyrrolidine analogs for dipeptidyl peptidase IV *(Journal of Enzyme Inhibition and Medicinal Chemistry,* vol. 26, nº 1) pp. 129-140

[13] Yong L, and W. Jian-ping W 2007 The Optimization Design of Measure & Payment Process in Construction by Using the Business Process Reengineering Technolog *(International Conference on Management Science and Engineering)*

[14] Moreno-Garcia R and Parra-Perez K 2012 La reingenieria de procesos una herramienta en la gestion de negocios: Caso Cereales Santiago (*Ciencia en su PC N° 4)* pp. 85-98

[15] Gómez-Palacios J and Noriega-Zambrano P 2018 *Reingeniería de Procesos Operativos y Administrativos en el Area Soporte de Ventas de Unidades Nuevas de un Concesionario Automotriz (http://www.dspace.espol.edu.ec/xmlui/handle/123456789/42369* Guayaquil-Ecuador)

[16] Troncoso-Palacio A, Neira-Rodado D, Ortíz-Barrios M, Jiménez-Delgado G, Hernández-Palma H 2018 Using Discrete-Event-Simulation for Improving Operational Efficiency in Laboratories: A Case Study in Pharmaceutical Industry. (Shangai China Advances in Swarm Intelligence. ICSI 2018. Lecture Notes in Computer Science, vol 10942. Springer, Cham) pp. 440-451

[17] V. Acosta Carpio, *Una metodologia de redise;o de procesos de negocios basada en la teoria de la reestructuraci[on de las organizaciones,* Santiago: Facultad Econom[ia y Negocios Universidad de chile, 2017.

[18] J. Osorio Acosta J and Paredes Alonso E 2001 Reingeniería de procesos en los hospitales públicos: ¿reinventando la rueda? (*Revista española de salud pública,* vol. 75, nº 3) pp. 193-206

[19] Conde Paucar I O and Luján Campo L A 1998 *Reingeniería de procesos corporativos para una compañía de seguros*

[20] Coley T 1994 Learning to Change in SmithKline Beecham *(Business Change & re-engineering,* vol. 1, nº 4) pp. 10-19

[21] M. Sadiq Sohail M, Daud S and Rajadurai J 2006 Restructuring a higher education institution: A case study from a developing country (*International Journal of Educational Management,* vol. 20, nº 4) pp. 279-290

[22] Choi C J 1995 Samsung: Re-engineering Korean style *(Long Range Planning,* vol. 28, nº 4) pp. 6-7, 74-80

[23] Hill F M 2010 Total quality management and business process re-engineering: A study of incremental and radical approaches to change management at BTNI *(Total Quality Management,* vol. 10, nº 1) pp. 37-45

[24] Rotab Khan M 2000 Business process reengineering of an air cargo handling process *(International Journal of Production Economics,* vol. 63, nº 1) pp. 99-108

[25] Rio Belver R M and Contreras Romeral M T 2009 Re-Ingeniería en el proceso de gestión de reclamaciones: aplicación al sector aeronaútico *(DYNA,* vol. 84, nº 6) pp. 485-492

[26] De-La Hoz-Hernandez J, Troncoso Palacio A and De-La-Hoz-Franco E 2019 Implementation of the Eclipse Process Framework Composer Tool for the Documentation of Quality Management Systems: A Case Applied in Healthcare Services *(Lecture Notes in Computer Science,* vol. 11656) pp. 200-210

Student perception of the use of ICT on learning Morphophysiology

A Quintana[1]

[1] Surgical Instrumentation Program of Rafael Núñez University Corporation, Cartagena, Colombia.

Email: alvaro.quintana@curnvirtual.edu.co

Abstract. The possibility of working in Morphology using the Technologies of Information and Communication promoting an interactive process focused on the student, who does not leave dependence, sometimes excessive, of the teachers to achieve the objectives. The general objective was to determine the importance of the use of ICT in Morphophysiology learning. A cross-sectional descriptive study with a quantitative approach was carried out. The results of the application of a questionnaire that collected the information in several dimensions. A number of 90 people were surveyed, which accounts for 32% of the population of morphology and physiology basic science students of the Nursing, Surgical Instrumentation and Prehospital Care programs of the Rafael Núñez University Corporation of Cartagena, in the period between February and June 2018. The results show that the use of ICT to support Morphophysiology learning is very well received by students and favors the learning in health sciences. ICT are tools that validate and support the observation and experimentation in the laboratory simulation.

1. Introduction

Since the last decade of the twentieth century, worldwide it has tried to modernize the educational methodology, so that it gradually incorporates the use of specialized software and some information and communication technologies (ICT). One of the changes proposed at the educational level has been to move from the classroom as an exclusive form of teaching, to a mixed model that combines the lessons in the classroom with the use of digital material accessible on the Internet. Education, at all levels, has been in need of change because the very society in which it is inevitably is no longer the same society in which it was raised. The current postmodern society is far, too, from the curricular programs that were once designed. In this stage of transition, education has tried to renew each other towards digital technology under an integrative-educational conception.

Current education aims at the use of information and communications technologies (ICT) as a fundamental tool to strengthen the teaching-learning process. The increasingly frequent use of simulators and other computer aids in medical education is evident, which allows the student to repeat the training as many times as necessary until acquiring the expected skills. Recent studies show that learning curves based on computer media are better than learning curves based on classical training [1].

For authors such as Jean Piaget, all people act when it comes contact with the environment. Their initial actions are open and quickly coordinated in series, the " cognitive structures of learning." these series of coordinated actions are what he calls " schemes" for then divers schemes will the body has developed at a particular time to form a "structure". Then as the organism develops, its cognitive structures change from the "instinctive" through the "sensory-motor" to the "operational" structure of adult thinking, is what Piaget calls "cognitive structures for learning [2].

Piaget explains in the exposition of his theory, how cognitive development is the result of the active adaptation of the organism to the environment, through assimilation and accommodation. As a result, variable cognitive structures are developed, which are, in turn, coordinated schemes, and which, at the same time, combine to form intelligence.
For Vigotsky, with his constructivist theory based on social interaction; The learning-development process expresses a complex unit that seeks to boost development from learning, which implies the use of tools, of pedagogical strategies constructed and innovative, in a logic of accompaniment by the teacher, academic peers, and even the same students among them, which would lead to the production of the combined action between individual and collective work [3].

For Bruner, the assimilation of information by humans is four different ways: visual, auditory, read/write and kinesthetic (VARK: *Visual Auditory Reading Kinesthetic*), and that such information requires a scaffold, which is a type of support directed by family members, teachers or experts that they carry out with the purpose of channeling and enriching the behavior of an apprentice [4].

The visual learner, you are easier to remember images, graphs, diagrams, photographs, etc., So to study morphology, anatomy, physiology, some authors recommend use texts such as atlas, structures, infograms, ideograms, maps, slides, photographs, visits to museums, use videos in which the visual material is accompanied by an explanation, or in laboratories where a teaching museum carefully observe the anatomical preparations there [5].

Kinesthetic student learns more through processes read and write, have an affinity for textbooks, the dictionaries, glossaries, notes written by teachers and websites with relevant information about Topics to study. Kinesthetics are also very good at using software and web resources that allow them to develop their skills and concerns [5].

From the cyber theory proposed by Norbert Wiener, it is assumed that the knowledge and results found both by associative theories (stimuli and responses) as well as from cognitive theories, can be used in association with computer science, providing the properties and characteristics of computerized system to the structuring of thought. This approach considers learning as a process of communication and control, a process constituted by a set of internal operations or events that are responsible for processing the information, and whose result is issued abroad in the form of observable behavior [6].

According to the UNESCO, in any educational area, strategic objectives should aim to improve the quality of education through diversified content and methods. This should promote experimentation, innovation, dissemination and sharing of information and good practices, the formation of learning communities and stimulate a fluid dialogue on the policies to follow [7].

Melo Santiesteban, Guadalupe; Velazquez Licea Eulalio; Pérez Morales, Alfonso; and cols, posed the research problem: How do ICTs help improve the teaching-learning process in health sciences, especially the morphology and physiology subjects? This study was carried out in the Faculty of Medicine of the Veracruzana University, Mexico; and the period August 2006 to February 2008. The aim this study was to find better strategies for teaching morphology in medical school. They concluded that ICTs contribute to providing information in a different way, which allows us to understand the axiological contents implicit in the subjects of morphology by strengthening the teaching and learning process of medical students at the University of Veracruz [8].

For McNulty, the students learn anatomy with the help of computer programs I webpages, as well as traditional methods such as dissection/prosection, get better test scores than those who never used these computer contents. Most students find computer programs useful for learning and using them themselves. Students who accessed computer resources more frequently scored higher on exams compared to students who never accessed them. Students using resources as web presentations, videos and others can be very useful and available at any time to be reviewed, given the advances in internet and MOs devices vile and their applications, students can save this information or Search for it immediately using "web browsing". While those who only study anatomy by dissection of corpses, they will forget the concepts and terms more frequently [9].

IOP Conf. Series: Materials Science and Engineering **844** (2020) 012053 doi:10.1088/1757-899X/844/1/012053

A study by Dobson, JL, between the fall of 200 8 and the spring of 2009, applying an online survey to a physiology undergraduate course at the University of Florida, with the purpose of investigating the relationship between: preferences and styles learning; gender and exam results.

In the survey, students were asked to provide descriptive information about themselves (for example, gender and older) and he asked them to self-assess their preferences sensory modality. Among the results achieved, a total of 901 students completed the questionnaire, 75% of which were women and 25% were men. The most predominant learning style was visual learning (46% for women and 49% for men) It was determined that the type of student learning is fundamental to the achievement of the objectives and the final impact of the course grades [10].

Burn, points out the variety of possibilities to teach with the new information and communication technologies, since these allow at a much larger scale, cheaper in time and more effective in the field of communication, perform the following processes that They can be used for teaching: Interaction (indefinitely review), feedback (deployment of the work process), convergence (integration of different authoring modes : video and audio), and inhibition (being able to deploy the work in different formats and platform for more audiences) [11].

Jenkins states in his studies that the use of ICT, allows in various subjects, "operations with knowledge" among which is the simulation, which allows interpreting and building dynamic models of real-world processes [12].

Algieri, Ruben Daniel Ferrante, Maria Soledad, Mazzoglio and Nabar, Martin Javier made a study whose main objective was to objectively evaluate you results showed that 86.9% (n = 133) of respondents believed necessary the use of other resources ICT during the practical work of Anatomy to complement the learning that is carried out with the cadaveric material. Among the possible pedagogical tools, 72.5% express preferences for classes with PowerPoint presentations; in the second place it was selected using 3D presentations containing reconstructions computer anatomical imaging (70.6%) the educational resource less useful from the viewpoint of students were mockups (16.3%) and classical classes with chalk-board (32%) 87.6% believe that the existence of virtual elements serve the educational process beyond the physical limits of the University [13].

Benvenuto, clarifies that technology will continue to provide new tools for teaching use and the answer is not only to provide computer resources to universities, but also to develop in teachers the skills and abilities in the use of ICT [14].

Fernandez, highlights the imperative need to train teachers in the use of ICT from well-founded methodological criteria so that they can select, use, design and produce the audiovisual and computer material that is required to guide student learning [15].

This study that considers the use of ICT by students of the Nursing, Surgical Instrumentation and Pre hospital Care programs of the Rafael Núñez University Corporation, to learn morphology and physiology, will allow the appropriation of new knowledge, the transformation of conceptions methodological, measurement of conceptual limits and improvement of the didactic strategies used, enriching the teaching and learning of morphology and physiology in students of health sciences.

The objective of this research was to determine the importance of the use of ICTs for the learning of morphology and physiology in the Nursing, Surgical Instrumentation and Prehospital Care programs of the Rafael Núñez University Corporation, during the period between February and June 2013.

2. Methodology

The research approach was quantitative, based on a non-experimental transversal design, with a descriptive methodology. For this purpose, a structured survey was designed, subjected to opinion of experts, to assess reliability and content validity of this instrument.

Respecting the ethical parameters for human studies, student participation was voluntary, respecting their right not to participate. There was no possibility of any type of damage as a result of student participation in the study. The information management was confidential requesting 90 students to answer a simple questionnaire delivered via the web through the institutional email.

A random sample of 90 study subjects was included on a population framework of 253 students, regardless of race, age or sex. A confidence level of 95% and a margin of error of 5% were considered for random sampling. Processing and results analysis was used descriptive statistics, from statistical processing with the program Microsoft Excel 2010, availing ourselves of the resources of graphic tabulation of data and application was made statistical formulas for the measurement of variables.

3. Results

When asked about the place where students access to ICT resources, most prefer doing at home (72%) and a smaller percentage in the library (16%) and other much less common spaces like Teachers room or computer rooms (Look figure 1).

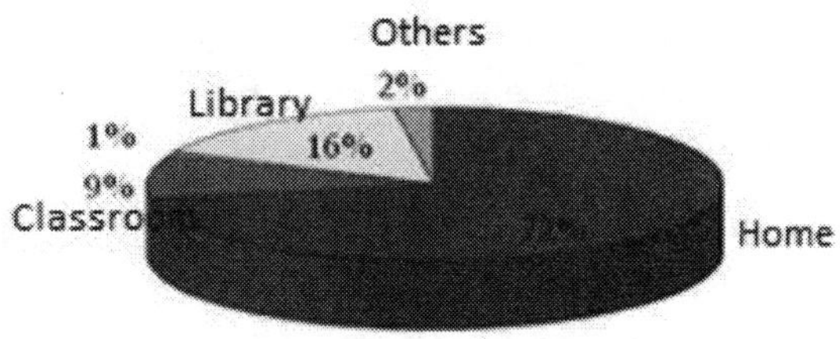

Figure 1. Place to access to the computer resources most frequently

Regarding access to computer resources, which ICT were most used to study morphology and physiology contents. Giving them from the question some ICT options more frequently known in the academic field. 51 students, that is, 57% of the sample, answered that if they use them very frequently, and 36 students (40%) say they use them but very little. While 3 students (3%) do not use them at all (Look figure 2).

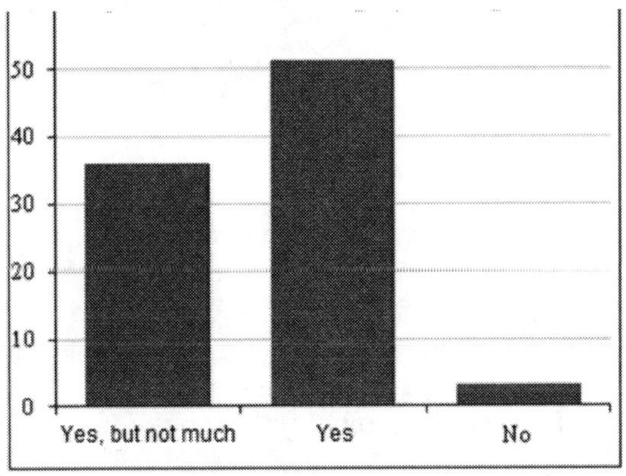

Figure 2. Frequency o use ICT to study morphology and physiology contents

When asked about the use of search engines such as Google, Wikipedia, among others to study Morphology and Physiology, 94% (85 students) state that they use them frequently for this purpose, compared to 4 % (4 students) They don't use them (Figure 3).

IOP Conf. Series: Materials Science and Engineering **844** (2020) 012053 doi:10.1088/1757-899X/844/1/012053

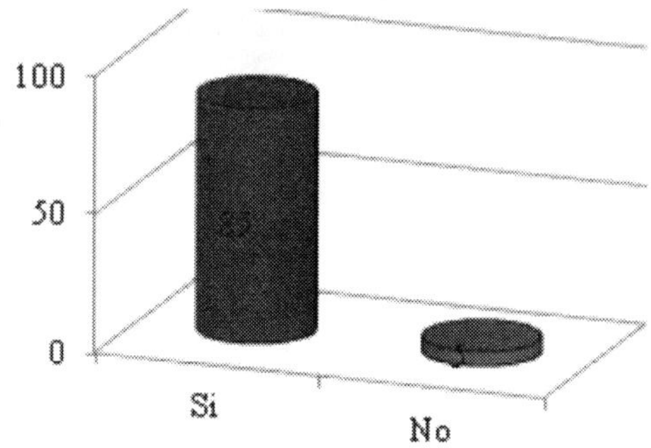

Figure 3. Use of Searching engines like Google, Wikipedia and others to study morphologhy and Phisiology

When asking the student whether or not he uses ICT in the morphology and physiology simulation laboratory, 62 students (70%) replied that if they used them and 27 (30%) who did not use them to support learning in The laboratory (figure 4).

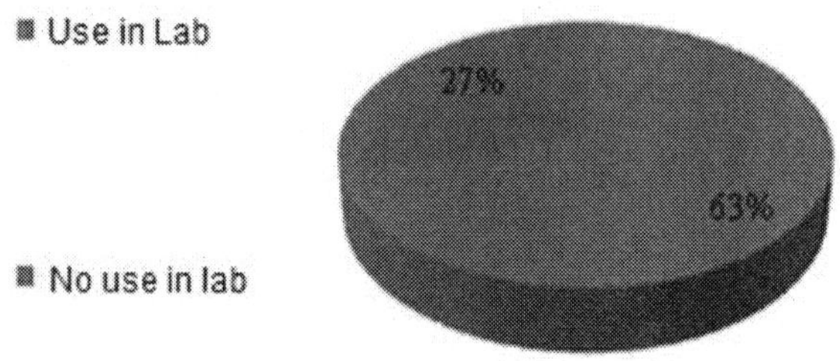

Figure 4. Use ICT in Laboratory spaces

As the student considers or not the Technologies Information and Communication have favored their teaching and learning morphology and physiology, the 99% state that ICT, if favored in the teaching process learning morphology and physiology, and only 1 student (1%) stated that these have not helped their teaching and learning of morphology and physiology (Figure 5).

Figure 5. Perception of improved knowledge with ICT

When inquiring about the characteristics or qualities of Information and Communication Technologies, which have been able to favor the teaching and learning processes of morphology and physiology; it was noted that one study, you believe these have favored the individualization of teaching. To 63 students the variety of codes information usually offered by ICT (text, sound, images, etc.) encourages them to prefer them to any other remedy. 5 students believe they favored autonomous learning. 8 students with the ICT give high motivation which can stimulate to use. For 12 students, what motivates them to rely on ICT when studying morphology and physiology is the ease of using them and the ease of access to them (figure 6).

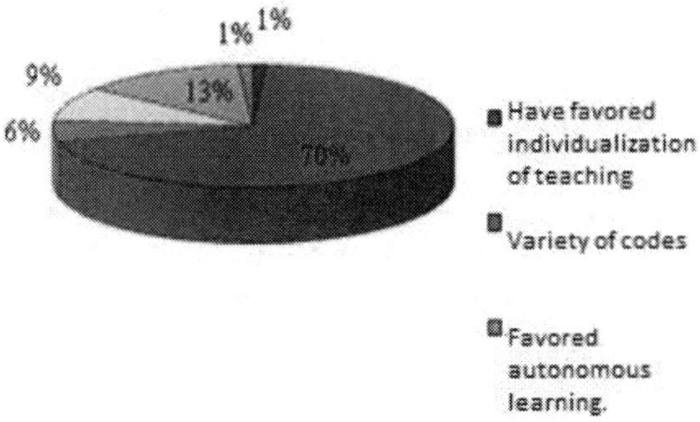

Figure 6. Features Technologies Information and Communication favors the processes of teaching and learning morphophyisiology

4. Considerations

The purpose of this study was to determine the importance of using ICT for learning morphology and physiology in nursing, Surgical instrumentation and Care Prehospital programs of the University Corporation Rafael Nunez, during the period from February to June 2018.

The findings made it possible to confirm specific aspects of the approaches of other authors such as Vázquez-Mata and Guillamet-Lloveras; who affirm that the use of ICT and computer tools of various kinds, favor the teaching-learning process of the students.

The use of ICT to study morphology and physiology is one of the many well-structured and innovative pedagogical tools that motivate the student to learn, either individually or collectively. This study allowed highlighting a series of interesting collaborative processes, which are being carried out in different scenarios such as the university classrooms, library, in a more autonomous place such as the student's house or in the simulation laboratories, aspects that also They have been proposed by Vigotsky, Jenkins, Burn among others.

As Martínez Marrero reflects, and given that some ICTs incorporate images, graphics, diagrams, photographs and videos; This study showed that these tools they are very useful to students couple studying morphology and f physiology.

Results show that students using recur sos ICT as presentations, video and others; They are motivated to use this type of resources because in some way and according to their learning style, they have the perception that they are given the appropriation of content, as McNulty, Sonntag, and Sinacore demonstrated.

According to the results of this study, the ICT tools most used by students to support the learning of anatomy and physiology were: ICT that use presentations, videos, interactive images, as they also observed in their Algieri study, Ferrante, Mazzoglio and Nabar.

5. Conclusions

The appropriation they make of ICT, the morphology and physiology students of the Rafael Núñez University Corporation is considerable high. They mostly use ICT tools that allow presentation resources, since they generally offer simple content and that by means of slides or slides, combine diversity in the presentation of information, whether in text, drawings, graphics or videos.

For students, the use of ICT to support simulation, is of great importance, because through these tools can understand better the tissues, organs, system as and physiological processes; tracking concepts, testing hypotheses, better detailing structures and facilitating them to propose better explanations. The interactivity that offer certain ICT tools in simulation labs, used simultaneously with a model or simulator itself, allows students cognitive exercises, imaging ideas and concepts more extended, thereby strengthening critical thinking when they compare simulators with s interactive content.

Universities must continue to promote institutional and academic actions from their management, with the purpose of identifying and strengthening in the students of higher education, skills and abilities necessary for the use of ICT.

In order to strengthen the teaching-learning processes of the sciences that require simulation components, it is convenient for universities to have specific virtual laboratories, which allow developing teaching objectives that are increasingly closer to the experimental. Morphology and physiology laboratories can be partially or totally considered in their infrastructure, such as virtual scenarios that use the so-called Virtual Learning Environments (VAS), which take advantage of the ICT functionalities, thus offering new environments for teaching and learning. freer from the restrictions imposed by time and space in face-to-face teaching and are able to ensure communication between students and teachers.

ICT are not intended to replace the use of simulators and anatomical and physiological models, but instead are tools that validate and support the observation and experimentation in the laboratory simulation, strengthening mental operations that by representations, positively influence the construction of interactive processes, focus on the student from the basic sciences of morphology and physiology.

The use of ICT in simulation exercises involves the incorporation of many resources such as internet, computers, electronic tablets, among others, which not all students can bring to the laboratory due to economic limitations, security conditions, among other factors; It is well worth considering for further studies.

It is necessary to expand the sample in subsequent studies, in order to make it more representative with respect to the total population, as well as making some kind of qualitative interpretation that yield deeper considerations and more comprehensive readings, than those offered by some quantitative studies.

References

[1] Vázquez-Mata, G., and Guillamet-Lloveras, A 2009 El entrenamiento basado en la simulación como innovación imprescindible en la formación médica Medical Education 12 149-155.
[2] Piaget J 1990. The child's conception of the World. New York: Littlefield Adams
[3] Vigotsky L 1979 The development of higher psychological processes Barcelona.
[4] Bruner J 1967 Toward a Theory of Instruction Cambridge Harvard Univ Press.
[5] Martínez Marrero E 2012 How to study anatomy Barranquilla Editorial Universidad del Norte pp 1-32.
[6] Wiener N 1948 The Cyber Theory of Learning Austin Texas the University of Texas.
[7] UNESCO 2004 Information and communication technologies in teacher training. Planning Guide Paris France.
[8] Melo Santiesteban, G and Velasquez Licea EP 2008 Didactics and technology in morphology of medical sciences.
[9] McNulty J Sonntag B and Sinacore J 2009 Evaluation of computer-aided instruction in a gross Anatomy course: a six-year study Anatomical Sciences Education pp 2-8
[10] Dobson J 2009 Learning style preferences and course performance in an undergraduate physiology class Advances in Physiology Education 33 pp 308-314.
[11] Burn A 2009 Making New Media Creative Production and Digital Literacies New York Peter Lang.

[12] Jenkins H 2006 Confronting the challenges of participatory culture: Media education for the 21st century Cambridge UK MacArthur Foundation.
[13] Algieri RD Ferrante MS and Mazzoglio and Nabar MJ 2008 August. http://diegolevis.com.ar
Retrieved September 3, 2012, from http://www.diegolevis.com.ar/secciones/Articulos/tic_medicina.pdf
[14] Benvenuto Vera A 2004 Information and communications technologies (ICT) in university teaching Theoria Vol 12 pp 109-118
[15] Fernández Muñoz R 2005 The university of Castilla la Mancha and the process of convergence to the EHEA. MEC ISBN 84-689-0273-X. Part II

IOP Conf. Series: Materials Science and Engineering **844** (2020) 012054 doi:10.1088/1757-899X/844/1/012054

Manufacture of hybrid pieces using recycled R-PET, polypropylene PP and cocoa pod husks ash CPHA, by pneumatic injection controlled with LabVIEW Software and Arduino Hardware

A D Rincón-Quintero[1,2], L A Del Portillo-Valdés[2], A Meneses-Jacomé[1], C G Cárdenas-Arias[1], C L Sandoval-Rodriguez[1], J G Ascanio-Villabona[1], B E Tarazona-Romero[1], J A Sandoval-Baldión[1], J C Ortiz-Ríos[1], R A Mantilla-Olejua[1]

[1] Design and materials research group (DIMAT), Automation and Control Energy Systems Research Group (GISEAC), Electromechanical engineering, Faculty of Natural Sciences and Engineering, Technological Units of Santander (UTS), Calle de los estudiantes No 9-82, 680005, Bucaramanga, Colombia.
[2] Energy in Building research group (ENEDI), Doctoral Program in Energy Efficiency and Sustainability in Engineering and Architecture, Department of Machines and Thermal Engines, University of the Basque Country (UPV/EHU), Engineer Torres Quevedo Plaza, 1, 48013, Bilbao, BI, España.

arincon@correo.uts.edu.co

Abstract. The research consists of the design and automation of a pneumatic injector, applying LabVIEW software through an integrated Arduino hardware module, programming through an interface, the pressure that guarantees the geometric quality in the models obtained. The final result shows a piece designed in mass fraction with 40% polypropylene resin (PP), 40% recycled polyethylene terephthalate in the form of scales (R-PET) and 20% from cocoa pod husks ash (CPHA). Likewise, the platform allows to define and control the conditions of generation of the product, defining a pattern according to the type and the mass proportion between the resin of PP, R-PET and CPHA. In addition, the design of the mechanical and thermal elements, consisting of the piston and the system of clamp-type resistors, the pneumatic injection means and the raw material feed gate, improve the efficiency and effectiveness of the equipment.

1. Introduction

Polymers are a widely used material in important sectors of the economies at different scales of the territory, they are the synthetic raw material par excellence used to manufacture products that meet the needs of the population[1], however, their impact on The natural environment has consequences that eventually lead to the degradation of natural resources[2].

The industrial processes where this material is handled, is aimed at reducing costs, and the search for methodologies where reuse is part of the process [3][4].

The manufacture of molds on fixed parts with varied industrial destinations demands high quantities of this material [5]. The processes of forming of plastics vary from product to product since they are designed to meet different characteristics such as the creation of geometric parts, parts of cars or bottles for liquid filling, which count from molding or blowing and more complex processes such as extrusion which allows creations of plastic films, profiles, plastic bags among others [6].

In order to have process efficiencies, thermoplastic injection is contemplated, which allows a molding of the material to meet the demands that arise, thus allowing the design of countless pieces [7].

The methodological process consists in the design of the mechanical, thermal and electronic elements. On the one hand, the injection pressure is determined that guarantees the geometric quality of the final model, applying the pascal law. Then, the design of the integral mechanical element is carried out that involves an efficient process of injection of the hybrid piece, based on the conduction of heat in transient regime and the natural convection [8]. Next, the plunger where the mixture of components with the clamp-type resistors that transfer heat by conduction and radiation will be deposited to carry out the fusion, also, the feeding gate, the pneumatic injection system. Next, an interface is developed through LabVIEW software and Arduino hardware for the control and visualization of variables: time, melting temperature and injection pressure, developing an algorithm for data acquisition [9]–[13].

2. Work development

This research integrates components and theoretical bases regarding heat transfer, the design of mechanical elements, the theory that governs the phenomena of pneumatics and process control, being essential to define these scientific areas in a brief but effective way.

As for heat transfer, this science predicts approximately the exchange of energy between two bodies or systems that are in a temperature differential, there are three (3) ways for this phenomenon to occur: Conduction, convection and radiation .

Automation consists of the programming of processes that are developed in a repetitive and precise way, optimizing them through interfaces that can control the variables of interest. All this, based on a control system, composed of elements that regulate and monitor these variables. A useful device in this area is the Arduino hardware, being a programmable device, where behavior or functionality can be changed, by means of commands in a specific language and compatible with certain Software. One of these is LabVIEW (Laboratory Virtual Instrument Engineering Workbench), which is a program with application in automation laboratories due to its versatile characteristics.

The methodology contemplates the definition and properties of the materials used in the research. Also, the procedure carried out to achieve the stated objectives is described step by step.

2.1. Synthetic polymers

These types of polymers result from polymerization processes, from raw materials with a relatively low molecular weight. An important factor in these materials is the melting point [14], represented in Table 1.

Table 1. Melting temperature (°C) of some polymers [14]

Polymers	Melting temperature (°C)
low density polyethylene LDPE	115
high density polyethylene HDPE	137
Polyvinyl chloride PVC	176 – 212
Polypropylene PP	168 – 176
Polystyrene	240

2.2. Polyethylene terephthalate PET

PET is an easy-molding thermosetting plastic when exposed to heat transfer [15][16]; It is commonly used in beverage container containers, which for this specific case represents the interest of this study. Well, it's about recycling bottles where soda and other food products are packed. This polymer was released in 1941 by Whinfield and Dickinson, after these, increased its production and application at the industry level [17].

PET is a product of terephthalic acid and ethylene glycol and its chemical formula is as follows:

$$[-CO-C_6H_6-CO-0-CH_2-0-]$$

It is a material that has a low molecular weight, its ability to withstand stress is high, has a good physical appearance after making the molds [18]. It is important for the food industry, as it does not affect the taste of food or beverages [19][20]. In addition, it turns out to be economical because it is 100% recyclable. The physical, mechanical, thermal and chemical properties of PET are shown below [21].

Table 2. Physical Properties PET[21]

Physical Properties PET	
Water Absorption - Balance (%)	< 0.7
Density (g/cm$^{3)}$	1.3 – 1.4
Refractive Index	1.58 – 1.64
Flammability	Auto extinguishable
UV resistance	Good

Table 3. Mechanical properties PET[21]

Mechanical properties	
Coefficient of friction	0.2 – 0.4
Rockwell Hardness	M94 - 101
Tensile strength (Mpa)	190 - 160
Impact resistence (1/J.m)	13 – 35

Table 4. Thermal Properties PET[21]

Thermal Properties	
Specific heat (KJ/Kg.K)	1.0 – 1.35
Coefficient of thermal expansion ($*10^{-6}$ K^{-1})	20-80
Thermal conductivity (W/K.m)	0.15 – 0.4
Maximum use temperature (°C)	115-170
Minimum use temperature (°C)	-40 - -60
Melting temperature (°C)	230 - 260

Table 5. Chemical properties PET[21]

Chemical properties	
Concentrated acids	Good
Alkalis	Bad
Alcohols	Good
Fats and oils	Good
Halogens	Good
Aromatic hydrocarbons	Acceptable

2.3. Polypropylene PP

It is a thermoplastic that brings together a series of properties that are difficult to find in other material such as: Its high thermal stability allows it to work for a long time at a temperature of 100°C in the air [22].

It is also resistant to boiling water and can be sterilized at temperatures up to 140 °C without fear of deformation.

Advantage:

- Light
- High tensile and compression resistance
- Excellent dielectric properties
- Resistance to most acids and alkalis
- Low moisture absorption coefficient

Structurally it is a vinyl polymer, similar to polyethylene, only one of the carbons of the monomer unit has a methyl group attached[23].

Industrially manufactured polypropylene is a linear polymer, whose backbone is a chain of saturated hydrocarbons. Every two carbon atoms of this main chain, a methyl group (CH_3) is branched. This makes it possible to distinguish three isomeric forms of polypropylene: Isotactic, Sindiotactic, Atactic. These are distinguished by the position of the methyl-CH_3 groups with respect to the spatial structure of the polymer chain.

Isotactic and syndiotactic forms, given their great regularity, tend to acquire an ordered, semi-crystalline spatial arrangement in the solid state, which gives the material exceptional physical properties. The atactic form, on the other hand, does not have any crystallinity. The most used industrial processes are directed towards the manufacture of isotactic polypropylene which is the one that has aroused the greatest commercial interest [24].

Polypropylene is the commercial polymer with the lowest density and ease of molding. It is used in a large number of sheets, fibers and filaments. Its properties include its high melting point (does not melt below 160 °C), high rigidity, high resistance to breakage and abrasion, dielectric properties, low friction, glossy surface and floating in water. It is resistant to acids, alkalis and many organic solvents. It is reheated close to 100 °C.

Polypropylene is sold with different molecular weights according to its purpose. In addition to polypropylene, a large number of copolymers of propylene exist on the market. The most important are those of propylene-ethylene [25].

2.4. Characterization of the Cocoa pod husks ashes

The characterization of this material follows an experimental procedure. First, the selection of the cocoa shell pods that are discarded during the pulping of the fruit is made. The drying is then carried out in a controlled atmosphere for 24 hours at a temperature of 65 °C, removing up to 50% of the humidity. Then, it is crushed and sieved in 2 mm meshes. The incineration of the sample is carried out in an isolated oven for 2 hours at an average temperature of 400 °C.

The ashes are sent to the X-Ray fluorescence laboratory of the National University of Colombia, where a semi-quantitative analysis is applied by means of a MagicPro PW-2440 Philips X-Ray fluorescence spectrometer. The results are shown in table 6.

Among the most relevant components we can see 72.08% potassium oxide (K2O), 11.57% Calcium Oxide (CaO), 5.42% Phosphorus Pentoxide (P_2O_5), 4.12% Magnesium Oxide (MgO), 3.09% Sulfur Trioxide and up to 2.15% Silicon Dioxide (SiO_2) commonly called silica. Each of these components is usable in different applications.

2.5. Mechanical design of the pneumatic injection machine using Autodesk Inventor Software.

The mechanical design includes the cylinder or shot sleeve with the respective closure cap, the feeding system of the material to be melted with the respective hopper, the injection nozzle holder of the injection mechanism and the structural base comprising all parts , mechanisms and systems of the machine.

Table 6. Characterization of the Cocoa pod husks ashes

Component	CVSC (% in weigh)
K_2O	72.08
CaO	11.57
P_2O_5	5.42
MgO	4.12
SO_3	3.09
SiO_2	2.15
AlO_3	0.39
MnO	0.26
Zn	0.24
Rb	0.23
Cl	0.14
Fe_2O_3	0.07
Cu	0.07
TiO_2	0.06
Cr	0.05
Sr	0.04
Na_2O	0.03

2.5.1 Power Unit, injection and fusion. The injection unit corresponding to a storage container or cylinder with elements to meet the process [26].

The injection cylinder or sleeve must be a completely smooth cylinder, so that no friction is not generated at the time of injection and return of the piston, considering a hole at the top, for feeding material to be processed. Measures taken are: length 40 cm, internal diameter 3.81 cm and 5.1 cm outer diameter. The injection nozzle is made conical so that it can give a lace possible molds that are designed for a total sealing.

To the feed hopper, it was considered an angle allowing the sliding material and no clogging thereof is generated before entering the cylinder.

The closure cap of the cylinder is designed with a center of flexible material that can generate wear on the rod of the pneumatic actuator, therefore, its role is to close the injection cylinder to prevent outgassing when melting occurs and also adjusts the tire stem actuator not to generate oscillating motion in the plane x in the injection.

2.5.2 Support of the injection system and structural base of the machine. The machine stand is designed with the characteristic that must be robust, for the subject to be compression and bending in the structure. The injection process generates different effort by pneumatic actuator. The piece is designed in resistant steel compression and bending, grub to give fastening the cylinder and the machine base. The pneumatic actuator bracket fixed to said support base of the machine and likewise helps give a parameter to center the machine or locate the injection cylinder in the same central axis of the actuator with the piston and rod. In part that is fixed to the base annealing is performed to fit the two parts and there is no possibility of movement.

The support base is designed in a rectangle 40 cm wide and 26 cm deep so that it can have a working area at the time of placing the molds for injection. To the support column of the machine, a steel is recommended seamless to generate a compressive strength greater degree, it is designed taking into account the height of the injection cylinder, with a length of 67.5 cm and an inner diameter of 3.81 cm.

Given the given specifications, modeling of machine support is performed, similarly a design on a table is performed to give a working height for the operator to perform a better process, all the above is shown in Figure 1.

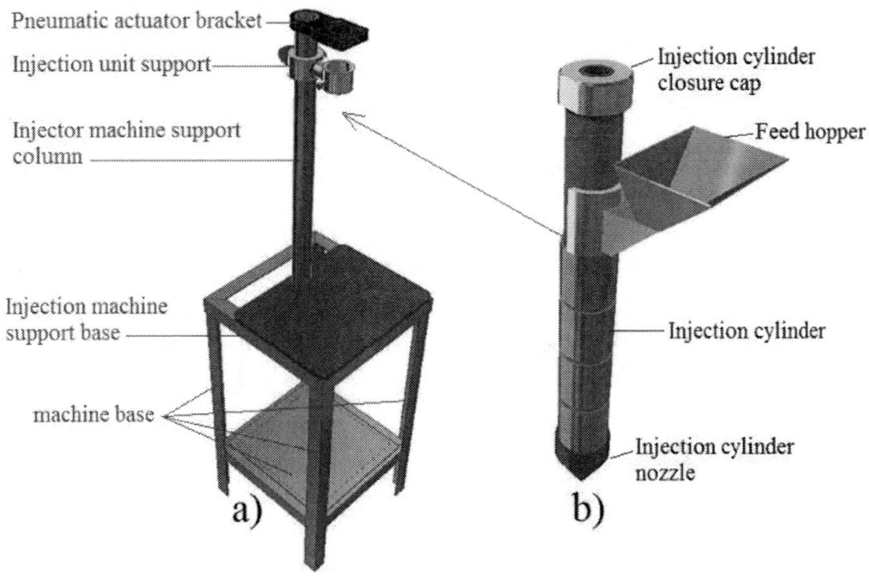

Figure 1. Mechanical design of the injection machine. a) support of the injection unit; b) Injection and fusion unit.

For this system, a piston design that will exert pressure on the material is also a patterned assembly actuator using the solenoid valves is made.

For a design part is made in nodular steel to be displaceable on the cylinder or sleeve of the fuel without causing any damage to the same surface, preventing leaks that may cause clogging in the injection.

Measurement was according piston into the cylinder injection in the case is 3.81 cm. For modeling, injection system takes into account the pneumatic actuator, which must meet a stipulated career to fulfill the operation, just as a solenoid valve is included with their respective assembly maintenance unit to perform an injection with a specified position. In Figure 2, the design is observed.

2.6. Pneumatic injection machine automation

The injection of plastic required a control system that allows closed loop to the automatic process. The process has the following steps, preheating cylinder, heating the plastic to melt it and then injected into a mold; in order to have a product with an established form and a semirecycled material.

2.6.1 Circuit electronic. Depending on the automatic process that you want to play on the injector, an electrical circuit is deployed that can take control of all the variables involved in the process.

2.6.1.1 Subsystems process. Subsystems used in the process are: Temperature (electric heaters), has five resistors distributed evenly to uniformly heat the liner (cylinder), also has three thermocouples which measure the temperature in real time of the same. Tire (piston) is formed by a pneumatic actuator which raises and lowers the injection piston, three position sensors on one side of the actuator for opening, closing and injection of the material and a directional solenoid 5 way three position closed center. Interactive Control Software (LabVIEW), the subsystem hardware consists of an Arduino UNO and their respective modules to receive information and broadcast it to a computer with LabVIEW software that controls the process.

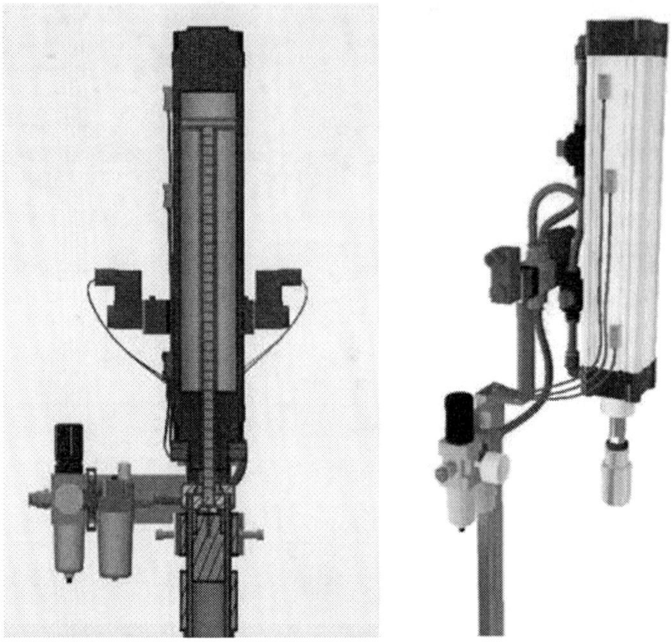

Figure 2. Injection system with pneumatic actuator

Table 7. Electronic components

Quantity	Component	Characteristic
3	Temperature sensor	Measuring range 0 ° to 350 ° C
5	Electric resistance	Power 280 W and 100 W
1	directional solenoid valve	Pneumatic valve 5 way / 3-position
3	Position sensors	Magnetic sensor
3	Isolating circuit coupled	Handle different voltages in a circuit
1	Arduino Uno	ADC 10 bits volt 0-5 v
1	2-channel relay module	Digital pins 120 v - 1 A
1	4 channel relay module	Digital pins 120 v - 1 A
3	Analog to digital converter for Thermocouple	Vcc: 5 v - Icc: 1.5 mA

2.6.2 Programming Software LabVIEW. It allows you to read signals, process and act in real time. The programming language compiler is the block diagram and also to create an interactive interface of the whole process. Here it is shown that the programming code for the injection molding process.

All programming is inside a while loop structure, so as to be continuous reading data. Inside is a structure "case" to have a manual option (False) and a sequential autonomous (true).

The interactive image program allows you to control all elements of the fuel through five (5) boolean switches within the program:

- Solenoid 1 directional valve (V1, Lower piston).
- Solenoid 2 directional valve (V2, Up piston).
- Heat resistance 1 (R1).
- Heat resistance 2 (R2).
- 3 Thermal resistance (R3).

The solenoids of the valves are conditioned so that only moving between the sensors S1 and S3. The only activate this contrary to the direction of the current position.

Additionally it has temperature readings of the three points (T1, T2 and T3), which have an ON-OFF system and a set temperature, which is not exceeded. It also has reading piston position (S1, S2, S3).

The interactive interface allows the observation made: the temperatures of the fuel, the piston position, the phase of the sequence found and the elapsed time of each phase. It also has interactive animations that make the process. All this is shown in Figure 3.

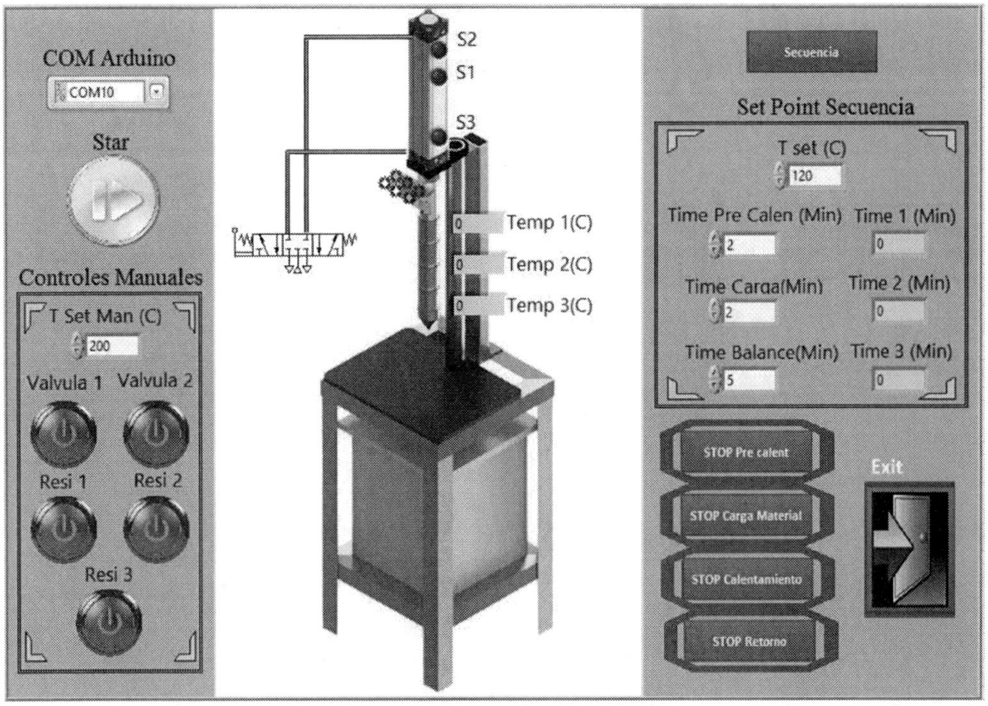

Figure 3. Interactive interface.

2.7. Construction of the automated pneumatic injection machine with LabVIEW Software.
It based on the mechanical design automation equipment, construction proceeds with recommendations for each element, system and mechanism of the machine materials; resulting in the equipment shown in Figure 4.

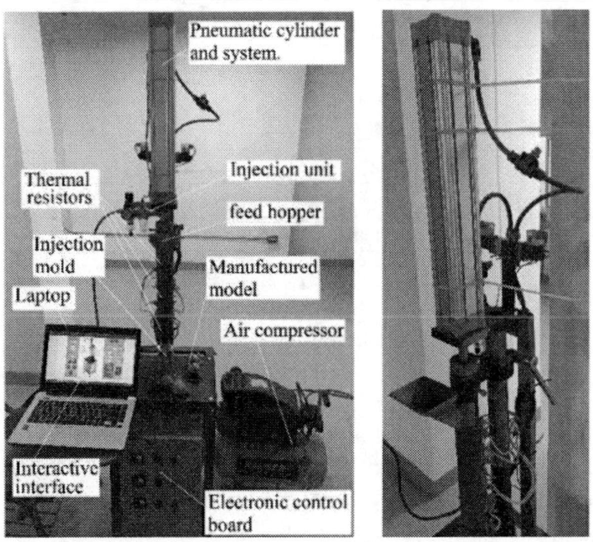

Figure 4. Final image of the automated pneumatic injector with all systems and mechanisms.

2.8 Application of experimental tests.
For this process, different mix designs mass fraction between polypropylene resin (PP), polyethylene terephthalate recycled in flake form (R-PET) and sheath ash cocoa shell (CPHA) are performed. These components can be seen in Figure 5.

Figure 5. Raw material. a) Polypropylene resin PP; b) Recycled polyethylene terephthalate in the form of scales R-PET; c) Cocoa pod husks ashes CPHA

Having defined the raw material, the testing for three (3) types of mix designs, recording the mass fractions, melting temperatures and times, the photographic record of the final piece and applied pressure injection are performed. All this is summarized in Tables 8,9 and 10.

Table 8. Design of components and manufacturing specifications of model A

Test Format: Model A						
 Photography model A	Test number:			1	Mold:	Trompo
	Component design					
	PP	80 g 100%	**R-PET**	--- 0%	**CPHA**	--- 0%
	Preheating (minutes)				10	
	Melting time of the mixture (minutes)				25	
	Maximum temperature (°C)				300	
	Injection time (seconds)				30	
	Injection pressure (psi$_a$)				80	
	Morphological appearance				95%	
	Observations: The manufactured piece has a homogeneous morphological appearance and an almost transparent color, the density remains stable. The molten raw material has a high viscosity, therefore the pressure and injection time increase.					

Table 9. Design of components and manufacturing specifications of model A

Test Format: Model B						
 Photography model B	Test number:			1	Mold:	Trompo
	Component design					
	PP	40 g 50%	**R-PET**	40 g 50%	**CPHA**	---g 0%
	Preheating (minutes)				10	
	Melting time of the mixture (minutes)				20	
	Maximum temperature (°C)				300	
	Injection time (seconds)				12	
	Injection pressure (psi$_a$)				70	
	Morphological appearance				96%	
	Observations: The manufactured part has a morphological appearance and a homogeneous color, although it is important to clarify that the color depends on the R-PET used and on which containers it was recycled. On the other hand, the viscosity of the molten mixture decreases with respect to the model with 100% PP, this due to the chemical and physical properties of R-PET, requiring a lower pressure and injection time.					

Table 10. Design of components and manufacturing specifications of model C

<table>
<tr><td colspan="7">Test Format: Model C</td></tr>
<tr><td rowspan="11">
Photography model C</td><td colspan="3">Test number:</td><td>1</td><td>Mold:</td><td>Trompo</td></tr>
<tr><td colspan="6">Component design</td></tr>
<tr><td rowspan="2">PP</td><td>32 g</td><td rowspan="2">R-PET</td><td>32 g</td><td rowspan="2">CPHA</td><td>16 g</td></tr>
<tr><td>40%</td><td>40%</td><td>20%</td></tr>
<tr><td colspan="4">Preheating (minutes)</td><td colspan="2">10</td></tr>
<tr><td colspan="4">Melting time of the mixture (minutes)</td><td colspan="2">19</td></tr>
<tr><td colspan="4">Maximum temperature (°C)</td><td colspan="2">280</td></tr>
<tr><td colspan="4">Injection time (seconds)</td><td colspan="2">10</td></tr>
<tr><td colspan="4">Injection pressure (psi_a)</td><td colspan="2">65</td></tr>
<tr><td colspan="4">Morphological appearance</td><td colspan="2">98%</td></tr>
<tr><td colspan="6">Observations: The manufactured piece has a homogeneous morphological appearance and the grayish color is due to the incorporation into the cocoa pod husks ash mixture. Also, it could be shown that the viscosity of the molten mixture decreases, therefore, a lower pressure and injection time is needed compared to the 100% PP model</td></tr>
</table>

3. Conclusions

An automated injection machine was designed and implemented to obtain a hybrid piece that takes advantage of the ashes of the cocoa shell pod and reuses PET bottles that are recycled and crushed in the form of scales. All this, in order to incorporate this waste into a thermoplastic element that can be used again. In addition, the control system operates on the LabView computer platform, through the Arduino architecture.

The final product is characterized in that the process is controlled in pressure limits and injection times. In addition to the maximum temperature of the mixture incorporated in the fusion cylinder, which are configurable in the application of a PC computer. Finally, and after several tests, three (3) types of models (A, B and C) are selected to which a mass-based mixture of each of its components (PP resin, R-PET and CPHA).

The morphological aspect of all models was homogeneous by up to 98%. Also, it was found that R PET is used up to 50% and even more.

By incorporating a greater amount of R-PET in the mixture (e.g. up to 50% on a mass basis), the viscosity decreases at the time of fusion, compared to the fusion of model A composed of 100% PP resin; needing a lower pressure and injection time.

The addition of CPHA in the mixture provides a model with a homogeneous and gray-gray appearance, but the most interesting thing is that the biodegrability properties increased by the addition of this compound, because it provides CaO, which in turn absorbs CO2 from the environment ,producing condensation on the surface of the final model.

References

[1] Karel Š and Achten H 2017 *Experimental development of a plastic bottle usable as a construction building block created out of polyethylene terephthalate : Testing PET (b) rick J. (Build Eng* vol 12 No May) pp 239–247

[2] Mwanza B G and Mbohwa C 2017 *Drivers to Sustainable Plastic Solid Waste Recycling : A Review* (*Procedia Manuf* vol 8 No October) pp 649–656

[3] Liu H Wu Q and Zhang Q 2009 *Bioresource Technology Preparation and properties of banana fiber-reinforced composites based on high density polyethylene (HDPE) / Nylon-6 blends* (*Bioresour Technol* vol 100 No 23) pp 6088–6097

[4] Palma-ramírez D Torres-huerta A M De Fuente D and Pla P E 2019 *Data supporting the morphological / topographical properties and the degradability on PET / PLA and PET / chitosan blends (Data in Brief* vol 25) pp 1-23
[5] Lahimer M C 2017 *Characterization of plastic packaging additives : Food contact , stability and toxicity* (*Arab J Chem* vol 10) pp S1938–S1954
[6] Yu L Dean K and Li L 2006 *Polymer blends and composites from renewable resources (Progress in Polymer Science* vol. 31) pp 576–602
[7] Sorgato M Masato D and Lucchetta G 2018 *Tribological effects of mold surface coatings during ejection in micro injection molding (J Manuf Process* vol. 36 No August) pp. 51–59
[8] Sorgato M Masato D Lucchetta G and Orazi L 2018 *Effect of different laser-induced periodic surface structures on polymer slip in PET injection moulding (CIRP Annals* vol 67 No 1) pp 575–578
[9] Sánchez-Alvarez A Luna-Moreno Q D Hernández-Morales J A Zaragoza-Zambrano J O and Castillo-Guerrero D H 2018 *Control of Stepper Motor Rotary Stages applied to optical sensing technique using LabView (Optik (Stuttg)* vol 164) pp. 65–71
[10] Zhang Y Li Y Gu X Liu H Zhang Y and Hu W 2019 *Laser spot image acquisition and processing based on LabVIEW (Optik (Stuttg)* vol 185 No November) pp 505–509
[11] Beyaz A 2018 *Harvest glove and LabView based mechanical damage determination on apples (Sci. Hortic (Amsterdam)* vol 228 No September) pp. 49–55
[12] Mahmoodi M James L A and Johansen T 2018 *Automated advanced image processing for micromodel flow experiments ; an application using LabVIEW (J Pet Sci Eng* vol 167 No November) pp 829–843
[13] Mitra A C Jagtap A and Kachare S 2018 *Development and Validation of Experimental Setup for Flexural Formula of Cantilever Beam Using NI-LabVIEW (Mater Today Proc* vol. 5 No 9) pp. 20326–20335
[14] La Mantia F P Morreale M Botta L Mistretta M C and Scaffaro R 2017 *Degradation of polymer blends : A brief review (Polym Degrad Stab* vol 145 pp 79–92
[15] Hahladakis J N and Iacovidou E 2018 *Closing the loop on plastic packaging materials: What is quality and how does it affect their circularity? (Sci Total Environ* vol 630) pp 1394–1400
[16] Kenny M Peponi L Puglia D Torre L and Valentini L 2014 *Processing of nanostructured polymers and advanced polymeric based nanocomposites* (*Mater Sci Eng R Reports* vol 85) pp 1–46, 2014.
[17] Siracusa V Dalla M Romani S and Rocculi P 2011 *Life Cycle Assessment of multilayer polymer film used on food packaging field (Ital Oral Surg* vol 1) pp 235–239
[18] Prambauer M Wendeler C Weitzenböck J and Burgstaller C 2019 *Geotextiles and Geomembranes Biodegradable geotextiles – An overview of existing and potential materials* (*Geotext Geomembranes* vol. 47 No 1) pp 48–59
[19] Taneli V 2017 *A review on new bio-based constituents for natural fiber-polymer composites (J Clean Prod* vol 149) pp 582–596
[20] Groh K J *et al 2019 Overview of known plastic packaging-associated chemicals and their hazards* (*Sci Total Environ* vol 651) pp 3253–3268
[21] Zhang Z Wang C and Mai K 2019 *Advanced Industrial and Engineering Polymer Research Reinforcement of recycled PET for mechanical properties of isotactic polypropylene (Adv Ind Eng Polym Res* vol 2 No 2) pp 69–76
[22] Coreño-Alonso J and Méndez-Bautista M T 2010 *Relación estructura - propiedades de polímeros (Educ Química* vol 21 No 4) pp 291–299
[23] Báez J E 2016 *El crecimiento en los extremos:Reactividad de grupos terminales en polímeros para la síntesis de copolímeros bloque* (*Educ Quim* vol 27 No 2) pp 97–104
[24] Serrano Hernández L Barceló Santana F H and Santos Espinoza A 2013 *Deflexión transversa de materiales alternativos a base de polímeros para fabricación de base de dentadura (Rev Odontológica Mex* vol 17 No 3) pp 146–151
[25] Eduardo B J Pedro R O Bogart M U Camille and Claudia Marcela G B 2013 *Desarrollo de un*

material compuesto ultraligero para construir la plataforma de una mesa vibradora (Ing Investig y Tecnol vol 14 No 4) pp 595–611
[26] Dahlbo H Poliakova V Mylläri V Sahimaa O and Anderson R 2018 *Recycling potential of post-consumer plastic packaging waste in Finland (Waste Manag* vol 71) pp 52–61

Incidence of corrosion in low voltage electrical conductor

J G Ascanio-Villabona[1], A D Rincón-Quintero[1,2], C G Cardenas-Arias[1] and C L Sandoval-Rodriguez[1]

[1] Automation and Control Energy Systems Research Group (GISEAC), Design and materials research group (DIMAT), Electromechanical engineering, Faculty of Natural Sciences and Engineering, Technological Units of Santander (UTS), Calle de los students No 9-82, 680005, Bucaramanga, Colombia .
[2] Energy in Building research group (ENEDI), Doctoral Program in Energy Efficiency and Sustainability in Engineering and Architecture, Department of Machines and Thermal Engines, University of the Basque Country (UPV/EHU), Engineer Torres Quevedo Plaza, 1, 48013, Bilbao, BI, España.

jascanio@correo.uts.edu.co

Abstract - This research used a salt spray booth to generate accelerated corrosion in a copper conductor, AWG 12-gauge purity of 95.95%, this caliber used in residential electrical installations according RETIE Chapter 3 Article 20 and NTC 2050 338. Additionally, to provide an enabling stage of corrosion, type mouse tail splices were made. The amount of corrosion is measured by weight loss, and incidence of corrosion in transport was evaluated from two perspectives: difference active at the beginning and end of the test power, and the change in resistance of the specimen measured at different times of the experiment. The test piece was subjected to a flow of alternating current into a low voltage system, while corrosion occurred. Using two single-phase two-wire electronic meters, the rate of power supply circuit test was recorded, to check for changes in the energy carried from exposure to the corrosive environment wire. The data collected showed that exposure times of the samples in the salt spray chamber involvement was not very noticeable in the copper wire, highlighting small corrosion spots. Similarly, the energy no changes were observed to induce corrosion, however, the resistance of the wire showed a total change of up to 2.4 [milliohms]. The data collected showed that exposure times of the samples in the salt spray chamber involvement was not very noticeable copper wire, highlighting small corrosion spots. Similarly, the energy no changes were observed to induce corrosion, however, the resistance of the wire showed a total change of up to 2.4 [milliohms].

1. Introduction

For the distribution of electricity above two basic steps to this, which are the generation and transport are needed [1] [2].In transport and distribution electrical conductors are needed, which are the main elements to conduct energy. These wires being exposed outdoors and often extreme conditions undergo alterations that cause damage to it causing power loss to the electrical system by corrosion effect on the driver.

Importantly in relation to wear from corrosion materials used in electrical systems in general they have been raised research focusing on the one hand to define the factors that influence corrosion of these materials And in another perspective fault analysis in electrical equipment caused by corrosion . Although, we evaluated the incidence of corrosion drivers employed in medium voltage systems, however, wear on drivers and joints due to corrosion in electrical low voltage installations is noticeable in the residential splices, boxes outlets and switches and electrical protection boxes. [3] [4] [5] [6]

Considering the above, this paper presents a study on low voltage, in order to establish the incidence of corrosion of the conductor in electrical characteristics (resistance-Transport power). For this type rat tail splices in the conductor connecting to an AC source and a load of 300 [W] nominal was used, and power were evaluated at different time instants. Then, the materials that were used and the methodology used to assess the changes in driver characteristics are described.

2. Materials and methods

2.1 Materials

To carry out tests cabin salt spray in the lab Technology unit Santander which has control for temperature and dosage of Sodium Chloride (NaCl) was used, and to moisture at room temperature 22 ° C. In Figure1. Salt spray booth (CNS) used is presented.

Figure 1. Employee Salt Spray booth at work.

HIKING class 1.0 wattmeters were used imitating the measurement system of the electric power distributor in the locality, which is in kWh, like the one shown in Figure 2 measuring the power consumption at two different sites before the cabin Wattmeter 1, and after the cab, Wattmeter 2 as shown in Figure 3.

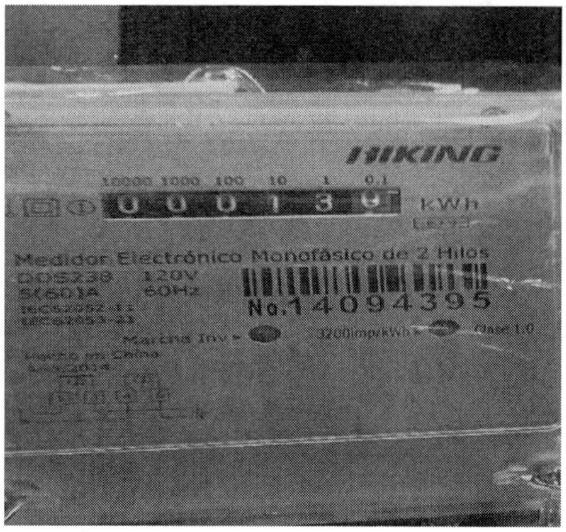

Figure 2. Wattmeter used for measurements.

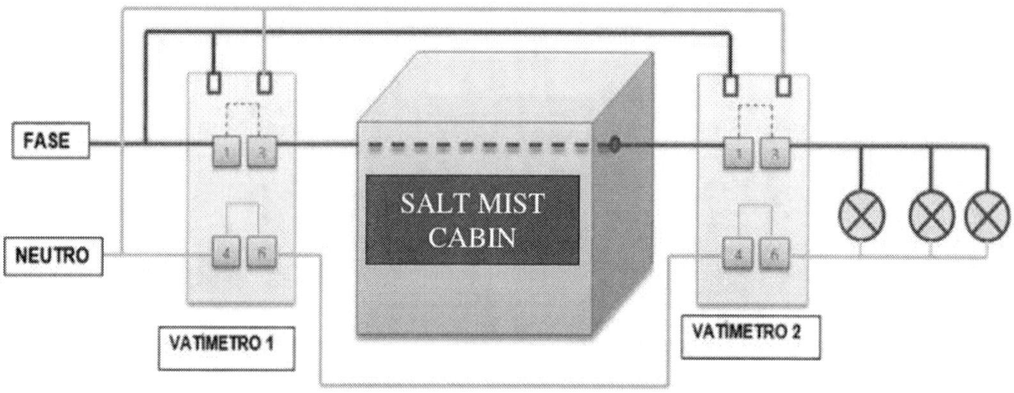

Figure 3. Connection diagram for measurements.

PM6306 impedances meter Fluke mark which has a measuring range 0.00005Ω to 50MΩ at ambient laboratory temperature of 22 ° C was used for resistance measurements. a scale model OHAUS PIONEER PA2014 0.1m was used [g] resolution for weight measurements. For loading three luminaires 100 [W] nominal was used, and the selected specimen was one AWG gauge number 12 copper since according RETIE chapter 3 article 20 and NTC 2050 section 310 and 338 it is convenient for electrical installations residential.

2.2 Methods

The research has two objectives: In the first instance define what type of connection cable corrosion is favored, if the simple connection (without joints) or complex connection (with connections) additionally if corrosion has any influence on the electrical characteristic's driver (Transmission power and resistance). Taking into account the above, research is descriptive because it attempts to measure the characteristics of the driver before a principal factor such as corrosion. Also, to establish the relationship between corrosion and electrical characteristics of the driver based on linear regression model is used [7] [7] [8]. In order to determine which was the type of connection where greater influence of corrosion is presented an experimental design was proposed treatments. 1. Type of connection and 2. Exposure time: Thus two factors are established. In this respect tests were performed with simple connection 8 hours exposure and 16 hours of no exposure to salt spray and the complex connection 8 hours exposure and 16 hours of non-exposure, completing this four treatments in each which the amount of corrosion, the energy transmitted and the strength was measured. 5 repetitions is recommended for experimental design with two factors were performed, as is the case in the present study; tests were relizadas in a period of 120 hours for each factor. It should be noted that the time intervals were chosen in this way because for making measurements, these should be made during working hours. [8] [9]

3. Works development

The tests were conducted in two stages with 12 AWG gauge copper number as according RETIE and NTC 2050 is suitable for residential electrical installations.

The first step was to expose completely smooth wire (single connection) to saltiness using the Salt Fog Cabinet (CNS) of Figure.1 which provide a special atmosphere to accelerate corrosion. The (CNS) is added a solution composed of 5 parts of sodium chloride ($NaCl$) and 95 parts of natural saline water, as indicated by ASTM B117 (ASTM B117), under the supervision of the CIC. To this was connected a completely resistive load of 300 [W] which is an average load of a joint or driver in a switch or home outlet, the supply of electricity from a source of 120 [V] AC. This step was performed in cycles 8 hours and 16 hours deenergized energized per day for 5 days, giving a total of 120 hours testing since the tests were done in days and working hours only. Also, the weight of the specimen after each cycle was measured and the electric power consumption in watts kilo-hour (kWh), in the entrance of the specimen as its output, as shown in Figure.3. Similarly, resistance in ohms (Ω) immediately terminated each cycle, the driver being at room temperature as it was not in overload was measured.

In test 2 The abov e procedure was replicated, but using complex connection, for this type joints used mouse tail as shown in Figure.5 and Figure.4.

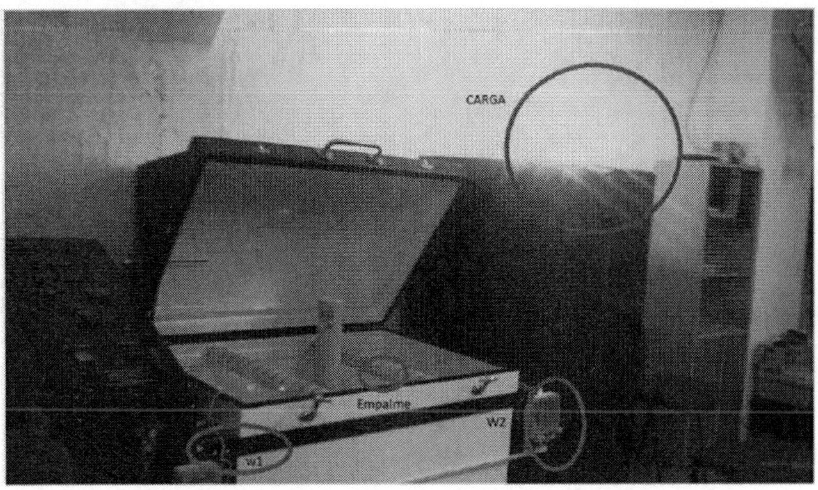

Figure 4. Load energy measurement and joints.

Figure 5. Rat Tail Sílice.

4. Results

To test 1 with the driver unsliced data obtained are shown in Table 1. The cycles are due to the possibility of laboratory measurement, cycle 1 being the initial state in the first test day. Is Day 1 Time 10:00 a.m. cycle 2 corresponds to the first day when they have passed 8 hours of operation and exposure is day 1 at 6:00 pm cycle 3 we correspond to measurements made past 16 hours, that is to second day 10: am , the fourth cycle is reading the same day to spend 8 hours and so on.

Table 1. Weight readings, energy in each meter and resistance in test 1 (connector without joints)

CYCLE	WEIGHT [G]	W1 [KWH]	W2 [KWH]	RESISTANCE [Ω]
1	55.3210	1.2	1.4	0.0154
2	55.3210	3.2	3.4	0.0154
3	55.3210	7.2	7.4	0.0154
4	55.3210	9.2	9.4	0.0155
5	55.3210	13.2	13.4	0.0156
6	55.3210	15.2	15.4	0.0157
7	55.3210	19.2	19.4	0.0157
8	55.3210	21.2	21.4	0.0158
9	55.3208	25.2	25.4	0.0163
10	55.3205	27.2	27.4	0.0177

In Figure 6 it can be seen the behavior of the resistance in the measuring time, which increases minimum values being assumed to microbiological corrosion, and in Figure.7 behavior driver's weight.

IOP Conf. Series: Materials Science and Engineering **844** (2020) 012055 doi:10.1088/1757-899X/844/1/012055

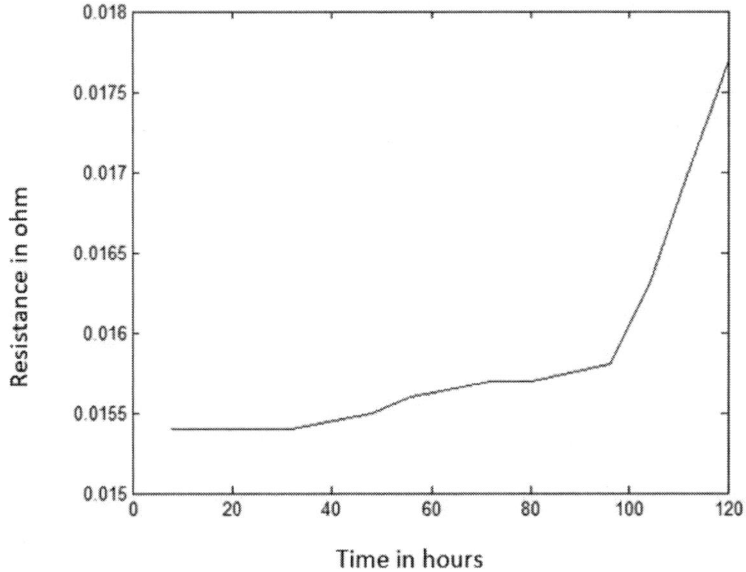

Figure 6. Behavior resistance over 120 hours of testing.

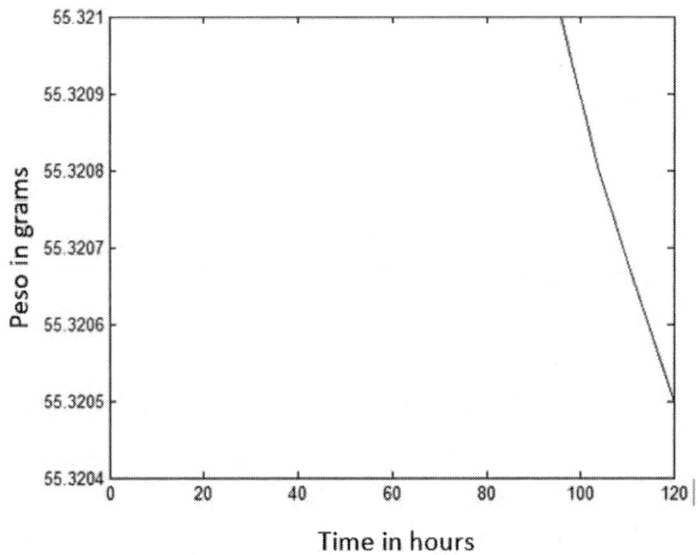

Figure 7. Behavior weight of the specimen over 120 hours of testing. Source: authors.

Similarly, in Figure 8 energy profiles are shown in both power consumption meter readings where w1 is the bottom line and reading w2 is the top line.

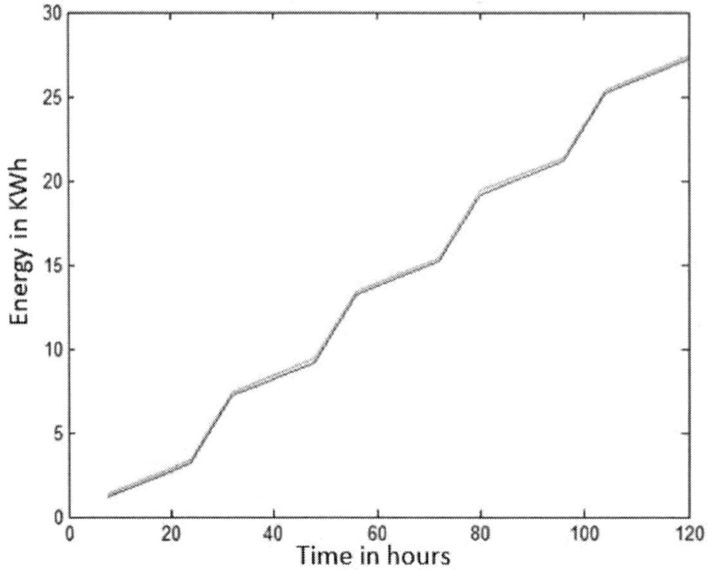

Figure 8. Behavior of energy transported. Source: authors.

To test 2 with the conductor splices using data obtained are shown in Table 2. The data presented there obey the procedure used for test 1

Table 2. Weight readings, energy in each meter and test Resistance 2 (connector joints)

CYCLE	WEIGHT [G]	W1 [KWH]	W2 [KWH]	RESISTANCE [Ω]
1	57.8265	27.2	27.4	0.0175
2	57.8265	29.8	30.1	0.0175
3	57.8265	35.0	35.3	0.0177
4	57.8265	37.6	37.9	0.0178
5	57.8263	42.8	43.1	0.0179
6	57.3260	45.4	45.7	0.0179
7	57.3258	51.0	51.3	0.0180
8	57.3253	53.6	53.9	0.0181
9	57.3248	58.8	59.2	0.0183
10	57.3240	61.4	61.8	0.0199

In Figure. 9 shows the behavior of the resistance along the test used 2 to 3 rat tail splices the same author. Also, in Wright Figure.10 behavior occurs.

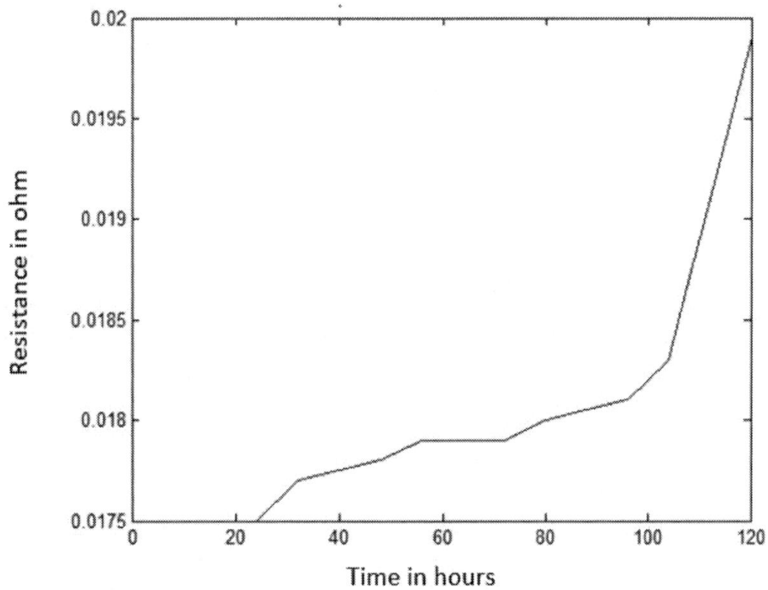

Figure 9. Behavior of resistance over 120 hours of test 2. Source: authors.

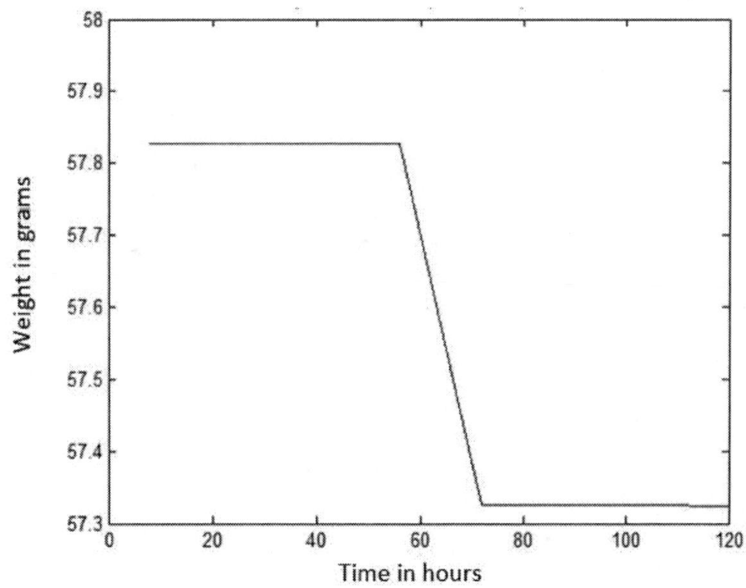

Figure 10. Behavior of the specimen weight over 120 hours of prueba2. Source: authors.

Similarly, in Figure.11 behavior of power transmission by reading W1 and W2 the upper curve the lower how.

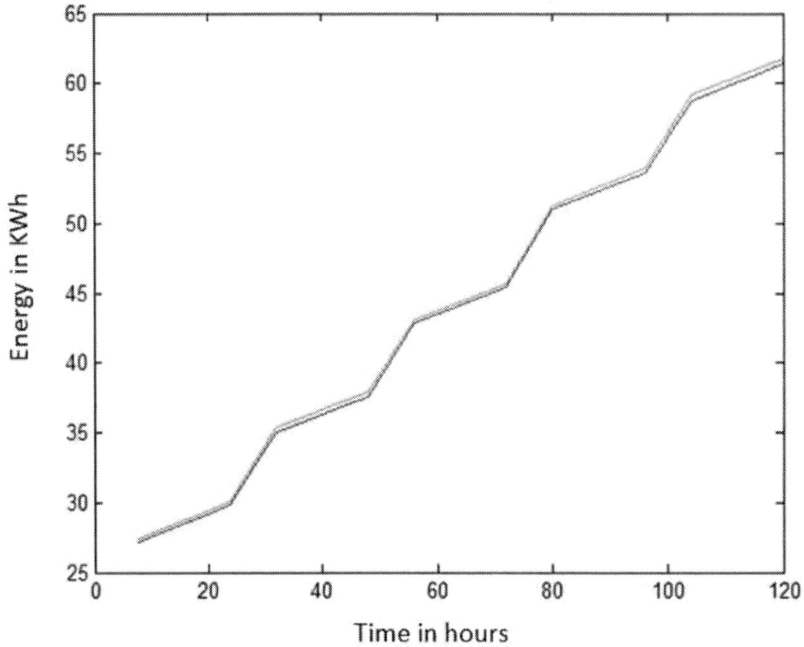

Figure 11. Behavior of energy transported in test 2. Source: authors.

From Figure 6 to 11 can be highlighted several observations. The first is that as shown in Figure. 7 and Figure 10, the weight loss was very small both in test 1 and the prueba2. Additionally, in Figure. 6 as in Figure.11 readings wattmeter 1 and 2 are very similar for each test. This indicates that exposure to corrosive environment did not allow a substantial material wear and did not affect the transport of energy. However, the conductor resistance in test 1 showed a total change of 2.3 milliohms and in test 2 2.4 m [Ω]. Whereby a linear regression model trying corrosion, represented by weight loss, as an independent variable and the dependent variable resistance in each of the tests sought.

For test 1 the model that best fits is an exponential model as shown in Equation 1.

$$R(t) = a * p(t)^b + c \tag{1}$$

Where:

R (t): Tensile test along
P (t): weight throughout the test
a: constant adjustment (-47.46)
b: adjustment constant (0.54689)
c: constant setting (426.1)

To validate this model the experimental data with the data model results were compared. The data are shown in Table 3.

Table 3. Validation data for the model resistance test 1 (driver without joints)

CYCLE	WEIGHT [G]	R MODEL OH	AS R [Ω]	ERROR %
1	55.3210	0.0144	0.0154	6.9515
2	55.3210	0.0144	0.0154	6.9515
3	55.3210	0.0144	0.0154	6.9515
4	55.3210	0.0144	0.0155	7.6460
5	55.3210	0.0144	0.0156	8.3405
6	55.3210	0.0144	0.0157	9.0350
7	55.3210	0.0144	0.0157	9.0350
8	55.3210	0.0144	0.0158	9.7294
9	55.3208	0.0152	0.0163	6.9449
10	55.3205	0.0165	0.0177	7.2393

To test 2 the same analysis was performed and the same model of equation 1. But with values a = -0.004401 was obtained, b = 0.6787 and c = 0.08679. Validation data shown in Table 4.

Table 4. Validation data for the model resistance test 2 (conductor joints)

CYCLE	WEIGHT [G]	R MODEL OH	AS R [Ω]	ERROR %
1	57.8265	0.0177	0.0175	1.0484
2	57.8265	0.0177	0.0175	1.0484
3	57.8265	0.0177	0.0177	0.0000
4	57.8265	0.0177	0.0178	0.6479
5	57.8263	0.0181	0.0179	1.2124
6	57.3260	0.0181	0.0179	1.0608
7	57.3258	0.0181	0.0180	.5089
8	57.3253	0.0181	0.0181	0.0000
9	57.3248	0.0181	0.0183	1.1447
10	57.3240	0.0181	0.0199	9.9840

5. Discussion

In this work two tests were to determine the influence of cable joints in the proliferation of corrosion. Similarly determine the type of incidence of corrosion in the electrical characteristics of the driver as the resistance and energy transport. On the one hand the total weight loss driver unspliced was 0.5m [g] and the conductor splices was 502.5m [g]. From this result it follows that there was greater weight loss in the conductor joints, which suggests that the presence of splices in a conductive favors outbreaks of corrosion. Furthermore, in the power transmission of the two conductor types that no changes can be inferred because of corrosion they occurred. In contrast to this, in resistance measurements if they every test these changes were generally 2.3m [Ω] and 2.4m [Ω] for the two tests respectively perceived. It is important to note that these values are in proportion to the average strength (15.85mΩ, 18.06mΩ, for test 1 and 2 respectively) of each test 14.51% and 13.29 respectively. This indicates that although there was a change in the resistance of the conductor, no changes were noted in the transmitted energy. This result may be limited by the accuracy of the wattmeter since this is the order of tenths.

Finally, the exponential model obtained represents an average approximate 6.98% and 1.7% error for the two tests, the resistance in the conductor based on the weight of the driver. This allows predicting changes resistance in the conductor when the weight has been affected by corrosion. It is important to note that changes in weight and resistance were evaluated for each test lasting 120 hours. In this sense it could be extrapolated to a longer obtaining the weight functions depending on the time and thus analytically meet resistance.

6. Conclusions

Descriptive analysis and experimental design allowed us to evaluate the incidence of two factors (connection type and exposure time) in induced corrosion in a copper conductor 95.95% purity used in residential electrical installations. The method of weight loss allowed to establish conditions altered corrosion element. With this it was possible to differentiate between the connection and without joints; latter showing further degradation by having a total weight loss was 100 times the driver without splicing. Additionally, according resistance profiles shown in Figures 6 and 9, changes manifest as time progresses test and turn corrosion; this shows that if the electrical characteristics vary with corrosion. But nevertheless, Finally, the methodology proposed by the experimental design provided results that were used to propose a model (equation 1) described with greater accuracy than 90% resistance behavior depending on the weight of the driver. It should be noted that the model parameters equation, according to this study, dependent on the structure of the conductor (with or without joints); in turn this evidence is important to adjust the parameters from experimental data if tests should do to other drivers of other features such as: other calibers, other arrangements, etc. However, to achieve a generalization of the model it should be extended to a dipping process of the specimen carried out directly in saline solution and no mist; It is also presumed that longer exposure times may exhibit changes larger magnitudes according to the trend shown in Figures. 6 and 9 because the behavior of the resistance in the measuring time increases in minimum values being assumed to microbiological corrosion.

Referencies

[1] Harper, G. E 2006 El ABC del alumbrado y las instalaciones electricas de baja tension, Mexico: Limusa.

[2] Molina, J. D. y Rudnick H 2010«Transmission of electric Energy: a Bibliographic Review,» *Latin America Transactions. IEEE-Revista IEEE America Latina Vol 8.,* pp. 245-258

[3] Blanco, M.C. 2008 «Corrosion En lineas Aereas,» Universidad Nacional Antonio Jose de Sucre, Venezuela

[4] Klengel, Robert. Klengel, Sandy. Stephan Tino and Petzold, Matthias, 2013 «Novel investigation of influencing factors for corrosive interface degradation in wire bond contacts,» de *EMPC 2013*, Grenoble: France

[5] Y. H. H. ,. G. H. W. J. .. Y. a. Z. Quin 2008 «Analysis and Treatment of transformer Faults Caused By Corrosive sulfur,»

[6] Cardenas, C.G, Pardo, O.M and Santos, A. 2016 «Incidence of Corrosion on Electric Power Losses in ACSR Cables,» *Tecciencia Vol 11,* pp. 27-33

[7] Hernandez Sampieri, Roberto. Fernandez C, Carlos y Baptista. L, Pilar 2006 Metodologia de la Investigacion, Mexico: McGraw-Hill

IOP Conf. Series: Materials Science and Engineering **844** (2020) 012055 doi:10.1088/1757-899X/844/1/012055

[8] Gutierrez P.Humberto y De La Vara .S. Roman 2008 Analisis y diseño de Experimentos, Mexico: McGraw-Hill Interamericana, 2008
[9] Montgomery, Douglas.C. 2004 Diseño y analisis de experimentos, Arizona: Limusa Wiley

Building a prototype for functional analysis of the energy potential of the water flow in pipe ½ "using microturbines applied to Unidades Tecnológicas de Santander

J G Ascanio-Villabona[1,], C L Sandoval-Rodriguez[1], A D Rincón-Quintero[1,2], B E Tarazona-Romero[1], and R E Paez-Castro [1]

[1]Automation and Control Energy Systems Research Group (GISEAC), Design and materials research group (DIMAT), Electromechanical engineering, Faculty of Natural Sciences and Engineering, Technological Units of Santander (UTS), Calle de los estudiantes No 9-82, 680005, Bucaramanga, Colombia.
[2] Energy in Building research group (ENEDI), Doctoral Program in Energy Efficiency and Sustainability in Engineering and Architecture, Department of Machines and Thermal Engines, University of the Basque Country (UPV/EHU), Engineer Torres Quevedo Plaza, 1, 48013, Bilbao, BI, España.

jascanio@correo.uts.edu.co

Summary: The project aims to build a working prototype for analysis of potential energy of water flow in pipe ½ "by using micro tangential turbines with a loss running low, constituting a small-scale model for the study power generation thereof, setting parameters related to a hydraulic distribution system operation residential drinking water. Taking advantage of the electric power, a microturbine is obtained by decanting flow through it. The research methodology is descriptive quantitative approach, to develop the project initially performed a literature review, followed by identification of the electrical variables and characteristics of the prototype to generate a simulation of a real hydraulic circuit, it followed this comes the recognition and selection of the elements for retrofitting. Thus, as looking through prototype and analysis of the energy potential, generating an academic, social and environmental impact on the institution; since being pioneers in this type of project it is expected to generate the conceptual foundations in the community, enabling understanding of the use of this resource and future implementation. social and environmental in the institution; since being pioneers in this type of project it is expected to generate the conceptual foundations in the community, enabling understanding of the use of this resource and future implementation. social and environmental in the institution; since being pioneers in this type of project it is expected to generate the conceptual foundations in the community, enabling understanding of the use of this resource and future implementation.

1. Introduction

Currently, one of the biggest concerns is the high demand for electricity [1]Due to population growth [2] and industrial development of the country [3] That have generated an energy deficit [4] considerable [5]. Which it has generated various analyzes of possible alternative energy solutions [6] not conventional [7] , Conducting studies [8] and looking that in everyday activities, you can generate electricity [9] needed to help combat this problem

Thus, as within the various alternative systems under study, there are micro turbines [10] they tend to operate in much the same compared as do large and expensive turbines [11] installed in conventional hydroelectric [12]Without having to be subject to multiple failures [13] and its devastating effects [14] Such as those who have registered several hydraulic generation plants [15].

One of the most innovative technologies [16] projected [17] today for power generation [18]Are the microturbines [10], Primarily because they tend to have a small size [19]. Because of its low-weight size allows these are easy to locate in a water system [20] or a process plant [21].

Contemplating other forms of clean energy generation and renewable [22] as produced by eoliths systems [23] and solar panels [24] A system of micro turbines installed in the sanitary water system can produce electricity [25] at times there is a consumer [26].

These micro turbines [10], They can be exploited at suitable locations where water circulates natural draft without [27] intervention pumping systems [28].

Given the above concept is intended to route the first phase of the investigation [29] to the nearest study area being the Technological Units of Santander. Accordingly to the above and according to the technical information provided by the Office Infrastructure Technology unit Santander (UTS), viability analyzes [30] to use this potential in the water system [31] Health Building B.

To this end it plans to build a prototype that simulates like the water system health conditions Building B, which will allow us to understand the behavior [32] the microturbine, his generation [33], Possible losses [34] and other data needed for evaluation, taking into account that is not being used efficiently, the excess energy generated in the sanitary water system.

2. Methodology used

For the project, the construction of a working prototype for analysis of potential energy of the water flow in pipe ½ "using microturbines applied to Technological Units of Santander, which perform arises through descriptive research methodology [35]With which it compile [36] information [37] with a quantitative approach [38] for further statistical analysis, through collection of electrical variables such as current, voltage and power and pressure so ON-LINE manually for OFF-LINE aftertreatment in the generator program trend curves [36] (GNU) in order, to perform the respective pressure comparison [37] input regarding the generation of micro turbine.

3. Used items

3.1 Hydraulic motor pump

The centrifugal motor pump is operating principle convert electrical energy into velocity and pressure of flow through the impeller or turbine rotating in the housing. The main purpose of the centrifugal pump is to move a volume of liquid from two different levels and / or use the pressure that it can supply. [38]

Rangel QB60 the pump seen in Figure 1, has a power half horsepower, a maximum flow rate of 35L / min, a dynamic head of 35 m. Table 5 shows all characteristics.

Figure 1. Moomba Ranger QB60

3.1.1. Performance curve motor pump Ranger QB60. The performance curves shown in Figure 2, contains information from two motor pumps Ranger mark and for selecting one considers the curve 1 as is corresponds to the motor pump of 0.5HP QB60.

This curve parameters are the flow rate (GPM) vs. Pressure (psi) for this model.

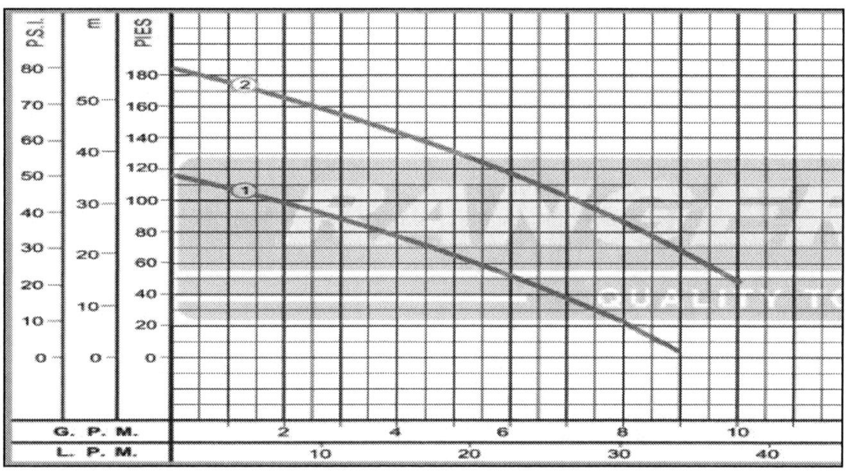

Figure 2. Motor Pump curve Ranger QB60

The motor pump Ranger QB60, generates a maximum pressure of 50 psi and a maximum flow of 35LPM, which are similar to the pressure handled in the water system because under NTC-1500 [37], We define the range of managed systems in conventional water pressures, which are 25 to 35 psi.

3.2 *Hydroelectric Generator Microturbine*

The AIYIMA brand in its product catalog, have a variety of hydraulic microturbines called HYDROELECTRIC GENERATOR MICROTURBINE [39], Which select the microturbine hydraulics model, shown in Figure 3, the main features are described below:

- Magnetic coupling clutch double single
- Inlet, outlet pipe are available to match a straight line or right angle mounting type L
- Ultra-Low Loss Flow

- Starting low water pressure
- isolation diode power directly to the rechargeable battery or power the capacitor 5F farads. [39]

Figure 3. Hydroelectric Generator Microturbine

The Hydraulic microturbine It is used in lighting shower and sound system, gas water heater with power and other electrical products. This high-power generator 60 can energize light emitting diodes (LED)

Table 1 shows the technical specifications such as voltage, pressure range, initial flow among others described.

Table 1. Micro sheet Hydroelectric turbine Generator

Diameter inlet and outlet	½ "
Output voltage	9.8V to 18.5V DC
Current Range	128mA to 260mA
Pressure range	0.08MPa to 0.55MPA
Maximum pressure	0.55MPA
Lost by traffic	3.6% (to 0.25MPa)
initial flow	1 L / min
Operating temperature range	5 ° C to 85 ° C
Dimensions	81.4mm x 43.8mm x80mm
Maximum power output	5W
Price	€ 10.41
Minimum Run	It does not require
Retail Distribution	YES
External diameter	20mm

With the yield curve you can know what the operating conditions of hydraulics, such as flow rate, minimum and maximum operating pressures are microturbine. Graphical representations voltage Vs Flow and Pressure Flow Vs is shown in Figure 4 and 5 respectively

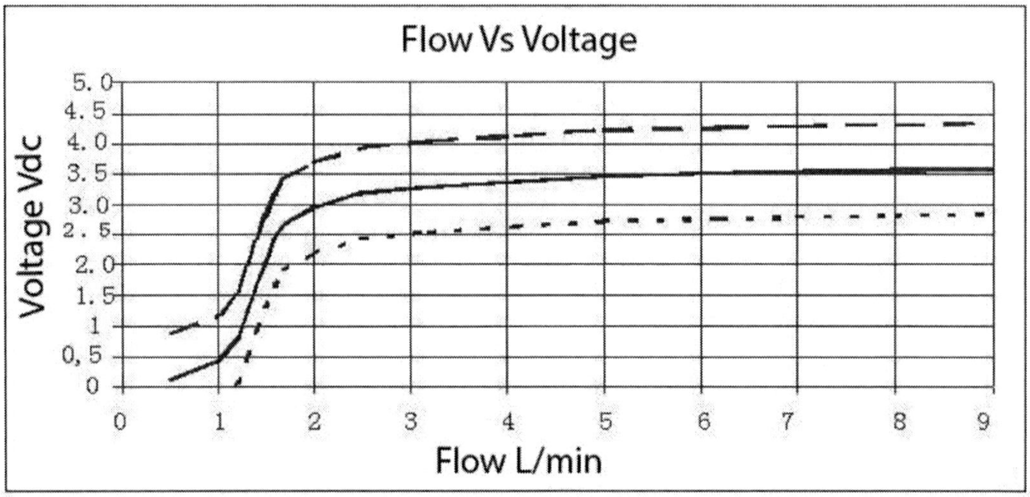

Figure 4. Flow Vs Voltage curve

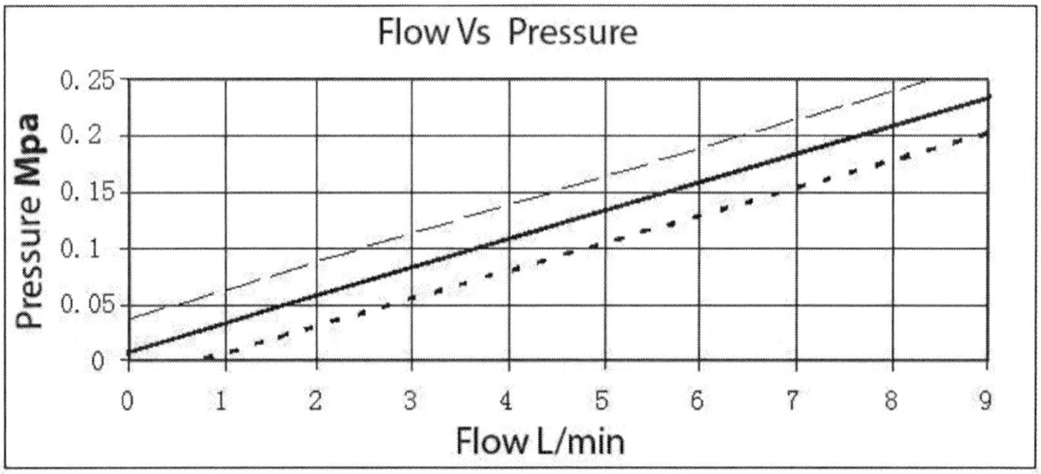

Figure 5. Flow Vs Pressure curve

3.3 *Pressure gauges*

A gauge is an instrument for measuring the pressure in fluid (Liquids and gases) in closed circuits. Measure the difference between the actual or absolute pressure and atmospheric, Calling this value, gauge pressure. This type of gauge is also known as "pressure gauges". [40]

The gauge shown in Figure 6, is used for this research project will be a range from 0 to 60 PSI. They will be used in order to measure outlet pressure of the motor pump, inlet pressure turbine and pressure loss of the fluid to decant it.

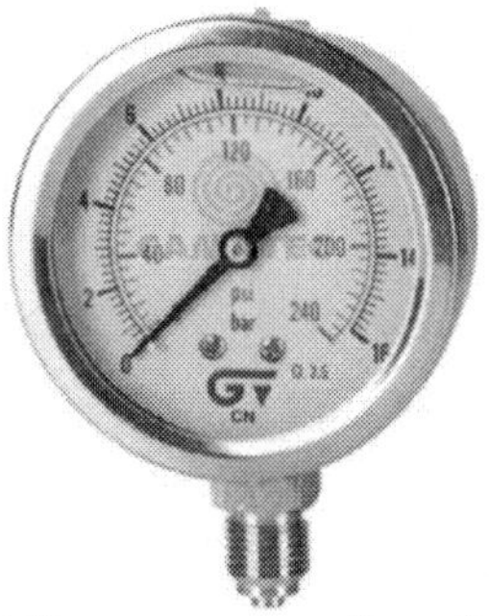

Figure 6. Pressure gauge from 0 to 60PSI

3.4 Check valve

The check valves or valves are used to keep back a fluid within a line. This implies that when the pumps are closed for any maintenance or just gravity does its work of returning the fluid downward, this valve closes instantaneously letting only the flow running in the right direction. That is also why they are called non-return valves. Obviously, it is a one-way valve that must be positioned properly to perform its function using the direction of traffic flow is correct. [41]. The check valve shown in Figure 7 will be used to avoid a possible shock at the motor pump Ranger QB60

Figure 7. Check Valve

4. Selecting the area to simulate the prototype

It is important that site selection is the most appropriate because it must take into account that water consumption caused by the high traffic of students in health services, will generate the flow and pressure of water that is required for there is a generation of effective and representative electrical energy, this can be captured, since this electric potential is being wasted.

In the study performed for selecting the right place, requested the Office of Infrastructure of Technological Units of Santander (UTS), the respective health plans designed for module Building B, so this study will focus on said location.

According to the design of Building B, it has 7 floors and a basement, where there are restrooms on each floor, distributed as follows:

From the first floor to the sixth, they are distributed four health Batteries that are intended for general use described as follows:

- Two sanitary units for men.
- Two health Battery for ladies.
- On the seventh floor they are distributed health services as follows:
- 5 private health services.
- A Cafeteria.
- Finally, in the basement, according to design drawings, these are not updated with the latest reforms made.

Given the foregoing, it has been determined for this study, the water distribution points, as described in the following items, are most suitable.

5. Inventory overall battery health building b.

Based on the above analysis, we found that building a sanitary battery B has made up a considerable amount of toilets, sinks and other type push.

Table 2 shows the number of individual services installed will be displayed and detailed for each floor. These are points potable water supply that may become points of facilities microturbines. These points are regulated by the NTC-1500 standard [37]Where defines minimum operating pressures and minimum pipe diameter distribution. For this research project, only the points having a pipe diameter of 1/2 inch is taken into account.

Table 2. Detailed inventory of health Battery Building B

	Gentlemen	**Ladies**	**Others**
Floor	Total	Total	Total
First	22	16	0
Second	22	16	0
Third	22	16	0
Quarter	22	16	0
Fifth	22	16	0
Sixth	22	16	0
Seventh	5	4	0
bathroom Vice	2	0	2
Private bathroom Rectory	5	4	0
Cafeteria	0	0	2
Total	141	100	4

6. Base standard NTC-1500

The committee hydraulic facilities in ISO-1500 standards in its third update defines the rules of facilities supply and distribution of water to plumbing devices that provide water for drinking, personal hygiene, preparation or food processing, medical products and pharmaceuticals It must be potable. Also it specifies parameters that must be considered for its design defining the minimum dimensions of the pipe to be selected such that under the conditions of peak demand cater to sanitary equipment for flow capacities and defines the minimum flow and minimum pressure [37] flow provided to the devices listed in Table 3. [37]. Based on this regulation, operating pressures and minimum diameters are considered for the selection of the microturbine are determined.

Table 3. minimum diameters for water supply pipes in the plumbing equipment

Apparatus	minimum pipe diameter (inches)
Bathtubs 1524 mm x 813 mm (60 inches x 32 inches) and	(1/2)
Bathtubs larger 60-inch x 32 inch	(1/2)
Bidet	(1/2)
Poceta combination and tray	(1/2)
Dishwasher, domestic	(1/2)
sprue	(1/2)
Hose taps	(1/2)
Dishwasher	(1/2)
Laundry 1, 2 or 3 compartments	(1/2)
Handwash	(1/2)
Shower or shower	(1/2)
Poceta, with overflow	(1/2)
Poceta Service	(1/2)
Urine discharge tank	(1/2)
Urinal, flushometer	(3/4)
Hydrant wall	(1/2)
Toilet flush tank	(1/2)
Odorless, wastegate	(1)
Toilet tank pressurization	(1/2)
Toilet, a piece	(1/2)

7. Count people entering and using the building battery health b

This count is done in the health batteries for women and men on the first floor of Building B, because for us it is the largest influx of students and outsiders, being located close to the goal of access to the institution, which concentrates most human activity.

Counting that takes place on Thursday 22 and 29, Friday 23 and November 30, 2018 between the hours of 6: 00a.ma 6: 00 p.m. give us as a result the number of people using health battery daily ladies and gentlemen Technology unit Santander, is plotted in figure 8.

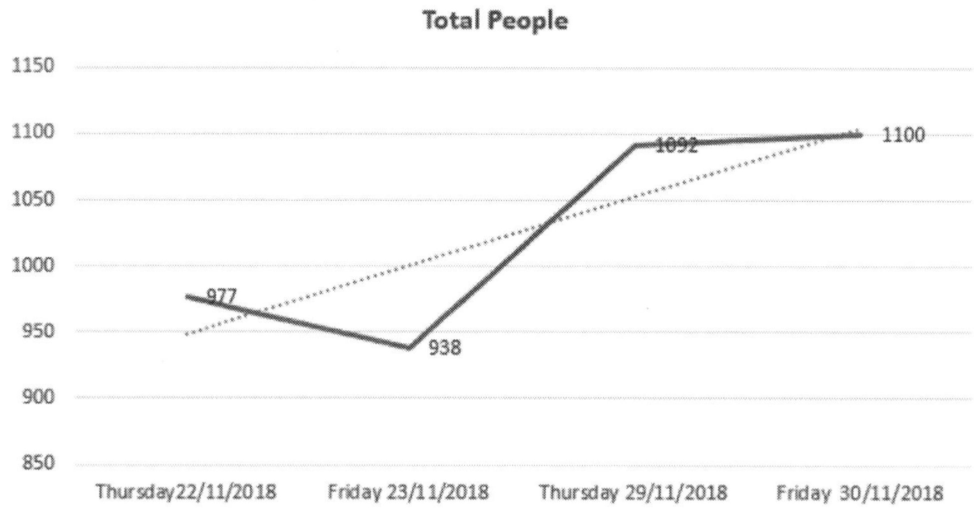

Figure 8. Graphical representation of the total number of people who entered and used battery health ladies and gentlemen

According to the data obtained in Figure 8, equation 1 raises to find the average people using these services daily.

$$\text{average (week)} = (D1+D2)/2 \quad (1)$$

Where:
D1: First day where the count is made
D2: Second day where the count is made

The data obtained with equation 1, are shown in Table 4

Table 4. Average number of people entering the building sanitary batteries B

Average (Week 1)	957 persons / day
Average (week 2)	1096 persons / day
Total average)	**1026 persons / day**

According to the results shown in Table 4 it is determined that the lavatories of ladies and gentlemen of the first floor of Building B, enter the first week 957 people a day in the second week fall 1096 people. Calculating the total average between Week 1 and Week 2, we need to such lavatory's ladies and gentlemen

IOP Conf. Series: Materials Science and Engineering **844** (2020) 012056 doi:10.1088/1757-899X/844/1/012056

of the first floor of Building B enter 1026 people a day, assuming that at least driven once sinks type push the bathrooms.

Accordingly, the average input is the time that the sink type pushes actuated.

In order to be able to measure the amount of water coming out at a given time sinks type push of the bathrooms on the first floor of the building B, the Laboratory of Environmental Resources Technological De Santander (UTS) Units provide us a suitable container for this they are as 600ml beaker and 100ml test tube, shown in figure 9.

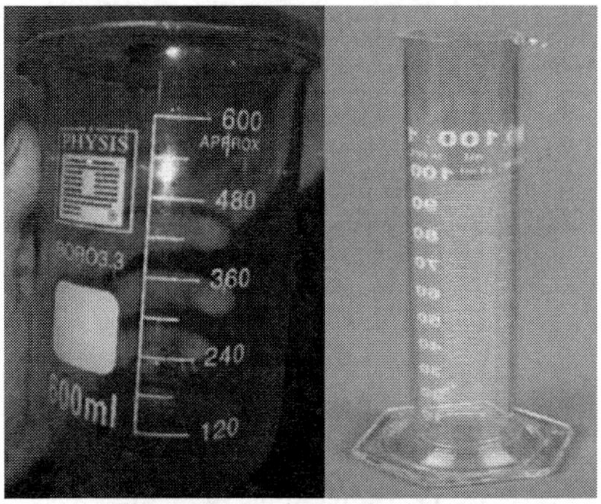

Figure 9. Beaker and test tube

From this test it was determined that a time 5.18seg, the beaker was filled to the level shown in Figure 10.

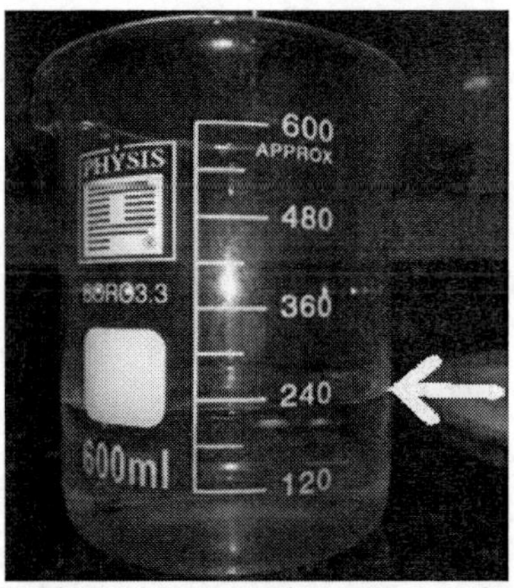

Figure 10. Beaker with water that came out in 5.18seg

To more accurately measuring the specimen 100ml in order to hear more about the amount of water that came out at this particular time is used.

The 100 ml graduated cylinder is completely filled twice and a third time reaching 14 ml partially, as shown in Figure 11. This tells us that in the water service 5.18seg gives us a flow rate of 214 ml

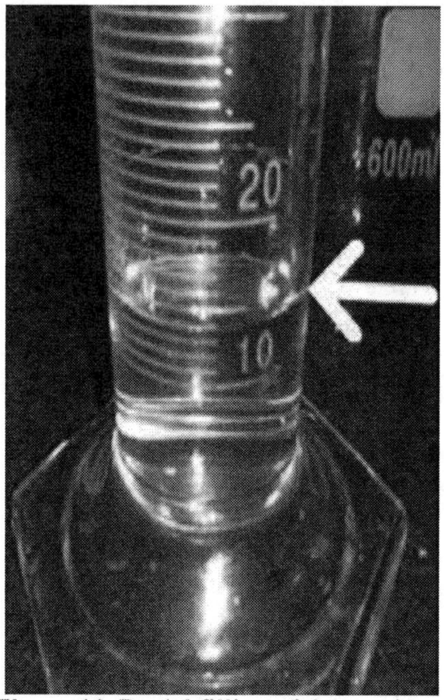

Figure 11. Partial filling of the specimen

In order to determine the microturbine generation, we must know approximately how much time in a day lasted keys sinks push open type.

Therefore, we perform equation 2.

$$TFGPD \cong T1P * CPP \quad (2)$$

Where:

TFGPD: Water flow time average day
T1P: Time 1 push
CPP: Average amount of push

Applying equation 2:

$$TFGPD \cong 5.18seg * 1026 \frac{drive}{day}$$

$$TFGPD \cong 5314,68 \frac{Driver*seg}{day}$$

knowing that one hour is 3600 seconds raised the equation 3:

$$TFGPD \ (hours) \cong 5314,68 \frac{Driver*seg}{day} * \frac{1H}{3600seg} \quad (3)$$

$$TFAPF \cong 1.5 \frac{Driver*seg}{day}$$

IOP Conf. Series: Materials Science and Engineering **844** (2020) 012056 doi:10.1088/1757-899X/844/1/012056

8. Comparison and Selection of the microturbine

Microturbine to select the most suited to water design, we consider the following parameters:

- Inlet and outlet diameter of ½ "
- Maximum working pressure greater than or equal to 20psi
- Less flow to start power generation
- It does not require a minimum operating time

With the above considerations the international market for microturbines these analyzes, finding few companies developing this technology. At European level there are two companies that develop these microturbines, which are:

- Tecnoturbines with Picoturbina model.
- Aiyima with model Micro Hydroelectric Generator turbine

Significantly, the domestic market is a limited supply of this type of hydraulic microturbines, since the use of these is still incipient in our country.

To make a correct selection of the microturbine to be used in prototype construction, a comparison of specifications between the Picoturbina and microturbine is Hydraulic. This comparison is carried out in Table 5.

Table 5. Comparative table between Picoturbina and microturbine Hydraulics

	Picoturbina	microturbine
Diameter inlet and outlet	½ "	½ "
Voltage range	105V to 12VDC	18.5 9.8V DC
Current Range	0A to 2A	120mA to 260mA
minimum operating pressure	0.045MPA	0.08MPa
Lost by traffic	not specified	3.6% to (0.25MPa)
Operating minimum flow	0.5 L / s (30LPM)	1LPM
Maximum power output	25W	5W
minimum operating time	20 minutes	It does not require

According to Table 5, the microturbine that is closest to the selection parameters, set in this item, is the hydraulic microturbine mark Aiyima since, unlike Picoturbina not required minimum operating time and operating pressure initial is the lowest.

9. Behavioral tests of hydraulic microturbine

To determine the characteristics of the prototype for simulation of water service, evidence which shows the operation and behavior of the hydraulic microturbine are made, as this gives us the idea of how you can be the design for this research project.

9.1 *Test 1*

In this first test the assembly of the microturbine and a manometer inlet, shown in Figure 12 is performed to make the measurements at different pressures and thus know the behavior of the microturbine, starting with the lower pressure to higher pressure in the water system of the selected place to test

Figure 12. Mounting conducted for test No. 1, with pressure from 0 to 60 psi

At a pressure of 10 psi, as shown in Figure 13, you are obtained one has a voltage of 13.80 volts, which is optimal for charging a 12V battery.

Figure 13. Evidence of voltage generation at a pressure of 10psi

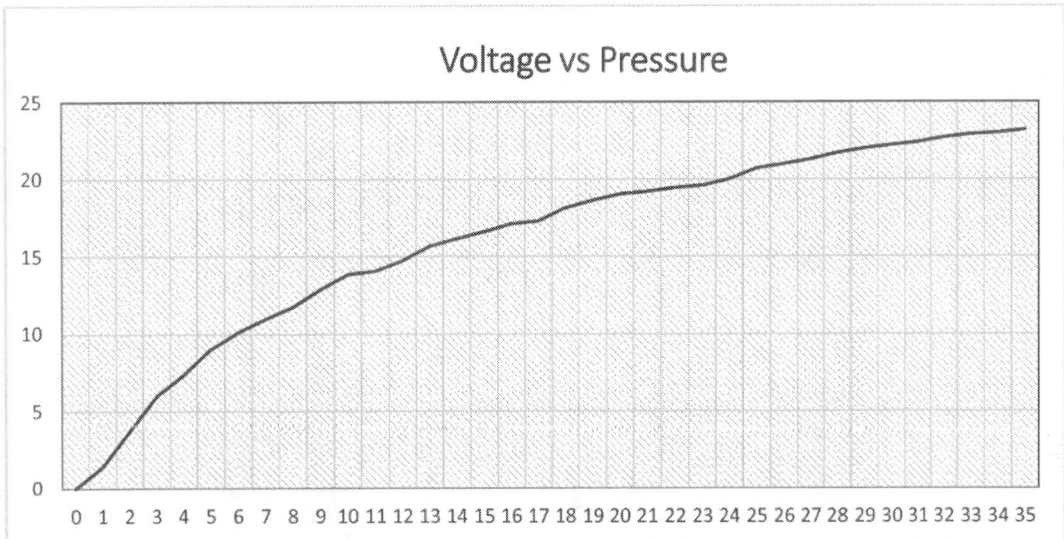

Figure 14. Graphing voltage Vs Pressure testing behavior

The graphical representation as that found in Figure 14 has the trend equation to the expression shown in equation 4:

$$y=-0.0195x^2+1.125665x+2.0699 \quad (4)$$

Where:
Y: Pressure Psi
X: Voltage in Volts
Trend equations fit to describe how the variable that is measured, changes over time.
The form of the equation adjusted trend depends on the model type is selected, as shown in Table 6 [42]

Table 6. Trend equations

Model type	Equation
Linear	yt = b0 + (b1 * t)
Quadratic	yt = b0 + b1 * t + (b2 * t2)
Exponential growth	yt = b0 + (B1T)
S curve (logistics Pearl-Reed)	t = (10a) / (b0 + b1 * B2T)

Where:
yt: The variable
b0: constant
b1 and b2: coefficients
t: Value of the unit time

9.2 *Test 2*

With this test installation of gauges and out of the microturbine is performed, as shown in Figure 15, in order to determine the losses caused by friction of the flow are, and to know whether these match losses they give us in the data sheet factory, which are 3.6% to 0.25MPa.

In our case, they will vary the results given by the manufacturer, since we work with a maximum pressure of 22psi. These tests for higher accuracy work, performed with air pressure in the Industrial Instrumentation Laboratory Technology unit of Santander, under the guidance of tutor's teachers.

Figure 15. Mounting conducted to determine traffic losses

In this test as shown in Figure 16, the inlet pressure takes 20psi to have a point of comparison with the output gauge.

IOP Conf. Series: Materials Science and Engineering **844** (2020) 012056 doi:10.1088/1757-899X/844/1/012056

Figure 16. Manometer measurement input

In Figure 17 note that the loss of output pressure is very low, therefore it can be concluded that the pressure does not affect the health system, and its output pressure is very close to 20psi.

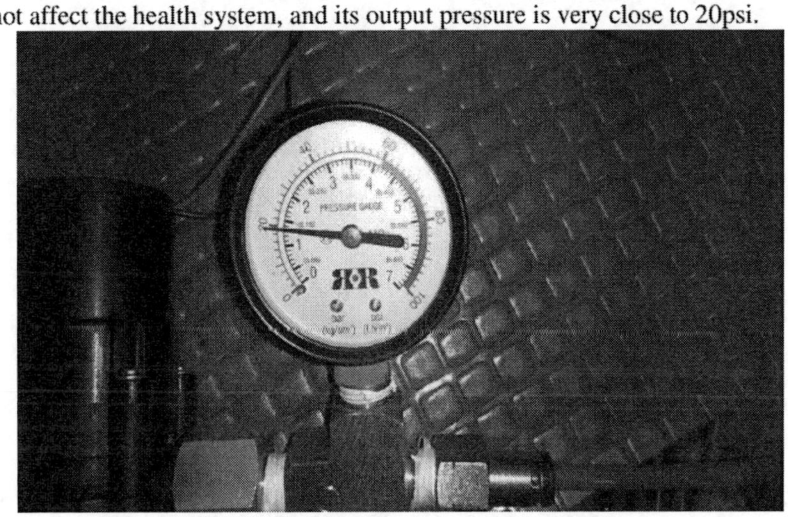

Figure 17. Measuring the output gauge

To allow comparison of the loss in output pressure, the respective theoretical calculations were performed to make a comparison with the actual data.

9.3. *Theoretical calculations*

Because the inlet pressure shown on the gauge is in the range of Psi, the conversion is performed megapascals, using equation 5:

$$P_{IN}(MPA)=\frac{1(MPA)*P_{IN}(PSI)}{145{,}038(PSI)} \quad (5)$$

Where:
Pin (mpa): found value for equation 5.
Pin (psi) inlet pressure supplied by the pressure gauge.

Applying equation 5:

$$Pin(mpa) = \frac{1(mpa)*20(psi)}{145{,}038(psi)}$$

$$Pin(mpa) = 0.137894897$$

Finally, to find the percentage of loss in the microturbine, we use equation 6.

$$X(\%) = \frac{3{,}6\%*Pin(mpa)}{0.25(mpa)} \tag{6}$$

Applying equation 6:

$$X(\%) = \frac{3{,}6\%*0{,}137894897(mpa)}{0.25(mpa)}$$

$$X(\%) \cong 2$$

From this we conclude that the loss is very low and therefore not affect the operation of the water system.

10. Prototype development

To perform this prototype takes into account multiple factors, such as flow rate and pressure of hydraulic system design of the building B Technology unit Santander (UTS), the pressures in and out of the microturbine, the outflow of the motor pump, recirculation water to avoid waste, the construction of the base or support for the installation of the tank and the motor pump and the pipes and the necessary accessories.

As the motor pump Ranger QB60 has an outlet diameter of 1 "and the microturbine has inlet and outlet ½", the corresponding line in PVC and accessories as ball valves, universal fittings, check valve, elbows, reducers, joints used, test.

also, it takes account of a bypass in order to decrease or increase the inlet pressure to the microturbine, since, if opened or ball valve (1) is closed, the pressures vary allowing to different measures them and obtain different voltages in power generation. Furthermore, in order to perform a purge upon completion of tests, and help us to drain the water in the upper pipeline.

To make a comparison of electrical variables, relative to the pressure, a pressure gauge at the inlet and outlet in the microturbine because with this variation values obtained support the calculation of loss of the microturbine is installed.

Based on the above considerations the final draft program performed in a computer assisted design, CAD, as shown in Figure 18 is made, all with the assistance of teacher tutors.

Significantly, with elaborate design with CAD, the losses are calculated by means of hydraulic calculation software, which is 2.0 Epanet General Public License (GNU)

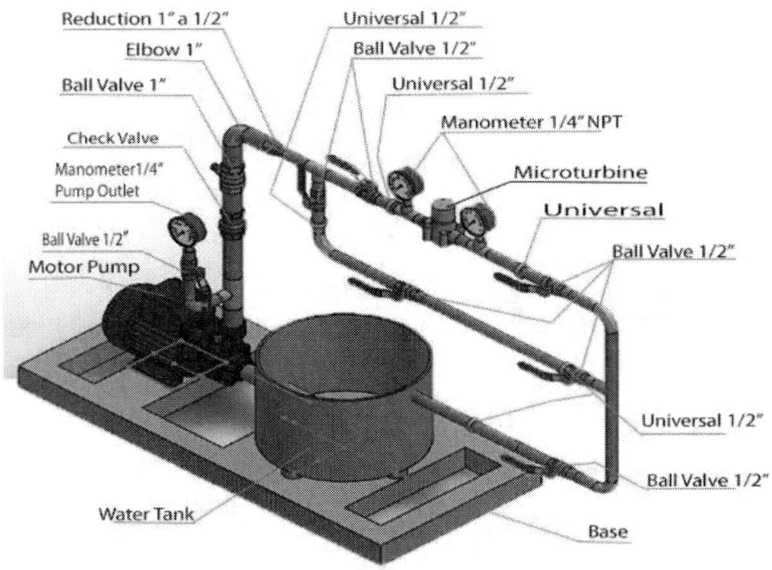

Figure 18. Sketch drawn in CAD

10.1 *Calculation of loss per charge*

To determine the feasibility of building sketch drawn in CAD, the respective calculations in order to interpret the losses that exist in this system to perform these calculations for hydraulic calculation software, Epanet V2.0 (GNU) is used it is made. This procedure is performed by pipe lengths according to their diameters and the results obtained are shown in a graph, generated by the program representation.

For these calculations the maximum flow of the motor pump QB60 Ranger, which is 35L / min is used. To develop these calculations, the constants shown in Table 7 are obtained.

Table 7. Constants water and PVC

Density of water at 25 ° C	997 kg / m3
Roughness PVC	0.005mm
PVC friction factor	0.0252

These constants together with the flow, are introduced into the Epanet V2.0 (GNU) software and we automatically generate the data shown in Table 8. These calculations are made as sections of pipe according to its length, the results vary.

Table 8. Data entered the software

Flow	35 l / min
Density	997 kg / m3
dynamic speed	0.00089 Pa * s
Inside diameter	35 mm
Speed	0.61 m / s
Reynolds number	23772
rugosity	0.005mm
Friction Factor	0.0252
Pressure drop per meter	0.01349m 0.01913627437 psi / m

10.2 *Calculation of load losses for section 1 "*

To prepare loss calculations for the pipe section 1 ", the length generated is used by the DAC, which is 0.6211m, this being introduced in the software, we generated the data shown in Figure 19.

Pipe Length	**0,6221**	m
Total, load loss in straight sections		
	0,0084	m
Loss of load (1)	**0,012**	**psi**

Figure 19. Head loss in straight sections

For this stretch of 1 ", it is necessary to know the constants accessories and the number of them, for the losses in accessories shown in Table 9.

Table 9. Losses in valves and fittings

Description	Quantity	K factor
T 1in, linear flow	one	0.2
Check valve	one	14.5
Ball valve 1in	one	0.05
90 ° elbow 1in	one	0.05
Load loss valves and accessories = 0.393 psi		

For total load losses, must have the elevation of the pipe, since the losses of total charge is the sum of the losses between the straight sections, fittings and elevation, these losses are shown in table 10.

Table 10. Total load losses

Pressure Difference by Pipeline Elevation	
elevation (h)	0,4858 m
Difference in Pressure Due to Unevenness	0,68876 **psi**
TOTAL LOAD LOSS (Straight sections + Accessories + Elevation)	
0,7715 m	
1,094 **psi**	

After typing all the above data, hydraulic calculations software, Epanet V2.0 (GNU), generated graphical representation of total losses shown in the red curve. The design uses a 1 "pipe SCH 40, which has an inner diameter of 26.2mm. This pipe section has a loss 0.741psi, as shown in Figure 21

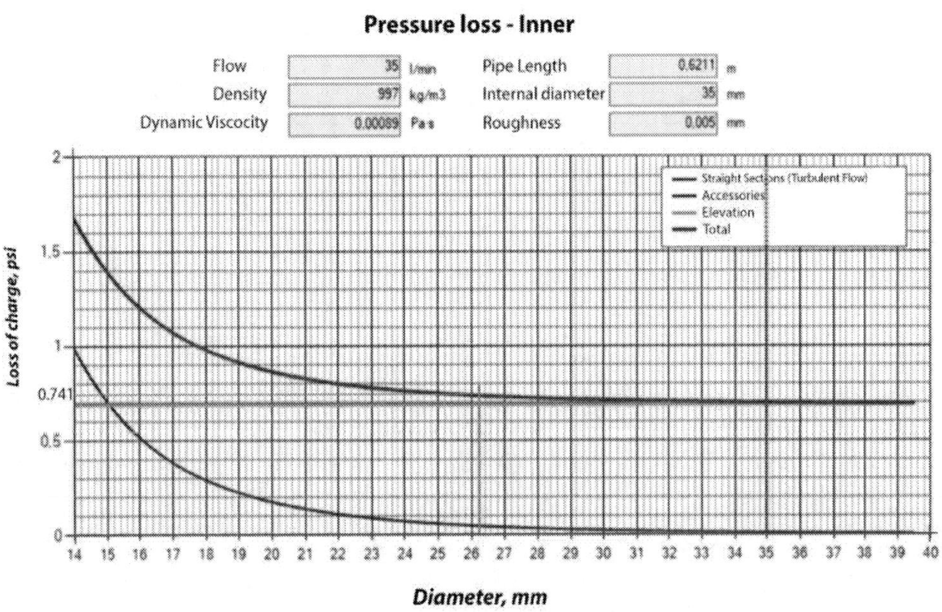

Figure 20. Graphical representation of pressure drop

10.3 *Calculation of losses to stretch ½ "*

For calculations of losses in the pipe section of 1/2 ", the total length of this stretch generated by the CAD, which is 1.749m, is to be introduced into the software, we generated data is used, shown in figure 22.

IOP Conf. Series: Materials Science and Engineering **844** (2020) 012056 doi:10.1088/1757-899X/844/1/012056

Loss of load in straight sections

pressure loss per meter:

	0,01349	m
	0,019136274	**psi/m**
Pipe Length	**1,749**	m

Total load loss in straight sections:

	0,0236	m
Loss of load (1)	0,033	**psi**

Figure 21. Losses in straight sections

For the section of 1/2 ", it is important to know the constants accessories and the amount of these in order to meet the losses in accessories that are reflected in Table 10.

Table 11. Lost minor

Description	Quantity	K factor
Reduction cone type 1in to 0.5in	one	0.37
Universal Valva	3	0.08
T linear 0.5in	3	0.2
90 ° elbow	two	0.3
Ball valve open 0.5in	3	0.05
Headloss in valve and accessories = 0.052psi		

To know the total load losses, we must have the elevation of the pipe as losses total load is the sum between losses of straight sections, accessories and elevation in this stretch of 1/2 "elevation is negative since this is in the upper circuit and carries the water back to the tank, the pressure losses are represented in figure 23.

After typing all the above data, hydraulic calculations software, Epanet V2.0 (GNU), generated graphical representation of total losses shown in the red curve. In the sketch tubing 1/2 "SCH 40, this has an internal diameter of 15.5mm and a loss of 0.980psi, as shown in Figure 24 is used.

Pressure Difference from Pipe Elevation

elevation (h)	-0,485	m
Difference in Pressure due to Unevenness	**-0,6878**	m

TOTAL LOAD LOSS (Straight sections + Accessories + Elevation)

-0,4247	m
-0,6	**psi**

Figure 22. Total load losses

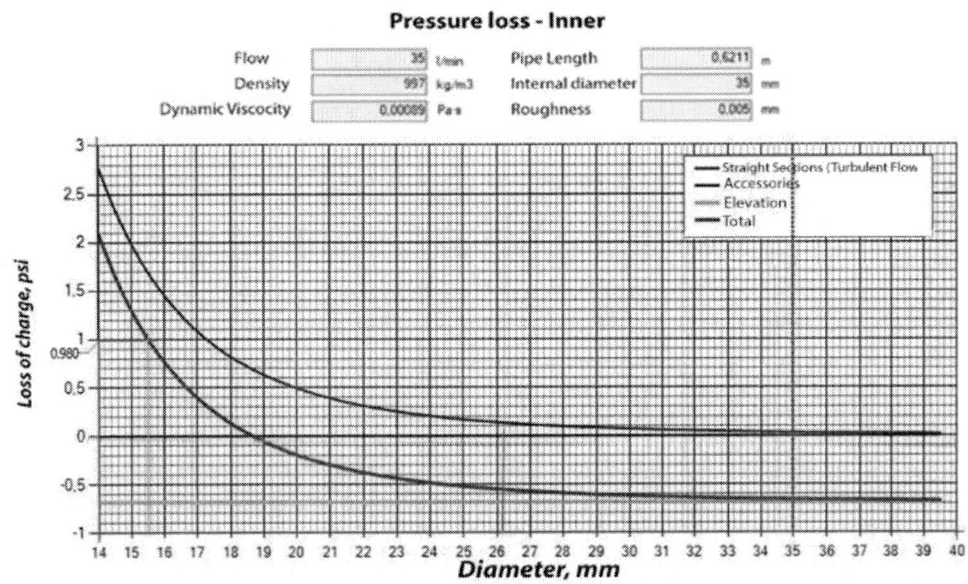

Figure 23. Graphical representation of total losses

11. Conclusions

- For this construction of a prototype for an analysis of energy potential it should be pressure and operating time to be handled in the prototype because it must be like to run in the bathrooms on the first floor of building B of Technological Units Santander, considering that:
  - The NTC-1500 standard set pressure and minimum flow according to the design water distribution, in the case of building design B are of 25psi and 1 GPM
  - Due to health battery is installed in the baths and then make a measurement operation time and operation duration time 1.5 hours / day distributed between 1092 drives set 5.18seg
- As shown in the development of this project to make the selection of the microturbine were considered, certain conditions and requirements, these lead us to choose the HYDROELECTRIC GENERATOR MICROTURBINE because it is the one that meets the parameters established both pressure

and their behavior. In paragraph 8 "comparison and selection of the microturbine" compared with other types of microturbines and define why your selection

- Because the microturbine will be in contact with potable water that meets the parameters set out in resolution 2115 of 2007. It is important to submit to the microturbine and water after trash to the respective studies such as:
  - Determine with certainty the material with which the microturbine is constructed, to know its behavior and degradation time.
  - Define if when in operation after a long time, degradation of the material occurs which is built the propeller of the microturbine and if some substance like mercury, selenium and other begin to exist in drinking water, being harmful these harmful to the human being
  - Determine when trash water by dissolved oxygen increases microturbine, this index is determined under the risk of water quality for human consumption - Irca-
- Because the market is booming microturbines is important to use other types of microturbines creating new analysis, measurements of electrical variables and thus experience different types of microturbines to take over this type of energy.

Referencias

[1] S. Clavijo, «La Republica,» 25 01 2019. [En línea]. Available: https://www.larepublica.co/analisis/sergio-clavijo-500041/dinamica-de-la-demanda-de-energia-en-2018-y-perspectivas-2019-2819927. [Último acceso: 27 07 2019].

[2] C. M. M. F.Oxenham, «Tasa de crecimiento de la población natural como medida paleodemográfica de crecimiento,» *Journal of Archaeological Science: Reports,* vol. 19, nº 19, pp. 352-356, 07 2018.

[3] CarolinaHenao-Rodríguez, «Reconversión industrial en Colombia necesaria para la adecuada inserción en el mercado internacional,» *Suma de Negocios,* vol. 8, nº 18, pp. 150-155, July-December 2018.

[4] D. A. S.-N. b. ,. N. C. B.-C. c. y. G. O.-P. Germán Alfonso Osma-Pinto a, «Consideraciones energéticas de las viviendas sociales en Colombia según el concepto NZEB,» *DYNA,* vol. 1, nº 1, p. 11, Agosto 2015.

[5] BP, «BP Statistical Review,» 2019. [En línea]. Available: https://www.bp.com/content/dam/bp/business-sites/en/global/corporate/pdfs/energy-economics/statistical-review/bp-stats-review-2019-electricity.pdf. [Último acceso: 23 Junio 2019].

[6] J. M. L. L. N. M. G. Fernando Villada Duque, «Effects of Incentives for Renewable,» vol. 01, nº 01, p. 16, 28 07 2017.

[7] M. V.-C. a. P. B. Z. Oscar Churio-Silvera, «Status of Non-Conventional Sources of Energy in Colombia: A Look at the Challenges and Opportunities of the Electric Sector,» *HIKARI Ltd,* vol. 11, nº 44, p. 10 , 2018.

[8] C. M. S. L. l. johnson, «Desarrollo y realización de estudios.,» *Principles and Practice of Clinical Research (Fourth Edition),* vol. 1, nº 15, pp. 203-2018, 2018.

[9] A. A. M. A. M. V. V. d. c. e. A. G. A. Kulagin, «Perspectivas de la industria de la energía eléctrica en medio de la transformación de los mercados globales de generación de energía.,» *springer link,* vol. 64, nº 10, p. 02, 15 03 2017.

[10] D. &. H. V. &. A. F. &. M. C. Biner, «Design & performance of a hydraulic micro-turbine with counter-rotating runners,» *researchgate,* vol. 1, nº 1, p. 11, 04 2015.

[11] P. C. H. P. S.L. Dixon B. Eng., «Capítulo 9 - Turbinas hidráulicas,» de *Mecánica de fluidos y termodinámica de turbomáquinas (Séptima edición)*, Madrid, Dossant, 2014, pp. 361-418.
[12] B. F. Towler, «Capítulo 10 - Hidroelectricidad.,» de *El futuro de la energía*, Brisbane, Academic Press, 2017, pp. 215-235.
[13] N. Nasim Uddin, «Lecciones aprendidas: falla de una represa del proyecto hidroeléctrico,» *Asce,* vol. 19, nº 1, p. 9, 02 2005.
[14] C. C. A. D. B. F. D. J. H. K. O. Winemiller, «Impactos de las represas hidroeléctricas en los peces y las pesquerías en los ríos tropicales a través de la lente de los rasgos funcionales,» *Opinión actual en sostenibilidad ambiental,* vol. 34, nº 1, pp. 20-40, 04 2019.
[15] C. Koç, «Problemas y soluciones relacionadas con centrales hidroeléctricas construidas en Buyuk Menderes y la cuenca del Mediterráneo occidental,» *Taylor Francis Online,* vol. 34, nº 15, pp. 1416-1425, 2012.
[16] C. J. F. I. D. Santiago Hoyos, «Integración de fuentes no convencionales de energía renovable al mercado eléctrico y su impacto sobre el precio,» *Ingeniería y Ciencia,* vol. 13, nº 26, p. 32, 2017.
[17] N. S. C. T. M. M. b. A. A. M. M. C. Felgueiras, «Nuevas tendencias en la producción y utilización de energía,» *Energía Procedia,* vol. 107, nº 1, pp. 7-14, 2019.
[18] L. L. Grigsby, Generación, transmisión y distribución de energía eléctrica., Boca raton: CRC Press, 2012.
[19] C. soares, Microtrubinas, Oxford: Butterworth-Heinemann, 2007.
[20] D. P. L. v. Beek, Water Resource Systems Planning and Management, Netherlands: Springer, 2017.
[21] L. H. y. J. Gomez, «cdtdegas.com,» 01 03 2019. [En línea]. Available: https://www.cdtdegas.com/images/Descargas/Nuestra_revista/MetFlu5/4_Microturbinas.pdf.
[22] L. H. L. Wenjia, «El análisis de los efectos de la generación de energía de energía limpia.,» *Energía Procedia,* vol. 152, nº 1, pp. 947-952, 2018.
[23] J. G. y. S.-C. R. González-Hernández, «Representación y estimación del coeficiente de potencia en sistemas de conversión de energía eólica. Revista Facultad de Ingeniería,» *Revista Facultad de Ingeniería,* vol. 28, nº 50, p. 13, 2019.
[24] «Propuesta de diseño de un sistema de energía solar fotovoltaica. Caso de aplicación en la ciudad de Bogotá,» *Salamanca-Ávila, Sebastián.,* vol. 1, nº 30, p. 15, 2017.
[25] M. H. R. Z. F. H. A. A. R. N. Abdullah, «Transmisión de energía eléctrica,» de *Manual de electrónica de potencia (Cuarta edición)*, Oxford, Butterworth-Heinemann, 2018, pp. 826-946.
[26] M. M. ,. C. F. ,. D. A. Vieux F, «Patrones de consumo de agua y bebidas entre niños de 4 a 13 años en el Reino Unido.,» *BMC Salud Pública.,* vol. 17, nº 1, p. 28, 2017.
[27] B. C. A. G. H. T. B. M. H. Wziontek, «Medición del efecto de los cambios en el almacenamiento de agua local en observaciones de gravedad in situ: estudio de caso del Observatorio Geodésico Wettzell, Alemania,» *Water resources research* , vol. 46, nº 1, p. 15, 2010.
[28] G. T. ,. M. C. F. T. Imene Yahyaoui, «Diseño y Evaluación de un Sistema de Bombeo de Agua Renovable.,» *IFAC-PapersOnLine,* vol. 48, nº 30, pp. 462-467, 2015.
[29] C. P. ,. S. I. ,. B. C. Hayden JA, «La identificación de las fases de la investigación ayuda a planificar, evaluar y aplicar los resultados de los estudios de pronóstico explicativo.,» *J Clin Epidemiol,* vol. 61, nº 6, p. 552–560, 06 2008.
[30] P. K. Dey, «Enfoque integrado para el análisis de viabilidad del proyecto: un estudio de caso,» *Evaluación de impacto y evaluación de proyectos* , vol. 19, nº 1, pp. 235-245, 2001.
[31] E. R. A. G. G. Alessandro Giacchino, «Soluciones híbridas para la Industria de Generación de Energía,» *Energía Procedia,* vol. 148, nº 1, pp. 814-821, 2018.
[32] Q. &. F. J. &. S. Liang, «nfluence of hydraulic characteristics on transient behavior of ultra-high head pump turbine,» *Qinghua,* vol. 1, nº 1, p. 240, 2019.

[33] O. Y. M. A. I. Y. M. A. Aly Hassan Elbatran, «Revisiones de sistemas hidroeléctricos y de turbinas,» *ResearchGate,* vol. 74, nº 5, pp. 89-90, 2015.

[34] I. Božić, «Determinación de pérdidas hidráulicas en el paso de flujo entre las paletas guía y el corredor de la turbina Kaplan,» *Journal of Hydraulic Research,* vol. 55, nº 3, pp. 349-361, 2015.

[35] E. d. l. F. D. M. Z. V. Jorge Veiga de Cabo, «Modelos de estudios en investigación aplicada: conceptos y criterios para el diseño,» *Medicina y Seguridad del Trabajo,* vol. 54, nº 210, pp. 81-88, 2008.

[36] R. P. McDonald, «El análisis informativo de las curvas de tendencia individuales,» *Investigación de comportamiento multivariable* , vol. 39, nº 3, pp. 517-563, 2004.

[37] INCONTEC, «NTC-1500 CODIGO COLOMBIANO DE INSTALACIONES,» de *INCONTEC,* Tercera ed., Bogota, Aprocog, 2017, p. 104 y 107.

[38] Intercal, «bombas-intercal.com,» 15 03 2019. [En línea]. Available: https://bombas-intercal.com/bombas-centrifugas-funcionamiento/. [Último acceso: 15 03 2019].

[39] Aiyima, «Aiyima,» 23 05 2018. [En línea]. Available: https://es.aliexpress.com/store/product/Turbina-generatore-idroelettrico-hydroelectric-Generator-micro-turbine-generator-water-LED/1897413_32692792829.html?spm=a219c.search0104.3.72.40a3157dUktgR3&ws_ab_test=searchweb0_0,searchweb201602_5_10065_10068_1054. [Último acceso: 02 02 2019].

[40] Tecnologia, «areatecnologia.com,» areatecnologia.com, - - -. [En línea]. Available: https://www.areatecnologia.com/herramientas/manometro.html. [Último acceso: 07 06 2016].

[41] S. Fluids, «SC Fluids,» SC Fluids, 06 06 2019. [En línea]. Available: http://www.valvulasymedidores.com/valvulas_check.html. [Último acceso: 06 07 2019].

[42] minitab, «support.minitab.com,» 15 03 2019. [En línea]. Available: https://support.minitab.com/es-mx/minitab/18/help-and-how-to/modeling-statistics/time-series/how-to/trend-analysis/interpret-the-results/all-statistics-and-graphs/. [Último acceso: 15 03 2019].

[43] mercadoherramientaonline, «mercadolibre,» 14 03 2019. [En línea]. Available: https://articulo.mercadolibre.com.co/MCO-474206991-motobomba-ranger-12hpq-max35lmin-qb60-_JM?quantity=1. [Último acceso: 14 03 2019].

[44] Aiyima, «Aiyima,» 23 05 2018. [En línea]. Available: https://es.aliexpress.com/item/New-Hydroelectric-power-Micro-hydro-generator-Portable-water-charger-Micro-hydro-generator-water-flow-generator-3/32847826663.html?spm=a219c.10010108.1000016.1.2ee7110fDmZZSQ&isOrigTitle=true. [Último acceso: 02 02 2019].

[45] laboratoriotienda, «labotienda.com,» 01 03 2019. [En línea]. Available: https://www.labotienda.com/es/productos-laboratorio/probeta-graduada-vidrio-100-ml/.

[46] Omega, «omega.com,» omega, - - -. [En línea]. Available: https://es.omega.com/prodinfo/galgas-de-presion.html. [Último acceso: 06 07 2019].

[47] Camatea, «camatec.es,» camatec.es, 06 07 2019. [En línea]. Available: https://camatec.es/manometros/1819-ca3814004.html. [Último acceso: 06 07 2019].

[48] Vidri, «Vidri,» Vidri, 06 07 2019. [En línea]. Available: https://www.vidri.com.sv/producto/116208/V%C3%A1lvula-check-de-retenci%C3%B3n-vertical-de-1-pulgada.html. [Último acceso: 06 07 2019].

[49] Evnas, «Evans.com,» 19 03 2015. [En línea]. Available: https://www.evans.com.mx/specs/AQUA30W. [Último acceso: 14 03 2019].

[50] Pediaa, «Pedia.com,» 21 06 1018. [En línea]. Available: https://pediaa.com/difference-between-descriptive-and-experimental-research/#Experimental%20Research. [Último acceso: 27 07 2019].

[51] ujaen, «ujaen.es,» 12 05 2019. [En línea]. Available: http://www.ujaen.es/investiga/tics_tfg/enfo_cuanti.html. [Último acceso: 27 07 2019].
[52] E. H. F. y. M. I. Z. GONZALEZ, «PREPARACIÓN DE UN PROYECTO DE INVESTIGACIÓN,» *Ciencia y enfermería,* vol. 9, nº 2, pp. 23-28, 2003.

IOP Conf. Series: Materials Science and Engineering **844** (2020) 012057 doi:10.1088/1757-899X/844/1/012057

Characterization of the mechanical vibration signals associated with unbalance and misalignment in rotating machines, using the cepstrum transformation and the principal component analysis

C L Sandoval-Rodriguez[1], J G Ascanio Villabona [1], C G Cárdenas-Arias[1], A D Rincon-Quintero[1-2] and B E Tarazona-Romero[1].

[1] Energy Systems, Automation and Control Research Group (GISEAC), Design and materials research group (DIMAT), Electromechanical Engineering, Faculty of Natural Sciences and Engineering, Technological Units of Santander (UTS), Road No 9-82 students, 680,005, Bucaramanga, Colombia.

[2] Energy in Building Research Group (ENEDI), Doctoral Program in Energy Efficiency and Sustainability in Engineering and Architecture, Department of Machines and Thermal Engines, University of the Basque Country (UPV / EHU), Engineer Torres Quevedo Square, 1, 48013, Bilbao, BI, Spain.

candoval@correo.uts.edu.co

Abstract. In the present document, the Cepstrum transform and the analysis of principal components were used to differentiate amplitudes in the mechanical vibrations produced by unbalance and misalignment with respect to a reference group. This document requires three stages. It begins with levelling in order to establish the control group. The unbalancing was carried out with a known mass located in the two radial distances of the first and second flywheels. The misalignment was made by running the sliding supports back 0.5, 1.0 and 1.5 degrees. In the second stage, Matlab algorithms were created for both cepsctrum and main component analysis. In the last stage the obtained data were analyzed identifying the differences that may exist in the analyzed records. The project focused on the use of Matlab to find differences at a frequency of 30 Hz. The results obtained made it possible to determine that it is possible to find differences with the proposed methodology.

1. Introduction.

Rotating machines have different operating states during its lifetime. These may be normal type or operation failures, which may be due to little study of the machine, and critical operation, which can affect permanently to different parts making up the machine [1]. Because of this industry began to adopt different types of maintenance on rotating machines in order to address the failures that may occur in these teams; preventively before the failure, correctively occurs when the failure occurs and is investigated because occurred; and fails when it seeks to replace damaged parts [1].

In rotary machines 2 important affectations studied in this project are presented. The first is due to improper mounting of the machine by poor manufacture of the bed or poor mechanical coupling between drive and driven shafts; this involvement is known as alignment fault [2].

On the other hand, the second involvement is concerned with wear, incorrect selection of a failed component, faulty manufacture of the components of the machine, some dynamic fracture machine part, among other causes; by which it can be produced by unbalance failure [3]. Although the type of failure can be detected in time, other inherent flaws in rotating machinery problems arise, which have to do with other factors such as; interpretation of curves, judgment and subjectivity of the analyst. The interpretation of the spectra in vibration analysis equipment or specialized software depends on the experience, knowledge and judgment on the analyst applied vibration; to determine whether visible sometimes audible failures are due to imbalance, misalignment, mechanical looseness, worn bearings, bent arrows or electrical problems [2]. In this sense, the solution to this problem focuses on the possibility of finding a method to moderate the interpretation of the information from the vibration, the analyst's subjectivity; and also serve to support maintenance professionals for a correct decision in determining faults in rotating machines.

Importantly, the scientific and academic community has been concerned about this problem and have generated around this, various alternatives both in terms of feature extraction [4-7], from mechanical vibrations, as in systems that allow discrimination of each operating mode [8-9] and focused data compression for the application of classification and troubleshooting rotating machinery [10]. In this regard, it has raised for this work to extract features related to the behavior of a rotating system affected by unbalance and misalignment independently.

The methodology used in the proposed solution is that the descriptive desired characterize faults; correlational because a reference group is set to analyze faults unbalance and misalignment in rotating machinery [11]. Similarly, the method used is the deductive because based on overall theories transformed cepstrum and principal component analysis in solving particular problems as proposed in this work [12]; on the other hand, the technique used to validate all results outlined in this research is the experimental [11] type.

Matlab is used to analyze three general types of samples; The control group, the group of imbalance and misalignment group. In Matlab different algorithms in order to select samples from the same variable that are grouped in reference to the amplitude at the fundamental frequency created. It also created algorithms to calculate the cepstrum transformed to perform principal component analysis 8 observations, to obtain the Euclidean distance and different characteristics in the analysis of 6 variables statistics.

2. Cepstrum Analysis and Principal Component Analysis

The transformed cepstrum Oppenheim and Schafer as [13], with Bogert emerged in 1963, referring to a way to analyze signals in the time domain, which did not belong to the frequency domain; cepstrum can be defined as the spectrum of the natural logarithm of the spectrum as shown in equation 1.

$$X[n] = IFFT\{log|FFT(x[n])|\} \tag{1}$$

In this equation the expression IFFT corresponds to the discrete inverse Fourier transform, the expression FFT is the discrete Fourier transform, the expression log is the logarithm base 10 and the expression x [n] is the signal in the time domain. With these coefficients aspects of the spectra such as the envelope or detailed characteristics [14] recovered.

It is clear that the drawback that occurs in the technique of feature extraction by applying transformations of the signal is the high dimensionality of the resulting space, which would form very large vectors description. Given this, the dimensionality reduction techniques help to preserve the most important information while the dimension of the transformed space is reduced [15-16]. The idea of the principal component analysis is to estimate the variability of each data set and retain greater variance components according to certain criteria [17] in order to use them as representative characteristics of each mode.

The first principal component is the linear combination of coefficients with a high variability; as shown in Equation 1. The second is due to the linear combination of the coefficients of the second greater variability and so on.

$$Xp_n = \sum Cp_i * \lambda p_i \qquad (2)$$

Where Xp is the main component, the subscript n indicates the order of the component, Cp is the coefficient related to the variance of the coefficient λ. After calculating the principal component s are retained only those showing greater variability using different criteria. For this work four criteria were implemented:

Jolife criteria: Only the components whose variance is greater than 0.7 are retained.
Kaiser criterion; Are retained components whose variance is greater than average.
standard criteria: The components are retained whose variance is below 2% (Criterion elbow)
Cumulative variance criterion: The components are retained whose variance is greater than 1% of the total cumulative variance of all components. With the application of the different criteria the components to be applied to the algorithm that allow differentiation are retained.

3. Sampling and observations

The calculation of the samples was conducted using Equ (3) in infinite sample size populations without known antecedent.; where Z is the confidence of the sample, P is the probability of success, Q is the probability of failure and E is the allowable error.

$$n = \frac{ZPQ}{E} \qquad (3)$$

For a population in the conditions described should be taken Z = 1.6448, the probability of success and failure should equal P = Q = 0.5, and the allowable error E = 0.01 was taken.

$$n = \frac{1.6448 * 0.5 * 0.5}{0.01} = 67.63 \cong 68$$

In Table 1, the variables used for performing bank failures in the vibration according to the limitations found in preliminary tests are described.

Table 1. Description of variables.

VARIABLE	DESCRIPTION
GCAB	Control group alignment and balancing
DA05G	Movement of the support for 0.5 ° degrees
DA10G	Movement supports to 1.0 ° degrees
DA15G	Movement supports to 1.5 ° degrees
DBV1R1	Place a mass m1 in the wheel 1 to the distance r1
DBV1R2	Place a mass m1 in the wheel 1 to the distance r2
DBV2R1	Place a mass m1 in the wheel 2 to the distance r1
DBV2R2	Place a mass m1 in the wheel 2 to the distance r2

Point for data collection, is located at the farthest support the engine in horizontal position to half the distance of the bearing and the signal acquisition was performed using the piezoelectric accelerometer Dytran. All samples were performed at 30 Hz, due to the limitations found in preliminary tests.

Installation and data acquisition was performed by means of a sensor Dytran brand, acquisition card National Instruments and Labview program [18]. In Fig.1 the hardware which was used for the tests shown.

Figure 1. Mechanical Hardware Test.

3.1. *Obtaining the characteristics using cepstrum coefficients into mel scale and principal components analysis.*

For comparison of characteristics comparison algorithm using Euclidean distance used by [19-20] it was used. This system is based on a priori knowledge base characterizing cases the required class. a comparison control group compared to the other seven groups shown in Table 2.

Table 2. Euclidean distance of CCM related variables

Variables Name	Relationship	no CCM	DE * e-3
CEPDA1	GCABvsDA05g	1	3.93322
CEPDA2	GCABvsDA10g	1	4.82292
CEPDA3	GCABvsDA15g	3	10.94881
CEPD1	GCABvsDBv1r1	4	11.80011
CEPD2	GCABvsDBv1r2	3	16.63780
CEPD3	GCABvsDBv2r1	2	17.28532
CEPD4	GCABvsDBv2r2	2	32.77406

DE: Euclidean distance.
CCM: Cepstrum coefficient on the Mel scale.

It should be noted that for the principal component analysis (PCA), the Fourier spectrum is calculated and spectral coefficients, applying the different criteria the components which in turn are the characteristics of each observation are retained. In this regard the components shown in Table 3 were obtained.

Table 3. Components retained on each criterion

Criterion	Retained components	GROUPS OBSERVED
Jolife	7	everyone
Kaiser	2	everyone
Standard	1	everyone
Cumulative variance	2	everyone

It is important to note that not necessarily the criterion to retain the minimum number of components will best performance features. To probe the above was necessary to apply the algorithm developed by [20] to give the results shown in Table 4 (case-based reasoning-RBC).

Table 4. Percentages classification and standard deviations for each criterion.

Criterion	Percentage Of Classification	Standard Deviations
Jolife	80.2	2
Kaiser	78.66	4,21
Standard	82	6,32
Cumulative variance	80.2	6.96

According to the above criteria Jolife (good classification rate and high concentration of data), for characterization as described in Table 5 was made.

Table 5. Statistical employees

Statistical	Calculation method	Description
E1: Average value	$T1 = \frac{\sum_{n=1}^{N} x(n)}{N}$	The average value in a set of data within a reasonable range represents a trend of values to a representative value for all [21]
E2: Standard Deviation	$T2 = \sqrt{\frac{\sum_{n=1}^{N}(x(n) - T1)^2}{N-1}}$	The standard deviation is the root of the sum of the squared difference values of the signal with the average value between the total data minus 1. The measure indicates how scattered the data are the mean value zero being an ideal value where the data signal and its average value agree [21]
E3: RMS root mean square	$T3 = \sqrt{\frac{\sum_{n=1}^{N}(x(n))^2}{N}}$	The root mean square or RMS is the root of the sum of the square of a number by the total amount of data. By relying heavily on the average value of the RMS data it is also susceptible to extreme outliers [21]
CV: Coefficient of variation	$CV = \frac{T2}{T1}$	the coefficient of variation which is defined as the standard deviation about the mean value of the signal [21]

3.2. Statistical results calculated

After applying statistical both mel cepstral coefficients in the scale and the main components for the criterion Jolife the results shown in Table 6. (Note that the vectors have compared the cepstral coefficients and the principal components or for each group were obtained observation)

Table 6. Results of the characterizations for the cepstral coefficients of Table 2.

Variable	statisticians			
	E1 * e4	E2 * e-4	CV%	E3 * e-4
GCAB	51.9257	3.0682	5.91	52.0149
DA05g	52.2840	3.2627	6.24	52.3842
DA10g	52.9456	4.3471	8.21	53.1211
DA15g	38.4259	3.6330	9.45	38.5947
DBv1r1m	62.8185	12.5058	19.91	64.0333
DBv1r2m	69.4527	2.8814	4.15	69.5116
DBv2r1m	77.7885	3.7489	4.82	77.8775
DBv2r2m	100.5636	5.0916	5.06	100.6905

E1: Average value
E2: Standard Deviation
E3: RMS root mean square
CV: Coefficient of variation

From the above it can be seen that the average value to differentiate each of the variables in Table 1, where the variables for the unaligned group are above and below the reference group and variables unbalanced group are above reference group. Variable statistical standard deviation E2 does not provide much information al1, but if the coefficient of variation is used, it can be concluded that the overall variables in Table 5, do not exceed 10% coefficient of variation; however DBv1r1 variable passes this value and is located in a coefficient of variation of approximately 20%; so in general all the variables do not have a too large standard deviation.

The variable statistic root mean square E3 allows differentiating each of the variables in Table 6, where the variables for the unaligned group are above and below the reference group and variables unbalanced group are above group reference.

4. Conclusions

The anomalies in rotary machines can be identified up to the fourth cepstrum coefficient in mel scale. The Euclidean distance made it possible to find differences between the Mel scale cepstrum coefficients for unbalance and misalignment conditions in rotary machines. On the other hand, the principal components that represent greater discrimination power were calculated taking into account the Jolife criterion, obtaining characteristics such as the mean value, standard deviation, variation coefficient and quadratic mean root. Additionally, the coefficient of variation allowed to determine that the variables studied for the unbalance and misalignment conditions do not have a very high standard deviation. Finally, it was evidenced that both the mean value and the root mean quadratic allow to differentiate all the groups when composing a characteristic vector containing the cepstrales coefficients in mel scale and the principal components obtained from the Fourier coefficients of the spectrum.

References

[1] Pérez González A, Rodríguez Cervantes P J, & Sancho Brú J L, (2007). Mantenimiento mecánico de máquinas. Universitat Jaume I, Castelló de la Plana, España..

[2] Gómez de León F C. (1998). Tecnología del mantenimiento industrial. Murcia: EDITUM.

[3] Rao S S, (2012). Vibraciones Mecánicas (5a edición) Pearson Educación., pp. 651-658.

[4] Li C, Zhang W E I, Peng G, & Liu S (2017). Bearing fault diagnosis using fully-connected winner-take-all autoencoder. *Access IEEE, 6, 6103-6115.*

[5] Feng Z, Zhang D, & Zuo MJ (2017). Adaptive mode decomposition methods and their applications in signal analysis for fault diagnosis machinery: a Review with examples. *IEEE access, 5, 24301-24331.*

[6] Wang D, Tsui KL, & Miao Q. (2017). Prognostics and health management: A review of bearing and gear vibration based health indicators. *Access IEEE, 6, 665-676.*

[7] Lu Y, Li Q, Pan Z, & Liang, SY (2018). Prognosis of bearing degradation factor using gradient Variable forgetting RLS combined time series model With. *Access IEEE, 6, 10986-10995.*

[8] Duan X, Zhao T, Li T, Liu J, Zou L, & Zhang L. (2017). Method for diagnosis of on-load tap changer based on wavelet theory and support vector machine. *The Journal of Engineering, 2017 (13)*, **2193-2197.**

[9] Liu S, Sun Y, & Zhang L. (2018). A novel fault diagnosis method based on noise-assisted DMEM and neural network for functional fuzzy rolling element bearings. *Access IEEE, 6*, 27048-27068.

[10] Ahmed H, & Nandi AK (2018). Compressive sampling and feature ranking framework for classification With bearing fault vibration signals. *Access IEEE, 6*, 44731-44746.

[11] Hernández Sampieri R, Fernández Collado C & Baptista Lucio P. (2010). Metodología de la investigación. pp. 102-160. 2006.

[12] Bernal, C. A. (2006). Metodología de la investigación (No. 001.42 B4564m Ej. 1 022619). Pearson, **pp. 56.** [Online]. Available: https://books.google.com.co/books?id=h4X_eFai59oC&pg=PA56&dq=metodo+deductivo&

hl=es-419&sa= X & SQI = 2 & pjf = 1 & ved = 0ahUKEwiNtKKFzYTUAhVDfiYKHUtAA10Q6AEIITAA # v = 1page & q = 20% deductive method & f = false.

[13] Oppenheim, A V & Schafer R W. (2004). From frequency to quefrency: A history of the cepstrum. *IEEE signal processing Magazine*, **21(5), 95-106**. [Online]. Available in: https://www.researchgate.net/ profile / Av_Oppenheim / publication / 3321562_From_Frequency_to_Quefrency_A_History_of_the_Cepstrum / links / 5519617d0cf2d241f3564c58.pdf

[14] Quiroga J., G. Trujillo, S. and Quintero, "Study of incipient bearing failure using the technique of the envelope and cepstrum" *Ingeniare, vol. 20*, no. 3, **pp. 350-359**. Nov. 2012 [Online]. Available in:http://www.scielo.cl/scielo.php?script=sci_arttext&pid=S0718-33052012000300009&lng=

[15] Smith LI (2002). A tutorial on principal components analysis.

[16] Rao CR (1964). The use and interpretation of principal component analysis in applied research. Sankhyā: The Indian Journal of Statistics, Series A, 329-358.

[17] PEÑA D. Multivariate analysis data, Chapter 5.McGraw-Hill. ISBN: 8448136101.

[18] Forero W, Lizcano D, & Colll J. "Implementation of a bank for the analysis of mechanical vibrations generated by unbalance and misalignment in rotating machines," Undergraduate Thesis, Technological Units of Santander, Bucaramanga, Colombia, 2016.

[19] Camacho Navarro, J; "Numerical adaptation and validation expert for structural health monitoring through pattern recognition system". Industrial University of Santander, Bucaramanga 2010.

[20] Sandoval Rodriguez, C L. Characterization of the dynamics of basic hand movements from the electromyography activity forearm. Industrial University of Santander 2013.

[21] Martinez C, statistics and sampling. 13th Edition. ECOE Editions, pp. 303-310. 2012

IOP Conf. Series: Materials Science and Engineering **844** (2020) 012058 doi:10.1088/1757-899X/844/1/012058

Development of an artificial vision system that allows non-destructive testing on flat concrete slabs for surface crack detection by processing of digital images in MATLAB

BE-Romero-Tarazona[1], CL-Rodriguez-Sandoval[1], JG-Villabona1-Ascanio[1] and A D Rincón-Quintero[1,2]

[1] Automation and Control Energy Systems Research Group (GISEAC), Design and materials research group (DIMAT), Electromechanical Engineering, Faculty of Natural Sciences and Engineering, Technological Units of Santander (UTS), Road No 9-82 students, 680,005, Bucaramanga, Colombia.
[2] Energy in Building research group (ENEDI), Doctoral Program in Energy Efficiency and Sustainability in Engineering and Architecture, Department of Machines and Thermal Engines, University of the Basque Country (UPV/EHU), Engineer Torres Quevedo Plaza, 1, 48013, Bilbao, BI, España

btarazona@correo.uts.edu.co

Abstract. A computing tool based on an intelligent algorithm generates new alternatives for an early crack detection on flat concrete slab in a surface testing. First, a system able to autonomously detect a surface crack, by means of a digital image processing system has been developed. This system carries on an effective selection of data by an algorism bound to a specialized MATLAB software that can assess in full detail the surface conditions of a concrete slab, and finding out any affected spot.

1. Introduction

Cracks are breaking sites occurring in concrete when the tensions stressing the concrete structures are superior to its resistance. Several conditions leading to this serious threat, some or chemical origin, other of physical origin. Depending on the time they first appear, it could be in the fresh stage or plastic stage, before the hardening has taken place, or in other instances, through the concrete hardening process [1].

It is here when the term durability arises. Durability may be defined as the concrete structure capacity to keep intact both their chemical and physical attributes throughout their usable life span when subjected to material break up as a result of multiple tensions these structures are designed to resist. Therefore, every potentially aggressive agent a concrete structure is exposed to, must be identified [2].

In turn, the first indications of structural degradation in particular are given in superfie thereof, evidenced as fissures and cracks [3], Locally reducing the stiffness and causing discontinuities in the material[4] [5], Which affects the durability of the structure and is critical for maintenance, because they remain in contact with the environment causing severe damage [3]. Several authors mention in their research technique crack detection by manual visual inspection, one says that it is the method most commonly used in practice, which is inefficient from the standpoint of cost, time, accuracy and safety[6]Another describes it as; the method acclaimed for inspecting cracks, stresses that the manual inspection, the outline of the crack or fissure is prepared manually and conditions of irregular observed, depending entirely on the knowledge and experience of the specialist, lacking a quantitative analysis[3].

But basically crack detection can be done in two ways, using destructive or non-destructive testing, for example, through the inclusion of visual tools and topographic examination assessing the deficiencies of surface [7] as a nondestructive method, as well as automatic detection, which is very effective [8], Which uses some techniques such as infrared and thermal test[9] [10], Ultrasonic testing[eleven], Laser test[12]and radiographic test[13]. Added to this, many detection methods through image processing[14]They have been highly proposed by automation, objectivity and efficiency. However, there are many difficulties in detection for various reasons[fifteen]. Methods based on image processing are summarized and divided into four categories[3] [15] : Integrated algorithm [16], Morphological approach[17]Method based on percolation[18], Technical practice, among others.

A promising alternative methods are based on computer vision that can automatically extract information from the cracks of images. Binarization image, developed for detecting text is appropriate for identifying cracks, since the texts and cracks are similar and consist of lines and curves distinguishable. However, standardizing identifying cracks using binarization image is a challenge because the binarization depends on the method and associated parameters[19].

Thus, it has experienced a growing interest in developing tools to analyze images (flat slabs) by nondestructive testing and Technological Units of Santander have a Laboratory of Industrial Control and Automation, which lacked a specialized software capable determining the existence of cracks on flat slabs, why this document presents the development of an algorithm capable of detecting independently the crack, by digital image processing using MATLAB. The purpose of the system is to allow detailed assessment of structures to analyze, locating the areas affected by the above-mentioned phenomenon.

2. Materials and Methods

2.1. *materials*

2.1.1. Camera. The camera used to capture the images before starting the process of analysis, brand Nikon D5300 camera is Wi-Fi + 18-55 mm lens, 24 megapixel.

2.1.2. MATLAB. To develop the algorithm will use the software MATLAB, which is a computing environment that enables the execution of numerical and symbolic calculations quickly and accurately, accompanied by graphic features and advanced visualizations Aptar for scientific work in engineering[20].

2.1.3. Computer. the use of a computer capable of running the MATLAB software for algorithm development and subsequent commissioning is required.

2.1.4. flat concrete slabs. It is required for the project, study material concrete, in Figure 1 the flat plates 6 Concrete used in the project proposed herein which were made and used in a previous project are evidenced by[21]With the following characteristics:

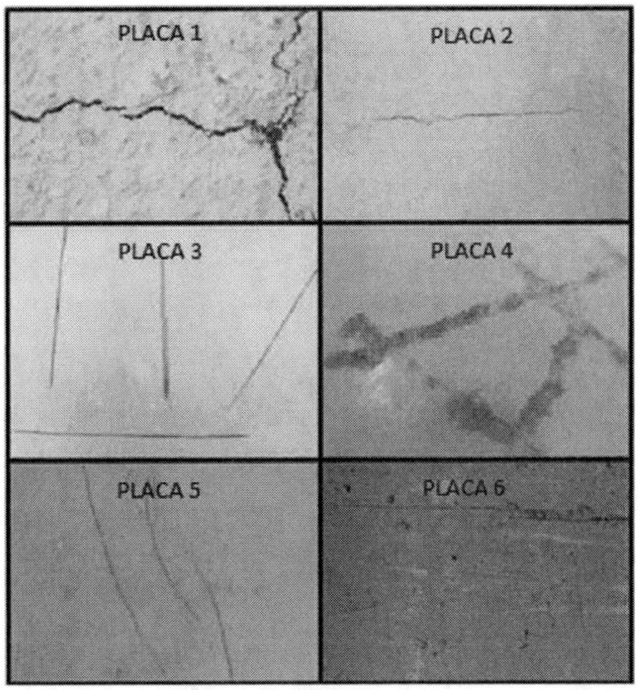

Figure 1. Flat plates Concrete

Mixture of sand, crushed, cement and water (3000 PSI Concrete), weight 65 kg.
Plate 1: Reflects minor cracks in the center. (Natural fissure)
PLATE 2: Reflects a small crack on the right side. (Natural fissure)
PLATE 3: 4 large cracks that reflects through the plate vertically and horizontally. (Created with a polisher)
Plate 4: Reflects fissure caused by chisel, with enamel finish.
PLATE 5: Reflects minor cracks. PLATE 6: Reflected multiple cracks, fissures and discontinuities mild throughout its structure. (Natural fissure)

2.2. Method

The methodology for the development of artificial vision system seeks to develop a tool capable of performing digital image processing in order to enhance images taken initially to flat concrete slabs and create a new one, in which the surface discontinuities is evident . The initial process requires a photographic record of the plates made with a NIKON D5300 described above, was considered lighting surrounding the plate in order to capture the image with adequate contrast, the process of making or photographic survey was conducted by a professional photographer. Subsequently, the algorithm can process images is developed for this, The experimental design is performed to determine how to treat the image properly with the digital image processing, in order to obtain an inspection or non-destructive testing of such structures hormigo with surface problems fissures type further determining area affectation. Figure 2 evidence the steps required for methodological development project.

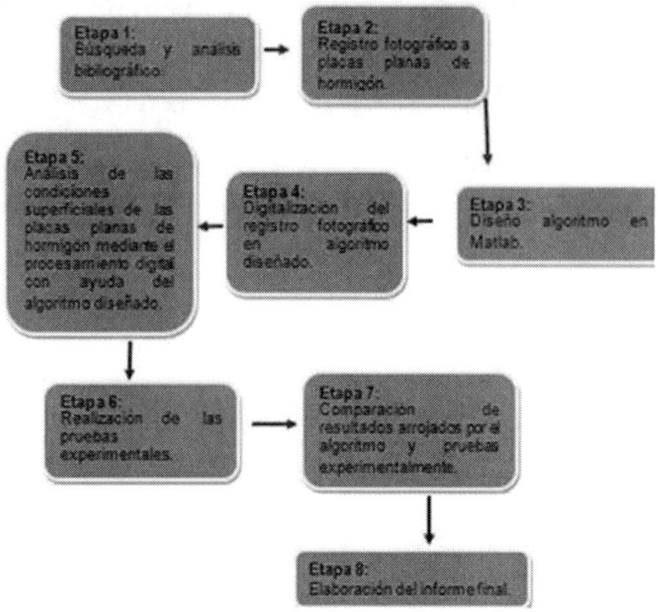

Figure 2. Flow diagram

3. Tests and results

a photographic record for digital processing and analysis of the images later, which is closely linked to the quality of the image capture rose. NIKON D5300 camera and technical services of a professional photographer was used for making initial images. In Figure 3, an unspecified images taken in the process prior to capture the same processing algorithm developed percentage is reflected.

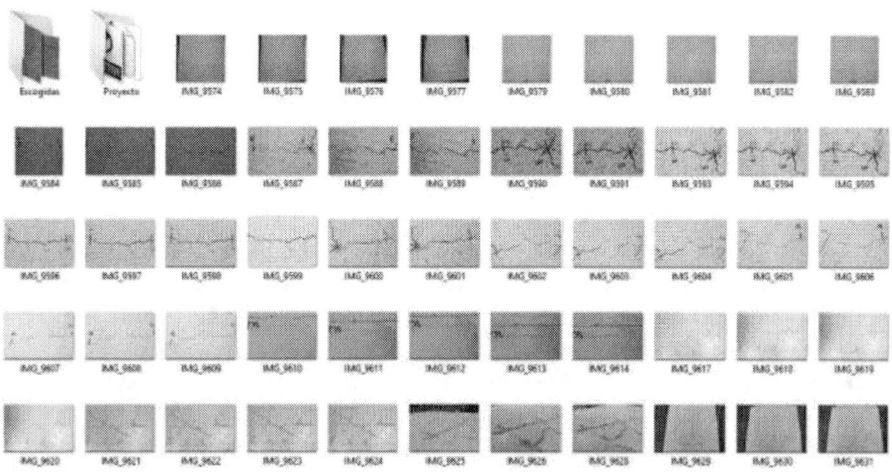

Figure 3. Photographic record

They took a total of 130 photos, taking into account three aspects to select appropriate: Sharpness, position (location and focus) and quality.

Subsequently, the algorithm is designed in MATLAB, is which performs an initial process of image recognition to pass to grayscale, apply a spatial filter, then converted to a binary image with a threshold and morphologic arrangements (Apertura and closing), preceded by process histogram, to finally have the binarized image arranged and the calculation of the impact area, figure 4 shows the methodological above mentioned process performed by the algorithm designed.

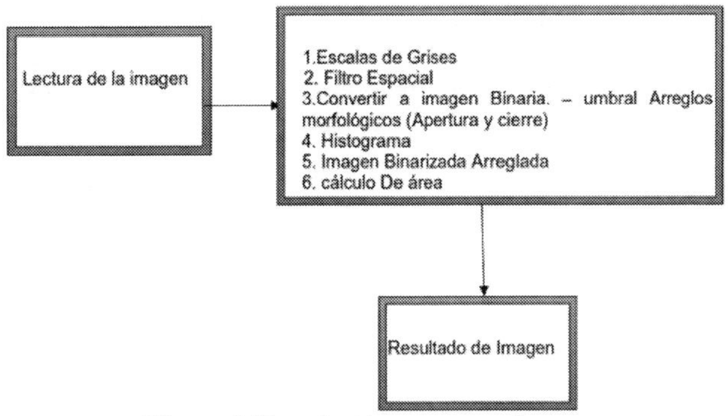

Figure 4. Flowchart image processing

Development work was important the realization of an experimental design in order to obtain information with the smallest number of experiments, based on the central objective was to measure how they influence K factors in identifying discontinuities in images by digital image processing in MATLAB.

For simplicity, an array of full factorial experiments provides the least number of tests which can be studied k factors in a full factorial design2^k[22].

Factorial design with the effect of two factors considered two levels in each study. Each replica of this design is 2x2 combinations or treatments that may denote different ways, reflected in Table 1.2^2

Table 1. How to write the treatment design

Factor A	Factor B	Reply y
-	-	$y_{11}(o)$
+	-	$y_{21}(a)$
-	+	$y_{12}(b)$
+	+	$y_{22}(ab)$

X = Alteration of structure-> "Independent".

Y = Process image for the alteración-> (filtering, morphological arrangements). "Dependent"

This design is the simplest of the series, considering two factors fl and f2, each two levels. It can be represented geometrically as a square with 4 trials.

F1 = morphologic Arrangement (opening and closing) = (square "square" Create a square structuring element whose width is w pixels, w should be a non-negative integer, disk "disk" Create a flat structuring element disk-shaped, where R specifies the radius. R must be a nonnegative integer.

F2 = spatial filter = (average "Average" Returns an averaging filter argument can be a vector that specifies the number of rows and columns in which case it is a square matrix, disk "Averaged circulate" Returns an averaging filter within circular the square matrix side) creates a predetermined two-dimensional filter

Algebraic signs to calculate design effects 2^2.

Tests were from four trials per plate for a total of 4x6 = 24 TESTS. Each test was performed with different MATLAB programming with different variations of filter, but with the same image processing on the four tests.

one. Y11 = 1 = PROTOTYPE TEST (Disk, square)
two. Y21 = PROTOTYPE TEST 2 = (Disk, disk)
3. Y12 = PROTOTYPE TEST 3 = (Average, square)
4. Y22 = PROTOTYPE TEST 4 = (Average, disk)

IOP Conf. Series: Materials Science and Engineering **844** (2020) 012058 doi:10.1088/1757-899X/844/1/012058

Table 2. Matrix factorial design experiments for 2 ^ 2

Combination treatments	Spatial filter (Average)	Spatial filter (Disk)	Spatial filter (Average)	Spatial filter (Disk)
Morphological Arrangements (Square)	1	0	0	0
Morphological Arrangements (Square)	0	1	0	0
Morphological Arrangements (Disk)	0	0	1	0
Morphological Arrangements (Disk)	0	0	0	1

4 programming prototypes were made in MATLAB, the result of experimental design to determine which facilitate the identification of the Fissure additionally calculating the area of the same in the study material. They were taken into account morphological and spatial filtering arrangements in experimental design.

Algorithm developed the prototype 1 is shown below:

```
close all
clear all
clc
%% Read  the image
[a b] = uigetfile({'*.jpg','All Files'});
I= imread([b a]);
%I = imread('fotos\Placa 1-7.jpg');
%% Convert the image to grayscale
Igray = rgb2gray(I);
%% Average spatial filter
H = fspecial('disk',5);
Ifilter1=imfilter(Igray,H);
imshow(Ifilter1)
%% Convert to binary image
imgBW=im2bw(Ifilter1,0.65); % Convert the image to a binary  image
using the threshold
se= strel('square',3);
I3 = imopen(imgBW,se);
I3 = imclose(I3,se);
I3=~I3;
I4= bwareaopen(I3,5000);
[L Ne]=bwlabel(I4);
RGB = label2rgb(L,'jet','k');
% Figures
figure(1),imshow(I),title('Original image');
figure(2),imshow(Igray),title('Grayscale image');
figure(3),imshow(Ifilter1),title('Filteres image');
```

```
figure(4),imhist(Ifilter1),title('Histogram');
figure(5),imshow(I3),title('Binarized Image');
figure(6),imshow(I4),title('Fixed Binarized Image');
figure(7),imshow(RGB),hold on;

prop = regionprops(L,'Centroid','Area');
areas=[prop.Area]';

for cnt=1:length(prop)
text(prop(cnt).Centroid(1),prop(cnt).Centroid(2),num2str(label(cnt)),'Fon
tSize',15,'color','white');
end
```

4 prototypes were made, each prototype performs tests on the 6 plates to determine which of the four allows to prove the existence of fissures in the plates, determining in turn the area of affectation of the plates. Figure 5 shows the result of one of the images initially taken from the Concrete 01 plate and the image after the revision by the artificial vision system with a numbering by zones evaluated by the algorithm in the image processing.

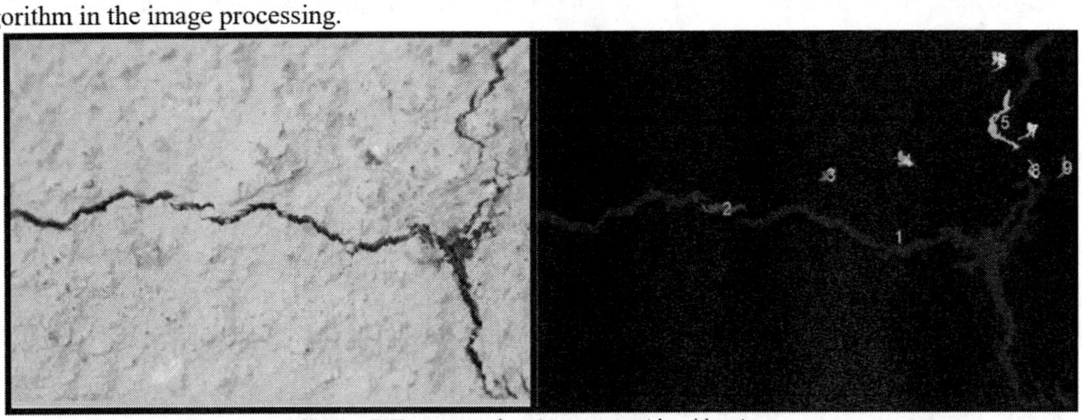

Figure 5. Treatment plate 1 prototype Algorithm 1

Table 3 reflects in column 2 the values of the pixel area corresponding to each of the zones evaluated by the algorithm, Figure 5 shows in the image processed by the algorithm the section corresponding to each number in the first column of the table, further in column three is the value of the centroid of the area pixelated by the vision system. with a numbering by zones evaluated by the algorithm in the image processing.

Table 3. Data plate processing 1 algorithm 1

Fields	Area	Centroid
1	741612	[3.4516e+03...
2	21701	[1.7783e+03...
3	7142	[2.7981e+03...
4	12954	[3.5437e+03...
5	36187	[4.4869e+03...
6	13451	[4.4601e+03...
7	8558	[4.7612e+03...
8	5377	[4.7791e+03...
9	6330	[5.0810e+03...

Figure 6 shows the result of one of the images taken from the Concrete plate 2 and the image after the revision by the artificial vision system with a numbering by zones evaluated by the algorithm in the processing of the image.

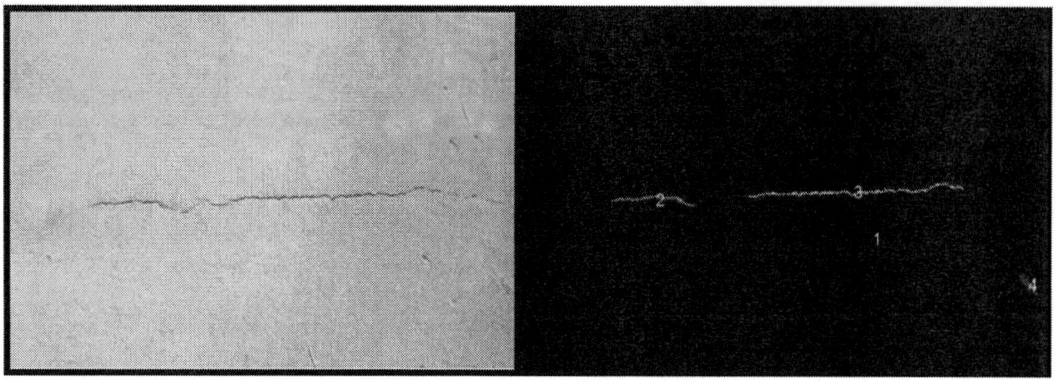

Figure 6. Treatment 2 plate prototype Algorithm 1

Table 4 reflects in column 2 the values of the pixel area corresponding to each of the zones evaluated by algorithm 1 on plate 2, Figure 6 shows in the image processed by the algorithm the section corresponding to each number in the first column of the table, additionally in column three is the centroid value of the pixelated area by the vision system.

Table 4. Data plate processing 2 algorithm 1

Fields	Área	Centroid
1	82116	[3.4959e+03...
2	15587	[1.3534e+03...
3	38118	[3.3073e+03...
4	9796	[5.0166e+03...

Figure 7 reflects the result of one of the images taken from the Concrete plate 3 and the image after the revision by the artificial vision system with a numbering by zones evaluated by the algorithm in the processing of the image.

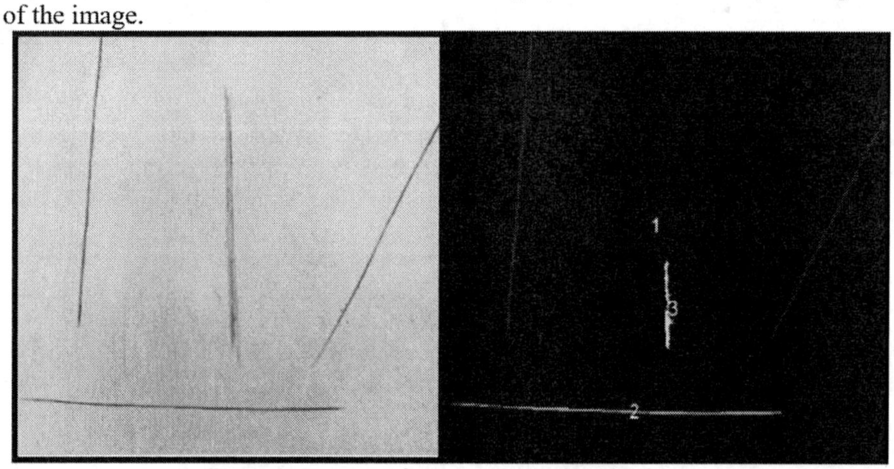

Figure 7. Treatment plate 3 prototype Algorithm 1

Table 5 reflects in column 2 the values of the pixel area corresponding to each of the zones evaluated by algorithm 1 on plate 3, Figure 7 shows in the image processed by the algorithm the section corresponding to each number in the first column of the table, additionally in column three is the centroid value of the pixelated area by the vision system.

Table 5. Data plate processing 3 algorithm 1

Fields	Área	Centroid
1	98767	[1.6544e+03…
2	47515	[1.4882e+03…
3	18035	[1.7870e+03…

Figure 8 reflects the result of one of the images taken from the Concrete plate 4 and the image after the revision by the artificial vision system with a numbering by zones evaluated by the algorithm in the processing of the image.

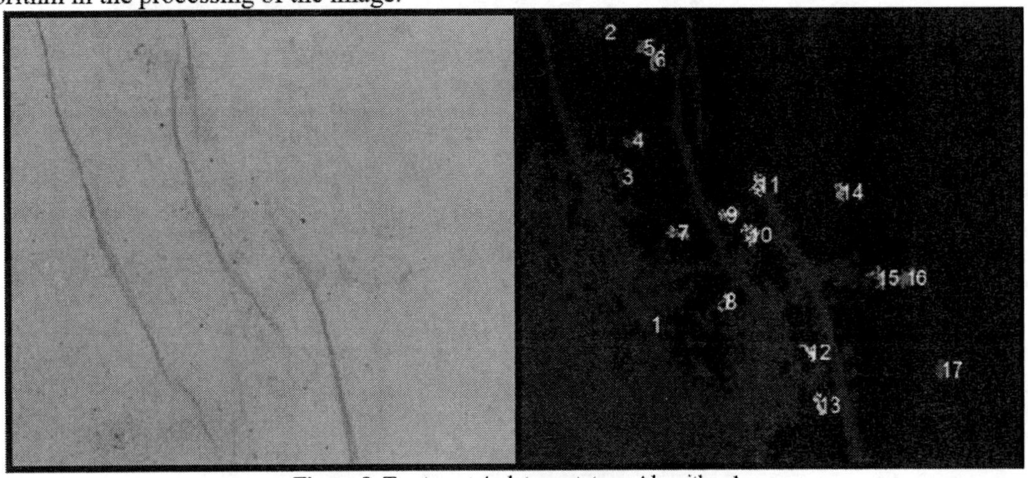

Figure 8. Treatment 4 plate prototype Algorithm 1

Table 6 reflects in column 2 the values of the pixel area corresponding to each of the zones evaluated by algorithm 1 on plate 4, Figure 8 shows in the image processed by the algorithm the section corresponding to each number in the first column of the table, additionally in column three is the centroid value of the pixelated area by the vision system.

Table 6. Data plate processing 4 algorithm 1

Fields	Área	Centroid
1	383602	[1.0626e+03…
2	5783	[721.1587 1…
3	6877	[855.7598 1…
4	6827	[938.4011 9…
5	5749	[1.0154e+03…
6	9068	[1.1087e+03…
7	6702	[1.2788e+03…
8	5910	[1.6573e+03…
9	5754	[1.6682e+03…
10	11106	[1.8514e+03…
11	8062	[1.9130e+03…
12	5342	[2.3167e+03…
13	6376	[2.3941e+03…
14	6738	[2.5701e+03…
15	8399	[2.8473e+03…
16	14721	[3.0804e+03…
17	5363	[3.3512e+03…

Figure 9 reflects the result of one of the images taken from the Concrete plate 05 and the image after the revision by the artificial vision system with a numbering by zones evaluated by the algorithm in the processing of the image.

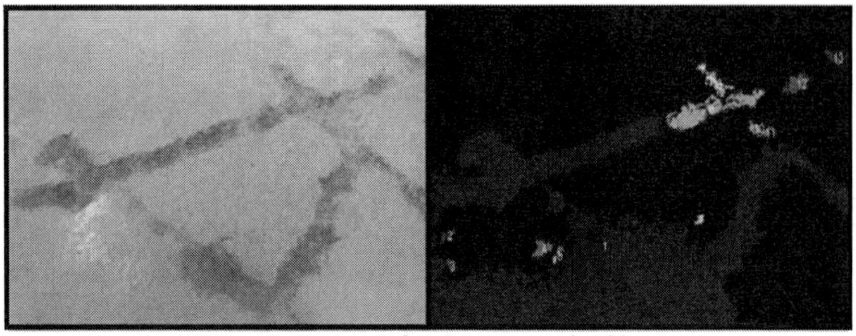

Figure 9. Treatment plate 5 prototype Algorithm 1

Table 7 reflects in column 2 the values of the pixel area corresponding to each of the zones evaluated by algorithm 1 on plate 5, in Figure 9, the image processed by the algorithm shows the section to which each number in the first column of the table corresponds, additionally in column three is the centroid value of the pixelated area by the vision system.

Table 7. Data plate processing 5 algorithm 1

Fields	Área	Centroid
1	4120220	[2.1330e+03...
2	10421	[217.5636 2...
3	6788	[257.3088 2...
4	21715	[1.3890e+03...
5	7849	[1.5792e+03...
6	150884	[3.4610e+03...
7	5979	[3.3304e+03...
8	5045	[3.3526e+03...
9	25299	[3.3525e+03...
10	9482	[4.0132e+03...
11	6174	[4.1553e+03...
12	32352	[4.5378e+03...
13	11408	[4.9983e+03...

Figure 10 reflects the result of one of the captured images of the concrete slab 06, the post review by the artificial vision system evaluated by the algorithm 01 image and no data values corresponding to specific areas.

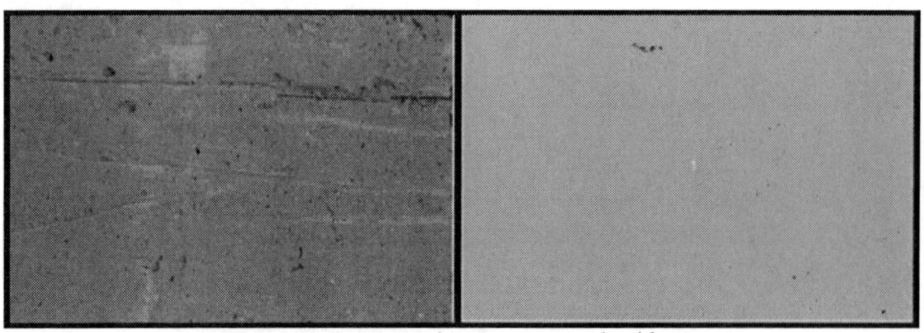

Figure 10. Treatment plate 6 prototype Algorithm 1

Next, the results of the other three algorithms performed in the process of executing the design of experiments, the algorithms are not included, only the images corresponding to those previously presented by algorithm 01, to facilitate the analysis of the results in this document. Figure 11 shows the result of one of the images initially taken from the Concrete plate 01 and the image after the revision by the artificial vision system with a numbering by zones evaluated by the algorithm in the image processing.

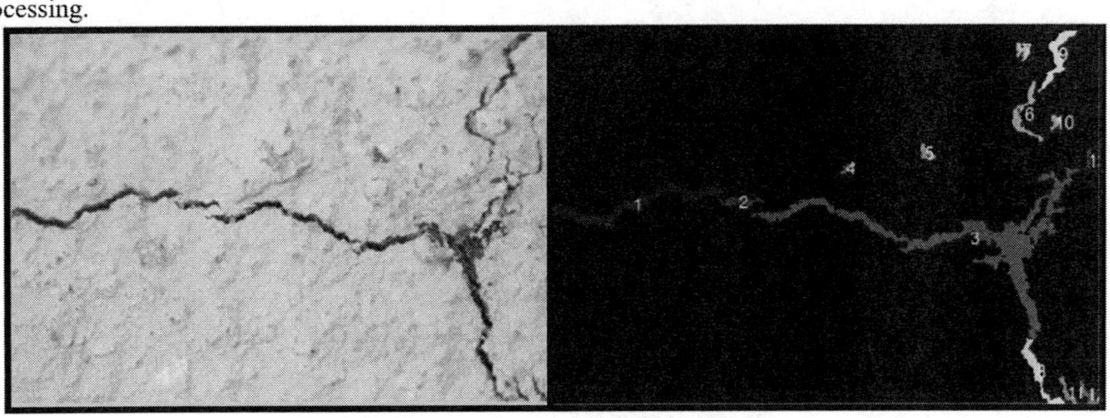

Figure 11. Treatment 1 prototype board Algoritmo2

Table 8 reflects in column 2 the values of the pixel area corresponding to each of the zones evaluated by algorithm 2 on plate 1, Figure 11 shows in the image processed by the algorithm the section corresponding to each number in the first column of the table, additionally in column three is the centroid value of the pixelated area by the vision system.

Table 8. Data plate processing 1 algorithm 2

Fields	Área	Centroid
1	129400	[782.9074 1...
2	21597	[1.7774e+03...
3	444908	[3.9764e+03...
4	7214	[2.7982e+03...
5	12809	[3.5443e+03...
6	358334	[4.4854e+03...
7	13211	[4.4613e+03...
8	44307	[4.6042e+03...
9	51704	[4.8009e+03...
10	8030	[4.7720e+03...
11	11215	[4.8830e+03...
12	12105	[5.0802e+03...
13	6195	[5.0821e+03...

Figure 12 reflects the result of one of the images initially captured from the Concrete plate 2 and the image after the revision by the artificial vision system with a numbering by zones evaluated by the algorithm in the image processing.

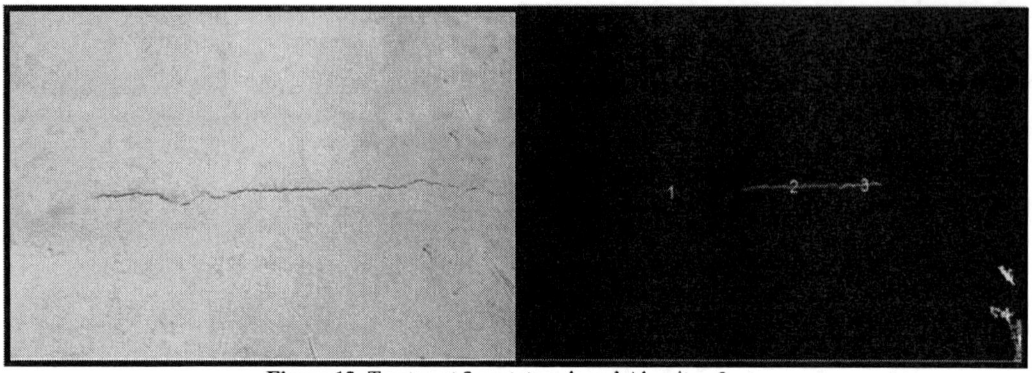

Figure 12. Treatment 2 prototype board Algoritmo2

Table 9 reflects in column 2 the values of the pixel area corresponding to each of the zones evaluated by algorithm 2 on plate 2, Figure 12 shows in the image processed by the algorithm the section corresponding to each number in the first column of the table, additionally in column three is the centroid value of the pixelated area by the vision system.

Table 9. Data plate processing 2 algorithm 2

Fields	Área	Centroid
1	7822	[1.5204e+03...
2	18128	[2.7843e+03...
3	8818	[3.5019e+03...
4	11453	[4.9627e+03...
5	10897	[5.0159e+03...
6	26020	[5.1376e+03...

Figure 13 reflects the result of one of the images initially captured from the Concrete plate 3 and the image after the revision by the artificial vision system with a numbering by zones evaluated by the algorithm in the image processing.

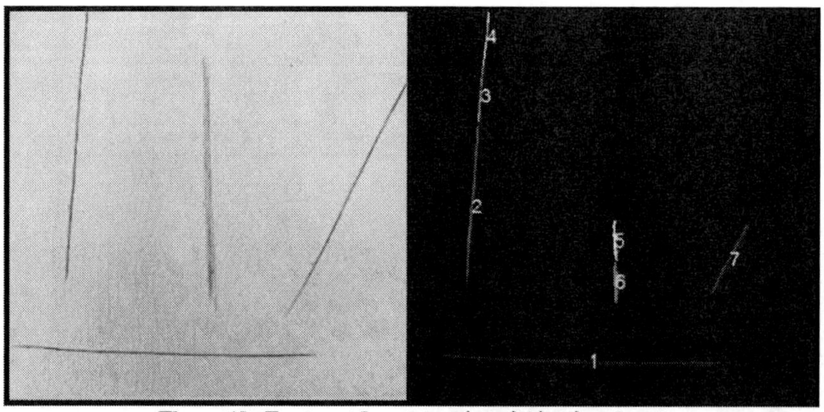

Figure 13. Treatment 3 prototype board Algoritmo2

Table 10 reflects in column 2 the values of the pixel area corresponding to each of the zones evaluated by algorithm 2 on plate 3, in Figure 13 the image processed by the algorithm shows the section corresponding to each number in the first column of the table, additionally in column three is the centroid value of the pixelated area by the vision system.

IOP Conf. Series: Materials Science and Engineering **844** (2020) 012058 doi:10.1088/1757-899X/844/1/012058

Table 10. Data plate processing 3 algorithm 2

Fields	Área	Centroid
1	44640	[1.5645e+03…
2	23627	[564.7436 1…
3	7037	[632.9156 7…
4	7793	[680.9493 2…
5	7607	[1.7826e+03…
6	10017	[1.7907e+03…
7	9977	[2.7607e+03…

Figure 14 reflects the result of one of the images initially captured from the Concrete plate 4 and the image after the revision by the artificial vision system with a numbering by zones evaluated by the algorithm in the image processing.

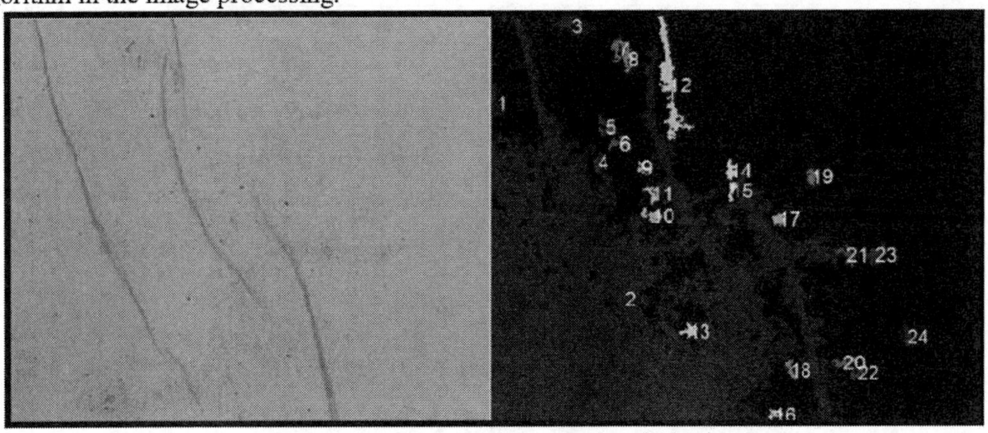

Figure 14. Treatment 4 prototype board Algoritmo2

Table 11 reflects in column 2 the values of the pixel area corresponding to each of the zones evaluated by algorithm 2 on plate 4, Figure 14 shows in the image processed by the algorithm the section corresponding to each number in the first column of the table, additionally in column three is the value of the centroid of the pixelated area by the vision system.

Figure 15 reflects the result of one of the images initially captured from the Concrete plate 5 and the image after the revision by the artificial vision system with a numbering by zones evaluated by the algorithm in the image processing.

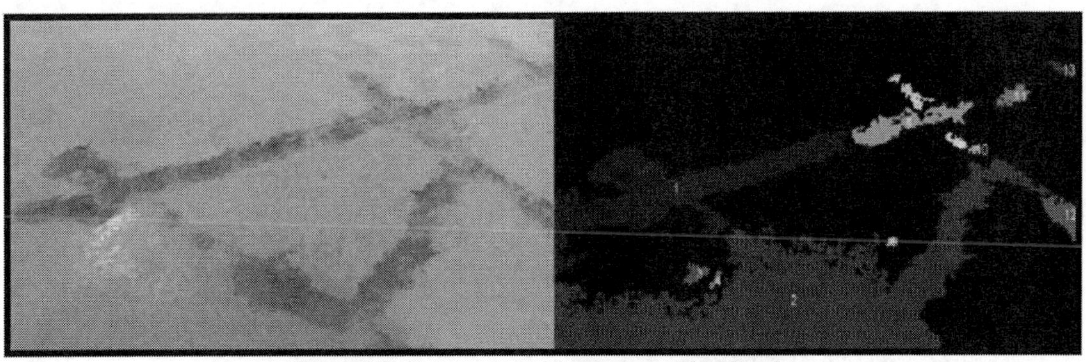

Figure 15. Plate treatment 5 prototype Algoritmo2

Table 11. Data plate processing 4 algorithm 2

Fields	Área	Centroid
1	18456	[42.1052 73...
2	4277608	[1.0742e+03...
3	24302	[652.8055 8...
4	9043	[857.6036 1...
5	1293	[922.6540 9...
6	9986	[1.0331e+03...
7	11151	[1.0242e+03...
8	8783	[1.0972e+03...
9	5134	[1.2110e+03...
10	10519	[1.2868e+03...
11	5678	[1.2815e+03...
12	72177	[1.4265e+03...
13	8276	[1.5806e+03...
14	9090	[1.9140e+03...
15	5760	[1.9199e+03...
16	6419	[2.2624e+03...
17	5936	[2.2952e+03...
18	7095	[2.3943e+03...
19	7868	[2.5711e+03...
20	8331	[2.8256e+03...
21	11176	[2.8455e+03...
22	5507	[2.9339e+03...
23	16717	[3.0797e+03...
24	7379	[3.3480e+03...

Table 12 reflects in column 2 the values of the pixel area corresponding to each of the zones evaluated by algorithm 2 on plate 5, Figure 15 shows in the image processed by the algorithm the section corresponding to each number in the first column of the table, additionally in column three is the value of the centroid of the pixelated area by the vision system.

Table 12. Data plate processing 5 algorithm 2

Fields	Área	Centroid
1	979627	[1.1568e+03...
2	3121096	[2.3252e+03...
3	24373	[1.3874e+03...
4	8102	[1.5758e+03...
5	165189	[3.4711e+03...
6	6453	[3.3298e+03...
7	5592	[3.3513e+03...
8	27349	[3.5586e+03...
9	15762	[3.9794e+03...
10	5610	[4.1573e+03...
11	35271	[4.5369e+03...
12	77571	[5.0315e+03...
13	14832	[5.0148e+03...

Figure 16 reflects the result of one of the captured images of the concrete plate 06, the post review by the artificial vision system evaluated by the algorithm 02 image and no data values corresponding to specific areas.

IOP Conf. Series: Materials Science and Engineering **844** (2020) 012058 doi:10.1088/1757-899X/844/1/012058

Figure 16. Treatment 6 prototype board Algoritmo2

Furthermore, the results of the image processing provided by Algorithm 03 begin in Figure 17, that evidences the result of one of the images initially captured from the Concrete plate 01 and the image after the revision by the system of artificial vision with a numbering by zones evaluated by the algorithm in the image processing.

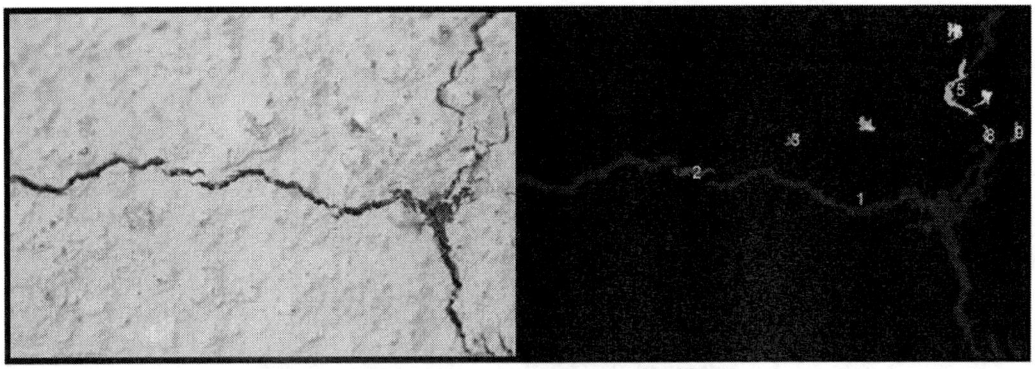

Figure 17. Treatment 1 plate prototype Algorithm 3

Table 13 reflects in column 2 the values of the pixel area corresponding to each of the zones evaluated by algorithm 3 on plate 1, Figure 17 shows in the image processed by the algorithm the section corresponding to each number in the first column of the table, additionally in column three is the value of the centroid of the pixelated area by the vision system

Table 13. Data plate processing 1 algorithm 3

Fields	Area	Centroid
1	753595	[3.4514e+03...
2	21824	[1.7777e+03...
3	7137	[2.7976e+03...
4	12959	[3.5432e+03...
5	36286	[4.4864e+03...
6	12461	[4.4598e+03...
7	8586	[4.7608e+03...
8	5421	[4.7787e+03...
9	6397	[5.0805e+03...

Figure 18 reflects the result of one of the images initially captured Concrete plate 02 and the image after the revision by the system of artificial vision with a numbering by zones evaluated by the algorithm in the image processing.

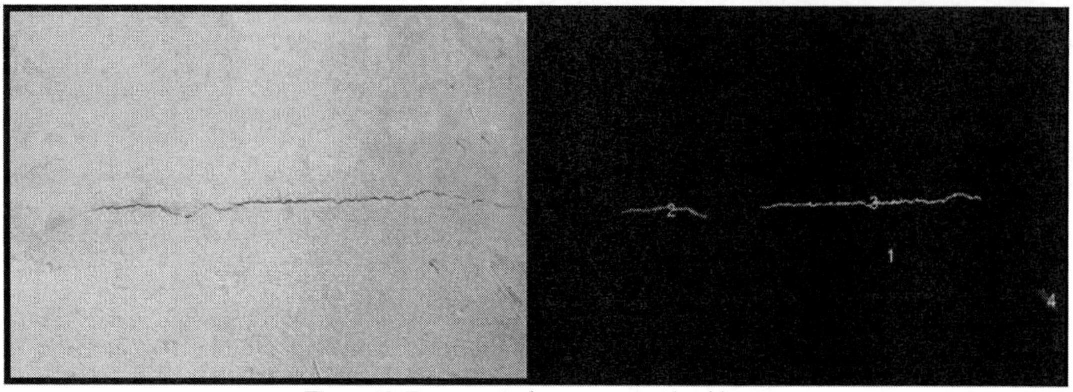

Figure 18. Treatment 2 plate prototype Algorithm 3

Table 14 reflects in column 2 the values of the pixel area corresponding to each of the zones evaluated by algorithm 3 on plate 2, Figure 18 shows in the image processed by the algorithm the section corresponding to each number in the first column of the table, additionally in column three is the value of the centroid of the pixelated area by the vision system

Table 14. Data plate processing 2 algorithm 3

Fields	Área	Centroid
1	88895	[3.4656e+03...
2	15973	[1.3523e+03...
3	39264	[3.3079e+03...
4	9795	[5.0160e+03...

Figure 19 reflects the result of one of the images initially captured Concrete plate 03 and the image after the revision by the system of artificial vision with a numbering by zones evaluated by the algorithm in the image processing.

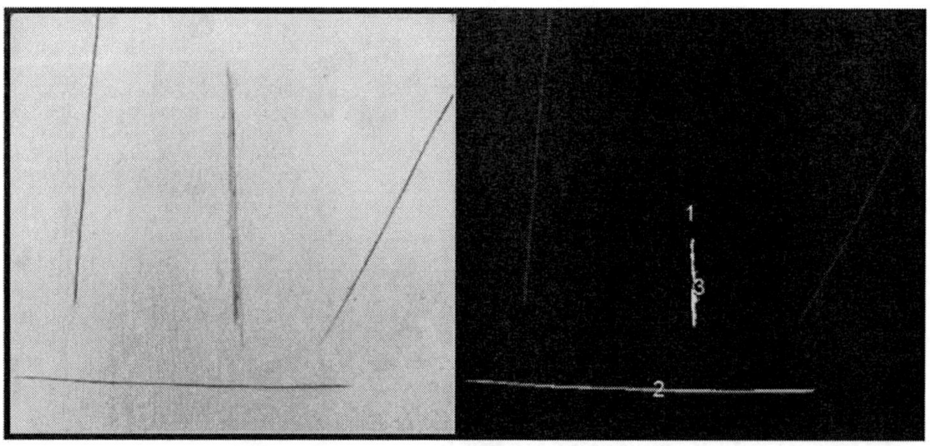

Figure 19. Treatment 3 prototype plate Algorithm 3

Table 15 reflects in column 2 the values of the pixel area corresponding to each of the zones evaluated by algorithm 3 on plate 3, Figure 19 shows in the image processed by the algorithm the section corresponding to each number in the first column of the table, additionally in column three is the value of the centroid of the pixelated area by the vision system

Table 15. Data plate processing 3 algorithm 3

Fields	Área	Centroid
1	105587	[1.7204e+03...
2	48706	[1.4867e+03...
3	18056	[1.7865e+03...

Figure 20 reflects the result of one of the images initially captured Concrete plate 04 and the image after the revision by the system of artificial vision with a numbering by zones evaluated by the algorithm in the image processing.

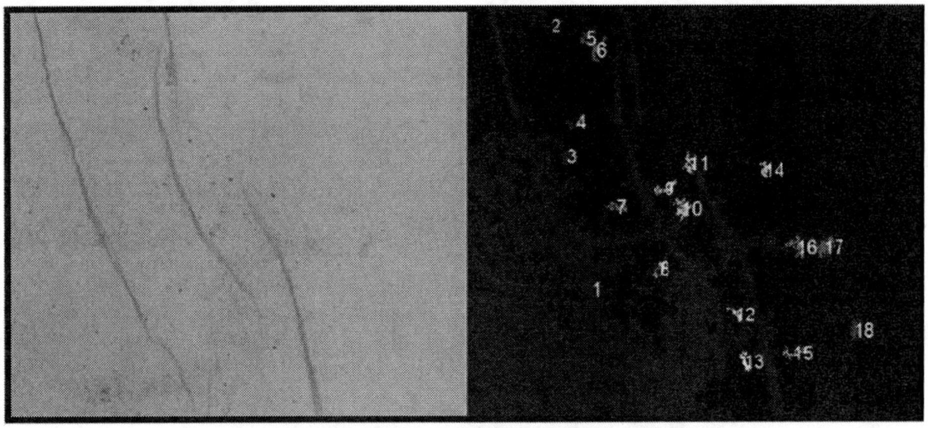

Figure 20. Treatment 4 plate prototype Algorithm 3

Table 16 reflects in column 2 the values of the pixel area corresponding to each of the zones evaluated by algorithm 3 on plate 4, Figure 20 shows in the image processed by the algorithm the section corresponding to each number in the first column of the table, additionally in column three is the value of the centroid of the pixelated area by the vision system.

Table 16. Data plate processing 4 algorithm 3

Fields	Área	Centroid
1	763595	[3.5714e+03...
2	11824	[1.7877e+03...
3	7937	[2.9976e+03...
4	12959	[3.9432e+03...
5	37786	[4.4334e+03...
6	12871	[4.4128e+03...
7	8766	[4.7008e+03...
8	5991	[4.7007e+03...
9	6897	[5.9805e+03...

Figure 21 reflects the result of one of the images initially captured Concrete plate 05 and the image after the revision by the system of artificial vision with a numbering by zones evaluated by the algorithm in the image processing.

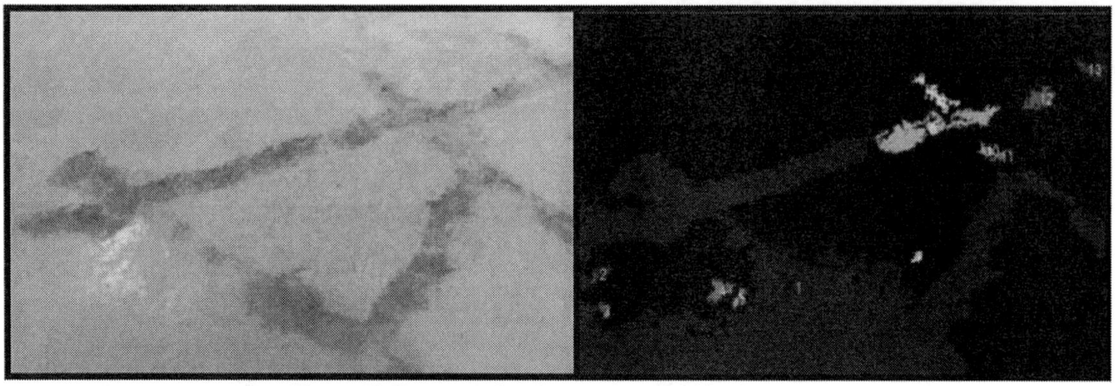

Figure 21. Treatment plate 5 prototype Algorithm 3

Table 17 reflects in column 2 the values of the pixel area corresponding to each of the zones evaluated by algorithm 3 on plate 5, Figure 21 shows in the image processed by the algorithm the section corresponding to each number in the first column of the table, additionally in column three is the value of the centroid of the pixelated area by the vision system.

Table 17. Data plate processing 5 algorithm 3

Fields	Área	Centroid
1	4130228	[2.1341e+03...
2	10418	[217.1615 2...
3	6792	[256.6637 2...
4	21886	[1.3881e+03...
5	7890	[1.5785e+03...
6	152097	[3.4617e+03...
7	6021	[3.3299e+03...
8	5094	[3.3519e+03...
9	24936	[3.5517e+03...
10	9572	[4.0125e+03...
11	6253	[4.1549e+03...
12	32716	[4.5377e+03...
13	13575	[5.0153e+03...

Figure 22 reflects the result of one of the captured images of the concrete plate 06, the post review by the artificial vision system evaluated by the algorithm 03 image and no data values corresponding to specific areas.

Figure 22. Treatment plate 6 prototype Algorithm 3

Finally, the results of image processing delivered by the algorithm 04 begins in Figure 23, which shows the result of one of the images initially captured from the concrete plate 01 and the image after the revision by the system of artificial vision with a numbering by zones evaluated by the algorithm in the image processing.

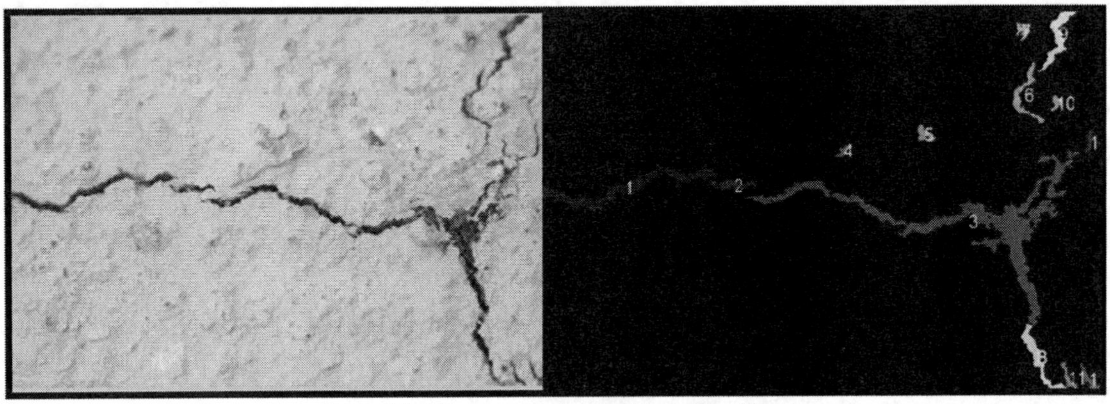

Figure 23. 1 Treatment 1 plate prototype Algorithm 4

Table 18 reflects in column 2 the values of the pixel area corresponding to each of the zones evaluated by algorithm 4 on plate 1, Figure 23 shows in the image processed by the algorithm the section corresponding to each number in the first column of the table, additionally in column three is the value of the centroid of the pixelated area by the vision system

Table 18. Data plate processing 1 algorithm 4

Fields	Área	Centroid
1	129900	[783.8090 1…
2	21737	[1.7770e+03…
3	446280	[3.9761e+03…
4	7114	[2.7976e+03…
5	12821	[3.5440e+03…
6	35846	[4.4849e+03…
7	13131	[4.4610e+03…
8	44500	[4.6040e+03…
9	51762	[4.8001e+03…
10	7991	[4.7716e+03…
11	11308	[4.8827e+03…
12	12222	[5.0800e+03…
13	6277	[5.0814e+03…

Figure 24 reflects the result of one of the images initially captured Concrete plate 02 and the image after the revision by the system of artificial vision with a numbering by zones evaluated by the algorithm in the image processing.

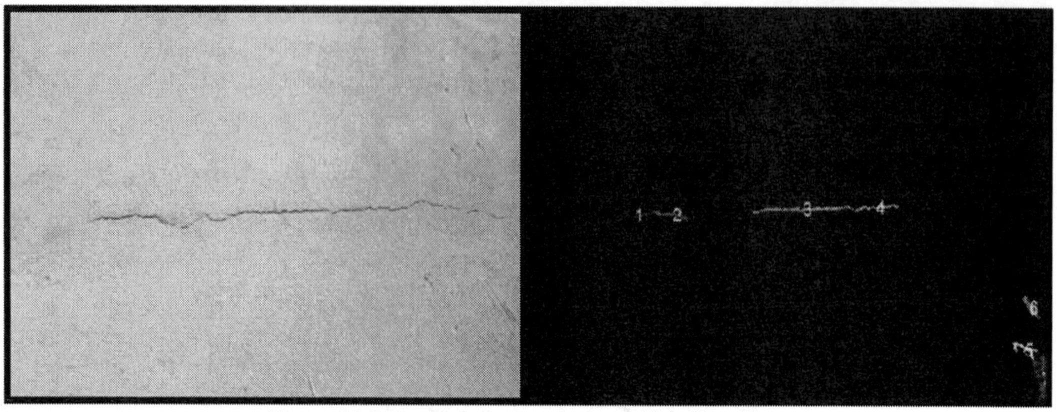

Figure 24. 2 Treatment 2 plate prototype Algorithm 4

Table 19 reflects in column 2 the values of the pixel area corresponding to each of the zones evaluated by algorithm 4 on plate 2, Figure 24 shows in the image processed by the algorithm the section corresponding to each number in the first column of the table, additionally in column three is the value of the centroid of the pixelated area by the vision system.

Table 19. Data plate processing 2 algorithm 4

Fields	Area	Centroid
1	6848	[1.1394e+03...
2	7986	[1.5189e+03...
3	18582	[2.7841e+03...
4	9008	[3.5020e+03...
5	9707	[4.9660e+03...
6	10743	[5.0147e+03...
7	25924	[5.1364e+03...

Figure 25 reflects the result of one of the images initially captured Concrete plate 03 and the image after the revision by the system of artificial vision with a numbering by zones evaluated by the algorithm in the image processing.

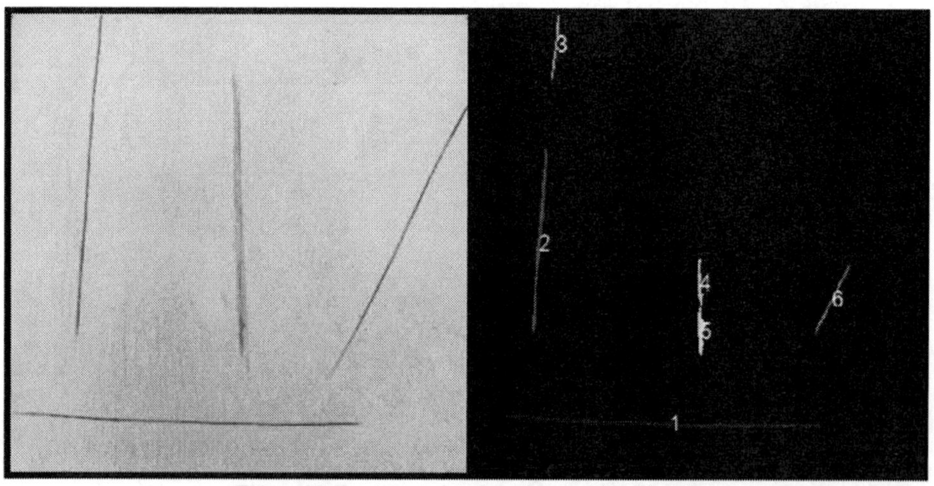

Figure 25. Treatment plate 3 prototype Algorithm 4

Table 20 reflects in column 2 the values of the pixel area corresponding to each of the zones evaluated by algorithm 4 on plate 3, Figure 25 shows in the image processed by the algorithm the section corresponding to each number in the first column of the table, additionally in column three is the value of the centroid of the pixelated area by the vision system

Table 20. Data plate processing 3 algorithm 4

Fields	Área	Centroid
1	45891	[1.5594e+03...
2	23263	[564.2397 1...
3	7686	[680.6746 2...
4	7590	[1.7819e+03...
5	10652	[1.7901e+03...
6	8887	[2.7807e+03...

Figure 26 reflects the result of one of the images initially captured Concrete plate 04 and the image after the revision by the system of artificial vision with a numbering by zones evaluated by the algorithm in the image processing.

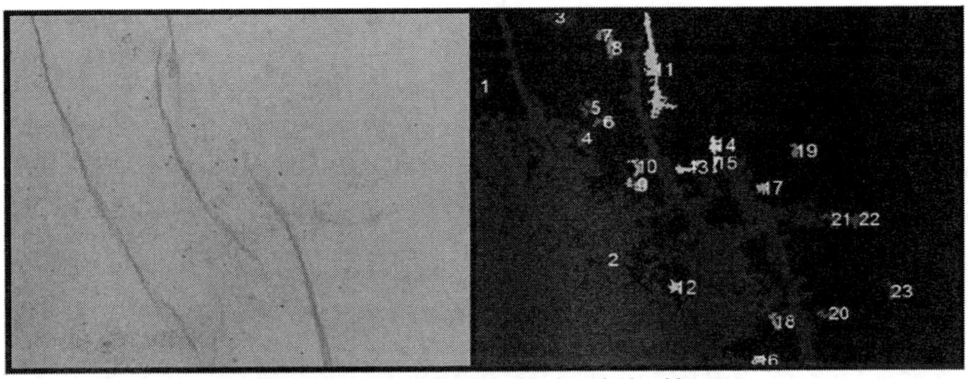

Figure 26. Treatment 4 prototype board Algorithm 4

Table 21 reflects in column 2 the values of the pixel area corresponding to each of the zones evaluated by algorithm 4 on plate 4, Figure 26 shows in the image processed by the algorithm the section corresponding to each number in the first column of the table, additionally in column three is the value of the centroid of the pixelated area by the vision system.

Figure 27 reflects the result of one of the images initially captured Concrete plate 05 and the image after the revision by the system of artificial vision with a numbering by zones evaluated by the algorithm in the image processing.

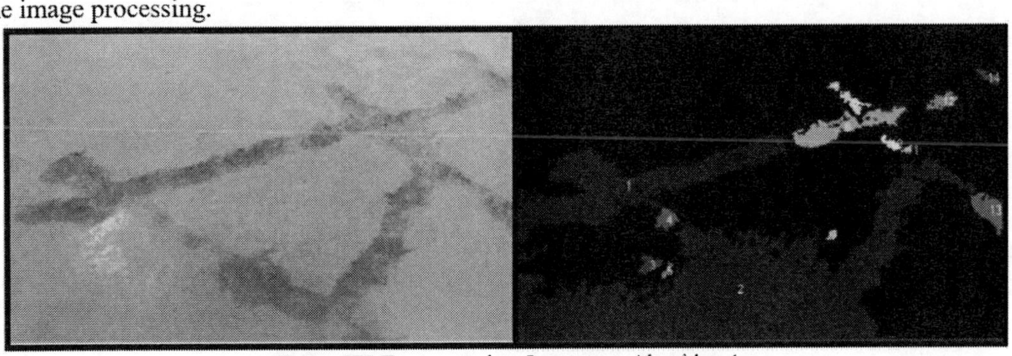

Figure 273 Treatment plate 5 prototype Algorithm 4

IOP Conf. Series: Materials Science and Engineering **844** (2020) 012058 doi:10.1088/1757-899X/844/1/012058

Table 21. Data plate processing 4 algorithm 4

Fields	Área	Centroid
1	16997	[40.3358 73...
2	4208663	[1.0683e+03...
3	23418	[650.2943 7...
4	8759	[856.9356 1...
5	10821	[929.6592 9...
6	9499	[1.0350e+03...
7	8269	[1.0276e+03...
8	8214	[1.0960e+03...
9	10150	[1.2868e+03...
10	5782	[1.2802e+03...
11	71969	[1.4260e+03...
12	6842	[1.5909e+03...
13	10382	[1.6994e+03...
14	9071	[1.9133e+03...
15	5510	[1.9195e+03...
16	6172	[2.2605e+03...
17	5801	[2.2949e+03...
18	6868	[2.3935e+03...
19	7730	[2.5705e+03...
20	7943	[2.8252e+03...
21	10750	[2.8450e+03...
22	15664	[3.0767e+03...
23	7051	[3.3490e+03...

Table 22 reflects in column 2 the values of the pixel area corresponding to each of the zones evaluated by algorithm 4 on plate 5, Figure 27 shows in the image processed by the algorithm the section corresponding to each number in the first column of the table, additionally in column three is the value of the centroid of the pixelated area by the vision system.

Table 22. Data plate processing 5 algorithm 4

Fields	Área	Centroid
1	15719	[3.9788e+03...
2	5520	[4.1582e+03...
3	34378	[4.5348e+03...
4	77714	[5.0312e+03...
5	14604	[5.0137e+03...

And Figure 28 reflects the result of one of the images captured from the concrete plate 06, the post-review by the artificial vision system evaluated by the algorithm 04 image and no data values corresponding to the specific areas.

Figure 28. Treatment plate 6 prototype Algorithm 4

Finally, Figure 29 reflects the graphical interface of algorithm management designed for the realization of the crack detection process in flat concrete structures.

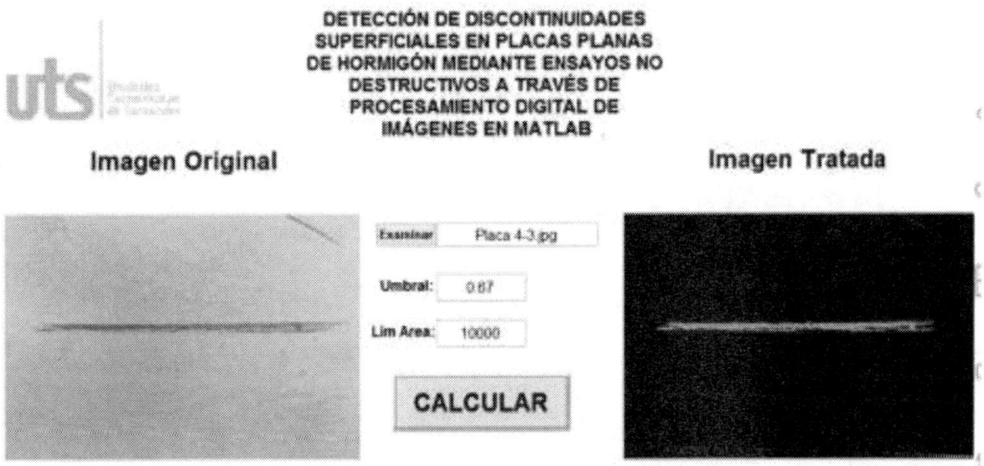

Figure 29. Interface artificial vision system

4. Analysis of the results

Image processing is reflected in the figures described in the previous paragraph (Tests and results), have a comparative as Figure 29, in which once the digital imaging procedure in MATLAB performed, the program to analyze the images with tests with the experimental design, identifying cracks and certain additional surface anomalies in the concrete slabs, noting these discontinuities, highlighting the area of the same (square pixels).

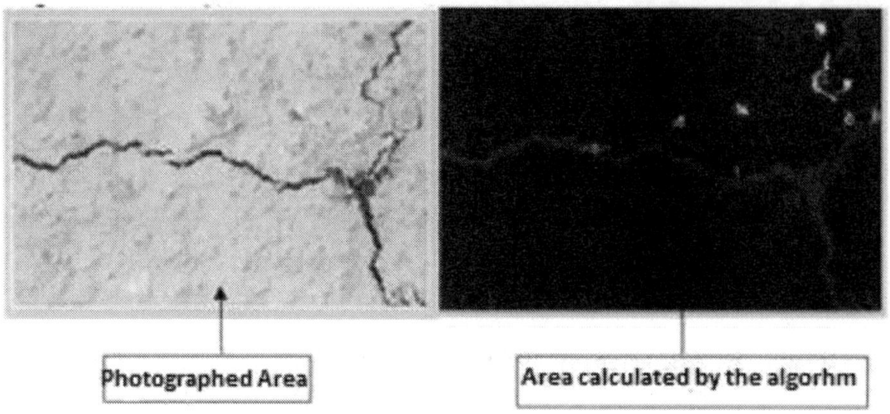

Figure 30. Data treatment

4 prototypes were made, each prototype test on 6 plates to determine which of the four allow detect is cracked plates, in turn determine the impact area thereof. Figures 5, 6,7, 8, 9 and 10 present the results of image processing of the plates 1, 2, 3, 4, 5 and 6 of the algorithm 01 designed, this analysis was replicated with 03 other algorithms. In Figures 5, 6 and 7 corresponding to the first three plates analyzed, images reflect some points marked in the fissures and a shade of blue in the discontinuities, the numbered points above allow calculation by the algorithm the approximate area of the crack. Furthermore, FIG 8 and 9 corresponding to the fourth and fifth plate analyzed, does not generate an image processing so clear, the result is a large percentage of affectation image processing which exceeds the naked eye the affected flat plate in the original photograph area finally figure 10 corresponding to the sixth plate, shows no visible result of cracks despite world. The phenomenon affecting Figures 8, 9 and 10 can occur because the plates have different areas with high concentrations of porosity which would affect the final outcome of digital image processing. shows no visible result of cracks despite world. The phenomenon affecting Figures 8, 9 and 10 can occur because the plates have different areas with high concentrations of porosity which would affect the final outcome of digital image processing. shows no visible result of cracks despite world. The phenomenon affecting Figures 8, 9 and 10 can occur because the plates have different areas with high concentrations of porosity which would affect the final outcome of digital image processing.

For analysis prototype algorithm 02, the results are reflected in Figures 11, 12, 13, 14, 15 and 16 respectively, for each in ascending order of one sixth plate. Figures 11, 12 and 13 correspond to the analysis in the plates 1, 2 and 3, reflecting an initial taking of the plate, a result of image processing by the algorithm image, and calculating the area of the crack, these results show a fissure type discontinuities visually visible in the initial photograph, but with a level of continuity reflected in the blueness of the result of digital processing inferior to those provided by the algorithm 01. Furthermore image, figure 14 and 15 corresponding to the fourth and fifth plate analyzed, does not generate an image processing so clear, the result is a large percentage of affectation image processing which exceeds the naked eye the affected flat plate in the original photograph area finally figure 16 corresponding to the sixth plate presents no visible result on cracks in spite of the world. The phenomenon affecting Figures 14, 15 and 16 can occur because the plates have different areas in high concentrations of porosity which would affect the final outcome of digital image processing.

The same phenomenon described above with the algorithm 02 applies to the 03, the figure they represent the first three tests with the first 3 plates, reflect a number of discontinuities fissure type visually visible on the original photograph, but with a level of continuity reflected in the blueness of the result of digital processing inferior to those provided by the algorithm 01. Furthermore image, figure 20 and 21 corresponding to the fourth and fifth plate analyzed, does not generate clear images processing and corresponding figure 24 the sixth plate shows no visible result of cracks. In turn, the figures refer to the analysis algorithm 04 is numbered 22, 23, 24, 25, 26 and 27,

Finally, after performing the process of analysis of the results the need to individually compare them arises, considering the algorithm 01 shows a result of scanning the image with a certain degree of accuracy greater naked eye. The plate 3, 4 and 5 did not allow for proper digitization of images due to the different types of discontinuities have cleft type, crack and high levels of porosity.

5. conclusions

The full factorial designs include optimal experiments to study which variables influence the process. In this project, the application only 4 per plate experiments it was determined the location of the main Fissures existing on flat concrete slabs used for the study.

major benefits of full factorial designs are generated when few variables studied are obtained. This is due to the number of experiments grows exponentially with the number of factors. For example, for six factors, the design comprises 64 experiments. These allow estimating 64 results, many of which will be significant. To study a large number of factors, it is more efficient to use only a fraction of a full factorial design. This is highlighted as it was essential for selecting the appropriate algorithm implementing this type of design of experiments and this foundation, helps explain why its use throughout the project.2^6

In each of the tests they were considered two parameters of great importance for identifying cracks as were the threshold and the boundary area, since such discontinuities are texture, size, shape, different color, therefore it performed with different values. This requires a GUIDE where management is facilitated these two variables was performed, generating the identification of the fissure with more continuity. That is, as it increases the threshold increases the chances would find points or objects that do not belong to the crack, so it is important to identify the closest point to the affected zone presenting the fissure, supported by the histogram of each figure.

In turn, when capturing images should take into account external factors since any abnormality present at the time of registration may impact upon scanning the image through artificial vision system, it is recommended to reduce the minimum noise, excess or lack of light, among others. Significantly, the texture and finish of the concrete (brightness) can influence when taking photos.

Finally, the area of the crack in flat slabs, comprising software (algorithm), performing four prototypes algorithm through an experimental design 24 tests on concrete slabs 6 present in the laboratory automation analyzed. Given the characteristics of the algorithm the results were satisfactory at the time of identifying areas with type discontinuities fissure present, generating a difference between the algorithms considered for the design of factorial experiments yielding as result the algorithm 01 4 analyzed, it generates a digital treatment better than other images,

References

[1] J. Toriac Corral, «Construction pathology, cracks and complications in hormigon works, origin and prevention,» *Ciencia y Sociedad,* vol. 29, nº 1, pp. 72-114, Marzo 2004.

[2] F. Muñoz Salinas y C. J. Mendoza Escobedo, «Durability in reinforced concrete structures from the perspective of the spanish standard force concrete structures» *Concreo y cemento. Investigacion y desarrollo,* vol. 4, nº 1, 2013.

[3] A. Mohan y S. Poobal, «Crack detection using image processing: A critical reniew and analysis,» *Alexandria Engineering Journal,* vol. 57, nº 2, pp. 787-798, 2017.

[4] J. Aboudi, «Stiffness reduction of cracked solids,» *Engineering Fracture Mechanics,* vol. 26, nº 5, pp. 637-650, 1987.

[5] B. Budiansky y R. J. O'connell, «Elastic moduli of a cracked solid,» *International Journal of Solids and Structures,* vol. 12, nº 2, pp. 81-97, 1976.

[6] H. Kim, E. Ahn, S. Cho, M. Shin y S.-H. Sim, «Comparative analysis of image binarization methods for crack identification in concrete structures,» *Cement and Concrete Research,* vol. 99, pp. 53-61, 2017.

[7] D. Dhital y J. Lee, «A Fully Non-Contact Ultrasonic Propagation Imaging System for Closed Surface Crack Evaluation,» *Experimental Mechanics,* vol. 52, nº 8, pp. 1111-1122, 2012.

[8] Y. Fujita y Y. Hamamoto, «A robust automatic crack detection method from noisy concrete surfaces,» *Machine Vision and Applications,* vol. 22, nº 2, pp. 245-254, Marzo 2011.

[9] Y. Qi, B. Wang, S. Li, X. Ren, J. Zhou, Y. Li y J. Mo, «Improved quantitative analysis method for evaluating fatigue cracks in thermal fatigue testing,» *Materials Letters,* vol. 242, pp. 115-118, 2019.

[10] G. Meneghetti, M. Ricotta y G. Pitarresi, «Infrared thermography-based evaluation of the elastic-plastic J-integral to correlate fatigue crack growth data of a stainless steel,» *International Journal of Fatigue,* vol. 125, pp. 149-160, 2019.

[11] N. Pal Kaur, J. K. Shah, S. Majhi y A. Mukherjee, «Healing and simultaneous ultrasonic monitoring of cracks in concrete,» *Materials Today Communications,* vol. 18, pp. 87-99, 2018.

[12] M. Pavana, D. Furfari, B. Ahmad, M. Gharghouri y M. Fitzpatrick, «Fatigue crack growth in a laser shock peened residual stress field,» *International Journal of Fatigue,* vol. 123, pp. 157-167, 2019.

[13] A. de Pannemaecker, J. Buffiere, S. Fouvry y O. Gratona, «In situ fretting fatigue crack propagation analysis using synchrotron X-ray radiography,» *International Journal of Fatigue,* vol. 97, pp. 56-69, 2017.

[14] M. R., A. S, Mohana y Y. Chitkara, *Inspection, Identification and Repair Monitoring of Cracked Concrete structure –An application of Image processing,* IEEE, 2018.

[15] P. Wang y H. Huang, *Comparison analysis on present image-based crack detection methods in concrete structures,* Yantai: IEEE, 2010.

[16] Y. Wang, J. Y. Zhang, J. Xin Liu, Y. Zhanga, Z. P. Chen, C. Guang Li, K. He y R. B. Yan, «Research on Crack Detection Algorithm of the Concrete Bridge Based on Image Processing,» *Procedia Computer Scienc,* vol. 154, pp. 610-616, 2019.

[17] E. Fesslera, E. Andrieu, V. Bonnand, V. Chiaruttini & S. Pierreta, «Relation between crack growth behaviour and crack front morphology under hold-time conditions in DA Inconel 718,» *International Journal of Fatigue,* vol. 96, pp. 17-27, 2017.

[18] P. Wang y H. Huan, *Comparison analysis on present image-based crack detection methods in concrete structures,* IEEE, 2010.

[19] h. Kim, S. Cho, M. Shin y e. Ahn, «Comparative analysis of image binarization methods for crack identification in concrete structures,» *Cement and Concrete Research* , 2017.

[20] C. Perez, MATLAB and its applications in science and engineering, pearson, Ed., PRENTICE HALL, 2002.

[21] B. Tarazona, J. Marin y M. Diego, *Desing and construction of a vibration analyzer for the detection of cracks and fissures in concrete structures,* Bucaramanga, 2015.

[22] P. D. Medina Varela y A. M. Lopez Reyes, «Critical analysis of the 2k factorial desing on applied cases» *Scientia Et Technica,* vol. XVIII, nº 47, pp. 101-106, Abril 2011.

[23] R. Hernandez Sampieri, C. Collado y P. Baptista, Investigation methodology, Sexta ed., McGRAW-HILL, 2014.

IOP Conf. Series: Materials Science and Engineering **844** (2020) 012059 doi:10.1088/1757-899X/844/1/012059

Elasticity modulus variation of the AISI SAE 1045 steel subjected to corrosion process by chloride using tension test destructive

C G Cárdenas-Arias[1*], A D Rincón-Quintero[1], A Santos-Jaimes[2], C L Sandoval-Rodriguez[1] D F Rojas-Gomez[1], S J Ardila-Galvis[1]

[1] Design and materials research group (DIMAT), Automation and Control Energy Systems Research Group (GISEAC), Electromechanical engineering, Faculty of Natural Sciences and Engineering, Technological Units of Santander (UTS), Calle de los estudiantes No 9-82, 680005, Bucaramanga, Colombia.
[2] Materials Research Group (GIM). Mechanical Engineering, Engineer School, Pontifical Bolivarian University (UPB), Kilometer 7 Highway Floridablanca – Piedecuesta, (057) (7) 6799220, Floridablanca, Colombia.

*ccardenas@correo.uts.edu.co

Abstract. The study of materials corrosion effect is a great importance topic for the manufacturing industry and also for the customer that requires this. That is why a research study was conducted that aimed to determine the elasticity modulus variation of AISI SAE 1045 steel subjected to corrosion. To achieve this, thirty-five specimens were used, which were exposed to chlorides by immersion processes and salt spray chamber in three periods of time and then subjected to stress test destructive, following the guidelines established under ASTM G-1 standards, ASTM G-31 and ASTM E-8. Variation was found in the elastic behavior of steel and curves that differ from the theoretical ones raised in the documents.

1. Introduction

Due to steel need and importance in the industry, a study has been carried out on AISI SAE 1045 steel in order to observe the elasticity modulus variation over time when subjected to corrosion through a series of tests, such as: immersion test and salt spray chamber. In order to analyze the behavior developed by the material corrosion and its condition, AISI SAE 1045 steel it has been decided to use because it is a material commonly used in the manufacture of parts for the automotive, agricultural, among others, where hardness and toughness is required in the final product. [1]

The AISI SAE 1045 steel study was experimental, and consisted of observing and analyzing the elasticity modulus variation subjecting 30 specimens of this material to corrosion, 15 of them intended for the immersion test method application using Sodium hypochlorite with demineralized water and with 15-day intervals, 5 specimens were removed. Likewise, the corrosion process of the specimens was carried out using the salt spray chamber, using sodium chloride with demineralized water and with the same 15-day interval between specimens, 5 specimens were extracted. To obtain the quantitative data of the elasticity modulus variation of the material, a destructive tensile test was carried out, which consisted of subjecting the specimens in axial direction, to an increasing effort until their breakage in order to determine one or various mechanical characteristics. [2]

In order to find out the corrosion incidence on the material properties, whether it was affected negatively or positively, depending on its nature and the surrounding conditions and passivation presence [3],[4], the study is performed to the elasticity modulus.

2. Work development

2.1. Test Tube Preparation

Compilation using bibliographic information in literatures and applying the standards ASTM G-1 [5], ASTM G-31 [6] and ASTM E-8 [7], the study of elasticity modulus variation proceeded of AISI SAE 1045 steel exposed to chloride ions to accelerate the corrosion process. The shape of the specimens was selected according to the ASTM E-8 standard for destructive stress testing, and for the dimensions the technical specifications were taken into account as maximum applied force capacity of the universal testing machine, the modeled in a CAD software Fig. 1 and manufactured by water jet cutting and grinding in CNC milling machine Fig. 2.

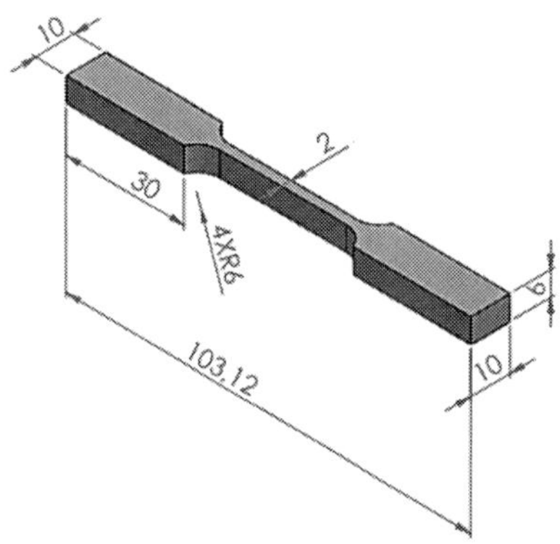

Figure 1. Tension test tube design.

Figure 2. Machined test tube.

2.2. Trial development

In order to standardize tests, chemical solutions used and exposure times in the test execution, the exposure time diagrams and respective identification of the specimens were made using letters from A to letter F, Table 1.

Table 1. Rehearsal and exposure time

TEST	TEST TUBE N°	AMOUNT TEST TUBE	EXHIBITION TIME (DAYS)
SALT IMMERSION (SODIUM HYPOCHLORITE + DEMINERALIZED WATER)	A	5	15
	B	5	30
	C	5	45
SALT FOG CHAMBER (SODIUM CHLORIDE + DEMINERALIZED WATER)	D	5	15
	E	5	30
	F	5	45

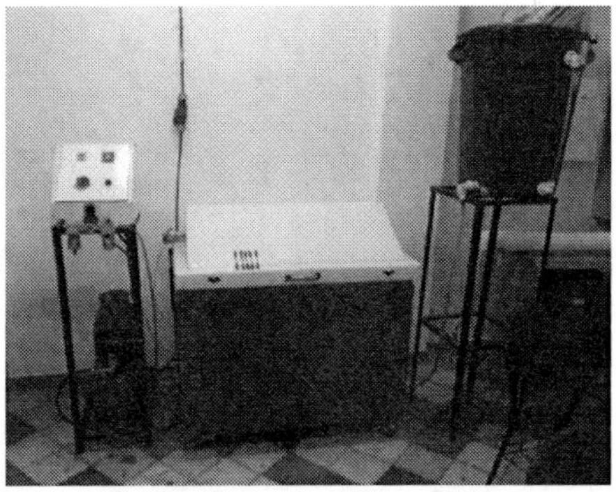

Figure 3. Salt spray chamber

In **Figure 3** is shown, the salt spray chamber where the specimens corrosion was carried out at the established times and under the standardized process of its use.

The term corrosion is used to describe the process of metallic materials deterioration (including both pure metals and their alloys), through chemical and electrochemical reactions[8]. Due the AISI SAE 1045 steel importance, it´s selectioned to observe its behavior when exposed in a corrosive environment.

Figure 4. Test tube positioned in salt spray chamber

When exposing the specimens in the salt spray chamber, **Figure 3** and **Figure 4** for the time established between them, the physical changes variation in the material was observed, one of them was the characteristic brown colour that increased its intensity with the passage of days and likewise the oxide was appreciated in large proportion when cleaning the specimens.

In the immersion test execution **Figure 5**, 15 specimens were subjected to sodium hypochlorite exposure with demineralized water and subsequently they were immersed under the established times, according to Table1.

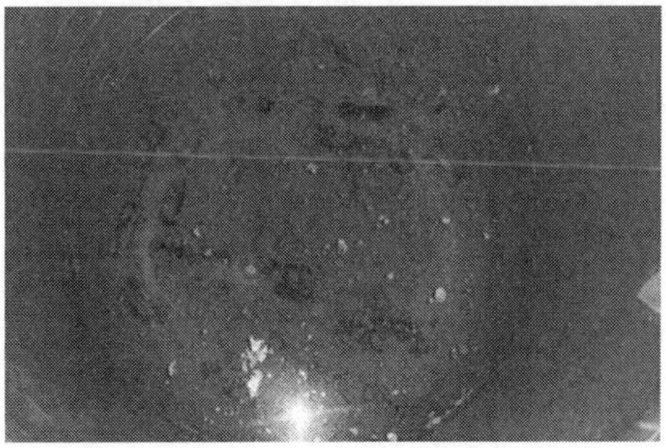

Figure 5. Test tube subjected to chloride ions immersion

IOP Conf. Series: Materials Science and Engineering **844** (2020) 012059 doi:10.1088/1757-899X/844/1/012059

Figure 6. Test tubes after 15 days to immersion corrosion exposure

Figure 7. Corroded specimens after 15 days

When extracting the steel specimens, the material surface variations are observed **Figure 6**, product of material irreversible interfacial reaction with its environment, which causes a consumption or dissolution of the material in the solution.

In Fig. 5, is seen the corrosion attack as small pitting holes, which are irregularly shaped and connected to each other. In the carbon steels case, the resulting corrosion compounds are deposited in these holes and, due to their porous nature, allow the attack to continue. [9]

3. Results

From the results obtained by submitting the specimens of the AISI SAE 1045 material in the universal testing machine, the representation of the stress curve vs. unit deformation Fig. 8 and Fig. 9 was obtained for immersion and fog chamber tests saline respectively.

The graphical representation shows how the average last effort of the specimens is affected with the passage of time and the elasticity modulus is also reflected, being explicit in the **Table 2**

Table 2. Elasticity modulus of the different specimens

TEST TUBE	TEST TYPE	ELASTICITY MODULUS E (MPA)
A	Immersion	196,469
B	Immersion	190,547
C	Immersion	186,459
D	Fog chamber	197,460
E	Fog chamber	192,721
F	Fog chamber	181,547

With immersion corrosion, an elastic modulus was obtained for test tube A of 196.47 GPa, as a result of exposing 5 test pieces to corrosion for 15 days and obtaining the representative value. 190.55 GPa by exposing the B specimens to corrosion, with a duration of exposure to corrosion for 30 days and finally the C specimens, obtained a representative sample of 186.46 GPa which allows to justify the effect of the elastic modulus of the material with the over time.

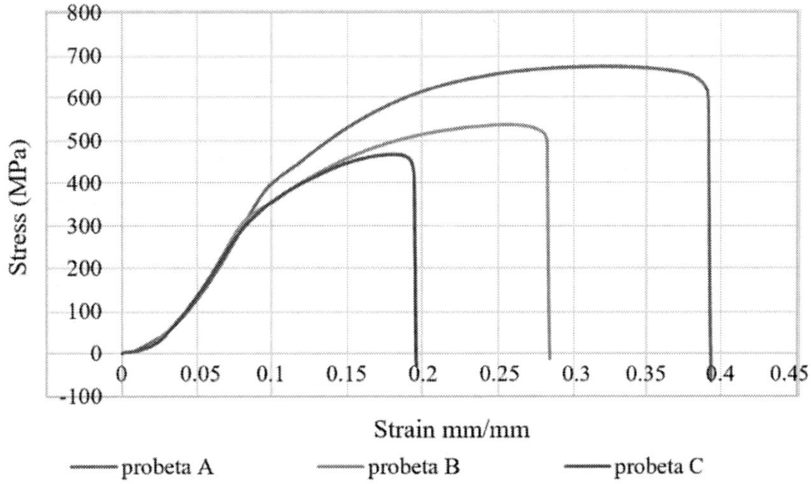

Figure 8. Stress – strain diagram test tubes A, B, C. Immersion test

With the data in the tests with immersion corrosion it was observed how the AISI SAE 1045 steel lost an average of 2% of the modulus of elasticity in the first series of specimens (A), 5% of the elastic modulus in the second series of specimens (B) and 7% of the elastic modulus in the third series of specimens (C). The loss of material and modulus of elasticity being remarkable. On average it lost 2.5% elastic property every 15 days under the immersion test.

With salt spray corrosion, it was established that corrosion directly affected the material, reflected in the decrease in the average elastic modulus of each cycle of the specimens, the average elastic modulus of the 5 specimens (D) being 197.46 GPa for the period of 15 days, test pieces (E) 192.72 GPa for the period of 30 days and test pieces (F) 181.54 GPa for the period of 45 days.

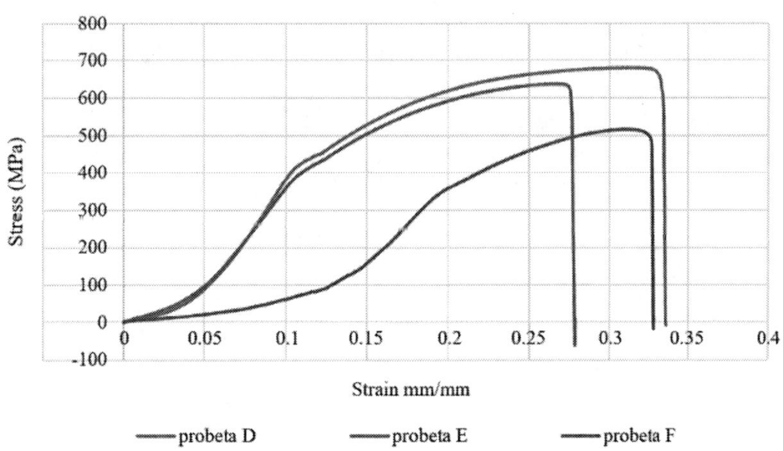

Figure 9. Stress – strain diagram test tubes D, E, F. Fog chamber test.

With the data obtained in the test with salt spray corrosion, it was observed how AISI SAE 1045 steel lost an average of 1.25% of the modulus of elasticity in the first series of specimens (D), 4% of the elastic modulus in the second series of test pieces (E) and 9% of the elastic module in the third series of test pieces (F).

In Figure 9, is see how the decrease obtained by the ultimate effort of the material with the passage of time and how the modulus of elasticity decreases considerably when exposed in an aggressive environment of corrosion. Based on information consulted on the 200 GPa (29000 KSI) elasticity modulus [10], conclusions could be made regarding the considerable decrease in the AISI SAE 1045 material module.

It's further observed that the curve shows some variation with the theoretical curve characteristic of ductile materials, showing considerable deformation with small efforts before presenting the slope corresponding to the elastic coefficient. The effect of stress reduction in the creep zone is not shown.

4. Conclusions

It is evidenced that corrosion actually affects the material in an ascending manner with the passage of time, allowing a notable decrease in the modulus of elasticity of the material to be observed when the AISI SAE 1045 steel is subjected to corrosion under the saline effect.

Comparing the corrosion procedures, the immersion is more aggressive, degrading the material at a higher initial speed, but in the long term, the effect on the fog chamber is more critical, which is observed in the results of the elastic constant.

The phenomenon of perfect plasticity, characteristic of ductile materials, does not occur.

References

[1] Saldaña A, Plascencia H, Aguilera E, Saldaña A, Márquez A and Diosdado J 2018 Influence of ball-burnishing on roughness, hardness and corrosion resistance of AISI 1045 steel *Surface & Coatings Technology* **339** 192

[2] Askeland D R, Fulay P P and Wright W J 2012 *Ciencia e Ingeniería de Materiales* (México D.F.: Cengage Learning Editoresx S. A.)

[3] Schweitzer P A 2009 *Fundamentals of Corrosion – Mechanisms, Causes and Preventive Methods* (New York: CRC Press Taylor & Francis Group)

[4] Revie R W 2011 *Uhlig's Corrosion Handbook* (New Jersey - Wiley & Sons Inc., Hoboken)

[5] ASTM International: Standard Practice for Preparing, Cleaning, and Evaluation Corrosion Test Specimens (G1-03). ASTM Int., West Conshonhocken, PA, USA, (2017)

[6] ASTM International: Standard Practice for Laboratory Immersion Corrosion Testing of Metals (G31-72). ASTM Int., West Conshonhocken, PA, USA, (2004)

[7] ASTM International: Standart Test Methods for Tension Testing of Metallic Materials (E8/E8M). ASTM Int., West Conshonhocken, PA, USA, (2016)

[8] Revie R W and Uhlig H H 2008 *Corrosion and Corrosion Control – An Introduction to Corrosion Science and Engineering* (New Jersey: Wiley–Interscience)

[9] Casallas S 2005 *Ingenio Libre* **32** 29

[10] Padzi M M, Abdullah S and Nuawi M Z 2014 On the need to decompose fatigue strain signals associated to fatigue life assessment of the AISI 1045 carbon steel *Materials and Design* **57** 408

IOP Conf. Series: Materials Science and Engineering **844** (2020) 012060 doi:10.1088/1757-899X/844/1/012060

Ultrasonic signal acquisition system for TOFD imaging

J Casagua[1], J Rodríguez[1], D Segura[1]

[1]Grupo de investigación MEM (Modelado Electrónica Monitoreo), Facultad de ingeniería Electrónica, Universidad Santo Tomás, Cra 9 No. 51-11, Bogotá, Colombia.

E-mail: johancasagua@usantotomas.edu.co

Abstract. This article describes the development of an application to capture and process ultrasonic signals that will lead to the formation of flight time diffraction images, TOFD. In this way the theoretical basis of operation of this technique is presented establishing concepts such as type A and type D scanning. In addition, the ultrasonic waveforms that are used in this technique are presented. Application development is carried out through layered programming, which consists of three modules: information capture, information storage and finally data processing and graphing. However, the ultrasonic signal acquisition system designed and implemented in the laboratories of the Santo Tomás University of Bogotá D.C is presented as an alternate ultrasound system. For the verification of the system an exploration is carried out on a steel specimen that intentionally contains a hole inside it and the system determines its existence. In addition, a scan is performed on a steel sample with a weld bead and defects within the bead are verified.

1. Introduction

Exploration techniques through non-destructive tests on materials such as steel play an important role in the industry, since in metallurgy, for example, it is necessary to establish techniques to locate discontinuities in welded joints; It is there where the ultrasonic systems come into operation. From the acquisition of the ultrasonic signals, images can be formed that determine the structural health of the materials, in addition to locating defects and defining their type. These systems provide assistance to operators, since it is not necessary to destroy a material or structure to examine its composition. That is why within the research group of Modeling, Electronics and Monitoring (MEM), of the School of Electronic Engineering of the University Santo Tomas Bogotá D.C the research has been carried out for some years since the processing of ultrasonic signals, to locate discontinuities in welded joints.

2. TOFD technique

Flight time diffraction is an ultrasonic technique that allows you to examine a metallic material non-invasively. Using this technique, you can perform different types of solid material exploration, of which Type A and Type B scans are highlighted. [1][2][4]

2.1. Type A Scan

Type A scan is an ultrasonic signal that contains information about the variation in amplitude with respect to time; that is waveforms that are generated when an ultrasound wave interacts with an eventual discontinuity located within a metallic part. [3]

IOP Conf. Series: Materials Science and Engineering **844** (2020) 012060 doi:10.1088/1757-899X/844/1/012060

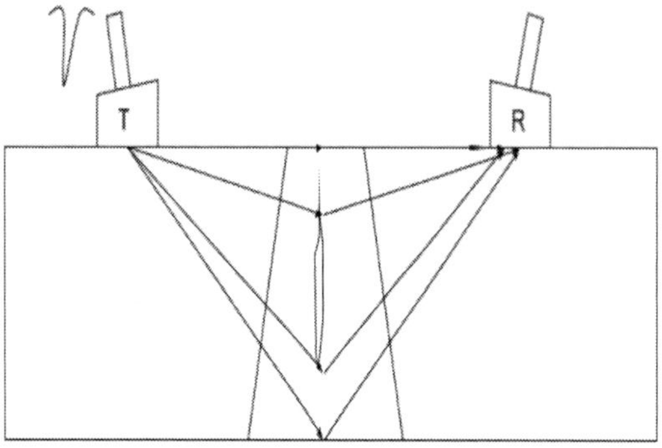

Figure 1. TOFD technique for detecting defects inside a steel specimen

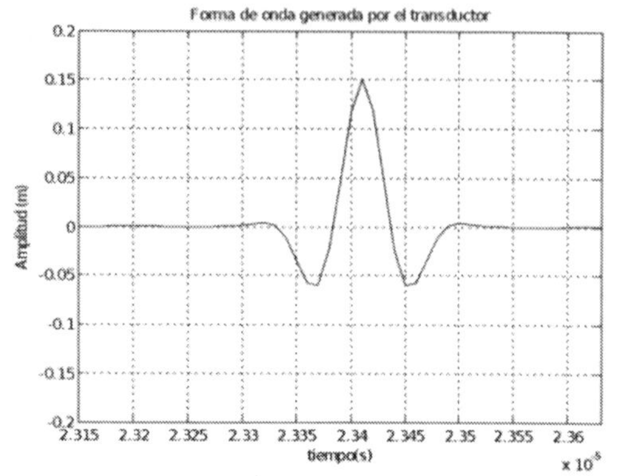

Figure 2. Waveform generated by the ultrasonic transducer. [1]

Examples of ultrasonic scanning signals type A are illustrated below considering a metallic part with and without discontinuity.

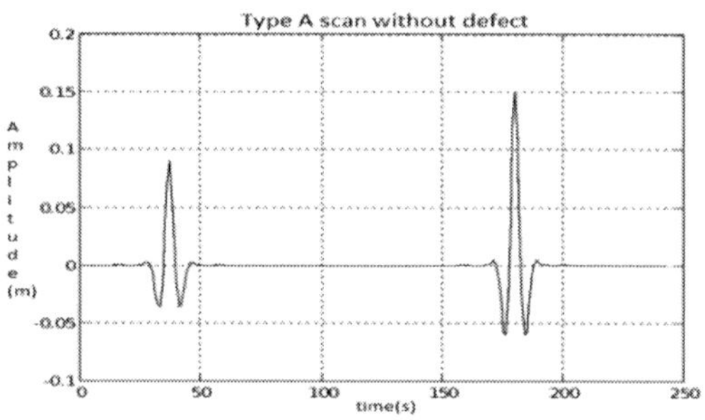

Figure 3. Type A ultrasonic scan signal without discontinuity in the metal part. [1]

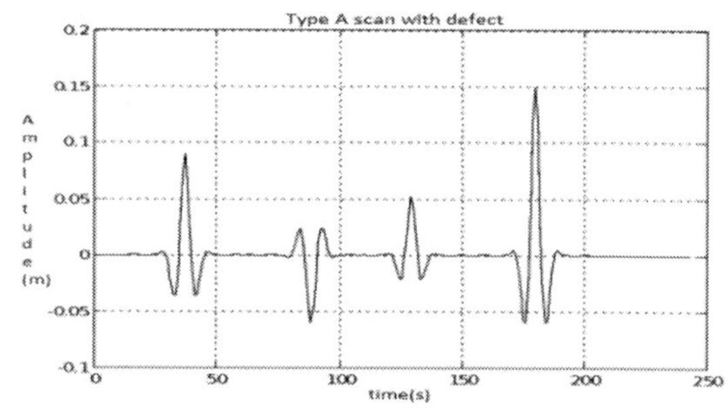

Figure 4. Type A ultrasonic scan signal with discontinuity in the metal. [1]

2.2. Type D Scan

An exploration D is a two-dimensional image constructed on the basis of a set of sweeps type A which are produced from the synchronous movement of a pair of transducers emitter-receiver along a certain direction of movement. Thus, for example, if said displacement is made with a spatial interval of one millimeter the generated sweeps A are stored each time the pair of transducers travels that distance. Subsequently, these sweeps are grouped to obtain a two- dimensional image scaled in shades of gray, in which each tone encodes amplitude values of type A signals. [3]

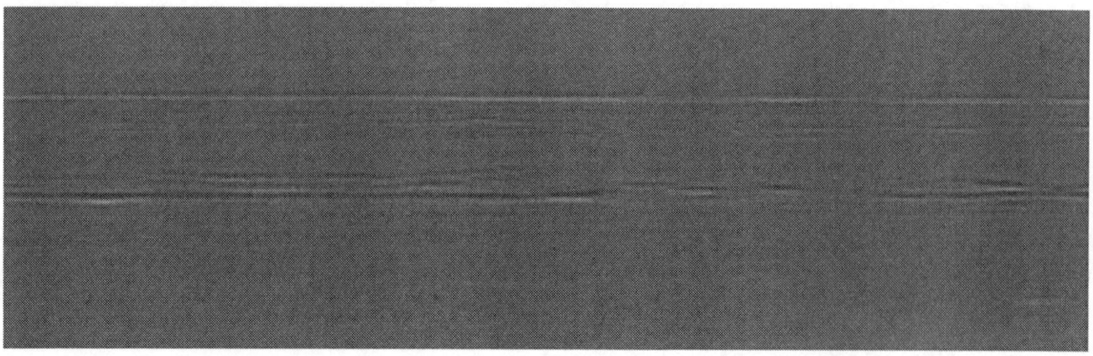

Figure 5. Type D scan captured by the implemented system.

2.3. *TOFD data interpretation*

The signal that results in the receiving transducer is:

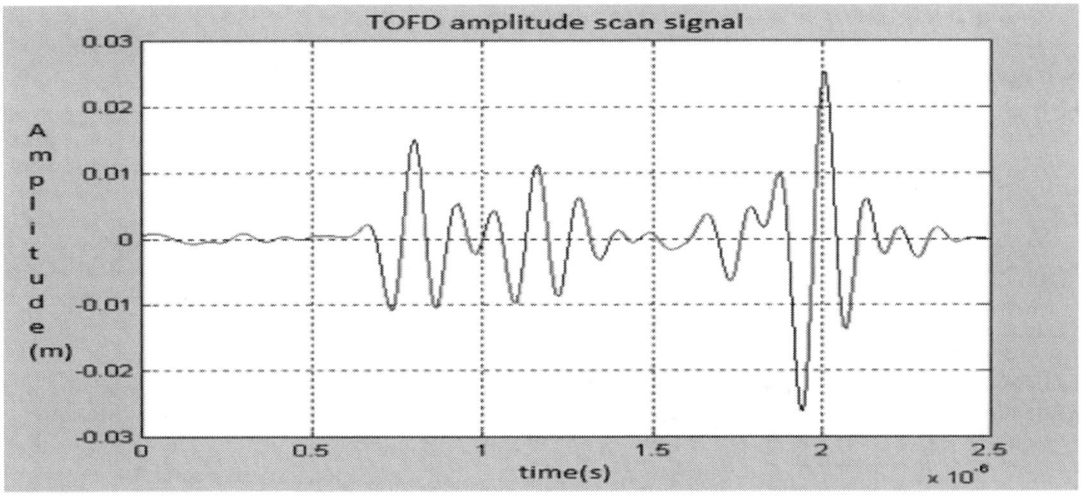

Figure 6. Signal obtained in the receiver transducer using the TOFD technique.

Considering the origin of the time t=0 at the moment in which the excitation of the transducer transmitter *T* occurs, and t_l in which the lateral wave arrives at the receiver of the transducer *R*. Furthermore, the delay time incorporated by the shoes must be considered; that is, $2t_o$, so that:

$$2t_0 + t_1 = 12.2\mu s \tag{1}$$

However, t_l is the period of time taken by the lateral wave to travel the distance from the edge of the transmitter transducer *T* to the edge of the transducer *R*, which is approximately:

$$t_1 = \frac{2S}{c} = \frac{45.42mm}{5920ms^{-1}} = 7.67\mu s \tag{2}$$

Therefore, the delay time entered by each shoe will be t_o=2.26μs. The arrival time interval used by the first diffracted wave must be:

$$2to + t2 = 12.2\mu s + (t2 - t1) \tag{3}$$

This last interval has a value of 0.3μs. This last interval has a value of 0.3μs; therefore t2, the duration of the first diffracted wave will be 7.98μs. From these figures the depth of the defect can be determined; this is:

$$d = \sqrt{\left(\frac{c}{2}\right)^2 t_2^2 - s^2} \tag{4}$$

For the case at hand, a depth of 6.5mm is obtained. Repeating a similar analysis for the second diffracted wave is that:

$$2to + t3 = 12.2\mu s + (t2 - t1) + (t3 - t2) = 12.2\mu s + 0.3\mu s + 0.7\mu s \tag{5}$$

IOP Conf. Series: Materials Science and Engineering **844** (2020) 012060 doi:10.1088/1757-899X/844/1/012060

From the above it obtains that t3 = 8.68μs, and therefore a distance travelled by the wave of 51.38mm. By Pythagoras we can determine the size of the defect; that is to say:

$$(51.38/2)_2 - (45.42/2)_2 = (6.5 + \mathrm{h})_2 \quad (6)$$

From the above it is obtained that h = 5.5mm. Finally, we will find the transit time for the echo from the back wall by the following procedure:

$$2t_0 + t_4 = 12.2\mu s + (t_2 - t_1) + (t_3 - t_2) + (t_4 - t_3) = 12.2\mu s + 0.3\mu s + 0.7\mu s + 0.28\mu s \quad (7)$$

Given the equation (7) it is had for t4 of 8.96μs, a result that supplies the distance travelled by the mentioned echo with a value of 53.04mm, which, after being confronted with the theoretical value of 51.75mm throws an error of 2.5%.

3. Process

As mentioned earlier, research work focused on the detection of defects in welded joints, through non-invasive procedures, has been carried out in the electronic engineering faculty. The aforementioned process consisted of a DSP card, of foreign manufacture, that stimulated a transducer at the output port by means of a tip signal; well known in the field of END. Subsequently, through the input port, it received the signal from another transducer, and then stored it, processed using a series of algorithms developed by the group [1]. Because the card broke down, as if the investigation was interrupted, the group of researchers found it necessary to look for alternatives that would lead to solving the problem. This is how one of the members of the MEM group presented the problem to one of the authors of this article, who using a "button", also developed by the group years ago managed to perform tests with the system and obtain images for processing. The "button" was developed by the seedbed of the MEM group, called Synapsis, which was in charge of another of its members.

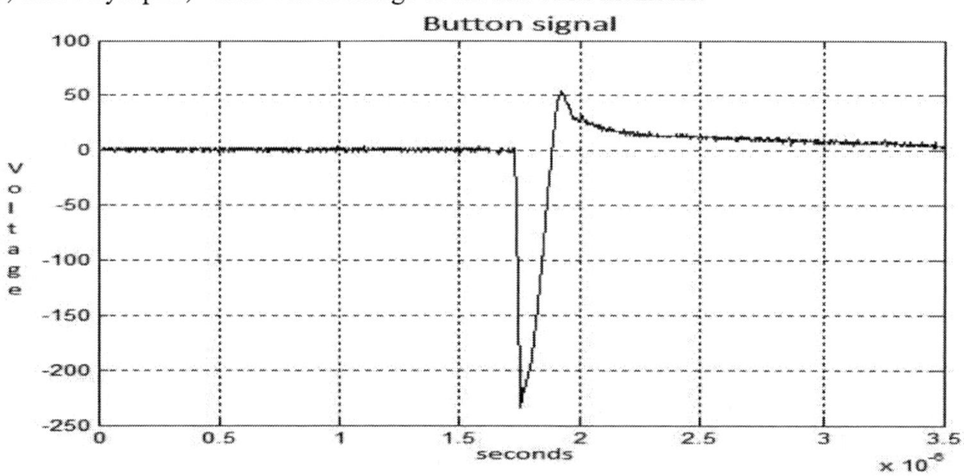

Figure 7. Tip type signal that drives the emitting transducer.

The signal presented in figure (5) is called a lace because of the great amplitude it presents in such a short time, which for this case is approximately -240 V in an approximate time of 0.3us.

3.1. *Computer connection interface with oscilloscope*

A system was designed and implemented that enables the TBS 1052B reference oscilloscope as an analog-to-digital converter, whose schematic diagram is illustrated below.

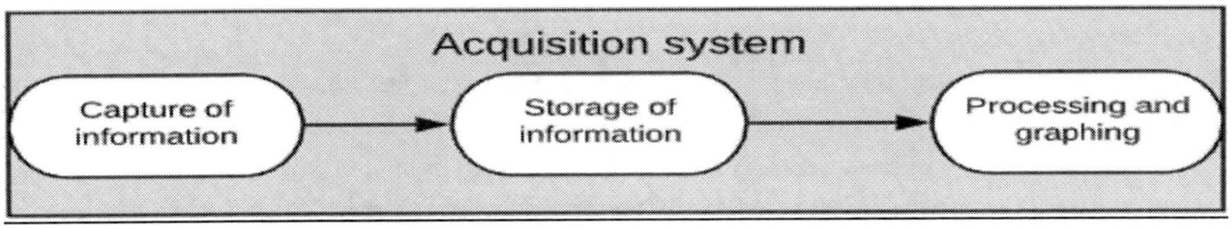

Figure 8. Schematic diagram of the system.

When observing figure (8) the implemented system is divided into three modules:

- Capture information.
- Storage of information.
- Processing and graphing.

The software was implemented using the .NET framework. The VISA library provided by the Tektronix firm is used to capture the information. Subsequently, communication is established between the oscilloscope and a computer through a USB port. For the storage of the information two files are implemented in the hard disk of the computer, from a specified address. The first file is of the ".bmp" format, and the second is of the ".txt" format. In both files the information of the set of sweeps type A is contained, only that in the first one, it can be seen visually, while in the second one there is the data of the sweeps type A. For the graphic representation, an interface was developed in which the necessary controls are arranged to start and stop the captures of sweeps type A, and to store them as shown below.

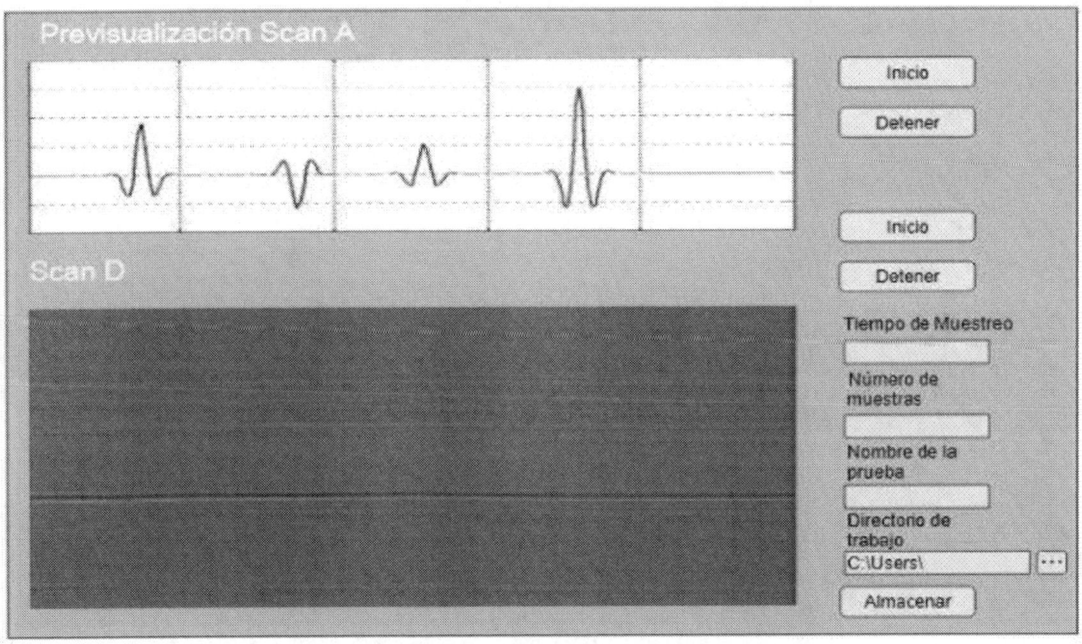

Figure 9. Graphical user interface (GUI) of the application.

The application allows a possible user to use it in a Windows environment, in addition to having a plotter that allows you to visualize the type A sweeps, and the grouping of these as a two-dimensional image composed of shades of gray, also type D sweeps This image is achieved assigning a gray scale, so that the points of the captured signal where there is greater amplitude is the darkest tone, and the points of the signal where there is less amplitude, the tone is lighter. In addition, the user can set the number of captures he wishes to obtain, as well as the sampling time, he can also stop the acquisition.

IOP Conf. Series: Materials Science and Engineering **844** (2020) 012060 doi:10.1088/1757-899X/844/1/012060

3.2. Acquisition system

To perform these captures, it is necessary to install the ultrasonic transducers, the implemented push button, and the oscilloscope that performs the function of analog-digital converter. According to figure (10), the ultrasonic system has a button that is responsible for stimulating the transmitter. In addition, through the oscilloscope the signal provided by the transducer receiver is acquired, this information is sent to the computer via USB where the algorithms are provided for further processing.

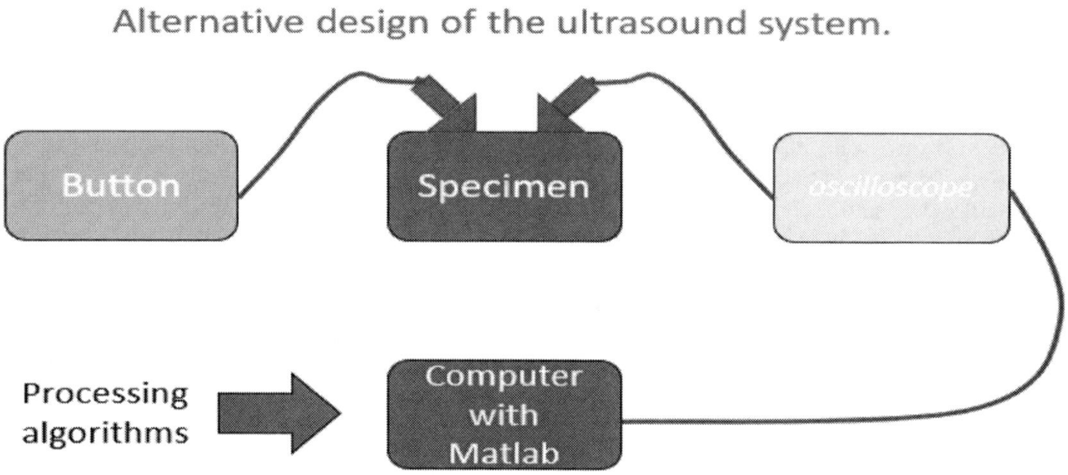

Figure 10. Ultrasonic information capture system.

4. Results

The exploration was carried out in a specimen that intentionally contains a hole in order to simulate a defect but this is not present in the whole specimen. For this capture the following enlarged image is obtained from the type D scan:

Figure 11. Scan D into a specimen with a hole.

The previous figure shows a type D image where the darkest tones represent the largest signals that travel through the edge and bottom of the sample. Between these two waveforms are the signals that provide the information. However, the image does not show variation of the signal in a section, this is because the hole is not found in the whole specimen.

Tests were also carried out on specimens attached with a welding bead. The results showed the following:

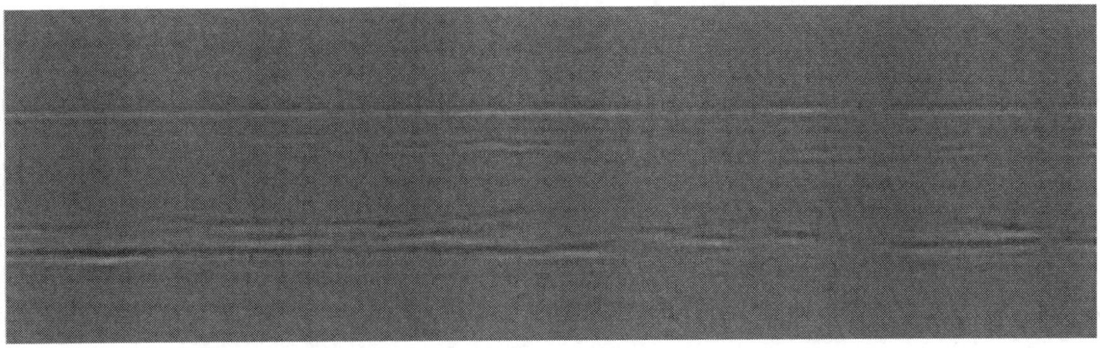

Figure 12. Scan D into a specimen attached with a welding bead.

In the same way as in the figure (11) a reference of greater amplitude (darker tone) is shown in the figure (12), and within this the information of the defects of the weld bead.

5. Conclusions

In this paper an ultrasonic signal acquisition system for TOFD imaging was presented. Based on the tests carried out, the social and industrial application of the system can be established, since by means of this the location of the defects present in metal structures can be determined without the need to destroy a piece or composition. The implemented system provides the necessary information of the captures made to the non-destructive tests, this plays an important role since if they want to export the data to another platform in which it is required to perform the image processing, for example to determine the type of defect in a welded joint, the thickness of the defect, its location etc, it can be done easily.

References

[1] Rodríguez J, Vitola J, Sandoval S, Forero E 2009 Design and construction of a system for non-destructive inspection of faults and defects in metals using ultrasonic signals **12** 9-21
[2] Segura D, Rodríguez J 2012 TOFD imaging system in nondestructive tests **17** 139-149
[3] Rodríguez J, Vitola J, Sandoval S 2009 Design and construction of an ultrasound system for detection of discontinuities in welding **41** 1
[4] Baby S, Balasubramanian T, Pardikar R J, Palaniappan M, Subbaratnam R 2002 Time of flight diffraction (TOFD) technique for accurate sizing of surface breaking cracks **12** 5-7
[5] Rodríguez M, Emeterio J L, Lázaro J C, Ramos A 2004 Ultrasonic flaw detection in NDE of highly scattering materials using wavelet and Wigner-Ville transform processing **42** 847-851
[6] Manjula K, Vijayarekha K, Venkatraman B, Karthik D 2012 Ultrasonic Time of Flight Diffraction Technique for Weld Defects **4** 5525-5533

Acknowledgments

This research was carried out within the framework of the FODEIN 2019 project entitled "Diseño de un sistema para la localización de daños en estructuras metálicas y de material compuesto expuestas a cambios de temperatura (Etapa I)", belonging to the Electronic Modelling and Monitoring (MEM) research group of the School of Electronic Engineering of the Santo Tomás University, which made possible the management of resources for the publication of this article.

A Novel Full Boundary Element Formulation for Harmonic Analysis of Elastic Membranes Coupled to Acoustics Fluids

A J Narváez-Cruz[1], **J F Useche-Vivero**[1], **J A Martínez-Trespalacios**[2,] **and J R Castro-Suarez**[3]

[1] Universidad Tecnológica de Bolívar, Cartagena de Indias, Colombia
[2] Institución Tecnológica Colegio Mayor de Bolívar, Cartagena de Indias, Colombia
[3] Universidad del Sinú, Cartagena de Indias, Colombia

E-mail: anarvaez@utb.edu.co

Abstract. *A novel full Boundary Element Formulation for the harmonic analysis of elastic membranes coupled to acoustics fluid is presented. The elastic membranes is modeled using the classical linear elastic pre-stretched membrane theory. The acoustic fluid is modeled using the acoustic-wave equation for homogeneous, isotropic, inviscid and irrotational fluids. Elastostatic fundamental solution is used in the boundary element formulation for the elastic membrane. The boundary element formulation for the acoustic fluid is based on the fundamental solution of three dimensional Poisson equation. Domain integrals related to inertial terms and those related with distributed pressure on membrane, were treated using the Dual Reciprocity Boundary Element Method. Fluid-structure coupling equations were established considering the continuity of the normal acceleration of the particles and dynamic pressure at fluid-structure interfaces. Results obtained shows the accuracy and efficiency of the proposed boundary element formulation.*

Nomenclature

$\ddot{p}$	second time derivative for pressure
$\ddot{w}$	second time derivative for displacement
Γ_f	fluid boundary surface
Γ_f^p	fluid boundary surface with stablished p
Γ_f^q	fluid boundary surface with stablished q
Γ_s	membrane boundary
Γ_s^f	membrane boundary with stablished f
Γ_s^w	membrane boundary with stablished w
Γ_{fs}	fluid-structure interface
$\hat{\dot{p}}_0(\mathbf{x})$	known initial first time derivative for pressure in Ω_f
$\hat{\dot{w}}_0(\mathbf{x})$	known initial velocity in Ω_s
$\hat{f}(\mathbf{x}, t)$	known normal traction on Γ_s^f
$\hat{p}(\mathbf{x}, t)$	known pressure on Γ_f^p

$\hat{p}_0(\mathbf{x})$ known initial pressure in Ω_f
$\hat{P}_j(\mathbf{x}',\mathbf{x})$ particular solutions for pressure as a poisson type equation
$\hat{q}(\mathbf{x},t)$ known normal pressure gradient on Γ_f^p
$\hat{Q}_j(\mathbf{x}',\mathbf{x})$ particular solutions for gradient pressures as a poisson type equation
$\hat{T}_j(\mathbf{x}',\mathbf{x})$ particular solutions for tractions as a poisson type equation
$\hat{w}(\mathbf{x},t)$ known displacement on Γ_s^w
$\hat{w}_0(\mathbf{x})$ known initial displacement in Ω_s
$\hat{W}_j(\mathbf{x}',\mathbf{x})$ particular solutions for displacements as a poisson type equation
κ bulk modulus
Ω_f fluid opened cavity domain
Ω_s flexible membrane domain
ρ_f fluid mass density
ρ_s membrane mass density
A_{ij} matrix of coefficients, obtained by taking the value of $p(t)$ at different DRM points
B_{ij} matrix of coefficients, obtained by taking the value of $\ddot{w}(t)$ at different DRM points
$c(\mathbf{x}')$ jump terms arising from the terms of $O(1/r)$ in the respective kernel
c_f fluid wave propagation velocity
c_w membrane wave propagation velocity
$f = f(\mathbf{x},t)$ normal membrane traction
F_{ij} matrix of coefficients, obtained by taking the value of $\ddot{p}(t)$ at different DRM points
$f_j(r)$ radial basis function
h membrane thickness
N number of boundary elements
n_i^w components of the outward normal vector
$NDRF$ number of total DRM collocation points used in the fluid
$NDRM$ number of total DRM collocation points used in the membrane
$P(\mathbf{x}',\mathbf{x})$ fundamentals solutions for pressure for three dimensional acoustic problems
$p = p(\mathbf{x},t)$ fluid pressure
p_w distributed pressure applied over the membrane
$p_{,\alpha\alpha}$ pressure laplacian
p_{max} distributed time harmonic pressure load amplitude
$Q(\mathbf{x}',\mathbf{x})$ fundamentals solutions for gradient pressures for three dimensional acoustic problems
$q = q(\mathbf{x},t)$ normal fluid pressure gradient
q_n pressure gradient acting on the fluid-structure interface
$T(\mathbf{x}',\mathbf{x})$ fundamentals solutions for tractions
T_0 unit length tension
$W(\mathbf{x}',\mathbf{x})$ fundamentals solutions for displacements
$w = w(\mathbf{x},t)$ membrane transversal displacement
$w_{,\alpha\alpha}$ displacement laplacian
α_j coefficients that relate A_{ij} with p_w
$\mathbf{A}$ fluid-membrane coupling sub-matrix of H

$\mathbf{C}_w$ and $\mathbf{C}_f$ connectivity matrices joining fluid and structural degree of freedom at fluid-structure interface
$\mathbf{H}$ and $\mathbf{G}$ global coupling influence matrices
$\mathbf{M}$ global coupling mass matrix
$\mathbf{n}_w$ outward normal vector
$\mathbf{S}$ fluid-membrane coupling sub-matrix of G
$\mathbf{x}$ field points
$\mathbf{x}'$ collocation points
$\ddot{\alpha}_j$ coefficients that relate F_{ij} with $\ddot{p}$
$\ddot{\beta}_j$ coefficients that relate B_{ij} with $\ddot{w}$
${}^f\mathbf{G}_{fs}$ and ${}^f\mathbf{G}_{ff}$ sub-matrices of ${}^f\mathbf{G}$ related with degrees of freedom defined on the interface Γ_{fs}
${}^s\mathbf{B}_{fs}$ sub-matrix of ${}^s\mathbf{B}$ related with pressure terms defined in Ω_s
t time
${}^f\mathbf{H}$ and ${}^f\mathbf{G}$ influence matrices for the fluid
${}^f\mathbf{M}$ fluid mass matrix
${}^s\mathbf{B}$ influence matrix related with distributed pressure $\mathbf{p}_w$ applied over the membrane
${}^s\mathbf{H}$ and ${}^s\mathbf{G}$ influence matrices for the membrane
${}^s\mathbf{M}$ membrane mass matrix

1. Introduction

Acoustic problems in real life applications involve considerations of structural and acoustic part together and thus calls for a coupled vibroacoustic treatment rather a pure acoustic approach. Thus, whenever an elastic structure is in contact with a fluid, the structural vibrations and the acoustic pressure field in the fluid are influenced by the mutual vibro-acoustic coupling interaction. The analysis of fluid-structure coupled systems is a challenging and complex task. In general, the use of experimentation or numerical methods represent the two uniques alternatives to obtain approximate solutions for these kind of problems. However, numerical methods based on domain discretization requires refined meshes for high frequency problems, since the length of the elements should be proportional to the size of the wavelength. This means a more time-consuming model. The boundary element method (BEM) is a modern numerical technique which has enjoyed increasing popularity over the last two decades, and is now an established alternative to traditional computational methods of engineering analysis [1, 2]. The main advantage of the BEM is its unique ability to provide a complete solution in terms of boundary values only, with substantial savings in modelling effort. Since consolidation of the boundary element method (BEM) as reliable numerical method for structural and fluid analysis, linear vibrations of structures coupled with an internal fluid has been carried out using hybrid BEM-FEM formulations. In these formulations BEM is used to model the fluid media and the FEM to model the structural response [3, 4, 5, 6, 7, 8, 9, 10, 11, 12, 13]. The main advantage of such formulations lies in a substantial reduction in the number of the degrees of freedom in the discretization of the fluid domain. However, in this formulations is necessary to discretize the entirely structure due to the use of the FEM. Few BEM-BEM coupled formulations for fluid structure-interaction has been published [14, 15, 16, 17]. However, despite the fact that the BEM has been used for dynamic analysis of membrane structures and for analyses of acoustic fluids, to the best of authors knowledge, these formulations do not have been used for the fluid-structure interaction problem analysis using a full boundary element formulation for such purpose. In this work, a new full boundary element formulation for the transient dynamic analysis of acoustic

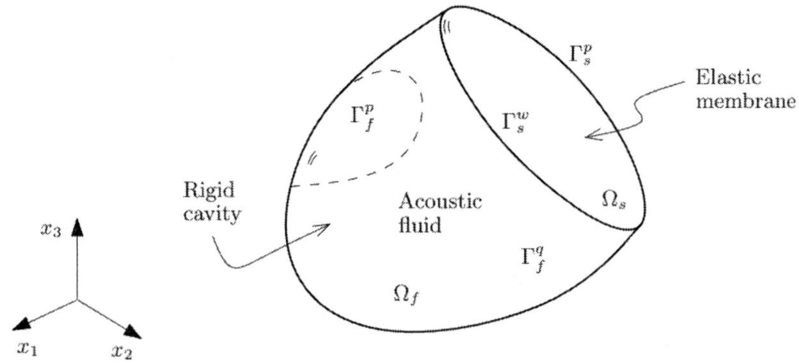

Figure 1. Fluid-structure interaction problem.

fluids coupled to elastic membranes is presented. Membranes are modeled using a boundary element formulation based on the linear elastic membrane theory under small deflection. The acoustic fluid is modeled using a boundary element formulation for the three dimensional acoustic wave equation. Fluid-structure coupling equations were established considering the continuity of the normal acceleration of the particles at fluid-structure interfaces. Domain integrals on both, fluid and structure equations, were treated using the Dual Reciprocity Boundary Element Method. The developed formulation was used to study the linear vibration response of elastic membranes coupled to acoustic fluids.

2. Structure subjected to a fluid pressure loading

Consider a partially opened cavity Ω_f with ridig walls and a flexible elastic membrane Ω_s with mass density ρ_s and thickness h (see Fig. 1). Cavity contains an homogeneous and isotropic acoustic fluid with mass density ρ_f. The membrane vibrations and the acoustic pressure field in the fluid are influenced by the mutual vibro-acoustic coupling interaction. In this work, the vibro-acoustic coupling interaction is modeled using an eulerian formulation where the acoustic response is described by the pressure, while the membrane response is described by the transversal displacement field.

2.1. Acoustic wave equation

The dynamic pressure of an ideal inviscid fluid under small perturbations in a spatial region Ω_f confined by the boundary surface $\Gamma_f = \Gamma_f^p \cup \Gamma_f^q$, is governed by the wave equation (see Fig. 2) [18]:

$$p_{,\alpha\alpha} = \frac{1}{c_f^2}\ddot{p} \tag{1}$$

In this equation p is the fluid pressure, $c_f^2 = \kappa/\rho_f$ stands for wave propagation velocity, κ is the bulk modulus. Above equation can be modified to include the effect of presence of an acoustic source. Double dot represents a second time derivative. Indicial notation is used throughout this work. Greek indices vary from 1 to 2 and Roman indices from 1 to 3. The boundary and initial conditions for equation (1) in the time interval $[0, t^*]$ are:

$$\begin{aligned} p(\mathbf{x}, t) &= \hat{p}(\mathbf{x}, t), \quad \mathbf{x} \in \Gamma_f^p, \quad t \in [0, t^*] \\ q(\mathbf{x}, t) = n_i^f p_{,i} &= \hat{q}(\mathbf{x}, t), \quad \mathbf{x} \in \Gamma_f^q, \quad t \in [0, t^*] \\ p(\mathbf{x}, 0) = \hat{p}_0(\mathbf{x}), &\quad \dot{p}(\mathbf{x}, 0) = \hat{\dot{p}}_0(\mathbf{x}), \quad \mathbf{x} \in \Omega_f \end{aligned} \tag{2}$$

where n_i^f are the components of the outward normal vector $\mathbf{n}_f$ at boundary. and t denotes time.

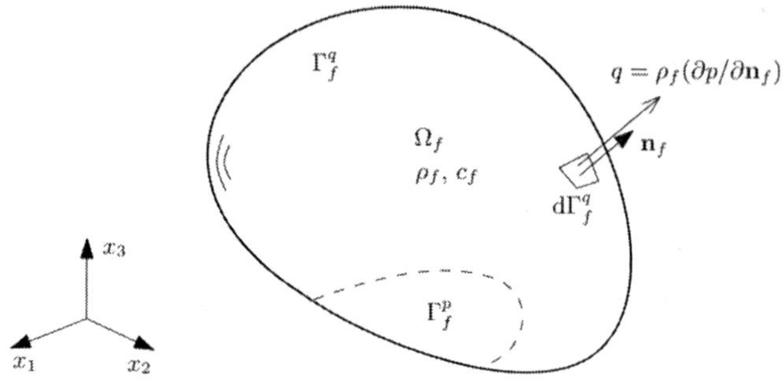

Figure 2. Acoustic fluid domain.

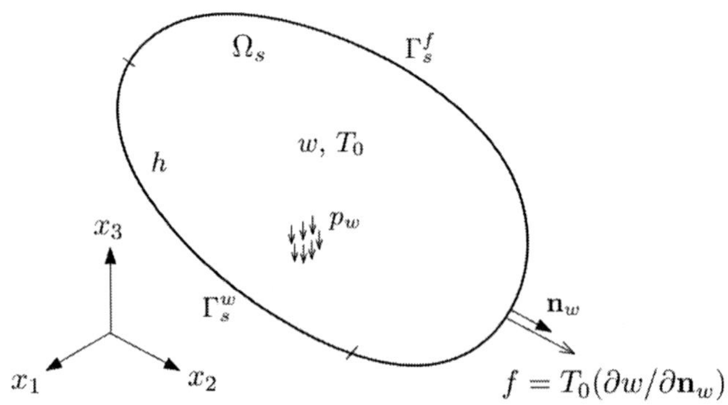

Figure 3. Elastic membrane domain.

2.2. Dynamic equation of an elastic membrane

Now consider a linear elastic membrane with thickness h ocuppying the spatial domain Ω_w confined by the boundary $\Gamma_s = \Gamma_s^w \cup \Gamma_s^f$ (see Fig. 3). An initial tension T_0 is uniformly applied to the membrane. In this work, the small deflection elastic membrane theory is considered. Thus, differential equation describing the trasversal displacement $w(\mathbf{x}, t)$ of this membrane in the time interval $[0, t^*]$ is given by:

$$w_{,\alpha\alpha} + \frac{p_w}{T_0} = \frac{1}{c_w^2}\ddot{w} \tag{3}$$

where $c_w^2 = T_0/\rho_s$ and $p_w(\mathbf{x}, t)$ is a distributed pressure applied over the membrane. The boundary and initial conditions for this equation are given by:

$$\begin{aligned} w(\mathbf{x}, t) &= \hat{w}(\mathbf{x}, t), \quad \mathbf{x} \in \Gamma_s^w, \quad t \in [0, t^*] \\ f(\mathbf{x}, t) &= n_i^w w_{,i} = \hat{f}(\mathbf{x}, t), \quad \mathbf{x} \in \Gamma_s^f, \quad t \in [0, t^*] \\ w(\mathbf{x}, 0) &= \hat{w}_0(\mathbf{x}), \quad \dot{w}(\mathbf{x}, 0) = \hat{\dot{w}}_0(\mathbf{x}), \quad \mathbf{x} \in \Omega_s \end{aligned} \tag{4}$$

In these expressions, n_i^w are the components of the outward normal vector $\mathbf{n}_w$ at boundary.

3. Boundary element equations

3.1. Boundary element equations for acoustic wave equation

The derivation of the integral formulation for equation (1) is based on application of the Boundary Element Method to the acoustic wave equation as presented in [2]. Thus, by using the weighted residual method and making use of the Green's identity, the following equation is obtained:

$$c(\mathbf{x}')p(\mathbf{x}') + \int_{\Gamma_f} Q(\mathbf{x}', \mathbf{x})p(\mathbf{x})\mathrm{d}\Gamma_f = \int_{\Gamma_f} P(\mathbf{x}', \mathbf{x})q(\mathbf{x})\mathrm{d}\Gamma_f + \frac{1}{c_f^2}\int_{\Omega_f} P(\mathbf{x}', \mathbf{X})\ddot{p}(\mathbf{X})\mathrm{d}\Omega_f \tag{5}$$

In this equation, $\mathbf{x}'$ and $\mathbf{x}$ represent collocation and field points, respectively; $P(\mathbf{x}', \mathbf{x})$ and $Q(\mathbf{x}', \mathbf{x})$ are fundamentals solutions for pressure and gradient pressures for three dimensional acoustic problems, respectively, as presented in [2]. The value of $c(\mathbf{x}')$ is equal to $\frac{1}{2}$ when $\mathbf{x}'$ is located on a smooth boundary.

In order to threat the domain integral, the Dual Reciprocity Boundary Element Method (DRM) is used as presented in [19]. In this way, equation (5) can be re-written as:

$$c(\mathbf{x}')p(\mathbf{x}') + \int_{\Gamma_f} Q(\mathbf{x}', \mathbf{x})p(\mathbf{x})\mathrm{d}\Gamma_f = \int_{\Gamma_f} P(\mathbf{x}', \mathbf{x})q(\mathbf{x})\mathrm{d}\Gamma_f +$$

$$+\frac{1}{c^2}\sum_{j=1}^{NDRF} \ddot{\alpha}_j(t)\Big[c_i\hat{P}_{ij} + \int_{\Gamma_f} Q(\mathbf{x}', \mathbf{x})\hat{P}_j\mathrm{d}\Gamma_f - \int_{\Gamma_f} P(\mathbf{x}', \mathbf{x})\hat{Q}_j\mathrm{d}\Gamma_f\Big] \tag{6}$$

In this equation, $NDRF$ represents the number of total DRM collocations points used in the fluid; $\hat{P}_j(\mathbf{x}', \mathbf{x})$ and $\hat{Q}_j(\mathbf{x}', \mathbf{x})$ are the particular solutions to equivalent homogeneous equation (1). These particular solutions were obtained considering the function $f_j(r) = 1 + r_j$ for the approximation of $\ddot{p}(t)$, as presented in [19]. Coefficients $\ddot{\alpha}_j$ are related to $\ddot{p}$ through: $\ddot{p} = F_{ij}\ddot{\alpha}_j(t)$, where F_{ij} is a matrix of coefficients, obtained by taking the value of $\ddot{p}(t)$ at different DRM points.

In order to discretize boundary surfaces of the acoustic medium, N boundary quadrilateral elements were used and $p(\mathbf{x})$ and $q(\mathbf{x})$ were assumed to be constant over each element and equal to their values at the mid-element node. Thus, the discretised form of equation (6) is given by:

$$c_i(\mathbf{x}')p(\mathbf{x}', t) + \sum_{k=1}^{N}\Big[\int_{\Gamma k} Q(\mathbf{x}', \mathbf{x})\mathrm{d}\Gamma\Big]p_k(t) - \sum_{k=1}^{N}\Big[\int_{\Gamma k} P(\mathbf{x}', \mathbf{x})\mathrm{d}\Gamma\Big]q_k(t) =$$

$$\frac{1}{c_f^2}\sum_{j=1}^{NDRF} \ddot{\alpha}_j\Big[c_i(\mathbf{x}')\hat{P}_{ij}(\mathbf{x}', \mathbf{x}) + \sum_{k=1}^{N}\int_{\Gamma k} Q(\mathbf{x}', x)\hat{P}_j(\mathbf{x}', \mathbf{x})\mathrm{d}\Gamma$$

$$-\sum_{k=1}^{N}\int_{\Gamma k} P(\mathbf{x}', \mathbf{x})\hat{Q}_j(\mathbf{x}', \mathbf{x})\mathrm{d}\Gamma\Big] \tag{7}$$

Applying this equation at each collocation point, the following linear system of equations is obtained:

$${}^f\mathbf{M}\ddot{\mathbf{p}} + {}^f\mathbf{H}\mathbf{p} = {}^f\mathbf{G}\mathbf{q} \tag{8}$$

where ${}^f\mathbf{M}$ is the fluid mass matrix, ${}^f\mathbf{H}$ and ${}^f\mathbf{G}$ are boundary element influence matrices; $\mathbf{p}$ and $\mathbf{q}$ are vectors of nodal pressures and normal derivative of pressure, respectively.

3.2. Boundary element equations for an elastic membrane

The derivation of the integral formulation for equation (3) is based on the application of the BEM to the membrane equation as presented in [2]. Thus, by using the weighted residual method, and making use of the Green's identity, the integral formulation for equation (3) is given by:

$$c(\mathbf{x}')w(\mathbf{x}', t) + \int_{\Gamma s} T(\mathbf{x}', \mathbf{x})w(\mathbf{x}, t)\mathrm{d}\Gamma_s - \int_{\Gamma s} W(\mathbf{x}', \mathbf{x})f(\mathbf{x}, t)\mathrm{d}\Gamma_s = -\frac{1}{T_0}\int_{\Omega s} W(\mathbf{x}', \mathbf{x})p_w(\mathbf{x}, t)\mathrm{d}\Omega_s$$

$$+\frac{1}{c_w^2}\int_{\Omega s} W(\mathbf{x}', \mathbf{x})\ddot{w}(\mathbf{x}, t)\mathrm{d}\Omega_s \tag{9}$$

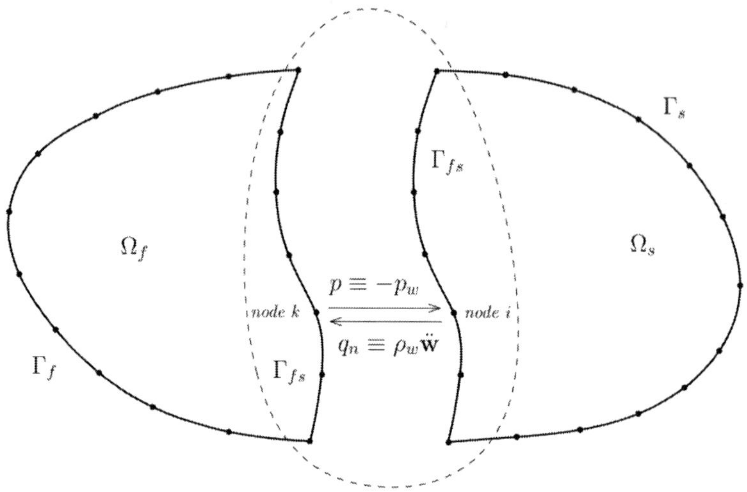

Figure 4. Discretized coupled problem using BEM.

where $\mathbf{x}$ and $\mathbf{x}'$ are field and collocation points respectively, $W(\mathbf{x}',\mathbf{x})$ and $T(\mathbf{x}',\mathbf{x})$ are fundamental solutions for displacement and traction, respectively as given in [1]. $c(\mathbf{x}')$ is the jump term arising from the terms of $O(1/r)$ in the kernel $T(\mathbf{x}',\mathbf{x})$.

In this work, the DRM was used to transform domain integrals related to inertial terms into boundary integrals. In this way, this equation (9) can be re-written as [19]:

$$c_i(\mathbf{x}')w(\mathbf{x}',t) + \int_{\Gamma s} T(\mathbf{x}',\mathbf{x})w(\mathbf{x},t)\mathrm{d}\Gamma_s - \int_{\Gamma s} W(\mathbf{x}',\mathbf{x})f(\mathbf{x},t)\mathrm{d}\Gamma_s = -\frac{1}{T_0}\sum_{j=1}^{NDRM} \alpha_j(t)\Big[c_i(\mathbf{x}')\hat{W}_j(\mathbf{x},t) + \int_{\Gamma s} T(\mathbf{x}',\mathbf{x})\hat{W}_j(\mathbf{x}',\mathbf{x})\mathrm{d}\Gamma_s - \int_{\Gamma s} W(\mathbf{x}',\mathbf{x})\hat{T}_j(\mathbf{x}',\mathbf{x})\mathrm{d}\Gamma_s\Big] + \frac{1}{c_w^2}\sum_{j=1}^{MDRM} \ddot{\beta}_j(t)\Big[c_i(\mathbf{x}')\hat{W}_j(\mathbf{x},t) + \int_{\Gamma s} T(\mathbf{x}',\mathbf{x})\hat{W}_j(\mathbf{x}',\mathbf{x})\mathrm{d}\Gamma_s - \int_{\Gamma s} W(\mathbf{x}',\mathbf{x})\hat{T}_j(\mathbf{x}',\mathbf{x})\mathrm{d}\Gamma_s\Big] \quad (10)$$

where $NDRM$ represent the total number of DRM collocation used in the membrane; $\hat{T}_j(\mathbf{x}',\mathbf{x})$ and $\hat{W}_j(\mathbf{x}',\mathbf{x})$ are the particular solutions to equivalent homogeneous equation (3). These particular solutions were obtained considering the function $f_j(r) = 1 + r_j$ for the approximation of $\ddot{w}(t)$ and $p_w(t)$ terms, as presented in [19]. Coefficients α_j are related to $p_w(t)$ through: $p_w = A_{ij}\alpha_j(t)$, where A_{ij} is a matrix of coefficients, obtained by taking the value of $p(t)$ at different DRM points. Similarly, coefficients $\ddot{\beta}_j$ are related to $\ddot{w}$ through: $\ddot{w} = B_{ij}\ddot{\beta}_j(t)$, where B_{ij} is a matrix of coefficients, obtained by taking the value of $\ddot{w}(t)$ at different DRM points.

Applying this equation at each collocation point, the following linear system of equations is obtained:

$$^s\mathbf{M}\ddot{\mathbf{w}} + {}^s\mathbf{H}\mathbf{w} = {}^s\mathbf{G}\mathbf{f} - {}^s\mathbf{B}\mathbf{p}_w \quad (11)$$

Here $^s\mathbf{M}$ is the membrane mass matrix, $^s\mathbf{H}$ and $^s\mathbf{G}$ are the influence matrices, $^s\mathbf{B}$ is the influence matrix related with distributed pressure $\mathbf{p}_w$ applied over the membrane.

4. Fluid-structure coupling equations

Fluid-structure coupling equations are given by compatibility considerations about normal pressure and dynamic pressure force acting at fluid-structure interface. Mathematically, these conditions can be written as follows [20]:

$$\mathbf{n}_f \cdot \nabla p = \mathbf{q}_n \equiv -\rho_w \mathbf{C}_w \ddot{\mathbf{w}} \tag{12}$$

$$\mathbf{p}_w \equiv -\mathbf{C}_f \mathbf{p} \tag{13}$$

That is, pressure gradient acting on the fluid-structure interface Γ_{fs} are related to normal acceleration of the plate and the acoustic pressure is equilibrated with pressure on the membrane (see Fig. 4). In these equations, $\mathbf{C}_w$ and $\mathbf{C}_f$ represent connectivity matrices joining fluid and structural degree of freedom at fluid-structure interface.

Replacing equations (13) and (13) into equations (8) and (11) we obtain the coupled fluid-structure equation problem:

$$\begin{bmatrix} {}^s\mathbf{M} & \mathbf{0} \\ \mathbf{S} & {}^f\mathbf{M} \end{bmatrix} \begin{Bmatrix} \ddot{\mathbf{w}} \\ \ddot{\mathbf{p}} \end{Bmatrix} + \begin{bmatrix} {}^s\mathbf{H} & -\mathbf{A} \\ \mathbf{0} & {}^f\mathbf{H} \end{bmatrix} \begin{Bmatrix} \mathbf{w} \\ \mathbf{p} \end{Bmatrix} = \begin{bmatrix} {}^s\mathbf{G} & \mathbf{0} \\ \mathbf{0} & {}^f\mathbf{G}_{ff} \end{bmatrix} \begin{Bmatrix} \mathbf{f} \\ \mathbf{q} \end{Bmatrix} \tag{14}$$

In this equation, the off-diagonal sub-matrices $\mathbf{A} = {}^s\mathbf{B}_{fs}\mathbf{C}_f$ and $\mathbf{S} = \rho_w {}^f\mathbf{G}_{fs}\mathbf{C}_w$ are fluid-membrane coupling matrices. ${}^f\mathbf{G}_{fs}$ and ${}^f\mathbf{G}_{ff}$ are sub-matrices of ${}^f\mathbf{G}$ related with degrees of freedom defined on the interface Γ_{fs}. ${}^s\mathbf{B}_{fs}$ is a sub-matrix of ${}^s\mathbf{B}$ related with pressure terms defined in Ω_s.

Equations (14) can be rewritten in a general way as:

$$\mathbf{M}\ddot{\mathbf{u}} + \mathbf{H}\mathbf{u} = \mathbf{G}\mathbf{r} \tag{15}$$

where, $\ddot{\mathbf{u}} = \{\ddot{\mathbf{w}}, \ddot{\mathbf{p}}\}^T$, $\mathbf{u} = \{\mathbf{w}, \mathbf{p}\}^T$ and $\mathbf{r} = \{\mathbf{f}, \mathbf{q}\}^T$.

5. Numerical examples

5.1. Box-shaped structure containing an acoustic fluid coupled to a membrane

Consider a partially opened box-shaped structure with ridig walls and a flexible elastic membrane with $c_w = 22.36$ m/s as presented in Fig. 5. The structure contains an acoustic fluid with mass density $\rho_f = 100$ kg / m^3 and a bulk's modulus $\kappa = 2.5 \times 10^5$ Pa. A distributed time harmonic pressure load $p_{max} = 1.0$ Pa is applied over the face of the structure located at $x_3 = 2$ m. Pressure gradient $q = n_i^f p_{,i} =$ is considered zero in the other faces.

Four meshes were used to show the convergence. Table 1 shows the respective meshes.

Table 1. Mesh characteristics for example 1

Meshes	Elements	Collocation Points
Mesh 1	54 (12)	125 (9)
Mesh 2	150 (20)	125 (25)
Mesh 3	294 (28)	343 (49)
Mesh 4	486 (36)	343 (81)

The collocation points are coincidents with fluid collocation points located in the fluid-membrane interface. A membrane tensile load $T_0 = 500$ N/m was considered in the analysis. The frequency-history response of the central deflection for frequency interval 0.039788 Hz. to 20.0 Hz. is shown in figure 6.

The collocation points are coincidents with fluid collocation points located in the fluid-membrane interface. A membrane tensile load $T_0 = 500$ N/m was considered in the analysis.

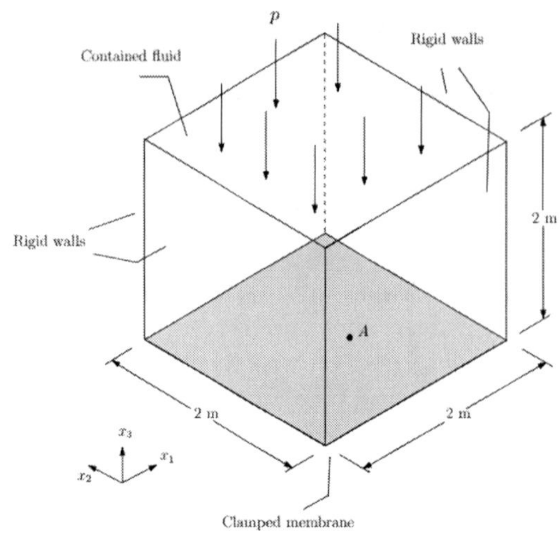

Figure 5. Discretized coupled problem using BEM.

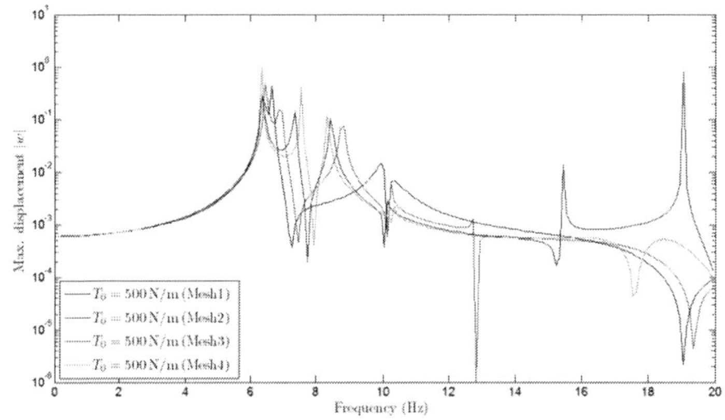

Figure 6. Maximum displacement frequency history at point $A(0.25,-0.25,-1)$ using DR-BEM

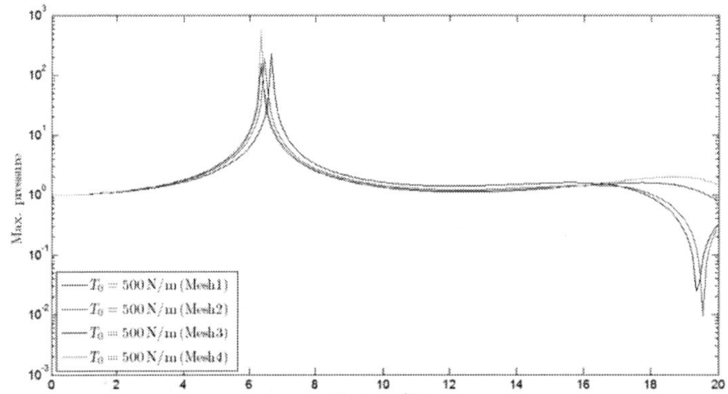

Figure 7. Pressure frequency history at point $A(0.25,-0.25,-1)$ using DR-BEM

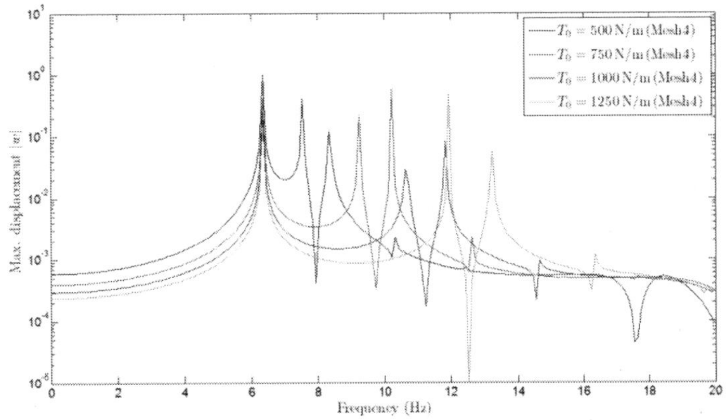

Figure 8. Maximum displacement frequency history at point $A(0.25,\text{-}0.25,\text{-}1)$ using DR-BEM

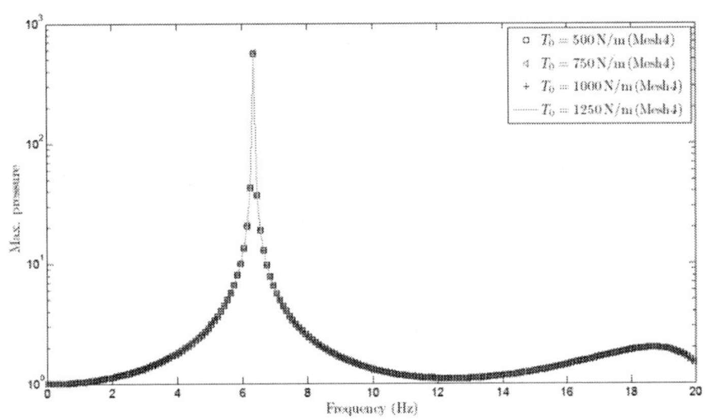

Figure 9. Pressure frequency history at point $A(0.25,\text{-}0.25,\text{-}1)$ using DR-BEM

5.2. Rigid channel containing an acoustic fluid coupled to a membrane

In this problem, a 10 m. long channel containing an acoustic fluid under a distributed time harmonic pressure load and coupled with a homogeneous elastic membrane with $c_w = 22.36$ m/s and density $\mu_w = 1.0$ kg/m^2 is analyzed (see Fig. 10). The fluid has a density ρ_f 100 kg/m^3 and a bulk's modulus of $\kappa = 2.5 \times 10^5$ Pa. The membrane is inclined 20.56° with respect to the vertical axis. A distributed time harmonic pressure load $p_{max} = 1.0$ Pa is applied on the membrane. Pressure gradient $q = n_i^f p_{,i} =$ is considered zero in the other faces.

Four meshes were used to show the convergence. Table 2 shows this meshes.

Table 2. Mesh characteristics for example 2

Meshes	Elements	Collocation Points
Mesh 1	162 (12)	90 (9)
Mesh 2	330 (20)	90 (25)
Mesh 3	658 (28)	171 (49)
Mesh 4	1098 (36)	171 (81)

IOP Conf. Series: Materials Science and Engineering **844** (2020) 012061 doi:10.1088/1757-899X/844/1/012061

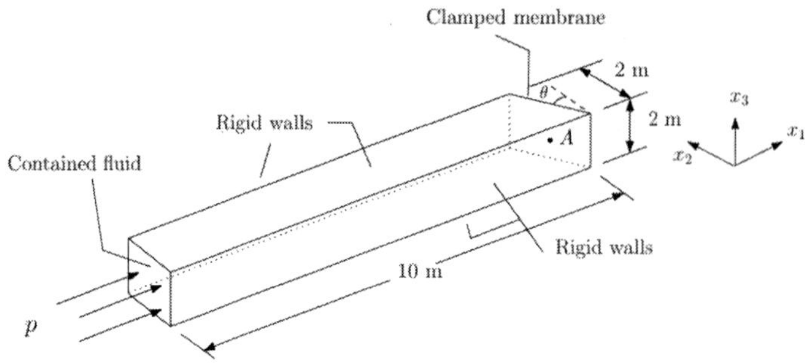

Figure 10. Discretized coupled problem using BEM.

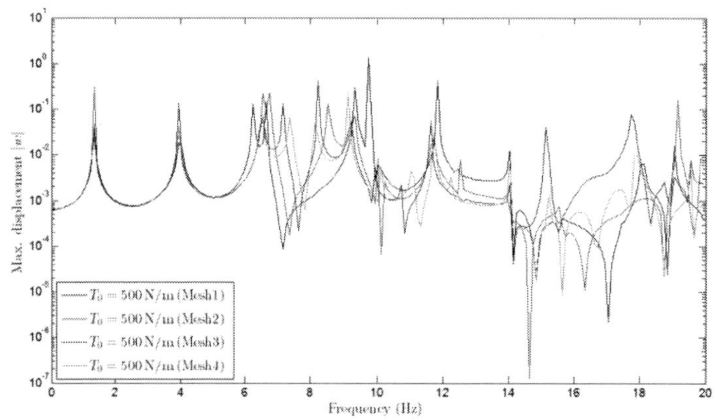

Figure 11. Maximum displacement frequency history at point A using DR-BEM

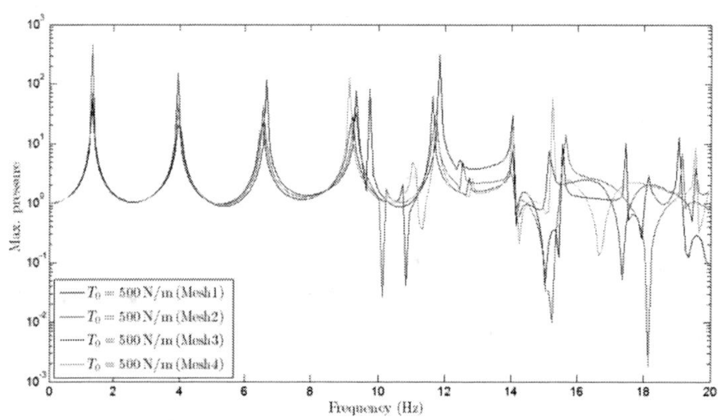

Figure 12. Pressure frequency history at point A using DR-BEM

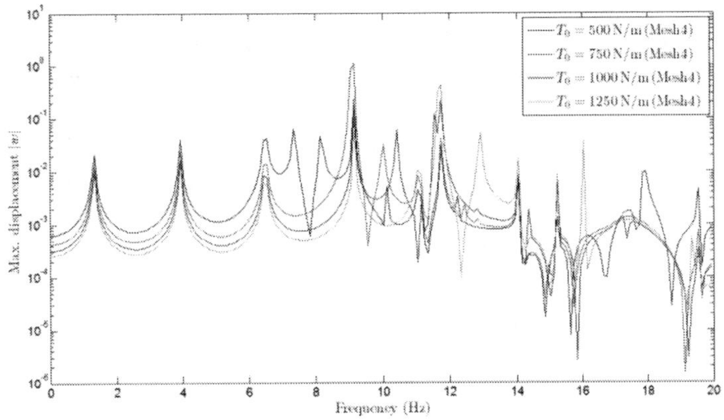

Figure 13. Maximum displacement frequency history at point A using DR-BEM

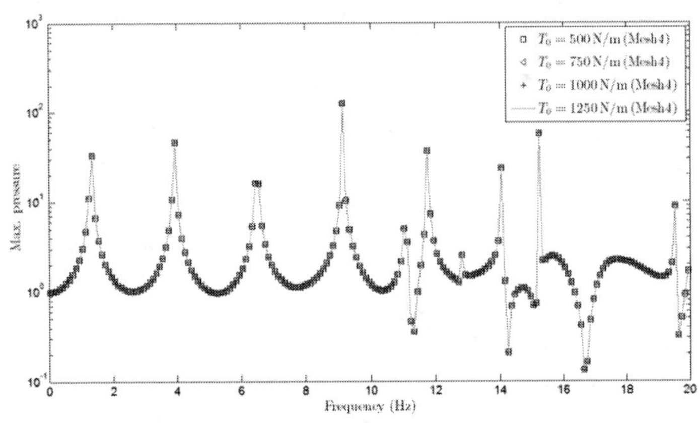

Figure 14. Pressure frequency history at point A using DR-BEM

IOP Conf. Series: Materials Science and Engineering **844** (2020) 012061 doi:10.1088/1757-899X/844/1/012061

6. Conclusions

A new full boundary element formulation for the harmonic dynamic analysis of acoustic fluids coupled to elastic membranes is presented. Membranes were modeled using a boundary element formulation based on the linear elastic membrane theory under small deflection. The acoustic fluid was modeled using a boundary element formulation for the three dimensional acoustic wave equation. Fluid-structure coupling equations were established considering the continuity of the normal acceleration of the particles at fluid-structure interfaces. Domain integrals on both, fluid and structure equations, were treated using the Dual Reciprocity Boundary Element Method. Results show good agreement with those obtained from finite difference models, turning proposed formulation a reliable and an alternative numerial engineering tool for the dynamic analysis of acoustic fluids coupled to flexible elastic membranes.

References

[1] Wrobel L C and Aliabadi M H 2002 *The Boundary Element Method Volume 2: Applications in Solid and Structures* (New York, NY: Wiley)
[2] Wrobel L C 2002 *The Boundary Element Method Volume 1: Applications in Thermo-Fluids and Acoustics* (New York, NY: Wiley)
[3] Mackerle J 1999 *Finite Elements in Analysis and Design Journal* **31** 231–240
[4] Shekari M R, NKhaji and Ahmadi M T 2009 *Journal of Fluids and Structures* **25** 567–585
[5] Naumenkob V, Strelnikovac E and Yeselevac E 2010 *Engineering Analysis with Boundary Elements Journal* **34** 856–862
[6] Gaul L and Wenzel W 2002 *Engineering Analysis with Boundary Elements* **26** 629–636
[7] Soares D 2009 *International Journal for Numerical Methods in Engineering* **78** 1076–1093
[8] Everstine G C and Henderson F M 1990 *The Journal of the Acoustical Society of America* **87** 1938–1947
[9] Fritze D, Marburg S and Hardtke H J 2005 *Computers and Structures Journal* **83** 143–154
[10] Citarella R, Federico L and Cicatielloa A 2007 *Engineering Analysis with Boundary Elements* **31** 248–258
[11] He Z, Liu G, Zhong Z, Zhang G and Cheng A 2011 *Engineering Analysis with Boundary Elements* **35** 140–147
[12] Chen P T, Ju S H and Cha K C 2000 *Journal of Sound and Vibration* **233** 407–422
[13] Marburg S and Nolte B 2010 *Computational Acoustics of Noise Propagation in Fluids - Finite and Boundary Element Methods* (New York, NY: Springer)
[14] Tanaka M and Masuda Y 1988 *Computer Methods in Applied Mechanics and Engineering Journal* **71** 225–234
[15] Jr D S and Mansur W 2006 *Journal of Computational Physics* **219** 498–512
[16] Burgschweiger R, Ochmann M and Nolte B 2008 *The Journal of the Acoustical Society of America* **123** 3757–3760
[17] Jr D S, von Estorff O and Mansur W J 2005 *International Journal for Numerical Methods in Engineering* **64** 1416–1431
[18] Kinsler L E, Freg A R, Coppens A B and Sanders J V 2005 *Fundamental of Acoustics* (London, UK: Jhon Wiley and Sons Inc.)
[19] Partridge P, Brebbia and Wrobel 1992 *The Dual Reciprocity Boundary Element Method* (Boston: Southampton)
[20] Morand P and Ohayon R 1995 *Fluid Structure Interaction* (New York, NY: John Wiley and Sons)

Business architecture for the management of urban incidents: believes the TI government for the legal office of the Sinu University

J C Arrieta[1], M C Bonfante[1], L A Blanquicett[1] and L F Murillo[2]

[1] Faculty of Exact Sciences and Engineering, Universidad del Sinú, Colombia
[2] Faculty of Engineering, Corporación Universitaria Rafael Núñez, Colombia

jarrieta@unisinucartagena.edu.co

Abstract. Currently, business models integrated with the IT architecture become a strategic plan to improve services and be more competitive in a globalized world. This work describes how, from the model-based architecture (MDA), such as the CIM, you can obtain the BPMN language software requirements, obtain the PIM component models, then generate the application code and, subsequently, perform functional tests of the product. The legal office of the Sinu University is presented as a case study, in which an integrated and automated business architecture is proposed.

1. Introduction

An organization can form its own business architecture, to respond to its strategic objectives aligned to its innovation projects, that is the case of the legal office of the Sinu University in Cartagena, which is a space of social practice and extension towards the community for the students of the last semester of the law degree.

The legal office has identified missional processes such as consultancy and legal representation that consists in the legal consultancy and representation before the court, which it's done with the supervision of the teachers.

The conciliation service aims to promote a culture of peace and co-existence and allows people to solve their conflicts by getting to agreement between the parts a new process is the one of urban incidents, management in which are solved cases related to unmet basic needs of the citizens, which serves like competences in the conflicts resolution. These cases are assigned according to the profiles of the students and they are evaluated by the teachers of the area, who finally measure their resolution capacity.

Due to the high volume of information that is handled, it is necessary to design a business architecture that can describe the business, information, applications and resources of computer technology that allow responding to the changing needs of the context. Models and tools will be used to describe the architecture, which will become a framework for those entities that provide social service to the community.

IOP Conf. Series: Materials Science and Engineering **844** (2020) 012062 doi:10.1088/1757-899X/844/1/012062

2. Related work

2.1. The model MDA in the lifecycle of software development
Model Driven Architecture (MDA) proposes the definition and use software development models different models of attraction, the same as the possibility of automatic generation codes from the defined models. The reference models are: Independent Computing Model (ICM), Platform Independent Model (PIM), Platform Specification Model (PSM).

2.1.1. Independent Computing Model (ICM). The fundamental objective of the models that compose this level is to achieve a characterization as real as possible of the environment where the software is going to be displayed. In order to carry out this task the business processes are represented by using BPMN that are developed in the application environment [1].

2.1.2. Platform Independent Model (PIM). Represents the models that describe a software solution which do not contain the details of the concrete platform where the solution will be implemented, hence its name platform independent models. These models arise as the result of the analysis and design [2]. The fundamental purpose is to show how the system components are related among them, in order to comply with the identified functionalities during the business process models.

2.1.3. Platform Specification Model (PSM). The specific models of the platform arise from the (PIM) and are created between the phases of design and coding.

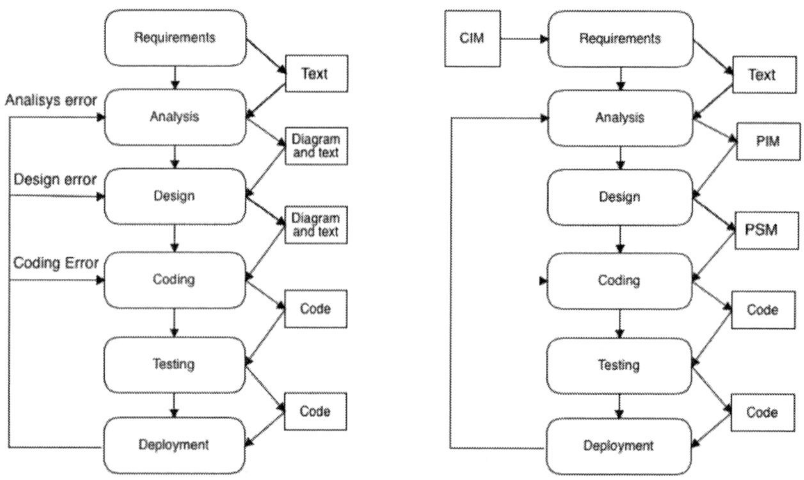

Figure 1. Traditional model and MDA MODEL. [3]

2.2. Business Architecture Design
A business architecture (BA) can be defined as a complete expression of the organization, which acts as a master plan in collaboration with business planning aspects, business operations, automation, business and technological infrastructure [4]. Another definition according to the IEEEE STD -1471 2000 is the fundamental organization of a system, composed of its components, the relationship between them and its environment and the principles that govern its designs and evolution [5]. The basic components of business architecture are:

- *Business process architecture.* It sees about the relationship with the strategy of the organization and the essential business processes. In this block is located the process architecture. Data architecture describes the physical and logical aspects of the data resources, as well as its corresponding management.

- *Interphases and integration architecture.* Provide the necessary basis in order to identify the apps that the business demands.
- *Technological architecture.* Provides the technological support required by apps, data, and identified processes in the other architectures.

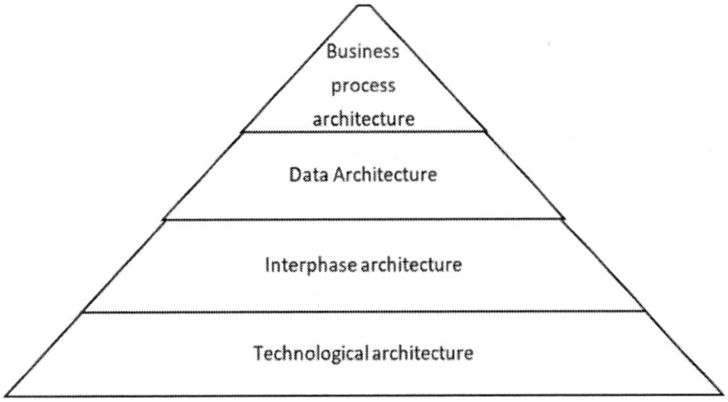

Figure 2. Basic components of business architecture. [5].

An organization has several architectures that models it, the relationship of architectural views that represent them, the whole of an organization, as shown in graph 3

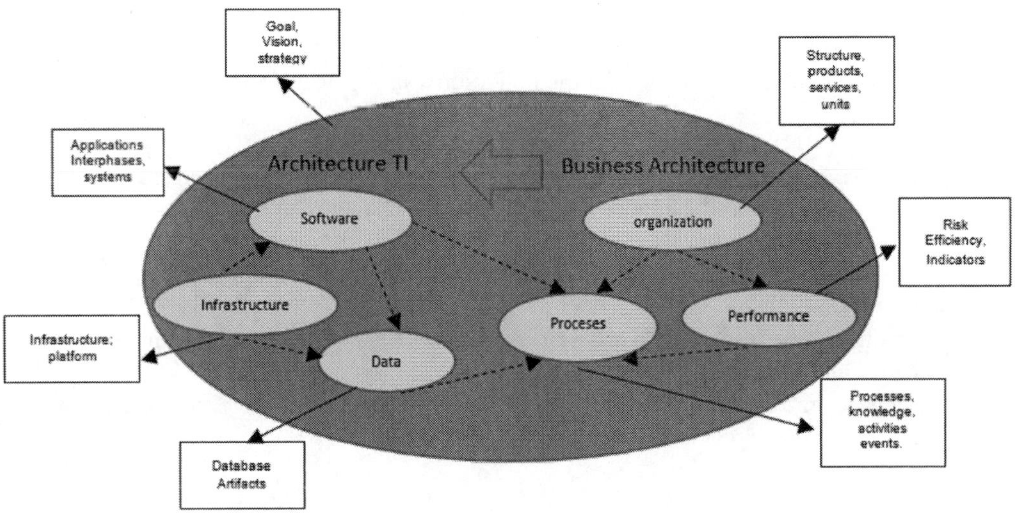

Figure 3. Scope of the different architectures [7]

The TI architecture is composed of three defined architectures: software applications, infrastructure networks, servers and data.

Besides there is a very strong relationship between them .and one depends on the other .business architecture is composed of three architecture or models which are: organizational, performance and, processes, where the lifecycle of architecture starts in the business it's complemented with the technological part [7].

Business architecture is conformed of some business processes (BP) that consists in the logically inter-related arranging of development task in time and space with begging and end, with defined input and output oriented toward the achievement of objective of the business, generating a value result for the owner of the process [8]. In addition, every business process is done through Business Process Model and Notation (BPMN) that consist in a standardized graphic notation that allows modeling of business processes in a workflow format [8].

The BPMN is related to software engineering requirements, since through business modeling we can obtain the requirements of the products, which are defined as the functionalities, features and restrictions that a software must have; and the requirements are defined as all the needs, or wishes ordered by the interested parties and involved in the software [9].

The Open Group gives the following definitions for each.

- Business architecture is the starting point for a software development, it is useful for the development team and analysts understand better the global requirement specification that the future system must satisfy.
- Data architecture is in charge of the interest of the database designers and administrators as well as the systems engineers responsible for developing and integrating the diverse system data components.
- Application architecture defines the necessary application for the management of the business information. The information system must support the availability and maintenance of the integrated information flow through the company, so it can be available in the quantity and quality necessary [10].
- System architecture oversees the buyer's interest of software and hardware, operation staff and system administrator [7].

In general, the Colombian government recommends six business architecture components: strategy, TI government, information systems, technology service, use and appropriation were adopted to describe this architecture.

3. Methods and Materials

For the realization of this work a series of methods, software, tools and materials were used.

3.1. Strategy

From the previous strategies, we can identify the participants, or the ones interested in the legal office their processes missions, and the applications and services that support those products.

- To advise the community in the different areas of the law: criminal, labor, civil, administrative and family law.
- To train the practitioner students who act as attorneys of the cases that arrive to legal office, personal and virtual.
- To connect and communicate the legal office with the urban environment where it operates and offers its academy services.

3.2. Business architecture

The diagram of the urban incident management process will be done as an example which is modeled in BPMN (Business process model notation) in which they involve the captures of an ordered sequences of the activities and information. Modeling a business process implies to represent how the company carries out its central objectives and activities [11]. The diagram shows which activities demand a web interface and the activities of delivering and receiving messages and the decisions made in the process.

The citizens can report urban incidents and problems of different kinds, collecting important information for his management and solutions. Such as geographic and geolocation, typology level of risk, graphs, videos and detailed description of the incident.

The system in turn assigns the incident automatically to a specialized teamwork (teachers and students of the law school) that is, availability to solve the incident, if the assignment is successful, it notifies the team and increases the incidents score. Every member of the teamwork must check the incident and report the relevant and necessary proceedings to solve the urban problem or incident, saving every evidence in the system (Documentary, photographic, videos, etc.) that allows later to do the tracking and traceability about the management of such solutions.

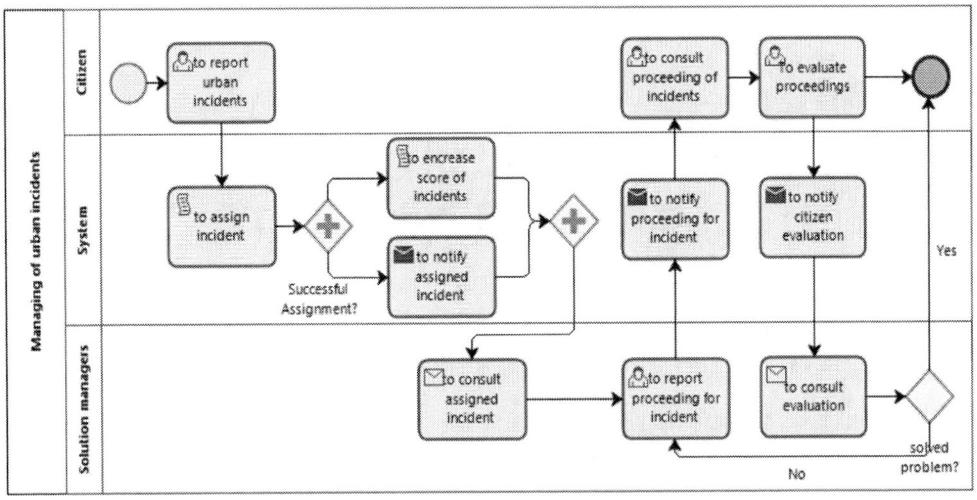

Figure 4. Diagram of the urban incident management process (CIM).

The citizens who report incidents receive notifications in real time about every one of the proceedings that the team-work carries out as an answer to the management done to solve the incident, and besides, they can consult and qualify the management done by the team-work assigned to solve the incident. Finally, the system generates statistical reports with the information generated by the citizen interaction through the legal office, around the problems of incidents that occur inside the urban core of the city, such information is of high importance now of making decisions of planning of the city.

3.3. Information Architecture

In order to represent the necessary information architecture to the legal office, a model entity relationship is represented, where the most important entities and their relationships between them: incidents, users, teams, typologies, incident files suggestions and answers to the suggestions.

4. Results

UniLab is the Digital Services Platform that supports all the mission processes of the Legal Office. It is integrated by the CityApp which is initially developed to be used from mobile devices with Android Operational Systems minimum 2.0 on. Its architecture is based on the integration of multiple articulation and communication online tools, in its great majority based on Restful Web Services technology, including: among them, Storage service on SQL Relational BD, Non SQL, unstructured and based in files, Post delivery service, Push Notification service in real time, geolocation and georeferencing service and integration with social networks.

The Register Incident option allows to report any anomalous or irregular situation that has occurred or potentially to occur within an urban or rural context, which can be classified in several classes or typologies depending on its nature or characteristics. These incidents are traffic problems, road accidents or buildings, with the roads, with street lighting, with the sewer system and public toilet, green zones, with historic cultural patrimony (heritage), safety and delinquency in public spaces, among others.

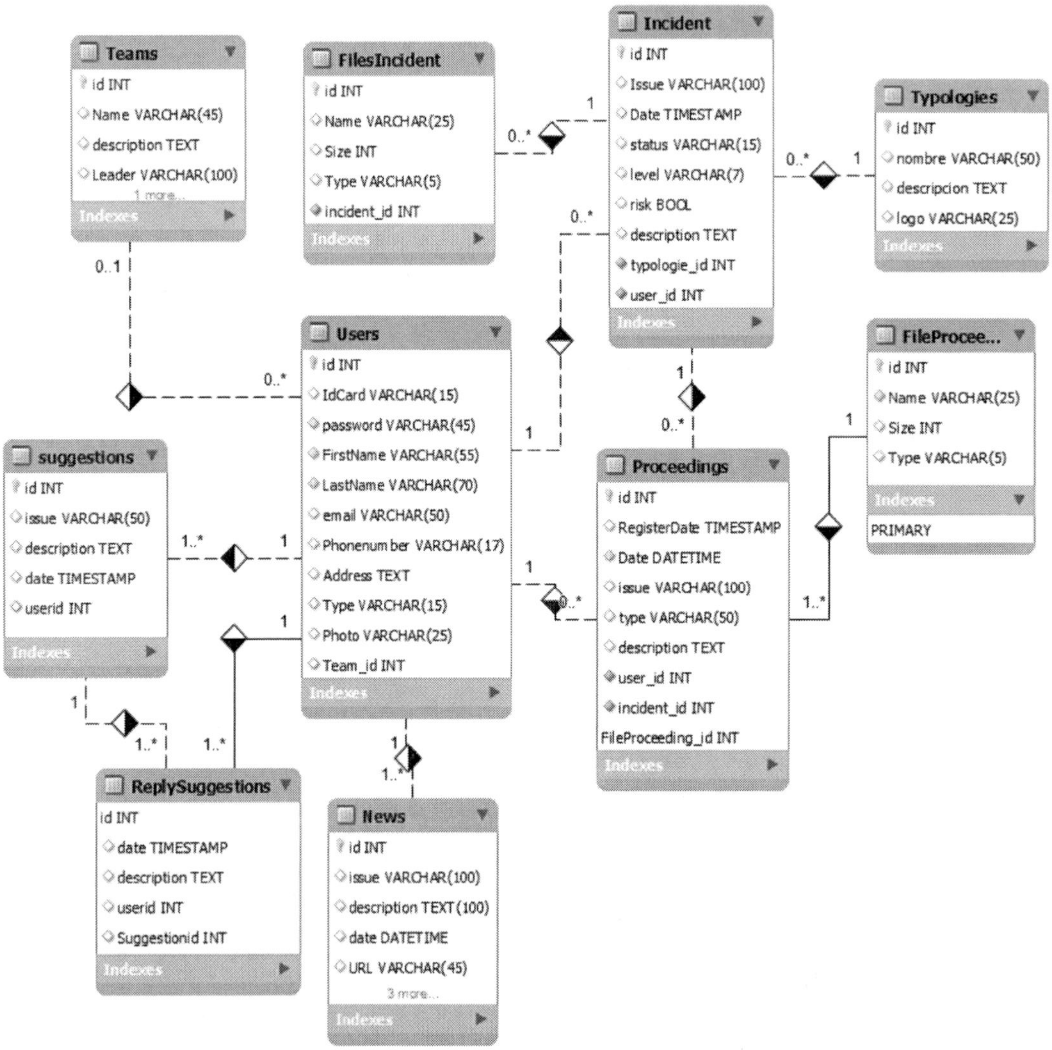

Figure 5. Data Model for Unilab (CIM).

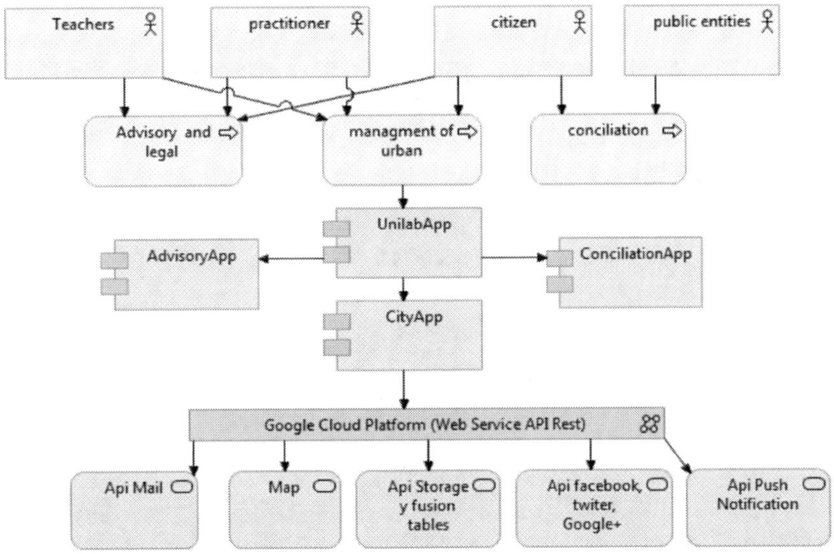

Figure 6. Company architecture legal office (PIM).

These incidents generally provide useful information at the time of carrying out the necessary procedures for their resolution, such as: information on the geographical location, a detailed description, images or videos that allow us to observe in detail and expand the information about the incident.

This information is sent to a work team willing to carry out the necessary measures that allow it to be resolved. If the incident user was reported by an Anonymous user or Citizen user, it must be first reviewed to determine its validity and relevance, this review or evaluation is carried out by user students, community leaders or teachers. In case of not considering that it is not relevant (as it may be false information or SPAM) this incident will be rejected and cannot be visualized in the system by any type of user. If approved, it can be consulted by any user, especially by the users who were assigned the responsibility of making the respective management that allows obtaining a solution.

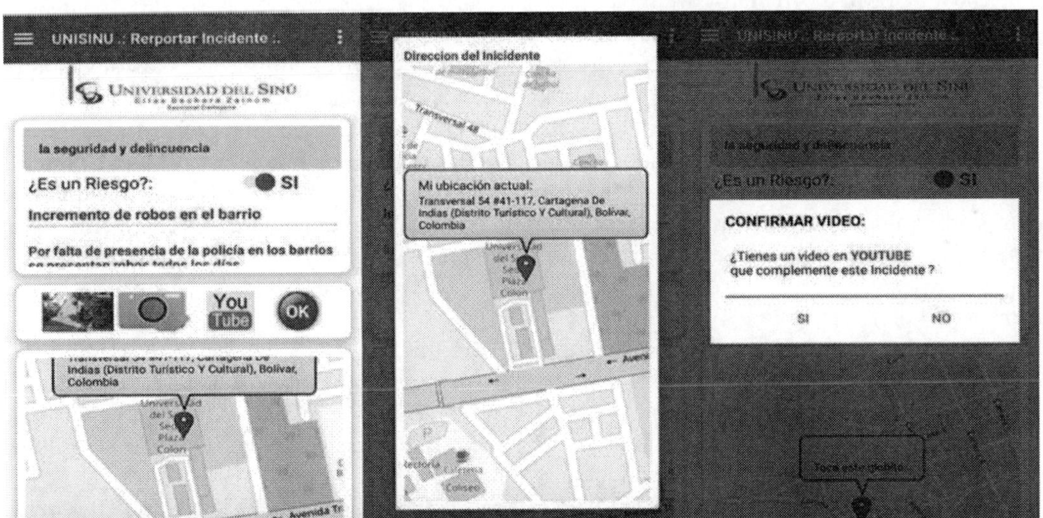

Figure 7. Unilab interface. CIM to PIM transformation.

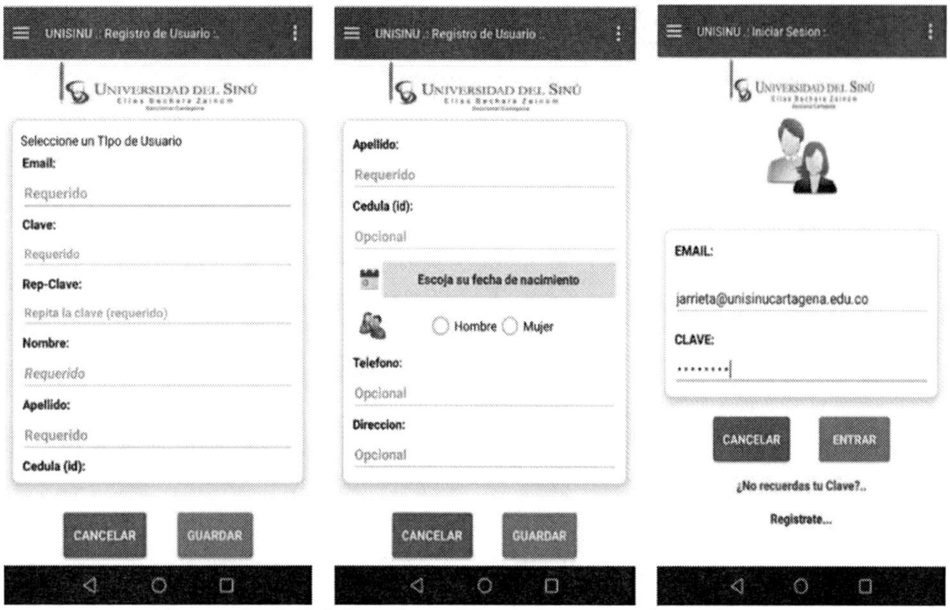

Figure 8. Unilab interface. CIM to PIM transformation.

5. Conclusions and recommendations

The selected architecture for the development of CityApp is through processes by using the notations of BPMN, which permits a standardized graphic notation of workflow. Easy to understand and to communicate among the users of every level of the office. The BPMN that allows to obtain elicitation of requirements in a much easier and efficient way. Through model allowing to pass from CIM to PIM for the development of computer product, using the philosophy of MDA; going through every one of the stages of the lifecycle of the software development from the moment of the capture of the requirement to the deployment of the app.

The development of the Cityapps, makes an integration between the different services, optimizing the resources that they have in the majority of mobile devices, in addition, it takes advantages the massive use at a global level, connectivity, portability, calculation capacity, integration to the context.

From the architecture directed by models, where applied in the development of software the models CIM and PIM, since from PSM cannot be done the transformation from the PIM to a block code (appinventor), because there is not a tool that can permit to generate the block code of the appinventor applications at the moment. It advises that a migration be made up to another tool that permit to apply all the MDA models for the development of a software product.

References

[1] Silega N, Macías D, Febles J and Noguera M. 2014 State of the arbitrary complexity and architecture Directed by models in the software development in Cuba," Rev. Cuba.sciense of computer , vol. 8, no. 1, pp. 87–96.

[2] Guzmán J, Losavio F and Matteo A. 2013 From Business Model to System Architecture Considering Goals, Aspects and Quality Standards.

[3] Quintero J and Anaya R. 2007 MDA y el papel de los modelos en el proceso de desarrollo de software. Revista EIA No. 8.

[4] Bueno H and Pineda W. 2015 Identification of company architeture principies for the management of impact factors in colombian public entities using TOGAF," INVENTUM,

vol. 10, no. 18, pp. 22–28.

[5] Gutiérrez A, Orantes S and López M. 2009 Arquitecturas empresariales: gestión de procesos de negocio vs. arquitecturas orientadas a servicios ¿se relacionan?, Tecnura, vol. 13, no. 25, pp. 136–144.

[6] Maldonado A and Velázquez A 2006 A method to define the process architecture. p. 12, 2006.

[7] Porras G 2008 Arquitectura Empresarial. Club de Investigaciones Tecnológicas, San José, Costa Rica.

[8] Jiménez S and Puldón J 2009 Procedures to obtain the funtional requirements from the analisys of bussiness processes. Eng. Ind., vol. 30, no. 1.

[9] A. Hossian, 2013 Modelo de Proceso de Conceptualización de Requisitos. Revista Latinoamericana de Ingenieria de Software.

[10] Cabrera A, Carrillo J, Abad M, Jaramillo D and Romero F, 2015 Design of architecture validations of company applications, RISTI - Rev. Ibérica Sist. E Tecnol. Informação, no. SPE4, pp. 79–91.

[11] Hitpass, B 2012 BPM: Business Process Management bases and Concepts of Implementation.

IOP Conf. Series: Materials Science and Engineering **844** (2020) 012063 doi:10.1088/1757-899X/844/1/012063

Discrete Events Simulation Method for Analyze Cycle Time: A Case Study in the Plastics Industry Sector

A Troncoso[1], A Sanchez[1] and J Gonzalez[2]

[1] Department of Industrial Management, Agroindustry and Operations, Universidad de la Costa CUC, Barranquilla, Colombia.

[2] Departament of Energy, Universidad de la Costa CUC, Barranquilla, Colombia.

*Corresponding author e-mail: atroncos1@cuc.edu.co

Abstract. This work shows the potential of the Discrete-Event Simulation -DES- in a software modeling process applied to business decision making, using a descriptive methodology in a productive process of a manufacturing plastic bags company. The analysis identified the process characteristics and built a simulated model in Arena® software. Simulation shown variability in the process greater than 50% in the cycle time, which negatively affects the economic interests of the company, impacting opportunities to supply the market risking its position with other companies in the sector. The method implemented in the present work allows to know the critical points of the process, evaluate them and project a path to obtain the optimal or most viable solution, and propose modifications rising up the probability of success of any improvement plan. This makes any economic investment or locative change a desition-making quantitative supported.

1. Introduction

Globally, the growing scientific and technological development and Free Trade Agreements (FTA) changes, allowed the entry of multinational companies to developing countries. This new scenario generated in some cases absorption of local companies due to a lack of competitive capacity, others closed, and others remain and undergo the transformation, adapting to the new challenges and demands of this increasingly competitive environment. This is why improvements should have, besides technological updates and innovation strategies, but assertive decision making that allows companies to accelerate their processes to become increasingly competitive. In the case of Colombia, according to [1] the increase in companies' productivity, must be one of the fundamental goals of competitiveness and productive development policies. As can be seen in [2], the manufacturing industry at least in Colombia can lose ground against other countries industry, showing the necessity to continue advancing in the strengthening of capacities in innovation and process improvement [3] even more due to events like gross domestic product (GDP) growing 2.2% in the second quarter of the first period of 2018 [4], which shows an alerts that should be attended. According to [5], the plastic sector has been negatively impacted in recent years. This sector has made specific requests to the Colombian government in tax, labor, and energy matters, due to the fierce competition with multinational companies, which have greater financial-economic muscle, better technology, and high-performance processes. These caused that the national companies have been acquired or absorbed by various multinationals, which makes the improvement in the productive and logistic systems of the national companies, a pressing need.

It's important to be aware, that companies must make strategic decisions to address the new market dynamics. This paper shows a case for process improvement decision making data-supported. For this, the production process of a company dedicated to the manufacture of plastic bags (polypropylene) was analyzed. Scientific literature shows that the simulation of discrete events (DES) is widely used quantitative processes in decision making. Works such as those carried out by [6], where a two-level remanufacturing system is modeled; or the one carried out by [7], which through simulation predicts the behavior of inputs and outputs of the material to designing a new warehouse. [8] Used simulation of discrete events to optimize resources in the emergency department in a hospital of the city of Cartagena. Also [9] used simulation of discrete events to assist supply chain managers in making decisions in the analysis of the inventory order history information for three different time periods. In accordance with [10], to carry out the simulation of discrete events it is necessary to perform a series of statistical tests, to establish confidence intervals so that the similarity between the simulated model and reality is identifiable.

2. Methodology

This work was developed following a three-phase methodology. The first phase documented the process, characterizing it and building a database with the results of the process. Second phase designed and evaluated the plastic bag production process through discrete events simulation, taking as a response variable the cycle time of three batches of different sizes. The third phase analyses and process the results for desition making related to raw material supply and resource allocation. An overview of the methodology is presented in Figure 1.

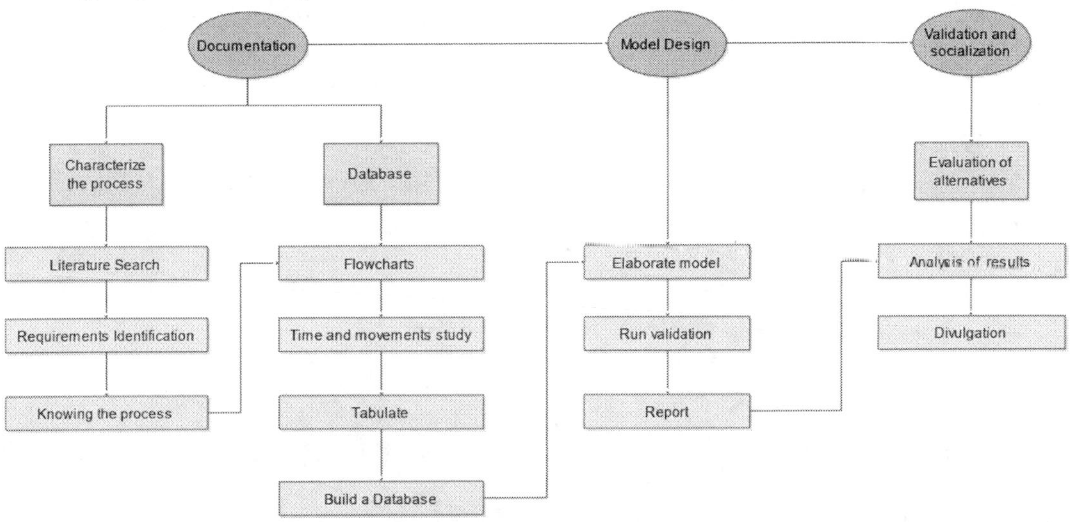

Figure 1. Applied methodology.

Phase 1 (Documentation) described the manufacturing process of plastic bags, in order to be able to characterize the process obtaining accurate and direct information from the personnel responsible for the production. Identifying at the same time the cost parameters for providing the financial analysis that complements the performance diagnosis. Also, data collected related to the variables and parameters in a company database helped to incorporate high-quality information into the simulated model. This information turned in flowcharts and SIPOC (Vendor-Entry-Process-Exit-Client) to perform the detailed model of the real system, following the methodology of [11].

Phase 2 (Model design) examined the data behavior by means of a series of statistical analyses, to determine if these present or don't abnormal system behavior. Then performed a heterogeneity test to check if data was homogeneous or not. In the event that the data is homogeneous, they should be represented by an existing probability distribution, and if not, each data set should be modeled separately. Then, the distribution parameters that should be entered in the simulation model should be calculated. This phase used a process simulation software for a virtual representation of the plastic bag manufacturing process.

Phase 3 (Validation and socialization) verify the reliability of the model. During this phase it performed 10 runs to estimate the size of the computational sample required. Once completed, a hypothesis testing is performed to determine if the simulated model is statistically equivalent to the actual manufacturing system. Finally, identified the critical points, bottlenecks or the rows and delays in each simulated station or process for strategies proposal that give solution to each specific point.

2.1. Case study analize approach

The problem approach selected to this work (Figure 2) was according to the contributions made by [12, 13]. Researchers selected 9 typical steps to analyze the problem through simulation of discrete events, described as next:

1. Problem formulation: includes the definition of the problem and set the objectives of the project.
2. Project planning: includes simulation programming tasks, assignment of roles and responsibilities.
3. System definition: identification of system components, input data, system processes, results, and performance measures.
4. Collection and analysis of input data: a collection of original data, use of existing data and statistical analysis.
5. Translation model: selection of the appropriate simulation, Software and model construction.
6. Verification & validation: Verification ensures that the model, works as intended and validation, determines whether the model represents reality.
7. Experimental design: determines which model alternatives will be beneficial to investigate.
8. Analysis: statistical comparisons between alternatives.
9. Presentation of conclusions and results.

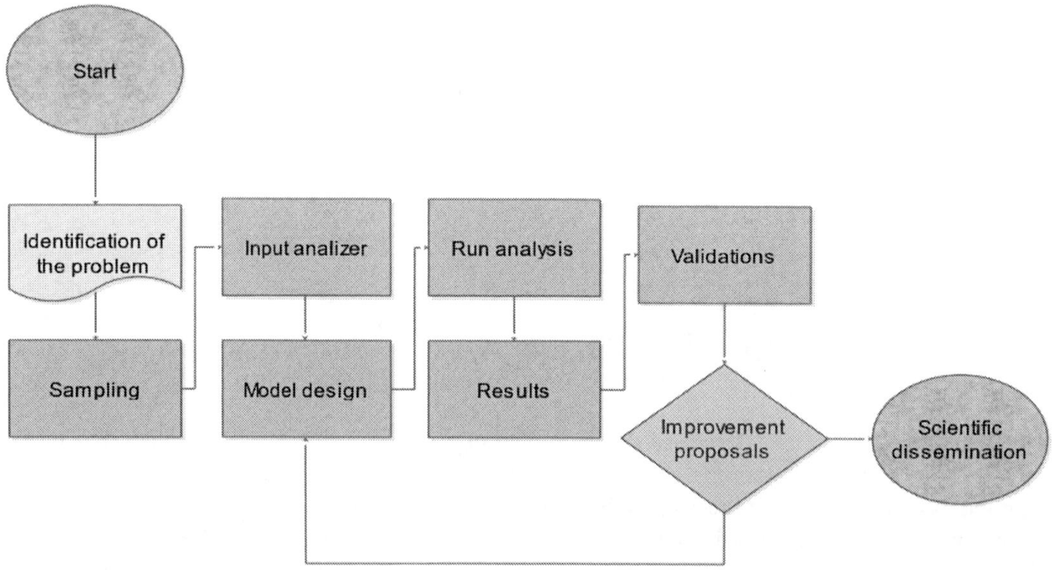

Figure 2. Steps for the development of the simulation model.

3. Case Study

The plastic bag manufacturing process for three types of bags consists of six main stages, namely:

1. Raw material supply process (High-density polyethylene and dyes)

2. Raw material processing
3. Extrusion (4 extruders of equal capacity 50-600 kg are used)
4. Sealing (1 sealer is used for each type of cut (3))
5. Coloration (an operator is used for product design)
6. Storage and dispatch

To identify the parties involved in this Process, researchers used the SIPOC diagram tool to identified input providers to the process, and what are the outputs of the process for customers (Figure 3) After that, the process diagram is carried out (Figure 4).

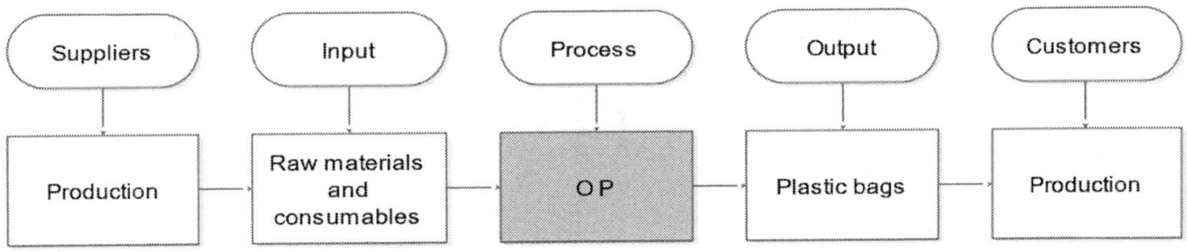

Figure 3. Application of the SIPOC tool

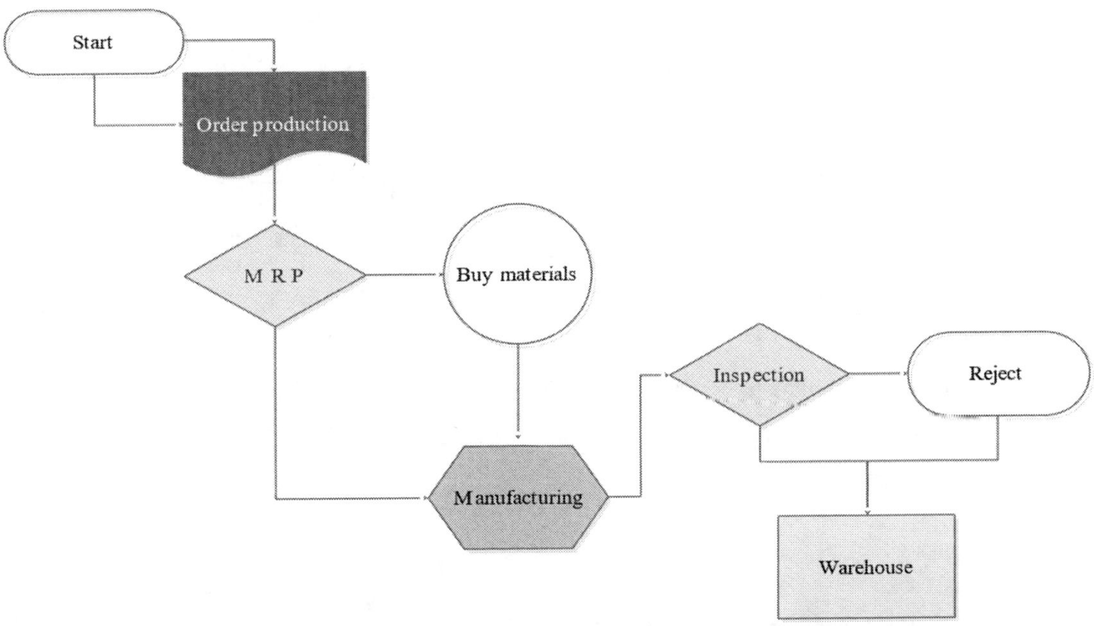

Figure 4. Diagram of the manufacturing process of plastic bags

The construction of the simulation model took into account the steps shown in the previous section: problem identification, data collection, model creation, validation, experimentation, and results. The plastic bag manufacturing process constructed in the Arena® software can be seen in Figure 5. On this, through statistical analysis determine which of the distributions of known probabilities resembles the process inputs associated with each activity. Once the distribution is identified and selected, a simulation of discrete events carried allowed the analysis of the cycle time of manufacturing plastic bags, to infer over it, and thus propose actions in order to improve the performance of the process.

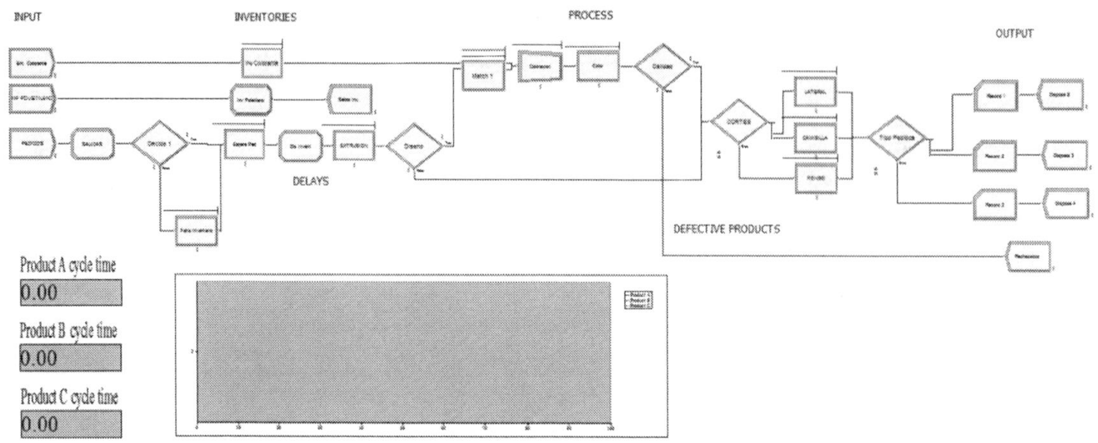

Figure 5. Simulation models of the manufacturing process of plastic bags, in Arena® software.

4. Results and discussions

Once the simulation was carried out, the resources that are critical within the process and the impact on the cycle times were identified, for example, in the manufacturing, the bottom cut activity has an average delay of 1.04 hours, while the shortage impacts on average 7.62 hours the cycle time. The results of the cycle times according to the order types are described below in Table 1:

Table 1. Cycle time by product type

Item	Average	Minimum value	Maximum value
Cycle time Product A	34,1263	33,0682	35,1844
Cycle time Product B	24,900	19,495	41,451
Cycle time Product C	10,961	7,267	14,861

At the same time, the statistic report yielded results about used resources in the process (Figure 6).

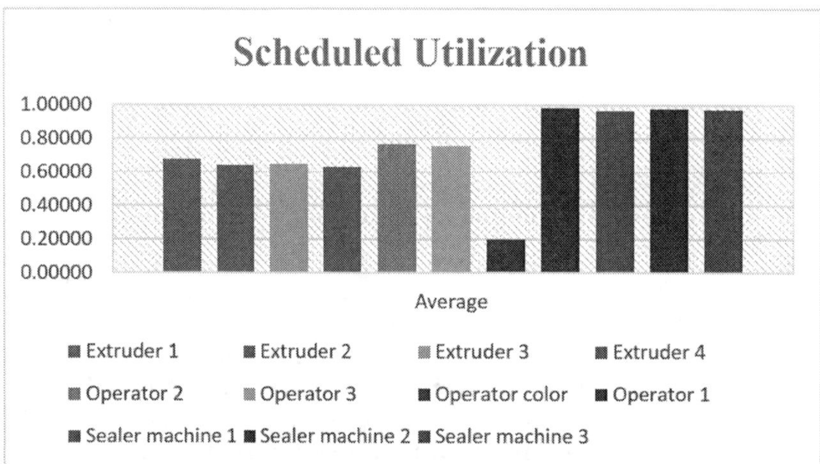

Figure 6. Resource utilization times

Homogeneity of cycle times can be observed in a pie chart shown in Figure 7.

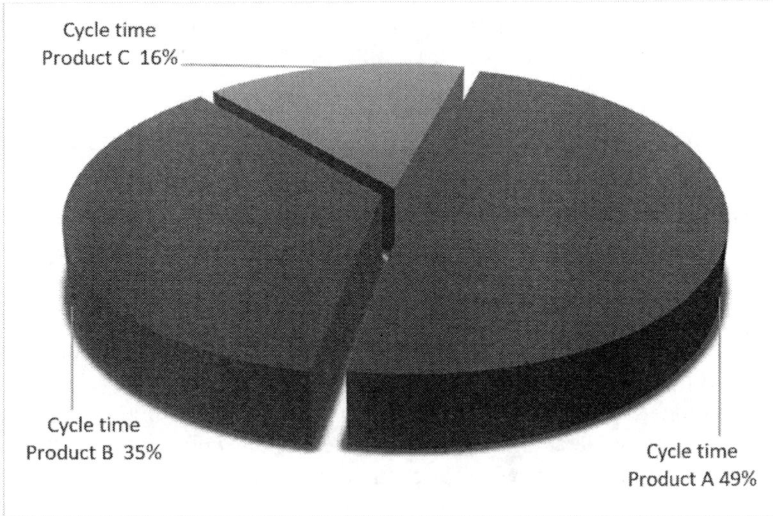

Figure 7. Percentage of time used for each product.

A deep homogeneity test is performed in Minitab® software using a box and whisker graph, to clearly visualize where the average of the data is, and check if the data follows the same sequence.

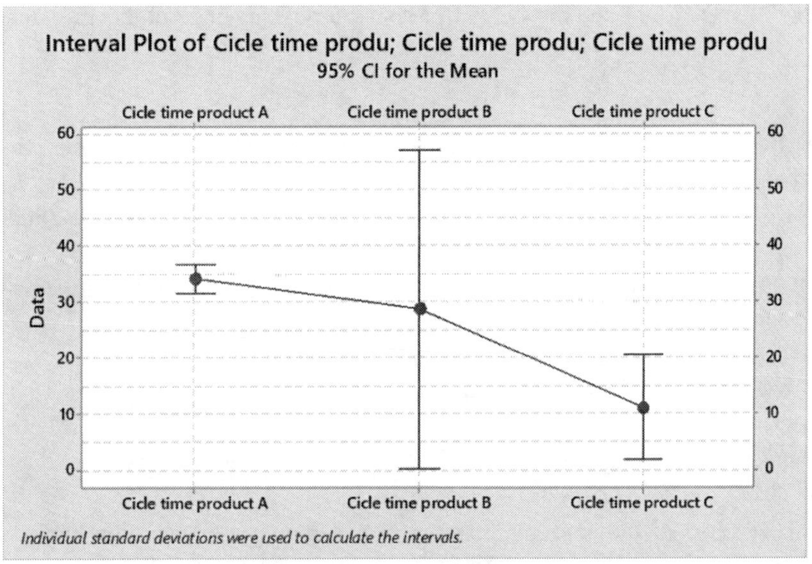

Figure 8. Homogeneity Test

The results are visible in Figure 8 where the high variability of the cycle time of product B is clearly evidenced, which is why this process should be intervened as soon as possible with the intention of reducing the cycle time. As can be seen, Discrete Event Simulation –DES in the Arena ® software, allows complex models analysis, due to its approach to processes with a flowchart type description, providing at the same time statistics data reports output based on different types of distribution, which is great for agile validations, [14, 15, 16]. Mixing a tool and techniques as made in the present work, show how useful is make quantitative process analysis before making any economic investment or location change into a process. Implementing tools for process analyzes techniques like DES allowed to make virtual adjustments and reevaluate it in order to find best or optimal strategies or improvement actions, that allow processes to generate true value to final clients and achieving productivity goals, improving company competitiveness, contributing business sustainability.

5. Conclusions

This case study demonstrated that there is the high potential of Discrete Events Simulation -DES- technique for process improvement decisions, but it should be noted that it is essential to have the truthful set of input data to make process modeling, which will give robustness to support desition making. Once this requirement is met, the descriptive methodology proposed in this article can be followed, using any process modeling tool. In this case study, the Arena® software.

The tool and methodology implemented for DES, allowed to detect in the case study a high variability in the process greater than 50% in the time cycle, which negatively affects the process performance and productivity goals of the company, cutting opportunity to supply the local market or compete with other companies in the sector. Also detected a high variability in the cycle time of one of the three different product of the process. Other advantages of using this technique were: observe the interactions between the units of the system, to know the behavior of the process at a certain moment of time, and estimate the probability of success and the percentage of occupation of the resources involved in the Process. Other advantages of using this technique are the observation of the interaction between the units of the System and gather knowledge of the behavior of the process in a certain moment of time.

Acknowledgements
Author's acknowledgment to the Paperthone-2018 program, also to the research groups: the LEAN DECISION, the PRODUCOM and the GINS, for it support in carrying out this research.

References

[1] Fedesarrollo 2014 Productividad y competitividad del Sistema de Ciudades (Bogotá).

[2] ANDI 2015. [En línea]. Available: http://www.andi.com.co/Uploads/Estrategia%20para%20una%20nueva%20industrializacion.pdf.

[3] ANDI 2017. [En línea]. Available: http://www.andi.com.co/Uploads/estrategia-para-una-nueva-industrializacion-ii.pdf.

[4] DANE 2018 dane.gov.co, Junio 2018. [En línea]. Available: http://www.dane.gov.co/files/investigaciones/boletines/pib/bol_PIB_IItrim18_producion_y_gasto.pdf.

[5] Revista Dinero 2018 Las reformas que pide el sector plástico al presidente Duque, *Industria,* 13 08 2018.

[6] Niranjan S 2017 A study of remanufacturing system in presence of unreliable supply of new products (*Las Vegas, NV, USA: Winter Simulation Conference WSC)* pp. 3459-3470

[7] Alvarez Rangel M A 2017 Simulación de un Sistema de Rollos para el Área de Corte (Lerma, Edo.de México)

[8] Medina P and Arrieta K Simulación de eventos discretos para optimizar recursos restrictivos del servicio de urgencias de un hospital (*Cartagena, Colombia Revista Científica Teknos,* vol. 13, nº 1) pp. 55-66, 2013

[9] Raznick E and Chatfield D C 2017 Information blackout in a multi-echelon supply chain simulation (*Las Vegas, NV, USA: Winter Simulation Conference WSC)*

[10] Herrera O J and Becerra L A 2014 Diseño General de las Etapas de Simulación de Procesos con Énfasis en el análisis de Entrada (Guayaquil Ecuador).

[11] Ortiz Barrios M.A., Escorcia Caballero J., Sánchez Sánchez F. 2015 A Methodology for the Creation of Integrated Service Networks in Outpatient Internal Medicine. In: Bravo J., Hervás R., Villarreal V. (eds) Ambient Intelligence for Health. AmIHEALTH 2015. Lecture Notes in Computer Science, vol 9456. Springer, Cham (Puerto Varas, Chile) pp. 247-257

[12] Shawki K, Kilani K and Gomaa M Analysis of earth-moving systems using discrete-event simulation (*Alexandria Engineering Journal)* pp. 533-540

[13] Troncoso-Palacio A, Neira-Rodado D, Ortíz-Barrios M, Jiménez-Delgado G, Hernández-Palma H 2018 Using Discrete-Event-Simulation for Improving Operational Efficiency in Laboratories: A Case Study in Pharmaceutical Industry. (Shangai China Advances in Swarm Intelligence. ICSI 2018. Lecture Notes in Computer Science, vol 10942. Springer, Cham) pp. 440-451

[14] Kelton W D, Sadowski R P and Sturrock D T 2008 Simulation with Arena (Nueva York: Editorial McGrawHill)

[15] Rabe M and Scheidler A A 2014 An approach for increasing the level of accuracy in Supply Chain simulation by using patterns on input data *(Savannah, GA Winter Simulation Conference WSC)*

[16] Kotachi M 2016 A Modeling and Simulation Platform for Evaluating Optimization Methods in Container Terminals *(Arlington, VA, USA: Winter Simulation Conference)*

IOP Conf. Series: Materials Science and Engineering **844** (2020) 012064 doi:10.1088/1757-899X/844/1/012064

The Reading Plan: A Strategy to support the Academic Performance of students of Systems Technology at Antonio de Arévalo University Foundation

N Sánchez-Castro[1]
[1]Fundación Universitaria Antonio de Arévalo-UNITECNAR-, Cartagena, Colombia

Nully.sanchez@unitecnar.edu.co

Abstract. This article is the result of the research: The reading plan at the Antonio de Arévalo University Foundation: a strategy to support the academic performance of students entering the first semester of higher education. This research is part of the linguistic, psycholinguistic and sociocultural approach of Daniel Cassany achieving results from a diagnostic test of reading levels and is based on three moments; the preparation, execution and evaluation of the project to the group under study. The conclusions indicate that the course oriented from this approach manages to develop reading and writing levels that allowed them to better understand the other subjects.

1. Introduction

The objective of this research was to design a strategy to support academic performance to improve reading comprehension of students of the Systems Technology Program of the Faculty of Engineering Sciences.The proposed reading plan is based on the description of a problem in the institution: the low results of the students in the national Saber Pro test, tests that the government applies to evaluate the quality of the educational service based on the competencies they possess Students related to the understanding, analysis and synthesis of written texts.

At Antonio de Arévalo University Foundation -Unitecnar- to improve reading levels, the development of the critical and analytical capacity of argumentation and the reading habit of students in the institution, is supported by different pedagogical strategies applied by the teacher in the course of Cátedra Tecnarista II. In that sense, the implementation of the reading plan through the Virtual Interactive Learning Platform System -SPLAVIA-, is a new training strategy that leads to the improvement of the academic process. With the purpose of analyze the results of the implementation of the project, a diagnostic test of reading and written composition was carried out, subsequently the investigation took three moments; the preparation, execution and evaluation of the project to the group under study.

Students who enter Unitecnar, come mostly from strata 1,2 and 3 and graduates of schools in these social sectors, for the most part, have a deficiency in reading and writing, To overcome these deficiencies, the institution guides its teachers so that in their teaching-learning process they rely on pedagogical strategies based on more structured and complex discursive genres based on the composition of essays, articles, reports, where the student is encouraged to take a critical stance in front of a text and appropriate the contents. Therefore, the academic process must develop a particular intellectual and social capacity that affects their understanding and writing.

For its part, the national government through the Ministry of Education has been implementing since 2010 the National Reading and Writing Plan, which contemplates in Reading is my story, that reading and writing are present in learning in all areas of knowledge.To achieve this, from the PNLE (National Reading and Writing Plan) reading and writing are appreciated as practices that are related to cognitive processes that are also closely related to the social dimensions of their use and as a basic right to which they must have access all the citizens. In this way, the PNLE contemplates three basic dimensions of reading and writing: its status as a right to have access to the exercise of full citizenship, its status as a tool for social insertion and the considerations of the cognitive field [1]. Despite the programs that the ministry has been implementing, these initiatives are not yet reflected in the skills of reading, writing and reading comprehension in student performance.

Educational institutions know and openly acknowledge that reading is the basis of knowledge. However, some have trouble finding the right way to overcome this low level of reader.

In this order of ideas, it is emphasized that the importance of this project lies in strengthening reading comprehension in the students of the institution, which includes processes of interpretation and development of reading levels. It is important to explain that teaching to read is part of the micro curriculum and the educational programs organized by the researcher. However, different authors have conceived different positions about reading and its conjugation in the student's cognitive process and the context in which learning relationships are established. Larrosa affirms that taking reading as training seriously can be, it seems to me, a way of breaking down borders and a way of affirming the formative and transformative (productive) power of the imagination [2] and focuses its attention on reading as an experience that transcends the reader's imagination in an analysis of disciplines, contexts and actions.

On the other hand, Gómez Soto concludes that there are two causes of the loss of interest in reading in today's society: one is the new leisure model that tends to dispersion, and the conditions imposed by the powers established for the development of culture in society [3].This author considers that reading habits have been lost with the lack of distribution of young people's free time. It is important to clarify the importance of generating spaces for reading and writing texts, not only in the classroom but (outside of it), for the academic and intellectual training of students. In that sense, Carlino, regarding the importance of training and teaching students in the writing of academic genres, affirms that it is common for teachers to insist that students should read, but it is more rare that we think they should also write (outside the exams) to learn our subjects. Writing is more careless than reading. And yet, writing is one of the most formative intellectual activities that exist. [4].

The research is based on the so-called linguistic, psycholinguistic and sociocultural approach, which for Cassany are not three ways of reading, but only three representations about reading, namely: linguistic, psycholinguistic and sociocultural.These approaches contribute to develop critical reading processes in students. In this way, those who are in the linguistic dimension, in this one the reader is able to recognize the explicit concepts in the text, this allows them to rank and organize ideas in the process of preparing a synthesis [5]. Similarly, to understand reading from the approach of psycholinguistics, Cassany affirms that reading does not only require knowing the combinatorial units and rules of the language. It also requires developing the cognitive skills involved in the act of understanding: provide prior knowledge, make inferences, formulate hypotheses and know how to verify or reformulate. Which means that, for this reader, the greatest effort is focused on the inference of the overall idea of reading, the communicative purpose, the formulation of hypotheses and verification of it. [6].

Consequently, for this author, Cassany, the sociocultural approach to read is not only a psychobiological process treated with linguistic units and mental abilities, it is also a cultural practice immersed in a particular society with a peculiar tradition, history, customs and forms of communication [7].That is, reading is a linguistic and cognitive process in which the social and cultural relations of the reader are implicit.

2. Reading Levels

Understanding is the process in which a relationship is established between the reader and the text. In this sense, there are three levels of thinking. The first is the literal level, Gordillo and del Pilar Flórez, points out that it focuses on the ideas and information that are explicitly stated in the text for recognition or evocation of facts [8]. It is possible to say that it is the basic level where the reader focuses on recognizing the ideas and information made explicit in text. The second is the inferential Castillo and Cool, as mentioned in Cisneros, Olave and Rojas states that in the inferential reading the possibility of relating information in the text to account for information that does not appear explicitly is explored. This type of reading implies a global understanding of the content of the text as well as the communication situation: to recognize the communicative information that underlies the texts, as well as the interlocutor and / or audience to whom the text is addressed. [9]. In fact we build inferences when we understand through relationships and associations of meanings. We seek relationships that go beyond what has been read, associating what has been read with our previous knowledge, formulating hypotheses and new ideas. The purpose of this level is to draw conclusions.

And the last is the critic Pérez says that the university student, and especially the future teacher, develop and exercise the ability to reason reading, critical, creative and autonomous thinking and the skills to understand, analyze and interpret what he reads, so that you can express and support clearly, orally or in writing, your ideas and opinions [10]. Indeed, it is of great value that the student and the teacher take distance from the content of the text they read, to assume a documented and sustained position. We make judgments about the text read, accept or reject it with arguments. This has an evaluative character where the training of the reader, his criteria and knowledge of what is read is involved.

3. Methodological framework

With the purpose of responding to the general objective, the research carried out falls within the mixed paradigmatic approach, which for Hernández, Fernández and Batista is a process that collects, analyzes and links quantitative and qualitative data in the same study or a series of investigations to respond to a problem [11]. The approach in question has a high qualitative component and the information was made through direct collection.

The type of research is descriptive-comparative. It is descriptive because the results of the diagnostic test were described and analyzed, the study of the results of each level of reading evaluated by academic cut and the survey applied to the focus group. In addition, the process and its results are clearly documentary; longitudinal, which is carried out through contracted measurements, with regard to the first, second and third academic period. The measurements must be periodic, trend and improvement options are established.

And it is comparative, because based on the information collected, the results of the implementation of the project in the students in the first period 2019 were compared, in order to measure the degree of reading comprehension. For this, an academic block of reading plan was designed on the SPLAVIA platform.

3.1. Instruments

The diagnostic test was designed to identify the level of reading that students present at the time of starting the course of Cátedra Tecnarista II. It is structured by eight questions (two literal questions, three inferential questions and three critical questions). The reading and writing test was applied to a group of 31 first semester students Technology in Systems in the first period of 2019. The Test was conducted individually through an argumentative text called "Sign of the Times" by Joao Anzanello Carrascoza. At the end of the semester a survey was prepared with the purpose of describing the expectations and opinions that students have when finishing the implementation of the reading plan. The instrument is structured by nine questions, students were asked to answer the questions as honestly as possible. The time to answer the survey was one hour.

IOP Conf. Series: Materials Science and Engineering **844** (2020) 012064 doi:10.1088/1757-899X/844/1/012064

3.2 Methodological strategies
The methodological deployment was approached from concrete actions, each of them responds to one of the objectives set. With respect to the first objective, the researcher carried out a diagnostic test of reading and written composition to identify the level of Interpretation and development of reading skills in the students of the Antonio de Arévalo University Foundation.

The design of the reading plan was developed in order to demonstrate the Second objective; after that to set the third and fourth objective, three phases were completed:
3.2.1.Awareness phase. Dissemination of the proposal of the Reading Plan to the directors of the institution, socialization to the students the training strategies framed in the reading plan that lead to the improvement of the teaching - learning process.

3.2.2. Structure phase. The reading plan will have an evaluation percentage within the evaluation process of the subject of Cátedra Tecnarista II, which will depend on the autonomy of the researcher and will be subject to their criteria, taking into account that this will apply by academic cut. The reading plan will have formats which show the different moments in which this strategy is implemented in the teaching - learning process. (Follow-up format Reader Plan).

3.2.3. Operationalization phase. In this phase the strategy of the reading plan is developed through the SPLAVIA platform in which the reading activity is designed, the competencies of the subject (interpretive, propositive and argumentative) and the level of reading to be strengthened in students by academic cut (literal, inferential and critical) in the development of this activity (see figure 1). For that, the researcher must fill in the rubric for the evaluation of the reading plan in her subject for which she will have the indicated formats, which must be available.

Figure 1. Course of Catedra Tecnarista II.

Therefore, it was necessary to design a block of reading plan in the subject that guides the researcher with methodological and theoretical activities of a linguistic, psycholinguistic and sociocultural approach that allow to remedy the problems that students bring when they enter higher education.

Consequently, in the first, second and third academic period, it is necessary to design activities aimed at developing the literal, inferential and critical levels in the course (see figure 2) requiring that in the first period methodological activities be designed to strengthen the level of literal reading for which students should read the narrative text called "The House Taken" by Julio Cortázar and summarize it, also make a concept map based on the explanation and examples given in class; activity that must be done individually. Regarding the categories of analysis used, the concept, the lexicon and the main idea were incorporated at this level.

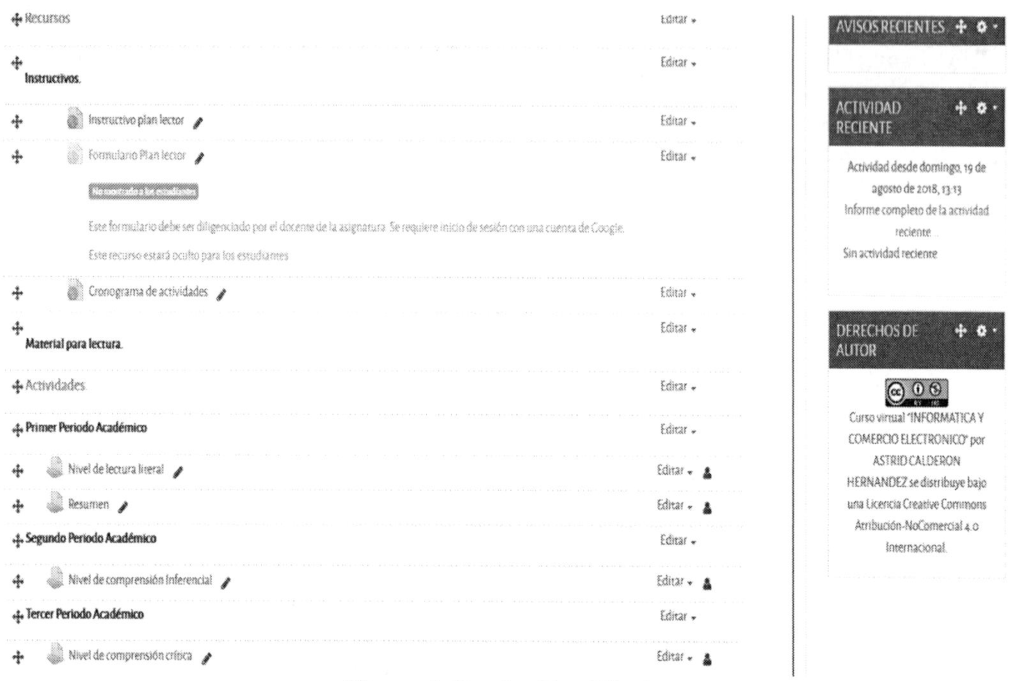

Figure 2. Reader Plan Block.

In the second academic period, the teaching-learning process was based on the level of inferential reading taking into account the local idea, the global idea, the author's purpose and the prediction of the events of Gabriel's "Third Resignation" story García Márquez, to achieve these purposes, students had to read the story and answer the question of inferential level elaborated by the researcher and for that they asked what the author refers to when referring to the third resignation in the story, both in the title and in the content. Also, to determine their level of written composition, they were asked to write some paragraphs of argumentative typology. They were also asked to write their personal opinion about the story.

Similarly, in the third academic period the level of critical reading was developed, the interest in addressing this level, was to analyze the ability to do essays and for the sustained and argued position in the issuance of value judgments on the story called "In the Right Hand of God the Father" by Tomás Carrasquilla.

The texts selected for the approach to the strategy were short stories to make it easier for students to develop reading skills at each reading level. In general, the activities were done in class and later the students uploaded it to the SPLAVIA platform for evaluation and qualification.

The fifth objective will take place by comparing and analyzing the results obtained with the focus group.

4. Results and Discussion

The results of the diagnostic test of reading and writing indicate that in the three competences that are evaluated, 55% of the students are located in the literal level are at the minimum level, 32% at an insufficient level and 4 % at a satisfactory level. Students recognize the local and global ideas of the text but have difficulty assuming a critical position towards reading. (See Figure 3)

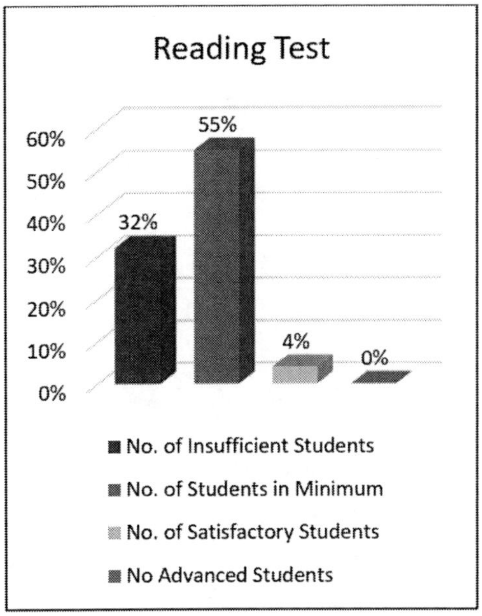

Figure 3. Reading test performance level.

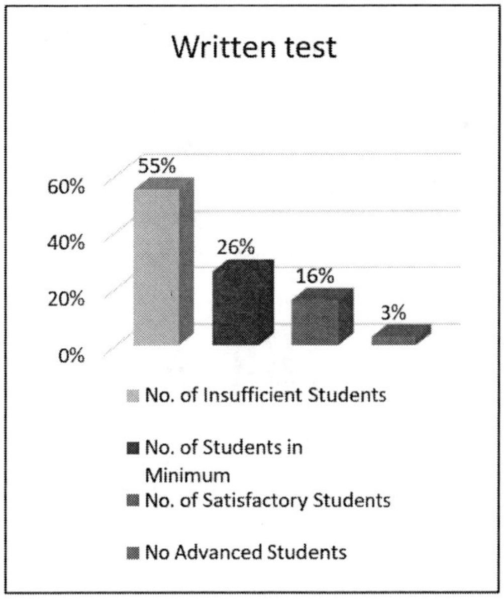

Figure 4. Written test performance level.

While proof of written composition, (See Figure 4) in relation to an inferential and critical level, 55% of the students are at an insufficient level; 26% at a minimum level, 16% at a high level and 3% at an advanced level. Students do not understand the information that is implicit in the text, for this reason they find it difficult to assume a critical posture and defend their arguments against reading.

The second moment of this investigation corresponds to the level of literal reading, (See Figure 5 y 6.) the 31 first semester students of Technology in Systems of the Faculty of Engineering Sciences, made conceptual maps and summaries of paraphrase concerning the level in question. The research showed that at the literal level, 90% and 84% of the students are at a minimum level, 10% and 13% at an insufficient level. Students recognize some important concepts in reading, what is difficult for them is to paraphrase what the text says.

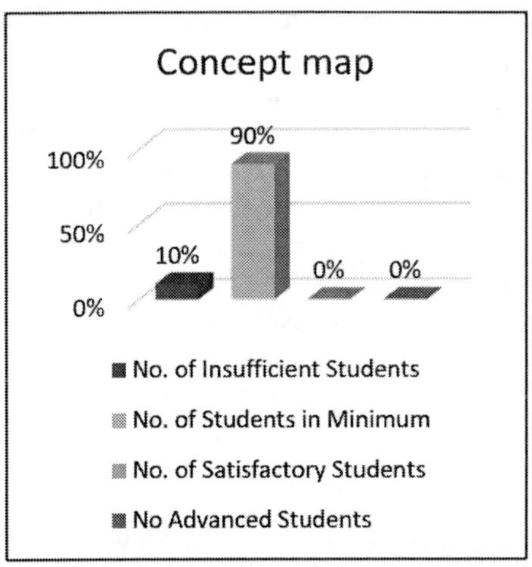

Figure 5. Literal level concept map.

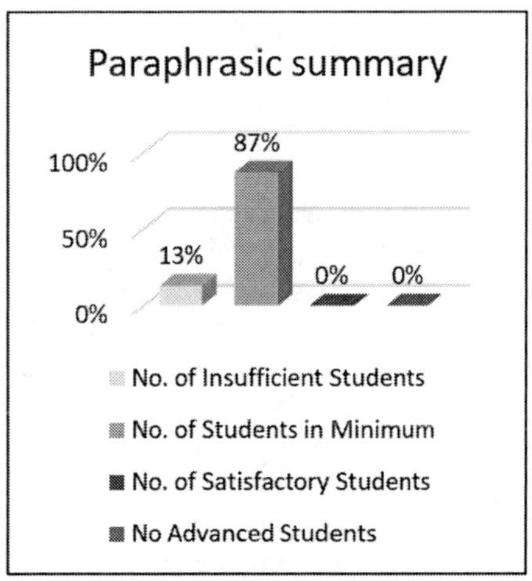

Figure 6. Literal level Paraphrasic summary.

In the third moment, the inferential reading level, 87% of the students are at a minimum level, 10% at a satisfactory level, and 3% at an insufficient level. Students infer the overall idea of reading, the subject and the communicative purpose of the author, but his writing is not organized, because it presents problems of coherence and cohesion that affects the meaning of the text. (See Figure 7.)

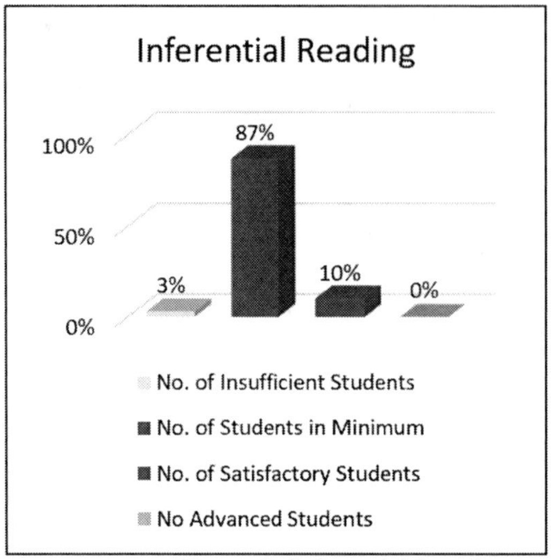

Figure 7. Inferential Level

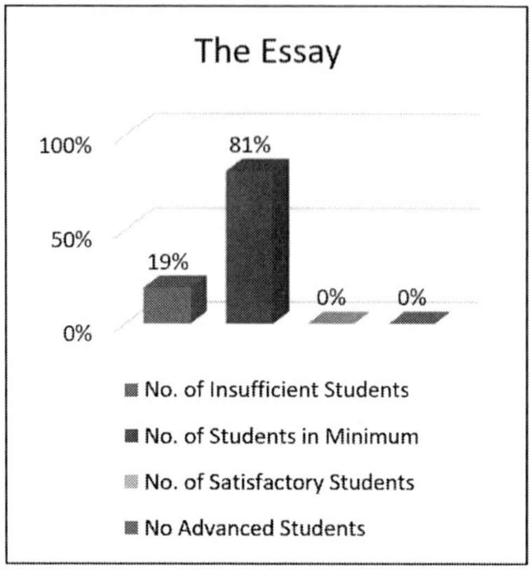

Figure 8. Critical reading level-the essay

On the other hand, with respect to the level of critical reading, (see Figure 8) 81% of the students are at a minimum level and 19% at an insufficient level. Students take a personal position on the subject, but it lacks the argument to support and defend their ideas. Also, his writing is not organized, because it presents problems of coherence and cohesion that affects the meaning of the essay.

Comparing the results by academic cut, the highest percentage of students are located at the literal level since they identify the information explicitly presented in the text. Similarly, respect the meaning of the text you read as it recognizes important information in reading. He also explains the reading in his own words to reconstruct the information contained in the text using different vocabulary, phrases or sentences, but equivalent in meaning. Although, the percentages of this level are not advanced, since students do not know some synonyms in the reading which makes it difficult for them to paraphrase what the text says.

Likewise, analyzing the results of the inferential level, a minimum achievement is evidenced because some students find it difficult to infer the information that is implicit in the text; surely due to lack of application of cognitive and metacognitive strategies consciously in the elaboration of sentences. On the other hand, the analysis of the essays indicates that, at the level of critical reading, the readers reach a minimum level, since they do not go to intertextuality, that is to say, they have difficulty in relating the content of the text with others that complement it. In this sense, readers partially understand the content of the text and support their ideas in the essay.

At the end of the course of the reading plan, a survey was carried out on 31 students of Technology in Systems, of which the following results were obtained, 84% students said they like to read literary, journalistic and academic texts and 16% do not like to read. Now, with respect to writing, 90% of students do not like to write and 10% indicate the pleasure of writing stories and reports. When asked how many texts they read in the semester, 65% indicated that they read 3 to 5 texts in the academic field, 44% read 1 to 2 texts in the workplace and intimate, and 7% read 5 to 6 texts in the family field .

At the end of the reading plan the students noticed progress in the level of literal and inferential reading since at the beginning of the course they had difficulties to understand the texts, especially to infer relationships of cause and effect. 65% manifest who has difficulty reading critical level, in this sense, when asked to perform an essay, they presented difficulty in assuming a critical position towards the text read. Consequently, student ratings indicate that 28% have difficulties in the process of written communication in punctuation and coherence, 23% in syntax, 18% in cohesion and accent, 16% in grammatical referents and connectors and 15% in communicative purpose, semantic and pragmatic resources.

The survey also obtained the following results that can be seen in the following figures. The evaluation carried out by the students in figure 9, indicate the expectations that students have with the implementation of the reading plan in the course, 41% indicate the ability to read critically, 47% clearly express their ideas through the writing, 6% in learning the contents of the subject and the other 6% personal improvement. Regarding the opinion of the students about the implementation of the reading plan at the end of the course in Figure 10, 48% consider that I create a reading habit, 32% indicate that it served to learn the contents and 19% improvement personal.

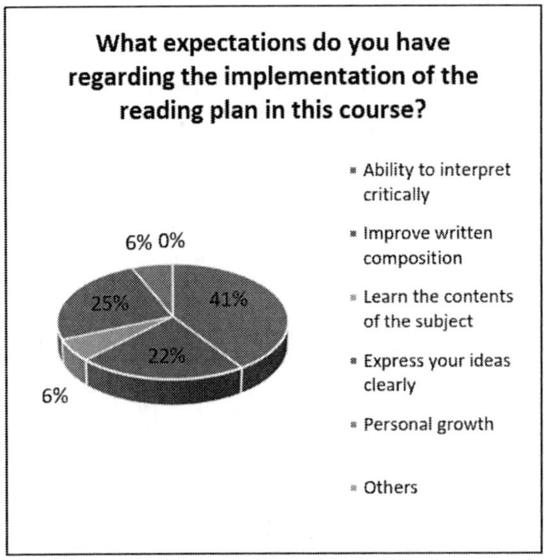

Figure 9. Expectation with the implementation.

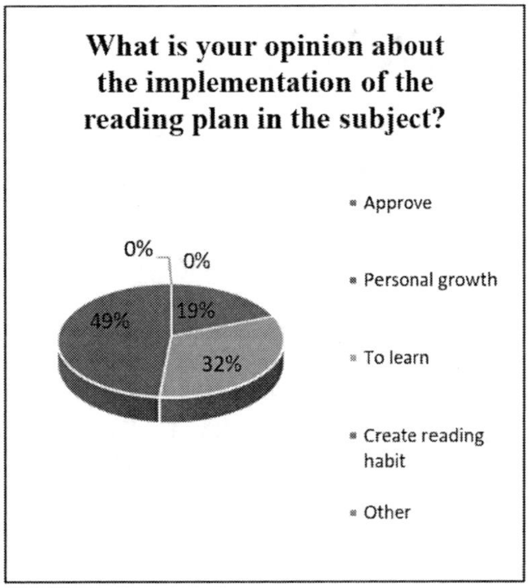

Figure 10. Opinion of the implementation.

Regarding the results of the survey at the end of the semester, it was carried out in order to evaluate the changes in the levels of reading and writing of the students through the implementation of the reading plan, 100% state that there were changes in the interpretation and understanding of texts and in the elaboration of his writings. (See Figure 11.)

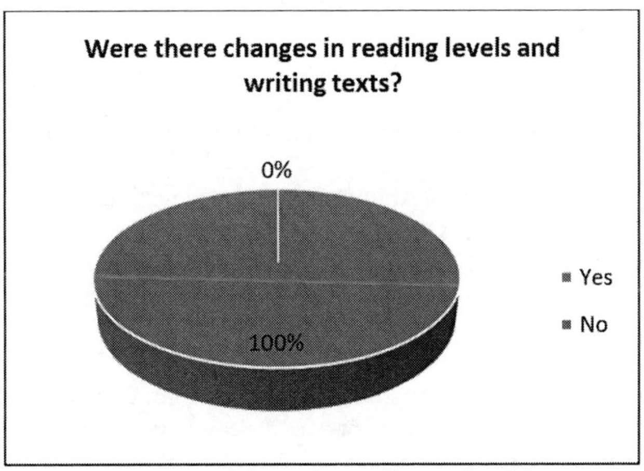

Figure 11. Changes in reading and writing.

5. Conclusion

In consideration of the results obtained in this research, it is necessary to continue with the implementation of the proposed reading plan aimed at training students of Systems Technology in reading and writing considering the diagnostic test, the analysis of the results implemented in each academic period and as expressed by the students in the survey:

- The development of skills and abilities in this course was allowing him to better understand the other subjects.
- At the end of the semester, the level of literal, inferential and critical reading was carried out at a minimum or acceptable level, this analysis was based on the reasons indicated above; they do not know some synonyms in the reading which makes it difficult for them to paraphrase what the text says. Similarly, analyzing the results of the inference level, a minimum achievement is evident because some students have failures to infer the information that is implicit in the reading; On the other hand, they take a personal position on the subject, but lack an argument to support and defend their ideas in the text.
- Likewise, when they were required to make a concept map, paraphrastic summary, paragraph of argumentative typology and essay, they performed it taking into account the linguistic, psycholinguistic and sociocultural approach without any inconvenience.
- On the other hand, The reading plan implemented through the Virtual Interactive Learning Platform System -SPLAVIA- in the subject of Cátedra Tecnarista II, It was a technological aid that facilitated the correction and evaluation of the exercises.

Therefore, the feedback from the workshops allowed the application of linguistic and psycholinguistic strategies leading to the production and interpretation of texts, the productive interaction between the teacher and the student. In this sense, it is important to consider the important and transformative role that the teacher plays as a counselor, motivator of reading and writing in the integral formation of students. Hence, the students are the reflection of the teacher's training, since they develop the skills and competencies that the teacher imparts in their pedagogical work.

References

[1] Ministry of National Education (PNLE) 2010 Colombia learns, the knowledge network. Retrieved on July 16, 2019. http://aprende.colombiaaprende.edu.co/es/content/leer-es-mi-cuento/w3-article-317423.php
[2] Larrosa, J. 1998. The Reading Experience. Literature studies and training. Editorial Laertes. p.28.
[3] Gómez Soto, I. 2002. Reading habits. Taken in LINUESA, María Clemente. Written Reading Morata editions. Madrid 2004. p. 49. D
[4] Carlino, Paula 2005. Writing, reading, and learning in college. An introduction to the academic literacy. Buenos Aires: Fund of. Economic Culture. p.16
[5] Cassany, D 2013. Behind the lines. About contemporary reading. Barcelona: Anagram. p.25
[6] Cassany, D 2013. Behind the lines. About contemporary reading. Barcelona: Anagram. p. 21-43
[7] Cassany, D 2013 Behind the lines About contemporary reading. Barcelona: Anagrama. p. 38
[8] Gordillo Alfonso A, Del Pilar Flórez, M. 2009. The levels of reading comprehension: towards a research and reflexive statement to improve reading comprehension in university students. Pedagogical News. p.97
[9] Cisneros-Estupiñán M, Olave-Arias, G, Rojas-García, I. 2012. How to improve inferential ability in university students. Education and Educators, vol. 15, not 1, p. 45-61.
[10] Pérez de Pérez, A. 2009. The development of critical reading: A path to media literacy in higher education. Lyrics vol. 51, n. 78, pp.309-356.
[11] Hernández Sampieri R, Fernández collado C, Batista Lucio P. 2006. Investigation methodology. Fourth edition. McGraw-Hill Interamericana Editores, S.A De C.V. Mexico, p. 755

Exploratory Analysis of the Market Trend of Power Transformers in Mexico for Manufacturing Sector for Exportation

M Rivera-Anaya [1], J R Corona-Armenta [2], E Oliva-López [1], F A Domínguez-Pacheco [1], J V Bernal [1]

[1] Department of Systems Engineering, Instituto Politecnico Nacional, Mexico
[2] Department of Organizational Systems, Universidad Autonoma del Estado de Hidalgo, Mexico

mayra_rivera10447@uaeh.edu.mx, m.r283@yahoo.com.mx

Abstract. The electric sector continues in constant development, the requirements that the market demands must be included in the design and manufacturing processes of the components that make up the grid, such as the Power Transformers. The objective of the research presented is to know the general characteristics that impact the design and manufacture of transformers in Mexico, for which, four stages were identified as part of the General Methodology, which includes the niche market, the companies that manufactures in Mexico, the volume of exported transformers, as well as the suppliers and raw materials that are necessary for their production. For which a literary research was carried out, also the consultation in secondary sources, field research in a manufacturing company, and analysis of statistical data through the Minitab program, and a comparison of the bill of materials (BOM) of the transformers with the highest export index. All this will allow us to know, influence and improve the design and manufacturing of transformers.

1. Introduction

The electricity sector is one of the fastest growing markets worldwide, factors such as population growth [1] and technological changes have influence the electricity networks [2].

Countries with the highest growth expectations such as China and the United States demand a future high investment in their infrastructure [3]. Are estimated the requirements of electric power in the United States at a consumption of 3,782,151 GWh, distributed in Residential 36.8%, followed by Commercial and Services 35.4%, Industry 22.4%, Farming 0.8%, Transportation 0.2% and Other Sectors 4.5% [4].

To guarantee efficiency in your electricity network and at the lowest possible cost, it is vitally important to include in the purchase and renovation of your infrastructure reliable components, among the most important is the transformer.

The transformer is considered by Hobday [5], as a complex product, among its most outstanding features are its high costs and customized production, which implies a small number of products manufactured with individual characteristics for each customer, which brings as consequently, the generic design requires a review during delivery to the final consumer [6].

The United States has global suppliers that contribute to meet its demand for Power Transformers (TP), among which companies established in Mexico such as GE / PROLEC, IEM, SIEMENS, WEG [3] stand out.

It is important to highlight the importance of this type of companies, given that 80.6% of the products manufactured in Mexico are sent to the United States, of this figure, 0.7% corresponds to the export of electrical transformers, which translates into 243,978 thousand dollars, according to figures reported by the National Institute of Statistics and Geography (INEGI Acronym in Spanish) [7]

Due to the importance of the TP, the general objective of this article is, to present the results achieved in the research of the transformer market for the manufacturing sector in Mexico; to know the general and material characteristics of power transformers designed and manufactured for export; through the identification of macroeconomic indicators, microeconomic analysis of the product trend and the Bill of Materials (BOM).

The scope of this article is an exploratory type, carried out in Mexico City, considering the description of the general methodology of literary search, field research in a manufacturing company and the analysis of the findings found; given that, in articles related to the manufacture of transformers and official reports issued by the competent agencies, a characterization of the TP manufactured in Mexico for export has not been issued

In the general search methodology, we consulted on pages such as: National Consortium for Scientific and Technological Information Resources (CONRICyt), SPRINGER, IEEE Xplore Digital Library, Electric Power Research, Taylor and Francis, Scopus, ScienceDirect, as well as reports on Organizations like International Energy Agency, OECD, US Department of Energy., The Word Economic Forum, and the Mexican Ministry of Energy and Federal Electricity Commission, National Chamber of Electrical Manufactures (CANAME acronym in Spanish), Directory of Transformer Factories in Mexico, and National Institute of Statistics and Geography (INEGI).

2. Research Development

The General research methodology, is supported by the systems approach, where a set of subjects, objects and concepts are conceptualized, which are related to fulfill a common mission, which allows us to identify a macro system with its different stages to the manufacture of a product, where it is located: Manufactured Niche, Manufactured Company, Manufactured Product, Manufacturing Supplier (NCPS) (See Figure 1)

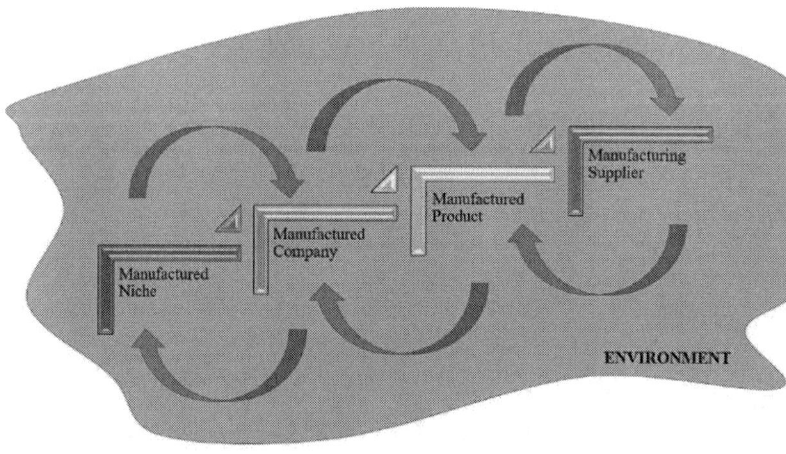

Figure 1. General Research Methodology

This leads us to conceptualize four stages:
A. Identification of the niche market.
B. Identification of transformer manufacturing companies.
C. Identification of products manufactured for export.
D. Identification of manufacturing suppliers.

A. Identification of the market niche.

The identification of the market niche, we carry out through the literary search in official reports such as International Energy Agency, World Economic Forum, among others, this allowed us to generate an article, in which the main socioeconomic trends of the electricity markets were identified, energy sources, generation and consumption capacity of countries with the greatest growth potential worldwide such as China and the United States [14].

This allowed us to know the expectation of growth in future PT requirements for the United States [3], and consider it as the appropriate target market for exports of transformer manufacturing companies and their components.

B. Identification of manufacturing companies.

For the identification of manufacturing companies in Mexico, we consult reports on pages of organizations such as the National Institute of Statistics and Geography (INEGI) [8], National Chamber of Electrical Manufactures (CANAME) [9], Directory of Transformer Factories in Mexico [10], Federal Electricity Commission (CFE), during the period July 2018 to December 2018. The keywords used: Manufacturing, Manufacturer, Company, Transformers; which allowed identifying related criteria such as Place of establishment and Products Offered.

1. Place of Establishment: The results obtained indicate that 63% are located in the center of the country. (See Figure 2)

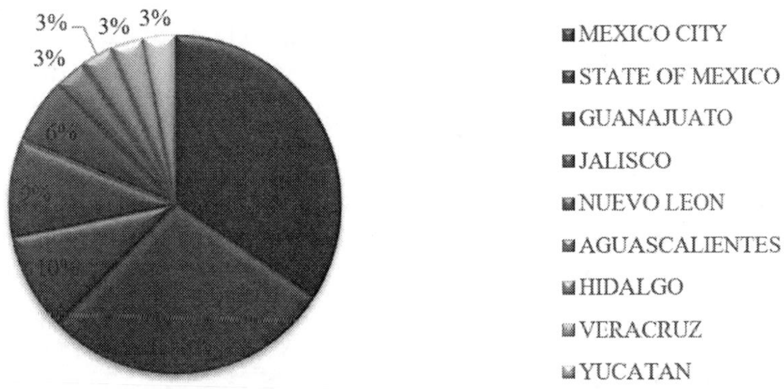

Figure 2. Establishment of Transformer Manufacturers by State

2. Products offered: The results found indicate that the products offered by the manufacturers are: 23 manufacture transformers type Distribution, 4 Distribution and Medium Power, 3 Maintenance Service (Workshops), 1 offers Distribution transformers, Medium Power and Power, and 1 company exports all its manufactured product. (See Figure 6). (The type of Transformer Distribution product, for the purposes of this study, refers to all those with a capacity of less than 10 MVA).

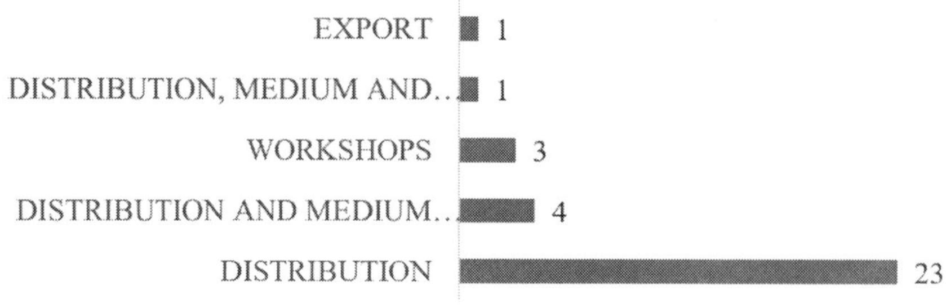

Figure 3. Products offered

C. Identification of products manufactured for export.

For the identification of manufactured products, the stages are required:
C.1 Analysis of sales order records.
C.2 Analysis of the sales trend.
C.3 Exploration of the list of materials, and identification of raw materials.

C.1 Analysis of sales order records.
For this stage, we conducted a field investigation, where we visited the manufacturing company "XYZ" located in the Center of Mexico, which has more than 500 employees, operating for more than 35 years in Mexico, offering products that range from Distribution type (Substation, encapsulated, dry, renewable, pole, with capacities less than 10 MVA), Medium Power and Power Transformers, ranging from 10 MVA to 350 MVA, to know the characteristics of the manufactured TP; for which, the registration of the orders executed during the period from the year 2000 to 2017 was carried out.
Due to the nature of this exploratory study, the collection and processing of the data, existing sources in the current market were considered, as well as some statistical analysis procedures described in the literature [11], considering the review of the orders Registered between 2000 and 2017, where Medium Power and Power Transformers are considered as objects of study, with a capacity of 10 MVA up to 230 MVA.

The total population is limited to 509 orders, where we considered the following seven variables:

1. Destination of the Transformer
2. Type of Client
3. Type of Industry
4. Transformer Capacity (MVA)
5. Transformer Type
6. Single phase - Three phase
7. Voltage Class in High Voltage (KV)

1. Destination of the Transformer: we found that the 95% are exported to United States, the remaining 5% to other countries. (See Figure 4)

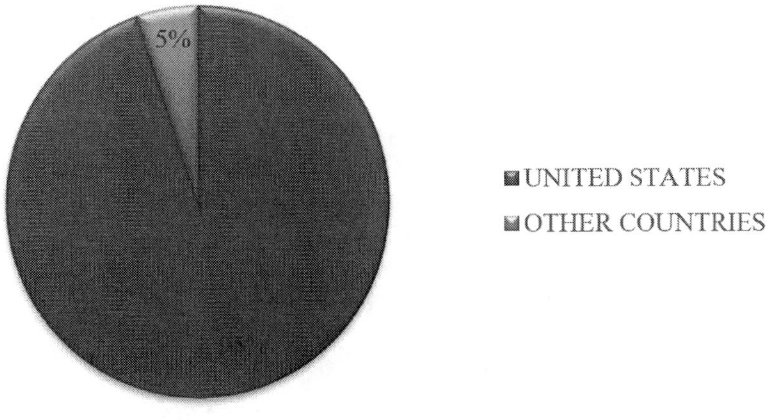

Figure 4. Exports

2. Type of Client: we found that are five types of clients (See Table 1), of which 54% belong to orders from Municipalities, followed by EPC, and IOUs [13] with 15% each. (See Figure 5)

Table 1. Kind of Client

Code	Definition
C&I	Industrial Client
EPC	Engineering Product Client
IOUs	Investor Owner Utilities
Municipals	Public Power Utilities
Renewable	Renewable

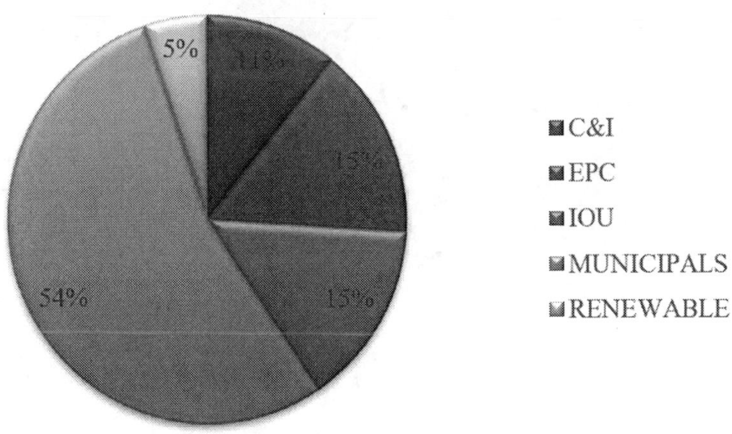

Figure 5. Kind of Client

3. Type of Industry: we found that the record of 8 Types of Industries, in which 70% exceeds that required in Utility, followed by Oil-Gas with 13%. (See Figure 6)

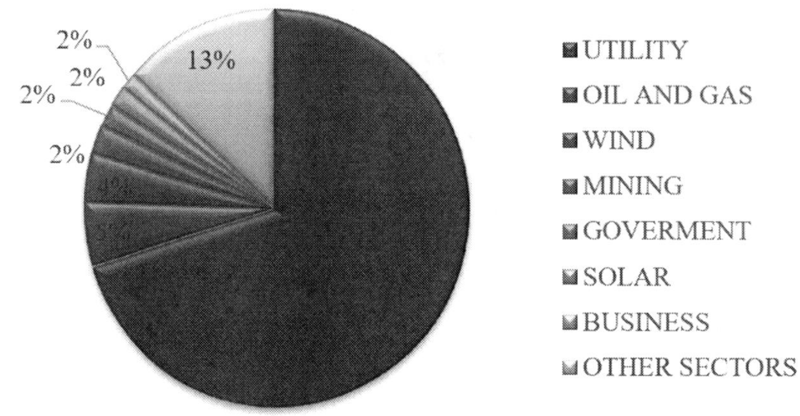

Figure 6. Kind of Industry

4. Capacity (KVA): we found 99 types of capacities, from10 to 224 MVA; of which, the largest number of orders is 20 MVA, followed by 30 and 25 MVA, which represents 45% of the requested orders. The representation of all types, we made consider a range of them, taking in account the ranges established in accordance with the Regulations established by IEEE for the Table of Preferred Ranges [12] (See Figure 7)

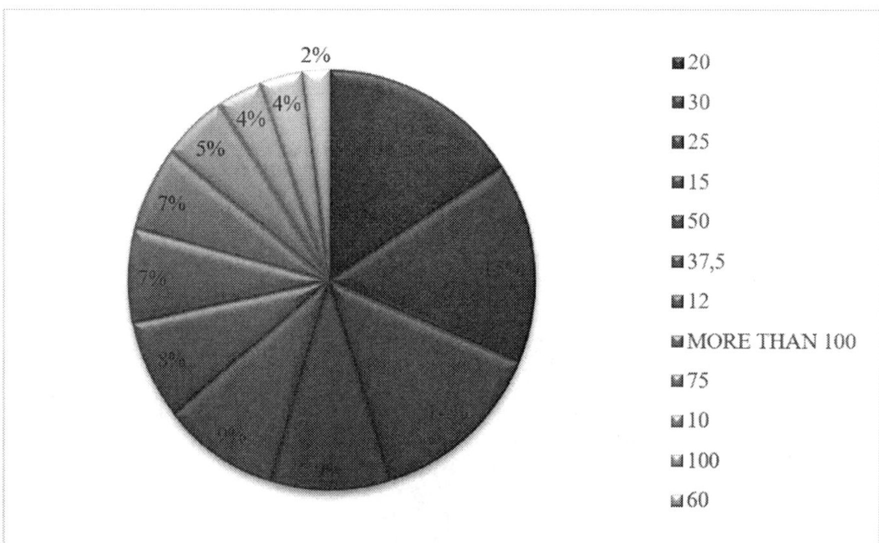

Figure 7. Capacity MVA

5. Transformer Type: we found that 86% are Substation Type and the rest of other Transformers type. (See Figure 8)

IOP Conf. Series: Materials Science and Engineering **844** (2020) 012065 doi:10.1088/1757-899X/844/1/012065

Figure 8. Kind of Transformer

6. Single phase - Three phase: we found that 97% of orders are required with 3 phases and 3% with one. (See Figure 9)

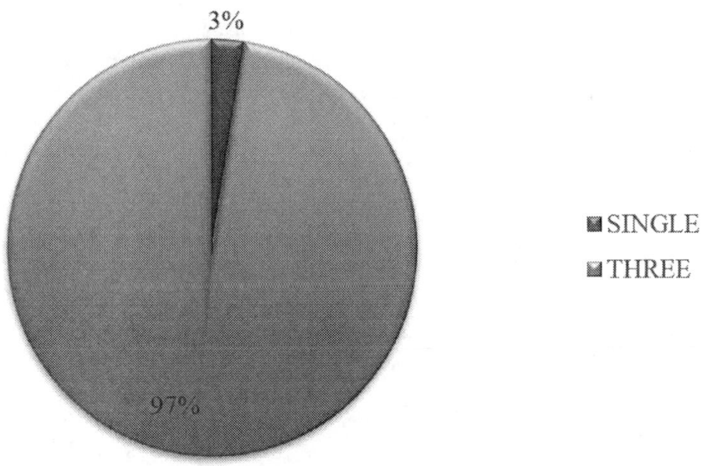

Figure 9. Number of phases

7. Voltage Kind: In this category, kind 138 and 69 KV in High Voltage stand out. (See Figure 10)

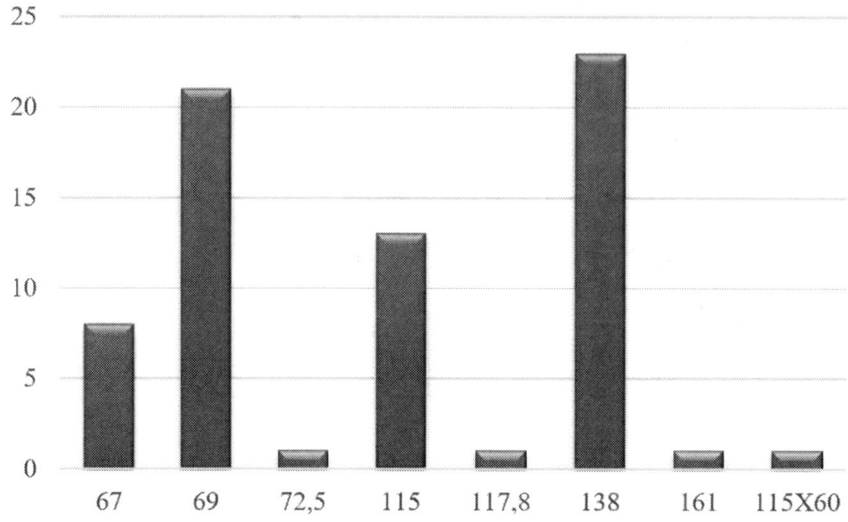

Figure 10. Export by voltage class

C.2 Sales trend analysis.

In the second stage, the analysis of the sales trend is carried out; Table 2 shows the quantities of units sold from the year 2000 to 2017 of kind 138 and 69 KV. It can be seen that, in some years, exports are zero, however, there are other periods in which it increases, this provides an unpredictable variability in demand, so a linear and quadratic trend adjustment is made, which These are the adjustments that can be made with the available data, to observe the behavior over time, as well as a forecast for the next three annual periods.

Table 2. Sold Units

Year	Model 69	Model 138
2000	0	0
2001	7	0
2002	0	0
2003	0	0
2004	9	4
2005	3	0
2006	0	0
2007	1	6
2008	0	6
2009	0	1
2010	0	4
2011	0	1
2012	0	1
2013	0	2
2014	0	1
2015	0	0
2016	1	0

We made adjustments to the sales of Model 138 through Minitab, it can be seen that sales are more adjusted to a linear trend, rather than the quadratic one, due to its average absolute percentage error (MAPE), also known as Deviation Percentage Absolute Average, which measures the accuracy of the method for the adjusted construction of time series values in statistics, is less, 59.7404 of the linear versus 71.7487, as shown in Figures 11 and 12, respectively. The forecasts for the following three periods, 1.75, 1.77 and 1.79, show a slight tendency to generate some export, although its slope, whose value is 0.025, is of minimum growth.

For Model 69, is presented the best fit in the quadratic, considering a lower MAPE value (61.7057) than the linear one (73.4477), which are presented in Figures 13 and 14 although the behavior in both is negative, that is to be considered for future planning. In this case, the forecasts do not indicate exports in the next 3 years, which leaves the propensity to manufacture in a position of extreme fragility. The MAD values, referring to the absolute deviation of the mean, and the MSD, the mean square deviation, in both cases, do not have a significant difference considering a 5% error.

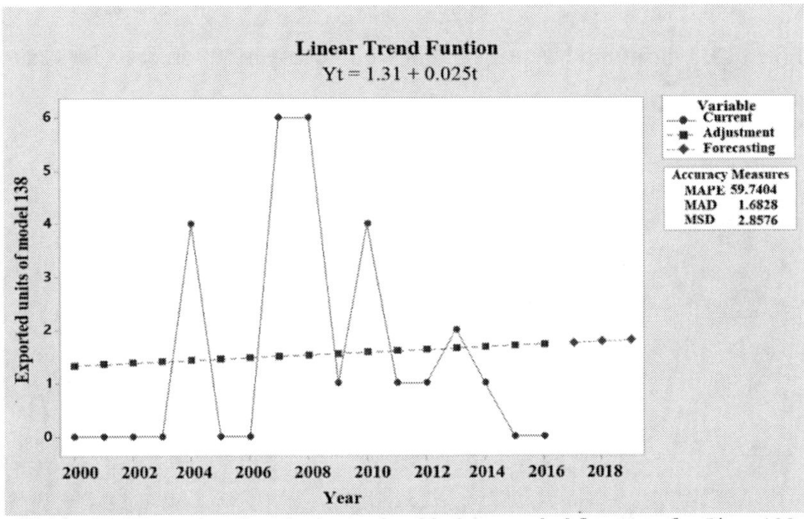

Figure 11. Linear trend analysis graph with three-period forecasts for Class 138

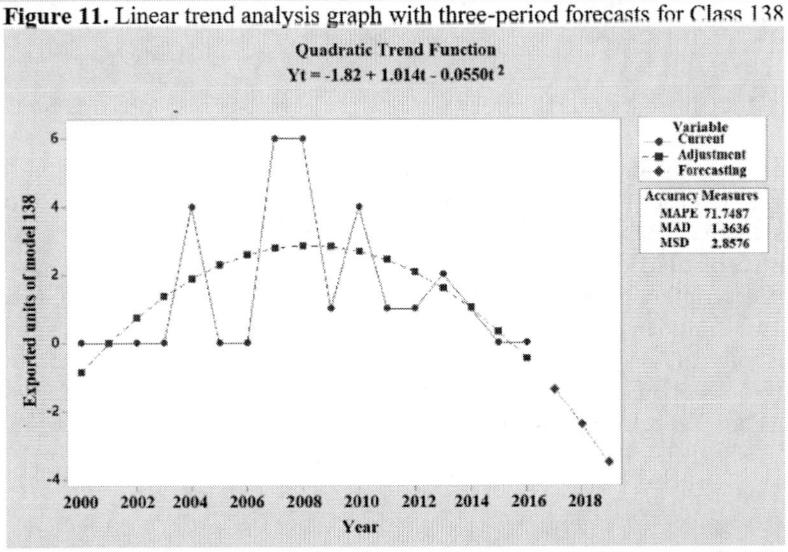

Figure 12. Graph of quadratic trend analysis with three-period forecasts for Class 138

IOP Conf. Series: Materials Science and Engineering **844** (2020) 012065 doi:10.1088/1757-899X/844/1/012065

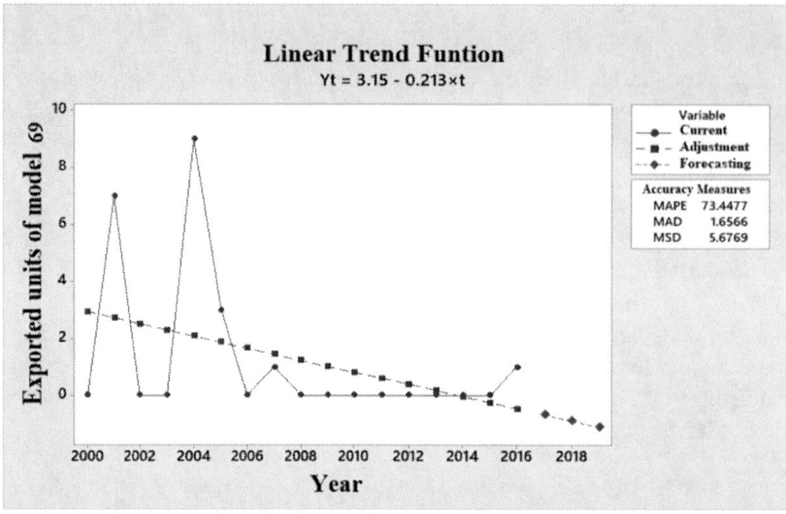

Figure 13. Linear trend analysis graph with three-period forecasts for Class 69

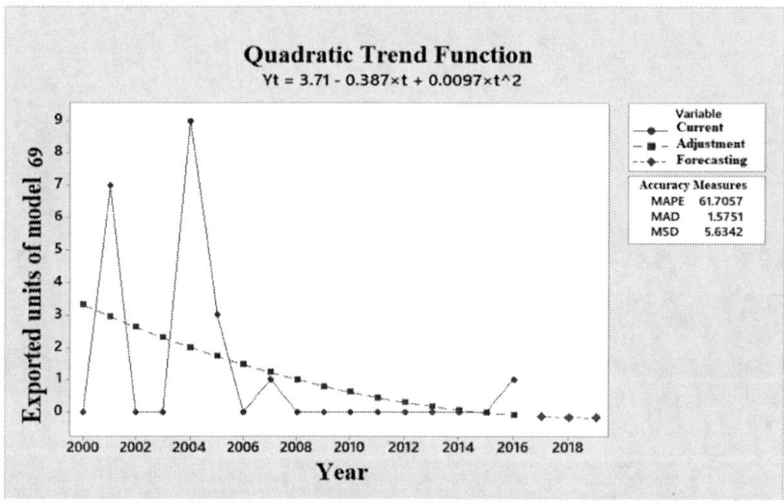

Figure 14. Graph of quadratic trend analysis with three-period forecasts for Class 69

C3. Materials list exploration and identification of raw materials

The exploration of the list of materials and identification of raw materials consisted of comparing the list of materials of three types of medium power transformers, randomly chosen from a population of 26 with characteristics of 20 MVA, Class 138, three phases.

We identify the description of the materials listed in each of the chosen models, (See Table 3, 4, 5), among which, we consider both the basic and raw materials; (oriented grain silicon steel, copper conductors, iron, oil, No Load Tap changer (NLTC), Load Tap changer (LTC) and insulations [3, 14]). This allowed us to count an average of 52 pieces or assemblies in each transformer model subdivided into 6 main parts, the Tank, Core Coil and Connections Assemblies (CCCA), Accessories, Nozzles and Connectors, Miscellaneous, and Shipping assemblies. (See Table 6)

Table 3. Model Standard IEEE 20 MVA Model 138 Kv

Assembly	Model 1
Tank	Steel A36
	L1500 Straight Radiator
	Butterfly Valve PN10 3
	Stainless Steel Conservator Tank
Assembly Core, Coils and Connections	Oriented Grain Silicon Steel Core M4
	Low Voltage Cable CTC 0.48 4.9X1.6 90
	High Voltage Cable 0.72mm 6.8X1.9mm T 90
	Round Copper 3/8X3"
	Thermostabilizer Copper Cable 14 3/0 AWG
	Paperboard PHP 3mmX2100mmX3200mm
	Insulating Paper Thermostabilizer 0.254X12
	Densified Wood 60mm
	No Load Tap Charger 550kV/300
	Paperboard Coils Assembly
	Assembly Iron Core
Accesories	Winding Thermometer
	Thermal plate
	Oil Thermometer
	Oil Gauche 152mm
	Pressure Valve
	Manovacometer
	Inertair Team
	Fan 0.5cv 4.01m^3/s
	Control Cabinet
	Wired Pipe List
	Current Transformer 1200/800/300/200:5
	Current Transformer 1000/5A C400 60hz
	Current Transformer 400/5 A C-400
Bushing and Connectors	Bushing Driver 650kV 800A
	Rigid Bushing Driver 15kV/900A/110kV
	Bushing Assembly
	Mho Connector
Miscellaneous	-40° Naftenic Insulation Oil
	Electronic Plug 1 and 2
	Transportation monitor
Shipment	Treated Wood 2.5X25X250 - 3.8X7.6X250 - 7.6X7.6X250
	1000 Liters Polyethylene Container
	Hard Plastic 1,200X1,050X1,500mm
	Gas Cylinder for High Pressure 9m^3

Table 4. Model Standard IEEE 20 MVA Model 138 Kv

Assembly	Model 2	
Tank	Steel A36	
	L1600 Straight Radiator	
	Butterfly Valve PN10 3	
	Stainless Steel Conservator Tank	
Assembly Core, Coils and Connections	Oriented Grain Silicon Steel Core M4	
	Low Voltage Regulation Cable CTC 0.6 6.2X1.6 90	
	Low Voltage Cable CTC 0.6 4.4X1.6 90	
	High Voltage Cable SRC 22HCC 0.84mm 6.1X2mm T 90	
	Thermostabilizer Copper Cable TMA14 500 MCM	
	Thermostabilizer Copper Cable TMA14 3/0 AWG	
	Paperboard PHP 3mmX2100mmX3200mm	
	Insulating Paper Thermostabilizer 0.254X12	
	Densified Wood 60mm	
	No Load Tap Charger 138kV/538A	
	Load Tap Charger 33 26,4 Kv	
	1.2 MVA Reactor	
	Paperboard Coil Assembly	
	Assembly Iron Core	
Accesories	Windig Thermometer	Pressure Valve
	Thermal Plate	Manovacometer
	Oil Thermometer	Air Dryer
	Oil Gauche 152mm	Inertarir Team
	Fan 0.5cv 4.01m³/s	
	Control Cabinet	Wired Pipe List
	Current Transformer 1200/800/300/200:5	
	Current Transformer 1200/800/300/200:5 A C-400	
	Current Transformer 400/5A C400 60hz	
	Tension Regulator	Adapter Panel 311X317.5X3mm
	Auxiliary Current Transformer	
Bushing and Connectors	Bushing Driver 650kV 800A	
	Rigid Bushing Driver 25kV/1315A/150kV	
	Bushing Assembly	
	Mho Connector	
Miscellaneous	-40° Naftenic Insulation Oil	
	Electronic Plug 1 and 2	Transportation Monitor
	Copper Terminal 375mm² 2 12.7mm	
	Copper Terminal 380mm2 32.9X26mm	
	Pipe Assembly 380mm²	
Shipment	Treated Wood 2.5X25X250 3.8X7.6X250	7.6X7.6X250
	1000 Lts Polyethylene Container	Hard Plastic 1,200x1,050x1,500mm

Table 5. Model Standard IEEE 20 MVA Model 138 Kv

Assembly	Model 3	
Tank	Steel A36	L2000 Radiator
	Butterfly Valve PN10 3	Stainless Steel Conservator Tank
Assembly Core, Coils and Connections	Oriented Grain Silicon Steel Core 23ZDKH85	
	Low Voltage Cable CTC 0.45 4.4X1.6 140	
	High Voltage Cable CTC 0.72 7.2X1.8 90	
	High Regulation Cable CTC 0.72 7.2X1.8 90	
	Thermostabilizer Copper Cable TMA14 600 MCM	
	Thermostabilizer Copper Cable TMA14 1000 MCM	
	Paperboard PHP 3mmX2100mmX3200mm	
	Insulating Paper Thermostabilizer 0.254X12	
	Densified Wood WOOD 60mm	
	Load Tap Charger LTC 72.5-10 19 W	
	Paperboard TX 10mmX2100mmX3200mm	
	Assembly Iron Core	
	Steel Core M4 for Reactor	
Accesories	Oil Thermometer	Relay Locker
	Thermal Plate	Relay
	Oil Thermometer	Rubber Bag 1X1140X2620mm
	Scada Electronic Level	Dehydrator
	Pressure Valve C-5M	Control Cabinet
	Pressure Relay	Wired Pipe List
	Current Transformer 1200/800/300/200:5	Current Transformer 3000/5A C400 60hz
	Current Transformer 1500/5A C400 60hz	Auxiliary Current Transformer
	Backup Control M-0329B BECK (1000400)	Parallelism Module
	Electric Regulator	Panel Adaptation
	Current Loop Internal Beck	Reley
Bushing and Connectors	Bushing 1200 a	Bushing Assembly
	Rigid Bushing Driver 15KV/3000A/110KV	
	Rigid Bushing Driver 15KV/1315A/110KV, 119T	
	Mho Connector	
Miscellaneous	FR3 Envirotemp Insulation Oil	
	Alarm Annunciator	Silica Gel Container 5kg
	Transformer Control	Wheel Assembly
	Transportation Monitor	Neutral Resistance 400 A, Y4.16/2.4 kV 60 seg.
	Copper Terminal 375mm² 2F12.7mm	380mm2 32.9X26mm
	Pipe Assembly 380mm²	
Shipment	Treated Wood 2.5X25X250	3.8X7.6X250 7.6X7.6X250
	1000 Lts Polyethylene Container	High Purity Gas Cylinder
	Hard Plastic 1,200X1,050X1,500mm	Pressurizing Device

IOP Conf. Series: Materials Science and Engineering **844** (2020) 012065 doi:10.1088/1757-899X/844/1/012065

Table 6. Assembly number

Assembly	Assembly - Pieces Number		
	Model 1	Model 2	Model 3
Tank	4	4	4
Assembly Core, Coils and Connections	11	14	13
Accessories	13	17	23
Bushings and Connectors	4	4	5
Miscellaneous	4	7	10
Shipment	6	7	7
Total	42	53	62

After counting the assemblies in each model, those considered critical were identified, this led us to verify that the greatest amount of raw materials are in the CCCA with 50% in Model 1, 50% in Model 2 and 46% in Model 3. (See Table 7)

Table 7. Raw Material

Assembly	Raw Material		
	Model 1	Model 2	Model 3
Tank	3	3	3
Assembly Core, Coils and Connections	10	12	11
Accesories	1	1	0
Bushings and Connectors	3	3	4
Miscellaneous	1	3	3
Shipment	3	3	3
Total	21	25	24

The exploratory study carried out, allows us to observe that the CCCA, composed of critical materials such as oriented grain silicon steel, Insulations, Copper, No Load Tap changer (NLTC), Load Tap changer (LTC), represents the highest percentage in the structure of the materials within the list of materials of the transformer models revised.

D. Identification of manufacturing suppliers.

For the identification of the suppliers of raw materials, we carried out the literary review of official reports issued by government agencies, we found that the US Department of Energy considers as official suppliers of silicon steel countries such as China, Czech Republic, Germany, Japan, Korea, Poland and Russia. [3, 15]

Mexico imports Insulations and Load Tap Changers from the United States and Germany [14], which represents an approximate lead-time of 8 months.

The lack of suppliers in Mexico, of the critical materials for the manufacture of the PT, is a factor that also significantly affect the cost of its manufacture. Therefore, part of the suggestions of the exploratory research carried out is to contextualize the importance of manufacturing companies of this type of materials in Mexico, given the future growth in the US electricity market.

3. Conclusions and Future work

The electricity sector in the United States will continue to grow [14], the result obtained from exploratory research, allows us to infer that, in the coming years, it will increase the export of power transformers for generation and transmission, 20 MVA Class 138 KV for the United States, manufactured in Mexico.

The results obtained from the trend analysis of the time series, sales for Class 69 KV, observe a decrease, however, they could be consider for future sales, when the United States government would authorize investments in Transmission sector.
The growth of the electricity sector and exports to the United States, offer the opportunity to open new distributors or manufacturers of critical materials in Mexico, and thereby reduce imports and manufacturing costs. The realization of a structured questionnaire or interview, to know the conditions by which the three models of transformers analyzed contain differences between their materials, is part of a future research that will allow a better analysis of the raw materials.
The exploratory research carried out, through the systems approach, allows generating a methodology that contributes to the identification of the most outstanding variables for the decision making of companies that intend to export different products manufactured in Mexico.

Acknowledgements
The authors are grateful for the support to the company "Transformers XYZ", for the facilities provided for conducting the research.

References
[1] United Nations 2015, World Population Prospects, p. 1
[2] World Economic Forum 2017, The Future of Electricity New Technologies Transforming the Grid Edge, p. 4-5
[3] U.S. Department of Energy 2014, Infrastructure Security and Energy Restoration, Office of Electricity Delivery and Energy Reliability. Large Power Transformers and the U.S. Electric Grind. (United States), pp. 3-28
[4] Secretaría de Energía 2015, Prospectiva del Sector Eléctrico 2015-2029. (México), p. 36
[5] Hobday M 1998, *Product complexity, innovation and industrial organization*. (England: Research Policy), pp. 689-710
[6] Vedman J & Alblas A 2012, *Managing design variety, process variety and engineering change: a case study of two capital good firms*. (Netherlands: SPRINGER), p.2.
[7] INEGI 2015, Balanza Comercial de Mercancías de México, (México), p. 13
[8] INEGI 2009. "Micro, Pequeña, Mediana y Gran Empresa: Estratificación de los Establecimientos: Censos Económicos 2009". (México)
[9] Cámara Nacional de Manufacturas Eléctricas 2018, Catálogo de Empresas y Productos 2018, (56ª Ed.) (México)
[10] Directorio de Fábricas de Transformadores ubicadas en México. Retrieved: https://www.directoriodefabricas.com/mexico/fabricantes-transformadores-mexico.html
[11] Anderson D, Sweeney D, and Williams T 2008, Statistics for Business and Economics, (United States10 th. Ed. Cengage Learning), p. 5-15
[12] IEEE Power and Energy Society 2010, IEEE Standard for General Requirements for Liquid – Immersed Distribution, Power and Regulating Transformers, C57.12.00. (United States) p.11
[13] U.S. Department of Energy 2015, United States Electricity Industry Primer, (United States) pp. 13, 27-28
[14] Rivera M., et. 2018. *General Context of the International Electricity Market for Power Transformers*. 16th LACCEI International Multi-Conference for Engineering, Education, and Technology: "Innovation in Education and Inclusion", (Peru), Retrieved: http://dx.doi.org/10.18687/LACCEI2018.1.1.335. p. 3
[15] U.S. Department of Commerce 2016, Global Steel Repot, (United States), p.5.
[16] Zaheer H and Saeed S, *Power transformers–global market analysis: Capacity additions and aging infrastructure driving the growth*, Transformers Magazine 5(1) (2018), (United States) p. 62–68.

IOP Conf. Series: Materials Science and Engineering **844** (2020) 012066 doi:10.1088/1757-899X/844/1/012066

Location of Quality Parameters in Small Smart Grid: Off-Grid Case Using Wavelet Transform

L. Rojas[1], C. Rojas[2], C. Pacheco[2], E. Caballero[3], M. Molina[3].

[1]Fundacion Universitaria Tecnológico Comfenalco, Cartagena, Colombia
[2]Universidad de Sucre, Sucre, Colombia
[3]Universidad Libre, Barranquilla, Colombia

[1]rojasml@tecnocomfenalco.edu.co

Abstract. In the following work the analysis made to an electrical signal by means of wavelet toolbox will be shown. The analysis will be done contemplating the use of the Discrete Wavelet Transform (DWT). This will show the multilevel decomposition taking into account different orders of daubechies. Taking into account the different levels it is expected to show the variants in terms of energetic quality such as sag, swell and harmonic detection. The final work will link the work of digital signal processing with previous work where the construction of a prototype solar system will be explained to obtain the electrical signals of voltage, current and power consumed. In this introductory stage will show the step-by-step development that was taken into account to carry out the design and construction of smart grid scale prototype with AC and DC load control through the HMI interface with LabVIEW software. The design process of the off-grid solar system was carried out through 7 stages that framed the scale of the solar installation. Once the design of the sizing was finished, programming was carried out in Human Machine Interface (HMI) graphics environment with LabVIEW software for optimal control of AC and DC variable loads. This control was established through current relays from 5 volts to 110 volts AC and 5 volts to 12 volts DC. The control interface was made through an embedded board with NI-VISA communication. It is important to note that the smart grid control prototype "Smart Grid" shows optimal results in its operation; denoting this that from the optics of the programming done in the graphics environment, the control of loads dependent on control canter's; act at disposition. Redundant this in improvements for the energy consumption derived from the storage system under the autonomy designed for three days.

1. Introduction

Smart grids. Are electrical networks that together can have information, communication and control technologies, in order to interconnect and operate a system efficiently. Inviting system users to be part of the production chain and not just as a consumer. If not as an entity capable of becoming self-generating and thus be able to provide their own energy and if possible their surplus inject them into the network if it is connected to the network. The matrix of change of the technology that the smart grids carry does not only contemplate the fact of connecting, monitoring and having the best domotic for the control through communications. It is also about the change of culture that we must have if we want to

IOP Conf. Series: Materials Science and Engineering **844** (2020) 012066 doi:10.1088/1757-899X/844/1/012066

disconnect from the network and stop consuming through common marketers or distributors. It is a change of mentality that will make you a citizen of the smart grids.

There are many theories, entities and authors that can emit a smart grid concept. According to Hassan Farhangin the electrical power industry is undergoing rapid change and he have talked about the path of smart grid in 2010 [1].

According to Vehbi C. Gungor, et al. The smart grid can be considered as a modern electric power grid infrastructure

for enhanced efficiency and reliability through automated control [2].

According to Xi Fang, Satyajayant Misra, Guoliang Xue and Dejun Yang. The idea was raised that the smart grid was considered as the network of the new generation and that it could use bi-directional flows of energy and information to create an automated electrical network [3].

Authors like Maria Lorena Tuballa and Michael Lochinvar Abundob, in the review of the development of smart grid technologies present an overview of the smart grid with its general features, functionalities and characteristics [4].

Currently Colombia has an energy matrix with generation of hydraulic type, thermal with a.c.p.m, thermal with carbon, thermal with combustoleo, gas, and thermal withJet-A1. _Co- generators with generation based on bagasse, carbon, gas, wind generation and solar generation. Showing a net effective capacity of 16856 MW.

Currently in Colombia the policies and regulatory changes have allowed to advance in types of initiatives that contemplate the use of photovoltaic solar energy. In reference [5], resolution issued by the commission of energy and gas regulation CREG 030 of 2018. By which the activities of small-scale self-generation and distributed generation in the National Interconnected System (SIN) are regulated.

Law 1715 of May 2014 [6]; in this law some guidelines about generalities and definitions are given as follows: In title 1 called "Generalities". And article 3 called "Definitions". It is called a small-scale auto-generator (AGPE); to the system that has power less than or equal to 1 MW. Similarly, the distributed generator (GD) is referred to as the entity that generates electricity near the consumption canter's, and is connected to the Local Distribution System and with installed power less than or equal to 0.1 MW. Article 3 mentions non- conventional sources of renewable energy and calls them (FNCER). In this resolution it is specified in addition in title 2 called "ıntegration to the network of self-generation and distributed generation", and article 5 called "technical standards of system availability at voltage level 1". That the sum of the installed power of the GD and AGPE must be equal to or less than 15 % of the nominal capacity of the circuit. In reference [6], the law 1715 of May 13, 2014 is decreed the law through which the integration of nonconventional renewable energies to the national energy system is regulated.

It is important to note that in Chapter 2, the provisions for the generation of electricity with non-conventional energy sources (FNCE) and the efficient management of energy are dictated. A call is made to Article 8, which specifies the promotion of self-generation to small, large-scale and distributed generation through excessive delivery mechanisms, bidirectional measurement systems, sale of energy by distributed generators, sale of credit energy, mass dissemination programs and targeted outreach programs.

It should be appreciated that the changes in the laws and energy regulations in Colombia are bringing about favourable change towards the initiatives of construction of photovoltaic solar systems disconnected from the network. It is interesting to note that at this time of regulatory changes in favor of non- conventional energies and in order to bring this technology to people and show how it works, the prototype construction of a solar photovoltaic system is presented. Which contemplates within its objectives, to establish a control of variable loads depending on relays of current, activated through programmed pulses with interface of communication between LabVIEW and Maker Hub library.

It is important to highlight that within the optimization of electrical systems, there is always the need to improve the parameters that affect the use of networks. These parameters affect the quality of energy in different scenarios. Such is the case of the scenarios that occur in photovoltaic electrical systems in terms of identification of sag, swell and detection of harmonics.

In this case of study, it is intended to show the analysis made to an electric signal of consumption in spaces from 0 to 1500 minutes. The processing of this signal will be done using the Discrete Wavelet Transform (DWT) and to dynamize this it will use the wavelet toolbox of MATLAB.

The following document will be presented as follows: As a first instance, an introduction will be presented, followed by a methodological stage with the design and calculation of all elements of the solar photovoltaic system at scale. Then we will present the design of the graphic control interface developed in LabVIEW software, presented the front panel and the block diagram. Following this, a description of the stage of physical construction with its parts will be presented.

To link the present work with the construction of the test prototype, an analysis of the discrete wavelet transform (DFT) will be presented, later a test case will be taken, the results will be analyzed and the conclusions will be stablished. Finally, acknowledgments and references will be presented.

2. Methodology design

In this section, the methodological design of this article will be presented.

2.1. Electric consumption

The first stage contemplated in the design of the solar photovoltaic system [7-10] to scale is framed in the dimensioning of the loads that will be part of the prototype. In the dimensioning, AC loads and DC loads were taken into account. For the purposes of designing and starting up the design of the prototype, low-wattage loads were taken into account. See Table 1. See Table 2.

Table 1. Consumption in direct current.

Consumption in direct current DC					
Description	Number	P (W)	Hours / day	Days of use / week	Energy (Wh / week)
Fans	4	1.8	1	7	50.4
Total energy per day					7.20
Total energy DC					50.4

2.2. Reference solar pico hours Formatting author names

Analyzing that, the solar system developed, takes as a base of location, the city of Cartagena in Colombia, it is noted that this city enjoys a high temperature during almost the twelve months of the year; with temperatures that vary from 30 to 40 degrees Celsius. and as it can be evidenced in the reference plane issued by the IDEAM, see Figure. 1, the radiation captured by cities in Colombia and in particular the solar irradiation of the city of Cartagena in the month of March, observing a variation in its space territorial between 4.5 and 6.5 Kwh / m. This data is important to be able to take the reference value of the peak solar hour HSP [11]. With which the solar photovoltaic system was designed to scale.

Now taking into account the above and taking as reference the irradiation of the month of March. See Figure. 2, and applying a correction factor k = 0.95 and additionally mention that the photovoltaic solar panel is under a fixed structure and this has a slope of 45 degrees. Which would indicate that in this month we would be taking advantage of its maximum energy according to the inclination. It was estimated that in this period there is an abundance of solar irradiation and the sunset can vary between 6 and 9 hours. For purposes of validation of the design, the peak solar hours of 6.16 HSP.

Table 2. Consumption in alternate current AC.

Consumption in alternate current AC					
Description	Number	P (W)	Hours / day	Days of use / week	Energy (Wh / week)
lights	4	0.7	1	7	19.6
Total energy per day					2.80
Total energy AC					19.6
Total energy system week					70
Total energy system per day					10

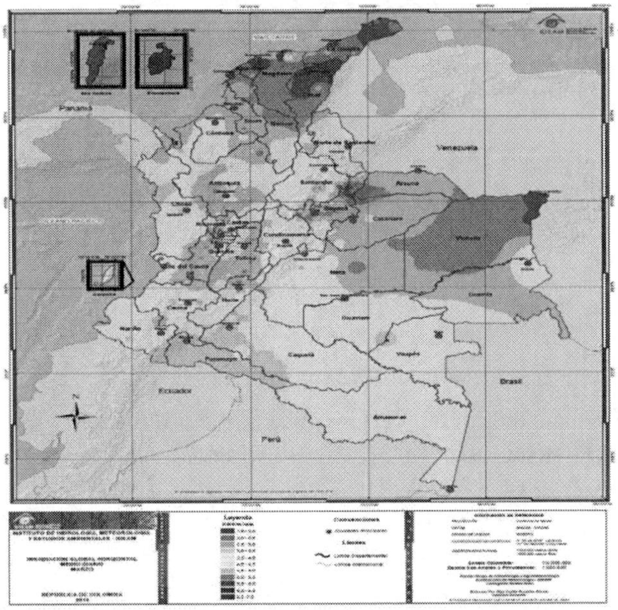

Figure 1. Radiation captured by cities in Colombia.

2.3. Global performance Installation

For the calculation of the overall performance [12-14] of the installation. The individual yields of each one of the elements that are part of the solar photovoltaic system at scale were taken into account. The elements to take into account can be evidenced in Table 3. And the data sheet of the photovoltaic module to be used can be evidenced in Table 4.

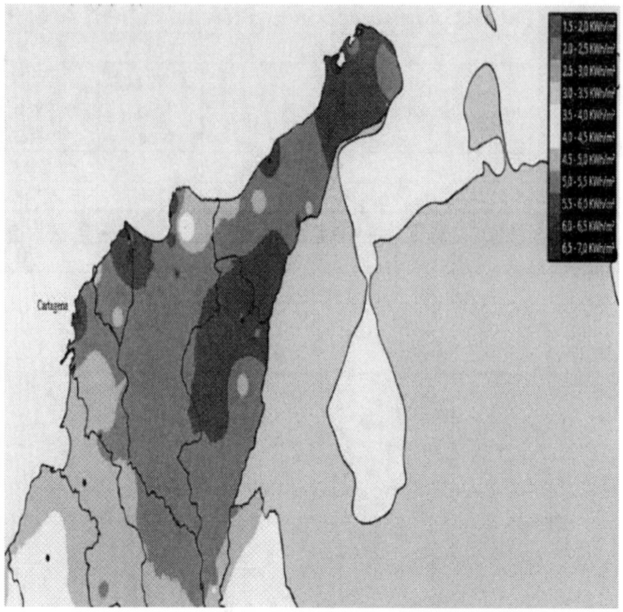

Figure 2. Radiation captured in Cartagena.

Using "(1)," we calculate the overall performance of the installation.

Taking into account the values in Table 3, it can be noted that the value of R = 0.6435 denotes an overall performance of the installation of 64.35%.

Table 3. Elements and characteristics

Value of chosen constants		
Description	Parameter	Value
New accumulations, without intense discharges	K_b	0,050
Low self-discharge batteries	K_a	0,002
Inverter performance	K_c	0,050
Efficient charge controller	K_r	0,100
Losses in wiring and equipment	K_v	0,150
Battery discharged up to 60%	P_d	0,600
Number of days of autonomy to ensure a no-load service	N	3

$$R = (1 - K_b - K_c - K_r - K_v) * \left((1 - K_a) * \left(\frac{N}{P_d} \right) \right) \tag{1}$$

$$R = 0{,}6435$$

Where:

K_b : Coefficient of losses for performance in the accumulator.

K_a : Fraction of energy that is lost by self-discharge.

K_c : Losses for the investor's performance.

K_r : Losses in the charge controller

K_v : Other losses not considered previously.

N : Number of days of autonomy to ensure a no-load service.

P_d : Maximum permissible discharge depth.

2.4. Solar panel calculation

In this stage of the design [15-17], the variables that we took into account and the calculation of the number of panels that are needed for the assembly of the solar photovoltaic system with an autonomy of three calendar days are presented. We needed to know the daily energy that the panels should provide, establishing a daily consumption and contrasting this with the overall performance of the installation. The daily energy that the panel was supposed to produce taking into account a selection was established.

Additionally, it was necessary to know, according to the selected panel. What was the energy produced by the system?

Using "(2)," we calculate the required energy that the panels have to supply.

Using "(3)," we calculate the energy produced per day by a photovoltaic panel.

$$EDS = \frac{CD}{RG}$$
$$EDS = \frac{10}{0.6435} \tag{2}$$
$$EDS = 15.54(Wh/dia)$$

$$EDP = I_{pm} * HSP$$
$$EDP = 0.56 * 6.16 \tag{3}$$
$$EDP = 3.45(Ah/paneldia)$$

Where:

CD : Total consumption per day
RG : Global performance
EDS : Daily energy to be produced by solar panels
EDP : Energy produced per day by a photovoltaic panel
HSP : Peak Solar Hour

Table 4. Data sheet of the photovoltaic module to be used

Description	Unit	Value
P_{max}	W	10
V_{nom}	V	12
V_{pm}	V	17.3
I_{pm}	A	0.56
V_{oc}	V	21.6
I_{sc}	A	6.13

For the smart grid scale model after the enhanced calculations, it is concluded that: The number of panels needed is equal to one. Taking into account the considerations of connection forms this can be series or parallel. Using "(4)," we calculate the number of panels in series. Using "(5)," we calculate the number of panels in parallel.

$$nps = \frac{V_s}{V_{nom}} = \frac{12}{12} = 1 \tag{4}$$

$$npp = \left(\frac{\frac{EDS}{V_s}}{I_{pm} * HSP} \right) = \left(\frac{\frac{15.54}{12}}{0.56 * 6.16} \right) = 0.37 \tag{5}$$

Where:

nps : Number of panels in series
npp : Number of panels in parallel

For the construction of the solar photovoltaic system to scale, a solar panel of 10 Watts was taken into account and the current that it supplies is 0.56 Amperes (A).

2.5. Regulator sizing
In this stage of the design we took into account that the charge regulator should work at the same voltage of the system and in theory should be able to support a current intensity greater than the intensity of the solar panel for assembly in an approximate value of 10 % [18-19].

Knowing that the panels provide a current intensity equal to 0.56 Amperes (A) and oversized this to 10%. It is obtained that the maximum intensity of the regulator is in the order of 0.62 Amperes (A). The regulator voltage is 12 volts (V).

2.6. Storage system calculation
At this stage of the design of the photovoltaic solar system, the storage system is presented [20-23]. At this point we had to calculate the capacity of the battery. For this purpose, we took into account data such as: Consumption, days of autonomy, voltage of the battery and depth of discharge of the battery selected for the assembly.

Knowing the Daily energy to be produced by solar panels (EDS), the days of autonomy (N), the nominal voltage (Vnom) of the battery equal to 12 volts and taking into account a discharge of 60% (Pd). Using "(6)," we calculate the accumulation capacity of the batteries in Ampere Hour (Ah).
In this stage of the design we took into account that the charge regulator should work at the same voltage of the system and in theory should be able to support a current intensity greater than the intensity of the solar panel for assembly in an approximate value of 10 % [18-19].

$$Abb = \left(\frac{\frac{EDS * N}{V_{nom}}}{\frac{P_d}{100}} \right) = 6(Ah) \tag{6}$$

Where:

Abb: Accumulation capacity battery bank

For the development of the assembly of the equipment according to the calculations established, it is concluded that one battery is needed and this can be connected in series or parallel. For connection in specific connection was made in series.

Taking into account the previous considerations for the assembly of the battery in the scale system, the battery under reference FL1290 was taken into account with the characteristics shown in Table 5.

Table 5. Specifications battery series FL1290

Description	Unit	Value
Cycle use	V	14.6 – 14.8
Standby use	V	13.7 – 13.9
Initial current Less Than	A	2.7
V_{nom}	V	12
Charge	Ah	9.0

2.7. Inverter sizing

In the final stage of the design of the solar photovoltaic system at scale, the calculation of the Inverter [24-25] was presented and for this purpose the total power of the consumptions in alternating current was considered, estimated a total close to 2.8 watts. Now taking into account the consumption and voltage of the 12 volt system. Using (7), the minimum power of the inverter can be set to work efficiently.

$$\begin{aligned} mpi &= p_{ac} * 1.2 \\ mpi &= 2.8 * 1.2 \ . \\ mpi &= 3.36(W) \end{aligned} \tag{7}$$

Where:

mpi: Minimum power of the inverter

To comply with the specifications of the design, when making the assembly of the solar photovoltaic system to scale. We took into account the inverter PI100LA. See Table 6.

Table 6. Specifications inverter series PI100LA

Description	Unit	Value
Max. Power	W	100
High Surge Peak	W	200
Input Voltage	V	12
Output Voltage	V	110
Frequency	Hz	60
Fuse	A	10
Output Wave Form		Modified Sine Wave
Power up To	A	0,45

3. Control load system

Finished the dimensioning of the solar system to scale. We present the design of the graphic programming for the control of variable loads. To give emphasis to this process, the Block Diagram is presented, followed by the Front Panel develop in LabVIEW [26-31]. The communications and control interface for the sending and receiving of data was made through LabVIEW with Arduino and Maker Hub [30, 32-33].

3.1. Block diagram

Finished the dimensioning of the solar system to scale. We present the design of the graphic programming for the control of variable loads. To give emphasis to this process, the Block Diagram is presented, followed by the Front Panel.

In the development of the programming for the control of variable loads AC and DC, it is established as a starting point. Use graphic programming blocks embedded with Maker Hub library and LINX communications. To start the programming cycle, an open point and a close point are established. To establish iterative cycles within the programming for load control, it is established to use a while loop structure.

On the other hand, taking into account that the real system includes 4 AC loads and four DC loads. A matrix of 8 inputs and 8 outputs was created that were controlled by a Digital Write N Chanel. It is made clear that a serial port reading input is included in the input of the programming cycle, which will be monitoring the signal coming from the board. This reading input is normally works at 9600 bauds per second and the reading input will be made through the ports com and lpt of the laptop that will act as HMI screen. The graphic programming established through the Block Diagram of LabVIEW software is presented in Figure. 3.

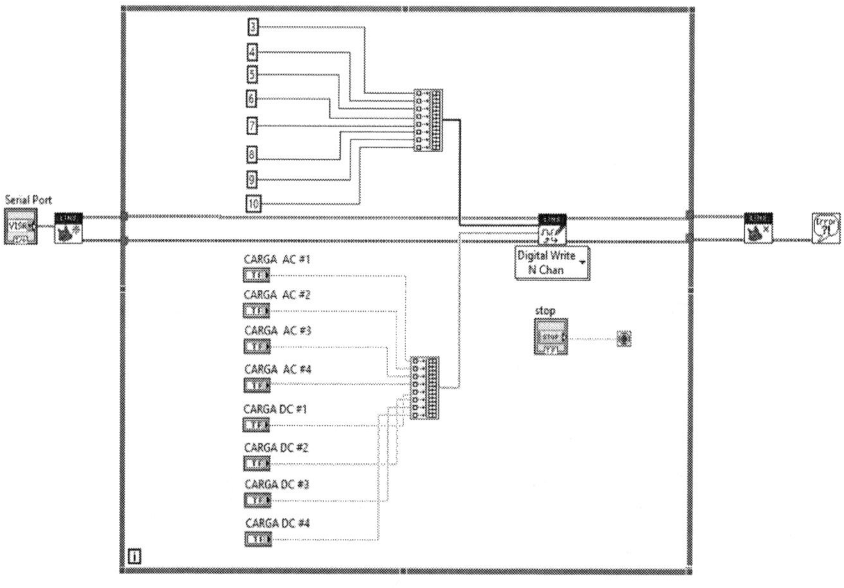

Figure 3. Block Diagram load control system.

3.2. Front panel

For the development of the control graphic environment, a man machine interface (HMI) was created in the LabVIEW programming software.

IOP Conf. Series: Materials Science and Engineering **844** (2020) 012066 doi:10.1088/1757-899X/844/1/012066

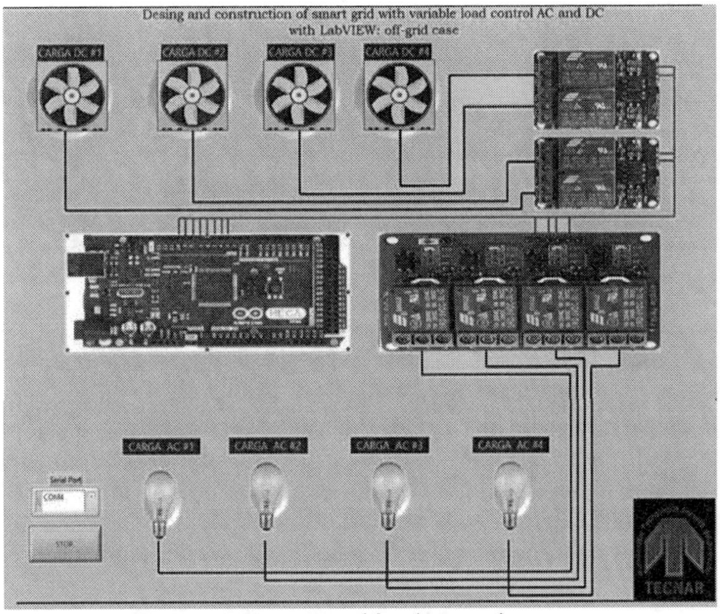

Figure 4. Front Panel load control system.

This environment allowed us to create a variable load control like the one observed in Figure. 4. In this we can observe an embedded board named Arduino Mega 2560, a relay system for AC load control (lighting) and a relay system for control of DC loads (fans). The graphical control system of loads to relays of current accentuates its serial communication with the PC through the COM4 port. In the same way an emergency stop stage like STOP is created.

This graphic design looks like the frontal image of the prototype already built, see Figure. 5.

4. Construction procedure

For the stage of physical construction of the prototype of solar photovoltaic system to scale, a rectangular container of 40 x 20 x 20 cm was built. On the upper base of the container, the 10-watt photovoltaic solar panel is supported at 45 degrees. On the front face of the model, is the control system. Where you can see the board, the AC loads (bulbs), the current relays, fuse protection system and wiring. On the left side face, there are DC loads (fans). On the back side is the regulator. Finally, in the internal part of the container, the connections with the battery and the inverter are immersed. See Figure. 5a, 5b, 5c and 5d.

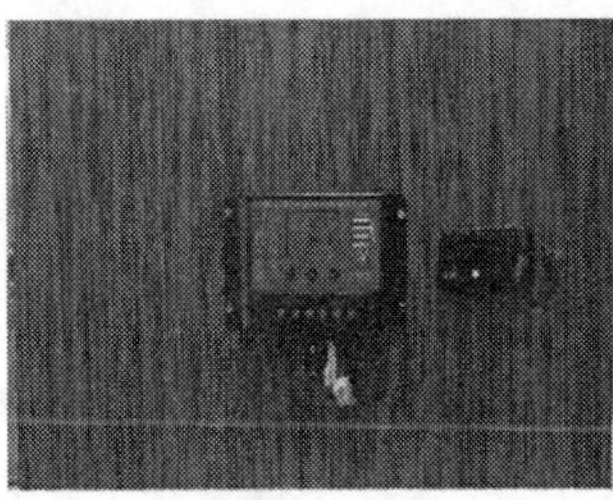

Figure 5a. Back prototype **Figure 5b.** Frontal prototype

IOP Conf. Series: Materials Science and Engineering **844** (2020) 012066 doi:10.1088/1757-899X/844/1/012066

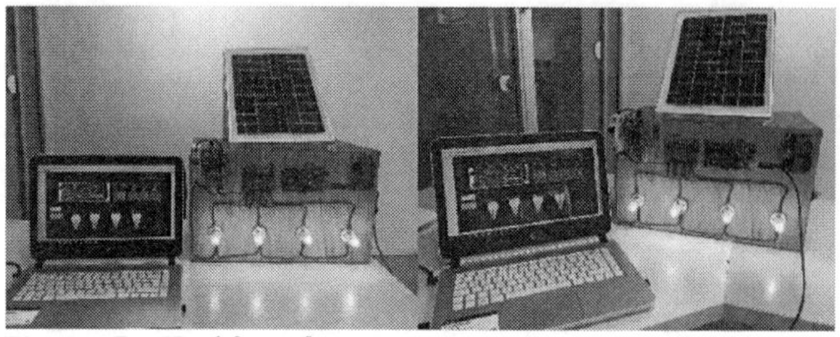

Figure 5c. Hmi interface **Figure 5d.** Hmi Interface

5. The wavelet transform

The wavelet transform (WT) in a mathematical processing technique commonly used in digital signal processing. This processing technique uses rectangular windows where the high and low frequencies are enclosed by applying them to scalar factors [34-36]. This method is mostly used for advanced frequency analysis over time. WT applications are included in ECG analysis, denoising signal, image filtering, image compression or as in this case is used for power quality analysis in electrical systems or electrical signals of voltage, current and power.

The wavelet transform can be found in the time domain. Wavelet Continuous Transform (CWT) and can be expressed mathematically using "8" of the following form.

$$CWT_{\psi}f(a,b)=\int_{-\infty}^{+\infty} f(t)\psi_{a,b}(t)dt$$
$$\text{where}$$
$$\psi_{a,b}(t)=|a|^{-\frac{1}{2}}\psi\left(\frac{t-b}{a}\right) \quad . \tag{8}$$

On the other hand, the wavelet transform can be found in the frequency domain. Discrete Wavelet Transform (DWT) [37] and can be expressed using "9," in the following way.

$$DWT_{\psi}f(m,n)=\int_{-\infty}^{+\infty} f(t)\psi_{m,n}(t)dt$$
$$\text{where} \quad . \tag{9}$$
$$\psi_{m,n}(t)=motherwavelet$$

Additional it must be mentioned that the DWT is used to design the multi resolution analysis (MRA). It should be clarified that the MRA represents the original signal in different resolutions after having applied two types of filters in each level. See Figure. 6. Each of these called low pass filter (LPF) and high pass filter (HPF). It is specified that the MRA performed a scaling of the function $\emptyset(t)$ using "10".

IOP Conf. Series: Materials Science and Engineering **844** (2020) 012066 doi:10.1088/1757-899X/844/1/012066

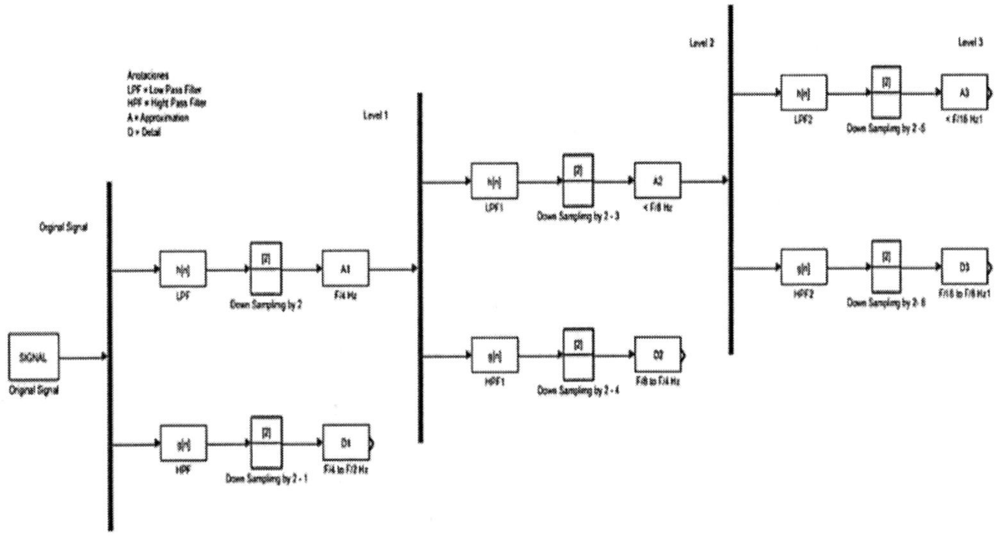

Figure 6. Schematic Level multiresolution signal analysis of DWT

$$\phi(t) = 2\sum_{n=-\infty}^{+\infty} c_n \phi(2t-n). \tag{10}$$

$$\psi(t) = 2\sum_{n=-\infty}^{+\infty} d_n \phi(2t-n). \tag{11}$$

Where d_n is also a squared summable sequence.

6. Analysis result

To start the results stage, the signal to analyze is shown as the first instance. See Figure. 7. Then a call will be made from the MATLAB Wavelet toolbox [38] and their respective perturbations will be focused.

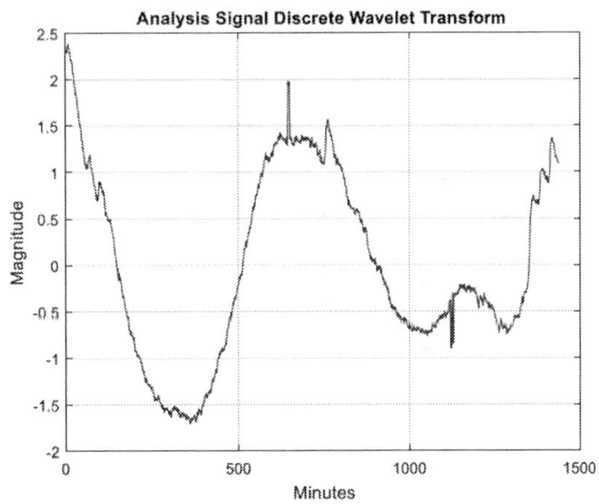

Figure 7. Analysis signal discrete wavelet transform

IOP Conf. Series: Materials Science and Engineering **844** (2020) 012066 doi:10.1088/1757-899X/844/1/012066

6.1. Sag detection

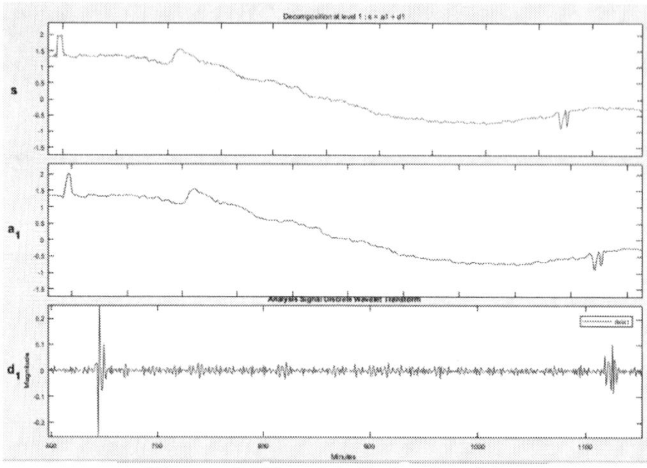

Figure 8. Wavelet db 4 at level 1

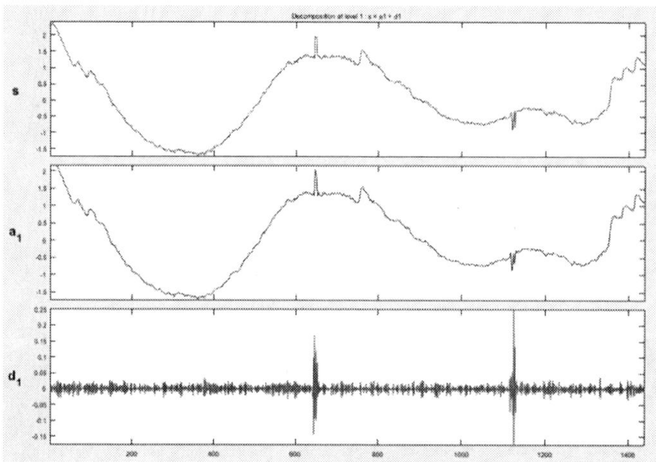

Figure 9. Wavelet db 8 at level 1

IOP Conf. Series: Materials Science and Engineering **844** (2020) 012066 doi:10.1088/1757-899X/844/1/012066

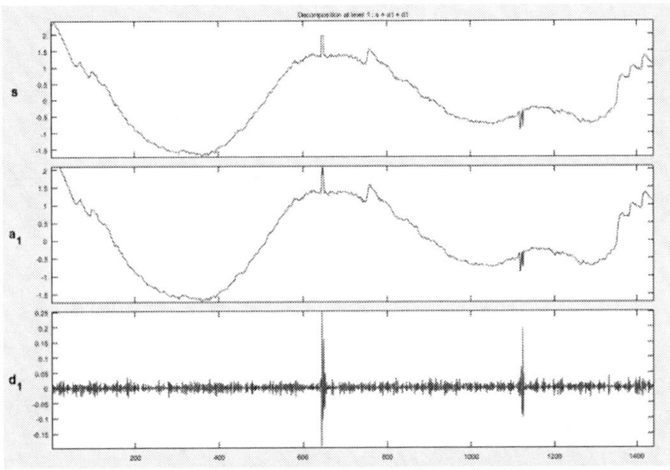

Figure 10. Wavelet db 10 at level 1

Taking into account the analysis made in level 1 with daubechies 4, 8 and 10 you can observe between the minutes 600 to 700 a behavior of sag voltage and likewise between the minutes 1100 to 1200. In each of these level daubechies One the behavior is similar. See Figure. 9, Figure. 10 and Figure. 10.

6.2. Harmonic detection

Taking into account the analysis done in level 4 with daubechies 4, a behavior similar to harmonics can be observed between 1300 and 1400 minutes. See Figure. 11.

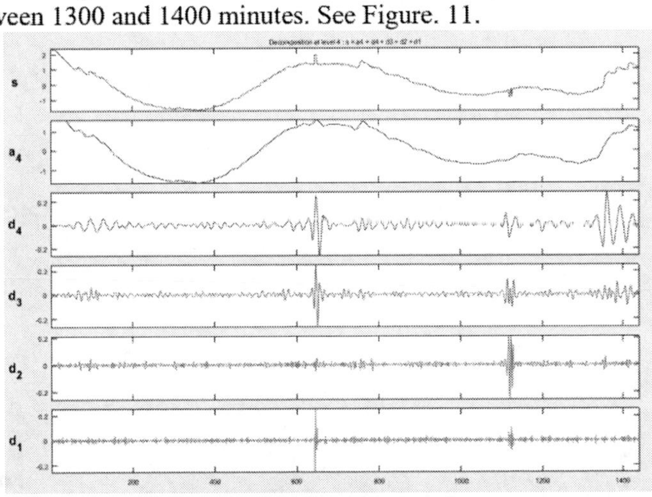

Figure 11. Wavelet db 4 at level 4

When the daubechies [39] db 8 family is used, the behavior to analyze the signal reveals a greater number of harmonics in the original signal. these behaviors are detected between minutes 700 and 800. See Figure. 12.

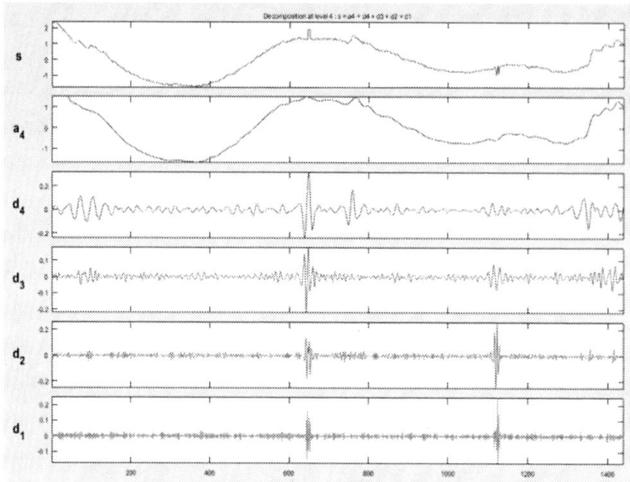

Figure 12. Wavelet db 8 at level 4

When the daubechies db 10 family is used, the behavior to analyze the signal reveals the same behaviors that db 8 and the harmonic are detected between minutes 700 and 800. And 1100 and 1200. See Figure. 13.

6.3. Swell detection

Taking into account the analysis done with the family db 8 in level 4, one can observe a swell behavior in d4 between the minutes 680 and 780, 850 and 900. Similarly, in db3 between minutes 760 and 840 but in a way a little less regular. See Figure. 14.

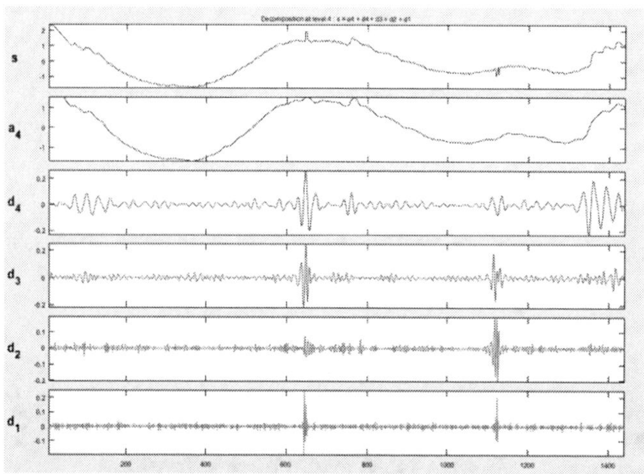

Figure 13. Wavelet db 8 at level 4.

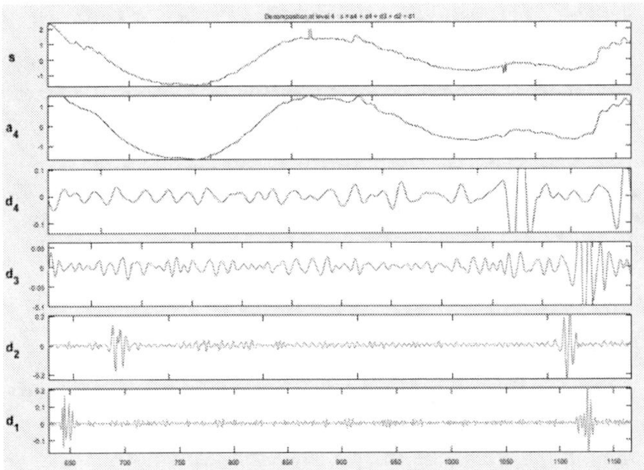

Figure 14. Wavelet db 8 at level 4.

7. Conclusions

In the present work the viability of using mathematical signal processing tools to analyze electrical signals is observed. This type of analysis is very helpful in the eventuality of wanting to implement filters that allow establishing or improving electrical signals due to signal fluctuations commonly experienced by medium and low voltage electrical distribution networks. The analysis through the wavelet transform and the wavelet mother through level daubechies allowed establishing that these small signals can be located. Future work is expected to establish digital filters that can suppress these small variations that affect electrical signals.

Acknowledgments

Acknowledgments are presented to the University Foundation Antonio de Arevalo who gave institutional endorsement for the present research project under contract project number 2018-00146.

References

[1] H. Farhangi, "The path of the smart grid," *IEEE Power and Energy Magazine,* vol. 8, no. 1, pp. 18-28, February 2010.

[2] V. C. Gungor, et al, "Smart Grid Technologies: Communication Technologies and Standards," *IEEE Transactions on Industrial Informatics*, vol. 7, no. 4, pp. 529-539, November 2001.

[3] X. Fang, S. Misra, G. Xue, D. Yang, "Smart Grid — The New and Improved Power Grid: A Survey", *IEEE Communications Surveys & Tutorials*, vol. 14, pp. 944-980, December 2011.

[4] M. L. Tuballa, M. L. Abundo, "A review of the development of Smart Grid technologies", *Renewable and Sustainable Energy Reviews*, vol. 59, pp. 710-725, June 2016.

[5] Ministry of Mines and Energy, CREG. (2010, Feb, 26). By which the activities of small-scale self-generation and distributed generation in the National Interconnected System are regulated [Online]. http://apolo.creg.gov.co/Publicac.nsf/1c09d18d2d5ffb5b05256eee00709c02/83b41035c2c4474f05258243005a1191?OpenDocument

[6] Congress of the Republic. (2014, May, 13). By means of which regulates the integration of non-conventional renewable energies to the National Energy System [Online]. http://www.secretariasenado.gov.co/senado/basedoc/ley_1715_2014.html

[7] M. R. Patel, *Wind and Solar Power Systems, Design, Analysis, and Operation,* 2nd. Ed. Florida: CRC Press Taylor & Francis Group, 2006.

[8] S. R. Wenhan, M. A. Green, M. E. Watt, R. Corkish, A. Sproul, *Applied Photovoltaics,* 3rd. Ed. London: Routledge, 2013.
[9] T. Markvart, L. Castañer, *Practical handbook of photovoltaics fundamentals and applications,* UK: Elsevier, 2003.
[10] C. Niucă, G. Chiriac, C. I. Gatu, D. Cuciureanu, I. Murgescu, "Laboratory Off-Grid Photovoltaic System with Two Axes Orientation", 2018 International Conference and Exposition on Electrical And Power Engineering (EPE), Romania, pp. 0846-0849, October 2018.
[11] M. P. Martinez, I. C. M. Rodriguez, E. Castro, "The hour equivalent solar pick: Definition and interpretation," *Ingenieria Energetica,* vol. 38, no. 2, pp.124-131, August 2017.
[12] T. P. Chang, "Performance evaluation for solar collectors in Taiwan," *Energy,* vol. 34, no. 1, pp. 32-40, January 2009.
[13] R. Rüther, M. M. Dacoregio, "Performance assessment of a 2 Kw grid-connected, building-integrated, amorphous silicon photovoltaic installation in Brazil," *Progress in Photovoltaics: Research and Applications,* vol. 8, no. 2, pp. 257-266, May 2000.
[14] D. L. King, M. A. Quintana, J. A. Kratochvil, D. E. Ellibe, B. R. Hansen, "Photovoltaic module performance and durability following long-term field exposure," *Progress in Photovoltaics: Research and Applications*, vol. 8, no. 2, pp. 241-256, May 2000.
[15] E. R. Zuhal, S. Marangozoglu, "New Design for Solar Panel Tracking System Based on Solar Calculations," 2018 IEEE 61st International Midwest Symposium on Circuits and Systems (MWSCAS), Canada, pp.1042-1045, August 2018.
[16] H. Terzioglu, F. A. Kazan, M. Arslan, "A New Approach to the Installation of Solar Panels," 2015 2nd International Conference on Information Science and Control Engineering, China, pp. 573-577, April 2015.
[17] K. U. Tahera, R. Fabiha, Md. Z. R. Khan, "Solar Photovoltaic System Design for a Residential Hall in BUET," 2018 10th International Conference on Electrical and Computer Engineering (ICECE), Bangladesh, pp. 1-4, December 2018.
[18] N. Efkarpidis, T. D. Rybel, J. Driesen, "Optimal placement and sizing of active in-line voltage regulators in Flemish LV distribution grids," 2015 International Conference on Renewable Energy Research and Applications (ICRERA), Italy, pp. 408-413, November 2015.
[19] H. Lee, P.K.T. Mok, K. N. Leung, "Design of low-power analog drivers based on slew-rate enhancement circuits for CMOS low-dropout regulators," *IEEE Transactions on Circuits and Systems II: Express Briefs,* vol. 52, no. 9, pp. 563-567, September 2005.
[20] J. Johnsson, "Sizing Batteries for Photovoltaic Systems," INTELEC '87 - The Ninth International Telecommunications Energy Conference, Sweden, pp. 139-142, June, 1987.
[21] S. Jogunuri, R. Kumar, D. Kumar, "Sizing an off-grid photovoltaic system (A case study)," 2017 International Conference on Energy, Communication, Data Analytics and Soft Computing (ICECDS), India, pp.2618-262, June 2018.
[22] Y. Yang, H. Li, A. Aichhorn, J. Zheng, M. Greenleaf, "Sizing Strategy of Distributed Battery Storage System With High Penetration of Photovoltaic for Voltage Regulation and Peak Load Shaving," *IEEE Transaction on Smart Grid,* vol. 5, no. 2, pp. 982-991, September 2013.
[23] P. G. Nikhill, D. Subhakar, "Sizing and Parametric Analysis of a Stand-Alone Photovoltaic Power Plant," *IEEE Journal Photovoltacis,* vol. 3, no. 2, pp. 776-784, March 2013.
[24] A. Khare, S. Rangnekar, "Optimal sizing of a grid integrated solar photovoltaic system," *IET Renewable Power Generation,* vol. 8, no. 1, pp. 67-75, January 2014.
[25] M. Bazrafshan, N. Gatsis, E. Dall'Anese, "Placement and Sizing of Inverter-Based Renewable Systems in Multi-Phase Distribution Networks," *IEEE Transactions on Power Systems,* vol. 34, no. 2, pp. 918-930, September 2018.
[26] A. Gasmelseed, "New Development Environment for Modern Bioelectromagnetics Signal Processing Applications [Application Notes]," *IEEE Microwave Magazine,* vol. 14, no. 5, pp.

134-1352, July 2013.

[27] F. J. Jimenez, F. R. Lara, M. D. Redel, "API for communication between Labview and Arduino UNO," *IEEE Latin America Transactions,* vol. 12, no. 6, pp. 971-976, September 2014.

[28] J. Chacon, H. Vargas, G. Farias, J. Sanchez, S. Dormido, "EJS, JIL Server, and LabVIEW: An Architecture for Rapid Development of Remote Labs," *IEEE Transactions on Learning Technologies,* vol. 8, no. 4, pp. 393-401, January 2015.

[29] T. R. de Melo, J. J. da Silva, J. S. da Rocha, "Implementation of a Decentralized PID Control System on an Experimental Platform Using LabVIEW," *IEEE Latin America Transactions,* vol. 15, no. 2, pp. 213- 218, February 2017.

[30] R. M. Shrenika, S. S. Chikmath, A. V. R. Kumar, Y. V. Divyashree, R. K. Swamy, "Non-contact Water Level Monitoring System Implemented Using LabVIEW and Arduino," 2017 International Conference on Recent Advances in Electronics and Communication Technology (ICRAECT), India, pp. 306-309, October 2017.

[31] A. El Hammoumi, S. Motahhir, A. Chalhl, A. El Ghzizal, A. Derouich, "Real-time virtual instrumentation of Arduino and LabVIEW based PV panel characteristics," IOP Conference Series: Earth and Environmental Science, vol. 161, no. 1, pp. 1-11, 2018.

[32] A. Tong-on, P. Saphet, M. Thepnurat, "Simple Harmonics Motion experiment based on LabVIEW interface for Arduino," Journal of Physics: Conference Series, vol. 901, no. 1, pp. 1-6, 2017.

[33] W. Alhalabi, "Patient monitoring at home using 32-channel cost- effective data acquisition device," *Telematics and Informatics,* vol. 35, no. 4, pp. 883-891, July 2018.

[34] S. Santoso, E. J. Powers and W. M. Grady, "Power quality assessment via wavelet transform analysis," Proceedings of the IEEE PES Summer Meeting, July 1995, Portland.

[35] D.C. Robertson, 0.1. Camps, J.S. Mayer, and W.B. Gish, "Wavelets and electromagnetic power system transients," Proceedings of the IEEE PES Summer Meeting, July 1995, Portland, USA

[36] F.A.C. Pires and N.S.D. Brim, "The analysis of transient phenomena using the wavelet theory," Pro- ceedings of the International Conference on Power Systems Transients, June 1997, Seattle, USA.

[37] A. M. Gaouda, M. A. Salama, M. R. Sultan, and A. Y. Chikhani, "Power Quality Detection and Classification Using Wavelet- Multi resolution Signal Decomposition," IEEE Trans. Power Delivery, Vol. 14(4), pp. 14691476, 1999.

[38] M. Misiti, Y. Misiti, G. Oppenheim and J.-M Poggi, Wavelet Toolbox For Use With Matlab: user's guide version 3., Mathworks Inc., 2006.

[39] I. Daubechies, Ten Lectures on Wavelets, SIAM, Philadelphia, Pennsylvania, 1992.

Effects of the implementation of a self-assessment model in the accreditation of programs in Higher Education

E Delgado-Florez[1], O Batista-Ayola[1] and Y Blanco-Jimenez[1]

[1] Fundación Universitaria Antonio de Arévalo-Unitecnar, Cartagena, Colombia

E-mail: emilio.delgado@unitecnar.edu.co

Abstract. The purpose of this article is to present the success of the Antonio de Arevalo University Foundation – Unitecnar, in the self-assessment process for undergraduate programs at the Faculty of Engineering Sciences, specifically the Technology in Systems program, describing the self-assessment model and the methodology applied in the program accreditation processes, which has resulted in three consecutive accreditation for the mentioned program. At present, the applied model has allowed the identification of strengths and aspects to be improved and the implementation of self-assessment and self-regulation processes in the different functional areas, from the generation of reflection spaces for a better contribution to the quality of the Administrative services and Academics, these excellent accreditation results have been obtained for nine (9) programs at the Unitecnar Faculty of Economics and Engineering Sciences.

1. Introduction

The National Accreditation Council - CNA (for its acronym in Spanish) is created by Law 30 of December 28, 1992, as an academic body, attached to the Ministry of National Education, with the main purpose of guaranteeing to the community that the Higher Education Institutions - IES (for its acronym in Spanish) will develop its educational function, offering programs with the highest criteria and quality standards [1].

The qualified registration system was installed in 2003, ten years later than the accreditation system. During this decade, the accreditation system was the only instrument for controlling the quality of higher education, but only for those institutions and programs that voluntarily accepted it. As of 2003, the qualified registration process was mandatory for all programs of this educational level [2].

The CNA defines Accreditation as a testimony given by the State regarding the quality of a program or institution based on a previous evaluation process, in which the institution, academic communities and the National Accreditation Council are involved, with the theme of the quality of higher education, some changes were introduced such as: the recognition of greater autonomy of educational institutions for the appointment of their highest managers and the creation of academic programs; the creation of a National Information System on Higher Education–SNIES to have information on educational offer and quality indicators of educational institutions; and, finally, the definition of a process of accreditation of educational institutions to guarantee students in particular and society in general that they meet the proposed objectives and the highest quality requirements [3].

The accreditation process is born initially, as a mechanism to strengthen the quality of Education and make a public recognition to the Institutions that will achieve high quality standards. In Colombia, as described by the CNA, the accreditation process does not arise within the framework of inspection and surveillance of the state, but in the promotion, recognition and continuous improvement of quality and also recognizes that the main and most effective investment made in the context of accreditation, is not properly the implementation of the model itself and of its evaluative processes, but the investment in the application of institutional improvement plans and programs, which have been designed by the institutions as a requirement for their entry into the system or as a result of the self-assessment and reports of academic peers [4].

From the perspective of interest groups, accreditation fulfills an important social function. In the case of students, they require a quality system that guarantees the provision of an adequate educational service so that graduates are highly likely to obtain employment and thus improve their personal and social life, the interest of employers is have a system that ensures the training of trained human resources so that it can perform a specific occupation [5] and the interest for IES, is a process of self-assessment of all the Administrative and Academic functional areas that allows a continuous improvement.

This article describes the effects of the implementation of the Self-Assessment Process for accreditation purposes, of Unitecnar based on the guidelines of the CNA, applied in the Undergraduate programs of the Faculty of Engineering Sciences. specifically Technicians and Technicians of the Systems area, which begins in 2007 when Unitecnar attending the projected opportunities, in areas such as: custom software development, computer application testing, and software adaptation to local markets and sectors , also, with the progressive deployment of telecommunications infrastructures and the massification of broadband, and opportunities related to management and production technologies in the cloud, the evolution of the knowledge society, its digital environment and greater interconnection and interactivity [6], begins with the review of all its processes to identify its strengths and weaknesses in these programs, which allowed it to be inferred that the demand for digital security technologies would be increasing, which made it necessary to offer an academic program with high quality standards that will guarantee a comprehensive professional that meets With the needs of the market.

For Unitecnar the Accreditation process is the highest recognition given by the MEN (Ministry of National Education) to the quality of academic programs of IES, which comply with specific guidelines determined by the evaluating entity, with a rigorous monitoring process and continuous improvement [7].

2. Self-Assessment Model

Unitecnar carries out its self-assessment processes, for accreditation purposes following a self-assessment model created by the Institution, which is based on the CNA guidelines for accreditation of Undergraduate Programs, which involves each of the elements of the Evaluation [8]. It determines or describes the level of proximity (with respect to an ideal), with which the different activities offered by the program are executed and the degree of compliance (compared to the ideal) of them is evaluated. See figure 1.

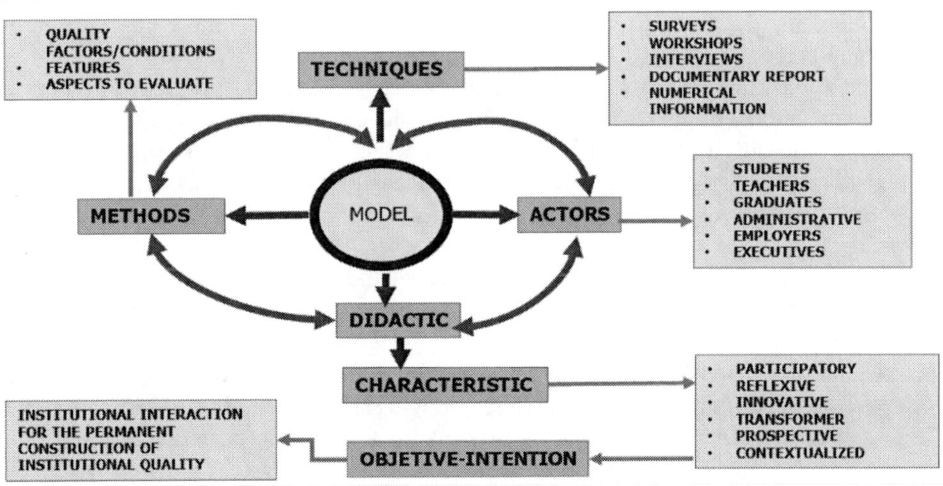

Figure 1. Unitecnar Self-Assessment Model.

3. Methodology of Application of the Self-Assessment Model

The self-assessment process is operationalized through the application of the model in four (4) stages, which together show the results of the final process, these are: First Stage: Pre evaluation, Second Stage: Evaluation, Third Stage: Post evaluation and Fourth Stage: Follow-up to Improvement Actions. Each of them contains the components evaluated by the CNA and described in the document Guidelines for the accreditation of Undergraduate programs, which are detailed below:

First Stage: Pre evaluation: In the first stage, the sensitization and socialization is carried out to the academic community about the self-assessment process to be carried out, the Quality Conditions Analysis is performed, the quality factors are analyzed, the Characteristics and Aspects to evaluate. The criteria for assigning weighting factors are defined, the criteria for qualifying self-assessment results, Discussion of the operational component of the process, the criteria, goals and activities are established in the short, medium and long term, the Work committees by factors, teachers' cloisters, discussion meetings and conceptualization are held to unify criteria.

Second Stage: Evaluation: In the second stage, pilot tests are developed to validate instruments, Collection of information through survey techniques, interviews, workshops and information gathering, Numerical and documentary, Consolidation of the information collected, Analysis and interpretation of the information about compliance with the quality conditions in the Programs, Drafting the diagnostic report with the results of the self-assessment, Conducting seminars for the socialization of the Results of the self-assessment.

Third Stage: Post evaluation: In the third stage the analysis of the results is addressed in the different committees in the direction of the diagnosis of the conditions, Current and to the proposal of improvement proposals, Preparation of the Improvement Plan for the program, by the Technical Committee.

Fourth Stage: Follow-up to Improvement Actions: In the fourth stage, the actions contemplated in the Improvement Plans, Preparation of Follow-up Reports of the improvement actions set out in the Improvement, Program and Institution Plans are monitored.

The Self-Assessment model is executed in a harmonized manner with what is requested by the MEN education authority and is pertinent from the point of view of the needs of the Institution, in order to self-evaluate its processes it carries out activities to assess the satisfaction of its entire academic community through periodic consultations such as, surveys of students and teachers that are carried out in order to propose improvement actions that favor academic and institutional quality.

The control and monitoring systems of the improvement plans and the achievements associated with the institutional project and its development plans are evidenced through the follow-up reports to the improvement plans. Each and every one of the self-assessments carried out in the Office of the Vice-Rector for Institutional Quality aims to generate an improvement plan and maintenance plans that are then part of the plans of action of teachers of the institution's plant to follow up at different times to verify their compliance.

Because the application of the self-assessment model to a specific program or to the institution generates a certain amount of data that is difficult to handle manually, and that in addition the collection of information in the different stages of the model is a fairly complex process to carry out, made it necessary to use a computer tool to manage information, easily accessible, that would provide a fast and reliable process of all types of data, with storage capacity, work automation, interactivity and digitization of all information [9] .

In response to the need to automate the execution of the self-assessment processes, SAIP for its acronym in Spanish for Institutional and Program Self-Assessment System was designed, developed and implemented. Since its inception, this system was conceived as a tool that greatly optimizes the operationalization of the application of the self-assessment model, managing in a better way the data and information that these processes handle, thus allowing the obtaining of results in a more efficient way. These self-assessments processes are necessary to request the accreditation and / or renewal accreditation of both a specific program and the institution before CNA. The whole system manages in a more efficient way self-assessment processes, because executing them manually implies a significant expense in time, in human, physical and technological resources.

SAIP was developed web-oriented, using the PHP programming language, with the Code Igniter framework [10] and the data is managed through a database in PostgreSQL [11]. The implementation of this tool generates the following benefits:

- Ease to collect all the information necessary for the generation of reports. Through the system the documentary / numerical information, strengths and weaknesses are entered.
- Be able to use the system at any time and place for being oriented to the Web.
- Survey application to the different sources and evaluators through the web, without having to move from the institution's facilities.
- Follow-up to the evaluators, allowing to identify who has carried out the survey in a very detailed way.
- Generation of reports which constitute a valuable input when it comes to constructing the documents presented to the CNA and in the improvement plan.
- Effective treatment of information, since being centralized all the information that is handled, you can access it and make the queries that are required to obtain specific information.
- Reduction of resources such as: time, human talent, among others.

Currently this system consists of two software: SAIP-Administration and SAIP-Survey. In SAIP-Administration, the entire self-assessment process is managed. This is linked in the institutional application platform called SIA (for its acronym in Spanish) and only those users to the platform can access it, logging in with their username and password, as shown in the figure 2. Through this software the following processes are handled:

1. *General data management*: Record, update and eliminate factors, characteristics, indicators or aspects to evaluate, questions and types of questions.
2. *Self-assessment processes*: Configure and manage self-assessments, defining the model to be worked on. This option defines the factors, characteristics, indicators and questions of a self-assessment model. In turn, the surveys that will be enabled for the different sources are configured.
3. *Institutional or Program Self-assessment*: Enable a self-assessment process for a specific program or institution.
4. *Registration of information (documentary / numerical)*: Record the weighting, documentary and numerical information, weaknesses and strengths to an institutional or program self-assessment
5. *Population and sample*: Identify the population to be evaluated, register parameters for sample calculation, register sample to an institutional/program self-assessment.
6. *Monitoring of evaluators*: List of evaluators of the self-assessment enabled, identifying who has carried out the survey, with the option of sending emails.
7. *Employer registration*: Register companies or employers.
8. *Definition of leaders*: Define self-assessment leaders and factor leaders.
9. *Download information*: Download list of evaluators, questionnaires according to type of source. In addition to all the information recorded in weaknesses, strengths and documentary and numerical information.
10. *Consult reports*: Consult the final report and the percentage self-assessment report of a process of an institutional or program self-assessment (see figure 2).

The user profiles that have been defined for access to the software are:

- *Administrator*: users with this profile are responsible for administering the entire system, having full access to all the options that are managed.
- *Self-assessment leader*: Profile that is enabled for those in charge of the program that is being self-assessed. The options to which you have access will allow you to monitor the self-assessment process applied to a program (see figure 2).
- *Leader factor*: Profile assigned to those responsible for leading a specific factor in a self-assessment process.

In detail, the processes that are enabled for each user profile are listed in Table 1.

Table 1. List of processes enabled for each user profile

	User Profiles		
Processes	**Administrator**	**Self-assessment leader**	**Leader factor**
1. General data management	X	-	-
2. Self-assessment processes	X	-	-
3. Institutional or Program Self-Assessment	X	-	-
4. Registration of information (documentary / numerical)	X	X	X
5. Population and sample	X	X	
6. Monitoring of evaluators	X	X	X
7. Employers Registry	X	X	-
8. Definition of leaders	X	X	X
9. Download information	X	X	X
10. Consult reports	X	X	X

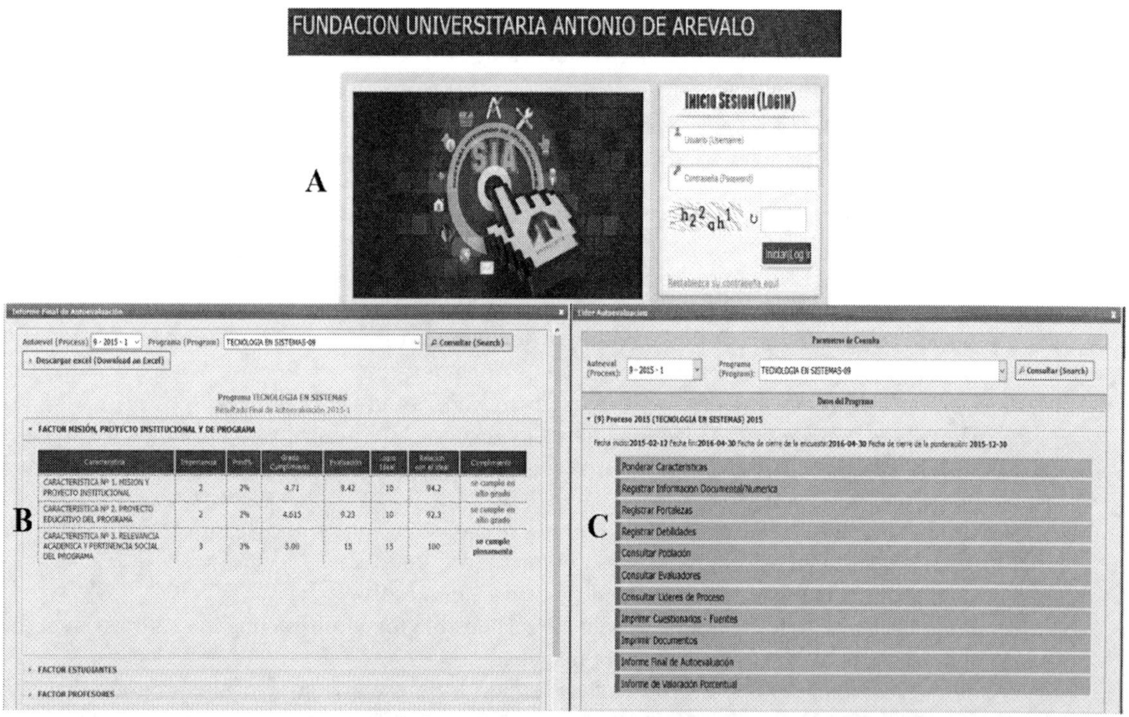

Figure 2. Views of SAIP-Administration. A) Login to SIA, B) Final self-assessment report and C) Self-assessment leader options.

On the other hand, SAIP-Survey is the software used to apply the surveys configured in each institutional or program self-assessment process. The evaluation of a program or the institution by the different sources is carried out through surveys, which are configured and enabled from SAIP-Administration. The sources are the actors or members of the university community those of the external sector, responsible and protagonists of the facts that are the object of evaluation and who will be able to identify and express strengths and weaknesses [12]. Therefore, the user profiles that will be handled in the software are the types of sources that participate in the process, such as professors, administrative staff, students, graduates and employers.

To carry out the survey, the evaluators enter the software with their identification and password, as shown in figure 3. The system identifies the types of sources it has handles and depending on which one you choose it will show you the surveys to be carried out (see figure 3).

Figure 3. Views of SAIP-Survey. A) Login in SAIP-Survey and B) Profiles enabled to conduct the survey.

IOP Conf. Series: Materials Science and Engineering **844** (2020) 012067 doi:10.1088/1757-899X/844/1/012067

4. Results

The Unitecnar self-assessment model, using SAIP software as a tool, has generated greater effectiveness in the self-assessment processes of the Systems Technology program, the result of which is the accreditation of the program for 6 years, where it stands out by the Ministry of Education that the program has achieved sufficient levels of quality so that in accordance with the rules governing the matter this fact is publicly recognized through a formal act of accreditation [13].

To achieve the aforementioned results, the self-assessment process was carried out, which consisted, first, in the formation of the committees by factor, consisting of: a teacher, a student, an employee, an entrepreneur, a graduate. The process leader acts as the leader of the factor within the committee. These committees are responsible for constructing the report of the assigned factor.

The Institutional Quality Vice-Rectory responsible for the institution's self- assessments processes assigns a team member as the leader of the self- assessment process, who will be in charge of programming the software for the program's self- assessment process defining the factors, characteristics, indicators, questions and questionnaires that will be applied for each type of source. In addition, it is established according to the population the selection of the sample of each of the actors participating in the process such as teachers, students, employees, entrepreneurs and graduates linked to the program. For the case study of the total population of the program, a sample was defined according to the statistical parameters of a confidence level of 90% and a margin of error of 10%. These parameters are configured in the software (see figure 4).

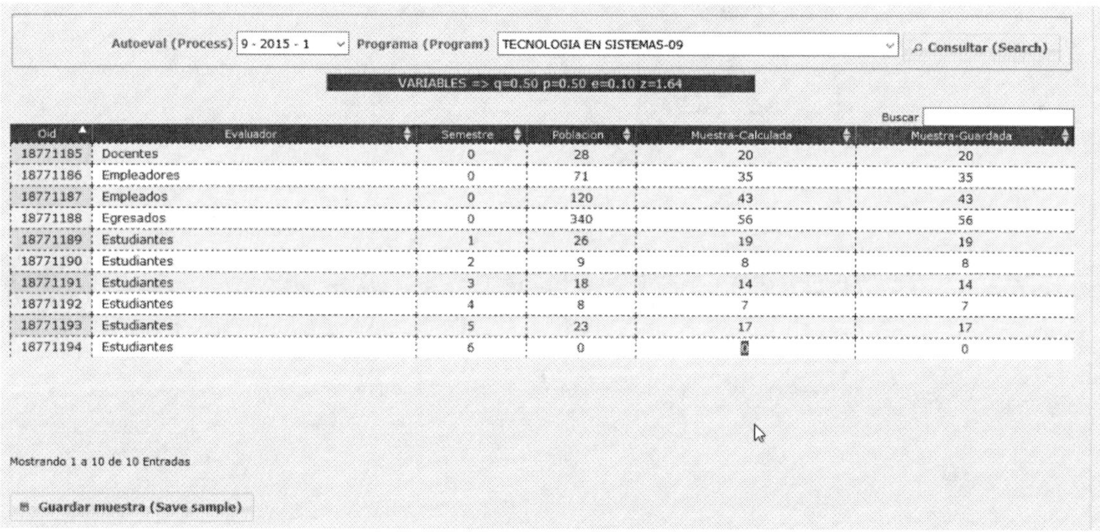

Oid	Evaluador	Semestre	Poblacion	Muestra-Calculada	Muestra-Guardada
18771185	Docentes	0	28	20	20
18771186	Empleadores	0	71	35	35
18771187	Empleados	0	120	43	43
18771188	Egresados	0	340	56	56
18771189	Estudiantes	1	26	19	19
18771190	Estudiantes	2	9	8	8
18771191	Estudiantes	3	18	14	14
18771192	Estudiantes	4	8	7	7
18771193	Estudiantes	5	23	17	17
18771194	Estudiantes	6	0	[illegible]	0

Figure 4. Sample values

Next, the survey is applied through the software, which can be tracked through a module in the application (see figure 5). This module also allows you to send reminder emails to the people involved who have not carried out the survey, guaranteed in this way that all the actors actively participate in the process.

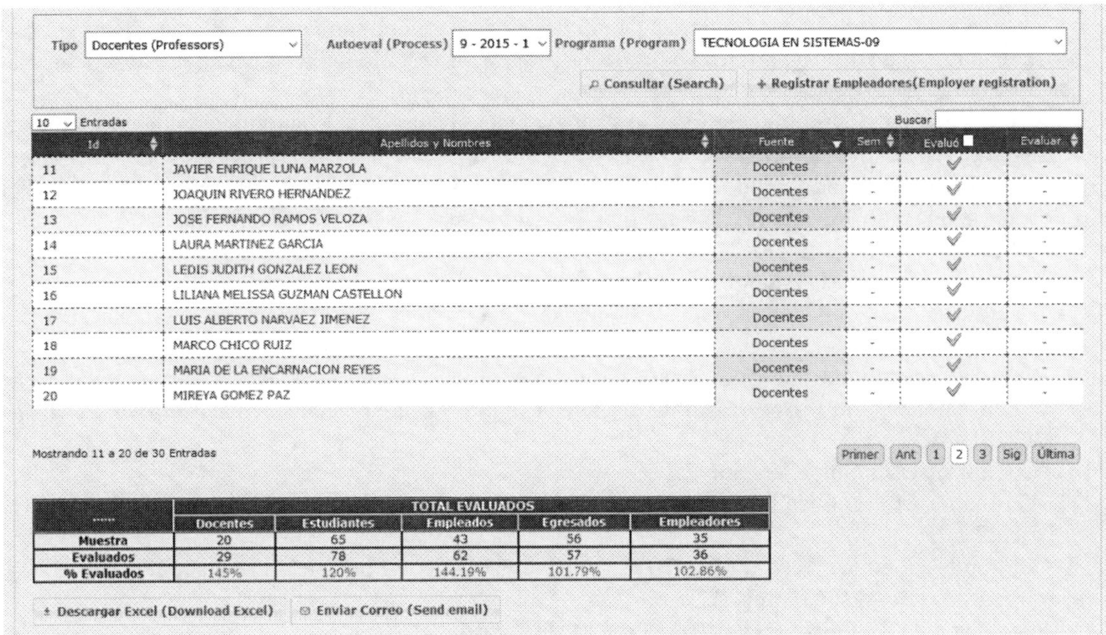

Figure 5. Monitoring of evaluators

Subsequently, the documentary and numerical information of each of the factors is entered. This information is recorded by the leaders of factors that have been previously assigned and defined in the software by the leader of the self-assessment process.

Finally, from the surveys carried out, the documentary and numerical information recorded and the weighting of each factor, the results of the process are generated, which are sent to the committees by factor for the realization of the report, highlighting the strengths and aspects to improve the program.

The process described above has been carried out three times for the Systems Technology program, achieving the accreditation for four years in the first self-assessment and presentation to the CNA; In the second renewal process, accreditation for four years was achieved in the same way, and for the third process of self-assessment and presentation before the CNA, the accreditation of the program for six years is achieved.

5. Conclusions

The methodology developed in the self- assessment process for accreditation purposes; applied to the Systems Technology program has allowed three consecutive accreditations to be obtained, which demonstrates that the efficiency and reliability of the Unitecnar self- assessment process.

The use of computer tools in organizations allows to improve their processes, as is the case of SAIP in Unitecnar, which has generated greater productivity in the self-assessment processes for the purpose of accreditation and renewal of accreditation of the Systems Technology program.

Conflict of Interest.

The authors report there are no conflicts of interest.

6. References

[1] Consejo Nacional de Acreditación 2013 (Bogotá: Colombia)
[2] Consejo Nacional de Acreditación 2013. Consejo Nacional de Acreditación. Bogotá, CNA.
[3] Consejo Nacional de Acreditación 2013 La evaluación externa en el contexto de la acreditación en Colombia (Bogotá: Colombia).

[4] Urbano N 2007 Efectos de la implementación del modelo colombiano de acreditación de programas académicos. Un análisis a partir del caso de los programas tecnológicos *Universitas Humanística* **64** 141-145.
[5] Consejo Nacional de Acreditación. 2013. Autoevaluación con fines de acreditación de programas de pregrado (Bogotá: Colombia)
[6] Correa G, Gaviria J and Giraldo R 2017 Revisión y diagnóstico sobre las condiciones de calidad en los programas de Ingeniería de Sistemas en Colombia *Revista iberoamericana de educación superior* **8**(22) 3-24.
[7] Fundación Universitaria Antonio de Arévalo 2019 Proceso de Acreditación.
[8] Consejo Nacional de Acreditación 2013 Lineamientos para la acreditación de programas (Bogotá: Colombia)
[9] Díaz J Pérez A and Florido R 2011 Impacto de las tecnologías de la información y las comunicaciones (TIC) para disminuir la brecha digital en la sociedad actual *Cultivos tropicales* **32**(1) 81-90
[10] CodeIgniter User Guide Version 2.1.3
[11] Mariuxi Z, Molina J and Redrovan F 2007 Administración de base de datos con PostgreSQL 3Ciencias **19**
[12] Consejo Nacional de Acreditación 2013 Autoevaluación con Fines de Acreditación de Programas de Pregrado (Bogotá: Colombia).
[13] Resolución de Acreditación no. 16738 del 24 de Agosto del 2017.

Development of a web system for the management of PQRS in higher education institutions

Y Blanco-Jimenez[1], L K Ahumada-Torres[1], J R Castro-Suarez[2] and M A Chico-Ruiz[1]
[1] Fundación Universitaria Antonio de Arévalo-Unitecnar, Cartagena, Colombia
[2] Fundación Universitaria Colombo Internacional, Unicolombo, Cartagena, Colombia

E-mail: coordinacion.software@unitecnar.edu.co

Abstract. The acquisition or development of a software in a company must always obey to meet a specific need, either to carry out a process in a better way or as an added value to a process that is currently managed efficiently. Such is the case of the web system for the management of requests, complaints, claims and suggestions (PQRS) of the Antonio de Arévalo University Foundation - Unitecnar, which emerges as a technological tool to systematize and optimize the current process of registration, consultation, management and monitoring PQRS. The article below details the development methodology implemented that for this case was the agile XP methodology, which allows iterative development, teamwork, and delivery in the shortest possible time. On the other hand, the technological tools used are described, which were: the CodeIgniter framework for application development in PHP, the Bootstrap framework for the creation of responsive web pages and PostgreSQL as a database management system. Finally, the results obtained that consist of two web applications will be explained. One to register, consult and evaluate PQRS. And the other is the tool through which the administration, configuration and monitoring of the PQRS is carried out.

1. Introduction

Technologies in recent times have had a vertiginous boom in all areas of society, and this includes organizations. We see today, how they are being integrated into companies through innovative computer systems. Therefore, its implementation is no longer a luxury or an investment but a fundamental need that allows large and small companies to be at the forefront of new times, with competitive processes both in the national and international market [1].

The acquisition or development of an information system in a company must always obey to meet a specific need, either to carry out a process in a better way or as an added value to a process that is currently managed efficiently.

On the other hand, the contribution that information systems have provided to companies today has been very important, which is why they are considered as one of the most relevant components of the current business environment, which offer great opportunities for success for companies, since they have the ability to collect, process, distribute and share data in a timely and integrated manner. In addition, they help narrow geographic gaps, allowing employees to be more efficient, which is reflected in an improvement of processes, management, and information management, resulting in a positive impact on productivity and competitiveness of the companies [2].

Additionally, the Internet, specifically the WEB, has become the means not only to disseminate information but also as a tool through which companies can carry out tasks or execute operations, with the ease of being able to be carried out from anywhere in the world. This is why computer-based information systems, particularly after the exponential relevance acquired by the web and the internet, have had a profound impact on how organizations, economies and societies in general carry out their activities. [3]

The use of technologies and information systems by organizations is indifferent to the sector to which they belong. This was evidenced in a recent study carried out by the company Fast Company in which it is stated that the most innovative companies are those found in the education sector [4]. This study clearly demonstrates that organizations dedicated to the education sector are increasingly involving information systems for the management of their academic and administrative processes, since having solid and reliable information systems contribute to the transformation and modernization of the sector [5].

All of the above has led the Antonio de Arévalo University Foundation - Unitecnar, making use of the internet and its institutional web platform - which generated competitive advantages in the market - to develop a web system to register, consult, evaluate and respond to requests, complaints, claims and suggestions made.

The purpose of this article is to show the results of the system developed and implemented for the management of PQRS, explaining the development methodology that for this case was XP, the use of the Codeigniter PHP and Bootstrap framework and the database managed with PostgreSQL.

2. Case study description

The Antonio de Arévalo University Foundation - Unitecnar is a higher education institution with headquarters in three of the main cities of Colombia, such as: Cartagena de Indias, Barranquilla and Santa Marta. This institution has established as a quality policy the continuous improvement in its academic and administrative processes, being the PQRS a tool to know the different concerns and situations that affect the users and interest groups that have some relationship with the institution.

The institution, in its quality management system, has a procedure to assess student satisfaction and a format for the registration and attention of PQRS. The only channel for the management of PQRS was email and its procedure was as follows:

- If a user had any request to submit, they should fill out and send the form by email to the Institutional Welfare Department (area in charge of this management).
- The Institutional Welfare Directorate, in turn, evaluated the case presented and sent it to the area responsible for providing a solution.
- The area in charge then performed the requested procedure. And the process ended by sending a message by email with the response to what was requested by the user.

As you can see, all the management of requests, complaints, claims and suggestions described above, was done through emails; thus becoming a slow and tedious process. Considering of the need to improve the process, it is decided by the development of an information system. This system, on the one hand, should allow the registration, consultation and evaluation of the PQRS. And on the other hand, it should allow the administration of registered PQRS, respond to requests, observe the traceability of a specific PRQS, consult user evaluations and generate reports for decision making.

3. Methodology

3.1. Agile XP methodology

There are different methodologies that can be applied in the field of software development, which detail the activities and processes to follow in the execution of projects of this type.

At present, the use of agile methodologies that are characterized by iterative and incremental development, frequent deliveries and simplicity of implementation has been imposed; prioritize requirements, continuous communication with the client and cooperation between developers [6]. It should be noted that the increasing use of this type of methodologies is due to the fact that they are the most appropriate in projects with system environments is very changing, and where it is required to drastically reduce development times, while maintaining high quality [7]. As was the case in Unitecnar, which required a customized information system and quality in the shortest possible time, to manage everything related to PQRS, so it is decided by the XP methodology (Extreme Programming). This methodology evidences principles such as incremental development, active customer participation, accepting change and simplicity [8].

The client requirements were defined through user stories, which is a technique to specify the software requirements [9]. Each story was described by the client in a simple language and with the help of the developers they are assigned a priority, estimating the time for its development [10].

Figure 1 shows the main user stories that make up the system.

No.	User stories	Priority (1-minima 5-maximum)	Estimated time (days)
1	PQRS management areas.	2	3
2	Types of PQRS, with their response times and transfers.	2	3
3	Registration of non-work days by headquarters.	2	2
4	Form to register a PQRS, including the different situations that may arise (sending mail)	5	10
5	PQRS consultation by an applicant. In addition you can evaluate the response received	4	5
6	PQRS administration.	4	7
7	PQRS management by the person in charge of an area (write draft, finalize PQRS, transfer, mail delivery).	5	12

Figure 1. User stories.

Subsequently, a series of deliveries or iterations defined between the client and the developers were established to develop a group of stories. You can create or modify stories at any time. Each delivery includes a new useful functionality within the system, which after being developed, is integrated, tested and evaluated to finish the cycle and then start with a new one, as seen in figure 2. Finally, when there are no more Cycles to execute, it is understood that the system has been fully developed

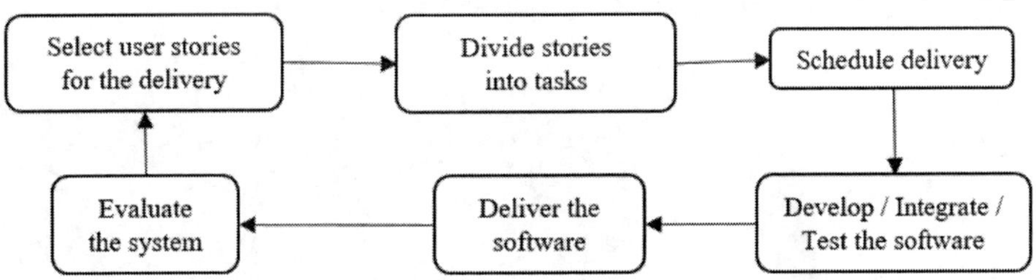

Figure 2. Delivery cycle in Extreme Programming. [8]

The lifecycle of this methodology is composed of the following phases:

3.1.1. Exploration phase: In this phase, the scope of the project was defined based on user stories. For this project, a time of 4 months was estimated. In addition, in this phase the developers became familiar with the technological tools that they would use in the project.

3.1.2. Planning Phase. User stories were prioritized, estimating the time that would be used for their development. In general, a work schedule was established, specifying four (4) iterations, each with its delivery times. Particularly, in this phase the stories that were part of the first installment were chosen.

3.1.3. Iterations Phase. Deliveries were executed, developing and testing the functionalities. As the project progresses, the client can add new stories, modify existing ones or even eliminate specific ones, so developers should adjust their next iterations. At the end of this phase the system is now ready to take to production.

3.1.4. Production Phase. In this phase, the system was enabled for operation and the new improvements that emerged were added to new iterations.

3.1.5. Project Death Phase. When there were no more iterations to perform, this phase has been reached, in which the system is documented, generating the different manuals necessary for the correct use of the system.

4. Technological tools for development

A non-functional requirement of the system is that it could be integrated into the institutional application platform called SIA (Integrated Application System), therefore, web technologies should be used, with the interaction of a database and free tools. Therefore, the technological tools used were:

4.1. Framework CodeIgniter

It is a framework for developing web applications in PHP, which allows the creation of applications quickly due to the set of libraries that it presents for common tasks, as well as a simple interface and a logical structure to access those libraries [11, 12]. The version that was used in this project was CodeIgniter 3.1.9, downloadable from its website, obtaining a folder structure that should be configured and adapted to the project to be developed, as shown in figure 3.

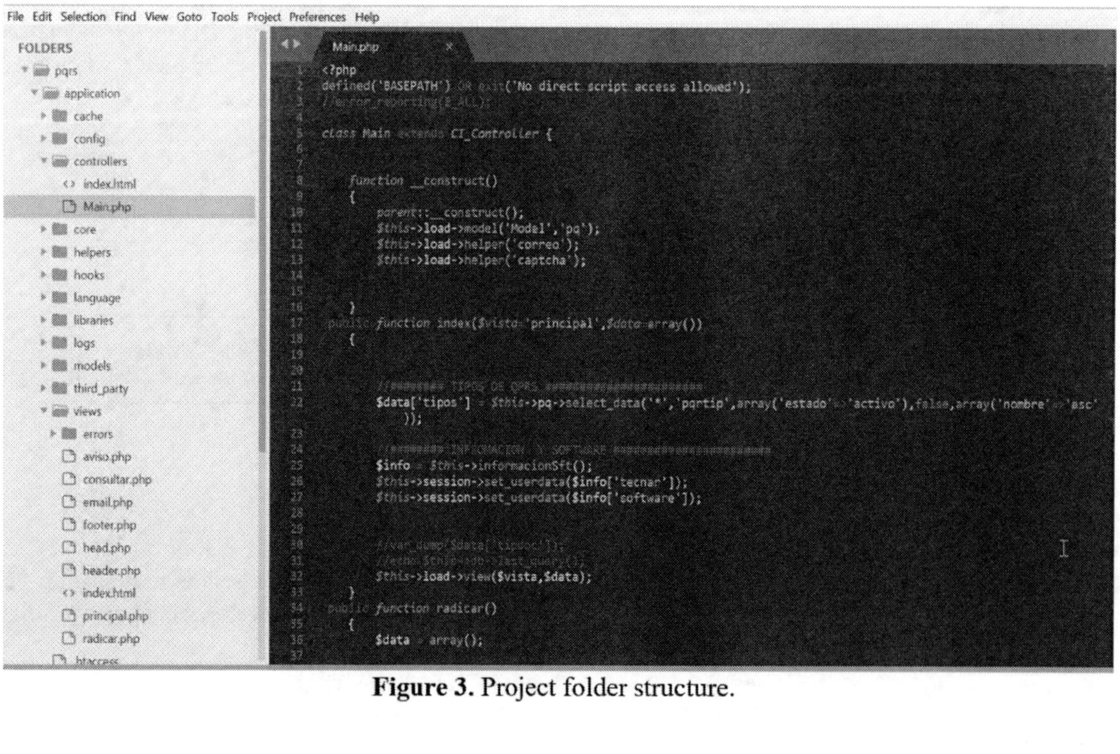

Figure 3. Project folder structure.

4.2. Database with PostgreSQL
It is an open source relational database management system [13], which offers features such as: different and very specific types of data, triggers, functions, views, etc. The version used was PostgreSQL 9.3.

4.3. Bootstrap
It is a library or set of open source tools for web application design, which includes design templates based on HTML and CSS for typography, forms and many others, as well as JavaScript add-ons, etc. [14]. Bootstrap allows you to create responsive, fast and simple web pages.

5. Results

5.1. Structure of the system database
The system has a structure of related tables as shown in figure 4. Likewise, triggers were developed for the execution of certain processes and views to generate queries with several tables, this in order to assign a certain workload to the base of data, thus saving lines of code and execution of processes on project pages.

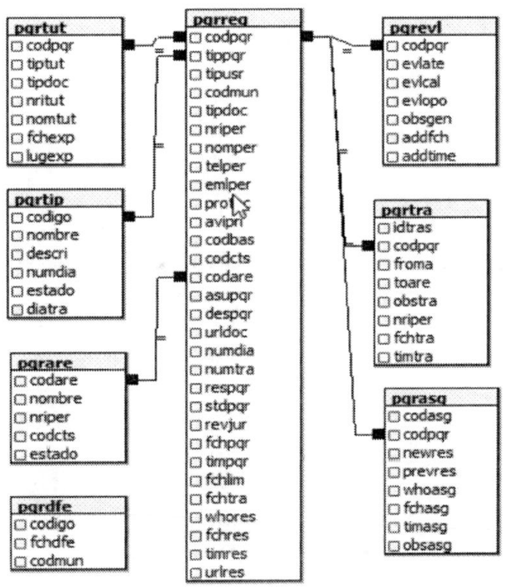

Figure 4. Table structure.

5.2. Description of the System developed
The system is fully responsive and easy to use, consisting of two web applications. One to register, consult and evaluate a PQRS. The other to carry out all administration, configuration, response and monitoring of the PQRS.

The stages of the process that are executed through the system are:

- Registration of a PQRS by the user.
- PQRS consultation by the area manager.
- Transfer to another area.
- Answer PQRS. When an answer is given, a message is sent to the request user.
- Consult a PQRS by the user.
- Evaluation of a PQRS by the user.
- Monitoring and reports.

5.2.1. Application for registration, consultation and evaluation of PQRS: This tool allows any user with access to the Web to make a request, complaint, claim or suggestion to the institution. The application is accessed through the URL (http://appz.unitecnar.edu.co/pqrs/), which will load a page like the one shown in figure 5-A. In addition, it has the options to consult the data of an already registered PQRS (see figure 5-B) and observe the response that has been received and finally evaluate it.

Figure 5. Application for registration, consultation and evaluation of PQRS. A) Main page. B) Form to file a PQRS.

IOP Conf. Series: Materials Science and Engineering **844** (2020) 012068 doi:10.1088/1757-899X/844/1/012068

5.2.2. Application for administration, configuration, response and monitoring of PQRS: This application is integrated into the institutional application platform - SIA. Only users authorized by the institution can enter (see figure 6), and have access to the options that are enabled.

Figure 6. Login to SIA

The user profiles that were defined for this system are: **Administrator** and **area managers**. Based on user stories and system design, **Administrators** can:

- Configure the locations that will handle PQRS.
- Configure the types of PQRS, defining response and transfer times.
- Configure non-work days by headquarters, as shown in Figure 7-A.
- Consult all PQRS, observing response times, delays (see figure 7-B).
- Transfer a PQRS to another area.
- Generate reports and download them in format for Excel and PDF.

Area managers are responsible for responding to a PQRS request. The options it handles are:

- Consult PQRS of the area.
- Give answer. The system allows saving a draft of it.
- Transfer a PQRS to another area, as long as it is on time.

5.3. Benefits of system implementation

The benefits that have been obtained are:

- More efficient management of the data, generating greater control and security in its management.
- Agile and secure access to information, with a tool that facilitates searches and report generation.
- Use of the system at any time and from any place since it was developed oriented to the web.
- Reduction of institutional costs in human talent, physical infrastructure, time spent in the execution of procedures and / or manual tasks, etc.
- Have a computer tool tailored to institutional needs.

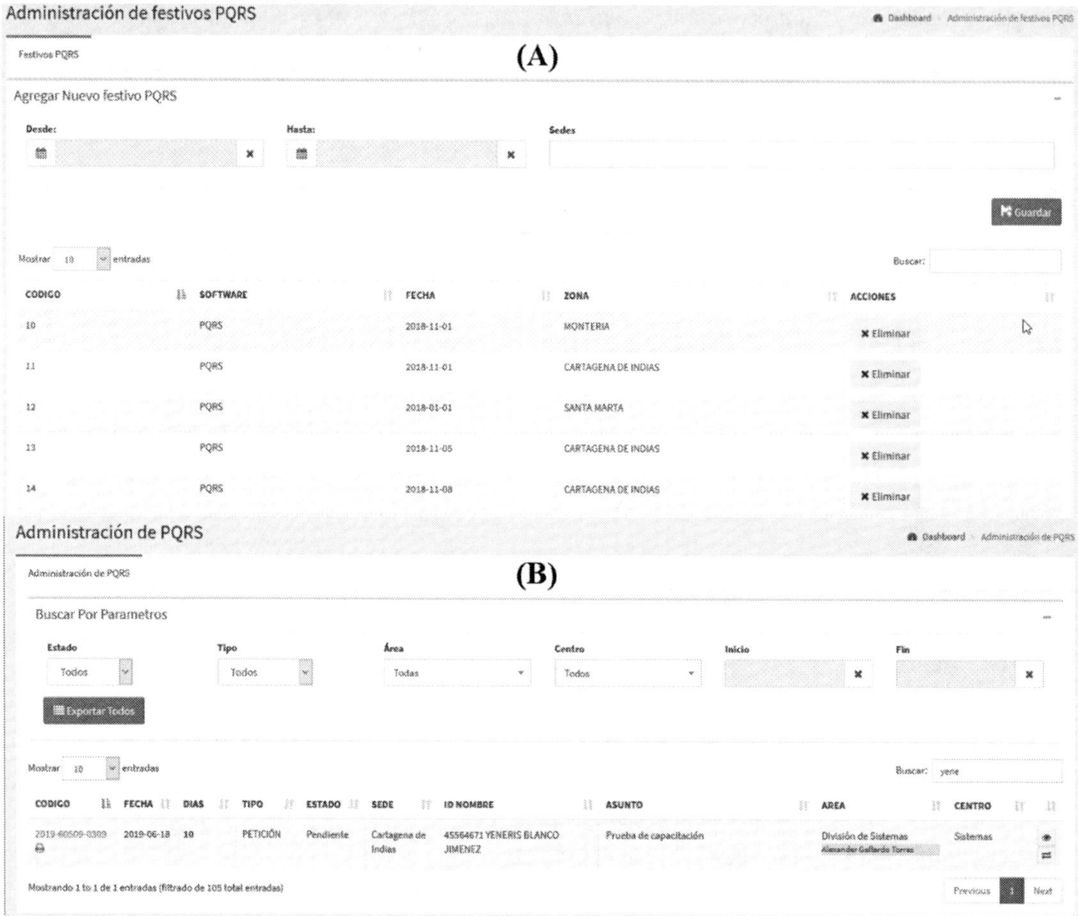

Figure 7. Application for administration, configuration, response and monitoring of PQRS. A) Management of non-work days by headquarters. B) Consult all PQRS.

6. Conclusions

Integrating technologies into the execution of activities carried out in organizations will always be of great benefit, contributing to the optimization of their processes, resulting in much more efficient management, reducing organizational costs, among others.

The implemented web system has allowed the community in general to have better communication channels to express their concerns and suggestions. On the other hand, the institution currently has a tool through which it obtains first-hand information on those situations that afflict users, in order to execute actions that allow improving processes, guaranteeing a better service and improving levels of satisfaction.

According to how the system was designed and developed, its implementation can be extended to other types of organizations, since it has the basic functionalities necessary to manage PQRS processes.

Conflict of Interest.

The authors report there are no conflicts of interest.

References

[1] Periodico ELTIEMPO 2004 https://www.eltiempo.com/archivo/documento/MAM-1532000
[2] Almazán D A, Tovar Y S and Quintero J M M 2017 Influencia de los sistemas de información en los resultados organizacionales *Contaduría y administración*, **62(2)** 303-320
[3] Proaño M, Orellana S and Martillo I 2018 Los sistemas de información y su importancia en la transformación digital de la empresa actual *Revista ESPACIOS 39*(**45**)

[4] Fast Company Staff 2018 The World's Most Innovative Companies *Fast Company* https://www.fastcompany.com/most-innovative-companies/2018/sectors/education
[5] Ministerio de Educación Nacional de Colombia 2008 Sistema de Atención al Ciudadano https://www.mineducacion.gov.co/1621/article-156303.html
[6] Navarro M E, Moreno M P, Aranda J, Parra L, Rueda J R and Pantano J C 2017 Selección de metodologías ágiles e integración de arquitecturas de software en el desarrollo de sistemas de información *XIX Workshop de Investigadores en Ciencias de la Computación (WICC 2017, ITBA, Buenos Aires)*
[7] Letelier P and Penadés M C 2012 Metodologías ágiles para el desarrollo de software: eXtreme Programming (XP)
[8] Pérez O A 2011 Cuatro enfoques metodológicos para el desarrollo de Software RUP–MSF–XP-SCRUM *Inventum* ***6*(10)** 64-78
[9] Medina I I S, Rojas F M and Medina, J M C 2018 Diseño de software para calcular la huella de carbono e hídrica durante la producción de café *Ingeniería Solidaria* ***14*(24)** 1-12
[10] Roger S P 2002 *Ingeniería de Software: Un enfoque práctico* (New York: McGraw Hill) pp. 84-88
[11] British Columbia Institute of Technology 2019 Guia de usuario de CodeIgniter https://codeigniter.com/user_guide/general/welcome.html
[12] Vidal C L, López, L L, Rojas J A and Castro M M 2017 Desarrollo de sistema web de reclutamiento y selección y de directivos por competencias mediante php codeigniter 3.0. *Información tecnológica*, ***28*(2)**, 203-212.
[13] The PostgreSQL Global Development Group 2018 PostgreSQL: The World's Most Advanced Open Source Relational Database https://www.postgresql.org/about/
[14] Refsnes Data 2019 Bootstrap 4 Get Started https://www.w3schools.com/bootstrap4/bootstrap_get_started.asp

IOP Conf. Series: Materials Science and Engineering **844** (2020) 012069 doi:10.1088/1757-899X/844/1/012069

Mechanical effort of ceramic particle-reinforced epoxy matrix used as a sealing layer on aluminum coatings by Arco Spray

E E Vergara[1,a], L M Castellanos[2,b], L M Diaz[2], A R Lucas[2] y B B De Avila[2]

[1] Naval Cadet School Almirante Padillas. Cartagena, Colombia
[2] Faculty of Engineering. Universidad Simón Bolívar. Barranquilla. Colombia

[a] edgar.vergara@armada.mil.co, [b] luis.castellanos@unisimonbolivar.edu.co

Abstract. The coating against the corrosion and wear of carbon steel with pure aluminum applied by thermal projection with electric arc (arc spray) presents a porous surface. This is the reason why adding a sealing with a polymer layer, regularly epoxy is necessary. However, epoxy-based sealing solutions usually do not have high performance against abrasive wear. In this research a new product has been developed in composite material with polymer matrix (PMCs), epoxy paint reinforced with ceramic particles, in order to improve the performance of this layer against abrasive wear. Preliminary runs were made with 3 variables: Epoxy matrix (2 types: epoxy S and epoxy C); Size of reinforcing ceramic particles (two levels: 1 μm 9 μm) and percentage of ceramic reinforcement (4 levels: 0%, 20 %, 33.3% and 50 %). Adhesion and abrasive wear tests were performed to show that the composite material with epoxy matrix type C, has better performance than the composite material with epoxy matrix type S. Increasing the percentage of reinforcement particles showed that the composite material with type C epoxy matrix maintained high levels of adhesion to the aluminum layer and showed an increase in abrasive wear resistance by about 60% compared to the resin without reinforcements.

1. Introduction

In recent years, surface treatment and corrosion control in structures is an issue that goes beyond protection and conservation, including in immersion materials such as: boats, ships, boats, tugs, barges, stilts, dock pilings, pipes and underwater platforms [1] metal coatings with Aluminum, Zinc and combinations thereof have proven to be a very good alternative for corrosion protection given the double protection of these metals, by barrier and by cathodic protection is also a clean, emission-free production that does not contaminate the environment . Worldwide, the coating with aluminum by Thermal Projection with electric arc is replacing the paints; However, the problem of abrasive wear, although improved with respect to paints, is an unresolved problem. During the last two years, a group of researchers from the Simon Bolivar University have worked and researched the optimization of aluminum coating, in order to achieve the double performance of the deposited layer against wear and corrosion [2] . Despite the high initial cost of applying aluminum coating, with arc spray technology, compared to high-performance paint coating, the group of researchers managed to estimate that the cost of the life cycle in 21 years of a Caribbean fuel transfer barge, coated with aluminum, is about 40% of the cost when coated with paint, and maintains structural integrity by extending its lifespan, while with painting processes at the end of the cycle in 21 years will be very deteriorated [3], this is due to the pores and conduction pathways that occur through the shrinkage of the coating during the curing process, causing the ease of corrosive electrolyte penetration into the film and access to the metal interface /

IOP Conf. Series: Materials Science and Engineering **844** (2020) 012069 doi:10.1088/1757-899X/844/1/012069

coating [4]; along with the effects of low draft rubbing on the rivers in the shipping sector, therefore a naval device coated with aluminum with the arc spray technique, will generate in the long term greater profitability, reducing maintenance costs [5].

The surface of aluminum coatings and composite materials with aluminum matrix applied by arc spray, are porous and require a sealing layer that covers such pores and maintains high corrosion resistance, usually sealing is performed with a layer of epoxy paint, but this does not perform high against abrasive wear.

According to a study aimed at the industrial applications of composite materials at a university in Caracas, Venezuela, the development of a coat of paint, "wash primer" which is used to generate a high resistance to corrosion and wear, increasing the life of the device by approximately 27 years; exposing it to a harsh environment, where corrosion and wear have little significance when exposed, it is also determined that once its contact with water is not consecutive, it recovers with the time it stays away from it [6] . Other studies have shown that the mechanical behaviour and wear of epoxy compounds, reinforced with particles such as Al, Ti, C, Si and Al_2O_3 , depend on the size of particles, the percentage of reinforcement and the preparation of the mixture [7] [8], in addition to identifying the appropriate type of polymer matrix, is very important for obtaining a good composite material. Among the most commonly used are the Epoxy resins, Polymethylmethacrylate and Polyeterimide [9].

In order to improve corrosion resistance, paints with high solid content have been developed; and its application requires an optimal coating design [10], for which a thorough analysis of the application process must be carried out; these advanced coatings allow good adhesion and extensive wear resistance, and are also made for specific ambient conditions and thus take advantage and achieve an optimal coating and be able to reduce costs for the mechanisms to be coated [11] [12].

This article presents the results of a preliminary study of the mechanical performance of new sealing products, developed by the GEMAS group of the Simon Bolivar University, made of composite material with reinforced polymer matrix (epoxy paint) with micro ceramic particles to increase wear performance in the shipping and maritime sector [13].

Two types of epoxy paint, red epoxy (S) and an epoxy with high solid (C), were used as a matrix; as a reinforcement, a high hardness ceramic powder was used in two particle sizes and in three percentage concentrations. Two types of mechanical tests were carried out after the specimens were obtained: adhesion, by the "pull off" method and to abrasive wear by the "pin on disk" method [14] [15], which serve as the basis for the development of new products in composite material for the Aluminum sealing by arc spray that ensure better performance as a solution to simultaneous wear and corrosion problems, mainly in the river shipping sector.

2. Materials and Methods

2.1. Materials.

In this study, two epoxy primers were used as a matrix of the composite coating reinforced with ceramic particles for the sealing of aluminum obtained by electric arc, one in red (referred to as S) and one in grey color (referred to as C), the latter with high aluminum metal solids. As a reinforcement, mixtures of metal oxides (Al, Ti and Zr) were used sifted with granulometry of 9 μm. As a substrate for the application of sealing in composite material, 80 mm x 80 mm specimens, with thickness of 3,175 mm (1/8 inch), coated with a layer of 200 μm of pure aluminum by arc spray were used, figure 1.

(a) (b) (c)
Figure 1. Samples coated (a) with aluminum, (b) and (c) with sealing layers S and C.

2.2. Methodology.
An experimental design is developed with two variables, one type of epoxy paint in levels (S or C) and the other the percentage of ceramic reinforcement in four levels (0%, 25%, 33.3% and 50%) see table 1, with two output factors that are: the degree of abrasive wear (see table 2), from the loss of layer thickness in the "pin on disk" (see figure 2 - (b)), as per ASTM G99-04 test and adhesion to the aluminum substrate (see table 3), evaluated by the "pull off" (see figure 2 - (a)), as say as per ASTM D4541. 4 replicates were made for each combination with a total of 12 specimens for each type of test.

The pin on disk test used a load of 2 N, abrasive paper No 600, rotation frequency of 342 RPM, turning radius 30 mm and the time that is limited by the thickness of the composite material, was estimated from a preliminary test, for this case with a thickness between 100 μm and 200 μm, the time taken was 30 s. Then, with a micrometer, the initial thickness of each sample is measured.

Using the ASTM G99-04 method, the procedure is performed to determine the wear of the materials using the pin on disk equipment, figure 2-(b), the thickness is measured again and the results tabulated in table 2 are obtained. Unreinforced epoxies are called (S-0) and (C-0), the following nomenclature is used to identify each specimen in composite material: the letter indicates the type of epoxy primer, the first number of ceramic particles in μm, and the last number is percentage of reinforcement in the matrix (see table 1). figure 2, shows the equipments used for this investigation.

Table 1. Samples obtained and their designation.

Percentage of reinforcement	Epoxy resin	
	S	C
0	S – 0	C – 0
20	S-9-25	C-9-25
33,3	S-9-33	C-9-33
50	S-9-50	C-9-50

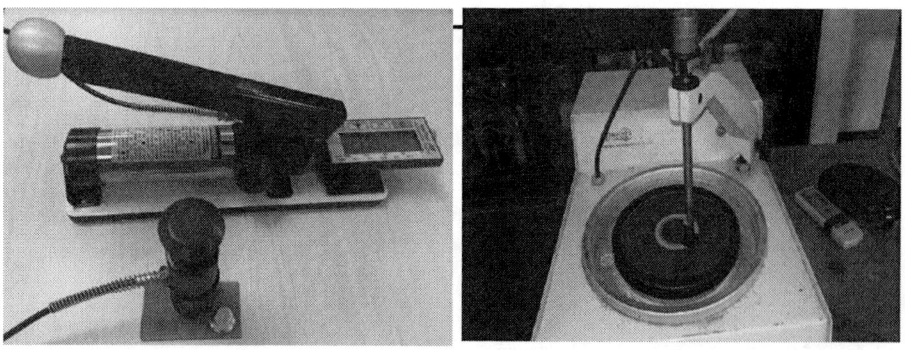

(a) (b)

Figure 2. Tests (a) Pull off adhesion and (b) Abrasive wear "pin on disk".

3. Conducting the experiment

Figure 1 shows the sequence of application of the two layers, (a) the base layer of 200 μm aluminum by arc spray on carbon steel A-36 to white metal, (b) and (c) the sealing layers with the composite materials obtained in the project. The mixtures of the ceramic reinforcement with the epoxy primer were made in parts by volume and dispersed by mechanical agitation in the hands of being applied with paint gun. They were dried in the environment for 48 hours before being cut with abrasive discs for laboratory testing. For each combination of variables, according to the experimental design 4 double-layered specimens were made, which were sectioned according to the test to be performed; for the wear test were obtained specimens of 1 cm^2 taken away from the edges, for the adhesion test the dimensions were 40 mm × 80 mm.

3.1. Experimental results

3.1.1. Abrasive wear testing

Table 2. Results. Pin on disk wear tests. Epoxy C and S reinforced with 9 μm ceramics.

Epoxy C	**Thickness loss (mm)**					
Specimens	**P1**	**P2**	**P3**	**P4**	**Average value**	**Error**
C- 0	0,10	0,11	0,08	0,09	0,10	0,011
C-9-25	0,05	0,07	0,06	0,06	0,06	0,007
C-9-33	0,04	0,03	0,04	0,03	0,04	0,005
C-9-50	0,03	0,03	0,04	0,03	0,03	0,004

Epoxy S	**Thickness loss (mm)**					
Specimens	**P1**	**P2**	**P3**	**P4**	**Average value**	**Error**
S- 0	0,10	0,12	0,11	0,11	0,11	0,007
S-9-25	0,10	0,05	0,06	0,05	0,07	0,021
S-9-33	0,07	0,05	0,06	0,05	0,06	0,008
S-9-50	0,06	0,07	0,09	0,06	0,07	0,012

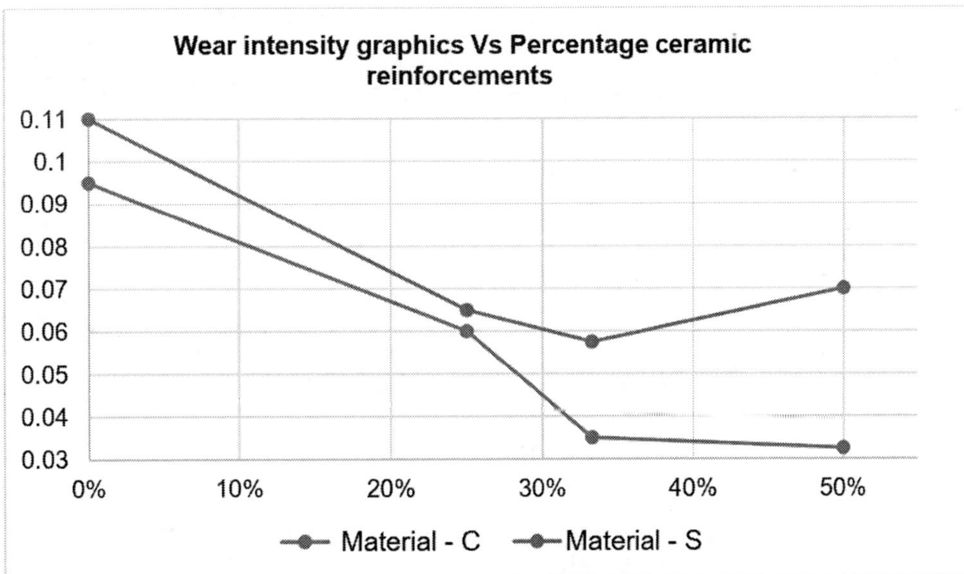

Figure 3. Wear Intensity Graphics Vs% Of Ceramic Reinforcements.

3.1.2. From the adhesion test. Table 3 shows the results of the adhesion tests by the "Pull off" method with the two epoxy paints as the matrix of the composite material. The adhesion values of each specimen represented by P_i are obtained from the average of three tests according to standard; The final mean value represents is the result of the 4 replicas.

Table 3. Results. Pull-off adhesion tests. Epoxy C and S reinforced with 9 μm ceramics.

Epoxy C	Pull Off Adhesion (Psi)					Error	Observation
Specimens	P1	P2	P3	P4	Average value		Separation mode
C- 0	980,6	879,0	990,4	979,0	957,3	45,4	Cohesive
C-9-25	986,0	821,0	688,0	845,0	835,0	105,7	Cohesive
C-9-33	750,3	710,8	680,0	740,1	720,3	27,4	Cohesive
C-9-50	545,0	571,0	560,0	571,3	561,8	10,7	Cohesive

Epoxy S	Pull Off Adhesion (Psi)					Error	Observation
Specimens	P1	P2	P3	P4	Average value		Separation mode
S- 0	601,2	580,7	595,7	593,0	592,7	7,5	adhesive with actuator
S-9-25	425,0	422,1	425,0	426,5	424,7	1,6	adhesive with actuator
S-9-33	134,0	134,6	138,0	137,4	136,0	1,7	adhesive with actuator
S-9-50	123,0	125,2	121,8	124,8	123,7	1,4	adhesive with actuator

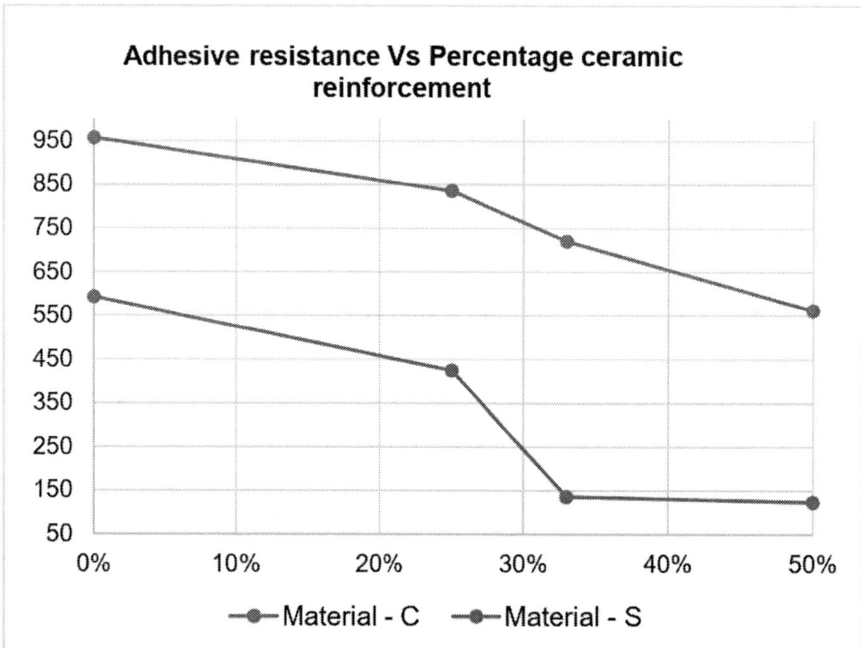

Figure 4. Adhesion test graphs (Psi) Vs the percentage of reinforcement.

Figure 5 shows some visual inspection images of tested specimens using the "Pull off" method, allowing you to determine whether the fault is adhesive, cohesive or mixed.

(a) (b)

Figure 5. Failure mode of tested specimens with "Pull off", (a) group C adhesive and (b) cohesive group S.

3.1.3. SEM/EDX analysis results. Figure 6(a) shows the three layers of a C-50 specimen, the base of carbon steel, then the middle layer of around 100 – 120 μm of aluminum per arc spray and the top layer thickened around 300 μm are shown. Some surface defects are noticed because it was manually paired where the gun did not give the required thickness; in real practice should not appear.

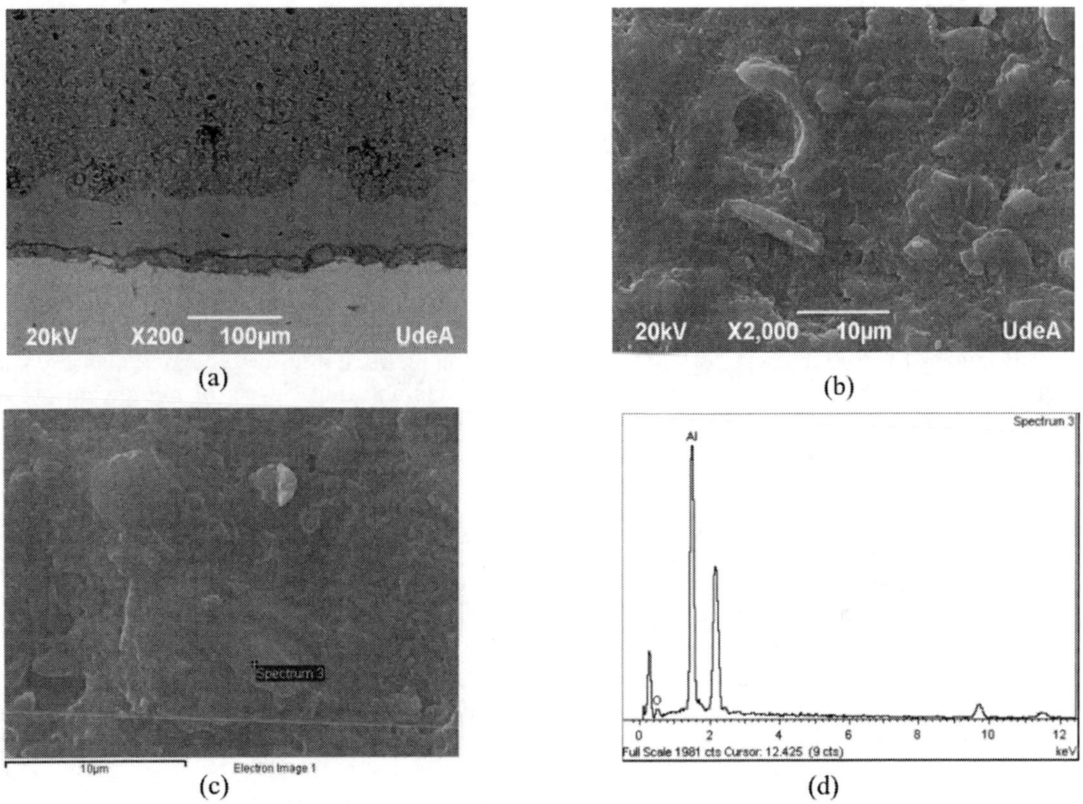

Figure 6. (a), (b) and (c) Images with different magnification obtained by SEM from a C-50, (d) Spectrun specimen obtained by EDX showing the peaks of Aluminium and oxygen of aluminium oxide shown in Figure 6 (c).

In images 6 (b) and (c) the high concentration of ceramic particles with good adhesion to the epoxy matrix is shown; there are also some pores. In Figure 6 (d) the spectrum obtained by EDX from the micro-analyzed ceramic particle in Figure 6 (c) shows the presence of alumina, with 79.35% Al by weight and 20.65% O by weight.

4. Discussion

Analysing mechanical performance, abrasive wear resistance and adhesion to the aluminum substrate, sealing layers with unreinforced epoxy matrix (S-0 and C-0) show that the red epoxy (S-0) without reinforcement has lower adhesion and less abrasive wear resistance than grey epoxy (C-0) as shown in figures 3 and 4. The use of these epoxy primers reinforced with ceramic particles of 9 μm, makes them a composite material with better performance against abrasive wear, in figure 3 it can be seen that with the increase of % reinforcement, the mechanical performance of the composite material is improved, being more relevant in the grey epoxy of high solids in metallic aluminum. According to the results obtained, it is observed that the worn thickness was reduced from 100 μm (without reinforcement) to 30 μm with a 33.3% reinforcement during test time: the increase to 50% reinforcement showed no significant change from 33.3%, indicating that attention should be paid to the interval 30% and 35 % of reinforcement to further investigate this material and analyse the rest of the properties of interest, such as resistance to bending, shock and adhesion.

The increase in the percentage of reinforcement, as expected, reduces the adhesion of the sealing layer, for this reason a maximum reinforcement limit must be found that guarantees improvements in wear resistance without detriment of the adhesion and without sacrificing resistance to impacts and mild bending of the substrate, the latter two properties have not been evaluated in this work. Figure 4 shows that in the grey epoxy (C) reinforced with ceramic the adhesion is maintained with high values with an increase of 50% reinforcement; for 33, 3% reinforcement the adherence was in the order of 560 Psi. The adhesion of the S compound turned out to be very low falling below 200 Psi for 33.3% reinforcement. AS illustrated in Figure 5, in the case of red epoxy (S) all failures in the adhesion test were of the cohesive type, indicating lack of cohesion inside the coating; in case of grey epoxy (C) all faults are mostly adhesive between the compound and the actuator, which shows strong adhesion between the aluminum and the sealing layer.

5. Conclusions

The study of the mechanical performance of the ceramic particle-reinforced epoxy matrix composite material has been carried out using two types of primers and three levels of reinforcement with an average size of 9 μm. The following conclusions are reached.

There is a significant difference in mechanical behavior between the two used epoxide primers, red (S) and grey (C), both unreinforced and reinforced with ceramic particles, being superior adhesion and resistance to abrasive wear of epoxy grey with high solids in aluminum metal, specially designed for porous aluminum sealing.

With the increase in the percentage of reinforcement both epoxies showed increased performance against abrasive wear and loss of adhesion to the aluminum substrate. Interestingly, these effects are interesting in grey epoxy, whose intensity to abrasive wear with reinforcements between 33% and 40 was reduced by 60% compared to the case without reinforcement, maintaining high levels of adhesion to the substrate. The adhesion values between grey epoxy and aluminum deposited by arc spray could not be specified, as more than 95% of the failures in the "pull off" test occurred between the actuator and the coating.

In general, a composite sealing product can be achieved with an epoxy matrix reinforced with ceramic particles with medium size of 9 μm, with a significant increase in abrasive wear resistance with reinforcement in the order of 33% and above, maintaining high levels of adherence to the substrate; above 33% reinforcement there was no significant increase in abrasive wear resistance, with no evidence of significant impact on adhesion to the aluminum substrate.

IOP Conf. Series: Materials Science and Engineering **844** (2020) 012069 doi:10.1088/1757-899X/844/1/012069

References

[1] De la Cruz P A 2017 Mecanismos de corrosión en agua de mar: Estructuras de acero
[2] Luddey L, Arévalo M, Pérez M D and Millan A R 2016. Resistencia a la corrosión en ambiente salino de un acero al carbono recubierto con aluminio por rociado térmico y pintura poli aspártica Corrosion resistance in saline environment of a carbon steel coated with aluminum by thermal spray and painting poly " vol. 30 no. 1 pp 21–31
[3] Morcillo M, Alcántara J, Díaz I, Chico B, Simancas J, and De la Fuente D 2015 "Marine atmospheric corrosion of carbon steels" Rev. Metal. vol. 51 no. 2 p 45
[4] Asghar J A, Naderia R, Ramezanzadeh B 2019 Epoxy-polyamide nanocomposite coating with graphene oxide as cerium nanocontainer generating effective dual active/barrier corrosion protection
[5] Campo M, Escalera M D, and Torres B 2007 "Comportamiento a desgaste de recubrimientos de material compuesto de matriz de aluminio fabricados por proyección térmica" vol. 43 no. 5 pp 359–369
[6] Maiz A D 2013 "Recubrimientos metálicos y cerámicos solucionan el problema económico y ambiental de países desarrollados y en desarrollo" vol. 25 pp 25–34
[7] Jamshidi R, Heidarpour A 2019 Improvement in the mechanical and tribological behavior of epoxy matrix with the inclusion of synthesized Ti3AlC2 MAX particles
[8] Arias A, Zaera R, López P J y Navaro C 2014 Fabricación y caracterización mecánica de un material compuesto de matriz polimérica y carga cerámica
[9] Frankel S G 1998 "Pitting Corrosion of Metals," J. Electrochem Soc vol. 145 no. 6 p 2186
[10] David R, Raja V S, Singh S K and Gore P 2018 "Development of anti-corrosive paint with improved toughness using carboxyl terminated modified epoxy resin" vol. 120 pp 58–70
[11] Gutiérrez Q W 2015 "Resistencia al desgaste abrasivo de recubrimientos producidos por proyección térmica para aplicaciones navales" p. 131
[12] Pi M M 2016 "Tratamientos Superficiales : Sistemas de aplicación de pintura utilizados en los astilleros"
[13] Hilerio I 2015 "Medición De Desgaste Abrasivo En Una Máquina De Pin on Disk (Perno Sobre Plato)"
[14] Fletcher J F and Barnes D J 2015 "Pull-Off Adhesion Testing of Coatings - Improve Your Technique"
[15] Evans B R, Lian J and Ji W 2018 "Evaluation of shielding performance for newly developed composite materials" Ann. Nucl. Energy vol. 116 pp 1–9
[16] Martinez X, Oller S, and Barbero E 2011 "Caracterización de la delaminación en materiales compuestos mediante la teoría de mezclas serie/paralelo," *Rev. Int. Metod. Numer. para Calc. y Disen. en Ing.* vol. 27, no. 3 pp 189–199
[17] García B E, De los Santos J, Soto T J, De los Santos V H and Abreu C E 2013 "Diseño de Elementos de una Máquina Pin-on-Disk" pp 14–16
[18] Garcia S R, y Real R J 2009 Comportamiento a fractura de resina epoxi cargadas con partículas de carburo de silicio: influencia de la proporción y silanización de las partículas Universidad Carlos III de Madrid
[19] Esteban J, Ospina C y Bedoya D C 2016 "Reingeniería en máquina para desgaste tipo pin sobre disco Reengineering in pin on disk wear machine" no. xx pp 1–8
[20] Coatings R 2013 "Standard Test Method for Pull-Off Strength of Coatings Using Portable Adhesion" pp 1–16
[21] Ojados S G 2013 Resistencia al desgaste y a la abrasión de polimeros técnicos y nanocomposites, Tesis doctoral Universidad Técnica de Cartagena España
[22] ASTM G99-04 2010 Standard Test Method for Wear Testing with a Pin-on-Disk Apparatus
[23] Douglas M 2007 *Diseño y analisis de experimento-segunda edición* (Mexico: Limusa Wiley) pp 60-86

Computational simulation of the temperature distribution of aerogel–based coated and uncoated multi–layer panels for thermal insulation

A Ángel[1,2], G Suárez[1,3] and J Niño [1,2]

[1] Grupo de Investigación en Matemáticas y Computación, Centro de Ciencia Básica
[2] Grupo de Investigación en Ingeniería Aeroespacial, Facultad de Ingeniería Aeronáutica
Universidad Pontificia Bolivariana, Medellín, Colombia

E-mail: [3]`gustavo.suarez@upb.edu.co`

Abstract. A computational simulation is developed to study the temperature distribution of two types of multi–layer panels, one of which utilizes an aerogel–based coating. The simulation considers properties such as layer thickness, thermal conductivity, specific heat and density of the materials. From a set of materials, a multi–layer is geometrically modeled to perform the simulation. The results obtained indicate the temperature distribution across the different layers of materials and evidence the presence of the aerogel material with respect to the uncoated model. This research aims to visualize the effect of whether or not having a special coating can benefit the thermal distribution in the panel, hence, the thermal control within enclosures and architectural structures.

1. Introduction

Various types of materials are being implemented to enhance thermal conditions within enclosures and architectural structures. One of the biggest costs that have to be considered in structural design is the thermal control in closed spaces, since these costs correspond to most of the energy consumption of buildings through wall thermal losses [1]. Even though there are various options of insulating coatings, there are few materials that out–stand in all of the thermo–physical properties needed to achieve an appropriate function as a thermal insulator [2]. Therefore, the development of new materials that are able to comply with the current needs of building insulation have been carried out, being aerogel–based materials some of the most promising for this task [3].

Kyriakidis et al [4] carried out comparisons through numerical modelling in which radiation within traditional and contemporary walling was studied. Also, both micro structural [5, 6] and composite panel studies [3, 7] have been made, both in which aerogel has been found to perform better subjected to various thermal conditions in comparison to different insulating materials alternatives.

In the present study, a comparison between an aerogel–coated and uncoated multi–layer panel was developed by means of computational simulations. Both panels, aerogel–coated and uncoated were evaluated through an external condition of a fixed 45 ^{o}C temperature.

IOP Conf. Series: Materials Science and Engineering **844** (2020) 012070 doi:10.1088/1757-899X/844/1/012070

2. Methodology

2.1. Layer properties and simulation conditions

The computational simulations were made on two multi–layer panels where the benefit of whether having or not an aerogel–based coating was studied by each sample's temperature distributions through the various layers of materials. The materials' sequence that make up the multi–layers was similar to those proposed in [8] and [9] and whose properties can be seen in Table 1. To evaluate the thermal conductivity of the multi–layer panels, a surface temperature of 45 °C was considered in the aerogel and concrete layers for the coated and uncoated cases, respectively.

Table 1: Thermo–physical properties of the multi–layer panel [8]

Material	Thickness (cm)	Thermal conductivity (W/(mK))	Specific heat (J/(kgK))	Density (kg/m^3)
Plaster (inside layer)	1.3	0.32	800	790
Glass Wool	16	0.041	840	12
Concrete	25	2.1	800	2400
Aerogel–based coating	4	0.027	1100	200

2.2. Mathematical modelling

The heat transfer equation in steady–state regime was established to model the thermo–conductive behavior through all of the layers of material that were implemented. The mathematical model is as proposed by Ángel, Suárez and Niño in [5].

$$k_c \nabla^2 T = -\rho h \in \Omega^2 \tag{1}$$

Where T corresponds to the temperature of the material on a specific point, k_c is the thermal conductivity of the aerogel SiO_2 which takes values between 13–20 mW/(mK), or each layer's thermal conductivity and ρh is the rate of heat generation or accumulation of internal energy within the material.

And the heat flow conditions or temperature that is applied over the external surface can be described through:

$$k_a \frac{\partial T}{\partial x} = h_c(T - T_\infty) \in \Omega^2 \tag{2}$$

$$T \in \partial\Omega \tag{3}$$

Where $\partial T/\partial x$ is the heat flow due to convection through the boundaries, h_c is the convective coefficient of the fluid to which the surface is exposed, k_a is the thermal conductivity of air in W/mK, T is the temperature along the boundary and T_∞ is the ambient temperature.

2.3. Geometrical and numerical modelling

A panel of 600×700 cm and of thickness as specified in Table 1, was developed. The aerogel–coated and uncoated panels are presented in Figure 1.

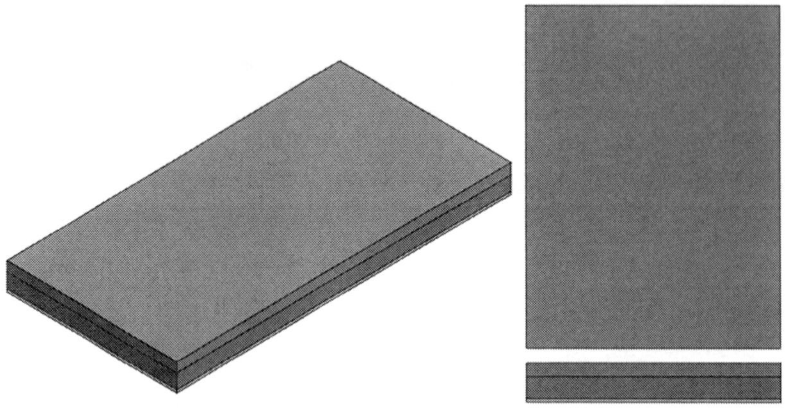

Figure 1: Geometrial modelling of the multi–layer uncoated panel.

The numerical meshing of both panels was developed through hexahedrical elements for each one of the layers, implementing finite element methods for the solution of the differential equations that describe the behavior of both composite panels [10]. The meshing of the multi–layer panels is shown in Figure 2.

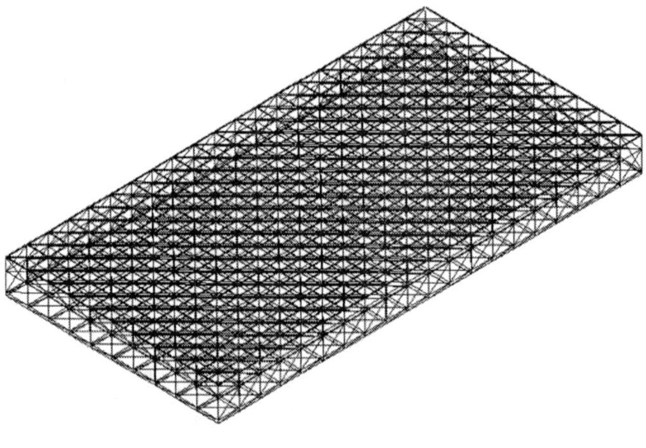

Figure 2: Multi–layer panel meshing

3. Results and discussion

The temperature distribution of the aerogel–based coating panel can be seen in Figure 3. Here, most of the layers have a constant temperature of around 30 ^{o}C and most of the heat exchange appears to happen in the outer layers, these being the aerogel and part of the concrete; where the biggest temperature differences occur, specially in the aerogel layer. The latter is due to the low thermal conductivity and high porosity of the aerogel material, which have been previously reported [11, 12] to be the major causes of its effectiveness retaining the energy that is trying to be conducted through the different layers. Also, within the aerogel layer there are some areas that have a greater temperature than the rest, this may be due to the heat transfer that may be easier to occur in certain directions rather than the others.

IOP Conf. Series: Materials Science and Engineering **844** (2020) 012070 doi:10.1088/1757-899X/844/1/012070

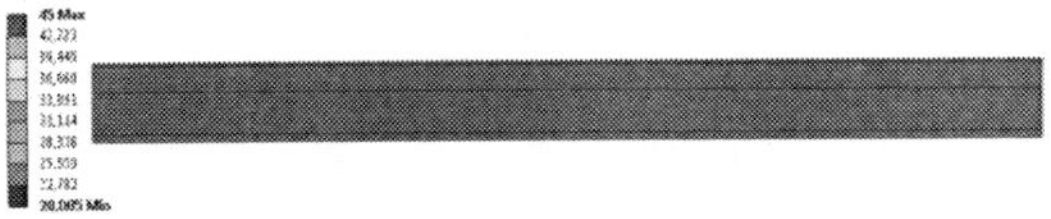

(a) Temperature distribution in °C.

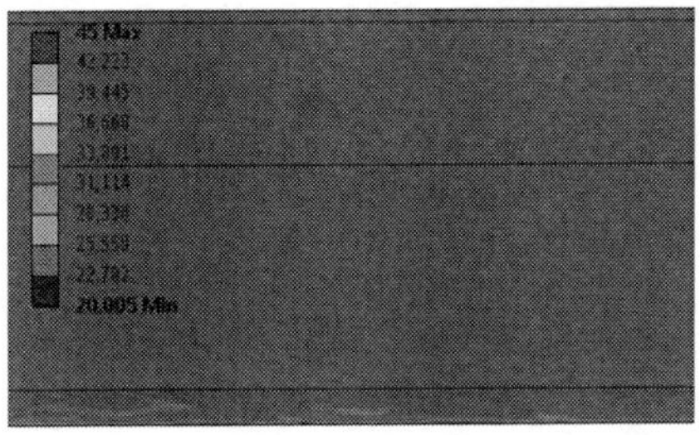

(b) Close up view of the temperature distribution.

Figure 3: Multi-layer panel with aerogel–based coating results.

In Figure 4 the temperature distribution of the uncoated panel can be seen. There is a greater area with a variable temperature, which corresponds to the concrete layer; nevertheless, the adjacent layers have a slight increase in temperature too. In general, there are greater temperature variations through the entire layers which clearly evidences the absence of aerogel. Although the temperature values are lower than in the aerogel–coated case, there are more temperature fluctuations.

(a) Temperature distribution in °C.

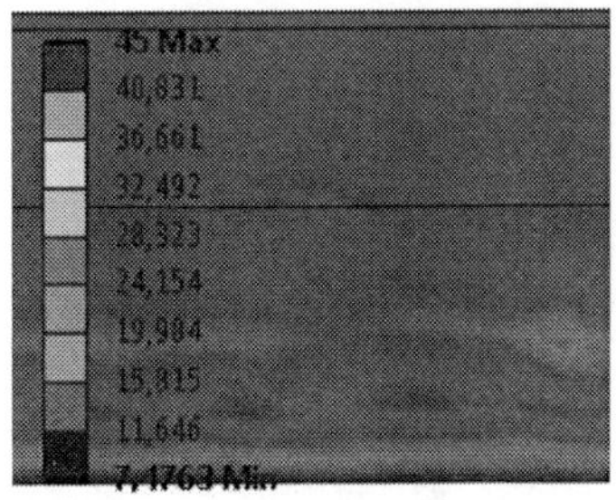

(b) Close up view of the temperature distribution.

Figure 4: Multi-layer panel without aerogel–based coating results.

4. Conclusions

Being thermal control one of the biggest costs that have to be taken into account for structural design, efforts have to be made in the understanding of the behavior of new insulating material alternatives. A computational simulation was made in which the ability of energy retention of aerogel was shown in comparison to a conventional multi–layer wall. Aerogel proved to make a difference in keeping a constant temperature through the entirety of the layers. A greater temperature was obtained in the over all aerogel–coated multi–layer relative to the uncoated multi–layer.

References

[1] Balaras C, Droutsa K, Argiriou A and Asimakopoulos D 2000 *Energy and Buildings* **31** 143–154
[2] Jelle B P 2011 *Energy and Buildings* **43** 2549–2563
[3] Yang J, Wu H, Xu X, Huang G, Xu T, Guo S and Liang Y 2019 *Renewable Energy* **138** 445–457
[4] Kyriakidis A, Michael A, Illampas R, Charmpis D C and Ioannou I 2018 *Energy* **161** 1016–1027
[5] Ángel A, Suárez G and Niño J 2019 *IOP Conference Series: Materials Science and Engineering* **519** 012008
[6] Qu Z, Fu Y, Liu Y and Zhou L 2018 *Applied Thermal Engineering* **132** 730–739
[7] Lucchi E, Becherini F, Di Tuccio M C, Troi A, Frick J, Roberti F, Hermann C, Fairnington I, Mezzasalma G, Pockelé L and Bernardi A 2017 *Building and Environment* **122** 258–268

[8] Ibrahim M, Biwole P H, Wurtz E and Achard P 2014 *Building and Environment* **81** 112–122
[9] Šimko M, Krajčík M, Šikula O, Šimko P and Kalús D 2018 *Energy and Buildings* **158** 135–146
[10] Nikishkov G 2010 Finite Element Equations for Heat Transfer *Programming Finite Elements in Java^{TM}* (London: Springer London) pp 13–19
[11] Cohen E and Glicksman L 2015 *Journal of Heat Transfer* **137** 081601
[12] Soleimani Dorcheh A and Abbasi M 2008 *Journal of Materials Processing Technology* **199** 10–26

The challenges of socio-legal research in law schools in Colombia and the use of new information and communication technologies

E Ramírez Llerena[1], M Chico Ruiz[2], Y Blanco Jiménez[3], J Luna Marzola[2], E Delgado Florez[2] and A Calderon Hernandez[2]

[1]Investigación Group FAGUCAR y/o CARTAGENA BAYKEEPER, Urbanización Buenavista Manzana A, Lote 5, Avenida Crisanto Luque and Investigación Group in Legal Sociology and Political Institutions. Universidad Libre Sede Cartagena. Pie de la Popa, Calle 30 No 20 - 177. Cartagena.

[2]Investigación Group Education and Educational Technology. Antonio de Arévalo University Foundation, Unitecnar. Cartagena, Avenida Pedro de Heredia, Calle 49A #31 – 45, Sector Tesca

[2]Investigación Group COMSIS. Antonio de Arévalo University Foundation, Unitecnar. Cartagena, Avenida Pedro de Heredia, Calle 49A #31 – 45, Sector Tesca

cartagenabaykeeper@gmail.com

elizabeth.ramirez@unilibrectg.edu.co,mchico@acm.org,yeneris.blanco@unitecnar.edu.co,javier.luna@unitecnar.edu.co,emilio.delgado@unitecnar.edu.co,astrid.tutora@gmail.com

Abstract. A correlational statistical analysis was carried out in two university institutions of Cartagena of India's, which allowed us to infer whether the educational entities envision the curricula of the Law program, the inquiry as a daily activity of students and professors and if they have socio-legal research center, with lines of research that place the development of investigative processes, with identity of managerial criteria, since it is the entity that designs and arranges the curriculum of this training program so that the lawyer who graduates, orients his daily legal practice in the investigative with the support of ICT. Surveys were applied to teachers of the law program, which allowed us to determine the strength of association or correlation between variables A and B through a population sample, to make inference to the formation of a lawyer, as well as a humanist, whose discipline be the socio-legal research so that, being a lawyer, you can devote yourself to observe, collect, select, describe, classify and analyse the elements that are in a social environment or space to frame it in the legal system with the support of legal computer systems . It was obtained as a result that (ICT) have evolved in a very fast way the daily life of teachers in the surveyed universities, since young teachers are not traumatized by the use of new technologies because they are daily familiar with computers, the internet and its communication tools, while teachers at the end of the century are considered digital illiterates.

1. Introduction

Today, Social Research according to Galindo [8] is part of the forms of information culture, since knowledge about the social is distributed among the population in a dissymmetric way, since some people get the information more than others on the other hand, some seek and need more information than others. This is a new starting point for the discussion about the use of new technologies in socio-legal research in Law.

As law researchers, teachers of the new century must have a comprehensive education: both humanistic, grounded in human rights; in the social, in the basic legal and in the scientific technical (in their own field); This implies being able to count on a law program, with a body of solid researchers [25] that can generate an institutional identity among its members, for a better training of the lawyer of the future, that of the digital era, by The Center for Socio-Legal Research, attached to the Law program in Cartagena universities, must be a reflective space on aspects related to epistemological proposals for social research and Law, stimulating science and aesthetics in the academic community; and, likewise, it must promote and increase the research carried out by managers, teachers and students in socio-legal knowledge with the support of legal information systems provided by educational technology, essential in the decision-making process that day demands the complexity of contemporary societies, giving life to a curriculum with a research approach, this is the thesis that will be developed in this scientific article.

The problem under investigation shows the refusal of the professors of the end of the 20th century, of the Law program of the Faculty of Social Sciences of the Antonio de Arevalo University Foundation, UNITECNAR regarding the use of the Moodle platform to promote research practices in the socio-legal research center as a true investigative entity, so that it has a positive impact on the development of the social space [17] as is the case of Cartagena de India's, in the Department of Bolívar in Colombia, as teachers assigned to these investigative units, in addition to lawyers, are the social actors who own the practices and ideologies whose description and analysis is the object of this investigation [19].

Likewise, the University - Research Center relationship must transcend the boundaries of everyday life in the teaching-learning processes, to reinvent other joints that under the theoretical and methodological parameters of social research - Participation Action Research - IAP, Ethnography, etc., allow to build an active and competitive model that allows the insertion of its results in the curriculum, therefore a quantitative correlational analysis was made to try to determine through the application of social surveys, if teachers of the Law Program of the Libre University in Cartagena are also reluctant to use the Moodle platform. Correlational research makes important contributions to scientific knowledge, since this research strategy helps scientists evaluate descriptive issues that may be the basis of subsequent socio-educational experiments.

2. Methodology

The situation of the culture of the use of ICTs in socio-legal research, particularly in the educational community of the Law program of the Faculty of Social Sciences of UNITECNAR, generated the question that teachers asked themselves: ¿Why use ICT when conducting a socio-legal research?, which leads us to establish as a general objective, to inquire through a correlational statistical survey, if the teachers of the Law programs of the Antonio de Arevalo University Foundation and the Libre University in Cartagena, are reluctant in the use of the Moodle platform for teaching - learning Law research; The hypothesis that was established was that the teachers of the subjects of the law program curriculum should generate research projects in the socio-legal classroom using new technologies because in this way they seek to create in the Law students, a profile of true researchers of this science, since, as Lawyers, they must not only know the theoretical and practical aspects of their discipline, but they must also have a vision of the social projection of the country, its reality and its historical, social processes, economic in association with the own legal science curriculum in its different manifestations: Criminal, Labor, Public, Private and Procedural.

For the Law, this is a quantitative Legal Partner investigation, that is, it has contributions from the quantitative research, which is one in which quantitative data on variables are collected and analysed.

3. Results

This was a correlational statistical investigation and a non-experimental research method was used in which the researchers measured two variables, that is, the purpose of our research was to detect which variables are connected to each other, when the teachers of the Law programs of the Antonio de Arevalo Tecnar University Foundation and the Libre University in Cartagena, use the Moodle platform for teaching - learning Law research, with which it was understood and evaluated, the statistical relationship between them, without influence of any strange variable .

The variable "A" was, the teachers of the Law programs of the Antonio de Arevalo Tecnar University Foundation use the Moodle platform for teaching - learning of the Law research and the variable "B" was, the teachers of the programs of Law at Libre University in Cartagena, and they also don't use the

Moodle platform for teaching - learning Law research. We wanted to determine if there is a positive or negative correlation or if there isn't correlation between the two variables, since teachers who teach socio-legal research in both law programs must use ICT to qualify the teaching-learning process.

The distinctive feature of this correlational investigation was that none of the variables involved were manipulated by the researchers, since a group of them applied a survey in the form of Google Doc, which in UNITECNAR was directed by the dean of the teachers who should use the SPLAVIA platform by institutional mail and at the UNILIBRE of Cartagena the same process was carried out by the dean.

The sampling was randomized, 8.3 teachers from professorship, 8.3 part-time teachers and 12.5 full-time teachers were surveyed. Teachers from UNITECNAR in Cartagena were surveyed in 16.7, from Barranquilla in 8.3% and from Sincelejo in 4.2%; 12.5% of full-time, part-time and full-time teachers were surveyed at UNILIBRE in Cartagena.

This section presents the results that were obtained after the statistical analysis performed, so we can say that correlational research allowed us to measure two variables and thus establish a statistical relationship between each of them:

Variable 1: Regarding the question about ¿whether you can connect a computer and its most common peripherals: printers, scanners, etc.? UNITECNAR teachers said yes in 95.8% and no in 4.2%. At UNILIBRE in Cartagena, 93.8 responded yes and 6.2% said no.

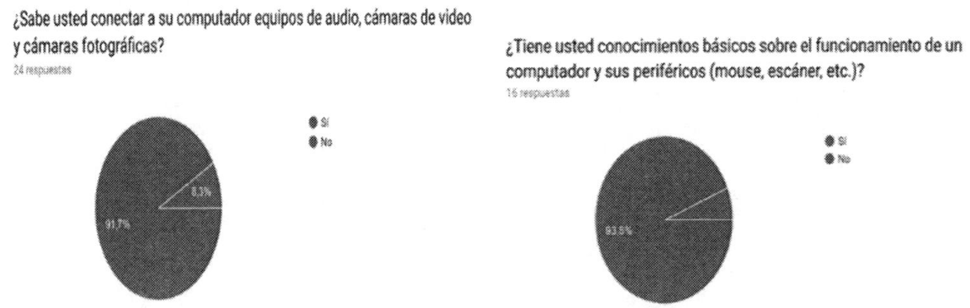

Figure. 1. whether you can connect a computer and its most common peripherals: printers, scanners.

Variable 2: About if ¿do you have basic knowledge about the operation of a computer and its peripherals (mouse, scanner, etc.)? UNITECNAR teachers said yes in 100% and no in 0%. In UNILIBRE Cartagena teachers said yes in 93.8% and no in 6.2%

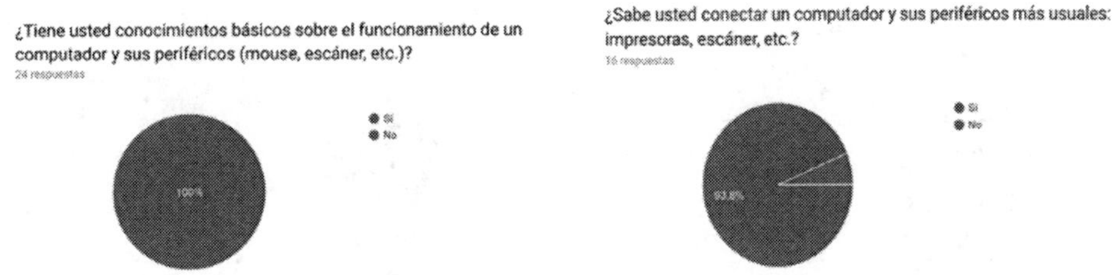

Figure 2. do you have basic knowledge about the operation of a computer and its peripherals (mouse, scanner, etc.)

Variable 3: As to whether ¿do you know how to connect audio equipment, video cameras and cameras to your computer? UNITECNAR teachers said yes in 97, 1% and no in 8.3%. In UNILIBRE Cartagena, teachers said yes in 62.5% and no in 37.5%.

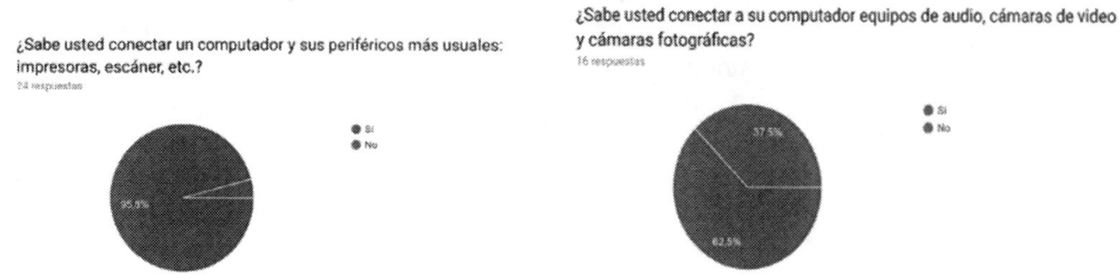

Figure 3. do you know how to connect audio equipment, video cameras and cameras to your computer

Variable 4: With reference to ¿if do you know how to install and uninstall computer programs on your computer? UNITECNAR teachers said yes in 62.5% and no in 37.5%. At UNILIBRE Cartagena they said yes at 50% and no at 50%.

Figure 4. do you know how to install and uninstall computer programs on your computer

Variable 5: If, ¿Do you know how to download programs, images, music, etc. from the internet? UNITECNAR teachers said yes in 91.7% and no in 8.3%. At UNILIBRE Cartagena they said yes at 75% and no at 25%.

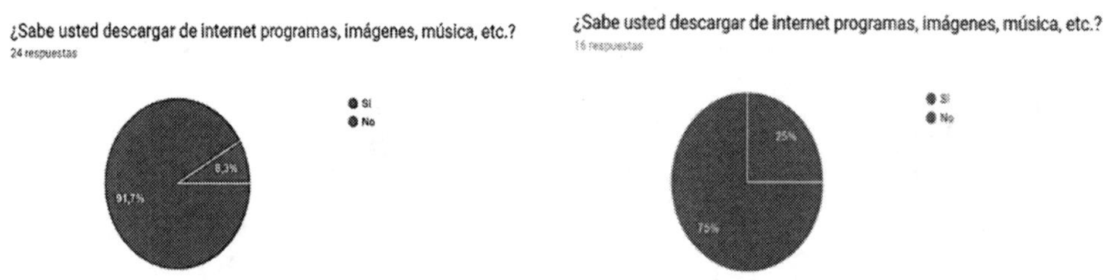

Figure 5. Do you know how to download programs, images, music, etc. from the internet

Variable 6: With reference to ¿do you understand the difference between hardware and software? UNITECNAR teachers said yes in 87.5% and no in 12.5%. At UNILIBRE Cartagena teachers said yes is 93.8% and 6.2% is not.

IOP Conf. Series: Materials Science and Engineering **844** (2020) 012071 doi:10.1088/1757-899X/844/1/012071

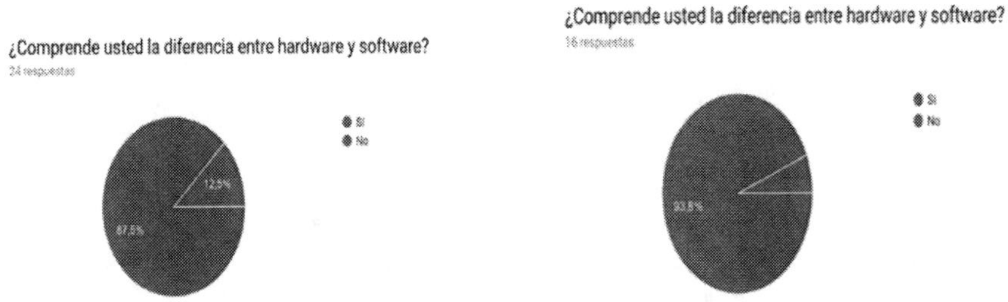

Figure 6. Do you understand the difference between hardware and software

Variable 7: Regarding whether ¿Do you know how to make inquiries in specialized databases such as Legis, E-book, EBESCO, etc.?, UNITECNAR teachers said yes in 70.8% and no in 29.2%. At UNILIBRE Cartagena they said yes at 87.5% and no at 12.5%.

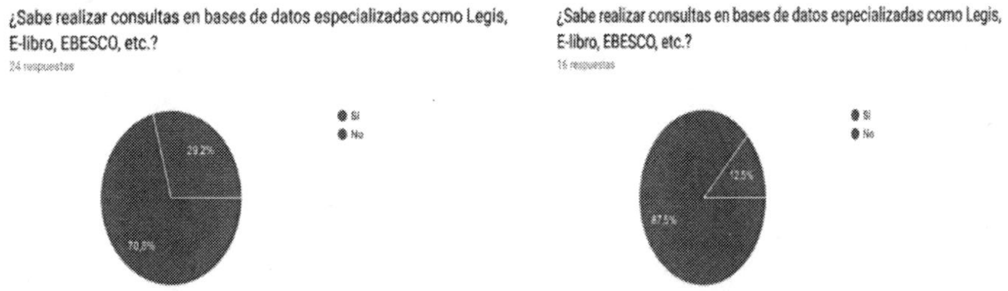

Figure 7. Do you know how to make inquiries in specialized databases such as Legis, E-book, EBESCO, etc.

Variable 8: With reference to whether ¿Do you know how to create and modify spreadsheets with a computer program (Excel, etc.) for specific purposes using functions such as formatting cells insert and hide rows, make dynamic tables ?, UNITECNAR teachers said that yes in 62.5% and no in 37.5%. In UNILIBRE Cartagena they said yes in 62.5% and no in 37.5%.

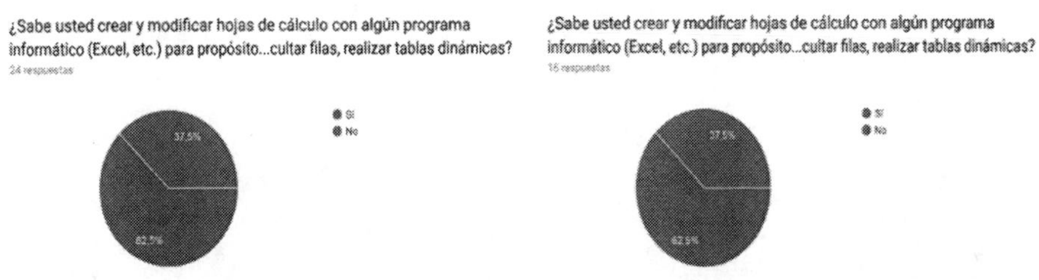

Figure 8. Do you know how to create and modify spreadsheets with a computer program (Excel, etc.) for specific purposes using functions such as formatting cells insert and hide rows, make dynamic tables

Variable 9: Regarding whether ¿Do you know how to use different programs that include tools to draw graphics such as Paint, Photoshop, etc.?, UNITECNAR teachers said yes in 58.3% and no in 41.7%. In UNILIBRE Cartagena they said yes in 56.3% and no in 43.8%.

IOP Conf. Series: Materials Science and Engineering **844** (2020) 012071 doi:10.1088/1757-899X/844/1/012071

Figure 9. Do you know how to use different programs that include tools to draw graphics such as Paint, Photoshop, etc

Variable 10: Regarding if ¿do you know how to create images and graphics through a computer problem?, UNITECNAR teachers said yes in 50% and no in 50%. At UNILIBRE Cartagena they said yes at 18.8% and no at 81.3%.

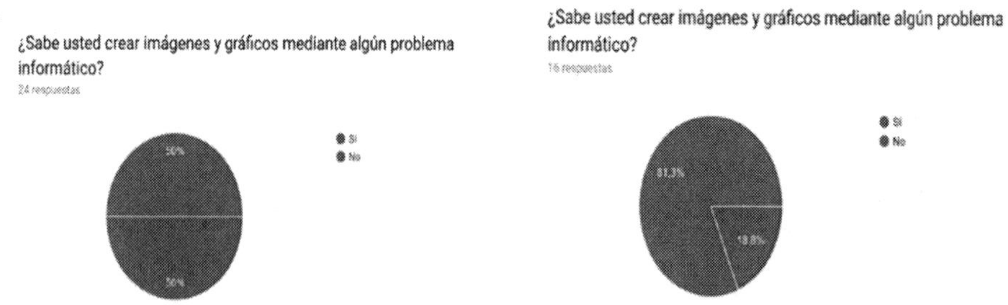

Figure10. do you know how to create images and graphics through a computer problem

Variable 11: On if ¿you can create or edit audio with a computer program?, UNITECNAR teachers said yes in 33.3% and no in 66.7%. At UNILIBRE Cartagena they said yes at 18.8% and no at 81.3%.

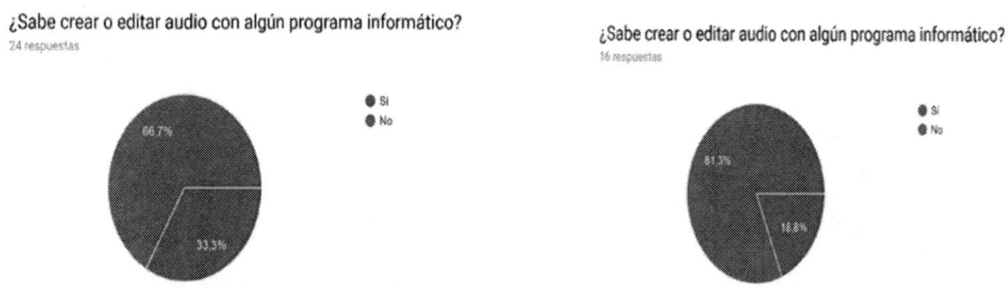

Figure 11. you can create or edit audio with a computer program

Variable 12: Asking for ¿if you know how to create a multimedia presentation through a program including static images, texts, audio and video graphic tracks?, UNITECNAR teachers said yes in 50% and no in 50%. At UNILIBRE Cartagena they said yes at 12.5% and no at 87.5%.

Figure 12. if you know how to create a multimedia presentation through a program including static images, texts, audio and video graphic tracks

Variable 13: If, ¿Do you know how to surf the Internet with browsers such as: Explorer, Chrome, Mozilla, etc.?, UNITECNAR teachers said yes in 100% and no in 0%. At UNILIBRE Cartagena they said yes at 18.8% and no at 81.3%.

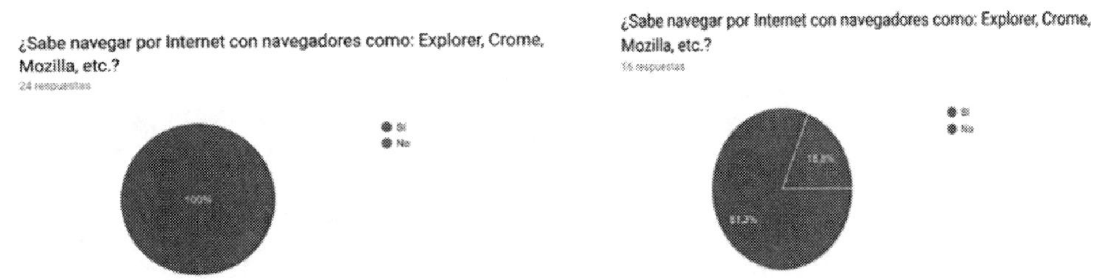

Figure 13. Do you know how to surf the Internet with browsers such as: Explorer, Chrome, Mozilla, etc.

Variable 14: In reference to if ¿do you know how to use Internet search engines such as: Google, Yahoo!, etc.? UNITECNAR teachers said yes in 100% and no in 0%. In UNILIBRE Cartagena they said yes in 93.8% and no in 6.2%.

Figure 14. Do you know how to use Internet search engines such as: Google, Yahoo!, etc.

Variable 15: About if ¿Do you know how to make videoconferences over IP with Messenger, Skype, NetMeeting, RENATA, etc.?, UNITECNAR teachers said yes in 83.3% and no in 16.7%. At UNILIBRE Cartagena they said yes at 31.3% and no at 68.8%.

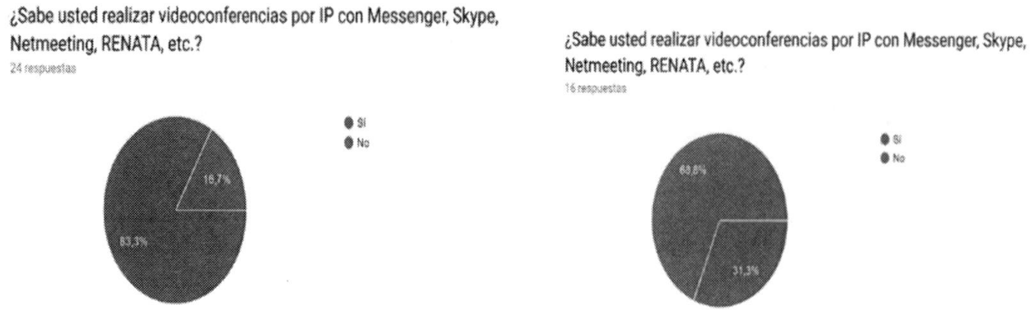

Figure 15. Do you know how to make videoconferences over IP with Messenger, Skype, NetMeeting, RENATA, etc.

Variable 16: Asking for ¿if you use email, instant messaging chat, forums, distribution list, that is, through the usual Internet communication tools?, UNITECNAR teachers said yes in 95.8% and no in a 4.2% At UNILIBRE Cartagena they said yes at 18.8% and no at 81.3%.

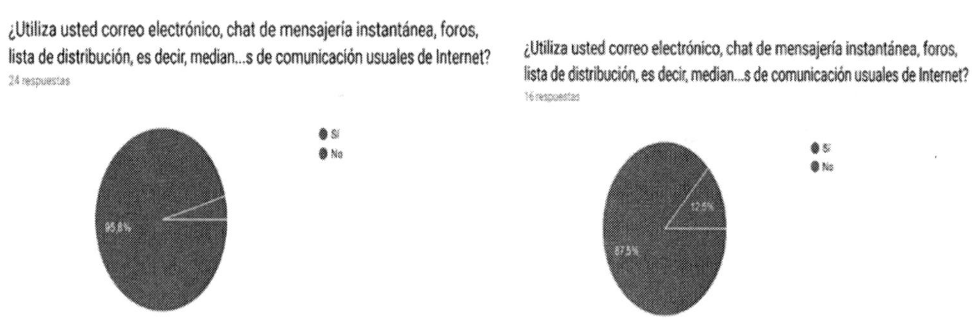

Figure 16. you use email, instant messaging chat, forums, distribution list, that is, through the usual Internet communication tools

Variable 17: Regarding if ¿do you know how to organize, analyse and synthesize the information through graph tables or diagrams?, UNITECNAR teachers said yes in 54.2% and no in 45.8%. At UNILIBRE Cartagena they said yes at 31.3% and no at 68.8%.

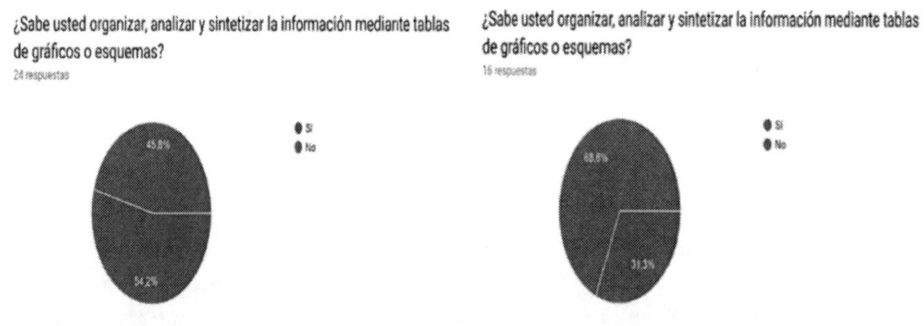

Figure 17. Do you know how to organize, analyse and synthesize the information through graph tables or diagrams

Variable 18: On ¿if you know how to use graphic organizers such as concept maps, mind maps, diagrams or diagrams to represent the relationships between ideas or concepts?, UNITECNAR teachers said yes in 75% and no in 25%. In UNILIBRE Cartagena they said yes in 48.3% and no in 56.3%.

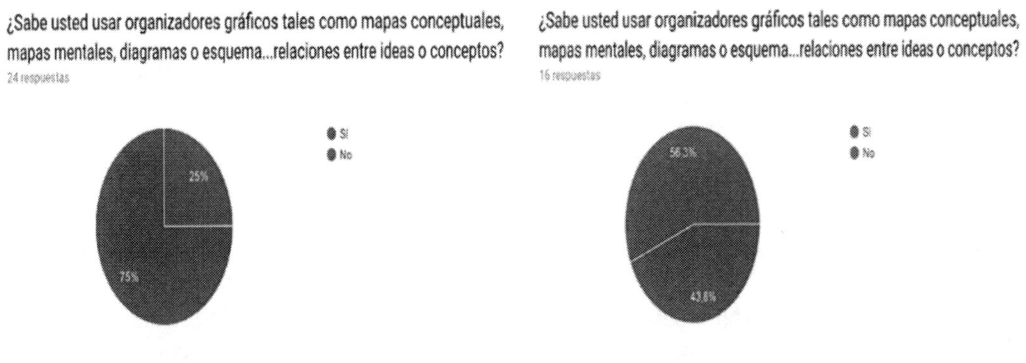

Figure 18. you know how to use graphic organizers such as concept maps, mind maps, diagrams or diagrams to represent the relationships between ideas or concepts

Variable 19: With reference to if, ¿Do you know significant methodological strategies supported by ICTs that are of valuable support in your face-to-face area using SPLAVIA - Moodle?, UNITECNAR teachers said yes in 70.8% and no in 29.2%. At UNILIBRE Cartagena, teachers do not work with the SPLAVIA platform, but they said that they know of significant methodological strategies supported by ICTs that are of valuable support in their face-to-face area using Platform-Moodle, blackboard, dokeos, yes in 31.3% and no in 68.8%

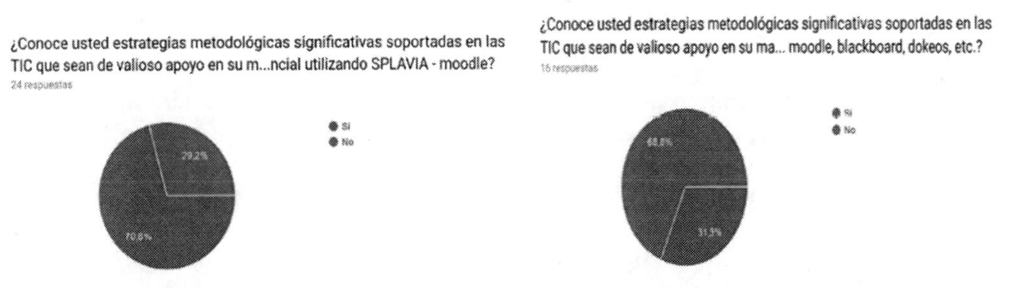

Figure 19. Do you know significant methodological strategies supported by ICTs that are of valuable support in your face-to-face area using SPLAVIA – Moodle.

Variable 20: If ¿you know the strengths and weaknesses of SPLAVIA - Moodle as an educational experience in TECNAR, obtained from various print and network sources?, UNITECNAR teachers said yes in 70.8% and no in 29.2% . At UNILIBRE Cartagena they said that they know the strengths and weaknesses of the platform - Moodle as an educational experience at the Libre University obtained from various print and network sources, yes in 18.8% and that no in 81.3.

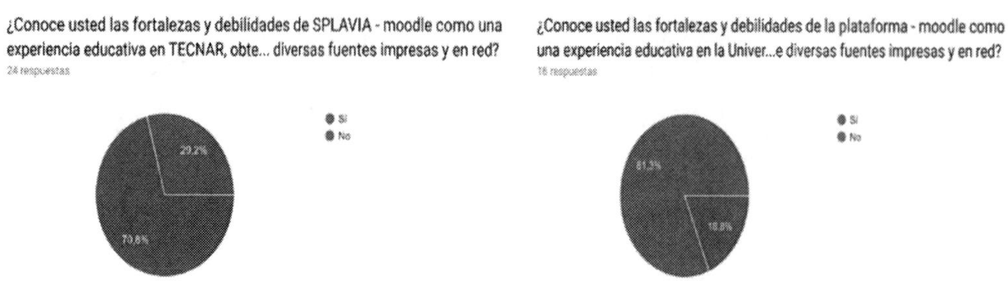

Figure 20. you know the strengths and weaknesses of SPLAVIA - Moodle as an educational experience in TECNAR, obtained from various print and network sources

Variable 21: About if, ¿Do you select learning strategies with ICT use to design a virtual learning environment with your students in the SPLAVIA - Moodle Platform?, UNITECNAR teachers said yes in 70.8% and no in 29.2%. At UNILIBRE Cartagena they said they select learning strategies with ICT use to design a virtual learning environment with their students in the Moodle Platform, yes in 25% and not in 75%.

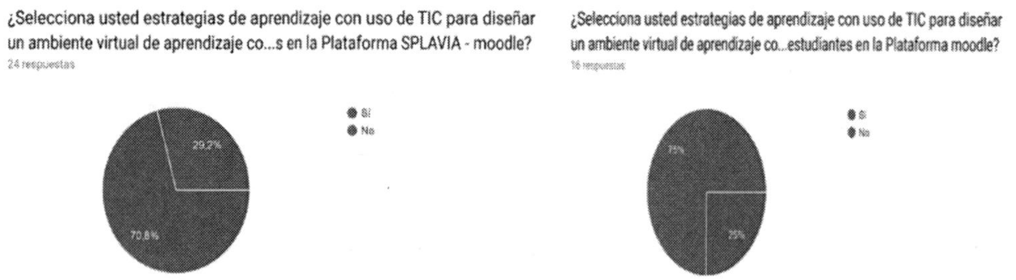

Figure 21. Do you select learning strategies with ICT use to design a virtual learning environment with your students in the SPLAVIA - Moodle Platform

Variable 22: Regarding whether, ¿Do you create computerized educational materials on the SPLAVIA - Moodle Platform as collaborative work among your students? UNITECNAR teachers said yes in 83.3% and no in 16.7%. At UNILIBRE Cartagena they said, they create computerized educational materials on the Platform - Moodle as collaborative work among their students, yes in 12.5% and that no in 87.5%

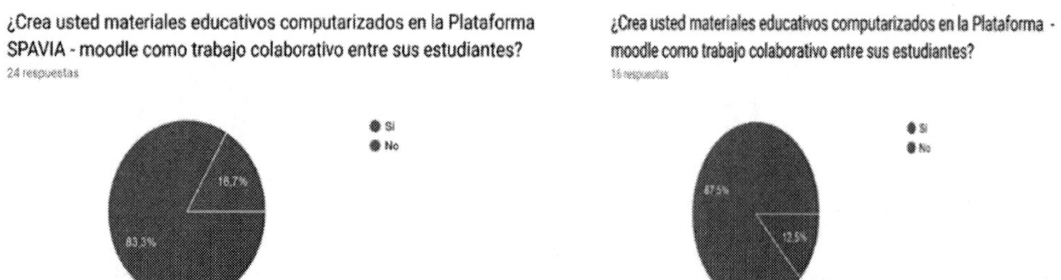

Figure 22. Do you create computerized educational materials on the SPLAVIA - Moodle Platform as collaborative work among your students

Variable 23: With reference to if, ¿do you create presentations to support teaching and learning is using text, graphic and multimedia elements in the SPLAVIA - Moodle Platform that use digital educational resources? UNITECNAR teachers said yes in 70.8% and no in 29.2%. At UNILIBRE Cartagena they said that they create presentations to support teaching and learning using textual, graphic and multimedia elements in the Moodle Platform that use digital educational resources, yes in 25% and not in 75%.

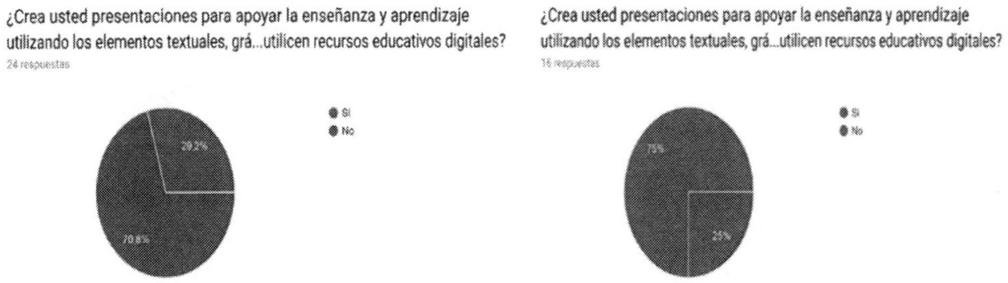

Figure 23. do you create presentations to support teaching and learning is using text, graphic and multimedia elements in the SPLAVIA - Moodle Platform that use digital educational resources

Variable 24: Asking for whether, if ¿it organizes groups of students, physical spaces, materials and tasks as pedagogical activities in the SPLAVIA - Moodle Platform ?, UNITECNAR teachers said yes in 58.3% and no in 41.7%. At UNILIBRE Cartagena they said that they organize groups of students, physical spaces, materials and tasks as pedagogical activities in the Moodle Platform or another, yes in 12.5% and that no in 87.5%

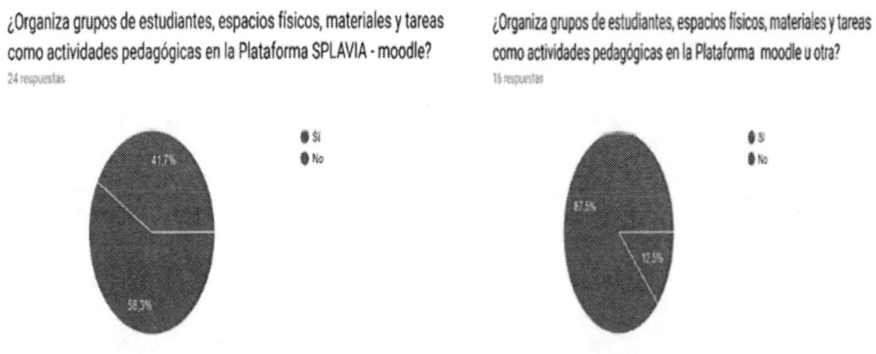

Figure 24. it organizes groups of students, physical spaces, materials and tasks as pedagogical activities in the SPLAVIA - Moodle Platform

Variable 25: About if, ¿Do you use technology to support pedagogical and didactic strategies that meet the diverse needs of your students in the SPLAVIA - Moodle Platform?, UNITECNAR teachers said yes in 79.2% and no in 20.8%. At UNILIBRE Cartagena they said that they use technology to support pedagogical and didactic strategies that meet the diverse needs of their students in the Moodle Platform or another, yes in 18.8% and that no in 81.3%.

Figure 25. Do you use technology to support pedagogical and didactic strategies that meet the diverse needs of your students in the SPLAVIA - Moodle Platform.

Variable 26: Regarding whether, ¿Do you implement pedagogical activities in the SPLAVIA - Moodle Platform in which you incorporate ICT resources as support elements for the learning sectors, using different proposals and methodological approaches such as: Webquest, collaborative work, micro projects, concept maps, etc. .?, UNITECNAR teachers said yes in 62.5% and no in 37.5%. In UNILIBRE Cartagena they said that they implement pedagogical activities in the Moodle Platform or another in which it incorporates ICT resources as support elements for the learning sectors, using different proposals and methodological approaches such as: Webquest, collaborative work, micro projects, concept maps, yes at 18.8% and that no 81.3%.

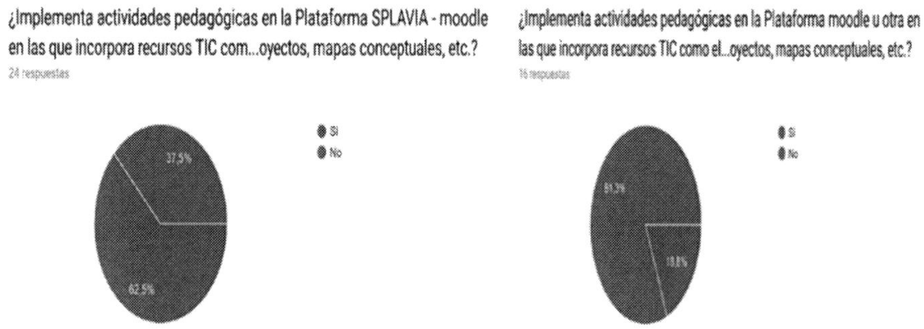

Figure 26. Do you implement pedagogical activities in the SPLAVIA - Moodle Platform in which you incorporate ICT resources as support elements for the learning sectors, using different proposals and methodological approaches

Variable 27: With reference to if ¿itself do you use pedagogical criteria to select computer applications and digital educational resources appropriate and possible to use in the SPLAVIA - Moodle Platform?, UNITECNAR teachers said yes in 75% and no in 25%. In UNILIBRE Cartagena they said that they use pedagogical criteria to select computer applications and digital educational resources suitable and possible to use in the Moodle Platform or another, yes in 12.5% and not in 87.5%.

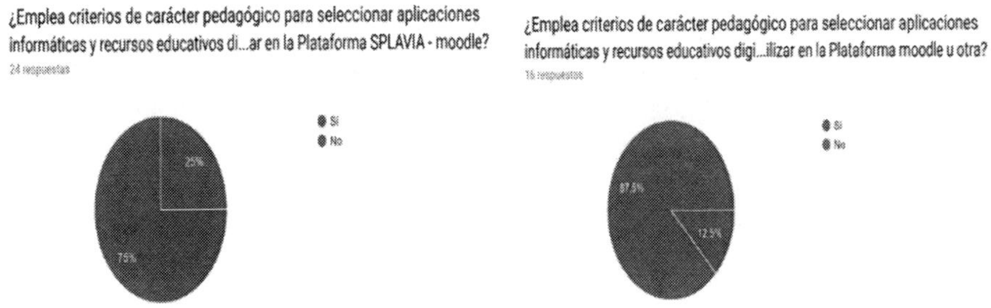

Figure 27. do you use pedagogical criteria to select computer applications and digital educational resources appropriate and possible to use in the SPLAVIA - Moodle Platform

Variable 28: About if, ¿do you design evaluation procedures and instruments for learning in ICT work environments such as the SPLAVIA - Moodle Platform?, UNITECNAR teachers said yes in 58.3% and no in 41.7%. At UNILIBRE Cartagena they said that they design procedures and assessment instruments for learning in ICT work environments such as the Moodle Platform or another, yes in 18.8% and no in 81.3%.

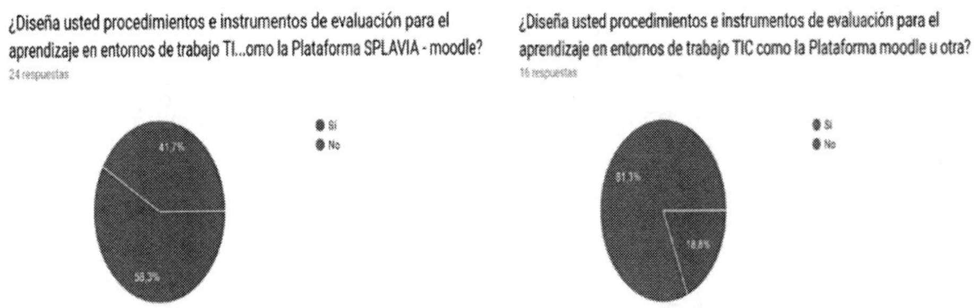

Figure 28. Do you design evaluation procedures and instruments for learning in ICT work environments such as the SPLAVIA - Moodle Platform

Variable 29: Asking for if, ¿do you know on-line training platforms other than the SPLAVIA-Moodle Platform and its use in the academic context? UNITECNAR teachers said yes in 66.7% and no in 33.3%. At UNILIBRE Cartagena they said that they know the On Line training platforms different from the Moodle Platform and its use in the academic context in 37.5% and that No in 62.5%.

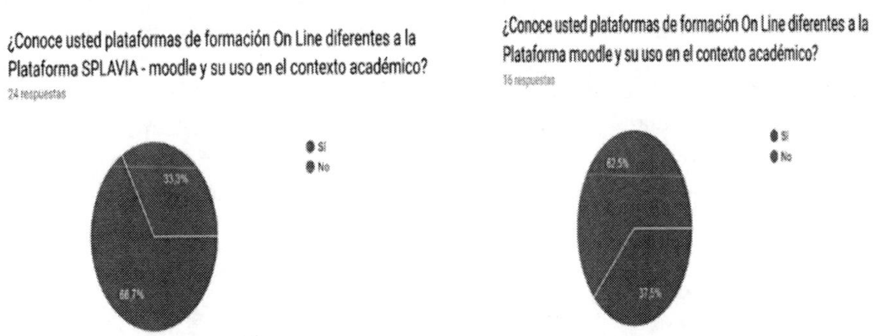

Figure 29. Do you know on-line training platforms other than the SPLAVIA-Moodle Platform and its use in the academic context

IOP Conf. Series: Materials Science and Engineering **844** (2020) 012071 doi:10.1088/1757-899X/844/1/012071

Variable 30: With reference to if, ¿do you know methodologies to support interaction and collaborative networking in the SPLAVIA - Moodle Platform?, UNITECNAR teachers said yes in 54.2% and no in 45.8%. At UNILIBRE Cartagena they said that they know the methodologies to support interaction and collaborative networking in the Moodle Platform, yes in 25% and that no in 75%.

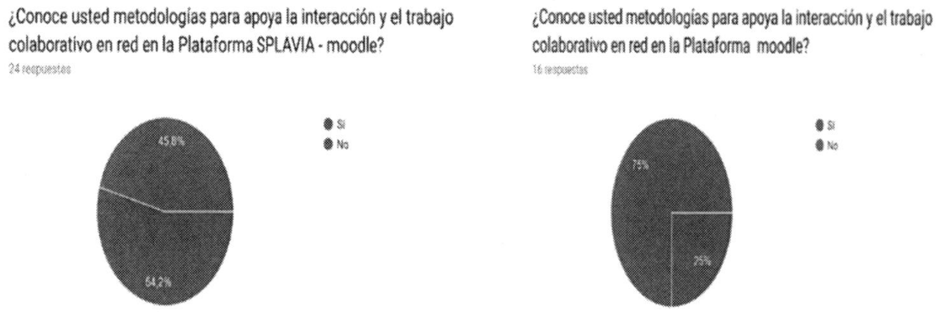

Figure 30. you know methodologies to support interaction and collaborative networking in the SPLAVIA - Moodle Platform

Variable 31: Regarding that if, ¿do you design online activities in the SPLAVIA - Moodle Platform that complement or support the face-to-face teaching-learning processes?, UNITECNAR teachers said yes in 66.7% and no in 33.3%. At UNILIBRE Cartagena they said that they design online activities in the Moodle Platform that complement or support the face-to-face teaching and learning processes, yes in 12.5% and not at 87.5%.

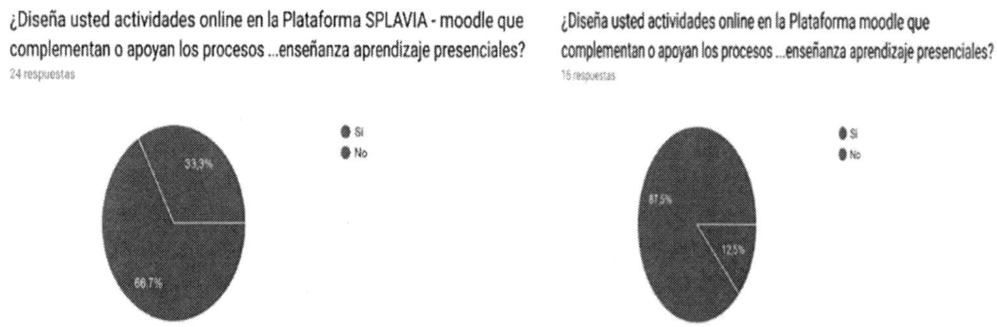

Figure 31. you design online activities in the SPLAVIA - Moodle Platform that complement or support the face-to-face teaching-learning processes

Variable 32: About if, ¿do you evaluate the impact of the online work of the SPLAVIA - Moodle Platform on learning processes?, UNITECNAR teachers said yes in 54.2% and no in 45.8%. At UNILIBRE Cartagena they said that they evaluate the impact of online work of the Moodle Platform in the learning processes, yes in 18.8% and that not in 81.3%.

IOP Conf. Series: Materials Science and Engineering **844** (2020) 012071 doi:10.1088/1757-899X/844/1/012071

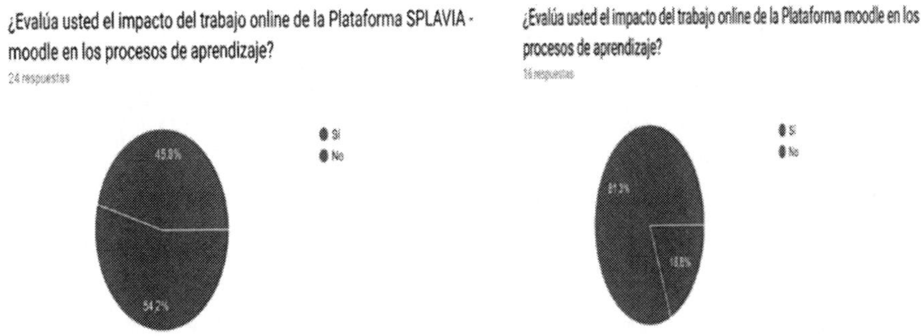

Figure 32. you evaluate the impact of the online work of the SPLAVIA - Moodle Platform on learning processes

Variable 33: Regarding that if, ¿do you recognize the ethical and legal aspects associated with digital information such as privacy, intellectual property, information security, etc.?, UNITECNAR teachers said yes in 87.5% and no in a 12.5% In UNILIBRE Cartagena they said yes in 75% and not in 25%.

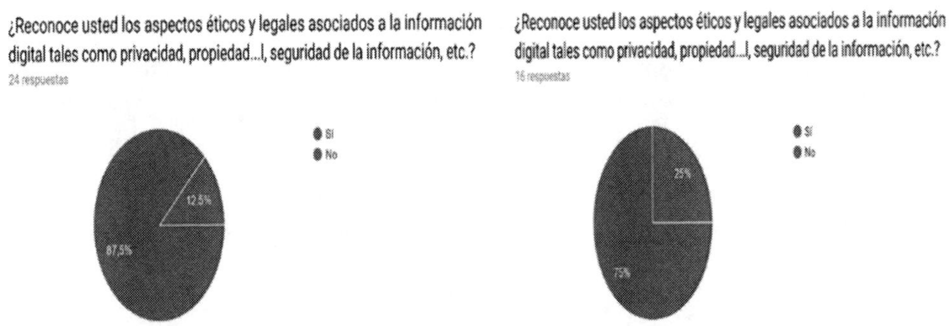

Figure 33. you recognize the ethical and legal aspects associated with digital information such as privacy, intellectual property, information security, etc

Variable 34: With reference, in ¿to the fact that if you use the Internet to support the tasks of your teaching work?, UNITECNAR teachers said yes in 100% and no in 0%. In UNILIBRE Cartagena they said yes in 75% and not in 25%.

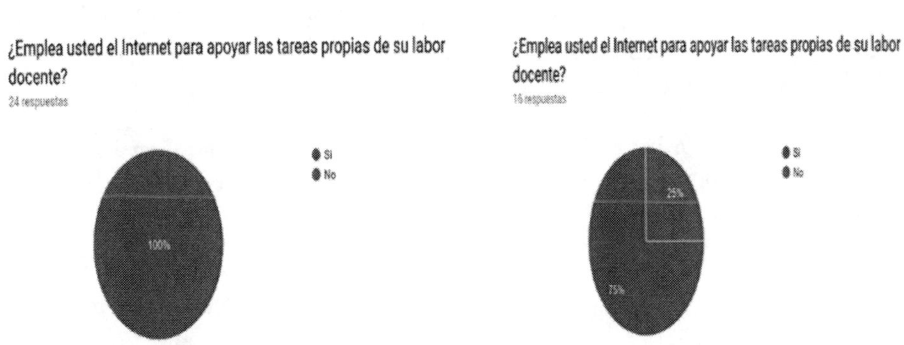

Figure 34. to the fact that if you use the Internet to support the tasks of your teaching work

Variable 35: On whether ¿do you use the communication resources provided by the technologies to establish a permanent contact with the students? UNITECNAR teachers said yes in 97, 1% and not in 8.3%. At UNILIBRE Cartagena headquarters they said yes at 62.5% and not at 37.5%.

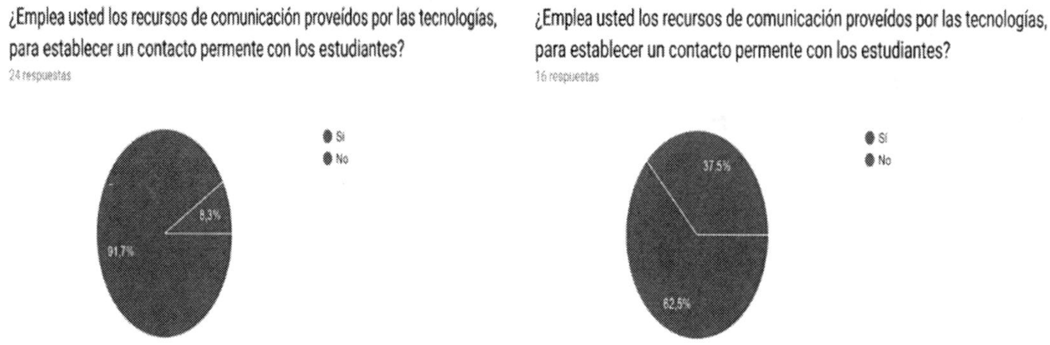

Figure 35. you use the communication resources provided by the technologies to establish a permanent contact with the students

Variable 36: Regarding whether, ¿do you create and maintain a list of sites relevant to your teaching work? UNITECNAR teachers said yes in 79.2% and no in 20.8%. At UNILIBRE Cartagena they said yes at 62.5% and not at 37.5%.

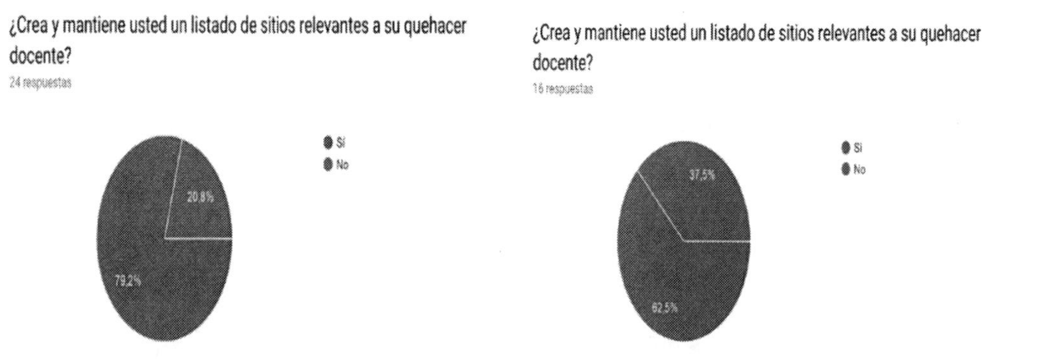

Figure 36. you create and maintain a list of sites relevant to your teaching work

Variable 37: As to whether do you access information sources for updating in educational technology, such as electronic journals, educational portals, edublogs, etc.?, UNITECNAR teachers said yes in 97, 1% and no in 8.3 %. In UNILIBRE Cartagena they said, yes in 56.3% and not in 43.8%.

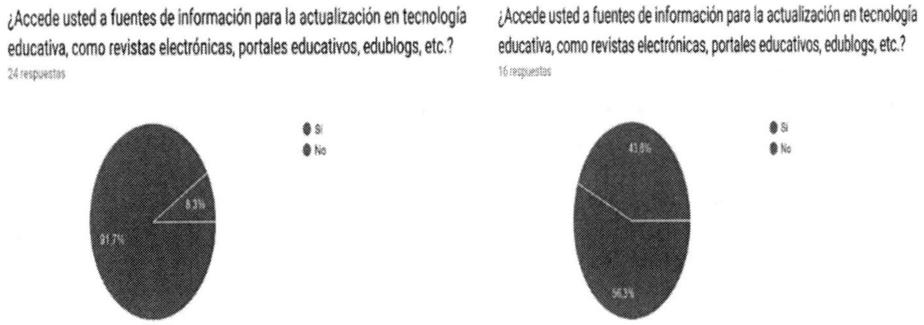

Figure 37. you access information sources for updating in educational technology, such as electronic journals, educational portals, edublogs, etc

Variable 38: With reference to that if ¿do you use the communication tools provided by the Internet, for the exchange of educational experiences with other professors and institutions of higher education?, UNITECNAR teachers said yes in 43.8% and not in 56.3% . In UNILIBRE Cartagena they said, yes in 56.3% and not in 43.8%.

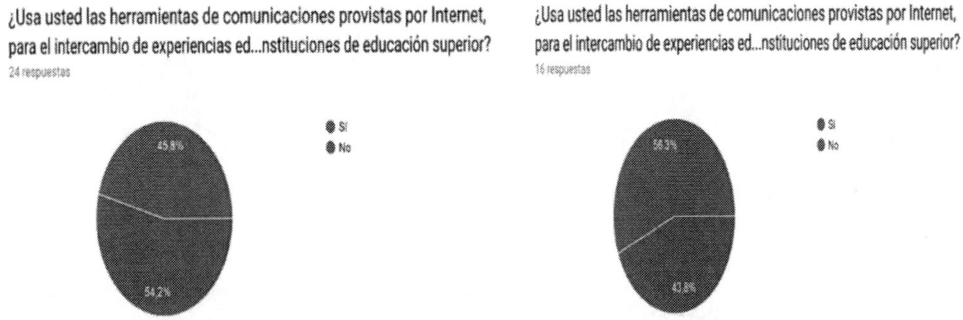

Figure 38. you use the communication tools provided by the Internet, for the exchange of educational experiences with other professors and institutions of higher education

Variable 39: On if¿ do you use educational portals as a place of access to communication spaces with peers that can support your teaching work?, UNITECNAR teachers said yes in 66.7% and not in 33.3%. At UNILIBRE Cartagena they said, yes at 50% and not at 50%.

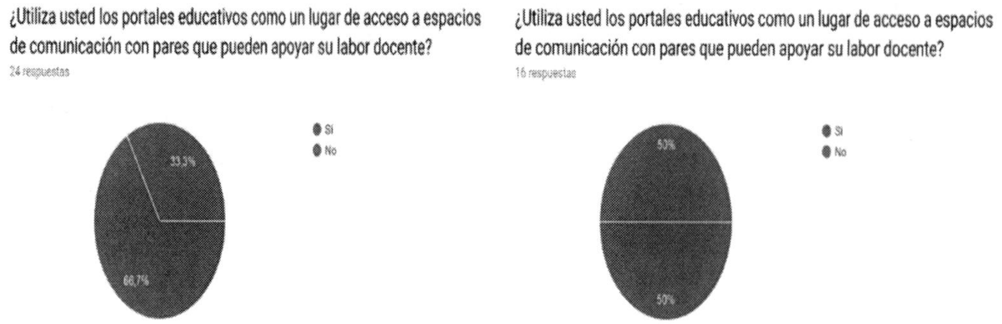

Figure 39. do you use educational portals as a place of access to communication spaces with peers that can support your teaching work

Variable 40: Asking if, ¿for whether do you participate in congresses, fairs, seminars, etc., related to the implementation of ICT in the classroom? UNITECNAR teachers said yes in 58.3% and not in 41.7%. At UNILIBRE Cartagena they said yes at 37.5% and no at 62.5%.

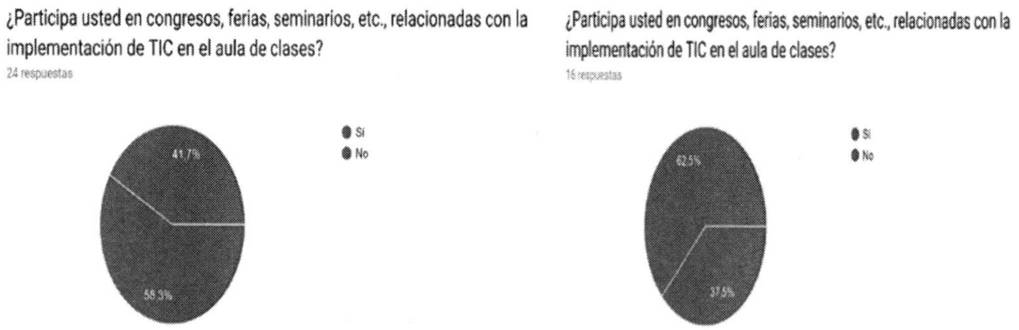

Figure 40. for whether do you participate in congresses, fairs, seminars, etc., related to the implementation of ICT in the classroom

The questions of the following questionnaire were of multiple choices, the teachers of UNITECNAR and UNILIBRE Cartagena, answered marking on the item of their choice.

Variable 1: On whether you have used the platform technical support service?, UNITECNAR teachers said that: Personal attention dare has done 83. 3%, through telephone service they do 12.5%, in an online chat they do 20.8%, and by email, 37.5%.

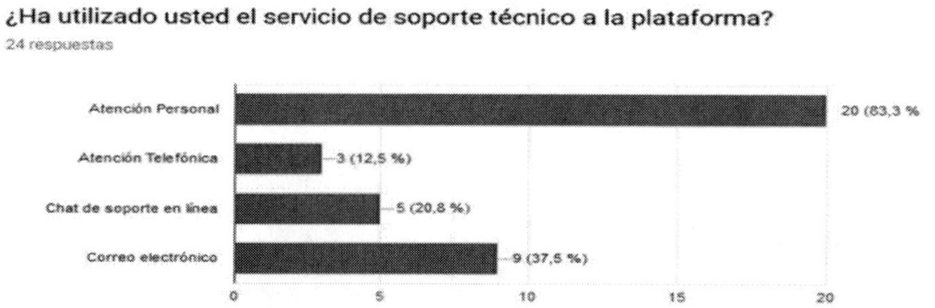

The teachers of UNILIBRE Cartagena said, Personal attention dare has done 62.5%, through telephone service they do 0%, In an online chat they do 0%, and By email, 43.8% does.

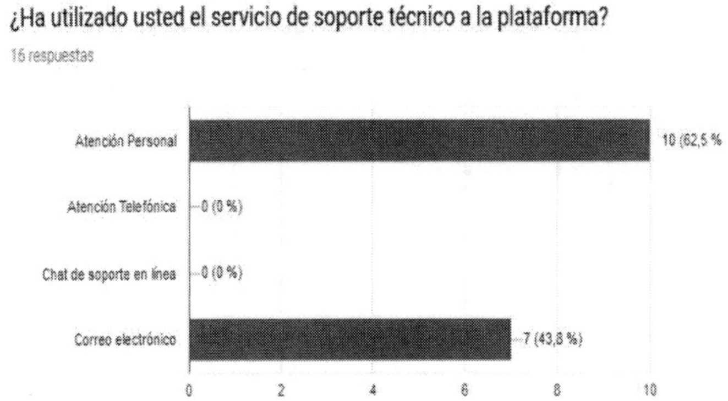

Variable 2: Asking for whether ¿the virtual classroom platform –SPLAVIA - Moodle- was available when needed? UNITECNAR teachers said that, Always, 54.2%, almost always 50%, rarely, 8.3%, and Never 0%.

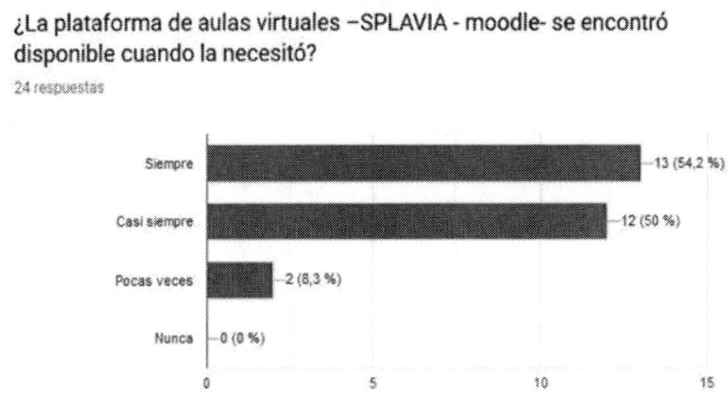

About that, the teachers of UNILIBRE Cartagena said, Always, 31.3% Almost always 12.5% Rarely, 12.5%, and Never un 43.8%.

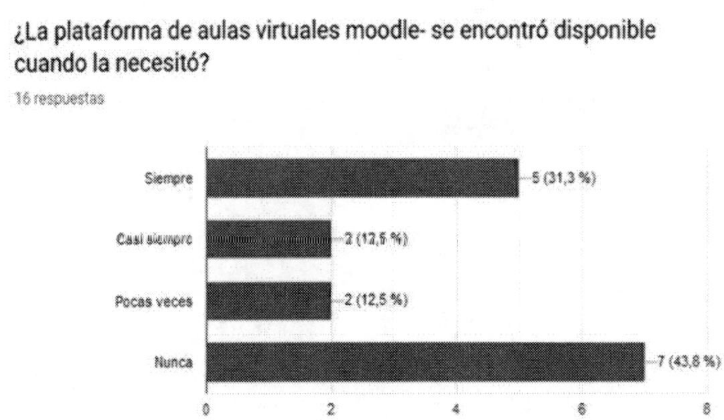

Variable 3: In relation to whether ¿the response time and performance of the virtual classrooms was adequate?, UNITECNAR teachers said that; Always, 54.2%, Almost always 41.7%, Rarely, 4.2%, and Never 0%.

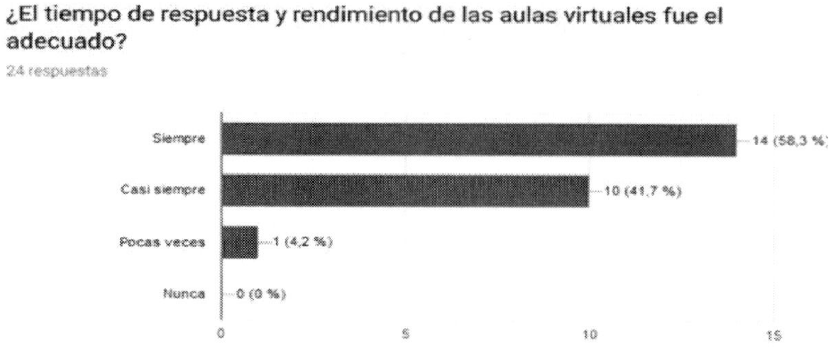

The teachers of UNILIBRE Cartagena said, about that; Always, 25%, almost always 25%, Rarely, 12.5%, and Never 37.5%

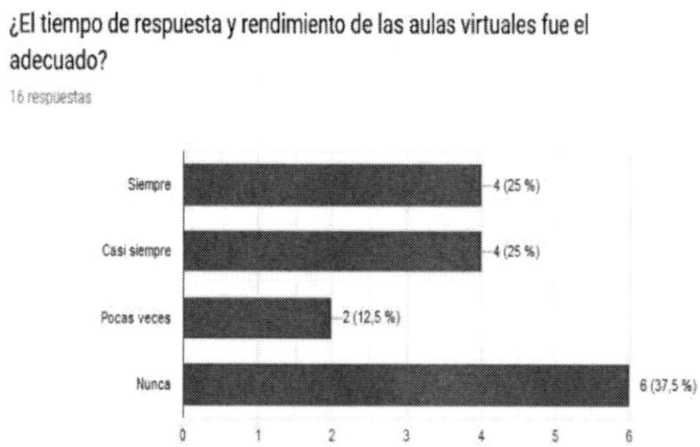

Variable 4: With reference to whether ¿it was possible to access the platform from any place and device?, UNITECNAR teachers said that; Always, 54.2%, Almost always 50%, Rarely, 4.2%, and Never 0%.

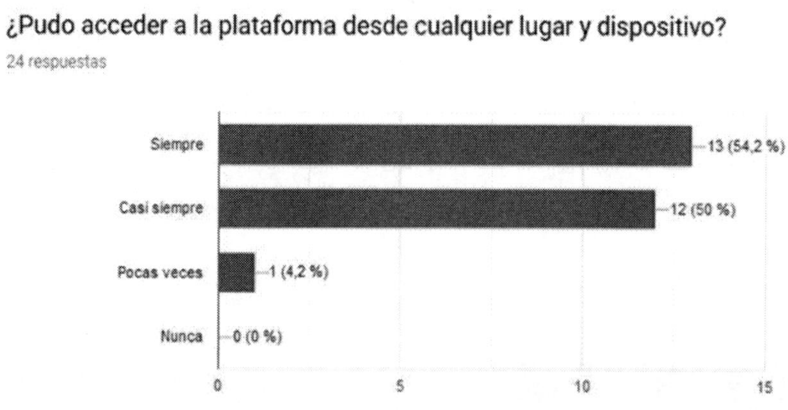

The teachers of UNILIBRE Cartagena said, about that; Always, 18.5%, Almost always 31.3%, Rarely, 6.3%, and Never 43.8%.

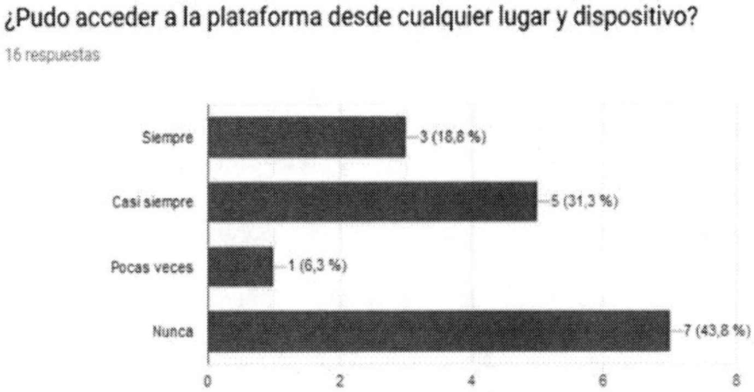

Variable 5: On whether ¿Do you consider the virtual classroom service with respect to the previous semester?, UNITECNAR teachers said that; Improved, by 70.8%, It remained the same, by 25%, It got worse, by 0%, and Don't know, 4.2%.

The teachers of UNILIBRE Cartagena said; Improved, by 56.3%, It remained the same, at 31.3%, It got worse, by 0%, and Don't know, at 12.5%.

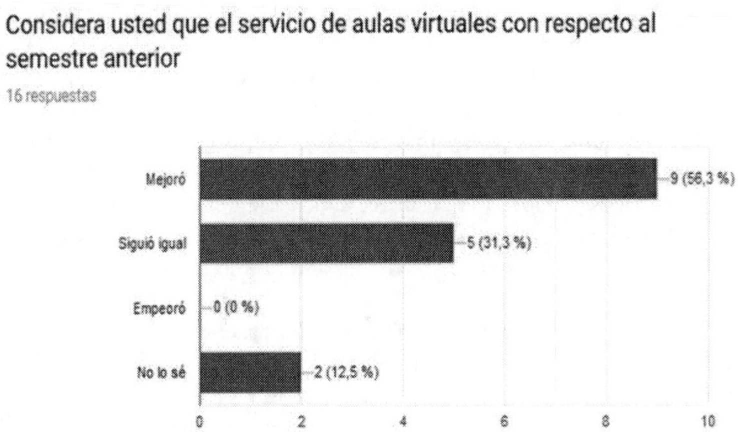

It can be inferred from the correlational analysis that the relationship between variable A and Variable B is negative since teachers of the Antonio de Arevalo University Foundation UNITECNAR, if they use the SPLAVIA platform for the training of investigating lawyers, only that teachers who trained as lawyers in the twentieth century are opposed to the use of new educational technologies, while teachers at the Libre University in Cartagena do not use any Moodle platform, although they know the basic concepts of the computer and the Internet.

4 Discussion.

A review of the state of the art that supports the problem investigated and justifies the holistic model proposed by the researchers was made, so it was necessary to use academic-scientific literature that explains why current curricular methods are deficient in the program of law and at the same time make possible the method proposed by the authors.

The Legal Partner Research in the Law Faculties. It is an interdisciplinary construction in which the social dimension challenges the spaces of juridicity, specific legal knowledge and legal practice. Taking into account; the social effects of a specific legal framework, and its correspondence with the foundations of a rule of law. The relations between the Law and other aspects that characterize the broad world of the social: customs, imaginary, social uses, etc.; The configuration of social and cultural practices of different human groups that affect the compliance or not of a legal system; The comparative analysis of contexts and the answers to their specific problems from the legal field, the Law and the knowledge of human behaviour; The analysis of the legal logic and the theory of argumentation that challenges its normative statements, in the face of a concrete reality molded by processes - of identity construction, constructed through long-term processes or in transition to modern forms thereof.

This should lead to the understanding that normative regulation must be understood with a non-ontological axiological content since it is aimed at regulating social and cultural phenomena, therefore the socio-legal researcher [18] can ask questions or build his hypotheses and assumptions from a theoretical and methodological reference but within a sociological conception of law.

Socio-legal research is aimed at creating Law, from regulating the cultural phenomena that are relevant to a State, at a historical moment in a given social space; taking place only when trying to give a sociological conception of Law, which is the one that assumes that it is born in relation to a cultural phenomenon, whose nature and modalities of empirical realization are necessary to know, in order to be able to issue viable norms to the conditions of time, mode and place where they must be fulfilled and legitimized before the community to the extent that they embrace the purposes for which the community created the institution. Example: When creating a new statute on the gay family in Colombia, it is necessary to start from a study on the different types of family that exist in this country, with ranges as heterogeneous as patriarchal, matriarchal and even polygamous forms; there are families that have a very precarious duration in time creating complex problems in the patrimonial relations of couples and in obligations in relation to progeny and spouses with each other, so it is necessary to know the characteristics and modalities of the social institution which is going to be regulated legally.

Teaching, Research and Extension; an integrated curriculum for a high quality law school or program in Colombia. The purpose of this article is to highlight the curricular position of the current education in the Law programs and its alternative of change [16] since it is important, to draw a route that allows the administration of a Law School, to be articulated from the Teaching, Research and Extension that allows achieving the high quality of the program. To achieve this goal, it is proposed to the Universities of Colombia (and the world) the adoption in their programs of a curricular model based on research processes. The research administration model proposed is Holistic, and it is intended to integrate the substantive axes of Higher Education: Teaching, Research and Extension, from the social, human and scientific dimensions, to develop gradual processes of change (visible and applicable research results) in the educational institution to obtain efficient results from it.

Kemmis [12] says that the forms of relationship between the theory and practice of education vary and today the millennium natives ask that the curriculum of law programs change, because they have understood that with the use of mobile phones, laptops, tablets, the internet and Moodle platforms leave the routine they live in classrooms with teachers who have become accustomed to conducting master classes, that is, they are dictatorial because they still teach classes to their students.

It is from the definition of a map, that in the institutional research route, that a series of events must be carried out that allow generating a research culture from the Law program, having as a link, the research curricular model that we propose as a holistic model, allowing the activities of the research work necessarily have implications with the effect of complementarity or sequence in the other axes of education such as research and extension, this is a systemic vision of the curriculum design process of the academic-pedagogical area of educational management [10].

Simply put, if the activity is teaching, its effects will replicate in research and extension, if the activity is research, it will replicate in teaching and extension and if the activity is extension, it will replicate in teaching and research. The integration of the substantive axes of education allows to have in an organized way all the events that are fulfilled both in one and another function of it and one or another academic and methodological - administrative unit; articulated with each other, so that all members of the academic community that belong to a training program produce tangible results of these substantive axes.

The operational phase of the Holistic Curricular Investigative Model (MOHID) proposed in this article is closely linked to the construction of curricular strategies that arise in the curriculum, such as a curricular design that consults the alternative nature of the process of building a new academic culture that allows the creation of new professional and pedagogical identities, such as the construction of a citizen culture for the care of the environment.

The interdisciplinary nature that the holistic research model (MOHID) pursues is not achieved with the current way in which knowledge is organized in the curricula in the faculties and law programs of Colombia, especially in the city of Cartagena. The subjects or subjects adapted as a curricular strategy have been subject to different analyses that call into question their insular, atomized, encyclopaedic and strongly rooted nature in their content mediating function, therefore, in this new curricular proposal, the students are considered thematic and problematic nuclei, as object of concretion of the proposed curricular model, from a holistic and integrative vision with the investigation.

The thematic and problematic core, as an interdisciplinary curricular strategy, is understood according to Nelson López, as: "The set of related knowledge, which makes it possible to define research practices and processes, around a problem object. This implies the construction of strategies that guarantee the theory - practice relationship and the construction of participatory actions, between individuals and groups in the strategy that allows integrating a field of problems; with a field of knowledge that becomes the training of a professional [13]

Working with the thematic and problematic core strategy, warns a direct correlation with the purpose of training the modern lawyer, which guides and determines the academic development of the training process in an academic program, to the extent that it becomes a device that facilitates the integration of training, research and social extension or projection, considered determining processes of the identity of institutions of higher education level, concatenated to the internationalization of the curriculum, to take it to the forefront of competitiveness and globalization.

The holistic research curricular model, based on the thematic and problematic core, is presented as a strategy that allows to overcome the split between research theory and practice, to the extent that it is the result of stressing a field of social reality problems [4] with a field of legal knowledge, in order to achieve a purpose of group-built and concerted training between the teacher and his students.

If there really is an intention to provide a comprehensive training of professionals (apprentices at the different levels of the curriculum, call semester or annuity), it is necessary to question the degrees of complementarity (or not) that are encouraged and encouraged to through the curricular structures in force when saying of Stenhouse [24] "To conceive the curricular process as an investigative process, it installs its development within the framework of critical rationality, which defines it as a custody and practice field that is interested in interaction of two large scenarios of meaning that have been given separately as differentiated concepts of conceiving the curricular process as an investigative process, installs its development within the framework of critical rationality, which defines it as a field of custody and practice that interested in the interaction of two large scenarios of meaning that have been given separately as concepts difference Two of Curriculum: the intentions for the school and the reality of the school; theory and ideas for practice and conditions of the reality of that practice"

The thematic and problematic nuclei mentioned above, given their complexity and extension can be broken down into programmatic blocks, understood as the integration of the subjects as interdisciplinary strategies of less complexity in the thematic nuclei. The problematic blocks can be considered as organizational units of subjects that allow the management and concrete development of the different dimensions, problems, objects and principles that constitute the thematic and problematic core. In turn, the subject blocks can lead to the construction and generation of specific research projects in the classroom, which refer to specific situations arising from the training, research and social projection from each of the subjects that make up the curriculum as Thematic axes, then these classroom projects end up in undergraduate work to complete the undergraduate law. As an example, we illustrate the experience of the Universidad del Quindío on the application of the PACA (Alternative Curriculum Action Program called PACA, which was proposed by NELSON LÓPEZ JIMÉNEZ at the Universidad del Quindío and with which we base our curriculum proposal), the following graphic illustrates the above:

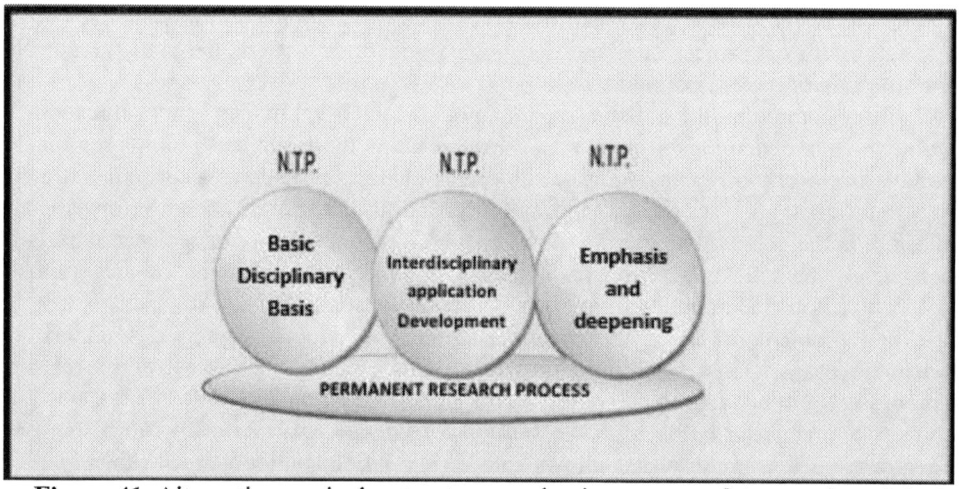

Figure 41. Alternative curricular structure, academic program of Accounting of the University of Quindío.

The thematic and problematic core strategy directly returns the nature of permanent and collective elaboration, which acquires the curricular process from the research on the conceptual horizon, which configures the field of action of the Curricular Investigative Model, MOHID. The research model that is presented and argued, reinstates in the deliberate intention to promote a substantive and structural change of the current curricular culture, while trying to weaken the hegemony of the current curricular forms in the perspective of building integrated, investigative and Socially and academically relevant, hypothetically it is assumed with this model, that what it is about is to create a new identity of the professional of the teaching and research of Colombian law in the faculties of Law. To operationalize this research model in the Schools or Law Program, managers, teachers and students should know about socio-legal research and the use of new information and communication technologies, ICTs.
Promotion of ICTs in the Law Faculties of the Universities of Cartagena. Today teachers must make use of Information Technology and Communications (ITC) that are defined by Law 1341 of 2009 as resources, tools, equipment, software, applications, networks and media that allow compilation, processing, storage and transmission of information, such as voice, data, text, video and images and that can be used in the classroom.

The Colombian Ministry of Education insists that teachers enrich their pedagogical processes with the use of ICTs, since by involving students in their use, they improve class productivity, because they educate their students in the search for information, focusing on the appropriation of technological tools that allow collaborative work and the exploration of learning objects such as socio-legal research in law programs.

The teacher who teaches socio-legal research can create small learning communities [20] integrated by him and his students through Moodle platforms; Therefore, information and communication technologies (ICTs) are electronic means of capturing, processing, storing and disseminating data and information, which facilitate the design of educational strategies in the classroom [2].

ICT can generate successful experiences such as the one proposed in this article, that of building a curriculum for a Law program based on research and that allows the articulation of from the Teaching, Research and Extension that allows to achieve not only the internationalization of the program but its high quality, as stated by Fuentelsaz, Maicas and Polo, [6]. In this sense, MOHID can be an educational innovation since it is the result of a process of integrating the technology existing today in the world of life to create a curriculum that allows to improve the processes of the socio-legal research activity as a product, which integrates a process of the Colombian innovation system. According to Medina and Espinosa [15] Innovation consists in the consolidation of a new product, or the process or system of an existing product in an improved version

5. Conclusion

On the improvement of socio-legal research with the use of new technologies, we can resume the words of Arbeláez [3] on this topic: "The use of new technologies has permeated the professional and educational fields to facilitate our performance in several areas, one of them has to do with access to information, another with data processing, and another, with immediate, synchronous and asynchronous communication, to disseminate information or to contact anyone anywhere in the world."

Certainly it is no longer necessary to share the same physical space to develop a work meeting in the research area, because we have computer systems such as Skype, Zoom, RENATA virtual office, or to develop some type of writing or project, and even, it is such the impact of the ICTs that more and more students get to pursue their undergraduate or postgraduate studies, without having attended the physical space of a university because these educational institutions have virtual platforms that allow distance education, without having to use the space and time.

This situation has not been alien to the field of scientific research of Law, where more and more academics use ICTs to solve problems of the personal management of their work or the management of research itself from the center of socio-legal research, which without the use of these technological resources would be costly in economic and time terms. However, there is a risk of confusing the medium with the end, that is, of thinking that the tools offered by the ICTs to the researcher are an end in themselves and not a means to solve situations of the research process, without taking into account the potentialities and risks that the researcher may run when he focuses his expectations on the use of technological tools, such as with a click, the entire research project or its results are deleted, checked, copied, etc.

Regarding the use of ICT for personal management, computer equipment and its various computer applications, if they have become an indispensable tool for the preparation and processing of information; the internet a means for the search of information that facilitates access to specialized sources of information and knowledge, with access to virtual and remote libraries, electronic journals, digital books, databases, in which you can consult a variety of articles of research, thesis, conference proceedings, among others; the physical and now virtual (in the cloud) information storage devices that allow not only to detach from the physical devices, but also to have information in any place where the researcher is without having his own computer, and share files with members of the same team or other places and universities to perform collaborative work, with tools such as Eduroam, videoconferences, VC Espresso, Google Drive, iCloud, Dropbox, among others.

On the other hand, ICTs facilitate the communication of research results and their accessibility to the community in general and to scientific communities in particular, which can be done through the publication of research results in personal or collective blogs of research groups, publication on the web or in electronic journals or in indexed journals that are found in the databases of universities or in academic search engines such as GOOGLE SCHOLAR, SCIELO, SNAAC, etc., located on the network.

In the very development of socio-legal research, technology has been making important developments to solve problems regarding the collection and analysis of information, as happened with the statistical survey applied to teachers of the law program of the Faculty of Science Social of UNITECNAR that showed that teachers of oscillating ages between 40 and 60 years do not use the SPLAVIA platform, are considered digital illiterates because they were formed with the use of the typewriter and are "puyographs", that is, they wrote with A single finger of each hand, at present they

IOP Conf. Series: Materials Science and Engineering **844** (2020) 012071 doi:10.1088/1757-899X/844/1/012071

have learned the basic use of the computer, but they do not know the processes of Hardware and other mobile applications and the improved use of the internet.

Young digital natives, on the other hand, play in classes with the various tools that have been developed for the advancement of a research class, such as online questionnaires, such as SurveyMonkey, which allow large population samples to be answered from anywhere in the world and download the results directly to quantitative analysis programs such as the SPSS (Statistical Package for the Social Sciences). Regarding the analysis of information, there is a fairly wide range of software for the analysis of quantitative and qualitative information; For the quantitative analysis, as already mentioned, there is the SPSS, R, Mintab, Stats and Excel, and for the analysis of qualitative information, not only of texts, but also of audios and videos, you can use Etnograf, Atlas Ti, Nud. IST Vivo 2, Aquad 5, Winmax and Maxqda. When measuring the same variable in the teachers of the law program of the Free University Cartagena Headquarters, we found a negative correlation (Hernández Sampier: 2004) since they do not use the Moodle platform in their classes, nor any legal computer system.

These computer programs that bring new technologies generate great expectation in terms of their potential, one of them is the speed in the handling, management, search and exposure of data; another is the improvement in the quality of research. This implies being able to handle a greater amount of information with greater quality in the analysis, and finally to improve transparency in the analysis process, due to its potential for standardization. While research cannot be abstracted from technological advances and get the most out of them, it is the research group who should be clear about why and for what to use a technological tool, and whether it will be beneficial for your project in terms of academic, economic and time, therefore as a strategy of socialization of the knowledge of the results of the research "Use of ICTs in the teaching - learning processes in the educational community of the Antonio de Arevalo Tecnar University Foundation and the Free University Headquarters Cartagena "and to motivate older teachers in the use of ICT, booklets, videos, podcast and posters were made that allowed them to understand the importance of using new technologies in the training processes of lawyers of the era of the new technologies.

Contextualizing this articulation of knowledge and technological mediations, it is considered that, in the field of Law, the great transformations, the social changes that are coming in Colombia, are allowing the development of legal sociology and its significant socio-legal research, especially in the Space of Human Rights, Armed Conflict, Forced Displacement, in the abuses of political power and abuses in the distribution of the economy in the environment, especially in terms of climate variations.

It is worth noting that, as long as there is no organized, homogeneous and disciplined movement around socio-legal research in Colombia, much of the work that critically addresses the problems that keep Colombian society in crisis will fail, hence the concern about the lack of good guidance in research in Law Faculties and Programs, so from the curriculum, articulating: Teaching, Research and Extension or Social Projection as substantive axes of Higher Education and the management of legal informatics, we find a holistic curriculum, and according to Garcés-Prettel [9] ICTs play a very important role in the discourses on pedagogical transformation, since they can contribute to the development of learning processes where knowledge is exercised by seeking solve functional, complex and everyday situations [21]

In terms of techno-pedagogical transformation, another essential component that higher education institutions can develop with the support of ICTs is to create tactics and strategies that engage students as actors in their own learning, which leads to good human development and with the transformations of the logic of power within the educational practice that should tend for horizontal relationships and for a coexistence centered on service, respect, participation, dialogue, and recognition of diversity [21] and to the other, as legitimate other [14].

In this regard Escontrela et al [5] suggest that each stage of humanity has had its technological tools and has also developed different models of use, and that in the early sixties and seventies, the pedagogical model of application is based on the individualization of learning or individual internalization of knowledge, in the eighties began to talk about the need to promote learning by discovery and highlight the benefits of group use, in the nineties, the cooperative model is the one that received the most attention, among other reasons, to the development of communications. Today we work under the collaborative scheme mediated by technologies that leads the individual to work assertively with the context, so it is expected that the new technologies will be incorporated into the socio-legal research of the law faculties in Colombian universities.

6. References

[1] Abdellah FG, Levine E. (1994) Preparing Nursing Research for the 21 st Century. Evolution. Methodologies, Chalges. Springer: New York.
[2] Alemna, A. A., & Sam, J. (2006). Critical issues in information and communication technologies for rural development in Ghana. Information Development, 22 (4), 236-239.
[3] Arbeláez Gómez, M C (2014). Information and communication technologies (ICT) an instrument for research. Andean Research, *16*(29), 997-1000. Retrieved October 31, 2019, from http://www.scielo.org.co/scielo.php?script=sci_arttext&pid=S0124-81462014000200001&lng=en&tlng=es.
[4] Berger L P y Luckmann T (1998) the social construction of reality. Amorrortu Editores.
[5] Escontrela Mao, R., & Stojanovic Casas, L. (2004). The integration of ICT in education: Notes for a relevant pedagogical model. Pedagogical Magazine. Caracas, vol.25, n.74, 481-502.
[6] Fuentelsaz, L., Maicas, J. P., y Polo, Y. (2005). Towards an effective management of information technologies. Universia Business Review, 6 (2), 40-53.
[7] Gálvez Albarracín E J (2014) Information and communication technologies, and innovation in MIPYMES of Colombia. Cuadernos de Administración. Universidad del Valle. Vol. 30 N° 51. Enero – junio.
[8] Galindo Cáceres J (1998) Técnicas de Investigación en Sociedad, Cultura y Comunicación. Pearson, Addison Wesley Longman. México.
[9] Garcés-Prettel, M; Ruiz-Cantillo R y Martínez Ávila D (2014) Pedagogical transformation mediated by information and communication technologies **(ICT)**. Revista Saber, Ciencia y Libertad. Volumen 9, No. 2, Julio - Diciembre.
[10] Guerra Montoya J W (2007) Curriculum design process. Teacher Training Project in Antioquia. SEDUCA – FUNLAM, 2006 - 2008
[11] Hernández Sampier, R (2004) Investigation and methodology. Editorial Felix Varela, La Habana.
[12] Kemmis, S (1993). The curriculum beyond the theory of Reproduction. 2ed, Madrid: Morata. 175 p.
[13] López Jiménez, N (2001) The curricular de-construction. Colección Seminarium Magisterio. Bogotá.
[14] Maturana, H (1999) Transformation in coexistence. Dolmen Ediciones. Santiago de Chile.
[15] Medina, C., y Espinoza, M. (1994). Innovation in modern organizations. Recuperado 12/06/2013 de http://www. azc.uam.mx/publicaciones/gestión/num5/doc06.htm.
[16] Mora Vargas A I (1998) Traditional curriculum approach Vs holistic approach. Revista Educación 22 (2) : 141 – 152.
[17] Moreno Durán Á (2003) Pierre Bourdieu. Elementary Introduction. Editorial Panamericana Formas e Impresos. S. A. Bogotá.
[18] Ramirez Llerena, E (2001) Socio-legal Research. Critical Approaches. Ediciones Doctrina y Ley Ltda. Bogotá D.C.
[19] Ramírez Llerena E (2015) Pierre Bourdieu's Theories and their impact on Law: Violence against legal operators, study of two homologous cases. Universidad Libre, Sede Cartagena. Programa de Derecho. Centro de Investigaciones.
[20] Ramírez Llerena E (2015) Research seedbeds: A significant learning experience in the Law program in the Libre University, in Cartagena. Universidad Libre, Sede Cartagena. Programa de Derecho. Centro de Investigaciones.
[21] Roegiers, X (2004) A pedagogy of integration. Deboeck Universidad. Bruselas.
[22] Rocha, C (2005) Communication and participation. En CEDAL, Participation is recognition (pags. 25-68). Audiovisual Educational Communication Center. Bogotá.
[23] Said Hung Elías (2015) Towards the promotion of ICT in the education sector in Colombia. Editor. Editorial North University. Barranquilla, Colombia.

[24] Stenhouse L (1984) Investigation and development of curriculum. Editorial Morata. Madrid.
[25] Torregrosa Jiménez N y Torregrosa Jiménez R (2012) Socio-legal Research a priority function in the training of 21st century lawyers in Colombia. Editorial. Verba Iuris 28. pp. 13-15. Julio - Diciembre 2012. Bogotá D.C. Colombia. ISSN: 0121-3474.

Vehicle Lateral Wander Measures Using Ultrasonic Sensors

E Oconitrillo-Varela, H Hernández-Vega and J P Aguiar-Moya

Laboratorio Nacional de Materiales y Modelos Estructurales, Escuela de Ingeniería Civil, Universidad de Costa Rica. Ciudad de la Investigación, Finca 2, Apartado Postal: 11501-2060

henry.hernandezvega@ucr.ac.cr

Abstract. This paper evaluates the measurement of vehicle lateral wander using ultrasonic sensors. Measurements were made on different sections of the Costa Rican National Road Network. The recorded data allowed to determine the cumulative frequency distribution, lateral wander data and a confidence interval for the measurements. In some routes, where the sample size is greater than 70 vehicles, the distribution of the lateral wander data adjusted to a normal distribution. However, in several cases the data does not resemble a normal distribution curve, or other distributions reported in the literature. Finally, an Android app was developed to automate the data collection process using the ultrasonic sensor setup, simplifying the measurement process. From the cumulative frequency results, obtained with the ultrasonic sensors, it was determined that most vehicles travel between 50 and 150 cm from the edge of the road. In order to validate the data, the results obtained with the ultrasonic sensor setup were compared with the results obtained by image analysis of video recordings. These image analysis is possible due to pavement markings every five centimeters on the pavement from the lane edge. The results from video recordings and ultrasonic sensors show a high coefficient of correlation.

1. Introduction

After the publication of the Mechanistic-Empirical Pavement Design Guide (MEPDG) [1], its implementation has been pursued in several Latin American countries [2]. One of the input parameters for pavement design is lateral displacement, which is not a parameter that has been extensively studied in the region. Consequently, lateral wander values taken from studies carried out in other countries are usually assumed. Lateral wander is the measurement of how channelized is the traffic traveling through a specific lane [3].

It is possible to adjust the pavement design to the through loading conditions when the real lateral displacement in a section of road is known. Also, the variability associated to the design models can be reduced by using representative field data, improving overall reliability. The accuracy of the measurements represents a key factor because lateral displacement is affected by many variables such as: geometric elements of the road, climatic conditions, traffic, speed, types of vehicles, drivers, among others [4].

It is necessary to determine average values of the lateral displacement in order to generate inputs that can be incorporated into future national pavement guidelines. Furthermore, this research seeks to be a reference for designers to use lateral wander values obtained in Costa Rica, in order to avoid the assumption of values from foreign research, since the University of Costa Rica is currently developing a pavement design guide, which requires adapted information that considers the specific conditions for the region [5].

In previous research work done in Costa Rica, it has been possible to measure lateral wander by using video recordings and pavement markings, as in the study conducted by [6]. This method is useful; however, it demands long time to process the data if no image analysis automation is used. Furthermore, the precision of this method is about ±6,25 centimeters.

In other countries, different methods have been developed. Lateral wander has been measured using a carpet of sensors placed on the pavement [3]. Another method consists of using three pneumatic sensors arranged on a Z pattern on the pavement [7]. The researchers were capable of measuring speed, lateral wander and count vehicles. Moreover, this displacement can be measured by video analysis to determine vehicle position on the lane [8]. Other methods consist of placing sensors embedded in the pavement, also in a Z configuration [9]. In general, literature has shown that wandering data follow a normal distribution. However, it was found that only in very specific cases the data collected adjusts to a normal distribution [6].

There is a need to develop new methodologies to obtain this type of data with greater precision, because of the absence of Latin-American studies related to vehicular displacement and the effects of this parameter in different fields of transport engineering. In this research, a method based on ultrasonic measurements is calibrated and validated. The method consists of an ultrasonic sensor controlled by an ArduinoUNO platform that measures distance of the vehicles from the lane edge.

2. Method

This section describes the sensor, how it works and the implementation of the tool.

2.1. Sensor Design

The Arduino tool has increased its popularity due to the benefits of open source hardware/software elements. There are many applications that can benefit from this tool, especially because it allows connecting many devices with relative ease. This application has not been widely applied in transport engineering, although smaller projects with distance sensors has been developed to recognize objects, specially, for home integration.

A description of every module of the prototype is presented below:

a) ArduinoUNO card: free software micro controller, based on the C++ programming language. The device is capable of controlling devices like sensors, lights, LCD screens, Bluetooth, clocks, lasers, GPSs, and others [10].

b) Ultrasonic Sensor HC-SR04: it is a sensor that sends a signal (TRIG) and when the signal hits an object, it is sent back to the sensor (ECHO). The sensor measures the time it takes the signal to return. Because the speed of the signal is known, it is possible to determine the distance from the object to the sensor [10].

c) Real Time Clock (RTC Clock): it works with the satellite time and date, with one calibration of the device it is possible to determine the exact time and date of each measurement.

d) Temperature and humidity sensor: refinements of the readings from the ultrasonic sensor´s can be adjusted according to the temperature and humidity recorded.

e) SD Card Module: this module allows the configuration of a SD card in order to store the measurements. The data are saved in a text file, which can be exported to a spreadsheet in order to post-process the results.

f) Bluetooth Module HC-06: this module sends and receives data from/to the ArduinoUNO card. This module was useful to this project because a tablet application was used in order to simultaneously capture the type of vehicle that is measured by the ultrasonic sensor.

Figure 1 shows the different electronic components that were used to produce the prototype used in this study to measure vehicle positioning in the lane.

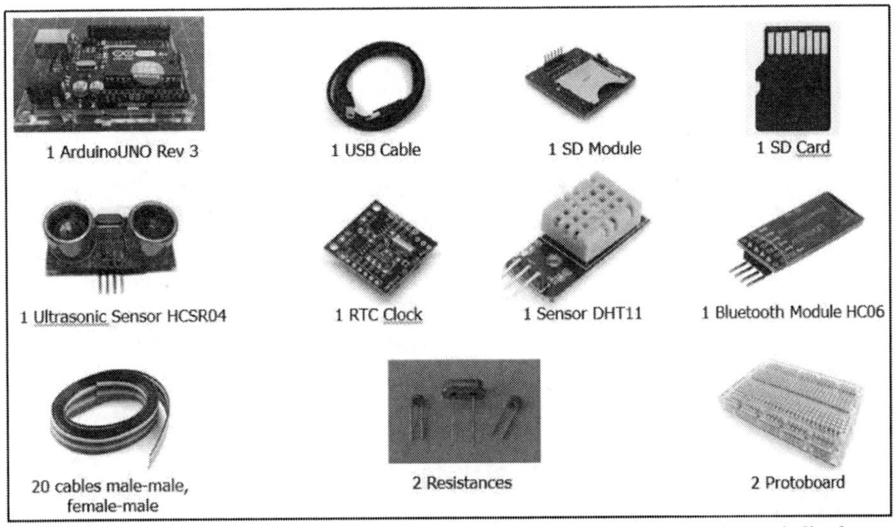

Figure 1. Electronic components necessary to elaborate the vehicular lateral displacement measurements

A code was developed in the Arduino interface using C++ language to control the connection among the devices (sensors, Real Time Clock, SD Card and Bluetooth). It is necessary to specify how the devices are connected to the ArduinoUNO card. The devices must be connected to the Arduino card; figure 2 shows the implemented setup.

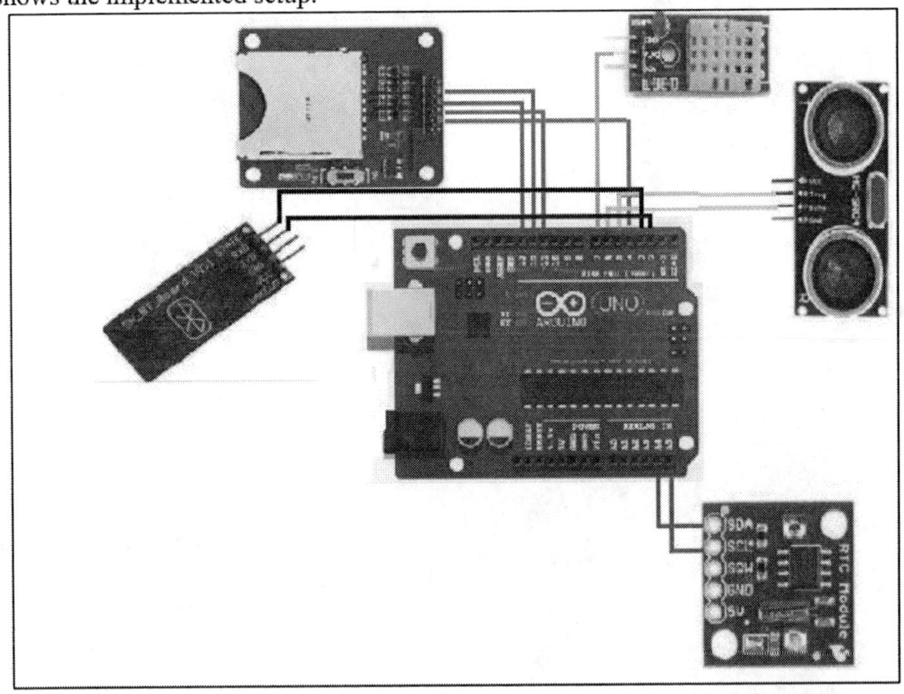

Figure 2. Diagram of modules and sensors connections to ArduinoUNO

2.2. How does it work?

Once the program has been coded and the connections between the different modules and sensors to the ArduinoUNO card are ready, the device is able to proceed performing distance measurements. To

IOP Conf. Series: Materials Science and Engineering **844** (2020) 012072 doi:10.1088/1757-899X/844/1/012072

measure distance of the vehicles from the lane edge by the ultrasonic sensor, the following procedure were followed:

a) Code the controller card: from Arduino Software it is necessary to connect Arduino card to the computer using a USB cable, to download the coded procedure.

b) Ultrasonic sensor placement: the device must be placed at the road shoulder. The distance between the lane edge and the sensor should be between 30 and 80 centimeters.

c) Bluetooth connection: in order to classify the vehicles passing through the sensor´s measure range, it is necessary to link Arduino to a vehicle classification application (Figure 3 shows the interface of the application developed in this project). The application also shows the distance measurement been recorded by the device.

d) Measure distance: the sensor is capable of registering every vehicle distance in an SD card. The data must then be downloaded to a computer for further processing.

The vehicles were classified as passenger cars, C2 Bus, C2+, C2, C3. C4, T3S2 and T3S3 as shown in figure 3. Figure 4 shows a schematic of the measuring process.

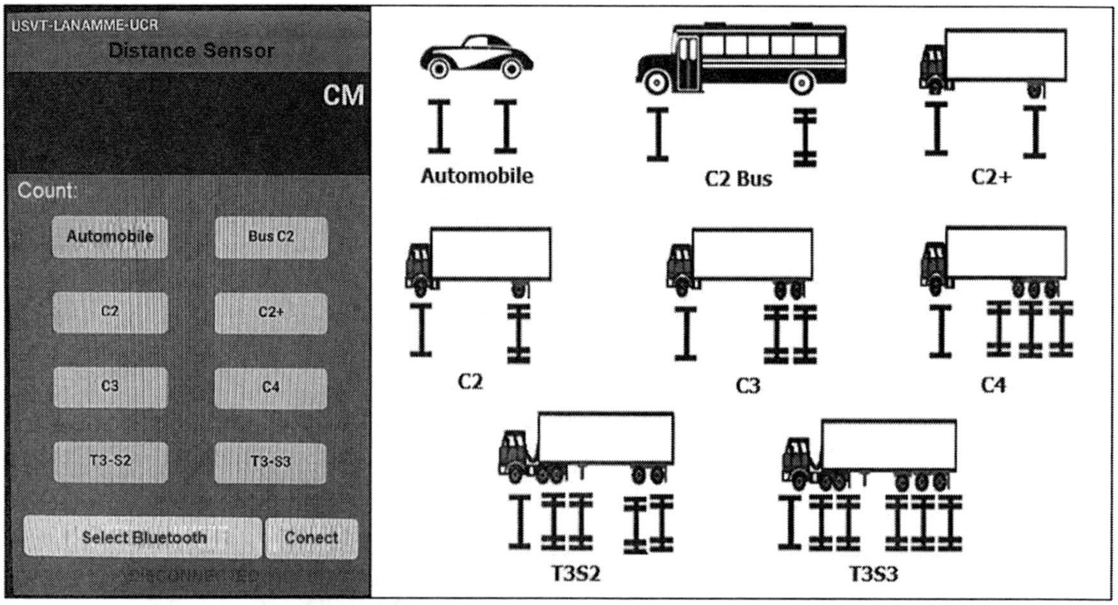

Figure 3. Android application interface (left) and vehicle classification scheme used (right).

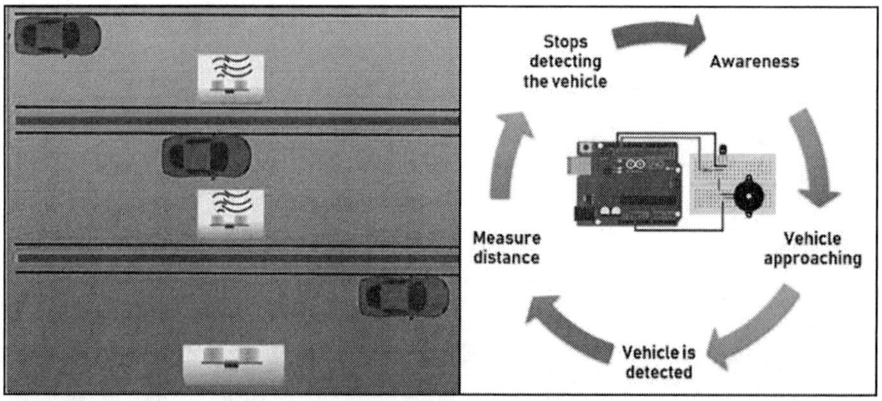

Figure 4. Illustrated process of measure (left) and schematic diagram of distance measure process by an ultrasonic sensor (right)

2.3. Tool implementation

Considering available data from a preceding study [6], six national routes were chosen to test the equipment capability. Every one of the sections had enough previous studies to guarantee a full characterization of traffic behavior in the route based on specific parameters. Some of the parameters are: lane width, shoulder width, operating speed, vehicular classification, average daily traffic, location and others environmental factors. Based on the previous requirements, measurements were made on six routes (see figure 5). Additionally, table 1 shows a summary for every test site.

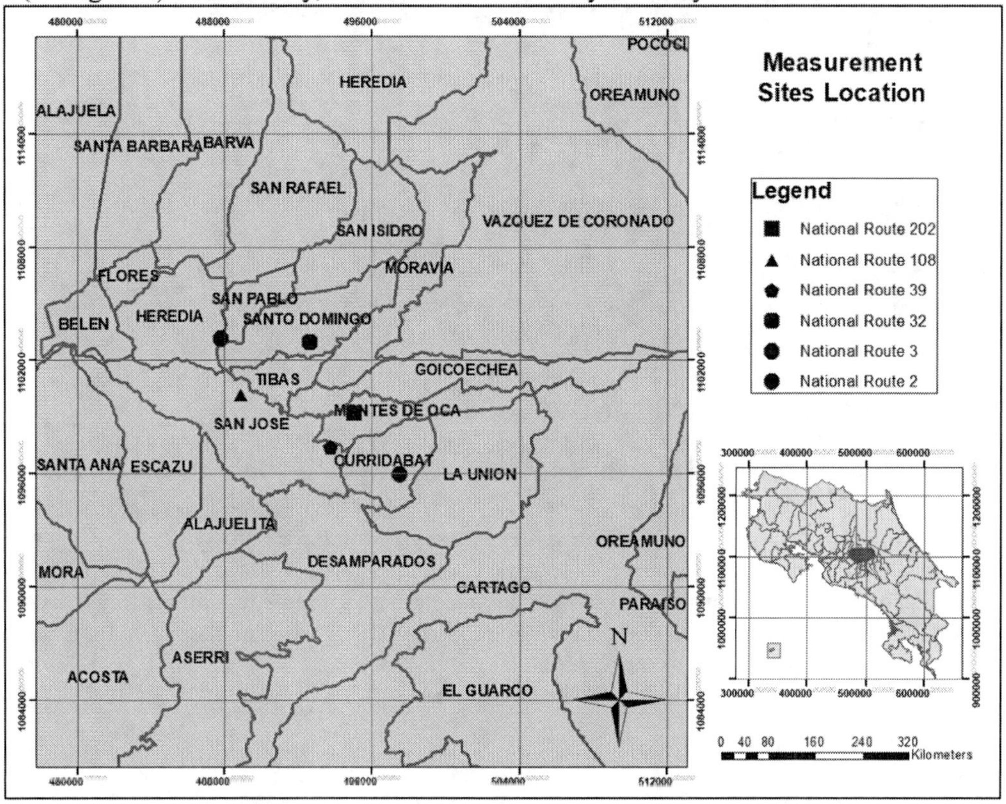

Figure 5. Ultrasonic sensor measurement sites location map

Table 1. Average daily traffic for every route and vehicular composition by vehicle class

Route	2	3	32	39	108	202
Average Daily Traffic (ADT) estimated for 2018	53001	32237	38784	66963	52065	23229
Percentage of passenger cars and motorcycles	68.71	78.79	76.34	84.43	68.61	86.53
Percentage of light Commercial Vehicles	14.65	9.84	11.03	10.51	16.52	8.89
Percentage of buses	4.38	5.38	2.69	0.78	4	2.05
Percentage of two axle trucks	6.65	4.52	3.67	3.22	6.61	2.23
Percentage of three axle trucks	3.23	0.87	0.83	0.51	1.39	0.2
Percentage of trucks with five or more axles	2.38	0.6	5.24	0.55	2.87	0.1

3. Results

3.1. Method validation

The validation of the method was performed by direct comparison to visual measurements of video recordings and the measurements made with the ultrasonic sensor. Figure 6 displays the markings used

IOP Conf. Series: Materials Science and Engineering **844** (2020) 012072 doi:10.1088/1757-899X/844/1/012072

for the visual analysis: marks were placed, every five centimeters from the edge of the lane mark, on the pavement. The data from the video recordings were processed by visual inspections in an office environment.

Figure 6. Marks placed on pavement. National Route 32

Figure 7 shows the results of the comparison between the measurements obtained by the ultrasonic sensor and the visual measurements from the video records. The figure demonstrates a high correlation between both methods when used to measure the lateral displacement.

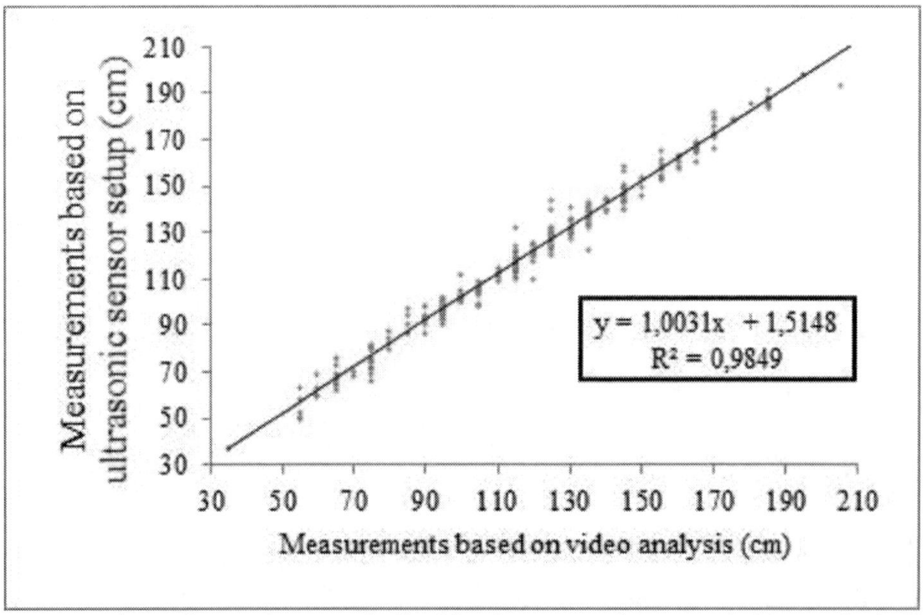

Figure 7. Comparison of measurements obtained by the proposed setup and video analysis (National Route 32, n=262)

The slope coefficient in a linear regression model suggests an almost perfect fit between the ultrasonic method and the visual method (t-stat = 127.02, p-value =5.40x10-227) indicating that both methods provide similar results. Once the method was validated from the ground truth method, measurements were performed in different sites.

3.2. Lateral wander measurements

Average lateral wander, considering the measurements on all analyzed national routes, is 28.9 cm with a standard error of 0.54. For light vehicles, lateral wander is 31.6 cm with a standard error of 0.68; and for heavy vehicles, is 28.4 cm with a standard error of 1.06. Lateral wander values obtained are consistent with other measurements made in the same routes, by [6]. Some vehicle types present very small sample sizes, considering their small share in total traffic composition and the time dedicated to collect the data for these routes. One device was used at the same time, and the duration of the individual tests varied between two and three hours at each site, and was defined by the run-time of the laptop´s battery (a laptop provided the electrical power to the device). Table 2 presents average values for all sites included in the present study.

Table 2. Average lateral wander (Lw) values with the sample size (n) for each vehicle type at each location

Vehicle type		Route						Average	Standard Error
		2	3	32	39	108	202		
Passenger Car	n	388	189	528	591	228	237	31.6	0.68
	Lw (cm)	30.8	32.9	34.9	36.4	24.0	30.6		
C2+	n	50	48	31	66	34	17	32.5	2.07
	Lw (cm)	37.9	34.6	35.7	37.5	20.2	29.0		
C2	n	35	57	37	44	10	14	38.8	2.76
	Lw (cm)	28.5	30.1	55.0	42.7	27.2	49.2		
C3	n	17	11	12	8	-	-	28.2	4.07
	Lw (cm)	28.7	19.6	38.6	25.7	-	-		
Bus	n	37	26	24	7	21	4	26.2	2.40
	Lw (cm)	27.1	37.2	32.1	14.2	21.8	24.8		
T3S2	n	22	6	36	7	10	-	24.3	2.70
	Lw (cm)	17.9	35.5	29.0	23.5	15.7	-		
T3S3	n	9	3	12	-	-	-	20.5	4.18
	Lw (cm)	13.2	21.4	26.8	-	-	-		

Figure 8 shows the lateral wander distribution for truck traffic on the analyzed routes. According to the data obtained, most of the routes exhibit a trend where 50% or more of the measures correspond to a distance less than 100 cm. Only measurements at National Route 202, show 70% of the heavy traffic driving from 100 cm or more from the lane edge. This route has a lane width of 3.7 meters, which is wider than the other analyzed routes where the lane with varies from 3.4 to 3.6 meters.

In order to capture the variability associated to lateral wander, box plots for light vehicles and trucks are shown in figure 9 and figure 10, respectively. Confidence interval is the range in which the average population on national routes for a 95 % level of confidence, it probably locates. For light vehicles, lateral displacement fluctuates from 0.3 to 0.4 meters; except for National Route 108, that shows a lower variation, possibly due to specific geometric and functional conditions of this road. In general, heavy vehicles have lower variability for each route; however, the overall average value differs for each route indicating that it should be considered as a section specific property.

To better characterize the lateral wander data, the distribution for each section was also analyzed. Fit to a given distribution or family of distributions can be made with statistical tests such as Anderson-Darling, Kolmogorov and Ryan Joiner. For each data set at a particular station, the adjustment to a probability distribution was verified. Although the tests can be done for all vehicle classes the analysis was performed for light and heavy vehicles in order to have an adequate sample size. From the tests, it was confirmed that most of the routes did not adjust to neither the normal or the Laplace probability distribution that were originally considered in this study. Only Route 3, Route 108 (heavy vehicles) and Route 202 (light vehicles) were confirmed that adjust to normal distribution (see table 3).

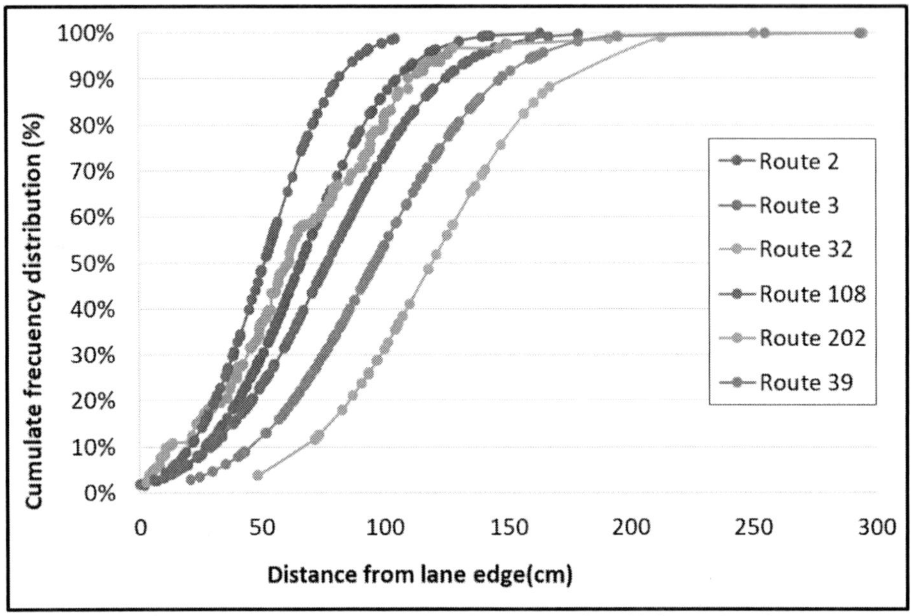

Figure 8. Heavy vehicular cumulate frequency for all the national routes

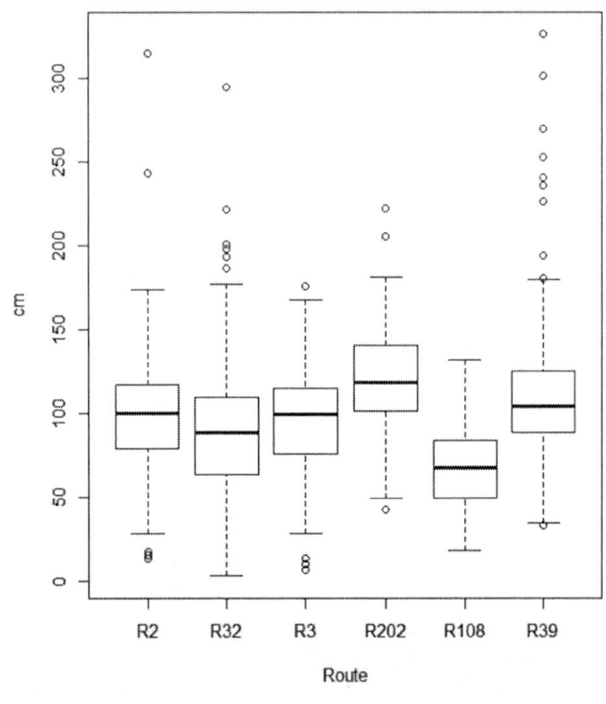

Figure 9. Distribution of passenger cars´ lateral wander by route

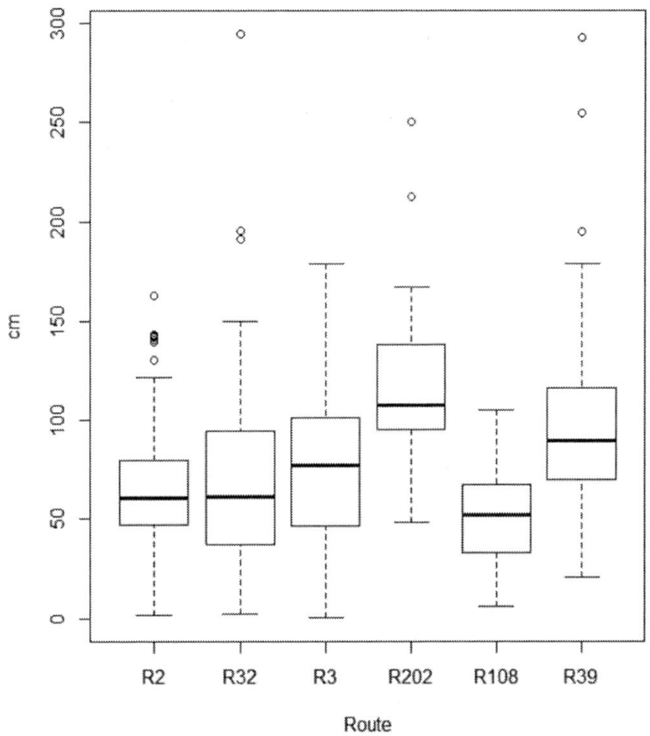

Figure 10. Distribution of heavy vehicles' lateral wander by route

Table 3. P-Values obtained from normality tests (Anderson-Darling)

Route	P-value	
	Non-Trucks	Heavy Vehicles
2	0.005	0.005
3	0.109	0.673
32	0.07	0.01
39	0.005	0.005
108	0.021	0.276
202	0.249	0.036

4. Conclusions

Based on the performed analysis, it is possible to obtain vehicular lateral displacement by ultrasonic sensors. The methodology was applied and validated in some of the most important routes of Costa Rica, such as National Route 2, National Route 3, National Route 32, and National Route 39.

Ultrasonic sensors controlled by Arduino are a useful tool to determine vehicular lateral displacement. It is necessary to visually inspect the type of vehicle passing through the measurement area, using a mobile application developed for this study.

The lateral wander distribution for some of the sections containing a significant sample size follow a normal distribution. When the route section data did not follow a normal distribution, it did not follow any other distribution tested by Minitab 18.

From cumulative frequency results, of the data obtained with the ultrasonic sensor, is determined that vehicles are concentrated between 50 and 150 cm from the edge of the road. Measurements at National Route 202, show 70% of the heavy traffic driving from 150 cm or more from the lane edge, which is sort of the center of the lane. So, this is a valuable input for pavement designers, because they can have a notion of the load distribution among the lane width.

Future research could compare results from the device proposed in the present study with those of the referenced papers and could discuss how the proposed solution could improve, or not, the state-of-the-art in terms of accuracy, precision, cost of the device, and others.

Acknowledgements
We would to express my gratitude to Jason Zamora and Jorge Mora, for their technical support during the design, developing and calibration of the method proposed in this paper.

References

[1] AASHTO 2008 *Mechanistic-Empirical Pavement Design Guide* (Washington D.C.: American Association of State Highway and Transportation Officials Executive Committee) pp 1-204
[2] Loría L, Badilla G, Jimenez M, Elizondo F and Aguiar J P 2011 *Experiences in the Characterization of Materials Used in the Calibration of the AASHTO Mechanistic-Empirical Pavement Design Guide (MEPDG) for Flexible Pavement for Costa Rica*. (Washington D.C.: Transportation Research Board 90th Annual Meeting) pp 11-3359
[3] Blab R and Litzka J 1995 *Measurements of the lateral distribution of heavy vehicles and its effects on the design of road pavements* (Michigan: Road Transport Technology) pp 389-395
[4] Stempihar J, Williams R and Drummer T 2005 *Quantifying the Lateral Displacement of Trucks for use in Pavement Design* (Washington D.C.:Transportation Research Board Preprint)
[5] Leiva F 2006 *Nueva guía mecanística-empirica para estructuras de pavimento* (San José: Programa Infraestructura del Transporte LanammeUCR)
[6] Morris-Mitchell V and Leiva-Villacorta F 2016 *Determinación del desplazamiento lateral de vehículos en carreteras urbanas de Costa Rica* (San José: Infraestructura Vial) vol 18 no 32 pp 5-11
[7] Erlingsson S, Said S and McGarvey T 2012 *Influence of heavy traffic lateral wander on pavement deterioration* (Sweden: Swedish National Road and Transport Research Institute)
[8] Lindheimer T, Fitzpatrick K, Avelar R and Miles J 2017 Effect of geometric factors on lateral position of vehicles in freeway buffer-separated managed lanes *Transportation Research Record: Journal of the Transportation Research Board No 2616* pp 10-18
[9] Timm D H and Priest A L 2005 *Wheel wander at the NCAT test track* (Auburn: National Center for Asphalt Technology) No 05-02 pp 28
[10] Fitzgerald S and Shiloh M 2013 *Arduino Projects Book* (Torino: Arduino LCC)

A Novel Full Boundary Element Formulation for Transient Analysis of Elastic Membranes Coupled to Acoustics Fluids

A J Narváez-Cruz[1], J F Useche-Vivero[1], J A Martínez-Trespalacios[1], and J R Castro-Suarez[2]

[1] Universidad Tecnológica de Bolívar, Cartagena de Indias, Colombia
[2] Universidad del Sinú, Cartagena de Indias, Colombia

E-mail: anarvaez@utb.edu.co

Abstract. A full time-direct Boundary Element Formulation for the dynamic analysis of elastic membranes coupled to acoustics fluid is presented. The elastic membranes is modeled using the classical linear elastic pre-stretched membrane theory. The acoustic fluid is modeled using the acoustic-wave equation for homogeneous, isotropic, inviscid and irrotational fluids. Elastostatic fundamental solution is used in the boundary element formulation for the elastic membrane. The boundary element formulation for the acoustic fluid is based on the fundamental solution of three dimensional Poisson equation. Domain integrals related to inertial terms and those related with distributed pressure on membrane, were treated using the Dual Reciprocity Boundary Element Method. Fluid-structure coupling equations were established considering the continuity of the normal acceleration of the particles and dynamic pressure at fluid-structure interfaces. The time integration is carried out using the Newmark method. Results obtained shows the accuracy and efficiency of the proposed boundary element formulation.

1. Introduction

Acoustic problems in real life applications involve considerations of structural and acoustic part together and thus calls for a coupled vibroacoustic treatment rather a pure acoustic approach. Thus, whenever an elastic structure is in contact with a fluid, the structural vibrations and the acoustic pressure field in the fluid are influenced by the mutual vibro-acoustic coupling interaction [1].

The analysis of fluid-structure coupled systems is a challenging and complex task. In general, the use of experimentation or numerical methods represent the two uniques alternatives to obtain approximate solutions for these kind of problems. However, numerical methods based on domain discretization requires refined meshes for high frequency problems, since the length of the elements should be proportional to the size of the wavelength. This means a more time-consuming model.

The boundary element method (BEM) is a modern numerical technique which has enjoyed increasing popularity over the last two decades, and is now an established alternative to traditional computational methods of engineering analysis [2, 3]. The main advantage of the BEM is its unique ability to provide a complete solution in terms of boundary values only, with substantial savings in modelling effort.

Since consolidation of the boundary element method (BEM) as reliable numerical method for structural and fluid analysis, linear vibrations of structures coupled with an internal fluid has been carried out using hybrid BEM-FEM formulations. In these formulations BEM is used to model the fluid media and the FEM to model the structural response [4, 5, 6, 7, 8, 9, 10, 11, 12, 13, 14, 15, 16, 17]. The main advantage of such formulations lies in a substantial reduction in the number of the degrees of freedom in the discretization of the fluid domain. However, in this formulations is necessary to discretize the entirely structure due to the use of the FEM.

Few BEM-BEM coupled formulations for fluid structure-interaction has been published [18, 19, 20, 21, 22, 23]. However, despite the fact that the BEM has been used for dynamic analysis of membrane structures and for analyses of acoustic fluids, to the best of authors knowledge, these formulations do not have been used for the fluid-structure interaction problem analysis using a full boundary element formulation for such purpose.

In this work, a new full boundary element formulation for the transient dynamic analysis of acoustic fluids coupled to elastic membranes is presented. Membranes are modeled using a boundary element formulation based on the linear elastic membrane theory under small deflection. The acoustic fluid is modeled using a boundary element formulation for the three-dimensional acoustic wave equation. Fluid-structure coupling equations were established considering the continuity of the normal acceleration of the particles at fluid-structure interfaces. Domain integrals on both, fluid and structure equations, were treated using the Dual Reciprocity Boundary Element Method. The developed formulation was used to study the linear vibration response of elastic membranes coupled to acoustic fluids.

2. Structure subjected to a fluid pressure loading

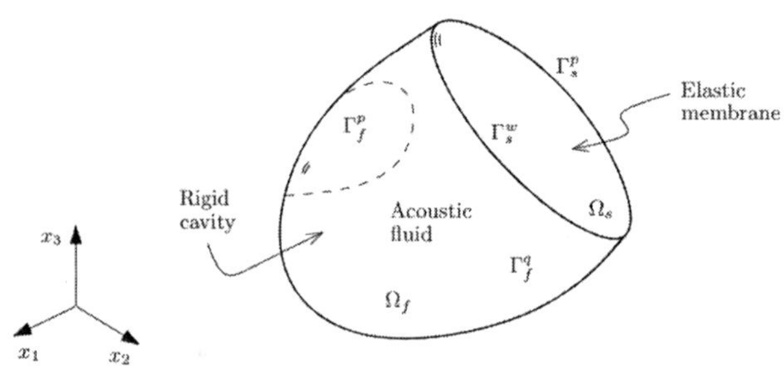

Figure 1. Fluid-structure interaction problem

Consider a partially opened cavity Ω_f with ridig walls and a flexible elastic membrane Ω_s with mass density ρ_s and thickness h (see figure 1). Cavity contains an homogeneous and isotropic acoustic fluid with mass density ρ_f. The membrane vibrations and the acoustic pressure field in the fluid are influenced by the mutual vibro-acoustic coupling interaction. In this work, the vibro-acoustic coupling interaction is modeled using an eulerian formulation where the acoustic response is described by the pressure, while the membrane response is described by the transversal displacement field.

2.1. Acoustic wave equation

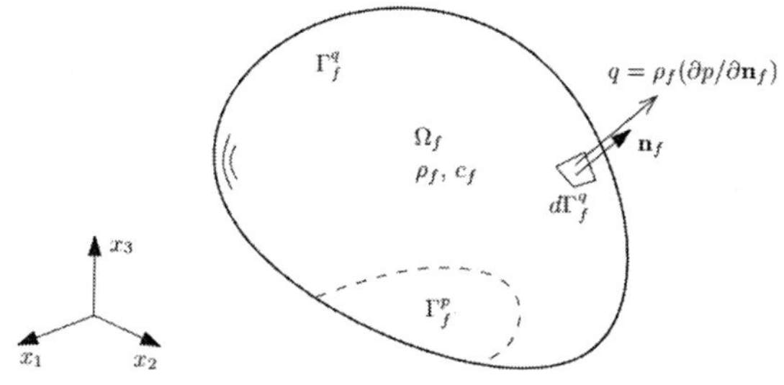

Figure 2. Acoustic fluid domain

The dynamic pressure of an ideal inviscid fluid under small perturbations in a spatial region Ω_f confined by the boundary surface $\Gamma_f = \Gamma_f^p \cup \Gamma_f^q$, is governed by the wave equation (see figure 2) [24]:

$$p_{,\alpha\alpha} = \frac{1}{c_f^2}\ddot{p} \tag{1}$$

In this equation p is the fluid pressure, $c_f^2 = \kappa/\rho_f$ stands for wave propagation velocity, κ is the bulk modulus. Above equation can be modified to include the effect of presence of an acoustic source. Double dot represents a second time derivative. Indicial notation is used throughout this work. Greek indices vary from 1 to 2 and Roman indices from 1 to 3. The boundary and initial conditions for equation (1) in the time interval $[0, t^*]$ are:

$$p(\mathbf{x}, t) = \hat{p}(\mathbf{x}, t), \quad \mathbf{x} \in \Gamma_f^p, \quad t \in [0, t^*]$$

$$q(\mathbf{x}, t) = n_i^f p_{,i} = \hat{q}(\mathbf{x}, t), \quad \mathbf{x} \in \Gamma_f^q, \quad t \in [0, t^*]$$

$$p(\mathbf{x}, 0) = \hat{p}_0(\mathbf{x}), \quad \dot{p}(\mathbf{x}, 0) = \hat{\dot{p}}_0(\mathbf{x}), \quad \mathbf{x} \in \Omega_f \tag{2}$$

where n_i^f are the components of the outward normal vector $\mathbf{n}_f$ at boundary. and t denotes time.

2.2. Dynamic equation of an elastic membrane

Now consider a linear elastic membrane with thickness h ocuppying the spatial domain Ω_w confined by the boundary $\Gamma_s = \Gamma_s^w \cup \Gamma_s^f$ (see figure 3). An initial tension T_0 is uniformly applied to the membrane. In this work, the small deflection elastic membrane theory is considered. Thus, differential equation describing the trasversal displacement $w(\mathbf{x}, t)$ of this membrane in the time interval $[0, t^*]$ is given by:

$$w_{,\alpha\alpha} + \frac{p_w}{T_0} = \frac{1}{c_w^2}\ddot{w} \tag{3}$$

where $c_w^2 = T_0/\rho_s$ and $p_w(\mathbf{x}, t)$ is a distributed pressure applied over the membrane. The boundary and initial conditions for this equation are given by:

$$w(\mathbf{x}, t) = \hat{w}(\mathbf{x}, t), \quad \mathbf{x} \in \Gamma_s^w, \quad t \in [0, t^*]$$

$$f(\mathbf{x}, t) = n_i^w w_{,i} = \hat{f}(\mathbf{x}, t), \quad \mathbf{x} \in \Gamma_s^f, \quad t \in [0, t^*]$$

$$w(\mathbf{x}, 0) = \hat{w}(\mathbf{x}), \quad \dot{w}(\mathbf{x}, 0) = \hat{\dot{w}}(\mathbf{x}), \quad \mathbf{x} \in \Omega_s \tag{4}$$

In these expressions, n_i^w are the components of the outward normal vector $\mathbf{n}_w$ at boundary.

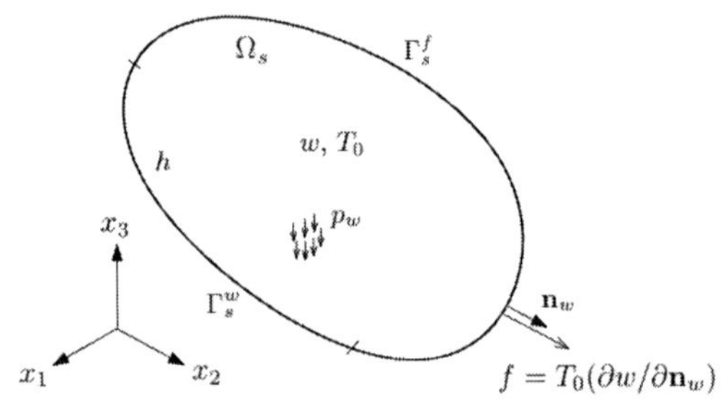

Figure 3. Elastic membrane domain

3. Boundary element equations

3.1. Boundary element equations for acoustic wave equation

The derivation of the integral formulation for equation (1) is based on the application of the Boundary Element Method to the acoustic wave equation as presented in [3]. Thus, by using the weighted residual method and making use of the Green's identity, the following equation is obtained:

$$c(\mathbf{x}')p(\mathbf{x}') + \int_{\Gamma_f} Q(\mathbf{x}', \mathbf{x})p(\mathbf{x})d\Gamma_f = \int_{\Gamma_f} P(\mathbf{x}', \mathbf{x})q(\mathbf{x})d\Gamma_f + \frac{1}{c_f^2}\int_{\Omega_f} P(\mathbf{x}', \mathbf{X})\ddot{p}(\mathbf{X})d\Omega_f \tag{5}$$

In this equation, $\mathbf{x}'$ and $\mathbf{x}$ represent collocation and field points, respectively; $P(\mathbf{x}', \mathbf{x})$ and $Q(\mathbf{x}', \mathbf{x})$ are fundamentals solutions for pressure and gradient pressures for three dimensional acoustic problems, respectively, as presented in [3]. The value of $c(\mathbf{x}')$ is equal to $\frac{1}{2}$ when $\mathbf{x}'$ is located on a smooth boundary.

In order to threat the domain integral, the Dual Reciprocity Boundary Element Method (DRM) is used as presented in [25]. In this way, equation (5) can be re-written as:

$$c(\mathbf{x}')p(\mathbf{x}') + \int_{\Gamma_f} Q(\mathbf{x}',\mathbf{x})p(\mathbf{x})\mathrm{d}\Gamma_f = \int_{\Gamma_f} P(\mathbf{x}',\mathbf{x})q(\mathbf{x})\mathrm{d}\Gamma_f$$
$$+\frac{1}{c^2}\sum_{j=1}^{NDRF}\ddot{\alpha}_j(t)\left[c_i\hat{P}_{ij} + \int_{\Gamma_f} Q(\mathbf{x}',\mathbf{x})\hat{P}_j\mathrm{d}\Gamma_f - \int_{\Gamma_f} P(\mathbf{x}',\mathbf{x})\hat{Q}_j\mathrm{d}\Gamma_f\right] \quad (6)$$

In this equation, $NDRF$ represents the number of total DRM collocations points used in the fluid; $\hat{P}_j(\mathbf{x}',\mathbf{x})$ and $\hat{Q}_j(\mathbf{x}',\mathbf{x})$ are the particular solutions to equivalent homogeneous equation (1). These particular solutions were obtained considering the function $f_j(r) = 1 + r_j$ for the approximation of $\ddot{p}(t)$, as presented in [25]. Coefficients $\ddot{\alpha}_j$ are related to $\ddot{p}$ through: $\ddot{p} = F_{ij}\ddot{\alpha}_j(t)$, where F_{ij} is a matrix of coefficients, obtained by taking the value of $\ddot{p}(t)$ at different DRM points.

To discretize boundary surfaces of the acoustic medium, N boundary quadrilateral elements were used and $p(\mathbf{x})$ and $q(\mathbf{x})$ were assumed to be constant over each element and equal to their values at the mid-element node. Thus, the discretized form of equation (6) is given by:

$$c_i(\mathbf{x}')p(\mathbf{x}',t) + \sum_{k=1}^{N}\left[\int_{\Gamma k} Q(\mathbf{x}',\mathbf{x})\mathrm{d}\Gamma\right]p_k(t) - \sum_{k=1}^{N}\left[\int_{\Gamma k} P(\mathbf{x}',\mathbf{x})\mathrm{d}\Gamma\right]q_k(t) =$$
$$\frac{1}{c_f^2}\sum_{j=1}^{NDRF}\ddot{\alpha}_j\left[c_i(\mathbf{x}')\hat{P}_{ij}(\mathbf{x}',\mathbf{x}) + \sum_{k=1}^{N}\int_{\Gamma k} Q(\mathbf{x}',\boldsymbol{x})\hat{P}_j(\mathbf{x}',\mathbf{x})\mathrm{d}\Gamma\right.$$
$$\left.-\sum_{k=1}^{N}\int_{\Gamma k} P(\mathbf{x}',\mathbf{x})\hat{Q}_j(\mathbf{x}',\mathbf{x})\mathrm{d}\Gamma\right] \quad (7)$$

Applying this equation at each collocation point, the following linear system of equations is obtained:

$$^f\mathbf{M}\ddot{\mathbf{p}} + {}^f\mathbf{H}\mathbf{p} = {}^f\mathbf{G}\mathbf{q} \quad (8)$$

where $^f\mathbf{M}$ is the fluid mass matrix, $^f\mathbf{H}$ and $^f\mathbf{G}$ are boundary element influence matrices; $\mathbf{p}$ and $\mathbf{q}$ are vectors of nodal pressures and normal derivative of pressure, respectively.

3.2. Boundary element equations for an elastic membrane

The derivation of the integral formulation for equation (3) is based on the application of the BEM to the membrane equation as presented in [3]. Thus, by using the weighted residual method, and making use of the Green's identity, the integral formulation for equation (3) is given by:

$$c(\mathbf{x}')w(\mathbf{x}',t) + \int_{\Gamma s} T(\mathbf{x}',\mathbf{x})w(\mathbf{x},t)\mathrm{d}\Gamma_s - \int_{\Gamma s} W(\mathbf{x}',\mathbf{x})f(\mathbf{x},t)\mathrm{d}\Gamma_s$$
$$= -\frac{1}{T_0}\int_{\Omega s} W(\mathbf{x}',\mathbf{x})p_w(\mathbf{x},t)\mathrm{d}\Omega_s + \frac{1}{c_w^2}\int_{\Omega s} W(\mathbf{x}',\mathbf{x})\ddot{w}(\mathbf{x},t)\mathrm{d}\Omega_s \quad (9)$$

where $\mathbf{x}$ and $\mathbf{x}'$ are field and collocation points respectively, $W(\mathbf{x}',\mathbf{x})$ and $T(\mathbf{x}',\mathbf{x})$ are fundamental solutions for displacement and traction, respectively as given in [2]. $c(\mathbf{x}')$ is the jump term arising from the terms of $O(1/r)$ in the kernel $T(\mathbf{x}',\mathbf{x})$.

In this work, the DRM was used to transform domain integrals related to inertial terms into boundary integrals. In this way, this equation (9) can be re-written as [25]:

$$c_i(\mathbf{x}')w(\mathbf{x}',t) + \int_{\Gamma s} T(\mathbf{x}',\mathbf{x})w(\mathbf{x},t)\mathrm{d}\Gamma_s - \int_{\Gamma s} W(\mathbf{x}',\mathbf{x})f(\mathbf{x},t)\mathrm{d}\Gamma_s =$$
$$-\frac{1}{T_0}\sum_{j=1}^{NDRM}\alpha_j(t)\Big[c_i(\mathbf{x}')\hat{W}_j(\mathbf{x},t) + \int_{\Gamma s} T(\mathbf{x}',\mathbf{x})\hat{W}_j(\mathbf{x}',\mathbf{x})\mathrm{d}\Gamma_s$$
$$-\int_{\Gamma s} W(\mathbf{x}',\mathbf{x})\hat{T}_j(\mathbf{x}',\mathbf{x})\mathrm{d}\Gamma_s\Big]$$
$$+\frac{1}{c_w^2}\sum_{j=1}^{MDRM}\ddot{\beta}_j(t)\Big[c_i(\mathbf{x}')\hat{W}_j(\mathbf{x},t) + \int_{\Gamma s} T(\mathbf{x}',\mathbf{x})\hat{W}_j(\mathbf{x}',\mathbf{x})\mathrm{d}\Gamma_s$$
$$-\int_{\Gamma s} W(\mathbf{x}',\mathbf{x})\hat{T}_j(\mathbf{x}',\mathbf{x})\mathrm{d}\Gamma_s\Big] \quad (10)$$

where $NDRM$ represent the total number of DRM collocation used in the membrane; $\hat{T}_j(\mathbf{x}',\mathbf{x})$ and $\hat{W}_j(\mathbf{x}',\mathbf{x})$ are the particular solutions to equivalent homogeneous equation (3). These particular solutions were obtained considering the function $f_j(r) = 1 + r_j$ for the approximation of $\ddot{w}(t)$ and $p_w(t)$ terms, as presented in [25]. Coefficients α_j are related to $p_w(t)$ through: $p_w = A_{ij}\alpha_j(t)$, where A_{ij} is a matrix of coefficients, obtained by taking the value of $p(t)$ at different DRM points. Similarly, coefficients $\ddot{\beta}_j$ are related to $\ddot{w}$ through: $\ddot{w} = B_{ij}\ddot{\beta}_j(t)$, where B_{ij} is a matrix of coefficients, obtained by taking the value of $\ddot{p}(t)$ at different DRM points.

Applying this equation at each collocation point, the following linear system of equations is obtained:

$$^s\mathbf{M}\ddot{\mathbf{w}} + {}^s\mathbf{H}\mathbf{w} = {}^s\mathbf{G}\mathbf{f} - {}^s\mathbf{B}\mathbf{p}_w \quad (11)$$

Here $^s\mathbf{M}$ is the membrane mass matrix, $^s\mathbf{H}$ and $^s\mathbf{G}$ are the influence matrices, $^s\mathbf{B}$ is the influence matrix related with distributed pressure $\mathbf{p}_w$ applied over the membrane.

4. Fluid-structure coupling equations

Fluid-structure coupling equations are given by compatibility considerations about normal pressure and dynamic pressure force acting at the fluid-structure interface. Mathematically, these conditions can be written as follows [26]:

$$\mathbf{n}_f \cdot \boldsymbol{\nabla} p = \mathbf{q}_n \equiv -\rho_w \mathbf{C}_w \ddot{\mathbf{w}} \quad (12)$$

$$\mathbf{p}_w \equiv -\mathbf{C}_f \mathbf{p} \quad (13)$$

That is, pressure gradient acting on the fluid-structure interface Γ_{fs} are related to normal acceleration of the plate and the acoustic pressure is equilibrated with pressure on the membrane (see figure 4). In these equations, $\mathbf{C}_w$ and $\mathbf{C}_f$ represent connectivity matrices joining fluid and structural degree of freedom at fluid-structure interface.

Replacing equations (12) and (13) into equations (8) and (11) we obtain the coupled fluid-

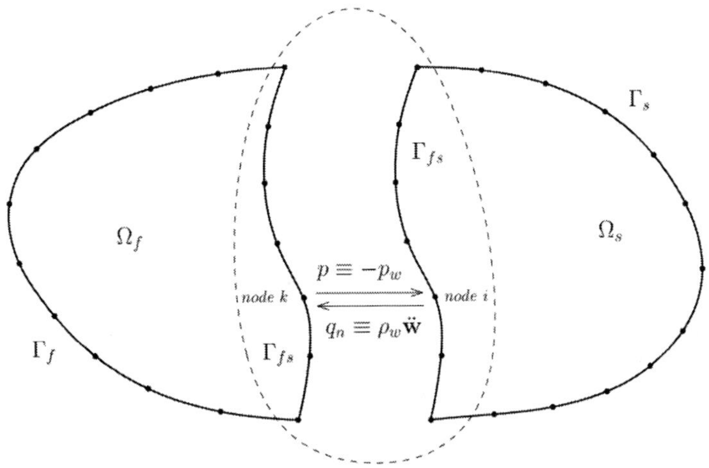

Figure 4. Acoustic fluid domain

structure equation problem:

$$\begin{bmatrix} {}^s\mathbf{M} & \mathbf{0} \\ \mathbf{S} & {}^f\mathbf{M} \end{bmatrix} \begin{Bmatrix} \ddot{\mathbf{w}} \\ \ddot{\mathbf{p}} \end{Bmatrix} + \begin{bmatrix} {}^s\mathbf{H} & -\mathbf{A} \\ \mathbf{0} & {}^f\mathbf{H} \end{bmatrix} \begin{Bmatrix} \mathbf{w} \\ \mathbf{p} \end{Bmatrix} = \begin{bmatrix} {}^s\mathbf{G} & \mathbf{0} \\ \mathbf{0} & {}^f\mathbf{G}_{ff} \end{bmatrix} \begin{Bmatrix} \mathbf{t} \\ \mathbf{q} \end{Bmatrix} \tag{14}$$

In this equation, the off-diagonal sub-matrices $\mathbf{A} = {}^s\mathbf{B}_{fs}\mathbf{C}_f$ and $\mathbf{S} = \rho_w {}^f\mathbf{G}_{fs}\mathbf{C}_w$ are fluid-membrane coupling matrices. ${}^f\mathbf{G}_{fs}$ and ${}^f\mathbf{G}_{ff}$ are sub-matrices of ${}^f\mathbf{G}$ related with degrees of freedom defined on the interface Γ_{fs}. ${}^s\mathbf{B}_{fs}$ is a sub-matrix of ${}^s\mathbf{B}$ related with pressure terms defined in Ω_s.

Equations (14) can be rewritten in a general way as:

$$\mathbf{M}\ddot{\mathbf{u}} + \mathbf{H}\mathbf{u} = \mathbf{G}\mathbf{r} \tag{15}$$

where, $\ddot{\mathbf{u}} = \{\ddot{\mathbf{w}}, \ddot{\mathbf{p}}\}^T$, $\mathbf{u} = \{\mathbf{w}, \mathbf{p}\}^T$ and $\mathbf{r} = \{\mathbf{f}, \mathbf{q}\}^T$.

5. Time integration

The two integration schemes which were tested to obtain the time response for the equation (15) are described below.

The Houbolt Method

The Houbolt integration scheme is an explicit unconditionally stable algorithm based on backward-type finite difference formula with error of order $O(\Delta\tau^2)$. The most important aspect of this method, when compared to other time integration methods based on central difference approximations or Newmark scheme, is the introduction of artificial damping which truncates the influence of higher modes in the response. Here the acceleration is approximated as:

$$\ddot{\mathbf{u}}^{\tau+\Delta\tau} = \frac{1}{\Delta\tau^2}(2\mathbf{u}^{\tau+\Delta\tau} - 5\mathbf{u}^{\tau} + 4\mathbf{u}^{\tau-\Delta\tau} - \mathbf{u}^{\tau-2\Delta\tau}) \tag{16}$$

where $\Delta\tau$ represents the time-step. Writing equation (15) at time $\tau + \Delta\tau$:

$$\mathbf{M}\ddot{\mathbf{u}}^{\tau+\Delta\tau} + \mathbf{H}\mathbf{u}^{\tau+\Delta\tau} = \mathbf{G}\mathbf{r}^{\tau+\Delta\tau} \tag{17}$$

and substituting equation (16) into equation (17), we have:

$$(2\mathbf{M} + \Delta\tau^2\mathbf{H})\mathbf{u}^{\tau+\Delta\tau} - \frac{1}{\Delta\tau^2}\mathbf{G}\mathbf{r}^{\tau+\Delta\tau} = \mathbf{M}(5\mathbf{u}^{\tau} - 4\mathbf{u}^{\tau-\Delta\tau} - \mathbf{u}^{\tau-2\Delta\tau}) \tag{18}$$

The above equation allows the calculation of the distribution of $\mathbf{u}$ at time $\tau + \Delta\tau$ by using boundary conditions at that time and information from three previous time steps. This algorithm requires a special starting procedure in which initial conditions for $\mathbf{u}$ are employed to calculate $\mathbf{u}^1$ and $\mathbf{u}^2$.

The Newmark Method

In this method the acceleration is approximated as:

$$\ddot{\mathbf{u}}^{\tau+\Delta\tau} = \frac{\mathbf{1}}{\beta\Delta\tau^2}\left[\mathbf{u}^{\tau+\Delta\tau} - \mathbf{u}^{\tau}\right] + \frac{1}{\beta\Delta\tau}\dot{\mathbf{u}}^{\tau} - \frac{(\mathbf{1} - \mathbf{2}\beta)}{\mathbf{2}\beta}\ddot{\mathbf{u}}^{\tau} \tag{19}$$

substituting equation (19) into equation (15), we have:

$$\left(\frac{\mathbf{M}}{\beta\Delta\tau^2} + \mathbf{H}\right)\mathbf{u}^{\tau+\Delta\tau} - \mathbf{G}\mathbf{r}^{\tau+\Delta\tau} = \mathbf{M}\left[\frac{\mathbf{1}}{\beta\boldsymbol{\Delta}\tau^{\mathbf{2}}}\mathbf{u}^{\tau} - \frac{\mathbf{1}}{\beta\boldsymbol{\Delta}\tau}\dot{\mathbf{u}}^{\tau} + \frac{(\mathbf{1} - \mathbf{2}\beta)}{\mathbf{2}\beta}\ddot{\mathbf{u}}^{\tau}\right] \tag{20}$$

6. Numerical examples

The results shown below for the different examples were obtained mainly from the Newmark method since the Houbold method, in general, does not converge. Hereinafter HBEM and NBEM will refer to the combination of BEM with the Houbolt and the Newmark method, respectively.

6.1. Box-shaped structure containing an acoustic fluid subjected a Heaviside load coupled to a membrane

The first example consists of a partially opened box-shaped structure with ridig walls and a flexible elastic membrane with $c_w = 0.7071$ m/s as presented in figure 5. The structure contains an acoustic fluid with mass density $\rho_f = 1.21$ kg / m^3 and a bulk's modulus $\kappa = 139876$ Pa. A distributed unit pressure load $p(t) = -1.0$ Pa is applied over the face of the structure located at $x_3 = 2$ m. Pressure gradients $q = n_i^f p_{,i}$ are considered zero at the other faces. Initial conditions for this problem are $\mathbf{u}(\mathbf{x}, 0) = \dot{\mathbf{u}}(\mathbf{x}, 0) = 0$.

To validate the BEM solution, a 3D finite element code (FEM) for this problem was developed. The meshes used for discretization are resumed at the table below.

These collocation points are coincidents with fluid collocation points located in the fluid-membrane interface.

6.2. Box-shaped structure containing an acoustic fluid subjected an harmonic load coupled to a membrane

For the second example, only the Heaviside pressure load was changed for a harmonic load defined as $p(t) = -sin(wt)$ Pa. Here both the frequency w and the amplitude was set to one.

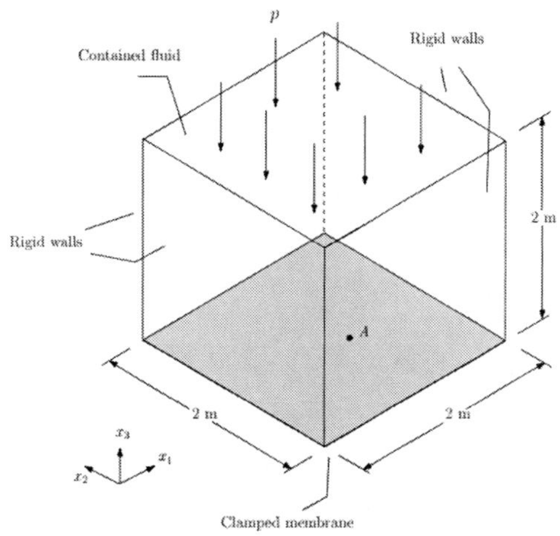

Figure 5. Box-shaped structure coupled to a membrane

Table 1. FEM Meshes

Mesh	Number of 3D elements	Number of fluid nodes	Number of 2D elements	Number of membrane nodes
FEM 6x6x6	216	343	36	49
FEM 8x8x20	1280	1701	64	81
FEM 12x12x20	2880	3549	144	169
FEM 16x16x20	5120	6069	256	289

Table 2. BEM Meshes

Mesh	Number of 2D quadrilateral elements	Number of interior collocation points	Number of 2D constant elements	Number of interior collocation points
BEM 5x5x5	125	125	20	25
BEM 7x7x7	343	125	28	49
BEM 9x9x9	729	125	36	81
BEM 11x11x11	1331	125	44	121
BEM 15x15x15	3375	125	60	225

6.3. Channel-shaped structure containing an acoustic fluid coupled to an inclined membrane

In this example lets to analyze a partially opened Channel-shaped structure with rigid walls containing an acoustic fluid coupled to a inclined flexible elastic membrane located at the right side of the channel. The membrane wave propagation velocity, the acoustic fluid mass density and the fluid compressibility are respectively $c_w = 0.7071$ m/s, $\rho_f = 1.21$ kg / m^3 and $\kappa = 139876$ Pa. A distributed harmonic pressure load $p(t) = -sin(wt)$ Pa is applied over the membrane, see the figure 1. Pressure gradients $q = n_i^f p_{,i} =$ considered zero at the rigid walls. Initial conditions for this problem are $\mathbf{u}(\mathbf{x}, 0) = \dot{\mathbf{u}}(\mathbf{x}, 0) = 0$.

The convergence in time for the displacement and pressure, using a refined mesh are shown below in figures 13 and 14. Figure 13 shows that the behavior of displacement over time as well as for pressure in figure 14 is periodic.

IOP Conf. Series: Materials Science and Engineering **844** (2020) 012073 doi:10.1088/1757-899X/844/1/012073

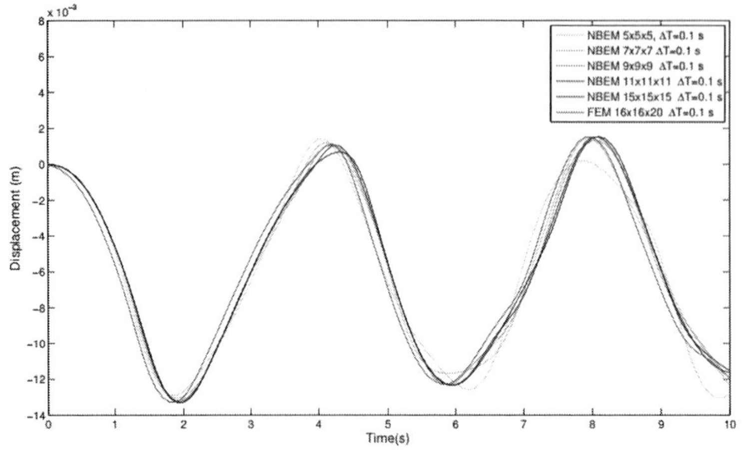

Figure 6. Displacement time history at point A(0.25,-0.25,-1) using DR-BEM with Newmark method

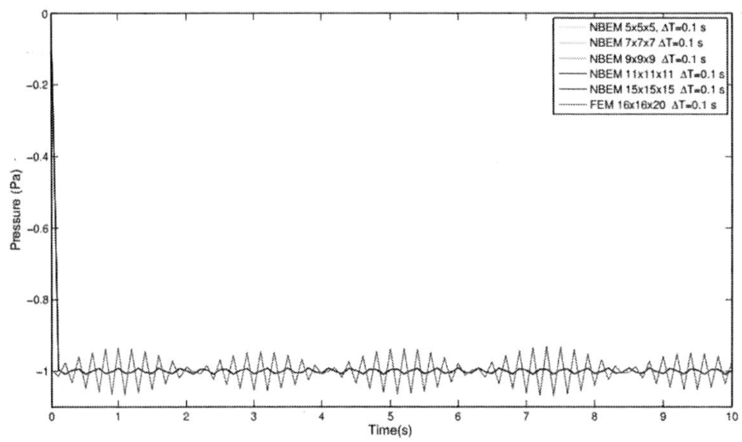

Figure 7. Pressure time history at point A(0.25,-0.25,-1) using DR-BEM with Newmark method

6.4. Channel-shaped structure containing an acoustic fluid coupled to a membrane

The last example consists of a partially opened Channel-shaped structure with rigid walls containing an acoustic fluid coupled to a flexible elastic membrane located at top face of the channel. All properties, pressure load and boundary conditions remain as before, see the figure 15.

The convergence in time for the displacement and presure, using a refined mesh are shown below in figures 16 and 17. It can be observed that the time history for displacement is lightly different compared with the inclined membrane, i.e., the shape of the wave is similar and only the amplitude changes, because the dimensions are differents, and as it was expected, the longer membrane has higher displacement at the center .

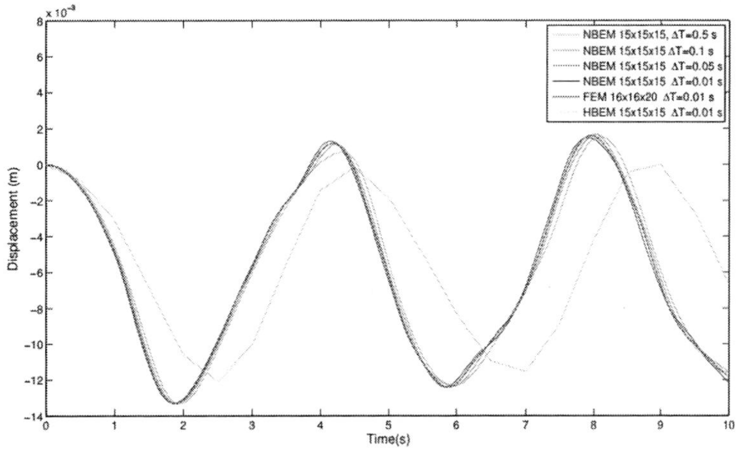

Figure 8. Displacement time history at point A(0.25,-0.25,-1) using DR-BEM with Newmark or Houbolt method

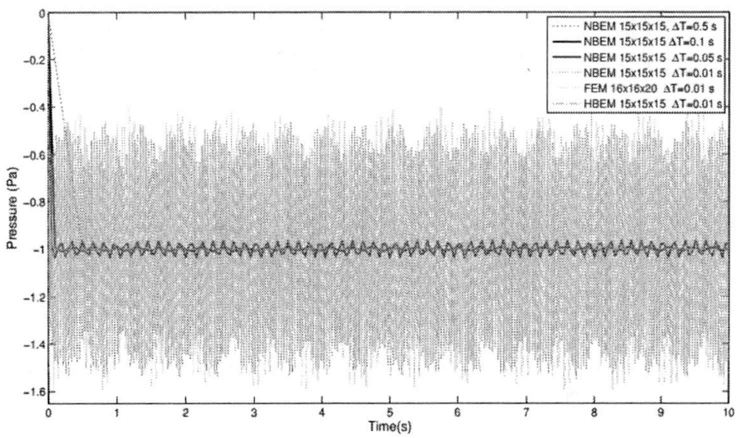

Figure 9. Pressure time history at point A(0.25,-0.25,-1) using DR-BEM with Newmark or Houbolt method

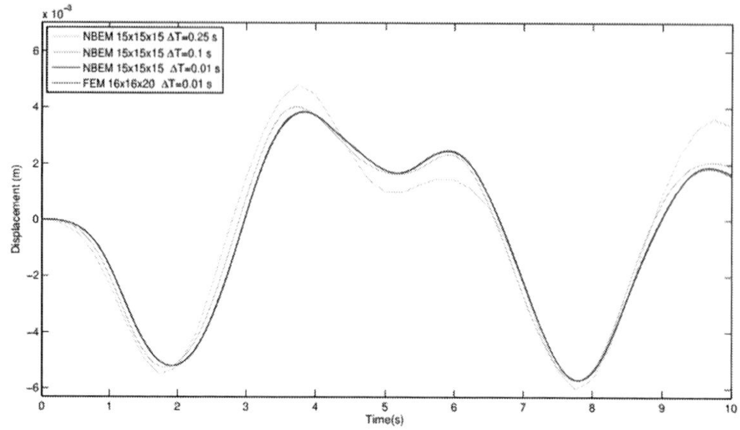

Figure 10. Displacement time history at point A(0.25,-0.25,-1) using DR-BEM with Newmark method

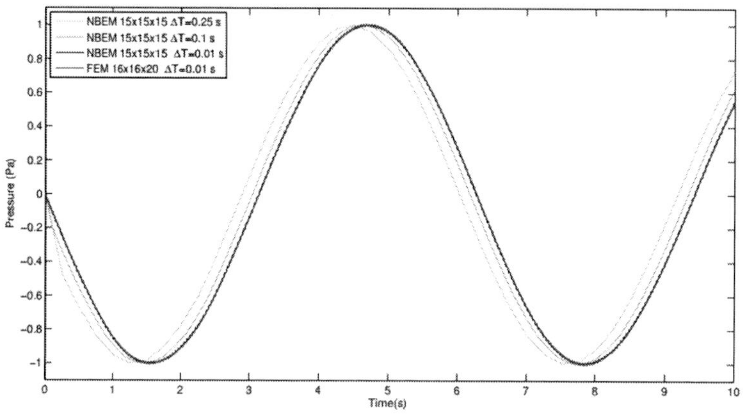

Figure 11. Pressure time history at point A(0.25,-0.25,-1) using DR-BEM with Newmark method

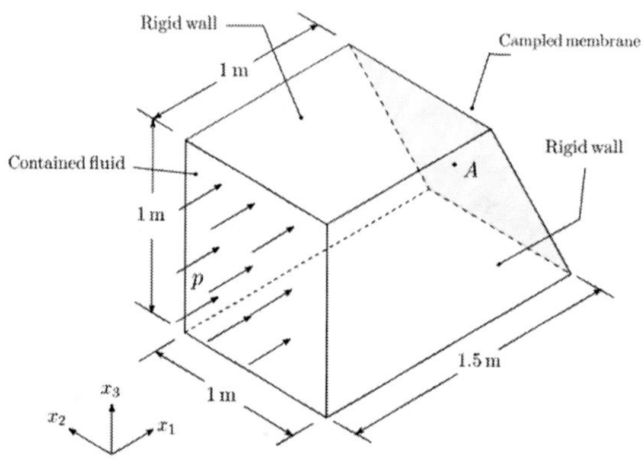

Figure 12. Channel 1

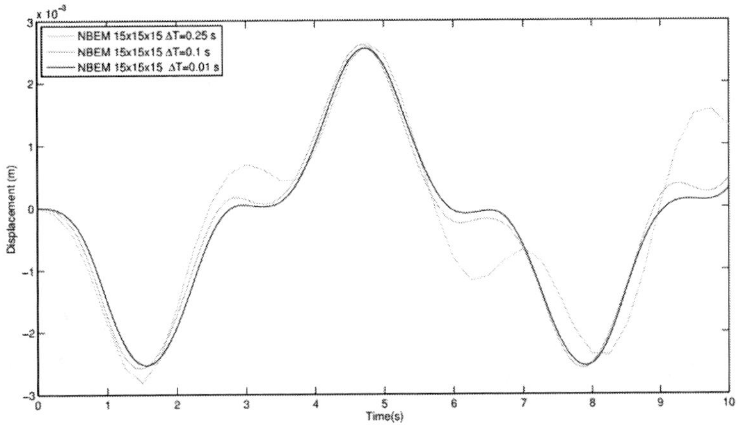

Figure 13. Displacement time history at point A(1.25,0.5,0.5) using DR-BEM and Newmark method

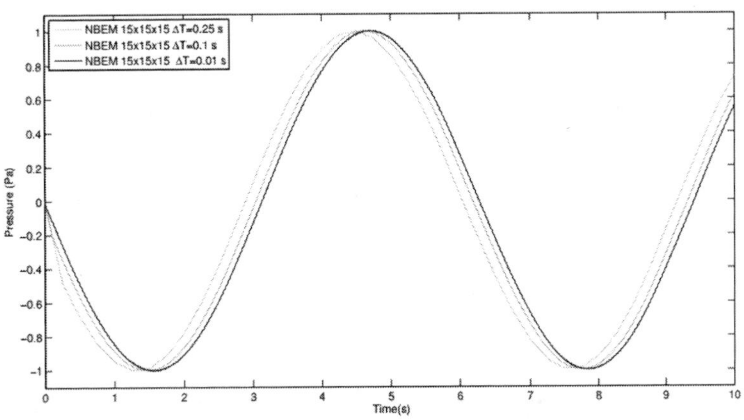

Figure 14. Pressure time history at point A(1.25,0.5,0.5) using DR-BEM and Newmark method

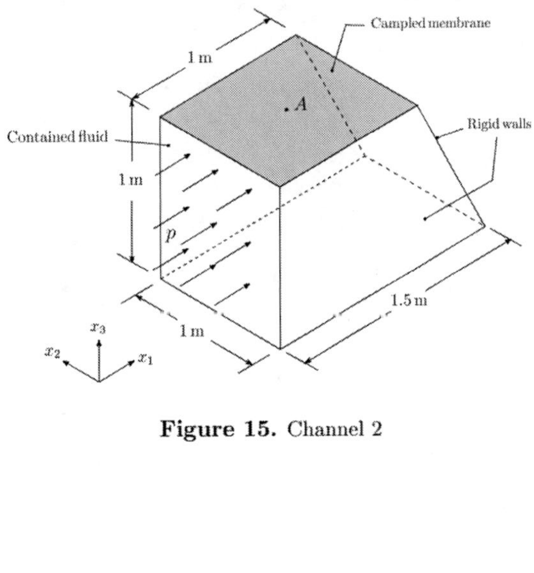

Figure 15. Channel 2

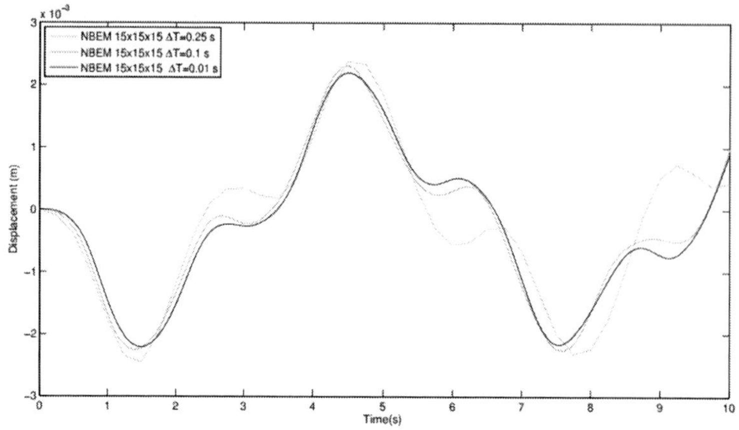

Figure 16. Displacement time history at point A(0.5,0.5,1) using DR-BEM and Newmark method

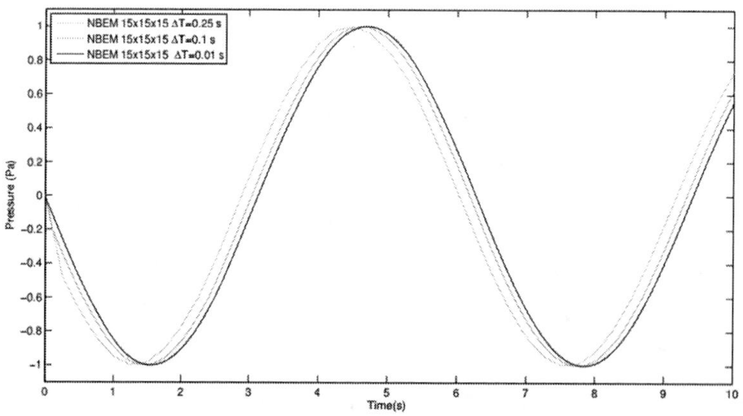

Figure 17. Pressure time history at point A(0.5,0.5,1) using DR-BEM and Newmark method

7. Conclusions

A new full boundary element formulation for the transient dynamic analysis of acoustic fluids coupled to elastic membranes is presented. Membranes were modeled using a boundary element formulation based on the linear elastic membrane theory under small deflection. The acoustic fluid was modeled using a boundary element formulation for the three dimensional acoustic wave equation. Fluid-structure coupling equations were established considering the continuity of the normal acceleration of the particles at fluid-structure interfaces. Domain integrals on both, fluid and structure equations, were treated using the Dual Reciprocity Boundary Element Method. Results show good agreement with those obtained from finite element models, turning the proposed formulation in an alternative numerical engineering tool for the dynamic analysis of acoustic fluids coupled flexible elastic membranes.

8. Acknowledgment

The authors are grateful to the Research Office of Technological University of Bolívar for supporting this research work.

References

[1] S. D. D handole, S. V. Modak, Review of Vibro-Acoustics analysis procedures for prediction of low frequency noise inside a cavity, *IMAC XXV A Conference and Exposition on Structural Dynamics*, 2005.

[2] L. C. Wrobel, M. H. Aliabadi, *The Boundary Element Method Volume 2: Applications in Solid and Structures*, Wiley, New York, 2002.

[3] L. C. Wrobel, *The Boundary Element Method Volume 1: Applications in Thermo-Fluids and Acoustics*, Wiley, 2002.

[4] I. T. Hwang, K. Ting, Boundary Element Method for Fluid-Structure Interaction Problems in Liquid Storage Tanks, *Journal of Pressure Vessel Technology*,111(4) (1989) 435–440.

[5] J. Mackerle, Fluid-Structure Interaction Problems, Finite Element and Boundary Element Approaches: A bibliography (1995-1998), *Finite Elements in Analysis and Design Journal*, 31 (1999) 231–240.

[6] M. R. Shekari, N.Khaji, M. T. Ahmadi, A coupled BE-FE study for evaluation of seismically isolated cylindrical liquid storage tanks considering fluid-structure interaction, *Journal of Fluids and Structures*, 25(3) (2009) 567–585.

[7] V. Naumenkob, E. Strelnikovac, E. Yeselevac, Free vibrations of shells of revolution filled with a fluid, *Engineering Analysis with Boundary Elements* 34(10) (2010) 856–862.

[8] L. Gaul, W. Wenzel, A coupled symmetric BEM-FEM method for acoustic fluid-structure interaction, *Engineering Analysis with Boundary Elements* 26 (2002) 629–636.

[9] Y. Young, Fluid-structure interaction analysis of flexible composite marine propellers, *Journal of Fluids and Structures* 24 (2008) 799–818.

[10] D. Soares, Acoustic modelling by BEM-FEM coupling procedures taking into account explicit and implicit multi-domain decomposition techniques, *International Journal for Numerical Methods in Engineering* 78 (2009) 1076–1093.

[11] G. C. Everstine, F. M. Henderson, Coupled finite element/boundary element approach for fluid-structure interaction, *The Journal of the Acoustical Society of America* 87 (1990) 1938–1947.

[12] D. Fritze, S. Marburg, H. J. Hardtke, FEM-BEM coupling and structural acoustic sensitivity analysis for shell geometries, *Computers and Structures* 83 (2005) 143–154.

[13] R. Citarella, L. Federico, A. Cicatiello, Modal acoustic transfer vector approach in a fem-bem vibro-acoustic analysis, *Engineering Analysis with Boundary Elements* 31 (2007) 248–258.

[14] G. Of, O. Steinbach, Coupled fe/be formulations for the fluid-structure interaction, in: Y. Huang, R. Kornhuber, O. Widlund, J. Xu (Eds.), *Domain Decomposition Methods in Science and Engineering* XIX, volume 78 of *Lecture Notes in Computational Science and Engineering*, Springer Berlin Heidelberg, 2011, pp. 293–300.

[15] Z. He, G. Liu, Z. Zhong, G. Zhang, A. Cheng, A coupled ES-FEM/BEM method for fluid-structure interaction problems, *Engineering Analysis with Boundary Elements* 35 (2011) 140–147.

[16] P. T. Chen, S. H. Ju, K. C. Cha, A symmetric formulation of coupled bem/fem in solving responses of submerged elastic structures for large degrees of freedom, *Journal of Sound and Vibration* 233 (2000) 407–422.

[17] S. Marburg, B. Nolte, Computational Acoustics of Noise Propagation in Fluids - Finite and Boundary Element Methods, Springer, 2010

[18] M. Tanaka, Y. Masuda, Boundary element method applied to certain structural-acoustic coupling problems, *Computer Methods in Applied Mechanics and Engineering* 71 (1988) 225–234.

[19] B. Nolte, *Fluid-structure interaction phenomena on the basis of a bem-bem coupling formulation,* pp. 705–710.

[20] B. Nolte, L. Gaul, *Fluid-structure interaction with the boundary element method,* volume 1, pp. 496–502.

[21] D. S. Jr., W. Mansur, Dynamic analysis of fluid-soil-structure interaction problems by the boundary element method, *Journal of Computational Physics* 219 (2006) 498–512.

[22] R. Burgschweiger, M. Ochmann, B. Nolte, Calculation of the acoustic target strength of elastic objects based on bem-bem coupling, *The Journal of the Acoustical Society of America* 123 (2008) 3757.

[23] D. Soares Jr, O. von Estorff, W. J. Mansur, Efficient non-linear solid-fluid interaction analysis by an iterative bem/fem coupling, *International Journal for Numerical Methods in Engineering* 64 (2005) 1416–1431.

[24] L. E. Kinsler, A. R. Freg, A. B. Coppens, J. V. Sanders, *Fundamental of acoustics*, Jhon wiley and sons Inc., 2005.

[25] P. Partridge, C. Brebbia, L. Wrobel, *The dual reciprocity boundary element method,* Southampton: Computational Mechanics Publications, 1992.

[26] P. Morand, R. Ohayon, *Fluid structure interaction,* John Wiley and Sons, 1995.

AUTHOR INDEX